Major Geological, Climatic, and Biological Events

Culmination of mountain-building followed by erosion and moderate, short-lived invasions of continental margins by the sea. Early warming trends were reversed by the middle of the period to cooler and finally to glacial conditions. Subtropical forests gave way to temperate forests and finally to extensive grasslands. Transition from archaic mammals to modern orders and eventually families. Evolution of humans during the last 5-8 million years.

Last great spread of epicontinental seas and shoreline swamps. At the end of the period extensive mountain-building cooled the climate worldwide. Angiosperm dominance began. Extinction of archaic birds and many reptiles by the end of the period.

Climate was warm and stable with little latitudinal or seasonal variation. Modern genera of many gymnosperms and advanced angiosperms appeared. Reptilian diversity was high in all habitats. First birds appeared.

Continents were relatively high with few shallow seas. The climate was warm; deserts were extensive. Gymnosperms dominated; angiosperms first appeared. Mammal-like reptiles were replaced by precursors of dinosaurs and the earliest true mammals appeared.

Land was generally higher than at any previous time. The climate was cold at the beginning of the period but warmed progressively. Glossopterid forests developed with the decline of the coal swamps. Mammal-like reptiles were diverse; widespread extinctions at the end of the period.

Generally warm and humid, but some glaciation in the Southern Hemisphere. Extensive coal-producing swamps with large anthropod faunas. Many specialized amphibians and the first appearance of reptiles.

Mountain-building produced locally arid conditions, but extensive lowland forests and swamps were the beginning of the great coal deposits. Extensive radiation of amphibians; extinction of some fish lineages and expansion of others.

The land was higher and climates cooler. Freshwater basins developed in addition to shallow seas. The first forests appeared and the first winged insects. There was an explosive radiation of fishes, followed by the disappearance of many jawless forms. The earliest tetrapods appeared.

The land was slowly being uplifted, but shallow seas were extensive. The climate was warm and terrestrial plants radiated. Eurypterid arthropods were at their maximum abundance in aquatic habitats and the first terrestrial arthropods appeared. The first gnathostomes appeared among a diverse group of marine and freshwater jawless fishes.

The maximum recorded extent of shallow seas was reached and the warming of the climate continued. Algae became more complex, vascular plants may have been present, and there was a variety of large invertebrates. Jawless fish fossils from this period are fragmentary but more widespread.

There were extensive shallow seas in equatorial regions. The climate was warm. Algae were abundant and there are records of trilobites and brachiopods. The first remains of vertebrates are found at the end of this period.

Changes in the lithosphere produced major land masses and areas of shallow seas. Multicellular organisms appeared and flourished—algae, fungi, and many invertebrates.

Formation of the earth and slow development of the lithosphere, hydrosphere, and atmosphere. Development of life in the hydrosphere.

VERTEBRATE LIFE

F. Harvey Pough

Arizona State University West

John B. Heiser
William N. McFarland

Cornell University

FOURTH EDITION

VERTEBRATE LIFE

Prentice Hall
Upper Saddle River, New Jersey 07458

Library of Congress Cataloging-in-Publication Data

Pough, F. Harvey.
  Vertebrate life / F. Harvey Pough, John B. Heiser, William N.
 McFarland,—4th ed.
    p. cm.
  Includes bibliographical references and index.
  ISBN 0–02–396370–0
  1. Vertebrates. 2. Vertebrates, Fossil. I. Heiser, John B. II. McFarland, William N. (William Norman). III. Title.
QL605.P68 1996
596—dc20   95–14458
                                                             CIP

Acquisitions Editor: *Sheri L. Snavely*
Editor in Chief: *Paul F. Corey*
Editorial Director: *Tim Bozik*
Assistant Vice President of Production and Manufacturing: *David W. Riccardi*
Executive Managing Editor: *Kathleen Schiaparelli*
Assistant Managing Editor: *Margaret Antonini*
Marketing Manager: *Kelly McDonald*
Manufacturing Buyer: *Trudy Pisciotti*
Creative Director: *Paula Maylahn*
Cover Designer: *DesignW, Inc./Wendy Helft*
Photo Editor: *Lorinda Morris-Nantz*
Photo Researchers: *Chris Migdoll, Diane Kraut*
Editorial Assistants: *Lisa Tarabokjia and Nancy Bauer*
Art Studio: *Academy ArtWorks, Inc.*
Copyediting and Text Composition: *Electronic Publishing Services Inc.*
Cover Art: © *RipTide, Inc. Maxine Fumagalli/Brendan Japantardi*

© 1996 by Prentice-Hall, Inc.
Simon & Schuster/A Viacom Company
Upper Saddle River, New Jersey 07458

Previous editions copyright © 1979, 1985, 1989 by Macmillan Publishing Company, a division of Macmillan, Inc.

Printed in the United States of America

10 9 8 7 6 5 4 3

ISBN 0-02-396370-0

Prentice-Hall International (UK) Limited, *London*
Prentice-Hall of Australia Pty. Limited, *Sydney*
Prentice-Hall Canada Inc., *Toronto*
Prentice-Hall Hispanoamericana, S.A., *Mexico*
Prentice-Hall of India Private Limited, *New Delhi*
Prentice-Hall of Japan, Inc., *Tokyo*
Simon & Schuster Asia Pte. Ltd., *Singapore*
Editora Prentice-Hall do Brasil, Ltda., *Rio de Janeiro*

PREFACE

The fourth edition of *Vertebrate Life* contains changes that reflect the extraordinary activity in vertebrate biology during the past two decades. The most pervasive innovations have resulted from the widespread adoption of phylogenetic systematics (cladistics) as the basis for determining the evolutionary relationships of organisms. The emphasis that this system of classification places on the importance of monophyletic groupings has ramifications in many areas of biology. As an objective (although frequently controversial) method that reflects information about the sequence of changes during evolution, cladistics provides an evolutionary framework in which ideas from other biological specialties can be accommodated. As a result, studies of behavior, physiology, and ecology are increasingly being placed in an explicitly evolutionary context, and this common ground has fostered increased interaction among those specialties.

We have retained the cladistic classification introduced in the third edition as the basis for the fourth edition of *Vertebrate Life*, and have included cladograms illustrating the postulated relationships of vertebrates. In doing so, we have tried to reconcile the views of various authorities and point out major areas of disagreement. The cladograms include synopses of the character states on which they are based and citations of the primary sources used. This information will facilitate exploration of different views, and will help faculty and students to modify the phylogenies presented here as new interpretations are published.

As a result of the cladistic perspective of this edition, we have reorganized the treatment of morphology and physiology to emphasize derived characters of vertebrates. Chapter 3 treats embryonic development and morphology, and Chapter 4 presents a parallel treatment of general aspects of vertebrate physiology and homeostasis. Topics unique to particular groups are highlighted in the chapters treating those groups.

Another important change in this edition is an emphasis on conservation, especially the application of basic biological information about organisms in programs of captive husbandry and management of threatened and endangered species. We believe that collaborative work by academic biologists with colleagues from zoos and conservation organizations, a synthesis we call applied organismal biology, offers the best hope for protecting biological diversity. We have provided examples of successes and failures of work of this sort, and have included admittedly speculative proposals for further applications in the hope that students will be attracted to this field.

Literature citations have been brought up to date, with many references from 1990 onward. As before, we have chosen citations on the basis of their helpfulness to students attempting to enter the literature of the subject; review articles are cited where possible, and recent references are used because students can trace earlier work through them.

The task of reviewing all of vertebrate biology is nearly overwhelming, and would have been impossible without the hours of time that colleagues spent helping us. We are exceedingly grateful to all of them.

Acknowledgments

Writing a book with a scope as broad as this one requires the assistance of many people. We are grateful to the following colleagues for their generous responses to our requests for information and their comments and suggestions:

Mary Allen (The National Zoo), John Baker (The Open University), Carol Beuchat (California State University at San Diego), the late Robert Bouma (Cornell University), Robert Carroll (McGill University), Mark Chappell (University of California at Riverside), Jennifer Clark (University Museum of Zoology, Cambridge University), Neil Clark (Hunterian Museum, University of Glasgow), Michael Coates (University Museum of Zoology, Cambridge University), Andres Collazo (California Institute of Technology), David Crews (University of Texas), Benjamin Dial (Chapman College), James Edwards (National Science Foundation), Carl Ferraris (California Academy of Sciences), Erik Gergus (Arizona State University West), Carola Haas (Virginia Polytechnic Institute and State University), Timothy Halliday (The Open University), David Hillis (University of Texas), Frank Gill (Academy of Natural Sciences, Philadelphia), Larry Herbst (University of Florida), Ronald Heyer (The National Museum of Natural History), William Hillenius (Oregon State University), James Hopson (The University of Chicago), Elliott Jacobson (University of Florida), Christine Janis (Brown University and The University of Chicago), William Layton (Dartmouth College), Amy McCune (Cornell University), Samuel McLeod (University of Southern California), Barbara Moore (Peabody Museum, Yale University), James Murphy (Dallas Zoo), Olav Oftedal (The National Zoo), Charles Oravetz (National Marine Fisheries Service), Gary Packard (Colorado State University), Alan Pooley (Rutgers University), Donald Prothero (Occidental College), David Roberts (Dallas Zoo), Alan Savitzky (Old Dominion University), Gordon Schuett (Arizona State University West), Donna Shaver (National Biological Survey), Barry Sinervo (University of Indiana), Joe Small (Bone Bug), Ellen Smith (University of Washington), J. A. van den Hoover (University of Stellenbosch), Kentwood Wells (University of Connecticut).

Brooks Burr (Southern Illinois University at Carbondale), Margaret Fusari (University of California at Santa Cruz), William Gutzke (Memphis State University), Christine Janis (Brown University and The University of Chicago), Fred Wasserman (Boston University), Jeffrey Carpenter

(Colorado State University), and Margaret Haag (University of Alberta) reviewed the entire text of the third edition and our plans for changes. Their suggestions have shaped nearly every aspect of this book, and we cannot sufficiently express our gratitude for their efforts.

F. Harvey Pough
John B. Heiser
William N. McFarland

BRIEF CONTENTS

CONTENTS

14 Geography and Ecology of the Mesozoic   445

15 The Lepidosaurs: Tuatara, Lizards, Amphisbaenians, and Snakes   451

16 Ectothermy: A Low-Cost Approach to Life   497

VERTEBRATE LIFE

PART I

VERTEBRATE DIVERSITY, FUNCTION, AND EVOLUTION

The 45,000 living species of vertebrates inhabit nearly every part of the Earth, and other kinds of vertebrates that are now extinct lived in habitats that no longer exist. Increasing knowledge of the diversity of vertebrates was one of the products of European exploration and expansion that began in the fifteenth and sixteenth centuries. In the middle of the eighteenth century the Swedish naturalist Carolus Linnaeus developed a binominal classification to catalog the varieties of animals and plants. The Linnean system remains the basis for naming living organisms today.

A century later Charles Darwin explained the diversity of plants and animals as the product of natural selection and evolution, and in the early twentieth century Darwin's work was coupled with the burgeoning information about mechanisms of genetic inheritance. This combination of genetics and evolutionary biology is known as the New Synthesis or Neo-Darwinism, and continues to be the basis for understanding the mechanics of evolution. Recent work has broadened our view of evolutionary mechanisms by suggesting, on one hand, that some major events in evolution may be the result of chance rather than selection, and, on the other hand, that natural selection can sometimes extend beyond individuals to related individuals, populations, or even to entire species. Methods of classifying animals also have changed their emphasis during the twentieth century, and classification, which began as a way of trying to organize the diversity of organisms, has become a way of generating testable hypotheses about evolution.

Vertebrate biology and the fossil record of vertebrates have been at the center of these changes in our view of life. Comparative studies of the anatomy, embryology, and physiology of living vertebrates have often supplemented the fossil record. These studies reveal that evolution acts by changing existing structures. All vertebrates have basic characteristics in common that are the products of their common ancestry, and progressive modifications of these characters can trace the progress of evolution. Thus, an understanding of vertebrate form and function is basic to understanding the evolution of vertebrates and the ecology and behavior of living species.

The Diversity, Evolution, and Classification of Vertebrates

Evolution is central to vertebrate biology because it provides a principle that organizes the diversity that we see among living vertebrates and helps to fit extinct forms into the context of living species. Classification, initially a process of attaching names to organisms, has become a method of understanding evolution. Current views of evolution stress natural selection operating at the level of individuals as a predominant mechanism that produces change over time. Additional mechanisms may involve selection that operates at higher and lower levels of biological organization and chance events. The processes and events of evolution are intimately linked to the changes that have occurred on Earth during the history of vertebrates. These changes have resulted from the movements of continents and the effects of those movements on climates and geography. In this chapter we present an overview of the scene, the participants, and the rules governing the events that have shaped the biology of vertebrates.

The Vertebrate Story

Mention "animal" and most people will think of a vertebrate. Vertebrates are often abundant and conspicuous parts of people's experience of the natural world. Vertebrates are also very diverse: The approximately 45,000 extant (= currently living) species of vertebrates range in size from fishes that weigh as little as 0.1 gram when they are fully mature to whales that weigh nearly 100,000 kilograms. Vertebrates live in virtually all the habitats on Earth: Bizarre fishes, some with mouths so large they can swallow prey larger than their own bodies, cruise through the depths of the sea, sometimes luring prey to them with glowing lights. Some 15 kilometers above the fishes, migrating birds fly over the crest of the Himalayas, the highest mountains on Earth. Birds can live at these high altitudes where mammals are incapacitated by lack of oxygen because the lungs of birds have a pattern of airflow that is different from that of mammals and bird lungs are more effective

than mammalian lungs at extracting oxygen from the air.

The behaviors of vertebrates are as diverse and complex as their body forms. Vertebrate life is energetically expensive, and vertebrates get the energy they need from food they eat. Carnivores eat the flesh of other animals and show a wide range of methods of capturing prey: Some predators search the environment to find prey, whereas others wait in one place for prey to come to them. Some carnivores pursue their prey at high speeds, others pull prey into their mouths by suction. In some cases the foraging behaviors that vertebrates use appear to be exactly the ones that maximize the amount of energy they obtain for the time they spend hunting; in other cases vertebrates can appear to be remarkably inept predators. Many vertebrates swallow their prey intact, sometimes while it is alive and struggling, but other vertebrates have very specific methods of dispatching prey: Venomous snakes inject complex mixtures of toxins, and cats (of all sizes from house cats to tigers) kill their prey with a distinctive bite on the neck. Herbivores eat plants: Plants cannot run away when an animal approaches, but they are hard to digest and they frequently contain toxic compounds. Herbivorous vertebrates show an array of specializations to deal with the difficulties of eating plants: These specializations include elaborately sculptured teeth and digestive tracts that provide sites in which symbiotic microorganisms digest compounds that are impervious to the digestive systems of vertebrates.

Reproduction is a critical factor in the evolutionary success of an organism, and vertebrates show an astonishing range of behaviors associated with mating and reproduction. In general, males court females and females care for the young, but these roles are reversed in many species of vertebrates. The forms of reproduction employed by vertebrates range from laying eggs to giving birth to babies that are largely or entirely independent of their parents (precocial young). These variations range across almost all kinds of vertebrates: Many fishes and amphibians produce precocial young and a few mammals lay eggs. In fact, only birds show no variation in their reproductive mode; all birds lay eggs. At the time of birth or hatching some vertebrates are entirely self-sufficient and never see their parents, whereas other vertebrates (including humans) have extended periods of obligatory parental care. Extensive parental care is found in seemingly unlikely groups of vertebrates—fishes that incubate eggs in their mouths, frogs that incubate eggs in their stomachs, and birds that feed their nestlings a fluid called crop milk that is very similar in composition to mammalian milk.

The diversity of living vertebrates is fascinating, but the species now living are only a small proportion of the species of vertebrates that have existed. For each living species there may be as many as ten extinct species, and some of these have no counterparts among living forms. The dinosaurs, for example, that dominated the Earth for 180 million years are so entirely different from any living animals that it is hard to reconstruct the lives they led. Even mammals were once more diverse than they are now: The Pleistocene saw giants of many kinds—ground sloths as big as modern rhinoceroses and raccoons and rodents as large as bears. Humans are great apes, close relatives of chimpanzees and gorillas, and much of the biology of humans is best understood in the context of our vertebrate heritage. The number of

species of vertebrates probably reached its maximum in the Pliocene and Pleistocene, and has been declining since then. Some parts of that decline can probably be attributed to the effects of humans on species they used for food or viewed as competitors, although changes in climate and vegetation have also been instrumental. In the modern world, however, the effects of human activities are paramount, and the fate of other species of vertebrates is very much affected, for good or ill, by human decisions. Our responsibilities to other vertebrates cannot be ignored.

The story of vertebrates is fascinating: Where they originated, how they evolved, what they do, and how they work provide endless intriguing details. In preparation to tell this story we must introduce some basic information: what the different kinds of vertebrates are called and how they are classified, how evolution works, and what the world within which the story of vertebrates unfolded was like. In this chapter we provide an overview of the vertebrates and the processes of evolution and environmental change that have shaped them.

The Different Kinds of Vertebrates

Describing and classifying the variety of vertebrates, living and extinct, has become more complicated recently than it used to be as a result of a change in the criteria used for recognizing natural groups of organisms. Most of us have grown used to classifying vertebrates as jawless fishes, cartilaginous fishes, bony fishes, amphibians, reptiles, birds, or mammals. Those names are familiar; each conjures up an image of a particular kind of animal. When someone said "reptile" we thought of turtles, alligators, crocodiles, lizards, and snakes. The term "reptile" communicated information about particular animals, and the same was true of the terms "jawless fishes" or "birds."

The animals that we recognized by those names still exist, of course, but some go by different names and are grouped differently now. The reason for the change is an increased emphasis on the proposition that *groups* of animals can be identified only if they share a common evolutionary lineage. Basically, the old method of classification (which can be called evolutionary systematics) lumped together animals that had very different evolutionary histories because they seemed similar in important respects. This process produced groups that contained unrelated evolutionary lineages. In contrast, the modern approach to classification, which is popularly known as cladistics, recognizes only groups of organisms that are related by common descent, or **phylogeny** (*phyla* = tribe, *genesis* = origin). The application of cladistic methods is making the study of evolution more rigorous than it has been heretofore. The natural groups recognized by cladistics are also easier to understand than the artificial groups we are used to, except that we are familiar with the names of the artificial groups and the names of the new groups are sometimes strange. At this point we need to establish a basis for talking about particular animals by naming them, and to relate the old, familiar names to the new, less familiar ones.

Figure 1–1 shows the major kinds of vertebrates and the relative numbers of living species and Table 1–1 compares the names of old and new groupings of vertebrates and shows the chapters in which the major discussions of the different groups will be found. In the following sections we describe briefly the different kinds of living vertebrates.

Hagfish and Lampreys (Myxinoidea and Petromyzontoidea)

Unique among living craniates because they lack jaws, the few species of hagfishes and lampreys occupy an important position in the study of vertebrate evolution. Hagfishes and lampreys have traditionally been grouped as agnathans (*a* = without, *gnath* = jaw) or cyclostomes (*cyclo* = round, *stoma* = mouth), but they probably represent two independent evolutionary lineages. Hagfishes appear to possess numerous ancestral vertebrate characteristics, whereas the evolutionary lineage of lampreys once had derived characters that are not present in living lampreys. The jawless condition of both lampreys and hagfish, however, is ancestral.

Lampreys and hagfishes are elongate, scaleless, and slimy and have no internal hard tissues. They are scavengers and parasites and are specialized for those roles. Hagfishes (about 40 species) are marine and occur on the continental shelf and open ocean at depths around 100 meters, whereas many of the 41 species of lampreys are migratory forms that live in oceans and spawn in rivers.

Sharks, Rays, and Ratfishes (Elasmobranchii and Holocephali)

Sharks have a reputation for ferocity that most of the 350 to 400 species would have difficulty living up to. Many sharks are small (15 centimeters or less), and the whale shark, which grows to 10 meters, is a filter feeder that subsists on plankton it strains from the water. The 450 species of rays are dorsoventrally flattened, frequently bottom dwellers that swim with undulations of their extremely broad pectoral fins. Ratfish are bizarre marine fishes with long, slender tails and buck-toothed faces that look rather like rabbits. The name Chondrichthyes (*chondro* = cartilage, *ichthyes* = fish) refers to the cartilaginous skeletons of these fishes.

Bowfin, Gars, and Others (Actinopterygii)

The fishes are so diverse that any attempt to characterize them briefly is doomed to failure. Two broad categories can be recognized, the ray-finned fishes (actinopterygians; *actino* = ray, *ptero* = wing or fin) and the lobe-finned fishes (sarcopterygians; *sarco* = lobe).

The actinopterygian fishes we have included in this category are called chondrosteans, holosteans, and neopterygians in various systems of classification. Most have cylindrical bodies, thick scales, and jaws armed with sharp teeth. These fishes seize prey in their mouths with a sudden rush or gulp; they lack the specializations of the jaw apparatus that allow derived bony fishes to use more complex feeding modes.

Teleost Fishes (Actinopterygii)

More than 20,000 species of fishes fall into this category and they cover every imaginable range of body sizes, habitats, and habits, from seahorses to the giant ocean sunfish. Most of the familiar fishes are in this category—the bass and panfish you have fished for in fresh water, and the sole (a kind of flounder) and redfish you have eaten in restaurants. Modifications of the jaw apparatus have allowed many teleosts to be highly specialized in their feeding habits.

Lungfishes and the Coelacanth (Actinistia and Dipnoi)

These are the sarcopterygian fishes. They are the living fishes most closely related to terrestrial verte-

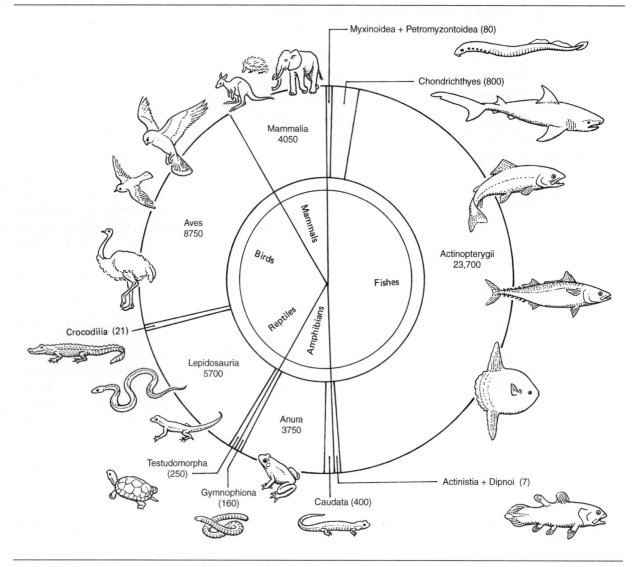

Figure 1–1 Diversity of vertebrates. The areas within the diagram correspond to the approximate numbers of living species in each group. Common names appear within the inner circle, and the formal names for the groups are on the outer parts of the diagram.

brates. The lobe-finned fishes are heavy-bodied and slow-moving. The four species of lungfishes live in fresh water, and the coelacanth is marine.

Salamanders, Frogs, and Caecilians (Caudata, Anura, and Gymnophiona)

These three groups of vertebrates are popularly known as amphibians (*amphi* = double, *bios* = life) in recognition of their complex life histories, which often include an aquatic larval form (the *larva* of a salamander or caecilian and the *tadpole* of a frog) and a terrestrial adult. All amphibians have bare skins (that is, lacking scales, hair, or feathers) that are important in the exchange of water, ions, and gases with their environment. Salamanders are elongate animals, mostly terrestrial and usually with four legs; caecilians are legless aquatic or burrowing animals; and anurans (frogs, toads, treefrogs) are short-bodied with large heads and large hind legs used for walking, jumping, and climbing.

Turtles (Testudomorpha)

Turtles are probably the most immediately recognizable of all vertebrates. The shell that encloses a turtle has no exact duplicate among other vertebrates, and the morphological modifications associated with the shell make turtles extremely peculiar animals. They are, for example, the only vertebrates with the shoulders (pectoral girdle) and hips (pelvic girdle) inside the ribs.

Tuatara, Lizards, and Snakes (Lepidosauria)

These three kinds of vertebrates can be recognized by their scale-covered skin as well as by characteristics of the skull. The two species of tuatara, stocky-bodied animals found only on some islands near New Zealand, are the sole living remnant of a lineage of animals called sphenodontids that were more diverse in the Mesozoic. In contrast, lizards and especially snakes are now at the peak of their diversity. Snakes are nearly as distinctive and widely recognized as turtles, but some lizards are legless like snakes.

Alligators and Crocodiles (Crocodilia)

These impressive animals (the saltwater crocodile has the potential to grow to a length of 7 meters) are in the same lineage (the Archosauria) as dinosaurs and birds. Crocodilians, as they are known collectively, are semiaquatic predators with long snouts armed with numerous teeth. Their skin contains many bones (osteoderms; *osteo* = bone, *derm* = skin) that lie beneath the scales and provide a kind of armor plating. Crocodilians are noted for the parental care they provide for their eggs and young.

Birds (Aves)

The birds are a lineage of archosauromorphs that evolved flight in the Mesozoic. The ability of birds to fly is based on feathers that provide the surfaces that create lift and propulsion, and feathers are the distinguishing characteristic of birds. Indeed, some fossils of *Archaeopteryx*, the earliest bird known, were originally classified as dinosaurs because no marks of the feathers were visible. Birds are conspicuous, often vocal, and active during the day (diurnal). As a result they have been studied extensively and much of our information about the behavior and ecology of terrestrial vertebrates is based on studies of birds.

Mammals (Mammalia)

The living mammals can be traced back through the synapsid lineage to an origin in the late Paleozoic from some of the earliest fully terrestrial vertebrates. Extant mammals include about 4050 species, most of which are placental (eutherian) mammals. Their name comes from the placenta, a structure that transfers nutrients from the mother to the embryo and removes the waste products of the embryo's metabolism. Most of the familiar animals of the world are placentals. Marsupials dominate the mammalian fauna only in Australia. Kangaroos, koalas, and wombats are familiar Australian marsupials. The strange monotremes, the duck-billed platypus and the echidnas, are mammals—they produce milk to feed their young—but the young are hatched from eggs, not born alive like marsupials and placentals.

Evolution

Evolution is the process that has shaped the vertebrate story, and it is the underlying principle of biology. An understanding of the principles and processes of evolution is essential to appreciating the diversity of vertebrates, because that diversity is the direct result of evolution.

Scientific ideas are shaped by the society and philosophical system in which they form, and in turn they may reshape society and philosophy (Box 1–1). That process has been a conspicuous part of the development of evolutionary theory in western societies, most recently in the conflicts over teaching of evolution in schools. Many parts of western thought had their origin in Greek philosophy, but evolution was not among them. The Greek conception of nature was a static one, and the idea of evolutionary change in species of plants and animals was foreign to Greek thought. Platonic philosophy was based on the concept of an abstract ideal; earthly manifestations of that archetype (individual

Table 1–1 **Comparison of the traditional classification of living vertebrates with phylogenetic systematics (cladistics). Extinct forms are omitted. This table does not reflect all of the complexities of the two classifications; see the chapters indicated for details.**

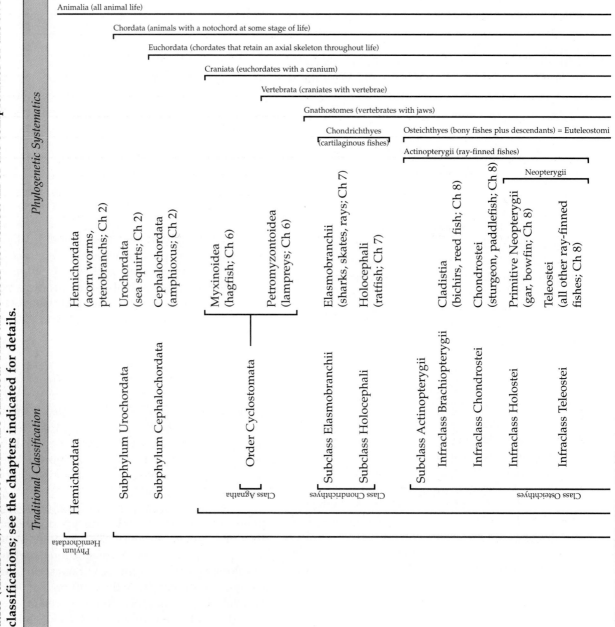

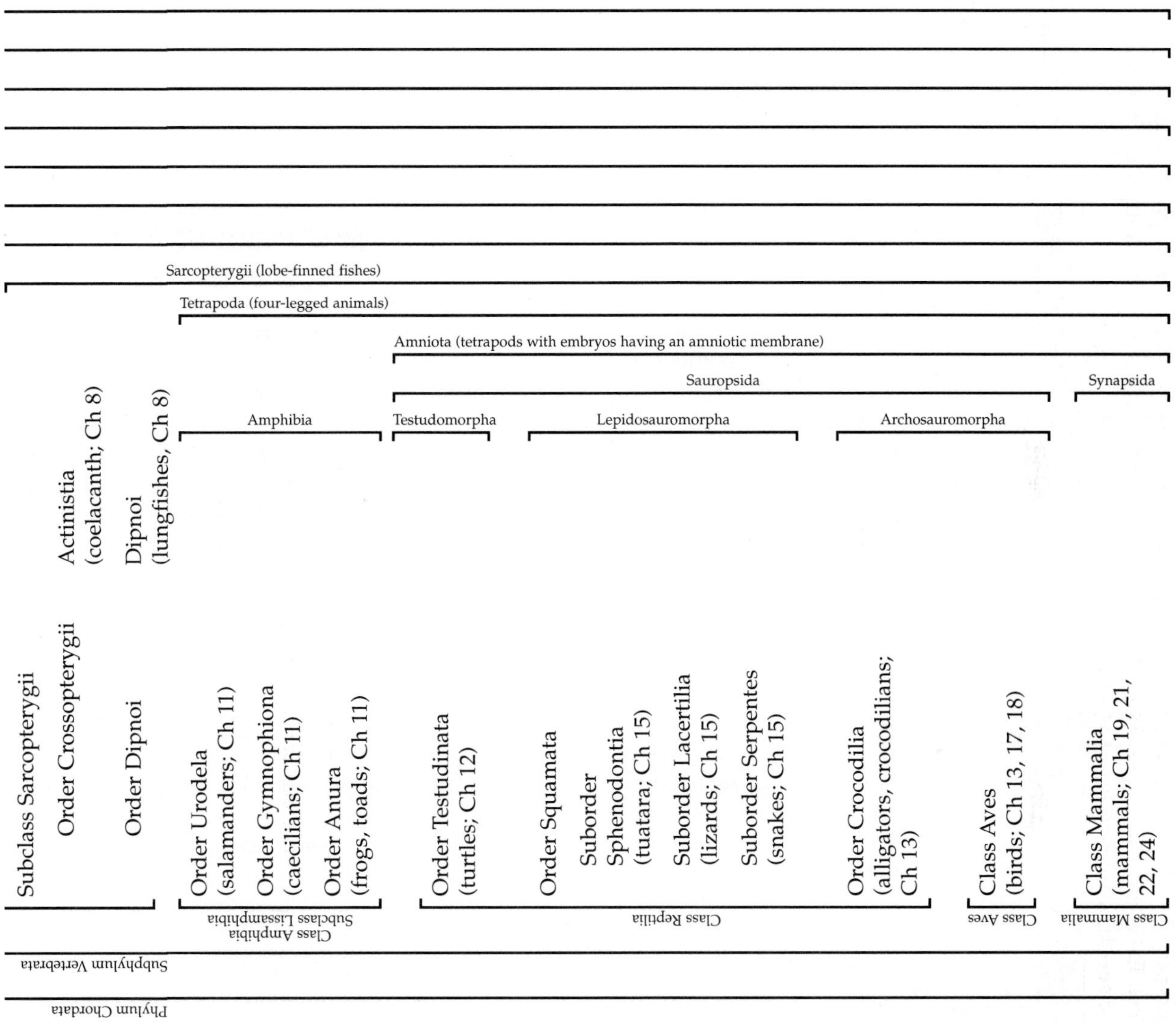

Sarcopterygii (lobe-finned fishes)

Tetrapoda (four-legged animals)

Amniota (tetrapods with embryos having an amniotic membrane)

Sauropsida

Synapsida

Amphibia

Testudomorpha

Lepidosauromorpha

Archosauromorpha

Actinistia (coelacanth; Ch 8)

Dipnoi (lungfishes, Ch 8)

Subclass Sarcopterygii

Order Crossopterygii

Order Dipnoi

Order Urodela (salamanders; Ch 11)

Order Gymnophiona (caecilians; Ch 11)

Order Anura (frogs, toads; Ch 11)

Order Testudinata (turtles; Ch 12)

Order Squamata

Suborder Sphenodontia (tuatara; Ch 15)

Suborder Lacertilia (lizards; Ch 15)

Suborder Serpentes (snakes; Ch 15)

Order Crocodilia (alligators, crocodilians; Ch 13)

Class Aves (birds; Ch 13, 17, 18)

Class Mammalia (mammals; Ch 19, 21, 22, 24)

Subclass Lissamphibia
Class Amphibia

Class Reptilia

Class Aves

Class Mammalia

Subphylum Vertebrata

Phylum Chordata

9

animals or plants, for example) were imperfect copies of its form. This attitude was incorporated into Christian theology as the view that an eternal, inviolable essence of every worldly object existed in the mind of God. A strict interpretation of the Bible insists that everything on Earth was created by God in its present form, and therefore nothing can change or has changed.

This view of life as unchanging was incorporated into biological thought and is still manifested in some biological practices. For example, the Rules of Zoological Nomenclature require that when a new species is named, one specimen of the new species must be designated the type specimen, or **holotype** of the species. The holotype is deposited in a museum, where it serves as a permanent record of the new species. The original role of the holotype was to show what the species was like. If you had another individual that you thought might be a member of the same species, you could compare it to the holotype and decide. Thus the holotype was a single example that was considered to define the essence of the new species. In effect, it was the designated representative of the Platonic ideal of the species.

Naming a new species still requires a holotype, but the role of the holotype has changed significantly, and in addition to the holotype it is usual to deposit several additional specimens called **paratypes** or the **paratypic series**. The paratypes are intended to show the range of variation of the species in such characteristics as body size, color, and pattern. The holotype now defines the species in those rare cases in which two unnamed species have been confused and both are represented in the paratypic series. The name remains with the species represented by the holotype, and the other species receives a new name.

The recognition that variation within species is biologically important is a major change from the idea that one specimen by itself could define a species, and it reflects the most significant change in biological thought that occurred between the writings of Plato (about 380 B.C.) and the work of Darwin and Wallace in the nineteenth century. In a Platonic view, the variations within a species are the imperfections of different copies of the archetype and they have no significance. To an evolutionary biologist, those variations are the raw material of evolution.

Box 1–1 Evolving Views of Evolution

Darwin's book, *The Origin of Species by Means of Natural Selection*, sold out in a single day when it was issued in November 1859. The fervor created by Darwin's theory of evolution continues to the present, and it has affected not just biology, but also the philosophical foundations of Western society. However, Darwin's realization that organisms are not immutable but instead change through time had been preceded by other concepts that the world was not static. The idea, which had prevailed through the dark ages, that Earth was an unchanging entity began to fall with the suggestions of Copernicus and Kepler that the Earth was not the center of the universe. The demonstrations by Newton and others that many physical events followed precise mathematical relationships and, particularly, the undeniable demonstration by paleogeologists that extinct organisms were recorded as

fossils in the rocks placed in question one of the fundamental tenets of Christianity—that all life originated from a divine creation and was consequently immutable.

Thus the idea that organisms were related and, perhaps, even descendents of earlier ancestors was not novel to many biologists in the middle of the nineteenth century. Earlier scientists had voiced the possibility that animals might change or evolve over time. Darwin's theory of evolution was accepted rapidly in the scientific community and within 20 years had become the major framework of biological thought. What features of Darwin's theory of evolution made it scientifically acceptable when earlier theories, such as that of Lamarck, had failed?

First, the evidence presented by Darwin in support of evolution was overwhelming. It covered fields as diverse as comparative anatomy, embryology, paleontology, geology, and animal and plant breeding. This time evolution could not be dismissed as a speculative concept. Darwin documented its reality with explicit logically argued examples. Descent from a common ancestor more readily explained the traits of organisms than a multitude of special creations. Second, to explain the origin of a new species Darwin provided a mechanism, **natural selection**, that was based on the inherent variation that he found so characteristic of species. In a natural population, selective forces, which might be environmental factors such as food availability, temperature, or moisture conditions or biotic factors such as predation or competition between individuals, would favor those individuals that were most fit over those individuals that were less fit in the conditions prevailing at the time. Darwin's concept of favored individuals was not so much survival of individuals as it was survival of the individual's progeny. He clearly saw that variation conferring a selective advantage would be passed on to future generations. Unfavorable variations would be diminished in future generations and ultimately eliminated.

Central to Darwin's theory was the idea that evolution proceeded by the accumulation of small, heritable changes, not large, sudden changes, and that selective forces acted on the individual. Furthermore, it was Darwin's contention that evolution acted without design—heritable traits accumulated randomly and natural selection depended on prevailing conditions.

Of course, Darwin was unaware of the mechanisms of genetic inheritance because Gregor Mendel's work was not reported until 1866 and was not widely disseminated until the early twentieth century. Nevertheless, Darwin realized that in some manner discrete changes or mutations that affected morphology and other aspects of an animal's biology, such as its behavior, occurred in individuals and were inherited. Darwin realized that variation within a species provided the framework on which natural selection could operate to produce new species. Evolution was seen to proceed not only by the loss or culling of unnecessary traits, but also by the selection of randomly accumulated variation in traits (the appearance of new traits by mutation and recombination). New attributes did not arise out of need for them, as postulated by Lamarckian evolution, but through the relentless operation of natural selection on the accumulation of variation among individuals of a species.

Although Darwin's voluminous works led to the rapid scientific acceptance of evolution, his theory of natural selection met with resistance. It was not until the 1920s that accumulating evidence, especially from the newly developing field of genetics, led the scientific community to become supporters of natural selection. The merger of Darwinian selection and genetic theory is referred to as **neo-Darwinism** or the **modern synthesis**, after a book by Julian Huxley entitled *Evolution: The Modern Synthesis*. Many books presented data showing that point mutations and genetic recombination were the source of variation and that evolution (changes in gene frequency) proceeded generally in small steps and resulted from natural selection acting on genetic variation. Such processes were considered sufficient to explain the origin of higher taxa if they acted over prolonged intervals. This view of the evolutionary process is now generally identified as **microevolution** or **phylogenetic gradualism**.

Recently, the hypothesis that evolution proceeds through a slow, constant rate of accumulation of small genetic mutations and/or gene recombinations has been challenged by several biologists who contend that speciation seen in the fossil record may not appear to be gradual but that new species may appear suddenly (Eldredge and Gould 1972, Stanley 1975, Gould and Eldredge 1977, Vrba 1980). Underlying their forcefully presented viewpoint is the fact that a gradual change or transition from one species to another is often missing in the fossil record. A gap frequently exists between recognizably related but distinct forms. Indeed, in the rare case when a species is represented by a long sequence of fossils, its characteristics usually show variation but no directional change of the sort expected if

Continued

Box 1–1 *(Continued)*

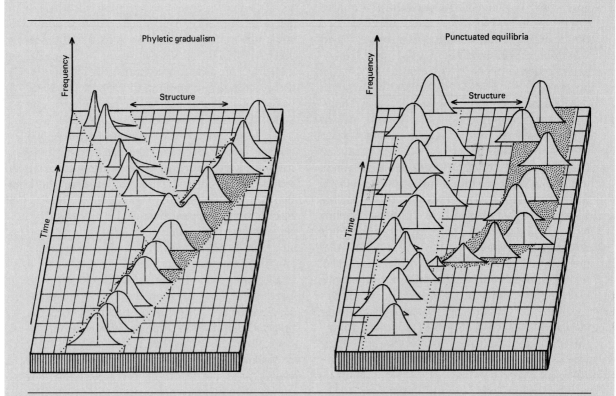

Figure 1–2 Examples of microevolution (or gradual changes) and of macroevolution (or punctuated changes) in species through time. The axes of the graphs plot a character (e.g., bill length) against time. The various curves show the frequency distribution of the character within the population at specific times. The tendency to speciate, or at least to diverge in that character between populations, is shown by the separation into two lineages. A speciation event occurs when the two populations become reproductively isolated as indicated by shading. (Modified after E. S. Vrba, 1980, *South African Journal of Science* 76:61–84.)

natural selection were operating. Rather than proceeding through a steady rate of accumulation of small changes in structure, physiology, and behavior, evolution appears to alternate between periods of rapid change and periods when little or no change occurs (Figure 1–2).

The competing theories of micro- and macroevolutionary processes of speciation have become popularly referred to as **gradualism** and **punctuated equilibrium**. Gradualists would expect a species to accumulate structural changes even in a more or less stable environment, whereas punctuationalists would expect a

species to remain in structural equilibrium unless the environment changed significantly. Central in both views is the fact that environment, although broadly stable over reasonable periods of time, does oscillate continually and therefore incessantly stresses each individual.

Punctuated equilibrium provides an explanation for the existence of a recognizable species through time. If species arise suddenly through rapid genetic structural adjustments and then remain in stable equilibrium until the next punctuation, they represent distinct entities with a proscribed structure and period of existence.

Biological Variation

The idea that no two people (except for identical twins) are exactly the same is a familiar one. Humans use individual variation every day in contexts that range from recognizing friends to fingerprinting criminal suspects. Individual variation is not confined to morphological variation like facial contours or fingerprints; it extends to any characteristic that can be measured.

These sorts of variation are examples of **phenotypic** variation. The phenotype of an organism means its form in a very broad sense. For example, phenotype can refer to color (pale skin versus dark skin), size or shape (tall versus short), or performance capacity (fast versus slow runners). Phenotypes can also be defined on the basis of molecular characteristics: Hemoglobin B is the normal form of the beta chain of the adult human hemoglobin molecule and hemoglobin S is the form of the beta chain that leads to sickle-cell anemia. The two forms differ in only one amino acid residue: A valine appears in hemoglobin S in place of a glutamine.

Behavioral phenotypes also can be defined: Guppies are most familiar as aquarium fish, but they occur wild in streams in the New World tropics. In Trinidad guppies occupy stream habitats with relatively few predators and other streams where predators are abundant, and populations in these habitats differ in male color and in the response of females to courtship by bright-colored and plain males (Breden and Stoner 1987). The bright colors of male guppies are part of the courtship display, but they also make males more visible to predators. In low-predation areas most male guppies are bright colored, but in areas of high predation they are plain. Experiments have shown a genetically determined difference in the responses of female guppies from streams with high and low predation to courtship by bright and plain males: Female guppies from low-predation habitats choose to mate with bright-colored males, whereas females from high-predation areas choose plain males. This response is apparently mediated by the chances that the offspring will survive. Bright-colored males sire bright-colored male offspring. In low-predation habitats, bright-colored male offspring are advantageous because they are attractive to females and run little risk of being eaten by a predator. However, in high-predation habitats bright-colored male offspring are vulnerable to predation, and females with the genetically determined behavioral trait of mating with plain males are more successful.

These types of phenotypic variation are manifestations of genetic variation. Evolution is defined as change through time in the frequencies of different forms of a gene in the gene pool of a population. Different forms of a gene are called **alleles** and different alleles produce different phenotypes. The **genotype** of an individual is its genetic composition. Remember that in sexually reproducing organisms an individual receives one allele from its mother and one from its father. To use the example of sickle-cell anemia, there are three possible genotypes: If an individual receives the normal form of the beta-chain allele from both parents it will be $Hb_B Hb_B$, that is, a **homozygous genotype**, and it produces the normal beta-chain phenotype. An individual that receives one allele from its mother and a different allele from its father would be $Hb_B Hb_S$, that is, a **heterozygote**. Heterozygotes for the sickle-cell allele have an advantage in areas where malaria is endemic, apparently because red blood cells that contain hemoglobin S collapse (sickle) when a malaria parasite enters them, and this sickling causes a decrease in the concentration of potassium inside the red cell and the eventual death of the parasite (Friedman 1978). An individual that receives the sickle-cell allele from both parents will have the genotype $Hb_S Hb_S$: These homozygous individuals are at a disadvantage because the hemoglobin molecules sickle whenever the oxygen concentration in the blood drops, as it may, for example, during physical exercise.

Evolution depends on the existence of variation in genotypes among individuals that is (1) manifested in the phenotype, (2) inherited, and (3) associated with differences in reproductive success. Natural selection changes the relative frequency of different alleles in a population, and these changes in alleles are reflected in changes in the frequencies of different phenotypes. It is easier to think of phenotypes in discussions of evolution because phenotypes are visible, whereas alleles are not. However, it is alleles that are inherited.

Variation in genotypes results from the combined actions of several genetic processes in sexually reproducing organisms. During meiosis and the formation of gametes (sperm and ova) the pairs of chromosomes that make up the genome of an individual are first duplicated and then separated into gametes that contain only one of each pair of chromosomes. (This is the **haploid** chromosomal complement.) When two gametes unite to form a zygote (fertilized egg) the **diploid** chromosome number is regained: Two chromosomes of each pair are present, one from the father and one from the mother. Alleles are not entirely independent entities; their function is affected by other alleles present in the genotype. Consequently, the genetic **rearrangement** that accompanies sexual reproduction is an important source of phenotypic variation. **Mutation** is another source of variation, of smaller magnitude than rearrangement but nonetheless important. Mutation can result from errors in copying the genetic code during the initial duplication of chromosomes during meiosis, or it can be caused by events that precede meiosis, such as exposure of the sex cells to ionizing radiation. Other cellular events can cause changes in the linear sequence of genetic loci on chromosomes. Because the action of alleles is influenced by neighboring alleles, these changes in sequence can affect the phenotype.

Natural Selection

Evolution results from the action of natural selection on different phenotypes, and natural selection works through **differential reproductive success**, that is, the contribution of different phenotypes to the gene pool of succeeding generations. The term **fitness** is used to describe the relative contribution of different individuals to future generations. If individual A produces 100 offspring that survive to reproduce and individual B produces only 90 offspring, individual B is only 90 percent as successful as individual A. The fitness of the most successful individual (A in this example) is defined as 1.00, and the fitness of other individuals are defined in relation to A. Thus the fitness of B would be 0.90. That means that the alleles represented in the genotype of individual B would be underrepresented in the next generation by 10 percent compared to alleles in the genotype of A. In this example, selection would be said to select against B and the selection coefficient would be 0.10.

Modes of Selection

Natural selection can affect the distribution of phenotypes in a population in three different ways (Figure 1–3). **Directional selection** discriminates against individuals at one extreme of the variation in a phenotypic character. For example, small individuals might be at a disadvantage compared to individuals of normal or larger-than-normal size. The effect of directional selection is to move the mean value of the character in the direction of the most fit phenotypes from generation to generation.

Stabilizing selection discriminates against individuals that have extreme variation in the phenotypic character in either direction. It favors individuals that have values that lie close to the mean, and it can result in a reduction of the amount of variation within a population, but it does not change the mean value.

Disruptive selection is the reverse of stabilizing selection. That is, individuals at both extremes of the range of variation are favored and individuals near the mean are at a disadvantage. Disruptive selection increases the variation but does not change the mean value.

Note that these descriptions have assumed that only one mode of selection affects the trait being considered. In fact, more than one mode of selection can occur simultaneously, and both the mean value of the trait and the amount of variation can change simultaneously. Natural selection is difficult to demonstrate under natural conditions because many different and sometimes conflicting selective forces operate simultaneously. Nonetheless, some examples are known, and John Endler (1986) has reviewed the evidence for natural selection in the wild.

Levels of Selection

One of the active areas of discussion and research in evolution is the question of the levels at which selection acts. Evolution can be defined as a change in the frequencies of alleles, but does selection act directly on alleles, or on phenotypes (that is, individuals) that have those alleles as part of their genotype, or at still higher levels of biological organization?

The best-known examples of selection act through differential reproduction of individuals, but other kinds of selection are possible. **Genic selection** describes selection for individual alleles. Normally, a heterozygous individual produces equal numbers of gametes with each of its two alleles. However, situations can occur in which one allele is transmitted to more than half of the viable gametes of a heterozygous individual. This phenomenon is called **segregation distortion** or **meiotic drive**. In that situation the allele that is overrepresented in the viable

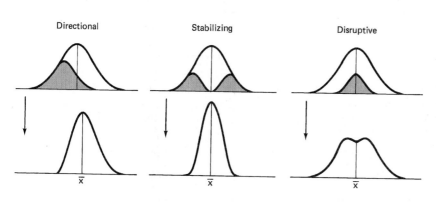

Figure 1–3 Three kinds of selection: directional, stabilizing, and disruptive. In each case the vertical axis is the proportion of individuals and the horizontal axis shows the range of variation. The outlines of the upper diagrams show the initial variation within the population. The individuals in the shaded areas are at a disadvantage relative to the rest of the population. The lower diagram shows the variation in the populations after selection ($\bar{x}$, average value).

zygotes is more fit (leaves more offspring) than the underrepresented allele.

Interdemic selection is another possibility. A **deme** is a local population of a wide-ranging species. Usually, the frequencies of alleles differ slightly from deme to deme. If populations that differ in allelic frequency become extinct at different rates or give rise to new populations at different rates, selection might act on all the alleles in a deme.

That idea can be extended to consider that different species might have characteristics that make them more or less likely to become extinct. For example, species with small geographic ranges may be wiped out by local changes in the environment, such as floods or droughts, whereas species with wide geographic ranges are less vulnerable to the effects of local changes. The same line of reasoning can be applied to other characteristics of a species—narrow versus broad choice of habitats, for example, or specialized versus generalized feeding habits. The rate of speciation might also be subject to selection. The assumption here is that an evolutionary lineage that forms new species rapidly is less likely to become extinct than is a lineage that forms new species slowly. If those characteristics of species are genetically determined, they might be subject to **species level selection**. This is still a controversial hypothesis; further discussion can be found in Jablonski (1986).

Evolution of Complex Systems

Organisms are extraordinarily complicated entities, and that complexity is one of the principal reasons they are so fascinating. We try to illustrate that perspective throughout this book by emphasizing the ways in which ecology, behavior, morphology, and physiology interact in living vertebrates, and by considering how similar interactions in the past may have influenced the evolution of vertebrates.

How do such complex systems evolve? Three general explanations have been proposed: saltation, mosaic evolution, and correlated progression. The **saltation** hypothesis suggested that the mixture of characters that distinguish a mammal from a reptile, for example, could have been acquired all at once as a result of the random effects of genetic mutation and recombination. This hypothesis, which is associated with the early part of the twentieth century, is not supported by either our current understanding of developmental mechanisms or the fossil record.

The hypothesis of **mosaic evolution** and the related hypothesis of **correlated progression** are more consistent with what we see during the embryonic development of vertebrates and in the fossil record. It is a commonplace observation that organisms are mosaics of ancestral and derived characters. For example, your hand (= forefoot) is very much like the forefoot of a lizard, but your foot (= hind foot) is quite different from a lizard's. Similarly, your eyes are essentially the same as those of a lizard, but your brain shows tremendous elaboration of some regions compared to the brain of a lizard. The distinction between mosaic evolution and correlated progression focuses on how much interaction occurred among the blocks (i.e., forefoot, hind foot, eye, brain) during evolution. Independent evolution of the blocks because of coincidental response to the same new environment is mosaic evolution, whereas an interaction in which changes in one block influence changes in other blocks is correlated progression.

Mosaic evolution and correlated progression are two ends of a continuum rather than being mutually exclusive alternatives, and particular examples may fall at various points along the line. A particularly good example, and one that is treated in more detail later in this book, is the evolution of mammals from early reptilelike animals. That transition involved a host of changes among interdependent systems. Endothermy (i.e., the regulation of body temperature by heat produced by an animal's metabolism), for example, requires a physiological mechanism for heat production (high metabolic rate) and a morphological mechanism for retaining heat (an insulating covering of hair), and neither feature is advantageous unless the other one is already present. This catch-22 situation is discussed in Chapter 4, and the sequence of appearance of the changes in skeletal structure, dentition, maternal care, and brain development that distinguish mammals from reptiles is described in Chapter 19. In the case of mammals, we clearly see a process of correlated progression, in which changes in skeletal structure were associated with changes in locomotion, physiology, reproduction, and parental care.

Inclusive Fitness

We have defined fitness as the relative genetic contribution of different individuals to future generations, and described it in terms of the number of offspring produced by an individual. That is an oversimplification in the sense that individuals of a species have alleles in common, and the more closely related two individuals are, the higher will be the proportion of their shared alleles. Siblings, for example, have an average of 50 percent of their alleles in common and half-siblings (one parent the same and one different) share 25 percent of their alleles. Remember that it is alleles that are transmitted from generation to generation, and you can see that it is possible for an individual to increase its own fitness (that is, to transmit alleles identical to its own) by assisting a related individual to reproduce successfully. For example, when your sibling reproduces, he or she transmits half the alleles you share to that offspring (which is your niece or nephew), and 25 percent of the alleles in the genotype of the offspring are the same as alleles in your genotype (50 percent of alleles shared between siblings × 50 percent of the genotype contributed by each parent = 25 percent of the alleles in the genotype of the offspring). Your

own offspring has 50 percent of the alleles in your genotype, twice as many as your niece or nephew. All else being equal, then, two nieces or nephews are the equivalent of one offspring of your own in terms of transmitting your alleles to future generations. The same calculations can be applied to increasingly distant genetic relationships. The important point is that reproduction of relatives contributes to the fitness of an individual. In some situations helping relatives to reproduce successfully may be the best way for an individual to increase its own fitness. This principle of **inclusive fitness** is believed to underlie many of the altruistic behaviors of birds and mammals (Chapters 18 and 23).

Variation and Evolution

Phenotypic and genotypic variation is the material on which natural selection operates, but not all kinds of variation are subject to selection, and characteristics of the biology of certain species can increase or decrease the importance of selection. In the first place, variation must be **heritable** if selection is to act on it. That is, variation in phenotypic characters must have an underlying genetic basis, and parents that manifest a particular character must produce offspring that also show that character. Not all variation does have a genetic basis, and many characters that do have a genetic basis are also affected by the environment. The body size of turtles at hatching is an example of this interaction between genetic and environmentally induced variation. Newly hatched turtles receive no parental care: They must dig their way out of nests in the soil, find their way to water, and begin to catch their own food entirely by their own efforts. Probably it is desirable for a female turtle to produce the largest hatchlings she can, because larger hatchlings can dig more strongly, move to water faster, and capture a wider variety of prey than can small individuals. The size of a hatchling turtle is correlated with the size of the egg it came from, and some female turtles lay larger eggs than others. All else being equal, female turtles that lay large eggs should produce larger hatchlings and be more fit than females that lay smaller eggs. However, the availability of water in the nest also affects the body size of hatchlings—nests in moist soil give rise to larger hatchlings than do nests in dry soil (Figure 1–4). Furthermore, hatchlings from moist nests can also crawl and swim faster than hatchlings from dry

Figure 1–4 These two hatchling snapping turtles (*Chelydra serpentina*) emerged from eggs laid by the same female turtle. The eggs were the same size when they were deposited, and the difference in the size of the hatchlings represents an environmental effect. The turtle on the left hatched from an egg that was incubated in moist substrate, whereas the one on the right is from an egg incubated in dry substrate. (Photograph courtesy of Gary C. Packard.)

nests. This component of variation in body size of hatchlings has no genetic basis and cannot be acted on directly by natural selection. If selection does act on this type of variation (and this sort of selection has not been demonstrated) it would have to be indirect, via a reduction in the fitness of female turtles that chose to nest in dry sites.

Variation That Is Subject to Natural Selection

Some female turtles have a genotype that results in their putting a large quantity of yolk into each egg so

that they produce large eggs, that is, **individual variation**. Individual variation is the type of variation one usually thinks of in terms of natural selection—some individuals are larger than others, or faster, or more colorful, or more aggressive. The list of characters can be extended indefinitely, and those kinds of differences result from the luck of the draw when gametes combine to form a zygote. The effects of genetic recombination and mutation lead to different genotypes and different phenotypes, and selection can act on those differences.

Individual variation is usually continuous. That is, some animals are small, others are large, and most are somewhere between those extremes. Continuous variation is the situation illustrated in Figure 1–3. A second type of variation is discontinuous: Instead of having a bell-shaped curve of frequencies of different forms, discontinuous variation can be sorted out into discrete categories. **Polymorphism** (*poly* = many, *morph* = form) is a common type of discontinuous variation. Figure 1–5 illustrates pattern polymorphism in a Puerto Rican frog, the coquí. Some individuals are uniformly colored, some have stripes along the sides of the body, and some have a mottled pattern of dark blotches. These different patterns are genetically determined and offspring inherit the patterns of their parents. One can find individuals with all those patterns in any population of the frogs, but some patterns are more common in one kind of habitat than another, probably because certain patterns are especially cryptic (hard for predators to detect) in particular habitats. In grasslands, for example, the striped patterns occur in high frequency, whereas in forests the unicolored and mottled patterns are more common. In grassy areas predators see the frogs against a background of grass stems, and the striped pattern may blend well with the straight, light-colored stems. In forests predators often see the frogs on a background of fallen leaves; there are few sharp, straight lines in a forest and the mottled and solid patterns of the frogs are probably more cryptic than stripes. The differences in frequencies of the patterns in different habitats indicate that natural selection has acted on the gene pool to change the frequencies of alleles producing the patterns.

Figure 1–5 Polymorphic patterns of the Puerto Rican coquí. These are four of the more than 20 different patterns that have been described: (a) mottled; (b) unicolor; (c) broad lateral stripes; (d) broad dorsal stripe. Striped patterns (such as c and d) are more common in grassland habitats than they are in the forest. (Photographs by Margaret M. Stewart [a, b, d] and F. Harvey Pough [c].)

Sexual dimorphism (*di* = two) is a special case of polymorphism in which males and females of a species differ in secondary sexual characters such as size or color. Sexual dimorphism results from the different roles that males and females play in reproduction and the differences in the selective pressures that affect the fitness of individuals of the two sexes. The difference in sexual roles begins with the investment of males and females in gametes: Males produce sperm, females produce ova. Sperm are small, energetically cheap, and quick to produce, whereas ova may contain large quantities of yolk and require a substantial investment of time and energy by the female.

As a result of this asymmetry of investment, the reproductive strategies that increase the fitness of males and females can be quite different. Because a male can readily produce more sperm he can usually increase his fitness by mating with as many females as possible—each mating has only a small cost and he can mate repeatedly with little delay between matings. The situation for the female is quite different: By the time she is ready to mate she has invested substantial energy and time in producing mature ova. Furthermore, in many species the female provides most of the parental care. As a result, it is costly in time and energy for a female to breed repeatedly. Instead, her fitness is increased by mating with the best possible male.

Sexual dimorphism often reflects these differences in the selective forces acting on males and females: In *The Descent of Man and Selection in Relation to Sex* Darwin pointed out that males are often larger than females and more aggressive. They may engage in ritualized or actual combat with other males, and may form dominance hierarchies in which high-ranking males have opportunities to court females and low-ranking males do not. Darwin noted that females often must be courted by a male before they will mate. Many secondary sexual characteristics of males, such as bright colors, the presence of horns or antlers, and vocalization, are examples of sexual dimorphism that are employed in combat or courtship. The role of female choice in determining the reproductive success of individual males has received increased attention recently. The bright-colored and plain-colored Trinidad guppies discussed in the preceding section provide an example of the role of female choice in determining the reproductive success of males, and additional examples are found in subsequent chapters. Eberhard (1985) has suggested that the enormous elaboration of male genitalia of vertebrates and invertebrates reflects sexual selection operating via female choice for males that are especially good at stimulating females during courtship.

Factors Promoting Speciation

Evolution consists of changes in frequencies of alleles within populations. If two populations of a species accumulate more and more differences in allelic frequencies over time, they become more and more distinct. Some alleles may be lost from one population but retained in the other population, and mutation may introduce new alleles into each population that are not represented in the other. How long can this process continue before the populations are no longer one species but two?

To answer that question we need a definition of a species, and that is a contentious issue among biologists. For sexually reproducing organisms, which includes the overwhelming majority of vertebrates, the definition of a **biological species** that was formulated by Ernst Mayr (1942) has been remarkably durable: "Species are groups of interbreeding natural populations that are reproductively isolated from other such groups." Critics of Mayr's biological species definition have pointed out that it does not encompass unisexual organisms, or geographically isolated populations that are prevented from interbreeding by a barrier such as a river or a mountain range. Also, the biological species definition does not include the idea of genealogical continuity over time.

The **evolutionary species** concept, first defined by George Gaylord Simpson in 1961 and modified by E. O. Wiley in 1978, defines a species as "a single lineage of ancestral-descendant populations which maintains its identity from other such lineages and which has its own evolutionary tendencies and historical fate." Frost and Hillis (1990) reviewed concepts of species and proposed a modification of the evolutionary species concept. They proposed that defining species as "the largest entities that have evolved whose parts, if distinguishable, are not likely to be on different phylogentic trajectories" would facilitate studies of biogeography and the evolution of characters, ecological specializations, and reproductive incompatibility.

All of these definitions of species emphasize that a species is the expression of a genetic system that

Figure 1–6 Examples of dimorphism. (a) A mating pair of European toads (*Bufo bufo*). Males of this species, and of many other amphibians, are substantially smaller than females. (b) African lions (*Panthera leo*). Among mammals, males are usually larger than females. (Photographs: [a] © K. G. Vock/OKAPIA, [b] © David Hosking.)

evolves independently from other species. This view of species stresses the fact that members of a species not only evolve independently of other species but also must be able to survive in the face of competition from members of other species.

The Role of Isolation in Speciation As long as populations are in contact, alleles can move among them. Even in species with large geographic ranges, **gene flow** maintains genetic continuity between populations at the extremes of the range of the species via intervening populations. A new allele created by a mutation in one part of the species range can eventually spread through the entire species, and an allele that is lost in one part of the range will eventually be replaced by gene flow from the gene pool of the species. Gene flow is both a constraining and a creative force in evolution (Slatkin 1987). On one hand, gene flow prevents local populations from accumulating enough genetic differences to evolve into different species, but at the same time gene flow allows superior alleles and combinations of alleles to spread through an entire species.

Allopatric speciation describes a situation in which a physical barrier such as a river or a mountain range prevents gene flow and allows two populations to diverge. **Sympatric speciation** describes the separation of a single population into two species without such a physical barrier. In this situation, ecological or behavioral differences between subdivisions of the population restrict or eliminate gene flow.

The Importance of Population Size in Evolution How speciation might proceed in large interbreeding populations has perennially puzzled evolutionists. A large population that occupies extensive areas can usually be divided into subpopulations, or demes. Large continental areas are seldom homogeneous, but rather are a series of patchy habitats because of variation in geology and vegetation. Members of a deme are more likely to breed with members of their own deme than with members of adjacent demes, if for no other reason than proximity. The result is restricted gene flow within the species and, presumably, a reduction in heterozygosity or genetic variation in the deme. If the deme is large the effect of interbreeding will be slight, but if it is small the reduction in genetic variation may be significant. Occasional outbreeding with individuals from other demes will tend to maintain heterozygosity within the deme: An average of one individual exchanged between populations per generation is sufficient to prevent divergence.

Superimposed on this concept of semi-isolated demes with limited outbreeding is the amount or the intensity of natural selection. If, for example, environmental change, predation, or competition for resources is intense within a deme, individuals with the most appropriate phenotypes for the local conditions will be favored and their genotypes will increase within the population. If gene flow between demes remains low, genetic divergence of demes can occur.

Perhaps more critical is the mechanism of **genetic drift** and its relationship to evolutionary processes. Genetic drift is a random process that can lead to the loss or the fixation of alleles in any population. The process is more rapid in small populations than in large ones. Genetic drift can lead to the loss of genetic variation in populations and divergence between populations by chance alone. Evolution, therefore, represents a delicate balance between mechanisms—if a deme is small genetic drift can reduce variation, as can intense selection. Genetic drift, however, is random with respect to any allele and natural selection is not.

Gene flow prevents a species composed of freely interbreeding populations from accumulating enough genetic differences among populations to speciate. Some kind of block to the free exchange of genes must be interposed between populations of a species before it can split into daughter species. In other words, some kind of spatial or **geographic isolation** is generally required before **reproductive isolation** can evolve between segments of a population. This process is called **allopatric speciation** (*allo* = different, *patris* = native country).

The rate of evolution is most likely to be rapid in small populations, and small breeding populations can result from various circumstances including dispersion of a few adults to a new location, the sudden local isolation of a new deme by some environmental event, or a sudden reduction in the size of a population (a population crash). These periods of diminished population size are referred to as **bottlenecks**, and they potentially stimulate rapid changes in the frequencies of alleles in the population. Furthermore, the small population resulting from the bottleneck will contain only those alleles present in the individuals that pass through the bottleneck, and these are the only alleles that will be represented in the descendant population. A similar phenomenon, called the **founder effect**, describes a situation in which a new isolated population is founded by a few individuals. For example, a flock of birds might be blown out to sea by a storm and ultimately land on an island. If conditions on the island are suitable, the waifs could breed and establish a new population that will contain only the alleles represented in the genotypes of the founding individuals. This is the sequence of events that is thought to have produced the bird faunas of many oceanic islands, including Hawaii and the Galapagos Islands.

A Classic Example of Evolution: The Galapagos Finches How geographic isolation promotes speciation is emphasized on oceanic islands, as Darwin himself observed in his examination of the endemic forms of animal life in the Galapagos Islands, which lie about 1000 kilometers off the coast of Ecuador. Darwin's finches (subfamily Geospizinae) have become the classic example of evolution at the species level (Figure 1–7). These 14 species have evolved rapidly from a single, ancestral population of seed-eating birds that arrived from South America in relatively recent geological time—perhaps less than 500,000 years ago.

From time to time individuals spread to other islands in the Galapagos from the original, founding population. In these new environments, with ecological opportunities for which no competition from other species existed, these new founders were able to repeat the process of establishment and eventual adaptation to local conditions. In time many of these isolated populations accumulated sufficient genetic differences so that, when secondary contact was established on islands that had already been colonized, they proved to be reproductively isolated from one another and could coexist as sympatric species, each with an ecological niche of its own (Grant 1986). Today, each island has as many as three to ten species, depending on the diversity of the vegetation and ecological opportunities.

Studies of diversity among the finches reveal that morphological variation has emphasized change in body size and in the shape of the bill. Three primary subgroups are recognized: the warbler-like finches, the tree finches, and the ground finches. Differences in beak size and shape represent specialization for different diets and ways of obtaining food.

Investigations of genetic differences among the finches produce species relationships that are basically similar to the arrangement proposed by David Lack in 1947. Although evolution has been rapid, allopatric speciation seems the simplest explanation for the diversity of Darwin's finches, especially when one considers the relatively high chances of dispersal of a founder population to a nearby island. A study by Gibbs and Grant (1987) revealed how rapidly the morphology of a species of finch can change in response to changes in the environment. The Galapagos Islands are characterized each year by wet and dry seasons. In exceptional years, however, the wet season can be prolonged and bring heavy rain. When that happens the plants flourish and food becomes abundant. These exceptional years are the result of climatic effects that reach far beyond the Galapagos—they are El Niño years, the name given to the periodic invasion of the eastern South Pacific by warm water masses. In 1982–83, a decade-long drought in the Galapagos was broken by a record El Niño year, the most intense in recent history. Gibbs and Grant had been monitoring a population of the medium-size ground finch *Geospiza fortis* on the small island of Daphne Major. The morphology of the birds changed between 1976–77 and 1984–85. Before the El Niño event, the population consisted of large-bodied and heavy-billed birds, but during and following the wet season both body size and bill size diminished. The differences in body and bill size have a genetic basis, and they also relate directly to success in feeding. In a drought, when food is scarce, the larger birds apparently are favored because they can crack the large, hard seeds that are the most abundant food for the finches. During wetter times the small birds are apparently successful because small, soft seeds are abundant. Here is a clear example of changing environmental conditions leading to the selection of genetically controlled morphologies that favor survival. The genotype of these finches must carry enough heterozygosity to express either phenotype.

For land-based animals such as finches, oceanic islands are obvious geographic features for the isolation and speciation of populations. The ocean is a barrier to dispersal from a continental source, but it is not an absolute barrier. Habitats on mountaintops are isolated from each other in much the same way as islands. It is easy to see how a surrounding lowland forest or desert may be an effective barrier to frequent dispersal of animals that are adapted to live only in the highland situation.

The rapid evolution of numerous related species, such as Darwin's finches, each of which represents a different adaptive type, is a recurrent process in the evolution of living organisms. It has been documented in all classes of vertebrates from fishes to mammals. Usually, the rapid evolution of groups of species results from new environmental

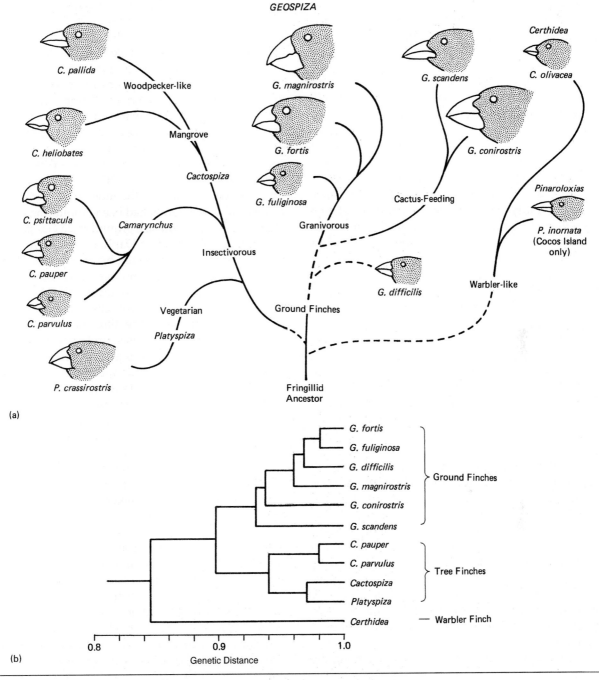

Figure 1–7 Differences among Darwin's finches and the likely rate and sequence of evolution. (a) The evolutionary sequence postulated by David Lack (1947, *Darwin's Finches*, Cambridge University Press, Cambridge, UK) was based mostly on morphological criteria. It is very similar to the phylogeny for ten of those species based on a genetic analysis of similarity (S. Y. Yang and J. L. Patton, 1981, *Auk* 98:230–242) as shown in (b). Differences occur only within the ground finches.

Box 1–2 Tempo in Evolution: The Molecular Clock Hypothesis

The fossil record of vertebrates is incomplete, especially in older deposits, because time and geologic events have folded, metamorphosed, and eroded away many of the fossil-bearing rocks. Of necessity, therefore, estimates of the tempo of vertebrate evolution have been based largely on groups for which the geological record is most complete. In spite of discontinuities, the fossil evidence documents that some kinds of vertebrates have evolved in spurts. Some paleontologists describe vertebrate evolution as a dynamic process in which speciation proceeds rapidly at some times and slowly at others. This oscillation in the rate of evolution is the theory of punctuated equilibrium.

It was surprising, therefore, when Zuckerkandl and Pauling (1962) postulated that substitutions in the amino acid sequences of certain proteins of vertebrates occurred at a constant rate. If this hypothesis is correct, it could be an important tool for evolutionary research because it would be possible to estimate the time that the phylogenetic lineages of two groups of animals separated by measuring the number of amino acid substitutions in their proteins. For example, if the amino acid composition of hemoglobin is compared among different taxa, the number of amino acid differences between any pair of them should be proportional to the time of evolutionary divergence if the differences have accumulated at a constant rate. The greater the amino acid differences, the farther back in time the taxa separated. Furthermore, because changes in amino acid sequences in proteins reflect changes in the nucleotide sequences of DNA, they indi-

rectly measure occurrences of point mutations in the gene complex. This hypothesis, called the **molecular clock**, has subsequently been extended to include the rates of change in the base pairs of DNA from the nucleus and from mitochondria.

If the molecular clock hypothesis is correct, a close correlation should result when the number of amino acid substitutions for proteins of different vertebrates are regressed against their time of divergence. This relationship is evident for some species: The blood serum albumins of numerous pairs of mammals have been tested for immune cross reactions to estimate the number of amino acid substitutions among the albumin molecule's approximately 580 amino acids. When the proteins from different pairs of mammalian carnivores and ungulates are examined for immune response and the immunological differences are plotted against divergence time of each pair of taxa, as estimated from the fossil record, a good correlation is obtained (Figure 1–8).

However, further studies have shown that the molecular clock does not necessarily keep good time: Different proteins evolve at different rates, and nuclear and mitochondrial DNA have different rates of evolution (Vawter and Brown 1986). Histones and structural proteins like collagen are conservative and show only low rates of substitution, whereas proteins involved in maintaining homeostasis, like hemoglobin, and in providing protection, such as the immunoglobulins, which serve as antibodies against foreign antigens, tend to evolve very rapidly. A

---

opportunities, each of which requires a different set of adaptations for successful exploitation. This phenomenon is called **adaptive radiation**. However, evidence from studies of protein evolution, which reflects genetic change, lends support to the view that molecular evolution proceeds at a fairly constant rate, especially when considered over long periods of time (Box 1–2).

Isolation on Continents It has not been so easy to explain how speciation has occurred on continents, particularly in regions where environmental conditions remain uniform over vast areas, such as in the tundra and boreal forests of the Northern Hemisphere or the Amazonian forest of South America. It now appears that a mechanism similar to isolation on islands has been at work repeatedly and that

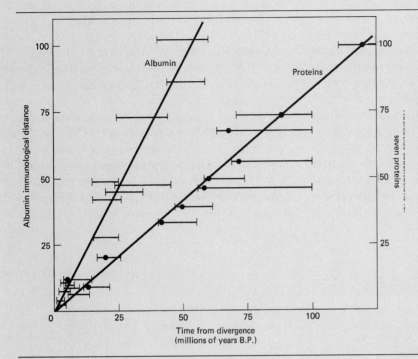

Figure 1–8 Constancy in the rate of molecular evolution of mammalian proteins. Left: Rates of albumin evolution in carnivorous and hoofed mammals. Right: Nucleotide substitutions since the time of divergence in living pairs of mammals. Closed symbols are the mean values for seven different proteins. Horizontal lines through the closed symbols show the estimated ranges of the times of divergence for each species pair. (Modified from C. H. Langley and W. M. Fitch, 1974, *Molecular Evolution* 3:161–177, and A. C. Wilson et al., 1977, *Annual Review of Biochemistry* 46:573–639.)

heavy-chain immunoglobulin with 500 amino acids residues could substitute as many as six to seven residues per million years. Furthermore, the rate of change in DNA varies by a factor of 5 among different phylogenetic groups, and within a single group, the rate of DNA evolution changes with time (Britten 1986). For example, birds have lower rates of DNA evolution than do rodents, and the rate of change in DNA among primates has decreased during their evolution.

Thus, the molecular clock hypothesis has not entirely fulfilled the expectations of its early proponents. There is no general rate of change in DNA that can be applied to all phylogenetic lineages over the entire course of evolution. However, measurements of biochemical differences within particular lineages over shorter periods have been immensely valuable in providing information about specific evolutionary relationships. Some of the most intriguing and controversial examples come from the application of molecular techniques to studies of the origin of humans (Lewin 1988a,b).

much speciation of land vertebrates on continents can be accounted for by geographic isolation.

The Pleistocene period—approximately the last 2 million years of geological history—has been characterized by changes in mean annual temperature and rainfall. Geological and biogeographic evidence supports the conclusion that island-like refugia of the tundra and taiga ecosystems persisted in northern latitudes even during maximum glaciations, whereas during interglacials these refugia have merged as continuous habitats of great expanse. Thus, species populations that were widely distributed in these uniform habitats during interglacials became fragmented and isolated in refugia during glaciations. This is precisely the sequence of events needed for speciation, and the present distributions of a number of bird and mammal species in boreal North America indicate that they have evolved in this way.

The changes in temperature and precipitation during the Pleistocene have been great enough to have effected changes in most biomes of the world, including those in the tropics. The Amazonian forest has apparently been broken up repeatedly into patches that persisted for a time as isolated forests and then coalesced to form a continuous biome again (Figure 1–9). Periods of drought have long been considered the cause of this disruption of continuous forest, but a recent interpretation suggests that floods were a major factor in the evolution of the Amazonian forest (Räsänen et al. 1987). Similar shifts in the distribution of savanna and forest biomes in Australia, also associated with long-term climatic changes, may account for much recent speciation among birds on that continent. Most biogeographers and paleoclimatologists now agree that climate and the major biotic associations of plants and animals have fluctuated repeatedly on a worldwide scale during the Pleistocene. The result has been the repeated fragmentation of biomes, alternating with re-formation of more continuous, continental distributions, plus some change in species composition of the biomes. Whereas geographic isolation on continents, as a requisite for speciation, was once difficult to explain, the Pleistocene and earlier cyclic events may now provide explanations. A few years ago ornithologists thought that most bird species dated from the Pliocene and were

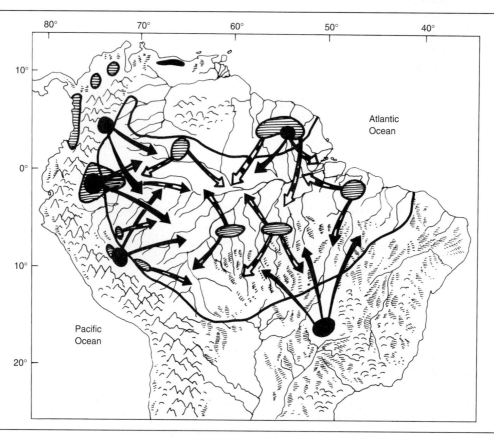

Figure 1–9 Proposed effects of changes during the Pleistocene on lowland forest vertebrates of the Amazon Basin. The black line shows the area within which forests are thought to have been fragmented. Within this region birds (black areas) and lizards (shaded areas) are thought to have been isolated in forest remnants. The arrows show postulated paths of reinvasion as the forests returned. (Based on P. E. Vanzolini, 1972, in *Tropical Forest Ecosystems in Africa and South America: A Comparative Review*, edited by B. J. Meggers et al., Smithsonian Institution Press, Washington, DC.)

several million years old. Now most ornithologists feel that speciation has been very rapid during the Pleistocene, especially among small birds with limited dispersal abilities—characteristics equally applicable to many other vertebrates. However, some aspects of refuge theory are currently being questioned, especially in the case of tropical species (Connor 1986).

Earth History and Vertebrate Evolution

The phenomena of isolation and secondary contact that are important in the evolution of individual species like Galapagos finches and eastern warblers have also been important in the larger story of vertebrate evolution. The world in which vertebrates have been evolving has changed enormously and repeatedly since the origin of vertebrates in the Cambrian, and these changes have affected vertebrate evolution directly and indirectly. An understanding of the sequence of changes in the positions of the continents and the significance of those positions in terms of climates and interchange of faunas is a central part of understanding the vertebrate story. A summary table of these events is presented in the front of the book, and details are given in Chapters 5, 9, 14, and 20. The history of life on Earth has occupied four **geological eras**: the Precambrian, Paleozoic, Mesozoic, and Cenozoic. These eras are divided into **periods**, and within the Cenozoic the periods are divided into **epochs**. The history of vertebrates extends from the Cambrian to the present, a time span that is sometimes called the **Phanerozoic**.

Movements of land masses have been a feature of Earth's history, at least since the Precambrian. The direction of vertebrate evolution has been in large part molded by continental drift. By the early Cambrian a recognizable scene had appeared: The seas had formed, continents floated atop the Earth's mantle, life had become complex, and an atmosphere of oxygen had formed, signifying that the photosynthetic production of food resources had become a central phenomenon of life.

The continents still drift today—North America is moving westward and Australia northward at approximately 4 centimeters per year. Because the movements are so complex, their sequence, their varied directions, and the precise timing of the changes are difficult to summarize. By viewing the movements broadly, however, a simple pattern unfolds during vertebrate history: *fragmentation–coalescence–fragmentation* (Figure 1–10).

Continents existed as separate entities over 500 million years ago. Some 300 million years ago all of these separate continents combined to form Pangaea, birthplace of the terrestrial vertebrates. Persisting and drifting northward as an entity, this single huge continent began to break apart about 100 million years ago. Its separation occurred in two stages: first into Laurasia and Gondwana, then into a series of units that have drifted and become the continents we know today.

The complex movements of the continents through time have had major effects on many phenomena significant to the evolution of vertebrates. The most obvious is the relationship between the location of land masses and their climates. At the start of the Paleozoic much of Pangaea was located on the equator and this situation persisted through the middle of the Mesozoic. Solar radiation is most intense at the equator, and climates at the equator are correspondingly warm. During the Paleozoic and much of the Mesozoic large areas of land enjoyed tropical conditions, and terrestrial vertebrates evolved and spread in these tropical regions. By the late Mesozoic much of the land mass of the Earth had moved out of equatorial regions, and most climates in the Northern and Southern Hemispheres are now temperate instead of tropical.

A less obvious effect of the position of continents on terrestrial climates comes about through changes in patterns of ocean circulation. For example, the Arctic Ocean is now largely isolated from the other oceans and it does not receive warm water via currents flowing from more equatorial regions. High latitudes are cold because they receive less solar radiation than do areas closer to the equator, and the Arctic Basin does not receive enough warm water to offset the lack of solar radiation. As a result, the Arctic Ocean is permanently frozen and cold climates extend well southward across the continents. The cooling of climates in the Northern Hemisphere at the end of the Mesozoic that had such a drastic effect on the dinosaurs is partly the result of the changes in oceanic circulation at that time.

Another factor that influences climates is the relative level of the continents and the seas. At some periods of the history of the Earth, most recently in the late Mesozoic and again in the first part of the

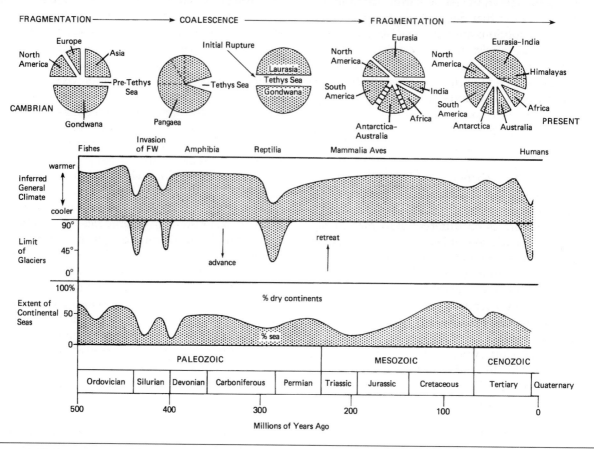

Figure 1–10 A summary of continental drift, climate, and the extent of epicontinental seas from the Ordovician to the present.

Cenozoic, large parts of the continents have been flooded by shallow seas. These **epicontinental seas** extended across the middle of North America and the middle of Eurasia in the Cretaceous and early Cenozoic. Water has a great capacity to absorb heat as environmental temperatures rise and to release that heat as temperatures fall. The heat capacity of water buffers temperature change in areas of land near bodies of water. Areas with **maritime climates** do not get very hot in summer or very cold in winter, and they are usually moist because water that evaporates from the sea falls as rain on the land. **Continental climates**, which characterize areas of land far from the sea, are usually dry with cold winters and hot summers. The draining of the epicontinental seas at the end of the Cretaceous was a second factor that probably contributed to the demise of the dinosaurs by making climates in the Northern Hemisphere more continental.

On a smaller scale, the upheaval of mountain chains also affects climates through a phenomenon known as the **rain shadow**. A range of mountains near the coast, such as the Sierra Nevadas in California, intercepts moist air as it moves inland and forces it upward. As the air rises it becomes cooler and the water vapor it contains condenses into drops of water and falls as rain on the windward sides of the mountains. When the air passes over the crests of the mountains it has already lost much of the moisture it contained, and the leeward side of the mountains is in a rain shadow and receives little moisture. In addition, as the air mass descends in altitude on the lee side of the mountain range, it becomes warmer. Warm air can hold more water vapor than cold air, but there is no source of water on the lee side of the mountains to replace the water that fell as rain on the windward side. As a result, the relative humidity of the air becomes progressively lower as the air streams down the lee side of the mountains. The hot, dry wind that descends from a mountain range when conditions are right is called a *chinook* in North America, a *foehn* in Europe, and a *simoom* in Asia. Mountain ranges are uplifted when continental plates collide, and the effects on local climates can affect animals. For example, the radiation of large grazing mammals in South America in the Cenozoic followed the formation of the Andes Mountains and the replacement of forest by grassland in their rain shadow.

In addition to changing climates, continental drift has formed and broken land connections between the continents. Isolation of different lineages of vertebrates on different land masses has produced dramatic examples of the independent evolution of similar types of organisms. These are well shown by mammals in the Cenozoic, a period when the Earth's continents were more widely separated than they ever have been before or since. Saber-tooth carnivores evolved among placental mammals in the Northern Hemisphere and among marsupials in South America, and grazing animals were represented by entirely different lineages of placental mammals in North America and South America and by kangaroos and wombats (both marsupials) in Australia.

Much of evolutionary history appears to depend on whether or not a particular lineage was in the right place at the right time. This **stochastic** (random, chance) element of evolution is assuming increasing prominence as more detailed information about the times of extinction of old groups and radiation of new groups is suggesting that competitive replacement of one group by another is not the usual mechanism of large-scale evolutionary change. The movements of continents and their effects on climates and the isolation or dispersal of animals are taking an increasingly central role in our understanding of vertebrate evolution.

Classification of Vertebrates

The diversity of vertebrates (45,000 living species and perhaps ten times that number of species now extinct) combines with the prevalence of convergent and parallel evolution to make the classification of vertebrates an extraordinarily difficult undertaking. Yet classification has long been at the heart of evolu-

tionary biology. Initially, classification of species was seen as a way of managing the diversity of organisms, much as an office filing system manages the paperwork of the office. Each species could be placed in a pigeonhole marked with its name, and when all species were in their pigeonholes, one could comprehend the diversity of vertebrates. This approach to classification was satisfactory as long as species were regarded as static and immutable: Once a species was placed in the filing system it was there to stay.

Acceptance of the fact of evolution has made that kind of classification inadequate. Now it is necessary to express evolutionary relationships among species by incorporating genealogical information in the system of classification (de Queiroz 1988). Ideally, a classification system should not only attach a label to each species, it should also encode the phylogenetic (*phyl* = tribe, *genesis* = origin) relationship between that species and other species. Modern techniques of **systematics** (the phylogenetic classification of organisms) are moving beyond filing systems and have become methods for generating testable hypotheses about evolution.

Classification and Evolution

Our system of classifying organisms is pre-Darwinian. It traces back to methods established by the naturalists of the seventeenth and eighteenth centuries, especially those of Carl von Linné, a Swedish naturalist, better known by his Latin pen name, Carolus Linnaeus. The Linnaean system employs **binominal nomenclature** to designate species, and arranges species into hierarchical categories (**taxa**) for classification. This system is incompatible in some respects with evolutionary biology (de Queiroz and Gauthier 1992), but it is still widely used. Thus, you must understand the traditional classification because it forms the basis for much of the literature of vertebrate biology, but you should also understand the shortcomings of the traditional system to appreciate how dramatically our view of systematics has changed.

Binominal Nomenclature The scientific naming of species became standardized when Linnaeus's monumental work, *Systema Naturae*, was published in sections between 1735 and 1758. He attempted to give an identifying name to every known species of plant and animal. His method involves assigning a generic name, which is a Latin noun, Latinized Greek, or a Latinized vernacular word, and a second species name, usually a Latin adjective, or similar derivative. Some familiar examples include *Homo sapiens* for human beings (*homo* = human, *sapiens* = wise), *Passer domesticus* for the house sparrow (*passer* = sparrow, *domesticus* = belonging to the house), and *Canis familiaris* for the domestic dog (*canis* = dog, *familiaris* = of the family).

Why use Latin words? Aside from the historical fact that Latin was the early universal language of European scholars and scientists, it has provided a uniform usage that has continued to be recognized worldwide by scientists regardless of their vernacular language. The same species may have different colloquial names, even in the same language. For example, *Felis concolor* ("the uniformly colored cat") is known in various parts of North America as cougar, puma, mountain lion, American panther, painter, and catamount. In Central and South America it is called león colorado, onça-vermelha, poema, guasura, or yaguá-pitá, but mammalogists of all nationalities recognize the name *Felis concolor* as referring to a specific kind of cat.

Hierarchical Groups: The Higher Taxa Linnaeus and other naturalists of his time developed what they called a natural system of classification in which all similar species are grouped together in one **genus** (plural **genera**), based on characters that define the genus. The most commonly used characters were anatomical, because they are the ones most easily preserved in museum specimens. Thus all dog-like species—various wolves, coyotes, and jackals—were grouped together in the genus *Canis* because they all share certain anatomical features, such as an erectile mane on the neck and a skull with a long, prominent sagittal crest on which massive temporal muscles originate for closure of the jaws. Linnaeus's method of grouping species was functional because it was based on anatomical (and to some extent on physiological and behavioral) similarities and differences. Although Linnaeus lived before there was any knowledge of genetics and the mechanisms of inheritance, he used taxonomic characters that we understand today are genetically determined biological traits that, within limits, express the degree of genetic similarity or difference among groups of organisms.

Subsequent development of biological classification has employed seven basic taxonomic cate-gories, listed below in decreasing order of inclusiveness:

Kingdom
  Phylum (= Division in plants)
    Class
      Order
        Family
          Genus
            Species

These are the taxonomic levels (categories) that are employed in most of the primary and secondary literature of vertebrate biology.

Traditional and Cladistic Classifications

All methods of classifying organisms are based on similarities among the included species, whether or not the similarities reflect common ancestry. **Phenetic** methods, such as classical numerical taxonomy, do not explicitly address evolutionary relationships. Pheneticists group organisms strictly by the number of characters they have in common. Characters are not weighted to reflect their relative importance, and as a result the presence of two arches in the skulls of lizards and birds would be treated equally to the presence of the oxygen-transport pigment hemoglobin in both groups. Phenetic classification is ideal for some types of data, such as biochemical information about the similarities and differences in the proteins or DNA sequences of species. With these sorts of data there is no basis for assuming that one difference is more important than another and assigning greater weight to it.

Morphological, behavioral, and physiological characteristics of organisms provide a different sort of information: It is possible to make a judgment that some differences are more significant than others. For example, only a few kinds of vertebrates have skulls with a pattern of two arches, but nearly all vertebrates have hemoglobin. Consequently, knowing that the species in question share the two-arched skull pattern tells you more about the closeness of their relationship than knowing that they have hemoglobin, and you would give more weight to the skull characters than to the hemoglobin. Today, two major evolutionary phylogenetic systems are in use for these sorts of data: the older systematics, which we will refer to as **traditional** or **evolutionary systematics** following Charig's (1982) terminology, and the newer **phylogenetic systematics**, also known as **cladistics**. The goal of each of these approaches to systematics is to establish the pattern of evolutionary descent of organisms, but the methods differ.

Phylogenetic Systematics

In 1966, Willi Hennig introduced phylogenetic systematics by forcefully emphasizing that a phylogeny can be reconstructed only on the basis of shared derived characters. "Derived" in this sense means "different from the ancestral condition." For example, the limbs of terrestrial vertebrates have a single bone in the upper limb and two bones in the lower limb. This arrangement of limb bones is different from the ancestral pattern seen in lobe-finned fishes, and is a shared derived character of terrestrial vertebrates. In cladistic terminology, shared derived characters are called **synapomorphies** (*syn* or *sym* = together, *apo* = away from, i.e., derived from, *morph* = form).

Of course, organisms also have shared characters that they have inherited from their ancestors.

These are called **plesiomorphies** (*plesi* = near). Terrestrial vertebrates have a vertebral column, for example, that was inherited essentially unchanged from lobe-finned fishes. Hennig called shared ancestral characters **symplesiomorphies**, and a species that retains many ancestral characters is sometimes described as "plesiomorphic." Plesiomorphies tell us nothing about degrees of relatedness. They are useful only in highlighting what characters are apomorphies. It is the insistence that *only* shared derived characters can be used to determine genealogies that particularly characterizes cladistics.

The jargon associated with cladistics makes it hard for newcomers to appreciate its importance. To clarify the meaning of the major terms, consider the examples presented in Figure 1–11. Any one of the three cladograms represents a possible genealogical relationship (i.e., phylogeny) for the three taxa identified as 1, 2, and 3. The letters A/a, B/b, and C/c represent characters—a morphological structure, for example, or an enzyme, a physiological process, or a behavior. To make the example a bit more concrete, we can say that A/a represents the toes on the front foot, B/b represents the skin covering, and C/c represents the tail. A, B, and C are the ancestral character states, that is, the conditions of each character that

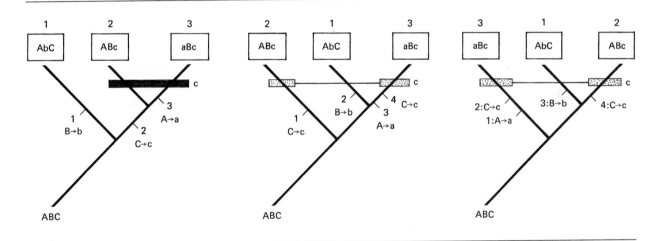

Figure 1–11 Cladograms showing the distribution of three character states in three taxa (1, 2, and 3). The three ancestral states are represented by the capital letters A, B, and C, and the derived conditions by a, b, and c. Bars connect derived characters (apomorphies). The dark bar shows a shared derived character (a synapomorphy) of the lineage that includes taxa 2 + 3. The stippled bars represent the two independent origins of the same derived character state that must be assumed to have occurred if the apomorphy was not present in the most recent common ancestor of taxa 2 and 3. The numbers identify changes from the ancestral character state to the derived condition.

was inherited from the most recent common ancestor of the lineages we are examining, and a, b, and c are the derived character states. For the purpose of this example, let's say that the ancestral character state for A/a is to have five toes on the front foot (= A) and the derived state is four toes (= a). We'll say that the ancestral state of character B/b is a scaly skin (= B), and the derived state is lacking scales (= b). And for C/c, the ancestral state is tail present (= C) and the derived state is tail absent (= c).

Figure 1–11 shows the distribution of those three character states in the three groups. The animals in group 1 have five toes on the front feet (A), lack scales (b), and have a tail (C). Animals in group 2 have five toes, scaly skins, and no tails, and animals in group 3 have fours toes, scaly skins, and no tails.

How can you use this information to decipher the genealogical relationships of the three groups of animals? Notice that the ancestral character state C occurs only in taxon 1 and the derived condition c is found in both taxa 2 and 3. The most *parsimonious* phylogeny (i.e., the phylogeny that requires the fewest number of evolutionary changes) is represented by the left diagram in Figure 1–11. It requires only three changes to produce the derived character states a, b, and c from the a starting point with the ancestral character states A, B, and C:

1. In the evolution of group 1, scales are lost (B → b).
2. In the evolution of group 2 + group 3, the tail is lost (C → c).
3. In the evolution of group 3, a toe is lost from the front foot (A → a).

The other two phylogenies are possible, but they would require that tail loss (the derived condition c) originated independently in taxon 2 and in taxon 3. Both of these phylogenies require four evolutionary changes, so they are less parsimonious than the first phylogeny we considered. Usually we consider that the most parsimonious phylogeny is most likely to be correct, and that's the one we would use. But a

phylogeny of this sort (called a **cladogram**) is a hypothesis about the genealogical relationships of the groups included. Like any scientific hypothesis, it can be tested with new data, and if it fails that test, it is rejected and a different hypothesis (cladogram) takes its place.

So far we have avoided one of the central issues of phylogenetic systematics: How do you know which character state is ancestral (plesiomorphic) and which is derived (apomorphic)? That is, how do you determine the direction (**polarity**) of evolutionary transformation of the characters? For that, you need additional information. Increasing the number of characters you are considering can help, but comparing the characters you are using with an outgroup that consist of the closest relatives of the ingroup (i.e., the organisms you are studying) is the preferred method. A well-chosen outgroup will consistently possess unequivocal ancestral character states compared to the ingroup. For example, lobe-finned fishes are an appropriate outgroup for terrestrial vertebrates. Sometimes the fossil record provides important information, but it's not the complete solution because it preserves only certain characters (mostly hard tissues such as bone). An examination of the embryonic development of the structures might be helpful, because embryonic stages often reveal similarities that are not seen in adults. (The development of gill pouches at some stage of development of all chordates is a good example of this phenomenon.)

A classification of the vertebrates by phylogenetic systematics (Figure 1–12) and a classification by evolutionary systematics (Figure 1–13) are only vaguely similar, although each is based on similar analyses. If both evolutionary systematists and cladists use basically the same characters, why are their results so different?

The two methods differ because of the criteria used to define the taxa that make up the cladogram, and that difference is fundamental. Cladists establish and name a taxonomic group within a cladogram

Figure 1–12 Phylogenetic relationships of extant vertebrates. This diagram shows the probable relationships among the major groups of vertebrates. The boxes across the top of the diagram show the traditional groupings of the taxa, and the names along the right side show how the lineages are grouped by phylogenetic systematics. Note that these groupings are nested progressively; that is, all sarcopterygians are osteichthyans, all osteichthyans are gnathostomes, all gnathostomes are vertebrates, all vertebrates are craniates, and so on. The numbers indicate derived characters that distinguish the lineages.

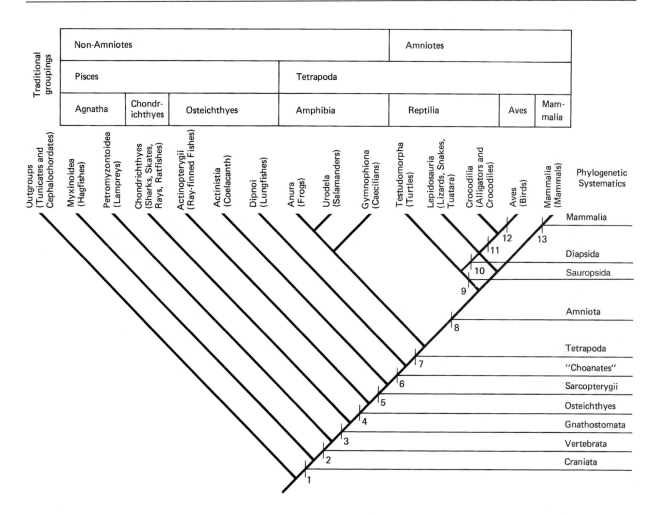

1. Distinct head region skeleton incorporating anterior end of notochord, one or more semicircular canals, brain consisting of three regions, paired kidneys, gill bars, neural crest tissue. 2. Arcualia or their derivatives form vertebrae, two or three semicircular canals. 3. Jaws formed from mandibular arch, teeth containing dentine, three semicircular canals, branchial skeleton internal to gill membranes, branchial arches contain four elements on each side plus one unpaired ventral median element, paired fins with internal skeleton and muscles supported by girdles in the body wall. 4. Presence of lung or swimbladder derived from the gut, unique pattern of dermal bones of shoulder, unique characters of jaw and branchial muscles. 5. Unique supporting skeleton in fins. 6. Choanae* present, derived paired appendage structure, conus arteriosus of heart partly divided, unique dermal bone pattern of braincase, loss of interhyal bone. 7. Distinctive paired pectoral and pelvic limbs characterized by a single bone in the proximal portion of the limb and two bones in the distal portion. 8. A distinctive arrangement of extraembryonic membranes (the amnion, chorion, and allantois). 9. Tabular and supratemporal bones small or absent, simple coronoid, centrum of atlas and intercentrum of axis fused, medial centrale of ankle absent. 10. Skull with a dorsal temporal fenestra, upper temporal arch formed by triradiate postorbital and triradiate squamosal bones. 11. Presence of a fenestra anterior to the orbit of the eye, orbit shaped like an inverted triangle. 12. Feathers, loss of teeth, metabolic heat production used to regulate body temperature (endothermy). 13. Hair, lower jaw formed only by dentary bone, mammary glands, independent development of endothermy.

*The homology of the choanae of dipnoans with those of tetrapods is questioned by many authorities, thus the name "Choanates" may not be appropriate but the grouping appears to be valid.

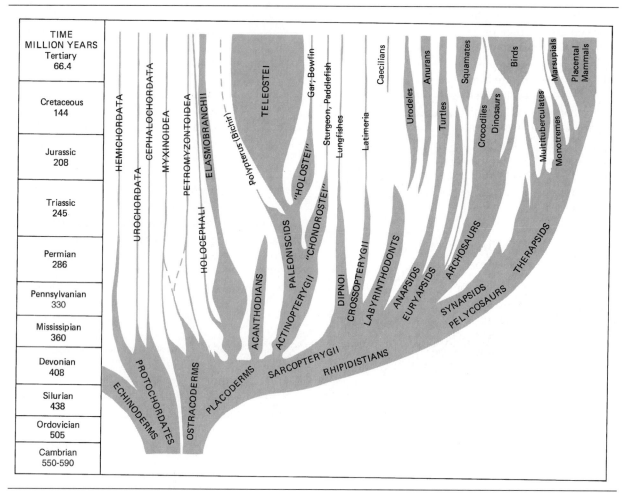

TIME MILLION YEARS

Period	Million Years
Tertiary	66.4
Cretaceous	144
Jurassic	208
Triassic	245
Permian	286
Pennsylvanian	330
Mississippian	360
Devonian	408
Silurian	438
Ordovician	505
Cambrian	550–590

Figure 1–13 Traditional evolutionary tree of the vertebrates. Width of branches indicates the relative number of recognized genera for a given time level on the vertical axis (time in millions of years indicates beginning of geological periods).

solely on the basis of **monophyly** (single origin). For example, every vertebrate depicted in Figure 1–12 that is ascendant from a chosen branch point (i.e., a clade or lineage of ascent) along the cladogram is related by the shared derived characters that diagnose that lineage. The name assigned to this lineage identifies all members in the clade, and only monophyletic groups can receive names. Evolutionary systematists are happy to accept monophyletic taxa, but they also are willing to include groups that are not monophyletic within a named taxon, or exclude a group from a monophyletic lineage and assign it to another taxon.

An example emphasizes the differences between the two schemes. The cladogram depicted in Figure 1–12 is our hypothesis of the evolutionary relation-

ships of the major living groupings of vertebrates. There are 13 dichotomous branches leading from the origin of the vertebrates from other chordates to the mammals. A cladist ranks the organisms by sequencing the cladogram from ancestral to derived dichotomies, and assigning names to the lineages originating at each branch point. This process produces a nested series of groups, starting with the most inclusive. Thus, the name Gnathostomata includes all vertebrate animals that have jaws; that is, every taxon to the right of the number 3 in Figure 1–12 is included in the Gnathostomata, every taxon to the right of number 4 is included in the Teleostomii, and so on.

In contrast, the traditional classification shown in Figure 1–13 is not based on nested series. As you move to the right in the diagram, one group is replaced by another of the same taxonomic level. Evolutionary systematists often elevate taxonomic rankings above their position in the phylogeny. For example, they elevate birds to a taxonomic ranking (Class Aves) equal to the other groups of terrestrial vertebrates (Class Amphibia, Class Reptilia, Class Mammalia) because they attribute great significance to the presence in birds of feathers and endothermy. There are proponents of both methods of classification (see Charig 1982), but evolutionary systematics has been almost entirely replaced by phylogenetic systematics.

Interpretation of Phylogenetic and Evolutionary Systematics

Because traditional taxonomy predated Darwin's explanation of evolution, it did not attempt to represent genealogical relationships. In contrast, phylogenetic taxonomy attempts to represent an evolutionary system based on genealogical continuity (de Queiroz 1988). That may not seem to be a big difference, but it has important implications. Traditional classification was based on similarity—animals that are alike were placed in the same taxonomic group, or at least in groups that were close to each other. Those animals might or might not have any genealogical connection; that was not important for the classification. A classification of this sort is functional, in the sense that it organizes information and attaches names to species, but it does not allow us to make predictions about unknown aspects of the biology of the species.

In contrast, phylogenetic systematics is based on the hypothesis that organisms that are grouped together share a common genealogy, and that a common heritage accounts for their similarity. If that hypothesis is correct, we can use phylogenetic systematics to ask questions and draw conclusions about evolution. We can examine the historical origin and functional significance of characters of living animals, and make inferences about the biology of extinct organisms. For example, the phylogenetic relationship of crocodilians, dinosaurs, and birds is shown in Figure 1–14a. We know that crocodilians and birds care for their eggs and young. Fossilized dinosaur nests, some of which contain remains of baby dinosaurs, suggest that at least some dinosaurs cared for their young. Is that a plausible inference?

Those dinosaurs are extinct, so we cannot hope to find any direct evidence of their reproductive behavior, but the phylogenetic diagram in Figure 1–14a shows that both of the closest living relatives of the dinosaurs, crocodilians and birds, do have parental care. Looking at living representatives of more distantly related lineages, we see that parental care is not universal among fishes, amphibians, or reptiles other than crocodilians. With that information, we can conclude that parental care is a derived character state of the evolutionary lineage that contains crocodilians + dinosaurs + birds (Archosauria). The most parsimonious explanation of the occurrence of parental care in both crocodilians and birds is that it had evolved in that lineage before the crocodilians separated from dinosaurs + birds. (We could postulate that parental care evolved separately in crocodilians and in birds, but that would be a less parsimonious hypothesis.) So we are probably correct when we interpret the fossil evidence as showing that dinosaurs had parental care.

How is that example related to taxonomy? Compare the taxonomic schemes in parts (b) and (c) of Figure 1–14. Phylogenetic taxonomy (Figure 1–14b) shows crocodilians, dinosaurs, and birds as a nested hierarchy within an evolutionary lineage called Archosauria. Birds and dinosaurs are in a more restricted lineage called Dinosauria, and birds alone are in a lineage called Aves. That is exactly the information encoded in the phylogenetic diagram (Figure 1–14a), and it is the information we need to decide if we are correct in our interpretation of dinosaurs' parental care. In contrast, traditional taxonomy (Figure 1–14c) places crocodilians and dinosaurs as orders within the Class Reptilia, and birds as the Class Aves. It provides no useful information about the genealogical relationship of the groups. In fact, the traditional classification misleads us by suggesting that the Class Aves is in some way larger, more important, or more inclusive than the Order Dinosauria, when genealogically Aves is a subset of Dinosauria.

Figure 1–14a also illustrates one of the ways that cladistics has made talking about restricted groups of animals more complicated than it used to be. Suppose you wanted to refer to just the two lineages of animals that are popularly known as dinosaurs— what could you call them? Well, if you call them dinosaurs, you're not being PC (phylogenetically correct), because the taxon Dinosauria includes birds. So if you speak of dinosaurs, you are including ornithischians + saurischians + birds, even though any seven-year-old would understand that you are

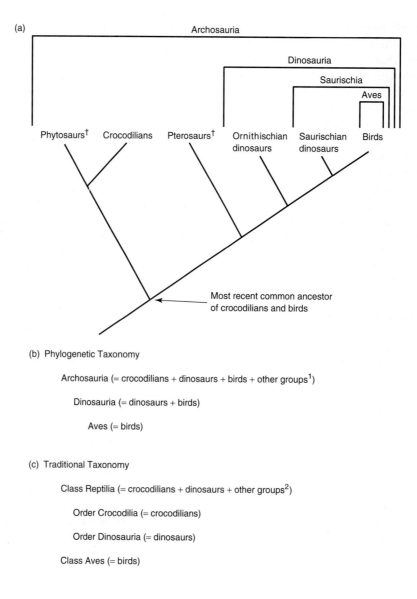

(a)

Archosauria

Dinosauria

Saurischia

Aves

Phytosaurs†   Crocodilians   Pterosaurs†   Ornithischian dinosaurs   Saurischian dinosaurs   Birds

Most recent common ancestor
of crocodilians and birds

(b) Phylogenetic Taxonomy

Archosauria (= crocodilians + dinosaurs + birds + other groups[1])

Dinosauria (= dinosaurs + birds)

Aves (= birds)

(c) Traditional Taxonomy

Class Reptilia (= crocodilians + dinosaurs + other groups[2])

Order Crocodilia (= crocodilians)

Order Dinosauria (= dinosaurs)

Class Aves (= birds)

Figure 1–14 A comparison of the information provided by phylogenetic taxonomy and traditional taxonomy. (a) A cladogram showing the relationships of the Archosauria, the evolutionary lineage that includes living crocodilians and birds. (The symbol † indicates lineages that are entirely extinct. Phytosaurs were crocodile-like animals that disappeared at the end of the Triassic, and pterosaurs were the flying reptiles of the Jurassic and Cretaceous.) (b) The phylogenetic taxonomy of the Archosauria shows the same nested arrangement of taxa as the cladogram. ([1] "Other groups" in this example refers to phytosaurs and pterosaurs.) (c) Traditional taxonomy provides no information about the evolutionary relationships of the taxa, and suggests that birds have a status equivalent to that of all the other archosaurians. ([2] "Other groups" in this example includes turtles, lizards, snakes, and tuatara as well as the crocodilians and dinosaurs listed as orders of reptiles.)

trying to restrict the conversation to extinct Mesozoic animals.

In fact, in cladistic terminology there *is* no correct taxonomic name for just the animals popularly known as dinosaurs. That's because cladistics recognizes only monophyletic lineages, and a monophyletic lineage, or clade, includes an ancestral form *and all its descendants*. The most recent common ancestor of ornithischians, saurischians, and birds in Figure 1–14a lies at the intersection of the lineage of ornithischians with saurischians + birds, so Dinosauria is a monophyletic lineage. But if you leave out birds, you no longer have all the descendants of the common ancestor, so ornithischians + saurischians minus birds does not fit the definition of a monophyletic lineage. (It would be called a **paraphyletic** group by cladists.)

Biologists who are interested in how organisms live often want to talk about paraphyletic groups. After all, the dinosaurs (in the popular sense of the word) differed from birds in many ways. The only correct way to refer to the animals popularly know as dinosaurs is to call them nonavian dinosaurs, and you will find that and other examples of paraphyletic groups later in the book. Sometimes even this construction does not work because there is no appropriate name for the part of the lineage you want to distinguish. In this situation we will use quotation marks (e.g., "ostracoderms") to indicate that the group is paraphyletic.

One other useful bit of terminology is **sister group**. The sister group is the monophyletic lineage that is most closely related to the monophyletic lineage you are considering. For example, the lineage that includes crododilians and phytosaurs is the sister group of the lineage that includes ornithischians, saurischians, and birds.

Determining Phylogenetic Relationships

We've established that the derived characters systematists used to group species into higher taxa must be inherited via common ancestry, that is, they must be **homologous** similarities (Hall 1994). Unfortunately, the word "homologous" has at least three distinct meanings in evolutionary biology, only one of which is the same as its use in everyday speech, and these different meanings can cause confusion. (1) In the narrowest sense, homologous means "descended from the same derived structure." (2) In a slightly

broader sense, homologous can mean "descended from the same ancestral structure." (3) In the form employed by molecular geneticists (most of whom are not familiar with evolutionary biology), homologous means "similar" without any implications of evolutionary descent. The first definition is the one employed by phylogenetic systematists, but the second definition is widespread in the literature of vertebrate biology. The third definition is not widely used in evolutionary biology, although it does appear in cases where DNA sequences are used to infer evolutionary relationships. A reader must be careful to determine which of the possible meanings of homology the author intended (Gould 1988).

We can turn to the limbs of terrestrial vertebrates again, this time to illustrate the difference between the first and second definitions of homologous: The forelimb of tetrapods consists of one bone (the humerus) in the upper limb and two bones (the radius and ulna) in the lower arm. This is the ancestral condition for terrestrial vertebrates, and all terrestrial vertebrates have forelimbs with this structure. By definition 2, all these forelimbs are homologous structures. That includes forelimbs as different as those of humans, horses, and birds. Birds have extensively modified the ancestral forelimb and hand to form wings. Wings are a derived character of birds, and no other vertebrates have exactly the same modifications of the forelimb bones. Different kinds of birds have wings of different shapes—short and broad, long and narrow, and so on. By the most restrictive definition of homology (definition 1), the wings of one species of bird are homologous to the wings of another species of bird because both are descended from the wing (derived forelimb) of an ancestral bird.

Structures that perform the same function or the same appearance without being homologous are called **homoplastic** (*plast* = form, shape). The wings of birds and bats are both forelimbs specialized for flight (Figure 1–15). By definition 1 they are not homologous because they are not derived from a common ancestor of birds and bats that had a forelimb specialized for flight; therefore, the wings of birds and bats would be called homoplastic. By definition 2, the wings of birds and bats would be considered homologous because they are derived from the ancestral vertebrate limb.

Homoplastic similarities include **convergence** and **parallelism**. It is well known that distantly related species with similar modes of life may have

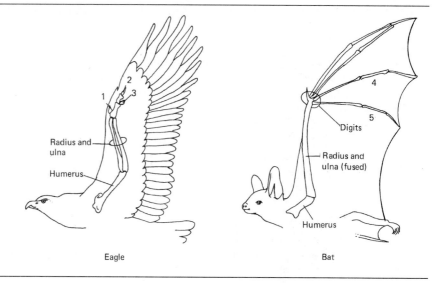

Figure 1–15 Wing bones of a bird and a bat showing the similarities and differences.

Radius and ulna

Humerus

Eagle

Digits

Radius and ulna (fused)

Humerus

Bat

superficially very similar morphological structures and body shapes as the result of the same selective forces acting on different genotypes (convergence). Examples of convergence among vertebrates include the similarity in body shape of a pelagic fish (shark), an aquatic lepidosauromorph (ichthyosaur), and an aquatic mammal (porpoise). The fusiform body shapes of these animals have been derived independently, beginning from very different ancestral body forms (fish-like, lizard-like, and mammalian, respectively). Parallelism describes the evolution of similar structures from a derived starting point. For example, some lineages of desert rodents, such as the jerboas of the Old World and the kangaroo rats of the New World, have long hind limbs, a long, bare tail with a tuft of hair on the end, and enlarged middle ears. Jerboas and kangaroo rats each originated from the rodent lineage (that is, from an already derived mammalian body form), and they have developed their specializations in parallel.

The Time Course of Vertebrate Diversity and the Effect of Human Population Growth

Starting from the appearance of the earliest vertebrates in the Ordovician, the diversity of vertebrates appears to have increased slowly through the Paleozoic and early Mesozoic, and then rapidly during the past hundred million years. This overall increase has been interrupted by eight periods of extinction for aquatic vertebrates and six for terrestrial forms (Benton 1990). The number of species of vertebrates reached a peak in the Pliocene or Pleistocene, and has declined since then (World Conservation Monitoring Centre 1992).

Extinction is as normal a part of evolution as species formation, and the duration of most species in the fossil record appears to be from one million to ten million years. Periods of major extinction (a reduction in diversity of 10 percent or more) are associated with changes in climate and the consequent changes in vegetation. (Impacts of extraterrestrial objects have clearly occurred many times during the history of vertebrates, but they have played only a minor role in extinctions.)

Changes in that pattern of reasonably long-lived species and extinctions that are correlated with changes in climate and vegetation appear about the time that humans became a dominant factor in many parts of the world (Wilson 1992, Steadman 1995). Hawaii probably had more than 100 species of native birds when Polynesian colonists arrived about 300 A.D. (Olson and James 1991a,b). By the time European colonists reached Hawaii in the late eighteenth century, that number had been reduced by half, and in the past two centuries another third of the native Hawaiian birds have become extinct. The arrival of humans in Australia (30,000 years ago), North America (10,000 years ago), Madagascar (1500 years ago), and New Zealand (1000 years ago) coincided in

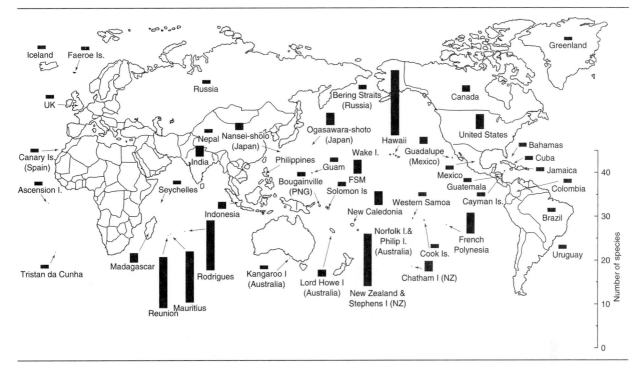

Figure 1–16 The numbers of confirmed extinctions of species of birds since 1600. Islands have suffered more extinctions than continental areas. (*Source*: World Conservation Monitoring Centre, 1992, *Global Biodiversity: Status of the Earth's Living Resources*, Chapman & Hall, London, UK.)

every instance with waves of extinction. These extinctions were selective—large mammals and flightless birds disappeared, whereas small mammals and most flying birds were not affected. Hunting was probably the principle cause of these extinctions, because the species that disappeared were those that human hunters concentrated on (Diamond 1989).

Animal species continue to become extinct today (Figure 1–16), and a worldwide survey of extinctions since the start of European colonization reveals two trends: Island extinctions began almost two centuries earlier than continental extinction, and both island and continental extinctions have increased rapidly from the early or mid nineteenth century through the twentieth century (World Conservation Monitoring Centre 1992).

It is not easy to calculate the rate at which extinctions are occurring, and different assumptions can produce very different values. Recent estimates of the number of species of birds that would be committed to extinction by 2015 range from 450 to 1350 (Heywood et al. 1994). (Committed to extinction means

that a species' populations in the wild are no longer viable, and the species will inevitably become extinct unless major conservation actions reverse the current trend.) What is clear is that destruction of habitat is the major threat, affecting more than 60 percent of threatened species of birds and nearly 80 percent of threatened species of mammals (Figure 1–17).

An especially ominous trend has become apparent recently: the decline of populations and the disappearance of species from causes that seem to be worldwide and indirect. Consider these examples:

1. Populations of frogs and salamanders are declining or disappearing entirely, and examples have been reported from all continents (Wake 1991, Phillips 1994, Blaustein et al. 1994b). Ultraviolet B radiation, which kills the eggs and embryos of some species of amphibians, may be one cause of these declines (Blaustein et al. 1994a). The amount of ultraviolet B radiation reaching the Earth's surface is increasing as a result of depletion of ozone in the stratosphere by chlorofluorocarbons released by human activities, and this phenomenon is not

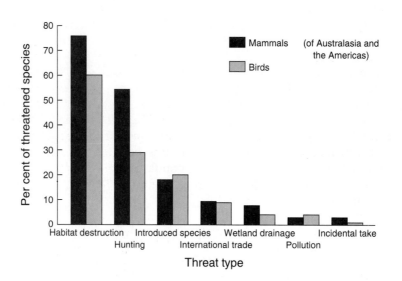

Figure 1–17 The major threats affecting birds (on a worldwide basis) and mammals (Australasia and the Americas). Habitat destruction is the single most important threat for both kinds of animals, affecting 60 percent of birds and 76 percent of mammals. Hunting (for food and sport) is a greater threat to mammals than to birds. Introduced species may be predators or competitors, and international trade refers to commercial exploitation for fur, feathers, and the pet trade. Incidental take is the term used to designate accidental mortality, such as dolphins that are drowned by boats fishing for tuna. (*Source*: World Conservation Monitoring Centre, 1992, *Global Biodiversity: Status of the Earth's Living Resources*, Chapman & Hall, London, UK.)

confined to the well-known hole over the Antarctic. Measurements of ultraviolet B radiation at Toronto, Canada, since 1989 show an increase of 35 percent per year in winter and 7 percent per year in summer (Kerr and McElroy 1993).

2. Populations of migratory birds in North America declined 50 percent from the 1940s to the 1980s. Destruction of forest habitats in North America (where the species breed) and in Mexico, the West Indies, and Central and South America (where they spend the winter) appears to have been an important part of these declines (Terborgh 1989).

3. Some birds in European forests are producing eggs with thin, porous shells that break easily. The problem is a lack of calcium in the birds' diets because acid in rain and snow has dissolved calcium from the soil. These depleted soils no longer support the populations of snails that once supplied the calcium the birds need to form eggshells (Graveland et al. 1994).

4. An average of 82 percent of alligator eggs from Lake Apopka in Florida fail to hatch, and about half of the alligators that do emerge from these eggs die within two weeks. In testimony before the Subcommittee on Health and the Environment of the Committee on Energy and Commerce of the United States House of Representatives in October 1993, Louis Guillette of the University of Florida reported that

examination of hatchlings and young alligators revealed that males had been feminized: They had estrogen concentrations typical of females and virtually no testosterone, the phallus was one-half to one-third the normal size, and their gonads had the structure of ovaries rather than testes (Committee on Energy and Commerce 1994). Female alligators from Lake Apopka had about twice the normal concentration of estrogen and the eggs in their ovaries were both abnormal and too numerous. The pesticide dicofol and its contaminants DDT, DDE, and DDD are probably responsible for these abnormalities. When alligator eggs from uncontaminated sites were painted with dicofol, the hatchlings had hormone levels that were nearly identical to hatchlings from Lake Apopka. Dicofol is just one example of an environmental hormone that acts like a synthetic estrogen. Similar effects are produced by other pesticides, including kepone, heptachlor, dieldrin, mirex, and toxaphene, as well as by polychlorinated biphenyls (PCBs), dioxin (a compound that is a contaminant of some herbicides and is produced by incinerating some plastics), and alkylphenol polyethoxylates (APEs, a group of surfactants used in many dishwashing liquids). In addition to alligators, the list of vertebrates that have probably been feminized by these synthetic estrogens includes fishes (trout, carp, sturgeon), birds (gulls, birds of prey), and mammals (the Florida panther) (Raloff 1994). Humans are probably not immune from the effects of environmental estrogens. Between 1938 and 1991 the average sperm densities of human males in Europe and North America declined from 113 million per milliliter of seminal fluid to 66 million per milliliter, and the average volume of seminal fluid decreased from 3.40 milliliters to 2.75 milliliters. Feminization by environmental estrogens has been suggested as the cause of this phenomenon (Carlsen et al. 1992).

The magnitude and complexity of the problems facing conservation biologists are daunting, and the scale on which remedial efforts must be attempted is nearly beyond comprehension. The growth of human populations is at the root of the crisis in biodiversity. The United Nations Population Fund estimated that in 1994 the human population of the Earth was 5.7 billion, and that it will reach 8.3 billion by 2025 and 9.8 billion by 2050. These increases will occur despite the fact that the total fertility rate (births per woman) is declining because the base population (to which the fertility rate applies) keeps growing (Bongaarts 1994). More people means more agriculture, more pollution, and a reduction in habitat available for other species.

Reviews of the biodiversity crisis agree that an essential first step in coming to grips with the problem is a clear understanding of what species exist, where they are, and what are the critical elements in their survival (Wilson 1992, World Conservation Monitoring Centre 1992). This is an enterprise that must enlist biologists from specialties as diverse as systematics, ecology, behavior, physiology, genetics, nutrition, and animal husbandry. Academic biologists have sometimes stood aloof from colleagues who work in applied areas of biology, but that kind of elitism hinders progress toward the synthesis we need. Throughout this book, we have noted areas in which a combination of basic and applied information about vertebrates has the potential to contribute to conservation of biodiversity.

Summary

The 45,000 species of living vertebrates span a size range from less than a gram to more than 100,000 kilograms and live in habitats from the bottom of the sea to the tops of mountains. This extraordinary diversity is the product of 500 million years of evolution. Evolution means a change in the relative frequencies of alleles in the gene pool of a species. Heritable variation among the individuals of a species is the raw material of evolution, and natural selection is the mechanism that produces evolutionary change. Natural selection works by differential reproduction, and fitness describes the relative contribution of different individuals to future generations. Most selection probably operates at the level of individuals, but it is possible that selection also operates at the level of alleles, populations, and even species.

In addition to individual variation, species often exhibit sexual dimorphism and geographic variation. Sexual dimorphism reflects the different selective forces that act on males and females of a species as a result of the asymmetry of reproductive investment by the two species. In most cases males can maximize their fitness by mating with as many females as possible, whereas females should attempt to choose the best possible mate. When a local population of a species is isolated from the rest of the species, genetic differences accumulate. If genetic differences become extensive, individuals of the isolated population may be unable to breed with individuals of the main population when they reestablish contact. This process is called allopatric speciation.

The Earth has changed dramatically during the half billion years of vertebrate history. Continents were fragmented when vertebrates first appeared; coalesced into one enormous continent, Pangaea, about 300 million years ago; and began to fragment again about 100 million years ago. This pattern of fragmentation–coalescence–fragmentation has resulted in isolation and recontact of major groups of vertebrates on a worldwide basis. On a continental scale the advance and retreat of glaciers in the Pleistocene caused homogeneous habitats to split and merge repeatedly, isolating populations of widespread species and leading to the evolution of new species.

Phylogenetic systematics, usually called cladis-tics, classifies animals on the basis of shared derived character states. Natural evolutionary groups can be defined only on the basis of these derived characters; retention of ancestral characters does not provide information about evolutionary lineages. Application of this principle produces groupings of animals that reflect evolutionary history as accurately as we can discern it, and forms a basis for making hypotheses about evolution.

The diversity of vertebrates has increased steadily (albeit with several episodes of extinction) for the past 500 million years, and peaked in the Pliocene or Pleistocene. Much of the decline in diversity of vertebrates (and other forms of life) since then can be traced to the direct and indirect effects of humans on the other species with which we share the planet. The major threats to the continued survival of species of vertebrates include habitat destruction, pollution, and hunting. At the base of all these phenomena is the enormous increase in human population size. Knowledge of vertebrates and other organisms is an essential part of the conservation of biodiversity, but control of human population growth is the only solution to the biodiversity crisis.

References

Benton, M. J. 1990. Patterns of evolution and extinction in vertebrates. Pages 218–241 in *Evolution and the Fossil Record*, edited by K. Allen and D. Briggs. Smithsonian Institution Press, Washington, DC.

Blaustein, A. R., P. D. Hoffman, D. G. Hokit, J. M. Kiesecker, S. C. Walls, and J. B. Hays. 1994a. UV repair and resistance to solar UV-B in amphibian eggs: a link to population declines. *Proceedings of the National Academy of Sciences, USA* 91:179.

Blaustein, A. R., D. B. Wake, and W. P. Sousa. 1994b. Amphibian declines: judging stability, persistence, and susceptibility of populations to local and global extinctions. *Conservation Biology* 8:60–71.

Bongaarts, J. 1994. Population policy options in the developing world. *Science* 263:771–776.

Breden, F., and G. Stoner. 1987. Male predation risk determining female preference in the Trinidad guppy. *Nature* 329:831–833.

Britten, R. J. 1986. Rates of DNA sequence evolution differ between taxonomic groups. *Science* 231:1393–1398.

Carlsen, E., A. Giwercman, N. Keiding, and N. E. Skakkebaek. 1992. Evidence for decreasing quality of semen during the past 50 years. *British Medical Journal* 305:609–619.

Charig, A. 1982. Systematics in biology: a fundamental comparison of some major schools of thought. Pages 363–440 in *Problems of Phylogenetic Reconstruction*, edited by K. A. Joysey and A. E. Friday. Academic, New York, NY.

Committee on Energy and Commerce, U. S. House of Representatives. 1994. *Health Effects of Estrogenic Pesticides*. Serial No. 103-87. U. S. Government Printing Office, Washington, DC.

Connor, E. F. 1986. The role of Pleistocene forest refugia in the evolution and biogeography of tropical biotas. *Trends in Ecology and Evolution* 1:164–168.

de Queiroz, K. 1988. Systematics and the Darwinian revolution. *Philosophy of Science* 55:238–259.

de Queiroz, K., and J. Gauthier. 1992. Phylogenetic taxonomy. *Annual Review of Ecology and Systematics* 23:449–480.

Diamond, J. 1989. Quaternary megafaunal extinctions: variations on a theme by Paganini. *Journal of Archaeological Science* 16:167–175.

Eberhard, W. G. 1985. *Sexual Selection and Animal Genitalia*. Harvard University Press, Cambridge, MA.

Eldredge, N., and S. J. Gould. 1972. Punctuated equilibria: an alternative to phyletic gradualism. Pages 82–115 in *Models in Paleontology*, edited by T. J. M. Schopf. Freeman, Cooper, San Francisco, CA.

Endler, J. A. 1986. *Natural Selection in the Wild*. Monographs in Population Biology, No. 21. Princeton University Press, Princeton, NJ.

Friedman, M. J. 1978. Erythrocyte mechanism of sickle cell resistance to malaria. *Proceedings of the National Academy of Sciences (USA)* 75:1994–1997.

Frost, D. R., and D. M. Hillis. 1990. Species in concept and practice: herpetological applications. *Herpetologica* 46:87–104.

Gibbs, H. L., and P. R. Grant. 1987. Oscillating selection on Darwin's finches. *Nature* 327:511–513.

Gould, S. J. 1988. The heart of terminology. *Natural History* 97(2):24–31.

Gould, S. J., and N. Eldredge. 1977. Punctuated equilibrium: the tempo and mode of evolution reconsidered. *Paleobiology* 3:115–151.

Grant, P. R. 1986. *Ecology and Evolution of Darwin's Finches*. Princeton University Press, Princeton, NJ.

Graveland, J., R. van der Wal, J. H. van Balen, and A. J. van Noordwijk. 1994. Poor reproduction in forest passerines from decline of snail abundance on acidified soils. *Nature* 368:446–448.

Hall, B. K. (editor). 1994. *Homology, The Hierarchical Basis of Comparative Biology*. Academic, New York, NY.

Hennig, W. 1966. *Phylogenetic Systematics*. University of Illinois Press, Urbana, IL.

Heywood, V. H., G. M. Mace, R. M. May, and S. N. Stuart. 1994. Uncertainties in extinction rates. *Nature* 368:105.

Jablonski, D. 1986. Background and mass extinctions: the alternation of evolutionary regimes. *Science* 231:129–133.

Kerr, J. B., and C. T. McElroy. 1993. Evidence for large upward trends of ultraviolet-B radiation linked to ozone depletion. *Science* 262:1032–1034.

Lack, D. 1947. *Darwin's Finches*. Cambridge University Press, Cambridge, MA.

Lewin, R. 1988a. Conflict over DNA clock results. *Science* 241:1598–1599.

Lewin, R. 1988b. DNA clock conflict continues. *Science* 241:1756–1759.

Mayr, E. 1942. *Systematics and the Origin of Species*. Columbia University Press, New York, NY.

Olson, S. L., and H. F. James. 1991a. Descriptions of thirty-two new species of birds from the Hawaiian Islands. Part 1: Nonpasseriformes. *Ornithological Monographs* 45:1–88.

Olson, S. L., and H. F. James. 1991b. Descriptions of thirty-two new species of birds from the Hawaiian Islands. Part 2: Passeriformes. *Ornithological Monographs* 46:1–88.

Phillips, K. 1994. *Tracking the Vanishing Frogs*. St. Martin's Press, New York, NY.

Raloff, J. 1994. The gender benders. *Science News* 145(2):24–27.

Räsänen, M. E., J. S. Salo, and R. J. Kalliola. 1987. Fluvial perturbance in the western Amazon Basin: regulation by long-term sub-Andean tectonics. *Science* 238:1398–1401.

Simpson, G. G. 1961. *Principles of Animal Taxonomy*. Columbia University Press, New York, NY.

Slatkin, M. 1987. Gene flow and the geographic structure of natural populations. *Science* 236:787–792.

Stanley, S. M. 1975. A theory of evolution above the species level. *Proceedings of the National Academy of Sciences* (USA) 72:646–650.

Steadman, D. W. 1995. Prehistoric extinctions of Pacific island birds: biodiversity meets zooarchaeology. *Science* 267:1123–1131.

Terborgh, J. 1989. *Where Have All the Birds Gone?* Princeton University Press, Princeton, NJ.

Vawter, L., and W. M. Brown. 1986. Nuclear and mitochondrial DNA comparisons reveal extreme rate variation in the molecular clock. *Science* 234:194–196.

Vrba, E. S. 1980. Evolution, species and fossils: how does life evolve? *South African Journal of Science* 76:61–84.

Wake, D. B. 1991. Declining amphibian populations. *Science* 253:860.

Wiley, E. O. 1978. The evolutionary species concept reconsidered. *Systematic Zoology* 27:17–26.

Wilson, E. O. 1992. *The Diversity of Life*. Norton, New York, NY.

World Conservation Monitoring Centre. 1992. *Global Biodiversity: Status of the Earth's Living Resources*. Chapman & Hall, London, UK.

Zuckerkandl, E., and L. Pauling. 1962. Molecular disease, evolution and genetic heterogeneity. Pages 189–225 in *Horizons in Biochemistry*, edited by M. Kasha and B. Pullman. Academic, New York, NY.

2

THE ORIGIN OF VERTEBRATES

The principles of evolution and natural selection that we described in the preceding chapter apply to all forms of life, not just to vertebrates. What are the derived characters that make vertebrates unique? In this chapter we explain what characteristics are necessary to diagnose a vertebrate, and describe the systems that make a vertebrate self-sustaining. We also address the question of the origin of vertebrates: What does a consideration of the derived characters of vertebrates tell us about the phylogenetic relationships of vertebrates? If we could identify the sister group of vertebrates, would that information help us to explain the process by which vertebrates arose from invertebrate ancestors? We compare the most widely accepted view of the origin of vertebrates with several alternatives that may provide different insights about the events and mechanisms responsible for vertebrate evolution.

The Significance of Similarity and Differences

Each group of vertebrates differs from all others in some fundamental way, but all share the common chordate–vertebrate characters. What is the meaning of the underlying similarity? Since Darwin's *Origin of Species* we have understood that the sharing of fundamental similarities (now understood to be of identical genetic and evolutionary origin), or homologies, among widely different groups of species indicates that they evolved from a common ancestor that possessed the same features. In general, the more homologies two species share, the more closely related they are.

A relationship between different taxonomic groups can be established only on the basis of shared derived characters. Ancestral character states, in contrast, are features that are shared by all members of a taxonomic grouping. The notochord, for example, is an ancestral character state for vertebrates because all chordates share the trait. The notochord is a shared derived character that separates chordates from other deuterostomes, such as the echinoderms, which do not possess a notochord.

What is the meaning of the diversity within a lineage of related groups and species? The differences relate to adaptation to different environmental conditions or opportunities. Each species exists in an ecological niche that is different from all others and that is often expressed by specialized body form and function. The diversity of species is indicative of the genetic responsiveness of each group to environmental differences.

Evolution and adaptation are major themes of this book. We shall therefore direct attention to the following sorts of questions. What were the historical, ancestral precursors of any structure or behavior under consideration? How does the structure or behavior promote survival and reproduction of the organism in its natural environment? Was the original function of the structure or behavior the same as its current function?

Adaptations can be tricky subjects to write about (Burian in Keller and Lloyd 1992), because an adaptation, presumably, has some useful function in the life of the organism possessing it. Because humans are purposive animals, perceive the means to ends, and anticipate results prior to their achievement, some philosophers and scientists of an earlier era ascribed a divine purpose as the cause for useful adaptations and for evolution (Singer 1959). This philosophy is called **teleology**.

When biologists discuss adaptations they refer to alterations in structure or function that result from natural selection operating on the genetic variability of organisms. These alterations increase the fitness of the individuals affected. Adaptations emerge, therefore, without the existence of a prior purpose. This scientific explanation of adaptations has been called **teleonomy** (Williams 1966, West–Eberhard in Keller and Lloyd 1992).

Some Familiar Facts About Vertebrates

Vertebrates derive their name from the serially arranged vertebrae, the axial endoskeleton that vertebrates share as a common diagnostic character (Figure 2–1). Skeletal elements have been elaborated into a cranium or skull, which houses various sense organs and a complex brain. An older name for the vertebrates is Craniata. In fact, the distinctive vertebrate cranium and tripartite brain evolved before the vertebral column and they are, perhaps, more characteristic of vertebrates than is the backbone. **Craniata** is currently used to include with the vertebrates those ancient forms that lack vertebrae but have cephalization and a protective box around the brain.

Vertebrates also share some fundamental morphological features with certain marine invertebrates, and are classified with them in the phylum **Chordata**. The common chordate structures are the **notochord, muscular postanal tail, dorsal hollow nerve cord**, and **pharyngeal slits** (at least in the majority of forms: see the discussion of *Pikaia* and conodontids below). Only the nerve cord remains as a functional entity in the adult stage of many vertebrates, but all four chordate features are evident at some stage in the development of all vertebrates. This is generally true also of the urochordates (tunicates or ascidians) and cephalochordates (acraniates or lancelets).

We now have the minimum information needed to define a **vertebrate**. A vertebrate is a chordate ani-

mal that has a cartilaginous or bony endoskeleton. The axial components of this endoskeleton consist of a cranium housing a three-part brain and a vertebral column through which the nerve cord passes. No other animals possess this constellation of fundamental characters. Bone is also a unique vertebrate tissue, which evolved as early as the late Cambrian. So far we have identified only the fundamental features of vertebrate morphology. Other organs and structures are required to make even the simplest vertebrate into a self-maintaining reproducing organism. There are ten organ systems involved in the vital functions of all vertebrates. These systems are listed in Table 2–1 and are described more completely in Chapter 3.

The Basic Vertebrate Body Plan and a Search for the Relatives of the Vertebrates

What is the minimum functional organization that we can call vertebrate? Do any vertebrates actually conform to this generalized plan? Or is it an abstraction based on an a priori assumption that evolution has proceeded from simple to complex organization?

The symmetry of a hypothetical ancestral vertebrate would certainly be bilateral with definite head and tail ends. The basic internal organization would be a tube-within-a-tube arrangement with the major internal organs lying within a body cavity or coelom (Figure 2–2). In these respects the ancestral vertebrate is not unlike many derived invertebrate animals, which also possess these general body features —mollusks, annelids, and arthropods.

Which animals are most closely related to vertebrates? The fact that no fossils are clearly intermediate between the earliest vertebrates known, the "**ostracoderms**" (a paraphyletic group) and any other group of animals presents difficulties for a biologist interested in the origin of the vertebrates. Indirect evidence must be used to infer evolutionary relationships between the vertebrates and other animals.

Lophophorates, Deuterostomes, and Chordates

Almost every invertebrate phylum has been suggested as the closest relative of the chordate line.

Figure 2–1 Examples of bony structures typical of vertebrates: (a) vertebra; (b) skull.

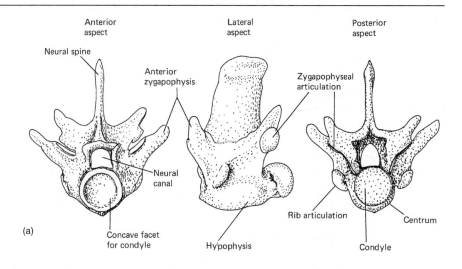

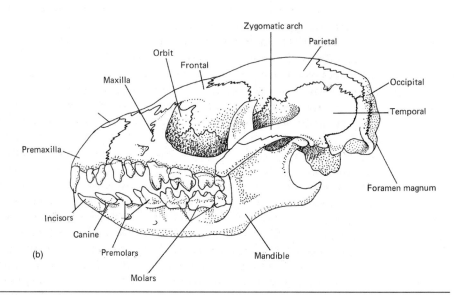

Affinities have been postulated with annelids, arachnids, and arthropods. However, developmental patterns in various coelomate animals reveal a different and rather surprising conclusion. The Chordata and Hemichordata, which are bilaterally symmetrical, are related to the Echinodermata, marine forms with radial symmetry as adults.

Coelomates evolved two different developmental patterns for forming the early embryonic and larval stages (Nielsen 1985, Schaeffer 1987). The Mollusca, Arthropoda, and Annelida are grouped as the **protostomes** (*proto* = first, *stoma* = mouth). The other phyla, which share a distinctive embryology, are placed in the **deuterostomes** (*deuter* = second). The blastopore of protostomes ultimately forms the mouth of the organism, whereas the blastopore of deuterostomes forms the anus. Table 2–2 summarizes the main differences between these superphyla. The filter-feeding Lophophorates (Phyla Phoronida, Ectoprocta, and Brachiopoda) do not fit this scheme because they possess developmental features of both protostomes and deuterostomes. Like the echinoderms, the lophophorates have free-swimming ciliated larvae. Quite probably the ancestor of echinoderms, chordates, and related phyla was a generalized sessile (fixed in one place) or semisessile lophophore-bearing animal as an adult with a free-swimming ciliated larval stage for dispersal.

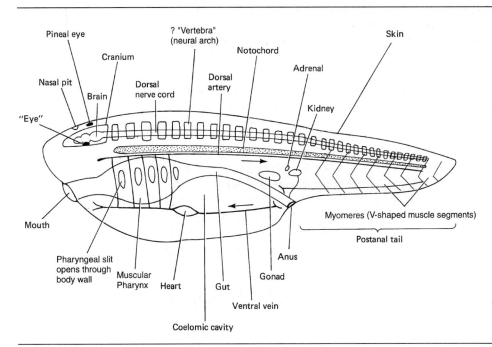

Figure 2–2 Body plan of a hypothetical vertebrate.

Pineal eye
? "Vertebra" (neural arch)
Cranium
Notochord
Skin
Adrenal
Nasal pit
Dorsal artery
Dorsal nerve cord
Brain
Kidney
"Eye"
Mouth
Myomeres (V-shaped muscle segments)
Postanal tail
Anus
Pharyngeal slit opens through body wall
Muscular Pharynx
Heart
Gut
Gonad
Ventral vein
Coelomic cavity

One existing genus of pterobranch hemichordate, *Cephalodiscus*, has a single pair of pharyngeal slits in addition to ciliated tentacles on arms, showing the potential for transitional stages between two modes of feeding. Changes may have occurred in the mobile larvae of prechordates that improved locomotor and feeding mechanisms, and in the sessile adult stage in which the feeding apparatus altered radically. From external food collection on arms or tentacles where cilia moved food particles to the mouth, development of slits in the pharynx permitted internal collection of food particles from a stream of water driven by cilia.

An ancestral tornaria-like larva may have gradually changed into an elongate, prototadpole larva, propelled by bands of cilia and with slits in the pharynx and a means to entrap food particles. These transformations produced the typical chordate tadpole larva. Emancipated from the severe limitations imposed by ciliary locomotion, such a larva could increase in size, and if it could become reproductive various evolutionary paths would become possible.

Examination of the comparative anatomy and embryology of living forms helps to bridge the morphological gap between vertebrates and other metazoa. New fossils are continually changing our perspectives on the origin of vertebrates. The closest relatives of vertebrates appear to lie among the invertebrate chordates—the **Urochordata** (tunicates or sea squirts and their kin) or the **Cephalochordata** (amphioxus or lancelets)—which possess a notochord, dorsal hollow nerve cord, pharyngeal slits, and postanal tail. No other group of living animals shows closer structural and developmental affinities with vertebrates.

The larvae of tunicates (**Urochordata**) are tadpole-like animals with pharyngeal slits, notochord, and dorsal hollow nerve tube fully developed. In addition, they possess a muscular, postanal tail, which moves in a fish-like swimming pattern (Figure 2–3). The larvae are sufficiently generalized to suggest that they represent only slightly modified living forms of chordates related to the vertebrate line of evolution. However, tunicate larvae, after a brief free-swimming existence generally of no more than a few minutes to a few days, metamorphose into adult animals that bear almost no resemblance to vertebrates. Adult tunicates are anchored or free-living animals with body plans that are so divergent from those of vertebrates that it is hard to imagine an evolutionary transformation from one to the other.

By what evolutionary process could the larval traits of the tunicate tadpole be passed on to succeeding generations of adult animals to set the stage for the evolution of the cephalochordates and vertebrates? A scenario was proposed by W. Garstang in

Table 2-1 **Basic vertebrate systems.**

System	Basic Functions	Major Components
Integumentary	1. Protect underlying tissues from injury. 2. Prevent excessive loss or absorption of water and the consequent effect on tissues. 3. Aid excretion and absorption of specific metabolites and ions. 4. Almost all sense organs are derived in part from the integument.	Skin: Composed of epidermis above and the dermis below (Greek, *epi* = upon; *derma* = skin) and the derivatives of these two layers, such as scales, feathers, and hair.
Skeletal	1. Provide a framework for all body systems. 2. Provide attachments for muscles, tendons, and fascia. 3. Enclose and protect vital organs. 4. Serve as a reserve storehouse for minerals.	Bones, cartilage, and ligaments. These tissues are divided into: 1. Axial skeleton: skull, vertebral column, and (when present) ribs. 2. Appendicular skeleton: pectoral and pelvic girdles and limbs (when present).
Muscular	1. Movement of body and body parts. 2. Maintenance of posture. 3. Internal transport and expulsion (movement of food through digestive tract, blood through vessels, germ cells through reproductive tract, bile from gallbladder, urine from kidneys, feces from alimentary canal). 4. Homeostatic adjustments such as size of opening of the pupil of the eye, the pylorus, the anus, blood vessels; heat production in some vertebrates.	Smooth (nonstriated) muscles of involuntary control found primarily in wall of digestive tract, genital ducts, and blood vessels. Cardiac muscle of involuntary control restricted to the heart. Striated muscles generally under voluntary control found attached to the skeleton so intimately that the name "musculoskeletal system" is often applied; tendons (the connective tissue bands that bind striated muscle to bone).
Digestive	1. Capture and physical/chemical disintegration of food. 2. Absorption, detoxification, alteration, storage, and controlled release of the products of digestion and metabolism.	Alimentary canal: mouth and oral cavity with associated teeth, tongue, and jaws (when present). Pharynx (associated intimately with the respiratory system). Esophagus. Stomach (when present). Intestine divided and specialized in various ways. Accessory glands: salivary (when present). Liver [responsible for most of the functions listed in (2)]. Pancreas.
Circulatory	1. Transport of materials to and from cells. 2. Transport, formation, and storage of blood cells for oxygen transport, defensive, and immunogenic functions. 3. Drain fluid from between cells and return it to the regular circulatory system from which it leaked.	Heart. Arteries (from the heart to the tissues). Arterioles (small arteries). Capillaries (extremely small vessels connecting arterioles and venules). Venules (small veins). Veins (from tissues to the heart). Spleen (and other sites in various vertebrates, but always intimately associated with the digestive tract and/or skeletal system). Lymphatic system.

1928. He suggested that vertebrates originated from tunicate-like larvae that failed to metamorphose but nevertheless developed functional gonads and reproduced: The genotype could be passed on without the necessity of a separate adult morph in the life cycle (Figure 2–4).

Garstang termed this evolutionary process **paedomorphosis** (*paedo* = child, *morph* = form). The

Table 2-1 *(Continued)*

System	Basic Functions	Major Components
Respiratory	1. Exchange of gases (primarily intake of oxygen and discharge of carbon dioxide) between the organism and its environment (water or air). 2. Various accessory functions from production of sound to nest building.	Lungs, gills, and/or skin, depending on which groups of vertebrates are under discussion; lungs and gills are derived from and intimately connected with the pharyngeal region of the digestive system.
Excretory	Chemical (and to a lesser extent physical) homeostasis or maintenance of constant internal environment by (1) excreting toxic and metabolic waste products, especially those containing nitrogen; (2) maintaining proper waste balance; (3) maintaining proper concentration of salt and other substances in the blood; (4) maintaining proper acid–base equilibrium in body fluids.	Kidneys and excretory ducts (tube-like passages) variously aided by the gills, lungs, skin, and/or intestines. The mode of development and use of common ducts makes this and the reproductive system inseparable morphologically so that the two are often referred to as the urogenital system.
Reproductive	Formation of zygotes by the union of two gametes to produce new individuals of the same biological variety.	Primary sex organs in the form of male (testes) or female (ovaries) gonads. Secondary sex organs concerned with transport of gametes from their site of formation to their site of union. Accessory sex organs assuring union of gametes, such as glands and external genitalia.
Endocrine	Regulation and correlation/integration of body activities through chemical substances (*hormones*) carried by the blood. The endocrine system acts more slowly than the nervous system, but it is capable of continuous action.	A large number of cell types discharge secretions that have regulatory effects on other cells. In plesiomorphic vertebrates, these cells tend to be widely scattered in other tissues. Derived vertebrates have discrete aggregations of these cells (endocrine glands).
Nervous	Regulation and correlation/integration of body activities through conduction within and between individual nervous cells or neurons, which eventually cause a response in some other system (e.g., muscular contractions). The nervous system is fast acting, and conduction velocity may exceed 90 meters per second.	Central nervous system (CNS): brain, spinal cord. Peripheral nervous system (PNS): craniospinal nerves, which exit from the protective skeletal sheath of the cranium and vertebrae and may be voluntary (to striated muscles) or involuntary (to smooth muscles); involuntary nerves are often referred to collectively as the autonomic nervous system. Sensory nerves from either complex organs (e.g., eye, ear) and simple receptors (e.g., cutaneous sensory nerves) enter the CNS via the craniospinal nerves.

attainment of sexual maturity in an arrested larval animal has also been called paedogenesis, progenesis, and neoteny, with the result that confusion exists about two rather different evolutionary–developmental processes (Gould 1977). **Neoteny** (*neo* = new, i.e., juvenile; *ten* = hold) is the retention of one or more larval or embryonic traits in the adult body. Neoteny results from slowing down or stopping the

Table 2-2 **Some basic differences between the protostomes and deuterostomes.**

Feature Compared	Protostomes	Deuterostomes
Cleavage	Spiral and determinate	Radial and indeterminate
Blastospore	Forms the mouth	Forms the anus
Coelom formation	Schizocoelous (a rift or split within the mesoderm)	Enterocoelous (outpocketing from the dorsolateral gut wall); *except vertebrates*
Type of larva	Trochophore	Tornaria or bipinnaria *except chordates*

development of the traits affected. Some familiar neotenic traits include the external gills of certain adult aquatic salamanders in which the larval gills are only slightly modified, and the small face and expanded brain case of human beings, traits that are widespread in mammalian embryos but are retained in the postfetal stages of humans. Although neoteny is characterized by the retention of embryonic or larval characteristics in an otherwise typically adult body, **progenesis** (*pro* = forward, i.e., early; *genesis* = birth, i.e., reproduction) describes the early development of the gonads in an otherwise larval body. The developmental process that leads to progenesis is thus the opposite of the one leading to neoteny, but both processes are a part of the process of paedomorphosis, whereby ancestrally juvenile, larval, or embryonic traits are incorporated in the reproductive stage of the species.

Both neoteny and progenesis are examples of a process called **heterochrony** (*hetero* = different, *chron* = time), that is, a change in the timing of events during development. For example, a vertebrate lineage that retained juvenile characteristics when it was sexually mature would have a proportionally larger head and eyes than normal for its sister group. Modifications of growth and ontogenetic development are usually not so exaggerated, but over evolutionary time heterochronic changes can have spectacular effects. Paedomorphosis of a larval tunicate-like organism, for example, may have initiated the entire vertebrate lineage.

Several additional hypotheses define vertebrate affinities with invertebrate taxa (Box 2–1). Some represent slight modifications of Garstang's original speculations (Figure 2–5b), whereas others differ markedly and, particularly, do not invoke paedomorphosis as a necessary step in the origin of vertebrates (Figure 2–5, c–e). The precise phylogenetic affinities between chordates and the other metazoa remain uncertain (Dzik 1993). Clearly, the weight of evidence favors deuterostomes as the group from which chordates arose, but no living adult or larval deuterostome is a strong candidate for closest living nonchordate relative.

Amphioxus as a Model Prevertebrate

The marine **Cephalochordata**, particularly the well-studied amphioxus or lancelet (*Branchiostoma lanceolatum*), possess the basic chordate characters in diagrammatic form. Significantly, their mode of filter feeding is similar to that of the larval stage of lampreys (Chapter 6) except that they use ciliary, not muscular, pumping. These characters point to a relationship with vertebrates and suggest that the morphology of cephalochordates might provide hints about the chordate relatives of the vertebrates.

The Cephalochordata consist of some 22 species, all of which are small, fusiform, superficially fish-like, marine animals usually less than five centimeters long. Lancelets are widely distributed in oceanic waters of the continental shelves and are burrowing, sedentary animals as adults. The main features of the lancelet's morphology relate to its methods of locomotion and feeding (Figure 2–6).

A significant characteristic of amphioxus is its fish-like locomotion, resulting from the contraction of serially arranged **myomeres**, blocks of striated muscle fibers arranged dorsolaterally along both sides of the body and separated by sheets of connective tissue. Contractions of the myomeres bend the body in a way that results in forward propulsion. (For a discussion of swimming, see Chapter 8.) An incompressible, elastic rod, the notochord, extends the full length of the cephalochordate body and prevents the body from shortening when the myomeres contract. As a result, the myomeral contractions bend the body. The notochord of amphioxus extends from the tip of the snout to the end of the tail, projecting well beyond the region of the myomeres at

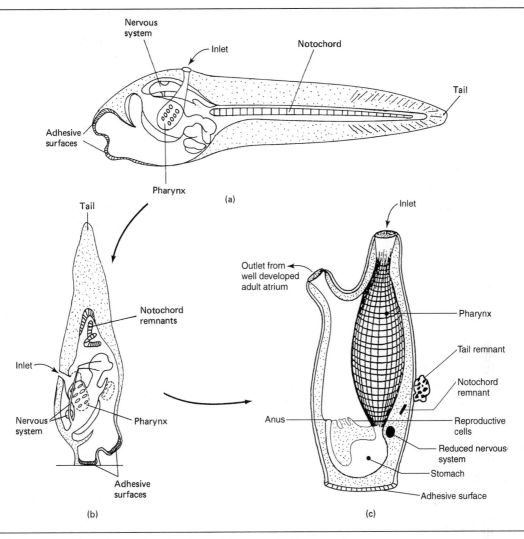

Figure 2–3 Tunicates: (a) the free-swimming larva, (b) an attached larva undergoing metamorphosis, (c) and the sessile adult.

both ends. This elongation of the notochord apparently aids in burrowing.

Amphioxus filters small food particles from a stream of water drawn through its mouth and pharynx by the movements of cilia. As in all animals that use cilia for filtering, a large surface area is required to obtain sufficient food. The pharyngeal apparatus occupies more than half of the body length. Its walls are perforated by up to 200 oblique, vertical slits on which cilia are located. The slits are separated by bars with internal skeletal rods. The sidewall of the perforated pharynx is covered by the protecting walls of the atrium. Filtered water exits to the outside of the animal through a posterior atriopore. Solid particles

entrained in the water current encounter a variety of organs that separate edible from inedible matter and conduct the entrapped food to the digestive tract. Buccal cirri, attached to the margin of the oral hood in front of the mouth, possess sensory cells and form a funnel-like sieve that prevents the entry of large particles. The velum presumably functions as an additional screening mechanism against unwanted items. Food particles are caught partly on a complex set of ciliated tracts known as the wheel organ. Mucus from Hatschek's pit is discharged onto the wheel organ, and food particles caught in this material are swept by cilia toward the mouth and enter with the main stream of water.

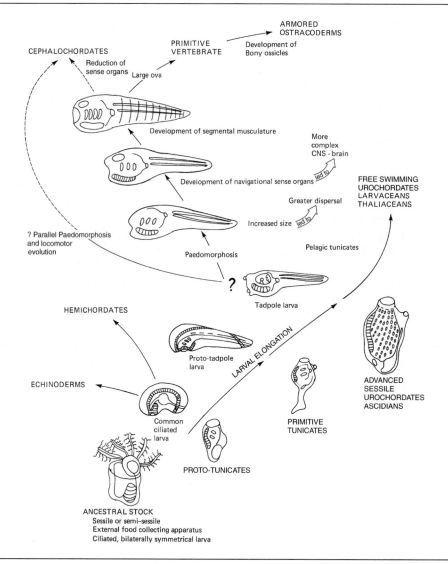

Figure 2–4 Garstang's hypothesis of the origin of vertebrates by paedomorphosis from an ancestral lophophorate.

The endostyle, a mucus-secreting organ on the floor of the pharynx, consists of ciliated cells alternating with mucus-secreting cells and produces sticky threads of mucus in which food particles become entangled. Ciliary currents then draw the food-carrying mucus strands up the sides of the pharynx and into a median, dorsal epipharyngeal groove in which cilia drive the material into the midgut. (A similar feeding mechanism exists among vertebrates—the food-filtering mechanism of the larva of lampreys—but a muscular pump is used to move the water instead of cilia.)

Amphioxus shows other fundamental similarities to the plan of vertebrate organization. It has a closed circulatory system in which slow waves of contraction drive the blood forward in the ventral blood vessels and posteriorly in the dorsal ones. Below the hind end of the pharynx there is a large sac, the sinus venosus, that collects blood from all parts of the body, like the vertebrate heart, although it is not the exclusive or even primary pumping organ. Amphioxus reveals no indication of a complex brain, possessing only an anterior expansion of the dorsal nerve cord into a cerebral vesicle capped

Box 2–1 Multiple Views of the Origin of Craniates

The origin of craniates continues to attract the attention of scientists, and new hypotheses are proposed every few years. Techniques such as DNA hybridization have been combined with new analyses of fossils and cladistic methods of classification to produce insights about the relationships between the craniates and their possible relatives. Garstang's original hypothesis (Figure 2–5a) involved paedomorphosis, the process of evolutionary selection acting on larval characters (Garstang 1928, Berril 1955, Tarlo 1960). Although Garstang's hypothesis fits the embryological evidence, the transformation of a tadpole-like tunicate larva into a vertebrate remains speculative. In the words of a recent critic, "neoteny of such a tadpole in the origin of craniates has the scientific status of a creation myth" (Jeffries 1986, p. 350). Nevertheless, paedomorphosis and the resulting transformations of neoteny and paedogenesis are common biological phenomena and Garstang's concept has not been falsified.

The phylogeny of the deuterostomes proposed by Schaeffer (1987) is based mostly on early embryonic characteristics that have classically been used to define the deuterostomes (Table 2–2). The question of paedomorphosis is bypassed in Schaeffer's suggested phylogeny and an amphioxus-like chordate is considered the sister group of the craniates (Figure 2–5b). Northcutt and Gans (1983) consider the origin of craniates from a functional viewpoint (Figure 2–5c). They largely avoid the question of the closest living relative, although they identify the prevertebrate as probably similar to the cephalochordates. They stress the development of the head anterior to the notochord as a structure unique to craniates. The head includes paired sense organs, a muscular water pumping pharynx, and respiratory gills. The functional innovation of craniates in their view was a shift from filter feeding to a predatory existence. This transition involved several phases (Figure 2–5c). A new embryonic tissue unique to craniates, the **neural crest** (Chapter 3), initiates the development of cranial nerves, branchial muscles, pharyngeal cartilage, and bone in the head region. The suggestions of Northcutt and Gans are compelling because they place the rapidly expanding information about structural affinities between the deuterostomes in an ecological and evolutionary framework. Unfortunately, just which fossils might be the "protovertebrates" is not established.

In 1977, Soren Lovtrup published *The Phylogeny of Vertebrates*, a work based on what he considered a rigorously objective and nonspecialized analysis of vertebrate affinities. The phylogeny harks back to older theories that considered that the roots of craniates were anchored in the protostomes (Figure 2–5d). Although Lovtrup's objectivity is to be admired, the ancestry of craniates that results ignores basic morphology, embryology, and fossil evidence, even when the evidence could illuminate relationships among living taxa. Thus, the hollow nerve cord, the notochord, the pharyngeal region, and most other common chordate characters are omitted. Instead, Lovtrup's phylogeny relies on physiological, chemical, and histological characters without weighing their relative importance, because he considers them more evolutionarily stable than morphological characters. If, for example, two out of three taxa share more biological chemicals than the third taxon, they are deemed more closely related. Although it is objective, his analysis denies the possibility of convergence between unrelated groups.

Another attempt to unravel vertebrate origins is a voluminous book by R. P. S. Jeffries (1986) entitled *The Ancestry of the Vertebrates*. Jeffries proposes that vertebrates and other living deuterostomes each originated from a widely distributed group of Paleozoic fossils—the calcichordates (Figure 2–5e and f). First formulated in detail by Torsten Gislen in 1930, the hypothesis was largely ignored because of the appeal of Garstang's theory of paedomorphosis.

Continued

Box 2–1 *(Continued)*

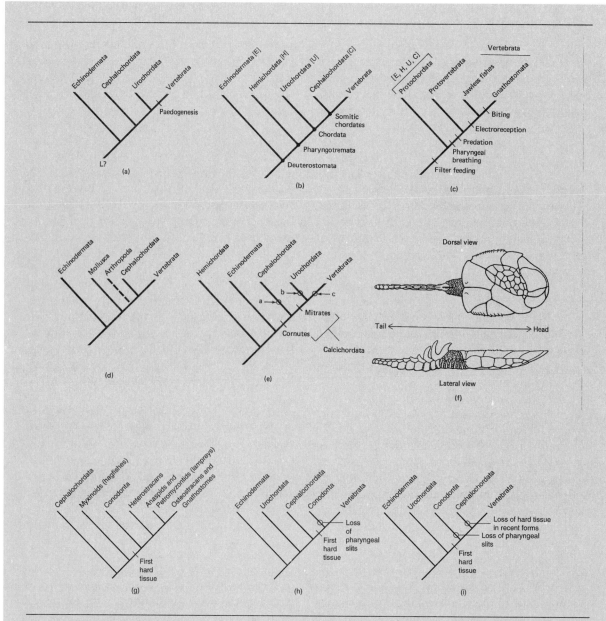

Figure 2–5 Suggested patterns of the phylogenetic relationship of vertebrates to other taxa: (a) Garstang's hypothesis of an origin by paedomorphosis; (b) Schaeffer's hypothesis, which places chordates as the sister group of vertebrates; (c) Northcutt and Gans's hypothesis based on an analysis of some functional characteristics of vertebrates; (d) Lovtrup's hypothesis based on nonmorphological characters; (e) Jeffries' hypothesis, which derives vertebrates from a calcichordate, such as that depicted in (f); the letters in (e) refer to genera of calcichordates: a, *Lagynocystis*; b, *Peltocystis*; c, *Chianocarpus*; (g) Briggs's inclusion of conodont animals; (h and i) other possible placement of conodontophorids.

Calcichordate fossils have usually been classified as echinoderms, but Jeffries contends that they are chordates. Unusual in this phylogeny is the assignment of specific fossil calcichordates as possible direct ancestors of each group of chordates. In Jeffries' scheme, tunicates (not cephalochordates) are the sister group of the craniates.

Jeffries' strict cladistic interpretation of the evidence is commendable. In question, however, are his interpretations of the fossil evidence and, especially, of the internal soft anatomy of the calcichordates. In a refreshing manner even Jeffries invites readers to be skeptical of his interpretations! Jeffries argues that calcichordates possess the requisite characters—a notochord, a segmented brain and trigeminal cranial nerves, a muscular postanal tail, and other structures shared with the chordates. Jeffries's hypothesis, like those of Schaeffer and of Northcutt and Gans, does not involve a paedomorphic transformation to explain the origin of craniates. The embryonic neural crest tissue that Northcutt, Gans, and Schaeffer designate as uniquely vertebrate, Jeffries suggests was present in the advanced calcichordates, because vertebrate structures derived from this tissue were present—a head already existed in the calcichordates, as did a tail. In his view, it was the development and elongation of the trunk from the posterior region of the calcichordate head that led to the craniates. This proposal is the opposite of the view of Northcutt and Gans, which stresses the origin of the head as critical to the evolution of predatory craniates.

The addition of the conodont animals has hardly had a clarifying effect on vertebrate phylogeny. D. E. G. Briggs (1992) helped muddy the ancestral waters with his cladogram (Figure 2–5g), because he interprets conodontophorans to be more closely related to the craniates than are hagfish. Another positioning for these curious beasts might be as the sister group of the craniates (Figure 2–5h), more closely related to them than are the cephalochordates, or as the sister group of the cephalochordates plus the craniates (Figure 2–5i). The possibility of dermal mineralization in the fossil cephalochordate *Palaeobranchiostoma* (Figure 2–5b) offers a resolution of evolutionary continuity of dermal skeleton if the latter scenario is accepted.

None of these hypotheses is unambiguously supported by data, and there is no definitive answer. What can be learned from the various phylogenies proposed for vertebrate origins is the importance of combining information from a wide variety of sources—embryology, molecular biology, morphology, physiology, chemistry, histology, modern classification techniques, and paleontology. It is crucial that any phylogeny be based on as much evidence as possible.

off anteriorly by a median, unpaired photosensitive pigment spot. Numerous other lines of evidence relate amphioxus to the craniates, including details of cranial and spinal nerve organization (Fritzsch and Northcutt 1993), muscle proteins, mitochondrial gene arrangements, DNA hybridization, and homeobox gene similarity (Gee 1994).

Despite these similarities, basic differences indicate that an amphioxus-like organism may not have been in the direct evolutionary lineage of the vertebrates. For example, the nerve cord of amphioxus has a superficially craniate-like series of paired spinal nerves that contain both sensory fibers (carrying information from sense organs) and motor fibers (sending impulses to stimulate muscles). Structures that were formerly called ventral spinal nerves are really specialized muscle fibers. Amphioxus differs from vertebrates in these respects, but its nerve pattern can be interpreted as an ancestral condition for vertebrates. Similarly, the excretory system of amphioxus is unlike that of vertebrates, but could be ancestral (Gans 1989). It consists of short excretory tubules that terminate in flask-shaped cells that drain body wastes from the body fluids bathing them. The open end of each tubule drains into the atrium. These structures have been called protonephridia, but they are much smaller than the nephrons in the kidneys of vertebrates.

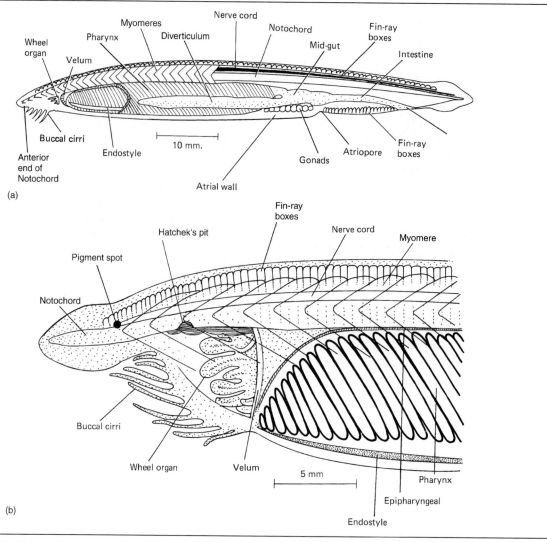

Figure 2–6 Cephalochordates: (a) lancelet, amphioxus—a longitudinal parasagittal section with the posterior myomeres removed; (b) detail of the anterior end of amphioxus showing the structures involved in filter feeding.

Amphioxus has no homologs of the paired eyes, ears, nose, or other cephalic sense organs of vertebrates. Some of the genes that are common to mice and amphioxus (homeobox genes) direct formation of the posterior part of the brain. The various members of this gene cluster are associated with distinct morphological landmarks along the anterior region of amphioxus, suggesting that this portion of an amphioxus is equivalent to the head of a craniate (Gee 1994). In the final analysis, most biologists believe that the cephalochordates are the closest living relatives of the vertebrates.

The middle Cambrian Burgess Shale in British

Columbia has yielded about 60 specimens of what has been called the world's first known chordate, the 40-millimeter *Pikaia* (Gould 1989). Although details of its anatomy remain to be described, it has frequently been compared to the cephalochordates (Figure 2–7a). A notochord running along the posterior two-thirds of the body and segmental, V-shaped muscle bands (myomeres) are the characters that most paleontologists note as obvious features. In addition, *Pikaia* had a small bilobed head with a pair of short, slender tentacles, perhaps one of the reasons it was originally described as a polychaete worm. Behind the head, bilaterally paired rows of about 12

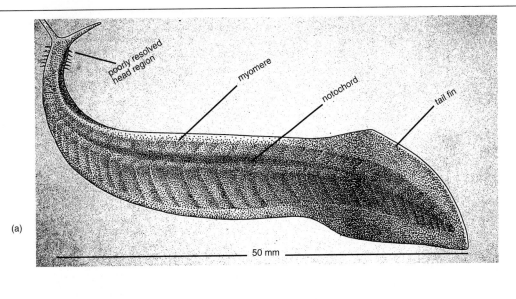

(a)

50 mm

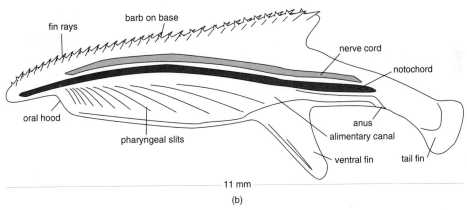

(b)

Figure 2–7 Fossil organisms thought to be related to cephalochordates: (a) *Pikaia* from the Middle Cambrian Burgess Shale of British Columbia. (Illustration by Marianne Collins from Stephen Jay Gould, 1989, *Wonderful Life: The Burgess Shale and The Nature of History*, Norton, NY. Reproduced with the permission of Marianne Collins and W. W. Norton & Company, Inc., copyright © 1989 by Stephen Jay Gould.) (b) *Palaeobranchiostoma* from the Early Permian of South Africa. Note the barbs on the dorsal fin, which may be dermal mineralized tissue not known in extant cephalochordates. (Modified from A. Blieck, 1992, *Geobios* [Lyons] 25[1]:101–113.)

tufts may represent exterior evidence of gill slits. The median fins show no signs of fin rays, but their presence and the lateral flattening of the body lead one to suspect that *Pikaia* swam above the sea floor (S. Conway Morris, unpublished).

In contrast, a well-developed branchial region characterizes the single early Permian specimen of *Palaeobranchiostoma*, which is clearly a relative of today's amphioxus (Figure 2–7b). It has dermal skeletal elements in the form of inverted V-shaped barbs studding its dorsal fin (Blieck 1992). Unfortunately, this South African fossil is only an 11-

millimeter imprint on shale and is much more recent than the critical time of craniate origins. It thus does little to help understand the origin of such craniate characters as the dermal skeleton. Cephalochordates provide insights about the likely ancestral craniate body plan. Cephalochordates and vertebrates represent two paths of evolution from a common ancestry far back in time.

Changes Leading to the First Vertebrates

Whether the process leading to the appearance of the vertebrates involved paedomorphosis or not, one can readily imagine how slight morphological changes could lead from sexually competent, tunicate-like larvae to organisms similar to amphioxus and the larvae of lampreys. The evolution of the first full vertebrates with crania and backbones, an increase in body size, the development of cephalic sense organs, an anterior brain, and extension of segmental muscles onto the trunk would have allowed the prevertebrates to reach a new level of organization. These new characteristics permitted successful life as swimming filter feeders in the sea, and a more active way of life could clearly follow as body size increased, the senses became keener, and prey capture mechanisms evolved (Northcutt and Gans 1983, Gans 1989, Forey and Janvier 1994).

Only the origin of bone remains a puzzle. Bone cannot easily be homologized with any skeletal tissues found among living invertebrate phyla, but it has some common features with the mineralized tissues of other animals. Like the exoskeletons of arthropods, shells of molluscs, and hard parts of other animals, bone results from the deposition of minerals in an organic matrix produced by special cells. The cells involved in the formation of calcified tissues derive from embryonic ectoderm and mesoderm, and the matrices produced usually contain polysaccharide and protein. The minerals deposited generally consist of calcium carbonate or calcium phosphate either in amorphous form or as crystals. Although bone is unique in its combination of characteristic cells, matrices, and minerals, all the components of the vertebrate hard skeleton exist in other animals. Thus, bone did not originate in early vertebrates as a completely novel tissue. Rather, it was a new elaboration of phenotypic features already well established in many metazoan lineages.

Curious microfossils known as conodonts are widespread and abundant in marine deposits from the late Cambrian to the late Triassic. These are small (generally less than one millimeter long) spine- or comb-like fossils of apatite. Their structure indicates that they were deposited in soft tissues by the apposition of concentric layers, exactly as were ancestral fish scales. Conodonts have a close similarity in chemical composition and microstructure to vertebrate bony scales. They have been variously described as skeletal parts of marine algae, as phosphatic annelid jaws, fish teeth, gill rakers, gastropod radulae, nematode copulatory spicules, arthropod spines, and denticles of free-swimming lophophorate animals (Higgins 1983). Conodont fossils of Silurian age began a new era in the group's taxonomic peregrinations—they have been claimed to be vertebrates akin to the jawless myxinoids or hagfishes (Briggs et al. 1983, Mikulic et al. 1985, Benton 1987).

When at last the fossil impressions of the soft tissues of the 40-millimeter-long conodont-bearing animal were discovered in North America and Europe, the chordate affinities of the Conodontophora were apparent (Figure 2–8). They seem to have had a notochord, myomeres, and perhaps even fin rays and two eyes protected by sclerotic rings composed of cartilage. Although the pharynx may have been muscular (to operate the conodont apparatus), there are no signs of pharyngeal slits. A close relationship between the Conodontophora and vertebrates has been suggested (Sansom et al. 1992, Briggs 1992), but this hypothesis is not completely accepted (e.g., Gans in Hanken and Hall 1993). Whether some or all of the conodont animals are eventually accepted as craniates, the sister group of craniates, or the sister group to cephalochordates plus craniates, they now seem firmly planted in the vertebrate ancestry (Gabbott et al. 1995, Purnell 1995).

Environment in Relation to the Origin of Bone

What selective advantage could mineralized tissues have had? Dentine and enamel-like tissues (which may predate bone in their evolution) may have originated as a protective and electrically insulating coating around the electroreceptors in the head of prevertebrates that enhanced electrosensitive detection of prey (Northcutt and Gans 1983, Gans in Hanken and Hall 1993).

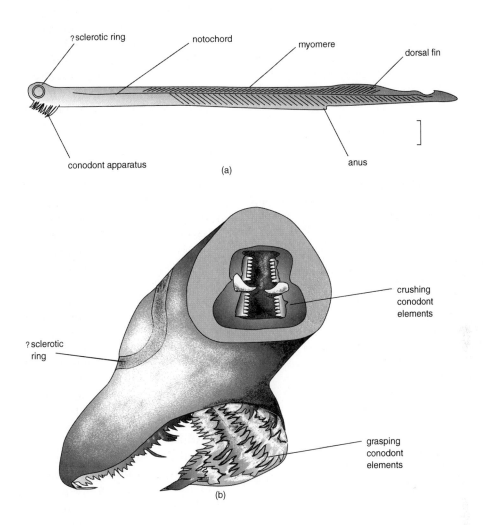

Figure 2–8 Conodont-bearing animals. (a) In lateral view and (b) close-up, cut-away view of head to show the position of the conodonts. (Modified from J. Dzik, 1993, *Evolutionary Biology* 27:339–386.)

Subsequently, the need for phosphorus, which is a relatively rare element in natural environments, may have been one of the early selective forces involved in the evolution of bone. Although we think of bone as primarily supportive or protective, it also serves as a store of calcium and phosphate. Mineral regulation involves deposition and mobilization of calcium and phosphate. High levels of activity by vertebrates usually employ anaerobic metabolism, which produces lactic acid as a metabolic product, and bone may be important in buffering this acid. Once mineralized tissues had evolved, they could have been transformed by other selective forces into the dermal teeth and armor plates of the "ostracoderms" (a paraphyletic group of early vertebrates).

Certainly the plates must have functioned as protective armor to thwart the attacks of powerful predators. The body design of early "ostracoderms" indicates that they were sluggish creatures and probably swam only for short distances before coming to rest on the bottom. They had no true paired appendages, and probably little maneuverability to avoid predators. Lacking jaws, they had little in the way of defense mechanisms. Their only defense against predation would be protective armor to discourage the onslaughts of grasping or biting feeders.

The "ostracoderm" fossils are frequently found in association with fossil eurypterids, giant scorpion-like arthropods. Eurypterids were dominant predators of the Cambrian, Ordovician, and Silurian. They first were marine, but became abundant in fresh waters during the Silurian. They were found in the same habitats as the ostracoderms, and it is likely that they were prominent among the predators that exerted the selective forces responsible for the evolution of extensive dermal bony armor in these small, vulnerable fishes.

Did Vertebrates Evolve in Marine or in Freshwater Habitats?

By the late Silurian, armored "ostracoderms" and early jawed fishes were abundant in both freshwater and marine environments. For years students of evolution have argued whether the vertebrates had a freshwater origin or a marine origin. Under what conditions did the first vertebrates evolve?

Suggestions of a Freshwater Origin

The Cambrian and Ordovician seas were rich environments and fossils of most major groups of animals have been found only in marine deposits of these periods. The first land plants appear in the Silurian, and the first land animals in the Devonian. Geologists have found few rocks earlier than those of the Silurian that were formed in freshwater deposits. Consequently, we know little about the freshwater biota of the Ordovician period or of earlier times. Nonetheless, the influential Alfred Sherwood Romer, a paleontologist, and Homer Smith, a renal physiologist, were staunch advocates of a freshwater origin of vertebrates (Romer 1967, Smith 1953).

Romer's proposal of a freshwater origin of vertebrates was based on the fossil record. "Ostracoderm" fragments found in the middle Ordovician Harding Sandstone of Colorado occur in association with fossils of marine origin. Their location in the sediments and their worn condition were interpreted by Romer to indicate that these fragments had washed downstream from rivers or lakes into estuaries and were fossilized in marine sediments. Romer was impressed by another aspect of the fossil record. At the time he formulated his ideas, vertebrate remains were unknown from the rich fossil-bearing rocks of the Cambrian and were rare in Ordovician rocks. If vertebrates arose in the sea, Romer wondered why they left no significant fossil record before the Silurian. If, however, vertebrates evolved in fresh water, Romer reasoned, their absence from the exclusively marine Cambrian and Ordovician formations is explained. Their sudden appearance in a variety of forms in the late Silurian and Devonian corresponds with the earliest occurrence of abundant freshwater sediments in the geological record. Romer thus proposed that active swimming by means of a muscular postanal tail evolved in the early vertebrates in response to stream currents.

Homer Smith's studies of vertebrate kidney structure and function also led him to support a freshwater origin hypothesis. The blood and body fluids of most marine organisms contain salts at concentrations similar to those of the seawater that bathes them. Any alteration of this concentration equilibrium will produce a flow of water from the more dilute to the more concentrated fluid—a process of **osmosis** (see Chapter 4). In bony fishes, body fluid concentrations are very dilute compared to seawater. In freshwater habitats this lowering of body fluid concentration, although reducing the inward osmotic flow of water, nevertheless cannot stop hydration of the body completely. A freshwater fish therefore takes up water and must have a mechanism for excreting the excess water and for retaining needed ions. The vertebrate glomerular kidney is well suited for these functions, and Smith proposed that dilute body fluids and the glomerular kidney evolved as adaptations to freshwater conditions.

Recent work has carefully examined the details of one of the sites where the American Harding Sandstone Formation contains numerous Ordovician vertebrate fragments (Graffin 1992). If the interpretation of these deposits is correct and if the study of further localities of the earliest vertebrate fragments yield similar results, these early ancestors must be considered as restricted to flowing, freshwater streams. At exactly what stage in the evolution of vertebrates from a craniate or precraniate form freshwater was invaded remains a tantalizing question.

Evidence for a Marine Origin

In spite of Romer's and Smith's arguments, evidence for a marine origin of vertebrates is widely accepted. All protochordate and deuterostome invertebrate phyla are exclusively or ancestrally marine forms. Because of chordate affinities to these phyla, there must have been a marine transitional form in the evolutionary line leading to the vertebrates. Most of the known Cambrian and Ordovician vertebrate fossils are thought to be marine fossils. The Harding Sandstone Formation, which Romer and others interpreted to be estuarine, is now known to extend over many thousands of square kilometers and is a mixture of offshore shelf, shore, beach, delta, and freshwater stream deposits. Denticles from early Ordovician rocks are generally thought to have come from a type of sandstone that is formed only under true marine conditions, and the same is claimed for all of the more recently discovered fragments of early ostracoderms from North American locations (Repetski 1978).

The exclusively marine hagfishes (*Myxiniformes*) have body fluids that are similar in total ion concentration to seawater, as do the tunicates and other deuterostomes. Nevertheless, the hagfishes have a well-developed glomerular kidney and so do the marine cartilaginous fishes, which are also osmotically similar to seawater. These kidneys remove divalent cations. Thus, a functional glomerulus is not necessarily associated only with osmotic regulation as Smith assumed. The glomerulus produces a fluid from which red blood cells and blood proteins have been excluded. The tubule cells can return water, salts, glucose, and other useful materials to the circulatory system, while allowing toxic substances, nitrogenous wastes, and excess salts to remain in the urine. The kidney system therefore separates certain ions and molecules, including water, from others. Thus kidney function can also be considered as excretory, not only osmoregulatory, and the kidney would be valuable to either a freshwater or a marine organism. In this view, whether a vertebrate is adapted to the sea, to fresh water, or to life on land, the glomerular kidney is valuable and could have evolved in a marine vertebrate or in a freshwater one. Once a powerful and efficient glomerular filtration system had been achieved, however, only slight modifications of the kidney would be needed to produce an efficient water-excreting device for osmotic regulation in rivers and lakes.

Summary

The history of vertebrates covers a span of more than 500 million years. A basic body plan consists of a bilateral, tubular organization, with such characteristic features as the notochord, pharyngeal slits, dorsal hollow nerve cord, vertebrae, and cranium. One of the protochordates, amphioxus, and the ammocoete larva of lampreys provide glimpses of what the earliest vertebrates may have been like.

The earliest known vertebrates are the "ostracoderms," a paraphyletic group of jawless, aquatic animals that first appear in the late Cambrian and Ordovician as fragmentary fossils. They ultimately evolved heavy bony armor. No fossils have been found that are intermediate between "ostracoderms" and any of the presumed invertebrate progenitors of the first vertebrates, but some idea of the possible origin of vertebrates from invertebrates can be inferred by comparison of living forms. The chordates are more closely allied to the Echinoderms and perhaps to certain lophophorate groups than to any other invertebrate phyla. The notochord, pharyngeal slits, and dorsal hollow nerve cord are shared with certain protovertebrate animals, and it is most probable that vertebrates arose from chordates with the general features of amphioxus. Garstang's hypothesis of vertebrate evolution from a tunicate larval stage remains unproven, although it is widely accepted. The first animals that can be called vertebrates probably evolved in Cambrian seas. Although this sequence of events is not proven, it is a good example of how biologists build evolutionary hypotheses from comparative studies of living and fossil animals.

References

Benton, M. J. 1987. Conodonts classified at last. *Nature* 325:482–483.

Berril, N. J. 1955. *The Origin of Vertebrates*. Oxford University Press, Oxford, UK.

Blieck, A. 1992. At the origin of chordates. *Geobios* (Lyons) 25(1):101–103.

Briggs, D. E. G. 1992. Conodonts: a major extinct group added to the vertebrates. *Science* 256:1285–1286.

Briggs, D. E. G., E. N. K. Clarkson, and R. J. Aldridge. 1983. The conodont animal. *Lethaia* 26:275–287.

Dzik, J. 1993. Early metazoan evolution and the meaning of its fossil record. *Evolutionary Biology* 27:339–386.

Forey, P. and P. Janvier. 1994. Evolution of the early vertebrates. *American Scientist* 82:554–565.

Fritzsch, B., and G. Northcutt. 1993. Cranial and spinal nerve organization in amphioxus and lampreys: evidence for an ancestral craniate pattern. *Acta Anatomica* 148:96–109.

Gabbott, S. E., R. J. Aldridge, and J. N. Theron. 1995. A giant conodont with preserved muscle tissue from the Upper Ordorician of South Africa. *Nature* 374:800–803.

Gans, C. 1989. Stages in the origin of vertebrates: analysis by means of scenarios. *Biological Reviews* 64:221–268.

Garstang, W. 1928. The morphology of the Tunicata and its bearing on the phylogeny of the Chordata. *Quarterly Journal of the Microscopical Society* 72:51–87.

Gee, H. 1994. Return of the amphioxus. *Nature* 370:504–505.

Gould, S. J. 1977. *Ontogeny and Phylogeny*. Belknap, Cambridge, MA. (But see B. A. Pierce and H. M. Smith, 1979, Neoteny or paedogenesis? *Journal of Herpetology* 13(I):119–121 for continuing problems with terminology.)

Gould, S. J. 1989. *Wonderful Life*. Norton, New York, NY.

Graffin, G. 1992. A new locality of fossiliferous Harding Sandstone: evidence for freshwater Ordovician vertebrates. *Journal of Vertebrate Paleontology* 12(1):1–10.

Hanken, J., and B. K. Hall. 1993. *The Skull*, volume 2. University of Chicago Press, Chicago, IL.

Higgins, A. 1983. The conodont animal. *Nature* 302(5904):107.

Jeffries, R. P. S. 1986. *The Ancestry of the Vertebrates*. British Museum of Natural History, Dorset Press, Dorchester, UK.

Keller, E. F., and E. A. Lloyd, (editors). 1992. *Keywords in Evolutionary Biology*. Harvard University Press, Cambridge, MA.

Lovtrup, S. 1977. *The Phylogeny of the Vertebrates*. Wiley, London, UK.

Mikulic, D. G., D. E. G. Briggs, and J. Kluessendorf. 1985. A Silurian soft-bodied biota. *Science* 228:715–717.

Nielsen, C. 1985. Animal phylogeny in the light of the trochlea theory. *Biological Journal of the Linnean Society* 25:243–249.

Northcutt, R. G., and C. Gans. 1983. The genesis of neural crest and epidermal placodes: a reinterpretation of vertebrate origins. *Quarterly Review of Biology* 58:1–28.

Purnell, M. A. 1995. Microwear on conodont elements and macrophagy in the first vertebrates. *Nature* 374:798–800.

Repetski, J. E. 1978. A fish from the Upper Cambrian of North America. *Science* 200:259–531.

Romer, A. S. 1967. Major steps in vertebrate evolution. *Science* 158:1629–1637.

Sansom, I. J., M. P. Smith, H. A. Armstrong, and M. M. Smith. 1992. Presence of the earliest vertebrate hard tissues in conodonts. *Science* 256:1308–1311.

Schaeffer, B. 1987. Deuterostome monophyly and phylogeny. *Evolutionary Biology* 21:179–235.

Singer, C. J. 1959. *A History of Biology*, 3d revised edition. Abelard–Schuman, London, UK.

Smith, H. 1953. *From Fish to Philosopher*. Little, Brown, Boston, MA.

Tarlo, L. B. H. 1960. The invertebrate origins of the vertebrates. *21st International Geological Congress* 22:113–123.

Williams, G. C. 1966. *Adaptation and Natural Selection, a Critique of Some Current Evolutionary Thought*. Princeton University Press, Princeton, NJ.

VERTEBRATE ORGAN SYSTEMS AND THEIR EVOLUTION

We saw in the first chapter that evolution is a process of change in the frequencies of alleles in the gene pool of a species. The frequencies of different alleles determine the relative numbers of various genotypes, and the genotypes are expressed as phenotypes. This relationship of gene to structure is expressed during development. Derived structures do not appear de novo, they are produced by modifications of ancestral patterns of embryonic development. This mechanism imparts continuity to the body forms of vertebrates. During the life of an individual a fertilized egg (a single cell) grows into the billions of cells that form an adult vertebrate. In the process the cell differentiates into a wide variety of specialized cells combined in an equally wide variety of ways to form tissues and organs. An understanding of embryonic development and the development and function of specialized organs is needed to appreciate the changes that have occurred during the evolution of vertebrates.

The Unity of Vertebrate Structure

The well-known physiologist Homer Smith summed up a century of accumulation of detailed morphological and physiological research by scores of biologists with the title of his 1953 book, *From Fish to Philosopher*. The continuity of vertebrate organ systems and their functional attributes had indeed been fully substantiated from their origins in some creature we would all recognize as fish-like to the condition in the human organism. It is important to understand how profound these similarities are in order to appreciate the importance of the differences between one vertebrate clade and its sister taxon. Without appreciation for the basic structure and function of the organ systems, the variations often seem trivial, when in fact they separate the lineages in question in ecological space and evolutionary time. An appreciation of the vertebrate organ systems requires basic

knowledge of their development in the individual from the **zygote** (fertilized egg) to the adult form. This requires examination of the sequential processes of **embryogenesis** and **organogenesis**. Next must come an understanding of the basic functions of organ systems and how the associated organs carry out those functions. Finally, we would benefit from understanding the evolutionary changes in each system. Unfortunately, only a few of the organ systems have left any fossil record. Our only recourse is to examine the character states of the various organs of extant vertebrates. Here we tread on thin ice, for any evolutionary hypotheses generated by this process can be only as good as our hypothesis of phylogenetic relationships. For taxa with only a few extant members we face the additional challenge of distinguishing whether we are dealing with ancestral states of organ systems or with derived specializations. Mindful of these problems, let us examine the basic patterns of development of vertebrates, the differentiation of organ systems, and their function.

Genes, Development, and Evolution

Advances in genetics are currently shedding light on the direct connections between genes and development. Evolution as viewed from the perspective of the paleontologist has often been isolated from other fields of biology. Now that perspective is beginning to interweave with the previously barely related perspectives of developmental, cellular, and molecular biologists (Marx 1992). Although it has long been understood that changes seen in the evolution of adult organisms result from changes in the developmental processes that made the adults, fossils of embryos are very rare and never sufficient to yield anything like a developmental sequence.

Genes and groups of structurally and positionally related genes (gene clusters or families) seem to have key regulatory roles in embryonic development, and may continue to be active in adult tissues. The differences in these gene clusters from organism to organism can be viewed as a record of the evolution of the developmental process and the changes responsible for biodiversity.

Newly discovered genes of this type are being described regularly. A partial list and the known effects of each gene cluster on vertebrate development include the following:

Hox or homeobox, a gene cluster affecting regionalization of the nervous system, head and neck (Keynes and Krumlauf 1994), and the limbs and digits (Tabin 1992, Tabin and Laufer 1993)

Pax, a gene cluster affecting cells derived from the neural crest, and controlling the dorsal–ventral organization of the neural tube, the developing brain from its anteriormost region, the developing eye and nose (Gruss and Walther 1992), the early segmental nature of the central nervous system, and the limbic system of the adult brain stem (Stoykova and Gruss 1994)

Wnt, a gene cluster affecting cell-to-cell communication (Nusse and Varmus 1992) and body axis formation (Marx 1991), the central nervous system (especially from the midbrain and posteriorly), and the normal growth and differentiation of the mammary gland where inappropriate expression of *Wnt* appears to contribute to tumor formation (McMahon 1992)

Many other vertebrates, including teleosts, amphibians, birds, mammals, and probably lampreys, have at least one of the gene clusters that occur in amphioxus (Garcia–Fernandez and Holland 1994, Gee 1994). The DNA base sequence and linear arrangement on the chromosome of these genes is the same in amphioxus and humans, and the genes are expressed during development in that sequence. The antiquity of the *Hox* cluster among multicellular animals is emphasized by its presence in fruit flies (*Drosophila*), a nematode, amphioxus, and vertebrates. Vertebrates have four sets (each slightly different) of the specific gene family, whereas the non-craniates have single sets. Thus the quadruple state of the *Hox* gene cluster is a derived character for craniates.

Embryogenesis

The vertebrate zygote is a typical eukaryotic cell. One difference from most other eukaryotic cells sets the stage for subsequent development: A zygote has a specialized, inert nutritive material called **yolk**. The large eggs of birds have an enormous quantity of concentrated yolk, whereas the tiny (150 micrometers in diameter) human ovum has a sparse, dilute yolk. The 2- or 3-millimeter-diameter eggs of many amphibians, lungfish, and plesiomorphic bony fish (e.g., sturgeon) have a dilute yolk that is more concentrated at one end of the zygote than the other. The eggs of many are less than 1 millimeter in diameter and have a concentrated yolk.

Eggs and zygotes are classified as **oligolecithal** (*oligo* = little, *lekithos* = egg yolk), as in cephalochordates and marsupial and placental mammals; **mesolecithal** (*meso* = intermediate), as in lampreys, sturgeon, amia, lungfish, and amphibians; or **macrolecithal** (*macros* = large) as in hagfish, sharks, rays, teleosts, turtles, lizards, snakes, crocodiles, birds, and monotremes. All macrolecithal eggs are also **telolecithal** (*teleos* = end), their yolk being segregated from the egg cytoplasm and at one end of the egg (Soin 1981). Oligolecithal eggs are also **isolecithal** (*isos* = equal), the thin yolk being evenly distributed throughout the egg and not segregated from the egg cytoplasm. Eggs with a higher yolk to cytoplasm ratio, whether meso- or macrolecithal, have a concentration of yolk at one end of the zygote. The end with the higher yolk content is known as the **vegetal pole**; the opposite end is called the **animal pole** for reasons that do not become obvious until the zygote begins division.

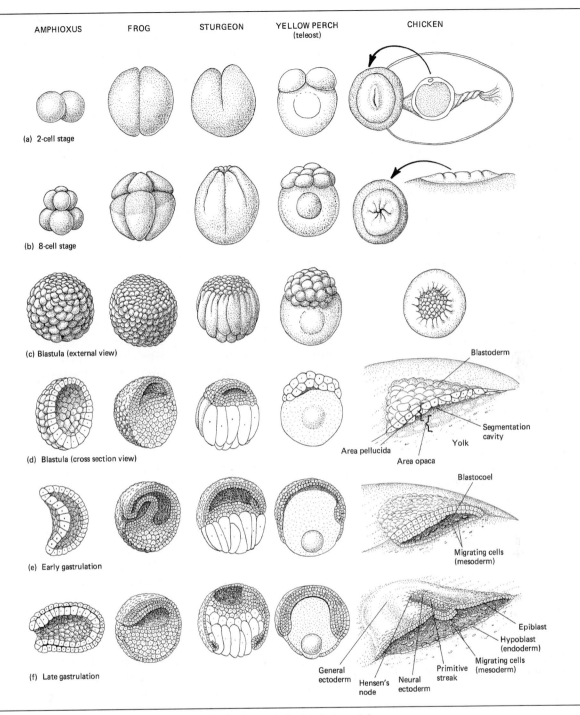

AMPHIOXUS   FROG   STURGEON   YELLOW PERCH (teleost)   CHICKEN

(a) 2-cell stage

(b) 8-cell stage

(c) Blastula (external view)

(d) Blastula (cross section view)

Blastoderm

Segmentation cavity

Area pellucida

Area opaca

Yolk

(e) Early gastrulation

Blastocoel

Migrating cells (mesoderm)

(f) Late gastrulation

General ectoderm

Hensen's node

Neural ectoderm

Primitive streak

Migrating cells (mesoderm)

Epiblast

Hypoblast (endoderm)

Figure 3–1 Cleavage (a–d) and gastrulation (e, f) of a cephalochordate and four vertebrate zygotes formed from eggs with scant, evenly distributed yolk (left) through those from eggs with a large amount of concentrated yolk at one end (right).

The early division of a vertebrate zygote occurs without growth changing the zygote—which is a single, large cell—to a multicellular embryo with smaller cells (Figure 3–1). This cell division without embryo growth is called **cleavage**, and it results in an embryonic stage called the **blastula** (*blastos* = a germ or bud). However, cleavage is more than just cell division. Rapid and unabated synthesis of nuclear material without cell growth occurs between successive cell divisions, and the nuclear-to-cytoplasmic ratio typical of adult cells is eventually reached. The high nuclear-to-cytoplasmic ratio of late cleavage cells may be important for effective genetic control of cellular differentiation, because embryonic gene expression first occurs at this time. The sequence and spatial arrangement of cell divisions of many chordates appear regular, especially early in development. This suggested to early researchers that individual cells may have distinct fates at this stage in the development of a vertebrate as they appeared to have in many invertebrate embryos. Such is not the case, however; development proceeds normally even if many cells are removed from the early cleavage stages. Hans Spemann found in the early decades of this century that a single cell removed from the early blastula could produce an entire embryo capable of successful development. Cleavage in vertebrates is thus said to be **indeterminate** and results in **totipotent** cells for a significant number of cell divisions.

The blastulae are quite different in various lineages of vertebrates because the quantity of yolk in an egg affects the process of cleavage. Apparently, the division of yolk-free cytoplasm is readily accomplished, but the presence of yolk slows cleavage in proportion to its concentration. Thus the oligolecithal zygotes of cephalochordates and mammals cleave evenly and completely to yield a blastula of nearly even-sized cells. Complete cleavage of the zygote is termed **holoblastic** cleavage (*holos* = entire). Mesolecithal zygotes cleave more quickly at the animal pole than at the vegetal pole, resulting in a blastula composed of many small cells at one end and fewer and larger yolk-filled cells at the opposite (vegetal) pole. This pattern is a direct reflection of the distribution of yolk in telolecithal eggs, but because the entire zygote cleaves, the process is properly called holoblastic. More dramatic still is cleavage of telolecithal zygotes. Here cell division within the highly concentrated and segregated yolk is inhibited completely. Cleavage occurs only in the cap of egg cytoplasm at the animal pole leading to a thin disk of cells (**blastodisc**) atop an enormous uncleaved yolk, a process known as **meroblastic** cleavage (*meros* = a part). The embryo is represented initially by the blastodisc. By processes that differ in various macrolecithal groups of vertebrates (indicating indepen-

dent evolution), the blastodisc proliferates an **extraembryonic membrane**. This membrane forms the **yolk sac**, which surrounds the inert yolk with blood vessels that transport nutrients to the developing embryo atop the yolk mass. It is now apparent why the cleaving animal hemisphere acquired that name —the definitive animal appears here. The yolk of the vegetal hemisphere is slowly absorbed to provide the energy and structural materials for the development of the embryo.

The blastula of a vertebrate may be a hollow sphere of cells, the central cavity of which is known as the **blastocoel** (*koilos* = hollow), or an unequally divided sphere with a restricted blastocoel in the animal hemisphere. The blastocoel of meroblastic vertebrates is represented by a space where the blastodisc has lifted off the undivided yolk. The importance of this hollow becomes evident as the vertebrate embryo enters its next stage of development, the formation of the **gastrula** (= belly or stomach).

Although the details of gastrulation vary, fundamentally the process involves the transformation of the blastula to a structure with three layers known as **germ layers**. The process of **gastrulation** is the time that the basic vertebrate body plan is laid down and the three axes of bilaterality, dorsal–ventral, and anterior–posterior are established. At this point many cells acquire developmental fates and positional information (Stern and Ingham 1992). Much has been made of the distinctions among the three cell layers differentiated during gastrulation—ectoderm, mesoderm, and endoderm. Their names suggest that their roles in embryogenesis have been fixed, but experimental manipulations show that there is little or no restriction on what tissue any cell from any layer in the gastrula can participate in forming. This apparent fixity is the result of later interactions determined by physical position within the embryo. For us what is significant are the normal fates of these cell layers in a naturally developing vertebrate embryo (Figure 3–2) as they interact, influence, and are influenced by the cell layers surrounding them. The fates of germ layers have been very conservative throughout vertebrate evolution. In all vertebrates the outermost layer of the gastrula, the **ectoderm** (*ecto* = outside; *derm* = skin), forms the adult superficial layers of skin, the most anterior and most posterior parts of the digestive tract, and the nervous system, including most of the sense organs, such as the eye and the ear. The innermost layer, the

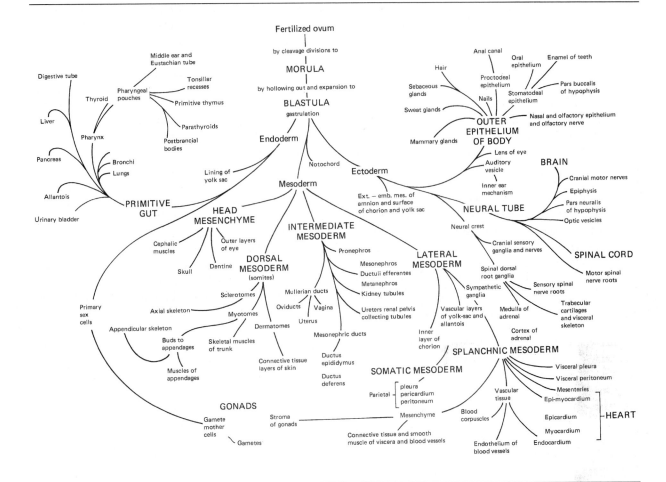

Figure 3–2 Schematic diagram showing basic vertebrate organ structures as derived from embryonic endoderm, mesoderm, or ectoderm. The diagram indicates origin of epithelial part of organ only. Most organs have supporting structures of mesodermal origin. Neural crest contributes to many more structures than can be illustrated here; see Table 3–1. (Modified after B. M. Patten, 1964, *Foundations of Embryology*, McGraw-Hill, New York, NY.)

endoderm (*endo* = within), forms the rest of the lining of the digestive tract, as well as the lining of the glands associated with the gut, and most respiratory surfaces of vertebrates. A middle layer, the **mesoderm** (*mesos* = middle), which is usually the last of the three layers to differentiate, forms the muscles, skeleton, connective tissues, and circulatory and urogenital systems.

Most germ layer formation occurs at a single spot, the **blastopore**, where layers of proliferating cells buckle and fold inward into the blastocoel. In meroblastic gastrulae, surface cells stream into the pocket-like blastocoel. The blastopore forms, or is in the vicinity of, the adult anus and thus indicates that the axes of the vertebrate body plan seem to have been established (Marx 1991). Vertebrates and other chordates share the blastopore-to-anus developmental pathway with echinoderms and a few minor invertebrate phyla, all of which are called deuterostomes (*deutero* = secondary; *stoma* = mouth). It is on the basis of this shared character of development that echinoderms are linked to chordates.

Although gastrulation sets the stage for further development, it remains for the subsequent stage to start the processes of growth, cellular differentiation, and organogenesis that lead to the formation of complex structures that begin to look like adult anatomy. **Neurulation**, the formation of the nervous system, is associated with major changes in superficial and internal cell arrangements. A population of mesodermal cells known as **chordamesoderm** aggregates early in development to form an elongate rod, the **notochord**, that defines the axis of the embryonic body. The ectoderm overlying the forming notochord begins to form two longitudinal folds with a mid-dorsal furrow between them. The crests of the ectodermal folds grow toward one another, forcing the furrow, its floor, and its sidewalls into the dorsal mesoderm flanking the notochord. When these **neural folds** meet, they fuse, sealing a hollow tube of now-isolated ectoderm below the surface of the embryo. This **neural tube** will become the central nervous system of the adult. Thus, neurulation produces two diagnostic characteristics of chordates: the notochord and dorsal hollow nerve tube. In the absence of chordamesoderm the neural tube will not form. Embryologists describe such a relationship by saying that chordamesoderm induces formation of the neural tube. However, we clearly have more to learn about this process, because it appears that ectoderm in general may differentiate into neural tissue unless it is signaled *not* to do so by adjacent mesoderm (Hemmati–Brivanlou and Melton 1994, Hemmati–Brivanlou et al. 1994).

The mesoderm remaining on either side after the differentiation of the notochord becomes divided into three morphologically distinct cell populations. These cell populations enter into complex interactions best understood by viewing an embryo stripped of its ectoderm along the dorsolateral aspect of the embryo (Figure 3–3a). Immediately flanking the notochord and neural tube is mesoderm that in the trunk becomes segmented into bilaterally paired blocks called **somites**. Lateral and ventral to the somites to the intermediate mesoderm is the sheet-like expanse of **lateral plate mesoderm**.

On closer examination of a cross section through a vertebrate embryo (Figure 3-3b), we see some of the internal morphogenesis of these mesodermal populations of cells. The entire mesoderm is hollow: Each somite has a cavity and the more extensive and continuous space in the lateral plate defines a superficial or **somatic** mesoderm layer immediately beneath the ectoderm and a deep or **splanchnic** mesoderm layer (*splanchnos* = viscera) surrounding the gut. The space in the lateral plate is the body cav-

ity or **coelom**. Soon after the somites form, three distinct regions can be identified in each somite. The dorsal portion, just beneath the ectoderm, forms the **dermatome** (*tomos* = a slice or section), which will produce the deep portions of the skin. Deep to the dermatome and across the slit-like cavity of the somite is the forming **myotome** (*myo* = muscle), which will form adult voluntary muscles. Cells located next to the notochord on the inner face of the somite form the **sclerotome** (*sclero* = hard), which will become vertebrae. The mesoderm intermediate between the somite and the lateral plate is also segmented and eventually separates and thickens into a **nephric ridge** (*nephros* = kidney), which develops into the kidney. Much later a second ridge forms from the medial border of the kidneys. This is the **genital ridge** and will form the gonads. The ducts serving these organs are often shared between the excretory and reproductive systems.

Organogenesis

Cells begin to differentiate during neurulation, losing a part of their totipotency as they assume more specific roles. In general, the cells of the endoderm form a close-knit lining for the hollow gut and the tubular structures that develop from the gut. Most endodermal cells are specialized for secretion or absorption. Cells of the ectoderm cover the outside of the body, forming a protective layer. Both the endoderm and the ectoderm are characterized as epithelia because of their tight intercellular junctions, sheet-like structure, and the fact that they generally flank an open space (e.g., the lumen of the gut). The rest of the cells, the mesodermal derivatives, form the inner matrix of the organism, a highly varied task accomplished by an enormous diversity of cells. Mesodermal cells that form the linings of the walls of the coelom, the blood vessels and heart, or the kidney and reproductive ducts resemble epithelial cells of the skin surface or gut linings. However, mesodermal cells that form blood, bone, or muscle, the general structural scaffolding of organs, or the deep layers of the skin have a very different morphology.

The specialized roles of cells are reflected in the organelles that are present and the extent of organelle elaboration and in the relationship of the cell to its neighbors. Coherent groups of cells with specialized functions form **tissues**. There are five

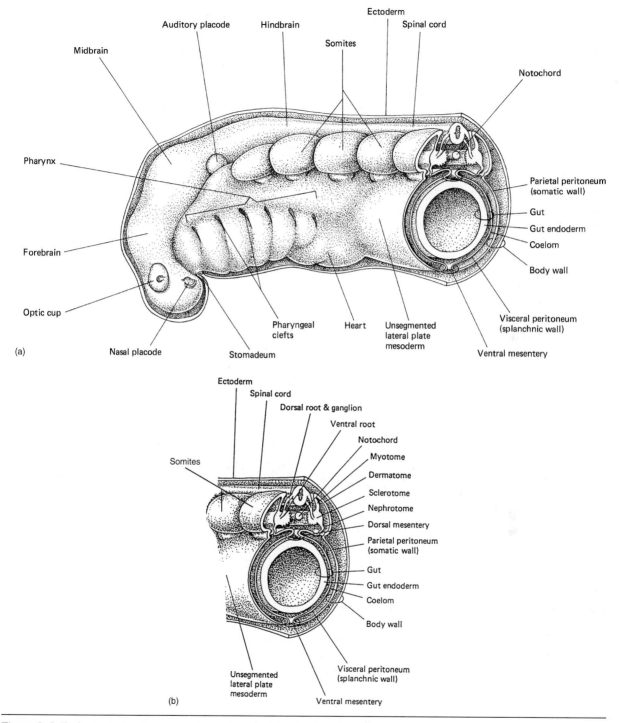

Figure 3–3 Embryonic development of a vertebrate. (a) Diagrammatic three-dimensional view of a portion of a generalized vertebrate embryo showing segmentation of the mesoderm in the trunk region and pharyngeal development. (b) Enlargement of cross section in trunk region. (After E. S. Goodrich, 1930, *Studies on the Structure and Development of Vertebrates*, Macmillan, London, UK.)

basic tissues in vertebrates: *epithelial*, *connective*, *blood*, *muscular*, and *nervous*. These tissues are combined in a wide assortment of permutations to form larger units called **organs**. Organs often have most or all of the five basic tissues. The basic functions of vertebrate life are supported by groups of organs united into one of the ten vertebrate organ systems (Table 2–1).

Wandering Cell Populations

Some of the most important and diagnostic tissues of vertebrates are formed by cells that break free of their parental germ layer and migrate as individual cells through the embryo to remote sites. These wandering cells, as well as any other cells that are not arranged in an epithelium, are called **mesenchyme**. (This term refers to the morphology of the cells, and it must not be confused with *mesoderm*, which is a germ layer that is one source of mesenchymal cells.) Because of the difficulty of following individual cells, we are still learning about the behavior and significance of these wandering cells.

Cells from many regions of the mesoderm become mesenchymal, migrate, and contribute to components of locally forming tissues, including cartilage, bone, smooth and some voluntary muscle, blood cells and vessels, and lymph vessels and glands. Many of these cells persist long into posthatching or postnatal life as undifferentiated progenitors of connective tissue cells. The **skeletogenous septum** is an important early site of migration and differentiation of mesenchymal cells (Figure 3–4). For example, in gnathostomes, but not jawless vertebrates, a horizontal skeletogenous septum develops, which runs laterally from the notochord to intersect the dermis of the skin along a midlateral line on each side of the body. Many segmental elements of the vertebrate endoskeleton form at the intersections of these thin sheets of connective tissue. When they are present, the sensory lateral line and its specialized scales form where the horizontal septum and the dermis meet; the neural spines form where the myosepta meet the dorsal septum; hemal spines arise where myosepta meet the ventral septum; dorsal or intermuscular ribs form where myosepta cross the horizontal septum; and ventral or subperitoneal ribs are produced where the lateral wall of the coelom and myosepta intersect. Around the notochord the sclerotomic mesoderm condenses and forms a vertebra at the intersection of every myoseptum with the horizontal, dorsal, and ventral septa. The myosepta are intersegmental, and the sclerotomal condensation and the resulting vertebra are also intersegmental. Thus, vertebrae alternate with developing muscle masses. The muscles eventually insert on the vertebrae and form the flexible spinomuscular basis of lateral undulatory locomotion. Thus, further unique euchordate features (muscles separated by myosepta into segmental myomeres) and vertebrate characters (vertebrae or their anlages) are defined by derived embryonic processes.

The **primitive sex cells** attain a distinctively large size, break loose from the other cells of the yolk sac, and migrate dorsally around the gut wall and up the newly formed dorsal mesenteries. From here they spread laterally to invade the genital ridges on either side of the body. They become the eggs or sperm of the adult. The origin and migratory behavior of these cells vary among vertebrates, but the evolutionary significance of that variation is unknown.

Perhaps the most significant of the wandering cell lineages in vertebrate development are the **neural crest cells** derived from the ectoderm. These cells are unique to craniates and are key players in the formation of nearly every derived character that sets vertebrates apart from other organisms (Table 3–1). The defining evolutionary step in the origin of the craniates was the evolution of neural crest cells. Le Douarin (1982) lists 42 different structures to which neural crest cells contribute. Neural crest cells become distinct at the time of neural tube formation, arising from cells at the angle between the developing neural tube and the margin of the remaining ectoderm, which is closing over the neural tube (Figure 3–5). They are first clearly visible lying above and to either side of the neural tube. From here they disperse laterally and ventrally, ultimately settling and differentiating in locations throughout the embryo. Derivatives of the neural crest form almost all of the peripheral nervous system (including the cranial nerve ganglia, the autonomic system, and the Schwann cells, which make the myelin sheath of peripheral nerves), several endocrine glands, pigment cells, and perhaps most remarkably, the skeleton and connective tissues of the head. The complexities of their movement and their interactions with other cells are still not thoroughly understood, and remain a focus of intense interest, although their

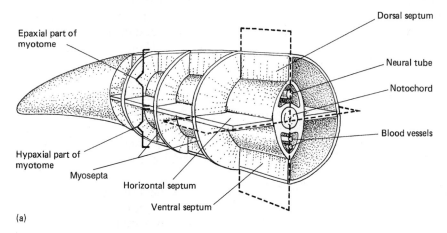

Epaxial part of myotome

Hypaxial part of myotome

Myosepta

Horizontal septum

Ventral septum

Dorsal septum

Neural tube

Notochord

Blood vessels

(a)

Figure 3–4 The skeletogenous septa of gnathostomes. (a) Diagrammatic representation of the orientation of the various mesenchymal septa in the tail of a generalized gnathostome. (b) Differentiation of the septa and the endoskeletal elements that grow at the various intersections in the trunk region of a generalized fish-like vertebrate. (c) Human thoracic vertebra and rib, which develop in the same skeletogenous septa intersections.

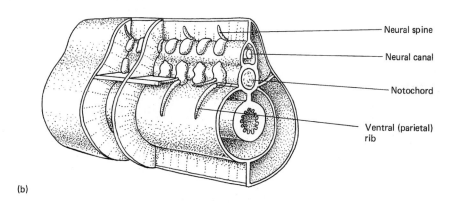

Neural spine

Neural canal

Notochord

Ventral (parietal) rib

(b)

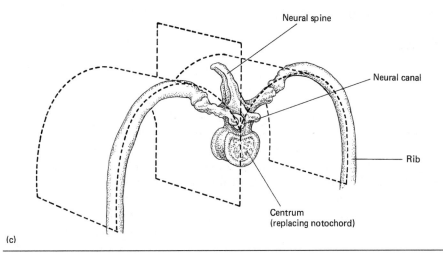

Neural spine

Neural canal

Rib

Centrum (replacing notochord)

(c)

contribution to skeletogenesis was first described in 1898 by Julia Platt.

A series of neurogenic epidermal thickenings called **placodes** are found only in the head region. Although the placodes are not part of the neural crest, it has been suggested that they and the neural crest cells evolved from a common ancestral cell line present in a prevertebrate ancestor. Some placodes

Table 3–1 Embryonic origins of shared derived characters of vertebrates.

Character	Embryonic Origin
Integument and skeleton	
Skin of face and ventral neck (dermis, smooth muscle, and adipose tissue)	Neural crest
Pigment cells	Neural crest
Cephalic armor	Neural crest
Tooth papillae	Neural crest
Anterior neurocranium, sensory capsules (sclera of the eye), and fragments of the cranial vault	Neural crest
Dorsal fin mesenchyme	Neural crest
Nervous system	
Cranial nerves with sensory ganglia, including satellite cells	Neural crest, epidermal placodes
Trunk nerves with sensory ganglia, including satellite cells	Neural crest
Peripheral motor ganglia, including sympathetic and parasympathetic ganglia and plexuses	Neural crest
Schwann cells (myelin sheath cells) of peripheral nerves	Neural crest
Meninges of prosencephalon and part of mesencephalon	Neural crest
Sense organs	
Nose	Epidermal placodes
Eyes (sclera?, lens, ciliary muscles, cornea)	Neural crest, epidermal placodes
Ears	Epidermal placodes
Lateral-line mechanoreceptors and electroreceptors	Neural crest, epidermal placodes
Gustatory organs (taste buds, olfactory receptors)	Epidermal placodes, endoderm
Pharynx and digestive tract	
Derivatives of visceral arch skeleton	Neural crest
Pharyngeal muscle	?Paraxial mesoderm
Connective tissues of pharyngeal muscles	Neural crest
Smooth muscle of gut	Lateral plate mesoderm
Calcitonin cells of ultimobranchial bodies and thyroid gland	Neural crest
Chromaffin cells of interrenal bodies and adrenal medulla	Neural crest
Connective tissue component of the pituitary, lacrymal, salivary, thyroid, parathyroid, and thymus glands	Neural crest
Circulatory system	
Gill capillaries	Lateral plate mesoderm
Muscularized aortic arches and their connective tissues components	Neural crest
Carotid sensory bodies in aortic arch	Neural crest
Muscular heart	Lateral plate mesoderm

Source: C. Gans and R. G. Northcutt, 1983, *Science* 220:268–274; N. Le Douarin, 1982, *The Neural Crest,* Cambridge University Press, Cambridge, UK.

are far dorsal, adjacent to the neural crest, whereas others are more ventral, just above the pharyngeal pouches. The cells of the placodes contribute to the sense organs (nose, eye, ear, lateral-line mechano- and electroreceptors, taste buds) and portions of the cranial sensory ganglia. In addition, portions of the placodes may migrate. The entire lateral line system of the trunk is formed by migration of cells from the head that are derived from both neural crest and placodes.

The Pharyngula

We have emphasized the profound similarity of developmental phenomena in vertebrates. Discovery of these common features occupied the entire careers of many of the best biologists of the nineteenth century. But developmental differences between the major clades of vertebrates are also important both because of uniquely derived patterns within each clade and because of the effects on cell division and migration of the quantity and location of yolk. However, the next stage of embryogenesis, the **pharyngula**, has a universality that transcends differences in cleavage, gastrulation, extraembryonic membrane formation, and neurulation. The stage is named for the characteristic development of **pharyngeal pouches**. The ancestral vertebrate feature of pharyngeal (or gill) slits makes at least a fleeting appearance in the embryos of all vertebrates. At this stage vertebrate embryos have their greatest similarity. After the pharyngula stage the distinctive characters of the adult organism begin to establish themselves.

The most conspicuous new feature of the pharyngula stage is the sculpturing of the sides of the posterior portion of the head into six to nine columns of solid tissue separated by deep furrows. The columns are the **pharyngeal** or **visceral arches**; the furrows, **pharyngeal grooves**. They form when endoderm of the pharyngeal lining evaginates in a series of pouches toward the exterior and visceral grooves of the ectoderm form indentations directly opposite the pouches on the external body wall. When the evaginations of endoderm meet the invaginations of the ectoderm, they unite to form thin partitions. In nonamniotes the partitions perforate to become the gill slits. Most are transient, and they are obliterated in adult tetrapods, but the most anterior maintains its thin membranous nature as the eardrum or **tympanum**. The linings of the pharyngeal pouches give rise to half a dozen or more glandular structures often associated with the lymphatic system, including the thymus gland, parathyroid glands, carotid bodies, and tonsils.

Each pharyngeal arch contains an arterial blood vessel carrying blood from the ventral aorta and developing heart. The vessel passes dorsally through the arch to unite with the other arch arteries to form the main conduit of blood to the body, the dorsal aorta. In fishes the arch vessels supply the gills; in tetrapods one arch is modified to supply the lungs. Each pharyngeal arch also contains the visceral (or gill) skeleton, typically consisting of an articulated crescent of pharyngo-, epi-, cerato-, and hypobranchial elements on each side and a midventral basibranchial (Figure 7–1). These elements support the gills of fishes and larval amphibians. In gnathostomes the skeletal elements of two anterior pharyngeal arches form the jaws, the defining character of the group and a key feature in vertebrate evolution. Finally, each arch contains pharyngeal (or bran-

chiomeric) muscle. In nonamniotes this muscle is used for respiration as well as for feeding, whereas in amniotes it is concerned primarily with food processing, head movement, and, in mammals, facial expression.

Experimental studies of chickens suggest that all voluntary musculature, including that associated with the pharynx, is derived from mesoderm immediately flanking the notochord and neural tube (Noden and deLahunta 1985). This result conflicts with the traditional view that pharyngeal muscle, like all other musculature associated with the gut, is derived from splanchnic mesoderm of the lateral plate (Figure 3–2). If this result is confirmed in studies of other vertebrates, as seems likely to be the case, the traditional distinction between the visceral (gut-associated) and somatic (notochord-associated) organ systems of vertebrates will require reevaluation. In either case, it is clear that the strong muscularization of the vertebrate pharynx is associated with an evolutionary switch from ciliary movement of feeding currents through the gill clefts to muscular pumping of feeding currents. The elastic gill skeleton and the increased vascularization are also pharyngeal arch correlates of this functional shift.

The principal shared derived characters of adult vertebrates all have their origins in shared derived characters of vertebrate embryos—the neural crest, epidermal placodes, and muscular pharynx. The vertebrates entered a new adaptive realm with the novel feeding shifts that occurred at their origin. These shifts had profound effects on food detection and ingestion and put new selective pressures on metabolic functions, especially food processing and gas exchange.

Protection, Support, and Movement

Exceptional mobility characterizes vertebrates, and the ability to move requires muscles and an endoskeleton. Mobility brings vertebrates into contact with a wide range of environments and objects in those environments, and a vertebrate's external protective covering must be tough but flexible. Bone, the mineralized tissue that we consider characteristic of the skeletons of living vertebrates, had its origin in this protective integument. The history of the shift of bone from superficial to internal skeleton

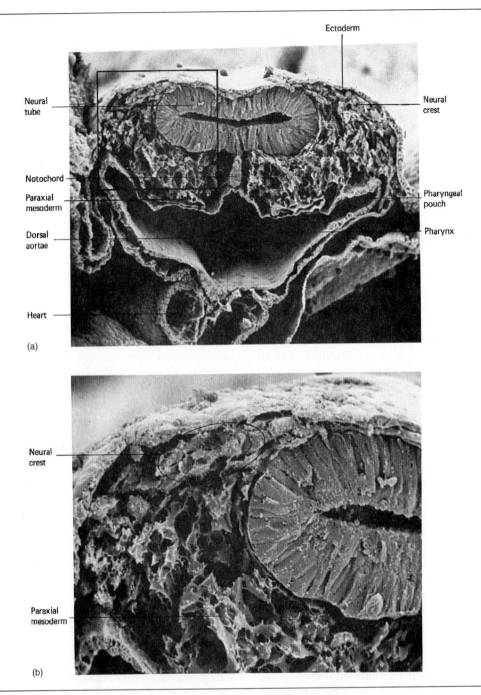

Figure 3–5 Formation and migration of neural crest from the neural tube to contribute to several unique vertebrate structures. (a) Quail embryo sectioned through the head at the level of the mesencephalon as indicated in (c). Note the epithelial nature of the ectoderm and pharyngeal lining and the mesenchymal nature of the neural crest and paraxial mesoderm. (b) Higher magnification of (a). (c) Lateral view of early embryo. (d) Lateral view of more advanced embryo showing some neural crest derivatives (all black and hatched structures). ([a] and [b] Courtesy of K. Z. Reiss.)

Figure 3–5 *(Continued)*

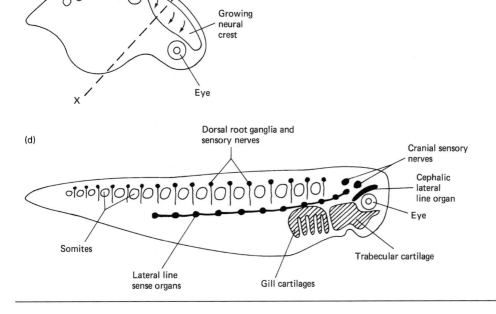

(c)

Somites

X

Growing neural crest

Eye

X

(d)

Dorsal root ganglia and sensory nerves

Cranial sensory nerves

Cephalic lateral line organ

Eye

Trabecular cartilage

Somites

Lateral line sense organs

Gill cartilages

parallels the history of increasing complexity of the vertebrate body plan, especially that of the organ systems involved in protection, support, and movement.

The Integument

The integument is a single organ, one of the largest of the body, making up 15 to 20 percent of the body weight of many vertebrates and much more in armored forms. It includes a series of derivatives of the skin, such as glands, scales, feathers, hair, spines, horns, and hoofs. The skin protects the body and receives information from the outside world (Bereiter–Hahn et al. 1986). The major divisions of the vertebrate skin (Figure 3–6) are the **epidermis** (the superficial cell layer derived from embryonic ectoderm) and the unique vertebrate **dermis** (the deeper cell layer of mesodermal and neural crest origin). The dermis extends deeper into a purely mesodermally derived subcutaneous tissue (**hypodermis**) that overlies the muscles and bones.

Epidermis The boundary between a vertebrate and the environment is the epidermis. As such, it is of paramount importance in protection, exchange, and

sensation. Nevertheless, it may be only a few cells thick in fishes. In nonamniotes (except for adult amphibians) the entire epidermis, even the surface cells, is alive and metabolically active. It often has secretory unicellular or multicellular glands.

In adult amphibians and other tetrapods the epidermis is an avascular cell layer with active and proliferating cells only in its deepest part and dead cells near the surface that are regularly shed. The layer of deep, active cells (**stratum germinativum**) obtains its metabolic needs from the underlying vascular dermis. Epithelial cell divisions at this interface force daughter cells toward the surface. These daughter cells produce and accumulate protein granules (keratohyalin) and related substances, which slowly change as the cells die and are forced nearer the skin's surface. By the time cells reach the surface they are adherent sacs of **soft keratin**, a tough, pliable hydrophobic protein.

The cohesiveness between cells of the tetrapod epidermis during all stages of development and degeneration imparts marvelous characteristics to the tissue: it is living, growing, regenerative—and at the same time dry, abrasive, resistant, and expendable. Nonamniotes, especially fishes and aquatic amphibians, are subject to osmotic movement of

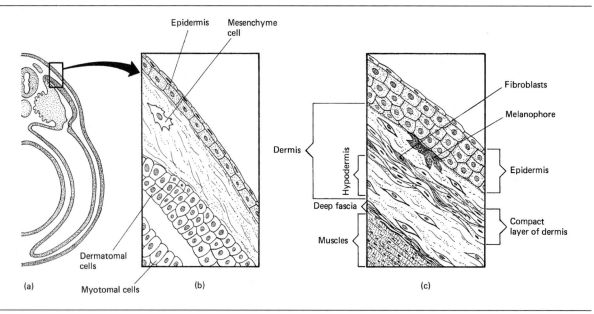

Figure 3–6 Development and differentiation of the vertebrate integument. (a) The ectoderm and somite contribute to the adult skin. (b) In addition, wandering cells from the neural crest and later the mesenchyme cells contribute to the differentiation of the integument. (c) The differentiated skin as described in the text. (After M. H. Wake, editor, 1979, *Hyman's Comparative Vertebrate Anatomy*, 3d edition, University of Chicago Press, Chicago, IL.)

water across the epidermis, and the living cells of the epidermis may play an active role in minimizing cutaneous water flux in conjunction with specialized cells of the gills (discussed in detail in Chapter 4). Terrestrial vertebrates are faced with evaporative water loss. Deposition of extracellular and intracellular lipids in the keratinized (dead) layers of the epidermis appears to establish a water barrier that greatly reduces cutaneous water loss. The details of the formation of an epidermal lipid water barrier differ in lizards, birds, and mammals, indicating multiple separate evolutions of cutaneous barriers. Epidermal derivatives are enormously varied among vertebrates—poison glands, feathers, hair, hoofs, horns, and so on. The interaction of epidermis and dermis appears to be essential for formation of these complex appendages of the integument. An example is the production by the epidermis of **enamel**, the hardest tissue evolved by vertebrates (Figure 3.7a). Mature enamel consists almost entirely of calcium phosphate–fluoride crystals in closely packed rods. The rods are secreted by basal epidermal cells and may attain considerable lengths in thick enamels. In living vertebrates enamel is produced only by epidermal cells sharing a basement membrane complex with dermis cells that are producing **dentine**, another mineralized tissue. No other region of epidermis in living vertebrates produces anything similar to enamel in spite of the basic similarity of most epidermal cells. However, this has not always been true in the history of the craniates. In early forms nearly the entire body surface demonstrated the capacity to form enamel and dentine (Smith and Hall 1993).

Dermis   The dermis, unlike the epidermis, is unique to vertebrates and is composed primarily of extracellular products. An extracellular basement membrane welds epidermis to dermis and is important in the relationship between the two layers. The dermis is elastic, loose, and thin over joints and in areas of mobility. In areas of continual friction with the outside world (the soles of feet, prehensile tails, flippers), the dermis is thick, firm, and immobile. Its deepest layer of interlaced bundles and sheets of collagen fibers is produced by sparsely scattered cells (fibroblasts). The interweaving and the amount of elastic fibers determine the tensile strength of skin. In sharks and many mammals this strength is extraordinary, and it is primarily this layer that is used to make leather. Occasionally, smooth muscle fibers occur in the dermis. Their contraction in mammals produces skin wrinkling such as that of the nipple of the mammary glands or the scrotum.

Blood vessels and nerves course through the lower dermis to their destination in the superficial dermis. Here arterioles break up to form a complex web of capillaries closely applied to the dermis–epidermis junction. These fuse again as venules. In birds and mammals venules exit from the superficial dermis, often in parallel with an incoming arteriole. Deep to the web of plexus of capillaries there may be direct connections between arterioles and venules (arteriovenous anastomoses) and two or more flat networks of vessels may occur at different depths within the dermis and hypodermis parallel to the body surface. These vascular networks are best developed in mammals lacking thick fur (for example, swine and humans), in naked regions of hairier mammals (the perineum, the scrotum, the nostrils), or in the egg-incubating brood patches of many birds. Lizards also have well-developed vascular networks in the dermis. The amount of blood and rate of flow within these vessels can be varied by central neural and hormonal control or directly by local temperature. Under conditions of excess internal heat, the vessel plexus dilates and warm blood is cooled by its close contact with the skin surface. When the skin is chilled, most of the capillaries constrict, minimizing blood flow to the skin. The close juxtaposition of arterioles and venules, which continue to carry small amounts of blood to the epidermis, conserves heat by countercurrent exchange.

In tetrapods the dermis houses the majority of sensory structures and nerves associated with the sensations of temperature, pressure, and pain. Some free nerve endings penetrate the epithelium, but the majority terminate in the dermis as specialized end organs.

Some neural crest cells come to rest in the dermis, in the interface between the dermis and epidermis, and in the deepest layers of the epidermis. These cells differentiate into **melanocytes** and produce granules of melanin that may be yellow, rusty red, brown, or black in color. Melanin is injected into adjacent cells that lack melanin-producing enzymes through long, narrow, cytoplasmic extensions of the melanocytes. Additional hues may be produced by structures overlying the melanin that refract and reflect light; by **chromatophores** in addition to

melanin-containing ones; and by vascularization of the skin, where the blood provides shades from pink to scarlet. Skin color that incorporates the vascular system has the advantage of being rapidly variable so that red display areas may become brilliant in excited individuals. Skin color that involves the nervous system and cells where pigments rapidly aggregate or spread out under neural influence are the ultimate in color-changing mechanisms. Many fish and some amphibians and lizards are capable of such exceptional change.

Hypodermis The subcutaneous tissue or hypodermis is not functionally a part of the skin but lies between the skin and the muscles and bones. This region contains collagenous and elastic fibers, and in several lineages of vertebrates it contains a major energy store of the body: subcutaneous fat or oils. Fat storage reaches its maximum development in mammals and birds from cold environments. Penguins, seals, sea lions, porpoises, and whales have hypodermal fat in the form of insulating blubber. A unique hypodermis characteristic of mammals is the extensive system of subcutaneous muscles that move the skin relative to underlying tissues. The fly-disturbing jiggle of a horse's skin is an example of its function. In many carnivores, ungulates, and especially in humans and the other derived primates, subcutaneous muscle from the muscles of the pharyngeal region reaches exceptional development in the facial region. It produces facial expressions, movement of the external ears, opening and closing the eyelids, and the ability to purse the lips and suck—a vital attribute to early mammalian life.

Mineralization The vascular and collagenous networks of the dermis play an important role in **mineralization**. Many fiber-rich connective tissues such as the dermis induce crystallization of calcium carbonates or calcium phosphates within the fibers of their extracellular matrix. Mineralization of connective tissue provides mechanical support and a store of essential minerals, but inhibits diffusion of substances through the mineralized extracellular matrix. This leaves the connective tissue cells embedded in mineralized matrix, and therefore isolated from blood vessels. They begin to starve and suffocate from lack of nutrients and oxygen, and are slowly poisoned by their own metabolic wastes. Vertebrates have two methods of achieving the advantages of mineralization while maintaining the vitality of the cells responsible for calcification: **dentine formation** and bone formation or **ossification** (Figure 3–7). Hydroxyapatite, a specific crystalline form of calcium phosphate, is involved in both processes.

Dentine formation (Figure 3–7a) is restricted to the dermis and usually to the region of the basement membrane complex at the interface with the epidermis. Mineralization occurs along narrow cytoplasmic extensions on the side of a cell away from the nearest blood vessel. As crystals of hydroxyapatite form along these cytoplasmic extensions, the dentine-forming cells retreat toward blood vessels. They elaborate longer cytoplasmic extensions to maintain contact with the dentine, and these extensions become embedded in the dentine. The resulting tissue is dense and hard, and contains nonmineralized

regions only in the thin tubules where the cytoplasmic extensions continue to deposit hydroxyapatite. These living projections maintain metabolic processes through their continuity with the rest of the cell. The cell bodies eventually retreat to clump around the blood vessel in a cavity within the mineralized tissue, the **pulp cavity**. Although dentine formation is usually thought of only in conjunction with teeth, the placoid and cosmoid scales of gnathostome fishes, fossil thelodont scales, and the heterostracan and osteostracan head shields and scales all contain dentine tissue. Thus dentine appears to have been a general characteristic of the earliest vertebrates, and may be the earliest vertebrate hard tissue to have evolved.

Ossification is a second mineralization process unique to vertebrates that allows continued vitality

Figure 3–7 Organization of vertebrate mineralized tissues: (a) enamel and dentine as seen developing in a tooth; (b) bone from a section of the shaft of a long bone of a mammal.

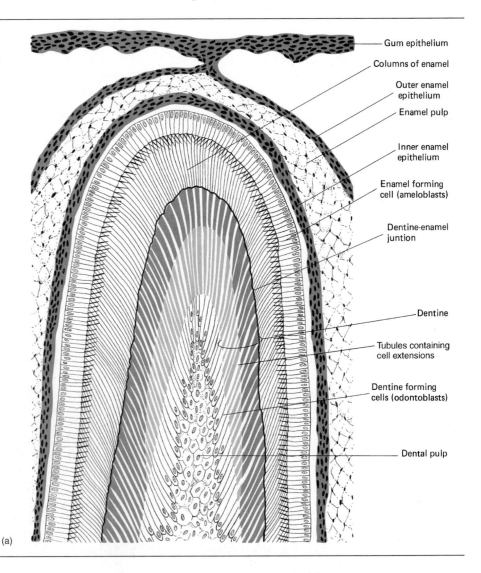

Gum epithelium

Columns of enamel

Outer enamel epithelium

Enamel pulp

Inner enamel epithelium

Enamel forming cell (ameloblasts)

Dentine-enamel juntion

Dentine

Tubules containing cell extensions

Dentine forming cells (odontoblasts)

Dental pulp

(a)

of the mineralizing cells. Ossification begins with cells forming a specialized network throughout a nonmineralized matrix. Cytoplasmic extensions make contact with those of other cells, thus indirectly joining to cells in perivascular (= close to a blood vessel) positions. Directly or indirectly every cell has cytoplasmic channels for metabolic exchange no matter how distant it may be from regional vascularization. Mineralization proceeds as bone matrix is deposited by the cells approximately midway between neighboring blood vessels. These sites of mineralization eventually form a latticework of branching and anastomosing needles of hydroxyapatite interwoven with the blood vessels (Fawcett and Raviola 1994). Much of the toughness and resiliency of bone is explained by the structure of the mineralized elements. The calcium phosphate crystals are aligned differently in alternate layers of the protein matrix, much like the structure of plywood. This irregular structure halts the spread of cracks and

fractures. The interconnected cells stimulate continued mineral deposition on the initial latticework until they become incarcerated in the hard matrix. These bone cells are able to remain metabolically active via their cytoplasmic supply lines to adjacent populations of cells. The unmineralized pockets in which the cells reside are the *lacunae* and the channels for the cell extensions the *canaliculi*. The spaces are evident in well-preserved fossils.

The metabolic activity of a bone cell continues through the period of **intramembranous ossification** (the building up of flat plate-like bones in the dermis as just described) and extends to reabsorption and remodeling of the internal microstructure of the bone throughout life. This ability to absorb and redeposit bone allows vertebrates to grow, to mobilize stores of minerals, and to heal damage to the mineralized tissue. When such **membrane bones** are restructured, the mineralized matrix and bone cell networks form layers and the blood vessels

Figure 3–7 *(Continued)*

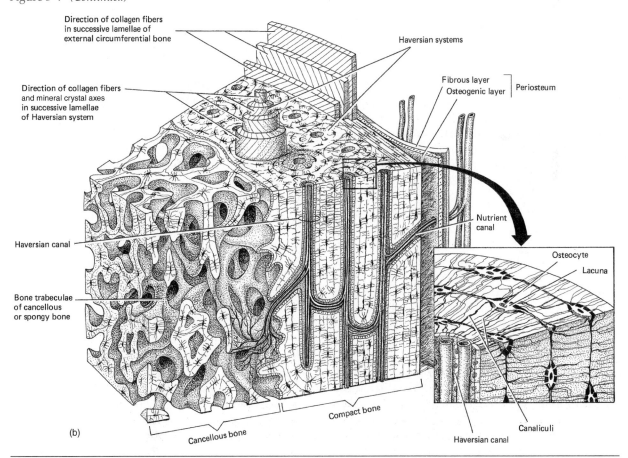

(b)

straighten out (Figure 3–7b). The resulting **lamellar bone** is arranged in concentric layers around vascular channels to form cylindrical units called **Haversian systems**. Intramembranous bone was characteristic of the superficial armor of the Paleozoic jawless vertebrates. Unlike dentine formation, however, ossification invaded deep-lying connective tissues to form the endoskeleton of later vertebrates.

The Endoskeleton

An endoskeleton is one of the basic characteristics of chordates. The **notochord** is the basic structural element of internal support for vertebrates. However, it is often obliterated during development by more familiar elements of the internal skeleton, the **vertebrae**. Phylogenetically older than vertebrae is the **cranium**, or skull, and appearing with it at the dawn of vertebrate life was the **visceral (pharyngeal) skeleton** of the gill arches and their derivatives.

Later contributions to the endoskeleton came with the evolution of the skeleton of the appendages and the ribs. The cranium, visceral skeleton, notochord, vertebrae, and ribs are together called the **axial skeleton**. The paired fin or limb skeletons and girdles make up the **appendicular skeleton**. The capacity to form mineralized tissues, the essence of the vertebrate endoskeleton, shows a clear historic pattern of step-by-step penetration of deep-lying connective tissues, beginning with dermal armor and progressing to cranial ossification, then to vertebral and finally to rib and appendicular ossification. Skeletal ossification may ancestrally have been limited to superficial tissues containing neural crest cells, later extending to deeper tissues of presumed neural crest origin, and finally to tissues of presumed mesodermal origin. Mineralized elements along the notochord were the first indications of vertebrae. Lateral plate skeletal derivatives (ribs, paired appendages) also acquired the capacity to support ossification of connective tissues. Each region of the axial and appendicular skeleton became more and more completely ossified during the evolution of the vertebrates, producing the bony skeletons of amniotes as we see them today.

One of the processes that becomes increasingly important phylogenetically is that of **endochondral ossification**. In this ontogenetic process a cartilaginous model of the adult structure develops and is subsequently invaded by blood vessels and bone-forming cells. The cartilage is eroded and replaced by bone. Ultimately, the entire structure, which retains the general shape of the original cartilage, is ossified. This process is quite obviously different from the formation of dermal bone or intramembranous ossification. Although these processes are distinct and appear to have separate phylogenetic histories, the final result, the bony tissue of the dermal or endochondral skeleton, is identical. Only developmental studies can determine how a particular bone was formed.

Skull The **skull** or cranium, which houses the brain and sense organs, is the most complicated portion of the skeleton. It is also rich with information, more often than not cryptically encoded, on the unique origins, phylogeny, ontogenetic processes, and adaptational success of vertebrate life (Hanken and Hall 1993). The cranium of vertebrates has evolved through additions from multiple ontogenetic sources, often repeatedly modified, to the point that recognition of homology is difficult at best. The skull is formed by three basic components: the dermatocranium, the chondrocranium, and the visceral arches or splanchnocranium.

The **dermatocranium** includes the roof of the skull, the superficial area around the orbits, the gill covers when they are present, the roof of the mouth (the palate and portions of the floor of the skull). The dermatocranium also contributes to the jaws, which in their most derived character state are formed primarily (e.g., birds) or entirely (mammals) by dermatocranium. The dermatocranium is in many ways the simplest of the skull's components. It is made up of dermal or intramembranous bones of the head integument (including the oral and pharyngeal lining), and it covers the other portions of the skull. Ossification first appears in the fossil record in these superficial tissues, and their record is better than that for the other ancestrally unmineralized skull elements. The dermatocranium appears to have been secondarily abandoned in living jawless fish and possibly chondrichthyans, but in other vertebrates it surrounds, unites with, or replaces portions of the other components of the skull so closely as to have usurped the functions of chondrocranial and visceral elements.

The **chondrocranium** is the deepest lying and probably phylogenetically the oldest portion of the skull. As its name implies, the entire structure begins phylogenetically and embryologically as cartilage that subsequently is replaced, partly or entirely, by endochondral ossification. A few living taxa retain a cartilaginous chondrocranium in adult life. Because it is often not formed by cartilage in adult vertebrates, some authors prefer to call it the **neurocranium**. The chondrocranium (Figure 3–8) is made up of portions of the floor of the cranial cavity that begin as a pair of cartilages flanking the anterior tip of the notochord (the **parachordals**). Other contributions come from the posterior wall of the brain cavity surrounding the spinal cord and from a pair of longitudinal cartilaginous bars (the **trabeculae**) located anteriorly and derived from the neural crest. The chondrocranium is completed by the cartilaginous capsules that form around two of the sense organs, the olfactory and otic capsules. The third pair of sense organs derived from placodes, the eyes, remain mobile and their connective tissue capsule becomes the sclera of the eyeball. These contributions to the chondrocranium by so many of the unique derived characters of vertebrates indicate the antiquity of the structure and its essential place in the origin of the vertebrates.

The final component, the **splanchnocranium**, comes from an unusual set of skeletal elements, the only hard ones to form in the wall of the gut (embryonic splanchnopleure), and are again derived from neural crest. Differentiating late during the pharyngula stage, the cartilaginous elements are all clearly related to the basic series of fleshy septa between pharyngeal gill slits or pouches. Each of these skeletal visceral arches is made up of several elements, usually five in each arch on each side of the head. In gnathostomes the anterior two pairs of arches become the jaws and their supporting structures (Figures 3–9 and 7–1), often undergoing partial or complete endochondral ossification. Where functional gills exist in adult vertebrates, they are supported by the more posterior arches. When gills are lost and respiration is via lungs, the posterior arches contribute to the larynx and trachea. The contribution of the visceral skeleton to the evolutionary success of the vertebrates cannot be overstated: The primary derived characters of gnathostomes originate from the visceral arch skeleton.

Notochord The notochord, long recognized as one of the most constant and characteristic features of ver-

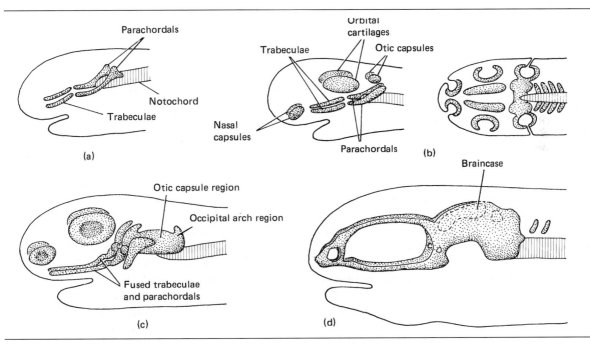

Figure 3–8 Early development of the chondrocranium. (a) through (d) are progressively more advanced stages of development. (b) Lateral view on left, dorsal view on right. The trabecular and parachordal cartilages are usually formed first, followed by the sensory capsules and the occipital arch. These centers of chondrification fuse to form the braincase. (Modified after W. W. Ballard, 1964, *Comparative Anatomy and Embryology*, Ronald, New York, NY.)

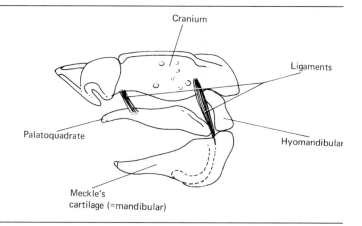

Figure 3–9 Relationships of the hyoid arch in supporting the jaw hinge as seen in the shark *Scyllium*. The hyomandibular, which acts as a strut to the jaw hinge, allows the jaw to be thrust forward for grasping and to enlarge the gape.

Cranium

Ligaments

Palatoquadrate

Hyomandibular

Meckle's cartilage (=mandibular)

tebrates, is an outwardly simple, stiff but flexible rod. It resists shortening of the body during segmental muscle contraction and translates unilateral contractions into the graceful bending curves of sinusoidal lateral undulations—the ancestral pattern of vertebrate locomotion. The notochord is a unique structure among the tissues of vertebrates. In some vertebrates it begins as a row of flat circular cells stacked like coins for the length of the body. At maturity it is made up of large, closely packed cells distended with fluid-filled vacuoles. This core of notochordal tissue is wrapped in a complex fibrous sheath that accounts for the rigidity of the rod. The physical characteristics of the notochord result from the incompressible nature of the fluids bound within the fibrous sheath. The sheath also represents the attachment surface between the notochord and the myomere connective tissue.

The notochord of all craniates ends just posterior to the bud of the pituitary gland. Posteriorly the notochord continues to the tip of the fleshy portion of the tail. In many fossil and extant fishes a notochord as just described (called an unrestricted notochord) appears as the endoskeletal body axis in the adult. In other fishes and in tetrapods the sheath of the notochord and the adjacent skeletogenous septa become involved in the development of structures that pinch and otherwise distort the notochord. In several clades of vertebrates the notochord becomes segmentally constricted into an axial series of nubbins (portions of the intervertebral disks in mammals) or obliterated altogether by derived axial endoskeletal elements, the vertebrae.

Vertebrae and Ribs In the majority of gnathostomes the axial skeleton is made up of a series of interseg-

mentally arranged **vertebrae**. These dense, usually mineralized units offer strength and a greater number and variety of attachment points for muscles than does the simple sheath of the notochord. The series of vertebrae from the skull to the tip of the tail constitutes the **vertebral column** or **spine**, an organ of complex function. The spine provides rigidity to the body, especially in large terrestrial vertebrates, where it becomes disproportionately massive. In swimming vertebrates the vertebral column retains the ancestral function of resisting body shortening during lateral undulation. To do this and yet permit flexibility, the spine is not made up entirely of rigid mineralized vertebrae. The vertebrae alternate with flexible cartilaginous **intervertebral disks**. These disks, which are partly remnants of the notochord, are bound between vertebrae by muscle and ligament attachments running from vertebra to vertebra. In terrestrial vertebrates the compressional forces of bilateral muscle contractions and the complex interlocking of their vertebrae often restrict the number of planes of movement possible in the spine. These interlocking surfaces provide a rigid support for the suspended weight of the body musculature and viscera.

Vertebrae are composite structures made up of mineralized cartilages or of both endochondral and intramembranous bony elements formed in the various skeletogenous septa intersecting at the central body axis. Six major components are usually recognizable: The **centrum** is a solid cylindrical spool that surrounds and often completely replaces or incorporates the notochord and makes up the body of the vertebra. A **neural arch** grows dorsally to cover the neural (spinal) cord like a rigid tent. A **hemal arch** grows ventrally (on postanal centra only) and similarly encloses the principal caudal blood vessels. **Neural** and **hemal spines** are bony blades that project into the dorsal and ventral skeletogenous septa, respectively, and provide sites for attachment of muscles and ligaments. Finally, a variable number of bilaterally symmetrical **apophyses** project from the vertebrae and attach via muscles and ligaments to other skeletal elements.

Ribs are functionally and anatomically related to the vertebral apophyses. They provide sites for muscle attachment and strengthen the body wall. Ribs form in intersecting skeletogenous septa, generally articulating with vertebral apophyses that develop in the same septa. Myosepta are one of the skeletogenous septa participating in rib formation, and thus ribs are intersegmental structures. **Intermuscular** or **dorsal ribs** form at the myoseptal and horizontal skeletogenous septum intersection. Where the myosepta intersect the connective tissue surrounding the coelom, **subperitoneal** or **ventral ribs** form between the coelomic wall and the hypaxial muscles (those ventral to the horizontal skeletogenous septum). The ventral ribs are thought to be phylogenetically older than the dorsal ribs, but both types occur together in many bony fishes along with additional rib-like intermuscular bones peculiar to these fishes.

Vertebrae and ribs are only moderately specialized in most fishes, usually in conjunction with unique derived characters of a particular taxon such as hearing in minnows and their relatives (Figure 8–29). The trend in the vertebrae and ribs of tetrapods has been toward differentiation into five regions from anterior to posterior: **cervical, thoracic, lumbar, sacral**, and **caudal**. Enhancing mobility of the head while continuing protection of the spinal cord is the function of the cervical vertebrae. The two most anterior cervical vertebrae are the highly differentiated **atlas** and the **axis**. The sacral vertebrae fuse with the pelvic girdle and transfer force to the appendicular skeleton. The trunk vertebrae are differentiated into two regions. The anterior region is the thoracic group with articulated ribs that function in respiratory movements in most amniotes as well as in support and protection for the lungs and heart. The **sternum** is a midventral, usually segmental structure, common to all tetrapods, that links the distal ends of right and left thoracic ribs. It ossifies well only in birds and mammals, and its evolutionary origins are obscure. The next posterior trunk vertebrae are the lumbar vertebrae, with abbreviated ribs fused to the centra and supporting an extensive muscular sling for the pliant, mobile viscera of the posterior coelom. Posterior to the sacral region, the caudal vertebrae are generally much like those of fishes but with short spinous processes.

The phylogenetic trend toward increased specialization of the vertebrae has proceeded independently in different lines as shown by the variable number of vertebrae of each region in closely related taxa. The fact that a taxon as diverse as the mammals almost invariably has seven cervical vertebrae is surprising, and is the only notable exception to a general pattern of variation seen in tetrapods. Regional specialization of the axial endoskeleton reflects the appearance of specialized subdivisions of the dorsal musculature, for the skeleton is closely linked to the muscles as the skeletomuscular complex.

Appendicular Skeleton Many of the evolutionary changes seen in elements of the axial skeleton of vertebrates are reflections of the vast reorganizations that have occurred in several stages of the evolution of the appendicular skeleton. Although median fins, especially the caudal fin, are essential to fishes, only the paired pectoral and pelvic appendicular structures have maintained a continuity of function across the boundary from fish to tetrapod.

Paired appendages appear in all of the early clades of jawed fishes. Their anatomical origins remain obscure, with no clear embryological clues such as those to the origin of jaws. In jawed fishes, especially derived ray-finned forms, the pectoral girdle is robust and articulates sturdily with the axial skeleton via the posterior part of the skull. The pelvic appendages retain the ancestral role of stabilizers but gain mobility often in conjunction with forward migration and attachment to the pectoral girdle.

In tetrapods the pelvic girdle becomes fused directly to modified sacral vertebrae and the hind limbs function as the primary propulsive mechanism. The tetrapod pectoral girdle is most often free of direct articulation with the axial skeleton and is often less robust than the pelvic girdle.

Homologies of elements of the appendicular skeletons of living vertebrates are obscure, although there is a high degree of early embryonic similarity. Note that the similarity between the elements of the anterior and posterior appendages (Figure 3–10) is not due to homology in the usual sense of that term. The appendages may, however, be considered as **serially homologous** structures, derived from similar genetic mechanisms acting in different body

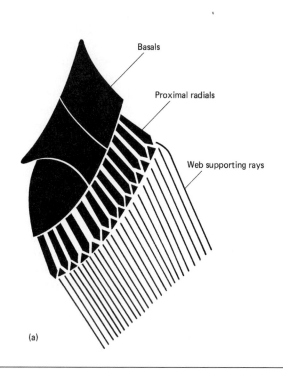

Basals
Proximal radials
Web supporting rays

(a)

APPENDICULAR SEGMENTS		PECTORAL	PELVIC
GIRDLE		Scapula (shoulder blade)	Pelvis (hip bone)
PROPODIUM		Humerus (upper arm)	Femur (thigh)
EPIPODIUM		Ulna/Radius (forearm)	Tibia/Fibula (lower leg)
MESOPODIUM		Carpus (wrist)	Tarsus (ankle)
METAPODIUM		Metacarpus (hand)	Metatarsus (foot)
PHALANGES		Phalanges (fingers)	Phalanges (toes)

(b)

Figure 3–10 Basic divisions of the endoskeletal supports of vertebrate paired appendages: (a) most fishes; (b) tetrapods as exemplified by the primitive mammalian condition with common and anatomical terminology compared.

segments. Although advances have been made in elucidating these genetic mechanisms (Tabin 1992, Tabin and Laufer 1993), consensus on the subject is not in sight (Coates 1993, Kessel 1993, Thorogood and Ferretti 1993).

The ancestral form of the endoskeletal elements of the paired fins of most fishes is a semicircular girdle embedded in the ventral musculature providing articulation for a small number of fanlike **basal elements** (Figure 3–10a). The basals support one or more ranks of cylindrical **radials**, which usually articulate with one or more ray-like structures that support most of the surface of the fin web. The thickness of this web and the development of intrinsic fin musculature determines the flexibility and mobility of the fin surface. These characteristics are generally poorly developed in plesiomorphic vertebrates such as sharks and well developed in rays and some derived bony fishes.

The basic tetrapod limb (Figure 3–10b) is made up of the limb girdle and five segments articulating

end to end: the propodium (humerus in the forelimb or femur in the hind limb), the epipodium (radius plus ulna or tibia plus fibula), the mesopodium (carpus or tarsus), the metapodium (metacarpus or metatarsus), and the phalanges (fingers or toes).

The Muscular System

The skeleton's functions of support and locomotion would be impossible without the muscular system with which the endoskeleton has so intimately coevolved. The musculature of vertebrates is generally divided into **involuntary** (= **smooth**) muscles derived from the splanchnic mesoderm and **voluntary** (= **striated** = **skeletal**) muscles, of which most, if not all, are derived from the myotomes of the mesoderm.

The gross morphology of most of the muscle of fishes is similar, but it is radically different from that of tetrapods (Figure 3–11). Fishes retain obvious axial segmentation of the musculature, reflecting that of the embryonic somites. The generally short

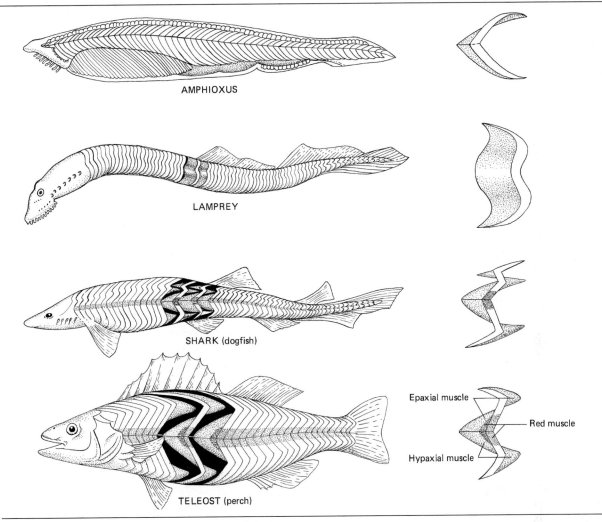

Figure 3–11 Myotomes. Acraniates like amphioxus have relatively simple myotomal form that is only slightly more complex in jawless fishes such as the lamprey. The myotomes are more complexly folded in advanced chondrichthyans and osteichthyans. Unshaded: surface appearance below skin; light shading: deep portions of myotome associated with axial skeleton but not visible from the surface; dark shading: areas of concentration of red muscle fibers. (Modified from A. G. Kluge et al., 1977, *Chordate Structure and Function*, 2d edition, Macmillan, New York, NY; and Q. Bone and N. B. Marshall, 1982, *Biology of Fishes*, Blackie, Glasgow, UK.)

muscle fibers of fishes run from the posterior face of one myoseptum to the anterior face of the next posterior myoseptum. Thus the muscles seem to be in simple blocks. However, in adult fishes each myomere and its adjacent and well-developed myosepta are complexly folded fore and aft so that a given myomere extends anteriorly and posteriorly over several body segments (and thus vertebrae). As a result, each intervertebral joint is acted upon by several sequentially arranged and innervated myomeres. The functional outcome is a smoothly graded bending of the spine as each motor nerve from head to tail fires down the sides of the swimming fish. The superficial midlateral muscle fibers in such an arrangement are nearly parallel to and at the level of the vertebral column. The deeper fibers, although running from myoseptum to myoseptum, may be strongly and regularly angled with respect to the longitudinal body axis due to the complex folding of the myosepta. The complex three-dimensional arrangement of fibers matches the amount of contraction of fibers close to the vertebral column (and thus near the center of curvature) to the amount of contraction of those fibers farthest from it, with a resultant efficiency not possible in an all-fibers-parallel arrangement.

Differences in the three-dimensional arrangement of muscle fibers also occur in tetrapods. In terrestrial vertebrates individual muscles are likely to differ in fiber orientation from neighboring muscles rather than having substantial differences of fiber orientation within the same muscle. One reason for differences in tetrapod muscle fiber orientation is the variable way the muscles relate to the skeleton. Muscles may attach directly to the connective tissue sheath of a bone, but it is more common for muscles to attach to the skeleton through **tendons**, tightly packed bundles of the long fibrous protein collagen. Tendons, whether in fishes or tetrapods, allow the force of muscle contraction to be applied at restricted insertions on a bone so that many muscles, through their tendons, can act on a small skeletal element. **Ligaments**, which run from one bony or cartilaginous element to another, are often elastic so they store energy when they are stretched and can return a joint to its original position without additional energy when the muscular contraction ceases.

Skeletomuscular Function

The skeleton, activated by muscles that may work at a distance through tendons, is essentially a system of levers. The skeletomuscular apparatus is subject to the same physical constraints as simple levers: Levers that maximize power act over short distances and are slow, whereas levers that maximize speed act over long distances and are weak. Sometimes the skeletomuscular systems are arranged into complex, interconnecting lever units known as **kinematic chains**. Prime examples of kinematic chains are the jaw-opening and protrusion mechanisms of fishes (Figure 8–22) and the kinetic head morphology of snakes. If one link in such a chain is fixed (by muscle contractions that brace it immovably against the axial skeleton), it limits the range of movements possible by other elements in the kinematic chain. When the fixed element is freed, however, the motions of the remaining parts of the kinematic chain of levers may be very different. By fixing in position different elements of such a chain at different times an animal can obtain a very wide range of motions. The force output from a limited number of skeletal and muscular elements is thus greatly diversified.

Energy Acquisition and Support of Metabolism

Vertebrates are among the most intensive consumers of energy on Earth. The energy vertebrates use is gleaned from the environment as food that must be processed to release energy and nutrients. This processing is the primary function of the digestive system.

After food has been digested and assimilated into the body, it must be transported to the tissues where the energy it contains is released and where some of its chemical constituents may be incorporated into the body tissues of the animal. Oxygen is required for the process of energy release, and the functions of gas exchange surfaces and the circulatory system are closely intertwined with those of the digestive system.

Feeding and Digestion

The trophic process of vertebrates is divisible into feeding and digestion. Feeding includes getting food into the oral chamber, some oral or pharyngeal processing (i.e, chewing in the braod sense), and swallowing. Digestion includes the breakdown of complex compounds into small molecules that are absorbed across the wall of the gut. Both feeding and digestion are two-part processes; each has a physical component and a chemical component, although the physical components dominate in vertebrate feeding, and the chemical ones dominate in digestion.

Most vertebrates feed on large particles relative to the size of their heads. In aquatic environments suction feeders take advantage of the viscous properties of water by creating a flow of water that carries food into the mouth. No similar mechanism exists among terrestrial vertebrates. The anterior teeth or the beaks of most vertebrates are used to capture prey and to sever pieces of food. In mammals a battery of specialized posterior teeth act in concert with jaws, cheeks, and tongue to chew food (Chapter 21). Most vertebrates do little or no mechanical processing of food in the mouth before swallowing. However, some chemical processing may take place.

Terrestrial vertebrates have salivary glands that secrete a watery lubricant that probably aids swallowing. The salivary secretions of many vertebrates contain enzymes that begin the process of digestion. Insectivorous mammals, certain lizards, and several lineages of snakes have elaborated these secretions into venoms that kill prey. A variety of venoms are known in fishes from many evolutionary lineages, but they all appear to be used defensively. None is known to be used in subduing potential prey, and venoms are rarely associated with the teeth of fishes, perhaps because fishes do not have salivary glands.

Swallowing is accomplished by contraction of the splanchnic musculature of the gut wall. These muscles, the mechanism of swallowing, and peristalsis (traveling waves of circular muscle contraction in the gut) are derived characters of the vertebrates. Other chordates use ciliary action throughout the gut to move food, digestate, and solid wastes.

The **stomach**, long thought to be a derived character of gnathostomes but now known to occur in some fossil jawless fish as well (see Chapter 6), is in general a rather simple organ in gross anatomy. Its lining is often abruptly different from that of the **esophagus**, from which it is separated by a muscular sphincter. The stomach of most vertebrates is capable of considerable distention. This is carried to an extreme in some deep-sea fishes. The swallowers (Elopomorpha) have a pharynx, esophagus, and stomach of such distensibility that they can swallow fishes twice their own length.

The size and structure of the stomach is related to dietary habits. Vertebrates that feed on fluids have reduced or nonexistent stomachs; these include jawless fishes such as lampreys and nectar-feeding birds and mammals. Herbivores, which feed primarily on the digestion-resistant leaves and stems of plants, often have greatly enlarged and frequently multichambered stomachs. As a rule, herbivores also have elongate intestines that accommodate the slow release of nutrients from plant digestion.

A glandular epithelium lines the stomach of vertebrates. The glands in the stomach lining secrete hydrochloric acid and the proteolytic enzyme pepsin, which acts only in an acid medium. The muscles of the stomach wall churn and mix the food, forming a particle-laden solution known as chyme. Chyme passes to the small intestine through the pyloric sphincter or valve for further digestion and absorption. The small intestine secretes mucus and probably enzymes that break down the cell walls of microorganisms as well as further digesting the chyme.

The intestine is straight only in living jawless fishes. The intestines of most vertebrates are many times longer than the body and are coiled or folded. Herbivores have especially long intestines. Sharks, skates, plesiomorphic bony fishes, including lungfishes, and at least one fossil vertebrate (a placoderm) illustrate an alternative method of increasing the area of intestinal surface in contact with the chyme—the **spiral valvular intestine**. This distinctive structure is apparently ancestral, at least within gnathostomes. The spiral valve consists of a broad sheet of the secretory and absorptive lining of the intestine that protrudes into the lumen (cavity) of the gut; the base of the fold extends from near the pylorus to near the end of the gut in a tight spiral path along the wall of the intestine. Both long intestines and spiral valves increase the surface area of the intestine, augmenting its function of absorption. Additional specializations, especially well developed in mammals, further increase the surface area of the intestine by large- and small-scale foldings of the lining. In birds and mammals finger- or leaflike projections composed of epithelial cells around a core of mesoderm form **villi**. The gut epithelial cells themselves have closely packed **microvilli** on the lumenal surface of each cell. These structures increase the surface area of the intestine greatly: The total surface area of the human intestine is about 300 square meters, about the size of a regulation basketball court.

Accessory digestive organs adjacent to the gut produce enzymes and related substances that break chyme down to the simple sugars and amino acids that can be absorbed by the gut. The **liver** is the largest gland in the body of any vertebrate, reaching a maximum of 25 percent of the body mass in some sharks. In these fishes it is a storage reservoir for oils in addition to its normal functions. In all vertebrates the liver produces digestive secretions, processes absorbed nutrients and metabolites, and detoxifies harmful substances. These functions are facilitated by the position of the liver in the circulatory system: It receives blood coming from intestinal walls carrying the molecules absorbed from the digestate. The secretory function of the liver is twofold: It produces **endocrine** secretions for release into the blood and **exocrine** secretions (defined as those carried to the outside of the body—in this case to the lumen of the gut). The exocrine secretions are combined in the bile, which is part waste (neutralized toxins) and part digestive juice. The digestive components are not enzymatic, but rather fat-emulsifying organic molecules that act like detergents upon entering the gut. They form submicroscopic droplets of fats a few molecules in size, allowing fats to be absorbed by the intestinal epithelium. Without these bile secretions, fats and oils in foods cannot be absorbed and would simply pass through the tract. Bile enters the intestine via the bile duct just posterior to the pyloric valve. A gallbladder may or may not occur as a blind storage sac along the bile duct. The distribution of the gallbladder among vertebrates, its size, and the overall rate of bile secretion, correlate positively with the amount of fats consumed in a species' natural diet.

The **pancreas** is the second of the accessory digestive glands along the intestine. Although not involved in as many complex biochemical processes as the liver, the pancreas (when present as a definitive gland) has both endocrine and exocrine functions segregated quite sharply by cell type. Lampreys and most bony fishes lack a discrete pancreas, but cells with pancreatic function are scattered in the mesenteries or embedded in the wall of the intestine or the liver. Pancreatic enzyme secretions are essential for the final breakdown of carbohydrates to simple sugars and fats to glycerol and fatty acids. Most important are as many as five potent pancreatic protease and nuclease enzymes that complete the digestion begun in the stomach. These enzymes would, of course, digest the pancreas if they were not secreted as inactive precursor molecules that are activated upon mixing with the lumenal contents. The bile and pancreatic juice are also rich in bicarbonate that neutralizes the stomach acid; thus, digestion in the gut (in contrast to that in the stomach) occurs at neutral pH. A final suite of digestive enzymes is incorporated in the membranes of the microvilli of the intestinal epithelial cells. Digestion occurs on the surfaces of these microvilli, which also possess the carrier molecules necessary for active transport of glucose and amino acids into the cells and eventually to the circulatory system of the gut. The products of fat digestion are biochemically reconfigured and shunted to the lacteals, specialized branches of the lymphatic system of the gut wall. From there fats are introduced to the general circulation, bypassing the liver.

One (in mammals) or a pair (in birds) of **ceca** (singular, cecum) are found at the junction of the large and small intestine. Some bony fishes have similar ceca in the region of the pyloric valve. Ceca appear to be fermentation chambers for gut microorganisms that digest cellulose. These symbiotic microorganisms break down plant cell components, especially carbohydrates, which may then be available for absorption. The complex symbioses between gut microorganisms and vertebrates from fishes to humans have been surprisingly little studied. The **large intestine** or **colon**, a unique character of tetrapods, is primarily a region of absorption, mucus being the only important secretion. Its surface is less complicated than that of the small intestine. The colon ends in a short **rectum** that generally has an involuntary sphincter formed by smooth muscle and a voluntary sphincter formed by striated muscle. The rectum may open to the outside directly through the anus (derived bony fishes and most mammals), but usually opens into a small chamber, the **cloaca**, shared with the opening of the urogenital ducts. A cloaca is found in the embryos of all vertebrates, and it is considered a distinctive ancestral character of vertebrates. **Feces** are the result of removal of water, ions, and many of the digested organic molecules from the chyme. The remaining undigested food residue, bile secretions, microorganisms, and actively secreted wastes such as heavy metals are passed to the outside of the body by the action of the muscles of the rectum. Anal glands frequently add lubricants as well as pheromones to the numerous volatile substances already present in the feces. Thus, feces are used in communication, especially to mark territorial boundaries.

Respiration and Ventilation

Energy is obtained from the metabolic oxidation of nutrients absorbed across the gut wall. This process requires the delivery of oxygen to every living cell. Organismal respiration is a multistage process that requires ventilation of respiratory surfaces where oxygen is absorbed from the environment, transportation of oxygen in the blood, and sometimes storage of oxygen in the tissues.

Ancestral chordates probably relied on **cutaneous respiration**—oxygen absorption across a thin skin—a mechanism that continues to be important for many vertebrates. The pharyngeal gill slits, initially part of the feeding system, had a large surface area and were constantly exposed to flowing water. Early in vertebrate evolution the endodermal lining of the pharyngeal arches developed a complex highly folded surface, the **gill lamellae**, richly invested with blood vessels. The gills of living jawed fishes are ectodermal variations on this theme (Chapter 7). Within the bony fishes the relative breadth of the lamellar attachment to the fleshy gill septum distinctly diminishes from the condition seen in the sharks and skates (the elasmobranchs, "plate gills"). An intermediate degree of attachment is seen in plesiomorphic ray-finned bony fishes; complete loss of the interlamellar septum and development of long, free gill filaments is characteristic of most teleost bony fishes.

In addition to functional gills, many bony fishes have accessory respiratory organs that are derived from the gut. The most common and phylogenetically oldest of these structures appears to be the ancestral swim bladder, the homolog of the tetrapod lung. These blind-ended, thin walled vascular sacs arise from outpocketings of the gut just posterior to the pharyngeal pouches. Fishes and amphibians use the hyoid apparatus to ventilate the lungs. The oral cavity is expanded, sucking air into the mouth, and then the floor of the mouth is raised, squeezing the air into the lungs. This method of lung ventilation is called a force pump or positive pressure mechanism.

Amniotes use a suction pump or negative pressure mechanism of lung ventilation. Movement of the ribs (in birds and squamates), internal organs (turtles and crocodilians), or a muscular diaphragm (mammals) increases the volume of the **pleural cavity**, which houses the lungs. This expansion creates a negative pressure (that is, below atmospheric pressure) in the pleural cavity and sucks air into the lungs. Air is expelled (exhaled) by compressing the pleural cavity, primarily through an elastic return of the rib cage to a resting position of smaller volume and a contraction of the elastic lungs. The lungs of most amniotes are divided into numerous blind chambers (**alveoli**) where gas exchange occurs. Alveoli are best developed in mammalian lungs, but some squamates also have well-developed alveoli. The posterior lung of some squamates is undivided and nonvascular. Similar nonvascular structures are highly developed in birds and are known as air sacs, some of which penetrate hollow bones or adhere closely to flight muscles. The respiratory portion of the lung of birds is also distinct. A complex system of channels allows air to flow in a single direction past the exchange surfaces of bird lungs.

Vocalization The regular large tidal flow of air through the pharynx and trachea to the lungs has set the stage for tetrapod vocalizations. The flow of respiratory air is forced through openings occluded by thin but often muscular membranes that vibrate and produce sounds. The usual site of sound production is the **larynx** (a derived character of tetrapods that is primarily derived from pharyngeal arch elements) at the junction of the pharynx and trachea. A second site, the **syrinx**, has taken over sound production in birds. The syrinx is formed by tracheal cartilages, muscles, diverticula of air sacs, and membranes. It is located where the trachea bifurcates into the right and left lungs.

Cardiovascular System

The cardiovascular system transports oxygen and nutrients to the living cells of the body, removes carbon dioxide and other metabolic waste products from the cells, and participates in maintaining the internal environment. In addition, the cardiovascular system carries hormones from their sites of release to their target tissues and is a second line of defense against pathogens and other foreign substances.

The Vessels The blood of vertebrates is contained in a **closed circulatory system** (Figure 3–12). That is, specialized vessels and organs contain the blood throughout the body. (This situation contrasts with the open circulatory systems of many invertebrates in which the blood percolates freely through nonvascular tissues.) Even blood sinuses, the expanded areas where blood accumulates within the vertebrate circulatory system, are lined by specialized tissue like the rest of the cardiovascular system. Thus, the cardiovascular system of vertebrates is composed of a muscular pump (the heart, which is a derived character of craniates) and interconnecting vessels. Arteries carry blood away from the heart in a ramifying series of vessels of progressively decreasing diameter, and blood returns to the heart in veins. Interposed between the smallest arteries (arterioles) and the smallest veins (venules) are the **capillaries**, which are the sites of exchange between the blood and tissues.

Capillaries pass close to every living cell and provide an enormous surface area for the exchange of gases, nutrients, and waste products. The surface area of the capillaries of a human is about 700 square

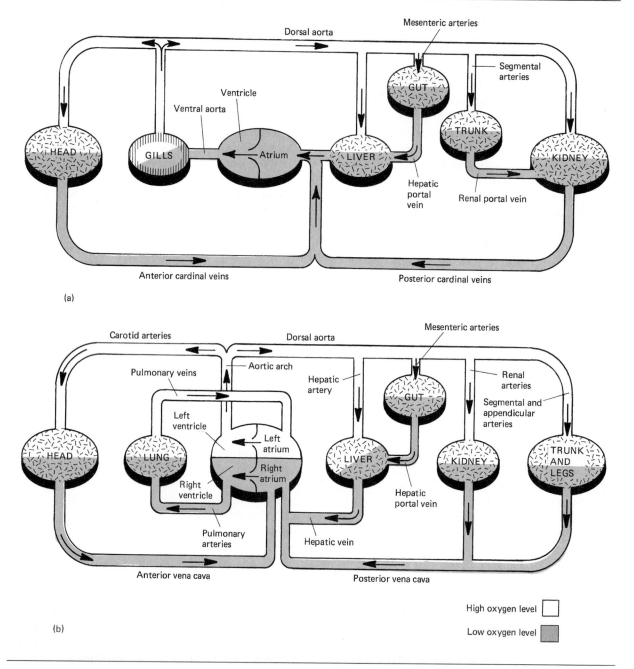

Figure 3–12 Basic plans of vertebrate cardiovascular circuits as seen in (a) a single-circuit fish-like pattern and (b) a double-circuit tetrapod-like pattern exemplified by that of a bird or mammal. Dark shading, venous blood.

meters, or nearly the area enclosed by the bases of a regulation baseball diamond.

Capillaries form dense capillary beds in metabolically active tissues and are sparsely distributed in tissues with low metabolic activity. Blood flow through a capillary bed is regulated by the opening and closing of precapillary sphincter muscles. Normally at any given time, only a fraction of the capillaries in a tissue have blood flowing in them; the rest are stagnant or empty. **Arteriovenous anastomoses** connect some arterioles directly to venules, allowing blood to bypass a capillary bed. When the metabolic activity of a tissue increases—when a muscle becomes active, for example—waste products of metabolism stimulate precapillary sphincters to dilate, increasing blood flow to that tissue.

The *rete mirabile* ("marvelous net," plural retia mirabilia) is a widely occurring vascular structure in vertebrates. In a rete mirabile the afferent vessel (an artery carrying blood from the heart to an organ or to a structure such as a limb) breaks down into a network of small, parallel vessels. The corresponding efferent vessel (a vein carrying blood from the organ back toward the heart) forms a similar series of small vessels. These vessels have thin walls and lie side by side, facilitating the transfer of heat or dissolved substances between them. Because the blood flows in opposite directions in the afferent and efferent vessels, a rete is a countercurrent exchange system.

Retia are commonly found in the limbs of tetrapods, where they help to conserve heat. Heat is rapidly lost from the limbs, which have a small volume and a large surface area, and especially in cold climates the temperature of the limbs of a bird or mammal may be many degrees lower than the core body temperature. Blood flowing into the limbs is chilled and would cool the body when it returned if it did not pass through a rete mirabile. In the rete the warm arterial blood transfers heat to the cold venous blood, warming it nearly to core temperature before it returns to the body and conserving heat.

Retia are not limited to limbs. Retia retain metabolic heat in the swimming muscles, sense organs, and/or parts of the brain of warm-bodied fishes such as tunas and some sharks, and additional retia allow tunas to keep their digestive organs and livers warm. Retia are also used to retain high concentrations of substances in restricted parts of the body. High concentrations of oxygen are kept in the swim bladders of fishes by a rete (see Chapter 8) and the high concentrations of sodium and chloride in mammalian kidneys depend in part on the presence of an elaborate rete (see Chapter 4).

The walls of capillaries are generally one cell thick. The cells making up the capillary walls and lining the interior of the rest of the circulatory system are called **endothelial** cells. Some endothelial cells in the capillaries of the intestine, pancreas, and endocrine glands are perforated by small holes, each closed by a thin diaphragm through which dissolved substances are believed to pass. The majority of endothelial cells have vesicles that transport dissolved substances across the capillary wall.

In addition to capillaries, the smallest veins (venules) may have walls thin enough to be involved in exchange with surrounding tissues. All the other vessels function only in the distribution and collection of blood. Arteries have walls made up of three layers: an internal endothelium, a middle layer of involuntary (smooth) muscle cells, and an outer layer of fibrous connective tissue. The muscle cells in the middle layer are usually arranged circumferentially and the connective tissue fibers in the outer walls are oriented longitudinally. The muscle layer is thick in arteries, whereas the connective tissue layer is better developed in veins.

Arteries and veins usually run side by side in a supply trunk related to a particular organ or set of tissues. Veins are usually more numerous than arteries, and the volume of the venous system is larger than the volume of the arterial system. Blood pressure in the venous system is lower than arterial pressure, partly because of this greater volume and partly because of the pressure drop that occurs as the blood passes through the resistance created by the narrow capillaries. Blood flows more slowly in veins than in arteries, and some veins have valves that prevent back flow.

Arteries are more elastic than veins because the rubbery, straw-colored protein elastin is present in the middle layer of the artery wall. Indeed, the main arteries of large mammals contain so much elastin that they appear yellow. Near the heart these vessels distend during ventricular contraction (systole) and elastically recoil during ventricular refilling (diastole). This expansion and contraction of the arteries smooths out the pulsatile flow of blood produced by the pumping of the heart. Elastin is unknown among invertebrates and has not been found in hagfish or lampreys, but it occurs in all gnathostomes (Sage and Gray 1979, 1981a,b). Sharks have elastin with a presumably ancestral amino acid composition, and ray-finned fishes have an elastin that is different from the elastins of sharks and also from that of lungfishes, coelacanths, and tetrapods. These changes mirror the basic pattern of vertebrate phylogeny and chronicle the increase in systemic blood pressure in gnathostome evolution.

Some vessels, known as **portal vessels**, are interposed between two capillary beds. In those instances arterial blood passes through the first capillary bed and is collected in a vein that subsequently breaks down into a second capillary bed in a different tissue or organ. Finally, the blood is collected in a systemic vein and carried to the heart. Portal vessels promote sequential processing of blood. The hepatic portal vein, for example, is interposed between the capillary bed of the gut and the sinusoids of the liver. Substances absorbed from the gut are transported directly to the liver where toxins are rendered harmless and some nutrients are further processed or removed for storage. The hepatic portal system appears in all living vertebrates and is considered an ancestral character for vertebrates. A hypophyseal portal system exists between the brain and the pituitary (Figure 3–25). A renal portal system between the gut and the kidneys occurs in some nonamniote gnathostomes and amniotes (except mammals) and is apparently a derived character of gnathostomes.

Lymphatic vessels resemble veins but have thinner walls and are not part of the closed blood circulatory system. The lymphatic system is a one-way system of vessels that largely parallels the blood vessels. Lymphatic vessels originate as blind-ended tubes in the tissues and connect to the major veins. Plasma that is forced out of capillaries into the tissues by hydrostatic pressure is collected by the lymphatic system and channeled back into the circulatory system. Specialized lymph drainage in the intestine via the lacteals has already been mentioned in the context of fat absorption. The lymphatic systems of most nonamniotes have muscular areas that contract and keep the lymph flowing. In amniotes lymph is moved by the compression of lymphatic vessels as muscles and tissues contract and move, and a series of valves keeps the flow moving toward the circulatory system. **Lymph nodes** lie along the course of the lymphatic vessels of mammals and a few birds. Lymph in the nodes is exposed to phagocytic cells that cleanse it of foreign material, and lymphocytes, part of the immune system, are added to the lymph and thus distributed via the circulatory system to all parts of the body.

Basic Vascular Circuits The ancestral vertebrate circulatory plan consists of midline vessels with the major arteries in a dorsal position and the veins ventral to them (Figure 3–13a). The heart pumps blood anteriorly in the **ventral aorta** (the only major ventral artery) to gill capillaries, where it is oxygenated. The blood then enters the **dorsal aorta** and flows posteriorly into arteries, arterioles, and finally capillaries. Blood leaving the capillaries collects in venules and

then in veins of increasing size, returning to the heart via the **posterior cardinal veins**. A fish-like or **single circulation** of this sort is a low-pressure system because of the drop in blood pressure across the resistance provided by the gill capillaries. The entire output of the heart passes through the gills, and by the time blood returns to the heart it may have passed through as many as three capillary beds (i.e., gills, general body tissues, and either the liver or kidneys after passing through a portal system). Even so, in some fishes blood pressure is higher than might be expected because of the vessel squeezing action of their trunk muscle contractions during swimming.

With the advent of lungs vertebrates evolved a **double circulation** in which the **pulmonary** circuit supplies the lungs with deoxygenated blood and the **systemic** circuit supplies oxygenated blood to the body. The single heart is the pump for both circuits, and is divided into right (pulmonary) and left (systemic) halves. The division of the heart into right and left sides is an actual morphological separation in archosaurs (crocodilians and birds) and mammals. In lungfishes, amphibians, squamates, and turtles there is no permanent morphological separation, but complex interactions between heart morphology and flow patterns maintain the separation of oxygenated and deoxygenated blood. The double circulation of tetrapods can be pictured as a figure eight with the heart at the intersection of the loops. One loop is the pulmonary circuit, and the other is the systemic circuit. This morphology permits the blood pressure to be low in the pulmonary circuit (where the blood flows through delicate capillaries during gas exchange) and high in the systemic circuit (where blood must be pumped long distances and through muscles that are contracting and squeezing the blood vessels).

The Heart As one would expect, the transition from a single circulation to a double circulation required major changes in the morphology of the heart and major vessels. The heart of fishes is a muscular tube folded upon itself and constricted into four sequential chambers: the sinus venosus, the atrium, the ventricle, and the conus arteriosus (or bulbus cordis). The sinus venosus and conus arteriosus are reduced or absent in the hearts of tetrapods and their prominence in fishes probably relates to the low peripheral pressure of the single circulation. The atrial and ventricular chambers are extensively modified in tetrapods.

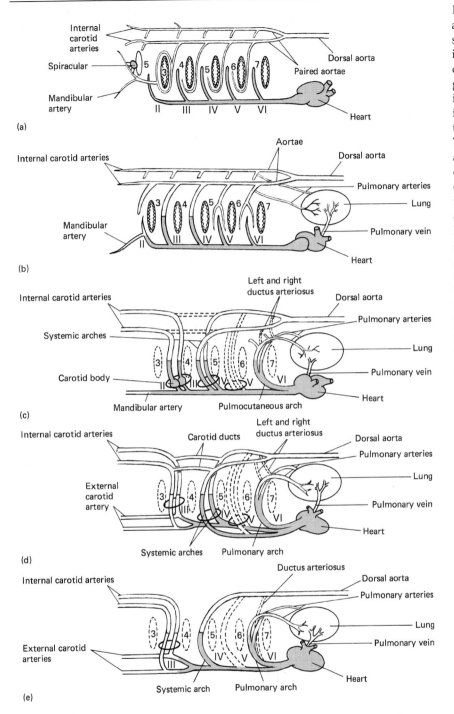

(a)

(b)

(c)

(d)

(e)

Figure 3–13 Evolution of the aortic arches. (a) Generalized scheme for a gill-breathing primitive vertebrate. Chondrichthyans and actinopterygians have numerous independently derived specializations that do not depart drastically from this basic plan. (b) The lungfish and (c) the adult anuran demonstrate variations on the double circulation and dual respiratory adaptations of these vertebrates. (d) Primitive amniote condition as found in a lizard. (e) Advanced amniote condition as found in a mammal. Arabic numbers, gill clefts/visceral pouches; Roman numerals, aortic arches. Numberings represent presumptive primitive condition for vertebrates. (See R. Lawson in M. H. Wake, 1979, *Hyman's Comparative Vertebrate Anatomy*, University of Chicago Press, Chicago, IL.)

The anatomical limits of the heart can be defined by the presence of **cardiac muscle**. Although the heart is controlled by the autonomic nervous system like most organs that contain smooth muscles, the fibers of cardiac muscle do not resemble smooth muscle but are striated like skeletal muscle. Unlike skeletal muscle, however, the cells of cardiac muscle appear to have one nucleus and distinct terminal junctions of the cells (the intercalated disks) that are visible in histological preparations.

Blood enters the tubular fish heart via multiple right and left systemic veins and usually by a hepatic vein from the liver. The **sinus venosus** is a thin-walled sac with a few cardiac muscle fibers. It is filled by pressure in the veins and pulsatile drops in pressure in the pericardial cavity as the heart beats. Suction produced by muscular contraction draws blood anteriorly into the **atrium**, which has valves at each end that prevent backflow. It thus acts as a one-way pre-pump that fills the main pump, the **ventricle**. A cross section of the ventricle of a fish shows thick, muscular walls and a spongy interior with a small lumen. Contraction of the ventricle forces the blood into the **conus arteriosus**. Teleosts lack a conus arteriosus formed by cardiac muscle, but have an elastic bulbus cordis that is formed by smooth muscle and is perhaps better regarded as a modified part of the ventral aorta. Either the conus or the bulbus serve as elastic reservoirs that damp the pulses of blood pressure that are produced by contractions of the ventricle, and they have a valvular construction that prevents blood from flowing back into the ventricle.

Evolution of Vertebrate Cardiovascular Systems The circulatory systems in lungfish and amphibians are quite similar. The resemblances probably reflect the presence in both taxa of dual respiratory mechanisms that vary in their dominance as environmental conditions change. The lungfish has both gills and lungs; adult amphibians have lungs and important cutaneous and mouth (buccopharyngeal) respiratory surfaces. Details for individual genera vary, but the trends in structural evolution of lungfish and amphibian cardiovascular systems include (1) evolution of a double circulation with systemic and pulmonary circuits, (2) alteration of the ancestral gill arch complement involving reduction in number of vascular arches and specialization of individual aortic arch blood distributions, and (3) morphological and functional division of the heart to serve the two circuits and the increasingly specialized aortic arches.

The atrium of the heart is divided into right and left chambers, either anatomically by an interatrial septum or functionally by flow patterns. Blood from the systemic veins flows into the right side of the heart and that from the lungs to the left. The ventricle shows a variable subdivision that correlates with the physiological importance of lung respiration to the species (Figure 3–14). In frogs the ventricle is undivided, but the position within the ventricle of a particular parcel of blood before ventricular con-

traction appears to determine its fate upon leaving the contracting ventricle. The short conus arteriosus contains a spiral valve of tissue that differentially guides the blood from the left and right sides of the ventricle to the aortic arches. The anatomical relationships are such that blood from the pulmonary veins enters the left atrium, which injects it on the left side of the common ventricle. The spongy muscular lumen minimizes the mixing of right- and left-side blood. Contraction of the ventricle tends to eject blood in laminar streams that spiral out of the pumping chamber carrying the left-side blood into the ventral portion of the spirally divided conus. This half of the conus is the one from which the carotid (head supplying) and systemic aortic arches arise.

Thus, when the lungs are actively ventilated, oxygen-rich blood returning from them to the heart is selectively distributed to the tissues of the head and body. Systemic blood entering the right atrium is directed into the dorsal half of the spiral-valved conus. It goes to the pulmonary arch (pulmocutaneous arch in amphibians), destined for oxygenation in the lungs. However, when the skin is the primary site of gaseous exchange, as it is when a frog is underwater, the highest oxygen content is in the systemic veins that drain the skin. The lungs may actually be net users of oxygen, and due to vascular constriction little blood passes through the pulmonary circuit. Because the ventricle is undivided and the majority of the blood is arriving from the systemic circuit, the ventral section of the conus receives blood from an overflow of the right side of the ventricle. The scant unoxygenated left atrial supply to the ventricle also flows through the ventral conus. Thus, the most oxygenated blood coming from the heart flows to the tissues of the head and body during this shift in primary respiratory surface, a phenomenon possible only because of the undivided ventricle. The variability of the cardiovascular output in dipnoans and amphibians is an essential part of their ability to use alternative respiratory surfaces effectively.

Although they generally use only lung respiration, lizards, snakes, turtles, and crocodilians also have a wide variety of cardiopulmonary morphologies. A small sinus venosus supplies the right atrium, but the conus region is absent in adults and the aortic arches arise directly from the heart. The atria are completely separate, but the ventricular part of the heart may be three incompletely separated chambers (turtles, lizards, and snakes) or two

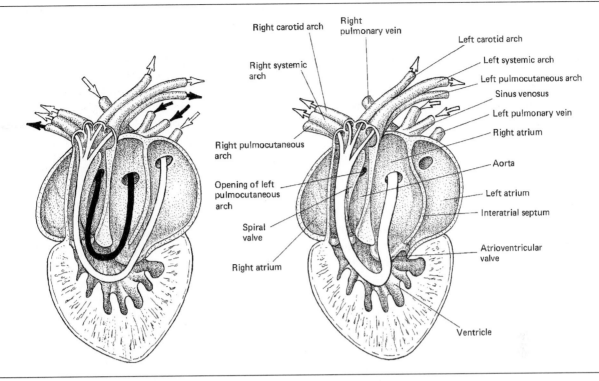

Figure 3–14 Blood flow through the heart of a frog. Left, pattern of flow when lungs are being ventilated; right, flow when only cutaneous respiration is taking place. Dark arrows, blood with low oxygen content; light arrows, most highly oxygenated blood.

completely separated chambers whose blood contents can mix via a window-like confluence of the arteries carrying blood from each ventricle (crocodilians). As in lungfishes and amphibians, these varied morphologies permit the selective shunting of blood between systemic and pulmonary circuits. The shifts are not in response to different respiratory surfaces but to the intermittent nature of lung ventilation in these vertebrates. Although obvious in diving forms, such as many turtles and the crocodiles, long periods of **apnea** (nonventilation of the lungs) are also characteristic of lizards and snakes, especially when they are cool and their metabolic rates correspondingly low. The ability, usually mediated by one or more vascular muscle sphincters on the pulmonary arteries, to shut off the blood flow to nonventilated lungs is likely to have several physiological advantages (Burggren 1987). Bypassing the pulmonary circuit may save energy when no oxygen is being extracted and reduce plasma leakage into the lungs. During lung ventilation, controlled mixing of oxygenated and venous blood

may result in higher rates of carbon dioxide elimination than are otherwise possible.

Birds and mammals have continual high-oxygen demands and thus experience apnea only in special circumstances such as diving. Their hearts are divided into two atria and two ventricles. They are specialized as a high-pressure systemic pump on the left and a lower-pressure pulmonary pump on the right. The two vascular circuits are also completely separated. Embryonic stages in heart and aortic arch development resemble the adult stages of more plesiomorphic cardiovascular systems. Avian and mammalian cardiovascular systems are convergent, not inherited from a common ancestor, and differ in details of their morphology. The most obvious difference is the reduction of the bilateral systemic arches characteristic of all other tetrapods. Birds retain the right side of the ancestral systemic arch pair, and mammals retain the left side. The hearts of birds and mammals are larger relative to body size than the hearts of other vertebrates, the terminal vessels are more numerous, distribution vessel walls are thicker, and the entire system, although under greater pressure, leaks less plasma into the interstitial space. The cardiovascular systems of birds and mammals are capable of high rates of cardiac contraction, high systemic pressures, rapid circulation, making sudden and profound adjustments in performance, and the highest degree of homeostatic control among vertebrates.

Homeostasis

The concept that organisms tend to maintain their internal environment (*milieu interieur*) within relatively narrow limits in spite of changes in the external environment was introduced by French physiologist Claude Bernard one year after Charles Darwin published *The Origin of Species*. Today, the concepts of organic evolution and an observed trend of increasing capacity of derived organisms to maintain a constancy of the *milieu interieur* are profoundly intertwined, especially in the study of vertebrates. If there is any general trend through time in the evolution of vertebrates, it is one of more and more powerful homeostasis. Vertebrate homeostasis is the topic of Chapter 4. The blood and the kidney are two vertebrate systems intimately associated with homeostasis.

Blood

Blood is a fluid tissue composed of liquid plasma and cellular constituents known as red blood cells (erythrocytes) and white blood cells (leucocytes). Blood transports oxygen, nutrients, carbon dioxide, nitrogenous wastes, hormones, and heat (to mention only its primary functions) and is responsible for the dynamic reactions of thermoregulation and of the immune system to invasion by foreign bodies.

Plasma is the water-based fluid that remains when all cellular components are removed from the blood. The water content of blood is especially important to thermal homeostasis. Because of its high specific heat, water is slow to warm and to cool. The large volume of blood that flows through metabolically active tissues transfers heat to the rest of the body. So efficient is blood in transferring heat that specialized circulatory structures allow certain vertebrates to maintain **regional heterothermy** (different temperatures in different parts of the body). The

most common of these specialized structures are arteriovenous anastomoses and retia.

Blood plasma contributes to other important fluids. The **interstitial fluid** is essentially blood plasma minus the large proteins, which do not pass through the capillary endothelium. **Lymph** is derived from interstitial fluid and is augmented by the addition of some of the larger proteins characteristic of circulating plasma and by certain white blood cells from the lymph nodes (lymphocytes). Cerebrospinal fluid and the similar aqueous humor of the eye have a low content of protein and different proportions of small organic compounds and inorganic salts compared to blood plasma. The cerebrospinal fluid is the result of active and highly selective transport of portions of the blood plasma by the vascular membranes that cover the brain.

From 20 percent to nearly half of the volume of the blood (depending on the species) is made up of erythrocytes. These red blood cells owe their red color to high concentrations of **hemoglobin**, the iron-containing globin protein that acts as a carrier for oxygen. The structure of erythrocytes varies among vertebrates. Most mammals have eliminated the nucleus and many other cell organelles from mature erythrocytes, so that technically they are not living cells. Other vertebrates retain nucleated erythrocytes but the nuclei seem functionally inert. Some fishes (e.g., the Antarctic crocodile ice fishes, Chaenichthyidae) have eliminated erythrocytes altogether, transporting oxygen exclusively in plasma. The sites of erythrocyte formation differ interspecifically and also change during the ontogeny of an individual. Adult sites of **hematopoiesis** (blood cell formation) include the blood vessels of fishes, the gut wall, kidneys, and liver of most vertebrates, and a few specialized organs especially characteristic of tetrapods: the **spleen**, the **thymus**, the **lymph nodes**, and the **red bone marrow**. In adult mammals erythrocytes are produced almost exclusively in the red bone marrow. Mature vertebrate erythrocytes appear incapable of replication or even self-repair, and there is a considerable turnover of these cells. New ones are produced and released into the circulation and old or damaged ones are scavenged from the blood, generally by the same hemopoietic tissues.

Erythrocytes contribute to homeostasis in a myriad of ways. The affinity of hemoglobin for oxygen (that is, how tightly oxygen molecules are held by hemoglobin) is changed by genetic, physical, and chemical effects. The most familiar illustration of the contribution of the oxygen affinity of hemoglobin to homeostasis is the way in which oxygen release is adjusted to match the metabolic requirements of different tissues.

The oxygen affinity of hemoglobin is reduced by acidity and increased by alkaline conditions. In other words, hemoglobin releases oxygen more readily at low pH than at higher pH. Carbon dioxide, which is produced by cellular metabolism, diffuses into the blood plasma and from there into the erythrocytes. Erythrocytes are rich in the enzyme **carbonic anhydrase**, which catalyzes the reaction between water and carbon dioxide that yields carbonic acid (H_2CO_3). Carbonic acid then dissociates to form a proton (H^+) and a bicarbonate ion (HCO_3^-). The protons liberated by the dissociation increase the acidity of the blood, and the lower pH reduces the affinity of hemoglobin for oxygen, resulting in the release of oxygen from the hemoglobin. This mechanism adjusts the delivery of oxygen to the metabolic needs of different tissues because carbon dioxide production is proportional to oxygen consumption.

In the gills or lungs the process is reversed: Hemoglobin that has released its oxygen in the tissues is exposed to high oxygen pressure and binds oxygen, releasing protons in the process. The increased concentration of protons drives the reaction in the reverse direction, recombining the protons with bicarbonate ions to form carbonic acid. Carbonic anhydrase in the erythrocytes catalyzes the formation of carbon dioxide and water from the carbonic acid, and the carbon dioxide diffuses across the wall of the capillary into the water or air.

Cells specialized to promote clotting of blood (**thrombocytes**) are present in all vertebrates except mammals, where they are replaced by noncellular **platelets**. These blood elements react to surfaces they do not normally encounter by adhering to the unfamiliar surface and to each other. Specific blood proteins react to damaged tissues as well, polymerizing as a network of fibers that enmesh cellular components of blood to form a wound-plugging clot.

White blood cells (**leucocytes**) are less abundant than erythrocytes. There are at least five types of white blood cells in humans. Leucocytes form the first line of defense against invasion by pathogenic bacteria and other foreign particles, they are involved in inflammation responses, and they support antigen–antibody activities of the immune system. Unlike erythrocytes, which normally remain entirely within the vascular channels, leucocytes squeeze between the endothelial cells of capillaries and venules and spend most of their lives moving freely through the loose connective tissues.

Immune System A significant aspect of the survival of an organism is the ability to distinguish between *self* and *nonself* so that invasion by foreign substances can be prevented. Although all animals show some capacity to recognize self at a molecular level, only the vertebrates possess an integrated cell and plasma–interstitial fluid **antibody immunity** that resists almost all organisms and toxins that damage tissues and does so with great *specificity* and *memory* (Cooper 1985).

By chemically tagging an **antigen** (*anti* = against, *genos* = birth) or foreign substance that invades the body with fluorescent dyes it has been possible to discover the cells in birds and mammals responsible for producing the antibodies that are the source of the specific and remembered immunity. They are white blood cells known as **B lymphocytes** because they were first discovered in the bursa of Fabricius, a lymphoid outgrowth of the cloaca of birds. B cells are widely dispersed in the lymphoid tissue of vertebrates. They produce immunoglobulins that are secreted into the body fluids. Some B lymphocytes transmit to their daughter cells precise information about the antigen, for the clone derived from them becomes the basis for immune system memory. The cells of the clone disperse widely in the lymphatic tissue of the body and when they are again stimulated by the same antigen rapidly begin to produce their special antibody.

Another class of lymphocytes, the T lymphocytes, so called because in mammals they are resident for a portion of their life in the **thymus gland**, can be distinguished only by their behavior. One group of T cells reproduce entire clones of cytotoxic cells. These killer T cells travel via the circulatory system and bind tightly to their specific antigen, injecting digestive enzymes into it. A second group of T cells, the helper T cells, stimulate the activity of B cells, other T cells, and phagocytic white blood cells. Suppressor T cells are believed to function as a negative feedback on other aspects of the immune reaction, keeping it from becoming excessively severe. These cells are also thought to be significant in helping the immune system to recognize self, and their malfunction is implicated in many autoimmune diseases.

Although both immunoglobulin antibodies and probable cytotoxic cells are found in jawless fishes, there has been no evidence until recently of the high level of differentiation of cell function characteristic of the system in birds and mammals. It is now known that the shark *Heterodontus* has genes nearly identical to those associated with T cells in birds and mammals (Rast and Litman 1994). Nevertheless, the degree of differentiation of these lymphocytes in ectothermic gnathostomes is not yet known, and the evolution of the unique vertebrate immune system remains uncertain.

Kidneys

Maintaining a narrow range of solute concentrations and low levels of accumulated toxic wastes are functions that go hand in hand, but for many years evolutionary physiologists have argued about which was the ancestral function of the kidney. If early craniates lived in a marine environment, kidneys are unlikely to have originated as water-excreting organs. Removal of nitrogenous and other wastes is likely to have been the kidney's primordial function. A freshwater origin for the first craniate would reverse the argument.

Two sets of anatomical terms, which unfortunately are not mutually exclusive, have been applied to the kidneys of vertebrates (Table 3–2). One set describes an ontogenetic (developmental) series of kidney types: **pronephros, mesonephros**, and **metanephros**, in order of their appearance in the embryonic development of amniotes. Another set of terms for kidneys is descriptive of their arrangement, drainage, and internal structure as seen in a phylogenetic (evolutionary) series of the adults of different vertebrate groups: **holonephros** (hypothetical, not actually known in any group), **opisthonephros** (nonamniotes), and **metanephros** (amniotes). Ontogenetically, and presumably phylogenetically, the earliest kidney tubules were segmentally arranged and opened through funnel-like ciliated mouths to the coelom, from which they drained fluid that was derived from the interstitial fluid (Figure 3–15). This fluid was conveyed along an **archinephric duct**, which connected each segment's tubule to the region of the anus. Here the ducts from each side opened to release the urine to the exterior. A later stage involved association of the individual tubules with circulatory capillary beds where wastes passed directly to the tubules. As this system became more and more efficient, the open connections to the coelom were lost but more numerous tubules crowded into each section. This stage is essentially that of the mesonephros in ontogeny and the opisthonephros of adult nonamniotes. Finally, the tubules became highly compacted and numerous, acquired a new draining mechanism that formed as an outgrowth of

Table 3–2 **Types of kidneys of vertebrates.**

Ontogenetic Classification	Phylogenetic Classification
Pronephros	Holonephros
Develops in anteriormost portion of the nephrogenic tissue in all vertebrates, segmental organization, drained by the archinephric duct, functional in embryos of fishes and amphibians, present but not functional in early embryos of amniotes.	A hypothetical structure, not found in any extant adult vertebrate. A single tubule per body segment for the entire length of the nephrogenic tissue, drained by the archinephric duct.
Mesonephros	Opisthonephros
Develops in the middle portion of the nephrogenic tissue of all vertebrates, drained by the archinephric duct, has no obvious segmentation, tubules are more numerous than segments, functional in embryos of all gnathostomes, but only briefly in embryos of amniotes.	Develops in middle and subsequently in the posterior portion of the nephrogenic tissue, drained by the archinephric duct, and additional posterior accessory ducts of nephrogenic origin, no segmentation, numerous tubules, functional in adults of all nonamniotes.

Metanephros

Develops in the posterior portion of the nephrogenic tissue of amniotes, drained by a cloacal outgrowth (the ureter), very large number of tubules with no evidence of segmentation, functional in late embryos and adults of all amniotes.

the cloaca, and achieved the capacity to concentrate the urine. These features are characteristic of amniote metanephric kidneys. The mammalian kidney is further specialized in having acquired an organized structure permitting an exceptionally dynamic range of urine concentrations. In spite of these evolutionary changes, the basic unit of all adult vertebrate kidneys is essentially the same—the tubular **nephrons**

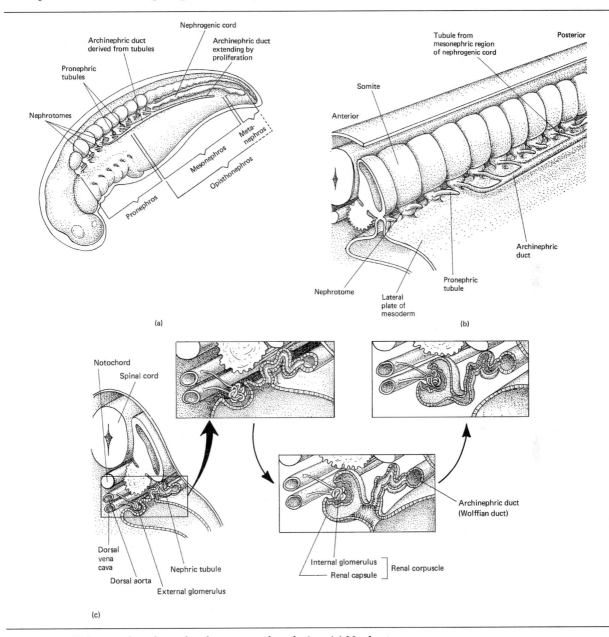

Figure 3–15 Kidney and nephron development and evolution. (a) Nephrotome regions in a generalized amniote embryo with region of adult nonamniote opisthonephros origin superimposed (see Table 3–2). (b) The relationship of the nephrotome and its developing tubules and archinephric duct to the somites, lateral plate of mesoderm, and coelom. (c) The presumed condition in primitive vertebrates (left) was replaced by an intermediate condition seen in the pronephri of the larvae of some living fishes (center), and finally by the condition seen in all living adult vertebrates (right).

(Figures 3–15c and 4–1 through 4–6). The function of the vertebrate kidney is an area of active research (Brown et al. 1993). The basics of kidney function (Chapter 4) are central to understanding the diversity in adaptability of vertebrate life.

Coordination and Integration

Physiological homeostasis at the level of individual organs is only part of the organismal homeostasis. Organ functions must be coordinated to work in concert if the action of one is not to cancel the action of another or extend an adjustment beyond the narrow tolerances of the tissues. Signals transmitted by the nervous system and endocrine secretions distributed to target organs by the cardiovascular system are significant coordinators in this regard.

The Nervous System: Brain Anatomy, Evolution, and Size

The basic unit of neuroanatomy is the **neuron**. Neurons are made up of nerve cell bodies with long, thin processes, the **dendrites** and **axons**, that extend from the cell body and transmit impulses. Axons are generally the longest and least branched of these extensions. In gnathostomes they are encased in a fatty insulating coat, the **myelin sheath**, which increases the conduction velocity of the nerve impulse. These sheaths are derived from Schwann cells that originate from the neural crest and thus are unique to vertebrates. An axon carries impulses over long distances to another nerve cell or population of nerve cells. Generally, axons in transit from one population of cells to another are collected together like wires in a cable. Such collections of axons in the **peripheral nervous system** (PNS) are called nerves; within the **central nervous system** (CNS) they are called tracts and compose most of the white matter (so called because of the color of the myelin). Nerve cell bodies are often clustered together, usually in groups with similar connections or functions, and are called ganglia (in the PNS) and nuclei (not to be confused with the cellular nucleus) that make up most of the gray matter (in the CNS). Nerve cells are embedded in generally nonconducting but physically and physiologically significant glial cells of a variety of distinct types.

The nerves of the PNS are segmentally arranged, exiting from the spinal cord between the vertebrae.

Each spinal nerve complex is made up of somatic sensory fibers from the body wall, visceral sensory fibers from the gut wall, visceral motor fibers to the muscles and glands of the gut, and somatic motor fibers to the muscles and glands of the body wall. The nerve cell bodies of both types of sensory fibers are collected in a segmental series of spinal ganglia adjacent to the spinal cord. They too are derived from neural crest cells and thus are absent from the segmental nerves of cephalochordates. These various types of fibers sort themselves out into the peripheral nerves in distinctive combinations that are consistent with our hypotheses of phylogenetic sequence. Lampreys have separate motor and sensory nerves that do not unite to form a complex spinal nerve. The most derived condition, special branches of combined fibers destined for a topographic area, is characteristic of amniotes. The cranial nerves are a special case of spinal nerve modification, including the isolation of fibers relating to different anatomical regions into separate nerves. The spinal nerve fibers that relate exclusively to the involuntary functions of the vertebrate body form two motor divisions that, together with their corresponding sensory nerves, are known as the **autonomic nervous system**. However, the autonomic system is not clearly separated from the CNS or the PNS. The **sympathetic motor division** of the autonomic system generally stimulates an organ to react in a manner appropriate to a stressful set of circumstances (e.g., to prepare for rapid energy expenditures). The **parasympathetic division** is antagonistic in its effect, causing the organs innervated to function appropriately for peaceful conditions (e.g., rest or digestion).

The CNS, too, shows evidence early in development of segmentation, although it does not correspond directly with somatic segments but, like the vertebrae, appears to be intersomitically segmented (Fraser 1993, Keynes and Krumlauf 1994). The **spinal cord** is the simplest part of the CNS, yet its cellular interconnections and functional relationships are complicated. The gross structure is that of a hollow tube with an inner core of gray matter and an outer layer of white matter. The spinal cord receives sensory inputs, transmits and integrates them with other portions of the CNS, and sends motor output as appropriate.

Ancestrally, the spinal cord had considerable autonomy, even in such complex movements as swimming. Fish can continue to produce coordinated

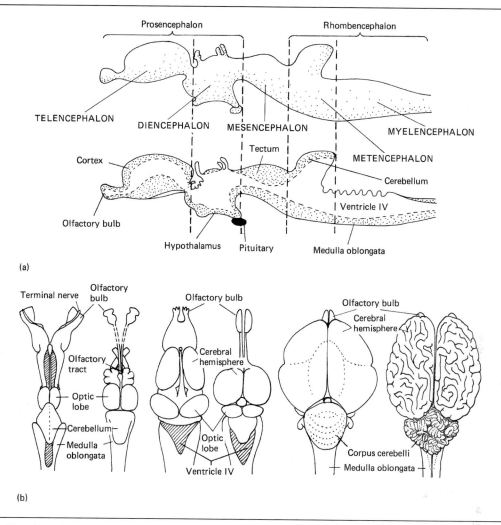

Figure 3–16 Vertebrate brain. (a) Diagrammatic developmental stage in generalized vertebrate brain showing principal divisions and structures described in the text. Lateral view above; sectioned view below. (b) Representative vertebrate brains seen in dorsal view. All brains are drawn to approximately the same total length, which emphasizes relative differences in regional development. From left to right: *Scymnus*, a shark; *Gadus*, a teleost; *Rana*, a frog; *Alligator*, a crocodilian; *Anser*, a goose; and *Equus*, a horse, as an example of an advanced modern mammal.

swimming movements when the brain is severed from the spinal cord. The trend in vertebrate evolution has been toward more complex circuits within the spinal cord and between the spinal cord and the brain. With these connections has come an ever-increasing dependence of spinal cord functions on control by higher CNS centers.

The neuroanatomy of the brain is exceedingly complex (Figure 3–16). Ancestrally (and embryologically in derived vertebrate clades) the brain was tri-

partite: One part related to olfaction, a second to vision, and the third to balance/vibration detection (the inner ear). The brain of living vertebrates is composed of five distinct regions, containing both gray and white matter, each region serving a different basic function.

Posteriorly, two regions differentiate from the embryonic brain region associated with the developing ear. The most posterior, the **myelencephalon** or **medulla oblongata**, is primarily an enlarged anterior extension of the spinal cord. The nuclei of the medulla oblongata form synapses primarily with the sensory organs of the muscles and skin of the head plus a distinctive set of nuclei, those associated with the sense organs of the inner ear. All impulses from the receptor cells of the balance (vestibular) and the hearing (cochlear) regions of the mammalian ear synapse first in these medulla oblongata nuclei.

The anterior portion of the posterior region of the embryonic brain, called the **metencephalon**, develops an important dorsal outgrowth, the **cerebellum**. The cerebellum coordinates and regulates motor activities whether they are reflexive, such as maintenance of posture, or directed, such as escape movements. The nuclear or gray matter of the cerebellum receives nerve impulses from the acoustic area of the myelencephalon (in particular, impulses relayed from the vestibular nuclei), impulses from the complex system of muscle and tendon stretch receptors, and indirectly, impulses from the skin, optic centers, and other coordinating brain centers.

The central embryonic brain region, the **mesencephalon**, develops in conjunction with the eye. The roof of the mesencephalon is known as the **tectum** and receives input from the optic nerve. The floor of this region contains fiber tracts that pass anteriorly and posteriorly to other regions of the brain as well as to nuclei concerned with eye movements.

A rather small region, the **diencephalon**, develops from the most anterior of the embryonic brain regions and is segmented in early development. In amniotes it is a major relay station between sensory areas and the higher brain centers. The eyes develop from stalk-like outgrowths of the diencephalon, the stalks remaining as the optic nerves (cranial nerve II, Table 3–3). A ventral outgrowth of the diencephalon contributes to the formation of the dominant endocrine organ, the pituitary gland, or hypophysis, which together with the floor of the diencephalon (hypothalamus) forms the primary center for neural–hormonal coordination and integration. Another endocrine gland, the pineal organ, is a dorsal outgrowth of the diencephalon. Originally it was a median photoreceptor (light-sensitive organ) of plesiomorphic vertebrates.

Finally, the most anterior region of the adult brain, the **telencephalon**, develops in association with the olfactory capsules and as the first nuclei of olfactory synapse. The telencephalon of plesiomorphic vertebrates also coordinates inputs from other sensory modalities. In mammals it becomes the primary seat of sensory integration and nervous control. The neuronal cells in the anterior olfactory nuclei of the telencephalon are variously enlarged into masses of gray matter that are often differentiated into distinctive layers and subdivisions with complex interconnections (Figure 3–17). Phylogenetic differences include the increase in size of the telencephalon relative to the rest of the brain, differentiation of nuclear (gray matter) areas, and the appearance of new nuclear types; corticalization, in which the gray matter emerges on the surface of the telencephalon, forcing the white matter to a central position; and finally, the convolution of the most recently evolved portion of the brain in derived mammals.

In plesiomorphic tetrapods (Figure 3–17a) the telencephalon is a small portion of the total brain. In ancestral amniotes (Figure 3–17b) expansions and migrations of some portions have forced others to an internal position. In a more derived stage (Figure 3–17c) the telencephalon constitutes the single major portion of the brain. A new nuclear region, the **neopallium**, has made its appearance and corticalization is well advanced. In birds (Figure 3–17d) the area associated with olfaction, is very small but the nuclei associated with innate, reflexive behaviors are enormous. These dominant nuclei are covered with a relatively thin coat, some of which is thought to be neopallium.

Two phenomena in the evolution of the mammalian brain make it unique among vertebrates. First is the development of an external matlike covering of gray matter from the ancestral central nuclear condition. This change places the nerve cell bodies on the outside of the brain, where additions to their number will not cause disruption of the tracts leading to them. Second is the domination of a region of gray matter distinctive in its cell types and cellular organization, the neopallium. Birds, although nearly as brainy as many mammals on a ratio of brain weight to body weight, have comparatively little of this type of neural tissue. The mammalian neopallium is a site of sensory projection, of integration, of memory, and of the construction and reconstruction of past or even hypothetical patterns of sensory input. It is ultimately the seat of intelligence. The neopallium and older olfactory nuclei of plesiomorphic mammals (Figure 3–17e) are the two major divisions of the telencephalon. The majority of connections to the brain stem run directly to the neopallium, unlike the condition seen in derived

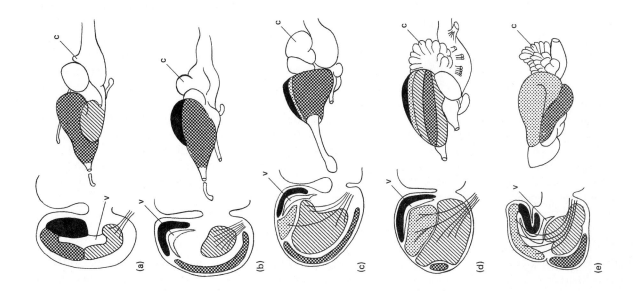

Figure 3–17 Various stages in the differentiation of the telencephalon of tetrapods in cross-sectional half-brain and lateral full-brain diagrams: (a) primitive tetrapod with relatively large contribution of nontelencephalon brain regions to total brain weight; (b) later stage, probably represented by primitive amniotes; (c) advanced stage, probably arrived at by several amniote lineages; (d) in birds the telencephalon and cerebellum make up the major portions of the brain; (e) primitive mammals; (f) advanced mammals. Lines representing cut nerve tracts indicate an association center's connections with the brain stem.

nonamniotes, plesiomorphic tetrapods, and birds, where the other nuclei serve as major association areas. In derived mammals (Figure 3–17f) the neopallium dominates the entire telencephalon and becomes highly folded. It communicates not only with the brain stem but also directly with its contralateral cerebral hemisphere via the corpus callosum.

Most vertebrate clades have evolved species of increasing body size during their histories. Numerous locomotory, homeostatic, and defensive benefits derive from large size. However, large size increases the demands on integrative systems. In large animals the tension and position of the additional muscle fibers must be monitored and their firing controlled, and the increased skin surface brings with it more sensory organs. These demand more nerve cells in the central nervous system, especially in the brain. Nevertheless, in vertebrates brain size does not increase in exact proportion to body mass. Rather, as body mass doubles, brain mass increases by a factor of about 1.6.

To compare the brain sizes of vertebrates, which vary from the teleost fish *Schindleria praematurus*, weighing only 2 milligrams, to the blue whale (*Balaenoptera musculus*), weighing over 160,000 kilograms, it is appropriate to compare brain mass per unit of body mass. When examined in this way, the brains of living birds and mammals are approximately 15 times larger than those of other vertebrates of the same body size. To visualize the relationship between various vertebrate taxa, including extinct groups, the data for each taxon can be enclosed in a polygon (Figure 3–18). Even the early mammals had four to five times as much brain tissue as other vertebrate taxa of similar body size. Obviously, an increase in relative brain-to-body mass ratios occurred in the evolution of modern mammals.

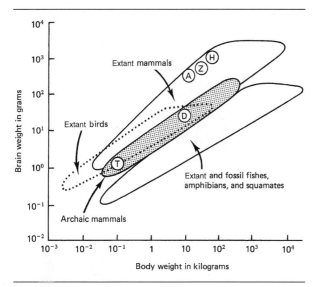

Figure 3–18 Relative brain weight of vertebrates. Brain weight plotted against body weight (both on logarithmic scales) based on measurements of 198 living vertebrate species and endocranial cast volumes of some fossils. T, *Triconodon* of the Jurassic; D, *Didelphis* the living Virginia opossum; A, *Australopithecus africanus*; Z, *Australopithecus boisei*; H, *Homo sapiens*. (Modified after H. J. Jerison, 1973, *Evolution of the Brain and Intelligence*, Academic, New York, NY.)

Spinal and Cranial Nerves

From the spinal cord in each body segment issue two nerves that supply motor (efferent) and sensory (afferent) fibers to the **somatic** and **visceral** components of that segment. Somatic components include the skeletal muscles, visceral components, the gut, and circulatory musculature. Although these nerves vary among vertebrates, a ventral and a dorsal root tract are always recognizable. In the ancestral condition the ventral root supplied only motor nerves to the myotomes or trunk muscles of each body segment (somatic components). The dorsal root was a mixed nerve that contained both visceral motor and sensory nerves.

Ventral root nerves are conservative and have retained throughout vertebrate evolution the metameric relationship between the muscle-forming myotome (somite) and the nerve supply to that somite. If a particular somite is lost (e.g., some head somites), so is the ventral nerve; if a myotome is modified and moved to serve other functions (e.g., the appendicular or limb muscles of tetrapods compared with the paired fin muscles of fishes), the

nerve preserves the metamerism and exits from the segment of the cord embryonically associated with that myotome. The dorsal root nerves appear to be more plastic; intersegmental capture is common. Perhaps this is a result of the dorsal root nerve containing the visceral motor fibers of the gut and the sensory nerves of the skin, two parts of the body that lack segmentation.

The cranial nerves are believed to represent independently derived serial homologs of elements of an amphioxus-like ancestral pattern of spinal nerves. They provide additional evidence that at least part of the vertebrate head and jaws were derived from a fusion and modification of anterior body segments (Table 3–3).

The Sense Organs

Sense organs are classified on the basis of the type of stimulus to which they react. **Chemoreceptors** are stimulated by chemical substances. **Mechanoreceptors** are sensitive to mechanical deformation even at the level of distortion of a single cell's stereocilia. **Radioreceptors** respond variously to electromagnetic disturbances of a broad spectral range, including thermal radiation. In mammals, the major senses include the classic five senses of taste, touch, sight, smell, and hearing. Most vertebrates share these and a few have additional senses or have greatly refined one or more of the familiar five.

Taste Taste buds made up of specialized sensory epithelial cells and associated nerves are restricted to the moist membranes of the walls of the mouth, the throat, and especially the tongue (Figure 3–19) of tetrapods, but many osteichthyans have taste buds on their heads, fins, or specialized oral structures called barbels (e.g., catfishes). Taste buds are stimulated when dissolved molecules contact their specialized surfaces. Although there are many similarities between the senses of taste and smell, they have different embryonic origins, innervations, sensitivities, and types of molecules to which they respond.

Touch At least four differently structured nerve ending–epithelium–connective tissue combinations found in tetrapods are sensitive to stimuli ranging from light touch to heavy pressure. In mammals touch receptors may also be combined with specialized hairs, the vibrissae, which grow on the muzzle, around the eyes, or on the lower legs. A specialized set

Table 3–3 **Cranial nerves and their primary function. Numbers refer to the anterior-to-posterior sequence of exit of the nerves from the mammalian brain, names broadly describe the nerve function (e.g., oculomotor = eye mover) or its appearance (e.g., trigeminal = a nerve with three major branches). ** = position of foramen magnum in nonamniotes.**

| | | Function | | Mixed Nerves | |
Number	Name	Somatic Motor	Somatic Sensory	Visceral Motor	Visceral Sensory
0	Terminalis		+?		
I	Olfactory		+		
II	Optic		+		
III	Oculomotor	+			
IV	Trochlear	+			
V	Trigeminal		+	+	
VI	Abducens	+			
VII	Facial		+	+	+
VIII	Statoacoustic		+		
IX	Glossopharyngeal		+	+	+
X	Vagus		+	+	+
**					
XI	Accessory			+	
XII	Hypoglossal	+			

of mechanoreceptors, the lateral-line system, occurs in aquatic nonamniotes (discussed in Chapter 8).

Vision Visual systems are considered the distance sense *par excellence*, sensitive to those wavelengths of electromagnetic radiation that reach the surface of the Earth with the least interference by the atmosphere and hydrosphere. Radiation of these wavelengths, which we call light, has several properties that make it a superior transmitter of information over distance. Light travels great distances in air with little attenuation (absorption). It travels in a straight line in homogeneous air or clear water. Light of different wavelengths is differentially absorbed, transmitted, and reflected so that the spectral (wavelength) distribution of light reflected from an object is altered. These properties of light give nearly unambiguous cues about the structure and texture of an object. Another significant attribute of light is its speed, which is instantaneous compared to the speed of transmission and action of the vertebrate neuromuscular system.

The receptor field of the vertebrate eye is arrayed in a hemispherical sheet, the **retina**. Each point on the retina corresponds to specific neural connections and a different visual axis in space. Thus, a vertebrate can determine where a target is in at least two-dimensional space and whether it is stationary or moving. Because of neuronal interactions, vertebrate eyes can also detect sharp beginnings and endings of visual stimuli (ons, offs, and edges) with great precision.

The eyes of vertebrates (Figure 3–20) consist of a light-shielding container, the sclera, and choroid coats. These prevent stimulation of the eye by light from multiple directions. The eye has a variable entrance aperture, the pupil surrounded by the iris, to control the amount of light that enters. A focusing system, the cornea, lens, and ciliary body, converges the rays of light on the photosensitive cells of the retina.

The retina (Figure 3–21) originates as an outgrowth of the diencephalon of the brain. The exceptional nature of the eye as an extension of the brain is not as generally appreciated as it deserves to be. For example, amphibians process distant stimuli in the elaborate neural network of the retina. The optic nerve sends signals to the brain with already encoded distinctions between predators, prey, and stationary objects in the environment. The structure of the mammalian optic nerve indicates that considerable processing takes place in the retina. There are

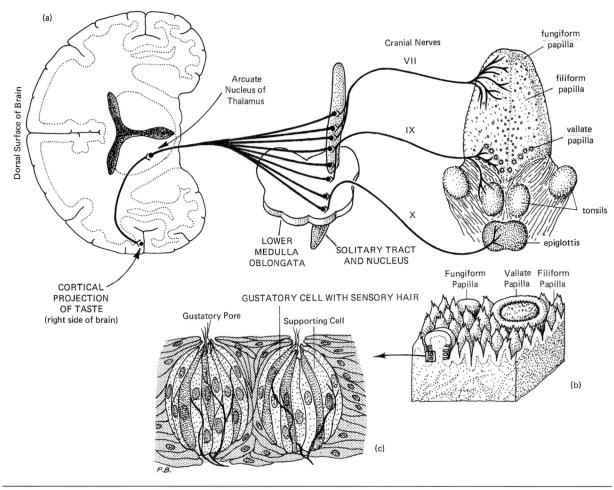

Figure 3–19 Sense of taste as illustrated by mammals. (a) Tongue of a human showing the topography and types of lingual papillae. The distribution of innervation and route of projection onto the border of the lateral fissure of the cerebrum are shown for one side of the tongue. (b) Enlargement of region of tongue illustrating details of papillae structure and position of taste buds. (c) Further enlargement of taste buds on a fungiform or vallate papilla showing cellular organization and sensory hairs, a common morphology of vertebrate taste buds. (Modified after F. H. Netter, 1962, *The CIBA Collection of Medical Illustrations*, volume I, *Nervous System*, CIBA Publications, Summit, NJ.)

approximately 100 million photoreceptor cells in the retina, but only one million axons in the optic nerve. Thus, each axon integrates information from many photoreceptor cells.

Lizards and birds have better visual acuity than most fishes and mammals. In part, their ability to distinguish between two close points in space, which would appear to vertebrates with lesser acuity as a single point, is due to their nearly pure cone retinas. Mammals approach this acuity only in one small region of the retina, the all cone **fovea**. **Cones** are one of two types of light-sensing cells in the vertebrate retina and are distinguished from the other type, the **rods**, by morphology, photochemistry, and neural connections. In addition to high visual acuity, cones are the basis for color vision. Cones have relatively low photosensitivity, however; they are at least two orders of magnitude less sensitive than rods and

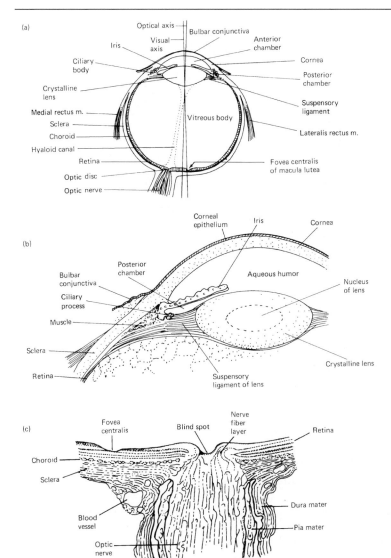

(a)

(b)

(c)

Figure 3–20 Vertebrate eye as illustrated by mammals. (a) The eye sectioned horizontally in the plane of the optic axis and optic nerve. In addition to the structures discussed in the text, (b) details structures of the anterior portion of the eye primarily responsible for image focus (cornea, lens, ciliary body) and quantitative control of light (iris). (c) Enlargement of the fundus of the eye showing the specialized sensory and neural regions found there. Note especially the continuity of the brain covering meninges (pia mater, arachnoid, and dura mater) to the protective and nutritive coats of the eye (sclera and choroid). (Modified after J. E. Crouch, 1969, *Text-Atlas of Cat Anatomy*, Lea & Febiger, Philadelphia, PA.)

thus fail to function at night and in other dim-light conditions.

Rods are morphologically distinct from cones and differ physiologically—they are exceptionally sensitive to dim light. Several rods synapse with a single neural element, increasing the chances that dim light will stimulate the neuron, but they achieve this increased sensitivity at the cost of visual acuity. An analogy can be made with photographic films: Increased sensitivity to dim light is accompanied by increased graininess of the image.

The loss of acuity in a nocturnal eye is considerable. Whereas the diurnal pigeon can distinguish between two lines producing retinal images less than 1 micrometer apart, the nocturnal rat requires a separation of more than 20 micrometers. Try to concentrate on the form or details of objects seen at the edge of the visual field of your eye—the peripheral rod-rich areas of the human retina—and the poor acuity of rod vision will be apparent.

The eyes and retina produce a two- or three-dimensional representation of the world (Figure 3–22). Each cone or closely adjacent group of rods on the retina corresponds precisely to a point or series of adjacent points in space. The muscles controlling the orientation of the eye and the accommodation of the

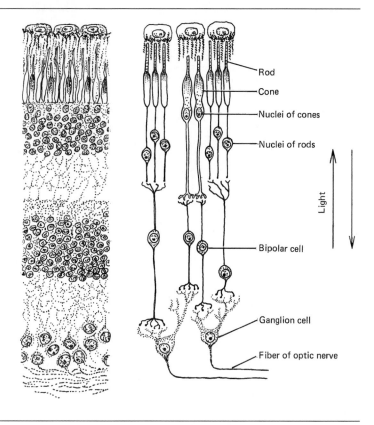

Figure 3–21 The retina. Histological (left) and diagrammatic (right) structure of the mammalian retina as seen in humans. The direction of light passage and neural conduction are indicated. The different types, great number, and interconnections of nerve cells in the retina are the basis for the considerable information processing that occurs within the eye. (See text for further discussion.)

Labels in figure: Rod, Cone, Nuclei of cones, Nuclei of rods, Light, Bipolar cell, Ganglion cell, Fiber of optic nerve

lens give sufficient information for very accurate neural reconstruction of the visually perceived world.

Olfaction The olfactory receptor apparatus is basically the same in all macrosmatic (strongly olfactory) vertebrates. In fishes the ectodermally derived olfactory organs are usually independent of the pharynx; separate incurrent and excurrent canals open on the exterior of the snout. The nasal epithelium is folded into a complex rosette of great surface area but does not have bony support. In mammals the sensory cells reside on the turbinals and nasal septum high in the nasal cavity (Figure 3–23). The lower turbinals and nasal epithelium are concerned with temperature and water regulation (Chapter 23).

The total area of the olfactory surfaces may exceed that of the entire body surface in olfactory-oriented vertebrates. A single molecule can excite a receptor cell, and only a few molecules are needed to elicit a behavioral response. A molecule of odorant probably fits more or less tightly into one or more of the different sensory pits that occur on the surface of olfactory epithelia. Once seated, a molecule lingers, producing prolonged stimulation that allows perception of the relative concentrations of stimulating substances. However, the lingering nature of the stimulus limits rapid temporal or spatial perception of the olfactory surroundings. Many mammals snort or sneeze during olfactory sensing, perhaps to flush air as completely as possible from the upper regions of their nasal cavities. The subtle differences in odor that mammals can detect are probably due to an ability to distinguish the hundreds of combinations possible from no more than a dozen basic odors. Such combinations would produce the individual differences in body odor that are important in maintenance of many mammalian social groups. The physiological state of an individual can often be determined from the odors it produces, especially in the feces and urine. Excretions are used by a wide variety of vertebrates to mark their territories or home ranges. One large group of freshwater fishes, the Ostariophysi (minnows and their 6000 relatives; Chapter 8) have a unique type of epidermal cell that exudes an alarm substance when the skin is abraded. Other fishes smell the substance and scatter, diving for the bottom in apparent antipredator behavior in direct response to the chemical signal.

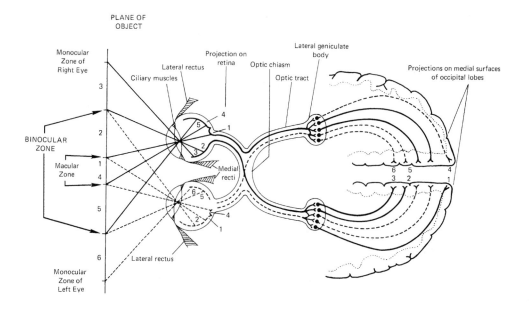

Figure 3–22 Neural connections and point-for-point correspondence between visual field and cerebral projections in the human brain. The one-dimensional representation of the line shown, as received by the brain, may be translated into a three-dimensional reconstruction of the visual world (depth perception). This is accomplished by information from the extrinsic ocular muscles on the degree of convergence or divergence of the eyes (such as from the lateral and medial rectus muscles shown) or by similar information from the ciliary muscles on the degree of lens distortion (accommodation) necessary for focusing. Many vertebrates lack significant binocular vision and have fewer visual system cues for depth perception.

Pheromones are chemical signals produced by an individual that affect the behavior and/or physiology of conspecifics. Minnow alarm substances are an example. The odors of feces and urine undoubtedly have such effects. Many male ungulates (hoofed mammals) sniff or taste the urine of females. This behavior is usually followed by *flehmen*, a behavior in which the male curls the upper lip and often holds his head high, probably inhaling. A specialized structure of the anterior palate, called the **vomeronasal**, contains branches of the olfactory nerve and is used during *flehmen*. Probably this behavior allows males to determine the stage of the reproductive cycle of a female.

Audition In contrast to the structural similarity in most of the sensory organs of touch, taste, sight, and smell of most fishes and the tetrapods, their organs of hearing are in many ways distinctive, especially those of mammals (Figure 3–24). The auditory apparatus of a typical mammal has three major functional units. One unit, the external ear, enhances the reception and directional characteristics of sound stimuli. The second unit, the middle ear, amplifies and transduces sound stimuli into fluid-borne analogs of the original sound. The third unit, the inner ear (some form of which is found in all craniates), discriminates the frequency and intensity of vibrations it receives and transmits this information to the central nervous system in the form of neurally encoded firing patterns.

The **pinna** (external ear) and narrowing external auditory meatus of mammals concentrate sound from the relatively large area encompassed by the external opening of the pinna to the small, thin, tympanic membrane. The pinna is unique to mammals, although it has a feathery analog in certain owls. The auditory sensitivity of a terrestrial mammal is reduced if the pinnae are removed.

The middle ear, found in all tetrapods, greatly increases auditory sensitivity. Under most conditions the sounds tetrapods attend to are airborne, but the inner ear, where sound stimuli become encoded into neural impulses, is a series of fluid-filled, membrane-

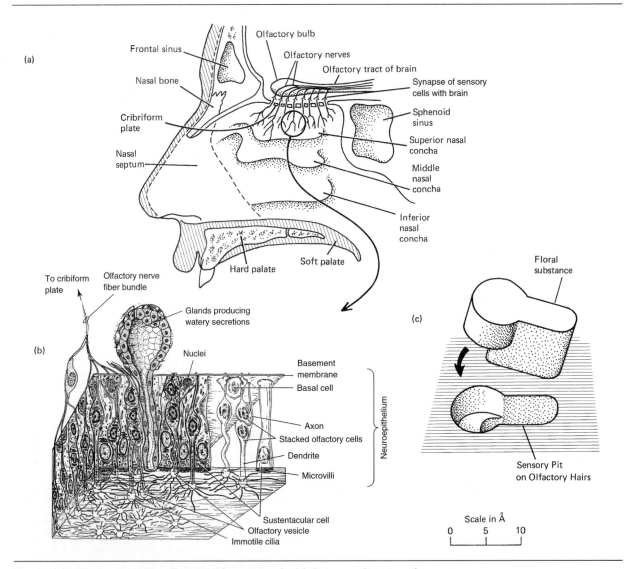

Figure 3–23 Sense of smell as illustrated by mammals. (a) Cutaway diagram of sagittal (medial) section of the human nasal cavity showing the distribution of the olfactory nerves and the nonolfactory nasal conchae (=turbinals). (b) The cellular organization of the olfactory epithelium is enlarged. Note that the sensory cells are the actual nerve cells that penetrate the epithelium with a modified dendrite. (c) Diagram of a hypothetical basis for sensitivity in olfaction. The three-dimensional shape of a substance determines whether or not it will lock into a sensory pit on the immotile cilia's surface. A sufficient number of fits results in sensory nerve firing. Each basic odor is associated with a different three-dimensional shape and set of sensory pits. ([c] Modified after J. Amoore, 1962, *Proceedings of the Toilet Goods Association, Scientific Section Supplement* 37:1–20.)

lined chambers. It requires considerably more energy to set the fluids of the inner ear in motion than most airborne sounds could impart directly. The middle ear receives the relatively low energy of airborne sound waves on its outer membranous end, the **tympanic membrane** or **eardrum**, and produces analogous vibrations in the fluids of the inner ear.

Two devices that enhance sound detection are employed in the mammalian middle ear, which is the most highly derived middle ear known. First, the area of the eardrum is about 20 times greater than that of the oval window, which separates the air-filled middle ear from the fluid-filled inner ear. In all tetrapods the gap between the tympanum and the oval window is bridged by bony elements, the ear ossicles. If mammals had a single bridging element, such as the columella of birds, we would expect something on the order of a 20-fold increase in the force per unit area between sound reception at the tympanum and vibrational input at the oval window. However, instead of a single bone, mammals have a chain of three ossicles linking the tympanum with the oval window. The **malleus, incus**, and **stapes** are interconnected and controlled by a series of fine muscles. Although there has been much debate about the exact function of the ossicles, they seem to have a mechanical advantage that boosts the force of vibration on the oval window and provides a much broader range of frequency sensitivity than a single transducing element.

If the middle ear were an airtight cavity and the pressure in the environment was different from that of the middle ear, the tympanum of tetrapods would be distended and put under increased

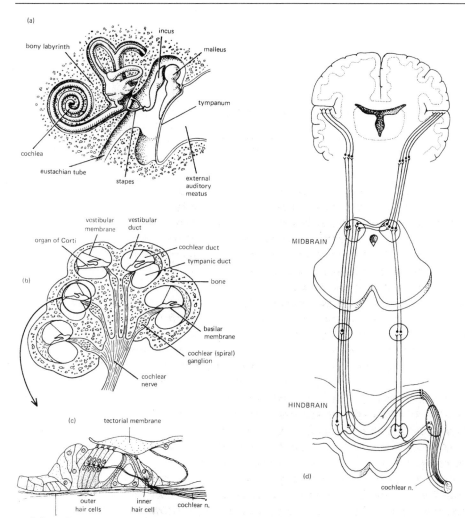

Figure 3–24 Sense of hearing as illustrated by mammals: (a) diagram of the auditory apparatus; (b) diagrammatic section of cochlea; (c) cellular structure of the organ of Corti; (d) neural pathways in audition show numerous synapses, decussations (crossing overs to the contralateral side), and final dual representation in the cerebral hemispheres, which facilitate comparison of inputs from each ear. ([a] Modified from A. S. Romer and T. S. Parsons, 1977, *The Vertebrate Body*, 5th edition, Saunders College, Philadelphia; [b] from A. G. Kluge et al., 1977, *Chordate Structure and Function*, 2d edition, Macmillan, New York, NY; [d] from F. H. Netter, 1962, *The CIBA Collection of Medical Illustrations*, volume I, *Nervous System*, CIBA Publications, Summit, NJ.)

tension, losing some of its responsiveness to sound pressure waves. The auditory tube, derived from the first embryonic visceral pouch, connects the mouth with the middle ear and alleviates this problem. Air flows in or out of the middle ear as air pressure changes. (These tubes sometimes become blocked, and when that happens changes in external air pressure can produce a painful sensation in addition to reduced auditory sensitivity.)

The tympanum is forced into the middle ear cavity when a sound wave impinges on it, momentarily compressing the air in the cavity. As the tympanum moves outward it temporarily creates reduced pressure in the middle ear. The pressure differences hinder free vibration of the tympanum and its attached ossicles by damping tympanic oscillations, and movement of air through the auditory tube is too slow to alleviate this problem. Mammals have minimized these damping pressure differences by an enlargement of the middle ear cavity into a **tympanic bulla** (plural: *bullae*). This hollow bubble of bone creates a middle ear volume so large that the volume changes produced by oscillations of the tympanum result in insignificant pressure variations. Auditory bullae are especially well developed in some desert rodents. The kangaroo rats of North America and the convergent but only distantly related jerboas of Africa and Asia have bullae that may constitute one-third of the length of the skull and have a volume greater than that of the braincase. These nocturnal rodents are particularly sensitive to low-frequency sounds, such as those made by owls in flight or by snakes moving across sand. The form of these bullae enhances the transmission of low-frequency, low-amplitude sounds from the tympanum to the inner ear (Webster 1966).

The inner ear is phylogenetically the oldest of the three divisions. However, the complexity of the mammalian inner ear is considerably greater than that found in other vertebrates. The inner ear of a mammal is a cavity within extremely dense bone, the periotic or petrous region of the temporal bone. The periotic bone forms from endochondral ossification of the chondrocranial otic capsules. Its central cavity is filled with a watery fluid, the perilymph, and has two flexible membrane-covered windows that face the middle ear cavity. Suspended in the inner ear cavity between the windows is a membranous sac with walls that are lined with sensory epithelia and invested with fibers of the eighth cranial nerve (the acoustic nerve, Table 3–3). This sac is filled with another fluid, the endolymph, and fits loosely within the bony cavity.

Both the cavity and the membrane sac are complex three-dimensional structures, as their names, the **bony labyrinth** and the **membranous labyrinth**, testify (Figure 3–24). Three general regions of these labyrinths can be identified in mammals: the **vestibule**, adjacent to the oval window; a posterodorsal extension of curving tubes known as the **semicircular canals**, three in gnathostomes, two in lampreys, and a distinctly different single one in hagfish; and from the floor of the labyrinth the **lagena**, which in mammals is much elaborated and called the **cochlea**. The mammalian cochlea is an anteroventral spirally coiled tube of bony labyrinth. It is divided into upper and lower perilymph-filled canals by a loose-fitting internal sleeve or tube of membranous labyrinth and endolymph.

The function of the various sac-like chambers and semicircular canals is essentially identical in all vertebrates: the detection of acceleration in three-dimensional space. The lagena (or cochlea) is the site of origin of neurally encoded auditory signals. Vibrations in the perilymph initiated by the columella (or the stapes of mammals) at the oval window of the vestibule pass into the lagena and travel all the way to the tip of the membranous labyrinth sleeve, around the blunt end of the sleeve, and back to the round window. In this long path the pressure wave deforms the flexible endolymphatic channel, especially at points where pressure waves traveling to the tip are in phase with those returning from it. These sites of maximum deformation occur at different points along the length of the lagena or the cochlea, depending on the frequency of the vibrations. Arrayed along the sleeve of the lagenar membrane is a band of hair cells innervated by twigs of the auditory nerve. Above the hair-like extensions of these cells a long shelf-like flap of tissue extends the length of the sleeve. When the membranous duct containing the hair cells is deformed by pressure waves in the perilymph, the sensory hairs are bent by contact with the tissue flap and cause firing in their respective neural connections. The location of the firing(s) along the length of the membranous labyrinth encodes the frequency of the original sound.

Endocrine System

The responses of vertebrates to their surroundings and to their internal needs are controlled by the nervous systems and the endocrine system. Both neurobiology and endocrinology deal with the study of the reception of stimuli, their transmission, and transfer from one area of the body to another. In nerve cells this process is achieved by electrical discharges that propagate along the nerve cell axon and are transferred to adjacent cells, generally by diffusion of chemical signals, for further neural processing. The neural transmission of stimuli from sensory receptor cells to effector cells (e.g., muscle) is very rapid, usually measured in fractions of a second. The endocrine system transfers information from one area to another via the release from a cell or organ complex (endocrine gland) of a chemical messenger, a **hormone**, that produces a response in the target cells.

The distinction between neural and endocrine processes has classically been defined as transmission of electrical versus chemical information and in terms of time, nervous activity being rapid (seconds or less) and endocrine activity slow (minutes to hours). The release of epinephrine (adrenalin) from the adrenal gland is followed by a multitude of responses in different target areas—increase in the strength of heartbeat, increased blood pressure, erection of hair (piloerection) in mammals, and inhibition of smooth muscle activity in the gut. Critical to the length of delay in response is the distance of the target organs from the site of hormone release. There are many examples of these kinds of hormones (Table 3–4).

Neurobiology and endocrinology have merged considerably in recent decades, however, with the discovery of **neurohormones**, hormones that act as chemical intermediates in the transfer of electrical activity from cell to cell. The classic example is acetylcholine, which is released from many nerve cell axonal terminals into a synapse (the junction between two nerve cells or a nerve cell and a muscle cell), and diffuses across the synapse to produce a response, usually a depolarization or hyperpolarization in the next nerve cell. Thus, in most cases neural activity in vertebrates and other metazoans involves both electrical transmission and the release of a neurohormone, or neurotransmitter substance, to pass information from cell to cell. Neurohormones cross synaptic junctions by chemical diffusion, which is slow relative to nerve conduction. The width of the synaptic junction is on the order of micrometers, and the synaptic time delay of neurohormonal diffusion from the presynaptic release point to the postsynaptic target sites can be measured in milliseconds. Hormones can therefore no longer be considered solely chemical messengers to distant target organs.

The **pituitary gland** and **hypothalamus**, often called the **pituitary axis**, are present in all living vertebrates (Figure 3–25). This complex is essentially where nervous activity from the peripheral and central nervous systems converges to provide major regulation of hormone release. Nerve activity directed into the hypothalamus causes the release of peptides (releasing factors) that are transported by the blood to distinct regions of the pituitary. They cause pituitary secretion of **tropins** (hormones that induce the release of the unique hormones of other endocrine glands). The release of thyroid hormone is a model (see Figure 3–25). Once thyroid hormone concentration rises in the blood it begins a feedback loop, thus inhibiting the secretion of more thyrotropins and regulating the level of thyroid hormone in circulation. Most hormones involve negative feedback loops of this sort.

Endocrine secretions are predominantly involved in the control and regulation of energy use, storage, and release, as well as energy allocation to special functions at critical times (Table 3–4). As a reflection of the role that hormones play in homeostasis, the majority of hormones that act in general metabolism, nutrition, or osmoregulation are counteracted by another hormone that acts in exactly the opposite direction. Often both hormones are the secretions of the same endocrine gland, but usually from distinctive cell types.

The existence of endocrine hormones predates the origin of vertebrates. In fact, hormones from invertebrates as well as vertebrates often can be classified as belonging to chemical families, such as the catecholamines (epinephrine, norepinephrine, and dopamine) and the neurohypophysial physins (oxytocin, vasopressin, and their various homologs). Hormone families share a similar chemical structure, but subtle chemical differences cause them to have different effects on their target organs. Within the vertebrates the structures of hormones have changed through time. The trend in the evolution of vertebrate endocrine glands has been consolidation either from the widely scattered clusters of cells or very small, poorly understood organs of fishes to larger, more definitive vascularized endocrine organs characteristic of amniotes.

Table 3–4　**Some examples of functions of the vertebrate endocrine system.**

Source of Hormone and Abbreviation	Effects
Hypothalamus	
Gonadotropic releasing hormone, GnRH	Stimulates FSH and LH secretion
Thyrotropin releasing hormone, TRH	Stimulates TSH secretion
Corticotropin releasing hormone, CRH	Stimulates ACTH secretion
Prolactin inhibiting factor, PIF	Suppresses prolactin secretion
Pituitary gland	
Posterior lobe	
Oxytocin	Stimulates milk secretion, uterine contraction
Vasopressin (arginine vasopressin), AVP (antidiuretic hormone), ADH	Stimulates renal water absorption
Intermediate lobe	
Melanocyte stimulating hormone, MSH	Stimulates melanin synthesis in skin
Distal lobe	
Follicle stimulating hormone, FSH	Female: Stimulates ovarian follicle growth, estradiol synthesis Male: Stimulates spermatogenesis
Luteinizing hormone, LH	Female: Stimulates ovulation, ovarian estradiol and progesterone synthesis Male: Stimulates testicular androgen synthesis
Prolactin, PRL	Stimulates milk synthesis, synthesis of progesterone by the corpus luteum in some species
Thyrotropin, TSH	Stimulates synthesis and release of thyroid hormone
Corticotropin, ACTH (adrenal cortical stimulating hormone)	Stimulates adrenal steroidogenesis
Somatotropin, STH (growth hormone, GH)	Stimulates somatomedin synthesis by the liver
Thyroid gland	
Thyroxine, T_4	Stimulates growth, differentiation, metabolism, heat production
Thyroid parafollicular cells or ultimobranchial glands	
Calcitonin	Lowers blood calcium ion concentration
Parathyroid glands	
Parathormone, PTH	Increases blood calcium ion concentration, lowers blood phosphate
Adrenal glands	
Adrenal steroidogenic tissue	
Cortisol, corticosterone	Stimulates carbohydrate metabolism
Aldosterone	Stimulates sodium retention by kidney

Table 3–4 *(Continued)*

Source of Hormone and Abbreviation	Effects
Adrenal chromaffin tissue	
Epinephrine, E	Multiple stimulatory and inhibitory effects on nerves, muscles, cellular secretions, and metabolism
Norepinephrine, NE	Generally the same as epinephrine
Ovary	
Preluteal follicle	
Estradiol (estrogen)	Stimulates female sexual development and behavior
Corpus luteum	
Progesterone	Stimulates growth of uterus and mammary glands, maternal behavior
Relaxin	Causes relaxation of pubic symphysis and dilation of cervix of uterus
Placenta	
Chorionic gonadotropin, CG	Stimulates progesterone synthesis by corpus luteum
Testes	
Testosterone (Leydig cells)	Stimulates male sexual development and behavior
Mullerian inhibiting hormone, MIS (Sertoli cells)	Causes degeneration of the oviducts, uterus, and the proximal part of the vagina in male mammalian embryos
Pineal (epiphysis)	
Melatonin	Suppresses gonadal development
Thymus gland	
Thymic hormones	Stimulates proliferation and differentiation of lymphocytes
Pancreatic islets	
Insulin	Lowers blood glucose; increases synthesis of protein, fat, and glycogen
Glucagon	Increases blood glucose
Gastrointestinal tract (only a few of the hormones secreted are listed)	
Gastrin	Stimulates hydrochloric acid secretion
Secretin	Stimulates secretion of bicarbonate by acinar cells of pancreas
Cholecystokin, CCK	Stimulates secretion of enzymes by acinar cells of pancreas, stimulates release of bile by gallbladder
Kidney	
Erythropoietin, TP	Stimulates erythropoiesis

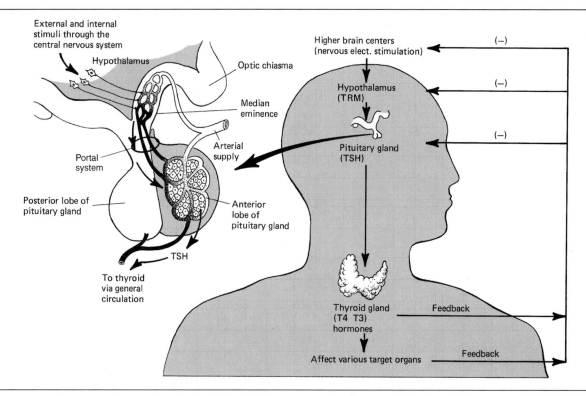

Figure 3–25 Pituitary axis and the release of thyroid hormones as an example of negative feedback to control hormonal release. High circulating levels of thyroid hormones in blood inhibit the release of thyroid releasing hormone from the hypothalamus and inhibit the release of thyroid stimulating hormone from the anterior lobe of the pituitary gland (indicated by minus signs), thereby lowering the rate of secretion of thyroid hormones from the thyroid gland.

Continuity of Life: The Reproductive System

Reproduction is vital to sustaining vertebrate life through the continuity and survival of genes over the long term, but it is subject to powerful environmental influences affecting gene frequencies and determining which genes remain in a population. Reproduction shows great variation among vertebrates. Nevertheless, a few aspects of the reproductive biology of vertebrates hold true throughout the group, have special generality, or illustrate trends consistent with phylogenetic hypotheses of relationships.

Gametogenesis and Gamete Conduits

During embryogenesis, paired genital ridges differentiate along the length of several somites from the medial border of intermediate mesoderm next to the nephrotome; primitive sex cells migrate from the gut wall into the ridges and the stage is set for differentiation of the gonads (Figure 3–26). Of the three cell types in the primordial gonad, only the **primordial sex cells** (probably derived from the migrant cell population) will produce gametes. Whether the gametes will be ova or sperm depends on the differentiation of the other two cell types. One of these is the epithelial lining of the coelomic cavity, and the second is composed of cells at least some of which had already differentiated and functioned in the Bowman's capsule of the embryonic mesonephros. These nephron cells de-differentiate, becoming less specialized and more like cells in early stages of embryonic development. De-differentiation is an uncommon phenomenon in vertebrate development. The coelomic epithelium organizes and thickens upon the arrival of the primordial sex cells and

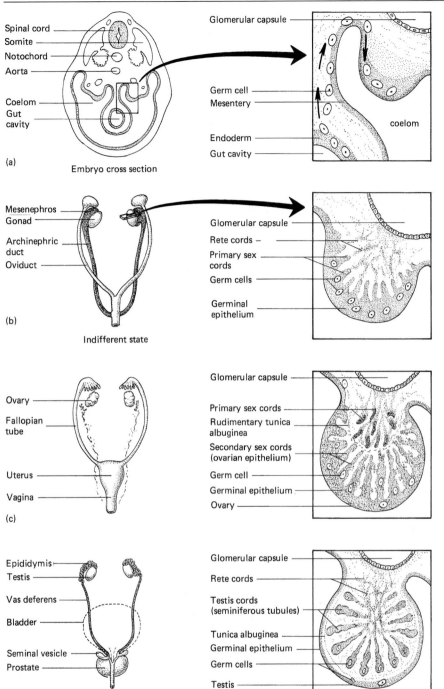

Spinal cord
Somite
Notochord
Aorta

Coelom
Gut cavity

(a)

Embryo cross section

Glomerular capsule

Germ cell
Mesentery

Endoderm

Gut cavity

coelom

Mesenephros
Gonad

Archinephric duct
Oviduct

(b)

Indifferent state

Glomerular capsule
Rete cords
Primary sex cords
Germ cells

Germinal epithelium

Ovary

Fallopian tube

Uterus

Vagina

(c)

Glomerular capsule

Primary sex cords
Rudimentary tunica albuginea
Secondary sex cords (ovarian epithelium)

Germ cell
Germinal epithelium
Ovary

Epididymis
Testis

Vas deferens

Bladder

Seminal vesicle
Prostate

(d)

Glomerular capsule

Rete cords

Testis cords (seminiferous tubules)

Tunica albuginea
Germinal epithelium
Germ cells

Testis

Figure 3–26 Generalized scheme of modification of the indifferent vertebrate gonad into an ovary or a testis. The gonadal structure, whether primary cortex or medulla, is derived from embryonic mesoderm. (a) The primordial germ cells, which give rise to eggs or sperm, are initially located in embryonic endoderm and migrate through the mesenteries during development into the indifferent gonad. (b) The germ cells arrange themselves between the medullary and cortical tissues of the indifferent gonad. (c) In formation of the ovary the cortical tissue predominates. (d) In the testis the medullary tissue predominates.

forms the gonadal cortex. The primordial sex cells multiply as well and with the de-differentiated nephric cells form a primordial gonad medulla. The relative number of each type of cell is considered to be under the control of sex-determining genes because it is the relative activities of these two cell populations that determine gonadal sex. Cortical cells attract migrating sex cells but inhibit cell division, whereas medullary cells attach cell processes to sex cells and stimulate both their migratory activity and cell division. The influence of cortex results in the formation of an ovary, and that of the medulla forms a testis. The definitive gonad, whether male, female, or both (see below) is an endocrine gland producing circulating hormones *and* a specialized exocrine gland that produces an unusual product: highly specialized cells—sperm or ova.

Sex Determination

Although there is a growing body of evidence that the social and physiological contexts of reproduction may bias the normal male to female sex ratio of vertebrate offspring (Ridley 1993), the actual mechanism of sex determination is usually less accessible to modulation. The gender of most mammals is expressed by genitalia and other secondary sex characteristics and is generally obvious from birth. In other classes of vertebrates, however, genitalia and sexually dimorphic structures are absent or not expressed until maturity is achieved (van Tienhoven 1983). To discuss sex determination in all vertebrates, we must bear in mind the basic definition of sex: The female sex produces eggs; the male sex produces sperm (Naftolin 1981). Egg production requires a functional ovary and sperm production requires a functional testis; each structure has a distinct cytological appearance in its mature form. In many kinds of vertebrates both an ovary and a testis are present in the same individual. By definition these individuals are both male and female, a seemingly bizarre condition, for in mammals individual sex is segregated into strong female and male roles. If some vertebrate individuals can function as mixed sexes, what is the purpose of the rigid sex segregation found in other vertebrates?

Sex and Sex Chromosomes In all vertebrates the gonad is initially indifferent and capable of producing an ovary or a testis (Figure 3–26). Individuals with both types of gonads present and functional are called **hermaphrodites**; individuals with either ovaries or testes are **gonochorists**. Hermaphroditism is common in nonamniotic vertebrates, especially fishes, but virtually absent among amniotes (Chapter 8).

In amphibians and fishes distinctive sex chromosomes have seldom been demonstrated. Breeding experiments clearly indicate, however, that male- and female-determining genes are distributed over several chromosomes (Bull 1983). Because of their morphological compatibility these chromosomes of many fishes and amphibians can cross over and exchange genes. As a result, intersexuality and several types of hermaphroditism, including functional sex reversal, are widespread in nonamniotes (van Tienhoven 1983).

In amniotes, however, sex reversals are atypical but sex determination may be affected by environmental temperature during development (Chapters 8 and 12). The sex of a mammal or bird is genetically prescribed by the sex chromosome pairs (Jablonka and Lamb 1990, Morell 1994). In humans, for example, zygotes of females contain 23 similar (homomorphic) chromosome pairs, but in males one chromosome (Y) is much smaller than its mate (X). Males therefore are chromosomally XY, a condition called heteromorphic. In males the Y chromosome carries a gene dominant for maleness (Gordon and Ruddle 1981, Haseltine and Ohno 1981). Sex chromosomal anomalies are well studied in humans. Individuals with only one X chromosome (XO) are female, although infertile. Individuals with sex chromosomal constitutions of XXY, XXXY, XXXXY, and XXXYY are phenotypic males, but sterile. Males of XYY are fertile. In mammals, the presence of the Y chromosome directs the medullary layer of the indifferent gonad to differentiate as a testis. In contrast, its absence and the presence of an X chromosome causes the cortical layer to develop as an ovary.

Sex chromosomes influence only the sexual development of the gonads, that is, the primary sex characters. Precisely how the sex genes act is unclear. Differentiation of testes in mammals, seems to be controlled by a gene located on the male Y chromosome, and its absence leads to the development of female gonads. In humans, for example, females with an XY chromosomal complement exist because they lack the male-determining portion of the Y chromosome. And males with an XX chromosomal complement are also known because they have the appropriate male-determining portion of the Y chromosome translocated onto one of the two X chromosomes (Kolata 1986). What seems to be consistent is that a gene located on the Y chromosome initiates male gonadal development, and female gonadal development results from its absence. Once a gonad has had its primary sex declared as female or male, the sex hormone estrogen or testosterone (and at least in mammals, a Mullerian-inhibiting hormone) is produced. These hormones affect the development of the secondary sex characters—all the structural and behavioral differences that exist between male and female (Ehrhardt and Meyer–Bahlburg 1981, Wilson et al. 1981). In humans the genitalia, breasts, hair patterns, and differential growth patterns are secondary sex characters. Horns, antlers, differences in plumage length, and dimorphic color patterns are familiar differences that we associate with sex in other vertebrates.

The development of so-called freemartins in cattle provides a natural example of the phenomenon of hormonal control of secondary sex characters. Twinning sometimes results in the mixing of fetal blood supplies. If one twin is male (XY) and the other female (XX), the female embryo is influenced by the testosterone secretions of the developing male. At birth the female calf (a freemartin) shows mixed male–female traits and is infertile. The genetic male calf is normal. The evidence suggests that in mammals testosterone leads to maleness, even in potentially genetic females (Bardin and Catterall 1981). Experimental evidence supports this conclusion. Castrating an early embryo of a placental mammal results in the development of female secondary sex characteristics, whether the embryo is XX or XY. Administration of testosterone to the castrated embryo, however, induces the development of male secondary sex characteristics (Bull 1983). Gonadal differentiation is not directly influenced by the hormone treatment, however, for XX embryos develop ovaries and XY embryos testes regardless of the hormonal effects on secondary sex characteristics.

Female spotted hyenas (*Crocuta crocuta*) exhibit male-like genitalia and dominance over males. The hyena placenta has been shown to convert high circulating concentrations of a steroid hormone precursor into testosterone, which results in the masculinization of the fetal external genetalia and perhaps affects the developing nervous system but lacks sterilizing action on the gonads (Yalcinkaya et al. 1993).

Heteromorphic sex chromosomes occur in birds, where the female is the heteromorphic sex and the male the homomorphic sex. To distinguish this difference, geneticists use alternative symbols: ZW for female and ZZ for male. The dissimilar chromosome W is dominant and directs the indifferent gonad to develop as an ovary. In the shelled (cleidoic) egg equivalent of freemartinism—double-yolked eggs where the embryonic blood supplies mix—ZW females are normal and ZZ males are sterile intersexes. The conclusion is that estrogen effects are dominant over the effects of testosterone in birds.

Why has sex determination been fixed rigidly in birds and mammals by heteromorphic sex chromosomes? Why is the heterogametic sex female in birds and male in mammals? In birds and mammals many species form large groups in which individual behaviors often involve distinct male and female roles. In mammals, fixed sex determination may have been essential for the exploitation of the behavioral complexity that was made possible by the increased integrative capacities associated with larger brain size. If, for example, male baboons wavered in fending off predators while females hesitated to herd the young to safety, the result could reduce reproductive success. Thus, although it is unlikely that any single factor explains fixation of sex determination, we suspect that the behavioral complexity of birds and mammals is a significant component.

The opposite sexual expression of heterogametic individuals in mammals on one hand and birds on the other is intriguing. In gonochoristic vertebrates, the heterogametic sex is physiologically dominant. Carrying young in utero as placental mammals do, however, places strictures on the determination of sex. Ursula Mittwoch (1973) pointed out that during pregnancy maternal estrogens can cross the maternal–fetal barrier. For a female fetus (XX) this presents no problem, but a male (XY) would be swamped by estrogenic effects (femaleness) unless a strong masculinizing agent existed. Mittwoch has suggested that this crossing of the placental barrier by maternal hormones explains the dominance of XY masculinizing effect in directing the secondary sex differentiation of mammals. Monotremes, mammals that lay eggs, are believed to lack a Y chromosome: Females are XX and males are X. In birds the egg isolates the embryo, whether male or female, from direct maternal influence. Genetic sex, expressed in the relative safety of the egg, could equally well be determined by female heteromorphy (WZ) or by male heteromorphy (XY). Female heteromorphy in birds is quite possibly a matter of evolutionary chance. Although there are some unique selective pressures associated with the evolution of female sex chromosome heteromorphy (Jablonka and Lamb 1990), they may not be as potent as those operating in the mammalian situation.

Female and Male Reproductive Organs Ovaries differentiate as relatively simple organs with an undifferentiated connective tissue **stroma** in which are embedded many **follicles**. Follicles begin as large primary sex cells completely surrounded by a covering of epithelial cells from the germinal cortex. As they mature the follicular cell layer becomes much larger, often forming a spherical, sometimes fluid-filled organ within the ovary nurturing the developing egg and producing the hormone estrogen. The follicle stimulates the development of yolk in the egg. Egg yolk constitutes a major energy investment in reproduction for many vertebrates, either because of the number of eggs produced or because of the amount of yolk in each egg. When the eggs mature the follicle ruptures, releasing the completed egg (**ovulation**). In most vertebrates the eggs are released into the coelomic cavity. Derived ray-finned fishes have very large, hollow sac-like ovaries and follicular rupture is toward the interior, so that the ovary becomes distended with ovulated eggs. The epithelial cells of the collapsed follicle remain in the stroma of the ovary after ovulation. In many vertebrates these cells form a second hormone-producing gland, the **corpus luteum** (plural copora lutea). The corpus luteum produces the hormone progesterone, which in mammals stimulates changes in the relevant portions of the female for retention of the developing young. In birds where no retention of ova occurs, corpora lutea do not develop. The production of eggs by the ovary may be almost continuous (humans), seasonal (the vast majority of vertebrates), or occur only once in a lifetime (semelparous fishes such as the eel, and certain salmon native to the North Pacific).

The testes differentiate as compact organs made up of interconnecting **seminiferous tubules** where sperm develop. Sperm are among the most highly specialized of animal cells. No cytoplasm in the conventional sense remains in a sperm cell; it is specialized to deliver nuclear information to the ovum. Sperm differentiate from prolifically dividing sex cells and are supported, nourished, and conditioned by cells that remain permanently attached to the tubule walls, the **supporting** or **Sertoli** cells. Unlike the follicular cells of the ovary, the testicular supporting cells are not endocrine but strictly sperm-nurturing cells. The hormone testosterone is produced by clusters of **interstitial cells** between the seminiferous tubules in tetrapods.

The reproductive tracts of vertebrates are the gamete conduits to the external environment. Lampreys and hagfish lack any conduits, sperm or eggs erupt from the gonad and move through the coelom to pores that open to the outside near the cloaca. The ancestral gnathostome condition involves two sets of ducts that carry the gametes to the exterior. One set of conduits are the **archinephric (Wolffian) ducts** that form on each side of the coelom by the fusion and posterior growth of the embryonic pronephric tubules (Figure 3–15). No duct transports sperm in jawless vertebrates. In male gnathostomes the archinephric duct transports sperm. In a few species (plesiomorphic ray-finned fishes, gymnophionans, and some other amphibians) sperm and urine are carried to the exterior through the archinephric duct. However, in the majority of gnathostomes the kidney tubules involved are modified exclusively for sperm transport. The archinephric duct also repeatedly has evolved toward exclusive sperm transport functions as the opisthonephric kidney develops accessory urinary ducts (cartilaginous fishes, most nonamniote tetrapods) or when the metanephric kidney evolved with its unique ureter (amniotes). The lungfishes and derived ray-finned fishes have very posteriorly positioned testes, which thus have connections with the urine-carrying archinephric duct only very posteriorly (lungfish) or not at all (teleosts). Short new ducts, unrelated to those of any other vertebrate gonad, transport sperm in these fishes. Like other vertebrates, bony fishes thus establish separation of sperm and urine. The terminus of the exclusively sperm-carrying archinephric ducts (or in some fishes of their distinct sperm ducts) is often incorporated into specialized **intromittent organs** that permit sperm to be directly introduced into the female reproductive tract. These organs may be derived from pelvic fins (claspers of cartilaginous fishes), anal fins (the gonopodia of fishes like guppies, mollies, and swordtails), gill covers (the gonopodia of some ray-finned fishes), or the walls of the cloaca (the penises and hemipenes of tetrapods).

The female reproductive tract develops in parallel to that of the archinephric duct, except in jawless fishes in which both sexes lack conduits. The distinctive paired ducts of females, called the **oviducts** or **Mullerian ducts**, form by a longitudinal splitting of the archinephric duct (in chondrichthyans and salamanders) or by an invagination of the peritoneum over the kidney (in plesiomorphic bony fishes and the remaining tetrapods). Teleost fishes form analogous but not homologous ducts by posterior extension of the sac-like ovaries, thus producing a conduit system that has no open coelomic segment. The Mullerian ducts form early in embryonic development and are generally located beside the archinephric ducts before the gonads differentiate. Thus, one or the other set of ducts degenerates during sexual differentiation.

The anterior end of the oviduct opens into the coelom, except in teleosts. Ova enter the **ostium** of the oviduct after they have ruptured from the ovary into the coelom. The oviducts are variously specialized in different groups to accommodate the great range of reproductive patterns seen in vertebrates. In forms with simple external fertilization of extruded eggs (**oviparous** species) the oviducts are simple tubes. The oviducts of most amniotes are specialized to produce albumen, and shells (cartilaginous fishes, most amniotes) or jelly coatings (amphibians, some bony fishes) for eggs that are to be extruded. (The term "oviparity" is also correctly applied to these species.) A wide variety of vascular and secretory specialization is found in the **uteri** (singular uterus). Uteri are usually derived from specializations of the oviducts in forms in which both fertilization and development are internal (**viviparity**). The degree of development and type of uterine specializations depend on whether additional energy, over and above that contained in the yolk, is provided to the embryo (Chapter 7).

Summary

The complex activities of vertebrates are supported by an equally complex morphology. Interactions between different tissues and structures are central to the embryonic development of a vertebrate and to its function as an organism. Patterns of embryonic development are generally phylogenetically conservative, and many of the shared derived characters of vertebrates can be traced to their origins in the early embryo. In particular, the neural crest cells, which are unique to vertebrates, participate in the embryonic origin of a majority of derived characters of vertebrates.

An adult vertebrate can be viewed as a number of systems that interact continuously. The integument separates a vertebrate from its environment and participates in regulating the exchange of matter and energy between the organism and the environment.

Support and movement are the province of the skeletomuscular system, and both are necessary for the effective function of the food-gathering and food-processing systems. Respiration and circulation carry metabolic substrates and oxygen to the tissues and remove wastes. The nitrogenous waste products of protein metabolism are eliminated by the integument of plesiomorphic forms and by the renal system of derived vertebrates. The renal system also participates in regulating salt and water balance. Coordination of these activities is accomplished by the nervous and endocrine systems, and the reproductive system transmits genetic information from generation to generation. In this chapter we have described the phylogenetic and embryonic origins of the major structures and systems of vertebrates, and discussed their modifications during vertebrate evolution.

References

Bardin, C. W., and J. F. Catterall. 1981. Testosterone: a major determinant of extragenital sexual dimorphism. *Science* 211:1285–1294.

Bereiter–Hahn, J., A. G. Matoltsy, and K. S. Richards (editors). 1986. *Biology of the Integument 2: Vertebrates*. Springer, New York, NY.

Brown, J.A., R. J. Balment, and J.C. Rankin. 1993. *New Insight in Vertebrate Kidney Function*. Cambridge University Press, Cambridge, UK.

Bull, J. J. 1983. *Evolution of Sex Determining Mechanisms*. Benjamin-Cummings, Menlo Park, CA.

Burggren, W. W. 1987. Form and function in reptilian circulations. *American Zoologist* 27:5–19.

Coates, M. 1993. *Hox* genes, fin folds and symmetry. *Nature* 364:195–196.

Cooper, E. L. 1985. Comparative immunology. *American Zoologist* 25:649–664.

Ehrhardt, A. A., and H. F. L. Meyer–Bahlburg. 1981. Effects of prenatal sex hormones on gender related behavior. *Science* 211:1312–1318.

Fawcett, D. W., and E. Raviola 1994. *A Textbook of Histology*, 12th edition. Chapman & Hall, New York, NY.

Fraser, S. 1993. Segmentation moves to the fore. *Current Biology* 3:787–789.

Garcia–Fernandez, J., and P. W. H. Holland. 1994. Archetypal organization of the amphioxus *Hox* gene cluster. *Nature* 370:563–566.

Gee, H. 1994. Return of amphioxus. *Nature* 370:504–505.

Gordon, J. W., and F. H. Ruddle. 1981. Mammalian gonadal determination and gametogenesis. *Science* 211:1265–1271.

Gruss, P., and C. Walther. 1992. *Pax* in development. *Cell* 69:719–722.

Hanken, J., and B. K. Hall (editors). 1993. *The Skull*, volumes 1, 2 and 3. University of Chicago Press, Chicago, IL. An excellent compendium of contemporary knowledge and hypothesis on this most distinctive character of vertebrates.

Haseltine, F. P., and S. Ohno. 1981. Mechanics of gonadal differentiation. *Science* 211:1272–1278.

Hemmati–Brivanlou, A., and D. A. Melton. 1994. Inhibition of activin receptor signaling promotes neuralization in *Xenopus*. *Cell* 77:273–281.

Hemmati–Brivanlou, A., O. G. Kelly, and D. A. Melton. 1994. Follistatin, an antagonist of activin, is espressed in the Spemann organizer and displays direct neuralizing activity. *Cell* 77:283–295.

Jablonka, E., and M. J. Lamb. 1990. The evolution of heteromorphic sex chromosomes. *Biological Reviews* 65(3):249–276.

Kessel, M. 1993. *Hox* genes, fin folds and symmetry. *Nature* 364:197.

Keynes, R., and R. Krumlauf. 1994. *Hox* genes and regionalization of the nervous system. *Annual Review of Neuroscience*. 17:109–132.

Kolata, G. 1986. Maleness pinpointed on Y chromosome. *Science* 234:1076–1077.

Le Douarin, N. 1982. *The Neural Crest*. Cambridge University Press, Cambridge, UK.

Marx, J. 1991. How embryos tell heads from tails. *Science* 254:1586–1588.

Marx, J. 1992. Homeobox genes go evolutionary. *Science* 255:399–401.

McMahon, A. P. 1992. The *Wnt* family of developmental regulators. *Trends in Genetics* 8(7):236–242.

Mittwoch, U. 1973. *Genetics of Sex Determination*. Academic, New York, NY.

Morell, V. 1994. The rise and fall of the Y chromosome. *Science* 263:171–172.

Naftolin, F. 1981. Understanding the basis of sex differences. *Science* 211:1263–1264.

Noden, D. M., and A. deLahunta. 1985. *The Embryology of Domestic Animals*. Williams & Wilkins, Baltimore, MD.

Nusse, R., and H. E Varmus. 1992. *Wnt* genes. *Cell* 69:1073–1087.

Rast, J. P., and G. W. Litman. 1994. T-cell receptor gene homologs are present in the most primitive jawed vertebrates. *Proceedings of the National Academy of Sciences USA*. 91:9248–9252.

Ridley, M. 1993. A boy or a girl: is it possible to load the dice? *Smithsonian* 24(3):113–123.

Sage, H., and W. R. Gray. 1979. Studies on the evolution of elastin, I: phylogenetic distribution. *Comparative Biochemistry and Physiology* 64B:313–327.

Sage, H., and W. R. Gray. 1981a. Studies on the evolution of elastin, II: histology. *Comparative Biochemistry and Physiology* 66B:13–22.

Sage, H., and W. R. Gray. 1981b. Studies on the evolution of elastin, III: the ancestral protein. *Comparative Biochemistry and Physiology* 68B:473–480.

Smith, H. W. 1953. *From Fish to Philosopher*. Little, Brown, Boston, MA.

Smith, M. M., and B. K. Hall. 1993. A developmental model for evolution of the vertebrate exoskeleton and teeth: the role of cranial and trunk neural crest. *Evolutionary Biology* 27:387–448.

Soin, S. G. 1981. A new classification of the structure of mature eggs of fishes according to the ratio of yolk to ooplasm. *Soviet Journal of Developmental Biology* (Translation of *Ontogenez*) 12:13–17.

Stern, C. D., and P. W. Ingham (editors). 1992. Gastrulation. *Development*, 1992 supplement. The Company of Biologists Limited, Cambridge, UK.

Stoykova, A., and P. Gruss. 1994. Roles of *Pax*-genes in developing and adult brain as suggested by expression patterns. *The Journal of Neuroscience* 14(3):1395–1412.

Tabin, C. J. 1992. Why we have (only) five fingers per hand: *Hox* genes and the evolution of paired limbs. *Development* 116:289–296.

Tabin, C., and E. Laufer 1993. *Hox* genes and serial homology. *Nature* 361:692–693.

Thorogood, P., and P. Ferretti 1993. *Hox* genes, fin folds and symmetry. *Nature*. 364:196.

van Tienhoven, A. 1983. *Reproductive Physiology of Vertebrates*, 2d edition. Cornell University Press, Ithaca, NY.

Webster, D. 1966. Ear structure and function in modern mammals. *American Zoologist* 6:451–466.

Wilson, J. D., F. W. George, and J. E. Griffin. 1981. The hormonal control of sexual development. *Science* 211:1278–1284.

Yalcinkaya, T. M., P. K. Siitori, and J.-L. Vigne. 1993. A mechanism for virilization of female spotted hyenas in utero. *Science* 260:1929–1931.

HOMEOSTASIS AND ENERGETICS: WATER BALANCE, TEMPERATURE REGULATION, AND ENERGY USE

In the preceding chapter we described some of the structural complexities of vertebrates, and here we consider the way those structures work. Vertebrates are complicated organisms: In particular they maintain very different conditions inside their bodies from the conditions in the environment immediately external to them. The concentrations of water, ions, and molecules inside the body of a vertebrate have profound effects on biochemical and physiological mechanisms, and they must be regulated within specific limits. Temperature affects both the biochemical processes of life and the cellular environment within which those processes take place. Regulation of their internal conditions (homeostasis) is a central part of the biology of vertebrates. Structure and function are usually tightly coupled—changing the form of a structure is likely to change its function as well. Thus, the evolution of vertebrate morphology has been accompanied by changes in the ways that vertebrates work. Some kinds of specializations of vertebrates are mutually exclusive, whereas others are mutually compatible and can be combined in the same organism. The relationships between structure and function are often reflected in broad aspects of the biology of vertebrates and directly affect their ecology and behavior. These relationships are also intimately related to the characteristics of the environments in which vertebrates live: For example, many of the problems faced by terrestrial and aquatic vertebrates are quite different. In this chapter we describe the basic aspects of homeostasis of vertebrates and how they have changed in major evolutionary steps such as the transition from aquatic to terrestrial life or from ectothermy to endothermy.

The Internal Environment of Vertebrates

Seventy to eighty percent of the body mass of most vertebrates is water, and many of the chemical reactions that release energy or synthesize new compounds take place in an aqueous environment containing a complex mixture of ions and other solutes. Some ions are cofactors that control the rates of metabolic processes; others are involved in the regulation of pH, the stability of cell membranes, or the electrical activity of nerves. Metabolic substrates and products must move from sites of synthesis to the sites of utilization. Almost everything that happens in the body tissues of vertebrates involves water, and maintaining the concentrations of water and solutes within narrow limits is a vital activity.

Temperature, too, is critical in the function of organisms. Biochemical reactions are temperature-sensitive. In general, the rates of reactions increase as temperature increases, but not all reactions have the same sensitivity to temperature. Furthermore, the permeability of cell membranes and other features of the cellular environment are sensitive to temperature. A metabolic pathway is a series of chemical reactions in which the product of one reaction is the substrate for the next, yet each of these reactions may have a different sensitivity to temperature so a change in temperature can mean that too much or too little substrate is produced to sustain the next reaction in the series. To complicate the process of regulation of substrates and products even more, the chemical reactions take place in a cellular milieu that is itself changed by temperature. Clearly, the smooth functioning of metabolic pathways is greatly simplified if an organism can limit the range of temperatures its tissues experience.

From the perspective of environmental physiology, an organism can be described as a complex of self-sustaining exchanges with the environment. Energy is the basis of these exchanges. Vertebrates gain energy from the environment as food and as heat; they use energy for activity, growth, and reproduction; and they release energy as waste products and as heat.

In this chapter we describe some of the mechanisms that vertebrates use to regulate their exchanges of water and heat with the environment, and the significance of those mechanisms in the use of energy by vertebrates.

Exchange of Water and Ions

In a sense, an organism can be thought of as an aqueous solution of organic and inorganic substances contained within a leaky membrane, the body surface. Exchange of material and energy with the environment is essential to the survival of the organism, and much of that exchange is regulated by the body surface. The significance of differential permeability of the skin to various compounds is particularly conspicuous in the case of aquatic vertebrates, but it applies to terrestrial vertebrates as well. Both active and passive processes of exchange are used by vertebrates to regulate their internal concentrations in the face of varying external conditions.

The Vertebrate Kidney

The cells of an organism can exist over only a narrow range of solute concentrations of the body fluids.

Another area of narrow tolerance is in the accumulation of wastes. The small nitrogen-containing molecules that result from protein catabolism are toxic, and are an especially important category of wastes. The vertebrate kidney has evolved superb capacities for homeostatic control of water balance and waste excretion.

The adult kidney consists of hundreds to millions of tubular **nephrons**, each of which produces urine. The primary function of the nephron is removing excess water, salts, waste metabolites, and foreign substances from the blood. In this process, the blood is first filtered through the **glomerulus**, a structure unique to vertebrates (Figure 4–1). Each glomerulus is composed of a leaky arterial capillary tuft encapsulated within a sieve-like filter. Arterial blood pressure forces fluid into the nephron to form an **ultrafiltrate** composed of blood minus blood cells and larger molecules. The ultrafiltrate is then processed to return essential metabolites (glucose, amino acids, and so on) and water to the general circulation. When necessary, the walls of the nephron actively excrete toxins. These actions take place in the **proximal convoluted tubules** (PCT) and **distal convoluted tubules** (DCT) of the nephron (Figure 4–2). Finally, wastes are excreted as urine to the exterior.

Regulation of Ions and Body Fluids

The first vertebrates, the ostracoderms, probably had ion levels like those of their marine invertebrate ancestors, which, presumably, were like those of most living marine invertebrates. The solute concentrations in the body fluids of many marine invertebrates are similar to those in seawater, as are those of hagfishes (Tables 4–1 and 4–2). In contrast, solute levels are greatly reduced in the blood of all other vertebrates, a characteristic shared only with invertebrates that have penetrated estuaries, fresh waters, or the terrestrial environment.

The presence of solutes in seawater or blood plasma lowers the kinetic activity of water. Therefore, water flows from a dilute solution (high kinetic activity of water) to a more concentrated solution (low kinetic activity), a phenomenon called **osmosis**. The osmotic concentrations of various animals and of seawater are shown in Table 4–1. In most marine invertebrates and the hagfish, the body fluids are in osmotic equilibrium with seawater; that is, they are **isosmotic** to seawater (approximately 1000 milliOsmoles per liter [mOsm/L]). Body fluid concentra-

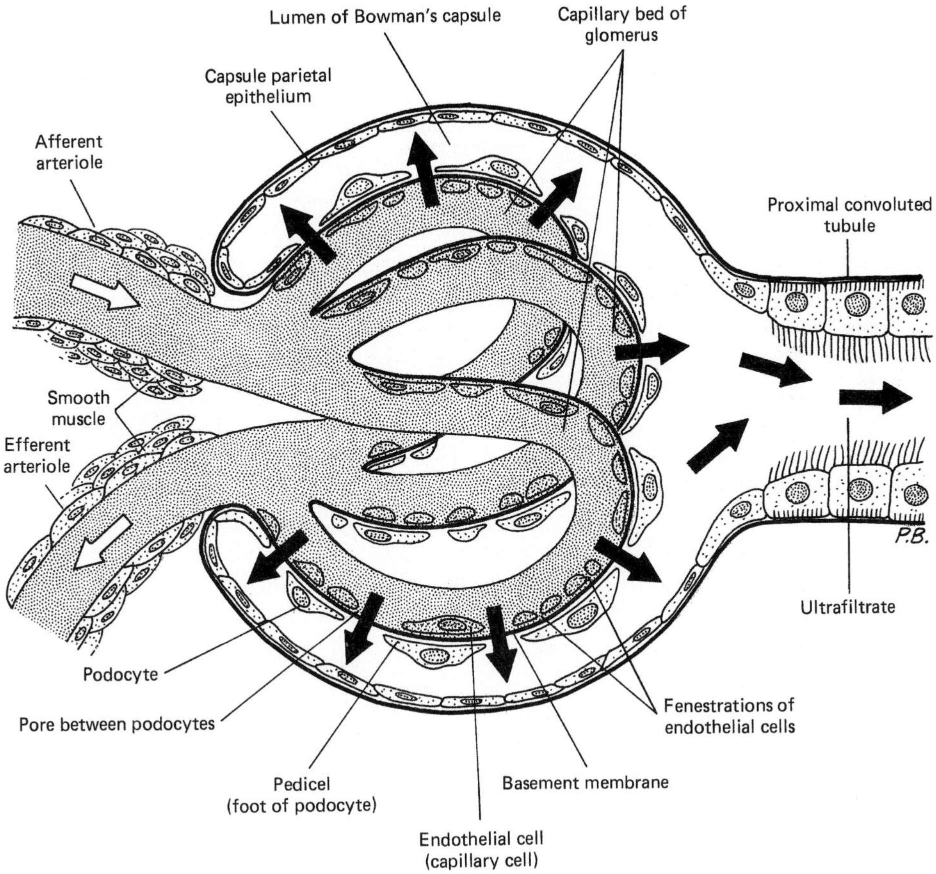

Figure 4–1 Detail of a typical mammalian glomerulus, illustrating the morphological basis for its function.

Labels in figure: Lumen of Bowman's capsule; Capillary bed of glomerus; Capsule parietal epithelium; Afferent arteriole; Proximal convoluted tubule; Smooth muscle; Efferent arteriole; P.B.; Ultrafiltrate; Podocyte; Pore between podocytes; Fenestrations of endothelial cells; Pedicel (foot of podocyte); Basement membrane; Endothelial cell (capillary cell)

tions in marine teleosts and lampreys are between 350 and 450 mOsm/L. Therefore, water flows outward from their blood to the sea. In chondrichthyans the osmolality of the blood is slightly higher than that of seawater, and water flows from sea to blood. These osmotic differences are specified by the terms **hyposmotic** (marine teleosts and lampreys) and **hyperosmotic** (coelacanth and chondrichthyans). Freshwater fishes are hyperosmotic to the medium, but through reduction in NaCl their blood osmolality is lower than that of their marine counterparts (Table 4–1). The water and salt balance of teleosts is constantly threatened by either (1) osmotic loss of water and salt gain when in seawater, or (2) osmotic gain of water and loss of salts when in fresh water. [For additional details see Evans (1980) and Nishimura and Imai (1982).]

Most fishes are **stenohaline** (*steno* = narrow, *haline* = salt): They inhabit either fresh water or seawater and survive only modest changes in salinity. Because they remain in one environment, the magnitude and direction of the osmotic gradient to which they are exposed is stable. Some fishes, however, are **euryhaline** (*eury* = wide): They inhabit both fresh water and seawater and tolerate large changes in salinity. In euryhaline species the osmotic and salt gradients are reversed as they move from one medium to the other.

Freshwater Organisms: Teleosts and Amphibians Several mechanisms are involved in the osmotic regulation of vertebrates that live in fresh water (Figure 4–2). The body surface of fishes has low permeability to water and to ions. However, fishes cannot entirely prevent osmotic exchange. Gills, which are permeable to gases, are also permeable to water. As a result, most water and ion movements take place across the gill surfaces. Water is gained by osmosis, and ions are lost by diffusion. To compensate for this influx of water, the kidney produces a large volume of urine. To reduce salt loss, the urine is diluted by actively reabsorbing salts. Indeed, urine processing in a freshwater teleost provides a simple model of vertebrate kidney function. A freshwater teleost does not

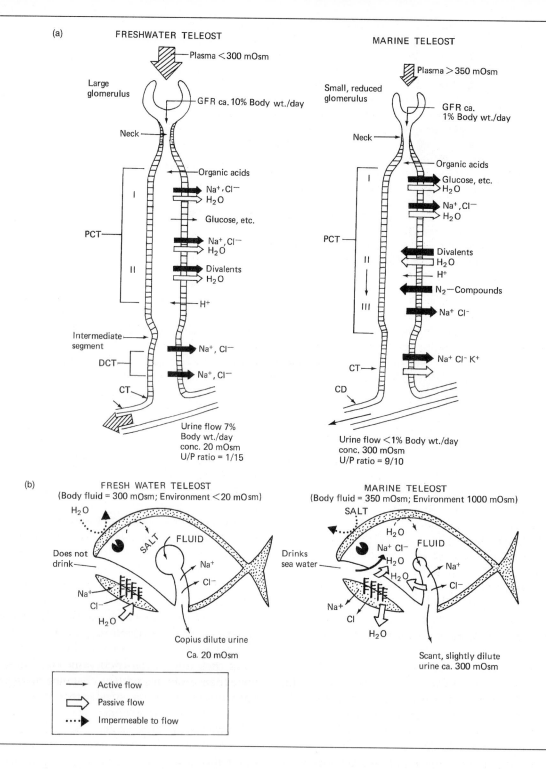

(a)

FRESHWATER TELEOST

Plasma <300 mOsm

Large glomerulus

GFR ca. 10% Body wt./day

Neck

Organic acids

Na⁺, Cl⁻
H_2O

Glucose, etc.

PCT

Na⁺, Cl⁻
H_2O

Divalents
H_2O

H⁺

Intermediate segment

Na⁺, Cl⁻

DCT

Na⁺, Cl⁻

CT

Urine flow 7%
Body wt./day
conc. 20 mOsm
U/P ratio = 1/15

MARINE TELEOST

Plasma >350 mOsm

Small, reduced glomerulus

GFR ca. 1% Body wt./day

Neck

Organic acids

Glucose, etc.
H_2O

Na⁺, Cl⁻
H_2O

Divalents
H_2O

H⁺

N_2—Compounds

Na⁺ Cl⁻

Na⁺ Cl⁻ K⁺

CT

CD

Urine flow <1% Body wt./day
conc. 300 mOsm
U/P ratio = 9/10

(b)

FRESH WATER TELEOST
(Body fluid = 300 mOsm; Environment <20 mOsm)

H_2O

Does not drink

SALT

FLUID

Na⁺

Cl⁻

Na⁺
Cl⁻
H_2O

Copius dilute urine
Ca. 20 mOsm

MARINE TELEOST
(Body fluid = 350 mOsm; Environment 1000 mOsm)

SALT

H_2O

Na⁺ Cl⁻
H_2O

Drinks sea water

H_2O

FLUID

Na⁺

Cl⁻

Na⁺

Cl

H_2O

Scant, slightly dilute urine ca. 300 mOsm

→ Active flow

⇒ Passive flow

····▶ Impermeable to flow

drink water, because osmotic water movement is already providing more water intake than it needs— drinking would only increase the amount of water it had to excrete via the kidneys.

The large glomeruli of freshwater teleosts produce a copious flow of urine, but the glomerular ultrafiltrate is isosmotic to the blood and contains essential blood salts. To conserve salt, ions are reabsorbed

Figure 4–2 Kidney structure and function of marine and freshwater teleosts. (a) Schematic comparison of nephron structure and functions in a freshwater and marine teleost fish. GFR is the glomerular filtration rate at which the ultrafiltrate is formed, expressed in percentage of body weight per day. PCT is the proximal convoluted tubule (in fishes sometimes referred to as the proximal tubule). Two segments (I and II) of the PCT are recognized in both freshwater and marine teleosts. Segment III of the PCT of marine teleosts is sometimes equated with the DCT (distal convoluted tubule) of freshwater teleosts. Darkened arrows represent active movements of substances, open arrows passive movement, and hatched arrows indicate by their size the relative magnitude of fluid flow. Note that Na^+ and Cl^- are reabsorbed in the PCT segment I and in the CT (collecting tubule) in both freshwater and marine teleosts; water flows osmotically across the PCT in both freshwater and marine teleosts but only across the CT of marine teleosts. Water permeability of the CT (and also the DCT) is therefore low in freshwater teleosts. U/P, ultrafiltrate to blood plasma concentration ratio—a measure of the concentrating power of a nephron. (b) General scheme of the osmotic and ionic gradients encountered by freshwater and marine teleosts.

across the proximal and distal convoluted tubules. Because the distal convoluted tubule is impermeable to water, the urine becomes less concentrated as ions are removed from it. Ultimately, the urine becomes hyposmotic to the blood. In this way the water that was absorbed across the gills is removed and ions are conserved. Nonetheless, some ions are lost in the urine in addition to those lost by diffusion across the gills. Salts from food compensate for some of this loss. In addition, freshwater teleosts have special cells located in the gills that absorb sodium and chloride ions from fresh water. These ions must be moved by active transport against a concentration gradient, and this requires energy (Kirschner 1995).

Freshwater amphibians face similar osmotic problems. The entire body surface of amphibians is involved in the active uptake of ions from the water. Like freshwater fishes, aquatic amphibians do not drink. Acidity inhibits this active transport of ions in both amphibians and fishes, and inability to maintain internal ion concentrations is one of the causes of death of these animals in habitats acidified by acid precipitation.

Marine Organisms: Teleosts and Other Fishes The osmotic and ionic gradients of vertebrates in seawater are largely the reverse of those experienced by freshwater vertebrates. Seawater is more concentrated than the body fluids of vertebrates, so there is a net outflow of water by osmosis and a net inward diffusion of ions.

Teleosts The integument of marine fishes, like that of freshwater teleosts, is highly impermeable, so that most osmotic and ion movements occur across the gills (Figure 4–2). The kidney glomeruli are small and the glomerular filtration rate is low. Less urine is formed, and the water lost in urine is reduced. Marine teleosts lack a water-impermeable distal convoluted tubule. As a result, urine leaving the nephron is less copious but more concentrated than that of freshwater teleosts, although it is always hyposmotic to blood. To compensate for osmotic dehydration, marine teleosts do something unusual —they drink seawater. Sodium and chloride ions are actively absorbed across the lining of the gut, and water flows osmotically into the blood. Estimates of seawater consumption vary, but many species drink in excess of 25 percent of their body weight per day and absorb 80 percent of this ingested water. Of course, drinking seawater to compensate for osmotic water loss increases the influx of sodium and chloride ions. To compensate for this salt loading, discrete cells, called **chloride cells**, located in the gills, actively pump sodium and chloride ions outward against a large concentration gradient.

Hagfishes and Chondricthyans Hagfishes minimize their problems with ion balance by regulating only divalent ions and reduce osmotic water movement by being nearly isosmotic to seawater. Chondrichthyans and coelacanths also minimize osmotic flow by maintaining the osmotic concentration of the body fluid close to that of seawater. These animals retain nitrogen-containing compounds (primarily urea and trimethylamine oxide) to produce osmolalities that are usually slightly hyperosmotic to seawater (Table 4–1). As a result, chondrichthyans

Table 4–1 Representative concentrations of the sodium and chloride and osmolality of the blood in vertebrates and marine invertebrates. Ion concentrations are expressed in millimoles per liter of water; all values are reported to the nearest 5 units. Osmolality reported in milliOsmoles (mOsm; 1 Osm = 1 mole dissolved particles per kilogram water).

Type of Animal	mOsm	Na^+	Cl^-	Other Major Osmotic Factor	Source
Seawater	~1000	475	550		
Fresh water	<10	~5	~5		
Marine invertebrates					
Coelenterates, mollusks, etc.	~1000	470	545		1
Crustacea	~1000	460	500		1
Marine vertebrates					
Hagfishes	~1000	535	540		2
Lamprey	~300	120	95		2
Teleosts	<350	180	150		3
Coelacanth	<1000 to 1180	180	200	Urea 375	4, 10
Elasmobranch (bull shark)	1050	290	290	Urea 360	5
Holocephalian	~1000	340	345	Urea 280	9
Freshwater vertebrates					
Polypterids	200	100	90		3
Acipenserids	250	130	105		3
Primitive neopterygians	280	150	130		3
Dipnoans	240	110	90		3
Teleosts	<300	140	120		3
Elasmobranch (bull shark)	680	245	220	Urea 170	5
Elasmobranch (freshwater rays)	310	150	150		6
Amphibians*	~250	~100	~80		7
Terrestrial vertebrates					
Reptiles	350	160	130		8
Birds	320	150	120		8
Mammals	300	145	105		8

*Ion levels and osmolality highly variable, but tend toward 200 mOsm in fresh water.

Sources: 1. W. T. W. Potts and G. J. Parry, 1964, *Osmotic and Ionic Regulation in Animals,* Macmillan, New York, NY; 2. J. D. Robertson, 1954, *Journal of Experimental Biology* 31:424–442; 3. M. R. Urist et al., 1972, *Comparative Biochemistry and Physiology* 42:393–408; 4. G. E. Pickford and F. G. Grant, 1964, *Science* 155:568–570; R. W. Griffith et al., 1975, *Journal of Experimental Zoology* 192:165–171; 5. T. B. Thorson et al., 1973, *Physiological Zoology* 46:29–42; 6. T. B. Thorson et al., 1967, *Science* 158:375–377; 7. P. J. Bentley, 1971, *Endocrines and Osmoregulation,* Zoophysiology and Ecology Series, volume 1, Springer, New York, NY; 8. C. L. Prosser, 1973, *Comparative Animal Physiology,* 3d edition, Saunders, Philadelphia, PA; 9. L. J. Read, 1971, *Comparative Biochemistry and Physiology,* 39A:185–192; 10. D. H. Evans, 1979, *Comparative Physiology of Osmoregulation in Animals,* edited by G. M. O. Maloiy, Academic, New York, NY.

gain water by osmotic diffusion across the gills and need not drink seawater. This net influx of water permits large kidney glomeruli, as in freshwater teleosts, to produce high filtration rates and therefore rapid cleansing of the blood. Urea is very soluble and diffuses through most biological membranes, but the gills of chondrichthyans are nearly impermeable to urea and the kidney tubules actively reabsorb it. With internal ion concentrations that are low relative to seawater, chondrichthyans experience ion influxes across the gills as do marine teleosts. Unlike the gills of marine teleosts, those of chondrichthyans have low ion permeabilities (less than 1 percent that of teleosts). Chondrichthyans generally do not have highly developed salt-excreting cells in the gills. Rather, they achieve ion balance by secreting from the rectal gland a fluid that is approximately isosmotic to body fluids and seawater, but contains higher concentrations of sodium and chloride ions.

The urea and trimethylamine oxide in the blood of chondrichthyans also contribute to buoyancy.

Chondrichthyans are more dense than seawater, so they sink when they are motionless. Sharks and rays lack the gas-filled swim bladders that bony fishes use to adjust their buoyancy, but many sharks have large, oil-filled livers. The oil is lighter than water, and the large size of the liver helps to make a shark buoyant. The Port Jackson shark (*Heterodontus portjacksoni*), however, has a small liver with a low oil content that does not contribute significantly to its buoyancy. Urea and trimethylamine oxide in the blood and muscle tissue of Port Jackson sharks provide positive buoyancy because they are less dense than an equal volume of water (Withers et al. 1994). Chloride ions also provide positive buoyancy, whereas sodium and protein are denser than water and are negatively buoyant. The net effect of these solutes is a significant positive buoyancy.

Freshwater Elasmobranchs and Marine Amphibians
Some elasmobranchs are euryhaline—sawfish, some sting rays, and bull sharks are examples. In seawater bull sharks retain high levels of urea, but in fresh water their blood urea levels decline. Sting rays in the family Potamotrygonidae spend their entire lives in fresh water and have very low blood urea concentrations. Their blood sodium and chloride ion concentrations are 35 to 40 percent below those in sharks that enter fresh water and only slightly above levels typical of freshwater teleosts (Table 4–1). The potamotrygonids have existed in the Amazon basin perhaps since the Tertiary, and their reduced ionosmotic gradient may reflect their long adaptation to fresh water. When exposed to increased salinity, potamotrygonids do not increase the concentration of urea in the blood as euryhaline elasmobranchs do, even though the enzymes required to produce urea are present. Apparently their long evolution in fresh water has led to an increase in the permeability of their gills to urea and reduced the ability of their kidney tubules to reabsorb it.

Most amphibians are found in freshwater or terrestrial habitats. One of the few species that occurs in salt water is the crab-eating frog, *Rana cancrivora*. This frog inhabits intertidal mudflats in southeast Asia and is exposed to 80 percent seawater at each high tide. During seawater exposure, the frog allows its blood ion concentrations to rise and thus reduces the ionic gradient. In addition, proteins are deaminated and the ammonia is rapidly converted to urea, which is released into the blood. Blood urea rises

Table 4–2 **Intracellular concentration of major inorganic ions in marine invertebrates. (Concentration values are in millimoles per liter; compare with Table 4–1.)***

	Na^{++}	Cl^-	K^+	Ca^{2+}	Mg^{2+}
Seawater	475	550	10	10	53
Marine invertebrates	54–325	54–380	48–175	3–89	8–96
Vertebrates					
Hagfish	122	107	117	2	13
All others	8–45	11–30	83–185	2–9	7–11

*Note that the monovalent ions Na^+ and Cl^- are reduced relative to seawater, and K^+ increased in all animals. For divalent cations, especially Mg^{2+}, a reduction is found in all vertebrates but not in all marine invertebrates.

from 20 to 30 millimoles per liter, and the frogs become hyperosmotic to the surrounding water. In this sense *Rana cancrivora* functions like an elasmobranch and absorbs water osmotically. Frog skin, unlike that of elasmobranchs, is permeable to urea and urea is rapidly lost. To compensate for this loss, the activity of the urea-synthesizing enzymes is very high. The tadpoles of *Rana cancrivora*, like most tadpoles, lack urea-synthesizing enzymes until late in their development. Thus, tadpoles of crab-eating frogs must use a method of osmoregulation different from that of adults. The tadpoles have extrarenal salt-excreting cells in the gills, and by pumping ions outward as they diffuse inward, the tadpoles maintain their blood hyposmotic to seawater in the same manner as marine teleosts.

Nitrogen Excretion by Vertebrates

Vertebrates require food in proportion to their activity. Metabolism of carbohydrates and fats (composed of carbon, hydrogen, and oxygen) produces carbon dioxide and water, which are easily voided. Proteins and nucleic acids are another matter, for they contain nitrogen. When protein is metabolized, the nitrogen is enzymatically reduced to ammonia through a process called deamination. Ammonia is very diffusible and soluble in water but also extremely toxic. Rapid excretion of ammonia is therefore crucial. Differences in how ammonia is excreted are partly a matter of the availability of water and partly the result of phylogeny. Nitrogen is eliminated by most vertebrates as ammonia, as urea, or as uric acid. Most vertebrates excrete all three of these substances, but the proportions of the three compounds differ among the groups of vertebrates (Figure 4–3).

Fishes and Amphibians Many aquatic invertebrates excrete ammonia directly, as do vertebrates with gills, permeable skins, or other permeable membranes that contact water. Excretion of nitrogenous wastes as ammonia is termed **ammonotelism**, excre-

tion as urea is **ureotelism**, and excretion as uric acid is **uricotelism**. Urea is synthesized from ammonia in a cellular enzymatic process called the **urea cycle**. Urea synthesis requires greater expenditure of energy than does ammonia production. The value of ureotelism, therefore, lies in the benefits derived from urea itself, not in energy economy.

Urea has two advantages. First, it is retained by some marine vertebrates to counter osmotic dehydration. A second function of urea synthesis is the

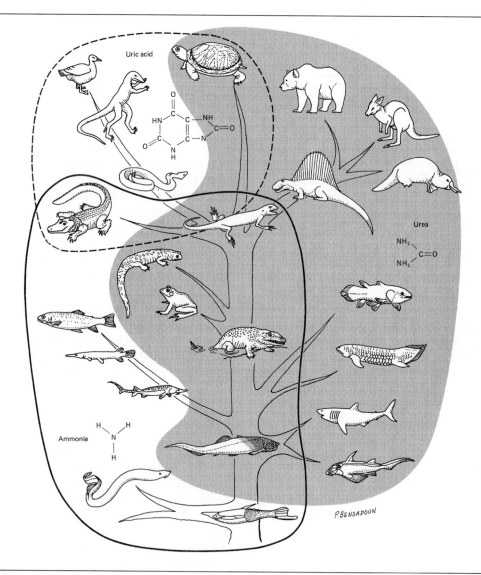

Figure 4–3 Phylogenetic distribution of the three major nitrogenous wastes in vertebrates. The types of wastes excreted by extinct vertebrates are unknown; examples merely provide visual continuity to the phylogeny. (Modified from B. Schmidt–Nielsen, 1972, pages 79–103 in *Nitrogen Metabolism and the Environment*, edited by J. W. Campbell and L. Goldstein, Academic, London, UK.)

detoxification of ammonia under environmental circumstances that prevent its rapid elimination. The less toxic urea can be concentrated in urine, thus conserving water.

Ureotelism probably evolved independently several times. Perhaps ureotelism developed in early freshwater fishes to avoid osmotic dehydration as they reinvaded the sea, as exemplified by chondrichthyans and coelacanths. In other instances ureotelism may have evolved in response to desiccation. Ureotelism was preadaptive for invasion of the land, because it allowed nitrogen to be retained in a detoxified state until sufficient water was available for excretion.

A fascinating example of this preadaptation was discovered in the ammonotelic African lungfish, *Protopterus*, by the late Homer Smith. During drought lungfishes estivate and slowly oxidize their carbohydrate, fat, and protein stores. The ammonia is detoxified and urea accumulates, increasing body fluid osmolality. This reduces the vapor pressure across the lung surface and water loss through evaporation. When rains return, the lungfishes rapidly take up water and excrete the accumulated urea. An analogous process takes place in some spadefoot toads that inhabit arid environments (see Chapter 16). Ureotelism may have been similarly advantageous to the first Devonian tetrapods.

Mammals The capacity of the mammalian kidney to conserve water, rid the body of nitrogenous and other wastes, and maintain a narrow ion and acid–base variation is essential to mammalian life. Only with the concentrating powers of the mammalian kidney could mammals have invaded so many diverse and severe environments. Understanding the mammalian kidney is a key factor in understanding the success of mammals.

The mammalian kidney is composed of millions of nephrons, the basic microanatomical units of kidney structure recognizable in all vertebrates. Each nephron is composed of a **glomerulus** that filters the blood and a long tubular conduit in which the chemical composition of the filtrate is altered. The mammalian kidney is capable of producing a urine more concentrated than that of any nonamniote and, in most cases, more concentrated than that of diapsids as well (Table 4–3). This ability greatly reduces water loss and the need for water, and is important for several lineages of mammals that live in arid habitats.

The basic mechanism of urine concentration is removal of water from an ultrafiltrate, leaving behind the concentrated excretory residue. Because cells are unable to transport water directly, they use osmotic gradients to manipulate movements of water molecules. In addition, the cells lining the nephron actively reabsorb substances important to

Table 4–3 **Maximum urine concentrations of tetrapods.**

Species	Maximum Observed Urine Concentration (mOsm/L)	Approximate Urine: Plasma Concentration Ratio
Shingle-back lizard (*Trachydosaurus rugosus*)	300	0.95
Pelican (*Pelecanus erythrorhynchos*)	700	approx. 2
Savanna sparrow (*Passerculus sandwichensis*)	2000	4.4
Human (*Homo sapiens*)	1430	4
Bottlenose porpoise (*Tursiops truncatus*)	1600–1800	approx. 5
Hill kangaroo (*Macropus robustus*)	2730	7–8
Camel (*Camelus dromedarius*)	2800	8
White rat (*Rattus norvegicus*)	2900	8.9
Cat (*Felis domesticus*)	3250	9.9
Pack rat (*Neotoma albigula*)	4250	11 (est.)
Marsupial mouse (*Dasycercus eristicauda*)	approx. 4000	12 (est.)
Kangaroo rat (*Dipodomys merriami*)	approx. 4650	12 (est.)
Vampire bat (*Desmodus rotundus*)	4650	14
Australian hopping mouse (*Notomys alexis*)	9370	22

Source: P. J. Bentley, 1959, *Journal of Physiology* 145:37–47; M. S. Gordon et al., 1982, *Animal Physiology*, 4th edition, Macmillan, New York; R. L. Malvin and M. Rayner, 1958, *American Journal of Physiology*, 214:187–191; R. E. MacMillen, 1972, *Symposium of the Zoological Society of London* 31:147–174; W. N. McFarland and W. A. Wimsatt, 1976, *Comparative Biochemistry and Physiology* 28: 985–1007; K. Schmidt-Nielsen, 1964, *Desert Animals*, Oxford University Press, Oxford, UK.

the body's economy from the ultrafiltrate and secrete toxic substances into it. The cells lining the nephron differ in permeability, molecular and ion transport activity, and reaction to the hormonal and osmotic environments in the surrounding body fluids.

The cells of the proximal convoluted tubule (PCT) have an enormous luminal surface area produced by long, closely spaced microvilli, and the cells contain many ATP-producing mitochondria (Figure 4–4). These structural features reflect the function of the PCT in rapid, massive transport of sodium from the lumen of the tubule to the peritubular space and capillaries; sodium transport is followed by passive movement of chloride to the peritubular space to neutralize electric charge (Figure 4–5). Water then flows osmotically in the same direction. Farther down the nephron, the cells of the thin segment of the loop of Henle are wafer-like and contain fewer mitochondria. The descending limb permits passive flow of sodium and water, and the ascending limb actively removes sodium from the ultrafiltrate. Finally, cells of the collecting tubule appear to be of two kinds. Most seem to be suited to the relatively impermeable state characteristic of periods of sufficient body water. Other cells are mitochondria-rich and have a greater surface area. They are probably the cells that respond to the presence of antidiuretic hormone (ADH) from the pituitary gland, triggered by insufficient body fluid. Under the influence of ADH, the collecting tubule actively exchanges ions, pumps urea, and becomes permeable to water, which flows from the lumen of the tubule into the concentrated peritubular fluids.

The nephron's activity may be considered a six-step sequential process, each step localized in regions having special cell characteristics and distinctive variations in the osmotic environment. The first step is production of an ultrafiltrate at the glomerulus (Figure 4–5). The ultrafiltrate is isosmotic with blood plasma and resembles whole blood after the removal of (1) cellular elements, (2) substances with a molecular weight of 70,000 or greater (primarily proteins), and (3) substances with molecular weights between 15,000 and 70,000, depending on the shapes of the molecules. An average filtration rate for resting humans approximates 120 milliliters of ultrafiltrate per minute. Obviously, a primary function of the nephron is reduction of the ultrafiltrate volume—to excrete 170 liters (45 gallons) of glomerular filtrate per day is impossible.

The second step in the production of the urine is the action of the **proximal convoluted tubule** (PCT, Figure 4–5) in decreasing the volume of the ultrafiltrate. The PCT cells have greatly enlarged lumenal surfaces that actively transport sodium and perhaps chloride ions from the lumen to the exterior of the nephron. Water flows osmotically through the PCT cells in response to the removal of sodium chloride. By this process about two-thirds of the salt is reabsorbed in the PCT, and the volume of the ultrafiltrate is reduced by the same amount. Although it is still very nearly isosmotic with blood, the substances contributing to the osmolality of the urine after it has passed through the PCT are at different concentrations than in the blood.

The next alteration occurs in the descending limb of the **loop of Henle**, where the thin, smooth-surfaced cells of this segment freely permit diffusion of sodium and water. Because the descending limb passes through tissues of increasing osmolality as it plunges into the medulla, water is lost from the urine and it becomes more concentrated. In humans the osmolality of the fluid in the descending limb may reach 1200 mOsm/L. Other mammals can achieve considerably higher concentrations. By this mechanism the volume of the forming urine is reduced to 25 percent of the initial filtrate volume, but it is still large. In an adult human, for example, 25 to 40 liters of fluid per day reach this stage, yet only a few liters will be urinated.

The fourth step takes place in the ascending limb of the loop of Henle, which possesses cells with large, numerous, densely packed mitochondria. The ATP produced by these organelles is utilized in actively removing sodium from the forming urine. Because these cells are impermeable to water, the volume of urine does not decrease and it enters the next segment of the nephron hyposmotic to the body fluids. Although this sodium-pumping, water-impermeable, ascending limb does not concentrate or reduce the volume of the forming urine, it sets the stage for these important processes.

The very last portion of the nephron changes in physiological character, but the cells closely resemble those of the ascending loop of Henle. This region, the **distal convoluted tubule** (DCT), is permeable to water. The osmolality surrounding the DCT is that of the body fluids, and water in the entering hyposmotic fluid flows outward and equilibrates osmotically. This process reduces the fluid volume to 5 to 20 percent of the original ultrafiltrate.

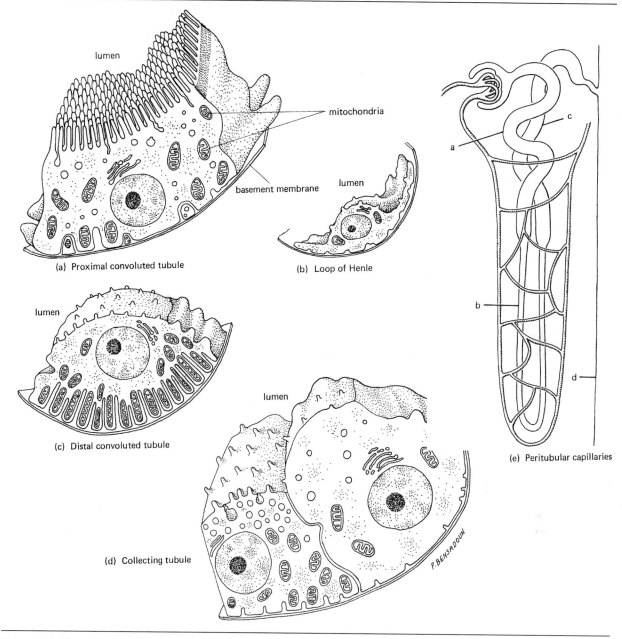

Figure 4–4 The structure of the cells lining the walls of the mammalian nephron:
(a) proximal convoluted tubule; (b) loop of Henle; (c) distal convoluted tubule;
(d) collecting tubule. In part (e) the letters correspond to the detailed drawings.

The final touch in the formation of a scant, highly concentrated mammalian urine occurs in the **collecting tubules**. Like the descending limb of the loop of Henle, the collecting ducts course through tissues of increasing osmolality, which withdraw water from the urine. The significant phenomenon associated with the collecting duct, and to a lesser extent with the DCT, is its conditional permeability to water. During excess fluid intake, the collecting duct demonstrates low water permeability: Only half of the water entering it may be reabsorbed and the remainder excreted. In this way a copious, dilute urine can be produced. When a mammal is dehydrated the collecting ducts and the DCT become very permeable to water and the

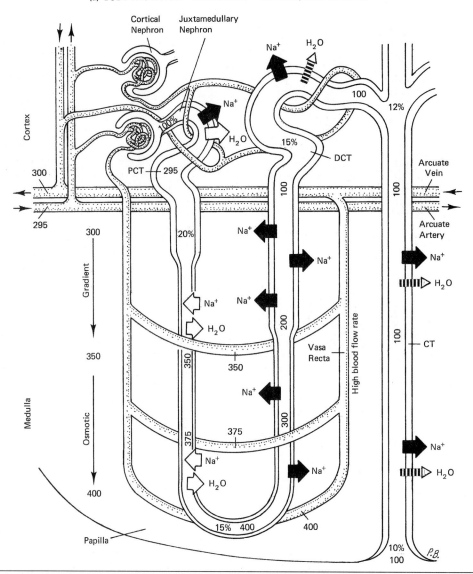

Figure 4–5 Diagram showing how the mammalian kidney produces dilute urine when the body is hydrated and concentrated urine when the body is dehydrated. Black arrows indicate active transport and white arrows indicate passive flow. The numbers represent the approximate milliosmolality of the fluids in the indicated regions. Percentages are the volumes of the forming urine relative to the volume of the initial ultrafiltrate. (a) When blood osmolality drops below normal concentration (about 300 mOsm/L), excess body water is excreted. (b) When osmolality rises above normal, water is conserved. (Based on F. H. Netter, 1973, *The CIBA Collection of Medical Illustrations*, volume 6, CIBA Publications, Summit, NJ.)

final urine volume may be less than 1 percent of the original ultrafiltrate volume. In certain desert rodents so little water is contained in the urine that it crystallizes almost immediately upon micturition.

A polypeptide called **antidiuretic hormone** (ADH; also known as **vasopressin**), is produced by specialized neurons in the hypothalamus, stored in the posterior pituitary, and released into the circula-

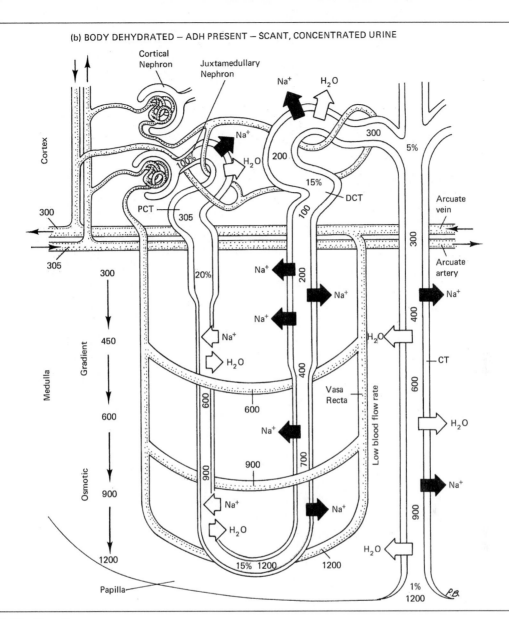

Figure 4–5 *(Continued)*

tion whenever blood osmolality is elevated or blood volume drops. When present in the kidney, ADH increases the permeability of the collecting duct to water and facilitates water reabsorption to produce a scant, concentrated urine. The absence of ADH has the opposite effects. Alcohol inhibits the release of human ADH, induces a copious urine flow, and this frequently results in dehydrated misery the following morning.

The key to concentrated urine production clearly depends on the passage of the loops of Henle and collecting ducts through tissues with increasing osmolality. These osmotic gradients are formed and maintained within the mammalian kidney as a result of its structure (Figure 4–6), which sets it apart from the kidneys of other vertebrates. Particularly important are the structural arrangements within the kidney medulla of the descending and ascending segments of the loop of Henle and its blood supply, the **vasa recta**. These elements create a series of parallel tubes with flow passing in opposite directions in adjacent vessels (countercurrent flow). As a result, sodium secreted from the ascending limb of the loop of Henle diffuses into the medullary tissues to increase their osmolality, and this excess salt is distributed by the countercurrent flow to create a steep osmotic gradient within the medulla (see Figure 4–5b). The final concentration of a mammal's urine is determined by the amount of sodium accumulated in the fluids of the medulla. Physiological alterations

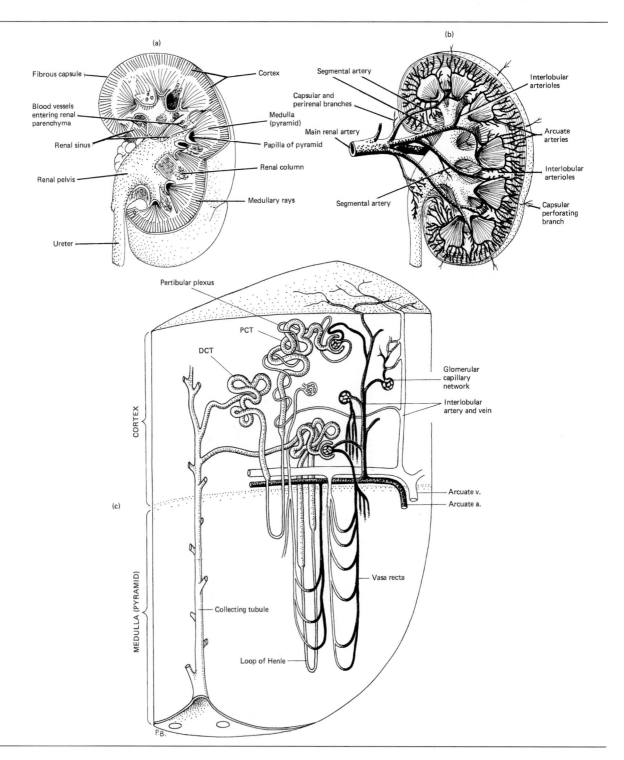

in the concentration in the medulla result primarily from the effect of ADH on the rate of blood flushing the medulla. When ADH is present, blood flow into the medulla is retarded and salt accumulates to create a steep osmotic gradient. Another hormone, aldosterone, from the adrenal gland increases the rate of sodium secretion into the medulla to promote an increase in medullary salt concentration.

In addition to these physiological means of concentrating urine, a variety of mammals have morphological alterations of the medulla. Most mammals have two types of nephrons: those with a cortical glomerulus and abbreviated loops of Henle that do not penetrate far into the medulla and those with juxtamedullary glomeruli, deep within the cortex, with loops that penetrate as far as the papilla of the renal pyramid (Figure 4–6c). Obviously, the longer, deeper loops of Henle experience large osmotic gradients along their lengths. The flow of blood to these two populations of nephrons seems to be independently controlled. Juxtamedullary glomeruli are more active in regulating water excretion; cortical glomeruli function in ion regulation. Finally, some desert rodents have exceptionally long renal pyramids. Thus, the loops of Henle and the vasa recta are extended and can produce large differences in osmolality from the cortical to the papillary ends. The maximum concentrations of urine measured from a given species of mammal correlate well with the length of its renal pyramids.

Fossils of early mammals suggest that they were primarily insectivores and carnivores with a high energy demand. This diet would be rich in protein, which, when metabolized, would produce large amounts of urea. Considerable water would be required to void this nitrogenous waste unless a means of concentrating urea was available. The unique concentrating power of the mammalian kidney may have been an early response to the accumulation of metabolic wastes from high-protein diets.

Diapsids and Turtles All of the living representatives of the diapsid lineage are uricotelic, and uric acid and its salts account for 80 to 90 percent of urinary nitrogen in most species. Turtles, also, excrete a variable proportion of their nitrogenous wastes as salts of uric acid (Table 4–4).

The kidneys of diapsids and turtles lack the long loops of Henle that allow mammals to reduce the volume of urine and raise its osmotic concentration to several times the osmotic concentration of the blood plasma. Urine from the kidneys of diapsids and turtles consists of a moderately dilute solution of uric acid and ions. It is isosmotic with the blood plasma, or even somewhat hyposmotic to the blood. However, uric acid differs from urea in being only slightly soluble in water. It will precipitate from a dilute solution, and that is what happens when urine from the ureters enters the cloaca or bladder. (Many diapsids lack a urinary bladder entirely; others have an ephemeral bladder that is lost shortly after they hatch; and some diapsids and probably all turtles have a functional bladder throughout life.) The uric acid combines with ions in the urine and precipitates as a light-colored mass that includes sodium, potassium, and ammonium salts of uric acid and also contains ions held by complex physical forces. When the uric acid and ions precipitate from solution, the urine becomes less concentrated. In effect, water is released and reabsorbed into the blood. In this respect, excretion of nitrogenous wastes as uric acid is even more economical of water than is excretion of urea, because the water used to produce urine is reabsorbed and reused.

Water is not the only substance that is reabsorbed from the cloaca, however. Many diapsids and turtles also reabsorb sodium ions and return them to the bloodstream. At first glance, that seems a remarkably inefficient thing to do. After all, energy was used to create the blood pressure that forced the ions through the walls of the glomerulus into the

---

Figure 4–6 Gross morphology of the mammalian kidney exemplified by that of a human. (a) Structural divisions of the kidney and proximal end of the ureter; (b) the renal artery and its subdivisions in relation to the structural components of the kidney. The renal vein (not shown) and its branches parallel those of the artery; (c) enlarged diagram of a section extending from the outer cortical surface of the apex of a renal pyramid, the renal papilla. The general relationship of the nephrons and blood vessels to the gross structure of the kidney can be visualized by comparing these diagrams. (Based on F. H. Netter, 1973, *The CIBA Collection of Medical Illustrations*, volume 6, CIBA Publications, Summit, NJ; and H. W. Smith, 1956, *Principles of Renal Physiology*, Oxford University Press, New York, NY.)

Table 4–4 Distribution of nitrogenous end products among diapsids and turtles.

Group	Percentage of Total Urinary Nitrogen		
	Ammonia	*Urea*	*Salts of Uric Acid*
Squamates			
Tuatara	3–4	10–28	65–80
Lizards and snakes	Small	0–8	90–98
Archosaurs			
Crocodilians	25	0–5	70
Birds	6–17	5–10	60–82
Turtles			
Aquatic	4–44	45–95	1–24
Desert	3–8	15–50	20–50

urine in the first place, and now more energy is being used in the cloaca to drive the active transport that returns the ions to the blood. The animal has used two energy-consuming processes and it is back where it started, with an excess of sodium ions in the blood. Why do that?

The solution of the paradox lies in a third water-conserving mechanism that is present in many diapsids and turtles, salt-secreting glands that provide an extrarenal pathway that disposes of salt with less water than the urine. In at least four groups of diapsids (lizards, snakes, crocodilians, and birds) some species possess glands specialized for the elective transport of ions out of the body (Peaker and Linzell 1975, Minnich 1982). Salt glands are widespread among lizards. In all cases it is the lateral nasal glands that excrete salt. The secretions of the glands empty into the nasal passages, and a lizard expels them by sneezing or by shaking its head. In birds, also, the lateral nasal gland has become specialized for salt excretion. The glands are situated in or around the orbit, usually above the eye. Marine birds (pelicans, albatrosses, penguins) have well-developed salt glands, as do many freshwater birds (ducks, loons, grebes), shorebirds (plovers, sandpipers), storks, flamingos, carnivorous birds (hawks, eagles, vultures), upland game birds, the ostrich, and the roadrunner. Depressions in the supraorbital region of the skull of the extinct aquatic birds *Hesperornis* and *Ichthyornis* suggest that salt glands were present in these forms as well.

In sea snakes (Hydrophiidae) and elephant-trunk snakes (Acrochordidae), the posterior sublingual gland secretes a salty fluid into the tongue sheath, from which it is expelled when the tongue is extended. In some species of homalopsines (a group of rear-fanged aquatic snakes from the Indoaustralian region) the premaxillary gland secretes salt. Salt-secreting glands on the dorsal surface of the tongue have been identified in several species of crocodiles, in a caiman, and in the American alligator.

The diversity of glands involved in salt excretion among diapsids indicates that this specialization has evolved independently in various groups. At least four different glands are used for salt secretion by diapsids, indicating that a salt gland is not an ancestral character for the group, and the differences between crocodilians and birds and between snakes and lizards suggest that salt glands are not ancestral either for archosaurs or for squamates.

Finally, in sea turtles and in the diamondback terrapin, a turtle that inhabits estuaries, the lachrymal gland secretes a salty fluid around the orbits of the eyes. Photographs of nesting sea turtles frequently show clear paths streaked by tears through the sand that adheres to the turtle's head. Those tears are the secretions of the salt glands.

Despite their different origins and locations, the functional properties of salt glands are quite similar. They secrete fluid containing primarily sodium or potassium cations and chloride or bicarbonate anions in high concentrations (Table 4–5). Sodium is the predominant cation in the salt gland secretions of marine vertebrates, and potassium is present in the secretions of terrestrial lizards, especially herbivorous species such as the desert iguana. Chloride is the major anion, and herbivorous lizards may also excrete bicarbonate ions.

The total osmotic concentration of the salt gland secretion may reach 2000 mOsm/L—more than six

Table 4–5 Salt gland secretions from diapsids and turtles.

Species and Condition	Ion Concentration (mmol/L)		
	Na$^+$	K$^+$	Cl$^-$
Lizards			
Desert iguana (*Dipsosaurus dorsalis*), estimated field conditions	180	1700	1000
Fringe-toed lizard (*Uma scoparia*), estimated field conditions	639	734	465
Snakes			
Sea snake (*Pelamis platurus*), salt-loaded	620	28	635
Homalopsine snake (*Cerberus rhynchops*), salt-loaded	414	56	—
Crocodilian			
Saltwater crocodile (*Crocodylus porosus*), natural diet	663	21	632
Birds			
Blackfooted albatross (*Diomeda nigripes*), salt-loaded	800–900	—	—
Herring gull (*Larus argentatus*), salt-loaded	718	24	—
Turtles			
Loggerhead sea turtle (*Caretta caretta*), seawater	732–878	18–31	810–992
Diamondback terrapin (*Malaclemys terrapin*), seawater	322–908	26–40	—

times the osmotic concentration of urine that can be produced by the kidney. This efficiency of excretion is the explanation of the paradox of active uptake of salt from the urine. As ions are actively reabsorbed, water follows passively, so an animal recovers both water and ions from the urine. The ions can then be excreted via the salt gland at much higher concentrations, with a proportional reduction in the amount of water needed to dispose of the salt. Thus, by investing energy in recovering ions from urine, diapsids and turtles with salt glands can conserve water by excreting ions through the more efficient extrarenal route.

Uricotelic Frogs Terrestrial amphibians have long been considered ureotelic, aquatic forms ammonotelic. However, J. P. Loveridge (1970) discovered that, during the dry season, a period when most frogs retire to a burrow and estivate, a South African frog, *Chiromantis xerampelina*, remains above ground.

Even more surprising, its excretions contain uric acid; biochemically, *Chiromantis* is like a lizard. It has subsequently been shown that the South American frog *Phyllomedusa sauvagei* responds in the same way to aridity (Shoemaker et al. 1972). There is a lesson in these unusual findings: Evolutionary convergence works on all levels of biological organization—anatomical, behavioral, physiological, biochemical—and its direction is determined by interactions with the environment.

Responses to Temperature

Vertebrates occupy habitats from cold polar latitudes to hot deserts. To appreciate this adaptability, we must consider how temperature affects a vertebrate such as a fish that has little capacity to maintain a difference between its body temperature and the temperature of the water around it (a poikilotherm).

Organisms have been described as "bags of chemicals catalyzed by enzymes." This view, although narrow, emphasizes that organisms are subject to the laws of physics and chemistry. Because temperature influences the rates at which chemical reactions proceed, temperature vitally affects the life processes of organisms. Most chemical reactions double or triple in rate for every rise of 10°C. We describe this change in rate by saying that the reaction has a Q_{10} of 2 or 3, respectively. A reaction that does not change rate with temperature has a Q_{10} equal to 1 (Figure 4–7).

The **standard metabolic rate** (SMR) of an organism is the minimum rate of oxygen consumption needed to sustain life. That is, the SMR includes the costs of ventilating the lungs or gills, of pumping blood through the circulatory system, of transporting ions across membranes, and of all the other activities that are necessary to maintain the integrity of an organism. The SMR does not include the costs of activities like locomotion or the cost of growth. The SMR is temperature-sensitive, and that means that the energy cost of living is affected by changes in

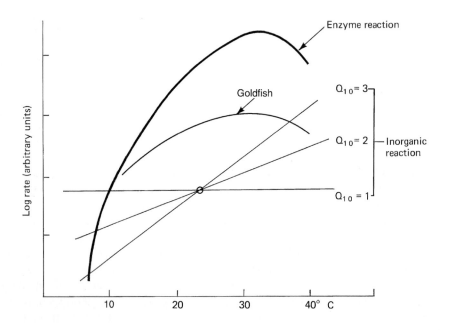

Figure 4–7 Rate-temperature responses in nonliving and living systems. The straight lines show typical Q_{10} responses of inorganic reactions. The curved lines show an enzyme-catalyzed organic reaction and the rate of oxygen consumption by a goldfish.

body temperature. If the SMR of a fish is 2 milliliters of oxygen per minute at 10°C and the Q_{10} response is 2, the fish will consume 4 milliliters of oxygen per minute at 20°C and 8 mL/min at 30°C.

Control of Body Temperature: Ectothermy and Endothermy

Because the rates of many biological processes are affected by temperature, it would probably be advantageous for any animal to be able to control its body temperature. However, the high heat capacity and heat conductivity of water make it difficult for most fishes or aquatic amphibians to maintain a temperature differential between their bodies and their surroundings. Air has both a lower heat capacity and a lower conductivity than water, and the body temperatures of most terrestrial vertebrates are at least partly independent of the air temperature. Some aquatic vertebrates also have body temperatures substantially above the temperature of the water around them. Maintaining these temperature differentials requires thermoregulatory mechanisms, and these are well developed among vertebrates.

The classification of vertebrates as poikilotherms (*poikil* = variable, *therm* = heat) and homeotherms (*homeo* = the same) was widely used through the middle of the twentieth century, but this terminology has become less appropriate as our knowledge of the temperature-regulating capacities of a wide variety of animals has become more sophisticated. Poikilothermy and homeothermy describe the variability of body temperature, and they cannot readily be applied to groups of animals. For example, mammals have been called homeotherms and fishes poikilotherms, but some mammals become torpid at night or in the winter and allow their body temperatures to drop 20°C or more from their normal levels, whereas many fishes live in water that changes temperature less than 2°C in an entire year. That example presents the contradictory situation of a homeotherm that experiences ten times as much variation in body temperature as a poikilotherm.

Because of complications of that sort, it is very hard to use the words "homeotherm" and "poikilotherm" rigorously. Some mammalogists and ornithologists still retain those terms, but most biologists concerned with temperature regulation prefer the terms **ectotherm** and **endotherm**. They are *not* synonymous with poikilotherm and homeotherm because, instead of referring to the variability of body temperature, they refer to the sources of energy used in thermoregulation. Ectotherms (*ecto* = outside) gain their heat largely from external sources—by basking in the sun, for example, or by resting on a warm rock. Endotherms (*endo* = inside) largely depend on metabolic production of heat to raise their body temperatures. The source of the heat used to maintain body temperatures is the major difference between ectotherms and endotherms. Terrestrial ectotherms like squamates and turtles and endotherms like birds and mammals all have activity temperatures in the range 30 to 40°C (Table 4–6).

Endothermy and ectothermy are not mutually exclusive mechanisms of temperature regulation, and many animals use them in combination. In general, birds and mammals are endothermal, but some species make extensive use of external sources of heat. For example, roadrunners are predatory birds that live in the deserts of the southwestern United States and adjacent Mexico. On cold nights roadrunners become hypothermic, allowing their body temperatures to fall from the normal level of 38 to 39°C down to 33 to 35°C. In the morning they bask in the sun, raising the feathers on the back to expose an area of black skin in the interscapular region. Calculations indicate that a roadrunner can save 132 joules per hour by using solar energy instead of metabolism to raise its body temperature. Snakes are normally ectothermal, but the females of several species of python coil around their eggs and produce heat by rhythmic contraction of their trunk muscles. The rate of contraction increases as air temperature falls, and a female Indian python is able to maintain her eggs close to 30°C at air temperatures as low as 23°C. This heat production entails a substantial increase in the python's metabolic rate—at 23°C, a female python uses about 20 times as much energy when she is brooding as she does normally. Thus, generalizations about the body temperatures and thermoregulatory capacities of vertebrates must be made cautiously, and the actual mechanisms used to regulate body temperature must be studied carefully.

Ectothermal Thermoregulation

From the time of Aristotle onward, lizards, snakes, and amphibians have paradoxically been called cold-blooded while they were thought to be able to tolerate extremely high temperatures. Salamanders

Table 4–6 **Representative body temperatures of vertebrates. Body temperatures are those that the animals maintain when they are able to thermoregulate normally.**

Group	Body Temperature
Primarily Ectothermal Groups	
Fishes	
Most fishes	Little different from water temperature
Warm-bodied fishes (tunas, some sharks)	About 30°C in water of 20°C
Amphibians	
Aquatic	Little different from water temperature
Terrestrial	Usually slightly below air temperature because of evaporative cooling; some amphibians raise their body temperatures 5–10°C above air temperature by basking
Amniotic ectotherms	
Turtles and crocodilians	From close to water temperature to about 35°C while thermoregulating
Squamates	From 20–25°C for forest-dwelling tropical species to 35–42°C for thermoregulating desert lizards
Primarily Endothermal Groups	
Birds	40–41°C
Mammals	
Monotremes*	28–30°C
Marsupials	33–36°C
Placentals	36–38°C

*Sloths, which are placentals, have body temperatures in this range.

frequently seek shelter in logs, and when a log is put on a fire, the salamanders it contains may come rushing out. Observations of this phenomenon gave rise to the belief that salamanders live in fire. In the first part of the twentieth century biologists were using similar lines of reasoning. In the desert lizards often sit on rocks. If you approach a lizard it runs away but the rock stays put, and touching the rock shows that it is painfully hot. Clearly, the reasoning went, the lizard must have been equally hot. Biologists marveled at the heat tolerance of lizards, and statements to this effect are found in authoritative textbooks of the period.

A study of thermoregulation of lizards by Raymond Cowles and Charles Bogert (1944) demonstrated the falsity of earlier observations and conclusions. They showed that reptiles can regulate their body temperatures with considerable precision, and that the level at which the temperature is regulated is characteristic of a species. The implications of this discovery in terms of the biology of amphibians and reptiles are still being explored.

Energy Exchange Between an Organism and Its Environment A brief discussion of the pathways by which thermal energy is exchanged between a living organism and its environment is necessary to understand the thermoregulatory mechanisms employed by terrestrial animals. An organism can gain or lose energy by several pathways, and by adjusting the relative flow through various pathways an animal can warm, cool, or maintain a stable body temperature (Tracy 1982).

Figure 4–8 illustrates pathways of thermal energy exchange. Solar energy can reach an animal in several ways. Direct solar **radiation** impinges on an animal when it is standing in a sunny spot. In addition, solar energy is reflected from clouds and dust particles in the atmosphere and from other objects in the environment and reaches the animal by these circuitous routes. The wavelength distribution of the energy in all these routes is the same—the portion of the solar spectrum that penetrates the Earth's atmosphere. About half this energy is contained in the visible wavelengths of the solar spectrum (400 to 700

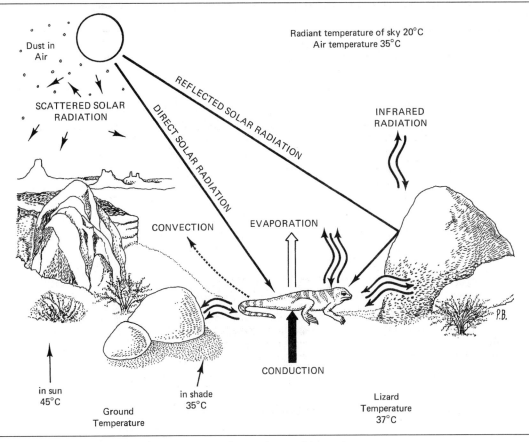

Dust in Air

SCATTERED SOLAR RADIATION

REFLECTED SOLAR RADIATION

DIRECT SOLAR RADIATION

Radiant temperature of sky 20°C
Air temperature 35°C

INFRARED RADIATION

CONVECTION

EVAPORATION

CONDUCTION

in sun
45°C

in shade
35°C

Ground Temperature

Lizard Temperature
37°C

P.B.

Figure 4–8 Energy is exchanged between a terrestrial organism and its environment by many pathways. These are illustrated in simplified form by a lizard resting on the floor of a desert arroyo. Small adjustments of posture or position can change the magnitude of the various routes of energy exchange and give a lizard considerable control over its body temperature.

nanometers) and most of the rest is in the infrared region of the spectrum (>700 nanometers).

Energy exchange in the **infrared** is an important part of the radiative heat balance. All objects, animate or inanimate, radiate energy at wavelengths determined by their absolute temperatures. Objects in the temperature range of animals and the Earth's surface (roughly –20 to +50°C) radiate in the infrared portion of the spectrum. Animals continuously radiate heat to the environment and receive infrared radiation from the environment. Thus, infrared radiation can lead to either heat gain or loss, depending on the relative temperature of the animal's body surface and the environmental surfaces and on the radiation characteristics of the surfaces themselves. In the example in Figure 4–8 the lizard is cooler than the sunlit rock in front of it and receives

more energy from the rock than it loses to the rock. However, the lizard is warmer than the shaded side of the rock behind it and has a net loss of energy in that exchange. The radiative temperature of the clear sky is about 20°C, so the lizard loses energy by radiation to the sky.

Heat is exchanged between objects in the environment and the air via **convection**. If an animal's surface temperature is higher than air temperature, convection leads to heat loss; if the air is warmer than the animal, convection is a route of heat gain. In still air, convective heat exchange is accomplished by convective currents formed by local heating; but in moving air, forced convection replaces natural convection and the rate of heat exchange is greatly increased. In the example shown, the lizard is warmer than the air and loses heat by convection.

Conductive heat exchange resembles convection in that its direction depends on the relative temperatures of the animal and environment. Conductive loss occurs between the body and the substrate where they are in contact. It can be modified by changing the surface area of the animal in contact with the substrate and by changing the rate of heat conduction in the parts of the animal's body that are in contact with the substrate. In this example the lizard gains heat by conduction from the warm ground.

Evaporation of water occurs from the body surface and from the pulmonary system. Each gram of water evaporated represents a loss of about 2450 joules (the exact value changes slightly with temperature). Evaporation almost always occurs from the animal to the environment, and thus represents a loss of heat. The inverse situation, condensation of water vapor on an animal, would produce heat gain, but it rarely occurs under natural conditions.

Metabolic heat production is the final pathway by which an animal can gain heat. Among ectotherms metabolic heat gain is usually trivial in relation to the heat derived directly or indirectly from solar energy. There are a few exceptions to that generalization, and some of them are discussed later. Endotherms, by definition, derive most of their heat energy from metabolism, but their routes of energy exchange with the environment are the same as those of ectotherms and must be balanced to maintain a stable body temperature.

Behavioral Control of Body Temperatures by Ectotherms
The behavioral mechanisms involved in ectothermal temperature regulation are quite straightforward and are employed by insects, birds, and mammals (including humans) as well as by ectothermal vertebrates (Avery 1979). Lizards, especially desert species, are particularly good at behavioral thermoregulation. Movement back and forth between sun and shade is the most obvious thermoregulatory mechanism they use. Early in the morning or on a cool day, lizards bask in the sun, whereas in the middle of a hot day they have retreated to shade and make only brief excursions into the sun. Sheltered or exposed microhabitats may be sought out. In the morning when a lizard is attempting to raise its body temperature, it is likely to be in a spot protected from the wind. Later in the day when it is getting too hot it may climb into a bush or onto a rock outcrop where it is exposed to the breeze and its convective heat loss is increased.

The amount of solar radiation absorbed by an animal can be altered by changing the orientation of its body with respect to the sun, the body contour, and the skin color. All of these mechanisms are used by lizards. An animal oriented perpendicular to the sun's rays intercepts the maximum amount of solar radiation, and one oriented parallel to the sun's rays intercepts minimum radiation. Lizards adjust their orientation to control heat gained by direct solar radiation. Many lizards are capable of spreading or folding the ribs to change the shape of the trunk. When the body is oriented perpendicular to the sun's rays and the ribs are spread, the surface area exposed to the sun is maximized and heat gain is increased. Compressing the ribs decreases the surface exposed to the sun and can be combined with orientation parallel to the rays to minimize heat gain. Horned lizards provide a good example of this type of control (Heath 1965). If the surface area that a horned lizard exposes to the sun directly overhead when it sits flat on the ground with its ribs held in a resting position is considered to be 100 percent, the maximum surface area the lizard can expose by orientation and change in body contour is 173 percent and the minimum is 28 percent. That is, the lizard can change its radiant heat gain more than sixfold solely by changing its position and body shape.

Color change can further increase a lizard's control of radiative exchange. (See the color insert.) Lizards darken by dispersing melanin in melanophore cells in the skin and lighten by drawing the melanin into the base of the melanophores. The lightness of a lizard affects the amount of solar radiation it absorbs in the visible part of the spectrum, and changes in heating rate (in the darkest color phase compared to the lightest) are from 10 to 75 percent.

Lizards can achieve a remarkable independence of air temperature as a result of their thermoregulatory capacities (Avery 1982). Lizards occur above the timberline in many mountain ranges, and during their periods of activity on sunny days are capable of maintaining body temperatures 30°C or more above air temperature. While air temperatures are near freezing, these lizards scamper about with body temperatures as high as those species that inhabit lowland deserts.

The repertoire of thermoregulatory mechanisms seen in lizards is greater than that of many other ectothermal vertebrates. Turtles, for example, cannot change their body contour or color, and their behavioral thermoregulation is limited to movements between sun and shade and in and out of water. Crocodilians are very like turtles, although young individuals may be able to make minor changes in body contour and color. Most snakes cannot change color, but some rattlesnakes lighten and darken as they warm and cool.

During the parts of a day when they are active, desert lizards maintain their body temperatures in a zone called the **activity temperature range**. This is the region of temperature within which a lizard carries out its full repertoire of activities—feeding, courtship, territorial defense, and so on. For many species of desert lizards the activity temperature range is as narrow as 4°C, but for other ectotherms it may be as broad as 10°C. Different species of lizards have different activity temperature ranges. The thermoregulatory activities of a lizard are directed toward keeping it within its activity temperature range, but the precise temperature it maintains within this range depends on a variety of internal and external conditions. For example, many ectotherms maintain higher body temperatures when they are digesting food than when they are fasting. Female lizards when they are carrying young may maintain different body temperatures than at other times, and ectotherms with experimentally induced bacterial infections show a fever that is achieved by maintaining a higher-than-normal body temperature by behavioral means (Kluger 1979).

Not all lizards regulate their body temperatures closely. Some lizards that live in the understory vegetation of tropical forests where sunlight does not penetrate do not raise their body temperatures above air temperatures. Differences in the intensity and availability of solar radiation in different seasons, different habitats, or even at different times of day can alter the balance of costs and benefits of thermoregulatory behavior (Huey 1982). These ecological aspects of thermoregulation are discussed in Chapter 15.

Physiological Control of the Rate of Change of Body Temperature by Ectotherms A new dimension was added to studies of ectothermal thermoregulation in the 1960s by the discovery that ectotherms can use physiological mechanisms to adjust their rate of temperature change (Bartholomew 1982). The original observations, made by George Bartholomew and his associates, showed that several different kinds of large lizards were able to heat faster than they cooled when exposed to the same differential between body and ambient temperatures. Subsequent studies by other investigators extended these observations to turtles and snakes. From the animal's viewpoint, heating rapidly and cooling slowly prolongs the time it can spend in the normal activity range.

Fred White and his colleagues have demonstrated that the basis of this control of heating and cooling rates lies in changes in peripheral circulation. Heating the skin of a lizard causes a localized vasodilation of dermal blood vessels in the warm area. Dilation of the blood vessels, in turn, increases the blood flow through them, and the blood is warmed in the skin and carries the heat into the core of the body. Thus, in the morning, when a cold lizard orients its body perpendicular to the sun's rays and the sun warms its back, local vasodilation in that region speeds transfer of the heat to the rest of the body.

The same mechanism can be used to avoid overheating. The Galapagos marine iguana is a good example (White 1973). Marine iguanas live on the bare lava flows on the coasts of the islands. In midday, beneath the equatorial sun, the black lava becomes extremely hot—uncomfortably, if not lethally hot for a lizard. Retreat to shade of the scanty vegetation or into rock cracks would be one way the iguanas could avoid overheating, but the males are territorial and those behaviors would mean abandoning their territories and probably having to fight for them again later in the day. Instead, the marine iguana stays where it is and uses physiological control of circulation and the cool breeze blowing off the ocean to form a heat shunt that absorbs solar energy on the dorsal surface and carries it through the body and dumps it out the ventral surface.

The process is as follows: In the morning the lizard is chilled from the preceding night and basks to bring its body temperature to the normal activity range. When its temperature reaches this level the lizard uses postural adjustments to slow the increase in body temperature, finally facing directly into the sun to minimize its heat load. In this posture the forepart of the body is held off the ground (Figure 4–9). The ventral surface is exposed to the cool wind blowing off the ocean, and a patch of lava under the animal is shaded by its body. This lava is soon cooled by the wind. Local vasodilation is produced by warming the blood vessels: It does not matter whether the heat comes from the outside (from the sun) or from inside (from warm blood). Warm blood circulating from the core of the body to the ventral

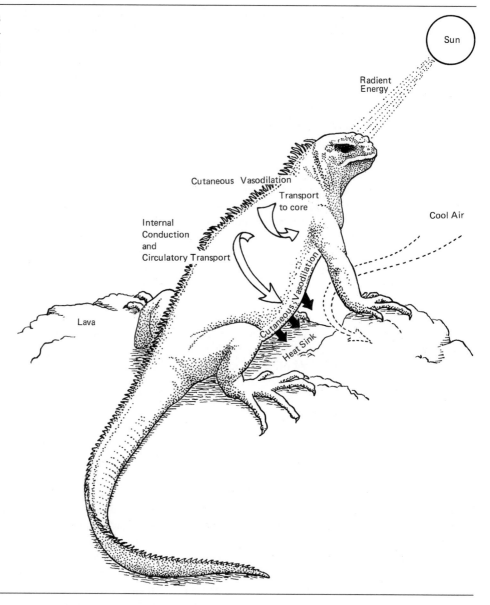

Figure 4–9 The Galapagos marine iguana uses a combination of behavioral and physiological thermoregulatory mechanisms to shunt heat absorbed by its dorsal surface out its ventral surface. (Modified from F. N. White, 1973, *Comparative Biochemistry and Physiology* 45A:503–513.)

skin warms it and produces vasodilation, increasing the flow to the ventral surface. The lizard's ventral skin is cooler than the rest of its body—it is shaded and cooled by the wind, and in addition it loses heat by radiation to the cool lava in the shade created by the lizard's body. In this way the same mechanism that earlier in the day allowed the lizard to warm rapidly is converted to a regulated heat shunt that rapidly transports solar energy from the dorsal to the ventral surface and keeps the lizard from overheating. In combination with postural adjustments and other behavioral mechanisms, such as the choice of a site where the breeze is strong, these physiological adjustments allow the lizard to remain on station in its territory all day.

The behavioral and physiological thermoregulatory mechanisms of ectotherms are intimately intertwined. Although we have tried to simplify our presentation by discussing them separately, it is essential to realize that neither behavioral, nor physiological, nor morphological thermoregulatory mechanisms function by themselves. They are used in combination, and they have evolved in combination. The thermoregulation of a lizard (and, as we will see, of a bird or mammal) involves all these mechanisms simultaneously.

Endothermal Thermoregulation

Birds and mammals are endotherms that regulate their high body temperatures by mechanisms that precisely balance metabolic heat production and heat loss to the environment. An endotherm can change the intensity of its heat production by varying metabolic rate over a wide range. In this way, an endotherm maintains a constant high body temperature by adjusting heat production to equal heat loss from its body under different environmental conditions.

Endotherms produce metabolic heat in several ways. Besides the obligatory heat production derived from the basal or resting metabolic rate, there is the heat increment of feeding, often called the **specific dynamic action** or **effect** of the food ration. This added heat production after ingestion of food apparently results from the energy requirements for assimilation and protein synthesis, and varies in amount depending on the type of foodstuff being processed. It is highest for a meat diet and lowest for a carbohydrate diet.

Activity of skeletal muscle produces large amounts of heat, especially during locomotion, which can result in a heat production exceeding the basal metabolic rate by 10- to 15-fold. This muscular heat can be advantageous for balancing heat loss in a cold environment, or it can be a problem requiring special mechanisms of dissipation in hot environments that approach or exceed the body temperature of the animal. **Shivering**, the generation of heat by muscle fiber contractions in an asynchronous pattern that does not result in gross movement of the whole muscles, is an important mechanism of heat production. Birds and mammals also possess mechanisms for nonshivering thermogenesis.

Because endotherms usually live under conditions in which ambient temperatures are lower than the regulated body temperatures of the animals themselves, heat loss to the environment is a more usual circumstance than heat gain, although heat gain can be a major problem in deserts. Balancing of heat loss is therefore one of the most important regulatory functions of an endotherm, and birds and mammals employ their plumage or hair in a very effective way as insulation against heat loss.

Any material that traps air is an insulator against conductive heat transfer. Hair and feathers provide insulation by trapping air, and the depth of the layer of trapped air can be adjusted by raising and lowering the hair or feathers. We humans have goose bumps on our arms and legs when we are cold because our few remaining hairs rise to a vertical position in an ancestral mammalian attempt to increase our insulation.

These physiological responses to temperature are controlled from neurons located in the hypothalamus of the brain. In some mammals, as in hibernators, the hypothalamic thermostat can be reset for a lower control temperature. In ectotherms the hypothalamic thermostat controls behaviors that place the animal in favorable circumstances (e.g., moving to a preferred ambient temperature away from excess heat, or orienting to maximize heat loss or gain).

Mechanisms of Endothermal Thermoregulation Body temperature and metabolic rate must be considered simultaneously to understand how endotherms maintain their body temperatures at a stable level in the face of environmental temperatures that may range from -70 to $+40°C$. Most birds and mammals conform to the generalized diagram in Figure 4–10.

Each species of endotherm has a definable range of ambient temperatures (T_1 to T_4) over which the body temperature can be kept stable by using physiological and postural adjustments of heat loss and heat production. This ambient temperature range is called the **zone of tolerance**. Above this range the animal's ability to dissipate heat is inadequate, and both the body temperature and metabolic rate increase as ambient temperature increases until the animal dies. At ambient temperatures below the zone of tolerance the animal's ability to generate heat to balance heat loss is exceeded, body temperature falls, metabolic rate declines, and cold death results. Large animals usually have lower values for T_1 and T_2 than small animals because heat is lost from the body surface, and large animals have smaller surface-to-mass ratios than those of small animals. Similarly, well-insulated species have lower values for T_1 and T_2 than those of poorly insulated ones, but thinly insulated species usually have higher values for T_3 than those of heavily insulated ones.

The **thermoneutral zone** (T_2 to T_3) is the range of ambient temperatures within which the metabolic rate of an endotherm is at its basal level and ther-

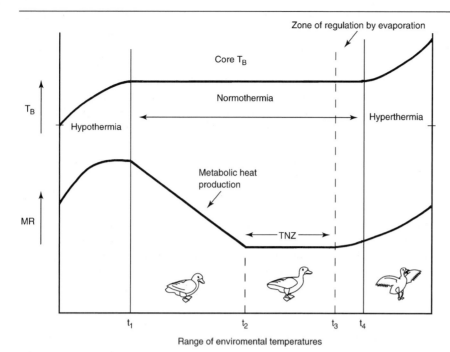

Figure 4–10 Generalized pattern of changes in body temperature and metabolic heat production of an endothermic homeotherm in relation to environmental temperature. Core T_B is the normal body temperature and varies somewhat for different mammals and birds. t_1, incipient lower lethal temperature; t_2, lower critical temperature; t_3, upper critical temperature; t_4, incipient upper lethal temperature; TNZ, thermoneutral zone.

moregulation is accomplished by changing the rate of heat loss. The thermoneutral zone is also called the zone of physical thermoregulation because an animal uses processes such as fluffing or sleeking its hair or feathers, postural changes such as huddling or stretching out, and changes in blood flow (vasoconstriction or vasodilation) to exposed parts of the body (feet, legs, face) to adjust its loss of heat.

The larger an animal is and the thicker its insulation, the lower will be the temperature it can withstand before physical processes become inadequate to balance its heat loss. The **lower critical temperature** (T_2) is the point at which an animal must increase metabolic heat production to maintain a stable body temperature. In the **zone of chemical thermogenesis**, the metabolic rate increases as the ambient temperature falls. The quality of the insulation determines how much additional metabolic heat production is required to offset a change in ambient temperature. That is, well-insulated animals have relatively shallow slopes for the graph of increasing metabolism below the lower critical temperature, and poorly insulated animals have steeper slopes (see Figure 22–3). Many birds and mammals from the Arctic and Antarctic are so well insulated that they can withstand the lowest temperatures on earth (about –70°C) by increasing their basal metabolism only threefold. Less well-insulated animals may be exposed to temperatures below their **lower lethal temperature** (T_1). At that point metabolic heat production has reached its maximum rate and is still insufficient to balance heat loss to the environment. Under those conditions the body temperature falls, and the Q_{10} effect of temperature on chemical reactions causes the metabolic rate to fall as well. A positive-feedback condition is initiated in which falling body temperature reduces heat production, causing a further reduction in body temperature. Death from hypothermia (low body temperature) follows.

Endotherms are remarkably good at maintaining stable body temperatures in cool environments, but they have difficulty at high ambient temperatures. The **upper critical temperature** (T_3) represents the point at which nonevaporative heat loss has been maximized by using all of the physical processes that are available to an animal—exposing the poorly insulated areas of the body and maximizing cutaneous blood flow. If these mechanisms are insufficient to balance heat gain, the only option vertebrates have is to use evaporation of water by panting, sweating, or gular flutter. The temperature range from T_3 to T_4 is the **zone of evaporative cooling**. Many mammals sweat, a process in which water is released from sweat glands on the surface of the body. Evaporation of the sweat cools the body surface. Other animals pant, breathing rapidly and shallowly so that evaporation of water from the respiratory system provides a cooling effect. Many birds use a rapid fluttering movement of the gular region to evaporate water for thermoregulation. Panting and gular flutter require muscular activity, and some of the evaporative cooling they achieve is used to offset the increased metabolic heat production they require.

At the **upper lethal temperature** (T_4) evaporative cooling cannot balance the heat flow from a hot environment, and body temperature rises. The Q_{10} effect of temperature produces an increase in the rate of metabolism, and metabolic heat production raises the body temperature, increasing the metabolic rate still further. This process can lead to death from hyperthermia (high body temperature).

The difficulty that endotherms experience in regulating body temperature in high environmental temperatures may be one of the reasons that the body temperatures of most endotherms are in the range 35 to 40°C. Most habitats seldom have air temperatures that exceed 35°C. Even the tropics have average yearly temperatures below 30°C. Thus the high body temperatures maintained by mammals ensure that in most situations the heat gradient is from animal to environment. Still higher body temperatures, around 50°C, for example, could ensure that mammals were always warmer than their environment. There are upper limits to the body temperatures that are feasible, however. Many proteins denature near 50°C. During heat stress, some birds and mammals may tolerate body temperatures of 45 to 46°C for a few hours, but only some bacteria, algae, and a few invertebrates exist at higher temperatures. This is another case in which the direction of vertebrate evolution has been established by a balance between biotic needs and physiochemical realities.

Advantages of a High Body Temperature Mammals and birds have high resting metabolic rates, at least six times the SMR of ectotherms. Although it is energetically expensive, there are benefits to regulating body temperature at a high rather than a low level that are independent of the mechanisms of thermoregulation discussed in the preceding section.

We have already noted that the biochemistry of vertebrates involves thousands of interacting enzyme-catalyzed reactions, most of which are temperature sensitive. A constant internal temperature is required to obtain maximum chemical coordination among these reactions. In addition, the higher the body temperature, the more rapid the response of cells to organismal needs. Although it is capable of acting at very cold temperatures (as in arctic fishes), the CNS functions more rapidly at high temperatures. As one example, neurotransmitters such as acetylcholine and norepinephrine act by diffusing from their site of release across the synaptic junction to the postsynaptic receptor surface. Because diffusion is a physical process, its rate increases as temperature increases. A high body temperature enhances the rate of information processing, a competitive advantage often neglected when considering the success of mammals and birds. Very rapid responses can be vital in catching prey and avoiding predators. Some ectotherms enjoy the same neurological benefits when they are warm, but very rapid responses on cool nights can occur only in endothermal homeotherms. In addition, muscle viscosity declines at high temperatures. This reduction in internal friction may result in more rapid, forceful contraction and faster response times.

Thus, endothermal homeothermy has some obvious advantages over ectothermy. (Ectothermy has its own advantages, which are discussed in Chapter 16.) Ectothermy is the ancestral condition for vertebrates. How did endothermy evolve?

The Evolution of Endothermy Endothermy has evolved from an ancestral ectothermal condition at least twice in the history of vertebrates—in birds and in mammals. Some evidence suggests that pterosaurs (flying archosaurs of the Mesozoic) might also have been endothermal, and if this is true, it would represent a third independent origin of endothermy. How would that transition occur?

The difference in the sources of heat used by ectotherms and endotherms creates a paradox when one tries to understand how an evolutionary lineage shifts from ectothermy to endothermy. Ectotherms rely on heat from outside their bodies, and the major specializations of ectothermal thermoregulation facilitate exchange of heat with the environment. The body surfaces of ectotherms have little insulation, probably because insulation would interfere with the gain and loss of heat. Metabolic rates of ectotherms are low, and ectotherms normally do not obtain sufficient heat from metabolism to warm the body significantly. (The warm-bodied tunas and sharks described in the next section and the sea turtles discussed in Chapter 13 are exceptions to this generalization because of their large body sizes, high levels of activity, and specializations of the circulatory system.) Thus, the thermoregulatory mechanisms of ectotherms are based on low metabolic rates, little insulation, and rapid exchange of heat with the environment.

Endothermal thermoregulation has exactly the opposite characteristics. The high metabolic rates of endotherms produce large quantities of heat, and that heat is retained in their bodies by the insulation provided by hair or feathers. Endothermal thermoregulation consists largely of adjusting the layer of insulation so that heat loss balances the heat produced by high rates of metabolism.

An evolutionary shift from ectothermy to endothermy appears to encounter a catch-22 situation: A high metabolic rate is of no use unless an animal has insulation to retain metabolically produced heat, because without insulation the heat is rapidly lost to the environment. However, insulation serves no purpose for an animal without a high rate of metabolism because there is little internally produced heat for the insulation to conserve. Indeed, insulation can be a handicap for an ectotherm, because it prevents it from warming up. Raymond Cowles demonstrated that fact in the 1930s when he made small fur coats for lizards and measured their rates of warming and cooling. The potential benefit of a fur coat for a lizard is, of course, its effect of keeping the lizard warm as the environment cools off. However, the lizards in Cowles' experiments never achieved that benefit of insulation, because when they were wearing fur coats they were unable to get warm in the first place.

Those well-dressed lizards illustrate the paradox of the evolution of endothermy: Insulation is ineffective without a high metabolic rate, and the heat produced by a high metabolic rate is wasted without insulation. By this line of reasoning, neither one of the two essential features of endothermy would be selectively advantageous for an ectotherm without the previous development of the other. So how did endothermy evolve?

Probably endothermy evolved as a by-product of selection for one or more other activities that involved some of the same characteristics that are needed by endotherms. Ideas about the evolution of endothermy are unavoidably speculative, but they can be tested to some extent by asking if the intermediate stages that are postulated could have been functional in thermoregulation. For example, feathers, which provide the insulating layer that permits birds to be endotherms, are derived from the scales that covered the bodies of Mesozoic archosaurs. One hypothesis for the origin of feathers suggests that an intermediate stage in their evolution consisted of elongate scales. An animal with scales of this type could orient its body while it was warming so that sunlight penetrated between the scales. When it was warm enough, a change of orientation would convert the scales to a series of parasols, blocking heat uptake (Regal 1975). This sort of scale probably would be suitable for the sort of ectothermal thermoregulation exemplified by the Galapagos marine iguana, and it might also have the basic features needed to provide insulation if the scales trapped a layer of air when they were lowered to press against each other. An animal with a body covering that had achieved some insulating capacity would be able to retain metabolically produced heat and might derive some benefit from a high metabolic rate (Regal 1985).

A different sequence of events, called the aerobic capacity model, has been suggested to account for the origin of mammalian endothermy. The synapsid lineage from which mammals are derived shows skeletal changes that seem to indicate a progressive increase in locomotor capacity (Chapter 19). If the animals were indeed becoming more active, it is plausible that the morphological changes were accompanied by an increasing aerobic metabolic capacity needed to sustain that activity, and that heat production by muscles during activity raised the body temperature (Bennett and Ruben 1979). In that situation, hair that provided insulation could help to maintain the elevated body temperature for some period after activity ceased.

Thus, in the evolution of mammalian endothermy metabolic heat production (as a by-product of locomotor activity) might have preceded insulation, whereas in birds insulation (a by-product of ectothermal thermoregulation) might have preceded high rates of metabolism. Other plausible evolutionary scenarios can be proposed to explain how endothermy could evolve in birds or mammals, and the ones suggested here may be incorrect. However, they do have the merit of illustrating an important point: The selective pressures that were responsible for the origin of a trait are not necessarily the ones that are responsible for its present value. In the case of the evolution of endotherms from ectotherms, the conflicting requirements of the two modes of thermoregulation are so different that some factors other than thermoregulation were almost certainly involved in the initial stages.

Regional Heterothermy: Warm Fishes

Endothermy and regulation of body temperature are not all-or-nothing phenomena for vertebrates. Regional heterothermy is a general term used to refer to different temperatures in different parts of an animal's body. Dramatic examples of regional heterothermy are found in several fishes that maintain some parts of their bodies at temperatures 15°C warmer than the water they are swimming in. That's a remarkable accomplishment for a fish, because each time the blood passes through the gills it comes into temperature equilibrium with the water. Thus, to raise its body temperature by using endothermal heat production, a fish must prevent the loss of heat to the water via the gills.

The mechanism used, as you will guess, is a counter-current system of blood flow in retia mirabile (Chapter 3). As cold arterial blood from the gills enters the warm part of the body, it flows through a rete and is warmed by heat from the warm venous blood that is leaving the tissue. This arrangement is found in some sharks, especially species in the family Lamnidae (including the mako, great white shark, and porbeagle), which have retia mirabile in the trunk. These retia retain the heat produced by the activity of the swimming muscles, with the result that those muscles are kept 5 to 10°C warmer than water temperature.

Scombroid fishes, a group of teleosts that includes the mackerels, tunas, and billfishes (swordfish, sailfish, spearfish, and marlin), have also evolved endothermal heat production. Tuna have an arrangement of retia that retains the heat produced by myoglobin-rich swimming muscles located close to the vertebral column (Figure 4–11). The temperature of these muscles is held near 30°C at water temperatures from 7 to 23°C. Additional heat exchangers are found in the brains and eyes of tunas and sharks, and these organs are warmer than water temperature, but somewhat cooler than the swimming muscles (Carey 1982).

The billfishes have a somewhat different arrangement in which only the brain and eyes are warmed, and the source of heat is a muscle that has changed its function from contraction to heat production (Block 1991). The superior rectus eye muscle of these billfishes has been extensively modified. Mitochondria occupy more than 60% of the volume of the cells, and changes in cell structure and biochemistry result in the release of heat by the calcium-cycling mechanism that is usually associated with contraction of muscles. A related scombroid, the butterfly mackerel, has a thermogenic organ with the same structural and biochemical characteristics found in billfishes, but in the mackerel it is the lateral rectus eye muscle that has been modified.

An analysis of the phylogenetic relationships of scombroid fishes by Barbara Block and her colleagues (Block et al. 1993) suggests that endothermal heat production has arisen independently three times in the lineage—once in the common ancestor of the living billfishes (by modification of the superior rectus eye muscle), once in the butterfly mackerel lineage (modification of the lateral rectus eye muscle), and a third time in the common ancestor of tunas and bonitos (involving the development of counter-current heat exchangers in muscle, viscera, and brain and development of red muscle along the horizontal septum of the body).

The ability of these fishes to keep parts of the body warm may allow them to venture into cold water that would otherwise interfere with body functions. Block has pointed out that modification of the eye muscles and the capacity for heat production among scombroids is related to the temperature of the water in which they swim and capture prey. The oxidative capacity of the heater cells of the butterfly mackerel, which is the species that occurs in the coldest water, is the highest of all vertebrates. Swordfish, which dive to great depths and spend several hours in water temperatures of 10°C or less, have better developed heater organs than do marlins, sailfish, and spearfish, which spend less time in cold water.

Energy Utilization: Patterns Among Vertebrates

Vertebrates usually require more energy than most metazoan animals of similar size, and those vertebrates that physiologically regulate their body temperature at high levels require much more energy. As

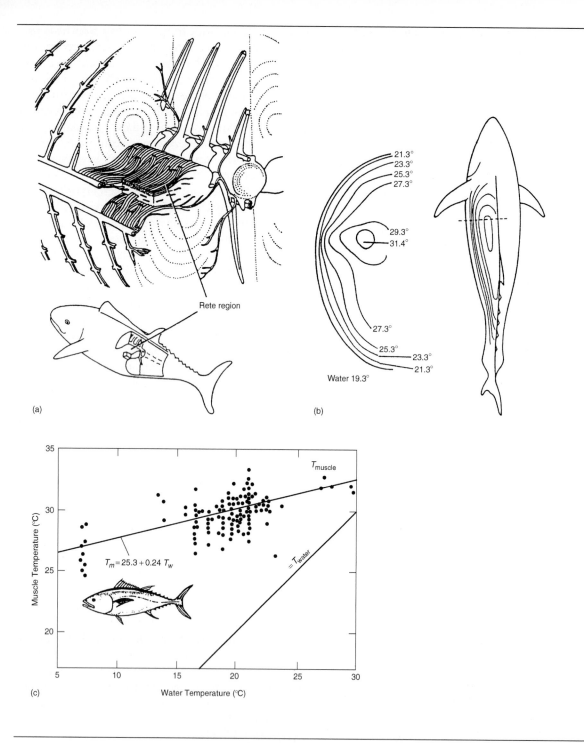

Figure 4–11 Details of body temperature regulation by the bluefin tuna. (a) The red muscle and *retia* are located adjacent to the vertebral column. (b) Cross-sectional views showing the temperature gradient between the core (at 31.4°C) and water temperature (19.3°C). (c) Core muscle temperatures of bluefins compared to water temperature. (Modified from F. G. Carey and J. M. Teal, 1966, *Proceedings of the National Academy of Sciences U. S. A.* 56:1464–1469.)

a result, vertebrates expend much of their active time in the search for food. As a consequence, the impact of vertebrates on the environment (that is, their demand for resources) is great in proportion to their abundance.

Aerobic and Anaerobic Metabolism

Most vertebrates require oxygen to live and oxidize their basic foods (carbohydrates, fats, and proteins) to carbon dioxide and water. To accomplish this task, oxygen must be supplied to the mitochondria of each cell, where the final stages of oxidation and ATP formation take place. All vertebrates accomplish the initial steps of the cellular breakdown of foods anaerobically (Figure 4–12). Thus, glucose is ultimately converted to pyruvate, and the energy released by this glycolytic process is converted in part into ATP. Generally, the pyruvate is decarboxylated and oxidized to carbon dioxide and water to yield additional ATP if oxygen is available. In active skeletal (striated) muscle, the pyruvate may accumulate faster than oxygen can be supplied for its complete oxidation. As a result, the tissue becomes anoxic (devoid of oxygen) and the accumulation of pyruvate inhibits further breakdown of glucose and thus slows ATP generation. By converting pyruvate to lactic acid, however, glycolysis can proceed until depletion of cellular energy stores causes fatigue. This process is called anaerobic metabolism because oxygen is not used as an electron acceptor.

A vertebrate can swim, fly, or run at maximum rate only for a limited time until it drops from fatigue. This is an important ecological point: There is a time limit as well as a mechanical limit placed on vertebrate mobility. Most behaviors take place at levels of energy expenditure only two to four times resting metabolic levels. For example, the lengthy migrations that many vertebrates undertake require much energy, but they are performed at speeds that do not exceed the limits of aerobic metabolism. In contrast, the brief pursuits of prey by predator, territorial defense, and nuptial gyrations between mates may drive oxygen consumption to ten times the resting rate and still require more ATP than can be synthesized by those aerobic metabolic pathways alone.

All vertebrates have the capacity for both aerobic and anaerobic metabolism, and the balance between the pathways differs in various species and situations. In general, endotherms have greater capacity for aerobic metabolism than do ectotherms, but even endotherms use anaerobic pathways to supplement aerobic production of ATP during intense activity. High levels of aerobic metabolism require high rates of transport of oxygen to active tissues, and animals with high aerobic metabolic capacities have large hearts, high rates of blood flow, high hematocrits, and high concentrations of hemoglobin in the blood (Table 4–7).

Physiological characteristics of muscles specialized for aerobic and anaerobic metabolism also differ. Red muscle takes its color from the myoglobin it contains. Myoglobin is a protein that binds oxygen and speeds its diffusion from blood to mitochondria, and muscles with high concentrations of myoglobin are well vascularized and have high activities of the enzymes associated with aerobic metabolic pathways. White muscle lacks myoglobin, is poorly vascularized, and has high activities of enzymes associated with anaerobic metabolic pathways. The distribution of red and white muscle in many vertebrates tells much about the sorts of activities those muscles support. For example, domestic chickens in a barnyard walk around all day, but they fly only if they are startled, and then for only a few seconds. The leg muscles of chickens are composed of red muscle (the dark meat) with high aerobic capacity. The flight (breast) muscles are white and are largely anaerobic. In tunas the swimming muscles are red (dark meat tuna), and other trunk muscles are white (light meat tuna).

The metabolic capacity of an animal is the sum of its aerobic and anaerobic energy production. A lizard (the desert iguana) and a mammal (the kangaroo rat) are both able to produce about 0.015 millimole of ATP per gram during 30 seconds of activity, but about 70 percent of that ATP comes from aerobic pathways in the mammal compared to 76 percent from anaerobic pathways for the lizard (Table 4–8). The ecological significance of that difference lies in the ability of the two animals to sustain activity. The oxygen and metabolic substrates used for anaerobic metabolism are carried to the muscles by the circulatory system. For anaerobic metabolism the substrate is glycogen, which is stored in the cell, and when that glycogen has been used up, anaerobic metabolism must stop. Thus, aerobic metabolism can continue nearly indefinitely,

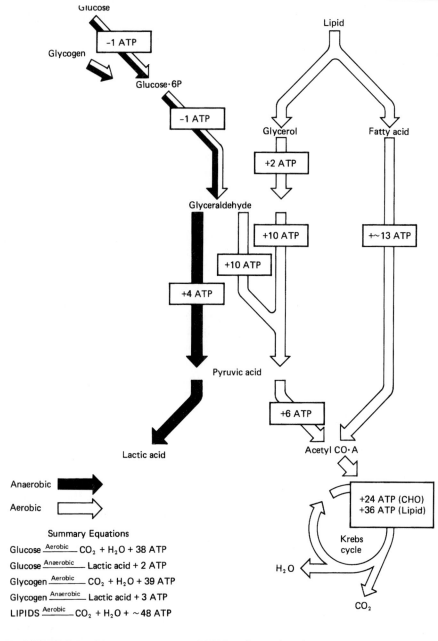

Figure 4–12 Schematic diagram of the main elements in aerobic and anaerobic energy metabolism. (From M. S. Gordon, G. A. Bartholomew, A. D. Grinnell, C. B. Jorgensen, and F. N. White, 1982, *Animal Physiology: Principles and Adaptations*, Macmillan, New York, NY.)

whereas anaerobic metabolism can produce large quantities of ATP, but for only a brief time before the substrate is depleted and the animal is exhausted. Animals with high aerobic capacities can be seen as specialized for sustained activity, whereas animals with low aerobic capacities are specialized for burst activity.

Metabolic Levels Among Vertebrates

The amount of energy that vertebrates use is reflected by the amount of oxygen they consume. In general, the minimum rate at which animals consume oxygen is set by the level necessary to support life. The maximum rate may be determined by the capac-

Table 4–7 Representative values of the heart and blood for vertebrates.

Species	Body Mass (kg)	Heart Mass (percentage of body mass)	Cardiac Output (mL/kg · min)	Blood Pressure (mm Hg)	Hematocrit (percent)	Hemoglobin (g/100 mL blood)
Fishes: carp	1	0.15	9	43	31	10.5
Lizards: iguana	1	0.19	58	75	31	8.4
Mammals: dog	14	0.65	150	134	46	14.8

ity of the respiratory and circulatory systems—that is, the rate at which oxygen and nutrients can be supplied to cells (Weibel 1984). The lower aerobic limit, often referred to as the **standard metabolic rate** (SMR), is set by the minimum energy required to maintain life in an organized state. As a rule, vertebrates seldom operate at their SMR. Therefore, it is necessary to define the conditions under which the SMR is measured. When the resting oxygen consumption is measured following a meal, the metabolism will be at least 5 to 30 percent higher than the SMR because of the costs of digestion. Other factors, such as visual or mechanical disturbance, cause significant increases in oxygen uptake by inducing stress, and low oxygen or high carbon dioxide levels have pronounced effects on energy metabolism. Because increasing temperature usually increases the rates of chemical reactions, including aerobic metabolism, it can have an overriding influence on organisms.

The Effect of Body Size SMR is reported in terms of the volume of oxygen consumed per unit of time at standard pressure and temperature (STP). Obviously, large animals consume more oxygen than small animals (Figure 4–13). To allow comparisons of animals of different sizes, the SMR is adjusted for body size to yield a mass-specific SMR (rate of oxygen consumption/body mass). It is immediately apparent that within vertebrates the mass-specific SMR decreases as body size increases. The slope of the regression of SMR on body mass is roughly similar for different kinds of vertebrates and also for a wide variety of invertebrates. This relationship has intrigued biologists for over 100 years, but a clear, unambiguous explanation has proved elusive.

Several biologists have pointed out that the increase in mass-specific SMR with decreased size might result from the relatively larger body surface area of small animals. Surface area is proportional to the 2/3 power (0.67) of the body mass. About 100 years ago Max Rubner showed that the heat loss across each unit of body surface of small and large dogs was the same (about 100 kilocalories per square meter), even though the mass-specific SMR was higher in smaller dogs. To explain this paradox he reasoned that small dogs must lose heat more rapidly than larger dogs because of their relatively high surface area. To maintain body temperatures small dogs compensate for their increased heat loss by an increase in mass-specific SMR. Rubner felt the regression of SMR on mass was sufficiently close to a slope of 0.67 that the phenomenon was explained by body geometry.

For many vertebrates this slope has a value near 0.75, that is, the SMR varies as the 3/4 power and not the 2/3 power of body mass (Kleiber 1961). Subsequently, theoretical arguments have been advanced to support Kleiber's rule. A mass exponent of 0.75 can be derived from the mechanics of locomotion (McMahon 1973) and from the geometry of four dimensions (Blum 1977). Recently the mass exponent of 0.67 has reemerged as the predicted value for comparisons of different-sized individuals of a single species (Heusner 1982). Feldman and McMahon

Table 4–8 Estimated aerobic and anaerobic contributions to 30 seconds of activity for a mammal (the kangaroo rat, *Dipodomys merriami*) and a lizard (the desert iguana, *Dipsosaurus dorsalis*).

| Species | Millimoles ATP per Gram Body Mass | | |
	Aerobic	Anaerobic	Total
Dipodomys	0.0098 (70%)	0.0043 (30%)	0.0141
Dipsosaurus	0.0044 (24%)	0.0142 (76%)	0.0186

Source: J. A. Ruben and D. E. Battalia, 1979, *Journal of Experimental Zoology* 208:73–76.

Figure 4–13 Comparison of body size and metabolic rate of ectotherms and endotherms. Upper graph: The total metabolism per hour plotted against the body mass (both on logarithmic scales) yields a straight line with a slope that usually lies between 0.65 and 0.85 and is conventionally considered to be equal to 0.75. Lower graph: Total metabolism converted to oxygen consumption per unit of body mass per hour to give a mass-specific metabolic rate (SMR). The plot of SMR against body mass (both on logarithmic scales) yields a slope that is conventionally considered to be −0.25.

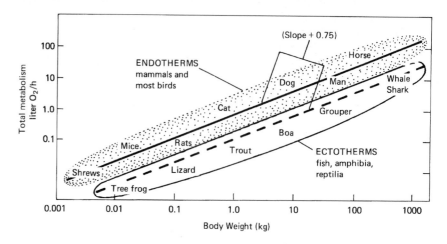

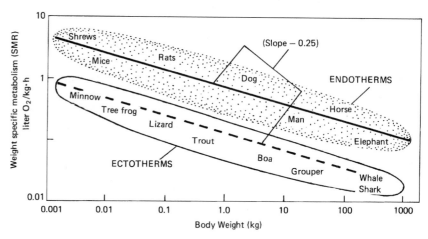

(1983) have suggested that both 0.75 and 0.67 are valid exponents, the first applying to comparisons of different species and the second being appropriate for comparisons of individuals within species. Although the mechanistic basis of the phenomenon remains unexplained, two points are clear: Metabolic rate is related to body size in all vertebrates, and that relationship has profound ecological and evolutionary consequences.

Although the exact values of the slopes relating metabolism to body mass are still subject to debate, it is clear that the slopes are less than 1. The metabolic rate can be thought of as the energy requirement of an animal. Because the slope of metabolism versus mass is less than 1, doubling the size of an animal does not double its energy requirement. To understand that relationship, assume that the mass exponent for metabolism is 0.75 and consider an animal weighing 2.5 kilograms and another animal of the same kind that weighs 5 kilograms. The metabolic rates of the two animals will be proportional to their body masses raised to the 0.75 power. Thus,

$$\text{MR of animal 1} = (2.5 \text{ kg})^{0.75} = 1.99$$

and

$$\text{MR of animal 2} = (5 \text{ kg})^{0.75} = 3.34$$

The energy requirements of the larger animal are only 1.68 times greater than those of the small animal. In ecological terms, that means that a small animal has a greater energy requirement *for its size* than does a large animal, although a large animal needs more energy *in total* than does a small animal.

At least two regression lines are required to fit the SMR data for all vertebrates (Figure 4–13). The lower SMRs of ectotherms result partly from their lack of internal heat generation to maintain a high body temperature. The SMR of an ectotherm averages one-sixth that of an endotherm of the same size. The cost of maintaining a high body temperature demands that birds and mammals consume more food than ectotherms of similar size. But vertebrates are mobile animals and, whether ectothermic or endothermic, all require more energy when active.

Summary

Vertebrates, like other organisms, are composed mostly of water. Inorganic and organic solutes are dissolved in the water, and the complex biochemical processes that make organisms self-sustaining require regulation of the water content and solute concentrations of their tissues and cells. Most vertebrates have osmotic concentrations between 250 and 350 milliOsmoles per kilogram of water, whereas fresh water is usually below 10 milliOsmolal and seawater is about 1000 milliOsmolal. Sodium and chloride are the major osmotically active compounds in seawater and in most vertebrates. The gills of fishes and skin of amphibians are permeable to water.

Freshwater teleosts and amphibians have osmotic and ion concentrations higher than their surroundings. Consequently, they must cope with an inward osmotic flow of water and an outward diffusion of ions. They produce copious, dilute urine to excrete water and expend energy to take up ions from the external medium. Marine teleosts are less concentrated than seawater; they lose water by osmosis and gain salt by diffusion. These fishes drink seawater and use active transport to excrete ions. Hagfishes, elasmobranchs, and coelacanths have osmotic concentrations close to that of seawater, but ionic concentrations that are different from those of their environment. As a result, osmotic water movement is low, but energy is used to regulate solute concentrations.

Deamination of proteins during metabolism produces ammonia, which is toxic. Ammonia is very soluble in water, and aquatic vertebrates excrete ammonia as their main nitrogen-containing waste product (ammonotelism). Terrestrial animals do not have enough water available to be ammonotelic. Mammals convert ammonia to urea, which is nontoxic and very soluble. The capacity of the mammalian kidney to produce concentrated urine allows mammals to excrete urea (ureotelism) without an excessive loss of water. The kidneys of diapsids and turtles do not have the urine-concentrating capacity of mammalian kidneys, and these animals convert much of the ammonia to uric acid (uricotelism). Uric acid is not very soluble, and it combines with ions to form urate salts that precipitate in the cloaca. As the salt precipitates, water is released, and uricotely is very economical of water. Some uricoteles save even more water by using extrarenal routes of salt secretion (salt glands) to eliminate sodium and chloride in solutions that may exceed 2000 milliOsmoles.

Temperature profoundly affects the biochemical processes that sustain vertebrates, and thermoregulatory mechanisms are widespread. Few fishes and amphibians can maintain a temperature difference between their bodies and the water around them, but some fast-swimming tunas and sharks have muscle temperatures that are 10°C or more above water temperature. Billfishes have specialized tissues that produce heat that warms the eyes and brain. Many terrestrial vertebrates have the capacity to regulate their body temperature. Ectotherms rely on sources of heat from outside the body for thermoregulation, balancing heat gained and lost by radiation, conduction, convection, and evaporation. This is a complex and effective process; many ectotherms maintain stable body temperatures substantially above ambient temperatures while they are thermoregulating. Endotherms use metabolically produced heat and manipulate insulation to balance the rates of heat production and loss. Endothermal thermoregulation confers considerable independence of environmental conditions but is energetically expensive. The mechanisms of ectothermal and endothermal thermoregulation are quite different, and an evolutionary transition from ectothermy to endothermy would be complex. Nonetheless, that transition occurred at least twice, once in the evolution of birds and once in the evolution of mammals. The evolutionary origins of the two essential components of endothermy—insulation and a high metabolic rate—were probably different from their current significance.

Vertebrates use two pathways of metabolic energy production: aerobic metabolism and anaerobic metabolism (glycolysis). Aerobic metabolism requires a circulatory system that can transport oxygen and metabolic substrates to active tissues, whereas anaerobic metabolism relies on the glycogen stores present in the cell. Both can produce ATP at high rates, but only aerobic metabolism can be sustained for long periods. Endotherms rely primarily on aerobic metabolism to sustain activity. As a result, they have high rates of oxygen consumption even when they are inactive. Ectotherms use anaerobic metabolism when they must produce ATP at high rates, and have low rates of oxygen consumption at rest—about 1/7 those of endotherms of the same body size. Large animals, whether endotherms or ectotherms, require more energy than small ones, but energy requirements increase more slowly than body mass. As a result, large animals require less energy per gram of body tissue than do small ones.

References

Avery, R. A. 1979. *Lizards: A Study in Thermoregulation*. University Park, Baltimore, MD.

Avery, R. A. 1982. Field studies of reptilian thermoregulation. Pages 93–166 in *Biology of the Reptilia*, volume 12, edited by C. Gans and F. H. Pough. Academic, London, UK.

Bartholomew, G. A. 1982. Physiological control of body temperature. Pages 167–211 in *Biology of the Reptilia*, volume 12, edited by C. Gans and F. H. Pough. Academic, London, UK.

Bennett, A. F., and J. A. Ruben. 1979. Endothermy and activity in vertebrates. *Science* 206:649–655.

Block, B. A. 1991. Evolutionary novelties: how fish have built a heater out of muscle. *American Zoologist* 31:726–742.

Block, B. A., J. R. Finnerty, A. F. R. Stewart, and J. Kidd. 1993. Evolution of endothermy in fish: mapping physiological traits on a molecular phylogeny. Science 260:210–214.

Blum, J. J. 1977. On the geometry of four dimensions and the relationship between metabolism and body mass. *Journal of Theoretical Biology* 64:599–601.

Carey, F. G. 1982. Warm fish. Pages 216–233 in *A Companion to Animal Physiology*, edited by C. R. Taylor, K. Johansen, and L. Bolis. Cambridge University Press, Cambridge, UK.

Cowles, R. B., and C. M. Bogert. 1944. A preliminary study of the thermal requirements of desert reptiles. *Bulletin of the American Museum of Natural History* 83:261–296.

Evans, D. H. 1980. Osmotic and ionic regulation by freshwater and marine fishes. Pages 93–122 in *Environmental Physiology of Fishes*, edited by M. A. Ali. Plenum, New York, NY.

Feldman, H. A., and T. A. McMahon. 1983. The 3/4 mass exponent for energy metabolism is not a statistical artifact. *Respiratory Physiology* 52:149–163.

Heath, J. E. 1965. Temperature regulation and diurnal activity in horned lizards. *University of California Publications in Zoology* 64:97–136.

Heusner, A. A. 1982. Energy metabolism and body size, I: Is the 0.75 mass exponent of Kleiber's equation a statistical artifact? *Respiration Physiology* 48:1–12.

Huey, R. B. 1982. Temperature, physiology, and the ecology of reptiles. Pages 25–91 in *Biology of the Reptilia*, volume 12, edited by C. Gans and F. H. Pough. Academic, London, UK.

Kirschner, L. B. 1995. Energetics of osmoregulation in fresh water vertebrates. *Journal of Experimental Zoology* 271:243–252.

Kleiber, M. 1961. *The Fire of Life: An Introduction to Animal Energetics*. Wiley, New York.

Kluger, M. J. 1979. Fever in ectotherms: evolutionary implications. *American Zoologist* 19:295–304.

Loveridge, J. P. 1970. Observations on nitrogenous excretion and water relations of *Chiromantis xerampelina* (Amphibia, Anura). *Arnoldia* 5:1–6.

McMahon, T. 1973. Size and shape in biology. *Science* 179:1201–1204.

Minnich, J. E. 1982. The use of water. Pages 325–395 in *Biology of the Reptilia*, volume 12, edited by C. Gans and F. H. Pough. Academic, London, UK.

Nishimura, H., and M. Imai. 1982. Control of renal function in freshwater and marine teleosts. *Federation Proceedings* 41:2355–2360.

Peaker, M., and J. L. Linzell. 1975. *Salt Glands in Birds and Reptiles*. Cambridge University Press, Cambridge, UK.

Regal, P. J. 1975. The evolutionary origin of feathers. *Quarterly Review of Biology* 50:35–66.

Regal, P. J. 1985. Commonsense and reconstructions of the biology of fossils: *Archaeopteryx* and feathers. Pages 67–74 in *The Beginnings of Birds, Proceedings of the International Archaeopteryx Conference, Eichstatt 1984*, edited by M. K. Hecht, J. H. Ostrom, G. Viohl, and P. Wellnhofer. Jura Museum, Eichstatt, West Germany.

Shoemaker, V. H., D. Balding, and R. Ruibal. 1972. Uricotelism and low evaporative water loss in a South American frog. *Science* 175:1018–1020.

Tracy, C. R. 1982. Biophysical modeling in reptilian physiology and ecology. Pages 275–321 in *Biology of the Reptilia*, volume 12, edited by C. Gans and F. H. Pough. Academic, London, UK.

Weibel, E. R. 1984. *The Pathway for Oxygen*. Harvard University Press, Cambridge, MA.

White, F. N. 1973. Temperature and the Galapagos marine iguana: insights into reptilian thermoregulation. *Comparative Biochemistry and Physiology* 45A:503–513.

Withers, P. C., G. Morrison, and M. Guppy. 1994. Buoyancy role of urea and TMAO in an elasmobranch fish, the Port Jackson shark, *Heterodontus portjacksoni*. *Physiological Zoology* 67:693–705.

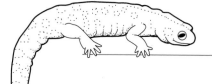

5 GEOGRAPHY AND ECOLOGY FROM THE CAMBRIAN TO THE MID-DEVONIAN

We have emphasized the importance of the stage on which the vertebrate story is played in shaping its plot, and in this chapter we illustrate the role of the environment by considering the conditions that prevailed from the early Paleozoic through the middle of the Devonian. The world then was very different from the one we know—the continents were in different places, climates were different, and there was little structurally complex life on land. All of these elements played a role in setting the stage for the origin and diversification of vertebrates.

Earth History, Changing Environments, and Vertebrate Evolution

Continents move because they float. The soil, rock, and pavements you walk on may not seem light, but they are not as dense as the material beneath them. The continents are formed of sedimentary and igneous rocks with an average density of 2.7 grams per cubic centimeter, whereas the mantle that lies beneath the continents consists of basaltic rocks with an average density of 3.0 grams per cubic centimeter. A continental block floats in the mantle, just as an ice cube floats in water.

Heat in the Earth's core produces slow convective currents in the mantle. Upwelling plumes of molten basalt rise toward the Earth's surface, forming **mid-ocean ridges** where they reach the top of the lithosphere and spread horizontally (Figure 5–1). The seafloor is criss-crossed by a chain of mid-oceanic ridges that extends around the globe. The youngest crust of the Earth is found in the centers of the ridges, and the seafloor becomes older as you move away from the axis of the ridge. **Subduction zones** form where the lithosphere sinks back down into the mantle, and the continents drift on **tectonic plates** formed by these processes. As a result of this cycle of upwelling at mid-oceanic ridges and sinking back into the mantle at subduction zones, rocks older than 200 million years do not occur anywhere on the seafloor.

Movements of the tectonic plates are responsible for the sequence of continental fragmentation, coalescence, and fragmentation that has occurred during the last 500 million years. Plants and animals were carried along as the continents drifted, collided, and separated. When continents moved toward the poles, the evolutionary lineages of plants and animals they carried with them encountered cooler climates and even glaciation. As once-separate continents collided, terrestrial and marine floras and faunas that had evolved in isolation were able to mix. The effect of these collisions may have been particularly acute for organisms in shallow marine habitats. When two continents coalesce, the total area of shallow marine habitats decreases because one large continent has less coastline in proportion to its land area than the two smaller continents had. Some lineages of both marine and terrestrial organisms became extinct when continents coalesced, and others extended their geographic ranges.

The position of continents affects patterns of oceanic circulation, and because ocean currents

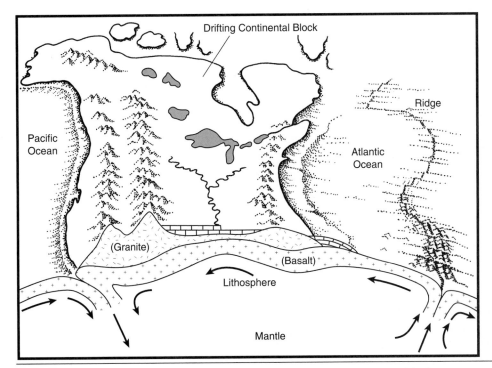

Figure 5–1 Generalized geological structure of a continent. The continental blocks float on a basaltic crust. Arrows show the movements of crustal elements and the interactions with the mantle that produce continental drift.

transport enormous quantities of heat those changes in water flow affect climates worldwide. During much of the Mesozoic, the position of the continents allowed free exchange of water at the North Pole with warmer water from the south, and the far north was warmer than it is now. Nonavian dinosaurs, crocodilians, and large turtles lived in forests of broad-leaved trees in Alaska. As the continents approached their present locations, the exchange of water between the Arctic Ocean and the Atlantic and Pacific oceans was reduced, and the Arctic Ocean cooled and then froze. Once the water had frozen, a permanent cover of ice and snow formed over the Arctic Ocean. This layer now reflects much of the solar energy that falls on the far north, producing the frigid Arctic that we know today.

Six ancient continent blocks existed from the Cambrian through the Silurian (Figure 5–2). A large block called **Laurentia** included most of modern North America, plus Greenland, Scotland, and part of northwestern Asia, and four smaller blocks contained other parts of what are now the Northern Hemisphere (**Baltica**—Scandinavia and much of central Europe, **Kazakhstania**—central southern

Asia, **Siberia**—northeastern Asia, and **China**—Mongolia, China, and all of Indochina). The sixth ancient continent, **Gondwana**, included most of what is now the Southern Hemisphere (South America, Africa, Antarctica, and Australia) plus India, Tibet, Iran, Saudi Arabia, Turkey, southern Europe, and part of the southeastern United States.

In the late Cambrian Gondwana, Laurentia, Siberia, and Kazakhstania straddled the equator, China lay to the north and Baltica to the south (Figure 5–2). Over the next one hundred million years, Gondwana drifted south and by the mid-Silurian it had reached the South Pole. In contrast, the remaining continents remained near the Equator (Veevers 1994).

The Environment of Early Craniate Evolution

The formation of continents during the early Proterozoic was accompanied by chemical changes that dramatically modified the composition of the oceans and the atmosphere (Holland 1984). Early seas were

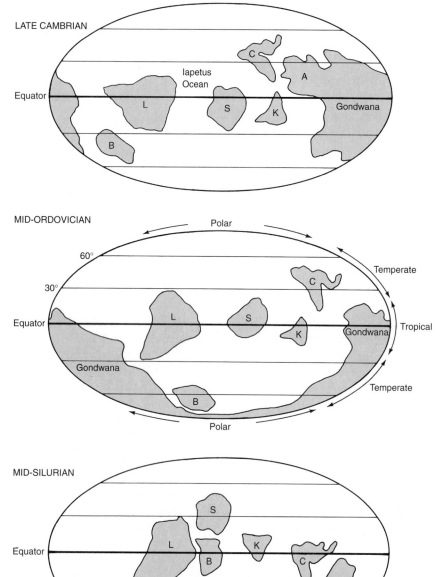

Figure 5–2 Location of continental blocks from the late Cambrian through the Silurian. Gondwana drifted southward, and the northern continents were located close to the equator.

probably acidic, with high levels of carbon dioxide and other acid-producing compounds. The ocean and the atmosphere were probably chemically reducing, due to the absence of free oxygen (Kempe and Degens 1985). The evolution of oxygen-producing and carbon dioxide-utilizing organisms during the two billion years of the Proterozoic resulted in the oceanic and atmospheric conditions we rec-

ognize today—chemically oxidative conditions and alkaline seas. The reductions in carbon dioxide and the acidity of the seas made possible the formation and precipitation of calcium carbonate-containing minerals. By the end of the Proterozoic a major biotic shift occurred—the evolution from soft-bodied organisms of forms capable of secreting articulated skeletal parts (McMenamin and McMenamin 1989).

The rapid radiation of animals with hard parts during the early Cambrian set the stage for the origin of craniates.

How well can we describe the habitat of these earliest craniates? The association of late Cambrian, Ordovician, and many of the Silurian chordates with brachiopods, crinoids, and corals—all marine invertebrates—attests to an origin in shallow, warm seas. During the late Cambrian and Ordovician, North America was largely covered by a shallow continental sea, apparently providing the necessary conditions for evolution of a chordate into a craniate. The general habitat of the first fossilized forms was probably the seafloor. Further, geochemists and geophysicists suggest that the Paleozoic sea was ionically much as it is today (Nicolls 1965). Early craniates therefore faced physiological problems similar to those faced by modern tunicates, echinoderms, and pterobranchs.

Heterotrophic animals like craniates ultimately depend on plants as a primary source of energy. What kinds of plants existed in the early Paleozoic? Simple, single-celled organisms originated in Precambrian seas—the cyanobacteria are examples (Rogers 1993). By the Ordovician, however, more complex multicellular green and red algae had evolved (Stewart 1983, Scagel et al. 1984). Phytoplankton such as diatoms and dinoflagellates, a major source of food in aquatic habitats today, was abundant in Ordovician seas. Most fossilized green and red algae were lime secretors, but fossils of non-lime-secreting algae suggest that noncalcareous multicellular algae were also important in these ancient shallow seas.

Terrestrial Ecosystems in the Paleozoic

The evolution of terrestrial plants and ecosystems can be traced from the Precambrian through the Tertiary (reviewed by Behrensmeyer et al. 1992). Terrestrial environments were nearly lifeless in the early Paleozoic. Mats of bacteria probably existed in wet terrestrial habitats from the Precambrian onward (Horodyski and Knauth 1994), but we have no direct evidence of land plants until the late Silurian. Fossilized soils from the Ordovician have mottled patterns that probably indicate the presence of bacterial mats, and traces of erosion suggest that some of the soil surface was covered by algae, but there is no evidence of rooted plants (DiMichele and Hook 1992). The first major radiation of plants onto land probably took place in the middle to late Ordovician. These pioneers included bryophytes (represented now by mosses, liverworts, and hornworts), lichens (symbiotic associations of algae and fungi), and fungi. The landscape would have looked bleak by our standards—mostly barren, with a few kinds of low-growing vegetation limited to moist areas.

The diversity of terrestrial life increased during the Silurian, and a rootless, leafless plant called *Cooksonia* is abundant in late Silurian fossil deposits. *Cooksonia*, which grew to a height of 10 to 15 millimeters, consisted of a group of unbranched stems topped by pinhead-size spore-producing structures. Fossils of a vascular plant with roots and leaves (*Baragwanathia*) have been found at late Silurian sites in Australia (White 1986). *Baragwanathia* has stems with leaves, and is very similar in structure to modern lycopods (club mosses). The reproductive leaves of modern club mosses have a different structure from the sterile leaves of the plant, but *Baragwanathia* did not show that distinction. Like its modern relatives, *Baragwanathia* grew to a height of several centimeters. *Rhynia* plants were leafless stems rising from horizontal branches that spread across the ground and absorbed water and nutrients. Cross sections of fossils reveal a core of tissues in the stems with channels that transported the water and nutrients from the site of uptake to the tissues in which photosynthesis took place. Lignin, a polymer that provides strength in modern plants, was present in these tubules (Niklas and Pratt 1980). Water impermeability in land plants is provided mainly by two waxy substances, cuterin on leaves and stems and suberin on roots, and these chemicals were also present in the earliest vascular land plants (Niklas 1979, Chapman 1985).

Invertebrates had invaded the land by the Silurian. Fossils of arachnids and scorpions are known, but we have no fossils of the animals those predators might have eaten. Probably they preyed on smaller arthropods that ate dead vegetation (called detritivores) or fungus, or on animals that grazed on microorganisms growing on dead vegetation. This simple food chain is very different from the complex webs of consumers and predators that began to develop in the Devonian and reached an essentially modern form by the Permian (DiMichele and Hook 1992).

Terrestrial ecosystems increased in complexity through the early and middle Devonian, but food webs remained simple. The land would still have looked barren, although the changes that had occurred since the Silurian would be apparent. Plants were limited to stream banks and other areas of nearly continuous moisture, and occurred in patches composed of single species, but the diversity of plant species was greater than it had been in the Silurian. Still more dramatic would have been the increased height that was possible for vascular plants (which could transport water from the site of uptake to other locations). By the middle Devonian, these plants probably attained heights of two meters, and the canopy they created would have modified microclimatic conditions on the ground.

Plants provide the basis for terrestrial animal life, but there is no evidence that Devonian invertebrates fed on living plants. Instead, detritivores consumed dead plant material. Millipedes were abundant, and springtails and mites were present. The oldest insects known come from early Devonian sediments in Canada. These detritivores were preyed upon by centipedes, scorpions, pseudoscorpions, and at least one species of spider.

Early Paleozoic Climates

Climate results from the interaction of sunlight, temperature, rainfall, evaporation, and wind throughout a year. Because climate profoundly affects the kinds of plants and animals that occupy an area, knowledge of paleoclimates helps us to understand the conditions under which plants and animals evolved. The primary factors that determine terrestrial climates of large areas like continents are latitudinal position (i.e., how far north or south of the equator, which affects the amount of solar energy received), proximity to an ocean (which buffers temperature change and provides water via evaporation and rainfall), and the presence of barriers like mountains that influence the movement of atmospheric moisture (Cox and Moore 1993).

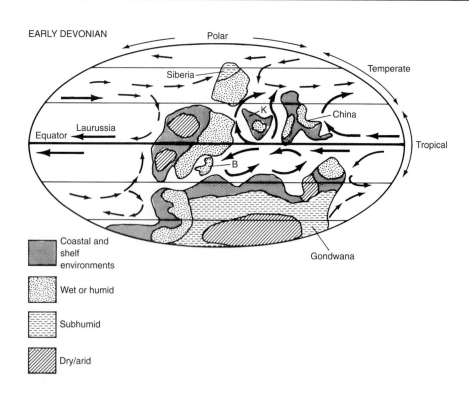

Figure 5–3 Location of continental blocks and probable climates and patterns of ocean circulation in the early Devonian. K, Kazakhstania; B, Baltica.

The early Paleozoic is characterized by thick limestone deposits along the edges and in the interior of the continents, suggesting that overall land profiles were low and that seas spread across parts of the continents (Bray 1985). Paleoclimatologists believe that the entire Earth may have been cool in the Cambrian. But throughout the last 500 million years or more the Earth has probably displayed generally the same type of climate as it does today, wherein the polar regions are cooler than equatorial locations, and tropical, temperate, and boreal climates can develop (Figure 5–3). Latitudinal drift of the continents (north or south of the equator) affected long-term continental climates. From the Ordovician and on into the Triassic the movements of the continents placed much of North America, and to a lesser extent western Europe, close to the equator. Over this long span of time these areas were exposed to a warm and stable year-round climate. Only areas outside the tropics would have been subjected to strong seasonal changes in climate, as they are today.

References

Behrensmeyer, A. K., J. D. Damuth, W. A. DiMichele, R. Potts, H. Dieter-Sues, and S. L. Wing (editors). 1992. *Terrestrial Ecosystems Through Time*. University of Chicago Press, Chicago, IL.

Bray, A. A. 1985. The evolution of the terrestrial vertebrates: environmental and physiological considerations. *Philosophical Transactions of the Royal Society of London* B309:289–322.

Chapman, D. J. 1985. Geological factors and biochemical aspects of the origin of land plants. Pages 23–45 in *Geological Factors and the Evolution of Plants*, edited by B. H. Tiffney. Yale University Press, New Haven, CT.

Cox, C. B., and P. D. Moore. 1993. *Biogeography: An Ecological and Evolutionary Approach*, 5th edition. Blackwell Scientific, Oxford, UK.

DiMichele, W. A., and R. W. Hook. (rapporteurs). 1992. Paleozoic terrestrial ecosystems. Pages 205–325 in *Terrestrial Ecosystems Through Time*, edited by A. K. Behrensmeyer, J. D. Damuth, W. A. DiMichele, R. Potts, H. Dieter-Sues, and S. L. Wing. University of Chicago Press, Chicago, IL.

Holland, H. D. 1984. *The Chemical Evolution of the Atmosphere and Oceans*. Princeton University Press, Princeton, NJ.

Horodyski, R. J., and L. P. Knauth. 1994. Life on land in the Precambrian. *Science* 26:494–498.

Kempe, S., and E. T. Degens. 1985. An early sodic ocean? *Chemical Geology* 53:95–108.

McMenamin, M. A. S., and D. L. McMenamin. 1989. *The Emergence of Animals: The Cambrian Breakthrough*. Columbia University Press, New York, NY.

Nicolls, G. D. 1965. The geochemical history of the oceans. Chapter 20 in *Chemical Oceanography*, volume 2, edited by J. P. Riley and G. Skirrow. Academic, New York, NY.

Niklas, K. J. 1979. An assessment of chemical features for the classification of plant fossils. *Taxon* 28:505–511.

Niklas, K. J., and L. Pratt. 1980. Evidence for lignin-like constituents in early Silurian (Llandoverian) plant fossils. *Science* 209:396–397.

Rogers, J. J. W. 1993. *A History of the Earth*. Cambridge University Press, Cambridge, UK.

Scagel, R. F., R. J. Bandoni, J. R. Maze, G. E. Rouse, W. B. Scholfield, and J. R. Stein. 1984. *Plants: An Evolutionary Survey*. Wadsworth, Belmont, CA.

Stewart, W. H. 1983. *Paleobotany and the Evolution of Plants*. Cambridge University Press, Cambridge, UK.

Veevers, J. J. 1994. Pangaea: evolution of a supercontinent and its consequences for Earth's paleoclimate and sedimentary environments. Pages 13–23 in *Pangaea: Paleoclimate, Tectonics, and Sedimentation During Accretion, Zenith, and Breakup of a Supercontinent*, edited by G. D. Klein. Geological Society of America, Special Paper 288.

White, M. E. 1986. *The Greening of Gondwana*. Reed Books, Frenchs Forest, Australia.

PART II

AQUATIC VERTEBRATES: CARTILAGINOUS AND BONY FISHES

Craniates originated in the sea, and more than half the living vertebrates are the products of evolutionary lineages that have never left an aquatic environment. Water now covers 69 percent of the Earth's surface (the percentage has been higher in the past) and provides habitats extending from deep oceans and lakes to fast-flowing streams and tiny pools in deserts. Fishes have adapted to all of these habitats, and the nearly 25,000 species of living fishes are the subject of this portion of the book.

Life in water poses challenges for vertebrates and offers many opportunities. Aquatic habitats are some of the most productive on Earth, and energy is plentifully available in many of them. Other aquatic habitats, such as the deep sea, have no in situ production of food and the animals that live in them depend on energy that comes from elsewhere. The physical structure of aquatic habitats has a similar range: Some aquatic habitats (coral reefs are an example) have enormous structural complexity, whereas others (like the open ocean) have virtually none. The diversity of fishes reflects specializations for this variety of habitats.

The diversity of fishes and the habitats they live in has offered unparalleled scope for variations in life history. Some fishes produce millions of eggs that are released into the water to drift and develop on their own, other species of fish produce a few eggs and guard both the eggs and the young, and many fishes give birth to precocial young. Males of some species of fish are larger than females, in others the reverse is true, some species have no males at all, and a few species of fish change sex partway through life. Feeding mechanisms have been a central element in the evolution of fishes, and the specializations of modern fishes extend from species that swallow prey longer than their own bodies to species that extend their jaws like a tube to suck up minute invertebrates from tiny crevices. In this part of the book we consider the evolution of this extraordinary array of vertebrates and the ecological conditions in the Devonian that contributed to the next major step of evolution, the origin of terrestrial vertebrates.

6

EARLIEST VERTEBRATES

The earliest craniates known were aquatic filter feeders, but represented an important advance over the protochordate filter feeders (Chapter 2). They used muscular contractions instead of cilia to move water. A muscular pump can move more water, all other things being equal, than a ciliated one, and the early craniates were able to grow larger than the protochordates. Bone, a distinctive form of mineralized tissue, was a second innovation of the earliest craniates. The bony armor that encased these animals probably gave some protection from predators and also may have been a store of minerals that contributed to homeostasis. These advances, combined with mobility, allowed the earliest craniates to radiate into adaptive zones that had previously been unoccupied. We know a remarkable amount about the anatomy of some of these early craniates because the internal structure of their bony armor reveals the positions and shapes of many parts of their soft anatomy. The brains and cranial nerves of these most plesiomorphic craniates were remarkably like the brain and cranial nerves of a living vertebrate, the lamprey. In this chapter we trace the earliest steps in the radiation of craniates some 500 million years ago.

The First Evidence of Craniates

A major advantage of the craniates was the evolution of the physiological ability to lay down a skeleton composed of calcium phosphate (bone). At first this ability seems to have been limited to superficial, dermal tissues but it soon extended to endoskeletal tissues of presumed mesodermal origin. Bony materials are more likely to be fossilized than soft tissues, and the fossil record of craniates subsequent to the evolution of bone is extensive (Forey and Janvier 1994).

The oldest fossil fragments believed to be from craniates occur in the late Cambrian and middle Ordovician, between 125 and 40 million years prior to the time when craniate fossils became abundant

(Repetski 1978). The best known of these early forms occur in the late Silurian to middle Devonian Old Red Sandstone in southwestern England and Wales and in similar rock found in Scotland, Norway, and Spitzbergen, dating from 400 million years ago. The microscopic structure of the early specimens is similar to that of the dermal bony plates of certain later unquestionably craniate forms (Figure 6–1) but it has also been compared to the remains of certain fossil arthropods (Blieck 1992). We can assume that these early animals were also characterized by laterally placed eyes, double nostrils, a complex of bone and bone-like tissues, and all the other features recognized as basic to the ancestral stock of craniates. These bony fragments also tell us that bone, at least on the surface of the body, evolved early in craniate history.

The fossil record of the first craniates reveals little about the course of evolution from the earliest

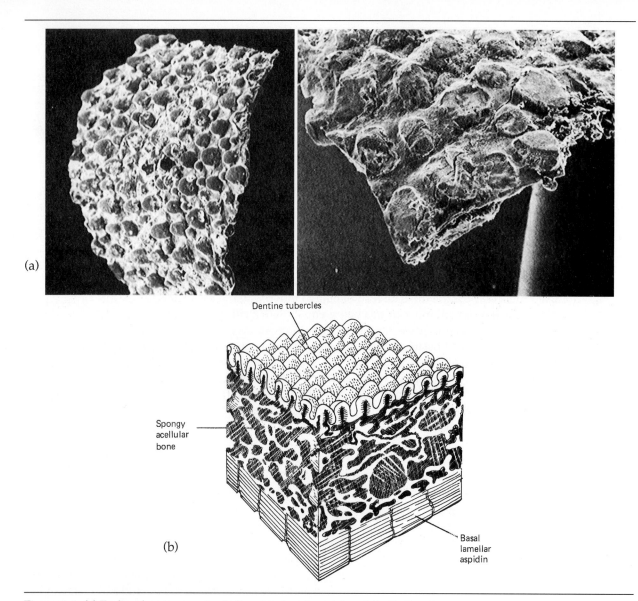

(a)

Dentine tubercles

Spongy
acellular
bone

Basal
lamellar
aspidin

(b)

Figure 6–1 (a) Earliest known remains believed to be of craniate origin. Scales of the pteraspid *Anatolepis* from the late Cambrian have definitely been identified from Oklahoma, Washington, and Wyoming. By the early Ordovician, *Anatolepis* was widely distributed in North America and to Spitzbergen, and related genera occurred in Australia. Two views of the same specimen show (left) tongue-like surface ornaments, which probably pointed posteriorly and (right) an edge showing a break through the plate revealing the solid surface layer, the cavity-rich middle layer, and the thick internal lamellar layer. (Scanning electron micrographs at approximately 220× courtesy of J. E. Repetski, U.S. National Museum.) (b) Three-dimensional block diagram of heterostracan dermal bone. (Based on B. Stahl, 1974, *Vertebrate History*, McGraw-Hill, New York, NY; and L. Halstead, 1969, *The Pattern of Vertebrate Evolution*, Oliver & Boyd, Edinburgh, UK.)

Cambrian craniates to the sudden appearance of a variety of agnathous and jawed forms in the late Silurian. It provides no clues about the evolution of craniate organization from an invertebrate progenitor. Nor does it shed light on the evolution of jawed vertebrates from their agnathous predecessors.

Many biologists conclude that vertebrates were the last major group of animals to evolve. Most of the other animal phyla appeared 50 million years before the first craniate fossils, in the oldest Cambrian rock sufficiently undistorted by geophysical forces to yield good fossil materials. The best guess is that the earliest craniates were rather uncommon, small, soft-bodied marine forms whose existence was unlikely to be recorded by the rare circumstances that lead to fossilization of soft parts.

The first fossil recognized craniates are encountered in the late Cambrian and Ordovician, and are unimpressive fragments. But when sectioned and examined microscopically (Figure 6–1), these fragments show an internal three-layered bony structure of considerable complexity—already as derived as that of craniate remains of a later date. If such histological complexity evolved gradually, this complexity suggests that these craniates had undergone considerable evolution *before* the earliest fossil evidence known was laid down. Bone formation requires the coordination of at least three types of tissues: fibroblast cells that lay down the highly organized collagen framework into which calcium phosphate is deposited as hydroxyapatite, cells that produce hormones that regulate the production of hydroxyapatite, and unique scleroblast cells that deposit the minerals in the collagen framework. It seems doubtful that these various tissues evolved in a sufficiently short time to produce the apparently sudden appearance of craniate fossils. If that is so, why are there no earlier fossils of craniates?

Perhaps conditions for fossilization simply did not exist during the earliest period of craniate evolution. That hypothesis seems unlikely, because delicate organisms such as jellyfish medusae were fossilizing during the Precambrian, and leaving such excellent impressions that their internal anatomy can be deciphered. Contemporaneous with fragments of the first craniates are brachiopods of enormous variety. Some of these invertebrates form their shell of hydroxyapatite, the same calcium phosphate salt found in bone, and they have fossilized perfectly.

Perhaps early craniates lived in a habitat that was not conducive to fossilization. Many students of vertebrate evolution, especially A. S. Romer (1966), have proposed that vertebrates evolved in freshwater streams and rivers (as we discussed in Chapter 2). These are areas of erosion, not deposition, and hence would not have provided a good fossil record. However, x-ray diffraction of the crystalline structure of the fragments and their surrounding matrices indicates that these specimens were fossilized where they are found today. Ecological reconstruction has been done using the techniques of stratigraphic analysis, which compare existing fossils in one formation with other and perhaps better understood assemblages of fossils elsewhere.

These studies have led to the conclusion that the earliest craniates lived, died, and were deposited in a marine environment. However, new localities for some of the strata that contain early craniates appear to be from freshwater streams and rivers (Graffin 1992). Whether freshwater or marine, some critical factor prevented the fauna from becoming abundant or diverse. Perhaps this critical factor was variable salinity. The most logical environment to fit all the evidence is a euryhaline estuarine zone. If so, some early craniates were capable of tolerating the stress of variable salinity, as are many modern fishes.

But why do craniates appear so late in the fossil record? The most likely hypothesis (but the least susceptible to verification) is that the earliest craniates and their evolution occurred in an environment that made the fossilization of soft tissues unlikely.

What Were the Earliest Craniates Like?

Two early sites have yielded most of the oldest craniate fossils (Repetski 1978, Romer 1968), but without microscopic examination the fragments would hardly have been recognized. One site is a small nearshore island in the Baltic Sea and the adjacent coast as far as Leningrad. The second site includes the Harding Sandstone Formation, extending from Arkansas to Montana. The more widespread Silurian fossils are similar to those of the Ordovician but are occasionally articulated. The most diverse assemblages of fossils so far unearthed are in the North American Silurian (Box 6–1).

Bony tissues greatly enhance the chances of fossilization, and our record of craniate life begins with bony remains. The initial position of bone was in a dermal exoskeletal armor. This encasing shell appears to have caused the early craniates much the

Box 6–1 Reconstructing the First Craniates

Although late Cambrian hydroxyapatite fragments are accepted as craniate remains, articulated pieces of early craniates that give us a basis for reconstructing the organisms are rare. No complete individuals are known until the Silurian/Devonian boundary. Only three geological formations, one each in Australia, North America, and South America, have yielded partial craniates of Ordovician age. The reasons seem clear. These heterostracan craniates were externally armored with a large number of small closely fitting, polygonal bony plates. Very special and rapid conditions of burial in beds destined to suffer minimal distortion in the subsequent 470 million years are required to keep such small plates articulated. These required conditions are understandably rare. It appears that a few shallow coastal marine, perhaps even tidal flat, environments provided the required conditions. Attempts at complete reconstructions of middle and late Ordovician craniates have been made (Ritchie in Rich and van Tets 1985, Elliott 1987, Gagnier 1989). Both *Arandaspis* from central Australia and *Astraspis* from the eastern slopes of the Rocky Mountains were 13 to 14 centimeters long

and had symmetrical tails (Figure 6–2). They were completely encased in small ornamented plates, each plate 3 to 5 millimeters in maximum dimension. Although the scales abut one another in the head and gill region, from about midbody posteriorly they may overlap as do scales in extant fishes. These bony plates show specializations for sensory canals, special protection around the eye, and in the reconstruction of the North American specimen, as many as eight gill openings on each side of the head.

If we take these characters to be ancestral for craniates, then the large head shield plates, single gill opening, and hypocercal tail of later heterostracans must be derived characteristics of this jawless fish radiation. Both reconstructions have well-developed eyes. If the eyes of hagfish are not secondarily degenerate, those paradoxical members of the modern fauna may have ancestors that predate the reconstructed Ordovician fishes. Full description of the gill region in the Australian and South American fossils and the rostral and mouth region of any of these Ordovician craniates will prove to be most interesting.

Figure 6–2 Reconstructions of Ordovician craniates: (a) *Arandaspis* from Australia; (b) *Astraspis* from North America; (c) *Sacabambaspis* from Bolivia. *Sacabambaspis* was twice as large as the others, averaging 35 centimeters. ([a] Modified from R. V. Rich and G. F. van Tets, 1985, *Kadimkara*, Pioneer Design Studio, Lilydale, Victoria, Australia; [b] from D. K. Elliott, 1987, *Science* 273:190–192; [c] from P. Y. Gagnier, 1989, *National Geographic Research* 5:250–253.)

same problem as the marine crustaceans experience: how to grow within a mineralized skin. The radiation of the ostracoderms is the history of evolutionary experiments to solve the growth-in-a-suit-of-armor dilemma. From the distinctive nature of the various armors, some paleontologists feel that bone probably evolved separately in each major lineage of "ostracoderms" (Carroll 1987).

In addition to bone, these Silurian organisms possessed another innovation, the switch from a cil-

iary mode to a muscular pumping mode of filter feeding. Ciliary filter feeding is common among invertebrates (Chapter 2). In these invertebrates (a great many of which are sessile or nearly so) water is wafted past food-snaring structures by the activity of great numbers of ciliated cells. Cilia are not especially effective at drawing water from a distance. Food-rich water must be close to a ciliated filter feeder. Living jawless craniates, however, suck water into the pharynx using muscles to contract the oral cham-

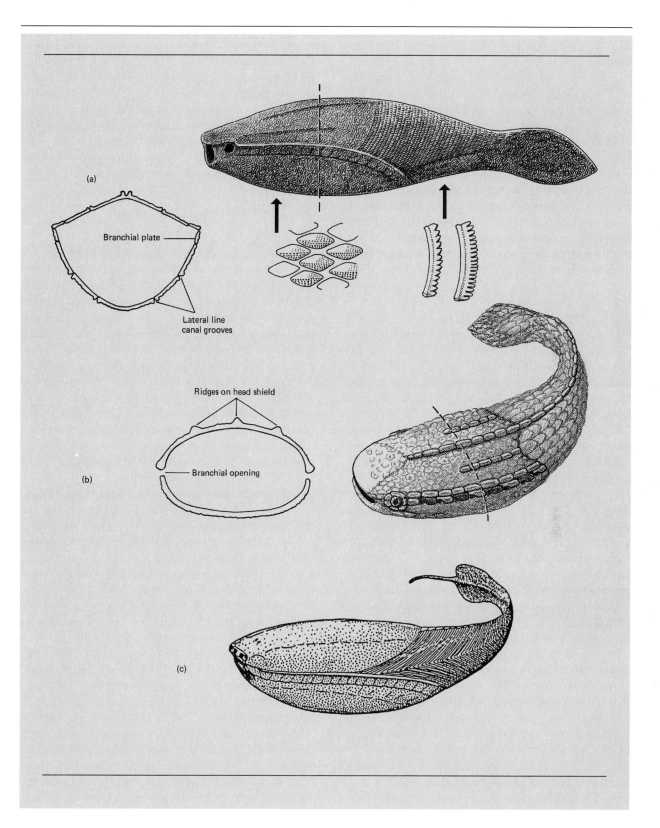

(a)

Branchial plate

Lateral line
canal grooves

Ridges on head shield

Branchial opening

(b)

(c)

ber or to move the tongue in a piston-like action. The structure of the pharynx and gills of Silurian craniates suggests that they did the same. This novel method of moving water for feeding as well as for respiration is not subject to the same restrictions as ciliary pumping. Water is moved in rapid, powerful pulses and considerable suction can be exerted to engulf large particles even if they are at some distance from the oral opening. This feeding mode may have permitted larger, more active organisms to evolve. Most early craniates were less than 20 centimeters long, but this represents a considerable size increase compared to the average invertebrate. An additional and significant difference between most invertebrate and vertebrate filter feeders is that motile vertebrates can move to areas, often ephemeral, where food is highly concentrated (Mallatt 1984).

Earliest Known Craniates

The earliest craniates were aquatic animals collectively called "**ostracoderms**" (*ostrac* = shell, *derm* = skin). The "ostracoderms" are a paraphyletic group containing at least two distinct lineages, each with recognizable subdivisions (Figure 6–3). The major "ostracoderm" groups are the **Pteraspida** (*ptera* = wing, *aspid* = shield) and the **Cephalaspida** (*cephal* = head).

Ranging from about 10 centimeters to more than 50 centimeters in length (Figure 6–3), the "ostracoderms" lacked jaws, although some of them may have had peculiar movable mouth parts not found in any other vertebrates. Typically, their mouths were fixed circular or slit-like openings that appear to have acted as intakes that filtered small food particles from the water or from bottom detritus. "Ostracoderms" also lacked paired appendages with any structural similarity to those of other craniates. Their respiratory apparatus consisted of a variable number of separate pharyngeal gill pouches that opened along the side of the head independently or through a common passage. The notochord was the main axial support throughout adult life. No vertebrae of "ostracoderms" have ever been found. Because "ostracoderms" and the living cyclostomes (hagfishes and lampreys) share these characteristics, they have long been placed together in the class Agnatha as representatives of a plesiomorphic level of craniate organization. However, because no useful information about interrelationships is contained in

shared ancestral characters, this grouping is a polyphyletic assemblage (i.e., it includes members of two or more lineages that are not sister groups) and as such is of no use in understanding evolutionary history. Many of the "ostracoderms" were specialized in the development of their bony armor, whereas the cyclostomes are specialized for burrowing or parasitic modes of existence. Figure 6–4 shows some possible interpretations of the relationships of early craniates. No resolution of these differing concepts of early craniate evolution is currently possible.

The Pteraspids and the First Craniate Life

The earliest jawless craniates, the Pteraspida, are known from the late Cambrian until the very end of the Devonian of North America, Europe, and Siberia (Figure 6–5). Because of their curious shelled appearance, they are called Heterostraci in some classifications (*hetero* = different, *strac* = shell). They varied in size from 10 centimeters to 2 meters and were encased anteriorly by bony articulating pieces that extended to the anus in some groups (Moy–Thomas and Miles 1971). Posterior to the anus was a short, probably mobile tail covered by smaller, protruding barb-like plates. The head shell had an ornamented dorsal plate, one or more lateral plates, and several large ventral elements. Thus the earliest recorded solution to the growth-in-a-suit-of-armor problem was that of the pteraspids, where either large articulating plates did not form until maximum size was attained, or numerous centers of bone formation enlarged circumferentially and fused into plates or a solid shield only as the animal reached maximum size. All these forms achieved an essentially solid carapace over the anterior one-third of their body, pierced by the mouth and a single pair each of eyes and external gill openings. No bone is known to have formed in endoskeletal structures and no bone cells are found within the osseous tissue, nor were true paired appendages or well-developed dorsal or anal fins present. The shell was composed of three distinct layers, and in all but two genera the shell appears to have continued to grow as the animal increased in size. Impressions on the inside of the dorsal plate suggest that the brain had two separate olfactory bulbs. Because it is assumed that these were connected with two separate nasal openings, these earliest craniates are grouped as the Diplorhina (*diplo* = two, *rhin* = nostril) in some early classifica-

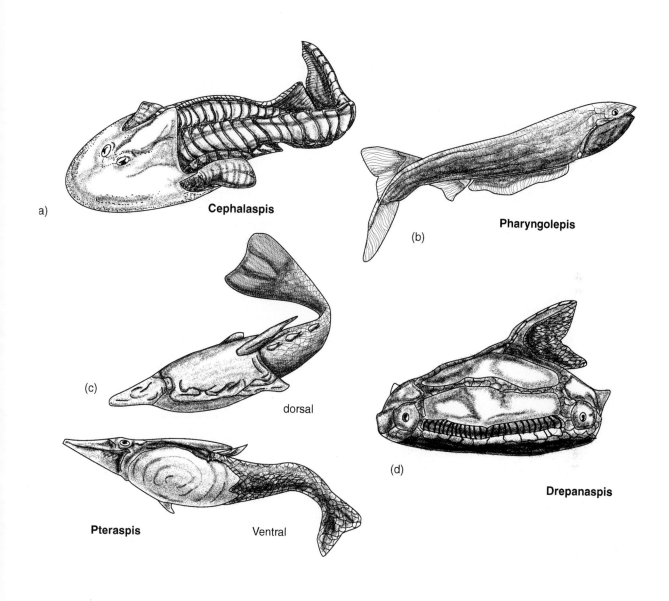

a)

Cephalaspis

(b)

Pharyngolepis

(c)

dorsal

Pteraspis   Ventral

(d)

Drepanaspis

Figure 6–3 Four representative "ostracoderms." Cephalaspids [=Monorhina, (a) and (b)]; Pteraspids [= Diplorhina, (c) and (d)].

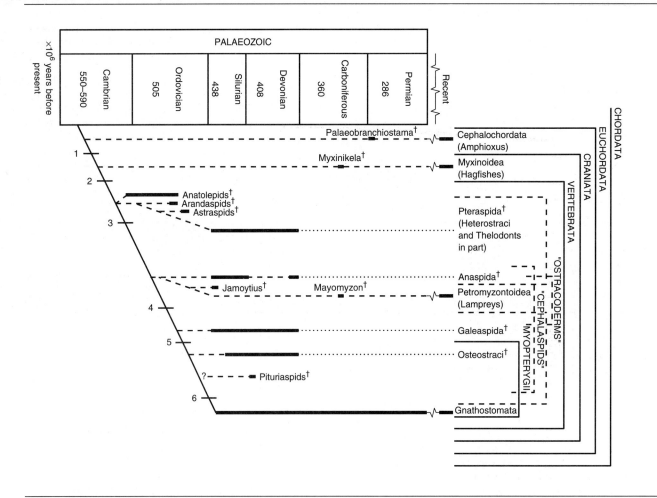

tions. Thus "Pteraspida," "Heterostraci," and "Diplorhina" are synonyms for these earliest known craniates.

Openings for two eyes are lateral, one on each side of the head shield. In the middle of the dorsal plate is a small opening for a third, median eye or **pineal organ**. The mouth is near the end of the body (terminal) but opens ventrally in many forms. Rimmed on its lower border by as many as two rows of small plates, the mouth is thought to have opened in a scoop-like manner of "V"-shaped cross section in the species best preserved (Soehn and Wilson 1990). The lower lobe of the tail is disproportionately large. This lower lobe contains the axial support element (the notochord), a tail fin construction called **hypocercal**. The body is generally round in cross section, like that of a tadpole, and early pteraspids show little sign of stabilizing projections. Possibly these fishes were erratic swimmers and resembled some tadpoles by swimming with something less than precisely controlled locomotion. While feeding they may have oriented head down and plowed their jawless mouth through the bottom sediments.

As might be guessed, evolutionary trends in the pteraspids led to the improvement of locomotor capabilities. During their later history, the cross section of benthic species flattened ventrally but remained arched or rounded dorsally (Figure 6–5). The head shield developed solid, lateral, wing-like stabilizing projections called **cornua** (horns). The head shield shortened, and except for a dorsal ridge of plates the bony covering became restricted to the anterior end. Although specialized edges around the mouth for biting and grasping did not evolve, some of the oral plates developed enlarged tooth-like projections that may have been used for scraping.

1. Craniata: Neural crest cells, coelom formed by split in unsegmented lateral plate, highly differentiated somites, gills supported by a distinctive skeleton, a distinct head region with the following characters: tripartite division of brain with cranial nerves differentiated from neural tube, segmental nerves, paired optic, auditory, and probably olfactory organs, one or more semicircular canals, cranium incorporating the anterior end of the notochord and enclosing brain and paired sensory organs. In other regions of the body are a system of distinctive endocrine glands, lateral line system, probable electrosensors, well-developed heart, paired kidneys, and at least 15 additional derived characters. 2. Vertebrata: presence of arcualia (unknown but presumed in Pteraspida, where known may fuse with additional elements in the adults of most forms to produce vertebrae), lateral line organs in a sensory canal, physiological capacity to form bone in the dermis, two or three semicircular canals, eyes well developed, and 20 additional derived characters. 3. True dorsal and anal fins with fin rays, asymmetrical tail shape. 4. Perichondral bone (at least in head), large orbits, large head vein (dorsal jugular). 5. Myopterygii: Cellular dermal bone, heterocercal tail, paired fins with internal musculature developed from a lateral plate that extends from behind the gills to the region of the cloaca, muscularized unpaired fins, large eyes with extrinsic musculature. If lampreys are the sister group to the Galeaspids +

Osteostracans + Anaspids and Gnathostomes, the similarities often drawn between lampreys and certain anaspids (i.e., *Jaymoytius*) are superficial. If *Jaymoytius* and lampreys are closely related (as shown here and in a different overall relationship to other fossil forms by Forey and Janvier, 1994, *American Scientist* 82:554–565), the character of paired lateral fins must have been lost by lampreys. Similarly, if anaspids and lampreys are the sister group of osteostracans, the evolution of these anaspid features must have paralleled their evolution in gnathostomes. 6. Gnathostomata: Jaws formed of bilateral palatoquadrate (upper) and mandibular (lower) cartilages of the mandibular visceral arch at some stage of development, teeth containing dentine, modified hyoid gill arch, branchial elements internal to gill membranes, branchial arches contain four elements on each side plus one unpaired ventral median element, paired trabeculae contribute to cranium, three semicircular canals, internal supporting girdles associated with pectoral and pelvic fins, myelinated nerve fibers, calcium carbonate statoconia or otoliths, lateral skeletogenous septum and the resultant unique trunk segmentation, endoskeletal fin radials and dermal fin rays at least in the tail, posterior paired fins in position of pelvic appendages, and over 30 additional derived characters. (Based primarily on J. G. Maisey, 1986, *Cladistics* 2:201–256, with consideration of P. Forey and P. Janvier, 1993, *Nature* 361:129–134.)

Figure 6–4 Phylogenetic relationships of the Craniata. This diagram depicts the probable relationships among the craniates. Extinct lineages are marked by a dagger (†). The numbers indicate derived characters that distinguish the lineages. Only the best corroborated relationships are shown. The quotation marks indicate that "ostracoderms" and "cephalaspids" are paraphyletic groups.

The pteraspid lineage also gave rise to species of bizarre appearance. Some developed enormous cornua that may have acted as water-planing surfaces (hydrofoils) that produced lift for the heavy head when swimming. In addition, some had two narrow sled-like runners on the ventral surface, which presumably held the head above the substrate. Several forms developed dorsally directed mouths and a much reduced skeleton. Perhaps these species fed at the surface by filtering large quantities of plankton-rich water. Forms are also known with a long tooth-bearing projection extending from the edge of the mouth, rather like that of the living sawfish. The function of this rostrum is difficult to understand because the mouth was dorsal to the saw, just opposite to morphologically similar forms among today's vertebrates. Perhaps the saw was used to stir up organisms from the bottom.

About the middle of the Silurian, when the diverse array of pteraspids entered the fossil record, another distinct but poorly known lineage of jawless craniates appeared. Rather than large plates, these small fish (Figure 6–6) were covered by numerous tiny **denticles**, small tooth-like structures not unlike those of living sharks. Isolated denticles from the Ordovician may belong to these fishes. No articulated specimens are known before the mid-Silurian, and by the mid-Devonian they were extinct. Several names have been given to this group of fishes, all referring to characteristics of the scales: **Thelodonti** (*thelo* = nipple, *dont* = tooth) and **Coelolepida** (*coel* = hollow, *lepida* = scale) are the names most commonly used. The phylogenetic relationships of thelodonts are uncertain. The distinctive small separate scales, each with a pulp cavity like a tooth, and the other characters established by the fossils have been con-

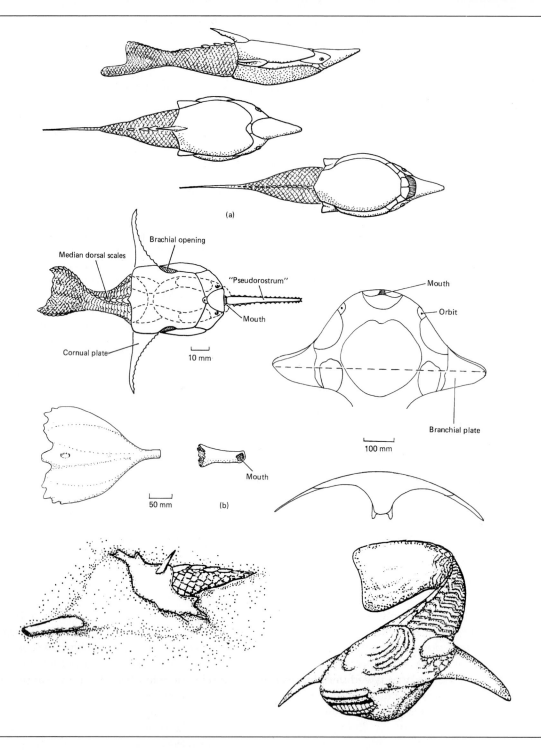

Brachial opening

Median dorsal scales

"Pseudorostrum"

Mouth

Cornual plate

10 mm

(a)

Mouth

Orbit

Branchial plate

100 mm

Mouth

50 mm

(b)

sidered plesiomorphic for craniates, and thus of no value in determining thelodont relationships.

Thelodonts were between 10 and 20 centimeters long. Most were dorsoventrally flattened anteriorly and laterally compressed posteriorly. Some were fusiform (torpedo shaped), with ridges that correspond to the dorsal and anal fins in more derived fishes and a hypocercal tail. In addition, broad-based flanges projected from their sides where anterior paired appendages occur in later craniates. Like the

Figure 6–5 Details of the bony armor of pteraspids. (a) Pteraspid *Pteraspis rostrata*, a typical, relatively unspecialized species. Reconstruction of an 18-centimeter specimen from the lower Devonian. Lateral view above, dorsal surface left, ventral surface right. (b) Specialized pteraspids (in dorsal and life views). Clockwise from the upper left: a sawfish-like form *Doryaspis*, lower Devonian; form with enlarged downcurved cornua (horns) and ventral runners (cross section of head region middle) and appearance in life, *Pycnosteus*, middle Devonian; and eyeless, tube-mouthed form *Eglonaspis*, lower and middle Devonian, which may have buried in the bottom, snout protruding. (Modified after J. A. Moy–Thomas and R. S. Miles, 1971, *Paleozoic Fishes*, Saunders, Philadelphia, PA; and P. Janvier in J. Hanken and B. K. Hall, 1993, *The Skull*, volume 2, University of Chicago Press, Chicago, IL.)

pteraspids, thelodonts had the ancestral craniate characters of lateral eyes, a pineal opening, and a jawless mouth.

Living fishes shaped like theledonts feed by swimming along the bottom of the sea. Thelodonts may have skimmed organic deposits off the bottom into their small mouths with the aid of muscular suction. Some of the thelodonts had internal denticles thought to have crushed prey and trapped the fragments as pharyngeal teeth and gill rakers do in extant fishes (Van der Brugghen and Janvier 1993).

Recent discoveries of articulated fossils of marine thelodonts from the Silurian and Devonian of northwestern Canada have greatly broadened our perception of these animals (Wilson and Caldwell 1993). All the newly discovered forms have bodies covered in typical thelodont scales, but the bodies are deep and laterally compressed with symmetrical and deeply forked tails. Most interesting is evidence of a well-developed stomach, a feature previously thought to be a derived character of jawed vertebrates.

Apparently, thelodonts were most numerous in coastal estuaries, but eventually also radiated into fresh water. Where they occurred together, the small, lightly armored thelodonts were probably behaviorally very different from the larger, armored, and heavy-bodied pteraspids. Their phylogenetic relationship to other early craniates is unknown, but the recent discovery of characters previously thought to be found only in jawed vertebrates puts these enigmatic craniates in the ranks of possible jawed vertebrate ancestors.

The Appearance of the Cephalaspids (Monorhina) and Vertebrate Radiation

A second group of craniates appeared at the same time as the thelodonts, and in the late Devonian they made a brief but impressive stand. Five distinct groups are often united as the **cephalaspids**. A single, large, slit-like nasal opening that lies in the center of the head anterior to the eyes and is usually associated with the pineal gland is a synapomorphy of the group. This structure has been the basis for another name widely used for these ancient craniates, the Monorhina (*mono* = one). The first group of cephalaspids is the **Osteostraci** (*osteo* = bone) of North America, Europe, and Siberia, a second the **Galeaspida** (*gale* = helmet, *aspid* = shield) of China and northern Vietnam, a third the recently described **Pituriaspida** (*pituri* is an Aboriginal word) from the mid-Devonian of Australia, a fourth the **Anaspida** (*an* = without) of North America, Europe, and China, and historically a fifth now generally considered polyphyletic group (Box 6–2), the still extant **Cyclostomata** (*cyclo* = round, *stoma* = mouth), including Carboniferous fossils such as *Hardistiella*, *Mayomyzon*, and *Myxinikela*.

Like pteraspids, **osteostracans** were heavily armored (Figure 6–8). However, the bone contained lacunae or spaces for bone cells and their head shield was a single, solid element devoid of sutures on its dorsal surface. Most pteraspids show evidence of periodic growth around the margins of their shield plates. The solid construction of the shield in osteostracans, however, and the absence of growth marks indicate that their head shield did not grow throughout life. Furthermore, all the individuals of a species of osteostracan are the same size. Perhaps osteostracans had a naked larval life, not unlike that of a lamprey ammocoete, and then metamorphosed into a stage where a head shield and other bony armor were deposited without further growth. Alternatively, all the different forms may not be species. They might represent stages in the life cycle of a few species. Between stages it would have been necessary for the dermal armor to be resorbed and the form changed before a new shield was laid down.

Like pteraspids, early osteostracans had extensive shields and no paired lateral stabilizers. Later types had short shields, movable paddle-like extensions of the body in the position of pectoral appendages, and horn-like extensions of the head shield just anterior to these paddles (Figures 6–5 and 6-8). Although the paddle-like extensions have been called paired pectoral fins and they seem to be muscular like those of jawed vertebrates, there is no evidence of an endoskeletal shoulder girdle, universal in the limbs of later vertebrates.

Unlike pteraspids, osteostracans had a **heterocercal** tail in which the lobe above the midline of the body was larger and stiffer than the lower lobe. Their heterocercal tail may have resulted in a locomotor system that provided considerable lift, which increased their overall mobility. Apparently osteostracans, like many pteraspids, fed by expanding the pharynx and sucking material from the bottom.

The osteostracans are referred to as models of early jawless vertebrates even though they appear later in the fossil record than pteraspids. Two features have made osteostracans better known than other jawless craniates of the Paleozoic. The first is their single-piece head shield, which resists disintegration better than a series of articulated plates. Second, within this shield is a braincase of cartilage with channels and foramina (small openings for the passage of nerves and blood vessels). The inner surface of the braincase and the channels are lined by thin layers of bone. These internal characters are preserved in sufficient detail to allow reconstruction of the soft anatomy of the head. Eric Stensio and his collaborators in Sweden and England have patiently polished and ground away layer after layer of the precious fossils, taking photographs of each successive layer until they had serial photographs through the complete head shield. From these they could trace in three dimensions the canals and cavities that in life had been paved with bone (Figure 6–8b). The resulting picture of the gill pouches in osteostracans is unique in that two or more pouches seem to have been positioned *anterior* to the otic capsule, and the rest were concentrated compactly in the posterior part of the head shield. Astonishingly, the internal

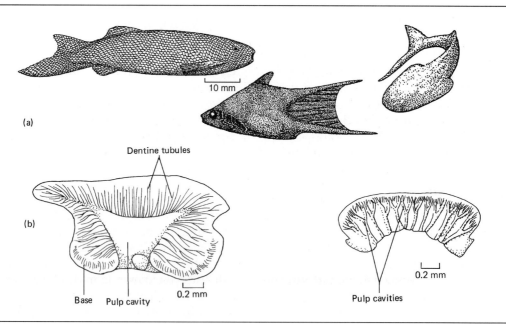

Figure 6–6 (a) Thelodonts, *Phlebolepis* (left), and *Loganellia* (middle), both from the upper Silurian, and (right) an unnamed Devonian species with deep body, forked tail, and a well-developed stomach; (b) cross sections of two types of thelodont scales. (Modified after J. A. Moy–Thomas and R. S. Miles, 1971, *Paleozoic Fishes*, Saunders, Philadelphia, PA; P. Janvier in J. Hanken and B. K. Hall, 1993, *The Skull*, volume 2, University of Chicago Press, Chicago, IL; and M. V. H. Wilson and M. W. Caldwell, 1993, *Nature* 361:442–444.)

anatomy of the brain and nervous system of osteostracans 425 million years old is very similar to that found in the modern lamprey.

Along the dorsolateral edges of the head shield, and sometimes in the center behind the pineal opening, are peculiar fields of thin, irregular small plates (Figure 6–8). These fields form depressions connected to the cranial cavity by huge canals that run through the shield and into the inner ear. Whether these were forerunners of the lateral line, electroreceptors, or even electrogenic organs is unknown. The osteostracans became abundant and diverse during the Devonian, even though competing with older groups and surviving in the presence of jawed vertebrates. In part, their success may relate to these mysterious, unique adjuncts to their nervous system.

An apparently geographically isolated group of early Devonian cephalaspids has come to light during the past two decades in southern China and northern Vietnam. A large diversity of these osteostracan look-alikes have been described as the **Galeaspida** (*gale* = a helmet). They differ from the osteostracans in lacking paired fins and in having a large slit-, bean-, or even heart-shaped opening on the dorsal surface of their head shield. This opening connected with the pharynx and may have acted as an inhalant canal. They apparently had paired nasal cavities in the single inhalant canal. The relationships of the galeaspids are not yet clear (Janvier 1984).

The most recently described major group of Paleozoic jawless fishes, the **Pituriaspids** (pituri is an Aboriginal word used in the region of the fossil discovery), has been found in the mid-Devonian of Australia (Figure 6–8). Superficially they resemble Galeaspids but are thought to have had paired pectoral appendages containing muscles like those of osteostracans. For this reason osteostracans and pituriaspids are grouped with the gnathostomes as **Myopterygii**.

In sediments from the late Silurian through the Devonian a fourth group of cephalaspids, the **anaspids** (*an* = without), are found. All about 15 centimeters long, these freshwater fishes had minnow-like body proportions (Figure 6–9) resembling the

Box 6–2 Interrelationships of Early Craniates

Our understanding of the early evolution and interrelationships of craniates is in a great deal of ferment. Two causes for argument are at the root of the many differing phylogenies proposed. First, the few extant forms are grossly different from the bulk of known fossil forms, especially in their lack of bony tissues, the very basis for fossil remains. This problem is compounded by the fact that the oldest vertebrate fossils are fragmented, incomplete, and rare. Neither of these difficulties is likely to change in the foreseeable future; thus it is useful to examine a few of the differing phylogenies, the basis upon which their authors have proposed them, and why subsequent authors have disagreed. Very few authors have simultaneously treated all the major fossil and living jawless fishes. The names used for the taxa have not been identical even when precisely the same animals were being discussed. For ease of comparison, let us hold the names and number of principal taxa in each clas-

sification more nearly constant than did the original authors and look at some of the cladograms resulting from various hypotheses of jawless fish interrelationships. The reader should realize that the cladograms presented here contain editorial inclusions and deletions that were not part of the originals.

Although periodically proposed throughout this century, the hypothesis that hagfish and gnathostomes are sister groups (Figure 6–7a) has little current support because of the lack of shared derived characters (Halstead 1982, Maisey 1986). A reexamination of the few hagfish embryos known (almost all of them were collected in 1896!) has found nothing inconsistent with a turn-of-the-century notion that the massive tongue-like structure in the floor of the hagfish oral cavity is homologous with the lower jaw of gnathostomes (Gorbman and Tamarin 1985). If other studies outlining extensive homologies in the feeding apparatus of hagfish

Continued

Box 6–2 (*Continued*)

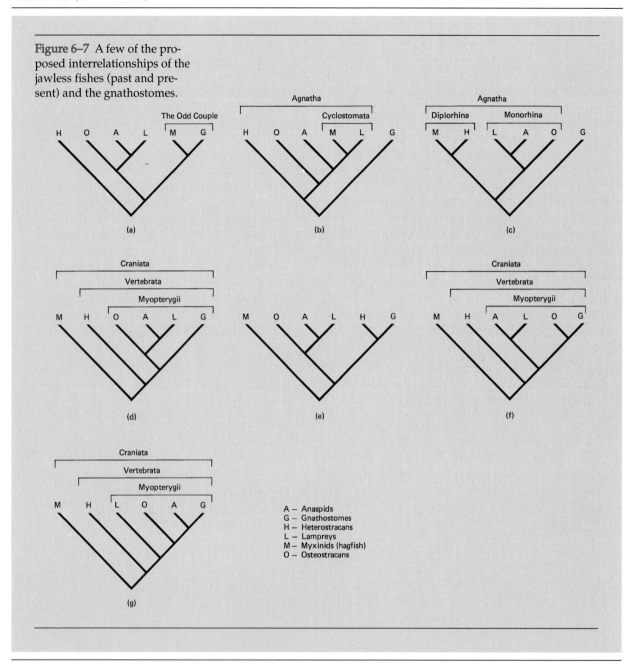

Figure 6–7 A few of the proposed interrelationships of the jawless fishes (past and present) and the gnathostomes.

A — Anaspids
G — Gnathostomes
H — Heterostracans
L — Lampreys
M — Myxinids (hagfish)
O — Osteostracans

probably unrelated thelodonts. Like their osteostracan relatives, the anaspids had a single median nasal opening anterior to the pineal foramen. Narrow scale rows (when they were present) covered the body in a manner similar to those along the posterior part of the osteostracans, but the flat scales were constructed of acellular layers. The head, however, was cov-

ered in most species by a complex of small plates or was naked. Anaspids also differed from osteostracans in having a hypocercal tail. They are considered to have been bottom-dwelling detritus feeders that fed in a head-down position reminiscent of that proposed for the heterostracans. Their stabilizing dorsal, anal, and lateral projections or folds, the spines and

and lampreys (Yalden 1985) prove to be correct, not only would the concept of a valid taxon Cyclostomata be supported (Figure 6–7b), but the sister group relationship of these living jawless vertebrates to the gnathostomes would be supported, a conclusion reached on other grounds by Schaeffer and Thomson (1980).

A very different view of the similarities between hagfish and lampreys has been reached by workers concentrating on the character states of fossil jawless fishes and attempting to interpret the anatomy of living forms in reference to these ancient extinct fishes. One view proposed in the 1920s but now not widely accepted is that of the hagfish as the sister group of heterostracans, and lampreys as the sister group of osteostracans (Figure 6–7c). This concept was based on extensive hypothetical reconstructions by Eric Stensio of the functional anatomy of heterostracans based on what was known of living hagfish (Olson 1971). The circularity of this reasoning has given way in most recent phylogenies to viewing similarities between the hagfish and heterostracans as being due to the retention of numerous ancestral characters in each taxon.

Proponents of the validity of a close relationship between lampreys and hagfish point to the differences between them as recent adaptations to anadromy by lampreys (e.g., osmoregulation) and to burrowing in the deep-sea floor by hagfish (e.g., degenerate not plesiomorphic structure of the eyes, inner ears, and lateral-line system). The majority of recent authors deny a close relationship on the grounds that there is little evidence that most hagfish characters are degenerate rather than ancestral. These authors derive hagfish from the base of the evolution of craniates, sometimes expressly identifying them as the sister group of vertebrates but not themselves vertebrates (Figure 6–7d).

Focusing on the lamprey side of living jawless fish relationships, most authors accept an anaspid/lamprey sister group arrangement but differ on whether these two are closely related to osteostracans (Figure 6–7d and e) or whether the osteostracans are closest to the gnathostomes (Figure 6–7f). This judgment depends entirely on which characters one values and which direction of the character state is considered to be derived in making interpretations. As an example, if acellular bone represented by gnathostome dentine is considered a condition derived from ancestral cellular bone, then, contrary to almost every other proposed arrangement, heterostracans become the sister group of gnathostomes since *all* of their mineralized tissue is acellular. Another entry into the your-guess-is-as-good-as-mine field of attempting to determine interrelationships of jawless fishes and gnathostomes (Figure 6–7g; Maisey 1986) sees anaspid/lamprey relationships as equivocal but offers the possibility of an anaspid/gnathostome sister relationship. We cautiously accept this view in our phylogeny (Figure 6–4). The most recently described groups of deep-bodied thelodonts, Chinese galeaspids and Australian pituriaspids, do nothing to clarify the picture.

In 1889, American paleontologist Edward Drinker Cope wrote, "We are embarrassed in the endeavor to present the relations of the earliest and lowest Vertebrata by want of knowledge of their structure" (see Forey 1984). The embarrassment continues (Forey and Janvier 1993, 1994).

scutes associated with these projections, and the compressed shape of their fusiform bodies probably allowed an agility and locomotor capacity not known in the heterostracans or osteostracans.

During the late Silurian and Devonian most major known groups of extinct jawless craniates coexisted (Figure 6–4). Muscular filter feeding, increased mobility, and the protection that dermal bone afforded were important characters of these animals. Together, these features triggered a proliferation of variations on the craniate theme that spread into the waters of the world. Wherever photosynthesis gave rise to small particulate matter capable of being sucked up and digested, vertebrates

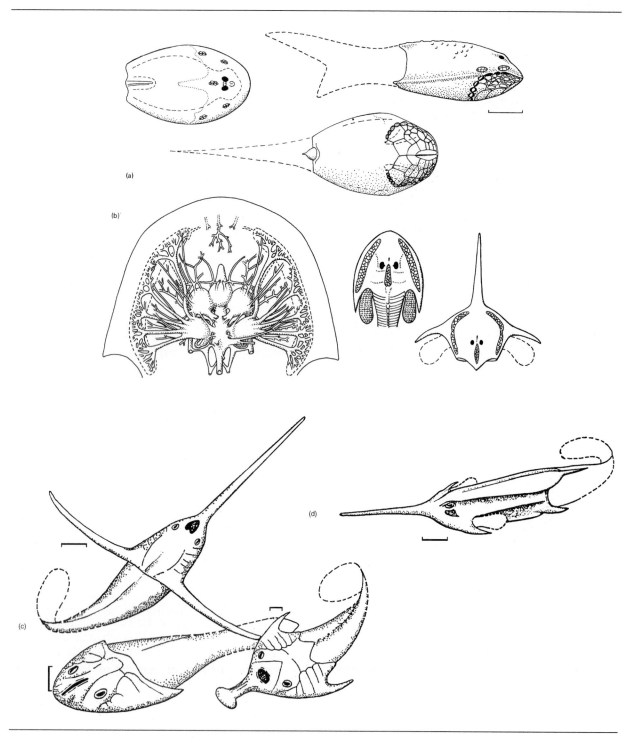

competed successfully with the established inverte-brate lineages. This basic agnathous body plan also gave rise to eyeless, tube-snouted heterostracans, and nearly naked, sucker-mouthed anaspids that left their marks on other organisms.

Here we observe for the first time a phenomenon repeated over and over again in the history of vertebrate life: A basic modification of the craniate framework appears, and a flood of forms using this new modification in conjunction with specializations

Figure 6–8 Details of osteostracans. (a) Plesiomorphic osteostracan, *Tremataspis*, upper Silurian, in lateral (right), dorsal (left), and ventral (below) reconstructions. (b) Derived osteostracans (l–r): reconstruction of the brain and cranial nerves of *Kiaeraspis*, lower Devonian, showing the detailed information obtainable from impressions left on the inner surface of the head shield; *Tyriaspis* of the upper Silurian and the long rostrum *Boreaspis* of the lower Devonian. (c) Diversity of morphology in galeaspids from the lower Devonian of China. (d) Pituriaspid from the lower Devonian of Australia. All size bars one centimeter. (Modified primarily after [a, b] J. A. Moy–Thomas and R. S. Miles, 1971, *Paleozoic Fishes*, Saunders, Philadelphia, PA; [c, d] P. Janvier in J. Hanken and B. K. Hall, 1993, *The Skull*, volume 2, University of Chicago Press, Chicago, IL.)

explodes onto the scene. From a generalized form, vertebrate life radiates in scores of directions to exploit the resources that this innovation has made available.

Extant Jawless Fishes

Two distinct groups of recent fishes, lampreys and hagfish, are jawless. Their fossil record is sparse. Lampreys are known from the Carboniferous—*Hardistiella* from Montana and *Mayomyzon* from Illinois. *Myxinikela*, an undisputed hagfish, and a second possible hagfish relative, *Gilpichthys* (Janvier 1981), have been found in the same deposits as *Mayomyzon* (Nelson 1994).

Extant "agnathans" (a paraphyletic group) have plesiomorphic characters that are present in the earliest craniates. They lack jaws and have no paired appendages to aid them in their locomotion. Some, however, are obligate ectoparasites of other vertebrates. Because they possess round, jawless mouths these fishes have often been combined in the Cyclostomata (*cyclo* = a circle, *stoma* = mouth), lampreys in the Petromyzontidae and hagfish or slimehags in the Myxinidae. The groups show such great differences in morphology as a result of their long phylogenetic separation and their different habits and habitats that they are not considered closely related by most systematists (Figure 6–4). Bone is not a universal characteristic of craniates, as the extant jawless craniates demonstrate.

Hagfishes (Myxinoidea)

The hagfishes (Figure 6–10) are entirely marine. The sister group of the Vertebrata, hagfish lack vertebrae. Over forty recognized species in six genera have nearly worldwide distribution, primarily on continental shelves (Brodal and Fange 1963). Hagfishes are never caught much above the bottom, often in deep regions of the shelf. Some live in colonies, each individual in a mud burrow marked in some species with a volcano-like mound at the entrance. Polychaete worms and shrimps are found in the guts of many species, and they probably live a mole-like existence, finding their prey beneath the ooze or at its surface. They must be active when out of their burrows, for they are quickly attracted to bait and moribund fishes caught in gill nets. Small morphological differences between populations indicate that hagfish are not wide ranging, but rather tend to live and breed locally.

Adult hagfish are generally under a meter in length. Elongated, scaleless, and pinkish to purple in color, hagfishes have a single terminal nasal opening that connects with the pharynx via a broad tube. The eyes are degenerate or rudimentary and covered with a thick skin. The mouth is surrounded by six tentacles that can be spread and swept to and fro by movements of the head when the hagfish is searching for food. Within the mouth two multicusped, horny plates border the sides of a protrusible tongue-like structure. These plates spread apart when protruded and fold together, the cusps interdigitating in a pincer-like action when retracted. The feeding apparatus of hagfish has been described as "extremely efficient at reeling in long worms," because the keratin plates alternately flick in and out of the oral cavity (Mallatt 1985). When feeding on something the size and shape of another fish, hagfishes concentrate their pinching efforts on surface irregularities, such as the gills or the anus, where they can more easily grasp the flesh. Once attached, they can tie a knot in their tail and pass it forward along their body until they are braced against their prey and tear off the flesh in their

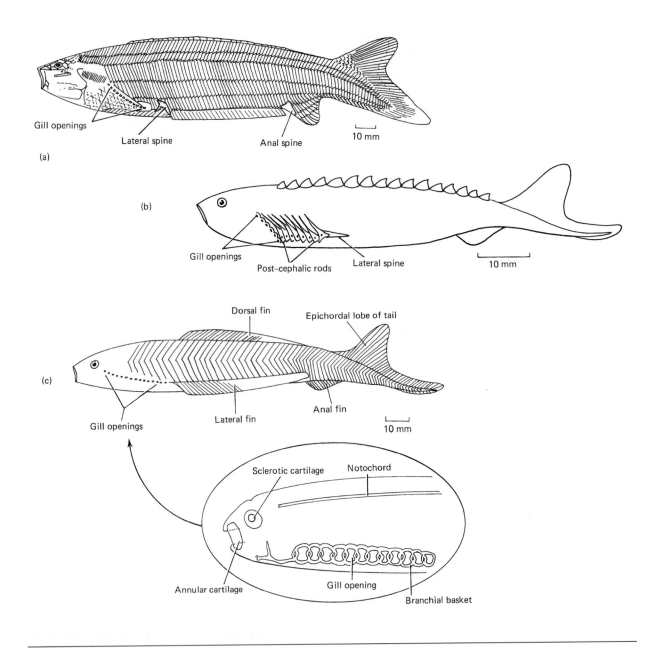

Figure 6–9 Reconstruction of upper Silurian fishes generally considered anaspids.
(a) *Pharyngolepis*; (b) *Lasanius*; (c) *Jaymoytius* showing (inset) internal structures
known from the head region. (Modified after J. A. Moy–Thomas and R. S. Miles,
1971, *Paleozoic Fishes*, Saunders, Philadelphia, PA.)

pinching grasp. Hagfishes take only dead or dying vertebrate prey, and they often begin by eating only enough flesh to enter the coelomic cavity, where they dine on soft parts. Once a food parcel reaches the hagfish's gut it is enfolded in a mucoid bag secreted by the gut wall. This membrane is permeable to digestive enzymes and digestate, and is excreted as a neat wrapper around the feces. The

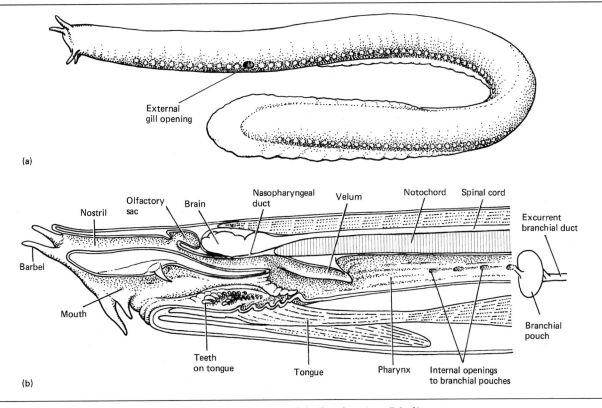

Figure 6–10 Hagfish: (a) lateral view; (b) sagittal section of the head region. (Modified from D. Jensen, 1966, *Scientific American* 214[2]:82–90.)

functional significance of this curious feature is unknown.

Different genera and species of hagfishes have variable numbers of external gill openings. From 1 to 15 openings occur on each side (Figure 6–10), but they do not correspond to the number of internal gills. The external openings occur as far back as the midbody, although the pouch-like gill chambers are just posterior to the head. The long tubes leading from the gills fuse to reduce the number of external openings to which they lead. The posterior position of the gill openings may be related to burrowing.

The internal anatomy of hagfishes is also peculiar. They have no vertebral anlage or plesiomorphic kidneys, and only one semicircular canal on each side of the head. This last character has been the subject of much technical debate. Northcutt (1985) pointed out that in spite of what the membranous structures look like, neurologically they are double sensory patches just as in lampreys, which have two semicircular canals. Hagfishes have long been thought to lack a lateral-line system. Recent work

suggests that *Eptatretus* has traces of the system, but whether this is an ancestral condition or a secondary reduction of ancestrally well-developed structures is not known.

Opening through the body wall to the outside are large mucous glands that secrete enormous quantities of mucus and tightly coiled proteinaceous threads. The latter straighten on contact with seawater to entrap the slimy mucus close to the hagfish's body. This obnoxious defense mechanism (Conniff 1991) is apparently a deterrent to predators. When danger is past, a hagfish draws a knot in its body and scrapes off the mass of mucus, then sneezes sharply to blow its nasal passage clear.

In contrast to all other craniates, hagfishes have accessory hearts in the caudal region in addition to the heart near the gills. They have capacious blood sinuses and very low blood pressure. The several hearts of the hagfishes are aneural, meaning that their pumping rhythm is intrinsic to the hearts themselves rather than coordinated via the central nervous system. The blood vascular system demonstrates few of the immune reactions characteristic of other craniates, and its osmotic concentration is approximately the same as that of seawater (see Chapter 4). Examination of the gonads suggests that at least some species are hermaphroditic, but nothing is known of mating. The eggs are oval and over a centimeter long. Encased in a tough clear covering, the yolky eggs are secured to the sea bottom by hooks and are thought to hatch into small, completely formed hagfish, bypassing a larval stage. Unfortunately, almost nothing is known of the embryology or early life history of any hagfish.

An increased economic interaction between hagfishes and humans over the past two decades is but one example of how vertebrate species are becoming more and more frequently threatened by burgeoning, technological, highly consumptive human society. Fishermen using stationary gear such as gillnets have been retrieving catches damaged beyond sale by scavenging hagfish. It is not surprising that they responded quickly when a commercial value equivalent to that of many food fishes was put on hagfish from a most unusual source: the specialty leather industry. Almost all so-called "eelskin" leather products are made via a proprietary tanning process from hagfish skin. Worldwide demand for this leather specialty led to the eradication of economically harvestable hagfish populations, first in Asian waters, then in some sites along the west coast of North America. Current fishing efforts are focusing on South American and North Atlantic hagfishes. It is typical of human exploitation of natural resources, such as fisheries, that are held in common by society that exploitation often depletes stocks because no attention is given to the biology of the resource and its renewable, sustainable characteristics. For example, we do not even know how long hagfish live, how old they are when they first begin to reproduce, exactly how, when, or where they breed, where the youngest juveniles live, what the diets and energy requirements of free-living hagfish are, or virtually any other of the prerequisite parameters for good management. As a result, "eelskin" wallets will probably become as rare as items made of whalebone (baleen) or ivory.

Lampreys (Petromyzontidae)

Although they are similar to hagfishes in size and shape, the 41 species of lampreys (Figure 6–11) are in other respects radically different from hagfish. They possess vertebrae, although these bones are minute. All but the most specialized lampreys are anadromous; that is, they ascend rivers and streams to breed. Some of the most specialized species are known only from fresh water. The adults neither feed nor migrate, and act solely as a reproductive stage in the life history of the species. Lampreys have a worldwide distribution except for the tropics and high polar regions. Anadromous species that spend some of their life in the sea attain the greatest size, although one meter is about the upper limit. The smallest species are less than one-fourth of this size.

Little is known of the habits of adult lampreys because they are generally observed only during reproductive activities or when captured with their host. Despite their anatomically well-developed senses, no clear picture has emerged of how a lamprey locates or initially attaches to its prey. In captivity, lampreys swim sporadically with exaggerated, rather awkward lateral undulations. They attach to the body of another vertebrate by suction, and rasp a shallow, seeping wound through the integument of the host. The round mouth and tiny esophagus are located at the bottom of a large fleshy funnel, the inner surface of which is studded with horny conical spines. The protrusible tongue-like structure is covered with similar spines, and together these structures allow tight attachment and rapid abrasion of the host's integument. An oral gland secretes an anticoagulant. Feeding is probably continuous when a lamprey is attached to its host.

Lampreys generally do not kill their hosts, but detach, leaving a weakened animal with an open wound. At sea, lampreys have been found feeding on several species of whales and porpoises in addition to fishes. Swimmers in the Great Lakes, after having been in the water long enough for their skin temperature to drop, have reported initial attempts by lampreys to attach to their bodies. The bulk of an adult lamprey's diet consists of body fluids of fishes. The digestive tract is reduced, as is appropriate for an animal that feeds on such a rich and easily digested diet as blood and tissue fluids.

The single nasal opening is high on the head, and continues as a blind-end tube beneath the brain in close proximity to the pituitary gland. The eyes are large and well developed, as is the pineal body, which lies under a pale spot just posterior to the nasal opening. In contrast to hagfishes, lampreys have two semicircular canals on each side of the head—a condition shared with the extinct osteostra-

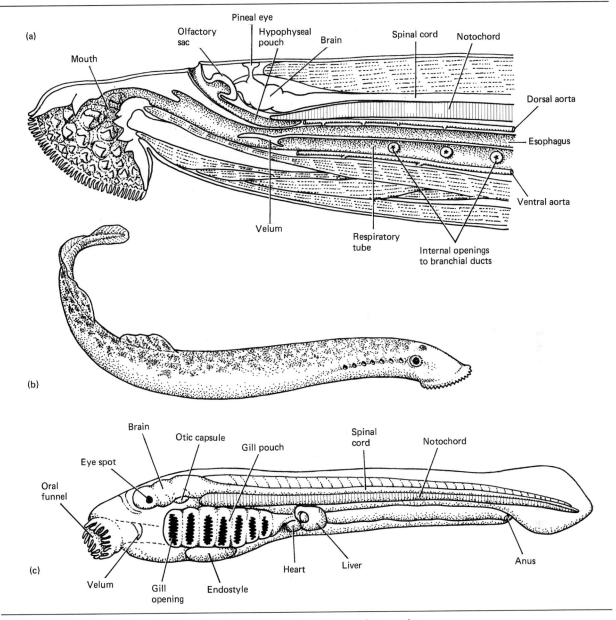

Figure 6–11 Lampreys: (a) sagittal section of the head region; (b) lateral view of an adult; (c) larval lamprey (ammocoete).

cans. The nerves, which exit segmentally along the length of the spinal cord, are completely separated into dorsal and ventral nerve roots, a lamprey peculiarity shared by no other living vertebrate. In addition, the heart is not aneural as in hagfishes, but is innervated by the parasympathetic nervous system. In lampreys these nerves cause cardiac acceleration and not deceleration as do the neural regulators of heart rate in all other craniates.

Seven pairs of gills open to the outside just behind the head. Chloride cells in the gills and well-developed kidneys regulate ions, water, and nitrogenous wastes and maintain the osmolality of the body fluids, allowing the lamprey to exist in a variety of salinities. Female lampreys produce hundreds to thousands of eggs about a millimeter in diameter and devoid of any specialized covering such as that found in the hagfish. Like the hagfish, however, lampreys have no duct to transport the specialized products of the gonads to the outside of the body. Instead, the eggs or sperm fill the coelom, and contractions of the body wall expel them from pores located near the openings of the urinary ducts. Fertilization is external.

Lampreys spawn after a temperature-triggered migration to the upper reaches of streams where the current flow is moderate and the streambed is composed of cobbles and gravel. They construct nests to receive the spawn. Males, later joined by females, select a site, attach themselves by their mouths to the largest rocks in the area, and thrash about violently. Smaller rocks are dislodged and carried a short distance by the current. The nest is complete when a pit is rimmed upstream by large stones and downstream by a mound of smaller stones and sand that producies eddies. Water in the nest is oxygenated by this turbulence, but does not flow strongly in a single direction. The weary nest builders spend the last of their energy depositing eggs and sperm—a process that may take two days. The female attaches to one of the upstream rocks laying eggs and the male wraps around her, fertilizing them as they are extruded. Adult lampreys die after breeding once.

The larvae hatch in about 2 weeks. Radically different from their parents, they were originally described as a distinct genus, Ammocoetes (Figure 6–11c). This name has been retained as a vernacular name for the larval form. A week to 10 days after hatching, the tiny 6- to 10-millimeter-long ammocoetes leave the nest. They are pink, worm-like organisms with a large fleshy oral hood and nonfunctional eyes hidden deep beneath the skin. Currents carry the ammocoetes downstream to backwaters and quiet banks, where they burrow into the soft mud and spend three to seven years as sedentary filter feeders. The protruding oral hood funnels water through the muscular pharynx where food particles are trapped in mucus and swallowed. An ammocoete may spend its entire larval life in the same burrow without any major morphological or behavioral change until it is 10 centimeters or more in length and several years old. Metamorphosis begins in midsummer, and produces a silver-gray juvenile ready to begin its life as a parasite. Downstream migration to a lake or the sea may not occur until the spring following metamorphosis. Adult life is usually no more than two years, and many lampreys return to spawn after one year. Some lamprey species lack parasitic

adults. The larvae metamorphose and leave their burrows to spawn immediately and die.

During this century humans and lampreys have increasingly been at odds. Although the sea lamprey, *Petromyzon marinus*, seems to have been indigenous to Lake Ontario, it was unknown from the other Great Lakes of North America before 1921. The St. Lawrence River flowing from Lake Ontario to the Atlantic Ocean was no barrier to colonization by sea lampreys, and the rivers and streams that fed into Lake Ontario held landlocked populations. During their spawning migrations, lampreys negotiate waterfalls by slowly creeping upward using their sucking mouth, but the 50-meter height of Niagara Falls (between Lake Ontario and Lake Erie) was too much for the most amorous lampreys. Even after the Welland Canal connected Lakes Erie and Ontario in 1829, lampreys did not immediately invade Lake Erie; it took a century for lampreys to establish themselves in Lake Erie's drainage basin.

Since the 1920s lampreys have expanded rapidly across the entire Great Lakes basin. The surprising fact is not that they were able to invade the upper Great Lakes, but that it took them so long to initiate the invasion. Environmental conditions that vary between the lakes may provide the answer to this curious delay. Lake Erie is the most eutrophic and warmest of all the lakes and has the least appropriate feeder streams. Many of these streams run through flat agricultural lands that have been under intensive cultivation since early in the nineteenth century. The streams are silty and frequently have had their courses changed by human activities. Because of the terrain, flow is slow and few rocky or gravel bottoms occur. Perhaps lampreys simply could not find appropriate spawning sites in Lake Erie to develop a strong population.

Once they reached the upper end of Lake Erie, however, lampreys quickly gained access to the other lakes. By 1946 they were known from all the Great Lakes. There they found suitable conditions and were able to expand unchecked until sporting and commercial interests became alarmed at the reduction of economically important fish species, such as lake trout, burbot, and lake whitefish. Chemical lampricides and electrical barriers and mechanical weirs at the mouths of spawning streams have been employed to bring the Great Lakes lamprey populations down to their present level. Although the populations of large fish species, including those of commercial value, are recovering, it may never be possible to discontinue these anti-lamprey measures, costly though they are. Human mismanagement (or initial lack of management) of the lamprey has been to our own disadvantage. The story of the demise of the Great Lakes fishery is but one of hundreds in the recent history of vertebrate life where human failure to understand and appreciate the interlocking nature of the biology of our nearest relatives has led to gross changes in our environment. Introduction of exotic (= not indigenous) species is a primary cause for the decline of many vertebrate species worldwide, especially in aquatic habitats (Allan and Flecker 1993).

Summary

Fossil evidence indicates that craniates probably evolved in a marine environment during the Cambrian. We know little about the group until some forms evolved bony dermal armor. The evolution of bone, muscular pump filter feeding, and increased motility led to at least two distinct groups of jawless craniates: first the Pteraspida, and later the Cephalaspida. The extensive radiation of these forms demonstrates the numerous successful solutions to the problem of growing inside an armored skin. Only two types of survivors from these early radiations exist today: the hagfishes and the lampreys. Nevertheless, extant jawless fishes illustrate the extreme specialization of which the ancestral craniate body plan is capable.

References

Allan, J. D., and A. S. Flecker. 1993. Biodiversity conservation in running waters. *Bioscience* 43(1):32–43.

Blieck, A. 1992. At the origin of chordates. *Geobios* (Lyon) 25:101–113.

Brodal, A., and R. Fange (editors). 1963. *The Biology of Myxine*. Universitetsforlaget, Oslo, Norway.

Carroll, R. L. 1987. *Vertebrate Paleontology and Evolution*. Freeman, New York, NY.

Conniff, R. 1991. The most disgusting fish in the sea. *Audubon* 93:100–118.

Elliott, D. K. 1987. A reassessment of *Astraspis desiderata*, the oldest North American vertebrate. *Science* 237:190–192.

Forey, P. L. 1984. Yet more reflections on agnathan–gnathostome relationships. *Journal of Vertebrate Paleontology* 4(3):330–343.

Forey, P., and P. Janvier. 1993. Agnathans and the origin of jawed vertebrates. *Nature* 361:129–134.

Forey, P., and P. Janvier. 1994. Evolution of the early vertebrates. *American Scientist* 82:554–565.

Gagnier, P. Y. 1989. A new image of *Sacabambaspis janvieri*, an early Ordovician jawless vertebrate from Bolivia. *National Geographic Research* 5:250–253.

Gorbman, A., and A. Tamarin. 1985. Early development of olfactory and adenohypophyseal structures of agnathans and its evolutionary implications. Pages 165–185 in *Evolutionary Biology of Primitive Fishes*, edited by R. E. Foreman et al. Plenum, New York, NY.

Graffin, G. 1992. A new locality of fossiliferous Harding sandstone: evidence for freshwater Ordovician vertebrates. *Journal of Vertebrate Paleontology* 12:1–10.

Halstead, L. B. 1982. Evolutionary trends and the phylogeny of the agnatha. Pages 159–196 in *Problems in Phylogenetic Reconstruction*, edited by K. A. Joysey and A. E. Friddy, Systematics Association Special Volume 21.

Janvier, P. 1981. The phylogeny of the Craniata, with particular reference to the significance of fossil "agnathans." *Journal of Vertebrate Paleontology* 1:121–159.

Janvier, P. 1984. The relationships of the Osteostraci and Galeaspida. *Journal of Vertebrate Paleontology* 4:344–358.

Maisey, J. G. 1986. Heads and tails: a chordate phylogeny. *Cladistics* 2:201–256.

Mallatt, J. 1984. Feeding ecology of the earliest vertebrates. *Zoological Journal of the Linnean Society* 82:261–272.

Mallatt, J. 1985. Reconstructing the life cycle and the feeding of ancestral vertebrates. Pages 59–68 in *Evolutionary Biology of Primitive Fishes*, edited by R. E. Foreman et al. Plenum, New York, NY.

Moy–Thomas, J. A., and R. S. Miles. 1971. *Paleozoic Fishes*. Saunders, Philadelphia, PA.

Nelson, J. S. 1994. *Fishes of the World*, 3d edition, Wiley, New York, NY.

Northcutt, R. G. 1985. The brain and sense organs of the earliest vertebrates: reconstruction of a morphotype. Pages 81–112 in *Evolutionary Biology of Primitive Fishes*, edited by R. E. Foreman et al. Plenum, New York, NY.

Olson, E. C. 1971. *Vertebrate Paleozoology*. Wiley–Interscience, New York, NY.

Repetski, J. E. 1978. A fish from the upper Cambrian of North America. *Science* 200:529–531.

Rich, P. V., and G. F. van Tets. 1985. *Kadimakara, Extinct Vertebrates of Australia*. Pioneer Design Studio, Lilydale, Victoria, Australia.

Romer, A. S. 1966. *Vertebrate Paleontology*, 3d edition. University of Chicago Press, Chicago, IL.

Romer, A. S. 1968. *Notes and Comments on Vertebrate Paleontology.* University of Chicago Press, Chicago, IL.

Schaeffer, B., and K. S. Thomson. 1980. Reflections on agnathan–gnathostome relationships. Pages 19–22 in *Aspects of Vertebrate History*, edited by L. L. Jacobs. Museum of Northern Arizona Press, Flagstaff, AZ.

Soehn, K. L., and M. V. H. Wilson. 1990. A compete articulated heterostracan from Wenlockian (Silurian) beds of the Delorme Group, Makenzie Mountains, Northwest Territories, Canada. *Journal of Vertebrate Paleontology* 10:405–419.

Van der Brugghen, W., and P. Janvier. 1993. Denticles in thelodonts. *Nature* 364:107.

Wilson, M. V. H., and M. W. Caldwell. 1993. New Silurian and Devonian fork-tailed 'thelodonts' are jawless vertebrates with stomachs and deep bodies. *Nature* 361:442–444.

Yalden, D. W. 1985. Feeding mechanisms as evidence for cyclostome monophyly. *Zoological Journal of the Linnean Society* 84:291–300.

THE RISE OF JAWED VERTEBRATES AND THE RADIATION OF THE CHONDRICHTHYES

7

Relatively soon after the first evidence of vertebrates in the fossil record, the next major step in vertebrate evolution appeared: jaws and internally supported paired appendages. Jaws play a variety of roles in the biology of vertebrates, but their major use is in feeding. Paired fins were a second important innovation because they gave a swimming vertebrate precise control of steering. The diversity of predatory specializations available to a vertebrate with jaws and precise steering is great, and the appearance of these characters signaled a new radiation of vertebrates. The cartilaginous fishes (sharks, rays, and ratfish) are the descendents of this radiation, and they combine derived characters such as a cartilaginous skeleton with a generally plesiomorphic anatomy. Sharks have undergone three major radiations, which can be broadly associated with increasingly specialized feeding mechanisms, and extant sharks are a diverse and successful group of fishes. In this chapter we consider the origin of jaws and paired appendages and the roles that these two innovations played in the success of cartilaginous fishes.

The First Appearance of Jaws and Unique Gnathostome Characters

Fossils of vertebrates with jaws are known from the mid-Silurian. It may seem strange that a major new morphological feature like jaws should arise before the extensive radiation of agnathous fishes occurred instead of arising from some later product of that radiation. This pattern of evolution, however, is seen over and over again in an examination of vertebrate life—major new innovations arise from less specialized members of a lineage.

Some ecological niches were so suited to the agnathous body plan that no evolutionary advance yet realized has been able to replace it. Lampreys and hagfishes provide evidence of this fact. The great majority of agnathous fishes, however, succumbed to what is generally thought to have been competition from jawed vertebrates. Actually, no one knows exactly what characteristics of the jawed vertebrates may eventually have led to the near exclusion of jawless fishes.

Albert Sherwood Romer (1962) suggested that "perhaps the greatest of all advances in vertebrate history was the development of jaws and the consequent revolution in the mode of life of early fishes." Jaws allow behaviors that otherwise would be difficult, if not impossible, to perform. The presence of jaws manipulated by muscles allows an organism to grasp objects firmly. When the jaws are armed with teeth their grip becomes surer. Teeth with sharp cutting edges reduce food particles to edible size and flat teeth grind hard foods. When vertebrates evolved jaws, therefore, new food sources became available. Jaws apparently placed the early gnathostomes in a commanding position, for many increased in size and gnathostomes appear to have replaced many lineages of jawless vertebrates during the Devonian.

The functions of jaws are not limited to capturing and chewing prey. A grasping, movable jaw permits a new behavior—manipulation of objects—that enters many aspects of the life of vertebrates. Jaws can be used to dig holes, or to carry pebbles or vegetation to build nests, or to grasp mates during courtship and juveniles during parental care. No wonder Romer placed so much importance on jaws.

Jawed fishes appear in the fossil record fully developed, without intermediates (Moy–Thomas and Miles 1971, Romer 1966). The Silurian fossils consist of detached spines, scales, teeth, and jaws. The first jaws in the fossil record give little insight into their evolutionary history. Our belief that jaws originated through modification of the gill arch skeleton comes from detailed studies of comparative anatomy and embryology. The neural crest seems to have been the key tissue in the evolution of jaws and of many of the other derived characters of gnathostomes (Chapter 3). Some neural crest cells migrate to the visceral arches to form the bilaterally paired upper palatoquadrate and lower mandibular cartilages that form the jaws (Figure 7–1). Between the jaws and the rest of the branchial skeleton lies the second visceral arch, known as the hyoid arch, which is ancestrally associated with supporting the jaws and bracing them against the cranium. The neural crest cells also contribute to the branchial skeleton, ancestrally consisting of five bilaterally paired series of hinged skeletal elements, four in each arch on each side extending from the vertebral column to the midventral line, where a single unpaired element unites the bilateral arches. This gill skeleton is entirely *internal* to the *ectodermal* gills, whereas the gill skeleton of jawless fish is *external* to the *endodermal* gills. Fossil forms on both sides of the jawless to jawed transition appear to have exactly the same dichotomy.

This confusing position and origin of the gills relative to their supports has long been a stumbling block to the otherwise reasonable hypothesis that some jawless vertebrate gave rise to jawed vertebrates. Careful reanalysis of the data (Mallatt 1984) generated a hypothesis to solve the problem. Extant sharks have extrabranchial cartilages in the locations of the branchial arch cartilages of lampreys, and lampreys may have both endodermal and ectodermal contributions to their gill surfaces. A common ancestor for those jawless fishes closest to gnathostomes

might have had skeletal support on *both* the internal and external sides of gills that included *both* ectodermal and endodermal components (Figure 7–1c). Such dual support could be explained by an increase in the size of the food items eaten by pre-gnathostomes. Perhaps connective tissue within the margin of the pharyngeal lumen chondrified, allowing pre-gnathostomes to trap larger particles without damaging the delicate respiratory lamellae. This new skeletonization would have set the stage for that later evolution of jaws.

A series of distinctive additional derived characters of the cranium, trunk, and mechanosensory apparatus is evident in fossil gnathostomes and is retained in extant forms (Table 7–1). The numerous traits distinguish extant gnathostomes from lampreys and especially from hagfish, although these are generally not detectable in fossils. They include the increased rotational action of the external ocular muscles; myelination of nerve fibers; and two types of the contractile protein actin, one specific to smooth muscle and the other to striated muscle (lamprey actins seem to be of one type only). The soft anatomy of gnathostomes includes several derived elements unknown in jawless fishes: a spiral valvular intestine, a renal portal venous flow, distinct oviducts and mesonephric ducts, a pancreas with both endocrine and exocrine functions, and a spleen. Several endocrine hormones, discrete endocrine glands, and storage/mobilization mechanisms for metabolites are derived characters of gnathostomes. All these distinguishing characteristics emphasize both the long evolutionary isolation of extant jawed and jawless fishes and the major innovations of the early gnathostome lineage.

Locomotor function was improved by a well-developed heterocercal tail and fin webs supported by collagenous fin rays. Undoubtedly overshadowing these is the most outstanding shared derived character of the gnathostomes besides the jaw—paired pectoral and pelvic appendages with internal supporting girdles.

The Origin of Fins

Jaws are an advantage only when applied to an object. Suction can draw objects into the mouth over modest distances, but generally the body must be guided to the graspable object. This sounds simple, but guidance of a body in three-dimensional space is complicated. Yaw (swinging to the right or left) com-

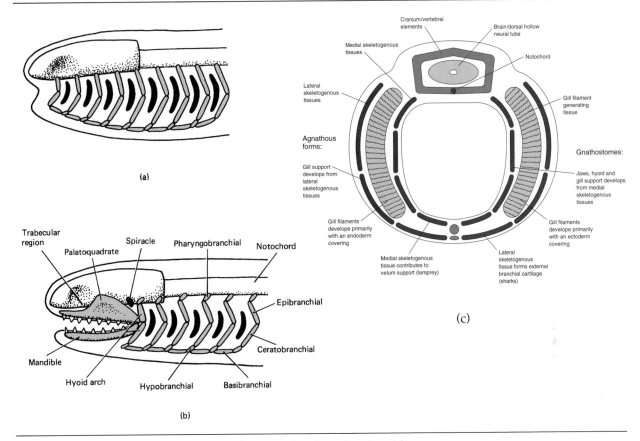

Figure 7–1 Evolution of the vertebrate jaw from anterior visceral arches: (a) agnathous condition; (b) gnathostomous condition; (c) diagrammatic cross section of a generalized craniate pharynx showing relationships between agnathous and gnathostome visceral arch components.

bines with pitch (tilting up or down) to make accurate contact with a target difficult. Roll (rotation around the body axis) must be countered for effective grasping. Especially if the target moves, perhaps evasively, quick adjustments of roll, pitch, and yaw are necessary. It is little wonder that the development of strong mobile fins was coincident with the evolution of jaws.

Fins act as hydrofoils, applying pressure to the surrounding water. Because water is practically incompressible, force applied by a fin in one direction against the water is opposed by an equal force in the opposite direction (Figure 7–2). Thus fins can resist roll if pressed on the water in the direction of the roll. Fins projecting horizontally near the anterior end of the body similarly counteract pitch. Yaw is controlled by vertical fins along the mid-dorsal and mid-ventral lines. Fins serve other functions as well.

They increase the area of the tail for greater thrust during propulsion. Presented at angles to a flow of water, they produce lift. Spiny fins are used in defense, and become systems to inject poison when combined with glandular secretions. Colorful fins are used to send visual signals to potential mates, rivals, and predators.

Fin structure of early fishes was variable. Agnathans had spines or enlarged scales derived from dermal armor that acted like fins. Osteostracans had paddle-like pectorally located structures without internal supports. Some anaspids had long fin-like sheets of tissue running along the flanks. The earliest paired fins, although a universal feature of gnathostomes, were dissimilar in details of structure. Two early groups of gnathostomes, the **acanthodians** and the **placoderms**, illustrate this. Acanthodians had a variable number of spines, sometimes with

Table 7–1 Derived characters of gnathostomes found in fossil and extant forms.

Cranial Characters

1. Paired trabeculae derived from the neural crest contribute to the floor of the cranium beneath the enlarged forebrain
2. Cranium enlarged anteriorly to end in a precerebral fontanelle
3. Incorporation of one or more occipital neural arches into the rear of the cranium
    3a. Cranium enlarged posteriorly, extending the position of the foramen magnum posteriorly
4. Addition of a third (horizontal) semicircular canal
5. Presence of calcium carbonate stataconia or otoliths
6. Development of a postorbital process on the chondrocranium separating the functions of supporting the jaws and enclosing the eye
7. Basicranial muscles that originate on the cranium and insert on the branchial arches

Trunk Characters

8. A horizontal septum of connective tissue dividing the trunk musculature into dorsal (epaxial) and ventral (hypaxial) units
9. Neural and haemal arches (usually mineralized) regularly appear along notochord

Sensory Characters

10. A unique and evolutionarily conservative pattern of cephalic lateral-line canals
11. Lateral line on the trunk region flanked or enclosed by specialized scales

Source: J. G. Maisey, 1986, Cladistics 2:201–256.

attached membranes, extending along the ventrolateral aspects of the trunk. The internal skeleton of acanthodian fins (Figure 8–17) was composed of at least two rows of rod-shaped cartilages associated with horny, thread-like rays that did not protrude far into the fin. Placoderm pectoral appendages generally had more numerous rod-shaped elements and horny fin rays extending far into the fin. Early fins, although always in approximately the same positions, were clearly different in internal and external construction and number.

There is little fossil evidence of the origin of fins, especially the paired pectoral and pelvic fins that were significant in later stages of vertebrate evolution. Although some early workers fancied a resemblance between the pectoral fins and the gill arches, a branchial origin is extremely doubtful. Fins are derived from mesoderm and have somatic, not visceral, musculature and innervation.

The fin-fold theory of the origin of paired appendages was long held in high regard. This theory was based on certain anaspids such as *Jaymoytius* (Figure 6–9) that had a pair of long-based lateral flaps extending from gills to anus. Early acanthodians had pairs of spines in the same region. By comparing this hypothetical series to the fin folds of amphioxus, discrete fins were pictured as originating from continuous fin folds. These fin folds were thought to be paired laterally but single dorsally and posteriorly. Broken into short segments and reduced in number, fin folds were said to have been the origin of the fins seen in today's fishes. But making a direct connection between the three constructions was merely an artificial assemblage of organisms with no direct phylogenetic relationship.

Because no fossil evidence to document these events has been uncovered, the fin-fold theory has become less appealing and it seems best to consider fins so beneficial that multiple evolutions have occurred. A multiple evolution with similar results is to be expected when only a limited variety of fin positions and shapes provides the hydrodynamic advantages we attribute to fins. Thus, the fins of the jawless ostracoderms may be convergent with those of gnathostomes. Early fossil gnathostomes may or may not have had strictly homologous paired appendages The girdles and basal elements of the paired appendages of extant gnathostomes seem to be homologous.

A well-developed heterocercal caudal fin is almost as universal a feature of all early jawed fishes as were paired fins. Comparably developed caudal fins were found in some jawless fish, although other jawless fish had hypocercal or other shaped tails. An abruptly up- or downturned notochord produces an increased depth of the caudal fin, a structure that is important in rapid acceleration (Webb and Smith 1980). In addition, the gnathostomes with their collagenous fin rays had a stiffened caudal web of considerable area, further enhancing acceleration. Burst swimming is important in predator avoidance, and provides significant economies in terms of locomotor energy when bursts of acceleration are alternated with glides. All fishes with a fin-strengthening upturned or downturned axial skeleton tip have a noncollapsible caudal fin that is effective for burst swimming.

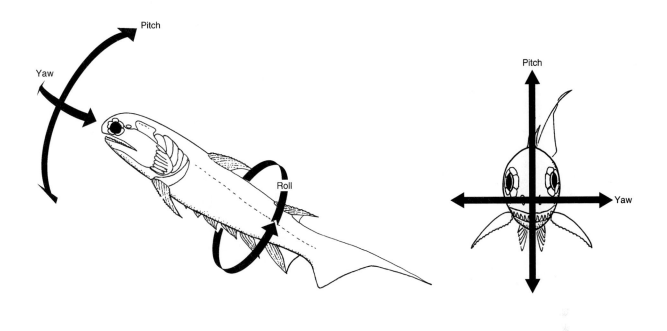

Figure 7–2 An acanthodian, *Climatius*, as seen in lateral and frontal views to show the orientation of pitch, yaw, and roll and the fins that counteract these movements.

The gnathostomes appear as four distinctive clades, all flowering in the Devonian—the age of fishes. One of these clades, the **placoderms**, was isolated from the other three (Chondrichthyans, Acanthodians, and Osteichthyans) despite sharing many derived gnathostome characters. The jaw muscles of placoderms are different from the jaw muscles of the other three groups, placoderms have nothing comparable to the teeth of the other gnathostomes, and the skeletal anatomy of the paired fins of placoderms lacks homologies with those of other gnathostomes. The placoderms seem to have left no descendants in the modern fauna. A second clade, the **Chondrichthyes**, clearly related to all other gnathostomes, evolved distinctive reductions and specializations of dermal armor, internal calcification, jaw and fin mobility, and reproduction. These chondrichthyans have successfully survived to the present.

The final two clades of fishes, the Acanthodians and Osteichthyes (together = Teleostomes), may be closely related and form the root of all subsequent vertebrate evolution (Chapter 8). Before turning to this majority of fish species past and present, we turn to the placoderms and chondrichthyans to examine the variety of early gnathostomes.

Placoderms: The Armored Fishes

Among the earliest gnathostomes in the fossil record is a confusingly diverse assemblage of generally heavily armored fishes, the placoderms (Figure 7–3). R. L. Carroll (1987) has pointed out that the placoderms are without modern analogs, and their massive external armor makes interpretation of their ways of life particularly difficult. Placoderms must have been primarily benthic fishes; their bodies were generally dorsoventrally depressed with flattened ventral surfaces. Although placoderms share an impressive list of derived characters with other gnathostomes, several elements of their morphology appear to isolate placoderms from all other jaw-

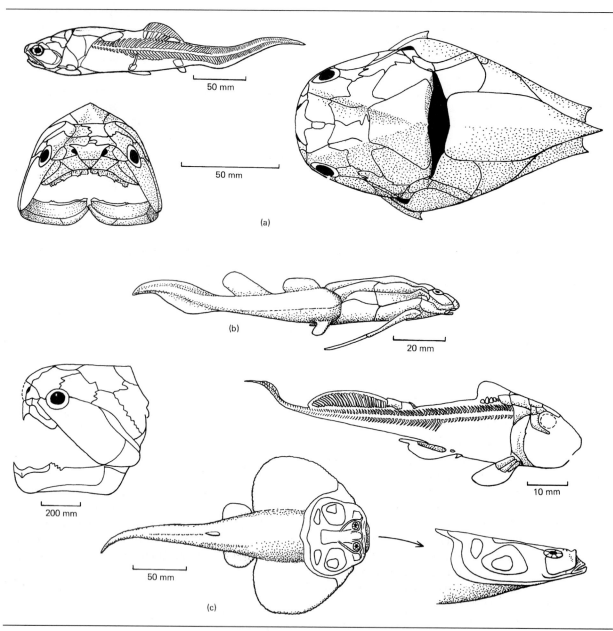

Figure 7–3 Placoderms: (a) lateral, frontal, and dorsal views of an arthrodire, *Coccosteus*, middle Devonian; (b) the peculiar placoderm *Bothriolepis*, with a jointed exoskeleton that supported pectoral appendages; (c) three widely varying types of placoderms: (left) the giant predator *Dunkleosteus*, upper Devonian; (right) the chimera-like *Rhamphodopsis* (see also Figure 7–11b); (below) the ray-like *Gemuendina*. (Modified after J. A. Moy–Thomas and R. S. Miles, 1971, *Paleozoic Fishes*, Saunders, Philadelphia, PA.)

bearing vertebrates. The most profound of these characters is the position of the jaw musculature. In all other gnathostomes the jaw muscles lie external to the jaw's skeletal elements. In those placoderms where it can be determined, it appears that the main mass of the jaw musculature is *medial* to the palato-quadrates. If this is true for placoderms in general, jaws may have evolved more than once among ancestral fishes, and the immediate common ancestor of placoderms and all other gnathostomes may not have had a functional jaw. Placoderms also lack teeth that correspond to those of any other gnathostome. They have a hyoid arch that is not clearly involved in the same suspensory function as in other gnathostomes and is distinctively different in arrangement and number of elements from that of other vertebrates.

Superficially the earliest placoderms, the arthrodires, resembled pteraspids and cephalaspids in appearance and habitat (Figure 7–3a). As the name placoderm (*placo* = plate, *derm* = skin) implies, they were covered with a thick, often ornamented bony shield over the anterior one-half to one-third of their bodies. By their extinction in the early Carboniferous, some placoderms had muscular, mobile pectoral fin structures that must have contributed to an active existence. Other placoderms, the antiarchs, developed pectoral appendages into stiff props by encasing all soft tissue in joined, bony tubes reminiscent of arthropod appendages (Figure 7–3b).

The head shield of early placoderms, which was formed by numerous large plates, was separated from the mosaic of small dermal bones that shielded the trunk by a narrow gap. A mobile connection between the anterior vertebrae and the skull allowed the head to be lifted. This craniovertebral joint permitted the mouth to be opened wider than would lowering only the mandible, or to be opened when the lower jaw was pressed against the seafloor as a placoderm quietly awaited the approach of its prey. During their evolution, predatory arthrodire placoderms developed a curious specialization further increasing the gape. The space between the head and the thoracic shields widened and a pair of joints evolved, one above each pectoral fin on a line that passed through the older craniovertebral joint of the axial skeleton. This arrangement provided great flexibility between the shields and allowed an enormous head-up gape. In addition, it probably increased respiratory efficiency and improved steering control. **Arthrodires**, the name given to this order of placoderms, pays tribute to this strange specialization: *arthros* (a joint) plus *dira* (the neck).

Placoderms were mostly creatures of the Devonian. During that period placoderms radiated into a large number of lineages and types. Ancestral placoderms were primarily marine but a great many lineages became adapted to freshwater and estuarine habitats. They nevertheless maintained their robust armor, indicating their ability to metabolize calcium phosphate in phosphate-poor environments. *Dunkleosteus* was a voracious, 10-meter-long, predatory arthrodire. *Bothriolepis*, an antiarch, supported itself on stilt-like pectoral fins. Another group of placoderms had the palatoquadrate firmly attached to the cranium and complex, solid tooth plates for crushing shellfish. In some placoderms, sexually dimorphic pelvic appendages suggest that internal fertilization occurred, probably coupled with complex courtship. Other groups show a tendency for exaggerated dorsoventral flattening, eyes on top of the head, and subterminal mouths, all indicating benthic specialization. *Gemuendina* bears a striking resemblance to modern skates, although it was completely armored with a mosaic of small plates and could not have used its broad pectoral fins in the skate's undulating manner of locomotion.

Placoderms lacked teeth. Slightly modified dermal bones lined the jaw cartilages of placoderms and, though they had long knife-like cutting edges and strong sharp points for slicing and piercing prey, they were subject to wear and breakage without replacement. The jaws of placoderms were often immovably bound to the cranium or tightly articulated to the rest of the head shield. This prevented their participation in any sucking action, a very successful feeding process as shown by its success in the jawless fishes and again in the vast majority of extant jawed fishes. The consensus among paleoichthyologists is that placoderms became extinct in the lower Carboniferous without giving rise to any surviving forms.

Chondrichthyes: The Cartilaginous Fishes

The sharks and their relatives first appear in the fossil record in the early Devonian. Since then, morphological refinements in some of their systems have evolved to levels surpassed by few other extant vertebrates; yet they retain many ancestral elements in their basic anatomy, and sharks have long been used to exemplify an ancestral vertebrate body form. Even at the molecular level these vertebrates appear to retain many ancestral character states (Martin et al. 1992). Identified by a cartilaginous skeleton, extant forms can be divided into two groups: those with a single gill opening on each side of the head and those with multiple gill openings on each side. The first group is called the **Holocephali** because of the undivided appearance of the head that results from having a single gill opening. Their common names of ratfish and chimaera come from their bizarre form: a long flexible tail, a fishlike body, and a head with big eyes and buckteeth that resembles a caricature of a rabbit. The second group is the **Elasmobranchii**, meaning plate-gilled. Elasmobranchs include the sharks, cylindrical forms with five to seven gill openings on each side of the head. They are also known as the Pleurotremata or Selachii. At least two distinct clades of extant sharks can be distinguished, the squalomorphs and the galeomorphs. A third group of elasmobranchs are dorsoventrally flattened with five gill slits on the ventral surface (the skates and rays, Hypotremata or Batoidea) (Greenwood et al. 1973).

Evolutionary Specializations of Elasmobranchs

In spite of a rather good fossil record, the phylogeny of cartilaginous fishes remains unclear. Early elasmobranchs, like extant species, were diverse in habits and habitats. Their initial radiation from a common ancestor emphasized changes in teeth, jaws, and fins. Apparently, the feeding and locomotor apparatus evolved at different rates within different lineages. In some lineages, derived dentition was coincident with ancestral fin structures, whereas the opposite combination is seen in others. As a result, fossil elasmobranchs display confusing mosaics of ancestral and derived characters.

Through time different lineages of elasmobranchs tended to accumulate similar but not identical modifications in their feeding and locomotor structures, presumably because of similar selective pressures. This pattern of similar adaptations in related lineages is an example of parallel evolution: When similar selective forces act on similar body forms and developmental mechanisms, certain modifications appear independently and often repeatedly in the course of time.

Adaptations found in several lineages are known as **general** or **broad adaptations**. Examples are paired appendages, jaws, and use of muscular pump filter feeding. Broad adaptations have penetrating effects on the organisms' integration of behavior, physiology, and morphology. Broad adaptations define an organizational level in a horizontal (not phylogenetic) classification. Species showing characters of a particular organizational level belong to the same **grade**. A grade may contain different phylogenetic lineages. Each phylogenetic lineage is called a **clade**. For elasmobranchs we can define three grades in their evolution, but at present it seems that only one of the clades was involved in the more recent evolutionary history of elasmobranchs (Figure 7–4) (Greenwood et al. 1973, Carroll 1987).

The Earliest Elasmobranch Radiation

The earliest elasmobranchs are identified by the form of the teeth common to the majority of the species— basically three-cusped with little root development (Figure 7–5). Although there is evidence of bone around their bases, the teeth are primarily dentine structures capped with an enamel-like coat. The central cusp is the largest in *Cladoselache*, the best known genus, and smallest in *Xenacanthus*, a more specialized form.

Cladoselache was shark-like in appearance, about 2 meters long when fully grown, with large fins and mouth and five separate external gill openings. The mouth opened terminally and the chondrocranium had several large areas for the tight ligamentous attachment of the upper jaw cartilage, the palatoquadrate. The jaw also obtained some support from

Figure 7–4 Phylogenetic relationships of placoderms, chondrichthyans, and teleostomes. This diagram shows the probable relationships among the major groups of basal gnathostomes. Extinct lineages are marked by a dagger (†). The numbers indicate derived characters that distinguish the lineages. Only the best-corroborated relationships are shown. The circle indicates that the relationships of the lineages at that node cannot yet be defined.

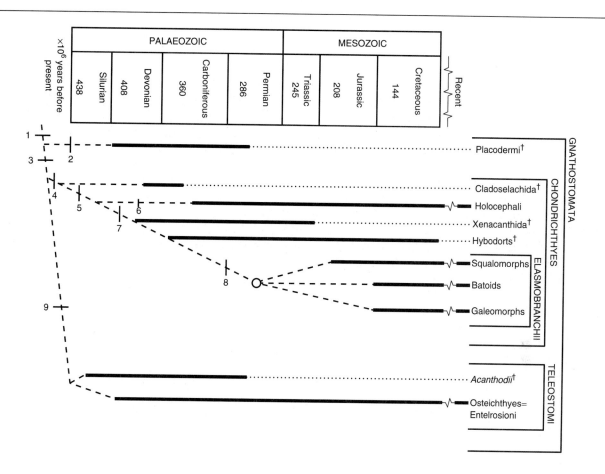

1. Gnathostomata: Jaws formed of bilateral palato-quadrate (upper) and mandibular (lower) cartilages of the mandibular visceral arch at some stage of development, teeth containing dentine, modified hyoid gill arch, branchial elements internal to gill membranes, branchial arches contain four elements on each side plus one unpaired ventral median element, paired trabeculae contribute to cranium, three semicircular canals, internal supporting girdles associated with pectoral and pelvic fins, myelinated nerve fibers, calcium carbonate statoconia or otoliths, and 34 additional derived characters. 2. Placodermi: Tentatively placed as the sister group of all other gnathostomes, but see Gardiner (1984) for a different view. Provisionally united by the following derived characters: a specialized joint in the neck vertebrae, a unique arrangement of dermal skeletal plates of the head and shoulder girdle, a distinctive articulation of the upper jaw, a unique pattern of lateral line canals on the head. 3. Chondrichthyes plus Teleostomi (Eugnathostoma): Epihyal element of second visceral arch modified as the hyomandibula, which is a supporting element for the jaw. 4. Chondrichthyes: Unique perichondral and endochondral mineralization as prismatic hydroxyap-atite tesserae, placoid scales, unique teeth and tooth replacement mechanisms, distinctive characters of the basal and radial elements of the fins, inner ear labyrinth opens externally via the endolymphatic duct, distinctive features of the endocrine system. 5. Claspers on male pelvic fins and at least four additional fin-support characters. 6. Holocephali: Hyostylic jaw suspension, gill arches beneath the braincase, dibasal pectoral fin, dorsal fin articulates with anterior elements of the axial skeleton, 24 additional derived characters of extant chimaeras. 7. Tribasal pectoral, shoulder joint narrowed, basibranchial separated by gap from basihyal. 8. Elasmobranchii: Pectoral fin with three basal elements, the anteriormost of which is supported by the shoulder girdle, characteristics of the nervous system, cranium, and gill arches, 25 additional derived characters. 9. Teleostomi: Hemibranchial elements of gills not attached to interbranchial septum, bony opercular covers, branchiostegal rays, five additional characters. (Based on G. V. Lauder and K. F. Liem, 1983, *Bulletin of the Museum of Comparative Zoology* 150:95–197; J. G. Maisey, 1986, *Cladistics* 2:201–256; and B. G. Gardiner, 1984, *Bulletin of the British Museum (Natural History) Geology* 37:173–427.)

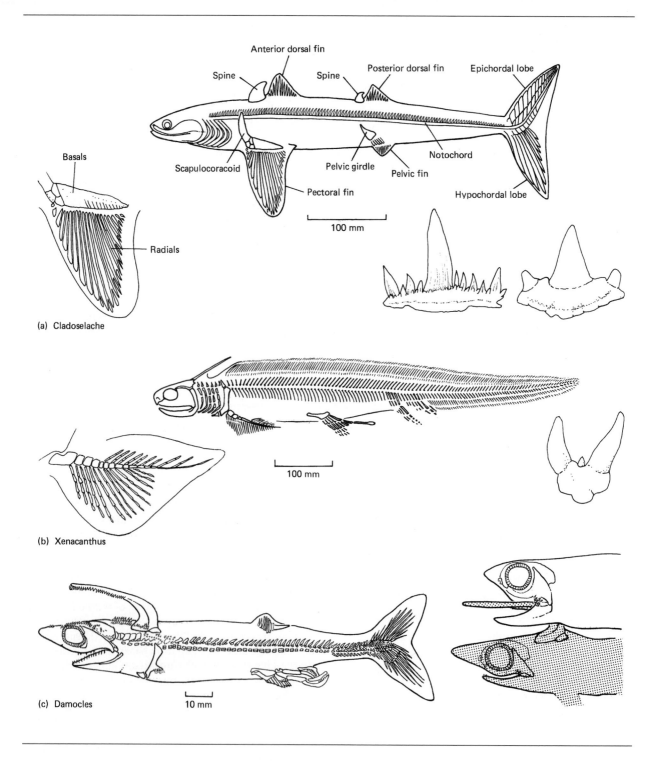

Basals

Radials

(a) Cladoselache

Anterior dorsal fin

Spine

Spine

Posterior dorsal fin

Epichordal lobe

Scapulocoracoid

Pelvic girdle

Notochord

Pelvic fin

Pectoral fin

Hypochordal lobe

100 mm

(b) Xenacanthus

100 mm

(c) Damocles

10 mm

the second visceral arch, the hyoid arch. The name **amphistylic** (*amphi* = both, *styl* = pillar or support) is applied to this mode of multiple sites of upper jaw suspension. The gape was large, the jaws extending well behind the rest of the skull. The three-pronged teeth were probably especially efficient for feeding on fishes that could be swallowed whole or severed by the knife-edge cusps.

Figure 7–5 Early elasmobranchs: (a) *Cladoselache* with details of the pectoral structure and teeth of the *"Cladodus"* type; (b) *Xenacanthus*, a freshwater elasmobranch
with details of its archipterygial pectoral fin structure and peculiar teeth; (c) left,
male *Damocles serratus*, a 15-centimeter shark from the late Carboniferous showing
sexually dimorphic nuchal spine and pelvic claspers; right, male (below) and
female (above) as fossilized, possibly in courtship position. (Modified after J. A.
Moy–Thomas and R. S. Miles, 1971, *Paleozoic Fishes*, Saunders, Philadelphia, PA;
and R. Lund, 1985, *Journal of Vertebrate Paleontology* 5:1–19, and 1986, *Journal of
Vertebrate Paleontology* 6:12–19.)

Wear of teeth, which renders them less functional, is a problem faced by all vertebrates. The earliest sharks, and also some acanthodians and early
bony fishes, solved this problem in a unique way.
Each tooth on the functional edge of the jaw was but
one member of a tooth whorl, attached to a ligamentous band that coursed down the inside of the
jaw cartilage deep below the fleshy lining of the
mouth. Aligned in each whorl in a file directly
behind the functional tooth were a series of developing teeth. In extant sharks essentially the same
dental apparatus is present. Tooth replacement is
rapid: Young sharks under ideal conditions replace
each lower jaw tooth every 8.2 days and each upper
jaw tooth every 7.8 days. If *Cladoselache* replaced its
teeth, as seems likely, a significant advantage in feeding mechanics is indicated for elasmobranchs compared to their placoderm contemporaries.

The body of *Cladoselache* was supported only by
a notochord, but cartilaginous neural arches gave
added protection to the spinal cord. The fins of *Cladoselache* consisted of two dorsal fins, paired pectoral
and pelvic fins, and a well-developed forked tail.
The first and sometimes the second dorsal fins were
preceded by stout spines, triangular in cross section
and thought by some to have been covered by soft
tissue during the life of the shark. The dorsal fins
were broad triangles with an internal structure consisting of a triangular basal cartilage and a parallel
series of long radial cartilages that extended to the
margin of the fin. The pectoral fins were larger but
similar in construction.

Among the early radiations of sharks, almost
every species seems to have had a different sort of
internal pectoral fin arrangement (Figure 7–5), but
all possessed basal elements that anchored the pectoral fins in place. From their structure the pectoral
fins appear to have been hydrofoils with little
capacity for altering their angle of contact with the
water. The pelvic fins were smaller, but otherwise
shaped like the pectorals. Some species in genera

other than *Cladoselache* show evidence of **claspers**:
male copulatory organs. No anal fin is known; it is
also lacking in many extant sharks.

The caudal fin of *Cladoselache* is distinctive (Figure 7–5). Externally symmetrical, its internal structure was asymmetrical and contained a band of sub-
chordal elements resembling the hemal arches that
protect the caudal blood vessels in extant sharks.
Long, unsegmented radial cartilages extended into
the hypochordal (lower) lobe of the fin. At the base
of the caudal fin were paired lateral keels that are
identifying characteristics of extant rapid pelagic
(open water) swimmers.

The integument included only a few scales, but
these resembled the teeth. Cusps of dentine were
covered with an enamel-like substance and contained a cellular core or pulp cavity. Unlike a tooth,
each scale had several pulp cavities to match its several cusps. These scales were limited to the fins, the
circumference of the eye, and within the mouth
behind the teeth. Their structural similarities leave
little doubt that the teeth of early sharks and other
vertebrates in general are derived from specialized
elements of the integument.

We can piece together the lives of many of the
early elasmobranchs from their morphology and fossil localities. Probably pelagic predators, most early
sharks, and *Cladoselache* in particular, would have
swum after their prey in a sinuous manner, engulfing fishes whole or slashing them with dagger-like
teeth. The lack of body denticles and calcification
suggests a tendency to reduce weight and thereby
increase buoyancy.

Reproduction of some forms involved internal
fertilization as evidenced by pelvic claspers, imply-
ing that a behavioral system existed to ensure successful mating. Also implied by claspers is a reproductive strategy of producing young potentially
retained within the protection of the body of the
mother for some period after fertilization. Since the
size and resources of the female are finite, such reten-

tion would result in a relatively small number of young. Thus as much as 350 million years ago, the elasmobranchs and possibly all chondrichthyans had evolved a reproductive life history with profound biological and evolutionary constraints on the population biology of sharks.

Descriptions of two species of 15-centimeter-long sharks from the early Carboniferous of Montana may indicate just how complex reproductive behavior was in early elasmobranchs. Males can be identified by pelvic claspers, sharp rostra, and an enormous forwardly curved middorsal spine firmly embedded just behind the head (Figure 7–5c). Richard Lund considered that the great degree of sexual dimorphism and the discovery of many more males than females of at least one of these species suggested that males displayed during courtship in a regular male display site (a lek). He also suggested that one of the specimens discovered might be a pair in a precopulatory courtship position (Figure 7–5c) with the blunt-snouted female grasping the male's nuchal spine in her jaws.

One of the score or so of genera produced in this early radiation of elasmobranch evolution is *Xenacanthus*, which had a braincase, jaws, and jaw suspension very similar to those of *Cladoselache*. But there the resemblance ends. The xenacanths were freshwater bottom dwellers. The idea that they were bottom dwellers derives from the similarity of their fins to those of extant Australian lungfish (which are bottom dwellers) and their heavily calcified cartilaginous skeleton that would have decreased their buoyancy. The xenacanths appeared in the Devonian and survived until the Triassic, when they died out without leaving descendants.

Another group of chondrichthyan fishes had a suite of morphological characteristics seen also among other fusiform, fast-swimming marine sharks. These characters included large pectoral fins and stiff, symmetrical, deeply forked tails. These sharks, the **edestoids**, were morphologically distinct from the main lines (clades) of elasmobranch evolution, especially because of their peculiar dentition (Figure 7–6). Most of the tooth whorls of edestoids were greatly reduced, but the symphysial (central) tooth row of the mandible was tremendously enlarged and each tooth interlocked with adjacent teeth at its base. Apparently, several members of this tooth whorl were functional at the same time. Blunt for crushing in some forms and compressed to create a series of knife-edge blades in others, the mandibular tooth row bit against small, flat teeth associated with a poorly developed palatoquadrate. Most edestoids replaced their teeth rapidly, the oldest worn teeth being shed from the tip of the mandible. In contrast, *Helicoprion* retained all of its teeth in a specialized chamber into which the life-long production of dentition spiraled (Figure 7–6b). Perhaps teeth no longer efficient in size or shape provided a solid foundation for the functional teeth.

The Second Elasmobranch Radiation

Further elasmobranch evolution involved reorganization in feeding and locomotor systems. These

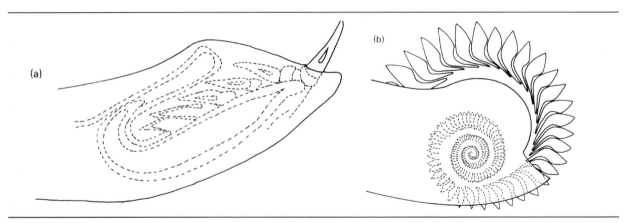

Figure 7–6 Tooth replacement by sharks. (a) Diagrammatic cross section of the jaw of an extant shark showing a single functional tooth backed by a band of replacement teeth in various stages of development; (b) lateral view of the symphysial (middle of the lower jaw) tooth whorl of the edestoid cladodont *Helicoprion*, showing the chamber into which the life-long production of teeth spiraled.

modifications began in the Carboniferous and lasted until the late Cretaceous. To identify these new adaptations, we shall describe a late and well-known genus of the Triassic and Cretaceous, *Hybodus*, which has left complete skeletons two meters in length (Figure 7–7a).

The diverse dentition of hybodont sharks seems pivotal to their success. The anterior teeth were sharp-cusped and appear to have been used for piercing, holding, and slashing softer foods. The posterior teeth were stout, blunt versions of the anterior teeth. Instead of becoming functional one at a time, they appeared above the fleshy lining of the mouth in batteries consisting of several teeth from each individual tooth whorl. An extant form with a similar dentition (Figure 7–7b) indicates how the mouth of a *Hybodus*-like shark must have looked: The extant horn sharks of the genus *Heterodontus*, which have *Hybodus*-like dentition, feed on small fish, crabs, shrimp, sea urchins, clams, mussels, and oysters. The sharp teeth near the symphysis seize and dispatch soft-bodied food, and shelled foods are crushed by pavement-like posterior teeth.

Also characteristic of the hybodonts were their fins. The pectoral girdle remained divided into separate right and left halves, but the articulation between girdle and the fin consisted of three narrowed plate-like basals instead of the long, often

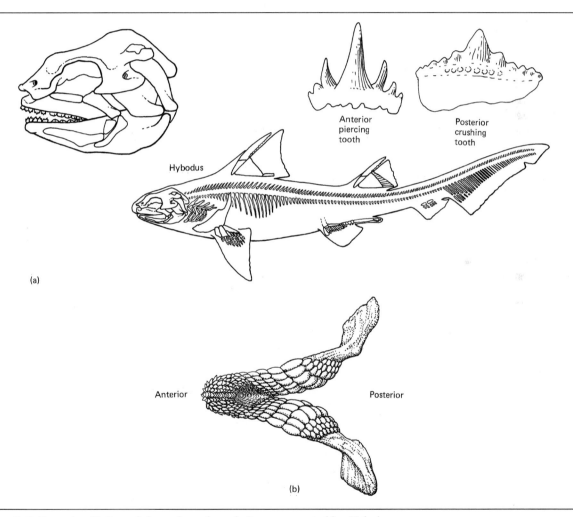

Figure 7–7 Second radiation of elasmobranch evolution represented by *Hybodus*. (a) Detail of head skeleton and teeth. (Modified after J. A. Moy–Thomas and R. S. Miles, 1971, *Paleozoic Fishes*, Saunders, Philadelphia, PA.) (b) Upper jaw of the extant hornshark, *Heterodontus*, with a dentition similar to that of many elasmobranchs in the second radiation during the late Paleozoic and early Mesozoic.

fused series seen in earlier sharks. This tribasal arrangement was also found in the pelvic fins, and both pairs of fins were supported on a narrow stalk composed of the three basals. Mobility of the distal portion of the paired fins was also increased. The cartilaginous radials did not extend to the fin margin and were segmented along their shortened length.

Proteinaceous, flexible structures called **ceratotrichia** extended from the outer radials to the margin of the fin. Intrinsic fin muscles arched the fin from anterior to posterior and along its long axis. This mobility allowed the paired fins to be used hydrodynamically in ways that seem impossible with the fin construction characteristic of *Cladoselache*. By assuming different shapes, the pectorals could produce lift anteriorly, aid in turning, or function as simple hydrofoils. Along with changes in the paired fins, the caudal fin assumed new functions and an anal fin appeared. Caudal fin shape was altered by reduction of the hypochordal lobe, division of its radials, and addition of flexible ceratotrichia. This construction is known as **heterocercal** (*hetero* = different, *kerkos* = tail). Undulated from side to side, the fin twisted due to water pressure so that the flexible lower lobe trailed behind the stiff upper one. This distribution of force produced forward thrust that, combined with the variable planing surfaces produced by flexible pectorals, could counter the shark's tendency to sink or could lift it from a benthic resting position. All gnathostomes from the early fossil record had similar, if stiffer, heterocercal caudal fins. The evolutionary success of the heterocercal fin compared to a straight or ventrally bent fin may relate to this generation of vertical forces, however slight they might have been (Webb and Smith 1980). Whatever mechanism the earlier sharks used to remain afloat, it is likely that the dynamic design of the fins in this second radiation of sharks allowed them more behavioral flexibility.

Other morphological changes in the sharks of the second major radiation include the appearance of a complete set of hemal arches that protected the arterial and venous trunks running below the notochord, well-developed ribs, and narrow, more pointed dorsal fin spines closely associated with the leading edge of the dorsal fins. These spines were deeply inserted in the muscle mass, exposed above the skin, ornamented with ridges and grooves, and studded with numerous barbs on the posterior surface, indicating defensive functions. Claspers are common to all species, leaving little doubt about the develop-ment of courtship and internal fertilization. In addition, male *Hybodus* had one or two pairs of hooked spines above the eye that may have been used as claspers during copulation.

Hybodus and its relatives resembled their presumed *Cladoselache*-like ancestors in having terminal mouths, amphistylic jaw suspension, unconstricted notochords, and multicusped teeth, but a direct line cannot be drawn between the two in time or in morphology. Some forms considered related to *Hybodus* had *Cladoselache*-like dentition combined with tribasal pectoral fins; others developed a very tetrapod-like support for highly mobile pectoral fins. Their caudal fin was reduced, and they probably moved around on the seafloor using their limb-like pectoral fins. Another form, known only from a five-centimeter juvenile, had a paddle-shaped rostrum one-third its body length. Other types were 2.5-meter giants with blunt snouts and enormous jaws. Despite their variety and the fact that they flourished during the Mesozoic, this second radiation of elasmobranchs became increasingly rare and disappeared from earth at the close of that era.

The Extant Radiation: Sharks and Rays

As early as the Triassic, and perhaps even in the late Carboniferous, fossilized indications of the extant radiation of elasmobranchs appear. By the Jurassic, sharks of extant appearance had evolved, and the Cretaceous contains genera that are still extant. Paleontologists are not in agreement about the origin of extant sharks. They may have evolved from *Hybodus*-like sharks, but a few details of the morphology of *Hybodus*-like sharks seem to preclude them from the lineage of extant sharks. Perhaps the evolution of the extant sharks, skates, and rays (the **Neoselachii**) was from a *Cladoselache*-like lineage that acquired characteristics in parallel with *Hybodus* and its relatives, or there may be a yet-unknown common ancestor for hybodonts and neoselachians. The most obvious difference between most members of the earlier radiations and extant sharks is the almost ubiquitous rostrum or snout that overhangs the ventrally positioned mouth in most extant forms. The technical characters distinguishing extant sharks are more subtle.

Extant elasmobranchs have an enlarged hyomandibular cartilage, which braces the posterior portion of the palatoquadrate and attaches firmly

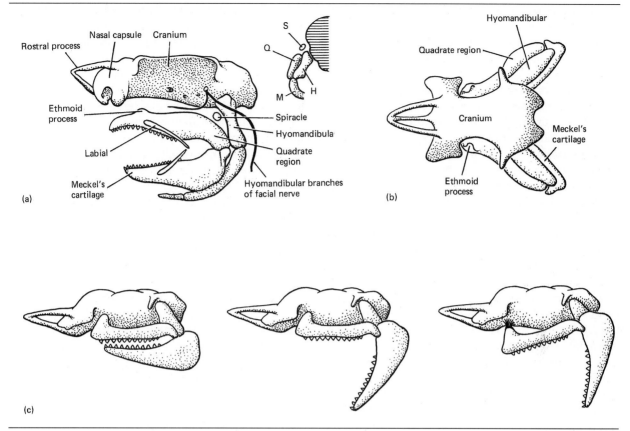

Figure 7–8 Anatomical relationships of the jaws and chondrocranium of extant hyostylic sharks based on *Scyllium* and *Carcharhinus*. (a) Lateral and cross-sectional views of the head skeleton of *Scyllium* with the jaws closed. (Modified after E. S. Goodrich, 1930, *Studies on the Structure and Development of Vertebrates*, Macmillan, London, UK.) (b) Dorsal view of *Carcharhinus*. (c) During jaw opening and upper jaw protrusion the hyomandibula rotates from a position parallel to the long axis of the cranium to a position nearly perpendicular to that axis. S, spiracle; Q, quadrate region of the palatoquadrate; H, hyomandibula; M, mandible. ([b, c] Modified after S. A. Moss, 1984, *Sharks: An Introduction for the Amateur Naturalist*, Prentice-Hall, Englewood Cliffs, NJ.)

but movably to the otic region of the cranium (Figure 7–8). A second connection to the chondrocranium is via paired palatoquadrate projections to either side of the braincase just behind the eyes and attached to it by elastic ligaments. Jaw suspension of this type is known as **hyostylic**. Hyostyly permits multiple jaw positions, each appropriate to different feeding opportunities (Moss 1984).

The right and left halves of the pectoral girdle are fused together ventrally into a single U-shaped scapulocoracoid cartilage. Muscles run from the ventral coracoid portion to the symphysis of the lower jaw and function in opening the mouth. The advan-

tages of the jaws of extant elasmobranchs are displayed when the upper jaw is protruded. Muscles swing the hyomandibula laterally and anteriorly to increase the distance between the right and left jaw articulations and thereby increase the volume of the orobranchial chamber. This expansion, which sucks water and food forcefully into the mouth, was not possible with an amphistylic jaw suspension because the palatoquadrate was tightly attached to the chondrocranium.

With hyomandibular extension, the palatoquadrate is protruded to the limits of the elastic ligaments on its orbital processes. This protrusion allows delicate plucking of benthic foodstuffs. Protrusion also drops the mouth away from the head to allow an extant shark to bite an organism much larger than itself despite its large, sensitive rostrum. The dentition of the palatoquadrate is specialized; the teeth are stouter than those in the mandible and often recurved and strongly serrated. When feeding on large prey a shark opens its mouth, sinks its lower and upper teeth deeply into the prey, and protrudes its upper jaw ever more deeply into the slash initiated by the teeth. As the jaws reach their maximum initial penetration, the shark throws its body into exaggerated lateral undulations, resulting in a violent side-to-side shaking of the head. The head movements bring the serrated upper teeth into action as saws to sever a large piece of flesh from the victim.

Mobility within the head skeleton, known as **cranial kinesis**, allows consumption of large food items. Cranial kinesis permits inclusion of large items in the diet of vertebrates, such as elasmobranchs, without excluding smaller, more diverse foodstuffs. Throughout their evolutionary history the Chondrichthyes have been consummate carnivores. In the third adaptive radiation of the elasmobranchs, locomotor, trophic, sensory, and behavioral characteristics evolved together in the mid-Mesozoic to produce forms still dominating the top levels of extant marine food webs. This position has gone hand-in-hand with the evolution of gigantism, one advantage of which is avoiding predation. The 360 species of sharks (Figure 7–9) and 456 species of skates and rays (Figure 7–13) known today are large organisms, even for vertebrates (Gilbert 1982). A typical shark is about two meters long, and a typical ray is half that length. Nevertheless a few interesting miniature forms have evolved and inhabit mostly deeper seas off the continental shelves (Box 7–1).

In spite of their enormous range in size, all extant elasmobranchs have common skeletal characteristics that earlier shark radiations lacked. The continuous notochord was replaced in the extant radiation by cartilaginous centra that calcify in several distinctive ways. Although centra occur in many other gnathostomes, those of extant sharks are so distinctive in morphology and fossil history that they are probably independently evolved. Between centra, spherical remnants of the notochord fit into depressions on the opposing faces of adjacent vertebrae. Thus, the axial skeleton is a laterally flexible structure with rigid central elements swiveling on ball-bearing joints of calcified cartilage and notochordal remnants. In addition to the neural and hemal arches, extra elements not found in the axial skeleton of other vertebrates (the intercalary plates) protect the spinal cord above and the major arteries and veins below the centra.

Shark scales also changed. The scales of extant sharks are single-cusped and have a single pulp cavity. Scales in more ancestral sharks may have begun development similarly but often fused to form larger scales later in life. Extant sharks add more individual scales to their skin as they grow, and new scales may be larger in proportion to the increase in size of the shark. The size, shape, and arrangement of these **placoid scales** reduce turbulence in the flow of water adjacent to the body surface and increase the efficiency of swimming. It seems that this simplification of denticles may relate to increased locomotor efficiency.

A characteristic that differentiates chondrichthyans from other extant jawed fishes is the absence of a gas-filled bladder. Elasmobranchs (and holocephalians) use their liver to counteract the weight of their dermal denticles, teeth, and calcified cartilages. The average tissue densities of sharks with their livers removed is 1.062 to 1.089 grams per milliliter. Because seawater has a density of about 1.030 grams per milliliter, one might conclude that a shark not swimming to stay afloat would sink. The liver of sharks, however, is well known for its high oil content, which gives shark liver tissue a density of about 0.95 grams per milliliter. The liver may contribute as much as 25 percent of the body mass. By adjusting the oil content and size of their livers through growth and reabsorption, sharks can adjust their buoyancy as well as store energy. A 4-meter tiger shark (*Galeocerdo cuvieri*) weighing 460 kilograms on land may weigh as little as 3.5 kilograms in the sea. Not surprisingly, benthic sharks have livers with fewer and smaller oil vacuoles in their cells.

Long before it was realized that sharks may weigh very little in their natural habitat due to liver buoyancy, a very different mechanism was credited with preventing sharks from sinking. The peculiar hydrodynamics of a laterally oscillated heterocercal tail were thought to produce an upward lift as well as a forward thrust. Anterior surfaces such as the pectoral fins and perhaps the snout, if properly oriented, will produce lift as the shark moves through the water. Thus, although gravity pulls a shark downward, two areas of lift—one anterior and one posterior—were considered to act on a swimming

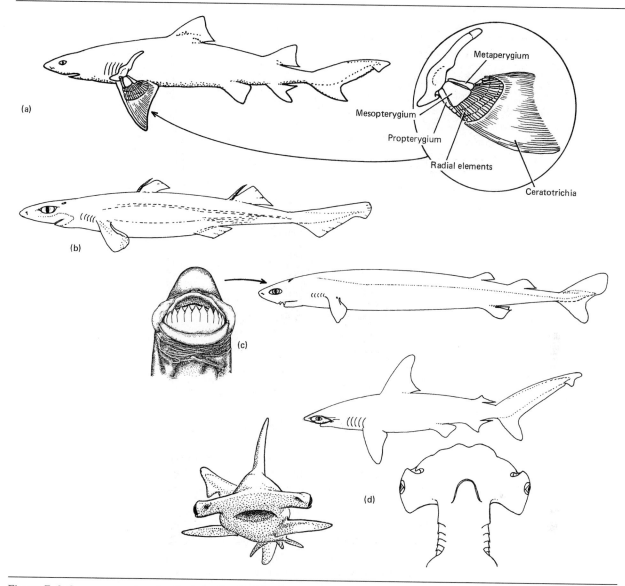

Figure 7–9 Some examples of extant sharks. (a) *Negaprion brevirostris*, the lemon shark, is widely used in elasmobranch research. It inhabits warm waters of the Atlantic frequented by bathers and divers and can attain a size sufficient to threaten humans. The internal anatomy of the pectoral girdle and fin are shown superimposed in their correct relative positions. (b) *Etmopterus vierens*, the green dogfish, is a miniature shark only 25 centimeters in length, yet it feeds on much larger prey items. (c) *Isistius brasiliensis*, the cookie-cutter shark, is another miniature species whose curious mouth (left) is able to take chunks from fish and cetaceans much larger than itself. (d) Hammerhead shark (*Sphyrna*) in lateral, ventral, and frontal views.

shark, permitting it to float. Two objections may be raised to this long-used explanation. First, because of the low density of its large liver a motionless pelagic shark is nearly neutrally buoyant. Any lift, either anterior or posterior, during forward locomotion would therefore cause a shark to rise in the water column—something we know does not necessarily occur. Second, the amount of lift resulting from the

Box 7–1 Food for Sharks

An interesting perspective on ecological principles that apply to many vertebrates can be obtained by examining the feeding habits of the smallest and largest elasmobranchs. The green dogfish, *Etmopterus virens*, attains a length of only 25 centimeters (Figure 7–9b). The bodies of these schooling sharks are punctuated with an elaborate pattern of green-glowing photophores. More than half of their food is squid and octopus, the eyes and beaks of which are sometimes so large that one wonders how they could have passed through the jaws and throat. Stewart Springer concludes that "green dogfish hunt in packs and may literally swarm over a squid or octopus much larger than they are, biting off chunks . . . and perhaps maintaining the integrity of their school visually through their distinctive lighting system."

Another miniature shark, *Isistius brasiliensis*, is even more bold in its feeding on large prey (Figure 7–9c). This brilliantly luminiscent shark lives in tropical deep waters in what is known as the **deep scattering layer** (DSL), a concentrated band composed by many species of vertically migrating animals. Many surface predators descend to the DSL in feeding forays; porpoises and tuna, which may weigh 400 kilograms, are among the voracious vertebrates that visit the DSL. These giant, fast-swimming predators often suffer wounds of curious origin, the shape of a silver dollar and about one centimeter in depth, reminiscent of the hole left in dough by a cookie cutter. These wounds have been matched in size and shape to the jaws of *Isistius*. Whether the photophores of a 40-centimeter *Isistius* play a part in some deception of the large predators on which it feeds or are used as social cues awaits observation from a deep submersible vehicle.

On the other end of the modern elasmobranch size spectrum are the largest extant fishes: the basking shark and the whale shark. The basking shark, *Cetorhinus maximus*, 10 meters long or more, lives in subpolar and temperate seas feeding exclusively on zooplankton, such as millimeter-long copepods. A feeding basking shark swims with its mouth wide open. Over a thousand erectile, whip-like denticles on the inner surface of the gill arches strain from the surface waters the hundreds of kilograms of food consumed each day. A typical adult basking shark filters about 1500 cubic meters of seawater per hour as it swims along with its mouth open (Priede 1984). Studies of the bioenergetics of basking sharks indicate that there is not enough food energy in the water column in winter to sustain these behemoths. During the winter months when plankton stocks are at a low, basking sharks disappear and are thought to rest on the bottom after shedding their gill rakers for the season. In contrast, whale sharks, *Rhiniodon typus*, which grow to 20 meters, are more tropical and able to feed on plankton and other small prey year around. Unique branchial specializa-

heterocercal tail and the pectoral fins is proportional to the forward velocity; the gravitational force is constant. Hence a shark would be neutrally buoyant and able to remain at the same level in the water column only over a narrow range of swimming speeds. If the shark swam too slowly, it would sink; if it swam too rapidly, it would rise.

K. S. Thomson has analyzed the swimming action of several extant shark species (Northcutt 1977, Moss 1984). Because dynamic adjustments of the muscles attached to the ceratotrichia of the hypochordal lobe can independently change the lobe's angle of attack, the heterocercal tail can deliver thrust over a wide range of angles, not simply forward and up as supposed previously. Thomson thinks that the highly controllable heterocercal tail of extant sharks allows them to develop extremely powerful dives and climbs in the water over a wide range of speeds, permitting sharks to make oblique attacks and shear off flesh from large prey. Students of fossil fishes have been slow to apply this functional analysis to the many forms with heterocercal tails.

tions have converted their gills to act as enormous sieves as well as respiratory surfaces. These solitary, white-spotted and striped Goliaths have been observed to feed head up and tail down at the surface. They lift above the water in the middle of a shoal of small fish until all the water in their orobranchial chamber has drained out the gill slits. With gills closed and mouth open the whale shark sinks tail down until surface water floods over the rim of its terminal mouth, bringing with it sustenance for the shark's 10,000-kilogram body.

In 1976 just outside the Hawaiian Islands and again in 1984 off southern California, two specimens of another very curious giant shark were inadvertently tangled in gear and hauled aboard ships. These 4.5-meter-long, rather soft-bodied, and apparently slow-moving sharks were completely unknown to science and have been named *Megachasma pelagios*, popularly known as megamouth. Megamouth has an enormous head and astonishingly wide terminal mouth filled with tiny teeth similar to the teeth of the basking and whale sharks. The cavernous mouth is lined with reflective crystals and may be bioluminescent. The stomach of the first specimen was filled with deepwater shrimp and that of the second with assorted plankton, including numerous remains of deepwater jellyfish. Although a few additional specimens have been reported, as yet it is not known how megamouth feeds, but one can imagine a glowing blue-green mouth cruising slowly through the gloom of the deep attracting the small life forms found there like moths to a flame, and then engulfing them.

Although it seems paradoxical that the largest vertebrates feed on tiny motes floating in the sea and the smallest species tackle organisms many times their own size, there are good ecological reasons behind these phenomena. Sunlight is the ultimate source of energy for life on earth, but no vertebrate has the ability to capture this energy directly. The enumeration of the indirect sources through which solar energy passes is called a **food chain** or, more properly because few vertebrates feed on a single food source, a **food web**. At each point of transfer in a web a great deal of energy is dissipated. On the average only about 10 percent of the energy in each step is passed to the next. Clearly, a vertebrate feeding on predatory tuna, as tiny *Isistius* does, is far removed from the primary source of energy. Because their individual and population requirements are comparatively low, the feeding methods employed by *Isistius* are effective. If a whale shark fed primarily on tuna, it would simply not be able to find enough food to maintain its enormous bulk. The solution for whale sharks and all extremely large vertebrates—nonavian dinosaurs, moas, elephants, blue whales—has been to feed near the sunny side of the food web by eating plants or organisms that eat plants.

Although not unique, the sensory systems of extant sharks, skates, and rays are certainly refined and diverse (Hodgson and Mathewson 1978, Sweet et al. 1983). Sharks may detect prey via **mechanoreceptors** of their **lateralis system**, an interconnected series of superficial tubes, pores, and patches of sensory cells distributed over the head and along the sides that respond to vibrations transmitted through the water. The basic units of mechanoreceptors are the **neuromast organs**, a cluster of sensory and supporting cells that are found on the surface and within the lateral-line canals. The anatomically related **ampullae of Lorenzini**, mucus-filled tubes with sensory cells and afferent neurons at their base, are exquisitely sensitive to electrical potentials and can even detect prey from their weak electrical fields (Box 7–2).

Chemoreception is another important sense. In fact, sharks have been described as swimming noses, so acute is their sense of smell. Experiments have shown that some sharks respond to chemicals in concentrations as low as one part in 10 billion!

Box 7–2 Electroreception by Elasmobranchs

The ability to detect electric fields is found in many fishes, especially elasmobranchs. On the heads of sharks, and on the heads and pectoral fins of rays, are structures known as the **ampullae of Lorenzini**. The ampullae are sensitive electroreceptors (Figure 7–10). The canal that connects the receptor to the surface pore is filled with an electrically conductive gel, and the wall of the canal is nonconductive. Because the canal runs for some distance beneath the epidermis, the sensory cell can detect a difference in electrical potential between the tissue in which it lies (which reflects the adjacent epidermis and environment) and the distant pore opening. Thus it can detect electrical fields, which are changes in electrical potential in space. The ancestral electroreceptor cell is a modification of the hair cells of the lateral line (Figures 8–10 and 8–15). Electroreceptors of elasmobranchs respond to minute changes in the electrical field surrounding an animal. They act like voltmeters, measuring a difference in electric potentials at discrete locations across the body surface. Voltage sensitivities are remarkable: Ampullary organs have thresholds lower than 0.01 microvolt per centimeter, a level of detection achieved only by the best voltmeters.

Elasmobranchs use their electric sensitivity to detect prey and possibly for navigation. All muscle activity generates electric potential: Motor nerve cells produce extremely brief changes in electrical potential, and muscular contraction generates changes of longer duration. In addition, a steady potential issues from an aquatic organism as a result of the chemical imbalance between the organism and its surroundings.

A. J. Kalmijn (1974) demonstrated that neither visual nor olfactory cues were necessary, and electrical activity in the absence of any other stimuli was sufficient to elicit an oriented feeding response (Figure 7–11). Subsequently, Kalmijn (1982) demonstrated electroreception by free-ranging sharks in the wild.

Figure 7–10 Ampullae of Lorenzini. (a) Distribution of the ampullae on the head of a spiny dogfish, *Squalus acanthius*. Open circles represent the surface pores and the black dots are the positions of the sensory cells. (b) A single ampullary organ.

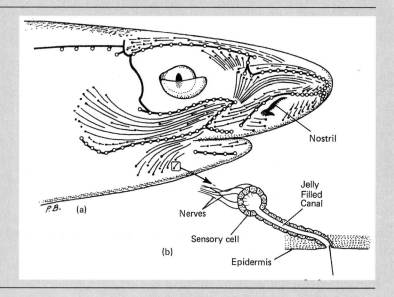

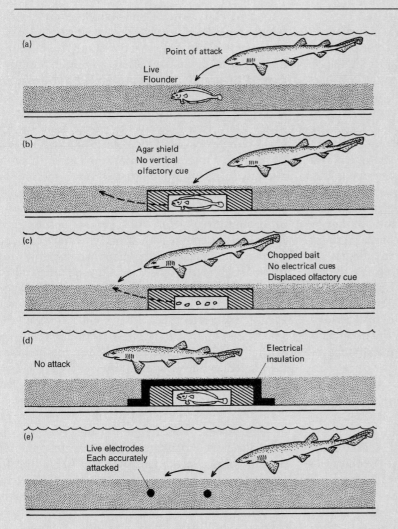

(a) Point of attack
Live
Flounder

(b) Agar shield
No vertical
olfactory cue

(c) Chopped bait
No electrical cues
Displaced olfactory cue

(d) No attack
Electrical
insulation

(e) Live electrodes
Each accurately
attacked

Figure 7–11 Kalmijn's experiments illustrating the electrolocation capacity of elasmobranchs. (Modified from A. J. Kalmijn, 1974, in *Handbook of Sensory Physiology*, volume 3, part 3, edited by A. Fessard, Springer, New York, NY.)

Electroreception might be used for navigation as well. The electromagnetic field at the Earth's surface produces tiny voltage gradients, and a shark swimming could encounter gradients as large as 0.4 microvolt per centimeter. In addition, ocean currents generate electric gradients as large as 0.5 microvolt per centimeter as they carry ions through the Earth's magnetic field. The use of these potentials for navigation has not been demonstrated. However, A. P. Klimley (1993) argues persuasively for hammerhead shark migration on and off of feeding grounds using electrodetection of geomagnetic gradients.

Hammerhead sharks of the genus *Sphyrna* (Figure 7–9) may have enhanced the directionality of their olfactory apparatus by placing the nostrils far apart on the odd lateral expansions of their heads.

Finally, vision is important to the feeding behavior of sharks. Especially well developed are mechanisms for vision at low light intensities at which humans would find vision impossible. This sensitivity is due to a rod-rich retina and cells with numerous plate-like crystals of guanine that are located just behind the retina in the choroid layer. Called the **tapetum lucidum**, the crystals act like microscopic mirrors to reflect light back through the retina and increase the chance that light will be absorbed. This mechanism, although of great benefit at night or in the depths, has obvious disadvantages in the bright sea surface of midday. To regulate the amount of bright light, cells containing the dark pigment melanin expand over the reflective surface to occlude the tapetum lucidum and absorb all light not stimulating the retina on first penetration. With so many sophisticated sensory systems, it is not surprising that the brains of many species of sharks are proportionately heavier than the brains of other fishes and approach the brain-to-body mass ratios of some tetrapods.

Anecdotal and circumstantial evidence suggests that sharks regularly use their various sensory modalities in an ordered sequence in locating, identifying, and attacking prey. Olfaction is often the first of the senses to alert a shark of potential prey, especially when the prey is wounded or otherwise releasing body fluids. A shark employs its sensitive sense of smell to swim up-current through an increasing odor gradient. Because of its exquisite sensitivity, a shark can use smell as a long-distance sense. Not as useful over great distances, but much more directional over a wide range of environmental conditions is another distance sense—vibration sensitivity. The lateralis system and the sensory areas of the inner ear are related forms of mechanoreceptors highly efficient in detecting vibrations such as those produced by a struggling fish. The vibration sense is effective in drawing sharks from considerable distances to a sound source. This has been demonstrated in macabre sea rescue operations in which sharks were apparently attracted by rescue helicopter rotor vibrations that fall into the same frequency range as those of a struggling fish. Under more controlled conditions, hydrophones can be used to attract sharks by broadcasting vibrations with frequencies like those produced by a struggling fish.

Whether by olfaction or vibration detection, once a shark is close to the stimulus source, vision takes over as the primary prey detection modality. If the prey is easily recognized visually, a shark may proceed directly to an attack. Unfamiliar prey is treated differently, as studies aimed at developing shark deterrents discovered. A circling shark may suddenly turn and rush toward unknown prey. Instead of opening its jaws to attack, however, the shark bumps or slashes the surface of the object with its rostrum. Opinions differ as to whether this is an attempt to determine texture through mechanoreception, to make a quick electrosensory appraisal, or to use the rough placoid scales to abrade the surface releasing fresh olfactory cues. Following further circling and apparent evaluation of all sensory cues from the potential prey, the shark may either wander off or attack. In the latter case the rostrum is raised, the jaws protruded, and in the last moments before contact many sharks draw an opaque eyelid across each eye to protect it. At this point it appears that sharks shift entirely to electroreception to track prey during the final stage of attack. This hypothesis was developed while studying the attacks by large sharks on bait suspended from boats and observed from submerged protective cages. After the occlusion of its eyes by the nictitating membrane, an attacking shark was frequently observed to veer from the bait and bite some inanimate, generally metallic object in the close vicinity (including the observer's cage, much to the dismay of the divers). Apparently, olfaction and vibration senses are of little use at very close range, leaving only electroreception to guide the attacking shark. The unnatural environment of the bait stations included stronger influences on local electric fields than the bait, and the sharks attacked these artificial sources of electrical activity.

Rare observations of sharks feeding under natural conditions indicate that these fishes are versatile and effective predators. The great white shark (*Carcharodon carcharias*) kills its mammalian prey by exsanguination—bleeding them to death (Klimley 1994). They hold a seal tightly in their jaws until it is no longer bleeding, then bite down removing an enormous chunk of flesh. The carcass floats to the surface and the shark returns to it for another bite. Attacks on sea lions, which have powerful front flippers probably used effectively in defense, often result in the prey being released before they are dead. But even sea lions, once attacked are repeatedly recaptured until they too are dead of blood loss. White sharks also prematurely release prey they find unacceptable after initial mouthing. Klimley suggested that lack of blubber is a primary rejection criterion. This behavior may explain why great white sharks seize and then release sea otters and humans along North America's West coast. A white shark may feed rather leisurely on an acceptable carcass and defend it from other white sharks with typical fish-like side-by-side tail slaps, as well as tail lobbing and breaching into the air—behaviors more familiar among dolphins and whales.

Much of the success of the extant grade of elasmobranchs may be attributed to their sophisticated breeding mechanisms. Internal fertilization is universal. The pelvic claspers have a solid skeletal structure that may increase their copulatory effectiveness. During copulation (Figure 7–12a) a single clasper is bent at 90 degrees to the long axis of the body, and a dorsal groove present on each clasper comes to lie directly under the cloacal papilla from which sperm exit. The single flexed clasper is inserted into the female's cloaca and locked there by an assortment of barbs, hooks, and spines near the clasper's tip. Male sharks of small species secure themselves *in copulo* by wrapping around the female's body. Large sharks swim side by side, their bodies touching or enter copulation in a sedentary position with their heads on the substrate and their bodies angled upward

(Tricas and LeFeuvre 1985). Some male sharks and skates bite the female's flanks or hold on to one of her pectoral fins with their jaws. In these species females may have skin on the back and flanks twice as thick as the skin of a male the same size. Whatever the position taken by the pair, when the single clasper is crossed to the contralateral side and securely inserted into the female cloaca, sperm from the genital tract are ejaculated into the clasper groove. Simultaneously, a muscular subcutaneous sac extending anteriorly beneath the skin of the male's pelvic fins contracts. This **siphon sac** has a secretory lining and is filled with seawater by the pumping activity of the male's pelvic fins before copulation. Seminal fluid from the siphon sac washes sperm down the groove into the female's cloaca, from which point the sperm swim up the female's reproductive tract.

With the evolution of internal fertilization elasmobranchs evolved a reproductive strategy favoring the production of a small number of offspring, retained, protected, and nourished for varying periods of time within the female's body. This requires a significant investment of her energy resources. The female has specialized structures at the anterior end of the oviducts, the **nidimental glands**, which secrete a proteinaceous shell around the fertilized egg. Most elasmobranch eggs are large (the size of a chicken yolk or larger) and contain a very substantial store of nutritious yolk. **Oviparous** (*ovum* = egg, *pario* = to bring forth) elasmobranchs have very large egg cases

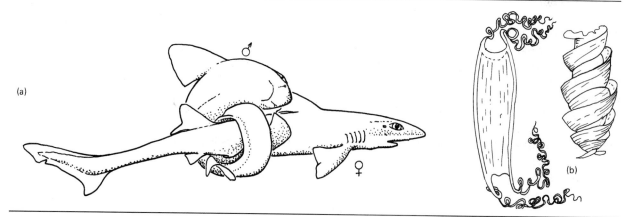

(a)

(b)

Figure 7–12 Reproduction by sharks. (a) Copulation in the smooth dogfish *Scyliorhinus*. Only a few other species of sharks and rays have been observed *in copulo*, but all assume postures so that one of the male's claspers can be inserted into the female's cloaca. (b) The egg cases of two oviparous sharks, *Scyliorhinus* (left) and *Heterodontus* (right). (Not to same scale.)

with openings for seawater exchange and protuberances to tangle or wedge themselves into protected portions of the substrate (Figure 7–12b). The zygote obtains nutrition exclusively from the yolk during the 6- to 10-month developmental period. Inorganic molecules including dissolved oxygen are taken from a flow of water induced by movements of the embryo. Upon hatching, the young are generally miniature replicas of the adults and seem to live much as they do when mature.

A most significant step in the evolution of elasmobranch reproduction was prolonged retention of the fertilized eggs in the reproductive tract. Many species retain the developing young within the oviducts until they are able to lead an independent life (ovoviviparity). Reduction in the nidimental gland's shell production and increased vascularization of the oviducts and yolk sacs are the only notable differences between oviparous and ovoviviparous forms. All nutrition comes from the yolk, and only inorganic ions and dissolved gases are exchanged between the maternal circulation and that of the developing young. This mode of reproduction is called **lecithotrophic viviparity** (*lecith* = egg, *troph* = nourishment). The eggs often hatch within the oviducts, and the young may spend as long in their mother after hatching as they did within the shell. As many as 100, but more often about a dozen, young are born at a time.

A natural step from the ovoviviparous condition is full viviparity or **matrotrophy** (*matro* = mother, *troph* = nourishment), the situation in which the nutritional supply is not limited to the yolk. Elasmobranchs have independently evolved matrotrophy several times. Some elasmobranchs develop long spaghetti-like extensions of the oviduct walls that penetrate the mouth and gill openings of the internally hatched young and secrete a milky nutritive substance. Other species simply continue to ovulate, and the young that hatch in the oviducts feed on these eggs. The most common and most complex form of viviparity among sharks is the yolk sac placenta, whereby each embryo obtains nourishment from the maternal uterine bloodstream via the highly vascular yolk sac of the embryo. This mode of reproduction is called **placentotrophic viviparity** (*placenta* = a placenta).

No matter which of the various routes of nourishment have brought young elasmobranchs to their free-living size, once the eggs are laid or the young born there is no evidence of further parental investment in them through care, protection, or feeding. In fact, elasmobranchs have long been considered solitary and asocial, but this view is changing. Accumulating field observations, often from aerial surveys (Kenney et al. 1985) or by scuba divers in remote areas, indicate that elasmobranchs of many species may aggregate in great numbers periodically, perhaps annually. More than 60 giant basking sharks were observed milling together and occasionally circling in head-to-tail formations in an area off Cape Cod in summer, and an additional 40 individuals were nearby. Over 200 hammerhead sharks have been seen near the surface off the eastern shore of Virginia in successive summers. Divers on seamounts that reach to within 30 meters of the surface in the Gulf of California have observed enormous aggregations of hammerheads schooling in an organized manner around the seamount tip. Some observations include behavior thought to be related to courtship (Moss 1984). More than 1000 individuals of the blue shark (*Prionace glauca*) have been observed near the surface over canyons on the edge of the continental shelf off Ocean City, Maryland. Fishermen are all too familiar with the large schools of spiny dogfish (genus *Squalus*) that seasonally move through shelf regions, ruining fishing by destroying gear, consuming bottom fish and invertebrates, and displacing commercially valuable species. These dogfish schools are usually made up of individuals that are all the same size and the same sex. The distribution of schools is also peculiar: Female schools may be inshore and males offshore or male schools may all be north of some point and the females south.

A similar sort of geographic sex segregation by hammerhead shark schools has been suggested to be the result of differential feeding preferences in habitats with different biological productivities (Klimley 1987). Females feeding on richer grounds grow faster, thus being larger at sexual maturity than males and better able to support embryonic young successfully. Such arguments do not explain why males should choose to live in less productive habitats, thus growing more slowly. Our understanding of these phenomena is slim, but it is clear that not all elasmobranchs are solitary all of the time.

No matter what the social behavior of the details of postcopulatory nutrition provided the embryo, sharks produce relatively few young during an individual female's lifetime. Although the young of sharks are large by most fishes' standards, they are subject to predation, especially from other sharks. Numerous species depend on protected nursery grounds—shallow inshore waters most subject to human disturbance and alteration. In addition, adult sharks are increasingly falling prey to humans. A rapid expansion in recreational and commercial shark fishing worldwide has threatened numerous species of these long-lived, slowly reproducing top predators. Internal fertilization and the life history characteristics that accompany it evolved in sharks fully 350 million years ago and have been a successful strategy throughout the world's oceans ever since. Now the alterations of habitat and heavy predation by humans threaten many species. The first United States shark fisheries management plan was established in 1993. Unfortunately, it applies only to national economic zone waters of the Eastern seaboard and the Gulf of Mexico. Whether or not such efforts to sustain populations of elasmobranchs will be successful will not be clear for many years.

For the most part we have been examining the characteristics of **pleurotremate** elasmobranchs—the sharks with gill openings (*trem*) on the sides (*pleur*) of the head. These forms number about 360 extant species. Although similar in overall appearance, the pleurotremate sharks actually come from two lineages. The more ancestral in their general anatomy (especially the smaller size of their brain) are the 80 species of **squaloid** sharks. Squaloids include the spiny and green dogfish, the cookie-cutter shark, and the basking and megamouth sharks. These species usually live in cold, deep water. The other 280 or so species of sharks are almost all members of the **galeoid** lineage that includes the hornshark, the nurse and carpet sharks, the whale shark, the mackerel sharks (including the great white shark), the carcharhinid sharks, and the hammerhead sharks. Galeoid sharks are the dominant carnivores of shallow, warm, species-rich regions of the oceans.

Surprising to many is the fact that the **hypotremate** elasmobranchs—the skates and rays—are more diverse than are the sharks. Approximately 456 extant species of skates and rays are currently recognized (Figure 7–13). These fishes have a long history of phylogenetic isolation from both clades of extant sharks. The suite of specializations characteristic of skates and rays relates to their early assumption of a benthic, **durophagous** (*duro* = hard, *phagus* = to eat) habit. The teeth are almost universally hard, flat crowned plates that form a pavement-like denti-

tion. The mouth may be highly and rapidly protrusible to provide powerful suction used to dislodge shelled invertebrates from the substrate.

Skates and rays are derivatives of the extant shark radiation adapted for benthic habitats. Radial cartilages extend to the tips of their pectoral fins, which are greatly enlarged. The anteriormost basal elements fuse with the chondrocranium in front of the eye and with each other in front of the rest of the head. Rays swim by undulating these massively enlarged pectoral fins. The flexibility of the fins results from a reduction in the number of placoid scales. The placoid scales so characteristic of the integument of a shark are absent from large areas of the bodies and pectoral fins of skates and rays. The few remaining denticles are often greatly enlarged to form sharp, stout bucklers along the dorsal midline. Many skates and rays cover themselves with a thin layer of sand. They spend hours partially buried and nearly invisible except for the dorsally prominent eyes and spiracles through which they survey their surroundings or take in oxygenated respiratory water. More derived forms, such as stingrays (family Dasyatidae), have a very few greatly elongate and venomous modified placoid scales at the base of the tail. The most highly specialized rays (of the family Mobulidae) are derived from these entirely naked types but spend little of their time resting on the bottom. Using powerfully extended pectorals, these devilfish or manta rays (up to six meters in width) swim through the open sea with flapping motions of their pectoral fins.

Skates and rays are primarily benthic invertebrate feeders (occasionally managing to capture small fishes), but the largest rays, like the largest sharks, are plankton strainers (Box 7–1). The dentition of many benthic rays is sexually dimorphic. Different dentitions coupled with the generally larger size of females might reduce competition for food resources between the sexes, but no difference in stomach contents has been found. Alternatively, the teeth of a male may be used to hold or stimulate a female during copulation.

Skates (family Rajidae) have specialized tissues in their long tails that are capable of emitting a weak electric discharge. Each species appears to have a unique pattern of discharge, and the discharges may identify conspecifics in the gloom of the seafloor. The electric rays and torpedo rays (family Torpedinidae) have modified gill muscles that produce electrical discharges of up to 200 volts that stun prey.

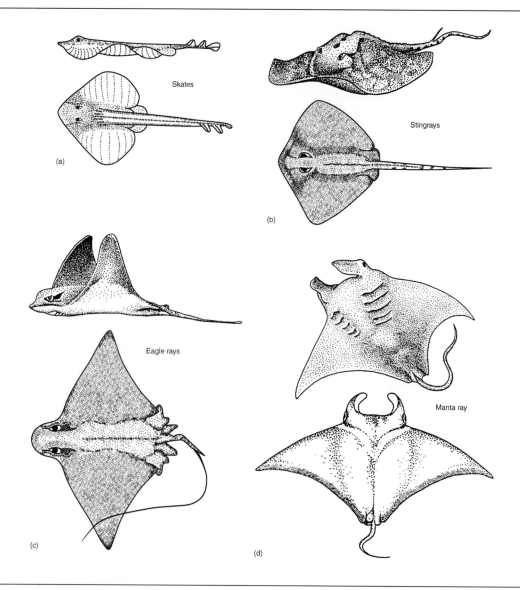

Figure 7–13 Some examples of extant skates and rays (Hypotremate elasmo-brachs): (a, b) benthic forms; (c, d) pelagic forms. (a) *Raja*, a typical skate, has an elongate but thick tail stalk containing electrogenic tissue and supporting two dorsal fins and a terminal caudal fin. Skates lay eggs enclosed in horny shells popularly called "mermaid's purses." (b) *Dasyatis*, a typical ray, has a whip-like tail stalk with fins represented by one or more enlarged, serrated, and venomous dorsal barbs. Rays bear living young; batoids display a reproductive diversity matching that of their shark kin. Pelagic batoids are closely related to the rays. (c) *Aetobatus* is representative of the eagle rays, wide-ranging shelled invertebrate predators. As befits their durophagus habit, they have batteries of broad, flat, crushing tooth plates. (d) *Manta* is representative of its family of gigantic sized fishes (up to 6 meters from wing tip to wing tip), which feed exclusively on zooplankton (Box 7–1). Extensions of the pectoral fins anterior to the eyes (the "horns" of these "devil rays") help funnel water into the mouth during filter feeding. (Modified in part from J. S. Nelson, 1994, *Fishes of the World*, Wiley, New York, NY; and P. B. Moyle, 1993, *Fish: An Enthusiasts Guide*, University of California Press, Berkeley, CA.)

A Second Radiation of Chondrichthyans: Holocephali

Most extant chondrichthyans are contained in the Elasmobranchii, but a small portion are grouped as ratfish or chimeras (Holocephali, Figure 7–14a). There is little agreement about holocephalian phylogeny. Two distinct fossil groups have been proposed as ancestors. A group of peculiar placoderms, the ptyctodontids, is known from the mid-Devonian. Rarely exceeding 20 centimeters in length, they showed a reduction in the extent and number of head and thoracic shield plates. A short palatoquadrate bound to the cranium carried a single pair of large upper tooth plates that were opposed by a smaller mandibular pair. The gills were covered by a single operculum. Postcranial characters, such as fin

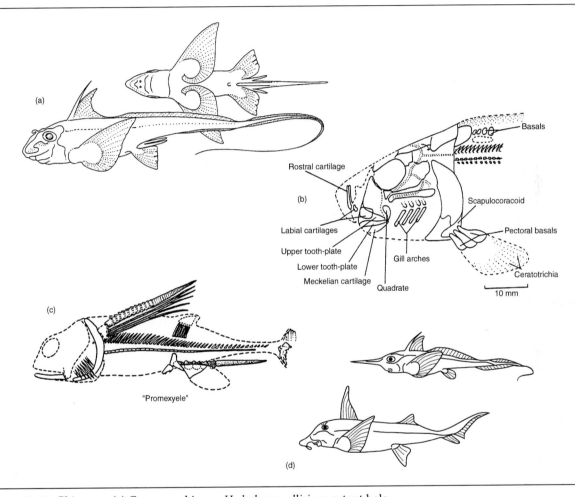

Figure 7–14 Chimeras. (a) Common chimera *Hydrolagus colliei*, an extant holocephalan, in lateral and ventral views (left) and representatives of two other groups: plownose chimera (lower right) and longnose chimera (upper right). (Modified in part after H. B. Bigelow and W. C. Schroeder, 1953, *Fishes of the Western North Atlantic*, part 2, Sears Foundation for Marine Research, Yale University, New Haven, CT.) (b) *Ctenurella*, an upper Devonian ptyctodont placoderm claimed by some paleontologists to be related to chimeras; see also Figure 7–3c. (Modified after J. A. Moy–Thomas and R. S. Miles, 1971, *Paleozoic Fishes*, Saunders, Philadelphia, PA.) (c) *Phomeryele*, a Carboniferous iniopterygian shark proposed by other paleontologists as a relative to the extant ratfishes. (Modified after P. H. Greenwood et al., 1973, *Interrelationships of Fishes*, Academic, New York, NY.)

spines, pared appendages, claspers, and caudal development, are strongly reminiscent of extant holocephalians (Figures 7–3c and 7–14a).

The first undoubted holocephalians are of Jurassic age, and the other fossils proposed as earlier members of the chimaera lineage do little to link the ptyctodont placoderms and extant forms. Ptyctodonts are more like extant holocephalians than the most ancestral holocephalians, which are in some cases rather shark-like. Since the 1950s Rainer Zangerl and co-workers have described a group of bizarre and obviously specialized forms (Figure 7–14c). These Iniopterygia (*inion* = back of the neck, *pteron* = wings) have characteristics that convinced Zangerl they were evidence for a link between the earliest sharks and holocephalans. As in extant holocephalians, the palatoquadrate is fused to the cranium (autostylic suspension), but the teeth, unlike those of extant chimeras, are in replacement families like those of elasmobranchs. The earliest holocephalans are thought to be of upper Carboniferous age, too old to be descendent from the contemporaneous iniopterygians. We currently think that both groups arose from ancestral sharks. If a connection is made between holocephalans and ancestral-grade elasmobranchs, the ptyctodont placoderms and chimeras will represent one of the most outstanding examples of convergent evolution in groups from different eras.

Whatever the origin of the chimeras, the approximately 30 extant forms (none much over a meter in length) have a soft anatomy more similar to sharks and rays than to any other extant fishes. They have long been grouped with elasmobranchs as Chondrichthyes because of their shared specializations. Generally found in water of deeper than 80 meters, the Holocephali move into shallow water to deposit their 10-centimeter horny shelled eggs from which hatch miniature chimeras. Several holocephalians have elaborate rostral extensions of unknown function (Figure 7–14a). From what is known, most species appear to feed on shrimp, gastropod mollusks, and sea urchins. Their locomotion is produced by lateral undulations of the body that throw the long tail into sinusoidal waves and by fluttering movements of the large, mobile pectorals. The solidly fused nipping and crushing tooth plates grow throughout life, adjusting their height to the wear they suffer. Of special interest are the armaments—a poison gland associated with the stout dorsal spine in some species, and mace-like cephalic claspers of males.

Summary

A major evolutionary innovation of the vertebrates was the appearance in the Silurian of fishes with jaws and paired appendages that had internal skeletal supports. The first of these jawed fishes, the acanthodians (Chapter 8), were soon joined by the diverse but phylogenetically isolated placoderms. Remains of these first gnathostomes yield few clues to the origins of jaws and paired appendages. However, patterns of embryonic development indicate that jaws and their supports evolved from anterior skeletal elements related to the visceral arches The extant fauna includes descendants of these early jawed fish: the Chondrichthyes, which include sharks, skates, rays, and chimeras. The evolution of chondrichthyans shows periods of relatively rapid changes in feeding and locomotor mechanisms followed by radiations and subsequent further morphological changes and reradiations, culminating in extant elasmobranchs and holocephalians.

From the array of adaptations present in extant chondrichthyans, it is not surprising that they have survived unchanged since the Mesozoic. The surprising fact is that they are not more diverse or found in a wider variety of habitats. To understand their limitations we must examine their competition—the bony fishes (Chapter 8).

References

Carroll, R. L. 1987. *Vertebrate Paleontology and Evolution*. Freeman, New York, NY.

Gilbert, P. W. (editor). 1982. *Oceanus* 24(4).

Greenwood, P. H., R. S. Miles, and C. Patterson (editors). 1973. *Interrelationships of Fishes*. Academic, New York, NY.

Hodgson, E. S., and R. F. Mathewson (editors). 1978. *Sensory Biology of Sharks, Skates, and Rays*. Technical Information Division, Naval Research Laboratory, Washington, DC.

Kalmijn, A. J. 1974. The detection of electric fields from inanimate and animate sources other than electric organs. Pages 147–

200 in *Handbook of Sensory Physiology*, volume 3, part 3, edited by A. Fessard. Springer, New York, NY.

Kalmijn, A. J. 1982. Electric and magnetic field detection in elasmobranch fishes. *Science* 218:916–918.

Kenney, R. D., R. E. Owen, and H. E. Winn. 1985. Shark distributions off the Northeast United States from marine mammal surveys. *Copeia* 1985:220–223.

Klimley, A. P. 1987. The determinants of sexual segregation in the scalloped hammerhead shark, *Sphyrna lewini*. *Environmental Biology of Fishes* 18:27–40.

Klimley, A. P. 1993. Highly directional swimming by scalloped hammerhead sharks, *Sphyrna lewini*, and subsurface irradiance, temperature, bathymetry, and geomagnetic field. *Marine Biology* 117:1–22.

Klimley, A. P. 1994. The predatory behavior of the white shark. *American Scientist* 82:122–133.

Mallatt, J. 1984. Early vertebrate evolution: pharyngeal structure and the origin of gnathostomes. *Journal of the Zoological Society, London* 204:169–183.

Martin, A. P., G. J. P. Naylor, and S. R. Palumbi. 1992. Rates of mitochondrial DNA evolution in sharks are slow compared with mammals. *Nature* 357:153–155.

Moss, S. A. 1984. *Sharks: An Introduction for the Amateur Naturalist*. Prentice-Hall, Englewood Cliffs, NJ.

Moy–Thomas, J. A., and R. S. Miles. 1971. *Paleozoic Fishes*. Saunders, Philadelphia, PA.

Northcutt, R. G. (editor). 1977. Recent advances in the biology of sharks. *American Zoologist* 17(2).

Priede, I. G. 1984. A basking shark (*Cetorhinus maximus*) tracked by satellite together with simultaneous remote sensing. *Fisheries Research* 2:201–216.

Romer, A. S. 1962. *The Vertebrate Body*, 3d edition, Saunders, Philadelphia, PA.

Romer, A. S. 1966. *Vertebrate Paleontology*, 3d edition. The University of Chicago Press, Chicago, IL.

Sweet, W. J. A. J., R. Nieuwenhuys, and B. L. Roberts. 1983. *The Central Nervous System of Cartilaginous Fishes*. Springer, New York, NY.

Tricas, T. C., and E. M. LeFeuvre. 1985. Mating in the reef whitetip shark *Triaenodon obesus*. *Marine Biology* 84:233–237.

Webb, P. W., and R. Smith. 1980. Function of the caudal fin in early fishes. *Copeia* 1980:559–562.

8 Dominating Life in Water: Teleostomes and the Major Radiation of Fishes

By the end of the Silurian the agnathous fishes (Chapter 6) had diversified and the cartilaginous gnathostomes (Chapter 7) were in the midst of their first radiation. The stage was set for the appearance of the final and most successful group of fishes, the bony fishes. The first fossils of bony fishes come from the late Silurian and their radiation was in full bloom by the middle of the Devonian. Once again, specialization of feeding mechanisms is one of the key features of the evolution of a major group of vertebrates. An increasing flexibility among the bones of the skull and jaws allowed the ray-finned fishes to exploit a wide range of prey types and predatory modes. Specializations of locomotion, habitat, behavior, and life histories have accompanied the specializations of feeding mechanisms, and bony fishes are the largest and most diverse group of living vertebrates. Their body forms, behaviors, and mechanical functions are intimately related to the characteristics of the aquatic habitat and to the properties of water as an environment for life.

Living in Water

Sixty-nine percent of the surface of the Earth is covered by fresh or salt water. Most of this water is held in the great ocean basins, which are populated everywhere by vertebrates, especially the bony fishes. The freshwater lakes and rivers of the planet hold a negligible percentage of all the water on earth—about 0.01 percent. This is much less than that tied up in the atmosphere, ice, and groundwater, but it is exceedingly rich biologically. Rivers and lakes are complex habitats with short histories on a geological time scale. Nevertheless, Earth's lakes and rivers have individual life spans long enough for evolutionary processes to occur within their isolated shores. Nearly 40 percent of all bony fishes, the largest taxon of vertebrates, live in fresh waters. Thus, watery environments gave rise to the first vertebrates and supported the first great steps in vertebrate evolution. Today they provide habitats for more kinds of vertebrates than any other environment.

Water is a demanding medium for vertebrate life. Even at saturation, water holds only about one-twentieth as much oxygen as an equal volume of air, and biological and chemical processes can lower the oxygen content of water to zero. Water is about 830 times as dense as air and 80 times more viscous, so movement in water is energetically expensive. Water's mass means that pressures increase significantly with increasing depth. Other physical characters limit the amount of light and affect the speed of sound transmission, imposing limits on sensory systems. Nevertheless, water is the ancestral medium of vertebrate life, and adaptation to it dominated the first 150 to 200 million years of vertebrate evolution.

Although each major clade of fishes seems to have solved environmental challenges in somewhat different ways, an examination of the specializations of the bony fishes for life in water illustrates one of their keys to evolutionary success—versatility. High rates of energy use, which are required to sustain the high activity levels characteristic of many fishes, depend on an efficient gas exchange in the gills. Most fishes swim with undulatory contractions of the body muscles. Guidance is usually visual, but other senses include the lateral line (for detection of low-frequency pressure waves) and electroreception. In this chapter we describe several basic specializations that have allowed fishes to become active and successful aquatic vertebrates. Many of the structures and functions of fishes set the stage for life on land.

Obtaining Oxygen in Water: Gills

Most aquatic vertebrates possess gills, evaginations from the body surface where respiratory gases are exchanged. Fish gills are enclosed in pharyngeal pockets (Figure 8–1). The flow of water is usually unidirectional—in through the mouth and out beneath the gill covers (operculae). Buccal flaps just inside the mouth and flaps at the margins of the operculae act as valves to prevent back flow. The respiratory surfaces of the gills are delicate projections from the lateral side of each gill arch (Figure 8–1b). Two columns of gill filaments extend from each gill arch. The tips of the filaments from adjacent arches meet when the filaments are extended. As water leaves the buccal cavity it passes over the filaments. Gas exchange takes place across numerous microscopic projections from the filaments, the secondary lamellae (Laurent and Dunel 1980).

The pumping action of the mouth and opercular cavities creates a positive pressure across the gills so that the respiratory current is only slightly interrupted during each pumping cycle (Figure 8–1c). Pelagic fishes, such as mackerel, sharks, tuna, and swordfish, have reduced or even lost the ability to pump water across the gills. A respiratory current is created by swimming with the mouth slightly opened, a method known as **ram ventilation**, and the fishes must perpetually swim. A great many other fishes switch to ram ventilation when they are swimming, and rely on buccal pumping when they are at rest.

The vascular arrangement in the teleost gill maximizes oxygen exchange. Each gill filament has two arteries, an **afferent** vessel running from the gill arch to filament tip and an **efferent** vessel returning blood to the arch (Farrell 1980). Each secondary lamella is a blood space connecting the afferent and efferent vessels (Figure 8–2a). The direction of blood flow through the lamellae is opposite to the direction of flow of water across the gill. This structural arrangement, known as a **countercurrent exchanger**, assures that as much oxygen as possible diffuses into the blood (Figure 8–2b). Pelagic fishes such as tunas, which sustain activity over extended periods of time, have reinforced gill filaments, large gill exchange areas, and a high oxygen-carrying capacity per milliliter of blood compared to sluggish benthic fishes, such as toadfishes and flatfishes (Table 8–1).

Terrestrial vertebrates seldom encounter low oxygen concentrations, for air contains 20.9 percent oxygen by volume. In aquatic habitats the amount of oxygen dissolved in water is much less than that present in a similar volume of air (a liter of air contains 209 milliliters of oxygen, whereas a liter of fresh water may contain only 10 milliliters of oxygen). Increasing temperature reduces the solubility of oxygen in water. High temperatures and the metabolism of microorganisms can produce anoxic ("without oxygen") conditions.

Fishes that live in these conditions cannot obtain enough oxygen via gills and have accessory respiratory structures that enable them to breath air. The electric eel, in addition to gills, has an extensive series of highly vascularized papillae in the pharyngeal region. The eels rise to the surface to gulp air, which diffuses across the papillae into the blood. Other fishes swallow air and extract oxygen through vascularized regions of the gut. The anabantid fishes of tropical Asia (e.g., the bettas and gouramies you see in pet stores) have vascularized chambers in the rear of the head, called labyrinths, which act in a similar manner. Many fishes are facultative air breathers; that is, oxygen uptake switches from gills to accessory organs when oxygen in the surrounding water becomes low. Others, like the electric eel and the anabantids, are obligatory air breathers. The gills alone cannot meet the respiratory needs of the fish even if the surrounding water is saturated with oxygen. These fishes drown if they cannot reach the surface to breathe air.

South American and African lungfish are obligate air breathers. Lungs are derived evaginations of the gut, and similar structures and responses can be found in the most plesiomorphic living ray-finned fishes, the polypterids. Air breathing is an ancestral characteristic of all osteichthyans, perhaps extending back to the early Devonian acanthodians. The early Devonian freshwater fishes probably encountered stresses like those faced by tropical freshwater fishes today. Early lungs may also have served as floats to produce a more nearly neutral density, as does the homologous swim bladder of teleosts.

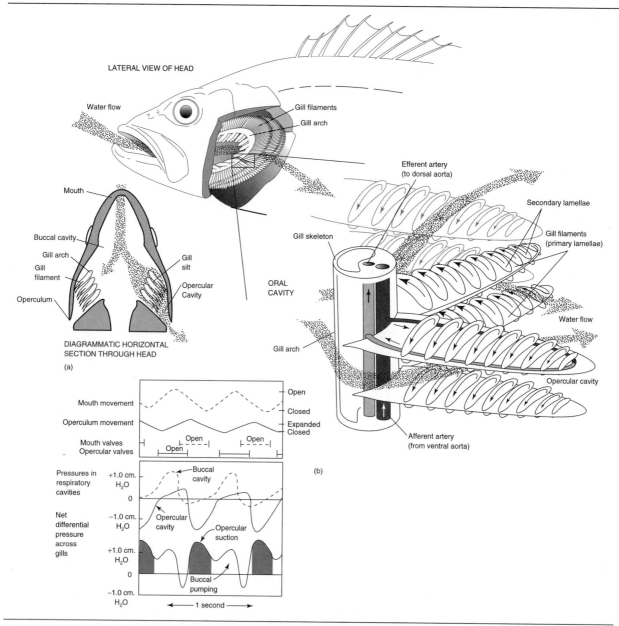

Figure 8–1 Anatomy and functional morphology of teleost gills: (a) position of gills in head and general flow of water; (b) water flow (shaded arrows) and blood flow (solid arrows) patterns through the gills; (c) water pressure changes across the gills during various phases of ventilation. (Modified after G. M. Hughes, 1963, *Comparative Physiology of Vertebrate Respiration*, Harvard University Press, Cambridge, MA.)

Locomotion in Water

Swimming results from sequential contractions of the muscles along one side of the body and simultaneous relaxation of those of the opposite side. Thus, a portion of the body momentarily bends and the bend is propagated posteriorly. When moving, therefore, a fish oscillates from side to side. These lateral undulations are most visible in elongate fishes, such as lampreys and eels (Figure 8–3). Most of the power for swimming comes from muscles in the posterior region of the fish (Rome et al. 1993).

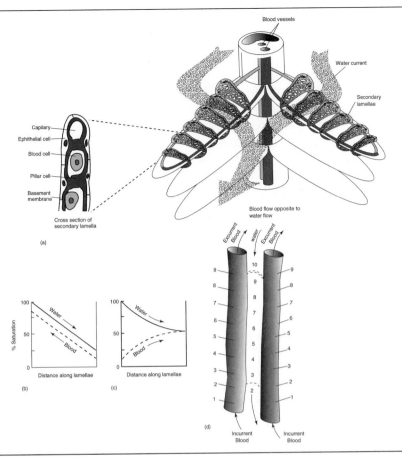

Figure 8–2 Countercurrent exchange in the gills of actinopterygian fishes. (a) The direction of water flow across the gill opposes the flow of blood through the secondary lamellae. Blood cells are separated from oxygen-rich water by the thin walls of the gill epithelium and the capillary wall (cross section of secondary lamella). (b) This results in a higher oxygen loading tension in the blood and lower oxygen tension in the water leaving the gills. (c) If water and blood flowed in the same direction over and within the secondary lamellae, an overall lower oxygen tension would occur in the blood leaving the gills. (d) Relative oxygen content of blood in secondary lamellae and the water passing over them. (Modified after G. M. Hughes, 1963, *Comparative Physiology of Vertebrate Respiration*, Harvard University Press, Cambridge, MA; R. E. Reinert in J. B. Gratzek and J. R. Matthews *Aquariology: The Science of Fish Health Management*, Tetra, Morris Plains, NJ; M. Hildebrand, 1988, *Analysis of Vertebrate Structure*, 3d edition, Wiley, New York, NY; and P. B. Moyle, 1993, *Fish: An Enthusiasts Guide*, University of California Press, Berkeley, CA.)

In 1926, Charles Breder classified the undulatory motions of fishes into three types: **anguilliform**—typical of highly flexible fishes capable of bending into more than half a sinusoidal wavelength; **carangiform**—undulations limited mostly to the caudal region, the body bending into less than half a wavelength; and **ostraciiform**—body inflexible, undulation of the caudal fin (Figures 8–3 and 8–6). These basic categories have been extensively subdivided and redefined since 1926 (Lindsey 1978, Webb 1984, Webb and Blake 1985) but are still useful for understanding locomotion in water.

Many of the specializations of body form, surface structure, fins, and muscle arrangement increase the efficiency of the different modes of swimming. A swimming fish must overcome the effect of gravity by producing lift and the drag of water by producing thrust (Figure 8–4).

Overcoming Gravity: The Generation of Vertical Lift

Because the average tissue density of fishes exceeds that of water, they sink unless their total density is reduced by a flotation device, or vertical lift is created by swimming. Many pelagic fishes, such as some tunas, mackerels, swordfishes (known collectively as scombroid fishes), and sharks, are negatively buoyant (i.e., weigh more than an equal volume of water). Tunas and many sharks must swim constantly to overcome gravity. In some cases lift is accomplished by extending large, wing-like pectoral fins at a positive angle of attack to the water. Many bony fishes, however, are neutrally buoyant (i.e., have the same density as water) and maintain position without body undulations. Only the pectoral fins backpaddle to counter a forward thrust produced by water ejected from the gills. Neutral buoyancy results from an internal float—a gas-filled swim bladder lying below the spinal column (Figure 8–5a). Fishes that are capable of hovering in the water usually have well-developed swim bladders.

The swim bladder is located outside the extraperitoneal cavity and ventral to the vertebral column. It is a gas-filled sac that arises as an evagination from the embryonic gut. The bladder occupies about 5 percent of the body volume of marine teleosts, and 7 percent of the volume of freshwater teleosts. The difference in volume corresponds to the difference in density of salt water and fresh water. The swim bladder wall, which is composed of interwoven collagen fibers, is virtually impermeable to gas diffusion. A teleost neither rises nor sinks, and needs to expend little energy to maintain its vertical position in the water column.

As a fish swims vertically up or down through the water column, its body is subjected to changing pressures from the overlying water column. Increased pressure at depth tends to compress the swim bladder, whereas decreased pressure toward the surface allows the bladder to expand. A mechanism is needed to maintain constant volume if the bladder is to act as a neutral buoyancy float. This is accomplished by moving gas into the bladder when a fish swims downward, and removing gas when it swims up. Plesiomorphic teleosts, such as bony tongues, eels, herrings, anchovies, and the minnows, salmons, and their kin, retain a connection, the **pneumatic duct**, between gut and swim bladder (Figure 8–5). These fishes, referred to as **physostomes** (*phys* = bladder, *stom* = mouth), can gulp air at the surface to fill the bladder, and can burp gas out to reduce the volume of the swim bladder. (Many physostomes also use the bladder as a lung. The pneumatic duct serves the same purpose as the windpipe or trachea of terrestrial vertebrates.)

The pneumatic duct is absent in adult teleosts from more derived clades, a condition termed **physoclistic** (*clist* = closed). Physoclists regulate the volume of the swim bladder by secreting gas into the bladder. Both physostomes and physoclists have a **gas gland**, which is located in the anterior ventral floor of the swim bladder (Figure 8–5b). Underlying the gas gland is the highly vascular **rete mirabile**—a complex, organized, overlapping array of blood vessels. This structure moves gas (especially oxygen) from blood to gas bladder, and is remarkably effective. Gas secretion occurs in many deep-sea fishes despite the enormous gas pressure within the bladder that is produced by the surrounding water pressure.

Table 8–1   **The relation between general level of activity, rate of oxygen consumption, respiratory structures, and blood characteristics in fishes of three activity levels.**

Activity Level	Species of Fish	Oxygen Consumption (mL O_2/g h^{-1})	No. Secondary Gill Lamellae mm^{-1} of Primary Gill Lamella	Gill Area (mm^2/g body mass)	Oxygen Capacity (mL O_2/100 mL blood)
High	Mackerel* (*Scomber*)	0.73	31	1160	14.8
Intermediate	Porgy (*Stenotomus*)	0.17	26	506	7.3
Sluggish	Toadfish† (*Opsanus*)	0.11	11	197	6.2

*Modified carangiform swimmers; swim continuously.
†Benthic fish

The gas gland releases lactic acid, which acidifies the blood in the rete mirabile. Acidification causes hemoglobin to release oxygen into solution (the Bohr effect). Because of the anatomical relations of the rete mirabile, which folds back upon itself in a **countercurrent multiplier** arrangement (Figure 8–5b), oxygen released from the hemoglobin accumulates and is retained within the rete mirabile until its pressure exceeds the oxygen pressure within the swim bladder. When this occurs, oxygen diffuses into the bladder, increasing its volume. The maximum multiplication of gas pressure that can be achieved is proportional to the length of the capillaries of the rete mirabile.

To compensate for gas expansion during ascent, physoclists open a muscular valve, called the **ovale**, located dorsally in the posterior region of the bladder adjacent to a capillary bed. The high internal pressure of gases in the bladder (especially oxygen) causes them to diffuse into the blood.

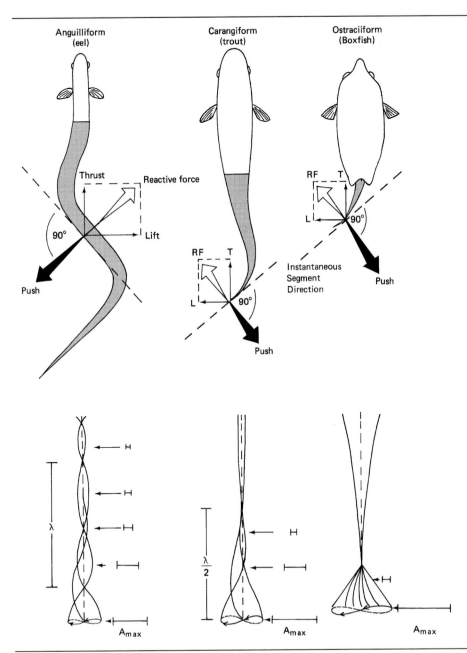

Figure 8–3 Basic movements of swimming fishes. Upper, outlines of some major swimming types showing regions of body that undulate (shaded). Note that the reactive force produced by one undulation's push on the water is canceled by that of the next, oppositely directed undulation. The thrust from each undulation is in the same direction and thus is additive; lower, diagrammatic waveforms created by undulations of points along the body and tail. A_{max} represents the maximum lateral displacement of any point. Note that A_{max} increases posteriorly; λ is the wavelength of the undulatory wave. Ostraciiform swimming, as initially defined by C. M. Breder, refers to a limited number of very specialized fishes, like box fishes and trunk fishes. A variety of fishes swim by propulsion from various fins rather than the body (see Figure 8–6).

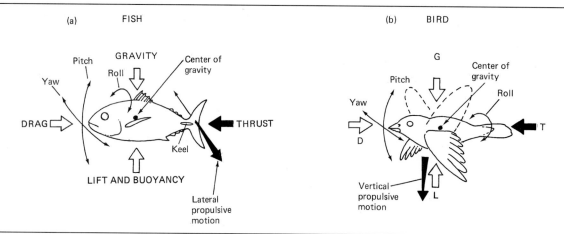

Figure 8–4 Comparison of forces associated with locomotion for a swimming and a flying vertebrate. (a) Motion of the caudal fin produces a lateral movement far from the center of gravity. The reactive force of the water to this movement causes the head to yaw in the same direction as the tail. (b) In a bird the major propulsive stroke of the wings is downward. Since both the propulsive stroke and the reactive force (lift) act near the center of gravity, the bird does not pitch. An up-and-down motion of the entire body occurs, however, because vertical lift is produced during the downward stroke only but gravity is constant.

Many deep-sea fishes have oily deposits of low density in the gas bladder, or have reduced or lost the gas bladder entirely and have lipids distributed in rich deposits throughout the body. These lipids provide static lift, as do the oils in shark livers. Because a smaller volume of the bladder contains gas, the amount of secretion required for a given vertical descent is less. Nevertheless, a long rete mirabile is needed to secrete oxygen at high pressures, and the gas gland in deep-sea fishes is very large. Mesopelagic fish that migrate large vertical distances depend more on lipids such as wax esters than on gas for buoyancy. Their close relatives that do not undertake such extensive vertical movements depend more on swim bladders for buoyancy. [For additional details, see Bone and Marshall (1982).]

Overcoming Drag: The Generation of Thrust

In general, fishes swim forward by pushing backward on the water. For every *active* force there is an opposite *reactive* force (Newton's third law of motion). Undulations produce an active force directed backward, and also a lateral force. The overall reactive force is directed forward and at an angle to the side.

Anguilliform and carangiform swimmers increase speed by increasing the frequency of their body undulations. Increasing the frequency of body undulations applies more power (force per unit time) to the water. Different fishes achieve very different maximum speeds—some (e.g., eels) are slow and others are very fast (e.g., tunas).

An eel's long body limits speed because it induces drag from the friction of water on the elongate surface of the fish. Fishes that swim rapidly are proportionately shorter and less flexible. Force from the contraction of anterior muscle segments is transferred through ligaments to the caudal peduncle and the tail. Morphological specializations of this swimming mode reach their zenith in fishes like tunas, where the caudal peduncle is slender and the tail greatly expanded vertically (Magnuson 1978).

Other fishes seldom flex the body to swim, but undulate the median fins (referred to as **balistiform** swimming). Usually, several complete waves are observed along the fin (Figure 8–6), and very fine adjustment in the direction of motion can be produced.

Many fishes (for example, ratfish, surf perches, and many coral reef fishes, such as surgeonfishes, wrasses, and parrot fishes) generally do not oscillate the body or median fins, but row the pectoral fins to

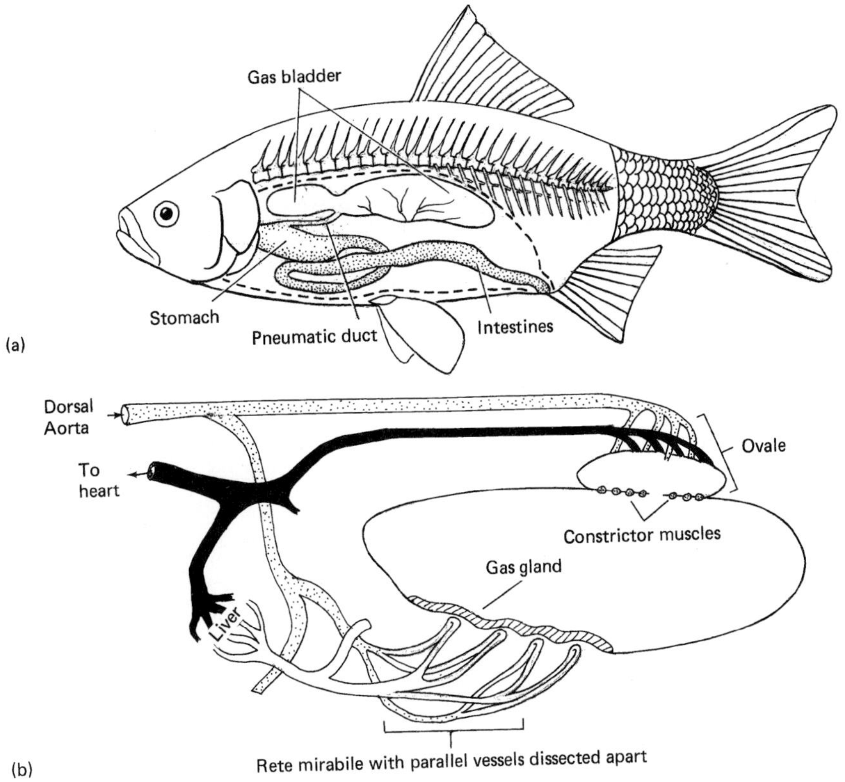

Figure 8–5 Swim bladder of actinopterygians: (a) the swim bladder is dorsal in the coelomic cavity just beneath the vertebral column; (b) vascular connections of a physoclistous swim bladder.

produce movement (**labriform** swimming). Most fishes show combinations of the basic swimming patterns. Intermediates between anguilliform, carangiform, ostraciiform, balistiform, and labriform movements exist, and most fishes can independently move paired fins to adjust body position. The muscles and nerves that move the fins are derivatives of the segmental muscles and spinal motor nerves that produce body undulations (Roberts 1981). In fact, the nerve–muscle complexes that move the limbs of tetrapods have the same origin. As different as swimming, crawling, running, and flying appear in vertebrates, the machinery that drives their locomotion has a common origin—the segmental neuromuscular complex.

Improving Thrust: Minimizing Drag

Drag is of two forms: **viscous drag** from friction between a fish's body and the water, and **inertial drag** from pressure differences created by the fish's displacement of water. Inertial drag is low at slow speeds, but increases rapidly with increasing speed. Viscous drag is relatively constant over a range of speeds. Viscous drag is affected by surface smoothness, inertial drag by body shape. Streamlined (teardrop) shapes produce minimum inertial drag when their maximum width is about one-fourth of their length and is placed about one-third of the length from the leading tip (Figure 8–7). The shape of many rapidly swimming vertebrates closely approximates these dimensions. A thin body has high viscous drag because it has a large surface area relative to its muscle mass, and a thick body induces high inertial drag because it displaces a large volume of water as it moves forward. Usually, fast-swimming fishes have small scales or are scaleless, with smooth body contours and low drag. (Of course, many slow-swimming fishes also are scaleless, showing that universal generalizations are difficult to make.) Mucus also contributes to the reduction of viscous drag.

Swimming movements in which only the caudal peduncle and fin undulate are usually called modified carangiform motion. Scombroids and many pelagic sharks have a caudal peduncle that is narrow in the vertical plane but is relatively wide from side-to-side. The peduncle of carangids is often studded

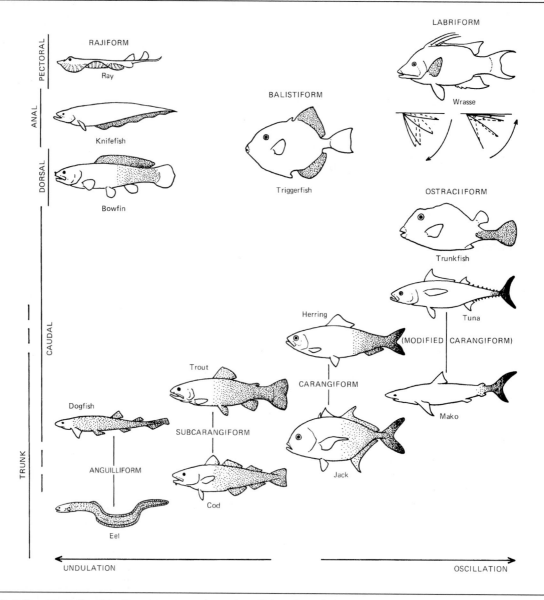

Figure 8–6 Location of swimming movements in various fishes. Stipple indicates areas of body undulated or moved in swimming. The names, such as carangiform, are used to describe the major types of locomotion found in fishes and are not a phylogenetic identification of all fishes using a given mode. (Modified after C. C. Lindsey, 1978, *Fish Physiology*, volume 7, Academic, New York, NY.)

laterally with bony plates called scutes. These structures present a knife-edge profile to the water as the peduncle undulates from side to side and contribute to the reduction of drag on the laterally sweeping peduncle. The importance of these seemingly minor morphological changes is underscored by the tail stalk of whales and porpoises, which is also narrow and has a double knife-edge profile. But the tail stalk of cetaceans is narrow laterally, whereas the caudal peduncle of scombroid fishes is narrow vertically. The difference reflects the plane of undulation—dorsoventrally for cetaceans and laterally for fishes. These strikingly similar specializations of modified carangiform swimmers—whether shark, scombroid,

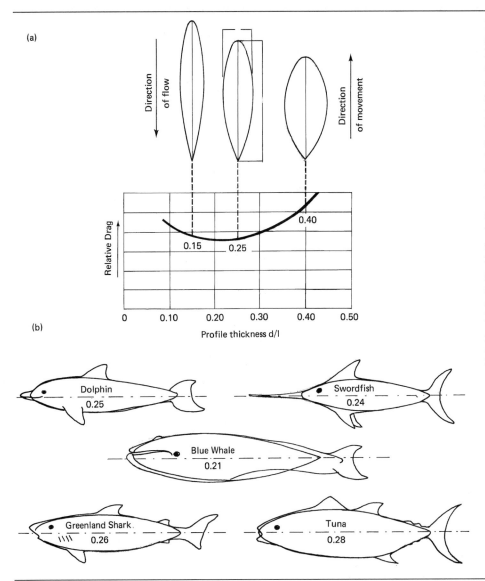

Figure 8–7 Effect of body shape on drag. (a) Stream-lined profiles with width (*d*) equal to approximately one-fourth of length (*l*) minimize drag. The examples are for solid, smooth test objects with thickest section about two-fifths of the distance from the tip. (b) Width-to-length ratios (*d/l*) for several swimming vertebrates. Like the test objects these vertebrates tend to be circular in cross section. Note that the ratio is near 0.25 and that the general body shape approximates a fusiform shape. (Modified after H. Hertel, 1966, *Structure, Form, Movement,* Krausskopf, Mainz, Germany.)

or cetacean—produce efficient conversion of muscle contractions into forward motion.

The tail creates turbulent vortices of swirling water in a fish's wake, which may be a source of inertial drag, or the vortices may be modified to produce beneficial thrust (Stix 1994). The total drag created by the caudal fin depends on its shape. When the **aspect ratio** of the fin (dorsal-to-ventral length divided by the anterior-to-posterior width) is large, the amount of thrust produced relative to drag is high. The stiff sickle-shaped fin of scombroids (mackerels, tunas) and of certain sharks (mako, great white) results in a high aspect ratio and efficient forward motion. Even the cross section of the forks of these caudal fins

assumes a streamlined teardrop shape, which further reduces drag. Many forms with these specializations swim continuously. But high-aspect-ratio fins do not present enough area to the water for a quick start.

The caudal fins of trout, minnows, and perch are not stiff and seldom have high aspect ratios. These subcarangiform swimmers change caudal fin area to modify propulsive thrust and to produce vertical movements of the posterior part of the body. The latter action is achieved by propagating an undulatory wave up or down the flexible caudal fin (Figure 8–6). In carangiforms propulsion often proceeds in bursts, usually from a standstill with rapid acceleration initiated by special neural systems, the Mauthner cells

Box 8–1 Mauthner Neurons and the Actinopterygian Brain

The complex behavioral adaptations shown by actinopterygians are associated with morphologically complex nervous systems. Studies of the nervous system of fishes are still in their infancy, but already clear is the distinctiveness of the central nervous system of ray-finned fishes.

Some especially well-developed actinopterygian neural adaptations are widespread in aquatic vertebrates. A neuro-locomotor specialization, the **Mauthnerian system**, present in many fishes and in tailed phases of amphibians, is well developed in the majority of teleosts, but absent in adult sharks and rays (Faber and Korn 1978, Eaton 1984). Located in the medulla oblongata of the brain are two giant nerve cells, one on either side of the midline. Each cell body is accompanied by two enlarged dendrites that synapse with cranial nerve VIII, the acoustic nerve. A single, heavily myelinated giant axon issues from each cell, crosses to the opposite side of the medulla, and then descends the full length of the spinal cord. Each giant axon synapses with motor neurons in every segment on one side of the body (Figure 8–8).

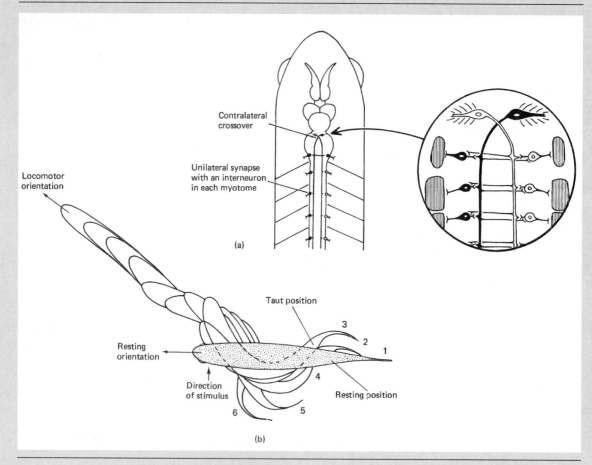

Figure 8–8 Mauthnerian system. (a) Anatomical relationships for the Mauthner neurons and associated structures in a teleost; (b) body snap or startle response of a teleost (trout) as seen from above in tracings from a motion picture.

Stimulation of one of the Mauthner cells results in a rapid, unilateral, forceful contraction of the trunk and tail myomeres. This reaction is similar and perhaps identical to the body snap or startle response that many teleosts exhibit when frightened by a sudden noise, mechanical shock, or change in illumination. The simultaneous, strong contraction of the muscles on one side of the body rapidly propels the fish forward. The value of such a startle response is obvious: A predator incautious enough to produce stimulae activating the Mauthnerian system may lose potential prey to this startle reaction.

Dinstinctive development of the forebrain (the telencephalon or anteriormost of the five basic craniate brain regions) is a shared derived character that is diagnostic of the ray-finned fishes (Nieuwenhuys 1982). In all other vertebrates the walls of the anterior end of the neural tube balloon outward on each side during development (Figure 8–9). The final result is the

Continued

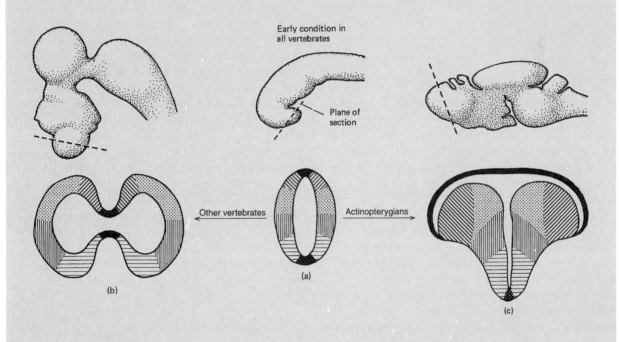

Figure 8–9 Unique development of the forebrain in actinopterygians. (a) The anteriormost portion of the neural tube (the future telencephalon of the brain) of actinopterygians early in its differentiation. (b) All other vertebrates, including sarcopterygians and their tetrapod descendents, undergo a rapid thickening of the side walls, which results in a ballooning laterally of the telencephalic walls and the formation of paired, hollow cerebral hemispheres. (c) In actinopterygians the walls also greatly thicken, but this results in eversion of the dorsal portions as the upper walls curl outward, stretching the primordial dorsal roof of the central cavity (the ventricle) to a thin covering over a ventricle of large surface area but narrow width.

Box 8–1 *(Continued)*

formation of hollow cerebral hemispheres. In actinopterygians the walls thicken, stretching the roof ever thinner as they evert. The result is solid cerebral hemispheres and an expansive hollow cloak of ventricle. Functional correlates of this structural difference are unknown. However, there is a marked increasing differentiation of the forebrain from *Polypterus* to sturgeons to *Amia* and the teleosts. Neural connections representing olfactory input become less important, structural complexity of the nerve cell architecture and interconnections increases, and connections to other distant regions of the brain increase. This pattern in actinopterygians is analogous to the evolution of the mammalian forebrain.

(see Box 8–1). The caudal peduncle of these fishes, unlike that of modified carangiform swimmers, is laterally compressed and deep. Under these circumstances the peduncle contributes a substantial part of the total force of propulsion.

Water and the Sensory World of Fishes

Water possesses properties that strongly influence the behaviors of fishes and other aquatic vertebrates. Light is absorbed by water molecules and scattered by suspended particles. Objects become invisible at a distance of a few hundred meters even in the very clearest water, whereas distance vision is virtually unlimited in clean air.

Fishes generally possess well-developed eyes. A major difference between aerial and aquatic vision relates to focusing light rays on the retina to produce a sharp image. Air, by definition, has an **index of refraction** of 1.00. The cornea in the eye of both terrestrial and aquatic vertebrates has an index of refraction of about 1.37, which is close to the index of refraction of water (1.33). Light rays are bent as they pass through a boundary between media with different refractive indices. The amount of bending is proportional to the difference in indices of refraction. Because the index of refraction of the cornea is substantially different from that of air, light rays are bent as they pass from air into the cornea, and the cornea is an important part of the focusing system of the eyes of terrestrial vertebrates. This relationship does not hold true in water because the refractive index of the cornea is too close to that of water to have much effect in focusing light. Terrestrial vertebrates rely on the lens, which often is flattened and pliable, for detailed focus. Fishes possess a less pliable spherical lens with high refractive power. The entire lens is moved toward or away from the retina to focus images. A similar lens has evolved in aquatic mammals, such as cetaceans. Otherwise, a fish's eye is similar to the eye of a terrestrial vertebrate.

Fishes have taste bud organs in the mouth and around the head and anterior fins. In addition, receptors of general chemical sense detect substances that are only slightly soluble in water, and olfactory organs on the snout detect soluble substances. Sharks and salmon are capable of detecting odoriferous compounds at less than 1 part per billion. Migrating salmon are directed to their stream of origin from astonishing distances by a chemical signature from the home stream permanently imprinted when they were juveniles. Plugging the nasal olfactory organs destroys their ability to home (Cromie 1982).

Mechanical receptors provide the basis for detection of displacement—touch, sound, pressure, and motion. Like all vertebrates, fishes possess an internal ear (the labyrinth organ, not to be confused with the organ of the same name that assists in respiration in some fishes), including the semicircular canals, which inform the animal of changes in speed and direction of motion. They also have gravity detectors at the base of the canals that allow them to distinguish up from down (Box 8–4). Most vertebrates also have an auditory region sensitive to sound pressure waves (Tavolga et al. 1981). These diverse functions of the labyrinth are dependent on basically similar types of sense cells, the **hair cells** (Figure 8–10). In fishes and aquatic amphibians clusters of hair cells and associated support cells form **neuromast organs** that are dispersed over the surface of the head and body. In jawed fishes neuromast organs often form a series of

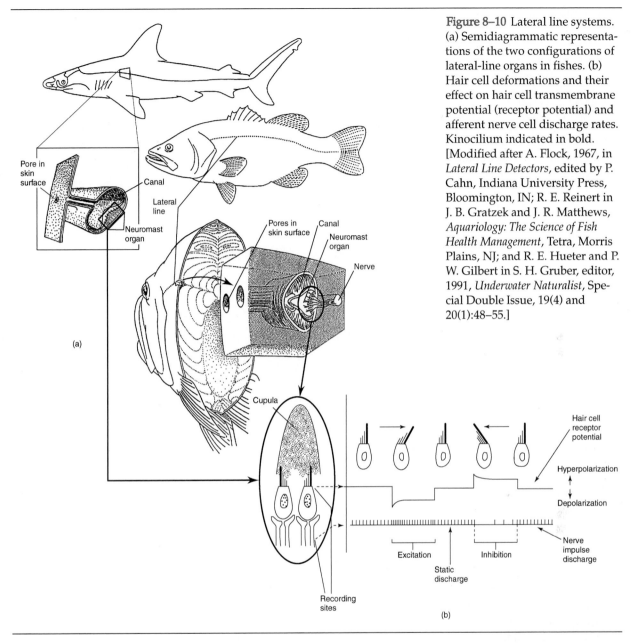

Figure 8–10 Lateral line systems. (a) Semidiagrammatic representations of the two configurations of lateral-line organs in fishes. (b) Hair cell deformations and their effect on hair cell transmembrane potential (receptor potential) and afferent nerve cell discharge rates. Kinocilium indicated in bold. [Modified after A. Flock, 1967, in *Lateral Line Detectors*, edited by P. Cahn, Indiana University Press, Bloomington, IN; R. E. Reinert in J. B. Gratzek and J. R. Matthews, *Aquariology: The Science of Fish Health Management*, Tetra, Morris Plains, NJ; and R. E. Hueter and P. W. Gilbert in S. H. Gruber, editor, 1991, *Underwater Naturalist*, Special Double Issue, 19(4) and 20(1):48–55.]

canals on the head, and one or more canals pass along the sides of the body onto the tail. This unique surface receptor system of fishes and aquatic amphibians is referred to as the **lateral-line system**.

Detection of Water Displacement: The Lateral Line

Neuromasts are distributed in two configurations—within tubular canals or exposed in epidermal depressions. Many kinds of fishes have both. Hair cells have a kinocilium placed asymmetrically in a cluster of microvilli and are arranged in pairs with the kinocilia positioned on opposite sides of adjacent cells. A neuromast contains many of these hair cell pairs. Each neuromast unit is innervated by two afferent lateral-line nerves: One transmits impulses from hair cells with kinocilia in one orientation, and the other carries impulses from cells with kinocilia positions reversed by 180 degrees.

All kinocilia and microvilli are embedded in a gelatinous structure, the **cupula**. Displacement of the cupula causes the kinocilia to bend. The resultant deformation either excites or inhibits the neuromast's nerve discharge. Each hair cell pair, therefore, encodes unambiguously the direction of cupula displacement. The excitatory output of each pair has a maximum sensitivity to displacement along the line joining the kinocilia and falling off in other directions. The net effect of cupula displacement is to increase the firing rate in one afferent nerve and to decrease it in the other nerve. These changes in lateral-line nerve firing rates inform a fish of the direction of water currents on different surfaces of its body.

Water currents of only 0.025 millimeter per second are detected by the exposed neuromasts of the aquatic frog, *Xenopus laevis*, with maximum response to currents of two or three millimeters per second. Similar responses occur in fishes. The lateral-line organs also respond to low-frequency sound, but controversy exists as to whether sound is a natural lateral-line stimulus. Sound induces traveling pressure waves in the water and also causes local water displacement as the pressure wave passes. It has been difficult to be sure whether neuromast output results from the accompanying water motions on the body surface or the sound's compression wave.

Several surface-feeding fishes and *Xenopus* provide vivid examples of how the lateral-line organs act under natural conditions. These species find insects on the water surface by detecting surface waves created by the prey's movements. In a series of clever experiments, E. Schwartz (see Fessard 1974) has demonstrated that each neuromast group on the head of the killifish, *Aplocheilus lineatus*, provides information about surface waves coming from a different direction (Figure 8–11). All groups, however, show stimulus field overlaps. Bilateral interactions between neuromast groups are indicated by the fact that extirpation of an organ from one side of the head disturbs the directional response to stimuli arriving from several directions.

The large numbers of neuromasts on the heads of some fishes might be important for sensing vortex trails in the wake of adjacent fishes in a school. Many of the fishes that form extremely dense schools lack lateral-line organs along the flanks (herrings, atherinids, mullets, and so on) and retain only the cephalic canal organs. The well-developed cephalic canal organs concentrate sensitivity to water motion in the head region, where it is needed to sense the degree of turbulence into which the fish is swimming, and the reduction of lateral elements would reduce

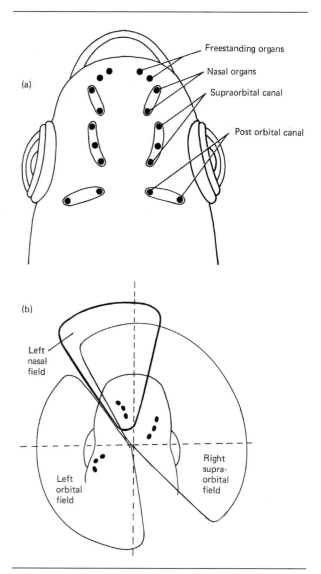

Figure 8–11 Distribution of the lateral-line canal organs on (a) the dorsal surface of the head of the killifish *Fundulus notatus*, and (b) the perceptual fields of the head canal organs in another killifish, *Aplocheilus lineatus*. The wedge-shaped areas indicate the relative directional sensitivity for each group of canal organs. Note that fields overlap on each side as well as on the same side of the body. (Modified after E. Schwartz, 1974, in *Handbook of Sensory Physiology*, volume 3, part 3, edited by A. Fessard, Springer, New York, NY.)

the constant noise from turbulence beside the fish. Over periods of time the reduction in drag through avoidance of turbulence could achieve a considerable metabolic swimming economy, as well as an even spacing between school members (Lindsey 1978).

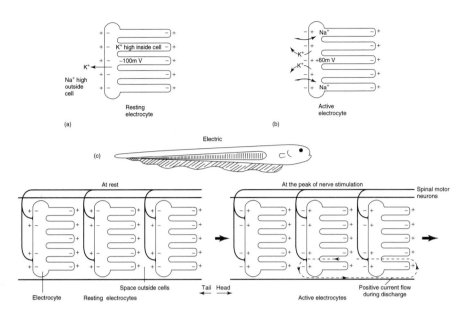

Figure 8–12 Use of transmembrane potentials of modified muscle cells by electric fishes to produce a discharge. K^+ (potassium ion) is maintained at a high and Na^+ (sodium ion) at a low internal concentration by the action of a Na^+/K^+ cell membrane pump. At rest, permeability of the membrane to K^+ exceeds the Na^+ membrane permeability. As a result, K^+ diffuses outward faster than Na^+ diffuses inward (arrow) and sets up the 100-millivolt resting potential. (a) At rest; (b) stimulated, sodium diffuses into the cell and potassium diffuses out; (c) differential movement of ions across the rough and smooth surfaces of electrocyte cells produces a directional current flow. By arranging electrocytes in series, some electric fishes can generate very high voltages. Electric eels, for example, have 6000 electrocytes in series and produce potentials in excess of 600 volts. (Modified in part from J. Bastian, 1994, *Physics Today* 47[2]:30–37.)

Electric Discharge and Electroreception

The torpedo ray in the Mediterranean, the electric catfish in the Nile, and the electric eel of South America can discharge enough electricity to stun other animals. The source of the electric current is modified muscle tissue. The cells of this modified muscle, called **electrocytes**, have lost the capacity to contract but are specialized for generating an ion current flow (Figure 8–12). When at rest the membranes of muscle cells or neurons are electrically charged, with the intracellular fluids about 100 millivolts negative relative to the extracellular fluids, primarily due to sodium ion exclusion (Figure 8–12a). A stimulus increases the permeability of the membrane to sodium ions, and a large, rapid local influx of sodium ions occurs (Figure 8–12b). This influx inverts the membrane potential and excites adjacent membranes to depolarize. As a result, the disturbance propagates over the cell surface. In an electrocyte, which is a modified noncontracting muscle cell, one cell surface is rough and the opposite surface smooth (Figure 8–12c). Innervation is associated with the smooth surface, and only the smooth surface depolarizes. The resulting sodium ion flux across the smooth surface into the cell and a potassium ion leakage across the rough surface and out of the cell yield a net positive current in one direction.

Because electrocytes are arranged in stacks or in series, like the batteries in a flashlight, the discharge potentials across each cell of the stack sum to produce high voltages. Synchrony of discharge is required, and this is produced by simultaneity of nerve impulses arriving at each electrocyte. The African electric catfish and the South American electric eel generate potentials in excess of 300 and 600 volts, respectively.

Most electric fish are found in tropical fresh waters of Africa and South America. Few marine forms are electric—only the torpedo ray (*Torpedo*), the ray genus *Narcine*, and some skates among elasmobranchs, and the stargazers (family Uranoscopidae) among the teleosts.

Unusual anatomical structures of electrocytes are present in several species of fishes that do not produce electric shocks (Figure 8–13). Because of their anatomical similarities to the electrocytes of strongly electric fishes, these organs were presumed to generate electric potentials, but not until the 1950s was their weak electric nature demonstrated (Bennett 1971a, Bass 1986). In these fishes, the discharge voltages are too small to be of direct defensive or offensive value. Instead, weakly electric fishes use their discharges for electrolocation and social communications (Bastian 1994). When a fish discharges its electric organ an electric field is established in its immediate vicinity (Figure 8–14). Because of the high energy costs of maintaining a continuous discharge, electric fishes produce a discontinuous, pulsating discharge and field. Some species produce a lifelong constant-frequency discharge, while others emit a variable-frequency discharge. Most weakly electric fishes pulse at rates between 50 and 300 cycles per second, but the sternarchid knifefishes of South America reach 1700 cycles per second, which is the most rapid continuous firing rate known for any vertebrate muscle or nerve.

The electric field from even weak discharges may extend outward for a considerable distance in fresh water because electric conductivity is low. This electric field will be distorted by the presence of both conductive and resistant objects. Rocks are highly resistive but other fishes, invertebrates, and plants are conductive. Distortions of the field cause a change in the distribution of electric potential across the fish's body surface. An electric fish detects the presence, position, and movement of objects by sensing where on its body maximum distortion of its electric field occurs.

The skin of weakly electric teleosts contains special sensory receptors: **ampullary** organs and **tuberous** organs (Fessard 1974). These organs detect tonic (steady) and phasic (rapidly changing) discharges, respectively. Electroreceptors are modifications of lateral-line neuromast receptors. Electroreceptors, like lateral-line receptors, have double innervation, an afferent channel sending impulses to the brain and an efferent channel that causes inhibition of the receptors (Bennett 1971b). Theodore Bullock and his collaborators have shown that during each electric organ discharge an inhibitory command is sent to the electroreceptors and the fish is rendered insensitive to its own discharge. Between pulses, electroreceptors sense distortion in the electric field or the presence of a foreign electric field. The electric organs and receptors of weakly electric fish provide a sixth sense. African and South American electric fishes are mostly nocturnal and usually inhabit turbid waters where vision is limited to short distances even in daylight.

Electric organ discharges vary with habits and habitat. Species that form groups or that live in shal-

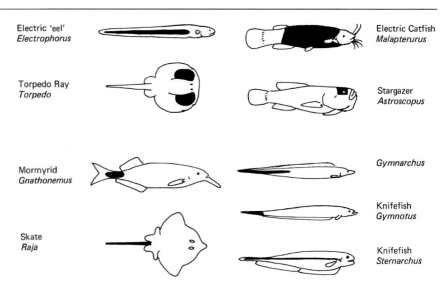

Figure 8–13 Electric fishes: morphology and position of electric organs and convergent aspects of body form are shown. Four strongly electric fishes are shown above and five weakly electric fishes below. Shaded areas indicate position of electrogenic organs. (Modified after M. V. L. Bennett, 1971, in *Fish Physiology*, volume 5, Academic, New York, NY.)

low, narrow streams generally have discharges with high frequency and short duration. These characteristics reduce the chances of interference from the discharges of neighbors. Territorial species, in contrast, have long electric organ discharges. Electric organ discharges vary from species to species. In fact, some species of electric fishes were first identified by their electric organ discharges. During the breeding sea-

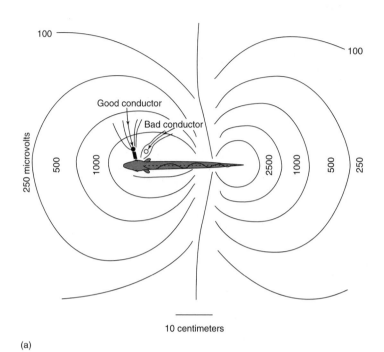

(a)

Figure 8–14 Electrolocation by fishes. (a) Distortion of the electric field (lines) surrounding a weakly electric fish by conductive and nonconductive objects. (Reference electrode 150 cm lateral to fish.) Conductive objects concentrate the field on the skin of the fish where the increase in electrical potential is detected by the electroreceptors. Nonconductive objects spread the field and diffuse potential differences along the body surface. (b) Electric field surrounding a weakly electric fish can be modulated by variations in the electric organ discharge for communication (left). When two electric fish come sufficiently close to one another significant interference can occur, requiring changes in the electric organ discharge (right).

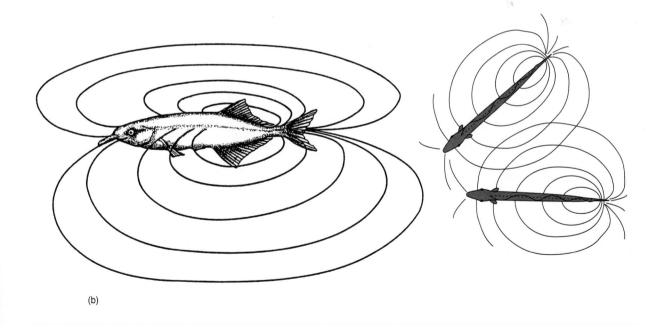

(b)

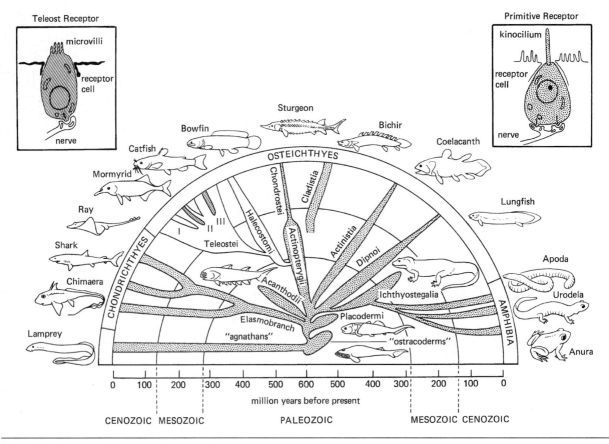

Figure 8–15 Phylogenetic distribution of electrosensitivity. Stippling and cross-hatching identify two classes of electroreceptors as indicated in the upper right and left panels. (Produced with the assistance of Carl Hopkins.)

son electric organ discharges distinguish immature individuals, ripe females, and sexually active males of some species.

Electrogenesis and electroreception are not restricted to a single group of aquatic vertebrates (Figure 8–15). Precisely localized activity occurs in the brain of the lamprey in response to electric fields (Bodznick and Northcutt 1981), and it seems likely that the earliest vertebrates also possessed electroreceptive capacity. All fish-like vertebrates of lineages evolved before the earliest neopterygians (represented by living gars and *Amia*) have electroreceptor cells. These cells have a prominent kinocilium, fire when the environment around the kinocilium is negative relative to the cell, and project to the medial region of the posterior third of the brain. The neopterygians lost electrosensitivity, and teleosts demonstrate at least two separate new evolutions of

electroreceptors, which are distinct from those of other vertebrates: They *lack* a kinocilium and fire when the environment is *positive* relative to the cell, and project to the *lateral* aspect of the rhombencephalon. At least one mammal, the duck-billed platypus, uses electroreception to detect prey (Scheich et al. 1986).

The Appearance of Teleostomes

The Devonian is known as the Age of Fishes because all major lineages of fishes, extant and extinct, coexisted in the fresh and marine waters of the planet during its 48-million-year duration. Most groups of gnathostomous fishes made their first appearance during the period including the most species-rich and morphologically diverse lineage of vertebrates, the Teleostomi (Figure 8–16), encompassing acanthodians and bony fish.

Acanthodians

The earliest jawed fishes in the fossil record are called acanthodians because of the stout spines (acanthi) anterior to their well-developed dorsal, anal, and often numerous paired fins (Denison 1979). The first fossil remains of acanthodians are isolated spines from the early Silurian. Later in the Silurian nearshore marine deposits contain more fin spines, scales, and teeth. Each of these elements is distinct from any comparable structures in other vertebrates. If they alone were available for study, the acanthodians would be considered as isolated from other vertebrate clades as are the placoderms. Such is not the case, however. When more complete specimens were discovered (early Devonian through the early Permian) they had the unmistakable spines, scales, and teeth of the Silurian fossils coupled with a mosaic of dermal, axial, and appendicular characters very like those of the majority of other gnathostomes and greatly different from those of any known jawless fish. Many of these characters seem to be ancestral for gnathostomes, as we might expect from such early forms. Nevertheless, these characters have been used to suggest relationships between acanthodians and (1) placoderms (on the basis of shared characters in the number of ossification centers in the jaws, gill arches, and shoulder girdles); (2) chondrichthyans (special border scales along the lateral line, details of the cranial and gill skeleton); and (3) osteichthyans (operculum, gill soft tissue and skeleton, branchiostegal rays, otoliths, and numerous other functional and morphological characters, especially of the cranium and jaw). In spite of considerable study, we must still wonder if these early gnathostomes had a cranium, jaw suspension, and dentition typical of other gnathostomes.

We follow Lauder and Liem (1983) in associating acanthodians as the sister group of the Osteichthyes. The **Teleostomi** (acanthodians + osteichthyans) are diagnosed by a unique mechanism of opening the mouth by lowering the mandible through movements of a hyoid apparatus transmitted to the lower jaw by ligaments, the presence of an ossified dermal operculum, a closely associated new element in the hyoid arch, the interhyal bone, and branchiostegal rays. All these elements become important in the later evolution of osteichthyan feeding mechanisms.

Acanthodians (Figure 8–17) were usually not more than 20 centimeters in length, although some 2-meter species are known. These marine and freshwater fishes were often clad in small, square-crowned scales, each of which grew in size as the animal grew. The head was large and blunt and housed large eyes. The mouth was large, and many species had jaws studded with teeth.

Unlike the teeth of all other gnathostomes, however, acanthodian teeth are not known to have been regularly replaced, and they lack enamel. Some acanthodians had a few enlarged scales that formed gill covers. In others, the head scales were lost completely except along the cephalic sensory canals. The brain was encased in a well-developed cartilaginous box, which also housed three semicircular canals. There are remains of neural and hemal arches, but no vertebral centra. The well-developed heterocercal tail, the fusiform body, and the arrangement of the fins indicate that acanthodians were good swimmers and probably were not bottom dwellers.

This basic body form lasted throughout acanthodian history, but several variations are evident. Some forms had robust spines in the position of the pectoral and pelvic fins. Between these were other pairs of spines of variable size. These spines were not embedded deeply in the body. Often the pectoral spines were associated with ventral dermal plates in a supportive girdle (Figure 8–17b). Other species lacked this dermal skeleton, and the spines were set more deeply in the body. The respiratory apparatus changed during the evolution of the acanthodian lineage and teeth were lost in more than one lineage. Jaw articulation and gill changes accompanying tooth loss indicate that toothless acanthodians were able to open their mouths and orobranchial chambers very widely. These species had long gill rakers, and probably were plankton feeders that swam with their enormous mouths open, straining small organisms from the water.

Such feeding specializations would have placed acanthodians at a key position in the aquatic and marine food webs of the late Paleozoic, transferring energy from tiny zooplankton into their own tissue and providing prey for larger fishes. Acanthodians may have been very abundant in Devonian seas, but during the early Permian the last forms disappeared, leaving the waters of the world to the Chondrichthyes and the most species-rich clade of vertebrates yet evolved—the bony fishes.

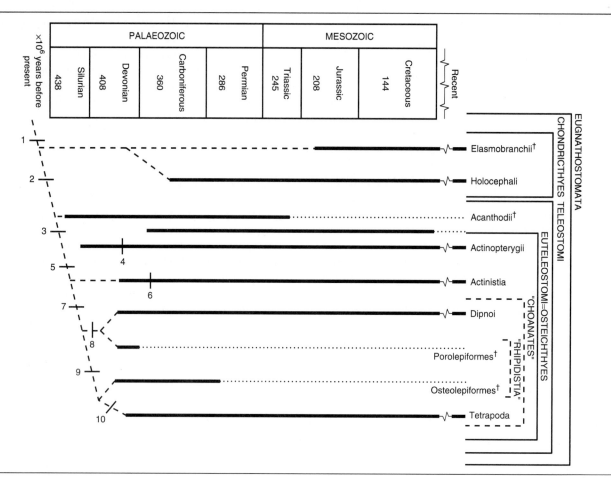

The Earliest Osteichthyes and the Major Groups of Bony Fishes

Fragmentary remains of bony fishes are known from the late Silurian. It is not until the Devonian that more complete remains are found. These animals resemble acanthodians in details of the head structure. The similarities may point to a common teleostome ancestor for acanthodians and osteichthyans in the early Silurian.

Remains of the bony fishes representing a radiation of forms already in full bloom appear in the lower to middle Devonian. Two major and distinctive osteichthyan (*osteo* = bone, *ichthys* = fish) types possessed locomotor and feeding characters that made them dominant fishes during the Devonian. Some evolved specializations that led to land vertebrates. Most lineages, however, retained fish-like characteristics and from among them rose the modern bony fishes, the largest group of living vertebrates (Table 8–2).

Fossils of the two basic types of osteichthyans are abundant from the middle Devonian onward. The Sarcopterygii (*sarcos* = fleshy, *pterygium* = fin; Figure 8–18a–d) and the Actinopterygii (*actinos* = stout ray; Figure 8–18e, f) are sister taxa. Synapomorphies include patterns of lateral-line canals, similar opercular and pectoral girdle dermal bone elements, and fin webs supported by bony dermal rays. A fissure allowed movement between the anterior and posterior halves of the neurocranium in many forms. The presence of bone is not a unifying osteichthyan characteristic because agnathans, placoderms, and acanthodians also possess true bone, and the bone loss of chondrichthyans is derived. The name Osteichthyes was coined before the occurrence of bones in other plesiomorphic vertebrates was recognized. Likewise, the various names long in use for the extant actinopterygian subgroups imply an increase in the ossification of the skeleton as an evolutionary trend (for example, chondrosteans, the "cartilaginous bony fishes," gave rise to a radiation

1. Chondrichthyes plus Teleostomi (Eugnathostoma): Epihyal element of second visceral arch modified as the hyomandibula, which is a supporting element for the jaw. **2.** Teleostomi: Hemibranchial elements of gills not attached to interbranchial septum, bony opercular covers, branchiostegal rays, five additional characters. **3.** Osteichthyes: Presence of lepidotrichia, differentiation of the muscles of the branchial region, presence of a lung or swim bladder derived from the gut, a unique pattern of ossification of the dermal bones of the shoulder girdle, medial insertion of the mandibular muscle on the lower jaw. **4.** Actinopterygii: Basal elements of pectoral fin enlarged, median fin rays attached to skeletal elements that do not extend into fin, single dorsal fin, scales with unique arrangement, shape, interlocking mechanism, and histology, and at least six additional derived characters. **5.** Sarcopterygii: Fleshy pectoral and pelvic fins have a single basal skeletal element, muscular lobes at the bases of those fins, enamel on surfaces of teeth, unique characters of jaws, articulation of jaw supports, gill arches, and shoulder girdles. **6.** Actinistia: Double jaw articulation, rostral organ, ossified swimbladder, loss of maxilla and branchiostegal. **7.** "Choanates": Internal oral openings from the olfactory passages. The homology of the choanae of dipnoans with those of tetrapods has not been established and porolepiformes may have lacked any such connection; thus the name "choanates" may be inappropriate, but the following shared derived characters indicate that the grouping of these lineages is valid: Pelvic girdle forms pubic and ischial processes, pectoral and pelvic appendages each have two primary joints (shoulder/hip and elbow/knee) with unique articulations, conus arteriosus of heart partly divided, unique dermal bone pattern of braincase, loss of interhyal bone, and at least 12 other characters (the exact number depends on which fossil forms are included). **8.** Reduced intracranial joint mobility, long leaf-shaped pectoral fin with central skeletal axis. **9.** True internal choanae present, a single pair of external nares. **10.** Tetrapoda: Digits on paired appendages, four limbs characterized by a single bone in the proximal portion and two bones in the distal portion. (Based on G. V. Lauder and K. F. Liem, 1983, *Bulletin of the Museum of Comparative Zoology* 150:95–197; and J. G. Maisey, 1986, *Cladistics* 2:201–256.)

Figure 8–16 Phylogenetic relationships of the Teleostomi. This diagram depicts the probable relationships among the major groups of teleostomes. Extinct lineages are marked by a dagger (†). The numbers indicate derived characters that distinguish the lineages. The quotation marks around "Choanates" and "Rhipidistia" indicate that these are paraphyletic groupings.

often called the holosteans, "entirely bony fishes," which culminated in teleosteans, "final bony fishes"). The fossil record indicates that a regular sequence of increasing ossification did not occur. To the contrary, a tendency to reduce ossification, especially in the skull and scales, is apparent when the full array of early Osteichthyes is compared with their derived descendents.

Early osteichthyans are not presently describable by a single widely acceptable phylogeny (Bemis et al. 1987). Although many relationships among ray-finned fishes are generally agreed upon because they are supported by several shared derived characters, the phylogenetic relationships of the Sarcopterygii are controversial. This controversy is important because from within the Sarcopterygii arose the tetrapods and with them terrestrial vertebrate life. For many years the sarcopterygians were thought to be composed of two sister groups: the lungfishes or Dipnoi (*di* = double, *pnoe* = breathing), and the Crossopterygii (*cross* = a fringe or tassel, *pterygium* = fin). Plesiomorphic Sarcopterygii have similar body shapes and sizes (20 to 70 centimeters), two dorsal fins, an epichordal lobe on the heterocercal caudal fin, and paired fins that were fleshy, scaled, and had a bony central axis. The paired fins' rays extend in a feather or compound leaf-like manner in contrast to the fan-like form of the fin rays in actinopterygian paired fins.

The jaw muscles of sarcopterygians were massive by comparison with those of actinopterygians, and the size of these muscles influenced the details of cranium, hyoid, and dermal skull characters that set sarcopterygians apart from the actinopterygians. Finally, the early sarcopterygians were coated with a peculiar layer of dentine-like material, **cosmine**, that spread right across the sutures between dermal bones and shows indications of being periodically reabsorbed. The very intricate interconnecting cavities characteristic of this cosmine are thought by some authorities to have contained an elaborate array of sensory organs, perhaps electroreceptors.

Despite these similarities only the Dipnoi are generally agreed to be monophyletic by most recent workers. The other Sarcopterygii have been variously combined as a single sister group, the **Crossopterygii**, now considered by most workers to be polyphyletic (i.e., including two or more separate lineages that are not sister groups); or as two equally ancient sister groups, **Rhipidistians** (entirely extinct forms) and **Actinistians** (for the sole surviving form *Latimeria* and its undisputed fossil relatives). Even this set of subdivisions has been brought into question by the revision of the rhipidistians into two or more separate clades. The names "Crossopterygii" and "Rhipidistia" have now lost their meaning. We shall take a further look at fossil sarcopterygians when we examine the origin of their sister taxon, the tetrapods.

The Evolution of the Actinopterygii

Basal actinopterygians include a variety of diverse taxa, formerly placed in an unnatural group of extinct fishes, the "paleoniscids," surrounding the origins of the polypterids or cladistians (a small group of living fishes that provide the best model for understanding the extinct forms), and the chondrosteans (living forms that are very different from the early fossil forms). Although fragments of late Silurian Actinopterygii exist, complete fossil skeletons are not found earlier than the middle to late Devonian. Early actinopterygians were small fishes (usually 5 to 25 centimeters long, although some were over a meter in length) with a single dorsal fin,

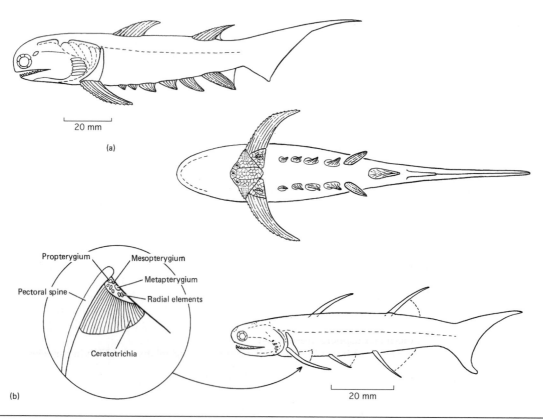

Figure 8–17 Reconstructions of acanthodians: (a) ancestral type: *Climatius*, lower Devonian, with multiple superficially attached spines (lateral view above, ventral view below); (b) more derived type: *Ischnacanthus*, lower Devonian, with fewer, more deeply embedded spines. (Inset) Detail of pectoral fin and spine of *Acanthodes*, lower Carboniferous.

a strongly heterocercal forked caudal fin with little fin web. Paired fins with long bases were common, but several taxa had lobate pectoral fins (Figure 8–18e, f). The interlocking scales, although thick like those of sarcopterygians, were otherwise distinct in structure and in growth pattern. Parallel arrays of closely packed radials supported the bases of the fins. The number of bony rays supporting the fin membrane was greater than the number of supporting radials and these rays were clearly derived from elongated scales aligned end to end. Two morphological aspects of the early ray-finned fishes deserve special attention: specializations for locomotion and for feeding. Unfortunately, the extant cladistians (polypterids) and chondrosteans are so specialized in these respects that they shed little light on the biology of early actinopterygians.

The lower jaw of early actinopterygians was supported by the hyomandibula, and in most forms was snapped closed in a scissors-like action by the adductor mandibulae muscles driving small conical teeth into prey (see Box 8–2, Figure 8–21a). This muscle originated in a narrow enclosed cavity between the maxilla and the palatoquadrate and inserted on the lower jaw near its articulation with the quadrate (Schaeffer and Rosen 1961). As a result, the bite was swift but the force created was low. The close-knit dermal bones of the cheeks permitted little expansion of the orobranchial chamber beyond that required for respiration. Some early actinopterygians did have a more specialized jaw apparatus. *Chirodus* (Figure 8–19e, left) had crushing tooth plates and a modified jaw to support them.

Table 8–2   **Classification and geographic distribution of Osteichthyes, the bony fishes.***

Osteichthyes (bony fishes), about 21,000 living species
    Sarcopterygii (fleshy-finned fishes), 7 living species
                  Dipnoi (lungfishes), 6 living species; Southern Hemisphere, fresh water
                  "Rhipidistia"[†]
                  Actinistia [Coelacanthiformes] (coelacanths), 1 living species; Indian Ocean islands, deep water marine
    Actinopterygii (ray-finned fishes), 20,850 living species
                  "Paleonisciformes"[†]
                  Polypteriformes [Cladistia] (bichirs), 11 living species; African, fresh water
                  Acipenseriformes (sturgeons and paddlefishes), 25 living species; Northern Hemisphere, coastal and fresh water
        Neopterygii,[‡] 20,814 living species
                  Lepisosteiformes [Ginglymodi] (gars), 7 living species; North and Central America, fresh and brackish water
                  Amiiformes (bowfins), 1 living species; North America, fresh water
      Teleostei (numbers of living species are conservative minima)
            Osteoglossomorpha (bony tongues), 206 living species; worldwide tropical fresh water
            Elopomorpha (tarpons and eels), 633 living species; worldwide, mostly marine
            Clupeomorpha (herrings and anchovies), 331 living species; worldwide, especially marine
            Euteleostei, 19,636 living species
                  Ostariophysi (catfish and minnows), 6050 living species; worldwide, fresh water
                  "Protacanthopterygii" (trouts and relatives), 320 living species; temperate Northern and Southern Hemisphere, fresh water
                  Scopelomorpha = "Myctophiformes" (lanternfishes and relatives), 677 living species; worldwide, mesopelagic (middle depth), marine
                  Paracanthopterygii (cods and anglerfishes), 1160 living species; Northern Hemisphere, primarily marine
                  Acanthopterygii (spiny-rayed fishes), 10,349 living species, including the Atherinomorpha (silversides), 1080 living species; worldwide, surface-dwelling, fresh water and marine; and Perciformes (perches), 7800 living species; worldwide, primarily marine

*Taxa in quotation marks are known to be artificial, but interrelationships are unresolved.
[†]Extinct.
[‡]The subdivision of the Neopterygii varies greatly from author to author.

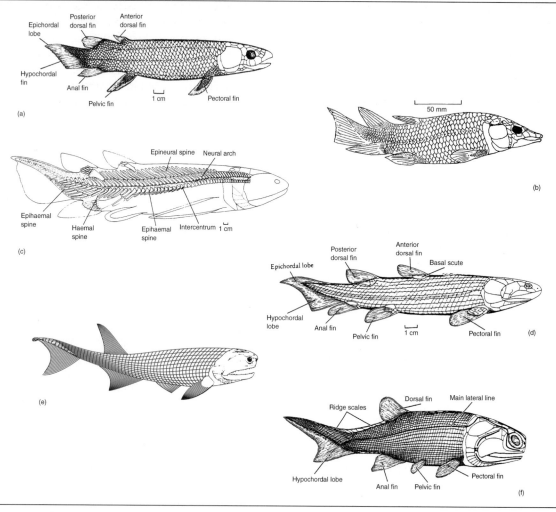

Figure 8–18 Plesiomorphic osteichthyans. (a, b) Dipnoans; (c, d) other Sarcopterygii; (e, f) Actinopterygians. (a) Relatively unspecialized dipnoan *Dipterus*, middle Devonian; (b) long-snouted dipnoan *Griphognathus*, late Devonian; (c) laterally compressed porolepiform *Holoptychius*, late Devonian to early Carboniferous; (d) cylindrical osteolepiform *Osteolepis*, middle Devonian; (e) fine-scaled actinopterygian *Cheirolepis*, middle to late Devonian; (f) typical early actinopterygian *Moythomasia*, late Devonian. (Modified from J. A. Moy–Thomas and R. S. Miles, 1971, *Paleozoic Fishes*, Saunders, Philadelphia, PA.)

These ancestral actinopterygians were successful in freshwater and marine habitats from 380 until 280 million years ago. No evidence supports a monophyletic origin of the fishes that have long been grouped together as "paleoniscoids." The different types share numerous characters (Figure 8–19), but these all may be ancestral. Near the end of the Paleozoic, the ray-finned fishes showed signs of change. The upper and lower lobes of the caudal fin were often nearly symmetrical, and all fin membranes were supported by fewer bony rays—about one for each internal supporting radial in the dorsal and anal fins. This morphological reorganization may have increased the flexibility of the fins.

The dermal armor of late Paleozoic ray-finned fishes was also reduced. The changes in fins and armor may have been complementary—more mobile fins mean more versatile locomotion, and increased ability to avoid predators may have permitted a reduction in heavy armor. This reduction of weight could have further stimulated the evolution of increased locomotor ability that was probably enhanced by perfection of the swim bladder as a delicately controlled hydrostatic device.

Better locomotion enhances predatory capacity. The food-gathering apparatus of several clades underwent radical changes to produce, in the upper Permian, a new clade of actinopterygians—the **neopterygians**. For over a century the most ancestral neopterygians have been called holosteans, but the characters they share may simply be ancestral for neopterygians as a whole. A new jaw mechanism was characterized by a short maxilla and a freeing of the posterior end of the maxilla from the other bones of the cheek (Figure 8–21b). Because the cheek no longer was solid, the nearly vertically oriented hyomandibula could swing out *laterally*, thus increasing the volume of the orobranchial chamber in a rapid motion to produce a powerful suction useful in capturing prey. The power of the sharply toothed jaw could be increased because the adductor muscle was not limited in size by a solid bony cheek. No longer enclosed, the jaw muscle mass expanded dorsally through the space opened by the freeing of the maxilla. In addition, an extra lever arm—the coronoid process—developed at the site of insertion of the adductor muscle, adding torque and thus power to the closure of the mandible of many of the more forceful biters.

The bones of the gill cover (operculum) were connected to the mandible so that expansion of the orobranchial chamber aided in opening the mouth. The anterior, articulated end of the maxilla developed a ball-and-socket joint with the neurocranium. Because of its ligamentous connection to the mandible, the free posterior end of the maxilla was rotated forward as the mouth opened. This directed the maxilla's marginal teeth forward, aiding in grasping prey. The folds of skin covering the maxilla changed the shape of the gape from a semicircle to a circular opening. The result was enhanced directionality of suction and elimination of a possible side-door escape route for small prey. These changes probably reduced the size of the opening of the expandable orobranchial cavity, and increased the suction produced at the mouth.

Thus, the first neopterygians had considerable trophic and locomotor advantages. The neopterygians first appeared in the Permian (Figure 8–20) and became the dominant fishes of the Mesozoic. During the Jurassic, and perhaps in the late Triassic, basal

neopterygians gave rise to fishes with further feeding and locomotor specializations. These fishes constitute the largest vertebrate radiation, the **Teleostei**. Although teleosts probably evolved in the sea, they soon radiated into fresh water. By the late Cretaceous, teleosts had replaced most of the more plesiomorphic neopterygians, and most of the more than 400 families of modern teleosts had evolved. Their first specializations involved changes in the fins.

The caudal fin of adult teleosts is supported by a few enlarged and modified hemal spines attached to the tip of the abruptly upturned vertebral column. Modified posterior neural arches—the uroneurals—add additional support to the dorsal side of the tail. These uroneurals are a derived character of teleosts (Lauder and Liem 1983). The caudal fin is symmetrical and flexible. This type of caudal structure is known as **homocercal**. In conjunction with a gas bladder, a homocercal tail allows a teleost to swim horizontally without using its paired fins as control surfaces. Drag is reduced by folding the fins close to the body. Relieved of a lift function, the paired fins of teleosts became more flexible, mobile, and diverse in shape, size, and position. Teleost fins have become specialized for activities from food getting to courtship and from sound production to walking and flying. As we saw among earlier groups, improvements in locomotion were accompanied by reduction of armor. Modern teleosts are thin-scaled by Paleozoic and Mesozoic standards, or lack scales entirely. The few heavily armored exceptions generally show a secondary reduction in locomotor abilities.

Teleosts, like most fishes that preceded them, also evolved improvements in their feeding apparatus. Trophic specializations in the earliest teleosts involved only a slight loosening of the premaxillae, so that they moved during jaw opening to accentuate the round mouth shape. One early clade of teleosts showed an enlargement of the free-swinging posterior end of the maxilla to form a nearly circular mouth when the jaws were fully opened (Figure 8–20). Later in the radiation of the teleosts distinctive changes in the jaw apparatus permitted a wide variety of feeding modes based on the speed of opening of the jaws and the powerful suction produced by the highly integrated jaw, gill arch, and cranium (Box 8–2).

In addition to speed and forceful suction, many teleosts have evolved a great deal of mobility in the skeletal elements that rim the mouth opening. This mobility allows the grasping margins of the jaws to be extended forward from the head, often at remark-

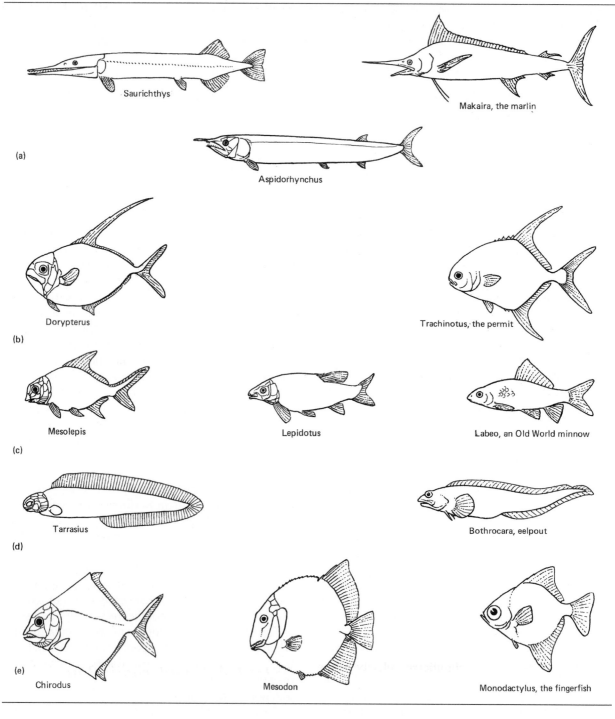

Saurichthys

Makaira, the marlin

(a)

Aspidorhynchus

Dorypterus

Trachinotus, the permit

(b)

Mesolepis

Lepidotus

Labeo, an Old World minnow

(c)

Tarrasius

Bothrocara, eelpout

(d)

Chirodus

Mesodon

Monodactylus, the fingerfish

(e)

able speed. The functional result is called the **protrusible jaw** and has evolved three or four times in different euteleost clades (Box 8–3). With so many groups converging on the same complex function, the adaptive significance of protrusibility would seem to be great. Surprisingly, there is no single hypothesis of advantage that has much experimental support (Lauder and Liem 1981). Protrusion may enhance the hydrodynamic efficiency of the circular mouth opening of ancestral teleosts, but this hypothesis seems insufficient to explain the complex anatomical changes necessary for protrusibility. Protrusion may

Figure 8–19 Convergence in specializations as indicated by general body form in actinopterygians from the earliest, neopterygian, and teleostean stages of evolution. Early ray-finned fishes of Carboniferous to early Triassic age are shown in the left column, Neopterygians of late Triassic to Cretaceous age in the middle column, and extant teleosts are shown on the right. No attempt to show the fishes to scale or to strictly match habitats (when they are known) has been made, but the general morphological convergence is readily apparent. (a) Piscivorous fishes with long bill-like rostra and/or jaws; (b) fork-tailed strong swimmers with trailing fins; (c) broad-finned bottom-feeding fishes; (d) eel-like fishes with confluent dorsal, caudal, and anal fins; (e) laterally compressed, maneuverable fishes.

aid in gripping prey, or the mouth's mobile jaws may be fitted to the substrate during feeding while the body remains in the horizontal position required for rapid escape from one's own predators. Protrusion also allows the mouth to close without reducing the volume of an expanded orobranchial cavity. This prevents prey from being flushed out of the mouth as the orobranchial cavity contracts at the end of an ingestion cycle. In many fishes with protrusible jaws, prey are clearly sucked into the mouth, but what part of the suction is produced by the protrusion itself depends greatly on the relative timing of the action of all parts of the system. No consistent pattern of steps in the feeding sequence indicates that protrusion is a significant contributor to suction efficiency. Another set of advantages of protrusible jaws may lie in the functional independence of the upper jaws relative to other parts of the feeding apparatus. Some fishes, such as the silversides and killifishes, can greatly protrude, moderately protrude, or not protrude the upper jaw while opening the mouth and creating suction. These modulations direct the mouth opening and major axis of suction either ventrally, straight ahead, or dorsally, allowing the fish to feed from substrate, water column, or surface with equal ease. Perhaps the most broadly applicable hypothesis for the strong selection for jaw protrusion is that shooting out the jaws in front of the head allows a predator to approach the prey with a portion of its feeding apparatus more rapidly than can fishes that lack protrusion. Protrusion can add significant velocity to the final moment of a predator's approach. The shooting out of protrusible jaws has been measured to increase the approach velocity of the predator by 39 to 89 percent in the crucial last instant.

Powerful mobile **pharyngeal jaws** also evolved several times (Figure 8–23). Ancestrally, ray-finned fishes have numerous dermal tooth plates in the pharynx. These plates are aligned with (but not fused to) both dorsal and ventral skeletal elements of the gill arches. A general trend of fusion of these tooth plates to one another and to a few gill arch elements above and below the esophagus can be traced in the Neopterygii. Ancestrally, these consolidated pharyngeal jaws are not very mobile and are used primarily to hold and manipulate prey in preparation for swallowing it whole. In the ostariophysan minnows and their relatives, the suckers, the primary jaws are toothless but protrusible and the pharyngeal jaws are greatly enlarged and can chew against a horny pad on the base of the skull. These feeding and digestive specializations allow extraction of nutrients from thick-walled plant cells and represent one of the largest radiations of herbivores among ray-finned fishes.

In the Neoteleostei the muscles associated with the branchial skeletal elements supporting the pharyngeal jaws have undergone radical evolution, resulting in a variety of powerful movements of the pharyngeal jaw tooth plates. Not only are the movements of these second jaws completely unrelated to the movements and functions of the primary jaws, but the upper and lower tooth plates of the pharyngeal jaws move quite independently of each other. With so many separate systems to work with, it is little wonder that some of the most extensive adaptive radiations among teleosts have been in fishes endowed with protrusible primary jaws and specialized mobile pharyngeal jaws.

Extant Actinopterygii: Ray-Finned Fishes

With an estimated 23,681 extant species (Nelson 1994), the extant actinopterygians present a fascinating, even bewildering, diversity of forms of vertebrate life. Because of their numbers we are forced to examine them in a disproportionately brief survey compared to other vertebrate taxa, focusing on the primary char-

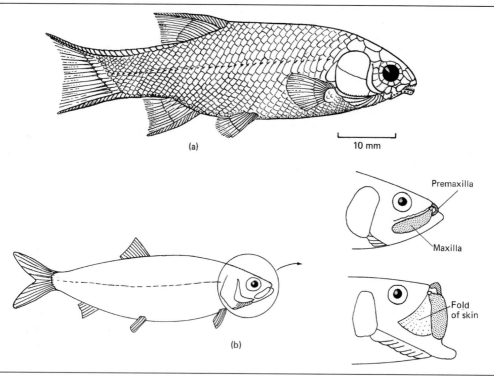

Figure 8–20 Early teleosts. (a) *Acentrophorus* of the Permian illustrates an early member of the Late Paleozoic neopterygian radiation. (b) *Leptolepis* an early Jurassic teleost with enlarged mobile maxillae, which form a nearly circular mouth when the jaws are fully opened. Membranes of skin close the gaps behind the protruded bony elements. Modern herrings have a similar jaw structure.

acteristics of selected groups and their evolution.

The study of the phylogenetic relationships of actinopterygians entered its current active state in 1966 with a major revised scheme of teleostean phylogeny proposing several new relationships. The following three decades have seen a phenomenal growth in our understanding of the interrelationships of ray-finned fishes (Rosen 1982). Nevertheless, some of the relationships indicated in Figure 8–24 are uncertain and should be considered as working hypotheses (Lauder and Liem 1983, Nelson 1994).

Polypterids and Chondrosteans

"Paleoniscoids" were replaced during the early Mesozoic by neopterygians, but a few skeletally degenerate or specialized forms have survived. The most plesiomorphic surviving lineage of actinopterygian fishes is the **Polypteriformes**. They are similar to paleoniscoids in many ways. However, polypteriforms have enough specializations to obscure their relationships to other fishes. Sometimes called the Cladistia, the Polypteriformes include 10 species of African bichirs and reedfish (Figure 8–25a), modest-size (less than a meter), slow-moving fishes with modified heterocercal tails. Polypteriforms differ from other extant basal actinopterygians in having well-ossified skeletons. In addition to a full complement of dermal and endochondral bones, polypteriforms are covered by thick, interlocking, multilayered scales. These **ganoid** scales are covered with a coat of ganoin, an enamel-like tissue characteristic of plesiomorphic actinopterygians. Larval bichirs (*Polypterus*) have external gills, possibly an ancestral condition for Osteichthyes. *Erpetoichthys*, the reedfish, is eel-like. All polypteriforms are predatory, and their jaw mechanics provide our best model of the original actinopterygian condition. Their peculiar flag-like dorsal finlets and the fleshy bases of their pectoral fins must be considered specializations, but so little is known about their natural history that the significance of these features cannot be appreciated.

Box 8–2 The Evolution of Jaw Mechanisms in Actinopterygians

Most of the main themes in actinopterygian evolution involve changes in the jaws from a simple prey-grabbing device to a highly sophisticated suction device. Suction is important in prey capture. A rapid approach by a predator toward prey pushes a wave of water in front of the predator. This diverts water around and away from the mouth. Potential prey thus can be deflected around the predator's body and away from the grasp of its jaws. If flow leading directly into the mouth could be created, prey would be drawn into the mouth. This is exactly what is done by neopterygians and some marine mammals. By rapidly increasing the volume of their orobranchial chamber, they create suction which draws a stream of water into the mouth.

The early Actinopterygii, such as *Moythomasia* (Figure 8–21a), were not much different from the acanthodians in the basic organization and function of their jaws during feeding. The jaws worked like a snap trap to grab prey. In water a significant force must be exerted to open the mouth against the resistance of the medium. Early actinopterygians achieved this in two simultaneous and complementary ways: (1) They lifted the cranium, to which the upper jaw was firmly attached by solid bony cheeks, by the action of epaxial muscles attached to the posterodorsal margin of the skull; and (2) they lowered the lower jaw by applying a posterodorsal force to a ligament that inserted on the lower jaw just posterior to its articulation with the upper jaw.

The force that rotates the lower jaw around its articulation and opens the mouth is generated in a peculiar, indirect way. Hypaxial muscles pull posteriorly on the pectoral girdle at the same time as similar bands of muscle between the anterioventral elements of the hyoid arch and the pectoral girdle contract. These actions pull posteroventrally on the anterior projections of the hyoid apparatus in the floor of the mouth. This causes the posterior portions of the hyoid to rotate dorsally. The ligament whose action opens the jaw is attached to these posterior hyoid elements and thus it is the movement of the hyoid that opens the lower jaw. Why such an indirect mechanism? The retraction of the hyoid has another component: It spreads the elements of the hyoid arch laterally, increasing the distance between the lateral walls of the orobranchial cavity and producing suction just as the mouth opens. Primitively, the extent of expansion was limited because the jaws, cheeks, and opercula of early actinopterygians were solidly encased in dermal bone with only a little flexibility between elements. The dorsal elements of the skull roof had little mobility.

The muscles that close early actinopterygian jaws are more direct and the action simpler than for the opening process. The mandibular adductor mass has several portions with fibers running at different angles providing different leverages. The muscle mass originates in the narrow space between the dermal cheek bones and **suspensorium** (the composite of hyomandibular and palatoquadrate derivatives that form the hinge joint from which the lower jaw is suspended from the cranium). The muscles thus have little space to expand as their fibers contract, leading to a fairly weak closure system. The adductor inserts on the lower jaw just anterior to its articulation with the suspensorium.

A major reorganization of the jaw mechanisms occurred with the evolution of the Neopterygii. The adductor muscles for the mandible became larger and more complex and inserted on a specialized coronoid process of the dentary, resulting in increased power in closing the jaws. Connections between the rigid cheek bones that had encased and limited the size of the adductor muscles in earlier ray-finned fishes were loosened. The stem neopterygians had an epaxial head lifting and a pectoral girdle/hyoid lower jaw opening mechanism very like that of their ancestors and retained by *Amia* (Figure 8–21b). In addition they had a second, independent jaw opening mechanism. Like the

Continued

Box 8–2 *(Continued)*

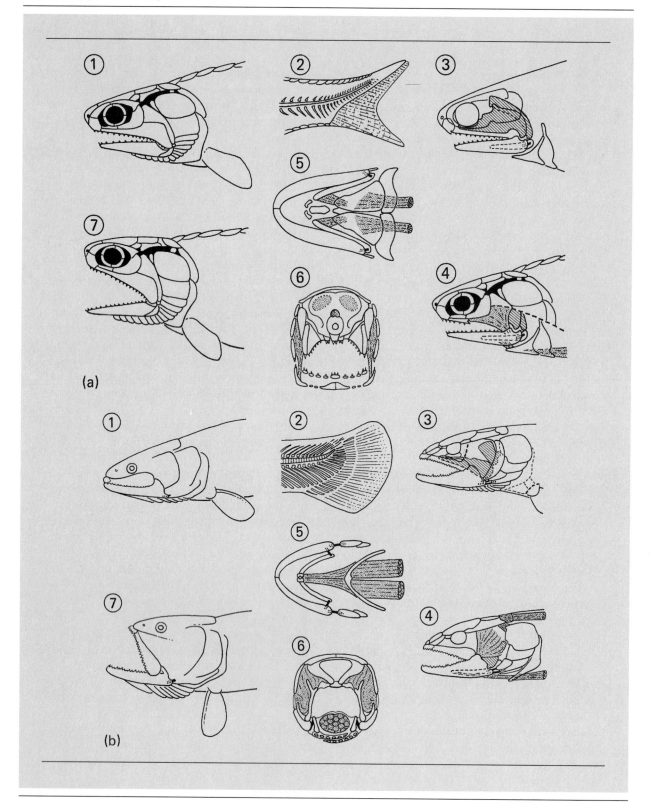

(a)

(b)

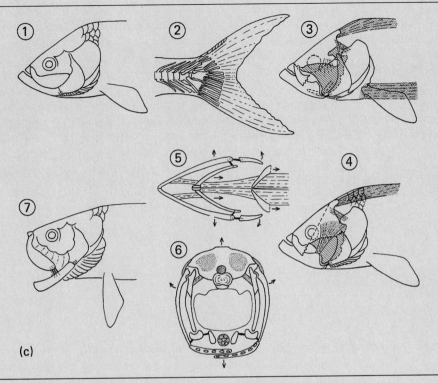

Figure 8–21 Morphological characters of three stages of actinopterygian evolution. (a) Early actinopterygian jaw size as indicated by gape (compare a1 and a7), hyomandibular support (a3), musculature (in ventral view [a5], cross section of the head [a6], and cutaway lateral view [a4]), and caudal construction (a2). Morphology is based on that reconstructed for *Moythomasia*, a "paleoniscid." (b) Neopterygian jaw size and partially circular gape (b1 and b7), nearly vertical hyomandibular orientation (b3), enlarged jaw musculature (ventral view [b5], cross section of head [b6], and cutaway view, [b4]), and caudal construction (b2). Morphology is that of *Amia*, the extant bowfin. (c) Teleostean jaw mobility to form a circular opening (c1 and c7), as illustrated in an elopomorph fish, the vertical (sometimes even anteriorly directed) hyomandibular (c3), the complex jaw musculature (ventral view [c5], cross section of head [c6], and cutaway view [c4]), and the homocercal caudal construction (c2). Morphology is based on *Megalops*, the extant tarpon.

pectoral girdle/hyoid mechanism, it is indirect and acts by way of another ligament attached to the lower jaw below and behind its articulation with the suspensorium. Many of the dermal bones of the cheek and operculum in neopterygians are free to move relative to others. This provides space for the enlargement of the jaw adductor musculature. The opercular series of bones rotates around a newly evolved socketlike articulation with the suspensorium. Rotation is powered by contraction of short opercular levator muscles originating on the back of the cranium and inserting on the dorsal border of the operculum. This rotation causes a posterodorsal movement of the ventralmost part of the opercular series. The force of this movement

Continued

Box 8–2 (*Continued*)

is transmitted via a ligament to the mandible, which swings open in response. Thus in the Neopterygii, two unrelated, biomechanically independent pathways for lowering the jaw permit the evolution of functional variation and specialization.

The teleosts (Figure 8–21c) were at first essentially identical in feeding mechanism to the earlier neopterygians with one exception. Instead of the maxilla alone rotating as the mouth opened, a new rotation site replaces that seen in

Amia: The premaxillae rotate on the tip of the skull, their toothed margins joining the maxilla in moving down and forward. This perfected the rounding of the open mouth even further. It also set the stage for at least two separate radiations of long sliding processes of the premaxillae that run up the midline of the snout when the jaws are closed. When the mouth is open the processes slide down, allowing the often tooth-bearing portions of the jaw to be greatly extended—a phenomenon known as **protrusibility**.

Box 8–3 The Protrusible Jaw

Although jaw protrusibility is generally associated with perciform fishes (Figure 8–22), it occurs in the related atherinid and paracanthopterygian fishes and in the unrelated cypriniform ostariophysans as well. Jaw morphology differs significantly among fishes with protrusible jaws, clearly indicating that some of these protrusibilities have evolved independently.

All protrusion mechanisms involve complex ligamentous attachments that allow the ascending processes of the premaxilla to slide forward on the top of the cranium without dislocation. In addition, since no muscles are in position to pull the premaxillae forward, they must be pushed by leverage from behind. Two sources provide

the necessary leverage. First, the premaxillae may be protruded by opening the lower jaw through ligamentous ties to the posterior tip of the premaxillae. Second, leverage can be provided by complex movements of the maxillae, which become isolated from the rim of the mouth by long, often toothed posterior projections of the premaxillae.

The independent movement of the protrusible upper jaw also permits closure of the mouth through maximum extension of the premaxillae while the orobranchial cavity is still expanded. Thus, engulfed prey are entrapped before the orobranchial cavity volume has been reduced through evacuation of water from the gill openings.

Figure 8–22 Jaw protrusion in suction feeding. (a) Top to bottom: The sequence of jaw movements in an African cichlid fish, *Serranochromis*. (Modified from K. F. Liem in A. G. Kluge et al., 1977, *Chordate Structure and Function*, 2d edition, Macmillan, New York.) (b) Muscle, ligament, and bone movements during premaxillary protrusion. (c) Skeletal movements and ligament actions during jaw protrusion. (After G. V. Lauder, 1980, *Biofluid Mechanics* 2:161–181.) (d) Frontal section (left) and cross section (right) of buccal expansion during suction feeding.

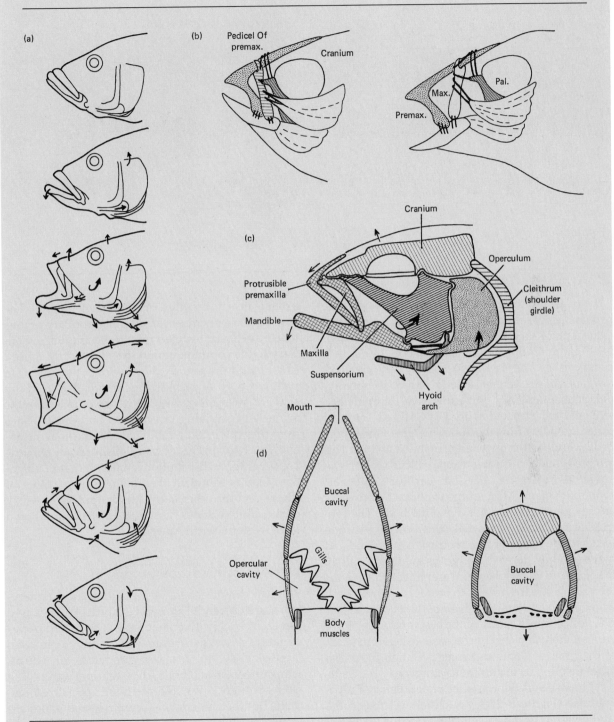

(a)

(b) Pedicel Of premax.   Cranium

Pal.

Max.

Premax.

(c) Cranium

Operculum

Protrusible premaxilla

Cleithrum (shoulder girdle)

Mandible

Maxilla

Suspensorium

Hyoid arch

(d) Mouth

Buccal cavity

Gills

Opercular cavity

Buccal cavity

Body muscles

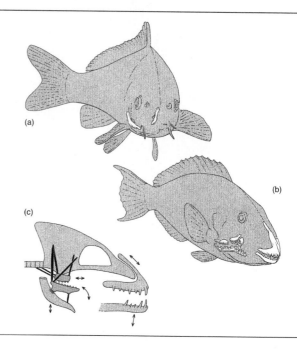

Figure 8–23 Position of pharyngeal jaws in a carp (a) and a wrasse (b). (c) Movements of the primary and pharyngeal jaws of a wrasse during feeding.

The **Chondrostei** or **Acipenseriformes** include two families of specialized plesiomorphic actinopterygians. The 24 species of sturgeon, family Acipenseridae (Figure 8–25b), are large (1- to 6-meter), active, benthic fishes. They lack endochondral bone and have lost much of the dermal skeleton of more plesiomorphic actinopterygians. Sturgeons have a strongly heterocercal tail armored with a specialized series of scales extending from the dorsal margin of the caudal peduncle along the upper edge of the caudal fin. This armored caudal cut-water (caudal fulcral scales) is an ancestral character of the earliest Paleozoic actinopterygians. Most sturgeons have five rows of enlarged armor-like scales along the body. The protrusible jaws of sturgeons make them effective suction feeders. The mode of jaw protrusion is unique and independently derived. Sturgeons are found only in the Northern Hemisphere and are either **anadromous** (ascending into fresh water to breed) or entirely fresh water in habit. Commercially important both for their rich flesh and as a source of the best caviar, they have been severely depleted by intensive fisheries in much of their range. Dams and river pollution have also taken their toll on anadromous sturgeon.

The two surviving species of paddlefish, Polyodontidae (Figure 8–25c), are closely related to the sturgeons but have a still greater reduction of dermal ossification. Their most outstanding feature is a greatly elongate and flattened rostrum, which extends nearly one-third of their 2-meter length. The rostrum is richly innervated with ampullary organs that are believed to detect minute electric fields. Contrary to the common notion that the paddle is used to stir food from muddy river bottoms, the American paddlefish is a planktivore. Paddlefish feed by actively swimming with their prodigious mouths agape and straining crustaceans and small fishes from the water, using modified gill rakers as strainers. The two species of paddlefish have a disjunct zoogeographic distribution similar to that of alligators: One is found in the Yangtze River valley of China, and the other in the Mississippi River valley of the United States. Fossil paddlefish are known from western North America.

Plesiomorphic Neopterygians

The two extant genera of plesiomorphic neopterygians are currently limited to North America and represent widely divergent types. The seven species of gars, *Lepisosteus* (Figure 8–25d), are medium- to large-size (1- to 4-meter) predators of warm-temperate fresh and brackish (estuarine) waters. The elongate body, jaws, and teeth are specialized features, but their interlocking multilayered scales are similar to those of many Paleozoic and Mesozoic actinopterygians. Gars feed on other fishes taken unaware when the seemingly lethargic and excellently camouflaged gar dashes alongside them and, with a sideways flip of the body, grasps them with needle-like teeth. Sympatric with gars is the single species of bowfin, *Amia calva* (Figure 8–25e). The head skeleton is not specialized by long prey-holding jaws like that of gar, but points toward the condition in more derived fishes in its modifications as a suction device (Lauder 1980a,b,c). *Amia*, which are 0.5 to 1 meter long, prey on almost any organism smaller than themselves. Scales of the bowfin are comparatively thin and made up of a single layer of bone as in teleost fishes; however, the asymmetric caudal fin is very similar to the heterocercal caudal fin of more plesiomorphic fishes. The interrelationships of gars, amia, and the teleosts are unresolved.

Teleosteans

Most extant fishes are teleosts. They share many characters of caudal and cranial structure and are grouped into four clades of varying size and diversity.

The **Osteoglossomorpha**, which appeared in Cretaceous seas, are now restricted to about 217 species in tropical fresh waters. *Osteoglossum* (Figure 8–26a) is a 1-meter-long predator from the Amazon. *Arapaima* is an even larger Amazonian predator (perhaps the largest strictly freshwater fish, reaching a length of at least 3 meters and perhaps to 4.5 meters), and *Mormyrus* (Figure 8–26b) is representative of the small African bottom feeders that use weak electric discharges to communicate with conspecifics. As dissimilar as they may seem, the osteoglossomorph fishes are united by unique osteological characters of the mouth and feeding mechanics.

The **Elopomorpha** (Figure 8–27) had appeared by the early Cretaceous. A specialized leptocephalous larva is a unique character of elopomorphs. These larvae spend a long time at the ocean surface, and are widely dispersed by currents. Elopomorphs include about 35 species of tarpons (Megalopidae), ladyfish (Elopidae), and bonefish (Albulidae) and 764 species of eels (Anguilliformes and Saccopharyngiformes).

Most elopomorphs are eel-like and marine, but some species are tolerant of fresh waters. The common American eel, *Anguilla rostrata*, has one of the most unusual life histories of any fish (Norman and Greenwood 1975). After growing to sexual maturity (perhaps 10 to 12 years) in rivers, lakes, and even ponds, the **catadromous** (downstream migrating) eels enter the sea. The North Atlantic eels migrate to the Sargasso Sea. Here they are thought to spawn and die, presumably at great depth. The eggs and newly hatched leptocephalous larvae float to the surface and drift in the current. Larval life continues until the larvae reach continental margins, where they transform into miniature eels and ascend rivers to feed and mature.

Most of the more than 350 species of **Clupeomorpha** are specialized for feeding on minute plankton gathered by a specialized mouth and gill straining apparatus. They are silvery, mostly marine schooling fishes of great commercial importance. Common examples are herrings, shad, sardines, and anchovies (Figure 8–27c). Several clupeomorphs are anadromous; the springtime migrations of American shad (*Alosa sapidissima*) from the North Atlantic into rivers in eastern North America involve millions of individuals, but they, too, have been greatly depleted by dams and pollution of aquatic environments.

The vast majority of extant teleosts belong to the fourth clade, the **Euteleostei**, which evolved before the upper Cretaceous (Figure 8–28). Their basal stock is represented today by the specialized ostariophysans and the generalized salmoniforms, but how these two monophyletic groups relate to each other and to the more derived (and monophyletic) neoteleostei is a matter of much dispute.

The **Ostariophysi**, the predominant fishes of the world's fresh waters, seem to be at the very base of the euteleostean radiation but have a distinctive derived character. Their name refers to small bones (*ostar* = a little bone) that connect the swim bladder (*physa* = a bladder) with the inner ear (Figure 8–29). Using the swim bladder as an amplifier and the chain of bones as conductors, this **Weberian apparatus** greatly enhances hearing sensitivity of these fishes. The ostariophysans are more sensitive to sounds and have a broader frequency range of detection than other fishes (Popper and Coombs 1980). Although all Ostariophysi have a Weberian apparatus, in other respects they are a diverse taxon of 6500 species and include the characins of South America and Africa, the carps and minnows (all continents except South America, Antarctica, and Australia), and the catfishes (all continents except Antarctica and in many shallow marine areas).

About 80 percent of the fish species in fresh water are ostariophysans. As a group they display diverse traits. For example, many ostariophysans have protrusible jaws and are adept at obtaining food in a variety of ways. In addition, pharyngeal teeth act as second jaws. Many forms have fin spines or special armor for protection, and the skin typically contains glands that produce substances used in olfactory communication. Although they have diverse reproductive habits, most lay sticky eggs or otherwise guard the eggs, preventing their loss downstream.

The esocid and salmonid fishes (Figure 8–30a) include important commercial and game fishes. These fishes have often been lumped into a taxon,

1. Actinopterygii: Basal elements of pectoral fin enlarged, median fin rays attached to skeletal elements that do not extend into fin, single dorsal fin, scales with unique arrangement, shape, interlocking mechanism, and histology, and at least six additional derived characters. 2. Cladistia plus Actinopteri (and presumably fossils such as *Moythomasia*): A specialized dentine (acrodin) forms a cap on the teeth, details of posterior braincase structure, specific basal elements of the pelvic fin are fused, and numerous features of the soft anatomy of extant forms that cannot be verified for fossils. 3. Polypteridae (Cladistia): Unique dorsal fin spines, facial bone fusion and pectoral fin skeleton and musculature. 4. Actinopteri plus fossils such as *Moythomasia* and *Mimia*: Derived characters of the dermal elements of the skull and pectoral girdle and fins. 5. Actinopteri: A spiracular canal formed by a diverticulum of the spiracle penetrating the postorbital process of the skull, other details of skull structure, three cartilages or ossifications in the hyoid below the interhyal, swim bladder connects dorsally to the foregut, fins edged by specialized scales (fulcra). 6. Chondrostei: Fusion of premaxillae, maxillae, and dermopalatines, unique anterior palatoquadrate symphysis. 7. Neopterygii: Dorsal and anal fins' rays reduced to equal the number of endoskeletal supports, upper lobe of caudal fin containing axial skeleton reduced in size to produce a nearly symmetrical caudal fin, upper pharyngeal teeth consolidated into tooth-bearing plates, characters of pectoral girdle and skull bones. 8. Lepisosteidae (Ginglymodi): Vertebrae with convex anterior faces and concave posterior faces (opisthocoel), toothed infraorbital bones contribute to elongate jaws. See character state "9." 9. "Halecostomi": Modifications of the cheek, jaw articulation, and opercular bones including a mobile maxilla. The relationships of the Lepisosteidae, Amiidae, their fossil relatives and the Teleostei are subject to many differing opinions at present with no clear resolution based on unique shared derived characters. More conservative phylogenies than presented here would represent them as an unresolved tricotomy. Others would unite the lepisosteids and the amiids as the monophyletic "Holostei." 10. Amiidae (Recent Halecomorphi): Jaw articulation formed by both the quadrate and the symplectic bones. 11. Teleostei: Elongate posterior neural arches (uroneurals), which contribute to the stiffening of the upper lobe of the internally asymmetrical caudal fin (the caudal is externally symmetrical = homocercal at least primitively in teleosts), unpaired ventral pharyngeal toothplates on basibranchial elements, premaxillae mobile, urohyal formed as an unpaired ossification of the tendon of the sternohyoideus muscle, details of skull foramina, jaw muscles, and axial and pectoral skeleton. 12. Recent Teleosts: Presence of an endoskeletal basihyal, four pharyngobranchials and three hypobranchials, median toothplates overlying basibranchials and basihyals. 13. Elopocephala: Two uroneural bones extend anteriorly to the second ural (tail) vertebral centrum, abdominal and anterior caudal epipleural intermuscular bones present. 14. Clupeocephala: Pharyngeal toothplates fused with endoskeletal gill arch elements, neural arch of first caudal centrum reduced or absent, distinctive patterns of ossification and articulation of the jaw joint. 15. Euteleostei: This numerically dominant group of vertebrates is poorly characterized with no known unique shared derived character present in all or perhaps even in most forms. Nevertheless, the following have been used as a basis for establishing monophyly: presence of an adipose fin posteriorly on the mid-dorsal line, presence of nuptial tubercles on the head and body, paired anterior membranous outgrowths of the first uroneural bones of the caudal fin. (These characters are usually lost in the most derived euteleosts.) The nature of these characters leads to a lack of consensus on the interrelationships of the basal clupeocephalids, although the group's monophyly is still generally accepted. (Based on G. V. Lauder and K. F. Liem, 1983, *Bulletin of the Museum of Comparative Zoology* 150:95–197; B. G. Gardiner, 1984, *Bulletin of the British Museum [Natural History] Geology* 37:173–427; J. G. Maisey, 1986, *Cladistics* 2:201–256; B. G. Gardiner and B. Schaeffer, 1989, *Zoological Journal of the Linnean Society* 97:135–187; P. E. Olsen and A. R. McCune, 1991, *Journal of Vertebrate Paleontology* 11:269–292; and J.S. Nelson, 1994, *Fishes of the World*, Wiley, New York, NY.)

Figure 8–24 Phylogenetic relationships of the Actinopterygii. This diagram depicts the probable phylogenetic relationships among the major groups of actinopterygians. A dagger (†) indicates extinct lineages. The numbers indicate derived characters that distinguish the lineages.

the "Protacanthopterygii" (Figure 8–28), but the basis for this classification may be no more than shared ancestral euteleostean characters that are not valid for determining phylogenetic relationships. The most plesiomorphic extant euteleosteans may be the esocids. These Northern Hemisphere temperate freshwater fishes include game species such as pickerel, pikes, and muskellunges and their relatives. The anadromous salmon usually spend their adult lives at sea, but the closely related trouts live in fresh

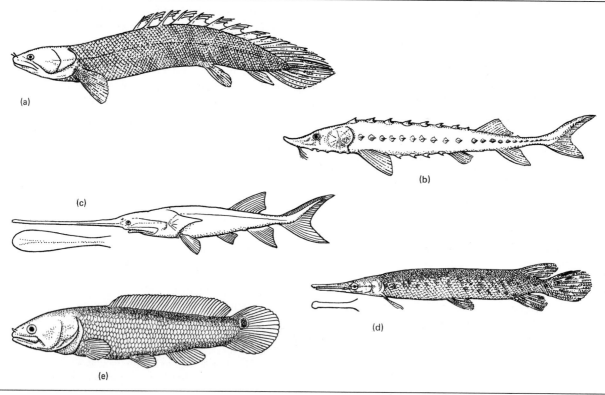

Figure 8–25 Extant nonteleostean actinopterygian fishes (a–c) and surviving plesiomorphic neopterygians (d, e). Not to scale. (a) *Polypterus*, the genus of bichirs; (b) *Acipenser*, one of the genera of sturgeons; (c) *Polyodon spathula*, one of two extant species of paddlefish; (d) *Lepisosteus*, the genus of gars; (e) *Amia calva*, the bowfin.

waters. The Southern Hemisphere galaxiids are also plesiomorphic euteleosts and live in habitats similar to those occupied by salmonids. An array of over 800 mesopelagic and deep-sea fishes in five orders, including deepwater lanternfishes, the myctophiforms (Figure 8–30b), resemble the salmonids but have derived characters of gill arch musculature, jaw structure, and fins. Tiny, light-producing organs called **photophores** are arranged on their bodies in species- and even sex-specific patterns. The light is produced by symbiotic species of *Photobacterium* and previously unknown groups of bacteria related to *Vibrio* (Haygood and Distel 1993). Some photophores probably act as signals to conspecifics in the darkness of the deep sea, where mates may be difficult to find. Others, such as those in the lure of anglerfish, probably attract prey.

Mobile jaws and protective, lightweight spines in the median fins have evolved in several groups of teleosts. About 1200 species of fishes, including cods and anglerfishes, are grouped as the **Paracanthopterygii**, although their similarities may represent convergence (Figure 8–31a). However, the **Acanthopterygii**, or true spiny-rayed fishes, whose 13,500 species dominate the surface and shallow marine waters of the world, appear to be monophyletic. Among the acanthopterygians, the atherinomorphs have protrusible jaws and specializations of form and behavior that suit them to shallow marine and freshwater habitats. This group includes the silversides, grunions, flyingfishes, halfbeaks, as well as the egg-laying and live-bearing cyprinodonts (Figure 8–31b). Killifish are examples of egg-laying cyprinodonts, and the live bearers include the guppies, mollies, and swordtails commonly maintained in home aquaria.

The majority of species of acanthopterygians and the largest order of fishes is the Perciformes, with 9293 extant species (Nelson 1994). A few of the well-known members are snooks, sea basses, sunfishes, perch, darters, dolphins (mahi mahi), snap-

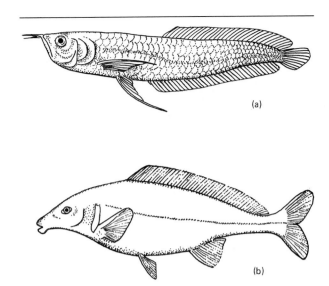

Figure 8–26 Extant osteoglossomorphs: (a) *Osteoglossum*, the arawana from South America; (b) *Mormyrus*, an elephant-nose from Africa. Not to scale.

pers, grunts, porgies, drums, cichlids, barracudas, tunas, billfish, and a majority of the fishes found on coral reefs.

Actinopterygian Reproduction and Conservation

Reproductive modes of actinopterygians show a greater diversity than is known in any other vertebrate taxon. Despite this diversity, the vast majority of ray-finned fishes are **oviparous** (producing eggs that develop outside the body of the mother). Within oviparous teleosts, marine and freshwater species show contrasting specializations.

Marine Teleosts

Most marine teleosts release large numbers of small, buoyant, transparent eggs into the water. These eggs are fertilized externally and left to develop and hatch while drifting in the open sea. The larvae are also small, and usually have little yolk reserve. They begin feeding on microplankton soon after hatching. Marine larvae are generally very different in appearance from their parents, and many larvae have been described for which the adult forms are unknown. Such larvae are often specialized for life in oceanic

plankton, feeding and growing while adrift at sea for weeks or months, depending on the species. The larvae eventually settle into the juvenile or adult habitats appropriate for their species. It is not yet generally understood whether arrival at the appropriate adult habitat (deep-sea floor, coral reef, or river mouth) is an active or passive process on the part of larvae. However, the arrival does coincide with metamorphosis from larval to juvenile morphology in a matter of hours to days. Although juveniles are usually identifiable to species, relatively few premetamorphosis larvae have successfully been reared in captivity to resolve unknown larval relationships.

This strategy of producing planktonic eggs and larvae exposed to a prolonged and risky pelagic existence appears to be wasteful of gametes. Nevertheless, complex life cycles of this sort are the principal mode of reproduction of marine fishes. Several hypotheses have been proposed to explain this paradox (Thresher 1984). One view involves the selective advantages gained by reduction of some types of predation on their zygotes that fishes may achieve by spawning pelagically. Predators that would capture the zygotes may be abundant in the parental habitat but relatively absent from the pelagic realm. Pelagic spawning fishes often migrate to areas of strong currents to spawn, or spawn in synchrony with maximum monthly or annual tidal currents, thus assuring rapid offshore dispersal of their zygotes. A second hypothesis explaining the advantages of pelagic spawning involves the high biological productivity of the sunlit surface of the pelagic environment. Microplankton of the appropriate size for food of pelagic larvae (bacteria, algae, protozoans, rotifers, and minute crustaceans) are abundant where sufficient nutrients reach sunlit waters. If energy is limiting to the adult population, a successful strategy might be to produce eggs with a minimum of energy reserve (yolk) that hatch into specialized larvae able to use this pelagic productivity. A final hypothesis involves species-level selection (Chapter 1). Floating, current-borne eggs and larvae increase the chances of young arriving at all possible patches of appropriate adult habitat. A widely dispersed species is not vulnerable to local environmental changes that could extinguish a species with a restricted geographic distribution. Perhaps the predominance of pelagic spawning species in the marine environment reflects the results of millions

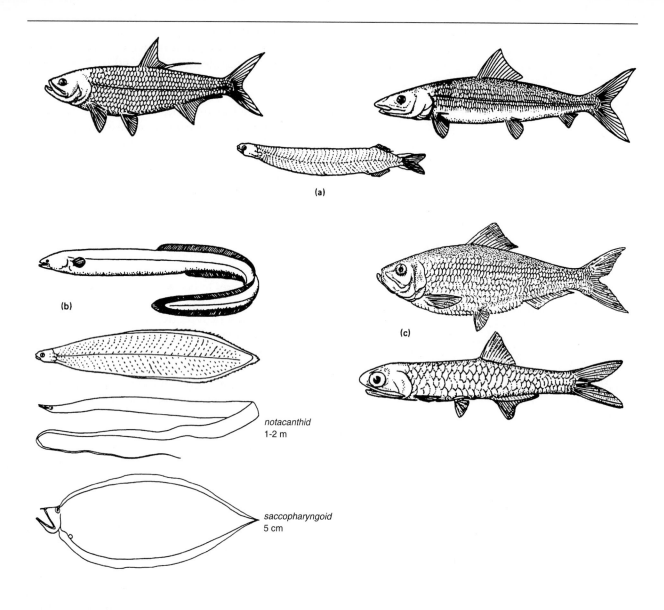

Figure 8–27 Extant teleosts of isolated phylogenetic position: (a) Elopomorpha represented by a tarpon (left) and a bonefish (right) and a typical fork-tailed lepto-cephalous larva (below); (b) anguilliform elopomorphs represented by the common eel, *Anguilla rostrata* (above), its leptocephalous larva (immediately below) and two other very different eel leptocephali; (c) Clupeomorpha represented by a herring (above) and an anchovy (below). Not to scale.

of years of extinctions of species with less dispersal-promoting habits.

Some marine species and the majority of fresh-water species lay adhesive eggs on rocks or plants or in gravel or sand, where they may guard the spawn. Nests vary from depressions in sand or gravel to elaborate constructions of woven plant material held together by parental secretions. Species of marine fishes that construct nests are smaller than species that are pelagic spawners, perhaps because small species are better able to find secure nest sites than are large species. Parental guarding of eggs may be unwittingly assisted by other organisms near, on, or in which the eggs are laid. Among these are stinging anemones, mussels, crabs, sponges, and tunicates. Perhaps the ultimate in protection of eggs is portage by one of the parents. Species are known that carry eggs on fins, under lips, in the mouth or gill cavities, or on specialized protuberances, skin patches, or even in pouches.

In spite of this diversity of reproductive habits among marine teleosts, the fact remains that the vast majority of species produce eggs that are shed into the environment with little further parental investment. These eggs tend to be small relative to the size of the adult fish, and they are produced in prodigious numbers. As a consequence of this reproductive characteristic of most marine and some freshwater teleosts, the number of individuals breeding in any given year (breeding stock size) bears no clear relationship to the number of individuals in the next generation (recruited stock year class strength). Thus, a breeding season rich with spawning adults may produce few or no offspring that survive to breed in subsequent seasons if environmental factors prevent the survival of eggs, larval, or juveniles. Conversely, a few breeding adults could, under exceptionally favorable circumstances, produce a very large number of offspring that survive to maturity. This is possible because of the large number of eggs a single female is capable of producing. The characteristic low predictability of future stock size based on current stock size has been a major stumbling block to effective fisheries management. Because so much of a population's future size depends on the environment experienced by eggs and larvae—conditions not usually obvious to fishermen or scientists—it is difficult to demonstrate the effects of overfishing in its early stages or the direct results of conservation efforts.

These inherent problems in fisheries management have resulted in destruction of commercial fish populations worldwide by overfishing. Many of the world's richest fisheries are on the verge of collapse. The Georges Bank, which lies east of Cape Cod, is an example of what overfishing can do. For years, conservation organizations called for a reduction in catches of cod, yellowtail flounder, and haddock. They were not heeded, and populations of those fish crashed dramatically in the 1990s. By October 1994, the situation was so bad that an industry group, the New England Fishery Management Council, directed its staff to devise measures that would reduce the

catch of those species essentially to zero. Many of those bottom-dwelling fish are caught unintentionally by vessels fishing for other species. If the moratorium on fishing recommended by the Fishery Management Council is implemented by the Department of Commerce, thousands of fishermen will be affected. Some of them will shift their boats to other heavily fished areas, such as the mid-Atlantic coast and the Gulf of Mexico.

Freshwater Teleosts

In contrast to their marine counterparts, freshwater teleosts generally produce and care for a small number of large, nonplanktonic eggs that hatch into adult-like body forms and behaviors. This reproductive strategy in fresh water has been related to the flowing and ephemeral characteristics of upland waters, which could easily flush a larval fish from its preferred habitat. Large yolk-rich demersal eggs may be ancestral for actinopterygians, and parental care may have evolved frequently throughout the evolution of ray-finned fishes. Producing pelagic eggs and larvae may be a derived characteristic of euteleosts. To understand more precisely the myriad variations on these two basic themes of actinopterygian reproduction, it will be necessary to know more about the early life history of fishes in the wild (Box 8–4).

Like their economically important marine kin, freshwater actinopterygians are threatened by human activities. However, the nearly 40 percent of fish species that live in the world's fresh waters are *all* threatened, regardless of their economic value. This universal threat is the result of global alteration and pollution of lakes, rivers, and streams (Warren and Burr 1994). Draining, damming, canalization, and diversion of rivers create habitats that no longer sustain indigenous fishes. In addition to the loss and physical degradation of fish habitat, much of the world—especially the Western nations and urbanized regions elsewhere—has fresh waters polluted by silt and toxic chemicals of human origin (Allan and Flecker 1993).

The United States has had, in recent years, over 2400 instances *annually* of beaches and flowing waterways closed because of pollution. These sites are too dangerous to human health for people to play in them, and those conditions are often lethal to organisms trying to live in them! Of the nearly 800 species of native freshwater fish in the United States,

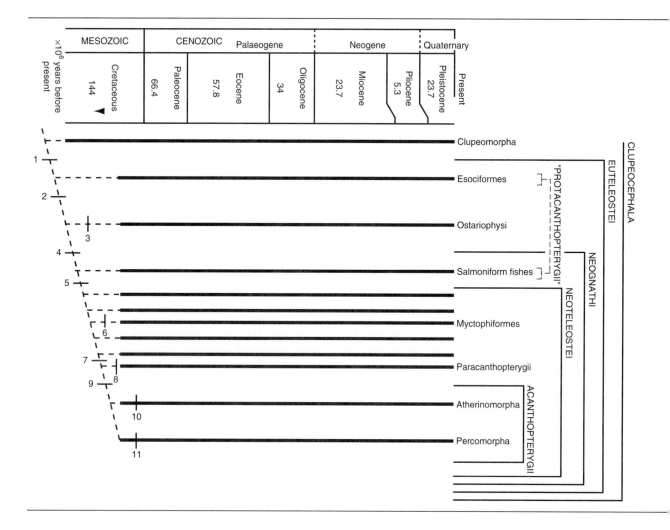

almost 20 percent are considered imperiled. As much as 85 percent of the fish fauna of some states is endangered, threatened, or of special concern (Figure 8–33).

Sex Reversal and Life History Strategies of Actinopterygians

The life histories of teleosts hold surprises for those who think of mammals as typical vertebrates. In at least a dozen separate lineages of teleosts, primarily among the Acanthopterygii, an individual's ultimate functional sex is not determined at the time of fertilization of the egg as it is in mammals. The functional sex of many teleost fishes changes as they grow, and some species are simultaneous hermaphrodites,

capable of producing eggs and sperm at the same time.

Whenever environmental factors affect the reproductive capacity of the sexes differently, the stage is set for natural selection to favor differentiation of one sex over the other (Turner 1983). There is an enormous size differential between female gametes (ova) and male gametes (sperm). A small female may be able to produce only a few eggs in her small ovaries, whereas even a small male has ample gonadal tissue to produce a prodigious quantity of sperm. Fishes are very small relative to adults when they hatch, but they continue to grow throughout life, often attaining sexual maturity long before reaching their largest known size. This means a female's egg production increases through her lifetime (Figure 8–34a). For example, female carp, *Cyprinus carpio*, hatch at a length of a few millime-

1. Euteleostei: This numerically dominant group of vertebrates is very poorly characterized. See Figure 8–24 for details. Presence of an adipose fin posteriorly on the mid-dorsal line, presence of nuptial tubercles on the head and body, paired anterior membranous outgrowths of the first uroneural bones of the caudal fin. (These characters are usually lost in the more derived euteleosts.) 2. Ostariophysi plus Neognathi: Loss of the toothplate on the fourth basibranchial. There is much disagreement about this region of the cladogram. The position of the esocids and the relationships of the ostariophysans being unresolved in the work of numerous authors. 3. Ostariophysi: Haemal spines of the anterior caudal vertebrae fused to the centra, presence of cells in the epidermis that produce a fright pheromone, reduction or loss of certain skull bones, most forms have a Weberian apparatus developed from modified ribs, vertebrae, and swim bladder. 4. Neognathi (Salmoniform fishes plus Neoteleostei): Increased contact of first vertebra with skull bones forming a tripartite occipital condyle, a rostral cartilage structure between the ethmoid and premaxillae bones. 5. Neoteleostei: Specialized tooth attachment to jaw bones, new muscles (retractors dorsalis) and muscle insertions associated with the upper branchial arch elements and upper jaw or operculum, characters of the jaw suspension, fusion of a toothplate to the third epibranchial, reduction of the caudal skeleton. 6. Myctophi-

formes plus four other superorders of primitive, stem neoteleosts living in the mesopelagic or deep sea. Some dispute on interrelationships exists. Trends within these taxa are the reduction of the fourth pharyngobranchial and its toothplate, third pharyngobranchial becoming the largest toothed element in the upper pharyngeal jaw and acquisition of a portion of the musculature of the fourth arch. 7. Paracanthopterygii plus Acanthopterygii (sometimes called the Holacanthopterygii): Expansion of ascending and articular premaxillary processes. 8. Paracanthopterygii: Caudal skeleton with a full neural spine on the second preural centrum followed by two epurals posteriorly. 9. Acanthopterygii: Insertion of the branchial retractor muscle on the third pharyngobranchial element only, characters of the pharyngobranchials and epibranchials, symphysial and toothed portions of the premaxilla capable of significant antero-ventral movement. Numerous taxa are omitted from the cladogram at this point. 10. Atherinomorpha: Unique jaw protrusion mechanism, reduction of infraorbital bones, loss of fourth pharyngobranchial. 11. Percomorpha: Pelvic girdle firmly joined to pectoral girdle, pelvic fins with one spine and five soft rays, a flange on the second circumorbital bone forms a shelf beneath the eye. (Based on G. V. Lauder and K. F. Liem, 1983, *Bulletin of the Museum of Comparative Zoology* 150:95–197; and J. S. Nelson, 1994, *Fishes of the World*, Wiley, New York, NY.)

Figure 8–28 Phylogenetic relationships of the Euteleostei. This diagram depicts the probable relationships among the major groups of modern teleosts. The numbers indicate derived characters that distinguish lineages. The quotation marks around "Protacanthopterygii" show that this is a paraphyletic grouping.

ters, reach sexual maturity in four to five years at about 430 millimeters, and are capable of laying at least 36,000 eggs. Carp may live 20 years, reaching 18 to 23 kilograms and a length of 1.2 meters. A 10-kilogram, 850-millimeter female carp contained 2,208,000 eggs. Thus a doubling of female length produced more than a 60-fold increase in gametes.

Environmental forces affect short-lived fish species differently than long-lived ones like carp. The Atlantic silverside, *Menidia menidia*, hatches in spring or summer, grows until conditions deteriorate in its habitat along the North American east coast in late fall, overwinters without growth, then spawns from May through July of its second year. Most individuals do not survive a second winter. A silverside hatched early in the season will have a long growth period and attain a length of 8 centimeters, whereas those hatched in August are much smaller when

growth ceases in winter. Females hatched from early spawnings have a reproductive advantage in the following year, because they are large and produce many eggs. The date of hatching, and hence size at reproduction, has no measurable effect on male silversides, which produce an excess of sperm no matter whether they are small or large. Silversides have a dual system of sex determination: primarily genetic sex determination in northern populations and primarily environmental sex determination in southern populations (Conover and Heins 1987). In southern populations, fishes that grow at cool temperatures (which correlate with spring season and indicate that there is a long growing period ahead) become females, whereas those growing at higher temperatures in mid to late summer become males. Environmental sex determination permits each individual to contribute a maximum of zygotes in the following year's breeding: Individuals that are likely to attain large size become females, and those likely to remain small become males. Farther north, genetic sex determination dominates, and the sex ratio of young silversides is 1:1 no matter at what temperature they grow. In the north the breeding season is so compressed that little difference in size separates early and late hatches.

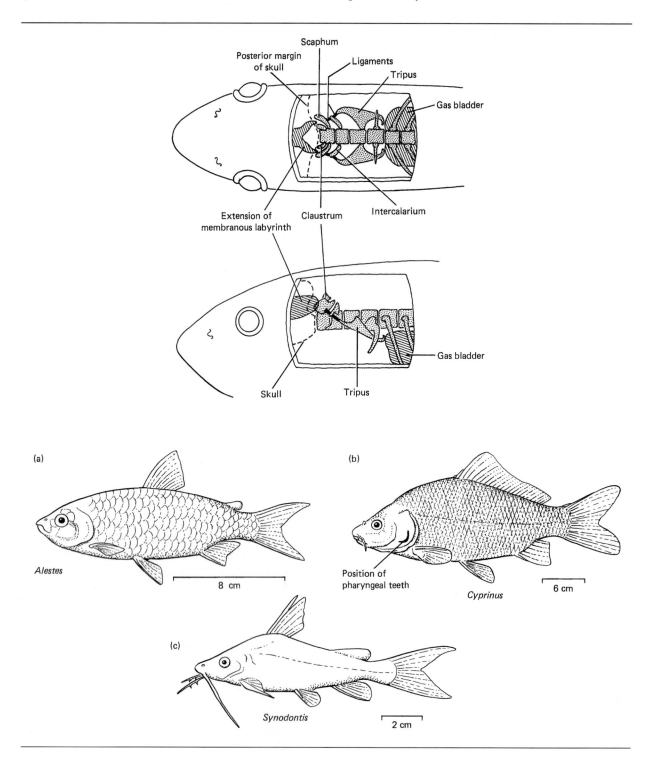

(a) *Alestes*

8 cm

(b) Position of pharyngeal teeth

Cyprinus

6 cm

(c) *Synodontis*

2 cm

Some species of pelagic spawning porgies are **sequential hermaphrodites**, meaning that small individuals are males and larger ones are females. Individuals change from male to female during their lifetime. This pattern of sex change is called **protandrous** hermaphroditism (*protos* = first, *andros* = male). The hypothesis developed to explain these sex changes is based on a **size advantage model**. As with silversides, a small fish is likely to produce more offspring as a male than it could as a female, and a large fish does best as a female. The exact size at which the sex change should take place is determined by the relative abilities of males and females to produce offspring.

Protogynous hermaphroditism (*protos* = first, *gyn* = female) occurs in species in which it is advantageous for males to be large. This often occurs in species with male–male competition, especially when a male dominates access to a limited resource. Females should spawn preferentially with the male that controls the resource. Thus, the dominant male's ability to contribute to zygotes is greater than that of lesser males or any individual female. Protogyny is widespread among coral reef fishes, where only a few spawning sites on the reef are situated in currents that will rapidly carry the fertilized eggs safely offshore for their pelagic phase. Females should become males at the size at which they can dominate the limiting resource. If they change too soon, they will lose chances to lay eggs without being able to fertilize eggs. If they delay changing, they miss a portion of the reproductive success that sex change might provide them.

A curious twist on this pattern is found among protogynous reef species with high population densities. Two types of males are found in these populations: large, sexually dimorphic males that have changed sex, and smaller males that look like females and do not change sex. The small males have much larger testes for their size than do the large males. These small males sneak in to fertilize eggs during spawning by females and territorial males.

Some fishes establish harems that consist of a single large male and several smaller females. Removal of the male initiates behavioral changes in the largest female: Within a few hours she begins behaving like a male, and in a week or two sex change is complete and the former female is producing sperm. (See the color insert.)

All-female species of fish also exist. They are produced initially by hybridization of two bisexual species, and use sperm from males (usually males of one of the parental species) to trigger zygote development. The sperm merely initiates embryonic development; its genetic material is not incorporated into the developing embryo. The best known of the all-female fishes are a half-dozen different forms of mollies, which are viviparous cyprinodonts from Mexico.

Extant Sarcopterygii: Lobe-Finned Fishes

Although they were abundant in the Devonian, the number of aquatic sarcopterygians dwindled in the late Paleozoic and Mesozoic. (All terrestrial vertebrates are sarcopterygians, so the total number of extant sarcopterygians is enormous.) The early evolution of sarcopterygians resulted in a significant radiation in fresh and marine waters. Today only four nontetrapod genera remain: the dipnoans (*Protopterus* in Africa, *Lepidosiren* in South America, *Neoceratodus* in Australia; Figure 8–35), and the actinistian *Latimeria* of waters 260 to 300 meters deep near the Comoro Archipelago off East Africa (Figure 8–36). We will discuss fossil sarcopterygians in more detail with their sister group, the tetrapods.

Figure 8–29 Ostariophysan fishes have a sound-detection system, the Weberian apparatus, that is a modification of the swim bladder and the first few vertebrae and their processes. Sound (pressure) waves impinging on the fish cause the air bladder to vibrate. The tripus is in contact with the air bladder; as the bladder vibrates the tripus pivots on its articulation with the vertebra. This motion is transmitted by ligaments to the intercalarium and scaphium. Movement of the scaphium compresses an extension of the membranous labyrinth against the claustrum, stimulating the auditory region of the brain. Typical ostariophysans include (a) characins, (b) minnows, and (c) catfish.

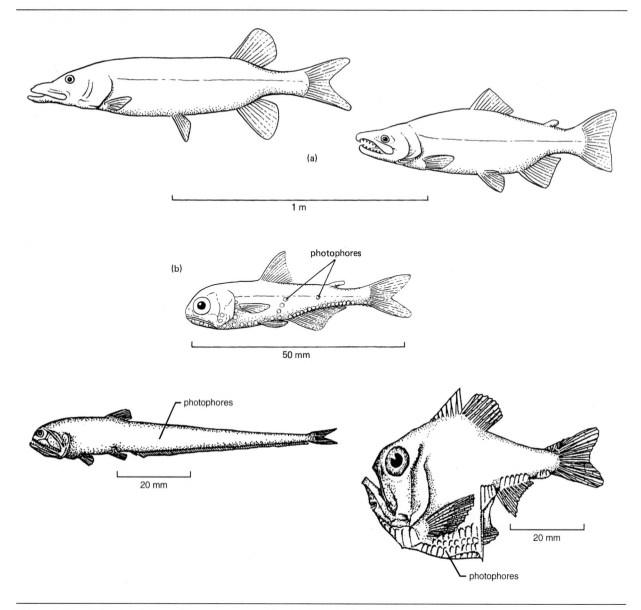

Figure 8–30 Plesiomorphic euteleosts represented by (a) the pike (left) and the salmon (right); (b) deep-sea fishes, represented by a bristlemouth (left), a lanternfish (center) and a deep sea hatchet fish (right). Species of each of these groups of deep-sea fishes differ in the number and arrangement of light-producing photophores concentrated on their ventral surface.

Dipnoans

The Dipnoi are distinguishable by the lack of articulated tooth-bearing premaxillary and maxillary bones and the fusion of the palatoquadrate to the undivided cranium. The teeth are scattered over the palate and fused into tooth ridges along the lateral palatal margins. Powerful adductor muscles of the lower jaw spread upward over the neurocranium. Throughout their evolution, this **durophagous** (feeding on hard foods) crushing apparatus has persisted. The earliest dipnoans were marine. During the Devonian, lungfishes evolved a body form quite distinct from the other Osteichthyes. The

Figure 8–31 (a) Paracanthopterygians represented by the cod (right) and the goosefish angler (left). Not to scale. (b) Atherinomorph fishes represented by (clockwise from upper left) an Atlantic silverside (*Menidia*), a flying fish, a half-beak, and live-bearing killifish, the male of which has a modified anal fin used in internal fertilization. Not to scale. The Atlantic silverside and some live-bearing killifish have unusual sex determination patterns (see pages 263–264).

anterior dorsal fin was lost, the remaining median fins fused around the posterior third of the body; the caudal fin, originally heterocercal, became symmetrical; and the mosaic of small dermal bones of the earliest dipnoan skulls (often covered by a continuous sheet of cosmine, an enameloid substance) evolved a pattern of fewer large elements without the cosmine cover. Since the Devonian, the dipnoan feeding apparatus has changed very little. Extant dipnoans, therefore, probably are not very different from their ancestors (Thomson 1969, Bemis et al. 1987).

Box 8–4 What a Fish's Ears Tell About Its Life

Following minute fish eggs or translucent larvae in the open sea or turbid rivers seems an insurmountable problem. However, there is an indirect method of tracing the details of the life history of an individual fish (Campana and Neilson 1985). A characteristic of Teleostomi is the presence of up to three compact mineralized structures suspended in the interior of each inner ear (Schultze 1990). These structures are especially well developed in the majority of teleosts, where these **otoliths** are often curiously shaped, fitting into the spaces of the membranous labyrinth very exactly and growing in proportion to the growth of the fish. They are important in orientation and locomotion, and they are formed in most teleosts during late incubation.

Otoliths grow in concentric layers much like the layers of an onion (Figure 8–32). Each of these layers reflects a day's growth. The relative width, density, and interruptions of the layers show the environmental conditions the individual encountered daily from hatching, including variations in temperature and food capture. Minute quantities of substances characteristic of the environment in which the fish has spent each day may also be incorporated in the bands further enhancing the log-like nature of the otolith. A day-by-day record of the individual is written in its otoliths. It is even possible to imprint a permanent code in the otoliths by subjecting young fish to a series of temperature increases and decreases (Brothers 1990). Fishery managers are using this method to mark juvenile fish before they are released. Years later, when the fish are recaptured as adults, the code embedded in the otoliths shows which brood they belong to. While still fraught with problems of interpretation, the study of otolith microstructure is significantly advancing our understanding of reproduction and early life history of fishes (Thresher 1988, Kingsmill 1993).

Figure 8–32 Scanning electron micrograph of an otolith from a juvenile French grunt, *Haemulon flavolineatum*. The central area represents the focus of the otolith from which growth proceeds. The alternating dark and light rings are daily growth increments. Note that the width of the growth increments varies, signifying day-to-day variation in the rate of growth. (Photograph courtesy of E. Brothers.)

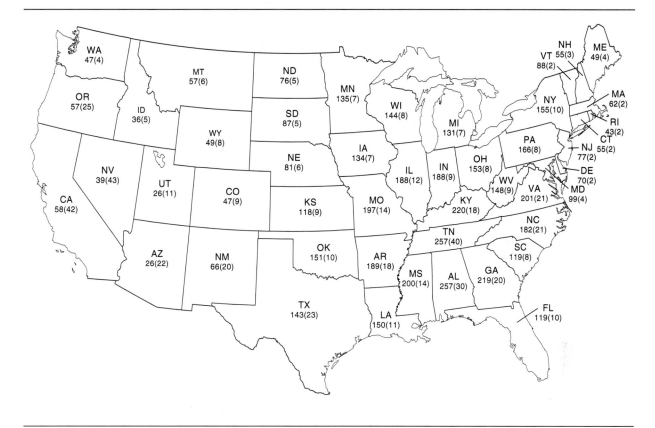

Figure 8–33 Number of known native freshwater fish species in each of the coterminous states of the United States. In parentheses, the number of species from that state considered by fisheries professionals as endangered, threatened, or of special concern. (From M. L. Warren, Jr. and B. M. Burr, 1994, "Status of fisheries of the United States: Overview of an imperiled fauna," *Fisheries* 19[1]:6–10.)

The monotypic Australian lungfish, *Neoceratodus forsteri*, is morphologically most similar to Paleozoic and Mesozoic Dipnoi. Like all other extant dipnoans, *Neoceratodus* is restricted to fresh waters; naturally occurring populations are limited to southeastern Queensland. The Australian lungfish may attain a length of 1.5 meters and a reported weight of 45 kilograms. It swims by body undulations or slowly walks across the bottom of a pond on its pectoral and pelvic appendages. Chemical senses seem important to lungfishes, and their mouths contain numerous taste buds. The nasal passages are located near the upper lip, with the incurrent openings on the rostrum just outside the mouth and the excurrent openings within the oral cavity. Thus, gill ventilation draws water across the nasal epithelium. *Neoceratodus* respires almost exclusively via its gills and uses its single lung only when stressed. Little is known of its behavior. Although they go through a complex courtship, which perhaps includes male territoriality and are very selective about the vegetation upon which they lay their adhesive eggs, no parental care has been observed after spawning. The jelly-coated eggs, 3 millimeters in diameter, hatch in three to four weeks, but the young have proven elusive and nothing is known of their juvenile life. Surprisingly little is known about the single South American lungfish, *Lepidosiren paradoxa*, but the closely related African lungfishes, *Protopterus*, with four recognized species, are better known. These two genera are distinguished by different numbers of weakly developed gills. Because their gills are very small, these lungfishes drown if they are prevented from using their paired lungs. Nevertheless, the gills are important in eliminating carbon dioxide. These thin-scaled, heavy-bodied, elongate fishes, 1 to 2 meters long, have unique filamentous and highly mobile paired appendages. For a time after their discovery 150

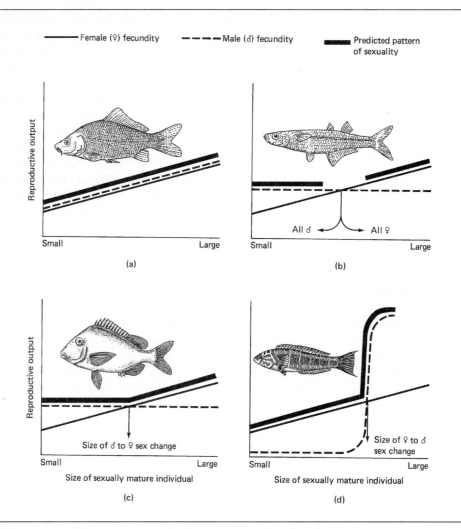

Female (♀) fecundity   ----- Male (♂) fecundity   Predicted pattern of sexuality

Figure 8–34 Size advantage hypothesis of sex determination and sexual patterns in organisms with labile sexuality, such as teleosts. Solid line, female fecundity; dashed line, male fecundity; shaded band, predicted pattern of sexuality. (a) Gonochorism (separate sexes) in iteroparous (multiple breeding season) species where males compete directly to fertilize eggs as in carp; (b) gonochorism in semelparous (single breeding season) species where sex determination complements size at reproduction as in some populations of silversides; (c) protandrous sequential hermaphroditism in iteroparous species with random mating as in some species of sea breams; (d) protogynous sequential hermaphroditism in iteroparous species with female choice for larger dominant males as in most wrasses. (Modified in part from R. R. Warner, 1984, *American Scientist* 72:128–136).

years ago lungfish were considered to be specialized urodele amphibians, which they superficially resemble. Although the skeletons of these lungfishes are mostly cartilaginous, their tooth plates are heavily mineralized and fossilize readily.

One habit of some species of African dipnoans, **estivation**, considerably increases the chance of fossilization. Similar in some ways to hibernation, estivation is induced by drying of the habitat rather than by cold. African lungfishes frequent areas that flood

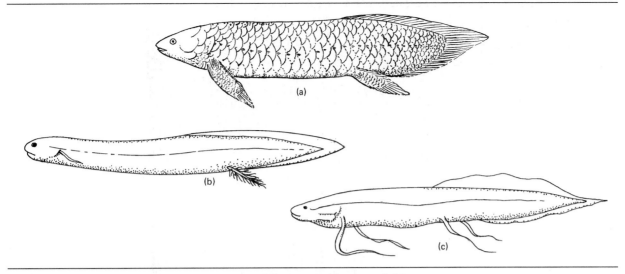

Figure 8–35 Extant dipnoans: (a) Australian lungfish, *Neoceratodus forsteri*; (b) South American lungfish, *Lepidosiren paradoxa*, male; note the specialized pelvic fins of the male during the breeding season; (c) African lungfish, *Protopterus*.

during the wet season and bake during the dry season—habitats not available to actinopterygians except by immigration during floods. The lungfishes enjoy the flood periods, feeding heavily and growing rapidly. When the flood waters recede, the lungfish digs a vertical burrow in the mud that ends in an enlarged chamber and varies in length in proportion to the size of the animal, the deepest being less than a meter. As drying proceeds, the lungfish becomes more lethargic and breathes air from the burrow opening. Eventually, even the water of the burrow dries up, and the lungfish enters the final stages of estivation. In the deep chamber, the lungfish remains folded into a U-shape with its tail over its eyes. Heavy mucus secretions the fish has produced since entering the burrow condense and dry to form a protective envelope around its body. Only an opening at its mouth remains to permit breathing.

Although the rate of energy consumption during estivation is very low, metabolism continues, using muscle proteins as an energy source. Lungfishes normally spend less than six months in estivation, but they have been revived after four years of enforced estivation. When the rains return, the withered and shrunken lungfish becomes active and feeds voraciously. In less than a month it regains its previous size.

Estivation is an ancient trait of dipnoans. Fossil burrows containing lungfish tooth plates have been found in Carboniferous and Permian deposits of North America and Europe. Without the unwitting assistance of the lungfishes, which initiated fossilization by burying themselves, such fossils might not exist.

Actinistians

Actinistians are unknown before the middle Devonian. Their hallmarks are an unlobed first dorsal fin and the unique, symmetrical, three-lobed tail with a central fleshy lobe that ends in a fringe of rays (Figure 8–36). Actinistians also differ from all other sarcopterygians in the head bones (they lack, among other elements, a maxilla), in details of the fin structure, and in the presence of a curious rostral organ. Following rapid evolution during the Devonian, the actinistians show a history of stability. Devonian actinistians differ from the more recent Cretaceous fossils mostly in the degree of skull ossification. Some early actinistians lived in shallow fresh waters, but the fossil remains of these and other osteichthyans during the Mesozoic are largely marine. The osteichthyans radiated into a variety of niches, but the actinistians retained their peculiar form. Fos-

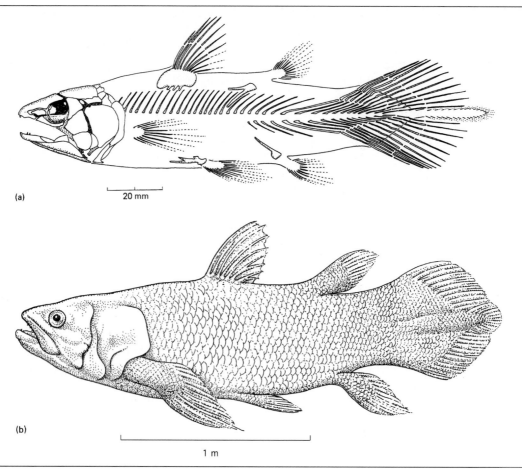

(a)

20 mm

(b)

1 m

Figure 8–36 Representative Actinistia (coelacanths): (a) *Rhabdoderma*, a Carboniferous actinistian; (b) *Latimeria chalumnae*, the extant coelacanth.

sil actinistians are not known after the Cretaceous, and until a little over 50 years ago they were thought to be extinct.

In 1938, an African fisherman bent over an unfamiliar catch and nearly lost his hand to its ferocious snap. Imagine the astonishment of the scientific community when J. L. B. Smith of Rhodes University announced that the catch was an actinistian! This large fish was so similar to Mesozoic fossil coelacanths that its systematic position was unquestionable. Smith named this living fossil *Latimeria chalumnae* in honor of Courtenay Latimer, who recognized it as unusual and brought the specimen to his attention (Smith 1956).

Despite public appeals, no further specimens of *Latimeria* were captured until 1952. Since then more than 150 specimens ranging in size from 75 centimeters to slightly over 2 meters and weighing from 13

to 80 kilograms have been caught, all in the Comoro Archipelago between Madagascar and Mozambique. Coelacanths are hooked near the bottom, usually in 260 to 300 meters of water about 1.5 kilometers offshore. Strong and aggressive, *Latimeria* is steely blue-gray with irregular white spots and reflective golden eyes. The reflection comes from a tapetum lucidum that enhances visual ability in dim light. A large cavity in the midline of the snout communicates with the exterior by three pairs of rostral tubes enclosed by canals in the wall of the chondrocranium. These tubes are filled with gelatinous material and open to the surface through a series of six pores. The rostral organ is almost certainly an electroreceptor (Bemis and Hetherington 1982). *Latimeria* is a predator—stomachs have contained fishes and cephalopods.

A fascinating glimpse of the life of the coelacanth

was reported by Hans Fricke and his colleagues who used a small submarine to observe the fish (Fricke et al. 1987; Fricke 1988). They saw six coelacanths at depths between 117 and 198 meters off a short stretch of the shoreline of one of the Comoro Islands. Coelacanths were seen only in the middle of the night and only on or near the bottom. Unlike extant lungfish, the coelacanths did not use their paired fins as props or to walk across the bottom. However, when they swam the pectoral and pelvic appendages were moved in the same sequence as tetrapods move their limbs.

The discovery of *Latimeria* has confirmed earlier reconstructions based on coelacanth fossils (Thomson 1986). A case in point is the mode of coelacanth reproduction. In 1927, D. M. S. Watson described two small skeletons from inside the body cavity of *Undina*, a Jurassic coelacanth, and suggested that coelacanths gave birth to their young. Because copulatory structures have never been found on any coelacanth fossil, some dismissed Watson's specimen as a case of cannibalism. Female *Latimeria* containing up to 19 eggs, each 9 centimeters in length, have now been captured. R. W. Griffith and K. S. Thomson sur-

mised that Watson was correct because of the small number of eggs and their lack of a shell or other coating to provide osmotic protection if released into the sea. While C. L. Smith and others at the American Museum of Natural History were dissecting a 1.6-meter specimen they discovered five young, each 30 centimeters long and at an advanced stage of embryonic development, in the single oviduct. Internal fertilization must occur, but how copulation is achieved is unknown, since males show no specialized copulatory organs. In spite of some excellent fossils and numerous specimens of the extant coelacanths, there has never been stable agreement about the relationships of the Actinistia with other gnathostomes. Workers have disagreed about the position of the coelacanths more than that of most other vertebrate taxa. In part this is due to the surprising combinations of morphology and physiology shown by *Latimeria*, which has many derived characters, some of which are most similar to Chondrichthyes, others to the Dipnoi, some to Actinopterygii, and also a curious collection of unique features (McCosker and Lagios 1979).

Summary

Because of its physical properties, water is a demanding medium for vertebrate life. Nevertheless, the greatest number of vertebrate species, the vast majority of them Osteichthyes, are found exclusively in the planet's oceans, lakes, and rivers. Highly efficient respiratory systems, the gills, facilitate rapid oxygen uptake to sustain the activity of fishes. Their locomotion is generally accomplished with undulations produced by contraction of the body muscles. Sensory guidance for activity is often visual, but other sensory systems also are highly refined. Fishes detect low-frequency pressure waves via the lateral-line system or navigate and communicate through electroreception.

At their first appearance in the fossil record, osteichthyans, the largest vertebrate taxon, are separable into distinct lineages. The Sarcopterygii (fleshy-finned fishes: lungfishes, actinistians, and other lobe-finned fishes) and the Actinopterygii (ray-finned

fishes) show indications of common ancestry. Extant sarcopterygian fishes offer exciting glimpses of adaptations evolved in Paleozoic environments. Actinopterygian fishes were distinct as early as the Devonian. Actinopterygians inhabit the 69 percent of the Earth's surface that is covered by water and are the most numerous and species-rich lineages of vertebrates. Several levels of development in food-gathering and locomotory structures characterize actinopterygian evolution. The radiations of these levels are represented today by relict groups: cladistians (bichirs and reedfish), chondrosteans (sturgeons and paddlefish), and the plesiomorphic neopterygians (gars and *Amia*). The most derived level—teleosteans—may number close to 24,000 extant species with two groups, ostariophysans in fresh water and acanthopterygians in seawater, constituting a large proportion of these species.

References

Allan, J. D., and A. S. Flecker. 1993. Biodiversity conservation in running waters. *Bioscience* 43:32–43.

Bass, A. H. 1986. Electric organs revisited: evolution of vertebrate communication and orientation organs. Pages 13–70 in *Electroreception*, edited by T. H. Bullock and W. Heiligenberg. Wiley, New York, NY.

Bastian, J. 1994. Electrosensory organisms. *Physics Today* 47(2):30–37.

Bemis, W. E., and T. E. Hetherington. 1982. The rostral organ of *Latimeria chalumnae*. Morphological evidence of an electroreceptive function. *Copeia* 1982:467–471.

Bemis, W. E., W. W. Burggren, and N. E. Kemp (editors). 1987. *The Biology and Evolution of Lungfishes*. Liss, New York, NY.

Bennett, M. V. L. 1971a. Electric organs. Pages 347–491 in *Fish Physiology*, volume 5, *Sensory Systems and Electric Organs*, edited by W. S. Hoar and D. J. Randall. Academic, New York, NY.

Bennett, M. V. L. 1971b. Electroreception. Pages 493–574 in *Fish Physiology*, volume 5. *Sensory Systems and Electric Organs*, edited by W. S. Hoar and D. J. Randall. Academic, New York, NY.

Bodznick, D., and R. G. Northcutt. 1981. Electroreception in lampreys: evidence that the earliest vertebrates were electroreceptive. *Science* 212:465–467.

Bone, Q., and N. B. Marshall. 1982. *Biology of Fishes*. Blackie & Son, Glasgow, UK.

Brothers, E. B. 1990. Otolith marking. *American Fisheries Society Symposium* 7:183–202.

Campana, S. E., and J. D. Neilson. 1985. Microstructure of fish otoliths. *Canadian Journal of Fisheries and Aquatic Science* 42:1014–1032.

Conover, D. O., and S. W. Heins. 1987. Adaptive variation in environmental and genetic sex determination in a fish. *Nature* 326:496–498.

Cromie, W. J. 1982. Born to navigate. *Mosaic* 13(4):17–23.

Denison, R. H. 1979. Acanthodii. In *Handbook of Paleoichthyology*, volume 5. Gustav Fischer, Stuttgart, Germany.

Eaton, R. C. (editor). 1984. *Neural Mechanisms of Startle Behavior*. Plenum, New York, NY.

Faber, D. S., and H. Korn (editors). 1978. *Neurobiology of the Mauthner Cell*. Raven, New York, NY.

Farrell, A. P. 1980. Vascular pathways in the gill of ling cod, *Ophiodon elongatus. Canadian Journal of Zoology* 58:796–806.

Fessard, A. (editor). 1974. Electroreceptors and other specialized receptors in lower vertebrates. Pages 59–124 in *Handbook of Sensory Physiology*, volume 3, part 3. Springer, New York, NY.

Fricke, H. 1988. Coelacanths, the fish that time forgot. *National Geographic* 173:824–838.

Fricke, H., O. Reinicke, H. Hofer, and W. Nachtigall. 1987. Locomotion of the coelacanth *Latimeria chalumnae* in its natural environment. *Nature* 324:331–333.

Haygood, M. G., and D. L. Distel. 1993. Bioluminescent symbionts of flashlight fishes and deep-sea anglerfishes form unique lineages related to the genus *Vibrio. Nature* 363:154–156.

Kingsmill, S. 1993. Ear stones speak volumes to fish researchers. *Science* 260:1233–1234.

Lauder, G. V. 1980a. Evolution of the feeding mechanism in primitive actinopterygian fishes: a functional anatomical analysis of *Polypterus, Lepisosteus* and *Amia. Journal of Morphology* 163:283–317.

Lauder, G. V. 1980b. On the evolution of the jaw adductor musculature in primitive gnathostome fishes. *Breviora* 460:1–10.

Lauder, G. V. 1980c. Hydrodynamics of prey capture by teleost fishes. *Biofluid Mechanics* 2:161–181.

Lauder, G. V., and K. F. Liem. 1981. Prey capture by *Luciocephalus pulcher*: implications for models of jaw protrusion in teleost fishes. *Environmental Biology of Fishes* 6:257–268.

Lauder, G. V., and K. F. Liem. 1983. The evolution and interrelationships of the actinopterygian fishes. *Bulletin of the Museum of Comparative Zoology* 150:95–197.

Laurent, P., and S. Dunel. 1980. Morphology of gill epithelia in fish. *American Journal of Physiology* 238:147–149 (R).

Lindsey, C. C. 1978. Form, function, and locomotory habits in fish. Pages 1–100 in *Fish Physiology*, volume 7, *Locomotion*, edited by W. S. Hoar and D. J. Randall. Academic, New York, NY.

Magnuson, J. J. 1978. Locomotion by scombroid fishes: hydromechanics, morphology, and behavior. Pages 239–313 in *Fish Physiology*, volume 7, *Locomotion*, edited by W. S. Hoar and D. J. Randall. Academic, New York, NY.

McCosker, J. E., and M. D. Lagios (editors). 1979. The biology and physiology of the living coelacanth. *California Academy of Sciences Occasional Papers* 134:1–175.

Nelson, J. S. 1994. *Fishes of the World*, 3d edition. Wiley, New York, NY.

Nieuwenhuys, R. 1982. An overview of the organization of the brain of actinopterygian fishes. *American Zoologist* 22(2):287–310.

Norman, J. R., and P. H. Greenwood. 1975. *A History of Fishes*, 3d edition. Ernest Benn, London, UK.

Popper, A. N., and S. Coombs. 1980. Auditory mechanisms in teleost fishes. *American Scientist* 68:429–440.

Roberts, B. L. 1981. The organization of the nervous system of fishes in relation to locomotion. Pages 11–136 in *Vertebrate Locomotion*, edited by M. H. Day. *Symposia of the Zoological Society of London*, volume 48. Academic, London, UK.

Rome, L. C., D. Swank, and D. Corda. 1993. How fish power swim. *Science* 261:340–343.

Rosen, D. E. 1982. Teleostean interrelationships, morphological function and evolutionary inference. *American Zoologist* 22:261–273.

Schaeffer, B., and D. E. Rosen. 1961. Major adaptive levels in the evolution of the actinopterygian feeding mechanism. *American Zoologist* 1:187–204.

Scheich, H. G. Langner, C. Tidemann, R. B. Coles, and A. Guppy. 1986. Electroreception and electrolocation in platypus. *Nature* 319:401–402.

Schultze, H.-P. 1990. A new acanthodian from the Pennsylvanian of Utah, U.S.A., and the distribution of otoliths in gnathostomes. *Journal of Vertebrate Paleontology* 10:49–58.

Smith, J. L. B. 1956. *Old Four Legs*. Longman, London, UK.

Stix, G. 1994. Robotuna. *Scientific American* 270(1):142.

Tavolga, W. N., A. N. Popper, and R. R. Fay (editors). 1981. *Hearing and Sound Communication in Fishes*. Springer, New York, NY.

Thomson, K. S. 1969. The biology of the lobe-finned fishes. *Biological Review* 44:91–154.

Thomson, K. S. 1986. Marginalia: a fishy story. *American Scientist* 74(2):169–171.

Thresher, R. E. 1984. *Reproduction in Reef Fishes*. T. F. H. Publications, Neptune City, NJ.

Thresher, R. E. 1988. Otolith microstructure and the demography of coral reef fishes. *Trends in Ecology and Evolution* 3:78–80.

Turner, B. J. (editor). 1983. *Evolutionary Genetics of Fishes*. Plenum, New York, NY.

Warren, M. L., Jr., and B. M. Burr. 1994. Status of fresh water fishes of the United States: overview of an imperiled fauna. *Fisheries* 19:6–18.

Webb, P. W. 1984. Form and function in fish swimming. *Scientific American* 251:72–82.

Webb, P. W., and R. W. Blake. 1985. Swimming. Pages 110–128 in *Functional Vertebrate Morphology*, edited by M. Hildebrand, D. M. Bramble, K. F. Liem, and D. B. Wake. Belknap, Cambridge, MA.

9 GEOGRAPHY AND ECOLOGY FROM THE MID-DEVONIAN TO LATE PERMIAN

Terrestrial environments in the mid-Devonian supported a substantial diversity of plants, some of which were as much as 2 meters tall. Probably plant life was confined to moist areas such as the edges of streams and lakes and low-lying swamps. The presence of plants creates local microclimates—shaded areas that are cool and humid. Those sheltered microclimates can be important to the plants themselves by providing favorable conditions for reproduction, and they increase the structural diversity of the habitat that is available to terrestrial animals. This was the environment that saw the appearance of terrestrial vertebrates.

Continental Geography in the Late Paleozoic

From the Devonian through the Permian the continents were drifting together (Scotese and McKerrow 1990). The continental blocks that correspond to parts of modern North America, Greenland, and western Europe had come into proximity along the equator (Figure 9–1). With the later addition of Siberia, these blocks formed a northern supercontinent known as **Laurasia.**

Most of Gondwana was in the far south, overlying the South Pole, but its northern edge was separated from the southern part of Laurasia only by a narrow extension of the **Tethys Sea.** The arm of the Tethys Sea between Laurasia and Gondwana did not close completely until the late Carboniferous (Figure 9–2). During the Carboniferous, the process of coalescence continued, and by the Permian most of the continental surface was united in a single continent, **Pangaea** (sometimes spelled Pangea). At its maximum extent, the land area of Pangaea covered 36 percent of the Earth's surface, compared to 31 percent for the present arrangement of continents (Veevers 1994). This supercontinent persisted for 160 million years, from the mid-Carboniferous to the mid-Jurassic, and profoundly influenced the evolution of terrestrial plants and animals.

Evolution of Terrestrial Ecosystems

The terrestrial environment of the mid-Devonian had substantially more different species of plants and animals than any previous period, but it differed in many respects from the modern ecosystems we are used to. In the first place, of course, there were no terrestrial vertebrates—the earliest of those appear in the late Devonian. Furthermore, plant life was limited to wet places—the margins of streams, rivers, and lakes and low-lying areas. In these areas plants grew in patches, each patch dominated by a single species, rather than in communities with a mixture of species as we see now. The tallest plants reached heights of 2 meters or so, and probably formed canopies dense enough to shade the ground and create diverse microclimates that animals could exploit.

Still more unusual to our eyes would be the absence of insects that fed on plants (DiMichele and

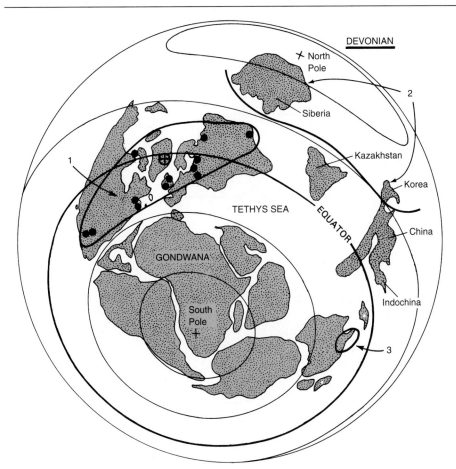

Figure 9–1 Location of continental blocks in the late Devonian. Laurentia, Greenland, and Baltica lie on the equator. An arm of the Tethys Sea extends westward between Gondana and the northern continents.

Hook 1992). It's hard to be sure whether the holes we see in fossil leaves were made while the leaf was still on the plant or after it had fallen to the ground, but the latter seems more likely. Probably terrestrial animal communities in the mid-Devonian were based on detritivores such as millipedes, springtails, and mites (Shear et al. 1989b). Those animals in turn were preyed upon by scorpions, pseudo-scorpions, and spiders (Shear and Kukalová–Peck 1990). The earliest evidence of silk-production by a spider is from the mid-Devonian (Shear et al. 1989a).

Terrestrial ecosystems became increasingly complex during the remainder of the Paleozoic. Plants diversified and increased in size, invertebrates with new specializations appeared (including insects that fed on living plants), and terrestrial vertebrates appeared and diversified. The late Devonian saw the spread of forests of the progymnosperm *Archaeopteris*, large trees with trunks up to a meter in diameter and reaching heights of at least 10 meters.

Most species of *Archaeopteris* put out horizontal branches that bore leaves, and stands of these trees would have created a shaded forest floor in low-lying areas. Some of these trunks show growth rings, indicating seasonal changes in the rate at which new tissue was deposited. Giant horsetails (*Calamites*), relatives of the living horsetails you can find growing in moist areas, reached heights of several meters. Many species of giant clubmosses (lycophytes) spent most of their lives as an unbranched trunk and looked rather like electric or telephone utility poles. Both male and female lycophytes produced large cones once in their lifetime. *Lepidodendron* grew as tall as 45 meters and produced spore-bearing cones 50 centimeters long (White 1986). Other areas were apparently covered by bush-like plants, vines, and low-growing ground covers.

The increasing diversity and habitat specificity of late Devonian floras continued and expanded in

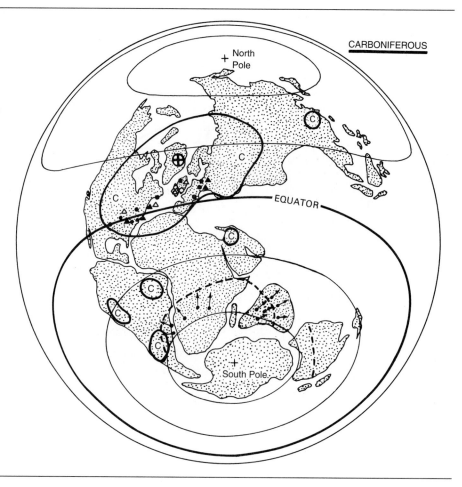

Figure 9–2 Location of continental blocks in the Carboniferous. This map illustrates an early stage of Pangaea. The location and extent of continental glaciation in the late Carboniferous is shown by the dashed lines and arrows radiating out from the South Pole.

the Carboniferous. Most of the major taxonomic groups of plants evolved during this time, and the land was partitioned among different ecological types corresponding to drainage patterns. Seed ferns (pteridosperms) and ferns lived in well-drained areas, and swamps were dominated by lyco-phytes, with horsetails, ferns, and seed ferns also present. An important innovation in the Devonian flora, the appearance of heterosporous plants (unequal spore sizes and incipient female–male differences), occurred in the early to middle Devonian. The first fossil seed is found in the late Devonian (Taylor 1981).

During the middle to late Devonian most plants were uniformly distributed between 30°S and 30°N latitude, where the climate was predominantly tropical. In Siberia, however, the flora of late Devonian age shows seasonal growth rings. Siberia was located between 30 and 60°N latitude in the Devonian

(Figure 9–1). This Siberian flora probably represents one of the first extensive terrestrial plant communities adapted to cooler conditions (Ziegler et al. 1981). In the Carboniferous, however, the distinction became blurred, possibly because Siberia drifted farther southward (Raymond et al. 1985).

Terrestrial invertebrates burgeoned during the Carboniferous. Millipedes, arachnids, and insects (especially roaches) were common. Detritivores continued to be an important part of the food web, but insect herbivory appears to have been well established by the end of the Carboniferous. Fossil leaves have ragged holes, seeds are penetrated by tunnels, and pollen is found in the guts and feces of fossilized insects. Dragonflies (one with a wingspan of 63 centimeters) flew through the air, and predatory invertebrates such as scorpions and arachnids prowled the forest floor. Arthropleurids were terrestrial predators, with body lengths that reached nearly 2 meters. In general, predatory arthropods were larger and more diverse in the Paleozoic than they are today, probably reflecting a different balance between invertebrates and vertebrates.

Terrestrial vertebrates appeared in the late Devonian and diversified during the Carboniferous. By the early Permian several vertebrate lineages had given rise to small insectivorous predators. These animals were similar in size and in general body form to modern salamanders and lizards, and probably lived much the same sorts of lives. Larger vertebrates (up to 1.5 meters in total length) were probably predators of these small species, and still larger predators topped the food web. An important development in the Permian was the appearance of herbivorous vertebrates. For the first time vertebrates were able to exploit the primary production of terrestrial plants directly, and the food web was no longer based solely on invertebrates. By the end of the Permian the structure and function of terrestrial ecosystems were essentially modern, although the kinds of plants and animals in those ecosystems were almost entirely different from the ones we know today (DiMichele and Hook 1992).

Devonian Climates

In Chapter 5 we suggested that the climate of equatorial latitudes in the early Paleozoic was tropical and reasonably uniform, and this pattern appears to have continued through the late Paleozoic. A single major climatic interruption that lowered temperatures worldwide by several degrees is indicated by extensive glaciation in the Ordovician. In the Silurian and Devonian, the climate was again warmer in the tropics and undisturbed by glaciation. Before the end of the Devonian, however, and again in the late Carboniferous major glacial events are indicated by the presence of glacial deposits. During this time Gondwana drifted about the Southern Hemisphere and on at least two occasions (late Ordovician and late Devonian) this huge landmass was centered over the South Pole.

References

DiMichele, W. A., and R. W. Hook. 1992. (rapporteurs). Paleozoic terrestrial ecosystems. Pages 205–325 in *Terrestrial Ecosystems Through Time*, edited by A. K. Behrensmeyer, J. D. Damuth, W. A. DiMichele, R. Potts, H. Dieter–Sues, and S. L. Wing, University of Chicago Press, Chicago, IL.

Raymond, A., W. C. Parker, and S. F. Barrett. 1985. Early Devonian phytogeography. Pages 129–167 in *Geological Factors and the Evolution of Plants*, edited by B. H. Tiffney. Yale University Press, New Haven, CT.

Scotese, C. R., and W. S. McKerrow. 1990. Revised world maps and introduction. Pages 1–21 in *Paleozoic Palaeogeography and Biogeography*, edited by W. S. McKerrow and C. R. Scotese. Geological Society Memoir No. 12.

Shear, W. A., and J. Kukalová–Peck. 1990. The ecology of Paleozoic terrestrial arthropods: the fossil evidence. *Canadian Journal of Zoology* 68:1807–1834.

Shear, W. A., J. M. Palmer, J. A. Coddington, and P. M. Bonamo. 1989a. A Devonian spinneret: early evidence of spiders and silk use. *Science* 246:479–481.

Shear, W. A., W. Schwaller, and P. Bonamo. 1989b. Record of Palaeozoic pseudoscorpions. *Nature* 341:527–529.

Taylor, T. N. 1981. *Paleobotany*. McGraw-Hill, New York, NY.

Veevers, J. J. 1994. Pangaea: evolution of a supercontinent and its consequences for Earth's paleoclimate and sedimentary environments. Pages 13–23 in *Pangea: Paleoclimate, Tectonics, and Sedimentation during Accretion, Zenith, and Breakup of a Supercontinent*, edited by G. D. Klein. Geological Society of America, Special Paper 288.

White, M. E. 1986. *The Greening of Gondwana*. Reed, Frenchs Forest, Australia.

Ziegler, A. M., R. K. Bambach, J. T. Parrish, S. F. Barrett, E. H. Gierlowski, W. C. Parker, A. Raymond, and J. J. Sepkoski. 1981. Paleozoic biogeography and climatology. Pages 213–266 in *Paleobotany, Paleoecology, and Evolution*, volume 2, edited by K. J. Niklas. Praeger Scientific, New York, NY.

TERRESTRIAL ECTOTHERMS: AMPHIBIANS, TURTLES, CROCODILIANS, AND SQUAMATES

The spread of plants and then invertebrates across the land provided a new habitat for vertebrates. The evolutionary transition from water to land is complex because water and air have such different properties: Aquatic animals are supported by water, whereas terrestrial animals need skeletons and limbs. Aquatic animals extract oxygen from a unidirectional flow of water across the gills, whereas terrestrial animals breathe air that they must pump in and out of sac-like lungs. Aquatic animals face problems of water and ion balance as the result of osmotic flow, whereas terrestrial animals lose water by evaporation. Even sensory systems like eyes and ears work differently in water and air. The transition from aquatic to terrestrial habitats must have been facilitated by characteristics of fishes that were functional both in water and in air, although the functions may not have been exactly the same in the two fluids.

Once in terrestrial habitats, vertebrates radiated into some of the most remarkable animals that have ever lived. The nonavian dinosaurs are the best known of these, but many smaller groups contained forms that were just as bizarre, although not as large as many of the dinosaurs. One contribution of phylogenetic systematics has been the emphasis it has placed on the relationship of birds, crocodilians, and dinosaurs. This perspective suggests that the complex behaviors we consider normal for birds might be ancestral characters of their lineage. Sure enough, living crocodilians display parental care that is quite like that of birds (allowing for the differences in the size and morphology of birds and crocodilians), and evidence is accumulating that at least some dinosaurs also showed extensive parental care and probably other behaviors we now associate with birds.

The distinction between ectotherms (animals that obtain the heat needed to raise their body temperatures from outside the body) and endotherms (animals that use metabolic heat production for thermoregulation) has important functional considerations that cut across phylogenetic lineages. In some respects crocodilians are more like turtles and squamates (lizards, snakes, and related forms) than like the birds that are their closest living relatives. The relationship between an ectothermal organism and its physical environment (solar radiation, air temperature, wind speed, and humidity) is often an important factor in its ecology and behavior. One consequence of relying on outside sources of energy for thermoregulation is efficient use of metabolic energy, and ectotherms transform a high proportion of the food they eat into their own body tissue. This characteristic gives them a unique position in the flow of energy through terrestrial ecosystems.

In this part of the book we describe the radiation of vertebrates into terrestrial habitats in the Paleozoic and Mesozoic, and various theories to account for the extinction of dinosaurs and for mass extinctions generally, and consider the advantages and disadvantages of being an ectotherm in the modern world.

10 ORIGIN AND RADIATION OF TETRAPODS IN THE LATE PALEOZOIC

By the middle Devonian the stage was set for the appearance of terrestrial vertebrates, and the first vertebrate to venture onto the land was a sarcopterygian. The sarcopterygians, introduced with the other bony fishes in Chapter 8, did not participate in the evolutionary success of the ray-finned fishes. Indeed, the only surviving aquatic sarcopterygians are the lungfishes and the coelacanth. However, all the terrestrial vertebrates are sarcopterygians.

The far-reaching structural changes required for life on land were barely complete when some lineages of tetrapods became secondarily aquatic, returning to freshwater habitats. However, other lineages became increasingly specialized for terrestrial life. Once again, a radiation of vertebrates was associated with progressive changes in the jaws that allowed new ways of feeding and simultaneous changes in the limbs that appear to have increased the agility of predators. These structural changes were widespread, but only one of the terrestrial lineages of Paleozoic tetrapods made the next major transition in vertebrate history with the appearance of the embryonic membranes that define the amniotic vertebrates.

The Earliest Tetrapods

Our understanding of the origin of terrestrial vertebrates (**tetrapods**) is advancing rapidly. New fossil material in Latvia, Scotland, Australia, and North America has supplemented information from additional specimens of *Acanthostega* and *Ichthyostega* found in the same deposits in East Greenland that first produced *Ichthyostega* in 1932. Analysis of the new specimens has focused on derived characters, and this perspective has emphasized the sequence in which the characteristics of tetrapods were acquired. The gap between fishes and tetrapods has narrowed; a newly defined group of sarcopterygian fishes is very like tetrapods, and the earliest tetrapods now appear to have been much more fish-like than we had previously realized. That information provides a basis for hypotheses about the ecology of animals at the transition between aquatic and terrestrial life.

The origin of tetrapods from sarcopterygian (lobe-finned) fishes between the middle and late Devonian can be inferred from features of the skull, teeth, and limbs. Cladistic analyses of the relationships of the sarcopterygians to tetrapods point to one of the sarcopterygian lineages as the most likely sister group of the tetrapods. This conclusion confirms and strengthens the classical view of tetrapod relationships that was based largely on ancestral characters shared by sarcopterygians and tetrapods.

The next stage in the history of tetrapods was their radiation into different lineages and different ecological types, during the late Paleozoic and Mesozoic. By the late Devonian and early Carboniferous, tetrapods had split into two lineages that are distinguished in part by the way the roof of the skull is fastened to the posterior portion of the braincase.

One of these lineages is the batrachomorphs, which were the largest and longest-lasting group of Paleozoic tetrapods. Some lineages of temnospondyls extended into the Cretaceous, and the living amphibians may be derived from temnospondyls.

The second early lineage of tetrapods, the reptilomorphs, contains a diverse array of animals. The diadectomorphs, large terrestrial reptilomorphs, appear to be the most likely candidates for the sister group of the Amniota. The first amniotic vertebrates appeared in the late Carboniferous. They were small, agile animals that show modifications of the skeleton and jaws, suggesting that they fed on terrestrial invertebrates. The diversity of nonamniotic tetrapods waned during the late Permian and Triassic. Simultaneously, amniotic tetrapods radiated into many of the terrestrial life zones that had been occupied by nonamniotes, and developed some specializations that had not previously been seen among tetrapods.

Tetrapod Relationships

Two groups of sarcopterygians have been considered alternatives as the closest relatives of tetrapods. The discovery of lungfishes seemed to provide an ideal model of a proto-tetrapod—what more could one ask than an air-breathing fish? However, lungfishes are very specialized animals, and Devonian lungfishes were scarcely less specialized than the living species. In contrast, a group of Devonian sarcopterygians, the osteolepiform fishes, were closer in many respects to what one would expect of a proto-tetrapod. The Osteolepiformes were cylindrical-bodied, large-headed fishes with thick scales, a single external nostril on each side, and variable caudal structures. Many osteolepiforms had ring-shaped vertebral centra with accessory ossifications associated with the spinal nerves; similar vertebrae are found in the earliest tetrapods. Although all osteolepiforms were basically free-swimming predators of shallow waters, some may have been specialized for life at the water's edge.

Most evolutionary biologists favored osteolepiforms as the sister group of tetrapods, but in 1981 Donn Rosen and his colleagues revived the argument in favor of lungfish (Rosen et al. 1981). Although this hypothesis was vigorously criticized (Jarvik 1981, Holmes 1985, Carroll 1987), it showed that the evidence in favor of osteolepiforms was not

entirely convincing in a cladistic analysis. As a result, recent studies have reexamined the question of the origin of tetrapods with cladistic methods, and it appears that neither the lungfishes nor the osteolepiforms is the sister group of tetrapods. Instead, a newly defined lineage of late Devonian sarcopterygians, the Panderichthyidae, seems the most likely candidate for this position (Vorobyeva and Schultze 1991, Ahlberg 1995).

The crocodile-like panderichthyids look like tetrapods with fins instead of limbs (Ahlberg and Milner 1994). Only two genera are known, *Panderichthys* and *Elpistostege*. Their bodies and heads were flattened, they lacked both dorsal and anal fins, their snouts were long, and their eyes were on the top of the head (Figure 10–1). In addition, panderichthyids and tetrapods have frontal bones and their ribs project ventrally from the vertebral column, whereas osteolepiforms lack frontal bones and have dorsally projecting ribs. These shared derived characters suggest that Panderichthyidae should be placed as the sister group of tetrapods, with osteolepiforms as the sister group of panderichthyids + tetrapods (Figure 10–2).

Early Tetrapods

The new fossils, especially specimens of *Acanthostega* from East Greenland, have shed light on the characteristics of early tetrapods, suggesting that they were more aquatic than we thought. In addition, one of the most widespread features of tetrapods, the pentadactyl (five-fingered) limb, turns out not to be an ancestral character.

The evidence for an aquatic way of life for early tetrapods comes partly from the presence of a groove on the ventral surface of the ceratobranchials (Coates and Clack 1991). The ceratobranchials are part of the branchial apparatus, which supports the gills of fishes. Elements of the branchial apparatus are retained in all tetrapods, including birds and mammals, so it is not merely the presence of ceratobranchials in *Acanthostega* that is important—it is the groove on their ventral surface. In derived fishes that groove accommodates the afferent branchial aortic arches, which carry blood to the gills. The presence of a similar groove on the ceratobranchials of *Acanthostega* strongly suggests that these tetrapods also

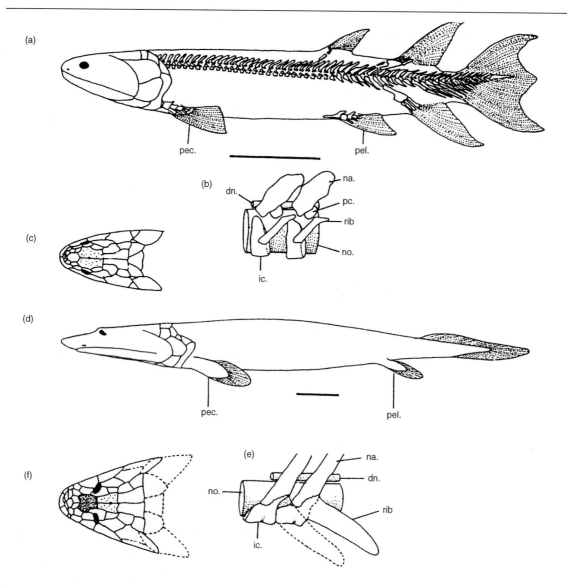

(a)

(b)
dn.
na.
pc.
rib
no.
ic.

(c)

(d)
pec.
pel.

(e)
na.
dn.
no.
rib
ic.

(f)

had gills. In addition, the cleithrum (shoulder girdle) of *Acanthostega* has a flange, the postbranchial lamina, on its anterior margin. In fishes, this ridge supports the posterior wall of the opercular chamber. The picture of *Acanthostega* that emerges from these features is an animal with fish-like internal gills and an open opercular chamber. That is, an animal with aquatic respiration, like a fish.

A second unexpected feature of *Acanthostega* is found in its feet—it had 8 toes on its front feet (Coates and Clack 1990). Furthermore, *Acanthostega* was not alone in its polydactyly (having more than five toes). *Ichthyostega* had seven toes on its hind foot, and *Tulerpeton*, a Devonian tetrapod from Russia,

had six toes (Figure 10–3). In fact, not one of the Devonian tetrapods yet known had five toes. These discoveries confound long-standing explanations of the supposed homologies of bones in the fins of sarcopterygian fins with those in tetrapod limbs, but they correspond beautifully with predictions based on a study of the embryology of limb development (Box 10–1).

These new discoveries leave us with a paradoxical situation: Animals with well-developed limbs and other structural features that suggest they were capable of locomotion on land appear to have retained gills that would function only in water. How does a land animal evolve in water?

Figure 10–1 A Devonian osteolepiform and panderichthyid. (a) The osteolepiform *Eusthenopteron* has a cylindrical body, and four unpaired fins (two dorsal fins, a caudal fin, and an anal fin) in addition to the paired pectoral and pelvic appendages. This was a large fish—the scale bar is 10 centimeters long. (b) A portion of the vertebral column showing the position of the notochord and dorsal nerve cord. The intercentra are ventral to the notochord and the pleurocentra are dorsal. The neural arches do not articulate, and the short ribs probably extended dorsally. (c) The skull roof of *Eusthenopteron*. The parietals are shaded; note that the area anterior to the parietals is filled by a mosaic of small bones. (d) The panderichthyid *Panderichthys rhombolepis* has a dorsoventrally flattened body with a long, broad snout, and eyes on top of the head. Note the absence of dorsal and anal fins. (Compare these features to *Ichthyostega,* Figure 10–6.) (e) A section of the vertebral column. There are no ossified pleurocentra, although cartilaginous pleurocentra might have been present. The ribs are larger than those of osteolepiforms, and project laterally and ventrally. (f) The skull roof of *Panderichthys*. Panderichthyids and tetrapods both have a single pair of large frontal bones (dark shading) immediately anterior to the parietals. dn, dorsal nerve cord; i, intercentrum; na, neural arch; no, notochord; pc, pleurocentrum; pec, pectoral appendage; pel, pelvic appendage. (From P. E. Ahlberg and A. R. Milner, 1994, *Nature* 368:507–514.)

Evolution of Tetrapod Characters in an Aquatic Habitat

Tantalizingly incomplete as the skeletal evidence is, it is massive compared to the information we have about the ecology of panderichthyids and early tetrapods. It is not possible even to be certain if the evolution of tetrapods occurred in purely freshwater habitats (Panchen 1977, Thomson 1980). Greenland and Australia, the two land masses from which Devonian tetrapods are known, are believed to have been separated by marine environments in the Devonian. Furthermore, *Tulerpeton* was found in a deposit that formed in a large, shallow marine basin that was a considerable distance from the nearest known land mass. Tetrapods may have evolved in brackish or saline lagoons, or even in marine habitats (Thomson 1993).

What sorts of lives did the panderichthyids lead? What was their habitat like? What were their major competitors and predators? What were the earliest tetrapods able to do that panderichthyids could not? This is the sort of information we need to assess the selective forces that shaped the evolution of tetrapods.

How Does a Land Animal Evolve in Water?

Certain inferences can be drawn from the fossil material available. We start from the basis that any animal must function in its habitat. If it does not, it does not leave descendants and its phylogenetic lineage becomes extinct. Evolutionary change occurs because the naturally occurring variation in organisms is subject to selection. Looking backward, we may see an evolutionary trend if progressive changes conferred an advantage on the individuals possessing them; but it is merely a fortunate coincidence if some features turn out to be advantageous for a different way of life. Tetrapod characteristics did not evolve because they would someday be useful to animals that would live on land; they evolved because they were advantageous for animals that were still living in water.

Panderichthyids were large fish, as much as a meter long, with heavy bodies, long snouts, and large teeth. They probably either stalked their prey or lay in ambush and made a sudden rush. One can picture a panderichthyid prowling through the dense growth of plants on the bottom of a Devonian pond or estuary. The flexible lobe fins would support it as it waited motionless for prey. A group of living fish, the frogfishes, provides a model for the usefulness of a tetrapod-like limb in water (Edwards 1989). The pectoral fins of frogfishes are modified into structures that look remarkably like the limbs of tetrapods (Figure 10–5), and are used to walk over the substrate. An analysis of frogfish locomotion revealed that they employ two gaits that are used by tetrapods—a walk and a (slow!) gallop.

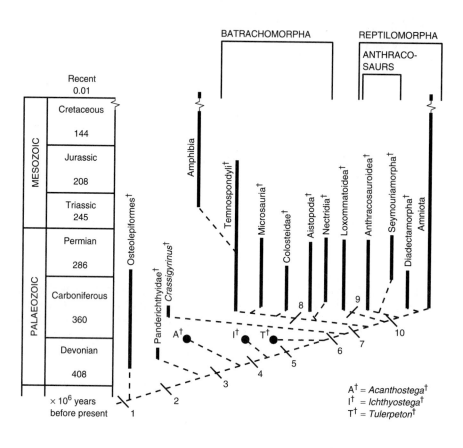

Figure 10–2 Phylogenetic relationships of sarcopterygians and tetrapods. This diagram shows the probable relationships among some of the sarcopterygian fishes and early tetrapods. Extinct groups are marked by a dagger (†). Numbers indicate derived characters that distinguish the lineages. **1.** Choana present, single pair of external nares. **2.** Flattened head with long snout, dorsal position of eyes, frontal bone present. **3.** Articulating surfaces on neural arches, characteristics of skull and limbs. **4.** Shape of ulna. **5.** Open lateral-line system on most or all dermal bones of the skull. Six or fewer digits. **6.** Absence of anocleithrum. **7.** Occipital condyles present, characters of skull and teeth. **8.** Skull roof attached to braincase via the exoccipital, loss of skull kinesis. **9.** Openings in the skull roof anterior to the eyes (antorbital vacuities) at least as large as the eyes. **10.** Several skull characters. (Modified from M. J. Benton (editor), 1988, *The Phylogeny and Classification of the Tetrapods*, Special Volume No. 35B, The Systematics Association, Oxford University Press, Oxford, UK; R. L. Carroll, 1988, *Vertebrate Paleontology and Evolution*, Freeman, New York, NY; M. J. Benton, 1990, *Vertebrate Paleontology*, Harper-Collins Academic, London, UK; M. J. Benton (editor), 1993, *The Fossil Record 2*, Chapman & Hall, London, UK; and P. E. Ahlberg and A. R. Milner, 1994, *Nature* 368:507–514.)

Although there is no direct fossil evidence of lungs in Devonian sarcopterygians, we believe they were present, because every related group (basal actinopterygians, dipnoans, actinistians, and tetrapods) is known to have lungs or the remnants of lungs. Panderichthyids presumably could breathe atmospheric oxygen by swimming to the surface and gulping air, or in shallow water, by propping themselves on their pectoral fins to lift their heads to the surface.

What Were the Advantages of Terrestrial Activity?

This question has fascinated biologists for a century or more, and there is no shortage of theories (Romer 1958, Szarski 1962, Schaeffer 1965, Bray 1985). These ideas can often be combined. The classic theory proposed is that the Devonian was a time of seasonal droughts. Shallow ponds that formed during the monsoon period often evaporated during the dry season, stranding their inhabitants in rapidly shrinking bodies of stagnant water. Fishes trapped in such situations are doomed unless the next rainy season begins before the pond is completely dry. We know that the living African lungfish copes with this situation by estivating in the mud of its dry pond until the rains return, but perhaps certain Devonian sarcopterygians had limbs that allowed them to crawl from a drying pond and move overland to larger ponds that still held water. Could millions of years of selection of the fishes best able to escape death by finding their way to permanent water produce a lineage showing increasing ability on land?

This theory has been criticized on several grounds. After all, a fish that succeeds in moving from a drying pond to one that still holds water has enabled itself to go on leading the life of a fish. That seems a backward way to evolve a terrestrial animal. Various alternative theories have been proposed that stress positive selective values associated with increasing terrestrial activity to sarcopterygians. One emphasizes the contrast between terrestrial and aquatic habitats in the Devonian. The water was swarming with a variety of fishes that had radiated to fill a multiplicity of ecological niches. Active, powerful predators and competitors abounded. In contrast, the land was free of vertebrates. Any sarcopterygians that could occupy terrestrial situations had a predator- and competitor-free environment at its disposal. The exploitation of this habitat

can be seen as proceeding by gradual steps.

The reinterpretation of early tetrapods as basically aquatic animals suggests that we should base speculations about selection for terrestrial activity on *Acanthostega* and *Ichthyostega*, rather than on panderichthyids. Juvenile *Ichthyostega* and *Acanthostega* might have congregated in shallow water, as do juveniles of living fishes and amphibians, to escape the attention of larger predatory fishes that are restricted to deeper water. At the edge of a lake or estuary many of the morphological and physiological characteristics of terrestrial vertebrates would have been useful to a still-aquatic tetrapod. Warm water holds little oxygen, and shallow pond edges are likely to be especially warm during the day. Thus, lungs are important to a vertebrate in that habitat, whether it be fish or tetrapod. Similarly, legs would have borne the weight of the animal in the absence of water deep enough to float in. In shallow water an air-breathing, upstanding tetrapod could have lifted its head above the water, and the change in the shape of the lens of the eye associated with the differences in the refractive indices of water and air could have started to occur. These behavioral and morphological features can be found among a number of living fishes, such as the mudskippers, climbing perch, and walking catfish that make extensive excursions out of the water, even climbing trees and capturing food on land.

Starting from aquatic tetrapods that snapped up terrestrial invertebrates that fell into the water, one can envision a gradual progression of increasingly agile forms capable of exploiting the terrestrial habitat for food as well as for shelter from aquatic predators. Terrestrial agility might have developed to the stage at which juvenile tetrapods moved overland from the pond of their origin to other ponds. Many vertebrates include a dispersal stage in their life history, usually in the juvenile period. In this stage individuals spread from the place of their origin to colonize suitable habitats, sometimes long distances from their starting point. This type of behavior is so widespread among living vertebrates that we may guess that it occurred in some form in the early tetrapods as well.

Perhaps the main selection for terrestrial life occurred in the juvenile stages of tetrapods' lives. The adults were large animals, and it is hard to imagine an adult *Ichthyostega* being sufficiently agile to capture a terrestrial invertebrate such as a scorpion scuttling about in its own terrestrial habitat. It is

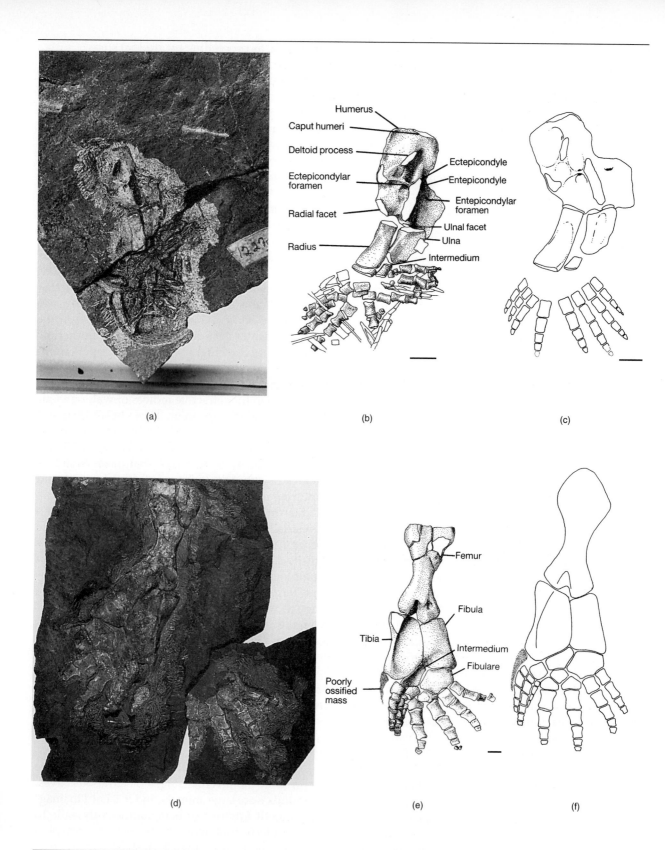

(a)

(b)

(c)

(d)

(e)

(f)

much easier to visualize a 15-centimeter-long juvenile following the scorpion under a log and grabbing it. A small body size would have greatly simplified the difficulties of support, locomotion, and respiration in the transition from an aquatic to a terrestrial habitat.

Devonian Tetrapods: *Acanthostega, Ichthyostega, Tulerpeton,* and *Hynerpeton*

Specimens collected mostly in the past decade have increased our knowledge of tetrapods in the late Devonian. It is clear that substantial diversity had evolved among tetrapods. They ranged from about 0.5 to 1.2 meters in length and, as we have noted, these animals have more than five digits: *Acanthostega* has 8 toes on both front and rear feet, *Ichthyostega* has seven digits on its rear foot (the front foot of *Ichthyostega* is unknown), and *Tulerpeton* has six digits on its front and rear feet.

Ichthyostega is the best known of the Devonian tetrapods, and an examination of its skeleton illustrates basic structural features associated with terrestrial life (Figure 10–6c). Most of the structures distinguishing aquatic from terrestrial animals are a consequence of the differences between water and air. Vertebrates never had to cope with being heavy until they emerged from the water. Water buoys an animal up, but air gives no such support. Resistance to gravity is of minor importance in the skeleton of a fish—the demands of locomotion are paramount. In contrast, resistance to the downward pull of gravity is a major factor shaping the skeleton of a terrestrial vertebrate.

When a terrestrial vertebrate stands, its body hangs from the vertebral column, which, like the arch of a suspension bridge, supports the weight of the trunk and transmits it to the ground by two sets of vertical supports, the girdles and legs. The vertebral column must be rigid and the girdles and legs sturdy and firmly connected with the vertebral column. Without all of these features, a terrestrial vertebrate cannot stand. In osteolepiforms such as *Eusthenopteron* (Figure 10–7a), each segment of the body contained two sets of bones that together formed the vertebral centrum. The anterior elements (intercentra) were wedge-shaped in lateral view and crescentic when viewed from the end. They lay beneath the notochord and extended upward around it. The posterior elements were a pair of small bones (pleurocentra) that lay on the dorsal surface of the notochord. The neural arches of adjacent vertebrae did not articulate. The structure of the vertebrae of *Ichthyostega* (Figure 10–7b) was similar to that of the osteolepiforms, but articulations between the neural arches transmitted forces from one vertebra to the next, providing some of the support needed by a terrestrial animal. These articulations were still more extensive in later forms such as the Permian temnospondyl *Eryops* (Figure 10–7c).

The broad, overlapping ribs of *Ichthyostega* probably added rigidity to its trunk. Coates and Clack (in press) pointed out that the ribcage of *Ichthyostega* is remarkably similar to that of an extant arboreal mammal, the two-toed anteater *Cyclopes didactylus* (Figure 10–6b). The anteater can extend itself horizontally from a tree limb by contracting its intercostal muscles to lock the overlapping ribs together while it anchors itself to the branch with its hind limbs and tail.

The posttemporal and supracleithral bones, which in fishes attach the cleithrum and clavicle to the skull, had disappeared completely from the pectoral girdle of *Ichthyostega*. A large interclavicle in the ventral midline braced the ventral part of the pectoral girdle, and the scapulocoracoid provided a strong dorsal attachment for the muscles that bound the pectoral girdle to the trunk. The forelimb of *Ichthyostega* was a strong prop, which was permanently bent at the elbow. In the pelvic region, the two bony ventral plates of fishes were greatly enlarged compared to their condition in osteolepiforms and formed a skid-like structure.

Figure 10–3 The polydactyl feet of *Acanthostega* (top row) and *Ichthyostega* (bottom row). (a) fossilized forelimb of *Acanthostega'*, (b) drawing of the fossil with the bones identified, (c) restoration of the forelimb of *Acanthostega*. (d) fossilized hindlimb of *Ichthyostega.* with its counterpart half, (e) drawing of the fossil with the bones identified, (f) restoration of the forelimb of *Ichthyostega*. The scale bars are 1 centimeter long. (From M. I. Coates and J. A. Clack, 1990, Nature 347:66-69. Courtesy of M. I. Coates and J. A. Clack.)

Box 10–1 Early Feet

How a tetrapod limb could evolve from the fin of a sarcopterygian fish has been hotly debated for more than a century. Until recently, the various theories were based on equating specific bones in the limbs of tetrapods with their presumed counterparts in the limbs of sarcopterygian fishes. Thus, Gegenbaur in the nineteenth century suggested that extension and additional segmentation of the radials seen in the limb of *Eusthenopteron* could produce a limb with digits like those seen in tetrapods (Figure 10–4a,b). The same speculation about equivalent structures has been applied to the fin skeletons of lungfish (Figure 10–4c,d).

A new perspective on the evolution of tetrapod limbs has been supplied by studies of the embryonic development of the limbs of derived vertebrates (Shubin and Alberch 1986). All tetrapod limbs have a common pattern of development, and the same sequence of events applies to the forelimbs and the hindlimbs. In the developing forelimb, the humerus branches to form the radius (anteriorly) and the ulna (posteriorly). The developmental axis runs through the ulna. Subsequent development on the anterior side of the limb occurs (1) by segmentation of the radius, which produces the radiale (shown in Figure 10–4e) and sometimes two additional segments, and (2) by both branching and segmentation of the ulna. A preaxial branch of the ulna produces the bones in the palm of your hand—the intermedium and centrales. The carpals (knuckles) and metacarpals (digits) result from postaxial branching. These postaxial digits are a new feature of tetrapods—nothing like them is known in any fish.

For pentadactyl tetrapods, the formation of digits starts with digit 4 and concludes with digit 1 (which is the thumb of humans). Digit 5 (our little finger) forms at different times in different lineages, and one of the attractive features of this developmental process is the ease with

Figure 10–4 Three hypotheses of the origin of tetrapod limbs. In every case the head of the animal is to the left. Thus the preaxial direction (i.e., anterior to the axis of the limb) is to the left. (a) The pectoral fin skeleton of *Eusthenopteron*. The limb has a longitudinal axis (solid line) and preaxial radials. (b) Gegenbaur's nineteenth century hypothesis of the origin of the vertebrate forefoot and fingers from preaxial radials. (c, d) The pectoral fin skeleton of the Australian lungfish *Neoceradotus*. (c) The fin axis and the preaxial radials that appear early in embryonic development by branching from the axis. The radials of *Eusthenopteron* may have developed in the same way. (d) The postaxial radials seen in the adult fin of *Neoceradotus* develop by condensation of tissue, not by branching. (e, f) A diagram of the forelimb skeleton of a mouse during early development. (e) The proximal (= nearest to the shoulder) parts of the limb skeleton consist of an axis with preaxial branches (radials) as seen in *Eusthenopteron* and *Neoceradotus* (compare the upper part of (e) to (a) and (c). The digits, however, are formed as postaxial branches. This pattern is unknown in fishes. (f) The embryonic limb skeleton of a mouse straightened for comparison with *Eusthenopteron* (a) and *Neoceradotus* (c). (g) The hindlimb of *Ichthyostega* showing the inferred position of the axis (solid line) and radials. Note that the tibia and fibula and the tarsal bones in the foot are formed by preaxial radials, whereas the digits originate as postaxial branches. (From P.E. Ahlberg and A. R. Milner, 1994, *Nature* 368:507–514.)

which more or fewer than five digits can develop. If the process of segmentation and branching continues, a polydactylous foot is produced with the extra digits forming beyond the thumb. If the developmental process is shortened, fewer than five digits are produced, and the thumb is the first digit to be lost.

Changes in the timing of development do not have to produce an all-or-none addition or loss of a digit; a reduction in size is common. Dogs, for example, have four well-developed digits (numbers 2 through 5), plus a vestigial digit (number 1) called a dewclaw. Many dogs are born without external dewclaws (a carpal may be present internally), and some breeds of dogs are required by their breed standards to have double dewclaws. This pattern of increase or reduction in the number of digits results from a change in the timing of development during evolution. Reduction in the number of digits in evolutionary lineages of birds and mammals is frequently associated with specialization for high-speed running—ostriches have three digits, artiodactyls (e.g., antelope) have two digits, and perissodactyls (horses) have a single digit. The embryonic process by which toes are formed explains why variation in the number digits is so widespread.

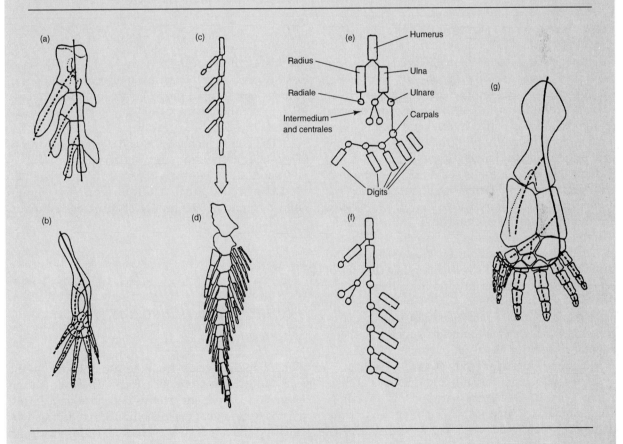

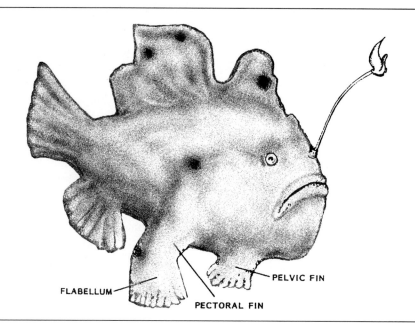

Figure 10–5 The frogfish *Antennarius pictus* in its typical posture, with its pectoral and pelvic fins planted firmly on the substrate. Only the right pectoral and pelvic fins can be seen in this view. The small pelvic fins are in an anterior position, but are not fused. When the animal walks, the left and right pelvic fins make contact with the substrate independently, allowing the gait of the fish to be compared to the gaits of tetrapods. (From J. L. Edwards, 1989, *American Zoologist* 29:235–254; courtesy of J. L. Edwards.)

FLABELLUM

PECTORAL FIN

PELVIC FIN

Acanthostega, Tulerpeton, and *Hynerpeton* are not as well known as *Ichthyostega,* but they differ enough to show that by the end of the Devonian, some 7 million years after their appearance, tetrapods had diversified into several niches. *Tulerpeton* was more lightly built than *Ichthyostega,* and had proportionally longer limbs, suggesting that it might have led a more active life. *Acanthostega* was probably less terrestrial than *Ichthyostega.* Its forelimb is less robust than that of *Ichthyostega* and more fin-like than its hind limb. The articulating surfaces of the vertebrae are small, the neural arches are weakly ossified, the ribs are short and straight, not broad and overlapping like the ribs of *Ichthyostega,* and the fin rays on the tail are longer than those of *Ichthyostega. Acanthostega* appears to have retained gills. *Hynerpeton* had lost the postbranchial lamina on the cleithrum, suggesting that it did not have an internal gill chamber like that of *Acanthostega* (Daeschler et al. 1994).

The combination of fish-like and terrestrial characters seen in Devonian tetrapods is puzzling. Coates and Clack (in press) called attention to similarities between *Ichthyostega* and extant pinnipeds, such as the elephant seal (*Mirounga leonina*). The forelegs of pinnipeds are props for the body on land, and the pelvic limbs are used as paddles and rudders in the water. The proportions of humerus and femur

of the elephant seal are very similar to those of *Ichthyostega,* and the permanently bent forelimb of *Ichthyostega* forms a prop like the forelimb of a seal. Figure 10–6c shows a restoration of *Ichthyostega* that reflects these seal-like features.

The mosaic of ancestral and derived features of Devonian tetrapods suggests that the origin of tetrapods and the origin of terrestrial life were two separate events. The first fully terrestrial vertebrates may be found in later groups of Paleozoic tetrapods.

The Radiation and Diversity of Nonamniotic Paleozoic Tetrapods

For 150 million years, from the late Devonian to the early Jurassic, nonamniotic tetrapods radiated into an unimaginable variety of terrestrial and aquatic forms. It is no wonder that the fossil record is confusing. Parallel and convergent evolution were widespread, and it is hard to separate phylogenetic relationship from convergent evolution. Large gaps in the fossil record, especially the lack of early representatives of many groups, still obscure relationships.

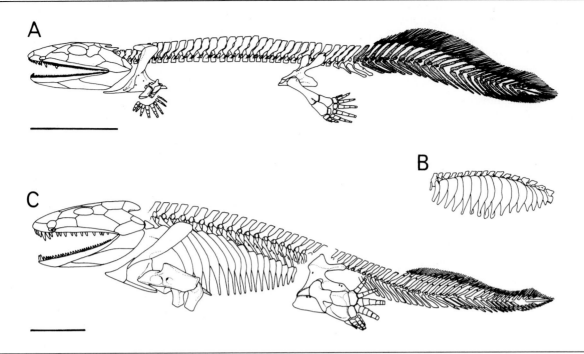

Figure 10–6 Skeletal reconstructions of Devonian tetrapods from east Greenland. (a) *Acanthostega gunneri:* provisional restoration omitting ribs and gastralia. Stippled regions of the tail are hypothetical, but the length of the fin rays in the tail is based on fossil material. (b) Expanded thoracolumbar ribs of an extant mammal, the two-toed anteater for comparison. (c) New restoration of *Ichthyostega;* the forefoot is unknown. (From M. I. Coates and J. A. Clack, in press, *Proceedings of the 7th International Sympsium on Lower Vertebrates.* Occasional Publications of the University of Kansas, Lawrence, KS.)

Crassigyrinus is a case in point. "Enigmatic" appears frequently in discussions of this early Carboniferous tetrapod from Scotland. It was a large animal (1.3 meters long), with tiny legs, and a massive skull with prominent teeth and fangs (Figure 10–8). Despite its relatively late appearance (in the Carboniferous), *Crassigyrinus* retained some ancestral characters, including limited ossification of the vertebral centra, the absence of an occipital condyle, and the presence of some skull bones that were reduced or absent in *Ichthyostega.* Clearly, *Crassigyrinus* was aquatic, but which way was it going? That is, was it a late survivor of a lineage that, like the Devonian tetrapods, had probably never become fully terrestrial? Or was it one of the earliest of the terrestrial vertebrates to return to the water and lose many of the characters associated with life on land? It is not yet possible to choose between those hypotheses

(Ahlberg and Milner 1994, Coates and Clack in press).

Temnospondyli

The temnospondyls were the most speciose nonamniotic tetrapods of the Paleozoic. Many temnospondyls were stocky, short-legged, heavy-bodied, large-headed semiaquatic predators that probably fed on fishes and other tetrapods (Figure 10–9). In general, temnospondyls had flat, akinetic skulls. *Eryops* of the late Paleozoic was about 1.6 meters long, and its large head comprised one-fifth of its total length. Its contemporary, *Cacops,* was only about 40 centimeters long, but its head was nearly one-third the total length of its body. The skeleton of *Cacops* indicates that it was quite terrestrial; the limb girdles are sturdy and the legs and toes are long in proportion to the body. Dissorophid temnospondyls such as *Cacops* appear to represent the height of temnospondyl terrestrial development.

Other temnospondyls were small, aquatic forms. The branchiosaurs (Figure 10–10) were paedomorphic temnospondyls less than 10 centimeters in total length. Branchiosaurs are probably the sister group of the living amphibians (Trueb and Cloutier 1991).

Many of the temnospondyls of the late Paleozoic and early Mesozoic were very large aquatic forms with short limbs and reduced ossification. The capitosaurs and metoposaurs of the Triassic had skulls nearly a meter long. *Cyclotosaurus* from the late Triassic was an aquatic form with small legs and a dorsoventrally flattened body (Figure 10–9c). These impressive temnospondyls may have represented the top of the aquatic food chain, feeding on their smaller relatives as well as any terrestrial animals incautious enough to try to swim in their vicinity.

Some of the most interesting ecological specializations among temnospondyls are found in aquatic forms. The Plagiosauridae (several genera, which are probably not as closely related as their inclusion in one family suggests) and the Brachyopidae are bizarre, flat-headed, short-snouted temnospondyls. *Gerrothorax*, a late Triassic plagiosaurid, was about 1

meter long (Figure 10–11). The body was very flat and armored dorsally and ventrally. The dorsal position of the eyes suggests that *Gerrothorax* lay in ambush on the bottom of the pond until a fish or another temnospondyl swam close enough to be seized with a sudden rush and gulp. The broad head and wide mouth would enable *Gerrothorax* to attack large prey. The broad, short skull and retention of external gills in at least some forms suggest that the brachyopids and plagiosaurids evolved by paedomorphosis from the larvae of temnospondyls. Because brachyopids and plagiosaurids were separated in both distance and time, it is likely that they represent at least two separate derivations, and quite possibly genera within the two families were independently derived.

The trematosaurids are among the most remarkable temnospondyls. As early as the Permian some temnospondyls reversed the trend to a broad, flat skull and evolved the elongate snout characteristic of specialized fish eaters (Figure 10–12). Forms such as *Archegosaurus* may later have given rise to even more specialized fish eaters including *Aphaneramma*, or the later forms may represent a convergent evolutionary lineage. The trematosaurids are found in early Triassic marine beds. Salt water is an unusual habitat for

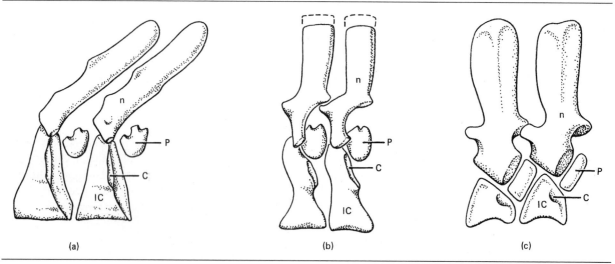

(a) (b) (c)

Figure 10–7 Vertebral structure of sarcopterygians and tetrapods (anterior is to the left): (a) the osteolepiform *Eusthenopteron*—the neural arches do not interlock; (b) *Ichthyostega*—zygapophyses form articulations between adjacent neural arches; (c) the terrestrial temnospondyl *Eryops*—zygapophyses are more fully developed. n, neural arch; pre, prezygapophysis; po, postzygaphphysis; p, pleurocentrum; ic, intercentrum; c, articulation for head of rib. (From A. S. Romer, 1966, *Vertebrate Paleontology*, 3d edition, University of Chicago Press, Chicago, IL.)

temnospondyls. The scaly covering that tremato-saurs might have retained from sarcopterygian ancestors could have helped to protect the adults from the osmotic stress of either fresh or salt water, but what of the presumed larval stages? Were trematosaurs viviparous like a few living amphibians, or could their larvae tolerate at least moderate salinity as do tadpoles of a few frogs and toads? Furthermore, if trematosaurs were able to invade the sea, why didn't other temnospondyls?

Microsauria

The microsaurs can be distinguished from members of the temnospondyl lineage by the modification of the first cervical vertebra into a one-piece atlas–axis complex that is unlike the multipiece atlas–axis complex of other tetrapods. The tuditanomorphs were almost exclusively terrestrial, but the micro-brachomorphs, which have cranial lateral line grooves and ossified supports for external gills, were probably aquatic. The name Microsauria (small reptile) is inappropriate, but this confusion with early amniotes is understandable. Some microsaurs had moderately elongate bodies with small but functional limbs and deep, sturdy skulls (Figure 10–13). Like the amniotes, they were probably small, active terrestrial predators that specialized in feeding on invertebrates. Some microsaurs had body proportions very like those of lizards and salamanders.

Colosteidae

These animals, represented by *Greererpeton*, *Pholidogaster*, and *Colosteus*, were probably secondarily aquatic. Like *Crassigyrinus*, they had elongate bodies, small limbs, and dorsoventrally flattened skulls with

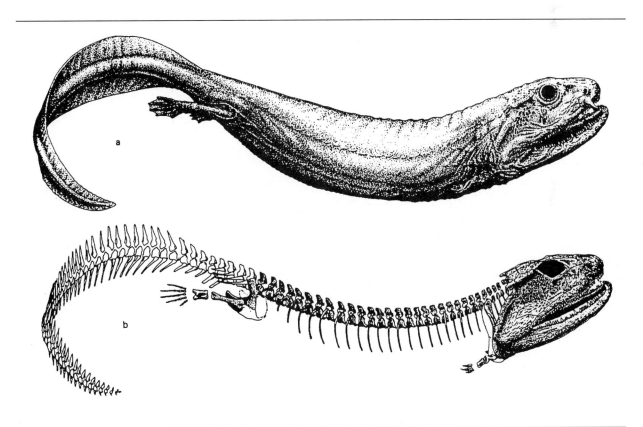

Figure 10–8 *Crassigyrinus* had a skull as big as that of a leopard, armed with lateral teeth that had sharp anterior and posterior keels and large fangs in the palate. (From A. L. Panchen and T. R. Smithson, 1990, *Transactions of the Royal Society of Edinburgh: Earth Sciences* 81:31–44.)

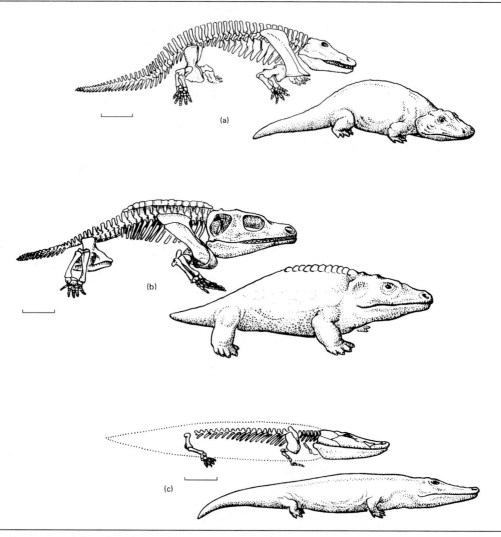

Figure 10–9 Temnospondyls radiated in both terrestrial and aquatic habitats. The dissorophids, (a) *Eryops* and (b) *Cacops,* both from the Permian, were relatively terrestrial, whereas capitosaurs, (c) *Cyclotosaurus,* from the late Triassic, were aquatic. The scale marks represent 10 centimeters. ([a] From E. A. Colbert, 1969, *Evolution of the Vertebrates,* Wiley, New York, NY; [b] from B. J. Stahl, 1974, *Vertebrate History: Problems in Evolution,* McGraw-Hill, New York, NY; [c] from C. L. Fenton and M. A. Fenton, 1958, *The Fossil Book,* Doubleday, New York, NY.)

lateral line grooves. *Greererpeton,* from the middle Carboniferous, was about 1.5 meters long.

Aïstopoda and Nectridia

The Aïstopoda is among the smallest groups of nonamniotic tetrapods of the Paleozoic, containing only six genera of legless forms (Figure 10–14). Most information comes from fossils from the late Carboniferous and early Permian. Aïstopods had very long bodies, some had more than 200 vertebrae in the trunk, and they lacked both limbs and limb girdles. The skulls of aïstopods had an open structure, like those of snakes and some lizards. The flexible skulls of snakes and lizards are associated with the ability to swallow large prey, and the same may have

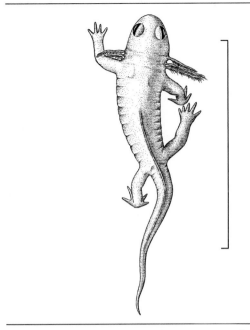

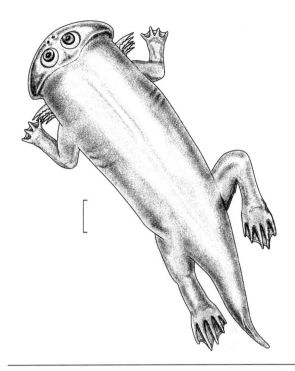

Figure 10–10 The paedomorphic temnospondyl *Branchiosaurus* was about the size of a salamander. The scale mark represents 5 centimeters.

Figure 10–11 *Gerrothorax*, a plagiosaur. The scale mark represents 10 centimeters.

been true of aïstopods. Flexible skulls are not very effective for digging burrows through soil, and fossorial snakes and lizards usually have rigid skulls. The open nature of aïstopod skulls suggests that these were not burrowing animals. They might have lived on the surface of the ground, foraging in leaf litter, or they might have been aquatic.

Nectrideans also were elongate aquatic animals, but in nectrideans the elongation occurred mainly by lengthening the tail rather than the trunk (Figure 10–14). Unlike aïstopods, nectrideans had only 15 to 26 trunk vertebrae. In *Urocordylus*, an early nectridean, the tail was twice as long as the body. The neural and hemal arches of the tail vertebrae were expanded and probably supported a flat muscular tail fin. In the urocordylids the skull was sharply pointed and the legs were small. In derived members of the lineage the ossification of the limbs and girdles was reduced.

A second lineage of nectrideans, represented by the family Keraterpetontidae, contained animals that were dorsoventrally flattened with broad skulls in which the tabular bones were elongated. In adult *Diploceraspis* the tabular bones formed crescent-shaped structures, often called horns, that at their greatest extent were more than five times the width of the anterior part of the skull. The ontogenetic development of the skull can be traced from the normal-looking oval skull of small individuals to fully developed horns in adults. The mouth was small and the eyes were far forward and directed upward, suggesting that the horned nectrideans may have lived on the bottoms of bodies of water. An imprint of the skin of a fossil nectridean shows that the horns were covered in life with a flap of skin that extended back to the shoulder. Was it a hydrofoil to help the animal swim, or a highly vascularized tissue that extracted oxygen from water? Possibly a ligament ran from the tabular horn to the pectoral girdle, stabilizing the head and reducing lateral oscillation during swimming in a manner analogous to the connection between the skull and pectoral girdle of fishes.

Loxommatoidea

Loxommatoids are known from the middle and late Carboniferous. The postcranial skeleton is nearly

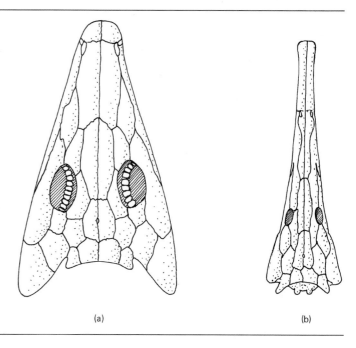

Figure 10–12 Specialized long-snouted, fish-eating dissorophids like *Archegosaurus* (a) appeared in the early Permian. The trematosaurs like *Aphaneramma* (b) were still more specialized marine forms of the Triassic.

(a)

(b)

unknown, but their skulls were elongate and crocodile-like with keyhole-shaped orbits and long, slender teeth that suggest a diet of fishes. Probably loxommatoids were aquatic predators that lived in large bodies of water.

Anthracosauroidea

The anthracosauroids were a less diverse group than the temnospondyls. Most anthracosauroids were terrestrial, and even the secondarily aquatic forms never attained the sizes reached by some of the huge aquatic temnospondyls. Anthracosauroids had deep skulls compared to the dorsoventrally flattened skulls of temnospondyls, and retained skull kinesis.

Gephyrostegus, from the late Carboniferous of Europe, represents a terrestrial anthracosauroid (Figure 10–15). It was about a meter long, scale-covered, and relatively lightly built. It lacked a lateral-line system and had large piercing teeth, which suggests that it fed on arthropods and small vertebrates. The animals called embolomeres include the families Proterogyrinidae, Anthracosauridae, Eogyrinidae, and Archeriidae. *Proterogyrinus*, from the middle Carboniferous of North America, appears to have been terrestrial; it had well-developed limbs and a relatively short trunk. In contrast, *Pholiderpeton* and *Archeria* were elongate aquatic animals that may have been as long as 4 meters.

Seymouriamorpha

The seymouriamorphs are known from the early Permain of North America. *Seymouria* (Figure 10–16c), which takes its name from the town of Seymour, Texas, was about 1 meter long and appears to have been a fully terrestrial animal. The limbs and girdles were well developed and the vertebral column was relatively short. However, lateral-line canals have been described in the skull of an adult *Seymouria*, suggesting the existence of an aquatic larval stage. Other seymouriamorphs were aquatic even as adults. Discosauriscids, which are found in Permian deposits in central and eastern Europe and China, are seymouriamorphs known only from aquatic larval or paedomorphic forms, and the kotlassiids from the late Permian and Triassic were also aquatic.

Diadectomorpha

The diadectomorphs form a monophyletic lineage that can be characterized by the presence of a prominent lateral shelf on the ilium, and are probably the sister group of the amniotes. Three families are included in the Diadectomorpha: the Limnoscelidae, Tseajaiidae, and Diadectidae. The limnoscelids and tseajaiids, had sharp-pointed teeth and were apparently carnivorous (Figure 10–16a). They were as large as 1.5 meters in length and appear to have been terrestrial or semiaquatic.

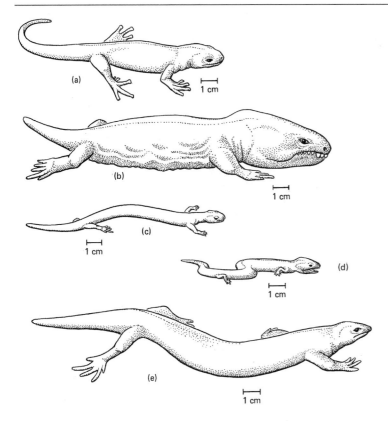

Figure 10–13 Microsaurs were a diverse group of tetrapods. Many were terrestrial, but the group also included aquatic and burrowing forms. (From R. L. Carroll and P. Gaskill, 1978, *Memoirs of the American Philosophical Society* 126:1–211.)

Diadectes was a stocky, heavy-bodied tetrapod of the early Permian about 2 meters in total length (Figure 10–16b). It had well-ossified limbs and girdles, and a short, sturdy vertebral column. *Diadectes* was unusual in having laterally expanded cheek teeth with broad, tricuspid grinding surfaces and chisel-shaped front teeth. This specialized dentition suggests that *Diadectes* was an herbivore. If so, it was possibly the first terrestrial herbivorous tetrapod. Other diadectids had conical teeth and were probably predatory.

Amniotes

A major difference between the groups we have discussed and the remaining tetrapods (**amniotes**) is the occurrence of an **amniotic egg** in the latter group. The amniotic or cleidoic (*cleido* = closed or locked) egg is sometimes referred to as the "land egg," but this is a misnomer. Many species of living amphibians and some fishes have nonamniotic eggs that develop quite successfully on land, and terrestrial invertebrates also lay nonamniotic eggs. Even the differences in moisture requirements of nonamniotic and amniotic eggs are not great. Both must have relatively moist conditions to avoid desiccation. Nonetheless, the amniotic egg is a derived character that distinguishes the two major groups of living tetrapods: amniotes (mammals and reptiles, including birds) and nonamniotes (amphibians).

The amniotic egg, as we know it, is characteristic of turtles, squamates, crocodilians, birds, monotremes, and, in modified form, of therian mammals. It is assumed to have been the reproductive mode of Mesozoic diapsids, and fossilized dinosaur eggs are relatively common in some deposits. An amniotic egg is a remarkable example of biological engineering (Figure 10–17). The shell, which may be leathery and flexible or calcified and rigid, provides mechanical protection while allowing movement of respiratory gases and water vapor. The albumin (egg

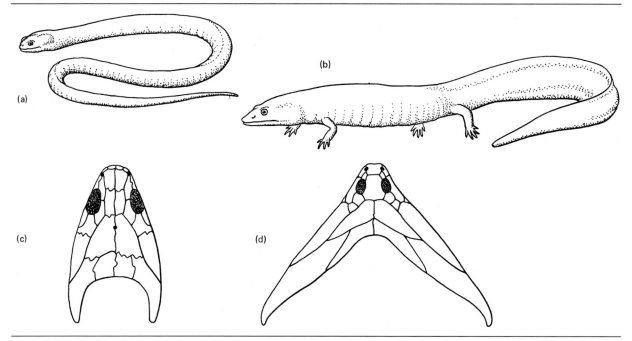

Figure 10–14 Nonamniotic tetrapods. (a) The legless aïstopod *Ophiderpeton* was about 75 centimeters long. (b) The aquatic nectrideans were more varied than the aïstopods. One type, illustrated by *Sauropleura,* had an elongated body and small legs, a laterally flattened tail, and a sharply pointed snout. The tabular bones of "horned nectrideans," (c) *Keraterpeton* and (d) *Diploceraspis,* were greatly elongated. ([a] and [b] From A. R. Milner and [c] from A. C. Milner, both in *The Terrestrial Environment and the Origin of Land Vertebrates,* edited by A. L. Panchen, 1980, Academic, London, UK; [d] from J. R. Beerbower, 1963, *Bulletin of the Museum of Comparative Zoology* 130:31–108.)

white) gives further protection against mechanical damage and provides a reservoir of water and protein. The large yolk is the energy supply for the developing embryo. At the beginning of embryonic development, the embryo is represented by a few cells resting on top of the yolk. As development proceeds these multiply, and endodermal tissue surrounds the yolk and encloses it in a yolk sac that is part of the developing gut. Blood vessels differentiate rapidly in the tissue of the yolk sac and transport food and gases to the embryo. By the end of development, only a small amount of yolk remains, and this is absorbed before or shortly after hatching.

In these respects the amniotic egg does not differ greatly from the nonamniotic eggs of amphibians and fishes. The significant differences lie in three extraembryonic membranes—the **chorion, amnion,** and **allantois.** The chorion and amnion develop from out-growths of the body wall at the ends of the embryo. These two pouches spread outward and around the embryo until they meet. At their junction, the membranes merge and leave an outer membrane, the chorion, which surrounds the embryo and yolk sac, and an inner membrane, the amnion, which surrounds the embryo itself. The allantoic membrane develops as an outgrowth of the hind gut posterior to the yolk sac and lies within the chorion. It is a respiratory organ and a storage place for nitrogenous wastes produced by the metabolism of the embryo. The allantois is left behind in the egg when the embryo emerges, and the nitrogenous wastes stored in it do not have to be reprocessed.

Early Amniotes

In the Late Carboniferous an evolutionary event occurred that, although it did not involve vertebrates,

Figure 10–15 Carboniferous tetrapods. (a) *Gephyrostegus*, a late Carboniferous anthracosauroid, was terrestrial. It had sturdy legs supporting a deep body. (b) *Pholiderpeton*, an embolomere of the same period, was aquatic. Its legs were small and its body was cylindrical. The scale marks represent 10 centimeters. ([a] From R. L. Carroll, 1972, *Handbuch der Palaeoherpetologie*, part 5B, pages 1–19 Gustav Fischer, Stuttgart, Germany; [b] from A. R. Milner, 1980, *The Terrestrial Environment and the Origin of Land Vertebrates*, edited by A. L. Panchen, Academic, London, UK.)

was to have a tremendous influence on the course of their evolution. This event was the radiation of insects in terrestrial habitats. Insects had appeared in the Devonian, and we have speculated about the role they might have played in the evolution of the earliest tetrapods. Through the Mississippian, however, the diversity of terrestrial insects appears to have remained limited, and the insects themselves may have been confined to areas near water. In the early Pennsylvanian the situation changed. The fossil record indicates that there was an abrupt expansion of many orders of insects, including dragonflies, stoneflies, and roaches. It is probable that the diversity of insects in the fossil record at this time reflects their spread into a variety of terrestrial habitats. The radiation of terrestrial insects was probably a response to the increasing quantity and diversity of terrestrial vegetation in the Carboniferous.

It is difficult to escape the conclusion that each of these groups, which today are often so interdependent (for example, bees seeking a food source and angiosperms requiring pollinators) rapidly exploited this new symbiotic relationship to their mutual benefit. How did these changes affect vertebrates? One need only think of how dependent many birds, mammals, lizards, and amphibians are on insects to realize that their evolution has profoundly influenced vertebrate evolution.

Most terrestrial vertebrates at that time were probably carnivorous. (No adult amphibian among living forms is herbivorous, and there is little evidence in the fossil record to suggest that Paleozoic tetrapods were herbivores. *Diadectes* is the only likely exception to this generalization.) Carnivorous vertebrates could not respond directly to the energy supply offered by terrestrial plants, but they could

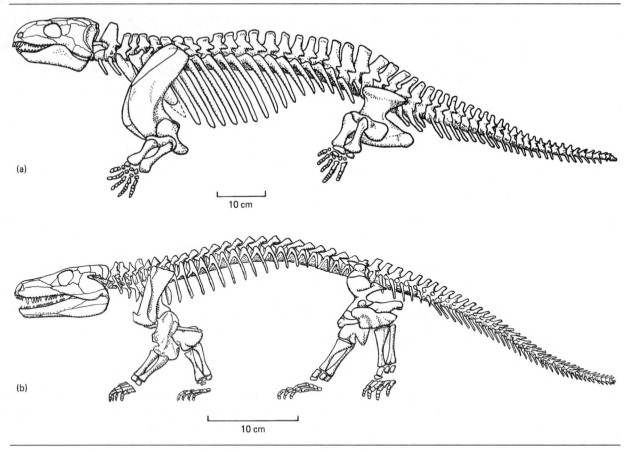

Figure 10–16 Terrestrial nonamniotic reptilomorph tetrapods of the late Paleozoic: (a) *Diadectes;* (b) *Seymouria.* (From R. L. Carroll, 1969, in *Biology of the Reptilia,* volume 1, edited by C. Gans, A. d'A. Bellairs, and T. S. Parsons, Academic, London, UK.)

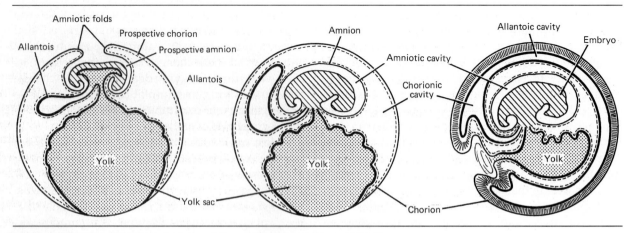

Figure 10–17 The distinctive features of the amniotic egg. (From T. W. Torrey, 1962, *Morphogenesis of the Vertebrates,* Wiley, New York, NY.)

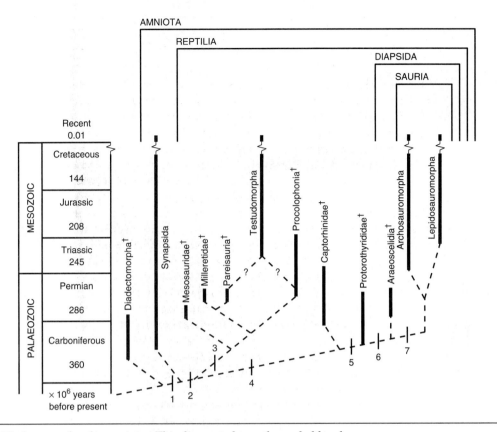

Figure 10–18 Phylogeny of early amniotes. This diagram shows the probable relationships among early amniotes. Extinct groups are marked by a dagger (†). Numbers indicate derived characters that distinguish the lineages. **1.** Amniota: Intertemporal absent, squamosal present in post-temporal fenestra, large exoccipitals contact medially above basioccipital, hemispherical and well-ossified occipital condyle, caniniform maxillary tooth. **2.** Reptilia: Suborbital foramen, anterior crista present on supraoccipital, enlarged postemporal fenestra, single coronoid, loss of ectopterygoid dentition, loss of medial centrale. **3.** Parareptiles: Loss of caniniform maxillary teeth, loss of supraglenoid foramen. **4.** Supratemporal small, parietal and squamosal broadly in contact, tabular not in contact with opisthotic, horizontal ventral margin of postorbital portion of skull, ontogenetic fusion of atlas pleurocentrum and axis intercentrum. **5.** Postorbital region of skull short, anterior pleurocentra keeled ventrally, limbs long and slender, hands and feet long and slender, metapodials overlap proximally. **6.** Diapsida: Upper and lower temporal fenestrae present, suborbital fenestra present, exoccipitals not in contact on occipital condyle; ridge-and-groove tibia–astralagal joint. **7.** Sauria: Dorsal origin of temporal musculature, quadrate exposed laterally, unossified dorsal process of stapes, loss of caniniform region in maxillary tooth row, sacral ribs oriented laterally, ontogenetic fusion of caudal ribs, modified ilium, short and stout fifth metatarsal, small proximal carpals and tarsals. (Modified from J. A. Gauthier, A. G. Kluge, and T. Rowe, 1988, pages 103–155 in *The Phylogeny and Classification of the Tetrapods,* Volume 1: *Amphibians, Reptiles, Birds,* edited by M. J. Benton, Clarendon Press, Oxford, UK; R. L. Carroll, 1988, *Vertebrate Paleontology and Evolution,* Freeman, New York, NY; M. J. Benton, 1990, *Vertebrate Paleontology,* HarperCollins Academic, London, UK; and M. J. Benton (editor), 1993, *The Fossil Record 2,* Chapman & Hall, London, UK.)

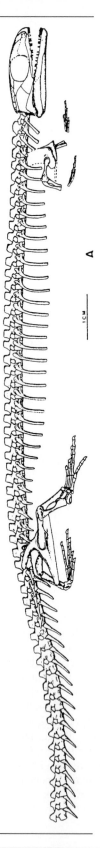

Figure 10–19 *Westlothiana lizziae*, from the early Carboniferous of Scotland, is considered to be earliest known amniote. (From T. R. Smithson, R. L. Carroll, A. L. Panchen, and S. M. Andrews, 1994, *Transactions of the Royal Society of Edinburgh: Earth Sciences* 84:383–412.)

and apparently did respond to the opportunities presented by the radiation of insects. Probably for the first time in evolutionary history there was an adequate food supply to support fully terrestrial vertebrate predators.

Two forces that shaped the radiation of amniotes were directly related to the exploitation of this new energy source. One was the evolution of a more effective jaw mechanism specifically adapted to feeding on insects. The evolution of more effective jaws was accompanied by changes in body structure that permitted more effective locomotion on land.

Carboniferous and Permian Tetrapods

Several groups of vertebrates, both nonamniotes and amniotes, evolved specializations for terrestrial life in the Carboniferous (Figure 10–18). We have already mentioned the microsaurs and the large dissorophid temnospondyls like *Cacops*. In the reptilomorph lineage the gephyrostegids and eoherpetontids were relatively small terrestrial predators, and the limnoscelids and tseajaiids were large predators. *Diadectes* may have been the first terrestrial herbivore.

Amniotes began to radiate into many of those same life zones in the Carboniferous and early Permian. Early amniotes such as *Westlothiana* (Figure 10–19) and *Hylonomus* (Figure 10–20a) had long skulls and short legs, whereas more derived forms had a short post-orbital region of the skull, and longer limbs and feet (Figure 10–20b,c). Protorothyridids like *Hylonomus* were the same size as early captorhinids, but probably much more agile. Their heads were small in proportion to their bodies, and they may have fed on small fast-moving prey such as insects. Protorothyridids survived into the middle Permian with only a slight increase in size.

The early diapsids (Figure 10–20c) were also about 15 centimeters in head-plus-body length, but they had longer legs and necks and were more lightly built than the former groups. They had a large number of small teeth, like the protorothyridids, and may have fed on insects. Diapsids changed little during the Paleozoic, but in the Mesozoic they radiated extensively (see Chapter 13).

Early synapsids (Figure 10–20b) were about the same size as the protorothyridids, but they had substantially larger heads and fewer but larger teeth that may have been specialized for piercing and tearing large prey animals. The change in the angle of the rear of the skull from vertical to sloping may have given the jaw muscles a mechanical advantage in resisting the struggles of prey held in the jaws. Later synapsids were much larger than the first representatives of the lineage, and the structure of the jaws became increasingly specialized (see Chapter 19). Mammals are derived from the synapsid lineage.

A group of animals that includes the pareiasaurs, milleretids, procolophonians, and mesosaurs are referred to as "Parareptiles." Although they appear to be linked by two shared derived characters, they may not represent a monophyletic lineage (Gauthier et al. 1988). Morphologically and ecologically, they are very diverse. Testudomorphs (turtles) group strongly with pareisaurs (Lee 1993) and almost as strongly with procolophonids (Reisz and Laurin 1991).

The mesosaurs (Figure 10–20e) are one of the classic pieces of evidence for continental drift. Their fossils are known only from middle Permian deposits in South Africa and from beds of the same age in South America. In the Permian, these areas were adjacent parts of Gondwana. Mesosaurs are the first amniotes known to have developed specializations for aquatic life; the large hind feet were probably webbed and the long, laterally compressed tail probably was used for swimming. The ribs are densely ossified and heavy; they may have increased the specific gravity of the animal, helping it to dive. The long snout was filled with slender teeth, and it is hard to picture that dentition being effective in capturing fish. Current theories speculate that mesosaurs used their teeth to strain from the water the small crustaceans that are abundant in the same deposits in which mesosaurs are found.

Milleretids, which lived in the middle and late Permian, were very like protorothyridids in body size and general form, although milleretids had longer legs (Figure 10–20f). Like the protorothyridids, milleretids had relatively small skulls and simple conical teeth that suggest a diet of insects.

The procolophonids of the late Permian and Triassic (Figure 10–20g) initially had a large number of small, peg-like teeth. The number of teeth was reduced in later members of this group, and each tooth was laterally expanded. That dentition appears to have been specialized for crushing or grinding, and procolophonids might have been herbivorous.

Pareiasaurs are known from the middle and late Permian of southern Africa, Europe, and China. In contrast to all the preceding groups, pariesaurs were large animals, approaching 3 meters in length (Figure 10–20h). The teeth were laterally compressed and leaf-shaped, like the teeth of derived herbivorous lizards.

The morphological and ecological diversity of the amniotes that had developed by the late Paleozoic testifies to the success of this clade. More-or-less coincident with the diversification of terrestrial amniotes was a contraction in the variety of terrestrial nonamniotes. Terrestrial batrachomorphs and reptilomorphs were at their peak in the Permian; the groups that survived through the Triassic were mostly aquatic forms such as the capitosaurs, trematosaurs, and metoposaurs. From the start of the Mesozoic onward, terrestrial habitats were dominated by a series of radiations of amniotic tetrapods.

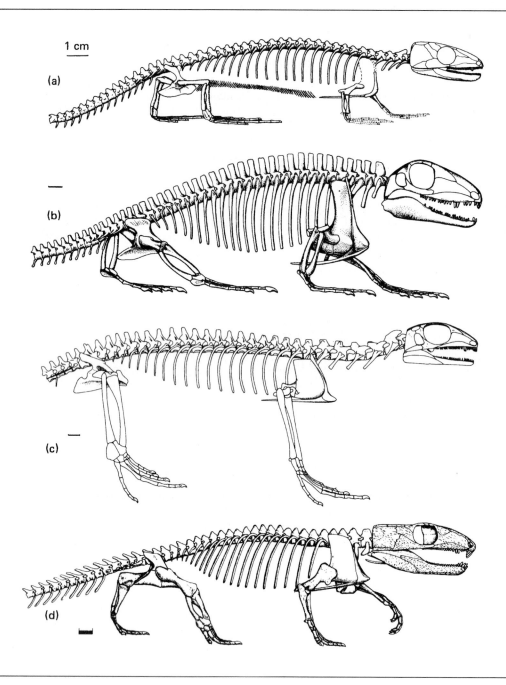

Figure 10–20 Early amniotes varied in size from a few centimeters long to a couple of meters, and their ecological roles were equally diverse. (a) *Hylonomus Iyelli*, a protorothyridid; (b) *Haptodus garnettensis*, a synapsid; (c) *Petrolacosaurus kansensis*, a diapsid; (d) *Eocaptorhinus laticeps*, a captorhinid; (e) *Mesosaurus brasiliensis*, a mesosaur; (f) *Milleretta rubidgei*, a millerosaur; (g) *Barasaurus besairiei*, a procolophonid; (h) *Pareiasaurus karpinskyi*, a pareiasaur. The scale mark is 1 centimeter. (From R. L. Carroll, 1982, *Annual Review of Ecology and Systematics*, 13:87–109.).

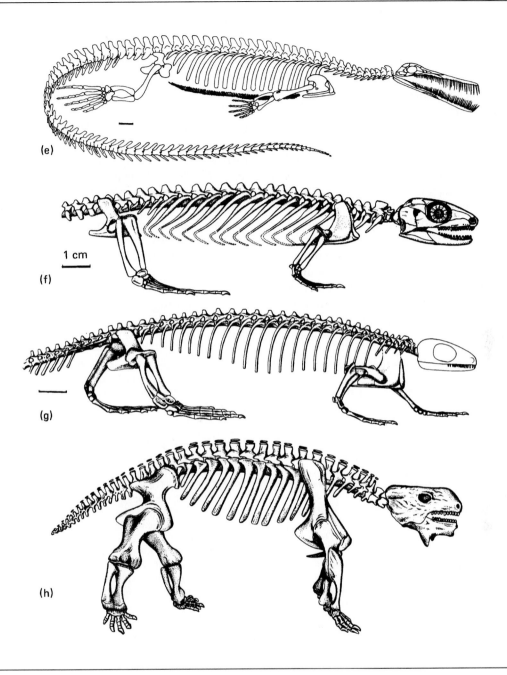

(e)

1 cm

(f)

(g)

(h)

Figure 10–20 *(Continued)*

Summary

The origin of tetrapods from panderichthyid fishes in the Devonian is inferred from similarities in the bones of the skull and braincase, vertebral structure, and limb skeletons. Paleozoic tetrapods can be divided into two clades, batrachomorphs and reptilomorphs. The temnospondyls (batrachomorphs) were predominantly aquatic, and some were as large as crocodiles. Temnospondyls radiated extensively in the late Carboniferous and Permian, and several lineages extended through the Triassic into the middle Jurassic. Modern amphibians—the salamanders, anurans, and caecilians—may be derived from the temnospondyl lineage.

The anthracosauroid lineage (reptilomorphs), which included terrestrial and aquatic forms that radiated during the Carboniferous, was never as diverse as the temnospondyls. Many of the anthracosaurs became extinct in the Permian, and only a few persisted into the Triassic. Amniotes are probably derived from the anthracosauroids.

The amniotic egg, with its distinctive extraembryonic membranes, is a shared derived character that distinguishes the nonamniotes (fishes and amphibians) from the amniotes (turtles, squamates, crocodilians, birds, and mammals). The earliest amniotes were small animals, and their appearance coincided with a major radiation of terrestrial insects in the Carboniferous. Progressive modifications of postcranial skeletons of early amniotes appear to show increased agility, and simultaneous changes in the jaws may be related to predation on insects. By the end of the Carboniferous, amniotes had begun to radiate into most of the terrestrial life zones that had been occupied by nonamniotes, and only the relatively aquatic groups of nonamniotic tetrapods maintained much diversity through the Triassic.

References

Ahlberg, P. E. 1995. *Elginerpeton panchei* and the earliest tetrapod clade. *Nature* 373:420–425.

Ahlberg, P. E., and A. R. Milner. 1994. The origin and early diversification of tetrapods. *Nature* 368:507–514.

Bray, A. A. 1985. The evolution of terrestrial vertebrates: environmental and physiological considerations. *Philosophical Transactions of the Royal Society of London* B309:289–322.

Carroll, R. L. 1982. Early evolution of reptiles. *Annual Review of Ecology and Systematics* 13:87–109.

Carroll, R. L. 1987. *Vertebrate Paleontology and Evolution.* Freeman, New York, NY.

Coates, M. I., and J. A. Clack. 1990. Polydactyly in the earliest known tetrapod limbs. *Nature* 347:66–69.

Coates, M. I., and J. A. Clack. 1991. Fish-like gills and breathing in the earliest known tetrapod. *Nature* 352:234–235.

Coates, M. I., and J. A. Clack. In press. Romer's gap: tetrapod origins and terrestriality. In *Proceedings of the 7th International Symposium on Lower Vertebrates,* edited by M. Arsenault and P. Janvier. Occasional Publications of the University of Kansas, Lawrence, KS.

Daeschler, E. B., N. H. Shubin, K. S. Thomson, and W. W. Amaral. 1994. A Devonian tetrapod from North America. *Science* 265:639–642.

Edwards, J. L. 1989. Two perspectives on the evolution of the tetrapod limb. *American Zoologist* 29:235–254.

Gauthier, J. A., A. G. Kluge, and T. Rowe. 1988. The early evolution of the Amniota. Pages 103–155 in *The Phylogeny and Classification of the Tetrapods,* volume 1, *Amphibians, Reptiles, Birds,* edited by M. J. Benton. Systematics Association Special Volume No. 35A, Clarendon, Oxford, UK.

Holmes, E. B. 1985. Are lungfishes the sister group of tetrapods? *Biological Journal of the Linnean Society* 25:379–397.

Jarvik, E. 1981. (Review of) Lungfishes, tetrapods, paleontology, and plesiomorphy. *Systematic Zoology* 30:378–384.

Lee, M. S. Y. 1993. The origin of the turtle body plan: bridging a famous morphological gap. *Science* 261:1716–1720.

Panchen, A. L. 1977. Geographical and ecological distribution of the earliest vertebrates. Pages 723–738 in *Major Patterns in Vertebrate Evolution,* edited by M. K. Hecht, P. C. Goody, and B. M. Hecht. NATO Advanced Study Series. Plenum, New York, NY.

Reisz, R. R., and M. Laurin. 1991. *Owenetta* and the origin of turtles. *Nature* 349:324–326.

Romer, A. S. 1958. Tetrapod limbs and early tetrapod life. *Evolution* 12:365–369.

Rosen, D. E., P. L. Forey, B. G. Gardiner, and C. Patterson. 1981. Lungfishes, tetrapods, paleontology, and plesiomorphy. *Bulletin of the American Museum of Natural History* 167:159–276.

Schaeffer, B. 1965. The rhipidistian–amphibian transition. *American Zoologist* 5:267–276.

Shubin, N. H., and P. Alberch. 1986. A morphogenetic approach to the origin and basic organization of the tetrapod limb. *Evolutionary Biology* 20:319–387.

Szarski, H. 1962. The origin of the amphibia. *Quarterly Review of Biology* 37:189–241.

Thomson, K. S. 1980. The ecology of Devonian lobe-finned fishes. Pages 187–222 in *The Terrestrial Environment and the Origin of Land Vertebrates,* edited by A. L. Panchen. Academic, London, UK.

Thomson, K. S. 1993. The origin of the tetrapods. *American Journal of Science* 293A:33–62.

Trueb, L., and R. Cloutier. 1991. A phylogenetic investigation of the inter- and intra-relationships of the Lissamphibia (Amphibia: Temnospondyli). Pages 223–313 in *Origins of the Higher Groups of Tetrapods: Controversy and Consenses,* edited by H.-P. Schultze and L. Trueb, Cornell University Press, Ithaca, NY.

Vorobyeva, E., and H.-P. Schultze. 1991. Description and systematics of panderichthyid fishes with comments on their relationship to tetrapods. Pages 68–109 in *Origins of the Higher Groups of Tetrapods: Controversy and Consenses,* edited by H.-P. Schultze and L. Trueb, Cornell University Press, Ithaca, NY.

11
Salamanders, Anurans, and Caecilians

P.B.

S hared derived characters of amphibians—especially a moist, permeable skin—have channeled their evolution in similar directions. Frogs are the most successful amphibians, and it is tempting to think that the variety of locomotor modes permitted by their specialized morphology may be related to their success: Frogs can jump, walk, climb, and swim. In contrast to frogs, salamanders retain the ancestral tetrapod locomotor pattern of lateral undulations combined with limb movements. The greatest diversity among salamanders is found in the Plethodontidae, many species of which project the tongue to capture prey on its sticky tip.

The range of reproductive specializations of amphibians is nearly as great as that of fishes, a remarkable fact when one remembers that there are more than five times as many species of fishes as amphibians. The ancestral reproductive mode of amphibians probably consisted of laying large numbers of eggs that hatched into aquatic larvae, and many amphibians still reproduce this way. An aquatic larva gives a terrestrial species access to resources that would not otherwise be available to it. Modifications of the ancestral reproductive mode include bypassing the larval stage, viviparity, and parental care of eggs and young, including females that feed their tadpoles.

Amphibians

At first glance, the three lineages of amphibians (Figure 11–1) appear to be very different kinds of animals: Frogs have long hind limbs and short, stiff bodies that don't bend when they walk, salamanders have forelimbs and hind limbs of equal size and move with lateral undulations, and caecilians are limbless and employ serpentine locomotion. These obvious differences are all related to locomotor specializations, however, and closer examination shows that amphibians have many characters in common (Table 11–1). We will see that some of these shared characters play important roles in the functional biology of amphibians. The moist, permeable skin in particular creates both limits and opportunities in terms of the things amphibians can do, and where and when they can do them.

The oldest fossils that may represent modern amphibians are isolated vertebrae of Permian age that appear to include both salamander and anuran types. The earliest relatively complete amphibian fossil is *Triadobatrachus* from the early Triassic (Figure 11–2). The skull is frog-like—broad, with the orbits enlarged into the cheek and temporal regions. The axial skeleton is less frog-like than the skull, but the vertebral column is short and the ilium extends forward. The posterior vertebrae do not form a urostyle, however, and the anterior end of the ilium is not fused to a sacral rib (compare to Figure 11–8). The absence of a long tail suggests that *Triadobatrachus* relied on limb movements for

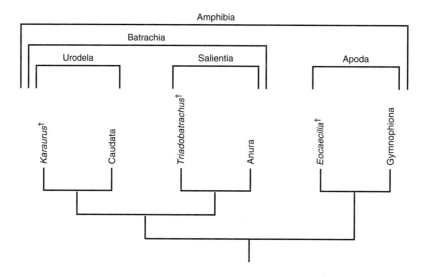

Figure 11–1 The major groups of extant amphibians. Extinct groups are marked by a dagger (†). (Based on W. E. Duellman and L. Trueb, 1985, *Biology of Amphibians*, McGraw-Hill, New York, NY; and David Hillis, personal communication.)

swimming and was probably able to jump. Fossils later than *Triadobatrachus* come from the Jurassic and Cretaceous and are as specialized as modern forms. In most cases they can be tentatively assigned to modern families.

All living adult amphibians are carnivorous, and relatively little morphological specialization is associated with different dietary habits within each group. Amphibians eat almost anything they are able to catch and swallow. The tongue of aquatic forms is broad, flat, and relatively immobile, but some terrestrial amphibians can protrude the tongue from the mouth to capture prey. The size of the head is an important determinant of the maximum size of prey that can be taken, and sympatric species of salamanders frequently have markedly different head sizes, suggesting that this is a feature that reduces competition. Frogs in the tropical American genus *Ceratophrys*, which feed largely on other frogs, have such large heads that they are practically walking mouths.

The anuran body form probably evolved from a more salamander-like starting point. Both jumping and swimming have been suggested as the mode of locomotion that made the change advantageous. Salamanders and caecilians swim as fish do by passing a sine wave down the body. Anurans have inflexible bodies and swim with simultaneous thrusts of the hind legs. Some paleontologists have proposed that the anuran body form evolved because of the advantages of that mode of swimming. An alternative hypothesis traces the anuran body form to selection for an amphibian that could rest near the edge of a body of water and escape aquatic or terrestrial predators with a rapid leap followed by locomotion on either land or water.

The earliest fossils of all the orders of modern amphibians are very like living forms (Carroll 1987). The oldest frog, *Vieraella*, from the early Jurassic of Argentina shows affinities to two modern families, the Ascaphidae and Discoglossidae. Salamanders and caecilians are known from the Jurassic. Clearly, the modern orders of amphibians have had separate evolutionary histories for a long time. The continued presence of such common characteristics as a permeable skin after at least 250 million years of independent evolution suggests that the shared characteristics are critical in shaping the evolutionary success of modern amphibians. In other characters, such as reproduction, locomotion, and defense, they show tremendous diversity.

Salamanders

The salamanders (Caudata) have the most generalized body form and locomotion of the living

Table 11–1   Shared derived characters of amphibians.

1. *Structure of the skin and the importance of cutaneous gas exchange:* All amphibians have mucous glands that keep the skin moist. A substantial part of an amphibian's exchange of oxygen and carbon dioxide with the environment takes place through the skin. All amphibians also have poison (granular) glands in the skin.

2. *Papilla amphibiorum:* All amphibians have a special sensory area, the papilla amphibiorum, in the wall of the sacculus of the inner ear. The papilla amphibiorum is sensitive to frequencies below 1000 hertz (cycles per second), and a second sensory area, the papilla basilaris, detects sound frequencies above 1000 hertz.

3. *Operculum–plectrum complex:* Most amphibians have two bones that are involved in transmitting sounds to the inner ear. The columella (plectrum) is derived from the hyoid arch and is present in salamanders and caecilians and in most frogs. The operculum develops in association with the fenestra ovalis of the inner ear. The columella and operculum are fused in anurans and caecilians and in some salamanders.

4. *Green rods:* Salamanders and frogs have a distinct type of retinal cell, the green rod. Caecilians apparently lack green rods, but the eyes of caecilians are extremely reduced and these cells may have been lost.

5. *Pedicellate teeth:* Nearly all modern amphibians have teeth in which the crown and base (pedicel) are composed of dentine and are separated by a narrow zone of uncalcified dentine or fibrous connective tissue. A few amphibians lack pedicellate teeth, including salamanders of the genus *Siren* and frogs of the genera *Phyllobates* and *Ceratophrys*, and the boundary between the crown and base is obscured in some other genera. Pedicellate teeth also occur in some actinopterygian fishes, which are not thought to be related to amphibians.

6. *Structure of the levator bulbi muscle:* This muscle is a thin sheet in the floor of the orbit that is innervated by the fifth cranial nerve. It causes the eyes to bulge outward, thereby enlarging the buccal cavity. This muscle is present in salamanders and anurans and in modified form in caecilians.

---

amphibians. Salamanders are elongate, and all but a very few species of completely aquatic salamanders have four functional limbs (Figure 11–3). Their walking gait is probably similar to that employed by the earliest tetrapods. It utilizes the lateral bending characteristic of fish locomotion in concert with leg movement. The 10 families (Figure 11–4), containing approximately 400 species, are almost entirely limited to the Northern Hemisphere; their southern-most occurrence is in northern South America (Table 11–2). North and Central America have the greatest diversity of salamanders—more species of salamanders are found in Tennessee than in all of Europe and Asia combined. Paedomorphosis is widespread among salamanders, and several families of aquatic salamanders are constituted solely of such paedomorphic forms (Table 11–2). These can be recognized by the retention of larval characteristics, including larval tooth and bone patterns, the absence of eyelids, retention of a functional lateral-line system, and (in some cases) retention of external gills.

The largest living salamanders are the Japanese and Chinese giant salamanders (*Andrias*), which reach lengths of 1 meter or more. The related North American hellbenders (*Cryptobranchus*) grow to 60 centimeters. All are members of the Cryptobranchidae and are paedomorphic and permanently aquatic. As their name indicates (*crypto* = hidden,

branchus = gill), they do not retain external gills, although they do have other larval characteristics. Another group of large aquatic salamanders, the mudpuppies (*Necturus*, Proteidae), consists of paedomorphic species that retain external gills. Mudpuppies occur in lakes and streams in eastern North America.

Several lineages of salamanders have adapted to life in caves (**troglodyty**). The constant temperature and moisture of caves makes them good salamander habitats, and food is supplied by cave-dwelling invertebrates. The brook salamanders (*Eurycea*, Plethodontidae) include a number of species that form a continuum from those with fully metamorphosed adults inhabiting the twilight zone near cave mouths to fully paedomorphic forms in the depths of caves or sinkholes. The Texas blind salamander, *Typhlomolge*, is a highly specialized troglodyte—blind, white, with external gills, extremely long legs, and a flattened snout used to probe underneath pebbles for food. The unrelated European olm (*Proteus*, Proteidae) is another troglodyte that has converged on the same body form.

Terrestrial salamanders like the North American mole salamanders (*Ambystoma*) and the European salamanders (*Salamandra*) have aquatic larvae that lose their gills at metamorphosis. The most fully terrestrial salamanders, the lungless plethodontids, include species in which the young hatch from eggs

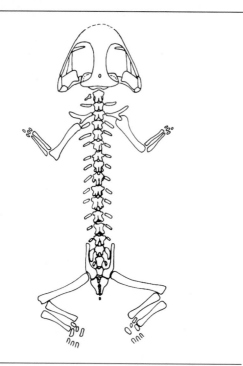

Figure 11–2 *Triadobatrachus*, a fossil from early Triassic sediments in Madagascar, shows some anuran-like characters. (From R. Estes and O. A. Reig, 1973, in *Evolutionary Biology of Anurans*, edited by J. L. Vial, University of Missouri Press, Columbia, MO.)

as miniatures of the adult and there is no aquatic larval stage.

Feeding Specializations of Plethodontid Salamanders
Lungs seem an unlikely organ for a terrestrial vertebrate to abandon, but among salamanders the evolutionary loss of lungs has been a successful tactic. The Plethodontidae is characterized by the absence of lungs and contains more species and has a wider geographic distribution than any other lineage of salamanders. Furthermore, many plethodontids have evolved specializations of the hyobranchial apparatus that allow them to protrude the tongue a considerable distance from the mouth to capture prey. This ability has not evolved in salamanders with lungs, probably because the hyobranchial apparatus in these forms is an essential part of the respiratory system.

Salamanders lack ribs, so they cannot use expansion and contraction of the rib cage to move air in and out of the lungs. Instead, they employ a buccal pump that forces air from the mouth into the lungs.

A sturdy hyobranchial apparatus in the floor of the mouth and throat is an essential part of this pumping system. The modifications of the hyobranchial apparatus that allow tongue protrusion are not compatible with its role in respiration, and reliance on the skin instead of the lungs for gas exchange may have been a necessary first step in the evolution of tongue protrusion by plethodontids.

The modifications of the respiratory system and hyobranchial apparatus that allow tongue protrusion appear to be linked with a number of other characteristics of the biology of plethodontids (Roth and Wake 1985, Wake and Marks 1993). These associations can be seen most clearly in the bolitoglossine plethodontids, which have the most specialized tongue-projection mechanisms (Figure 11–5). Bolitoglossine plethodontids (*bolis* = dart, *glossa* = tongue) can project the tongue a distance equivalent to their head plus trunk length and can pick off moving prey. This ability requires fine visual discrimination of distance and direction, and the eyes of bolitoglossines are placed more frontally on the head than the eyes of less specialized plethodontids. Furthermore, the eyes of bolitoglossines have a large number of nerves that travel to the ipsilateral (= same side) visual centers of the brain as well as the strong contralateral (= opposite side) visual projection that is typical of salamanders. As a result of this neuroanatomy, bolitoglossines have a complete dual projection of the binocular visual fields to both hemispheres of the brain and can make very exact and rapid estimation of the distance of a prey object from the salamander.

Tongue projection is reflected in diverse aspects of the life history characteristics of plethodontid salamanders. Aquatic larval salamanders employ suction feeding, opening the mouth and expanding the throat to create a current of water that carries the prey item with it. The hyobranchial apparatus is an essential part of this feeding mechanism, and some elements of the hyobranchial apparatus become well developed during the larval period. These larval specializations of the hyobranchial apparatus are quite different from those associated with the tongue-projection mechanism of adult plethodontids. Furthermore, larval salamanders have laterally placed eyes and the optic nerves project mostly to the contralateral side of the brain. Thus, the morphological specializations of aquatic plethodontid larvae that make them successful in that ecological role are different from the specializations of adults,

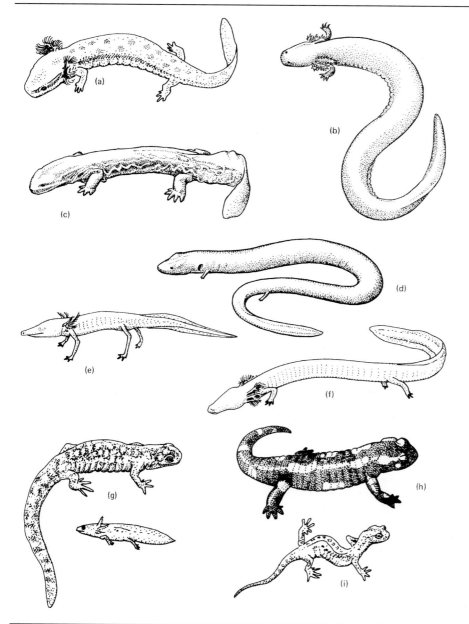

Figure 11–3 Salamander body forms reflect differences in life histories and habitats. Aquatic salamanders: (a) mudpuppy (*Necturus*); (b) siren (*Siren*); (c) hellbender (*Cryptobranchus*); (d) congo eel (*Amphiuma*). Specialized cave-dwellers: (e) Texas blind salamander (*Typhlomolge*); (f) olm (*Proteus*). Terrestrial salamanders: (g) tiger salamander (*Ambystoma*) and its aquatic larval form; (h) fire salamander (*Salamandra*); (i) slimy salamander (*Plethodon*).

and this situation may create a conflict of selective forces.

The bolitoglossines do not have aquatic larvae, and the morphological specializations of adult bolitoglossines appear during embryonic development. In contrast, the hemidactyline plethodontids have aquatic larvae and have considerable ability to project the tongues, but even as adults hemidactylines retain the large first ceratobranchial that appears in the larvae. This is a mechanically less efficient arrangement than the large second ceratobranchial of bolitoglossines, and the ability of hemidactylines to project their tongues is correspondingly less than that of bolitoglossines. Thus, the development of a specialized feeding mechanism by plethodontid salamanders has gone hand-in-hand with such diverse aspects of their biology as respiratory physiology and life history, and demonstrates that organisms evolve as whole functioning units, not as collections of independent characters.

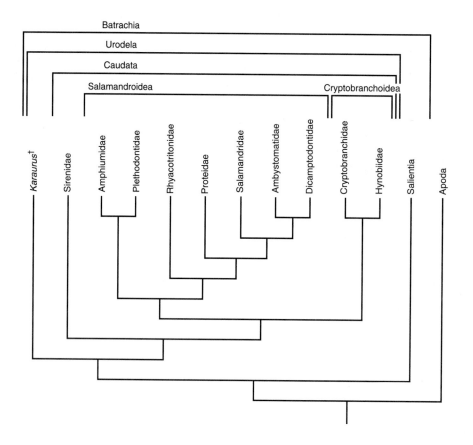

Figure 11–4 Lineages of salamanders. An extinct form is marked by a dagger
(†). (Based on W. E. Duellman and L. Trueb, 1985, *Biology of Amphibians*, McGraw-Hill, New York, NY; and David Hillis, personal communication.)

Social Behavior of Plethodontid Salamanders Plethodontid salamanders can be recognized externally by the paired nasolabial grooves that extend ventrally from the external naris to the lip of the upper jaw (Figure 11–6). These grooves are an important part of the chemosensory system of plethodontids. As a plethodontid salamander moves about, it repeatedly presses its snout against the substrate. Fluid is drawn into the grooves and moves upward to the external nares, into the nasal chambers, and over the chemoreceptors of the vomeronasal organ. Many plethodontid salamanders are territorial and use scent to mark their territories. Male salamanders of some species are aggressive toward conspecific males. Robert Jaeger has studied the territorial behavior of the red-backed salamander, *Plethodon cinereus*, a common species in woodlands of eastern North America. These small salamanders can be kept in cages in the laboratory and fed fruit flies.

Male red-backed salamanders establish territories in the cages. A resident male salamander marks the substrate of its cage with pheromones (chemical substances that are released by an individual and stimulate responses by other individuals of the species). A salamander can distinguish between substrates it has marked and those marked by another male salamander or by a female salamander. Male salamanders can also distinguish between the familiar scent of a neighboring male salamander and the scent of a male they have not previously encountered, and they react differently to those scents.

In laboratory experiments, red-backed salamanders select their prey in a way that maximizes their energy intake: When equal numbers of large and small fruit flies are released in the cages, the salamanders first capture the large flies. This is the most profitable foraging behavior for the salamanders because it provides the maximum energy intake per capture. In a series of experiments, Jaeger and his colleagues showed that territorial behavior and fighting can interfere with the ability of salamanders to select the most profitable prey (Jaeger et al. 1983). These experiments used "surrogate salamanders" that were made of a roll of moist filter paper the same length and diameter as a salamander. The surrogates were placed in the cages of resident salamanders to produce three experimental conditions: a control surrogate, a familiar surrogate, and an unfamiliar surrogate. In the control experiment, male red-backed salamanders were exposed to a surrogate that was only moistened filter paper; it did not carry any salamander pheromone. For both of the other groups the surrogate was rolled across the substrate of the cage of a different male salamander to absorb the scent of that salamander before it was placed in the cage of a resident male.

The experiments lasted 7 days; the first 6 days were conditioning periods and the test itself occurred on the seventh day. For the first 6 days, the resident salamanders in both of the experimental groups were given surrogates bearing the scent of another male salamander. The residents thus had the opportunity to become familiar with the scent of that male. On the seventh day, however, the familiar and unfamiliar surrogate groups were treated differently. The familiar surrogate group once again received a surrogate salamander bearing the scent of the same individual it had been exposed to for the previous 6 days, whereas the resident salamanders in the unfamiliar surrogate group received a surrogate bearing the scent of a different salamander, one to which they had never been exposed before. After a 5-minute pause, a mixture of large and small fruit flies was placed in each cage, and the behavior of the resident salamander was recorded.

The salamanders in the familiar surrogate group showed little response to the now-familiar scent of the other male salamander. They fed as usual, capturing large fruit flies. In contrast, the salamanders that were exposed to the scent of an unfamiliar surrogate began to give threatening and submissive displays, and their rate of prey capture decreased as a result of the time they spent displaying. In addition, salamanders exposed to unfa-miliar surrogates did not concentrate on catching large fruit flies, so the average energy intake per capture also declined. The combined effects of the reduced time spent feeding and the failure to concentrate on the most profitable prey items caused an overall 50 percent decrease in the rate of energy intake for the salamanders exposed to the scent of an unfamiliar male.

The ability of male salamanders to recognize the scent of another male after a week of habituation in the laboratory cages suggests that they would show the same behavior in the woods. That is, a male salamander could learn to recognize and ignore the scent of a male in the adjacent territory, while still being able to recognize and attack a strange intruder. Learning not to respond to the presence of a neighbor may allow a salamander to forage more effectively, and it may also help to avoid injuries that can occur during territorial encounters. Resident male red-backed salamanders challenge strange intruders, and the encounters involve aggressive and submissive displays and biting (Figure 11–7). Most bites are directed at the snout of an opponent, and may damage the nasolabial grooves. Salamanders with damaged nasolabial grooves apparently have difficulty perceiving olfactory stimuli. Twelve salamanders that had been bitten on the snout were able to capture an average of only 5.8 fruit flies in a 2-hour period compared to an average of 18.6 flies for 12 salamanders that had not been bitten. In a sample of 144 red-backed salamanders from the Shenandoah National Forest, 11.8 percent had been bitten on the nasolabial grooves, and these animals weighed less than the unbitten animals, presumably because their foraging success had been reduced (Jaeger 1981).

The possibility of serious damage to an important sensory system during territorial defense provides an additional advantage for a red-backed salamander of being able to distinguish neighbors (which are always there and are not worth attacking) from intruders (which represent a threat and should be attacked). The phenomenon of being able to recognize territorial neighbors has been called dear enemy recognition, and may be generally advantageous because it minimizes the time and energy that territorial individuals expend on territorial defense and also minimizes the risk of injury during territorial encounters. Similar dear enemy recognition has been described among territorial birds that show more aggressive behavior upon hearing the songs of

Table 11–2 **Characteristics of amphibians. The arrangement of the table follows that of Figures 11–1, 11–4, 11–8, and 11–14. The symbol † indicates a fossil. Numbers of species are based on McDiarmid (1993).**

BATRACHIA
Urodela
  Karaurus†: A fossil from the Jurassic of Kazakhstan with many ancestral characters.

Caudata
  Sirenidae: Small (15 cm) to large (75 cm) elongate aquatic salamanders with external gills and lacking the pelvic girdle and hindlimbs (3 species in North America).
  Amphiumidae: Very large (1 m) elongate aquatic salamanders lacking gills (3 species in North America).
  Plethodontidae: Tiny (3 cm) to medium-size (30 cm) aquatic or terrestrial salamanders, some with aquatic larvae, others with direct development (244 species in North, Central, and South America, plus 1 species in Europe).
  Rhyacotritonidae: Very small (<10 cm) semiaquatic salamanders with aquatic larvae. (4 species in North America).
  Proteidae: Paedomorphic aquatic salamanders with external gills (6 species in North America [*Necturus*] and Europe [*Proteus*]).
  Salamandridae: Small to medium-size terrestrial and aquatic salamanders (49 species in Europe and Asia, 6 in North America).
  Ambystomatidae: Small to large terrestrial salamanders with aquatic larvae (33 species in North America).
  Dicamptodontidae: Small to large (10–35 cm) semiaquatic salamanders with aquatic larvae (4 species in North America).
  Cryptobranchidae: Very large to enormous (>1.5 m) paedomorphic aquatic salamanders with external fertilization of the eggs (1 species in North America and 2 species in Asia).
  Hynobiidae: Small to medium-size terrestrial or aquatic salamanders with external fertilization of the eggs and aquatic larvae (35 species in Asia).

Salientia
  Triadobatrachus†: A fossil from the Triassic of Madagascar with a combination of ancestral and derived characters. See Figure 11–2.

Anura
  Ascaphidae: Small (5 cm) aquatic frog found in cold springs and torrential mountain streams; fertilization is internal (1 species in North America).
  Leiopelmatidae: Small semiaquatic or terrestrial frogs (3 species in New Zealand).
  Bombinatoridae: Small to medium-size semiaquatic frogs (7 species in Europe and Asia).
  Discoglossidae: Small to medium size terrestrial and semiaquatic frogs (9 species from western Europe and northern Africa).
  Pipidae: Specialized aquatic frogs; *Xenopus, Hymenochirus,* and some species of *Pipa* have aquatic larvae; other species of *Pipa* have eggs that develop directly into juvenile frogs (27 species in South America and Africa).
  Rhinophrynidae: A burrowing frog with aquatic larvae (1 species in Central America)
  Megophryidae: Terrestrial and semi-aquatic frogs; males of the Oriental genus *Leptobrachium* have a series of cornified spines on the upper lip that extends around the head; the spines are probably used in combat with other males (80 species on mainland Asia and the Indoaustralian Archipelago).
  Pelodytidae: Small terrestrial frogs with aquatic larvae (2 species in Europe and Asia).
  Pelobatidae: Short-legged terrestrial frogs with aquatic larvae (10 species in North America, Asia, Europe, and northern Africa).
  Allophrynidae: A small arboreal frog. (1 species in South America).
  Brachycephalidae: Very small (<16 mm) terrestrial frogs that probably have direct development (3 species in southeastern Brazil).
  Bufonidae: Small (20 mm) to enormous (25 cm) mainly terrestrial frogs; most have aquatic larvae, but some species of *Nectophrynoides* are viviparous (about 356 species in North, Central, and South America, Africa, Europe, and Asia).

Table 11–2 *(Continued)*

Centrolenidae: Mostly small arboreal frogs with aquatic larvae that live in streams (88 species in Central and South America).

Heleophrynidae: Medium-size frogs that live in mountain streams and have aquatic larvae (5 species in extreme southern Africa).

Hylidae: Mostly arboreal frog, but a few species are aquatic or terrestrial; most species have aquatic larvae, but the marsupial frogs (*Gastrotheca*) show several variations, including direct development of juvenile frogs in pouches in the female's skin (about 700 species in North, Central, and South America, Europe, Asia, and Australia).

Leptodactylidae: Very small (12 mm) to enormous (25 cm) frogs from all habitats and with diverse modes of reproduction (about 810 species in southern North America, Central and South America, and the West Indies).

Myobatrachidae: Small (20 mm) to large (12 cm) aquatic and terrestrial frogs with diverse reproductive modes (about 110 species in Australia, Tasmania, and New Guinea).

Pseudidae: Aquatic frogs with enormous tadpoles (25 cm for *Pseudis paradoxa*) that metamorphose into medium-size (to 7 cm) adults (4 species in South America).

Rhinodermatidae: Small terrestrial frogs that lay eggs on land; the tadpoles of *Rhinoderma rufum* are transported to water, whereas those of *R. darwini* complete development in the vocal sacs of the male (2 species in southern Chile and Argentina).

Sooglossidae: Small terrestrial frogs that lay eggs on land; the eggs hatch into juvenile frogs or into nonfeeding tadpoles that are carried on the back of an adult (3 species in the Seychelles Islands).

Arthroleptidae: Small to medium-size terrestrial frogs (73 species from sub-Saharan Africa).

Dendrobatidae: Small terrestrial frogs, many of which are brightly colored and extremely toxic; terrestrial eggs hatch into tadpoles that are transported to water by an adult (about 150 species in Central and South America).

Hemisotidae: Small burrowing frogs (8 species from sub-Saharan Africa).

Hyperoliidae: Small to medium-size mostly arboreal frogs with aquatic larvae (about 225 species in Africa, Madagascar, and the Seychelles Islands).

Microhylidae: Small to medium-size terrestrial or arboreal frogs; many have aquatic larvae, but some species have nonfeeding tadpoles and others have direct development (more than 300 species in North, Central, and South America, Asia, Africa, and Madagascar).

Ranidae: Medium-size to enormous (30 cm) aquatic or terrestrial frogs; most have aquatic tadpoles but several genera show direct development (more than 650 species in North, Central, and South America, Europe, Asia, and Africa).

Rhacophoridae: Very small to large mostly arboreal frogs; some species have filter-feeding aquatic larvae, whereas others lay eggs in holes in trees and have larvae that do not feed (about 200 species in Africa, Madagascar, and Asia).

APODA

Eocaecilia[†]: An early Jurassic fossil from the Kayenta formation of western North America with a combination of ancestral and derived characters. It has a fossa for a chemosensory tentacle, which is a unique derived character of apodans, but it also has four legs, whereas all living apodans are legless (Jenkins and Walsh 1993).

Gymnophiona

Rhinatrematidae: Small (to 30 cm) terrestrial gymnophionans believed to have aquatic larvae (9 species in South America).

Ichthyophidae: Moderately large (to 50 cm) terrestrial gymnophionans with aquatic larvae (36 species in southeast Asia).

Uraeotyphlidae: Small terrestrial gymnophionans, oviparous, perhaps with direct development (4 species in India).

Scolecomorphidae: Moderately large terrestrial caecilians, possibly viviparous (5 species in Africa).

Caeciliaidae: Very small (10 cm) to very large (1.5 m) terrestrial caecilians with both oviparous and viviparous species; no aquatic larval stage (89 species in Central and South America, Africa, India, and the Seychelles Islands).

Typhlonectidae: Large (to 75 cm) aquatic caecilians with viviparous reproduction (22 species in South America).

Figure 11–5 Prey capture by a bolito-glossine salamander, *Hydromantes*. (Courtesy of G. Roth.)

strangers than they do when hearing the songs of neighbors.

Anurans

In contrast to the limited number of species of salamanders and their restricted geographic distribution, the anurans (*an* = without, *uro* = tail) include 27 families, nearly 3750 species, and occur on all of the continents except Antarctica (Figure 11–8). Specialization of the body for jumping is the most conspicuous skeletal feature of anurans. The hind limbs and muscles form a lever system that can catapult an anuran into the air (Figure 11–9), and numerous morphological specializations are associated with this type of locomotion: The hind legs are elongate and the tibia and fibula are fused. A powerful pelvis

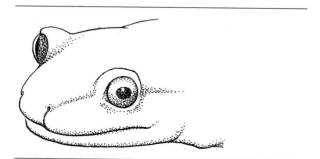

Figure 11–6 Nasolabial grooves of a plethodontid salamander.

strongly fastened to the vertebral column is clearly necessary, as is stiffening of the vertebral column. The ilium is elongate and reaches far anteriorly, and the posterior vertebrae are fused into a solid rod, the **urostyle.** The pelvis and urostyle render the posterior half of the trunk rigid. The vertebral column is short, with only five to nine presacral vertebrae, and these are strongly braced by zygapophyses that restrict lateral bending. The strong forelimbs and flexible pectoral girdle absorb the impact of landing. The eyes are large and are placed well forward on the head, giving binocular vision.

The hind limbs generate the power to propel the frog into the air, and this high level of power production results from structural and biochemical features of the limb muscles (Lutz and Rome 1994). The internal architecture of the semimembranosus muscle and its origin on the ischium and insertion below the knee allow it to operate at the length that produces maximum contraction force during the entire period of contraction. In addition, the muscle shortens faster and generates more power than muscles from most other animals. Furthermore, the intracellular physiological processes of muscle contraction continue at the maximum level throughout contraction, rather than declining as is the case in muscles of most vertebrates.

Specializations of the locomotor system can be used to distinguish many different kinds of anurans (Figure 11–10). The difficulty is finding names for

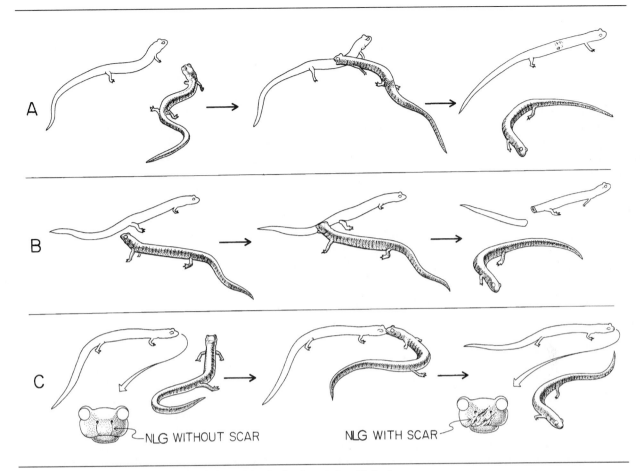

Figure 11–7 Aggressive behaviors of the red-backed salamander, *Plethodon cinereus*. The resident salamander is dark and the intruder is light in these drawings. (a) Resident bites the intruder on the body. Injuries in this region are unlikely to do permanent damage. (b) Bitten on the tail by the resident, the intruder autotomizes (breaks off) its tail to escape. Salamanders store fat in their tails, and this injury may delay reproduction for a year while the tail is regenerated. (c) Resident bites intruder on the snout, injuring the nasolabial grooves. The nasolabial grooves are used for olfaction, and these injuries can reduce a salamander's success in finding prey. (From R. Jaeger, 1981, *The American Naturalist* 117:968. Courtesy of Robert Jaeger. © 1981 The University of Chicago Press. All rights reserved.)

them—the diversity of anurans exceeds the number of common names that can be used to distinguish various ecological specialties (Figure 11–11). Animals called frogs usually have long legs and move by jumping. Many species of ranids have this body form, and very similar jumping frogs are found in other lineages as well. Semiaquatic forms are moderately streamlined and have webbed feet. Stout-bodied terrestrial anurans that make short hops instead of long leaps are often called toads. They usually have blunt heads, heavy bodies, relatively short legs, and little webbing between the toes. This body form is represented by members of the Bufonidae, and very similar body forms are found in other families, including the spadefoot toads of western North America and the horned frogs of South America. Spadefoot toads take their name from a keratinized structure on the hind foot that they use for digging backward into the soil with rapid movements of their hind legs. The horned frogs have extremely large heads and mouths. They feed on small vertebrates, including birds and mammals, but particularly on other frogs. The tadpoles of horned frogs also are carnivorous and feed on other tadpoles. Many frogs that burrow head-first have pointed heads, stout bodies, and short legs.

Arboreal frogs usually have large heads and eyes, and often slim waists and long legs. Arboreal frogs in many different families move by quadrupedal walking and climbing as much as by leaping. Many arboreal species of hylids and rhacophorids have enlarged toe disks and are called treefrogs. The surfaces of the toe pads consist of an epidermal layer with peg-like projections separated by spaces or canals (Figure 11–12). Mucus glands distributed over the disks secrete a viscous solution of polymers in water, and excess fluid drains away through the canals. Arboreal species of frogs appear to have at least two methods of sticking to the surfaces on which they climb: Interlocking of the projections of the toe disks with irregularities of rough surface appears to account for the ability of treefrogs to climb tree trunks, whereas capillary attraction allows the toe disks to stick to smooth surfaces like leaves. The mucus secreted by the glands on the disks wets the disk and the leaf surface and establishes a meniscus at the interface between air and fluid at the edges of the toes. Some arboreal frogs have such effective adhesion that they can walk upside down on the bottom surface of a sheet of glass. Expanded toe disks are not limited exclusively to arboreal frogs; some terrestrial species that move across smooth, slippery surfaces also have toe disks.

Specialized aquatic anurans in the Pipidae are dorsoventrally flattened and have thick waists, powerful hind legs, and large hind feet with extensive webs. Many aquatic anurans also have well-developed lateral-line systems and sensory structures at the tips of their fingers.

Several aspects of the natural history of anurans appear to be related to their different modes of locomotion. In particular, short-legged species that move by hopping are frequently wide-ranging predators that cover large areas as they search for food. This behavior exposes them to predators, and their short legs prevent them from fleeing rapidly enough to escape. Many of these anurans have potent defensive chemicals that are released from glands in the skin when they are attacked. Species of frogs that move by jumping, in contrast to those that hop, are usually sedentary predators that wait

in ambush for prey that passes their hiding places. These species are usually cryptically colored, and they often lack chemical defenses. If they are discovered by a predator, they rely on a series of rapid leaps to get away. Anurans that forage widely encounter different kinds of prey from those that wait in one spot, and differences in dietary habits may be associated with differences in locomotor mode. Aquatic species of anurans use suction feeding to engulf food in the water, but most semi-aquatic and terrestrial species have highly specialized sticky tongues that can be flipped out to trap prey and carry it back to the mouth (Figure 11–13).

Caecilians

The third group of living amphibians is the least known and does not even have an English common name (Figure 11–14). These are the caecilians (Gymnophiona), legless burrowing or aquatic amphibians that occur in tropical habitats around the world (Table 11–2). The eyes of caecilians are covered by skin or even by bone, but the retinae of many species have the layered organization that is typical of vertebrates and appear to be functional as photoreceptors. Conspicuous dermal folds (annuli) encircle the bodies of caecilians. The primary annuli overlie vertebrae and myotomal septa and reflect body segmentation. Many species of caecilians have dermal scales in pockets in the annuli; scales are not known in the other groups of living amphibians. A second unique feature of caecilians is a pair of protrusible tentacles between the eye and nostril. Some structures that are associated with the eyes of other vertebrates have become associated with the tentacles of caecilians. One of the eye muscles, the retractor bulbi, has become the retractor muscle for the tentacle; the levator bulbi moves the tentacle sheath; and the Harderian gland lubricates the channel of the tentacle. It is likely that the tentacle is a sensory organ that allows chemical substances to be transported from the animal's surroundings to the vomeronasal organ on the roof of the mouth. Caecilians feed on small or elongate prey—termites, earthworms, and larval and adult insects—and the tentacle may allow them to detect the presence of prey when they are underground. Females of some species of

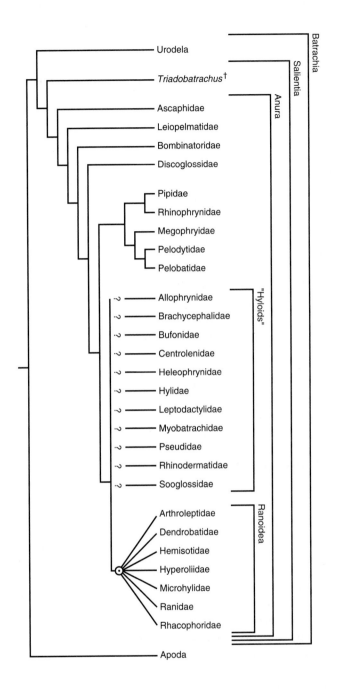

Figure 11–8 Lineages of frogs. The Ranoidea is believed to be a monophyletic group, but the branching sequence has not been resolved. The group marked "Hyloids" is probably not monophyletic. An extinct form is marked by a dagger (†). (Based on W. E. Duellman and L. Trueb, 1985, *Biology of Amphibians*, McGraw-Hill, New York, NY; and David Hillis, personal communication.)

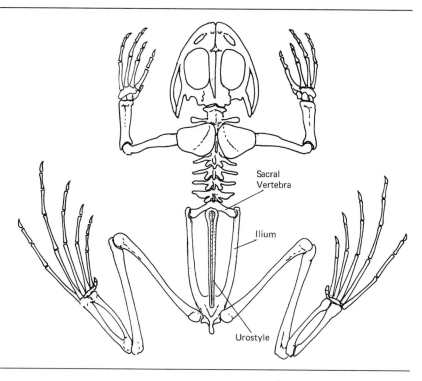

Figure 11–9 Anuran skeleton showing numerous specializations for saltatory locomotion.

Sacral Vertebra

Ilium

Urostyle

caecilians brood their eggs, whereas other species give birth to living young. The embryos of terrestrial species have long, filamentous gills and the embryos of aquatic species have sack-like gills.

Diversity of Life Histories of Amphibians

Of all the characteristics of amphibians, none is more remarkable than the variety they display in modes of reproduction and parental care. It is astonishing that the range of reproductive modes among the 4300 species of amphibians far exceeds that of any other group of vertebrates except for fishes, which outnumber amphibian species by 5 to 1. Most species of amphibians lay eggs. The eggs may be deposited in water or on land and they may hatch into aquatic larvae or into miniatures of the terrestrial adults. The adults of some species of frogs carry eggs attached to the surface of their bodies. Others carry their eggs in pockets in the skin of the back or flanks, in the vocal sacks, or even in the stomach. In still other species the females retain the eggs in the oviducts and give birth to metamorphosed young. Many amphibians have no parental care of their eggs or young, but in many other species a parent remains with the eggs and sometimes with the hatchlings, transports tadpoles from the nest to water, and in a few species an adult even feeds the tadpoles.

Caecilians

The reproductive adaptations of caecilians are as specialized as their body form and ecology. Internal fertilization is accomplished by a male intromittent organ that is protruded from the cloaca. Some species of caecilians lay eggs, and the female may coil around the eggs, remaining with them until they hatch (Figure 11–15). Viviparity is widespread in the order, however, and at least three of the lineages include species in which the eggs are retained in the oviducts and the female gives birth to living young. At birth young caecilians are 30 to 60 percent of their mother's body length. A female *Typhlonectes* 500 millimeters long may give birth to nine babies, each 200 millimeters long. The initial growth of the fetuses is supported by yolk contained in the egg at the time of fertilization, but this yolk is exhausted long before embryonic development is complete. *Typhlonectes* fetuses have absorbed all of the yolk in the eggs by the time they

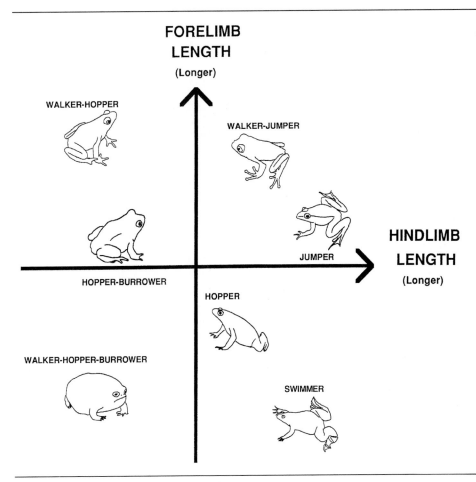

FORELIMB
LENGTH

(Longer)

WALKER-HOPPER

WALKER-JUMPER

JUMPER

HINDLIMB
LENGTH

(Longer)

HOPPER-BURROWER

HOPPER

WALKER-HOPPER-BURROWER

SWIMMER

Figure 11–10 The relation of body form and locomotor mode among anurans. (From F. H. Pough, 1992, Behavioral energetics, pages 395–436 in *Environmental Physiology of the Amphibia*, edited by M. E. Feder and W. W. Burggren, The University of Chicago Press, Chicago, IL. © 1992 The University of Chicago Press. All rights reserved.)

are 30 millimeters long. Thus, the energy they need to grow to 200 millimeters (a 6.6-fold increase in length) must be supplied by the mother. The energetic demands of producing nine babies, each one increasing its length 6.6 times and reaching 40 percent of the mother's length at birth, must be considerable.

The fetuses obtain this energy by scraping material from the walls of the oviducts with specialized embryonic teeth. The epithelium of the oviduct proliferates and forms thick beds surrounded by ramifications of connective tissue and capillaries. As the fetuses exhaust their yolk supply, these beds begin to secrete a thick, white, creamy substance that has been called uterine milk. When their yolk supply has been exhausted, the fetuses emerge from their egg membranes, uncurl, and align themselves lengthwise in the oviducts. The fetuses apparently bite the walls of the oviduct, stimulating secretion and stripping

some epithelial cells and muscle fibers that they swallow with the uterine milk. Small fetuses are regularly spaced along the oviducts. Large fetuses have their heads spaced at intervals, although the body of one fetus may overlap the head of the next. This spacing probably gives all the fetuses access to the secretory areas on the walls of the oviducts.

Gas exchange appears to be achieved by apposition of fetal gills to the walls of the oviducts. All the terrestrial species have fetuses with a pair of triple branched filamentous gills. In preserved specimens the fetuses frequently have one gill extending above the head and the other stretched along the body. In the aquatic genus *Typhlonectes*, the gills are sack-like but are usually positioned in the same way. Both the gills and the walls of the oviducts are highly vascularized, and it seems likely that exchange of gases, and possibly of small molecules such as metabolic substrates and waste

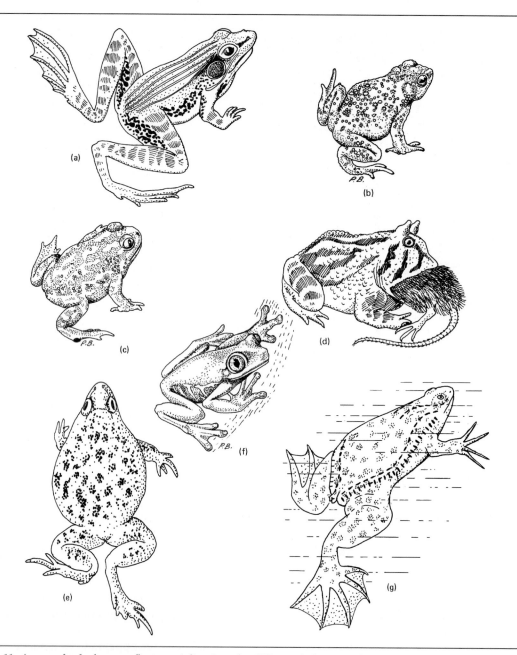

Figure 11–11 Anuran body forms reflect specializations for different habitats and different methods of locomotion. Semiaquatic form: (a) African ridged frog (*Ptychadena*, Ranidae). Terrestrial anurans: (b) true toad (*Bufo* Bufonidae); (c) spadefoot toad (*Scaphiopus* Pelobatidae); (d) horned frog (*Ceratophrys* Leptodactylidae). Burrowing species (e), African shovel-nosed frog (*Hemisus*, Hemisotidae). Arboreal frog (f Central American leaf frog (*Agalychnis*, Hylidae). Specialized aquatic frog: (g) African clawed frog (*Xenopus* Pipidae).

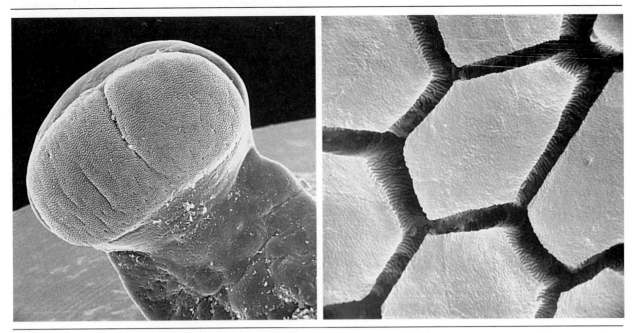

Figure 11–12 Toe disks of a hylid frog. [(a) Appeared in *Biological Journal of the Linnaean Society*, volume 13 (1980). Photographs courtesy of Sharon B. Emerson.]

products, takes place across the adjacent gill and oviduct. The gills are absorbed before birth, and cutaneous exchange may be important for fetuses late in development.

Details of fetal dentition differ among species of caecilians, suggesting that this specialized form of fetal nourishment may have evolved independently in different phylogenetic lines. Analogous methods of supplying energy to fetuses are known in some elasmobranch fishes.

Salamanders

Most groups of salamanders use internal fertilization, but the Cryptobranchoidea (Cryptobranchidae and Hynobiidae) and probably the Sirenidae retain external fertilization. Internal fertilization in salamanders is accomplished not by an intromittent organ but by the transfer of a packet of sperm (the **spermatophore**) from the male to the female (Figure 11–16). The form of the spermatophore differs in various species of salamanders, but all consist of a sperm cap on a gelatinous base. The base is a cast of the interior of the male's cloaca, and in some species it reproduces the ridges and furrows in accurate detail. Males of the Asian salamandrid *Euproctus* deposit a spermatophore on the body of a female

and then, holding her with their tail or jaws, use their feet to insert the spermatophore into her cloaca. Females of the hynobiid salamander *Ranodon sibiricus* deposit egg sacs on top of a spermatophore. In derived species of salamanders the male deposits a spermatophore on the substrate and the female picks off the cap with her cloaca. The sperm are released as the cap dissolves, and fertilization occurs in the oviducts.

Courtship Courtship patterns are important for species recognition, and they show great interspecific variation. Males of some species have elaborate secondary sexual characters that are used during courtship (reviewed by Halliday 1990). Pheromones (chemicals used for communication) are released primarily by males and play a large role in the courtship of salamanders: They probably contribute to species recognition, and may stimulate endocrine activity that increases the receptivity of females.

Pheromone delivery by most salamanders that breed on land involves physical contact between a male and female during which the male applies secretions of specialized courtship glands (hedonic glands) to the nostrils or body of the female. Several modes of pheromone delivery have been described. (1) Males of many plethod-

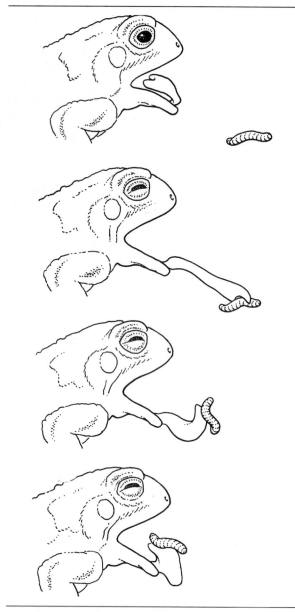

Figure 11–13 Prey capture by a toad.

the female with the pheromone (Figure 11–17c). (3) Males of two small species of *Desmognathus* use specialized mandibular teeth to bite and stimulate the female. (4) Male salamandrids (Salamandridae) rub the female's snout with hedonic glands on their cheeks (the red-spotted newt, *Notophthalmus viridescens*), chin (the rough-skinned newt, *Taricha granulosa* [Figure 11–17a]), or cloaca (the Spanish newt *Pleurodeles wall*).

Newts in the genera *Triturus* and *Cynops* transfer pheromones without physical contact between the male and female. The males of these species perform elaborate courtship displays in which the male vibrates its tail to create a stream of water that wafts pheromones, secreted by a gland in his cloaca, toward the female (Figure 11–17d).

Three groups of *Triturus*, probably representing evolutionary lineages, are tentatively recognized within the genus, and the evolution of courtship probably reflects this phylogenetic relationship (Arntzen and Sparreboom 1989, Halliday 1990, Halliday and Arano 1991). Two trends are apparent: an increase in diversity of the sexual displays performed by the male, and an increase in the importance of positive feedback from the female. The behaviors seen in *Triturus alpestris* may represent the ancestral condition. This species shows little sexual dimorphism (Figure 11–18c), and the male's display consists only of fanning (a display in which the tail is folded back against the flank nearest the female and the tail tip is vibrated rapidly). The male's behavior is nearly independent of response by the female—a male *T. alpestris* may perform his entire courtship sequence and deposit a spermatophore without active response by the female he is courting.

A group of large newts, including *Triturus cristatus* and *T. vittatus* (Figure 11–18a,b) show derived morphological and behavioral characters. They are highly sexually dimorphic, and males defend display sites. Their displays are relatively static, and lack the rapid fanning movements of the tail that characterize other groups of *Triturus*. A male of these species does not deposit a spermatophore unless the female he is courting touches his tail with her snout.

A group of small-bodied species includes *Triturus vulgaris* and *T. boscai* (Figure 11–18d,e). These newts show less sexual dimorphism than the large species, and have a more diverse array of behaviors that include a nearly static lateral display, whipping

ontids (e.g., *Plethodon jordani*) have a large gland beneath the chin (the mental gland), and secretions of the gland are applied to the nostrils of the female with a slapping motion (Figure 11–17b). (2) The anterior teeth of males of many species of *Desmognathus* and *Eurycea* (both members of the Plethodontidae) hypertrophy during the breeding season. A male of these species spreads secretion from his mental gland on the female's skin, and then abrades the skin with his teeth, inoculating

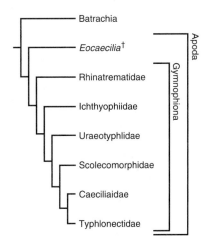

Figure 11–14 Lineages of caecilians. An extinct form is marked by a dagger (†). (Based on W. E. Duellman and L. Trueb, 1985, *Biology of Amphibians*, McGraw-Hill, New York, NY; and David Hillis, personal communication.)

the tail violently against the female's body, fanning with the tail tip, and other displays with names like "wiggle," and "flamenco" that occur in some of the species in the group. Response by the female is an essential component of courtship for these species—a male will not move on from the static display that begins courtship to the next phase unless the female approaches him repeatedly, and he will not deposit a spermatophore until the female touches his tail.

These trends to greater sexual dimorphism, more diverse displays, and more active involvement of the female in courtship may reflect sexual selection by females within the derived groups. Halliday (1990) has suggested that in the ancestral condition there was a single male display, and females mated with the males that performed it most vigorously. That kind of selection by females would produce a population of males, all of which displayed vigorously, and males that added new components to their courtship might be more attractive to females than their rivals.

Eggs and Larvae In most cases, salamanders that breed in water lay their eggs in water. The eggs may be laid singly, or in a mass of transparent gelatinous material. The eggs hatch into gilled aquatic larvae that, except in paedomorphic forms, transform into terrestrial adults. Some families, including the lungless salamanders (Plethodontidae), have a number of species that have dispensed in part or entirely with an aquatic larval stage. The dusky salamander, *Desmognathus fuscus*, lays its eggs beneath a rock or log near water, and the female remains with them until after they have hatched. The larvae have small gills at hatching, and may either take up an aquatic existence or move directly to terrestrial life. The redbacked salamander, *Plethodon cinereus*, lays its eggs in a hollow space in a rotten log or beneath a rock. The embryos have gills, but these are reabsorbed before hatching and the hatchlings are miniatures of the adults.

A few salamanders give birth to living young. The European salamander (*Salamandra salamandra*) produces 20 or more small larvae, each about one-twentieth the length of an adult. The larvae are released in water and have an aquatic stage that lasts about 3 months. The closely related alpine salamander (*S. atra*) gives birth to one or two fully developed young about one-third the adult body length. A female alpine salamander produces as many eggs as a European salamander, but only one egg in each oviduct develops. The remaining eggs break up into a mass that provides food for the developing embryo.

Paedomorphosis Paedomorphosis is the rule in families like the Cryptobranchidae and Proteidae and characterizes most troglodytes. It also appears as a variant in the life history of species of salamanders

Figure 11–15 Caecilians: (a) adult, showing body form; (b) a female coiled around her eggs. Embryos of terrestrial (c) and aquatic (d) species. [(b) Modified from H. Gadow, 1909, *Amphibia and Reptiles*, Macmillan, London; (c) and (d) modified from E. H. Taylor, 1968, *The Caecilians of the World*, University of Kansas Press, Lawrence, KS.]

(a)

(b)

(c)

(d)

that usually metamorphose, and can be a short-term response to conditions in aquatic or terrestrial habitats. The life histories of two species of salamanders from eastern North America provide examples of the flexibility of paedomorphosis.

The small-mouthed salamander, *Ambystoma talpoideum*, is the only species of mole salamander in eastern North America that displays paedomorphosis, although a number of species of *Ambystoma* in the western United States and in Mexico are paedomorphic. Small-mouthed salamanders breed in the autumn and winter, and during the following summer some larvae metamorphose to become terrestrial juveniles. These animals become sexually mature by autumn and return to the ponds to breed when they are about a year old. Ponds in South Carolina also contain paedomorphic larvae that remain in the ponds through the summer and mature and breed in the winter. Some of these paedomorphs metamorphose after breeding, whereas others do not metamorphose and remain in the ponds, becoming sexually mature while still in the larval body form.

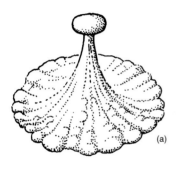 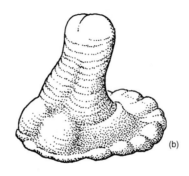 

Figure 11–16 Spermatophores from (a) red-spotted newt, *Notophthalmus viridescens;* (b) dusky salamander, *Desmognathus fuscus;* (c) two-lined salamander, *Eurycea bislineata.* (Modified from G. K. Noble, 1931, *The Biology of the Amphibia,* McGraw-Hill, New York, NY.)

Anurans

Anurans are the most familiar amphibians, largely because of the vocalizations associated with their reproductive behavior. It is not even necessary to get outside a city to hear them. In springtime a weed-choked drainage ditch beside a highway or a trash-filled marsh at the edge of a shopping center parking lot is likely to attract a few toads or treefrogs that have not yet succumbed to human usurpation of their habitat.

The mating systems of anurans can be divided roughly into *explosive breeding,* in which the breeding season is very short, sometimes only a few days, and *prolonged breeding* with breeding seasons that may extend for several months. Explosive breeders include many species of toads and other anurans that breed in temporary aquatic habitats such as vernal ponds or pools created in the desert. Because these bodies of water do not last very long, breeding congregations of anurans usually form as soon as the site is available. Males and females arrive at the breeding sites nearly simultaneously and often in very large numbers. Because the entire population breeds in a short time, the numbers of males and females present are approximately equal. Time is the main constraint on how many females a male is able to court, and mating success is usually approximately the same for all the males in a chorus.

In species with prolonged breeding seasons, the males usually arrive at the breeding sites first. Males of some species, such as green frogs (*Rana clamitans*), establish territories in which they spend several months, defending the spot against the approach of other males. The males of other species move between daytime retreats and nocturnal calling sites on a daily basis. Females come to the breeding site just to breed, and leave when they have finished. Just a few females arrive every day, and the number of males at the breeding site is greater than the number of females every night. Mating success may be very skewed, with many of the males not mating at all, and a few males mating several times. Males of anuran species with prolonged breeding seasons compete to attract females, usually by vocalizing. The characteristics of a male frog's vocalization (pitch, length, or repetition rate) might provide information that a female frog could use to evaluate his quality as a potential mate. This is an active area of study in anuran behavior.

Vocalizations Anuran calls are diverse; they vary from species to species, and most species have two or three different sorts of calls used in different situations. The most familiar calls are the ones usually referred to as mating calls, although a less specific term such as **advertisement calls** is preferable. These calls range from the high-pitched *peep* of a spring peeper to the nasal *waaah* of a spadefoot toad or the bass *jug-o-rum* of a bullfrog. The characteristics of a call identify the species and sex of the calling individual. Many species of anurans are territorial, and males of at least one species, the North American bullfrog (*Rana catesbeiana*) recognize each other individually by voice.

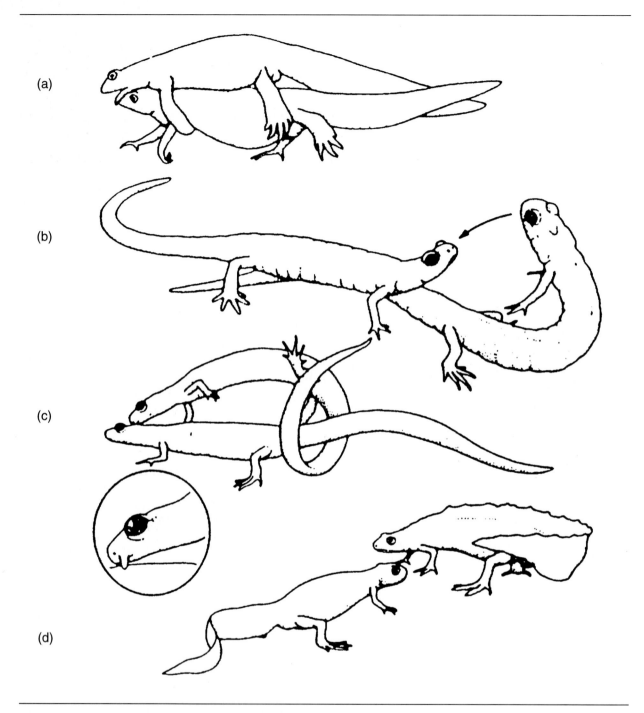

Figure 11–17 Transfer of pheromones by salamanders. (a) The rough-skinned, *Taricha granulosa.* (b) Jordan's salamander, *Plethodon jordani.* (c) The two-lined salamander, *Eurycea bislineata.* (d) The smooth newt, *Triturus vulgaris.* (From T. R. Halliday, 1990, *Advances in the Study of Behavior* 19:139. © 1990 Academic Press, Inc.)

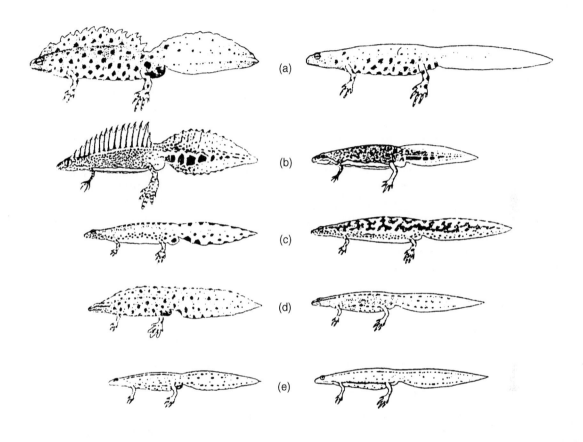

Figure 11–18 Dimorphism of secondary sexual characters of European newts (*Tritutus*). The male is on the left, the female is on the right. (a) The great crested newt, *T. cristatus.* (b) The banded newt, *T. vittatus.* (c) The alpine newt, *T. alpestris.* (d) The smooth newt, *T. vulgaris.* (e) Bosca's newt, *T. boscai.* The scale bar represents 5 centimeters. (From T. R. Halliday and B. Arano, 1991, *Trends in Ecology and Evolution* 6:114. © 1991 Elsevier Science Publishers Ltd.)

An advertisement call is a conservative evolutionary character, and among related taxa there is often considerable similarity in advertisement calls. Superimposed on the basic similarity are the effects of morphological factors, such as body size, as well as ecological factors that stem from characteristics of the habitat. Most toads (*Bufo*) have a trilled advertisement call that consists of a train of repeated pulses (a trill), but the pitch of the call varies with the body size. The oak toad (*B. quercicus*), which has a body length of only 2 or 3 centimeters, has a dominant frequency of 5200 hertz. The larger southwestern toad (*B. microscaphus*), which is 8 centimeters long, has a lower dominant frequency 1500 hertz, and the giant toad (*B. marinus*), with a body length of nearly 20 centimeters, has the lowest pitch of all, 600 hertz.

Female frogs are responsive to the advertisement calls of males of their species for a brief period when their eggs are ready to be laid. The hormones associated with ovulation are throught to sensitize specific cells in the auditory pathway that are responsive to the species-specific characteristics of the male's call. Mixed choruses of anurans are common in the mating season; a dozen species may breed simultaneously in one pond. A female's response to her own species' mating call is a mechanism for species recognition in that situation.

Costs and Benefits of Vocalization The vocalizations of male frogs are costly in two senses. The actual energy that goes into call production can be very large, and the variations in calling pattern that accompany social interactions among male frogs in a breeding chorus can increase the cost per call (see Box 11–1). Another cost of vocalization for a male frog is the risk of predation. A critical function of vocalization is to permit a female frog to locate a male, but female frogs are not the only animals that can use vocalizations as a cue to find male frogs—predators of frogs also find that calling males are easy to locate. The túngara frog (*Physalaemus pustulosus*) is a small terrestrial leptodactylid that occurs in Central America (Figure 11–19). Stanley Rand and Michael Ryan have studied the costs and benefits of vocalization for this species (Ryan 1985).

Túngara frogs breed in small pools, and breeding assemblies range from a single male to choruses of several hundred males. The advertisement call of a male túngara frog is a strange noise, a whine that sounds as if it would be more at home in an arcade of video games than in the tropical night. The whine starts at a frequency of 900 hertz and sweeps downward to 400 hertz in about 400 milliseconds (Figure 11–20). The whine may be produced by itself, or it may be followed by one or several *chucks* with a dominant frequency of 250 hertz. When a male túngara frog is calling alone in a pond it usually gives only the whine portion of the call, but as additional males join a chorus, more and more of the frogs produce calls that include chucks. By playing recordings of the whine calls to male frogs in breeding ponds, Rand was able to make them shift to giving calls that included chucks. That observation suggested that it was the presence of other calling males that stimulated frogs to make

Figure 11–19 Male túngara frog, *Physalaemus pustulosus*, vocalizing. Air is forced from the lungs (a) into the vocal sacs (b). (Photographs courtesy of Theodore L. Taigen.)

their calls more complex by adding chucks to the end of the whine.

What advantage would a male frog in a chorus gain from using a whine–chuck call instead of a whine? Rand suggested that the complex call might be more attractive to female frogs than the simple call. He tested that hypothesis by placing female túngara frogs in a test arena with a speaker at each side. One speaker broadcast a whine call and the second speaker broadcast a whine-chuck. Rand released female frogs individually in the center of the arena and noted which speaker they moved toward. As he had predicted, most of the female frogs (14 of the 15 he tested) chose the speaker broadcasting the whine–chuck call.

If female frogs are attracted to whine–chuck calls in preference to whine calls, why do male frogs give whine–chuck calls only when other males are present? Why not always give the most attractive call possible? One possibility is that whine–chuck calls require more energy than whines, and males save energy by using whine–chucks only when competition with other males makes the energy expenditure necessary. However, measurements of the energy expenditure of calling male túngara frogs showed that energy cost was not related to the number of chucks. Another possibility is that male frogs giving whine–chuck calls are more vulnerable to predators than frogs giving only whine calls. Túngara frogs in breeding choruses are preyed upon by frog-eating bats, *Trachops cirrhosus*, and the bats locate the frogs by homing on their vocalizations.

In a series of playback experiments Ryan and Merlin Tuttle placed pairs of speakers in the forest and broadcast vocalizations of túngara frogs. One speaker played a recording of a whine and the other a recording of a whine–chuck. The bats responded as if the speakers were frogs: They flew toward the speakers and even landed on them. In five experiments at different sites, the bats approached speakers broadcasting whine–chuck calls twice as frequently as those playing simple whines (168 approaches versus 81). Thus, female frogs are not alone in finding whine–chuck calls more attractive than simple whines—an important predator of frogs also responds more strongly to the complex calls. Predation can be a serious risk for male túngara frogs. Ryan and his colleagues measured the rates of predation in choruses of different sizes. The major predators were frog-eating bats, a species of opossum (*Philander opossum*), and a larger species of

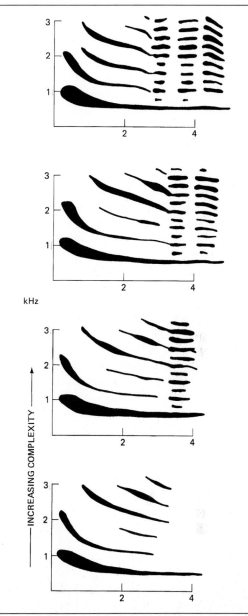

Figure 11–20 Sonograms of the advertisement call of *Physalaemus pustulosus*. The calls increase in complexity from bottom (a whine only) to top (a whine followed by three chucks). A sonogram is a graphic representation of a sound: Time is shown on the horizontal axis and frequency on the vertical axis. (Modified from M. J. Ryan, 1985, *The Túngara Frog*, University of Chicago Press, Chicago, IL.)

frog (*Leptodactylus pentadactylus*); the bats were the most important predators of the túngara frogs. Large choruses of frogs did not attract more bats than small choruses, and consequently the risk of

Box 11–1 The Energy Cost of Vocalization by Frogs

The vocalizations of frogs, like most acoustic signals of tetrapods, are produced when air from the lungs is forced over the vocal cords, causing them to vibrate. Contraction of trunk muscles provides the pressure in the lungs that propels the air across the vocal cords, and these contractions require metabolic energy. Measurement of the actual energy expenditure by frogs during calling is technically difficult because a frog must be placed in an airtight metabolism chamber to measure the amount of oxygen it consumes, and that procedure can frighten the frog and prevent it from calling. Ted Taigen and Kent Wells (1985) at the University of Connecticut overcame that difficulty in studies of the gray treefrog, *Hyla versicolor*, by taking the metabolism chambers to the breeding ponds. Calling male frogs were placed in the chambers early in the evening and then left undisturbed. With the stimulus of the chorus around them, frogs would call in the chambers. Their vocalizations were recorded with microphones attached to each chamber, and the amount of oxygen they used during calling was determined from the decline in the concentration of oxygen in the chamber over time (Figure 11–20).

The rates at which individual frogs consumed oxygen were directly proportional to their rates of vocalization (Figure 11–21). At the lowest calling rate, 150 calls per hour, oxygen consumption was barely above resting levels. However, at the highest calling rates, 1500 calls per hour, the frogs were consuming oxygen even more rapidly than they did during high levels of locomotor activity. Examination of the trunk muscles of the male frogs, which hypertrophy enormously during the breeding season, revealed biochemical specializations that appear to permit this high level of oxygen consumption during vocalization (Marsh and Taigen 1987).

The advertisement call of the gray treefrog is a trill that lasts from 0.3 to 0.6 second. During their studies, Wells and Taigen found that gray treefrogs gave short calls when they were in small choruses, and lengthened their calls when many other males were calling near them (Wells and Taigen 1986). It has subsequently been shown that long calls are more attractive to female frogs than short calls (Klump and Gerhardt 1987), but the long calls require more energy. A 0.6-second call requires about twice as much energy as a 0.3-second call, and the rate of

Figure 11–21 Gray treefrog (*Hyla versicolor*) in a metabolism chamber beside a breeding pond. A microphone in the chamber records the frog's calls, and a thermocouple measures the temperature inside the chamber. Gas samples are drawn from the tube for measurements of oxygen consumption. (Photograph courtesy of Theodore L. Taigen.)

oxygen consumption during calling increases as the length of the calls increases. That relationship suggests that, all else being equal, a male gray treefrog that increased its call duration to be more attractive to female frogs would pay a price for its attractiveness with a higher rate of energy expenditure.

Indirect evidence suggests that the energy cost of calling might limit the time a male gray treefrog could spend in a breeding chorus. The treefrogs call for only 2 to 4 hours each night, and stores of glycogen (the metabolic substrate used by calling frogs) decreased by 50 percent in that time. Wells and Taigen were able to simulate the effects of different chorus sizes by playing tape recordings of vocalizations to frogs. The frogs matched their own calls to the recorded calls they heard—short responses to short calls, medium to medium calls, and long responses to long calls. As the length of their calls increased, the frogs reduced the rate at which they called. The reduction in rate of calling approximately balanced the increased length of each call, so the overall calling effort (the number of seconds of vocalization per hour) was nearly independent of call duration. Thus, it appears that male gray treefrogs compensate for the higher energy cost of long calls by giving fewer of them. However, that compromise may not entirely eliminate the problem of high energy costs for frogs giving long calls. Even though the calling effort was approximately the same, males giving long calls at slow rates spent fewer hours per night calling than did frogs that produced short calls at higher rates.

The high energy cost of calling offers an explanation for the pattern of short and long calls produced by male gray treefrogs. The length of time that an isolated male can call may be the most important determinant of his success in attracting a female, and the trade-off, between rate of calling and call duration suggests that male frogs are performing at or near their physiological limits. Giving short calls and calling for several hours every night may be the best strategy if a male has that option avilable. In a large chorus, however, competition with other males is intense and giving a longer and more attractive call may be important, even if the male can call for only a short time.

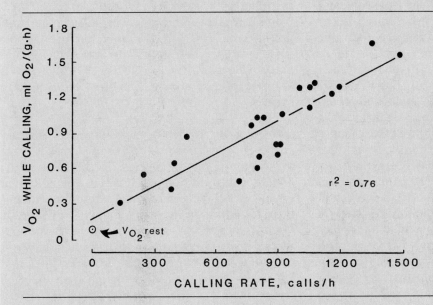

Figure 11–22 Rates of oxygen consumption of frogs calling inside metabolism chambers (on the vertical axis) as a function of the rate of calling (on the horizontal axis). The energy expended by a calling frog increases linearly with the number of times it calls per hour. (From T. L. Taigen and K. D. Wells, *Journal of Comparative Physiology,* B155:163–170.)

predation for an individual frog was less in a large chorus than in a small one. Predation was an astonishing 19 percent of the frogs per night in the smallest chorus and a substantial 1.5 percent per night even in the largest chorus. When a male frog shifts from a simple whine to a whine–chuck call, it increases its chances of attracting a female, but it simultaneously increases its risk of attracting a predator. In small choruses the competition from other males for females is relatively small and the risk of predation is relatively large. Under those conditions it is apparently advantageous for a male túngara frog to give simple whines. However, as chorus size increases, competition with other males also increases while the risk of predation falls. In that situation the advantage of giving a complex call apparently outweighs the risks.

Modes of Reproduction Fertilization is external in most anurans; the male uses his forelegs to clasp the female in the pectoral region (**axillary amplexus**) or pelvic region (**inguinal amplexus**). Amplexus may be maintained for several hours or even days before the female lays eggs. Males of the tailed frog of the Pacific northwest (*Ascaphus truei*) have an extension of the cloaca (the "tail" that gives them their name) that is used to introduce sperm into the cloaca of the female. Internal fertilization has been demonstrated for the Puerto Rican coquí (*Eleutherodactylus coqui*) and may be widespread among frogs that lay eggs on land (Townsend et al. 1981). Fertilization must also be internal for the few species of anurans that give birth to living young.

Anurans show even greater diversity in their modes of reproduction than salamanders. Similar reproductive habits have clearly evolved independently in different groups. Large eggs produce large offspring that probably have a better chance of surviving than smaller ones, but large eggs also require more time to hatch and are exposed to predators for a longer period. Thus, the evolution of large eggs and hatchlings has often been accompanied by the simultaneous evolution of behaviors that protect the eggs, and sometimes the tadpoles as well, from predation. A study of Amazon rainforest frogs revealed a positive relationship between the intensity of predation on frogs eggs in a pond and the proportion of frog species in the area that laid eggs in terrestrial situations (Magnusson and Hero 1991). Many arboreal frogs (rep-resented in Figure 11–23a by *Centrolenella*) lay their eggs in the leaves of trees overhanging water. The eggs undergo their embryonic development out of the reach of aquatic egg predators, and when the tadpoles hatch they drop into the water and take up an aquatic existence. Other frogs, such as *Physalaemus pustulosus* (Figure 11–23b), achieve the same result by constructing foam nests that float on the water surface. The female emits a copious mucus secretion during amplexus that the pair of frogs beat into a foam with their hind legs. The eggs are laid in the foam mass, and when the tadpoles hatch they drop through the foam into the water.

Although these methods reduce egg mortality, the tadpoles are subjected to predation and competition. Some anurans avoid both problems by finding or constructing breeding sites free from competition and predation. Some frogs, for example, lay their eggs in the water that accumulates in bromeliads—epiphytic tropical plants that grow in trees and are morphologically specialized to collect rainwater. A large bromeliad may hold several liters of water, and the frogs pass through egg and larval stages in that protected microhabitat. Many tropical frogs lay eggs on land near water. The eggs or tadpoles may be released from the nest sites when pond levels rise after a rainstorm. Other frogs construct pools in the mud banks beside streams. These volcano-shaped structures are filled with water by rain or seepage and provide a favorable environment for the eggs and tadpoles. Some frogs have eliminated the tadpole stage entirely. These frogs lay large eggs on land that develop directly to little frogs. This pattern is characteristic of about 20 percent of all anuran species.

Parental Care Adults of many species of frogs guard the eggs specifically. In some cases it is the male that protects the eggs, in others it is the female, and in most cases it is not clearly known which sex is involved because external sex identification is difficult for many anurans. Some of the frogs that lay their eggs over water remain with them. Some species sit beside the eggs, others rest on top of them. Many of the terrestrial frogs that lay direct-developing eggs remain with the eggs and will attack an animal that approaches the nest. Removing the guarding frog frequently results in the eggs desiccating and dying before hatching or

Figure 11–23 Reproductive modes of anurans; see the text for discussion. (a) Eggs laid over water, *Centrolenella;* (b) eggs in a nest of foam, *Physalaemus;* (c) eggs carried by the adult, *Rhinoderma;* (d) tadpoles carried by the adult, *Colostethus.* Eggs carried on the back of an adult: (e) *Hemiphractus;* (f) *Pipa.* [(c) From G. K. Noble, 1931, *The Biology of the Amphibia,* McGraw-Hill, New York, NY; (e) from W. E. Duellman, 1970, *The Hylid Frogs of Middle America,* Monograph of the Museum of Natural History, 1, The University of Kansas, Lawrence, KS; (f) from M. Lamotte and J. Lescure, 1977, *La Terre et la Vie* 31:225–311.]

being eaten by predators (Taigen et al. 1984, Townsend et al. 1984). Male African bullfrogs (*Pyxicephalus adspersus*) guard their eggs, and then continue to guard the tadpoles after they hatch. The male frog moves with the school of tadpoles, and will even dig a channel to allow the tadpoles to swim from one pool in a marsh to an adjacent one. Tadpoles of several species of the tropical American frog genus *Leptodactylus* follow their mother around the pond. *Leptodactylus* are large and aggressive frogs that are able to deter many potential predators.

Some of the poison-dart frogs of the American tropics deposit their eggs on the ground, and one of the parents remains with the eggs until they hatch into tadpoles. The tadpoles adhere to the adult and are transported to water (Figure 11–23d). Females of the Panamanian frog *Colostethus inguinalis* carry their tadpoles for more than a week and the tadpoles increase in size during this period. The largest tadpoles being carried by females had small amounts of plant material in their stomachs, suggesting that they had begun to feed while they were still being transported by their mother (Wells 1980). Females of another Central American poison-dart frog, *Dendrobates pumilio*, release their tadpoles in small pools of water, and then return at intervals to the pools to deposit unfertilized eggs that the tadpoles eat (Brust 1993).

Other anurans, instead of remaining with the eggs, carry the eggs with them. The male of the European midwife toad (*Alytes obstetricians*) gathers the egg strings about his hind legs as the female lays them. He carries them with him until they are ready to hatch, at which time he releases the tadpoles into water. The male of the terrestrial Darwin's frog (*Rhinoderma darwinii*) of Chile snaps up the eggs the female lays and carries them in his vocal pouches, which extend back to the pelvic region (Figure 11–23c). The embryos pass through metamorphosis in the vocal sacks and emerge as fully developed froglets. Males are not alone in caring for eggs. The females of a group of treefrogs carry the eggs on their back, in an open oval depression, a closed pouch, or individual pockets (Figure 11–23e). The eggs develop into miniature frogs before they leave their mother's back (Del Pino 1989). A similar specialization is seen in the completely aquatic Surinam toad, *Pipa pipa*. In the breeding season the skin of the female's back thickens and softens. In egg laying the male and female in amplexus swim in vertical loops in the water. On the upward part of the loop the female is above the male and releases a few eggs which fall onto his ventral surface. He fertilizes them and, on the downward loop, presses them against the female's back. They sink into the soft skin and a cover forms over each egg, enclosing it in a small capsule (Figure 11–23f). The eggs develop through metamorphosis in the capsules.

Tadpoles of the two species of the Australian frog genus *Rheobatrachus* are carried in the stomach of the female frog. The female swallows eggs or newly hatched larvae and retains them in her stomach through metamorphosis. This behavior was first described in *Rheobatrachus silus* and is accompanied by extensive morphological and physiological modifications of the stomach. These changes include distension of the proximal portion of the stomach, separation of individual muscle cells from the surrounding connective tissue, and inhibition of hydrochloric acid secretion, perhaps by prostaglandin released by the tadpoles (Tyler 1983). In January 1984, a second species of gastric brooding frog, *R. vitellinus*, was discovered in Queensland. Strangely, this species lacks the extensive structural changes in the stomach that characterize the gastric brooding of *R. silus* (Leong et al. 1986). The striking differences between the two species suggest the surprising possibility that this bizarre reproductive mode might have evolved independently. Both species of *Rheobatrachus* disappeared within a few years of their discovery, and there is no obvious explanation for the apparent extinction of these fascinating frogs. The repeated pattern of amphibian population declines, even for species from habitats that appear undisturbed, suggests that some of these declines may be produced by global-level effects of human activities.

Only a few anurans are viviparous. The females of some African bufonids retain the eggs in the oviducts and give birth to baby toads. Some of the energy and nutrients required for embryonic development are obtained from the oviduct. The golden coquí (*Eleutherodactylus jasperi*, a Puerto Rican leptodactylid) also gives birth to fully formed young, but in this case the energy and nutrients come from the yolk of the egg. (The golden coquí is yet another species of amphibian that has vanished and is presumed to be extinct.)

The Ecology of Tadpoles Although many species of frogs have evolved reproductive modes that bypass an aquatic larval stage, a life history that includes a tadpole has certain advantages. A tadpole is a completely different animal from an adult anuran, both morphologically and ecologically.

Tadpoles are as diverse in their morphological and ecological specializations as adult frogs and occupy nearly as great a range of habitats (Figure 11–24). Tadpoles that live in still water usually have ovoid bodies and tails with fins that are as large as the muscular part of the tail, whereas tadpoles that live in fast-flowing water have more streamlined bodies and smaller tail fins. Semiterrestrial tadpoles wiggle through mud and leaves and climb on damp rock faces; they are often dorsoventrally flattened and have little or no tail fin, and many tadpoles that live in bromeliads have a similar body form. Direct developing tadpoles have large yolk supplies, and reduced mouthparts and tail fins. The mouthparts of tadpoles also show variation that is related to diet (Figure 11–25). Filter-feeding tadpoles that hover in midwater lack keratinized mouthparts, whereas species that graze from surfaces have small beaks that are often surrounded by rows of denticles. Predatory tadpoles have larger beaks that can bite pieces from other tadpoles. Funnel-mouthed, surface-feeding tadpoles have greatly expanded mouthparts that skim material from the surface of the water.

Tadpoles of most species of anurans are filter-feeding herbivores, whereas all adult anurans are carnivores that catch prey individually. Because of these differences, tadpoles can exploit resources that are not available to adult anurans (Wassersug 1975). This advantage may be a factor that has led many species of frogs to retain the ancestral pattern of life history in which an aquatic larva matures into a terrestrial adult. Many aquatic habitats experience annual flushes of primary production when nutrients washed into a pool by rain or melting snow stimulate the rapid growth of algae. The energy and nutrients in this algal bloom are transient resources that are available for a brief time to organisms that are able to exploit them.

Tadpoles are excellent eating machines. All tadpoles are filter feeders, and feeding and ventilation of the gills are related activities. The stream of water that moves through the mouth and nares to ventilate the gills carries with it particles of food. As the stream of water passes through the branchial basket, small food particles are trapped in mucus secreted by epithelial cells. The mucus, carrying particles of food with it, is moved from the gill filters to the ciliary grooves on the margins of the roof of the pharynx and then transported posteriorly to the esophagus.

Although all tadpoles filter food particles from a stream of water that passes across the gills, the method by which the food particles are put into suspension differs among species. Some tadpoles filter floating plankton from the water. Tadpoles of this type are represented in several families of anurans, especially the Pipidae and Hylidae, and usually hover in the water column. Midwater-feeding

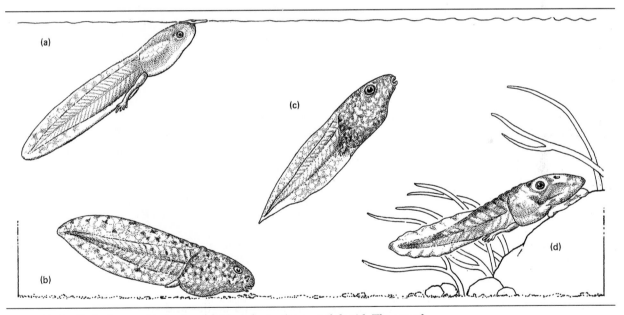

Figure 11–24 Body forms of tadpoles: (a) *Megophrys minor*, a pelobatid. The mouthparts unfold into a platter over which water and particles of food on the surface are drawn into the mouth. (b) *Rana aurora*, a ranid. A generalized feeder that nibbles and scrapes food from surfaces. (c) *Agalychnis callidryas*, a hylid. A midwater suspension feeder shows the large fins and protruding eyes that are typical of midwater tadpoles. It maintains its position in the water column with rapid undulations of the terminal part of its tail. (d) An unidentified species of *Nyctimystes*, a hylid. A stream-dwelling tadpole that adheres to rocks in swiftly moving water with a sucker-like mouth while scraping algae and bacteria from the rocks. The low fins and powerful tail are characteristic of tadpoles living in swift water.

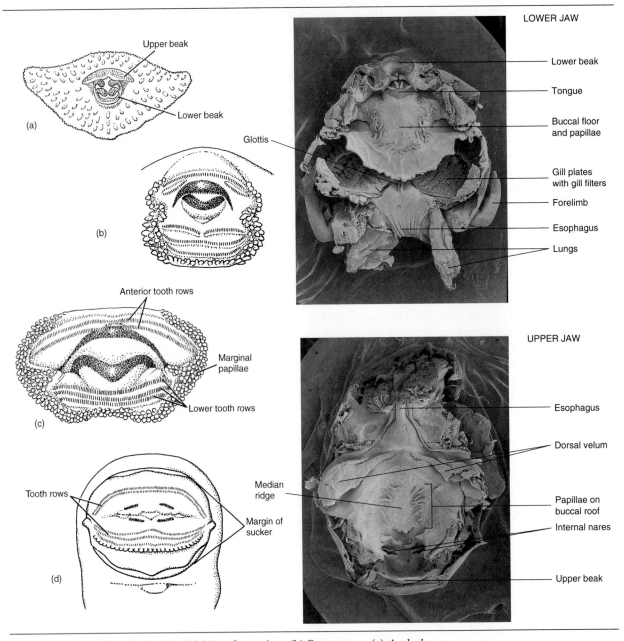

Figure 11–25 Mouths of tadpoles: (a) *Megophrys minor*; (b) *Rana aurora*; (c) *Agalych-nis callidryas*; (d) *Nyctimystes* sp.; (e) scanning electron micrograph of the inside of the mouth and buccal region of a tadpole (*Alsodes monticola*, a leptodactylid). [(e) courtesy of Richard J. Wassersug.]

tadpoles are out in the open, where they are vulnerable to predators and they show various characteristics that may reduce the risk of predation. Tadpoles of the African clawed frogs are nearly transparent, and they may be hard for predators to see. Some midwater tadpoles form schools that, like schools of fishes, may confuse a predator by presenting it with so many potential prey that it has difficulty concentrating its attack on one individual.

Many tadpoles are bottom feeders and scrape bacteria and algae off the surfaces of rocks or the leaves of plants. The rasping action of keratinized mouthparts frees the material and allows it to be whirled into suspension in the water stream entering the mouth of a tadpole, and then filtered out by the branchial apparatus. Some bottom-feeding tadpoles like toads and spadefoot toads form dense aggregations that create currents that lift particles of food into suspension in the water. These aggregations may be groups of siblings. Toad (*Bufo americanus*) and cascade frog tadpoles (*Rana cascadae*) are able to distinguish siblings from nonsiblings and they associate preferentially with siblings (Blaustein and O'Hara 1981, Waldman 1982). Recognition is probably accomplished by olfaction, and toad tadpoles can distinguish full siblings (both parents the same) from maternal half-siblings (only the mother the same), and they can distinguish maternal half-siblings from paternal half-siblings. Prior experience and diet also play a role in kin recognition by tadpoles, and kin recognition might be an artifact of habitat selection (Pfennig 1990a, Gamboa et al. 1991).

Some tadpoles are carnivorous and feed on other tadpoles. The large Central American form *Leptodactylus pentadactylus* that preys on smaller species of anurans such as the túngara frog has a carnivorous tadpole that preys on eggs and tadpoles of other species of anurans. Predatory tadpoles have large mouths with a sharp keratinized beak. Predatory individuals appear among the tadpoles of some species of anurans that are normally herbivorous. Some species of spadefoot toads in western North America are famous for this phenomenon. Spadefoot tadpoles are normally herbivorous, but when tadpoles of the southern spadefoot toad, *Scaphiopus multiplicatus*, eat the freshwater shrimp that occur in some breeding ponds they are transformed into the carnivorous morph (Pfennig 1990b). These carnivorous tadpoles have large heads and jaws and a powerful beak. In addition to eating shrimp, they prey on other tadpoles.

In an Amazonian rainforest, tadpoles are by far the most important predators of frog eggs. In fact, egg predation decreases as the density of fish increases, apparently because the fish eat tadpoles that would otherwise eat frog eggs (Magnusson and Hero 1991). Carnivorous tadpoles are also found among some species of frogs that deposit their eggs or larvae in bromeliads. These relatively small reservoirs of water may have little food for tadpoles. It seems possible that the first tadpole to be placed in a bromeliad pool may feed largely on other frog eggs—either unfertilized eggs deliberately deposited by the mother of the tadpole, as is the case for the poison-dart frog *Dendrobates pumilio,* or fertilized eggs subsequently deposited by unsuspecting female frogs.

The feeding mechanisms that make tadpoles such effective collectors of food particles suspended in the water allow them to grow rapidly, but that growth contains the seeds of its own termination. As tadpoles grow bigger they become less effective at gathering food because of the changing relationship between the size of food-gathering surfaces and the size of their bodies. The branchial surfaces that trap food particles are two dimensional. Consequently, the food-collecting apparatus of a tadpole increases in size approximately as the square of the linear dimensions of the tadpole. However, the food the tadpole collects must nourish its entire body, and the volume of the body increases in proportion to the cube of the linear dimensions of the tadpole. The result of that relationship is a decreasing effectiveness of food collection as a tadpole grows; the body it must nourish increases in size faster than its food-collecting apparatus.

The morphological specializations of tadpoles are entirely different from those of adult frogs, and the transition from tadpole to frog involves a very complete metamorphosis in which tadpole structures are broken down and their chemical constituents are rebuilt into the structures of adult frogs.

Amphibian Metamorphosis

The importance of thyroid hormones for amphibian metamorphosis was discovered quite by accident in the early twentieth century by the German biologist Friedrich Gudersnatch. He was able to induce rapid precocious metamorphosis in tadpoles by feeding them extracts of beef thyroid glands. Some of the details of the interaction of neurosecretions and endocrine gland hormones have been worked out, but no fully integrated explanation of the mechanisms of hormonal control of amphibian metamorphosis is yet possible.

The most dramatic example of metamorphosis is found among anurans, where almost every tadpole structure is altered. Anuran larval development is generally divided into three periods: (1) During premetamorphosis tadpoles increase in size with little change in form; (2) in prometamorphosis the hind legs appear and growth of the body continues at a slower rate; and (3) during metamorphic climax the forelegs emerge and the tail regresses. These changes are stimulated by the actions of thyroxine, and production and release of thyroxine is controlled by a product of the pituitary gland, thyroid stimulating hormone (TSH).

The action of thyroxine on larval tissues is both specific and local. In other words, it has a different effect in different tissues, and that effect is produced by the presence of thyroxine in the tissue; it does not depend on induction by neighboring tissues. The particular effect of thyroxine in a given tissue is genetically determined, and virtually every tissue of the body is involved (Table 11–3). In the liver, for example, thyroxine stimulates the enzymes responsible for the synthesis of urea (the urea cycle enzymes) and starts the synthesis of serum albumin. In the eye it induces the formation of rhodopsin. When thyroxine is administered to the striated muscles of a tadpole's developing leg, it stimulates growth; but when administered to the striated muscles of the tail, it stimulates the breakdown of tissue. When a larval salamander's tail is treated with thyroxine, only the tail fin disappears; but thyroxine causes the complete absorption of the tail of a frog tadpole.

Metamorphosis of salamanders is relatively undramatic compared to the process in anurans. Extensive changes occur at the molecular and tissue level in Salamanders, but the loss of gills and absorption of the tail fin are the obvious external changes. In contrast, the metamorphosis of a tadpole to a frog involves readily visible changes in almost every part of the body. The tail is absorbed and recycled into the production of adult structures. The small tadpole mouth that accommodated algae broadens into the huge mouth of an adult frog. The long tadpole gut, characteristic of herbivorous vertebrates, changes to the short gut of a carnivorous animal. Respiration is shifted from gills to lungs, and partly metamorphosed froglets can be seen swimming to the surface to gulp air.

Metamorphic climax begins with the appearance of the forelimbs and ends with the disappearance of the tail. This is the most rapid part of metamorphosis, occupying only a few days after a larval period that lasts for weeks or months. One reason for the rapidity of metamorphic climax may lie in the vulnerability of larvae to predators during this period. A larva with legs and a tail is neither a good tadpole nor a good frog: The legs inhibit swimming and the tail interferes with jumping. As a result, predators are more successful at catching anurans during metamorphic climax than they are in prometamorphosis or following the completion of metamorphosis. Metamorphosing chorus frogs (*Pseudacris triseriata*) were most vulnerable to garter snakes when they had developed legs and still retained a tail. Both tadpoles (with a tail and no legs) and metamorphosed frogs (with legs and no tail) were more successful than the metamorphosing individuals at escaping from snakes (Figure 11–26). Life-history theory predicts that selection will act to shorten the periods in the lifetime of a species when it is most vulnerable to predation, and the speed of

Table 11–3 **Some of the morphological and physiological changes induced by thyroid hormones during amphibian metamorphosis.**

Body form and stucture
  Formation of dermal glands.
  Restructuring of mouth and head
  Intestinal regression and reorganization
  Calcification of skeleton

Appendages
  Degeneration of skin and muscle of tail
  Growth of skin and muscle of limbs

Nervous system and sense organs
  Increase in rhodopsin in retina
  Growth of extrinsic eye muscles
  Formation of nictitating membrane of the eye
  Growth of cerebellum
  Growth of preoptic nucleus of the hypothalamus

Respiratory system
  Degeneration of the gill arches and gills
  Degeneration of the operculum that covers the gills
  Development of lungs
  Shift from larval to adult hemoglobin

Organs
  Pronephric resorption in the kidney
  Induction of urea-cycle enzymes in the liver
  Reduction and restructuring of the pancreas

Source: Based on B. A. White and C. S. Nicoll, 1981, in *Metamorphosis, a Problem in Developmental Biology*, edited by L. I. Gilbert and E. Freeden, Plenum, New York, NY.

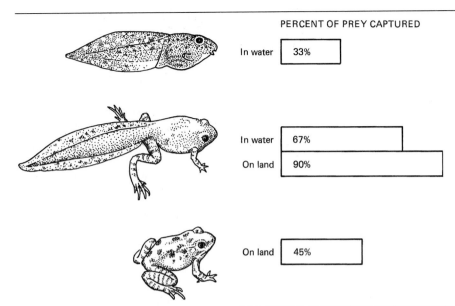

PERCENT OF PREY CAPTURED

In water	33%
In water	67%
On land	90%
On land	45%

Figure 11–26 Metamorphosing chorus frogs (*Pseudacris triseriata*) were more vulnerable to garter snakes than tadpoles or fully transformed frogs. In water, the snakes captured 33 percent of the tadpoles offered compared to 67 percent of the transforming frogs. On land the snakes captured 45 percent of the fully transformed frogs that were offered and 90 percent of the transforming ones. (Data from R. J. Wassersug and D. G. Sperry, 1977, *Ecology* 58:830–839.)

metamorphic climax may be a manifestation of that phenomenon.

Water Relations of Amphibians

Amphibians have a glandular skin that lacks external scales and is highly permeable to water. Both the permeability and glandularity of the skin have been of major importance in shaping the ecology and evolution of amphibians. Mucus glands are distributed over the entire body surface and secrete mucopolysaccharide compounds. The primary function of the mucus is to keep the skin moist and permeable. For an amphibian, a dry skin means reduction in permeability to water and gases. That, in turn, reduces oxygen uptake and the ability of the animal to use evaporative cooling to maintain its body temperature within equable limits. Experimentally produced interference with mucus gland secretion can lead to lethal overheating of frogs undergoing normal basking activity.

Permeability of Amphibian Skin

Both water and gases pass readily through amphibian skin. In biological systems, permeability to water is inseparable from permeability to gases, and amphibians depend on cutaneous respiration for a significant part of their gas exchange. Although the skin permits passive movement of water and gases, it controls the movement of other compounds. Sodium is actively transported from the outer surface to the inner, and urea is retained by the skin. These characteristics are important in the regulation of osmotic concentration and in facilitating uptake of water by terrestrial species.

The internal osmotic pressure of amphibians is approximately two-thirds that characteristics of most other vertebrates. The primary reason for the dilute body fluids of amphibians is low sodium content—approximately 100 milliequivalents compared to 150 milliequivalents in other vertebrates. Amphibians can tolerate a doubling of the normal sodium concentration, whereas an increase from 150 milliequivalents to 170 milliequivalents is the maximum humans can tolerate.

A watery animal with a permeable skin seems an unlikely candidate for success in an arid habitat, and most amphibians are restricted to moderately moist microhabitats. Anurans have been by far the most successful invaders of arid habitats. All but the harshest deserts have substantial anuran populations, and in different parts of the world, different families have converged on similar specializations. Avoiding the harsh conditions of the ground surface is the most common mechanism by which amphibians have managed to invade deserts and other arid habitats. Anurans and salamanders in deserts may spend 9 or 10 months of the year in moist retreat sites, sometimes more than a meter underground, emerging only during the rainy season and compressing feeding, growth, and reproduction into just a few months (Chapter 16).

Many species of arboreal frogs have skins that are less permeable to water than the skin of terrestrial frogs, and a remarkable specialization is seen in a few treefrogs. The African rhacophorid *Chiromantis xerampelina* and the South American hylid *Phyllomedusa sauvagei* lose water through the skin at a rate only one-tenth that of most frogs. *Phyllomedusa* has been shown to achieve this low rate of evaporative water loss by using its legs to spread the lipid-containing secretions of dermal glands over its body surface in a complex sequence of wiping movements. These two frogs are unusual also because they excrete nitrogenous wastes as salts of uric acid rather than as urea (see Chapter 4). This uricotelism provides still more water conservation.

Behavioral Control of Evaporative Water Loss

For animals with skins as permeable as those of most amphibians, the main difference between rain forests and deserts may be how frequently they encounter a water shortage. The Puerto Rican coquí lives in wet tropical forests; nonetheless, it has elaborate behaviors that reduce evaporative water loss during its periods of activity (Pough et al. 1983). Male coquís emerge from their daytime retreat sites at dusk and move 1 or 2 meters to calling sites on leaves in the understory vegetation. They remain at their calling sites until shortly before dawn, when they return to their daytime retreats. The activities of the frogs vary from night to night, depending on whether it rained during the afternoon. On nights after a rainstorm, when the forest is wet, the coquís begin to vocalize soon after dusk and continue until about midnight, when they fall silent for several hours. They resume calling briefly just before dawn. When they are calling, coquís

(a)

(b)

(c)

Figure 11–27 Male Puerto Rican coquí: (a) During vocalization nearly all the body surface is exposed to evaporation; (b) in the alert posture in which frogs wait to catch prey most of the body surface is exposed; (c) in the water-conserving posture adopted on dry nights half the body surface is protected from exposure. (Photographs by F. Harvey Pough.)

extend their legs and raise themselves off the surface of the leaf (Figure 11–27). In this position they lose water by evaporation from the entire body surface.

On dry nights the behavior of the frogs is quite different. The males move from their retreat sites to their calling stations, but they call only sporadically. Most of the time they rest in a water-conserving posture in which the body and chin are flattened against the leaf surface and the limbs are pressed against the body. A frog in this posture exposes only half its body surface to the air, thereby reducing its rate of evaporative water loss. The effectiveness of the postural adjustments is illustrated by the water losses of frogs in the forest at El Verde, Puerto Rico, on dry nights. Frogs in one test group were placed individually in small wire mesh cages that were placed on leaf surfaces. A second group was composed of unrestrained frogs sitting on leaves. The caged frogs spent most of the night climbing around the cages trying to get out. This activity, like vocalization, exposed the entire body surface to the air, and the caged frogs had an evaporative water loss that averaged 27.5 percent of their initial body mass. In contrast, the unrestrained frogs adopted water-conserving postures and lost an average of only 8 percent of their initial body mass by evaporation.

Experiments showed that the jumping ability of coquís was not affected by an evaporative loss of as much as 10 percent of the initial body mass, but a loss of 20 percent or more substantially decreased the distance frogs could jump (Beuchat et al. 1984). Thus, coquís use behavior to limit their evaporative water losses on dry nights to levels that do not affect their ability to escape from predators or to capture prey. Without those behaviors, however, they would probably lose enough water by evaporation to affect their survival.

Uptake and Storage of Water

The mechanisms that amphibians use for obtaining water in terrestrial environments have received less attention than those for retaining it. Amphibians do not drink water. Because of the permeability of their skins, species that live in aquatic habitats face a continuous osmotic influx of water that they must balance by producing urine. Species in arid habitats rarely encounter enough liquid water in one place to drink it, and if they should find a puddle they can quickly absorb the water they need through their skins. The impressive adaptations of terrestrial amphibians are ones that facilitate rehydration from limited sources of water. One such specialization is the **pelvic patch.** This is an area of highly vascularized skin in the pelvic region that is responsible for a very large portion of an anuran's cutaneous water absorption. Toads that are dehydrated and completely immersed in water rehydrate only slightly faster than those placed in water just deep enough to wet the pelvic area. In arid regions, water is frequently available as a thin layer of moisture on a rock, or as wet soil. The pelvic patch allows an anuran to absorb this water.

The urinary bladder plays an important role in the water relations of terrestrial amphibians, especially anurans. Amphibian kidneys produce urine that is hyposmotic to the blood, so the urine in the bladder is dilute. Amphibians can reabsorb water from urine to replace water they lose by evaporation, and terrestrial amphibians have larger bladders than aquatic species. Storage capacities of 20 to 30 percent of the body mass of the animal are common for terrestrial anurans, and some species have still larger bladders: The Australian desert frogs *Notaden nicholsi* and *Neobatrachus wilsmorei* can store urine equivalent to about 50 percent of their body mass, and a bladder volume of 78.9 percent of body mass has been reported for the Australian frog *Helioporus eyrei*.

Behavior is as important in facilitating water uptake as it is in reducing water loss. Leopard frogs, *Rana pipiens,* spend the summer activity season in grassy meadows where they have no access to ponds. The frogs spend the day in retreats they create by pushing vegetation aside to expose moist soil. In the retreats, the frogs rest with the pelvic patch in contact with the ground, and tests have shown that the frogs are able to absorb water from the soil. On nights when dew forms, many frogs move from their retreats and spend some hours in the early morning sitting on dew-covered grass before returning to their retreats. Leopard frogs show a daily pattern of water gain and loss during a period of several days when no rain falls: In the morning the frogs are sleek and glistening with moisture, and they have urine in their bladders. That observation indicates that in the morning the frogs have enough water to form urine. By evening the frogs have dry skins, and little urine in the bladder, suggesting that as they lost water by evaporation during the day they had reabsorbed water from the urine to maintain the water content of their tissues.

By the following morning the frogs have absorbed more water from dew and are sleek and well-hydrated again. Net gains and losses of water are shown by daily fluctuations in body masses of the frogs: In the mornings they are as much as 4 or 5 percent heavier than their overall average mass, and in the evenings they are lighter than the average by a similar amount (Dole 1965, 1967). Thus, these terrestrial frogs are able to balance their water budgets by absorbing water from moist soil and from dew to replace the water they lose by evaporation and in urine. As a result they are independent of sources of water like ponds or streams and are able to colonize meadows and woods far from any permanent sources of water.

Poison Glands and Other Defense Mechanisms

The mucus that covers the skin of an amphibian has a variety of properties. In at least some species it has antibacterial activity—for example, a potent antibiotic that may have medical applications has been isolated from the skin of the African clawed frog. It is mucus that makes some amphibians slippery and hard for a predator to hold. Other species have mucus that is extremely adhesive. Many salamanders, for example have a concentration of mucous glands on the dorsal surface of the tail. When one of these salamanders is attacked by a predator, it bends its tail forward and buffets its attacker. The sticky mucus causes debris to adhere to the predator's

snout or beak, and with luck the attacker soon concentrates on cleaning itself, losing interest in the salamander. When the California slender salamander is seized by a garter snake, the salamander curls its tail around the snake's head and neck. This behavior makes the salamander hard for the snake to swallow, and also spreads sticky secretions on the snake's body. A small snake can find its body glued into a coil from which it is unable to escape.

Although secretions of the mucous glands of some species of amphibians are irritating or toxic to predators, an amphibian's primary chemical defense system is located in the poison glands (Figure 11–28). These glands are concentrated on the dorsal surfaces of the animal, and defense postures of both anurans and salamanders present the glandular areas to potential predators. A wide variety of irritating and, in some cases, exceedingly toxic compounds are produced by amphibians. The compounds produced by different groups reflect their phylogenetic relationship, and skin toxins have been helpful in classification of some groups.

Many amphibians advertise their distasteful properties with conspicuous **aposematic**, or warning, colors and behaviors. A predator that makes the mistake of seizing one is likely to spit it out because it is distasteful. The toxins in the skin may also induce vomiting that reinforces the unpleasant experience for the predator. Subsequently, the predator will remember its unpleasant experience and avoid the distinctly marked animal that produced it. Some toxic amphibians combine a cryptic dorsal color with an aposematic ventral pattern. Normally, the cryptic color conceals them from predators, but if they are

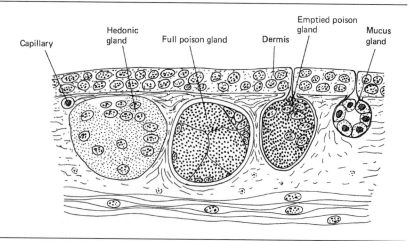

Figure 11–28 Cross section of skin from the base of the tail of a red-backed salamander, *Plethodon cinereus*. Three types of glands can be seen. (Modified from G. K. Noble, 1931, *The Biology of the Amphibia*, McGraw-Hill, New York, NY.)

attacked they adopt a posture that displays the brightly colored ventral surface (Figure 11–29).

Some salamanders have a morphological specialization that enhances the defensive effects of their chemical secretions. The European salamander *Pleurodeles waltl* and two genera of Asian salamanders (*Echinotriton* and *Tylotriton*) have ribs that pierce the body wall when a predator seizes the salamander. You can imagine the shock for a predator that bites a salamander and finds its tongue and palate impaled by a dozen or more bony spikes! Even worse, the ribs penetrate poison glands as they emerge through the body wall and each rib carries a drop of poison into the wound.

Anurans appear to have less complex defensive mechanisms than do salamanders. They do have skin toxins, and some frogs are lethally toxic (Box 11–2). Many anurans make long leaps to escape a predator, and others feign death. Some cryptically colored frogs extend their legs stiffly when they play dead. In this posture they look so much like the leaf litter on the ground that they may be hard for a visually oriented predator such as a bird to see. Very large frogs attack potential predators. They increase their apparent size by inflating the lungs, and hop toward the predator, often croaking loudly. That can be an unnerving experience, and some of the carnivorous species such as the horned frogs of South America

(*Ceratophrys*), which have recurved teeth on the maxillae and tooth-like serrations on the mandibles, can inflict a painful bite.

Red efts are classic examples of aposematic animals. (See the color insert.) They are bright orange and are active during the day, making no effort to conceal themselves. Efts contain tetrodotoxin, which is a potent neurotoxin. Touching an eft to your lips produces an immediate unpleasant numbness and tingling sensation, and the behavior of animals that normally prey on salamanders indicates that it affects them the same way. As a result, an eft that is attacked by a predator is likely to be rejected before it is injured. After one or two such experiences, a predator will no longer attack efts. Support for the belief that this protection may operate in nature is provided by the observation that 4 of 11 wild-caught bluejays (*Cyanocitta cristata*) refused to attack the first red eft they were offered in a laboratory feeding trial (Tilley et al. 1982). That behavior suggests that those four birds had learned to avoid red efts before they were captured. The remaining seven birds attacked at least one eft, but dropped it immediately. After one or two experiences of this sort, the birds made retching movements at the sight of an eft and refused to attack.

Of course, aposematic colors and patterns work to deter predation only if a predator can see the

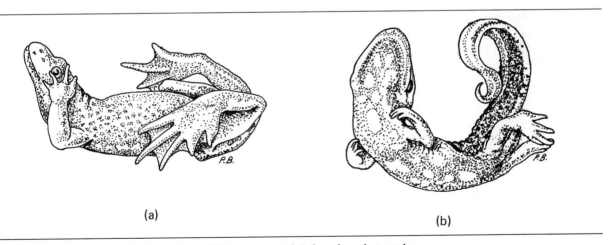

(a)   (b)

Figure 11–29 Aposematic displays of amphibians present bright colors that predators can learn to associate with the animals' toxic properties. (a) The European fire-bellied toad has a cryptically colored dorsal surface and a brightly colored underside that is displayed in the *unken* reflex when the animal is attacked. (b) The Hong Kong newt has a brownish dorsal surface and a mottled red and black venter that is revealed by its aposematic display. [(a) Modified from H. Gadow, 1909, *Amphibia and Reptiles*, London, UK; (b) from a photograph by E. D. Brodie, Jr.]

A great diversity of pharmacologically active substances has been found in the skins of amphibians. Some of them are extremely toxic and others are less toxic but capable of producing unpleasant sensations when a predator bites an amphibian. Biogenic amines such as serotonin and histamine, peptides such as bradykinin, and hemolytic proteins have been found in frogs and salamanders belonging to many families. Many of these substances, such as bufotoxin, physalaemin, and leptodactyline, are named for the animals in which they were discovered.

Cutaneous alkaloids are abundant and diverse among the poison-dart frogs, the family Dendrobatidae, of the New World tropics. More than 200 new alkaloids have been described from dendrobatids, mostly species in the genera *Dendrobates*, *Minyobates*, *Epidobates*, and *Phyllobates*. Most of these frogs are brightly colored and move about on the ground surface in daylight, making no attempt at concealment.

The name poison-dart frogs refers to the use by South American Indians of the toxins of some of these frogs to poison the tips of the blowgun darts used for hunting. The use of frogs in this manner appears to be limited to three species of *Phyllobates* that occur in western Colombia, although plant poisons like curare are used to poison blowgun darts in other parts of South America (Myers et al. 1978). A unique alkaloid, batrachotoxin, occurs in the genus *Phyllobates*. Batrachotoxin is a potent neurotoxin that prevents the closing of sodium channels in nerve and muscle cells, leading to irreversible depolarization and producing cardiac arrhythmias, fibrillation, and cardiac failure.

The bright yellow *Phyllobates terribilis* is the largest and most toxic species in the genus. The Emberá Choco Indians of Colombia use *Phyllobates terribilis* as a source of poison for their blowgun darts. The dart points are rubbed several times across the back of a frog, and set aside to dry. The Indians handle the frogs carefully, holding them with leaves—a wise precaution because batrachotoxin is exceedingly poisonous. A single frog may contain up to 1900 micrograms of batrachotoxin, and less than 200 micrograms is probably a lethal dose for a human if it enters the body through a cut. Batrachotoxin is also toxic when it is eaten. In fact, the investigators inadvertently caused the death of a dog and a chicken in the Indian village in which they were living when the animals got into garbage that included plastic bags in which the frogs had been carried. Cooking destroys the poison and makes prey killed by darts anointed with the skin secretions of *Phyllobates terribilis* safe to eat.

Skin secretions from another frog are used in the hunting magic of several Amazonian tribes. *Phyllomedusa bicolor*, a large, green hylid frog, secretes a variety of peptides including a hitherto unknown compound that has been named adenoregulin (Daly et al. 1992). Mucus scraped from a frog's skin is dried and stored for later use. When it is mixed with saliva and rubbed into areas of freshly burned skin, it induces a feeling of illness, followed by listlessness, and finally a profound euphoria. Adenoregulin, the constituent of the mucus that is responsible for these effects, apparently enhances binding of adenosine and its analogs to A_1 receptors of the brain. These receptors are distributed throughout the brain, and a way to modify their function could contribute to treatments for a variety of central nervous system disorders, including depression, seizures, and loss of cognitive function in conditions such as Alzheimer's disease. Adenoregulin may be yet another example of a medically important compound from the rapidly dwindling tropical forests.

aposematic signal. Nocturnal animals may have difficulty being conspicuous if they rely on visual signals. One species of dendrobatid frog apparently deters predators with a foul odor (Myers et al. 1991). The aptly named *Aromobates nocturnus* from the cloud forests of the Venezuelan Andes is the only nocturnal dendrobatid. It is an inconspicuous frog, about 5 centimeters long with a dark olive color. The frogs emit a foul, skunk-like odor when they are handled.

Mimicry

The existence of unpalatable animals that deter predators with aposematic colors and behaviors offers the opportunity for other species that lack noxious qualities to take advantage of predators that have learned by experience to avoid the aposematic species. In this phenomenon, known as **mimicry**, the mimic (a species that lacks noxious properties) resembles a noxious model and that resemblance causes a third species, the dupe, to mistake the mimic for the model. Some of the best known cases of mimicry among vertebrates involve salamanders (Pough 1988). One that has been investigated involves two color morphs of the common red-backed salamander, *Plethodon cinereus.*

Red-backed salamanders normally have dark pigment on the sides of the body, but in some regions an erythristic (*erythr* = red) color morph is found that lacks the dark pigmentation and has red-orange on the sides as well as on the back. These erythristic morphs resemble red efts, and could be mimics of efts. Red-backed salamanders are palatable to many predators, and mimicry of the noxious red efts might confer some degree of protection on individuals of the erythristic morph. That hypothesis was tested in a series of experiments (Brodie and Brodie 1980). Salamanders were put in leaf-filled trays from which they could not escape, and the trays were placed in a forest where birds were foraging. The birds learned to search through the leaves in the trays to find the salamanders. This is a very life-like situation for a test of mimicry because some species of birds are important predators of salamanders. For example, red-backed salamanders and dusky salamanders (*Desmognathus ochrophaeus*) made up 25 percent of the prey items fed to their young by hermit thrushes in western New York.

Three species of salamanders were used in the experiments, and the number of each species was adjusted to represent a hypothetical community of salamanders containing 40 percent dusky salamanders, 30 percent red efts, 24 percent striped red-backed salamanders, and 6 percent erythristic red-backed salamanders. The dusky salamanders are palatable to birds and are light brown; they do not resemble either efts or red-backed salamanders and they served as a control in the experiment. The striped red-backed salamanders represent a second control: The hypothesis of mimicry of red efts by erythristic salamanders leads to the prediction that the striped salamanders, which do not look like efts, will be eaten by birds, whereas the erythristic salamanders, which are as palatable as the striped ones but which do look like the noxious efts, will not be eaten.

A predetermined number of each kind of salamander was put in the trays and birds were allowed to forage for two hours. At the end of that time the salamanders that remained were counted. As expected, only 1 percent of the efts had been taken by birds, whereas 44 to 60 percent of the palatable salamanders had disappeared (Table 11-4). As predicted, the birds ate fewer of the erythristic form of the red-backed salamanders than they ate of the striped form.

These results show that the erythristic morph of the red-backed salamander does obtain some protection from avian predators as a result of its resemblance to the red eft. In this case the resemblance is visual, but mimicry can exist in any sensory mode to which a dupe is sensitive. Olfactory mimicry by amphibians might be effective against predators such as shrews and snakes, which rely on scent to find and identify prey. This possibility has scarcely been considered, but careful investigations may yield fascinating new examples.

Why Are Amphibians Vanishing?

The global decline of amphibian populations described in Chapter 1 is alarming, especially because we may have no idea *why* a species has disappeared from places in which it was formerly abundant. In some cases local events appear to provide an explanation. Habitat changes produced by logging are usually destructive to amphibians, for example, because frogs and salamanders depend on cool, moist microhabitats on the forest floor. When the for-

Table 11–4 Differential survival of salamanders exposed to foraging birds.

Experimental Design

Hypothesis: The erythristic morph of *Plethodon cinereus* is a mimic of the red eft.

Predictions:
1. Birds will not eat red efts because the efts are noxious.
2. Birds will readily eat dusky salamanders, which are palatable and not mimetic.
3. Birds will eat the striped *Plethodon,* which are also palatable and not mimetic.
4. Birds will mistake the erythristic *Plethodon* for efts and will not eat them.

Results

Percentage of Salamanders Gone From Trays

		Plethodon	
Red Efts	Dusky Salamanders	Striped	Erythristic
1.0	52.6	60.1	43.9

Interpretation

The predictions of the hypothesis were supported:
1. Birds did not eat the noxious red efts (prediction 1).
2. Birds did eat the palatable, nonmimetic dusky salamanders (prediction 2).
3. Birds ate the striped morph of *Plethodon* (prediction 3).
4. Birds ate significantly fewer of the mimetic morph of *Plethodon* than of the striped morph (prediction 4).

Source: Based on E. D. Brodie, Jr., and E. D. Brodie III, 1980, *Science* 208:181–182.

est canopy is removed, sunlight reaches the ground and conditions become too hot and dry for amphibians. The rock removed from mines releases toxic chemicals, cyanide used to extract gold poisons surface water, oil wells spread toxic hydrocarbons—the list of abuses is nearly endless.

Some local causes of amphibian mortality are not only obvious, they are positively undignified. Federal land in the western United States is leased for grazing, and cattle drink from the ponds that are breeding sites for anurans. As the ponds shrink during the summer, they leave a band of mud that cattle cross when they come to drink. The deep hoof prints the cattle make can be death traps for newly metamorphosed frogs and toads that tumble in and cannot climb out. Even worse, a few juvenile anurans that have the bad luck to pass behind a cow at exactly the wrong moment are trapped and suffocated beneath a pile of fresh manure!

But the truly disturbing questions involve species such as the Costa Rica golden toad (*Bufo periglenes*) and other high-altitude frogs that live in habitats where there is no sign of local environmental damage (Pounds and Crump 1994). The global scope of the problem suggests that we should look for global explanations. Two factors that have probably contributed to some of these declines are acid precipitation and increased ultraviolet radiation.

Precipitation (rain, snow, and fog) over large parts of the world, especially the Northern Hemi-

Figure 11–30 Effects of acidity on embryos of the spotted salamander, *Ambystoma maculatum*. Normal embryonic development is shown on the left, and abnormalities observed in acid conditions are on the right. Highly acidic conditions (pH values lower than about 5) kill embryos in early stages of development, whereas less acidic conditions (pH 5 to 6) produce abnormalities later in development. (From F. H. Pough, 1978, *NAHO* 11(1):6–9. Courtesy of NAHO, Published by the New York State Museum.)

DEVELOPMENT OF SPOTTED SALAMANDER EGGS

NORMAL	ABNORMAL

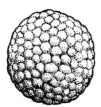

A. Cells divide evenly.

a. When exposed to highly acid conditions, cells divide unevenly.

B. Yolk plug retracts.

b. Yolk plug fails to retract fully and results in "c."

C. Embryo's development includes growth of its posterior portion.

c. Posterior portion of embryo is deformed. Deformities in "a" through "c" are always lethal to an embryo.

D. The gills and forelimbs develop as the embryo reaches a later stage.

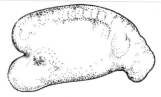

d. Less acid conditions produce damage at a later stage of development, including a swelling of the chest near the heart.

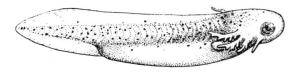

E. Gills become more lobed as the embryo nears hatching.

e. Less acidity also produces stunted gills. Deformities in "d" through "e" are frequently lethal to the embryo.

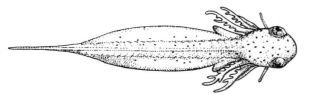

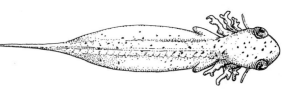

sphere, is at least a hundredfold more acidic than it would be if the water were in equilibrium with carbon dioxide in the air. The acidity is produced by nitric and sulfuric acids that form when water vapor combines with oxides of nitrogen and sulfur released by combustion of fossil fuels. Water in the breeding ponds of many amphibians in the Northern Hemisphere has become more acidic in the past 50 years, and this acidity has both direct and indirect effects on amphibian eggs and larvae (Dunson and Wyman 1992). Embryos of many species of frogs and salamanders are killed or damaged at pH 5 or less (Figure 11–30). Larvae that hatch may be smaller than normal, and sometimes have strange lumps or kinks in their bodies. Spotted salamander larvae grow slowly in acid water because their prey-capture efforts are clumsy and they eat less than do larvae at higher pH (Preest 1993).

Another global phenomenon is an increase in the amount of ultraviolet radiation reaching the Earth's surface as a result of destruction of ozone in the stratosphere by chemical pollutants. The effects are most dramatic at the poles, and are spreading into lower latitudes in both hemispheres. Ultraviolet light, especially the 290- to 320-nanometer UV-B band, kills amphibian eggs and embryos (Blaustein et al. 1994). Only 50 to 60 percent of the eggs of the Cascade frog (*Rana cascadae*) and the western toad (*Bufo boreas*) in ponds in the Cascade Mountains of Oregon survived when they were exposed to incident sunlight, but when a filter that blocked UV-B was placed over the eggs, survival climbed to 70 to 85%. Eggs of the Pacific treefrog (*Hyla regilla*) were not affected by unfiltered sunlight. The enzyme photolyase repairs UV-induced damage to DNA, and the different sensitivity of the three species of anurans to UV-B corresponded to differences in the activity of photolyase in their eggs.

Sensitivity of amphibian eggs to UV-B may be contributing to the decline of some species. Because ozone depletion is a global phenomenon, this mechanism might account for the species that are declining in habitats that show no evidence of local environmental degradation. High-altitude species, which represent some of the most puzzling examples of declines, may be especially vulnerable to thinning of the ozone layer because the intensity of ultraviolet

radiation normally increases with altitude. However, neither ozone depletion nor acid precipitation are yet severe at tropical latitudes, and other mechanisms—still unknown—must be responsible for the disappearance of golden toads and other tropical frogs.

Biologists from many countries met in England in 1989 at the First World Congress of Herpetology. In a week of formal presentations of scientific studies and in casual conversations at meals and in hallways the participants discovered that an alarmingly large proportion of them knew of populations of amphibians that had once been abundant and now were rare, or even entirely gone. Events that had appeared to be isolated instances turned out to be part of a global pattern. As a result of that discovery, David Wake of the University of California at Berkeley, persuaded the National Academy of Sciences to convene a meeting of biologists concerned about vanishing amphibians. Biologists from all over the world met at the West Coast center of the Academy in February 1990. All reported that populations of amphibians in their countries were disappearing, and often there was no apparent reason. Following that meeting, an international effort to identify the causes of amphibian declines was initiated by the Declining Amphibian Populations Task Force of the Species Survival Commission of the World Conservation Union (IUCN). This work is being conducted almost entirely by the voluntary efforts of concerned scientists. Regional Working Groups of the Task Force are composed of scientists who monitor the status of amphibian populations in their areas. Issue-Based Working Groups are responsible for assembling lists of chemical contaminants and climatic factors that are likely to affect amphibian populations and for coordinating the efforts of individual scientists.

Standard methods of surveying populations and marking individual animals are critical for an effort of this sort, which reaches across national boundaries and relies on the efforts of many investigators over periods of years. *Measuring and Monitoring Biological Diversity: Standard Methods for Amphibians* (Heyer et al. 1993) was compiled to provide a framework for studies of vanishing amphibian populations. *Tracking the Vanishing Frogs* (Phillips 1994) traces the discovery of the phenomenon of amphibian population declines, and examines the evidence for various possible causes. FROGLOG, the newsletter of the Task Force, is published at the Department of Biology of The Open University (Walton Hall, Milton Keynes, MK7 6AA, United Kingdom), and is partly funded by Frog's Leap Winery in St. Helena, California.

In 1995 the Declining Amphibian Populations Task Force established an amphibian conservation forum, AmphibianDecline, on the Internet. Topics include information about the status of particular species or populations as well as information about disease, ultraviolet radiation, population genetics, acid precipitation, pesticides, introduction of exotic species, and habitat destruction. Subscribers to the forum receive all e-mail sent to AmphibianDecline, and may also send messages that will automatically be distributed to other AmphibianDecline subscribers. (Anyone with e-mail access to the Internet can subscribe to AmphibianDecline. To subscribe, send a message to listproc@ucdavis.edu. The subject should be Subscribe and the message should read Subscribe AmphibianDecline Firstname Lastname. Use your own first and last name and do not include anything else in the message. You will receive confirmation that your subscription has been received as well as instructions for sending messages to AmphibianDecline and for ending a subscription.)

Probably most population declines represent the combined effects of several factors. Physical conditions, such as acidity or ultraviolet radiation, may interact with biological factors. Amphibians that are stressed by acidity or ultraviolet radiation may be susceptible to diseases or parasites that they would ordinarily be able to resist. Interacting mechanisms of that sort will be hard to identify, and harder still to ameliorate. The prospect for many species of amphibians appears bleak.

Summary

Locomotor adaptations distinguish the lineages of amphibians. Salamanders (Caudata) usually have short, sturdy legs that are used with lateral undulation of the body in walking. Aquatic salamanders use lateral undulations of the body and tail to swim, and some specialized aquatic species are elongate and have very small legs. Frogs and toads (Anura) are characterized by specializations of the pelvis and hindlimbs that permit both legs to be used simultaneously to deliver a powerful thrust used both for jumping and for swimming. Many anurans walk quadrupedally when they move slowly and some are agile climbers. The caecilians (Gymnophiona) are legless tropical amphibians; some are burrowers and others are aquatic.

The diversity of reproductive modes of amphibians exceeds that of any other group of vertebrates except the fishes. Fertilization is internal in derived salamanders, but most frogs rely on external fertilization. All caecilians have internal fertilization. Many species of amphibians have aquatic larvae. Tadpoles, the aquatic larvae of anurans, are specialized for life in still or flowing water, and some species of frogs deposit their tadpoles in very specific sites such as the pools of water that accumulate in the leaf axils of bromeliads or other plants. The specializations of tadpoles are entirely different from the specializations of frogs, and metamorphosis causes changes in all parts of the body. Direct development that omits the larval stage is also widespread among anurans and is often combined with parental care of the eggs. Viviparity occurs in all three orders.

In many respects the biology of amphibians is determined by properties of their skin. Hedonic glands are key elements in reproductive behaviors, poison glands protect the animals against predators, and mucus glands keep the skin moist, facilitating gas exchange. Above all, the permeability of the skin to water limits most amphibians to microhabitats in which they can control water gain and loss. That sounds like a severe restriction, but in the proper microhabitat amphibians can utilize the permeability of their skin to achieve a remarkable degree of independence of standing water. Thus, the picture that is sometimes presented of amphibians as animals barely hanging on as a sort of evolutionary oversight is misleading. Only a detailed examination of all facets of their biology can produce an accurate picture of amphibians as organisms.

An examination of that sort reinforces the view that the skin is a dominant structural characteristic of amphibians. This is true not only in terms of the limitations and opportunities presented by its permeability to water and gases, but also as a result of the intertwined functions of the skin glands in defensive and reproductive behaviors. The structure and function of the skin may be primary characteristics that have shaped the evolution and ecology of amphibians, and may also be responsible for some aspects of their susceptibility to pollution. All over the world populations of amphibians are disappearing at an alarming rate, and some of these extinctions may be caused by regional or global effects of human activities that are likely to affect other organisms as well.

References

Arntzen, J. W., and M. Sparreboom. 1989. A phylogeny for Old World newts, genus *Triturus:* biochemical and behavioural data. *Journal of Zoology* (London) 219:645–664.

Beuchat, C. A., F. H. Pough, and M. M. Stewart. 1984. Response to simultaneous dehydration and thermal stress in three species of Puerto Rican frogs. *Journal of Comparative Physiology* B154:579–585.

Blaustein, A. R., and R. K. O'Hara. 1981. Genetic control of sibling recognition? *Nature* 290:246–248.

Blaustein, A. R., P. D. Hoffman, D. G. Hokit, J. M. Kiesecker, S. C. Walls, and J. B. Hays. 1994. UV repair and resistance to solar UV-B in amphibian eggs: a link to population declines. *Proceedings of the National Academy of Sciences, USA* 91:179–1795.

Brodie, E. D. Jr., and E. D. Brodie III. 1980. Differential avoidance of mimetic salamanders by free-ranging birds. *Science* 208:181–182.

Brust, D. G. 1993. Maternal brood care by *Dendrobates pumilio:* a frog that feeds its young. *Journal of Herpetology* 27:96–98.

Carroll, R. L. 1987. *Vertebrate Paleontology and Evolution.* Freeman, New York, NY.

Daly, J. W., J. Caceres, R. W. Moni, F. Gusovsky, M. Moos, Jr., K. B. Seamon, K. Milton, and C. W. Myers. 1992. Frog secretions and hunting magic in the upper Amazon: identification of a peptide that interacts with an adenosine receptor. *Proceedings of the National Academy of Sciences, USA* 89:10960–10963.

Del Pino, E. M. 1989. Modifications of oogenesis and development in marsupial frogs. *Development* 107:169–187.

Dole, J. W. 1965. Summer movements of adult leopard frogs, *Rana pipiens* Schreber, in northern Michigan. *Ecology* 46:236–255.

Dole, J. W. 1967. The role of substrate moisture and dew in the water economy of leopard frogs, *Rana pipiens. Copeia* 1967:141–149.

Dunson, W. A., and R. L. Wyman (editors). 1992. Amphibian declines and habitat acidification. *Journal of Herpetology* 26:349–442.

Gamboa, G. J., K. A. Berven, R. A. Schemidt, T. G. Fishwild, and K. M. Jankens. 1991. Kin recognition by larval wood frogs (*Rana sylvatica*): effects of diet and prior exposure to conspecifics. *Oecologia* 86:319–324.

Halliday, T. R. 1990. The evolution of courtship behavior in newts and salamanders. *Advances in the Study of Behavior* 19:137–169.

Halliday, T. R., and B. Arano. 1991. Resolving the phylogeny of European newts. *Trends in Ecology and Evolution* 6:113–121.

Heyer, W. R., M. A. Donnelly, R. W. McDiarmid, L.-A. C. Hayek, and M. S. Foster (editors). 1993. *Measuring and Monitoring Biological Diversity: Standard Methods for Amphibians.* Smithsonian Institution Press, Washington, DC.

Jaeger, R. G. 1981. Dear enemy recognition and the costs of aggression between salamanders. *American Naturalist* 117:962–974.

Jaeger, R. G., K. C. B. Nishikawa, and D. E. Barnard. 1983. Foraging tactics of a terrestrial salamander: costs of territorial defense. *Animal Behaviour* 31:191–198.

Jenkins, F. A., Jr., and D. M. Walsh. 1993. An early Jurassic caecilian with limbs. *Nature* 365:246–250.

Klump, G. M., and H. C. Gerhardt. 1987. Use of non-arbitrary acoustic criteria in mate choice by female gray tree frogs. *Nature* 326:286–288.

Leong, A. S.-Y., M. J. Tyler, and D. J. C. Shearman. 1986. Gastric brooding: a new form in a recently discovered Australian frog of the genus *Rheobatrachus. Australian Journal of Zoology* 34:205–209.

Lutz, G. J., and L. C. Rome. 1994. Built for jumping: the design of the frog muscular system: *Science* 263:370–372.

Magnusson, W. E., and J.-M. Hero. 1991. Predation and evolution of complex ovoposition behaviour in Amazon rainforest frogs. *Oecologia* 86:310–318.

Marsh, R. L., and T. L. Taigen. 1987. Properties enhancing aerobic capacity of calling muscles in gray tree frogs, *Hyla versicolor. American Journal of Physiology* 252:R786–R793.

Myers, C. W., J. W. Daly, and B. Malkin. 1978. A dangerously toxic new frog (*Phyllobates*) used by Embera Indians of western Colombia, with discussion of blowgun fabrication and dart poisoning. *Bulletin of the American Museum of Natural History* 161:307–366.

Myers, C. W., A. Paolillio O., and J. W. Daly. 1991. Discovery of a defensively malodorous and noctural frog in the family Dendorbatidae: phylogenetic significance of a new genus and species from the Venezuelan Andes. *American Museum Novitates Number* 3002:1–33.

Pfennig, D. W. 1990a. "Kin recognition" among spadefoot toad tadpoles: a side-effect of habitat selection? *Evolution* 44:785–798.

Pfennig, D. W. 1990b. The adaptive significance of an environmentally-cued developmental switch in an anuran tadpole. Oecologia 85:101–107.

Phillips, K. 1994. *Tracking the Vanishing Frogs*. St. Martin's Press, New York, NY.

Pough, F. H. 1988. Mimicry of vertebrates: are the rules different? *American Naturalist* 131 (Suppl.):S67–S102.

Pough, F. H., T. L. Taigen, M. M. Stewart, and P. F. Brussard. 1983. Behavioral modification of evaporative water loss by a Puerto Rican frog. *Ecology* 64:244–252.

Pounds, J. A., and M. L. Crump. 1994. Amphibian declines and climate disturbance: the case of the golden toad and harlequin frog. *Conservation Biology* 8:72–85.

Preest, M. R. 1993. Mechanism of growth rate reduction in acid-exposed larval salamanders, *Ambystoma maculatum*. *Physiological Zoology* 66:686–707.

Roth, G., and D. B. Wake. 1985. Trends in the functional morphology and sensorimotor control of feeding behavior in salamanders: an example of the role of internal dynamics in evolution. *Acta Biotheoretica* 34:175–192.

Ryan, M. J. 1985. *The Túngara Frog: A Study in Sexual Selection and Communication*. University of Chicago Press, Chicago, IL.

Taigen, T. L., F. H. Pough, and M. M. Stewart. 1984. Water balance of terrestrial anuran eggs (*Eleutherodactylus coqui*): importance of parental care. *Ecology* 65:248–255.

Taigen, T. L., and K. D. Wells. 1985. Energetics of vocalization by an anuran amphibian (*Hyla versicolor*). *Journal of Comparative Physiology* B155:163–170.

Tilley, S. G., B. L. Lundrigan, and L. P. Brower. 1982. Erythrism and mimicry in the salamander *Plethodon cinereus*. *Herpetologica* 38:409–417.

Townsend, D. S., M. M. Stewart, F. H. Pough, and P. F. Brussard. 1981. Internal fertilization in an oviparous frog. *Science* 212:465–471.

Townsend, D. S., M. M. Stewart, and F. H. Pough. 1984. Male parental care and its adaptive significance in a neotropical frog. *Animal Behaviour* 32:421–431.

Tyler, M. J. (editor). 1983. *The Gastric Brooding Frog*. Croom Helm, Beckenham, Kent, UK.

Wake, D. B., and S. B. Marks. 1993. Development and evolution of plethodontid salamanders: a review of prior studies and a prospectus for future research. *Herpetologica* 49:194–203.

Waldman, B. 1982. Sibling association among schooling toad tadpoles, field evidence and implications. *Animal Behaviour* 30:700–713.

Wassersug, R. J. 1975. The adaptive significance of the tadpole stage with comments on the maintenance of complex life cycles in anurans. *American Zoologist* 15:405–417.

Wells, K. D. 1980. Evidence for growth of tadpoles during parental transport in *Colostethus inguinalis*. *Journal of Herpetology* 14:428–430.

Wells, K. D., and T. L. Taigen. 1986. The effect of social interactions on calling energetics in the grey treefrog (*Hyla versicolor*). *Behavioral Ecology and Sociobiology* 19:9–18.

12 TURTLES

Turtles provide a contrast to amphibians in the relative lack of diversity in their life histories. All turtles lay eggs and none exhibit parental care. Turtles show morphological specializations associated with terrestrial, freshwater, and marine habitats, and marine turtles make long-distance migrations that rival those of birds. Probably, turtles and birds use many of the same navigation mechanisms to find their way. Most turtles are long-lived animals with relatively poor capacities for rapid population growth, and many, especially sea turtles and large tortoises, are endangered by human activities. Some efforts to conserve turtles have apparently been frustrated by a peculiarity of the embryonic development of some species of turtles—the sex of an individual is determined by the temperature to which it is exposed in the nest. This experience emphasizes the critical importance of information about the basic biology of animals to successful conservation and management.

Everyone Recognizes a Turtle

Turtles found a successful approach to life in the Triassic and have scarcely changed since. The shell, which is the key to their success, has also limited the diversity of the group (Table 12–1). For obvious reasons flying or gliding turtles have never existed, and even arboreality is only slightly developed. Shell morphology reflects the ecology of turtle species: The most terrestrial forms, the tortoises of the family Testudinidae, have high domed shells and elephant-like feet (Figure 12–1a). Smaller species of tortoises may show adaptations for burrowing. The gopher tortoises of North America are an example. The forelegs are flattened into scoops and the dome of the shell is reduced. The Bolson tortoise of northern Mexico constructs burrows a meter or more deep and several meters long in the hard desert soil. These tortoises bask at the mouths of their burrows, and when a predator appears they throw themselves down the steep entrance tunnels of the burrows to

escape, just as an aquatic turtle dives off a log. The pancake tortoise of Africa is a radical departure from the usual tortoise morphology (Figure 12–1b). The shell is flat and flexible because its ossification is much reduced. This turtle lives in rocky foothill regions and scrambles over the rocks with nearly as much agility as a lizard. When threatened by a predator, the pancake tortoise crawls into a rock crevice and uses its legs to wedge itself in place. The flexible shell presses against the overhanging rock and creates so much friction that it is almost impossible to pull the tortoise out.

Other terrestrial turtles have moderately domed **carapaces** (upper shells), like the box turtles of the family Emydidae (Figure 12–1c). This is only one of several kinds of turtles that have evolved flexible regions in the **plastron** (lower shell) that allow the front and rear lobes to be pulled upward to close the openings of the shell. Aquatic turtles have low carapaces that offer little resistance to movement through water. The Emydidae and Bataguridae contain a large number of pond turtles (Figure 12–1d), including the painted turtles and the red-eared turtles often seen in pet stores and anatomy and physiology laboratory courses.

The snapping turtles (family Chelydridae) and the mud and musk turtles (family Kinosternidae) prowl along the bottoms of ponds and slow rivers and are not particularly streamlined (Figure 12–1f, g). The mud turtle has a hinged plastron, but the musk and snapping turtles have very reduced plastrons. They rely on strong jaws for protection. A reduction in the size of the plastron makes these species more agile than most turtles, and musk turtles may climb several feet into trees, probably to bask. If a turtle falls on your head while you are canoeing, it is probably a musk turtle.

The soft-shelled turtles (family Trionychidae) are fast swimmers (Figure 12–1e). The ossification of the shell is greatly reduced, lightening the animal, and the feet are large with extensive webbing. Soft-shelled turtles lie in ambush partly buried in debris on the bottom of the pond. Their long necks allow them to reach considerable distances to seize the invertebrates and small fish on which they feed.

The two lineages of living turtles can be traced through fossils to the Mesozoic (Gaffney et al. 1987, Gaffney and Kitching 1994). The **cryptodires** (*crypto* = hidden, *dire* = neck) retract the head into the shell by bending the neck in a vertical S-shape. The **pleurodires** (*pleuro* = side) retract the head by bending the neck horizontally. All of the turtles discussed so far have been cryptodires, and these are the dominant group of turtles. Cryptodires are the only turtles now found in most of the Northern Hemisphere, and there are aquatic and terrestrial cryptodires in South America and terrestrial ones in Africa. Only Australia has no cryptodires. Pleurodires are now found only in the Southern Hemisphere, although they had worldwide distribution in the late Mesozoic and early Cenozoic. *Stupendemys*, a pleurodire from the Pliocene of Venezuela, had a carapace more than 2 meters long. All the living pleurodires are at least semiaquatic, but some fossil pleurodires had high, domed shells that suggest they may have been terrestrial. The most terrestrial of the living pleurodires are probably the African pond turtles (Figure 12–1h), which readily move overland from one pond to another.

The snake-necked pleurodiran turtles (family Chelidae) are found in South America, Australia, and New Guinea (Figure 12–1i). As their name implies, they have long, slender necks. In some species the length of the neck is considerably greater than that of the body. These forms feed on fish that they catch with a sudden dart of the head. Other snake-necked turtles have much shorter necks. Some of these feed on mollusks and have enlarged palatal surfaces used to crush shells. The same specialization for feeding on mollusks is seen in certain cryptodiran turtles.

An unusual feeding method among turtles is found in a pleurodire, the matamata of South America (Figure 12–1j). Large matamatas reach shell lengths of 40 centimeters. They are bizarre-looking animals. The shell and head are broad and flattened, and numerous flaps of skin project from the sides of the head and the broad neck. To these are added trailing bits of adhering algae. The effect is exceedingly cryptic. It is hard to recognize a matamata as a turtle even in an aquarium; against the mud and debris of a river bottom they are practically invisible. In addition to obscuring the shape of the turtle, the flaps of skin on the head are sensitive to minute vibrations in water caused by the passage of a fish. When it senses the presence of prey, the matamata abruptly opens its mouth and expands its throat. Water rushes in carrying the prey with it, and the matamata closes its mouth, expels the water, and swallows the prey. The matamata lacks the horny beak that other turtles use for seizing prey or biting off pieces of plants.

Living marine turtles are cryptodires. The families Cheloniidae and Dermochelyidae show more extensive specialization for aquatic life than any freshwater turtle. All have the forelimbs modified as flippers. Cheloniids retain epidermal scutes on the shell (Figure 12–1k). The largest of the sea turtles of the family Cheloniidae is the loggerhead, which once reached weights exceeding 400 kilograms. The largest marine turtle, the leatherback, reaches shell lengths of more than 2 meters and weights exceeding 600 kilograms (Figure 12–1l). The dermal ossification has been reduced to bony platelets embedded in the skin. This is a pelagic turtle that ranges far from land, and it has a wider geographic distribution than any other ectothermal amniote: Leatherback turtles penetrate far into cool temperate seas, and have been recorded in the Atlantic from Newfoundland to Argentina and from Norway to South Africa, and in the Pacific from the USSR to Tasmania. In cold water leatherback turtles maintain body temperatures substantially above water temperature, using counter-

Figure 12–1 Body forms of turtles: (a) tortoise, *Testudo*; (b) pancake tortoise, *Malacochersus*; (c) terrestrial box turtle, *Terrapene*; (d) pond turtle, *Trachemys*; (e) softshell turtle, *Aplone*; (f) mud turtle, *Kinosternon*; (g) alligator snapping turtle, *Macroclemys*; (h) African pond turtle, *Pelusios*; (i) Australian snake-necked turtle, *Chelodina*; (j) South American matamata, *Chelys*; (k) loggerhead sea turtle, *Caretta*; (l) leatherback sea turtle, *Dermochelys*.

current heat exchangers to retain the heat released by muscular activity (Paladino et al. 1990). Leatherback turtles dive to depths of more than 1000 meters. One dive that drove the depth recorder off-scale is estimated to have reached 1200 meters, which exceeds the deepest dive recorded for a sperm whale (1140 meters). Leatherback turtles feed largely on jellyfish, whereas the smaller hawksbill sea turtles eat sponges that are defended by spicules of silica (glass) as well as a variety of chemicals (including alkaloids and terpenes) that are toxic to most vertebrates (Meylan 1988).

Phylogenetic Relationships of Turtles

Nearly 200 of the approximately 250 species of turtles are cryptodires (Figure 12–2). Turtles show a combination of ancestral features and highly specialized characters that are not shared with any other group of vertebrates, and their phylogenetic affinities are not clearly known. The turtle lineage probably originated among the early amniotes of the late Carboniferous. Like those animals, turtles have **anapsid** ("without an arch") skulls, but the shells and postcranial skeletons of turtles are unique. Two of the groups of parareptiles that were discussed in Chapter 10, the procolophonids and the parieasaurs, have been proposed as the sister group for turtles. At the moment the case for parieasaurs appears stronger, but a lively debate is in progress and the issue is far from resolved (Reisz and Laurin 1991, Lee 1993).

The earliest turtles are found in late Triassic deposits in Germany, Thailand, and Argentina. These animals had nearly all of the specialized characteristics of derived turtles, and shed no light on the phylogenetic affinities of the group. *Proganochelys* from the German Triassic beds was larger than most living turtles, with a high, arched shell nearly a meter long. The marginal teeth had been lost and the maxilla, premaxilla, and dentary bones were probably covered with a horny beak as they are in derived turtles. The skull of *Proganochelys* retained the supratemporal and lacrimal bones and the lacrimal duct, and the palate had rows of denticles; all these structures have been lost by derived turtles. The plastron of *Proganochelys* also contained some bones that have been lost by derived turtles, and the vertebrae of the neck lack specializations that would have allowed the head to be retracted into the shell. *Palaeochersis* (late Triassic of Argentina) and *Australochelys* (early Jurassic of South Africa) are probably the sister group of later turtles, including Cryptodira + Pleurodira (Gaffney and Kitching 1994, Rougier et al. 1995).

Turtles with neck vertebrae specialized for retraction are not known before the Cretaceous, but differences in the skulls and shells allow the pleurodiran and cryptodiran lineages to be traced back to the late Triassic (shells of pleurodires) and late Jurassic (skulls of cryptodires). The otic capsules of all turtles beyond the proganochelids are enlarged, and the jaw adductor muscles bend posteriorly over the otic capsule (Figure 12–3). The muscles pass over a pulley-like structure, the **trochlear process.** In cryptodires the trochlear process is formed by the anterior surface of the otic capsule itself, whereas in pleurodires it is formed by a lateral process of the pterygoid. Fusion of the pelvic girdle to the carapace and plastron distinguishes pleurodires from cryptodires, which have a suture attaching the shell to the girdle. The beginnings of these changes are seen in *Australochelys.*

Turtle Structure and Functions

Turtles are among the most bizarre vertebrates: Covered in bone, with the limbs inside the ribs, and with horny beaks instead of teeth—if turtles had become extinct at the end of the Mesozoic they would rival dinosaurs in their novelty. However, because they survived they are regarded as commonplace, and are used in comparative anatomy courses to represent ectothermal amniotes (inappropriately, because they are so specialized). In fact, much of the morphology and physiology of turtles is derived rather than ancestral, and turtles are ecologically quite different from the other ectothermal amniotes, the crocodilians (which are archosaurs) and the lepidosaurs (tuatara, lizards, snakes, and amphisbaenians).

Shell and Skeleton

The shell is the most distinctive feature of a turtle (Figure 12–4). The carapace is composed of dermal bone that typically grows from 59 separate centers of ossification. Eight plates along the dorsal midline form the neural series and are fused to the neural arches of the vertebrae. Lateral to the neural bones are eight paired costal bones, which are fused to the broadened ribs. The ribs of turtles are unique among tetrapods in being external to the girdles. Eleven pairs of peripheral bones, plus two unpaired bones in the dorsal midline, form the margin of the carapace. The plastron is formed largely from dermal ossifications, but the entoplastron is derived from the interclavicle, and the paired epiplastra anterior to it are derived from the clavicles. Processes from the hypoplastron fuse with the first and fifth pleurals, forming a rigid connection between the plastron and carapace.

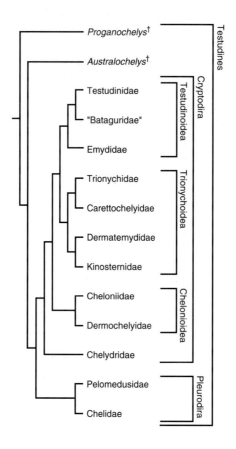

Figure 12–2 Phylogeny of turtles. The "Bataguridae" may not be monophyletic. Extinct forms are marked with a dagger (†). (Based on E. S. Gaffney and P. A. Moylan, 1988, pages 1157–1219 in *The Phylogeny and Classification of the Tetrapods*, Vol. 1, edited by M. J. Benton, Oxford University Press, Oxford, UK; and David Hillis and Alan Savitzky, personal communication.)

The bones of the carapace are covered by horny scutes of epidermal origin that do not coincide in number or position with the underlying bones. The carapace has a row of five central scutes, bordered on each side by four lateral scutes. Eleven marginal scutes on each side turn under the edge of the carapace. The plastron is covered by a series of six paired scutes.

Flexible areas, called hinges, are present in the shells of many turtles. The most familiar examples are the North American and Asian box turtles (*Terrapene* and *Cuora*) in which a hinge between the hyoplastral and hypoplastral bones allows the anterior and posterior lobes of the plastron to be raised to close off the front and rear openings of the shell. Mud turtles (*Kinosternon*) have two hinges in the plastron: The anterior hinge runs between the epiplastra and the entoplastron (which is triangular in kinosternid turtles rather than diamond-shaped) and the posterior hinge is between the hypoplastron and xiphiplastron. In the pleurodiran turtle *Pelusios* a hinge runs between the mesoplastron and the hypoplastron. Some species of tortoises have plastral hinges; in *Testudo* the hinge lies between the hyoplastron and xiphiplastra as it does in *Kinosternon*, but in another genus of tortoise, *Pyxis*, the hinge is anterior and involves a break across the entoplastron. The African forest tortoises (*Kinixys*) have a hinge on the posterior part of the carapace. The margins of the epidermal shields and the dermal bones of the carapace are aligned, and the hinge runs between the second and third pleural bones and the fourth and fifth costals. The presence of hinges is sexually dimorphic in some species of tortoises. The erratic phylogenetic occurrence of kinetic shells and differences among related species indicates that shell kinesis has evolved many times in turtles.

Asymmetry of the paired epidermal scutes is quite common among turtles, and modifications of the bony structure of the shell are seen in some families. Soft-shelled turtles lack peripheral ossifications and epidermal scutes. The distal ends of the broadened ribs are embedded in flexible connective tissue, and the carapace and plastron are covered with skin. The New Guinea river turtle (*Carretochelys*) is also covered by skin instead of scutes, but in this species the peripheral bones are present and the edge of the shell is stiff. The leatherback sea turtle (*Dermochelys*) has a carapace formed of cartilage with thousands of small polygonal bones embedded in it, and the plastral bones are reduced to a thin rim around the edge of the plastron. The neural and pleural ossifications of the pancake tortoise (*Malacochersus*) are greatly reduced, but the epidermal plates are well developed.

Derived turtles have only 18 presacral vertebrae, 10 in the trunk and 8 in the neck. The centra of the trunk vertebrae are elongated and lie beneath the dermal bones in the dorsal midline of the shell. The centra are constricted in their centers and fused to each other. The neural arches in the anterior two-thirds of the trunk lie between the centra as a result of anterior displacement, and the spinal nerves exit near the middle of the preceding centrum. The ribs are also shifted anteriorly; they articulate with the anterior part of the neurocentral boundary, and in the anterior part of the trunk, where the shift is most pronounced, the ribs extend onto the preceding vertebra.

Cryptodires have two sacral vertebrae (the 19th and 20th vertebrae) with broadened ribs that meet the ilia of the pelvis. Pleurodires have the pelvic girdle firmly fused to the dermal carapace by the ilia dorsally and by the pubic and ischial

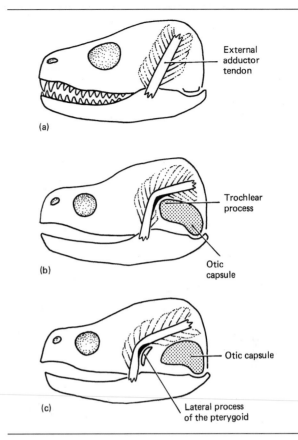

Figure 12–3 Position of the external adductor tendon in (a) the ancestral (parareptilian) condition, (b) cryptodiran turtles, and (c) pleurodiran turtles. (From E. S. Gaffney, 1975, *Bulletin of the American Museum of Natural History* 15:387–436.)

bones ventrally, and the sacral region of the vertebral column is less distinct. The ribs on the 17th, 18th, 19th, and sometimes the 20th vertebrae are fused to the centra and end on the ilia or the ilium–carapacial junction.

The cervical vertebrae of cryptodires have articulations that permit the S-shaped bend used to retract the head into the shell. Specialized condyles (ginglymes) permit vertical rotation. This type of rotation, **ginglymoidy**, is peculiar to cryptodires, but the anatomical details vary within the group. In most families the hinge is formed by two successive ginglymoidal joints between the 6th and 7th and the 7th and 8th cervical vertebrae. The lateral bending of the necks of pleurodiran turtles is accomplished by ball-and-socket or cylindrical joints between adjacent cervical vertebrae.

Figure 12–4 Shell and vertebral column of a turtle: (a) epidermal scutes of the carapace (left) and plastron (right); (b) dermal bones of the carapace (left) and plastron (right); (c) vertebral column of a turtle, seen from the inside of the carapace. Note that anteriorly the ribs articulate with two vertebral centra. (From R. Zangerl, 1969, in *Biology of the Reptilia*, volume 1, edited by C. Gans, A. d'A. Bellairs, and T. S. Parsons, Academic, London, UK.)

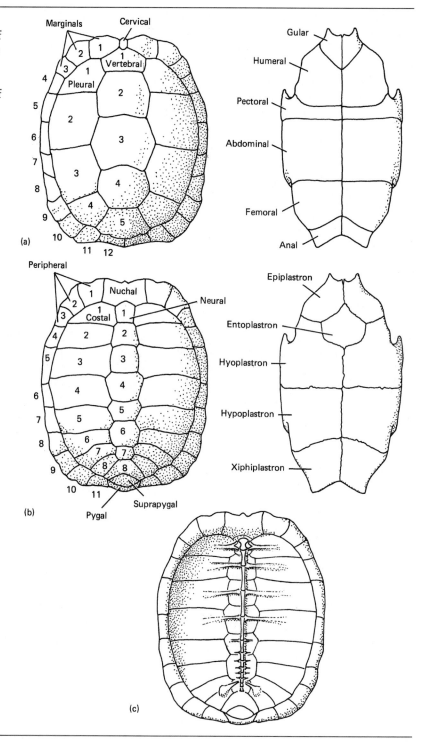

The Heart

The circulatory systems of tetrapods can be viewed as consisting of two circuits: The systemic circuit carries oxygenated blood from the heart to the head, trunk, and appendages, whereas the pulmonary circuit carries deoxygenated blood from the heart to the lungs. The blood pressure in the systemic circuit is higher than the pressure in the pulmonary circuit, and the two circuits operate in series. That is, blood flows from the heart through the lungs, back to the heart, and then to the body. The morphology of the hearts of birds and mammals makes this sequential flow obligatory, but the hearts of turtles, squamates, and amphibians have the ability to shift blood between the pulmonary and systemic circuits.

This flexibility in the route of blood flow can be accomplished because the ventricular chambers in the hearts of turtles and squamates are in anatomical continuity instead of being divided by a septum like the ventricles of birds and mammals. The pattern of blood flow can best be explained by considering the morphology of the heart and how intracardiac pressure changes during a heartbeat. Figure 12–5 shows a schematic view of the heart of a turtle or squamate. The left and right atria are completely separate, and three subcompartments can be distinguished in the ventricle. A muscular ridge in the core of the heart divides the ventricle into two spaces, the **cavum pulmonale** and the **cavum venosum.** The muscular ridge is not fused to the wall of the ventricle, and thus the cavum pulmonale and the cavum venosum are only partly separated. A third subcompartment of the ventricle, the **cavum arteriosum,** is located dorsal to the cavum pulmonale and cavum venosum. The cavum arteriosum communicates with the cavum venosum through an intraventricular canal. The pulmonary artery opens from the cavum pulmonale, and the left and right aortic arches open from the cavum venosum.

The right atrium receives deoxygenated blood from the body via the sinus venosus and empties into the cavum venosum, and the left atrium receives oxygenated blood from the lungs and empties into the cavum arteriosum. The atria are separated from the ventricle by flap-like atrioventricular valves that open as the atria contract, and then close as the ventricle contracts, preventing blood from being forced back into the atria. The anatomical arrangement of the connections between the atria, their valves, and the three subcompartments of the ventricle are crucial, because it is those connections that allow pressure differentials to direct the flow of blood and to prevent mixing of oxygenated and deoxygenated blood.

When the atria contract the atrioventricular valves open, allowing blood to flow into the ventricle. Blood from the right atrium flows into the cavum venosum, and blood from the left atrium flows into the cavum arteriosum. At this stage in the heartbeat the large median flaps of the valve between the right atrium and the cavum venosum are pressed against the opening of the intraventricular canal, sealing it off from the cavum venosum. As a result, the oxygenated blood from the left atrium is confined to the cavum arteriosum. Deoxygenated blood from the right atrium fills the cavum venosum and then continues over the muscular ridge into the cavum pulmonale.

When the ventricle contracts, the blood pressure inside the heart increases. Ejection of blood from the heart into the pulmonary circuit precedes flow into the systemic circuit because resistance is lower in the pulmonary circuit. As deoxygenated blood flows out of the cavum pulmonale into the pulmonary artery, the displacement of blood from the cavum venosum across the muscular ridge into the cavum pulmonale continues. As the ventricle shortens during contraction, the muscular ridge comes into contact with the wall of the ventricle and closes off the passage for blood between the cavum venosum and cavum pulmonale.

Simultaneously, blood pressure inside the heart increases, and the flaps of the right atrioventricular valve are forced into the closed position, preventing backflow of blood from the cavum venosum into the atrium. When the valve closes, it no longer blocks the intraventricular canal. Oxygenated blood from the cavum pulmonale can now flow through the intraventricular canal and into the cavum venosum. At this stage in the heartbeat, the wall of the ventricle is pressed firmly against the muscular ridge, separating the oxygenated blood in the cavum venosum from the deoxygenated blood in the cavum pulmonale.

As the pressure in the ventricle continues to rise, the oxygenated blood in the cavum venosum is ejected into the aortic arches. This system effectively prevents mixing of oxygenated and deoxygenated blood in the heart, despite the absence of a permanent morphological separation of the two circuits.

Respiration

Primitive amniotes probably used movements of the rib cage to draw air into the lungs and to force it out, and lizards still employ that mechanism. The fusion of the ribs of turtles with their rigid shells makes

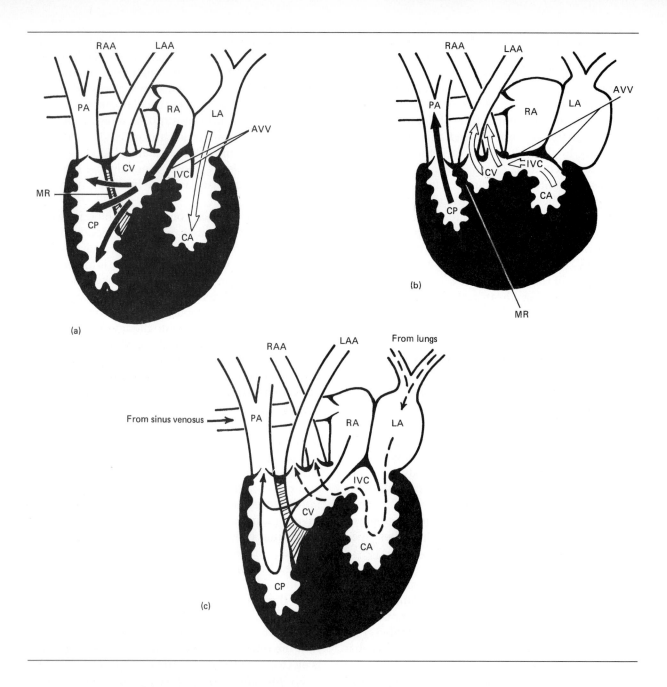

Part III   Terrestrial Ectotherms: Amphibians, Turtles, Crocodilians, and Squamates

that method of breathing impossible. Only the openings at the anterior and posterior ends of the shell contain flexible tissues. The lungs of a turtle, which are large, are attached to the carapace dorsally and laterally. Ventrally, the lungs are attached to a sheet of nonmuscular connective tissue that is itself attached to the viscera (Figure 12–6). The weight of the viscera keeps this diaphragmatic sheet stretched downward.

Turtles produce changes in pressure in the lungs by contraction of muscles that force the viscera upward, compressing the lungs and expelling air, followed by contraction of other muscles that increase the volume of the visceral cavity, allowing the viscera to settle downward. Because the viscera are attached to the diaphragmatic sheet, which in turn is attached to the lungs, the downward movement of the viscera expands the lungs, drawing in air. In tortoises both inhalation and exhalation require muscular activity. The viscera are forced upward against the lungs by

Figure 12–5 Blood flow in the heart of a turtle. (a) As the atria contract oxygenated blood (open arrows) from the left atrium (LA) enters the cavum arteriosum (CA) while deoxygenated blood (dark arrows) from the right atrium (RA) first enters the cavum venosum (CV) and then crosses the muscular ridge (MR) and enters the cavum pulmonale (CP). The atrioventricular valve (AVV) blocks the intraventricular canal (IVC) and prevents mixing of oxygenated and deoxygenated blood. (b) As the ventricle contracts the deoxygenated blood in the cavum pulmonale is expelled through the pulmonary arteries; the AVV closes, no longer obstructing the ICV; and the oxygenated blood in the cavum arteriosum is forced into the cavum venosum and expelled through the aortic arches. The adpression of the wall of the ventricle to the muscular ridge prevents mixing of deoxygenated and oxygenated blood. (c) Summary of the pattern of blood flow through the heart of a turtle. (Modified from N. Heisler et al., 1983, *Journal of Experimental Biology* 105:15–32.)

the contraction of the transverse abdominus muscle posteriorly and the pectoralis muscle anteriorly. The transverse abdominus inserts on the cup-shaped connective tissue, the **posterior limiting membrane,** that closes off the posterior opening of the visceral cavity. Contraction of the transverse abdominus flattens the cup inward, thereby reducing the volume of the visceral cavity. The pectoralis draws the shoulder girdle back into the shell, further reducing the volume of the visceral cavity.

The inspiratory muscles are the abdominal oblique, which originates near the posterior margin of the plastron and inserts on the external side of the posterior limiting membrane, and the serratus, which originates near the anterior edge of the carapace and inserts on the pectoral girdle. Contraction of the abdominal oblique pulls the posterior limiting membrane outward, and contraction of the serratus rotates the pectoral girdle outward. Both of these movements increase the volume of the visceral cavity, allowing the viscera to settle back downward and causing the lungs to expand. The in-and-out movements of the forelimbs and the soft tissue at the rear of the shell during breathing are conspicuous.

The basic problem of respiring within a rigid shell are the same for most turtles, but the mechanisms show some variation. For example, aquatic turtles can use the hydrostatic pressure of water to help move air in and out of the lungs. In addition, many aquatic turtles are able to absorb oxygen and release carbon dioxide to the water. The pharynx and cloaca appear to be the major sites of aquatic gas exchange. In 1860, in his "Contributions to the Natural History of the U. S. A.," Louis Agassiz pointed out that the pharynx of softshell turtles contains fringe-like processes and suggested that these structures are used for underwater respiration. Subsequent study has shown that when softshell turtles are confined under water they use movements of the hyoid apparatus to draw water in and out of the pharynx, and that pharyngeal respiration accounts for most of the oxygen absorbed from the water. The Australian turtle *Rheodytes leukops* uses cloacal respiration. Its cloacal orifice is as much as 30 millimeters in diameter, and the turtle holds it open. Large bursae (sacks) open from the wall of the cloaca, and the bursae have a well-vascularized lining with numerous projections (villi). The turtle pumps water in and out of the bursae at rates of 15 to 60 times per minute. Captive turtles rarely surface to breathe, and experiments have shown that the rate of oxygen uptake through the cloacal bursae is very high.

Patterns of Circulation and Respiration

The morphological complexity of the hearts of turtles and of squamates allows them to adjust blood flow through the pulmonary and systemic circuits to meet short-term changes in respiratory requirements. The key to these adjustments is changing pressures in the systemic and pulmonary circuits.

Recall that in the turtle heart deoxygenated blood from the right atrium normally flows from the cavum venosum across the muscular ridge and into the cavum pulmonale. As the ventricle contracts, the blood pressure inside the heart increases and blood is first ejected into the pulmonary artery, because the resistance to flow in the pulmonary cir-

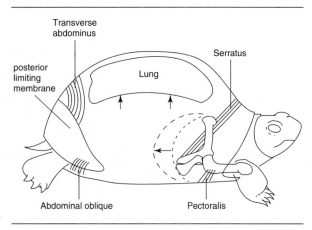

Figure 12–6 Schematic view of the lungs and respiratory movements of a tortoise. (Modified from C. Gans and G. M. Hughes, 1967, *Journal of Experimental Biology* 47:1–20.)

cuit is normally less than the resistance in the systemic circuit. However, the resistance to blood flow in the pulmonary circuit can be increased by muscles that narrow the diameter of blood vessels. When this happens, the delicate balance of pressure in the heart that maintained the separation of oxygenated and deoxygenated blood is changed. When the resistance of the pulmonary circuit is essentially the same as that of the systemic circuit, blood flows out of the cavum pulmonale and cavum venosum at the same time and some deoxygenated blood bypasses the lungs and flows into the systemic circuit (Figure 12–7). This process is called a **right-to-left intracardiac shunt.** Right-to-left refers to the shift of deoxygenated blood from the pulmonary into the systemic circuit, and intracardiac means that it occurs in the heart rather than by flow between the major blood vessels. (Left-to-right shunts also occur, but they are beyond the scope of this discussion. More information can be found in Burggren [1987].)

Why would it be useful to divert deoxygenated blood from the lungs into the systemic circulation? The ability to make this shunt is not unique to turtles—it occurs also among squamate reptiles and in crocodilians. The heart morphology of squamates is very like that of turtles and the same mechanism of changing pressures in the pulmonary and systemic circuits is used to achieve an intracardiac shunt. Crocodilians have hearts in which the ventricle is permanently divided into right and left halves by a septum, and they employ a different mechanism

to achieve a right-to-left shunt (Chapter 13).

The widespread occurrence of blood shunts among amniotic ectotherms suggests that the ability has important consequences for the animals, and one of these has already been discussed in Chapter 4: Lizards and crocodilians use a right-to-left shunt of blood for thermoregulation. By increasing systemic blood flow as they are warming, they increase the transport of heat from the limbs and body surface into the core of the body, thereby warming more rapidly.

The most general function for blood shunts may lie in the ability they provide to match patterns of lung ventilation and pulmonary gas flow (Burggren 1987). Squamates, crocodilians, and turtles normally breathe intermittently, and periods of lung ventilation alternate with periods of **apnea** (no breathing). Turtles are particularly prone to periods of apnea because their method of lung ventilation means they cannot breathe when they withdraw their heads and legs into their shells. Aquatic turtles, like other aquatic animals, are apneic when they dive, and right-to-left blood shunts may occur during diving. Limiting blood flow to the lungs during periods of apnea may permit more effective use of the pulmonary store of oxygen.

Temperature Regulation of Turtles

Turtles are ectotherms, and like lizards and crocodilians, they can achieve a considerable degree of stability in body temperature by regulating their exchange of heat energy with the environment. Turtles basking on a log in a pond are a familiar sight in many parts of the world, and this basking probably has primarily a thermoregulatory function. The body temperatures of basking pond turtles are higher than water and air temperatures, and these higher temperatures may speed digestion, growth, and the production of eggs. In addition, aerial basking may help aquatic turtles to rid themselves of algae and leeches. A few turtles are quite arboreal; these turtles have small plastrons that provide considerable freedom of movement for the limbs. The big-headed turtle (*Platysternon megacephalum*), a chelydrid from southeast Asia, lives in fast-flowing streams at high altitudes and is said to climb on rocks and trees to bask. In North America musk turtles (*Sternotherus*) bask on overhanging branches and drop into the water when they are disturbed.

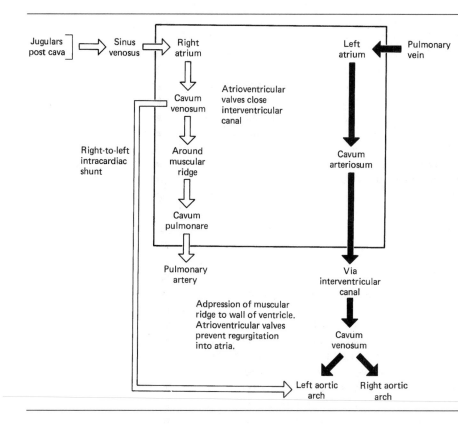

Figure 12–7 Diagram of the right-to-left shunt of blood in the heart of a turtle. Light arrows show deoxygenated blood and dark arrows show oxygenated blood. The box encloses the cycle of events during normal blood flow. (Compare to Figure 12–5.) When resistance in the pulmonary circuit increases, some deoxygenated blood from the cavum venosum flows into the left aortic arch instead of into the pulmonary arteries. (Modified from M. S. Gordon et al., 1982, *Animal Physiology: Principles and Adaptations*, 4th edition, Macmillan, New York, NY.)

Terrestrial turtles can thermoregulate by moving between sun and shade. Small tortoises warm and cool quite rapidly, and they appear to behave very much like other small ectotherms in selecting suitable microclimates for thermoregulation. Familiarity with a home range may facilitate this type of thermoregulation. A study conducted in Italy compared the thermoregulation of resident Hermann's tortoises (animals living in their own home ranges) with individuals that were brought to the study site and tested before they had learned their way around (Chelazzi and Calzolai 1986). The resident tortoises warmed faster and maintained more stable shell temperatures than did the strangers.

The large body size of many tortoises provides a considerable thermal inertia, and large species like the Galapagos and Aldabra tortoises heat and cool slowly. The giant tortoises of Aldabra Atoll (*Geochelone gigantea*) allow their body temperatures to rise to 32 to 33°C on sunny days and they cool to 28 to 30°C overnight. Aldabra tortoises weigh 60 kilograms or more, and even for these large animals, overheating can be a problem. The difficulty is particularly acute for some tortoises on Grande Terre (Swingland and Frazier 1979, Swingland and Lessells 1979). During the rainy season each year a portion of this population moves from the center of the island to the coast. The migrant turtles gain access to a seasonal flush of plant growth on the coast, and migrant females are able to lay more eggs than females that remain inland. However, shade is limited on the coast and the rainy season is the hottest time of the year. Tortoises must restrict their activity to the vicinity of patches of shade, which may be no more than a single tree in the midst of a grassy plain. During the morning tortoises forage on the plain, but as their temperatures rise they move back to the shade of the tree. Competition for shade appears to result in death from overheating of smaller tortoises, especially females.

Marine turtles are large enough to achieve a considerable degree of endothermy (Spotila and Standora 1985). A body temperature of 37°C was recorded by telemetry from a green turtle swimming in water that was 29°C. The leatherback turtle is the largest living turtle; adults may weigh more than 600 kilograms. It ranges far from warm equatorial regions, and in the summer can be found off the coasts of New England and Nova Scotia in water as cool as 8 to 15°C. Body temperatures of these turtles appear to be 18°C or more above water temperatures, and a countercurrent arrangement of blood vessels in the flippers may contribute to retaining heat produced by muscular activity.

Ecology and Behavior of Turtles

Turtles are long-lived animals. Even small species like the painted turtle do not mature until they are 7 or 8 years old, and they may live to be 14 or older. Larger species of turtles live longer. Estimates of centuries for the life spans of tortoises are exaggerated, but large tortoises and sea turtles may live as long as humans, and even box turtles may live over 50 years. These longevities make the life histories of turtles hard to study. Furthermore, a long lifetime is usually associated with a low replacement rate of individuals in the population, and species with those characteristics are at risk of extinction when hunting or habitat destruction reduces their numbers. Conservation efforts for sea turtles and tortoises are especially important areas of concern.

Social Behavior and Courtship

Tactile, visual, and olfactory signals are employed by turtles during social interactions. Many of the pond turtles of North America have distinctive stripes of color on their heads, necks, and on the forelimbs and hind limbs and the tail. These patterns are used by herpetologists to distinguish the species, and they may be species-isolating mechanisms for the turtles as well. During the mating season male pond turtles swim in pursuit of other turtles, and the color and pattern on the posterior limbs may enable males to identify females of their own species. Pheromones may also play a role in species identification. Long claws on the forefeet distinguish males of many species of pond turtles from females, and during courtship the male turtle swims backward in front of the female, vibrating his claws against the sides of her head (Figure 12–8).

Among terrestrial turtles, the behavior of tortoises is best known. Many tortoises vocalize during courtship; the sounds they produce have been described as grunts, moans, and bellows. The frequencies of the calls that have been measured range from 500 to 2500 hertz. Some tortoises have glands that become enlarged during the breeding season and appear to produce pheromones. The secretion of the subdentary gland found on the underside of the jaw of tortoises in the North American genus *Gopherus* appears to identify both the species and the sex of an individual. During courtship, the males and females of the Florida gopher tortoise rub their subdentary gland across one or both forelimbs, and then extend the limbs toward the other individual, which may sniff at them. Males also sniff the cloacal region of other tortoises, and male tortoises of some species trail females for days during the breeding season. Fecal pellets may be territorial markers; fresh fecal pellets from a dominant male tortoise have been reported to cause dispersal of conspecifics.

Tactile signals used by tortoises include biting, ramming, and hooking. These behaviors are used primarily by males, and they are employed against other males and also against females. Bites are usually directed at the head or limbs, whereas ramming and hooking are directed against the shell. The epiplastral region is used for ramming, and in some species the epiplastral bones of males are elongated and project forward beneath the neck. A tortoise about to ram another individual raises itself on its legs, rocks backward, and then plunges forward, hitting the shell of the other individual with a thump that can be heard for 100 meters in large species. During hooking the epiplastral projections are placed under the shell of an adversary, and the aggressor lifts the front end of its shell and walks forward. The combination of lifting and pushing hustles the adversary along and may even overturn it.

Movements of the head appear to act as social signals for tortoises, and elevating the head is a signal of dominance in some species. Herds of tortoises have social hierarchies that are determined largely by aggressive encounters. Ramming, biting, and hooking are employed in these encounters, and the larger individual is usually the winner, although experience may play some role. These social hierarchies are expressed in behaviors such as access to food or forage areas, mates, and resting sites. Dominance relationships also appear to be involved in determining the sequence in which individual tortoises move from one place to another. The social structure of a herd of tortoises can be a nuisance for zookeepers trying to move the animals from an outdoor pen into an enclosure for the night, because the tortoises resist moving out of their proper rank sequence.

(a)

(b)

Figure 12–8 Social behavior of turtles. (a) Male painted turtle (*Chrysemys picta*) courting a female by vibrating the elongated claws of his forefeet against the sides of her head. (b) The head-raising dominance posture of a Galapagos tortoise, *Geochelone*. (This behavior can sometimes be elicited by crouching in front of a male tortoise and raising your arm.) ([b] Modified from S. F. Schafer and C. O. Krekorian, 1983, *Herpetologica* 39:448–456.)

Little information is available about the social behavior of freshwater and marine turtles because they are harder to study than terrestrial species. However, underwater observations by Julie Booth of green turtles (*Chelonia mydas*) on the Great Barrier Reef of Australia revealed some of the social behaviors of these animals (Booth and Peters 1972). A small population of green turtles is resident in the lagoon at Fairfax Island, and this population is augmented by migrants that arrive in the breeding season. Male green turtles patrol the lagoon and attempt to court any female they encounter. The behavior of the female determines whether mating occurs. An unreceptive female turtle signals her unwillingness to mate by swimming away from a male. If the male pursues her, the female turns to face the male and assumes a refusal posture, hanging vertically in the water with her plastron facing the male and her limbs widespread. Male turtles usually swim off when a female adopts the refusal posture. Male turtles investigate other large objects in the water, including divers. Julie Booth noted that when male turtles swam toward her, she was able to discourage them by assuming the female turtle refusal position. The most remarkable aspect of the social behavior of these turtles is the presence of a section of the lagoon that appears to function as a female reserve. The bottom of the lagoon in this area is occupied only by female turtles resting on the sand. Male turtles patrol around the area and may swim across it at the surface of the water, but they do not attempt to initiate courtships with females in the reserve. As soon as a female leaves the reserve, however, she is courted by the waiting male turtles.

Nesting Behavior

All turtles are oviparous. Female turtles use their hind limbs to excavate a nest in sand or soil, and deposit a clutch that ranges from four or five eggs for small species to more than 100 eggs for the largest sea turtles. The period of embryonic development of turtles has been reported to extend from 28 days for an Asian species of softshell turtle to 420 days for the African leopard tortoise; 40 to 60 days is typical for many species. Turtles in the families Cheloniidae, Dermochelyidae, and Chelydridae lay eggs with soft, flexible shells, as do most species in the families Bataguridae, Emydidae, and Pelomedusidae. The eggs of turtles in the families Carettochelyidae, Chelidae, Kinosternidae, Testudinidae, and Trionychidae have rigid shells. In general, soft-shelled eggs develop more rapidly than hard-shelled eggs.

Environmental Effects on Egg Development Temperature, wetness, and the concentrations of oxygen and carbon dioxide can have profound effects on the embryonic development of turtles (Packard and Packard 1988). The temperature of a nest affects the rate of embryonic development, and excessively high or low temperatures can be lethal. The discovery that the sex of some turtles, lizards, and crocodilians is determined by the temperature they experienced during embryonic development has important implications for understanding patterns of life history and for conservation of these species. Temperature-dependent sex determination is widespread among turtles, apparently universal among crocodilians, and is known for tuatara and a few species of lizards. The effect of temperatures on sex determination is correlated with sexual size dimorphism of adults—high incubation temperatures produce the larger sex, which is usually females for turtles (Ewert and Nelson 1991). The switch from one sex to the other occurs within a span of 3 or 4°C (Figure 12–9). Male crocodilians, tuatara, and lizards are usually larger than females, and in these groups high temperatures during embryonic development produce males. Temperatures of natural nests are not completely stable, of course. There is some daily temperature variation superimposed on a seasonal cycle of changing environmental temperatures. In turtles the middle third of embryonic development is the critical period for sex determination; the sex of the embryos depends on the temperatures they experience during those few weeks. When eggs are exposed to a daily temperature cycle, the high point of the cycle is most critical for sex determination.

As a consequence of the narrowness of the thermal windows that are involved in sex determination and the variation that exists in environmental temperatures, both sexes are produced under field conditions, but not necessarily in the same nests. A nest site may be cooler in late summer when it is shaded by vegetation than it was early in the spring when it was exposed to the sun. Thus, eggs laid early in the season would produce females, whereas eggs deposited in the same place later in the year would produce males. Temperature can also differ between the top and the bottom of a nest. For example, temperatures averaged from 33 to 35°C in the top center of dry nests of American alligators in marshes, and males hatched from eggs in this area. At the bottom and sides of the same nests, average temperatures were from 30 to 32°C and eggs from those regions hatched into females.

Temperature-dependent sex determination has important implications for efforts to conserve endangered species. Nests of the American alligator in marshes are cooler than those built on levees; females hatch from marsh nests and males from nests built on levees. The sex ratio of hatchlings from 8000 eggs collected from natural nests was five females to one male, reflecting the relative abundance of marsh and levee nests. It is not clear whether this large imbalance of the sexes represents a normal condition for alligators, or if it is the result of recent changes in the availability of nest sites.

Some conservation efforts have been confounded by temperature-dependent sex determination in sea turtles. A number of programs have depended on collecting eggs from natural nests and incubating them under controlled conditions. Unfortunately, these unnaturally uniform conditions of incubation can result in producing hatchlings of only one sex (Mrosovsky and Yntema 1981). However, environmental sex determination can also be exploited by breeding programs (Box 12–1).

The amount of moisture in the soil surrounding a turtle nest is another important variable during embryonic development of the eggs. The wetness of a nest interacts with temperature in sex determination and also influences the rate of embryonic development and the size and vigor of the hatchlings that are produced (Miller et al. 1987). Dry substrates induced the development of some female painted turtles at low temperatures (26.5 and 27.0°C) that

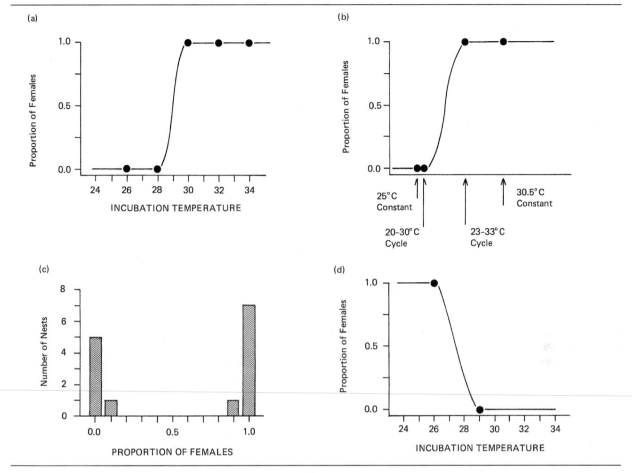

Figure 12–9 Temperature-dependent sex determination. (a) Eggs of the European pond turtle *Emys orbicularis* hatch into males when they are incubated at 26 or 28°C and into females at 30°C or above. (b) The North American map turtle *Graptemys ouachitensis* shows the same pattern. A temperature that cycles between 20 and 30°C produces males, whereas a temperature cycle of 23 to 33°C produces females. (c) Natural nests of map turtles produce predominantly males or females, depending on the nest temperature. (d) Eggs of the lizard *Agama agama* also show temperature-dependent sex determination, but the male and female determining temperatures are the opposite of those in turtles—for the lizard low temperatures produce females and high temperatures produce males. ([a] and [d] From J. J. Bull, 1980, *Quarterly Review of Biology* 55:13–21; [b] and [c] from J. J. Bull and R. C. Vogt, 1979, *Science* 206:1186–1188.)

would normally have produced only males. The wetness of the substrate did not affect the sex of turtles from eggs incubated at 30.5 and 32°C: All of the hatchlings from these eggs were females, as would be expected on the basis of temperature-dependent sex determination alone.

Wet incubation conditions produce larger hatchlings than do dry conditions, apparently because water is needed for metabolism of the yolk. When water is limited, turtles hatch early and at smaller body sizes, and their guts contain a quantity of yolk that was not used during embryonic development. Hatchlings from nests under wetter conditions are larger and contain less unmetabolized yolk. The large hatchlings that emerge from moist nests are able to run and swim faster than hatchlings from drier nests, and as a result they may be more successful at escaping from predators and at catching food.

Box 12–1 High-Tech Hatchlings

Temperature-dependent sex determination provides a way to manipulate the ratio of males to females produced by breeding programs for endangered species of turtles, lizards, or crocodilians. An excess of females would probably be desirable in most situations, because a single male can fertilize the eggs produced by many females. Thus, being able to skew the sex ratio of hatchlings toward females could provide a way to accelerate the production of animals that can be reintroduced into their native habitat.

Using the temperature sensitivity of reptilian sexual development to adjust the sex ratio of hatchlings is a complicated process, however, with both financial and technical difficulties. The eggs must be incubated under precisely controlled conditions, because as little as one degree separates the temperatures that produce all males from those that produce all females. Incubators capable of such precise temperature control cost several thousand dollars each, and it's hard enough to keep an incubator running properly for several months in a university laboratory—imagine trying to do it on a tropical beach where the electric power supply is uncertain at best!

An alternative method of controlling sex determination consists of administering the hormone estradiol-17ß to eggs during the middle third of their development. When it is administered at the right time in embryonic development, the exogenous (= originating outside the organism) hormone causes 100% of the treated eggs to develop as females regardless of the incubation temperature (Figure 12–10). This method of controlling the sex of hatchlings was patented in 1993 by Reptile Conservation International Inc., a nonprofit company. It could be a valuable addition to management techniques for some species of endangered reptiles because it is inexpensive ($20 worth of estradiol will treat 250,000 eggs) and noninvasive. The hormone can be applied to the eggshell and allowed to diffuse into the embryo. That method minimizes the risk of introducing bacteria into the egg, and estradiol-induced sex determination is now being tested in Mexico and Brazil with eggs that are left in natural nests.

Hatching and the Behavior of Baby Turtles

Turtles are self-sufficient at hatching, but in some instances interactions among the young may be essential to allow them to escape the nest. Sea turtle nests are quite deep, the eggs may be buried 50 centimeters beneath the sand, and the hatchling turtles must struggle upward through the sand to the surface. After several weeks of incubation the eggs all hatch within a period of a few hours, and a hundred or so baby turtles find themselves in a small chamber at the bottom of the nest hole. Spontaneous activity by a few individuals sets the whole group into motion, crawling over and under each other. The turtles at the top of the pile loosen sand from the roof of the chamber as they scramble about, and the sand filters down through the mass of baby turtles to the bottom of the chamber.

Periods of a few minutes of frantic activity are interspersed by periods of rest, possibly because the turtles' exertions reduce the concentration of oxygen in the nest and they must wait for more oxygen to diffuse into the nest from the surrounding sand. Gradually, the entire group of turtles moves upward through the sand as a unit until it reaches the surface. As the baby turtles approach the surface, high sand temperatures probably inhibit further activity, and they wait a few centimeters below the surface until night when a decline in temperature triggers emergence. All of the babies emerge from a nest in a very brief period, and all of the babies in different nests that are ready to emerge on a given night leave their nests at almost the same time, probably because their behavior is cued by temperature. The result is the sudden appearance of hundreds or even thousands of baby turtles on the beach, each one crawling toward the ocean as fast as it can.

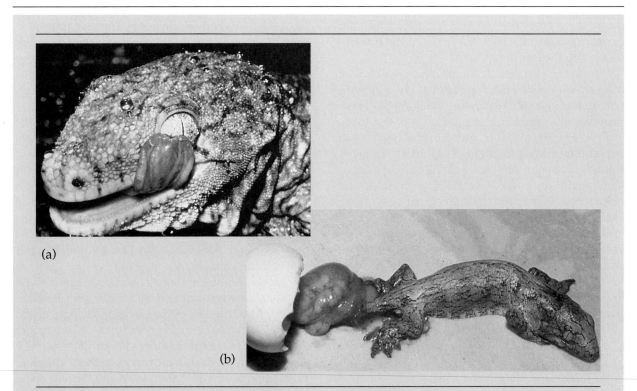

(a)

(b)

Figure 12–10 The Dallas Zoo is applying estradiol-17β to eggs of the New Caledonian giant gecko (*Rhacodactylus leachianus*). (a) The geckos, which reach an adult length of 35 centimeters, are threatened by logging, which is destroying their habitat. Most of the giant geckos in zoos are males, and breeding programs have produced an overwhelming proportion of male offspring. (b) By using hormone treatment of eggs, the Dallas Zoo has succeeded in producing female hatchlings. (Photographs courtesy of David T. Roberts, Dallas Zoo.)

Simultaneous emergence is an important feature of the reproduction of sea turtles, because the babies suffer severe mortality crossing the few meters of beach and surf. Terrestrial predators await their appearance—crabs, foxes, raccoons, and other predators gather at the turtles' breeding beaches at hatching time. Some of the predators come from distant places to prey on the baby turtles. In the surf, sharks and bony fish patrol the beach. Few, if any, baby turtles would get past that gauntlet if it were not for the simultaneous emergence that brings all the babies out at once and temporarily swamps the predators.

Turtles exhibit no parental care, and the long period of embryonic development renders their nests vulnerable to predators. Females of many species of turtles scrape the ground in a wide area around the nest when they have finished burying their eggs. This behavior may make it harder for predators to identify the exact location of the nest. Major breeding sites of sea turtles are often on islands that lack mammalian predators that could excavate the nests. Another important feature of a nesting beach is provision of suitable conditions for the hatchling turtles. Newly hatched sea turtles are small animals; they weigh 25 to 50 grams, which is less than 0.05 percent of the body mass of an adult sea turtle. The enormous disparity in body size of hatchling and adult turtles is probably accompanied by equally great differences in their ecological requirements and their swimming abilities. Many of the major sea turtle nesting areas are upstream from the feeding grounds, and that location may allow currents to carry the baby turtles from the breeding beaches to the feeding grounds.

We know even less about the biology of baby sea turtles than we do about the adults. The phenomenon of the "lost year" following hatching has been a long-standing puzzle in the life cycle of sea turtles (Carr 1987). For example, green turtles hatch in the late summer at Tortuguero. The turtles disappear from sight as soon as they are at sea, and they are not seen again until they weigh 4 or 5 kilograms. Apparently, they spend the intervening period floating in ocean currents. Material drifting on the surface of the sea accumulates in areas where currents converge, forming drift lines of flotsam that include sargassum (a brown algae) and the vertebrate and invertebrate fauna associated with it. These drift lines are probably important resources for juvenile sea turtles.

Navigation and Migration

Pond turtles and terrestrial turtles usually lay their eggs in nests they construct within their home ranges. The mechanisms of orientation they use to find nesting areas are probably the same ones they use to find their way among foraging and resting areas. Familiarity with local landmarks is an effective method of navigation for these turtles, and they may also use the sun for orientation. Sea turtles have a more difficult time, partly because the open ocean lacks conspicuous landmarks, and also because feeding and nesting areas are often widely separated. Most sea turtles are carnivorous. The leatherback turtle feeds on jellyfish, ridley and loggerhead turtles eat crabs and other benthic invertebrates, and the hawksbill turtle uses its beak to scrape encrusting organisms (sponges, tunicates, bryozoans, mollusks, and algae) from reefs. Juvenile green turtles are carnivorous, but the adults feed on vegetation, particularly turtle grass (*Thalassia testudinium*), which grows in shallow water on protected shorelines in the tropics. The areas that provide food for sea turtles often lack the characteristics needed for successful nesting, and many sea turtles move long distances between their feeding grounds and their breeding areas.

The ability of sea turtles to navigate over thousands of kilometers of ocean and find their way to nesting beaches that may be no more than tiny coves on a small island is a remarkable phenomenon. The migrations of sea turtles, especially the green turtle, have been studied for decades. Turtles captured at breeding sites in the Caribbean and Atlantic Oceans have been individually marked with metal tags since 1956, and tag returns from turtle catchers and fishing boats have allowed the major patterns of movements of the populations to be established (Figure 12–11). Similar studies by investigators in other parts of the world are beginning to shed light on movements of other species of turtles.

Four major nesting sites of green turtles have been identified in the Caribbean and South Atlantic: one at Tortuguero on the coast of Costa Rica, one on Aves Island in the eastern Caribbean, one on the coast of Surinam, and one on Ascension Island between South America and Africa. Male and female green turtles congregate at these nesting grounds during the nesting season. The male turtles remain offshore, where they court and mate with females, and the female turtles come ashore to lay eggs on the beaches. A typical female green turtle at Tortuguero produces three clutches of eggs about 12 days apart. About a third of the female turtles in the Tortuguero population nest in alternate years, and the remaining two-thirds of the turtles follow a 3-year breeding cycle. The coast at Tortuguero lacks the beds of turtle grass on which green turtles feed, and the turtles come to Tortuguero only for nesting. In the intervals between breeding periods, the turtles disperse around the Caribbean. The largest part of the Tortuguero population spreads northward along the coast of Central America. The Miskito Bank off the northern coast of Nicaragua appears to be the main feeding ground for the Tortuguero colony. A smaller number of turtles from Tortuguero swim south along the coast of Panama, Colombia, and Venezuela. Female green turtles return to their natal beaches to nest (Meylan et al. 1990), and the precision with which they home is astonishing. Female green turtles at Tortuguero return to the same kilometer of beach to deposit each of the three clutches of eggs they lay in a breeding season.

Probably the most striking example of the ability of sea turtles to home to their nesting beaches is provided by the green turtle colony that nests on Ascension Island, a small volcanic peak that emerges from the ocean 2200 kilometers east of the coast of Brazil. The island is less than 20 kilometers in diameter—a tiny target in the vastness of the South Atlantic. The Ascension Island population has its

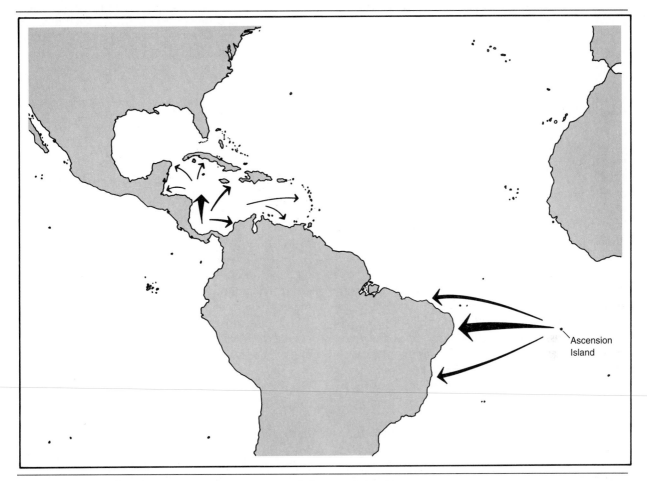

Figure 12–11 Migratory movements of green turtles (*Chelonia mydas*) in the Caribbean and South Atlantic. The population that nests on beaches in the Caribbean is drawn from feeding grounds in the Caribbean and Gulf of Mexico. The turtles that nest on Ascension Island feed along the coast of northern South America.

feeding grounds on the coast of Brazil. Migrating and homing birds use a variety of orientation mechanisms, including the ability to detect the magnetic field of the earth, to perceive polarized light, to use the sun and stars for orientation, and to hear very low frequency sounds. Sea turtles probably have a similarly wide repertoire of mechanisms for navigation.

A study of navigation by hatchling loggerhead turtles showed that they used at least three cues for orientation: light, wave direction, and magnetism (Lohmann 1991). These stimuli play sequential roles in the turtles' behavior. After hatching on beaches on the Atlantic coast of Florida, baby loggerheads enter the Gulf Stream which flows northward off the Florida Coast. They drift with the current along the coast of the United States, and then eastward across the Atlantic. Off the coast of Portugal, the Gulf Stream divides into two branches. One turns north toward England, and the other swings south past the bulge of Africa and eventually back westward across the Atlantic. It's important for the baby turtles to turn right at Portugal; if they fail to make that turn they are swept past England into the chilly North Atlantic, where they perish. If they do turn southward off the coast of Portugal, they are eventually carried back to the coast of tropical America—a round-trip that takes 5 to 7 years.

When they emerge from their nests, the hatchlings crawl toward the brightest light they see. Normally it is lighter over the ocean than over land, and this behavior brings them to the water's edge. (Shopping centers, street lights, even porch lights on beach-front houses can confuse these and other species of sea turtles and lead them inland, where they are crushed on roads or die in the sun the next day.)

In the ocean, the loggerhead hatchlings swim into the waves. This response moves them away from shore and ultimately into the Gulf Stream, but once the turtles are in the open ocean, waves may come from any direction and would no longer provide a reliable cue for navigation. Here magnetic orientation appears to take over, and it appears to tell the turtles when to turn right to catch the current that will carry them to the South Atlantic. We usually think of the Earth's magnetic field as providing two-dimensional information—north–south and east–west—but it's more complicated than that. The field loops out of the north and south magnetic poles of the Earth. At the Equator the field is essentially parallel to the Earth's surface (in other words, it forms an angle of 0 degrees), and at the poles it intersects the surface at an angle of 90 degrees. Thus, the three-dimensional orientation of the Earth's magnetic field provides directional information (Which way is magnetic north?) and information about latitude (What is the angle at which the magnetic field intersects the Earth's surface?).

When loggerheads in an enclosure on land that had no waves were exposed to an artificial magnetic field at the 57-degree angle of intersection with the Earth's surface that is characteristic of Florida, they swam toward artificial east even when the magnetic field was changed 180 degrees so that the direction they thought was east was actually west. That is, they were able to use a compass sense to determine direction. But that wasn't all they could do. When the angle of intersection of the artificial magnetic field was increased to 60 degrees, the turtles turned south. A 60-degree angle of intersection corresponds to the latitude where the Gulf Stream forks off the coast of Portugal, and where the turtles must turn south to reach the South Atlantic. Thus, it appears that loggerhead turtles can use magnetic sensitivity to recognize both direction and latitude.

Still more senses may be involved in navigation by sea turtles. An animal like a sea turtle that crosses thousands of miles of open ocean and can raise its head only a few centimeters above the water's surface, probably uses every cue it can. Olfaction may assist some kinds of orientation, such as location of a specific beach when a turtle has reached the breeding grounds, and a sun compass could be used for short-term movements between the beach and offshore areas, as well as for long-distance navigation. Sea turtles are inherently difficult animals to study, but evidence for these mechanisms is accumulating.

Conservation of Turtles

Many species of turtles have slow rates of growth and require long periods to reach maturity. These are characteristics that predispose a species to the risk of extinction when changing conditions increase the mortality of adults or drastically reduce recruitment of juveniles into the population. The plight of large tortoises and sea turtles is particularly severe, partly because these species are among the largest and slowest growing of turtles, and also because other aspects of their biology expose them to additional risk (Figure 12–12). The conservation of tortoises and sea turtles is a subject of active international concern (Swingland and Klemens 1989, Burke et al. 1993, Gibbons 1994).

The largest living tortoises are found on the Galapagos and Aldabra Islands. The relative isolation of these small and (for humans) inhospitable landmasses has probably been an important factor in the survival of the tortoises. Human colonization of the islands has brought with it domestic animals such as goats and donkeys that compete with the tortoises for the limited quantities of vegetation to be found in these arid habitats, and dogs, cats, and rats that prey on tortoise eggs and on baby tortoises.

The limited geographic range of a tortoise that occurs only on a single island makes it vulnerable to local habitat destruction. In 1985 and again in 1994 brush fires on the island of Isabela in the Galapagos Archipelago threatened the 20 surviving individuals of *Geochelone guntheri,* and emphasized the advantage of moving some or all of the turtles to the breeding station operated by the Charles Darwin Research Station on Santa Cruz Island. This station has a successful record of breeding and releasing another species of Galapagos tortoise, *Geochelone nigra hoodensis,* which is native to Espanola Island. In the early 1960s only 14 individuals of this form could be located. All were adults and apparently had not bred successfully for many years. All of the tortoises were moved to the Research Station, and the first babies were produced in 1971. Since then, nearly 400 tortoises have been raised, and half of them have been released on Espanola. This success story shows that

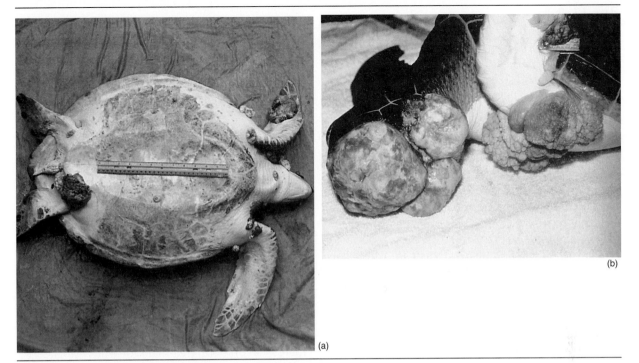

(a)

(b)

Figure 12–12 A green turtle with cutaneous papillomas. (a) These tumors, which are probably caused by a virus, have been found on most species of sea turtles and in most parts of the world (Herbst, in press). The tumors grow to more than 30 centimeters in diameter, and can appear on any skin-covered surface. (b) They are especially common on the conjuctiva of the eyes, and may grow over the cornea. Tumors were not recorded on green turtles in the Indian River Lagoon, Florida, until 1982, and by 1994 approximately 50 percent of the green turtles were affected. The first record of papilloma on green turtles in Kaneohe Bay, Hawaii, was in 1958, and since 1989 the incidence has ranged from 49 to 92 percent. The tumors can be lethal, and their increased frequency is an ominous development for species that were already endangered (Herbst, in press). (Photographs courtesy of Larry Herbst.)

carefully controlled captive breeding and release programs can be an effective method of conservation for endangered species of turtles. These programs carry with them the risk of introducing diseases into wild populations, however, and have the potential to do substantial harm (Box 12–2).

Conservation of sea turtles provides special challenges. The species range over thousands of square kilometers of ocean, including international waters and also coastal areas that are under the jurisdictions of many different nations. Protection of sea turtles is a complex undertaking, and the limited number of breeding sites used by many species adds to the difficulty. All of the seven species of sea turtles face threats, but the melancholy distinction of being the most endangered sea turtle goes to Kemp's ridley (*Lepidochelys kempi*, Figure 12–14). This species has

only one major nesting site, a 14-mile stretch of beach on the coast near Rancho Nuevo in Tamauliapas, Mexico. Kemp's ridley is unique among sea turtles in nesting by day, and in enormous numbers. The influx of female turtles to the beach is called an *arribada*, and a movie made in 1947 shows an *arribada* estimated to contain 47,000 turtles, all nesting at once. Although the turtles had probably been an important source of food for inhabitants of the region since pre-Columbian times, the location of the nesting beach was unknown to the scientific community until 1966. By then, the largest *arribadas* consisted of only 3000 to 5000 turtles, and their numbers have continued to decrease as adult females and eggs taken from the nesting beach have been used for food, and as adults and juveniles have drowned in fishing and shrimping nets in the Gulf of Mexico. By

Box 12–2 Sick Turtles

The desert tortoise (*Gopherus agassizi*) is one of the largest terrestrial turtles in North America. Its geographic range includes the southwestern corner of Utah plus the southwestern third of Arizona and adjacent parts of Nevada and California, and extends southward into Mexico. (Some of the behavioral and physiological mechanisms that allow these tortoises to survive in a harsh desert environment are discussed in Chapter 16.)

Populations of desert tortoises have declined since the 1950s as human activity has intruded on the desert habitat. Between 1979 and 1989, most tortoise populations in the Mohave and Colorado Deserts decreased by 30 to 70 percent (Berry 1989). The situation has become even more grave with the appearance of upper respiratory tract disease (URTD), which attacks desert tortoises, often with fatal results. Infected turtles first have a runny nose, which becomes progressively worse until the turtles exude foam from their nostrils, wheeze when they breathe, cease feeding, become listless, and ultimately die (Figure 12–13). In 1988 tortoises in the Desert Tortoise Natural Area in Kern County, California, first showed symptoms of URTD (Jacobson et al. 1991). In 1989, 627 dead tortoises were found and 43 percent of the live tortoises on the Natural Area showed symptoms of URTD.

A large variety of bacteria were cultured from the nasal passages of the sick turtles, including *Mycoplasma*, which has subsequently been shown to be the cause of the disease. Desert tortoises are popular pets in the desert southwest, and a high proportion of pet turtles are infected by *Mycoplasma*. The infection may have been introduced to the Desert Tortoise Natural Area when pet tortoises were released, and its spread may have been accelerated by the poor physical condition of the wild tortoises that resulted from habitat degradation and a prolonged drought. *Mycoplasma* infections are notoriously difficult to cure. Captive tortoises can be treated with a combination of antibiotics, but there is no practical treatment for wild tortoises.

URTD has now been reported in a population of the gopher tortoise (*Gopherus polyphenus*) on Sanibel Island off the coast of Florida (Jacobson 1993). Again, captive tortoises appear to have introduced the infection into a wild population: Until 1978 tortoises used in tortoise races in Fort Myers were released on Sanibel Island, and infected tortoises from the races may have carried *Mycoplasma* with them.

1994, despite conservation efforts, the entire population probably contained fewer than 800 females (Shaver 1990).

From 1978 to 1988 the Mexican Instituto Nacional de Pesca worked with the U. S. Fish and Wildlife Service, the National Marine Fisheries Services, the National Park Service, and the Texas Parks and Wildlife Department to establish a second breeding population of Kemp's ridley at Padre Island National Seashore in Texas. Each year about 2000 eggs were collected in Mexico as they were laid, and shipped to Padre Island where they were incubated and hatched. A total of 17,358 hatchlings were produced at Padre Island in those 11 years (Shaver 1992). A parallel effort incubated the majority of eggs from the species on the beach at Rancho Nuevo.

The goal of the project was to produce at least a 1:1 ratio of male and female hatchlings, and preferably a preponderance of females to establish a population of turtles that would return to Padre Island to breed. Several individuals, including National Parks Service employee Donna Shaver, cooperated in a study of the sex ratio of the hatchling turtles that had been produced between 1978 and 1984. Embryos that had failed to hatch had been preserved for later examination, and the sex of the dead turtles could be determined by histological exami-

Figure 12–13 A gopher tortoise with a runny nose—a sign that this tortoise is infected with the *Mycoplasma* that causes upper respiratory tract disease. (Photograph courtesy of Elliott Jacobson.)

These examples emphasize the risk of releasing animals that have been held in captivity into wild populations. Captive breeding programs must take extraordinary measures to ensure that the animals to be released are quarantined in a facility that is isolated from other animals. A breeding colony should be self-contained; once it is established, no outside animals should be introduced, and no equipment or containers should be moved in or out. Even the clothing of animal caretakers can carry pathogens, and a dressing room must be provided so they can wash and change their clothes when they enter or leave.

These precautions are time-consuming and expensive, but neglecting them can be disastrous.

The tremendous expenditure of time and money needed to breed healthy animals for release programs emphasizes the importance of applying information about the basic biology of animals to management plans. Estradiol may allow biologists to manipulate the sex ratios of hatchlings without removing the eggs from the nest (Box 12–1), and adjusting the sizes of eggs to produce hatchlings with the best chances of survival (Box 15–3) makes the best use of expensive captive-bred animals.

nation of their gonads. The results were discouraging (Figure 12–15). In three of the five years for which adequate samples were available only one-third of the hatchlings were females, and the highest proportion of females ever achieved was only 50 percent (Shaver et al. 1988). Those ratios were far from the project's goal of at least a 1:1 sex ratio, and the deviation was in the wrong direction—an excess of males was being produced.

The phenomenon of temperature-dependent sex determination in turtles had been described in the biological literature some 30 years earlier, but its implications for conservation of sea turtles had not been appreciated (Mrosovsky and Yntema 1981). The

Padre Island project, like many other sea turtle conservation projects around the world, incubated the eggs in moist sand in styrofoam boxes that were kept in a covered egg house on the beach (Figure 12–16). Shaver found that the temperature inside the boxes was slightly too cool, and this was why most of the eggs developed into male turtles.

From 1985 onward, temperatures in the egg houses at Rancho Nuevo and Padre Island were raised about 3°C. This small increase in incubation temperature was sufficient to shift sex determination in favor of females, and from 1985 through the end of the project in 1988 the proportion of female hatchlings increased dramatically (Figure 12–15).

Figure 12–14 An adult female Kemp's ridley sea turtle. (Photograph courtesy of Donna Shaver.)

This example illustrates the importance of applying basic biological information about organisms to management and conservation programs, and shows how effective even one person can be in applying that information. Recent studies of methods of manipulating the eggs of lizards and turtles during development may offer other ways to increase the effectiveness of conservation and management programs (see Boxes 12–1 and 15–3).

Attempts to save sea turtles are in progress all over the world, sponsored by a variety of governmental agencies, private organizations, and even dedicated individuals. The problems they face are massive, and the most effective methods to employ are still subject to disagreement (Pritchard 1980, Ehrenfeld 1981). For example, is controlled exploitation of sea turtles more feasible than an outright ban on the use of sea turtles and their products? Do turtle farming operations benefit conservation by producing captive-bred individuals, or do they indirectly harm natural populations by sustaining a demand for turtle products that would otherwise vanish?

Beyond those questions, which have their origins in the ways people in rich and poor nations respond to the often-conflicting demands of earning a living versus conserving natural resources, is another series of questions that arise from our inadequate knowledge of the biology of sea turtles. For example, is it a wise management practice to dig up nests of turtle eggs from the nesting beaches and incubate them in a protected area? Predation on eggs in natural nests can be high, but sea turtles display temperature-dependent sex determination, and the widespread technique of incubating eggs in plastic foam boxes appears to produce predominantly male hatchlings. Is the practice of "headstarting" sea turtles beneficial? In this technique, baby turtles are kept in captivity for some weeks or months and allowed to grow before they are released at sea. This method avoids the very high losses of baby turtles to predators that occurs when the newly hatched babies make their own way down the beach and into the sea. However, imprint-

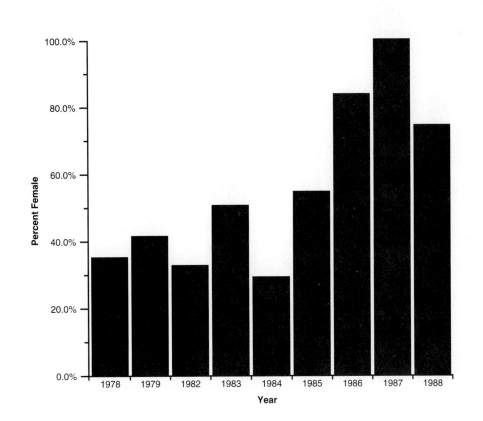

Figure 12–15 Proportion of female hatchlings produced in the egg house at Padre Island National Seashore from 1978 through 1988. Note the increase in the proportion of females from 1985 onward, after incubation temperatures were raised. (No reliable samples are available for 1980 and 1981.) (From D. Shaver et al., 1988, pages 103–108 in *Proceedings of the Eight Annual Workshop on Sea Turtle Conservation and Biology, Feb. 24–26, 1988, Fort Fisher, NC*, NOAA Technical Memorandum NMFS-SEFC-214, and Donna Shaver, personal communication.)

ing on the characteristics of the beach and the adjacent water might occur as the baby turtle makes its own way to sea. If that is the case, headstarting may prevent the imprinting that is essential for successful navigation of an adult turtle back to the nesting beach. We simply cannot evaluate the effects of these manipulations because we do not know enough about the biology of sea turtles.

These questions are a subset of a broader set of questions about the effectiveness of conservation methods as they have been applied to reptiles (Dodd and Siegel 1991, Burke 1991, Reinert 1991). The problems cited by these authors emphasize the central role of information about all aspects of the biology of organisms in successful conservation plans. This sort of information is not easy to obtain for any species of organism, and sea turtles are more difficult to study than most animals. Nonetheless, pitfalls lie in wait for even the best-intended management techniques, and conservation must incorporate an understanding of the biology of the organisms involved. We appear to be some distance from achieving that goal—a summary of conservation programs for sea turtles published by the National Academy of Sciences (National Research Council 1990) did not even mention temperature-dependent sex determination!

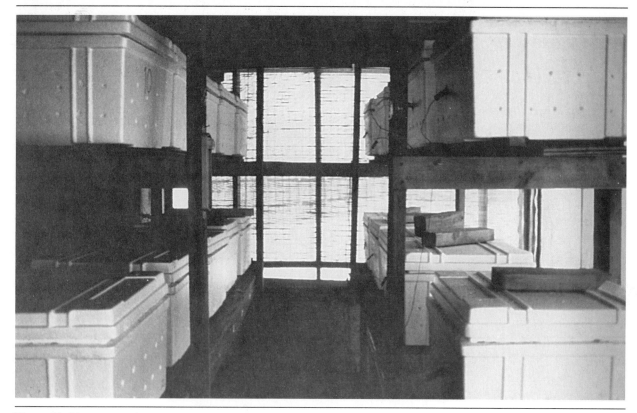

Figure 12–16 Interior of the egg house at Padre Island National Seashore. Sea turtle eggs are being incubated in sand-filled styrofoam boxes. (Photograph courtesy of Donna Shaver.)

Summary

The earliest turtles known, fossils from the Triassic, have nearly all the features of derived turtles. The first Triassic forms were not able to withdraw their heads into the shell, but this ability appeared in the two major lineages of living turtles, which were established by the late Triassic. The cryptodiran turtles retract the head with a vertical flexion of the neck vertebrae, whereas the pleurodires use a sideward bend.

Turtles are among the most morphologically specialized vertebrates. The shell is formed of dermal bone that is fused to the vertebral column and ribs. In most turtles the dermal shell is overlain by a horny layer of epidermal scutes. The limb girdles are inside the rib cage. Breathing presents special difficulties for an animal that is encased in a rigid shell: Exhalation is accomplished by muscles that squeeze the viscera against the lungs, and inhalation is accomplished by muscles that increase the volume of the visceral cavity, thereby allowing the lungs to expand. The heart of turtles (and of squamates as well) is able to shift blood between the pulmonary and systemic circuits in response to the changing requirements of gas exchange and thermoregulation.

The social behavior of turtles includes visual, tactile, and olfactory signals used in courtship. Dominance hierarchies shape the feeding, resting, and mating behaviors of some of the large species of tortoises. All species of turtles lay eggs, and none provides parental care. Coordinated activity by hatchling sea turtles may be necessary to enable them to dig themselves out of the nest, and simultaneous emergence of baby sea turtles from their nests helps them to evade predators as they rush down the beach into the ocean. Sea turtles migrate tens, hundreds, and even thousands of kilometers between their feeding areas and their nesting beaches, and use a large variety of cues for navigation.

The life history of many turtles makes them vulnerable to extinction. Slow rates of growth and long periods required to reach maturity are characteristic of turtles in general and of large species of turtles in particular. Tortoises and sea turtles are especially threatened, and conservation efforts are in progress in many countries. Recently discovered features of the basic biology of turtles have important implications for conservation efforts. For example, many species of turtles show temperature-dependent sex determination. That is, the sex of an individual turtle is determined by the temperature it experiences in the egg during embryonic development. Some conservation efforts undertaken before this phenomenon was appreciated resulted in the production and release of thousands of hatchling baby turtles, nearly all of which were probably males. Using information about basic aspects of the biology of turtles is a crucial part of efforts to sustain existing populations and to reestablish populations that have been lost.

References

Berry, K. 1989. *Gopherus agassizi*, desert tortoise. Pages 5–7 in The conservation biology of tortoises. *Occasional Papers of the IUCN Species Survival Commission*, No. 5, edited by I. R. Swingland and M W. Klemens. IUCN, Gland, Switzerland.

Booth, J., and J. A. Peters. 1972. Behavioural studies on the green turtle (*Chelonia mydas*) in the sea. *Animal Behaviour* 20:808–812.

Burggren, W. W. 1987. Form and function in reptilian circulations. *American Zoologist* 27:5–19.

Burke, R. L. 1991. Relocations, repatriations, and translocations of amphibians and reptiles: taking a broader view. *Herpetologica* 47:350–357.

Burke, V. J., N. B. Frazer, and J. W. Gibbons. 1993. Conservation of turtles: the chelonian dilemma. Pages 35–38 in *Proceedings of the 13th Annual Symposium on Sea Turtle Biology and Conservation*. U.S. Department of Commerce, National Oceanic and Atmospheric Administration, Jekyll Island, GA.

Carr, A. 1987. New perspectives on the pelagic stage of sea turtle development. *Conservation Biology* 1:103–121.

Chelazzi, G., and R. Calzolai. 1986. Thermal benefits from familiarity with the environment in a reptile. *Oecologia* 68:557–558.

Dodd, C. K. Jr., and R. A. Siegel. 1991. Relocations repatriation, and translocation of amphibians and reptiles: Are they conservation strategies that work? *Herpetologica* 47:336–350.

Ehrenfeld, D. 1981. Options and limitations in the conservation of sea turtles. Pages 457–463 in *Biology and Conservation of Sea Turtles*, edited by K. A. Bjorndal. Smithsonian Institution Press, Washington, DC.

Ewert, M. A., and C. E. Nelson. 1991. Sex determination in turtles: diverse patterns and some possible adaptive values. *Copeia* 1991:50–69.

Gaffney, E. S., and J. W. Kitching. 1994. The most ancient African turtle. *Nature* 369:55–58.

Gaffney, E. S., J. H. Hutchison, F. A. Jenkins, Jr., L. J. Meeker. 1987. Modern turtle origins: the oldest known cryptodire. *Science* 237:289–291.

Gibbons, J. W. 1994. Reproductive patterns of reptiles and amphibians: considerations for captive breeding and conservation. Pages 119–123 in *Captive Management and Conservation of Amphibians and Reptiles*, edited by J. B. Murphy, K. Adler, and J. T. Collins. Contributions to Herpetology, vol. 2, Society for the Study of Amphibians and Reptiles, Ithaca, NY.

Herbst, L. H. In press. Fibropapillomatosis of marine turtles. *Annual Review of Fish Diseases*.

Jacobson, E. R. 1993. Implications of infectious diseases for captive propagation and introduction programs of threatened/endangered reptiles. *Journal of Zoo and Wildlife Medicine* 24:245–255.

Jacobson, E R., J. M. Gaskin, M. B. Brown, R. K. Harris, C. H. Gardiner, J. L. LaPointe, H. P. Adams, and C. Reggiardo. 1991. Chronic upper respiratory disease of free-ranging desert tortoises (*Xerobates agassizi*). *Journal of Wildlife Disease* 27:296–316.

Lee, M. S. Y. 1993. The origin of the turtle body plan: bridging a famous morphological gap. *Science* 261:1716–1720.

Lohmann, K. J. 1991. Magnetic orientation by hatchling loggerhead sea turtles. *Journal of Experimental Biology* 155:37–49.

Meylan, A. 1988. Spongivory in hawksbill turtles: a diet of glass. *Science* 239:393–395.

Meylan, A. B., B. W. Bowen, and J. C. Avise. 1990. A genetic test of the natal homing versus social facilitation model for green turtle migration. *Science* 248:724–727.

Miller, K., G. C. Packard, and M. J. Packard. 1987. Hydric conditions during incubation influence locomotor performance of hatchling snapping turtles. *Journal of Experimental Biology* 127:401–412.

Mrosovsky, N., and C. L. Yntema. 1981. Temperature dependence of sexual differentiation in sea turtles: implications for conservation practices. Pages 271–280 in *Biology and Conservation of Sea Turtles*, edited by K. A. Bjorndal. Smithsonian Institution Press, Washington, DC.

National Research Council. 1990. *Decline of the Sea Turtles: Causes and Prevention*. National Academy Press, Washington, DC.

Packard, G. C., and M. J. Packard. 1988. Physiological ecology of reptile eggs. Pages 523–605 in *Biology of the Reptilia*, volume 16, edited by C. Gans and R. B. Huey. Alan Liss, Philadelphia, PA.

Paladino, F. V., M. P. O'Connor, and J. R. Spotila. 1990. Metabolism of leatherback turtles, gigantothermy, and thermoregulation of dinosaurs. *Nature* 344:858–860.

Pritchard, P. C. H. 1980. The conservation of sea turtles: practices and problems. *American Zoologist* 20:609–617.

Reinert, H. K. 1991. Translocations as a conservation strategy for amphibians and reptiles: some comments, concerns, and observations. *Herpetologica* 47:357–363.

Reisz, R. R., and M. Laurin. 1991. *Owenetta* and the origin of turtles. *Nature* 349:324–326.

Rougier, G. W., M. S. de la Fuente, and A. B. Arcucci. 1995. Late Triassic turtles from South America. *Science* 268:855–858.

Shaver, D. 1990. Kemp's ridley project at Padre Island enters a new phase. *Park Science* 10(1):12–13.

Shaver, D. 1992. Kemp's ridley research continues at Padre Island National Seashore. *Park Science* 12(4):26–27.

Shaver, D., D. W. Owens, A. H. Chaney, C. W. Caillouet, Jr., P. Burchfield, and R. Marquez M. 1988. Styrofoam box and beach temperatures in relation to incubation and sex ratios of Kemp's ridley sea turtles. Pages 103–108 in *Proceedings of the Eight Annual Workshop on Sea Turtle Conservation and Biology, Feb. 24–26, 1988, Fort Fisher, NC*. NOAA Technical Memorandum NMFS-SEFC-214.

Spotila, J. R., and E. A. Standora. 1985. Environmental constraints on the thermal energetics of sea turtles. *Copeia* 1985:694–702.

Swingland, I. R., and J. G. Frazier. 1979. The conflict between feeding and overheating in the Aldabran giant tortoise. Pages 611–615 in *A Handbook on Biotelemetry and Radio Tracking*, edited by C. J. Amlaner, Jr., and D. W. MacDonald. Pergamon, Oxford, UK.

Swingland, I. R., and M. W. Klemens (editors). 1989. The conservation biology of tortoises. *Occasional Papers of the IUCN Species Survival Commission*. No. 5. IUCN, Gland, Switzerland.

Swingland, I. R., and C. M. Lessells. 1979. The natural regulation of giant tortoise populations on Aldabra Atoll. Movement polymorphism, reproductive success and mortality. *Journal of Animal Ecology* 48:639–654.

MESOZOIC DIAPSIDS: NONAVIAN DINOSAURS, BIRDS, CROCODILIANS, AND OTHERS

13

At the same time that turtles were evolving in the Triassic, the most diverse lineage of amniotic verte-brates, the Diapsida, was starting its radiation. The most spectacular diapsids were the nonavian dinosaurs, but the lineage also gave rise to a majority of the species of extant terrestrial vertebrates. Birds are diapsids, as are the squamates (lizards, snakes, and amphisbaenians). A variety of other, lesser-known forms fills the roster of diapsids, including crocodilians, ichthyosaurs, and pterosaurs, to name only a few.

The nonavian dinosaur fauna of the Mesozoic was unlike anything that has existed before or since. Many dinosaurs were enormous, and it is difficult to recre-ate the details of the lives they led because we have no living models of truly large terrestrial vertebrates. Even elephants are only as large as a small to medium-size dinosaur.

A second group of diapsids, the lepidosauromorphs, radiated into a variety of animals in the Mesozoic and then had another radiation that produced somewhat different animals that survive today. The extant lepidosauromorphs are such suc-cessful animals in their own right that they will be discussed in the next chapter; however, a discussion of the Mesozoic world would be incomplete without con-sidering the secondarily aquatic marine forms (ichthyosaurs, plesiosaurs, pla-codonts, and the strange *Hupehsuchus*), and these were probably lepidosauro-morphs.

The remarkable success of large diapsids in the Mesozoic ended at the close of that era with the extinction of the nonavian dinosaurs. That mass extinction has attracted more than its share of attention because nonavian dinosaurs have great popular appeal; but as mass extinctions go, it was minor. However, it does pro-vide a good opportunity to consider in detail the merits of two types of expla-nations of mass extinctions, the gradualist versus the catastrophism schools of thought.

The Mesozoic Fauna

The Mesozoic Era, frequently called the Age of Reptiles, extended for some 180 million years from the close of the Paleozoic 245 million years ago to the beginning of the Cenozoic only 66 million years ago. Through this vast period evolved a worldwide fauna that diversified and radiated into most of the adaptive zones occupied by all the terrestrial vertebrates living today and some that no longer exist (for example, the enormous herbivorous and carnivorous tetrapods called nonavian dinosaurs). Although the nonavian dinosaurs are the most familiar representatives of the Age of Reptiles, they are only one of many groups.

Inevitably such a huge number of animals is complicated and confusing, not only on first acquaintance, but even after study. Parallel and convergent evolution were widespread in Mesozoic tetrapods. Long-snouted fish eaters evolved repeatedly, as did heavily armored quadrupeds and highly specialized marine forms. A trend to bipedalism was general, and a secondary reversion to quadrupedal locomotion is seen in many forms. Knowledge of phylogenetic relationships is in a state of flux, and the scheme outlined in Figure 13–1 will undoubtedly need revision as additional material is analyzed. Current views of the ecology of nonavian dinosaurs are likewise undergoing a radical revision.

This chapter commences with a brief review of the phylogenetic relationships of Mesozoic tetrapods and some aspects of their functional morphology and major evolutionary trends. More detailed information on these topics and additional illustrations of

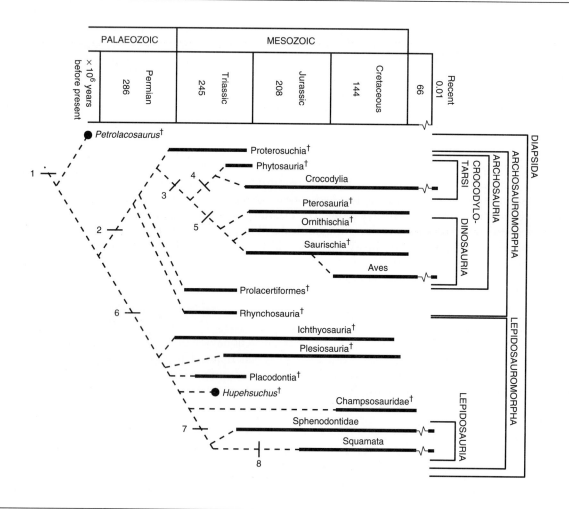

members of the groups discussed can be found in the references cited at the end of the chapter. Following a consideration of some aspects of the ecology of nonavian dinosaurs, we consider their disappearance at the end of the Cretaceous.

Phylogenetic Relationships Among Diapsids

Our understanding of the phylogenetic relationships of several groups of Mesozoic tetrapods has changed in the last decade. Many of these forms (thecodonts, crocodilians, pterosaurs, nonavian dinosaurs, squamates, and rhynchosaurs) had skulls with two temporal openings (a **diapsid** skull), and the Diapsida is considered a monophyletic lineage that includes most of the major groups of Mesozoic tetrapods as well as the living crocodilians, birds, tuatara (*Sphenodon*), and squamates (lizards, snakes, and amphisbaenians) (Figure 13–1).

The name diapsid means "two arches" and refers to the presence of an upper and a lower fenestra in the temporal region of the skull (Figure 13–2). More distinctive than the openings themselves is the morphology of the bones that form the arch separating

them. The upper temporal arch is composed of a three-pronged postorbital bone and a three-pronged squamosal. The lower temporal arch is formed by the jugal and quadratojugal bones. The lower arch has been lost repeatedly in the radiation of diapsids, and the upper arch also is missing in some forms. Living lizards and snakes clearly show the importance of those modifications of the skull in permitting increased skull flexion (**kinesis**) during feeding, and the same significance may attach to loss of the arches in some extinct forms. In addition to the two temporal fenestrae, derived diapsids have a suborbital fenestra on each side of the head anterior to the eye, and the presence of this fenestra modifies the relationships among the bones of the palate and the side of the skull.

The earliest diapsid known is *Petrolacosaurus* from late Carboniferous deposits in Kansas. It is a moderately small animal, 60 to 70 centimeters from snout to tail tip, with a long neck, large eyes, and long limbs (Figure 13–3). It gives the impression of having been an agile terrestrial animal that may have fed on large insects and other arthropods. The early diapsids were diverse—*Askeptosaurus* (Figure 13–28d) represents a lineage of Triassic marine diapsids known as thallatosaurs. The derived diapsids can be split into two groups, the **Archosauromorpha** and

1. Diapsida: Skull with a dorsal temporal fenestra, upper temporal arch formed by triradiate postorbital and triradiate squamosal, modifications of the snout, characteristics of the nervous system, including a true Jacobson's organ present at least in embryos and with olfactory bulbs anterior to the eyes. 2. Archosauromorpha: Characteristics of the snout, stapes lacking a foramen, characteristics of the vertebrae, humerus, and feet. 3. Archosauria: Presence of an antorbital fenestra, orbit shaped like an inverted triangle, teeth laterally compressed, fourth trochanter on femur. 4. Crocodylotarsi: Ankle (tarsus) in which the astragalus forms a distinct peg that fits into a deep socket on the calcaneum, plus characters of the cervical ribs and the humerus. 5. Dinosauria: Characteristics of the palate, pectoral and pelvic girdles, hand, hindlimb, and foot. 6. Lepidosauromorpha: Postfrontal enters border of upper temporal fenestra, characteristics of the vertebrae, ribs, and sternal plates. 7. Lepidosauria: Determi-

nant growth with epiphyses on the articulating surfaces of the long bones, characteristics of the skull, pelvis, and feet. 8. Squamata: Fusion of bones in snout region, characteristics of the palate and skull roof, vertebrae, ribs, pectoral girdle, and humerus. (Based on M. J. Benton, 1985, *Zoological Journal of the Linnean Society* 84:97–164; J. Gauthier, 1986, pages 1–55 in *The Origin of Birds and the Evolution of Flight*, edited by K. Padian, *Memoirs of the California Academy of Sciences*, Number 8; H.-D. Sues, 1987, *Zoological Journal of the Linnean Society* 90:109–131; M. J. Benton [editor], 1988, *The Phylogeny and Classification of the Tetrapods*, Special Volume No. 35B, The Systematics Association, Oxford University Press, Oxford, UK; R. L. Carroll, 1988, *Vertebrate Paleontology and Evolution*, Freeman, New York, NY; M. J. Benton, 1990, *Vertebrate Paleontology*, HarperCollins Academic, London, UK; and M. J. Benton [editor], 1993, *The Fossil Record 2*, Chapman & Hall, London, UK.)

Figure 13–1 Phylogenetic relationships of the Diapsida. This diagram shows the probable relationships among the major groups of diapsids. Extinct lineages are marked by a dagger (†). The numbers indicate derived characters that distinguish the lineages.

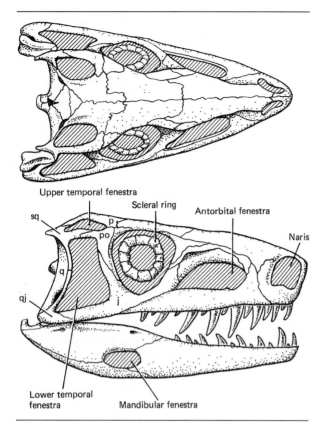

Upper temporal fenestra

Scleral ring

Antorbital fenestra

sq

p

po

Naris

q

qj

j

Lower temporal fenestra

Mandibular fenestra

Figure 13–2 The skull of *Euparkeria*, a primitive archosaur, shows the two arches in the temporal region that characterize the diapsid condition, and the antorbital fenestra. (From A. S. Romer, 1966, *Vertebrate Paleontology*, 3d edition, University of Chicago Press, Chicago, IL.)

the **Lepidosauromorpha** (Figure 13–1). The archosauromorphs include living crocodilians and birds, and the extinct rhynchosaurs, phytosaurs, nonavian dinosaurs, pterosaurs, and several late Permian and Triassic forms. The lepidosauromorphs include the extinct Younginiformes and the living tuatara and squamates plus their extinct relatives. In addition, four groups of specialized marine tetrapods (the placodonts, plesiosaurs, ichthyosaurs, and *Hupehsuchus*) are tentatively considered to be lepidosauromorphs. The skulls of these animals have a dorsal temporal opening, but lack a lower temporal fenestra, and the postorbital and squamosal bones do not have the three-pronged shape characteristic of diapsids. However, these patterns are within the range of modifications of the basic diapsid skull that is seen among other members of the clade.

The Archosauromorpha

The Archosauromorpha includes the most familiar of the Mesozoic diapsids, the nonavian dinosaurs, as well as their close relatives, the crocodilians, phytosaurs, birds, and pterosaurs (Figure 13–1). Less well-known groups of archosauromorphs are the rhynchosaurs and Prolacertiformes. The archosauromorphs are distinguished by several characteristics of the skull and axial skeleton (Figure 13–2).

Early archosaurs are grouped here as the Proterosuchia. This approach simplifies a complex area in which no one phylogenetic scheme is yet widely accepted. The proterosuchians in our classification include the Triassic forms known as thecodontians in earlier classifications such as that of Romer (1966). *Proterosuchus* was a quadrupedal, lizard-shaped carnivore, 2 or 3 meters long, that is known from South Africa. Related forms are known from deposits in China, Bengal, Eurasia, Australia, and Antarctica. *Erythrosuchus*, another Triassic quadruped, was twice the size of *Proterosuchus* and massively built, whereas *Euparkeria* was a lightly built animal about 150 centimeters long. Its hind limbs were half again as long as the fore limbs, suggesting that it was capable of bipedal locomotion (Figure 13–3).

Rhynchosauria

The rhynchosaurs were squat, heavily built tetrapods as much as 2 meters long (Figure 13–4). The distinctive specializations of rhynchosaurs lie in the structure of their jaws and teeth. In late Triassic forms the teeth in the upper jaw were borne on two maxillary tooth plates, each of which had several rows of teeth and was divided into two parts by a deep longitudinal groove. The lower jaw also bore two rows of teeth, one on the crest of the jaw and one lower down on the inner (lingual) side of the jaw. When the mouth was closed, the lower jaw fit snugly into the groove like the blade of a penknife folding into its handle. The premaxilla formed a heavy beak, and the jaw-closing muscles were powerful, giving the head a triangular shape.

Rhynchosaurs probably diverged from the diapsid stock in the Permian. The earliest rhynchosaurs known are small, lizard-like animals from the Triassic of South Africa. By the late Triassic, large rhynchosaurs were dominant members of many fossil faunas accounting for approximately half of all the terrestrial animals found (Benton 1983a). The abun-

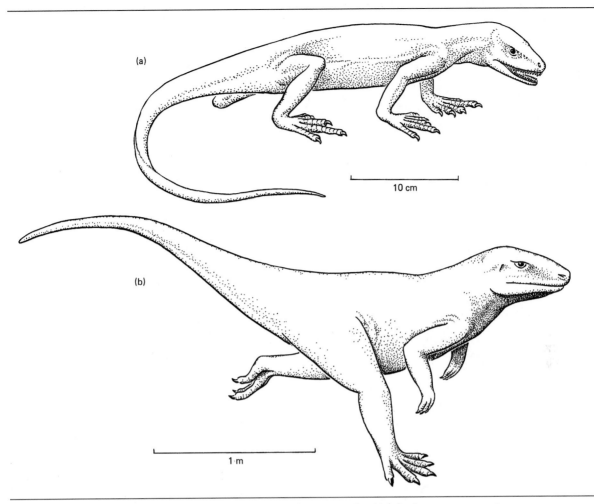

Figure 13–3 Early diapsids: (a) *Petrolacosaurus* from the late Pennsylvanian had forelimbs and hind limbs that were about the same length; (b) *Euparkeria* had hind limbs that were much longer than its forelimbs and was probably bipedal. ([a] From R. R. Reisz, 1977, *Science* 196:1091–1093.)

dance of rhynchosaurs in the late Triassic has been compared to that of antelopes in modern African savanna faunas.

Rhynchosaurs became extinct about 17 million years before the end of the Triassic, and a few million years after that nonavian dinosaurs were the dominant terrestrial animals. What could account for such a rapid (in geological time) disappearance of a successful group of animals with a worldwide distribution? Theories of competitive replacement of one group of animals by another better adapted group have long dominated the interpretation of paleontological information. However, as more detailed information becomes available this view is becoming less tenable as a general proposition, although it may apply to particular cases. Most of the time the temporal sequence of events appears to show that a once-dominant group waned, and shortly thereafter (in geological time) a new group burgeoned (Benton 1983b,c). That temporal mismatch appears to apply to the disappearance of rhynchosaurs (and the simultaneous disappearance of many kinds of synapsids—see Chapter 19) and the radiation of nonavian dinosaurs.

An event that does appear to correspond to the waning of the rhynchosaurs is a change in the climate and flora at the end of the Triassic. During the middle of the Triassic, when rhynchosaurs were dominant,

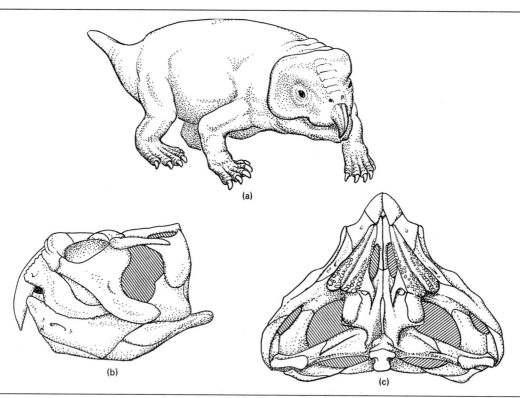

Figure 13–4 Rhynchosaurs: (a) reconstruction of the South American rhynchosaur *Scaphonyx;* (b) side view and (c) palatal view of the skull of the Scottish rhynchosaur *Hyperodapedon.* (From M. J. Benton, 1983, *Quarterly Review of Biology,* 58:29–55.)

most of the southern part of the world was covered by an assemblage of plants known as the *Dicroidium* flora, which consisted of seed ferns, horsetails, cycads, ginkgos, and conifers. The powerful and precise vertical chopping motion of the jaws of rhynchosaurs, which was guided by the groove in the tooth plates of the upper jaw, may have been suitable for cutting tough vegetation. Climatic conditions appear to have become more arid at the end of the Triassic, and the *Dicroidium* flora was replaced by a mixture of conifers and bennettitaleans (large, tree-like cycads). Rhynchosaurs probably could not rear up on their hind legs to browse on tall vegetation, and may have found their food supply greatly diminished as plants became taller and more widely spaced.

Prolacertiformes

The bar formed by the jugal and quadratojugal bones that closes the ventral side of the lower temporal opening in the diapsid skull was incomplete in the animals grouped as Prolacertiformes (Figure 13–5a). In living squamates this condition imparts a mobility to the quadrate bone that increases the efficiency of the lower jaw. Grouped among the Prolacertiformes are several medium-size tetrapods such as *Prolacerta* and *Protorosaurus* with lizard-like body proportions that appear to have been agile, terrestrial predators. Also included in the Prolacertiformes is the bizarre genus *Tanystropheus.* Several species of *Tanystropheus* are known, some of which were 6 meters in length. The body, limbs, and tail of *Tanystropheus* were of normal lizard-like proportions, but the neck, which contained 9 to 12 enormously elongate vertebrae, was as long as the body and tail combined. A small head, its jaws armed with conical teeth, perched on the end of this remarkable neck. The two parts of the body appear so different that when the first complete skeleton of this genus was discovered, it was found that bones from the front part of the animal had previously been described as belonging to a pterosaur and bones from the trunk had been identified as being from a nonavian dinosaur. Fossilized stomach contents have shown that *Tanystropheus* ate cephalopods (octopus and squid), but it's not clear how these strange animals captured their prey. The neck vertebrae did not have space for the attachment of strong muscles for bending, and thin ribs that extended backward from each vertebra would have stiffened the entire structure. Perhaps *Tanystropheus* cruised slowly in shallow water, probing in hollows in rocks or coral reefs in search of concealed prey.

Archosauria

The archosaurs are the animals most frequently associated with the great radiation of tetrapods in the Mesozoic. Nonavian dinosaurs and pterosaurs are distinctive components of many Mesozoic faunas, and other less familiar archosaurs were also abundant. The archosaurs are distinguished by the presence, in many forms, of an antorbital ("in front of the eye") fenestra. The skull was deep, the orbit of the eye was shaped like an inverted triangle rather than being circular, and the teeth were laterally compressed (Figure 13–5b). A trend toward bipedalism was widespread (but not universal) among archosaurs, and the ventral side of the shaft of the femur had a distinctive area with a rough surface, the fourth trochanter, which was the site of insertion of the powerful caudiofemoral muscle (Figure 13–5c).

Crocodylotarsi The archosaur stock gave rise to two lineages of aquatic fish eaters, the phytosaurs and crocodilians. The phytosaurs were the earlier radiation, and during the Triassic they were abundant and important elements of the shoreline fauna. In contrast to crocodilians, in which the nostrils are at the tip of the snout and a secondary palate separates the nasal passages from the mouth, phytosaur nostrils were located on an elevation just anterior to the eyes (Box 13–1). True crocodilians appeared in the Triassic and seem to have replaced phytosaurs by the end of that period. In most respects crocodilians conform closely to the skeletal structure of archosaurs, but the skull and pelvis are specialized. Crocodilians retained the nostrils at the tip of the snout and developed a secondary palate that carries the air passages posteriorly to the rear of the mouth. A flap of tissue arising from the base of the tongue can form a watertight seal between the mouth and throat. Thus, a crocodilian can breathe while only its nostrils are exposed without inhaling water. The increasing

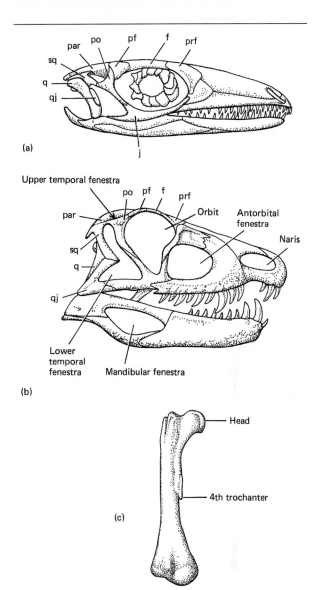

Figure 13–5 Morphological features of diapsids. (a) Skull of *Prolacerta* showing the incomplete lower temporal arch; (b) skull of the carnosaur *Ornithosuchus* showing the characteristic features of archosaurs: two temporal arches, a keyhole-shaped orbit, and an antorbital fenestra; (c) femur of *Thescelosaurus* showing the fourth trochanter. Key: f, frontal; par, parietal; pf, postfrontal; po, postorbital; prf, prefrontal; sq, squamosal; q, quadrate; qj, quadratojugal. ([a] and [b] From A. S. Romer, 1966, *Vertebrate Paleontology*, 3d edition, University of Chicago Press, Chicago, IL; [c] from A. S. Romer, 1956. *Osteology of the Reptiles*, University of Chicago Press, Chicago, IL.)

Box 13–1 Long-Snouted Fish Eaters

Aquatic and semiaquatic tetrapods with a crocodile-like body form have evolved repeatedly among tetrapods. The distinctive feature of this specialization is an elongate snout used to capture fish with a sideward sweep of the head. Among the earliest examples of this body form were the Triassic temnospondyl trematosaurs (Chapter 10). In the Diapsida, crocodile-like animals evolved in both the lepidosauromorph and the archosauromorph lineages. The champsosaurs (Figure 13–6a) are probably lepidosauromorphs. They are known from Cretaceous and early Tertiary fossil beds in North American and Europe. Champsosaurs were about 2 meters long. The posterior part of the skull was broad, suggesting the presence of strong jaw muscles.

Phytosaurs (Figure 13–6b) are the sister group of crocodilians and appear to have been abundant in the Triassic. Fossils are known from North America, Europe, and India. Crocodilians (Figure 13–6c) appeared in the Triassic and radiated extensively in the Jurassic and Cretaceous. The earliest crocodilians appear to have been terrestrial. *Protosuchus* ("first crocodile"), a late Triassic form, had long legs and a short, broad skull. Later crocodilians were more aquatic and had elongate snouts. The most specialized crocodilians were the thallatosuchians ("sea crocodiles"), marine forms that lacked dermal bony armor. Thallatosuchians had paddle-like limbs and a tail fin like that of ichthyosaurs in which the lower lobe was supported by the vertebral column and the upper lobe lacked skeletal support (Figure 13–6d).

In the ancestral diapsid skull the internal nares were located in the anterior part of the mouth, close to the external nares on the snout, and air passed through the length of the mouth as it was inhaled and exhaled. That arrangement is not effective for an aquatic animal, because the mouth is often full of water. (None of these animals had lips that could exclude water from the mouth.) Thus, they faced a problem in getting air from the nostrils into the trachea without inhaling water at the same time. Two solutions to the problem emerged: Phytosaurs shifted the position of the nostrils from the tip of the snout to a location just anterior to the eyes. In some phytosaurs the nostrils were located in a volcano-shaped elevation on the front of the skull. Champsosaurs and crocodilians, in contrast, evolved a secondary palate, a shelf of bone in the roof of the mouth that separates the nasal passages from the mouth itself. In champsosaurs the maxillary and ethmoid bones formed the secondary palate, whereas in crocodilians the maxillary and palatine bones formed most of the palate. Both solutions placed the internal nares at the rear of the mouth where a fleshy valve could keep water out of the air passages.

---

involvement of the premaxilla, the maxillae, and pterygoids in the secondary palate can be traced from Mesozoic crocodilians to modern forms.

Modern crocodilians are semiaquatic animals, but Triassic crocodilians were terrestrial. They were thin, slender animals about the size of a large cat, and give the impression of having been active hunters that probably preyed on smaller diapsids. Traces of this terrestrial origin persist in living crocodilians. They have well-developed limbs, and some species make extensive overland movements. Crocodilians can gallop, moving the limbs from their normal laterally extended posture to a nearly vertical position beneath the body.

The Cretaceous was the high point in crocodilian evolution. The extension of warm climates to land areas that are now in cool temperate climate zones favored both diversity and large size. *Deinosuchus* ("terrible crocodile") from the Cretaceous of Texas had a skull that was nearly 2 meters long. If this crocodilian had the same body proportions as extant forms, it would have had a total length of 12 to 15

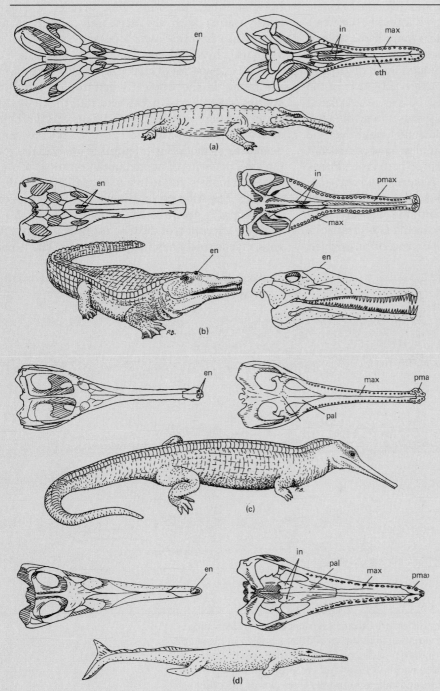

Figure 13–6 Convergent evolution of long-snouted diapsids: (a) champsosaur; (b) phytosaur; (c) crocodilian; (d) thallatosuchian. en, external nares; eth, ethmoid; in, internal nares; max, maxilla; pal, palatine; pmax, premaxilla. ([a] and [c] Skulls from A. S. Romer, 1966, *Vertebrate Paleontology*, 3d edition, University of Chicago Press, Chicago, IL; [b,d] skulls modified from W. K. Gregory, 1951, *Evolution Emerging*, Macmillan, New York, NY; [a–c] restorations from various sources; [d] restoration from J. Piveteau, 1955, *Traité de Paléontologie*, volume 5, Mason et Cie, Paris, France.)

meters, and might well have preyed on nonavian dinosaurs.

Enormous crocodilians persisted long after nonavian dinosaurs disappeared. A skull of the Miocene crocodile *Purussaurus brasiliensis* found in the Amazon Basin in 1986 is 1.5 meters long. If the animal that bore that skull had the same proportions as an alligator, it would have had a total lenth of 11 to 12 meters and have stood 2.5 meters tall—that is, the height of the ceiling in most houses. An isolated lower jaw in the paleontology museum at the Universidade do Acre is 30 centimeters longer than the jaw of the complete skull, and may have come from an animal 13 to 14 meters long. These crocodilians would have been as large as *Tyrannosaurus rex.*

A heavy, laterally flattened tail propels the body of a crocodilian in water, and the legs are held against the sides of the body. In the late Jurassic, a lineage of specialized marine crocodiles enjoyed brief success. These thallatosuchians had long skulls with pointed snouts. They lacked the dermal body armor typical of most crocodilians, and had developed a lobed tail very like that of the early ichthyosaurs, with the vertebral column turned downward into the lower lobe and the upper lobe supported by stiff tissue. The feet were paddle-like.

Only 21 species of crocodiles now survive. Most are found in the tropics or subtropics, but three species have ranges that extend into the temperate zone. In many respects crocodilians are the living animals most like Mesozoic forms, and they have been used as models in attempts to analyze the ecology and behavior of nonavian dinosaurs.

Systematists divide living crocodilians into three families: The Alligatoridae includes the two species of living alligators and the caimans (Figure 13–7). With the exception of the Chinese alligator, the Alligatoridae is solely a New World group. The Ameri-

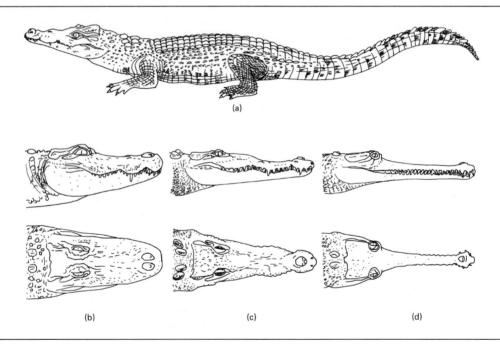

Figure 13–7 Modern crocodilians differ little from each other or from Mesozoic forms. The greatest interspecific variation in living crocodilians is seen in the shape of the head. Alligators and caimans are broad-snouted forms with varied diets. Crocodiles include a range of snout widths. The widest crocodile snouts are almost as broad as those of most alligators and caimans, and these species of crocodilians have varied diets that include turtles, fish, and terrestrial animals. Other crocodiles have very narrow snouts, and these species are primarily fish eaters. (a) Cuban crocodile; (b) Chinese alligator; (c) American crocodile; (d) gharial. (Modified from H. Wermuth and R. Mertens, 1961, *Schildkröten, Krocodile, Brückenechsen,* Gustav Fisher, Jena, East Germany.)

can alligator is found in the Gulf coast states, and several species of caimans range from Mexico to South America and through the Caribbean. Alligators and caimans are freshwater forms, whereas the Crocodylidae includes species such as the saltwater crocodile that inhabits estuaries, mangrove swamps, and the lower regions of large rivers. This species occurs widely in the Indo-Pacific region and penetrates the Indo-Australian archipelago to northern Australia. In the New World, the American crocodile is quite at home in the sea, and occurs in coastal regions from the southern tip of Florida through the Caribbean to northern South America.

The saltwater crocodile is probably the largest living species of crocodilian. Until recently, adults may have reached lengths of 7 meters. Crocodilians grow slowly once they reach maturity, and it takes a long time to attain large size. In the face of intensive hunting in the last two centuries, few crocodilians now attain the sizes they are genetically capable of reaching. Not all crocodilians are giants—several diminutive species live in small bodies of water. The dwarf caiman of South America and the dwarf crocodile of Africa are about a meter long as adults and live in swift-flowing streams.

The third family of crocodilians, the Gavialidae, contains only a single species—the gharial, which once lived in large rivers from northern India to Burma. This species has the narrowest snout of any crocodilian; the mandibular symphysis (the fusion between the mandibles at the anterior end of the lower jaw) extends back to the level of the 23rd or 24th tooth in the lower jaw. A very narrow snout of this sort is a specialization for feeding on fish that are caught with a sudden sideward jerk of the head. We have already called attention to the evolution of similar skull shapes in a variety of Mesozoic animals, including trematosaurs, phytosaurs, and the short-necked plesiosaurs.

Living crocodilians are ectotherms and small individuals bask in the sun to raise their body temperatures. A basking crocodilian can increase its rate of heating by using a right-to-left intracardiac blood shunt to increase blood flow in the peripheral circulation, just as lizards do. However, the structure of the crocodilian heart is different from that of the squamate and turtle heart, and the intracardiac blood shunt is achieved in a different way. Crocodilians, like birds and mammals, have a septum that separates the left and right sides of the ventricle. (It is the absence of that septum in the hearts of squamates and turtles that permits them to use pressure differentials to shift blood from the pulmonary [right] side of the ventricle across the muscular ridge to the systemic [left] side [see Figure 12–6].)

In the crocodilian heart, the right aortic arch opens from the left ventricle and receives oxygenated blood (Figure 13–8). The left aortic arch and the pulmonary artery both open from the right ventricle. Deoxygenated blood enters the pulmonary artery,

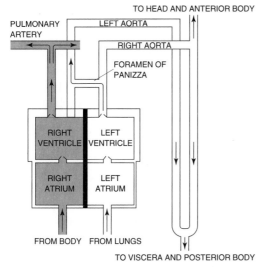

Figure 13–8 The relationship of the heart and major vessels of a crocodilian. The right aortic arch opens from the left ventricle and receives oxygenated blood, which flows to both the anterior and posterior parts of the body. The left aortic arch opens from the right ventricle. When pressure in the right ventricle equals or exceeds pressure in the left, the atrioventricular valve opens and deoxygenated blood flows into the left aorta, which carries blood only to the posterior part of the body. When pressure in the left ventricle exceeds pressure in the right ventricle, the right atrioventricular valve is held shut, and oxygenated blood flows via the Foramen of Panizza into the left aortic arch. (From A. G. Kluge (editor), 1977, *Chordate Structure and Function*, Macmillan, New York, NY.)

and you would expect that it would also flow into the left aortic arch, but studies of alligators have shown that the pattern of blood flow in the heart depends on what the alligator is doing (Jones and Shelton 1993). When an alligator is at rest, blood pressure is approximately the same in the right and left ventricles. In this situation, deoxygenated blood does flow from the right ventricle into the left aortic arch, (this is a right-to-left intracardiac shunt, because "right" and "left" refer to the ventricles, not to the aortic arches) and then posteriorly to the viscera. Deoxygenated blood contains hydrogen ions that are produced when carbon dioxide combines with the bicarbonate buffering system of the blood. The hydrogen ions that enter the left aortic arch may be used for the secretion of hydrochloric acid in the stomach during digestion. Note that the right aortic arch supplies the blood to the head, so the brain receives only oxygenated blood.

A different pattern of blood flow occurs when the alligator is active and pressure in the left ventricle rises above that of the right ventricle. The left and right aortic arches are connected via the Foramen of Panizza. When pressure in the right aortic arch exceeds that in the left, blood flows through this passage from the right aortic arch into the left. The increased pressure in the left aortic arch holds the ventricular valve closed, preventing entry of deoxygenated blood from the right ventricle, and both aortic arches receive oxygenated blood.

A third situation probably occurs when an alligator dives or is using the right-to-left shunt to increase blood flow to the limbs to accelerate heating. When blood vessels in the pulmonary circuit are constricted, pressure in the right ventricle rises and deoxygenated blood flows into the left aortic arch.

Pterosauria The archosaurs gave rise to two independent radiations of fliers. The birds are one of these lineages, and their similarity to archosaurs is so striking that had they disappeared at the end of the Mesozoic, they would be considered no more than another group of highly specialized archosaurs. The other lineage of flying archosaurs were the pterosaurs of the Jurassic and Cretaceous (Figure 13–9). They ranged from the sparrow-size *Pterodactylus* to *Quetzalcoatlus,* with a wingspan of 13 meters. The wing formation of pterosaurs was entirely different from that of birds. The fourth finger of pterosaurs was elongate and supported a membrane of skin anchored to the side of the body and perhaps to the hind leg. A small splint-like bone was attached to the front edge of the carpus and probably supported a membrane that ran forward to the neck. The rhamphorynchoid pterosaurs had a long tail with an expanded portion on the end that was presumably used for steering; the pterodactyloids lacked a tail.

Flight is a demanding means of locomotion for a vertebrate, and it is not surprising that pterosaurs and birds show a high degree of convergent evolution. The long bones of pterosaurs were hollow, as they are in birds and many other archosaurs, reducing weight with little loss of strength. The sternum, to which the powerful flight muscles attach, was well developed in pterosaurs, although it lacked the keel seen in birds. The eyes were large, and casts of the brain cavities of pterosaurs show that the parts of the brain associated with vision were large and the olfactory areas small, as they are in birds. The cerebellum, which is concerned with balance and coordination of movement, was large in proportion to the other parts of the brain as it is in birds.

Some pterosaurs had lost their teeth and evolved a bird-like beak. Others had sharp, conical teeth in blunt skulls reminiscent of those of bats. Some pterosaurs with elongate skulls and stout sharp teeth may have caught fish or small tetrapods. *Pterodaustro* had an enormously long snout lined with a comb-like array of fine teeth that may have been used for sieving small aquatic organisms. *Dsungaripterus* had long jaws that met at the tips like a pair of forceps. The tips of the jaws were probably covered with a horny beak, and blunt teeth occupied the rear of the jaw. These animals may have plucked snails from

Figure 13–9 Pterosaurs: (a) *Rhamphorhynchus* from the Jurassic; (b) *Pteranodon* from the Cretaceous. The skulls of pterosaurs suggest dietary specializations. (c) *Anurognathus* may have been insectivorous; (d) *Eudimorphodon* may have eaten small vertebrates; (e) *Dorygnathus* may have been a fish eater; (f) *Pterodaustro* had a comb-like array of teeth that may have been used to sieve plankton; (g) *Dsungaripterus* may have pulled mollusks from rocks with a horny beak and then crushed them with its molariform teeth. (Skulls modified from D. Norman, 1985, *The Illustrated Encyclopedia of Dinosaurs,* Salamander Books, London, UK.)

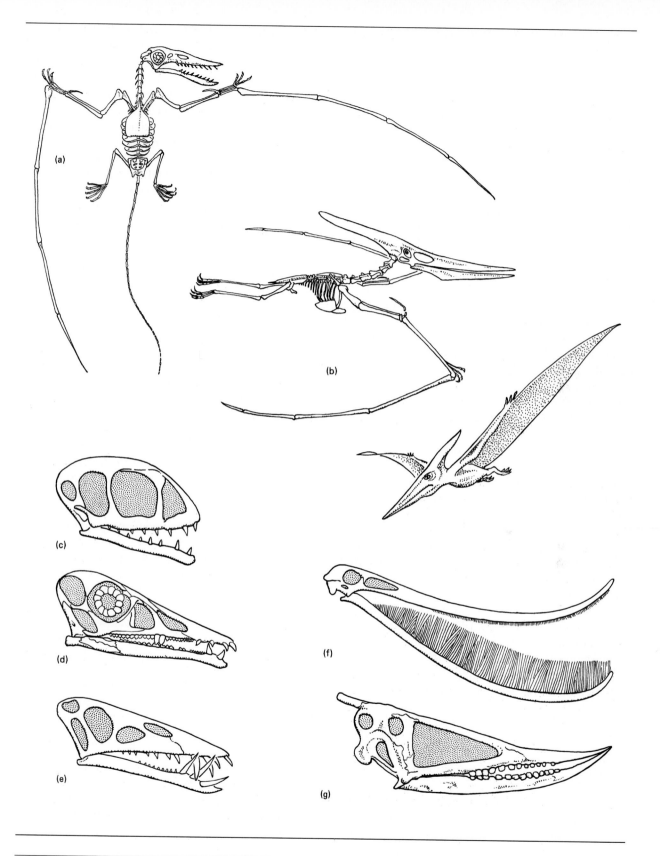

rocks with their beaks and then crushed them with their broad teeth.

The flight capacities of pterosaurs have long been debated, and most hypotheses about their ecology have been based on the assumption that they were weak fliers. That assumption has led to suggestions of restrictions of activities and habitats of pterosaurs that seem unlikely for a group of animals that was clearly diverse and successful through much of the Mesozoic. An aerodynamic analysis suggests that the flying abilities of pterosaurs have been underestimated (Hazlehurst and Rayner 1992). This view suggests that small pterosaurs were slow, maneuverable fliers like bats. The large pterosaurs appear to have been specialized for soaring like frigate birds and some vultures.

Speculations about the flying abilities of pterosaurs depend on what assumption one makes about the shape of the wing. It extended outward to the tips of the fourth finger, but where was it attached to the body? Did it stop at the waist, or extend onto the hind limbs as the wing does in bats? The structure of the wing may have varied among pterosaurs. A fossil of *Pterodactylus* shows the wing attached at least to the thigh (Padian and Rayner 1993), whereas an extremely well-preserved fossil of *Sordes pilosus* from Jurassic sediments in Kazakhstan shows that the hind legs were involved in the flight structures (Unwin and Bakhurina 1994). The wing of this species attached along the outside of the hind limb to the ankle, and another flight membrane, the uropatagium, stretched between the hind legs and was controlled by the fifth toe. This degree of involvement of the hind limbs with the wings would have limited their role in terrestrial locomotion (as is the case for bats), and *Sordes* may have been a clumsy walker on flat surfaces but a good climber on rocks and trees (Unwin 1987). Other pterosaurs, such as *Pterodactylus,* may have been capable of bipedal locomotion (Padian and Rayner 1993).

The new fossil of *Sordes* also corrects a mistaken interpretation of earlier material: Traces of thin fibers had been interpreted as hair, and had led to the suggestion that *Sordes,* and perhaps all pterosaurs, were endotherms. The new fossil shows that these fibers were part of a system that stiffened the outer part of the wing (Unwin and Bakhurina 1994).

Nonavian Dinosaurs By far the most generally known of the archosaurs are the Saurischia and Ornithischia. These groups are linked in popular terminolo-

gy as dinosaurs, but differ in the specializations they developed. Both groups appear to have been ancestrally bipedal and to have evolved some secondarily quadrupedal forms.

Many of the morphological trends that can be traced in archosaur evolution appear to be associated with increased locomotor efficiency. The two most important developments were the movement of the legs under the body and a widespread tendency toward bipedalism. Early archosauromorphs had a sprawling posture like that of many living amphibians and squamates. The humerus and femur were held out horizontally from the body, and the elbow and knee were bent at a right angle. Derived archosaurs had legs that were held vertically beneath the body.

Among early tetrapods, muscles originating on the pubis and inserting on the femur protract the leg (move it forward), muscles originating on the ischium abduct the femur (move it toward the midline of the body), and muscles originating on the tail retract the femur (move it posteriorly). The ancestral tetrapod pelvis, little changed from *Ichthyostega* through early archosauromorphs, was plate-like (Figure 13–10a). The ilium articulated with one or two sacral vertebrae, and the pubis and ischium did not extend far anterior or posterior to the socket for articulation with the femur (acetabulum). The pubofemoral and ischiofemoral muscles extended ventrally from the pelvis to insert on the femur. (The downward force of their contraction was countered by iliofemoral muscles that ran from the ilium to the dorsal surface of the femur.) As long as the femur projected horizontally from the body, this system was effective. The pubofemoral and caudiofemoral muscles were long enough to swing the femur through a large arc relative to the ground. As the legs were held more nearly under the body, the pubofemoral muscles became less effective. As the femur rotated closer to the pubis, the sites of muscle origin and insertion moved closer together and the muscles themselves became shorter. A muscle's maximum contraction is about 30 percent of its resting length; thus the shorter muscles would have been unable to swing the femur through an arc large enough for effective locomotion had there not been changes in the pelvis associated with the evolution of bipedalism (Charig 1972).

The bipedal ornithischian and saurischian dinosaurs carried the legs completely under the body and show associated changes in pelvic structure. The two groups attained the same mechanically advantageous end in different ways (Figure 13–

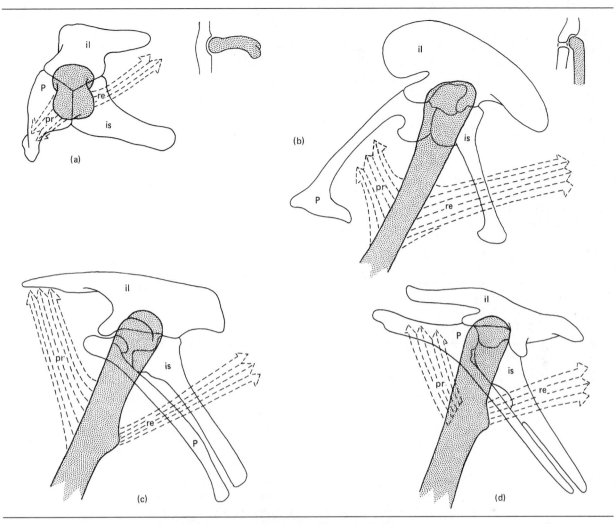

Figure 13–10 Functional aspects of the pelvises of nonavian dinosaurs. Pelvic morphology of an early archosaur (a, *Euparkeria*), a saurischian dinosaur (b, *Ceratosaurus*), and two ornithischian dinosaurs (c, *Scelidosaurus*; d, *Thescelosaurus*). The presumed action of femoral protractor muscles (pr) and retractors (re) is shown by arrows. Insets show an anterior view of the articulation of the femur with the pelvis. p, pubis; il, ilium; is, ischium.

10). In quadrupedal saurischians, the pubis and ischium both became elongated and the pubis was rotated anteriorly, so that the pubofemoral muscles ran back from the pubis to the femur and were able to protract it (Figure 13–10b). The pubis of early ornithischians did not project anteriorly (Figure 13–10c). Instead, the ilium was elongated anteriorly, and it appears likely that the femoral protractors originated on the anterior part of the ilium, from which they ran posteriorly to the femur. This condition is seen in the pelvis of ornithischians such as *Sceli-*

dosaurus (Figure 13–10c), and appears to be maintained in the ankylosaurs, a group of derived quadrupedal ornithischians. Other ornithischians developed an anterior projection of the pubis that ran parallel to and projected beyond the anterior part of the ilium (Figure 13–10d). This development occurred in both bipedal and quadrupedal lineages and provided a still more anterior origin for protractor muscles.

The trend toward bipedalism was important in terms of opening new adaptive zones to archosaurs.

A fully quadrupedal animal uses its forelegs for walking, and any changes in limb morphology must be compatible with that function. As animals become increasingly bipedal, the importance of the forelegs for locomotion decreases and the scope of the specialized functions that can develop increases. Many of the smaller carnivorous dinosaurs that were fully bipedal used their forelegs to seize prey. Specialization of forelimbs as wings occurred twice among diapsids, once in the evolution of birds and once in pterosaurs.

Bipedal animals have hind legs that are considerably longer than their forelegs, and the degree of disproportion between hind legs and forelegs is assumed to reflect the extent of bipedalism in a given species. The quadrupedal archosaurs had longer hind legs than forelegs, and this condition is thought by most paleontologists to indicate that they were secondarily quadrupedal, having evolved from bipedal ancestors.

The Saurischian Dinosaurs and the Origin of Birds

Two groups of saurischian dinosaurs are distinguished, the **Theropoda** and the **Sauropodomorpha**. Theropods, which include the extant birds, are carnivorous bipeds, whereas the extinct sauropodomorphs were quadrupedal herbivores. Ten shared derived characters unite saurischians (Gauthier 1986); the most obvious is an elongate, mobile, S-shaped neck. This character distinguishes birds among living amniotes. Other bird-like characters of saurischians are found in modifications of the hand, skull, and postcranial skeleton.

Sauropodomorph Dinosaurs

The earliest sauropod dinosaurs were the prosauropods, a group that was abundant and diverse in the late Triassic and early Jurassic (Galton 1990). Three types of prosauropods are known, differing in size and tooth structure. The anchisaurids ranged in size from *Anchisaurus* (2.5 meters) to *Plateosaurus* (6 meters). The anchisaurids had long necks and small heads (Figure 13–11a), and the teeth of the best-known forms had large serrations. Modern herbivorous lizards (iguanas) have teeth with very much the same form, and anchisaurids were probably herbiv-

orous. Supporting this view is the presence of **gastroliths** (*gast* = stomach, *lith* = stone) associated with some fossil prosauropods. These stones were probably swallowed by the dinosaurs and lodged in a muscular gizzard where they assisted in grinding plant material to a pulp that could be digested more readily; some birds use gastroliths in this manner. Prosauropods had cheeks that retained food in the mouth as it was processed by the teeth. The earliest prosauropods were small and bipedal. Later forms were larger, and their body proportions suggest that they could stand vertically on their hind legs, but probably employed a quadrupedal posture most of the time.

Derived prosauropods, such as the melanorosaurids were larger than the early prosauropods (*Riojasaurus* from the late Triassic of Argentina was 11 meters long). No skulls of melanorosaurids have been found, so the structure of their teeth is unknown. The long, slender neck of *Riojasaurus* suggest that the head was small, like that of early prosauropods. The yunnanosaurids were smaller than the melanorosaurids and more lightly built, and they differed from the earlier prosauropods in having teeth shaped like flattened cylinders with a chisel-shaped tip. This is the tooth structure seen in the giant sauropod dinosaurs, and is quite distinct from that of the laterally flattened, serrated teeth of early prosauropods such as *Plateosaurus*.

The long necks of all the prosauropods suggest that they were able to browse on plant material at heights up to several meters above the ground. The ability to reach tall plants might have been a significant advantage during the shift from the low-growing *Dicroidium* flora to the taller bennettitaleans and conifers that occurred in the late Triassic. (It would also have had a powerful influence on the evolution of plants—see Chapter 14.)

The derived sauropods of the Jurassic and Cretaceous were enormous quadrupedal herbivores. The sauropods were the largest terrestrial vertebrates that have ever existed, reaching lengths of 25 meters and weighing 20,000 to 50,000 kilograms. Three huge sauropods discovered in Colorado have not yet been fully described (Jensen 1985). One of these, *Supersaurus*, may have been 40 meters long and have weighed more than 100,000 kilograms, the equivalent of 20 elephants. A new fossil from New Mexico, popularly known as "*Seismosaurus*," may be even longer than *Supersaurus* (McIntosh 1989).

Figure 13–11 Sauropodomorph dinosaurs: (a) *Plateosaurus*; (b) *Camarasaurus*;
(c) *Diplodocus*.

Two major types of giant sauropods can be distinguished, the diplodocoids and camarasauroids. The diplodocoids include *Apatosaurus* (formerly known as *Brontosaurus*) and *Diplodocus* (Figure 13–11). These animals had long necks (15 cervical vertebrae) and long tails (up to 80 caudal vertebrae) that ended in a thin whiplash. Their front legs were relatively short, and the trunk slanted upward from the shoulders to the hips. Their skulls were elongate, teeth were limited to the front of the mouth, and the modest development of the bones of the lower jaw suggests that the jaw muscles were not particularly powerful.

Camarasauroids had necks with only 12 vertebrae, and their tails were shorter than those of diplodocoids (about 50 vertebrae), and lacked the whip-like extension that was characteristic of the diplodocoids. The forelimbs of camarasauroids were relatively long, and the vertebral column was nearly horizontal. Brachiosaurids had still longer front legs, and the trunk sloped steeply downward from the shoulders to the hips. Camarasauroids and brachiosaurids had compact skulls with stout jaws and large chisel-shaped teeth. The teeth of *Camarasaurus* and *Brachiosaurus* show evidence of heavy wear, suggesting that they fed on abrasive material.

Both kinds of sauropods were enormously heavy, and their vertebrae show features that helped the spinal column to withstand the stresses to which it was subjected (Figure 13–12). The vertebrae themselves were massive, and the neural arches were well developed. Strong ligaments transmitted forces from one arch to adjacent ones to help equalize the stress. The head and tail were cantilevered from the body, supported by a heavy spinal ligament. The sides of the neural arches and centra had hollows in them, possibly occupied by air sacs in life, that reduced the mass of the bones with little reduction in strength. The feet of these forms were elephant-like, and fossilized tracks indicate that the hind legs bore about two-thirds of the body weight. Some trackways show no tail marks, suggesting that the tails were carried in the air, not dragged along the ground in the manner shown in almost all illustrations of these dinosaurs.

Another mechanical problem that the sauropods would have faced was the difficulty of pumping blood to a head that was sometimes as much as 20 meters above the ground and 6 or 7 meters above the level of the heart (Lillywhite 1991). Blood is mostly water, and water is heavy. When their heads were raised to browse on trees, the tallest sauropods would have required ventricular blood pressures exceeding 500 millimeters of mercury to overcome the hydrostatic pressure of a 7-meter column of blood between the heart and the brain. A column of blood extending to a head 20 meters above the ground could have produced blood pressures as great as 1000 millimeters of mercury in the vessels of the legs and feet of a large sauropod. Pressures that high would have tended to force water across the walls of the capillaries, causing the legs and feet to swell. Muscular constriction of small arteries in the limbs could have reduced pressure in the capillaries, and prevented this leakage of blood. Giraffes are the closest living parallel of the long-necked sauropod dinosaurs, and they encounter similar problems in pumping blood to the head and preventing the accumulation of water in the legs. Their hearts generate pressures approaching 300 millimeters of mercury to

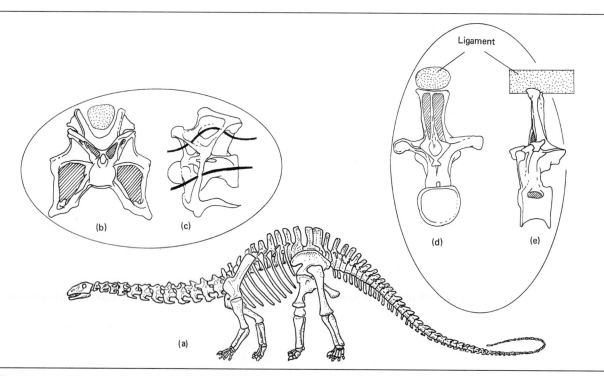

Figure 13–12 The skeletons of large diplodocid sauropods like *Apatosaurus* (a) combined lightness with strength. Vertebrae from the dorsal region (b, posterior view; c, lateral view) and neck (d, anterior view; e, lateral view) show the bony arches that acted like flying buttresses on a large building. (The black ribbons in [e] indicate the position of the arches.)

pump blood to the brain, and tight skin on the legs of giraffes prevents swelling.

Theropod Dinosaurs

The theropod dinosaurs include birds and all saurischian dinosaurs that are more closely related to birds than to sauropodomorph dinosaurs (Gauthier 1986). The four lineages of theropod dinosaurs shown in Figure 13–13 can be divided into three general types of animals: large predators (ceratosaurs and carnosaurs), fast-moving predators that attacked small prey (ornithomimids), and fast-moving predators that attacked prey larger than themselves (deinonychosaurs or dromeosaurs).

Ceratosaurs and Carnosaurs The carnosaurs are the impressive predators that form the centerpieces of paleontological displayed in many museums. Increasing body size among carnosaurs through the Mesozoic paralleled a similar trend in the herbivorous saurischians and ornithischians that were their prey. *Megalosaurus* of the late Jurassic was 6 meters long, and more fully bipedal than earlier forms (Figure 13–14). The head was large in proportion to the body, and the long teeth were fearsome weapons. The late Cretaceous tyrannosaurids such as *Tarbosaurus* and *Tyrannosaurus* were still longer, up to 15 meters in length, and stood some 6 meters high. The front legs of the most specialized of these giant carnosaurs seem absurdly small; they were too short even to reach the mouth and had only two small fingers on each hand. Instead of relying on the forelegs to capture prey as coelurosaurs and ornithomimids probably did, carnosaurs appear to have concentrated their weapons in the skull. The size of the head increased relative to the body, and the neck shortened. The head was lightened by the elaboration of antorbital and mandibular fenestrae that reduced the skull to a series of bony arches providing maximum strength for a given weight. The teeth were as much as 15 centimeters long, dagger-shaped with serrated edges. Experimental studies that used fossilized tyrannosaur teeth to bite meat showed that the serrations increased the cutting effect only slightly, but they trapped and retained meat fibers (Abler 1992). This debris would have supported the growth of bacteria, and a tyrannosaur bite would almost surely have become infected. Perhaps tyrannosaurs did not kill necessarily large prey such as sauropods in the initial attack, but relied on infection to weaken the victim and make it susceptible to a subsequent attack.

The spinosaurids, large (to 13 meters) carnosaurs known from the Cretaceous of North America and Africa, are distinguished by having trunk vertebrae with neural spines that projected upward as much as 1.8 meters above the back. These spines are assumed to have supported a dorsal crest. A huge sail-like structure of this sort might have had several functions. Dorsal crests composed of individual scales occur among lizards. They are sexually dimorphic—better developed in males than in females—and are used in social behavior. A function of this sort seems a plausible evolutionary origin for the sail of *Spinosaurus*, but was probably not its only function. The surface area of the sail of a *Spinosaurus* or an *Acrocanthosaurus* was a substantial fraction of the total surface area of the animal, and the sail would have affected energy exchange between the animal and the environment.

Whatever the selective forces that led to the first appearance of the sail, thermoregulation was almost surely one of its ultimate functions, because it was too large not to have an effect on energy exchange. If a *Spinosaurus* oriented its body perpendicular to the rays of the sun, the sail would absorb heat that could be transported into the body by blood flow. Large lizards and crocodilians use this mechanism to speed warming in the morning, although they lack specialized heat exchange structures. The sail of a spinosaurid could also have been used to dissipate heat. If the animal faced into the sun, only the front edge of the sail would receive direct solar radiation, and the sides of the sail could transfer heat from the body by convection and radiation.

Coelurosaurs The coelurosaurs include birds and all the theropods more closely related to birds than to carnosaurs. Several characters of living birds are seen in coelurosaurs (Gauthier 1986). The most interesting of these from the perspective of the origin of birds include features usually thought to be associated with powered flight, especially a fused bony sternum and a furcula (wishbone) formed by fusion of the clavicles.

Early coelurosaurs like the Triassic form *Coelophysis* (Figure 13–14a), which was about 3 meters in total length, were probably active, cursorial predators on small dinosaurs, lizards, and insects. The ostrich-like dinosaurs carried the characteristics of coelurosaurs to a greater degree of specialization.

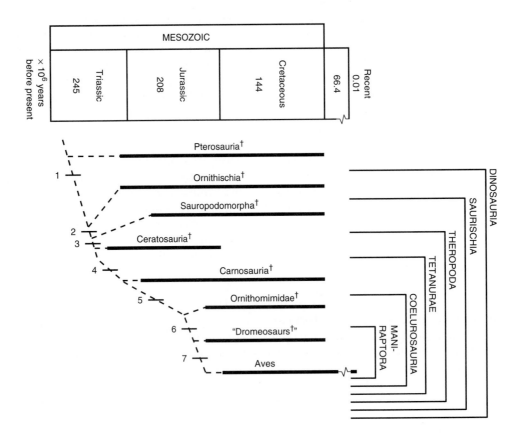

Ornithomimus of the late Cretaceous was ostrich-like in size, shape, and probably ecology as well. It had a small skull on a long neck, and its toothless jaws were covered with a horny bill. The forelegs were longer than those of *Coelophysis*, and only three digits were developed on the hands. The inner digit was opposable and the wrist was flexible, making the hand an effective organ for the capture of small prey. Like ostriches, *Ornithomimus* was probably omnivorous and fed on fruits, insects, small vertebrates, and eggs. Quite possibly it lived in herds, as do ostriches, and its long legs suggest that it inhabited open regions rather than forests.

Apparently not all ornithomimosaurs preyed on small animals. A theropod from the Gobi Desert, *Deinocheirus*, had hands more than 60 centimeters long that appear to have been used for grasping and dismembering large prey. The proportions of the hands and arms are like those of coelurosaurs, and if this theropod had the same body proportions as other coelurosaurs, it may have been more than 7.5 meters tall, exceeding the carnosaur *Tyrannosaurus rex*, previously the tallest theropod known.

Deinonychosaurs Deinonychus was unearthed by an expedition from Yale University in early Cretaceous sediments in Montana (Figure 13–14d). It is a small theropod, a little over 2 meters long. Its distinctive features are the claw on the second toe of the hind foot and the tail. In other theropods the hind feet are clearly specialized for bipedal locomotion and are very similar to bird feet. In these forms the third toe is the largest, the second and fourth are smaller, and the fifth has sometimes disappeared entirely. The first toe is turned backward, as in birds, to provide support behind the axis of the leg. The second toe of deinonychosaurs and especially the claw on that toe are enlarged (Figure 13–15). In its normal position the claw was apparently held off the ground and it could be bent upward even farther.

It seems likely that deinonychosaurs used these claws in hunting, disemboweling prey with a kick.

1. Dinosauria: Characteristics of the palate, pectoral and pelvic girdles, hand, hindlimb, and foot. 2. Saurischia: Construction of the snout, extension of the temporal musculature onto the frontal bones, elongation of the neck, modifications of the articulations between vertebrae, and modifications of the hand. 3. Theropoda: Construction of the lower jaw, bones of the skull roof and palate, fenestra in the maxilla, characters of the vertebrae, neural arches, and transverse processes lacking posterior to a transition point in the middle of the tail, modifications of the hand and foot, fibula and tibia closely adpressed, thin-walled (hollow) long bones. 4. Tetanurae: Large fenestra posteriorly located in the maxilla, large fanglike teeth absent from dentary, maxillary tooth row ends anterior to orbit of eye, transition point in tail is farther anterior than in theropods, expanded distal portion of pubis, characters of the foot. 5. Coelurosauria: Fenestra in roof of mouth, characters of cervical vertebrae and ribs, furcula (wishbone) formed by fused clavi-

cles, fused bony sternal plates, elongate forelimb and hand, characters of the foot. 6. Maniraptora: Prefrontal reduced or absent, characters of the vertebrae, transition point in tail vertebrae close to base of tail, characteristics of the feet and pelvis. 7. Aves: Progressive loss of teeth on maxilla and dentary, well-developed bill, feathers, characteristics of skull, jaws, vertebrae, and axial and appendicular skeleton. (Based on J. Gauthier, 1986, pages 1–55 in *The Origin of Birds and the Evolution of Flight*, edited by K. Padian, *Memoirs of the California Academy of Sciences*, Number 8; M. J. Benton (editor), 1988, *The Phylogeny and Classification of the Tetrapods*, Special Volume No. 35B, The Systematics Association, Oxford University Press, Oxford, UK; R. L. Carroll, 1988, *Vertebrate Paleontology and Evolution*, Freeman, New York, NY; M. J. Benton, 1990, *Vertebrate Paleontology*, HarperCollins Academic, London, UK; and M. J. Benton (editor), 1993, *The Fossil Record 2*, Chapman & Hall, London, UK.)

Figure 13–13 Phylogenetic relationships of the Saurischia. This diagram shows the probable relationships among the major groups of saurischian dinosaurs, including birds. Extinct lineages are marked by a dagger (†). The numbers indicate derived characters that distinguish the lineages.

The structure of the tail was equally remarkable. The caudal vertebrae were surrounded by bony rods that were extensions of the prezygapophyses (dorsally) and hemal arches (ventrally) that ran forward about 10 vertebrae from their place of origin. Contraction of muscles at the base of the tail would be transmitted through these bony rods, drawing the vertebrae together and making the tail a rigid structure that could be used as a counterbalance or swung like a heavy stick. Possibly the tail was part of the armament of *Deinonychus*, used to knock prey to the ground where it could be kicked, and it may have served as a counterweight for balance as *Deinonychus* made sharp turns.

The discovery of *Deinonychus* stimulated a reexamination of fossils of several other genera of small theropod dinosaurs from the Cretaceous, including *Dromeosaurus* and *Velociraptor*. All of these forms have an enlarged claw on the second toe of the hind foot, and they are now grouped with *Deinonychus* and birds in the Maniraptora. *Deinonychus*-like claws 35 centimeters long that were discovered in early Cretaceous sediments in Utah in the autumn of 1991 probably came from a previously unknown theropod (nicknamed Super-Slasher by paleontologists) that was nearly as large as a *Tyrannosaurus rex* and

had the speed, agility, and predatory behavior of *Deinonychus* (Figure 13–16).

Birds as Dinosaurs

The similarity of birds and nonavian dinosaurs has long been recognized. In the 1860s and 1870s Thomas Henry Huxley was an ardent advocate of that relationship, writing that birds are nothing more than "glorified reptiles." Huxley, in fact, was so impressed by their many similarities that he placed birds and reptiles together in his classification scheme as the Class Sauropsida, and birds are now viewed as the most derived theropod dinosaurs. The similarities of birds and theropods include the following derived characters:

1. Elongate, mobile S-shaped neck
2. Skull and neck joined by a single occipital condyle
3. Intertarsal ankle joint
4. Hollow, pneumatic bones

Birds differ from other theropods mainly in features directly associated with flight and endothermy, both of which are dependent on feathers. Perhaps in no

Figure 13–14 Theropod dinosaurs: (a) *Coelophysis*, a Triassic ceratosaur; (b) *Ornithomimus*, a Cretaceous ornithomimid; (c) *Megalosaurus*, a Jurassic form, probably a carnosaur. (d) *Deinonychus*, a Cretaceous deinonychosaur.

other major group of vertebrates has evolution been so determined by a single structural feature as in the case of birds with their highly modifiable and multifunctional feathers.

As we pointed out in Chapter 4, birds have nearly constant body temperatures of 40 to 41°C. These high temperatures depend on two characteristics: the high metabolic rates of birds and the excellent insulation provided by feathers. The paradox presented by the evolution of endothermy is that neither a high metabolic rate nor insulation is advantageous by itself; both are required for endothermal thermoregulation to be functional. Thus, the origin of feathers probably resulted initially from selection for some function other than endothermal thermoregulation.

One possibility is that elongate scales initially facilitated ectothermal thermoregulation as explained in Chapter 4.

The second major difference between birds and the other living archosaurs (the crocodilians) is, of course, the ability of birds to fly. Is it possible that feathers evolved initially because they conferred an advantage in locomotion, and subsequently took on a thermoregulatory role? To answer this question we must examine hypotheses about the origin of flight. Three types of locomotion have been suggested as being important in the evolution of flight by birds: running along the ground, leaping into the air, or gliding from trees. Hypotheses about the origin of flight can be divided into two categories: The "from

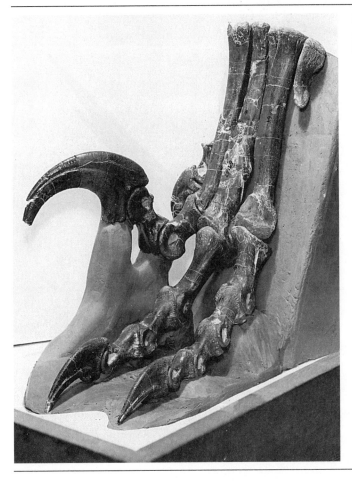

Figure 13–15 The foot of *Deinonychus*, showing the enlarged claw. (Courtesy of Barbara Moore, Peabody Museum, Yale University, New Haven, CT.)

the ground up" hypotheses picture flight as an outgrowth of selection for terrestrial locomotion (see Ostrom 1986), whereas the "from the trees down" ideas suggest that birds were initially arboreal animals that first took to the air as gliders (see Rayner 1988). We will examine the evidence for both categories of hypotheses, but first we must consider the earliest bird, *Archaeopteryx*.

Archaeopteryx and the Origin of Birds At least three different groups of diapsids developed gliding or powered flight in the Mesozoic. The kuehneosaurids of the late Triassic were the first tetrapods to become airborne. These were diapsids, up to a meter long, probably in the archosauromorph lineage. They are distinguished by extremely elongate ribs that probably supported a wing of skin like that seen in the living East Indian lizards of the genus *Draco*. *Draco* are arboreal lizards, and they use their wings to glide from tree to tree. A flight starts with a dive from an

elevated perch. The lizard descends at an angle of about 45 degrees, then levels out and uses the kinetic energy developed during the dive to glide nearly horizontally. A brief upward glide immediately precedes landing on another perch. Glides as long as 60 meters have been recorded with a loss of altitude of less than 2 meters.

The ribs of *Draco* can be folded back against the body when they are not in use. The skin of the wings is brightly colored, and the wings are spread in social displays as well as for flight. Many lizards flatten the body laterally by spreading the ribs, making themselves look bigger during social interactions. The elongate ribs and wings of *Draco* and the kuehneosaurids might have had their origin in that widespread behavior.

True wings and powered flight (as distinct from airfoils used for gliding) evolved twice among diapsids, in the pterosaurs and independently in the birds. The oldest known birds are from the Jurassic

Figure 13–16 Two large deinonychosaurs ("super slashers") attacking a brachiosaurid. (From *The New York Times*, illustration by Michael Rothman. Copyright © 1992 by The New York Times Company. Reprinted by permission.)

and are even more dinosaur-like than extant birds. Knowledge of these birds is currently based on one fossilized impression of a feather and on five (possibly six) fossilized skeletons, some with very distinct impressions of feathers (Figure 13–17). These fossil birds, which are about the size of a crow, have been given the name *Archaeopteryx* ("ancient wing"). Unquestionably, the most important features of these fossils are the feathers. Without the feather impressions, the fossils would be considered to be nonavian dinosaurs. In fact, that is exactly what happened to one *Archaeopteryx* fossil that was collected in Germany in 1951. It lacks obvious indications of feathers and was labeled as *Compsognathus* (a small coelurosaur) until 1973, when it was recognized as the fifth known specimen of *Archaeopteryx*.

Feathers were well developed in *Archaeopteryx*, indicating that they had long been present in the lineage. In addition to a presumed covering of body feathers, the wing feathers were differentiated into an outer series of primaries on the hand bones and an inner series of secondaries along the outer arm. This arrangement of flight feathers is essentially the same as that seen in extant birds, and the flight feathers on the wings of *Archaeopteryx* have asymmetrical vanes like those of flying birds, suggesting that they had been shaped by aerodynamic forces associated with flapping flight (Feduccia and Tordoff 1979). In contrast, the feathers of flightless birds serve mostly as insulation and have symmetrical vanes on each side of the rachis. The long tail of *Archaeopteryx* is not known in any other bird. The rectrices (tail feathers)

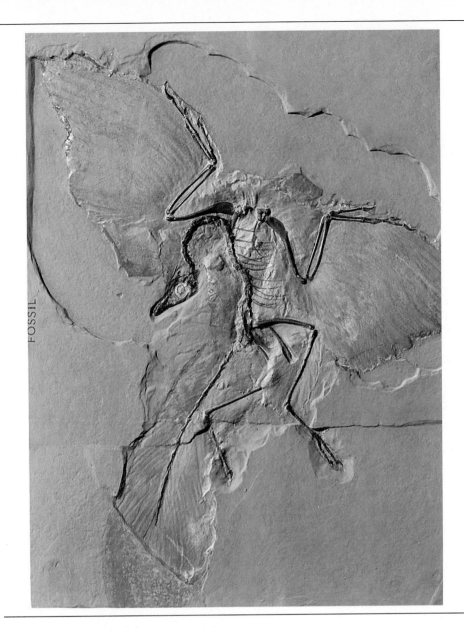

FOSSIL

Figure 13–17 *Archaeopteryx lithographica*. (From the Sanford Bird Hall, American Museum of Natural History #325097. Courtesy Department Library Services.)

are arranged in 15 pairs along the sides of the 6th through 20th caudal vertebrae. Feathers are the definitive character of birds, and are believed to have evolved from scales. The evidence for this homology lies in the biochemical similarities of scales and feathers, and the fact that feathers and scales develop from similar embryonic structures consisting of dermis and epidermis.

No intermediate fossils link *Archaeopteryx* with any of the groups from which it might have evolved, and the question of which theropods are the closest relatives of birds continues to be contro-versial. The generally accepted view places the closest relatives of birds among the deinonychosaurs. Support for this view, proposed in 1974 by John Ostrom of Yale University, rests on more than 20 features that *Archaeopteryx* shares with other Mani-raptora. The most striking of these are the remarkable similarities of the forelimbs, hands, hind limbs, and feet of *Archaeopteryx* to those of deinony-chosaurs (Figure 13–18). However, suggestions persist that birds were derived from archosaur lineages that separated from the saurischian stock earlier than the theropods (Feduccia and Wild 1993). If

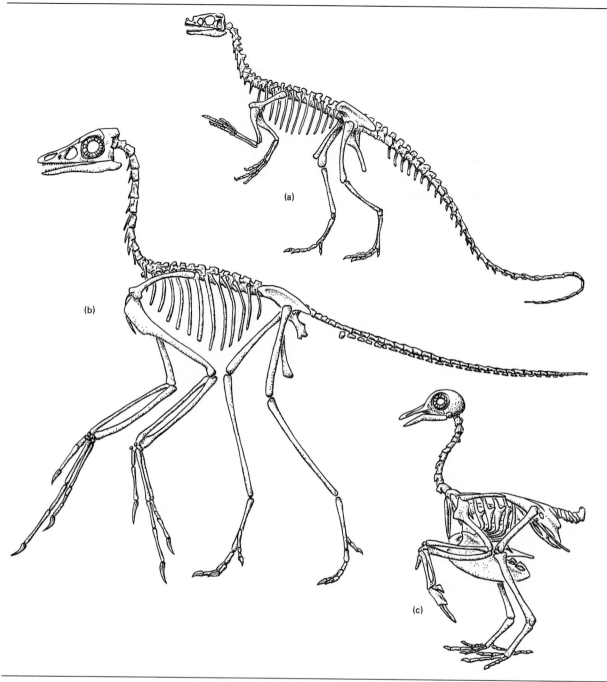

Figure 13–18 Skeleton of *Archaeopteryx* (b) compared to that of *Ornitholestes*, a maniraptor (a), and a modern bird (c). (Modified from D. Norman, 1985, *The Illustrated Encyclopedia of Dinosaurs*, Salamander Books, London, UK.)

these hypotheses are correct, the remarkable similarities of deinonychosaurs and birds noted in the caption of Figure 13–13 would be examples of convergence.

The Origin of Flight Flapping flight has evolved in three separate groups of vertebrates: pterosaurs, birds, and bats. The wings of these different vertebrates represent examples of convergent evolution,

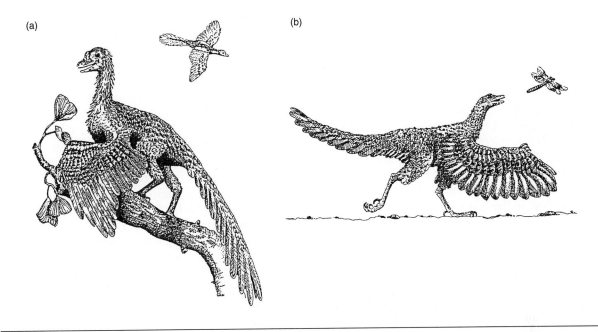

Figure 13–19 Two reconstructions of *Archaeopteryx*. (a) The from-the-trees-down hypothesis, showing *Archaeopteryx* as an arboreal climber and (rear) glider. (b) The from-the-ground-up hypothesis showing *Archaeopteryx* as a cursorial arboreal hunter. (From J. M. V. Rayner, 1988, *Biological Journal of the Linnean Society* 34:269–287.)

and the actual structural details of the wing design are quite different in the three groups. Only birds employ a complicated series of overlapping epidermal derivatives (feathers) as the main wing surface.

What were the selective advantages for the evolution of wings and flight in the proavian ancestors of birds? Two competing hypotheses have existed for a century—the arboreal theory and the terrestrial theory (Figure 13–19). To examine these hypotheses, we need to give some further consideration to *Archaeopteryx*.

What kind of life did *Archaeopteryx* lead? It has often been considered an arboreal climber, jumper, and glider with limited powers of flapping flight that allowed it to extend the distance it could travel through the air between trees. Yalden (1985) and Feduccia (1993) interpreted the morphology of *Archaeopteryx* as indicating that it climbed in trees like a squirrel, and this view finds a parallel in current ideas about the biology of pterosaurs. In this scenario it scampered about bipedally on the branches of trees, aided by grasping toes, a reversed hallux,

grasping claws on the leading edges of its wings, and a long tail that was used for balance. This conception finds a parallel among extant birds in the young hoatzin, which climbs about aided by functional clawed fingers on its wings.

The arboreal theory of the origin of avian flight has long dominated the field (Rayner 1988). According to this view, the proavian relatives of *Archaeopteryx* were tree climbers that jumped from branch to branch and from tree to tree much as some squirrels, lizards, and monkeys do. Under selective pressures favoring increased distance and accuracy of travel between trees, structures that provided some surface area for lift would be advantageous. A functional analogy can be made to gliding lizards such as *Draco*, although the morphological structures involved in the proavian model are the forelimbs, not the ribs. By this hypothesis, the evolution of flying forms passed from gliding stages through intermediate stages, such as *Archaeopteryx*, in which gliding was aided by weak flapping flight, to fully airborne flapping fliers. However, one problem that has not been satisfactorily explained by the arboreal theory is selection for bipedalism in an arboreal habitat (Gauthier and Padian 1985). Could a two-legged creature land upright on the branch of a tree without already possessing well-coordinated, aerodynamically controlled braking ability?

If, as seems reasonable from the fossil record of the coelurosaurs and the structure of *Archaeopteryx* itself, the lineage giving rise to birds consisted of bipedal, terrestrial forms, is it necessary to invoke arboreal selection pressures at all for the evolution of avian flight? The from-the-ground-up theory postulates that flapping flight evolved directly from ground-dwelling, bipedal runners.

According to the first version of this hypothesis (the cursorial theory), proavians were fast, bipedal runners that used their wings as planes to increase lift and lighten the load for running. In a later development, the wings were flapped as the animal ran to provide additional forward propulsion, much as a chicken flaps across the barnyard to escape from a dog. Finally, the pectoral muscles and flight feathers became sufficiently developed for full-powered flight through the air.

The cursorial theory in its original form failed as an explanation because in physical and mechanical terms flapping is not an effective mechanism to increase running speed. Maximum traction on the ground is required to achieve acceleration, and this traction can be provided only by solid contact of the feet with a firm substrate. Planing with wings would have reduced traction, and the push from the small surfaces of the protowings probably would not have compensated the loss in speed from the hind legs, much less added to acceleration.

A recently identified specimen of *Archaeopteryx*, misidentified as a coelurosaur for over 100 years, revealed some previously unknown details of the hand and led John Ostrom to modify the cursorial theory. Some elements of the manus are extremely well preserved in this specimen and show the actual horny claws on digits 1 and 3. These claws look like the talons of a predatory bird.

The similarities in morphology between the hand, metacarpus, forearm, humerus, and pectoral apparatus of *Archaeopteryx* and those of several coelurosaurs may be evidence of a similarity in biological roles for both—a grasping function for predation. Although bearing feathers, the forelimb and shoulder of *Archaeopteryx* have not been structurally

much modified from the skeletal condition of these small theropods, and they differ from all known birds in lacking several features that are critical for powered flight (fused carpometacarpus, restricted wrist and elbow joints, modified coracoids, and a plate-like sternum with keel) (Jenkins 1993). In fact, the only skeletal feature suggesting flight is the well-developed **furcula** (wishbone), which was present in coelurosaurs and is present in extant birds, although reduced or absent in flightless forms. Thus, the entire pectoral appendage (skeleton and muscles) of *Archaeopteryx* appears to have been as well adapted for predation as for flight. From these considerations, Ostrom postulated that the incipient wings of the proavians evolved first as snares to trap insects or other prey against the ground or to bat them down out of the air, making it easier for them to be grasped by the claws and teeth. The structures subsequently became further modified into flapping appendages capable of subduing larger prey and coincidentally aided in leaping attacks on that prey.

More recently, aerodynamic models have suggested that evolution of the avian wing could have assisted horizontal jumps after prey. By spreading or moving its forelimbs, the proavian cursor could control pitch, roll, and yaw during a jump and also maintain balance on landing (Caple et al. 1983).

Despite the details of anatomy detectable from the fossils, *Achaeopteryx* remains difficult to interpret in functional terms. The structural features of *Archaeopteryx* seem consistent with those of a basically ground-dwelling, running and jumping predator that was also capable of powered flight over short distances (Peterson 1985). A small cursorial predator of this sort fits well within the diversity of coelurosaurian dinosaurs, and emphasizes the remarkable similarities to birds that are seen in derived coelurosaurs.

A physiological perspective has enriched our interpretation of the biology of *Archaeopteryx* (Ruben 1991, 1993). Paleontologists have assumed that the energy requirements of take-off and powered flight are too great for an ectotherm. These assumptions are not supported by our understanding of the metabolic capacities of modern lizards. During burst activity the locomotor muscles of terrestrial lizards produce at least twice as much power (measured as watts per kilogram of muscle) as do the locomotor muscles of birds and mammals. This power is derived largely from anaerobic metabolic pathways, as described in Chapter 4, and high levels of power output cannot be sustained indefinitely. Nonetheless, some modern lizards are capable of substantial periods of rapid locomotion—the Komodo monitor lizard is reported to sprint for a kilometer at a velocity of 30 kilometers per hour. Ruben calculated that a similar power output would have allowed an *Archaeopteryx* to fly at least 1.5 kilometers at a velocity of 40 kilometers per hour.

An animal that could take off from the ground and fly rapidly for several hundred meters would be able to escape predators or fly up into trees. Many living birds, including cursorial predators such as the North American roadrunner and the African secretary bird, use flight in exactly this way. The Central American chachalaca is somewhat more arboreal than roadrunners and secretary birds, and may be a still better model for the biology of *Archaeopteryx*.

The Ornithischian Dinosaurs

The difference in the structure of the pelvis of saurischian and ornithischian dinosaurs indicates an early separation of the two groups. However, ornithischians and saurischians show parallels in body form that probably reflect the mechanical problems of being very large terrestrial animals. Ornithischians were herbivorous and radiated into considerably more diverse morphological forms than did the herbivorous sauropod saurischians. All ornithischians had cheeks, and horny beaks rather than teeth at the front of the mouth. The ornithischians were never as bipedal as the theropod saurischians, and the forelimbs were not greatly reduced.

Three groups of ornithischian dinosaurs can be distinguished (Figure 13–20):

Thyreophora: The armored dinosaurs. Quadrupedal forms including stegosaurs (forms with a double row of plates or spines on the back and tail) and ankylosaurs (heavily armored forms, some with club-like tails).

Ornithopoda: Bipedal forms, including the duck-billed dinosaurs.

Marginocephalia: The pachycephalosaurs (bipedal dinosaurs with enormously thick skulls) and ceratopsians (the quadrupedal horned dinosaurs).

Thyreophora: Stegosaurs and Ankylosaurs

The stegosaurs were a group of quadrupedal herbivorous ornithischians that were most abundant in the Jurassic, although some forms persisted to the end of the Cretaceous. *Stegosaurus*, a large form from the Jurassic of western North America, is the most familiar of these dinosaurs. It was up to 6 meters long, and its forelegs were much shorter than its hind legs (Figure 13–21). A double series of leaf-shaped plates were probably set alternately on the left and right sides of the vertebral column. Two pairs of spikes on the tail made it a formidable weapon. *Kentrosaurus*, an African species that was contemporaneous with *Stegosaurus*, was smaller (2.5 meters), and had a series of seven pairs of spikes that started near the middle of the trunk and extended down the tail. Anterior to these spikes were about seven pairs of plates similar to those of *Stegosaurus*, but smaller.

The function of the plates of *Stegosaurus* has been a matter of contention for decades. Originally they were assumed to have provided protection from attacks of carnosaurs, and some reconstructions have shown the plates lying flat against the sides of the body as shields. A defensive function is not very convincing, however. Whether the plates were erect or flat, they left large areas on the sides of the body and the belly unprotected. Another idea is that the plates were used for heat exchange (Farlow et al. 1976). Examination shows that the plates were extensively vascularized and could have carried a large flow of blood to be warmed or cooled according to the needs of the animal (Buffrénil et al. 1986). *Kentrosaurus*, the African counterpart of *Stegosaurus*, had much smaller dorsal plates than *Stegosaurus* and the plates on *Kentrosaurus* extended only from the neck to the middle of the trunk. Here the plates were replaced by a row of spikes that ran down the tail and appear to have had a primarily defensive function rather than a thermoregulatory one. It is frustrating not to be able to compare the thermoregulatory behaviors of the two kinds of stegosaurs in a controlled experiment.

The short front legs of stegosaurs kept their heads close to the ground, and their heavy bodies do not give the impression that they were able to stand upright on their hind legs to feed on trees as ornithopods and sauropods probably did. Stegosaurs may

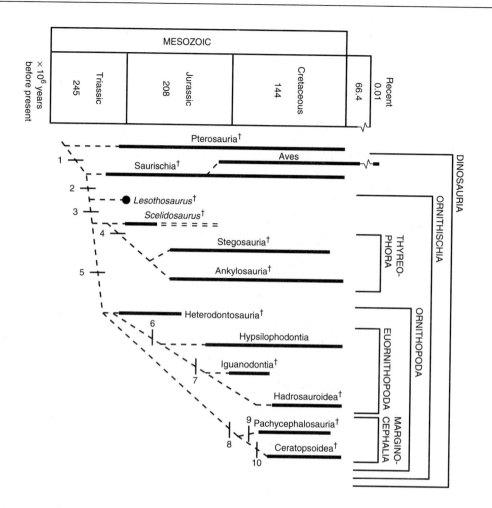

1. Dinosauria: Characteristics of the palate, pectoral and pelvic girdles, hand, hindlimb, and foot. 2. Ornithischia: Characters of the teeth, skull, palate, sacral vertebrae, pelvic girdle. 3. Genasauria: Characters of the jaws and teeth, including a toothless region on the anterior portion of the premaxilla. 4. Thyreophora: Characters of the orbit of the eye, rows of keeled scutes on the dorsal body surface. 5. Cerapoda: Five or fewer premaxillary teeth, a diastema between premaxillary and maxillary teeth, characters of the pelvis. 6. Euornithopoda: Ventral extension of the quadrate lowering the jaw articulation below the level of the tooth row, characters of the teeth and snout. 7. Iguanodontia: Premaxillary teeth absent, characters of the tooth surfaces and tooth enamel, lower jaw and skull. 8. Marginocephalia: A shelf formed by the parietals and squamosals extends over the occiput, char- acters of the snout and pelvis. 9. Pachycephalosauria: Thickened skull roof (frontals and parietals), other char- acters of the skull, dorsal vertebrae, forelimbs, and pelvis. 10. Ceratopsia: Head triangular in dorsal view; tall, narrow anterior beak; jugals flare beyond the skull roof; deep, transversely arched palate, immobile mandibular symphysis. (Based on P. C. Sereno, 1986, *National Geographic Research* 2:234–256; M. J. Benton (edi- tor), 1988, *The Phylogeny and Classification of the Tetrapods*, Special Volume No. 35B, The Systematics Association, Oxford University Press, Oxford, UK; R. L. Carroll, 1988, *Vertebrate Paleontology and Evolution*, Freeman, New York, NY; M. J. Benton, 1990, *Vertebrate Paleontology*, Harper- Collins Academic, London, UK; and M. J. Benton (edi- tor), 1993, *The Fossil Record 2*, Chapman & Hall, London, UK.)

Figure 13–20 Phylogenetic relationships of the Ornithischia. This diagram shows the probable relationships among the major groups of ornithischian dinosaurs. Extinct lineages are marked by a dagger (†). The numbers indicate derived charac- ters that distinguish the lineages.

Figure 13–21 Quadrupedal ornithischians. (a) *Stegosaurus* and (b) *Kentrosaurus* were stegosaurians, (c) *Ankylosaurus*, an ankylosaurian, and (d) *Styracosaurus*, a ceratopsoid.

have browsed on ferns, cycads, and other low-growing plants. The skull was surprisingly small for such a large animal, and had the familiar horny beak at the front of the jaws. The teeth show none of the specializations seen in hadrosaurs or ceratopsians, and the coronoid process of the lower jaw is not well developed. Unlike hadrosaurs and ceratopsians, which appear to have been able to grind or cut plant material into small pieces that could be digested efficiently, stegosaurs may have eaten large quantities of food without much chewing and relied largely on the fermentative activity of symbiotic bacteria and protozoans to aid digestion. Stegosaurs may also have used gastroliths in a muscular gizzard to pulverize plant material.

The ankylosaurs are a group of heavily armored dinosaurs that are found in Jurassic and Cretaceous deposits in North America and Eurasia. Ankylosaurs were quadrupedal ornithischians that ranged from 2 to 6 meters in length. They had short legs and broad bodies, with **osteoderms** (bones embedded in the skin) that were fused together on the neck, back, hips, and tail to produce large shield-like pieces. Bony plates also covered the skull and jaws, and even the eyelids of *Euoplocephalus* had bony armor. Ankylosaurs had short tails, and some species had a lump of bone at the end of the tail that could apparently be swung like a club. The posteriormost caudal vertebrae of these club-tailed forms have elongated neural and hemal arches that touch or overlap the arches on adjacent vertebrae and ossified tendons running down both sides of the vertebrae. Contraction of the muscles that inserted on these tendons probably pulled the posterior caudal vertebrae together to form a stiff handle for swinging the club head at the end of the tail. The tail of these animals resembles nothing so much as an enormous medieval mace. Other species had spines projecting from the back and sides of the body, and ankylosaurs must have been difficult animals for tyrannosaurids to attack. The brains of ankylosaurs appear to have had large olfactory stalks leading to complex nasal passages that probably had sheets of bone that supported an epithelium with chemosensory cells. If this interpretation is correct, ankylosaurs may have had a keen sense of smell.

Ornithopods

Ornithopods from the early Jurassic, the heterodontosaurids and related groups, were mostly small (1 to 3 meters long) and bipedal. They had four toes on the hind feet and, unlike the bipedal saurischians, retained five toes on the forefeet. Their cheek teeth were specialized for grinding plant material. Some heterodontosaurids had sharp tusks that may have been better developed in males than in females. The Cretaceous hypsilophodontids had horny beaks with which they may have cropped plant material that was subsequently ground by the high-crowned cheek teeth that gave the family its name ("high-ridged tooth").

The first nonavian dinosaur fossil to be recognized as such was an ornithopod, *Iguanodon,* found in Cretaceous sediments in England (Figure 13–22). Specimens have subsequently been found in Europe and Mongolia, and related forms have been discovered in Africa and Australia. *Iguanodon* reached lengths of 10 meters, although most specimens are smaller. Iguanodontids from the early Cretaceous had large heads and elongated snouts that ended in broad, toothless beaks. Their teeth, which were in the rear of the jaws, were laterally flattened and had serrated edges, very like the lateral teeth of living herbivorous lizards like *Iguana.*

The first digit on each forefoot of derived ornithopods was modified as a spine that projected upward. These spines show a striking resemblance to the spines on the forefeet of some frogs that are used as defensive weapons and during intraspecific encounters. *Ouranosaurus,* an ornithopod known from the early middle Cretaceous of Africa, had a large sail that was supported by elongated neural spines on the vertebrae of the trunk and tail. Behavioral and thermoregulatory functions can be postulated for the sail of *Ouranosaurus* as they were for the similar structures of the spinosaurid carnosaurs.

Hadrosaurs The derived ornithopods include several specialized forms of hadrosaurs (duck-billed dinosaurs). Hadrosaurs were the last group of ornithopods to evolve, appearing in the middle of the Cretaceous. As their name implies, some duck-billed dinosaurs had flat snouts with a duck-like bill (Figure 13–22). These were large animals, some reaching lengths in excess of 10 meters and weights greater than 10,000 kilograms. The anterior part of the jaws was toothless, but a remarkable battery of teeth occupied the rear of the jaws. On each side of the upper and lower jaws were four tooth rows, each containing about 40 teeth packed closely side by side to form a massive tooth plate. Several sets of replacement teeth lay beneath those in use, so a hadrosaur had several thousand teeth in its mouth, of which several hundred were in use simultaneously. Fossilized stomach contents of hadrosaurs consist of pine needles and twigs, seed, and fruit of terrestrial plants.

The rise of the hadrosaurs was approximately coincident with a change in the terrestrial flora during the middle Cretaceous. The bennettitaleans and seed ferns that had spread during the Triassic now were replaced by flowering plants (angiosperms). Simultaneous with the burgeoning of the angiosperms and hadrosaurian dinosaurs was a decline in the enormous sauropod dinosaurs such as *Diplodocus* and *Brachiosaurus.* Those lineages were most diverse in the late Jurassic and early Cretaceous, and only a few forms persisted after the middle of the Cretaceous.

Three kinds of hadrosaurs are distinguished: flat-headed, solid-crested, and hollow-crested (Figure 13–23). In the flat-headed forms (hadrosaurines) the nasal bones are not especially enlarged, although the nasal region may have been covered by folds of flesh. In the solid-crested forms (lambeosaurines) the nasal and frontal bones grew upward, meeting in a spike that projected over the skull roof. In the hollow-crested forms (saurolophines) a similar projection was formed by the premaxillary and nasal bones. In *Corythosaurus* those bones formed a helmet-like crest that covered the top of the skull, whereas in *Parasaurolophus* a long, curved structure extended over the shoulders. Although the crests of the lambeosaurines contained only bone, the nasal passages ran through the crests of the saurolophines. Inspired

Figure 13–22 Bipedal ornithischian dinosaurs: (a) *Iguanodon;* (b) *Hadrosaurus;*
(c) *Pachycephalosaurus;* (d) *Ouranosaurus.*

air traveled a circuitous route from the external nares through the crest to the internal nares, which were located in the palate just anterior to the eyes.

Perhaps these bizarre structures were associated with species-specific visual displays and vocalizations (Hopson 1975). The crests might have supported a frill attached to the neck, which could have been utilized in behavioral displays analogous to the displays of many living lizards that have similar frills. Possibly in the noncrested forms the nasal regions were covered by extensive folds of fleshy tissue that could be inflated by closing the nasal valves. Analogous structures can be found in the inflatable proboscises of elephant seals and hooded seals. The inflated structures of seals are resonating chambers used to produce vocalizations. The size and shape of the nasal cavities of lambeosaurine hadrosaurs suggest that adults produced low-frequency sounds, but juveniles would have had higher-pitched vocalizations (Weishampel 1981).

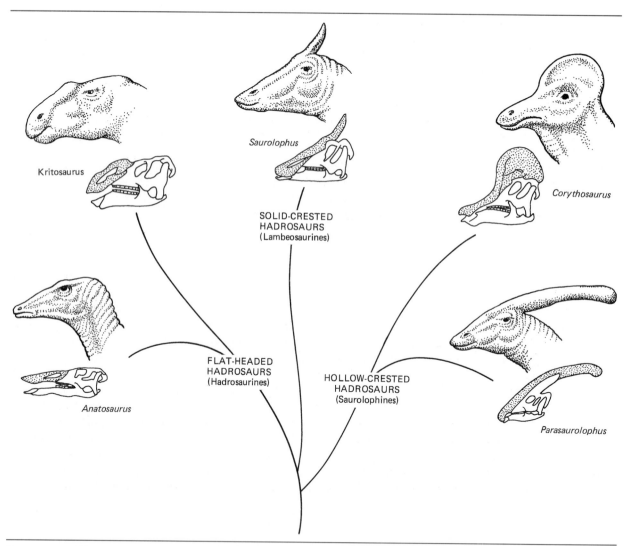

Figure 13–23 The bizarre development of the nasal and maxillary bones of some hadrosaurs gave their heads a superficially antelope-like appearance. In the flat-headed and solid-crested forms the nasal passages ran directly from the external nares to the mouth. In the hollow-crested forms the premaxillary and nasal bones contributed to the formation of the crests, and the nasal passages were diverted up and back through the crests before they reached the internal nares.

Marginocephalia

Pachycephalosaurs The pachycephalosaurs are among the most bizarre ornithischians known (Figure 13–22). The postcranial skeleton conforms to the bipedal pattern seen in ornithopods, but on the head an enormous bony dome thickens the skull roof. The bone is as much as 25 centimeters thick in a skull only 60 centimeters long. The angle of the occipital condyle indicates that the head was held so that the axis of the neck extended directly through the dome. The trunk vertebrae have articulations and ossified tendons that appear to have stiffened the vertebral column and resisted twisting. The pelvis was attached to at least six, and possibly to eight, vertebrae.

The thickened skull roof and the features of the postcranial skeleton have led some paleontologists to suggest that pachycephalosaurs used their heads like battering rams, perhaps for defense against carnivorous dinosaurs, or perhaps for intraspecific combat. An analogy has been drawn with goats and especially mountain sheep, in which males and females use head-to-head butting in social interactions. However, sheep and goats have horns that absorb some of the impact and they have protective air sacs at the front of the brain. Pachycephalosaurs had neither of these specializations, although stretchable ligaments in the neck may have helped to absorb the shock of impact. The Galapagos marine iguana may be a better model for the behavior of pachycephalosaurs. These lizards have blunt heads with spikes very like miniature versions of the heads of pachycephalosaurs. Male marine iguanas press their heads together, twisting and wrestling during territorial disputes. Perhaps pachycephalosaurs used their bony heads in the same way.

Ceratopsians The ceratopsians appeared in the middle Cretaceous. By this time the easy transit from one continent to another that had characterized much of the Mesozoic was coming to an end. Early ceratopsians are found in western North America and eastern Asia (which were connected across the Bering Sea), but they were apparently excluded from the rest of the world by the shallow epicontinental seas that covered the central parts of both North America and Eurasia in the late Mesozoic, and derived ceratopsians are known only from western North America.

The distinctive features of ceratopsians are found in the frill over the neck, which is formed by an enlargement of the parietal and squamosal bones, a parrot-like beak, and a battery of shearing teeth in each jaw (Figure 13–21). The earliest ceratopsians were the psittacosaurs from Asia. These were bipedal dinosaurs and they had no trace of a frill, but they did have a horny beak that covered a rostral bone at the front of the upper jaw. (The rostral bone is a distinctive feature of ceratopsian dinosaurs.) *Protoceratops*, one of the earliest of the quadrupedal ceratopsians, had developed a modest frill that extended backward over the neck and formed the origin for powerful jaw-closing muscles that extended anteriorly through slits at the rear of the skull and inserted on the coronoid process of the lower jaw. The teeth were arranged in batteries in each jaw somewhat like those of hadrosaurs, but with an important difference. The teeth of ceratopsians formed a series of knife-like edges rather than a solid surface like hadrosaur teeth. The feeding method of ceratopsians seems likely to have consisted of shearing vegetation into short lengths, rather than crushing it as hadrosaurs did.

Protoceratops had a simple frill, unadorned by spikes, and also lacked a nasal horn. Derived ceratopsians had both of these elaborations. Two groups can be distinguished: In the short-frilled ceratopsians (*Monoclonius*, *Styracosaurus*, and others) the frill extended backward over the neck, whereas in the long-frilled forms (*Chasmosaurus*, *Pentaceratops*, and others) the frill extended past the shoulders. Both short- and long-frilled ceratopsians had nasal and brow horns developed to varying degrees. The possible functions of the frills and horns of ceratopsians have attracted much speculation (Dodson 1993). Probably the initial stages in the evolution of the frill involved jaw mechanics and the importance of strong jaw muscles. Even in *Protoceratops*, however, males had larger frills than females, and that sexual dimorphism suggests that frills played a role in the social behavior of ceratopsians. Furthermore, the nasal and brow horns would have been formidable weapons for defense and for intraspecific combat. An analogy to the horns of antelope or the antlers of deer, which function in both defense and social behavior, seems appropriate for ceratopsians.

The Ecology and Behavior of Nonavian Dinosaurs

Many of our ideas about the sorts of lives nonavian dinosaurs had persisted with little change from the nineteenth century. Although the amount of information available about such topics as biomechanics, paleoclimatology, and the ecology and behavior of living archosaurs has increased enormously in the last hundred years, it is only recently that paleontologists have returned to an examination of nonavian dinosaur fossils armed with this new perspective (Charig 1979, Hopson and Radinsky 1980, Coombs 1990, Colbert 1993).

Classic views of the lives of nonavian dinosaurs were extrapolated from a superficial view of the biology of large living diapsids, especially crocodilians and large lizards. These animals are usually seen in zoos, where they are well fed and isolated from the stimuli they experience in their normal habitats. Zoo animals present an exaggerated impression of the lethargy of large reptiles, and they are unsuitable as models for nonavian dinosaurs.

The Mesozoic world was quite different from the modern world, and the roles of Mesozoic archosaurs were correspondingly different from those of modern forms. Diapsids were the dominant terrestrial vertebrates of the Mesozoic, and they filled more adaptive zones than modern diapsids do in the mammal-dominated terrestrial ecosystems of the Cenozoic.

Recently some paleontologists have begun to compare the anatomy of nonavian dinosaurs with the anatomy of large mammals and birds, and to speculate about ecological and behavioral similarities on that basis. At the same time, studies of the behavior of crocodilians under field conditions have shown that they have much more complex behavior patterns than had been suspected from observations of captive animals. These lines of reasoning have led to the conclusion that nonavian dinosaurs were probably considerably more active than had been suspected, and that they probably had an intraspecific social organization at least as complex as that of crocodilians and birds. Peripheral information about ecology and behavior of nonavian dinosaurs is provided by material such as fossilized stomach contents, fossilized footprints, and fossilized eggs, nests, and juveniles.

Crocodilians and Birds as Models for Nonavian Dinosaurs

One of the strengths of the cladistic approach to classification is the emphasis it places on shared derived characteristics of related organisms. Usually these are morphological characters, and they are em-ployed to draw inferences about phylogenetic relationships, but the process can be used in other ways. For example, if a phylogeny can be established by using morphological characters, other characteristics—ecological, behavioral, or physiological—can be superimposed on the phylogeny, and their evolution can be interpreted in a phylogenetic context.

A variety of morphological features common to crocodilians and birds places them both in the archosaur lineage, and this relationship can be combined with information about the social behavior of crocodilians and birds to draw inferences about the probable behavior of nonavian dinosaurs. The analysis of parental care is a good example of this approach. The extensive parental care provided to their young by many birds has long been known, partly because most birds are conspicuous animals that are relatively easy to study. Parental care by crocodilians is less well known, but it appears to be as extensive as that of most birds. All crocodilians probably protect their nests, and elaborate parental care has been described for some species (Pooley and Gans 1976, Ferguson 1985, Lang 1986).

Baby crocodilians begin to vocalize before they have fully emerged from the eggs, and these vocalizations are loud enough to be heard some distance away. The vocalizations of the babies stimulate one or both parents to excavate the nest, using their feet and jaws to pull away vegetation or soil (Figure 13–24). For example, the female American alligator bites chunks of vegetation out of her nest to release the young when they start to vocalize. Then she picks up the babies in her mouth and carries them, one or two at a time, to water where she releases them. This process is repeated until all of the hatchlings have been carried from the nest to the water. The parents of some species of crocodilians gently break the eggshells with their teeth to help the young escape. The sight of a crocodile, with jaws that could crush the leg of a cow, delicately rupturing the shell of an egg little larger than a hen's egg and releasing the hatchling unharmed is truly remarkable.

Young crocodilians stay near their mother for a considerable period—two years for the American alligator—and may feed on small pieces of food the female drops in the process of eating. Like many birds, baby crocodilians are capable of catching their own food shortly after they hatch and are not dependent on their parents for nutrition.

Crocodilians, like birds, are vocal archosaurs (Lang 1989). Male crocodilians vocalize to announce

(a)

(b)

(c)

Figure 13–24 Parental care by the mugger crocodile, *Crocodylus palustris*. The numbered tags allowed individual crocodiles to be recognized: (a) male parent picking a hatchling; (b,c) male parent carrying hatchlings to the water, 9 meters away, where the mother was waiting. (Photographs courtesy of Jeffrey W. Lang.)

their territorial status, and courtship is accompanied by vocalizations from males and females. Baby crocodilians begin to vocalize while they are still in the eggs, and hatchlings emit a distress squeak when they are frightened that stimulates adult male and female crocodilians to come to their defense. When staff members at a crocodile farm in Papua New Guinea rescued a hatchling New Guinea crocodile that had strayed from its pond, its distress call brought 20 adult crocodiles surging toward the hatchling (and the staff members!). The dominant male head-slapped the water repeatedly, and then charged into the chain link fence where the staff members were standing while the females swam about, gave deep guttural calls, and head-slapped the water.

The nesting behavior and parental care of crocodilians overlaps that of many birds. For example, the bush turkeys (megapods) of Australia bury their eggs in piles of soil and vegetation in craters they excavate in the ground and release their young at the end of incubation. The young of many birds, including bush turkeys, are well developed at hatching (**precocial**) and are able to find their own food. In these birds, as in crocodilians, the important function of parental care appears to be protection of the young from predators.

The most parsimonious explanation of the presence of well-developed parental care in crocodilians and birds is that it was present in the common ancestor of crocodilians and birds. In other words, parental care of young appears to be an ancestral character of the archosaur lineage, at least at the level of crocodilians. If that is the case, nonavian dinosaurs and pterosaurs would have inherited parental care as a part of their ancestral behavioral repertoire, and one would expect them to have exhibited the behavior seen in birds and crocodilians. Behavior is difficult to decipher from the fossil record, but evidence is accumulating to suggest that nonavian dinosaurs did, indeed, engage in parental care and other forms of complex social behaviors.

Ecology of Sauropods

The large sauropod dinosaurs, *Apatosaurus, Diplodocus,* and related forms, were the largest terrestrial animals that have ever lived. The largest of them may have reached body lengths exceeding 30 meters and weights of 100,000 kilograms. (For comparison, an elephant is about 5 meters long and weighs 5000 kilograms.) From the earliest discovery of their fossils, paleontologists doubted that such massive animals could have walked on land and felt they must have been limited to a semiaquatic life in swamps. Mechanical analysis of sauropod skeletons does not support that conclusion (Alexander 1989). The skeletons of the large sauropods clearly reveal selective forces favoring a combination of strength with light weight. The arches of the vertebrae acted like flying buttresses on a large building, while the V-shaped neural spines of diplodocoids held a massive and possibly elastic ligament that helped to support the head and tail. In cross section the trunk was deep, shaped like the body of a terrestrial animal such as an elephant rather than rounded like that of the aquatic hippopotamus. The tails of sauropods are not laterally flattened like tails used for swimming. Instead, they are round in cross section and, in diplodocoids, terminate in a long, thin whiplash. These structures look like counterweights and defensive weapons.

Fossil trackways of sauropods indicate that the legs were held under the body; the tracks of the left and right feet are only a single foot width apart. Analysis of the limb bones suggests that they were held straight in an elephant-like pose and moved fore and aft parallel to the midline of the body. This is what would be expected on mechanical grounds, because no other leg morphology is possible for a very large animal. Bone is far less resistant to bending forces exerted across its long axis than it is to compressional forces exerted parallel to the axis. As an animal's body increases in size, mass grows as the cube of its linear dimensions but the cross-sectional area of the bones increases as the square of their linear dimensions. The strength of bone is roughly proportional to its cross-sectional area. As a result, when the body size of an animal increases, the strength of the skeleton increases more slowly than the stress to which it is subjected. One evolutionary response to this relationship is disproportionate increase in the diameter of bones—an elephant skeleton is proportionally larger than a mouse skeleton. Another response is to transform bending forces to compression forces by bringing the legs more directly under the body. In a large animal, such as an elephant or a sauropod dinosaur, not only are the legs held under the body, but the knee joint tends to be locked as the animal walks. This morphology produces the ambling locomotion familiar in elephants, and it is likely that sauropods walked with an elephant-like gait, holding their heads and tails in the air.

Sauropod teeth are sometimes described as being small and weak. Certainly they were small in proportion to the size of the body, as was the entire skull. In absolute terms, however, they were neither small nor weak. They were larger than the teeth of browsing mammals, and there is no reason to believe that plant material was tougher in the Mesozoic than it is today. There were no flat (molariform) teeth to crush the ingested plant material. This function may have been served by gastroliths, stones that are deliberately eaten by an animal and retained in some part of the gut where they crush food, and the breakdown of plant material may have been aided by symbiotic microorganisms (Farlow 1987).

The fossilized stomach contents of a sauropod dinosaur, found in Jurassic sediments in Utah, includes pieces of twigs and small branches about 2.5 centimeters long and 1 centimeter in diameter. The fragmented and shredded character of the woody material indicates that even without molariform teeth the sauropod had some method of crushing its food. This discovery appears to confirm the view of sauropod ecology that was developed from study of the skeleton and analysis of plant fossils found in association with sauropod fossils: Sauropods probably occupied open country with an undergrowth of ferns and cycads and an upper story of conifers. They were preyed upon by the large theropod carnivores and sought escape in flight or defended themselves by whipping their tails. Their long necks probably enabled them to graze from tree tops, probably standing on their hind legs by using the tail as a counterweight. It may be significant that Mesozoic conifers bore branches only near the tops of the trees, far out of reach of any but a very long-necked dinosaur.

Sauropods might have been an important force shaping the landscape and preventing ecological succession from transforming open countryside to dense forest. As such, their presence could have been important in creating and preserving suitable habitat for other species, such as the cursorial ornithomimids. Like many herbivorous lizards, the sauropods may have been omnivorous and opportunistic in their feeding, taking whatever was readily available, including carrion. The fossilized stomach contents mentioned previously contain traces of bone as well as a tooth from the contemporary carnivorous dinosaur *Allosaurus*.

Fossil trackways reveal a few details from which glimpses of nonavian dinosaur behavior can be reconstructed (Thulborn 1990). A famous trackway found in Texas, parts of which are now on display at the University of Texas at Austin and in the American Museum of Natural History in New York City, shows the footprints of a sauropod trailed by a large theropod. In some cases the theropod had stepped into the footprints of its potential prey. Unfortunately, the trackway ended before the outcome of this Cretaceous drama was revealed. Some insight into the intraspecific behavior of sauropods may be revealed by a series of tracks found in early Cretaceous sediments at Davenport Ranch in Texas. These reveal the passage of 23 brontosaur-like dinosaurs some 120 million years ago. A group of individuals moving in the same direction at the same time would be remarkable for most living diapsids, but the brontosaur tracks suggest that this is what happened. Furthermore, the tracks may show that the herd moved in a structured fashion with the young animals in the middle, surrounded by adults.

Predatory Behavior of Theropods

The carnivorous saurischian dinosaurs, like the herbivorous forms, present an impression of relatively small morphological diversity. The major evolutionary lineages have already been traced—the ostrich-like ornithomimids, the giant theropods, and the deinonychosaurs that combined the speed of ornithomimids with the predatory habits of the theropods. The ornithomimids parallel flightless birds so closely in morphology that it seems reasonable to assume that they lived essentially the same life in the Cretaceous that ostriches, emus, and related forms live now. The appearance of increasingly cursorial forms that relied on running to escape

predators may indicate that habitats became increasingly open during the Mesozoic.

There was probably a trend to increasingly specialized methods of prey capture among carnosaurs. This hypothesis is suggested by the reduction of the size of the forelimbs of carnosaurs that presumably indicates an increased reliance on the teeth for both seizing and killing prey. The deinonychosaurs probably relied instead on fleetness of foot to capture active prey. The discovery of five *Deinonychus* skeletons in close association with the skeleton of *Tenontosaurus*, an ornithischian three times their size, might indicate that *Deinonychus* hunted in packs (Figure 13–25). Deinonychosaurs probably used their clawed forefeet to seize prey and then slashed at it with the sickle-like claws on the hind feet. This tactic appears to be illustrated by a remarkable discovery in Mongolia of a deinonychosaur called *Velociraptor*. It was preserved in combat with a *Protoceratops*, its hands grasping the head of its prey and its enormous claw embedded in the midsection of the *Protoceratops*.

Social Behavior of Ornithischians

The morphological diversity of the ornithischian dinosaurs suggests an equivalent diversity in behavior and ecology. Social interactions based on visual displays and vocalizations may have been well developed among hadrosaurs, and pachycephalosaurs may have engaged in shoving contests like those of ungulate mammals. Individual interactions of these sorts may have been integrated into group behaviors. Fossilized eggs of dinosaurs provide information about nesting behaviors (Box 13–2). Evidence of parental care may be revealed by a nest of 15 baby hadrosaurs (*Maiasaura*) in the late Cretaceous Two Medicine Formation in Montana (Horner and Makela 1979). The babies were about a meter long, approximately twice the size of hatchlings found in the same area, indicating that the group had remained together after they hatched. Furthermore, the teeth of the baby hadrosaurs showed that they had been feeding; some teeth were worn down to one-quarter of their original length. The object presumed to be a nest was a mound 3 meters in diameter and 1.5 meters high with a saucer-shaped depression in the center (Figure 13–25). Such a large structure would have made the babies conspicuous to predators, and it seems likely that a parent remained with the young. (*Maiasaura* can be translat-

Figure 13–25 Nonavian dinosaurs may have had a variety of social behaviors. (a) Deinonychosaurs may have hunted in packs to attack large prey such as ornithopods. (b) Ceratopsians (*Chasmosaurus*) may have formed a ring to confront predators. (c) Hadrosaurs (*Maiasaura*) may have nested in colonies. (Modified from D. Norman, 1985, *The Illustrated Encyclopedia of Dinosaurs*, Salamander Books, London, UK.)

ed as "good mother reptile.") The morphology of the inner ears of lambeosaurs suggests that adults would have been able to hear the high-pitched vocalizations of juveniles, strengthening the inference of parental care (Weishampel 1981). The association between adults and young appears to have lasted for a con-

siderable time. Fossils suggest that *Maiasauria* and the lambeosaur *Hypacrosaurus* grew to one-quarter of adult size before they left the nesting grounds, and a species of hypsilophodontid found at the same site grew to half its adult size (Horner 1994). Both vocal communication and prolonged association of parent dinosaurs and their young are plausible in the light of the behaviors known for crocodilians.

Additional fossils in the same formation suggest that the area contained nest sites of other species of hadrosaurs and of ceratopsians as well (Horner 1982, 1984). Eggshells and baby dinosaurs are abundant in the Two Medicine Formation but rare in adjacent sediments. A similar concentration of conspicuous nests, eggs, and juveniles of the small ceratopsian *Protoceratops* discovered in Mongolia suggests parental care in this species as well. The possibility of parental care and social behavior among sauropods is raised by the discovery of five baby prosauropod dinosaurs (*Mussaurus*) with the remains of two eggs in a nest in late Triassic deposits in Argentina (Bonaparte and Vince 1979).

Protection of nests and juveniles from predators is a plausible reason for parental care (Kirkland 1994), and Katherine Troyer has suggested a second important reason for an association between juveniles and adults of herbivorous dinosaurs such as the hadrosaurs and ceratopsoids (Troyer 1984). Working with iguanas, which are herbivorous lizards, she showed that hatchlings must ingest the feces of adult iguanas (or of other hatchlings that have themselves ingested feces from adults) to inoculate their guts with the symbiotic microorganisms that permit them to digest plant material. The microorganisms responsible for the fermentation of plant material are anaerobic and soon die when feces are exposed to air. Thus, a close association between hatchlings and adults is necessary to achieve the transfer of the microorganisms. Herbivorous dinosaurs probably relied on fermentation of plant material just as most extant herbivorous vertebrates do, and some contact between juveniles and adults would have been necessary to inoculate hatchling dinosaurs with the symbiotic microorganisms.

Were Nonavian Dinosaurs Endotherms?

The cladistic approach that was helpful for inferences about the social behavior of non-avian dinosaurs is less effective in deciding what thermoregulatory mechanisms they may have employed. Ectothermy is an ancestral characteristic in the archosaur lineage, and crocodilians are ectotherms. Clearly, the endothermy seen in birds is derived, but when did it appear? It could have been at any point between crocodilians and birds. That is, endothermy might be characteristic of pterosaurs + dinosaurs + birds, or of dinosaurs + birds, or it might be limited to birds. All the lineages between crocodilians and birds are extinct, so we cannot draw any direct evidence from them—paleophysiology is a speculative subject.

Much of the controversy about the temperature relations of nonavian dinosaurs stems from failure to distinguish between *homeothermy* and *endothermy*. Homeothermy means only that the body temperature of an organism is reasonably stable, whereas endothermy and ectothermy are mechanisms of temperature regulation. Either ectothermy or endothermy can produce homeothermy, as we saw in Chapter 4. Furthermore, ectothermy and endothermy are the ends of a spectrum, and many living animals occupy positions between those extremes.

The diversity of nonavian dinosaurs and the enormous size of many species complicate discussions of their thermoregulatory mechanisms. The great ecological and phylogenetic diversity of nonavian dinosaurs has largely been overlooked in the debate about their thermoregulatory mechanisms. Not all kinds of evidence apply to all species, and it is likely that there were substantial differences in the biology of different lineages. Quite apart from any other consideration, the difference in body size of an ornithomimid and a sauropod would make them very different kinds of animals (Farlow 1990).

Furthermore, we have no terrestrial animals that compare to nonavian dinosaurs. Even elephants (which weigh about 5000 kilograms) are only as large as medium-size sauropods, and the largest living reptiles (leatherback sea turtles and saltwater crocodiles) weigh about 1000 kilograms. Thus, we have no models of nonavian dinosaurs to work with, and must extrapolate observations of living animals far beyond the size ranges for which we have measurements.

What can elephants and leatherback turtles tell us about the metabolism of nonavian dinosaurs? First, they tell us that when you compare animals of large body size, there is very little difference between the metabolic rates of endotherms and ectotherms: The metabolic rate of a 4000 kilogram elephant is 0.07 liters of oxygen per kilogram per hour, whereas the rate for a 400 kilogram leatherback turtle is 0.06 liters of oxygen per kilogram per hour. Thus, the metabolic distinction between endotherms and ectotherms disappears at large body sizes, and discussions of whether large nonavian dinosaurs were endotherms or ectotherms are meaningless.

Box 13–2 Dinosaur Eggs and Nests

Fossils of dinosaur eggs have been found in late Cretaceous deposits in Mongolia, China, France, India, and the United States. Most of these fossils are fragments of eggshells, but intact eggs containing embryos have also been discovered (Figure 13–26).

Concentrations of nests and eggs ascribed to sauropods in Cretaceous deposits in southern France suggest that these animals had well-defined nesting grounds (Kerourio 1981). Eggs thought to be those of the large sauropod *Hypselosaurus priscus* are found in association with fossilized vegetation similar to that used by alligators to construct their nests. The orientation of the nests suggests that each female dinosaur probably deposited about 50 eggs. The eggs had an average volume of 1.9 liters, about 40 times the volume of a chicken egg. A total of 50 of these eggs would weigh about 100 kilograms, or 1 percent of the estimated body mass of the mother. Crocodilians and large turtles have egg outputs that vary from 1 to 10 percent of the adult body mass, so an estimate of 1 percent for *Hypselosaurus* seems reasonable. The eggs might have been deposited in small groups instead of all together, because 50 eggs in one clutch would have consumed oxygen faster than it could diffuse through the walls of the nest (Seymour 1979).

Nearly all fossil dinosaur eggs are attributed to herbivorous dinosaurs, and two patterns of egg-laying can be distinguished (Moratalla and Powell 1994). Some dinosaurs laid eggs in clutches within an excavation that might have been filled with rotting vegetation that would have provided both heat and moisture for the eggs. (This method of egg incubation is used by many of the living crocodilians, whereas other crocodilians bury their eggs in sandy soil.) Nests of *Protoceratops* fall in this category—the 30 to 35 eggs in each nest are arranged in concentric circles with their blunt ends up. Eggs of *Orodromeus makelai*, a hypsilophodontid, are also oriented vertically with the blunt end up, but are arranged in a spiral within the circular nest.

The fossil of a theropod dinosaur that apparently died while attending a nest of eggs was discovered in the Gobi Desert in 1923, but its significance was not recognized until 70 years later (Norell et al. 1994). The eggs, which were about 12 centimeters long and 6 centimeters in diameter, were thought to have been deposited by the small ceratopsian *Protoceratops andrewsi* because adults of that species were by far the most abundant nonavian dinosaurs at the site. The theropod was assumed to have been robbing the nest, and was given the name *Oviraptor philoceratops*, which means "egg seizer, lover of ceratops." In 1993, paleontologists from the American Museum of Natural History, the Mongolian Academy of Sciences, and the Mongolian Museum of Natural History discovered a fossilized embryo in

---

Leatherback turtles employ a form of thermoregulation that has been called "gigantothermy" (Paladino et al. 1990). The large body mass of these animals has two important consequences for temperature regulation. First, it produces enormous thermal inertia, so body temperatures change slowly. Second, it allows the animals to change the effective thickness of their insulation by changing the distribution of blood flow to the surface of the body versus the core. When a leatherback is in cold water, it can retain the heat produced by muscular activity in its body by restricting blood flow to the surface of the body and employing countercurrent heat exchangers in its limbs. In warmer water, when it needs to dissipate metabolic heat to avoid overheating, it can increase blood flow to the surface and bypass the countercurrent system in its limbs. These mechanisms allow leatherback turtles to make migratory journeys of 10,000 kilometers between arctic oceans and the tropics.

Metabolic rates of leatherback turtles were used as a basis for computer simulations of body temperatures of nonavian dinosaurs (Spotila et al. 1991). Three metabolic rates were compared: (1) a standard estimate of resting metabolism of reptiles; (2) the

---

Figure 13–26
Dinosaur eggs from
the Gobi Desert.
(a) Fossilized nest of
an oviraptorid dino-
saur (AMNH 410765).
(b) The fossilized
skeleton of an embryo
of an oviraptorid
dinosaur (AMNH
K17088). This is the
first embryo of a car-
nivorous dinosaur
ever found. (Courtesy
of the American
Museum of Natural
History.)

(a)

(b)

an egg identical to the supposed *Protoceratops* eggs. To their surprise, the embryo was an *Oviraptor* nearly ready to hatch. This discovery suggests that the adult *Oviraptor* found with a clutch of eggs in 1923 got a bum rap—it probably died while resting on its own nest.

measured resting rate for leatherback turtles, which is three times the standard reptilian rate; (3) the metabolic rate of leatherback turtles while they are digging nests. This is hard work, and raises the metabolic rate to about 10 times the standard reptilian rate. The model also incorporated two rates of blood flow (low = resting rate, high = five times resting) and two blood flow patterns (low = flow to the body surface negligible, high = flow to the body surface three times flow to the core).

Five nonavian dinosaurs were modeled: *Compsognathus* (2 kilograms), *Deinonychus* (75 kilograms), *Tenontosaurus* (600 kilograms), *Edmontosaurus* (3600 kilograms), and *Apatosaurus* (30,000 kilograms). The model predicts that the core body temperatures of these animals would be 0.3 to 100°C higher than air temperature (Figure 13–27). As you would expect, the larger the animal, the greater the predicted difference between core body temperature and air temperature. For the largest species, *Apatosaurus*, even the standard reptilian metabolic rate produces a body temperature 10°C above air temperature, and the highest metabolic rate would raise its body temperature above the boiling point of water!

Even a relatively small animal such as *Deinony-chus* would have a body temperature 10 to 11°C above air temperature if it had the same metabolic rate as a nesting leatherback turtle, a low rate of blood flow, and low blood circulation to the surface of its body. Thus, the metabolic rates observed for large reptiles appear to be sufficient to allow even a small nonavian dinosaur to have a body temperature well above air temperature, and larger animals with metabolic rates higher than those of extant reptiles

would have died from overheating. (This computer simulation did not include the substantial amount of heat that would be produced by fermentative digestion of plant material in the guts of the herbivores *Tenontosaurus, Edmontosaurus,* and *Apatosaurus* [Farlow 1987]. That additional source of heat would increase the difference between air temperature and body temperature predicted by the model.)

The smallest animal considered, *Compsognathus,* differs from the rest of the species. The maximum

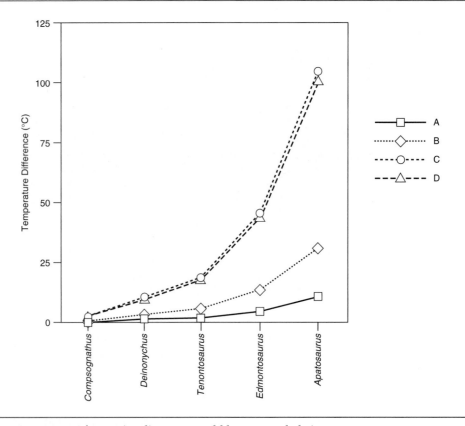

Figure 13–27 Body temperatures of nonavian dinosaurs could have exceeded air temperature by more than 100°C. The estimated difference between body temperature and air temperature depends on the body size of the animal being considered, and assumptions about metabolic rate and blood flow. The graph shows the elevation of body temperature calculated for dinosaurs from 10 kilograms to 30,000 kilograms. Four combinations of metabolic rate and blood flow are illustrated: A = standard reptilian metabolic rate, low cardiac output, low blood flow to the body surface. B = leatherback turtle resting metabolic rate (approximately 3 times reptilian standard rate), low cardiac output, low blood flow to the body surface. C = leatherback turtle nesting metabolic rate (approximately 10 times reptilian standard rate), low cardiac output, low blood flow to the body surface. D = leatherback turtle nesting metabolic rate (approximately 10 times reptilian standard rate), high cardiac output, low blood flow to the body surface. (Based on data in J. R. Spotila et al., 1991, *Modern Geology* 16:203–227.)

temperature differential predicted for *Compsognathus* is only 3°C above air temperature. *Compsognathus* was the same size as lizards such as green iguanas, and probably used the same kinds of behavioral thermoregulation that iguanas do—basking in the sun in the morning, moving to shade later in the day to avoid overheating, and retreating to sheltered sites at night.

The calculations we have described were based on the assumption of steady-state conditions, but the thermal environmental of an animal changes from hour to hour, from day to night, and from summer to winter. How would nonavian dinosaurs respond to these kinds of variation in the environment? Spotila and his colleagues used a second computer simulation to investigate that question. Once again, the results depend on body size, reflecting the importance of thermal inertia. The body temperature of *Compsognathus* would respond to changes in environmental temperature over periods of a few hours (like that of an iguana), whereas *Deinonychus* and *Tenontosaurus* would be affected by day–night changes. The body temperature of *Edmontosaurus* would change over a period of weeks, and *Apatosaurus* would require months to change temperature.

Fossils of nonavian dinosaurs from near the poles raise the question of migration. Although the climate in the Mesozoic showed less north–south variation in temperature than we see today, polar regions would have had cool, dark winters. Small nonavian dinosaurs, such as *Compsognathus,* could have retreated to burrows and hibernated, just as extant reptiles do, but what about *Deinonychus* and the large herbivores? Spotila and his colleagues calculated the migratory abilities of these species by extrapolating measurements of the energy used for locomotion by mammals and reptiles. They calculated that fat stores alone would have fueled journeys of 848 kilometers (for *Deinonychus*) to 7330 kilometers for *Apatosaurus*. (If you assume that the animals paused to feed during migrations, as birds do, they could have gone much farther.) Migrations take place at a leisurely pace, and leatherback turtles regularly migrate 10,000 kilometers from their feeding grounds in the arctic to the tropical beaches where they nest. Large nonavian dinosaurs may well have undertaken similar journeys. The main obstacle to their migration may have been geographic. Leatherback turtles are marine, and have unobstructed routes to travel, but that may not have been the case for terrestrial animals. The great inland seas that filled the center of North America and Eurasia during the first part of the Cretaceous had drained by the end of the Period. As the seas retreated, rivers cut valleys into what had been level terrain. These river valleys might have hindered migrations by nonavian dinosaurs.

The Lepidosauromorpha: Ichthyosaurs, Lizards, and Others

The second lineage of diapsids, the Lepidosauromorpha, is distinguished by derived characters of the skull and postcranial skeleton (Benton 1985a). These characters include entry of the postfrontal bone into the dorsal border of the upper temporal fossa, an additional articulating surface in the midline of neural arch of the vertebrae, cervical vertebrae with centra that are shorter than those of the trunk vertebrae, and single-headed ribs. The earliest lepidosauromorphs known are the Younginiformes of the late Permian and early Triassic. Terrestrial forms such as *Youngina* were about 50 centimeters long and very like *Petrolacosaurus,* but appear to have had shorter necks and limbs than that genus. Aquatic Younginiformes like *Hovasaurus* were approximately the same size as the terrestrial forms and had laterally flattened tails. The champsosaurs, *Champsosaurus* and *Simoedosaurus,* are problematic forms that may be lepidosauromorphs. They were aquatic animals, about the size of extant crocodilians, and showed many of the same specializations for aquatic life, including nostrils at the tip of the snout and a secondary palate that provided a passageway for air (Box 13–1). The champsosaurs appeared in the Cretaceous and, unlike many of the large Mesozoic diapsids, persisted past the end of that period into the early Tertiary.

Marine Diapsids: Placodonts, Nothosaurs, Plesiosaurs, Ichthyosaurs, and *Huphesuschus*

Several lineages of specialized marine tetrapods are tentatively grouped with the lepidosauromorphs (Figure 13–28).

Placodonts The Triassic placodonts ("flat teeth") were mollusk eaters specialized for crushing hard-

Figure 13–28 Reconstructions of Mesozoic aquatic reptiles. (All except [d] are probably lepidosauromorphs; [d] is an early archosauromorph.) (a) The Upper Jurassic ichthyosaur *Opthalamosaurus*, approximately 2.5 meters long. The body form parallels that of the fastest swimming extant fish. (b) The Upper Jurassic plesiosaur *Cryptoclidus*, approximately 3 meters long. Plesiosaurs are exceptional among diapsid reptiles in elaborating both the front and hind limbs as large paddles to "row" the body through the water in the manner of living sea lions. The trunk is dorsoventrally compressed and the tail serves as a rudder. (c) The Triassic nothosaur *Pachypleurosaurus*, approximately 1 meter long. The tail was long and laterally compressed. The forelimbs may have been used for propulsion, but the rear limbs served as rudders. (d) The middle Triassic thallatosaur *Askeptosaurus*. The body is elongate to facilitate anguilliform swimming. The size and degree of ossification of the limbs was reduced, but they are little modified for aquatic propulsion. Approximately 2 meters long. (e) The placodont *Placodus* from the middle Triassic. Approximately 1 meter long. Limbs were little modified for aquatic propulsion. (f) *Hupehsuchus* from the middle Triassic of China. Up to 2 meters long. (Figure and caption from R. L. Carroll and Z. Dong, 1991, *Philosophical Transactions of the Royal Society, London,* B 331:131–153.)

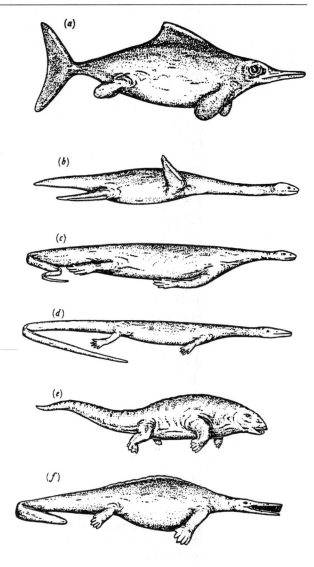

shelled food rather than for rapid pursuit of prey (Figure 13–29). *Placodus* had large, flat maxillary teeth and a heavy palate with enormous teeth on the palatine bones. The anterior teeth projected forward and might have been used to pull mussels or oysters off rocks. Another placodont, *Placochelys*, had a toothless beak instead of projecting teeth, but retained broad teeth in the rear of the mouth. In contrast, *Henodus* was almost toothless and may have crushed its food with horny plates like those of turtles. The similarity of some of these placodonts to turtles extended to their appearance as well. *Henodus* and *Placochelys* had a body armor that was as extensive as that of turtles, but it was composed of a mosaic of small polygonal dermal bones rather than of large plates like those of turtles. In *Henodus* the dermal bones were apparently covered with epidermal scutes as they are in turtles.

Nothosaurs and Plesiosaurs Nothosaurs are elongate marine reptiles from the middle Triassic with small heads, long necks, and paddle-like limbs. The tail was laterally flattened and was probably the main propulsive organ.

The plesiosaurs ("ribbon reptiles") appeared at the Triassic/Jurassic transition and persisted to the

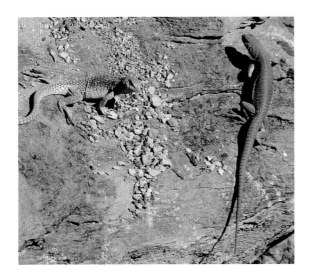 

Color change is a temperature-regulating mechanism used by lizards such as the desert iguana (*Dipsosaurus dorsalis*). (Left) When they first emerge in the morning, desert iguanas are dark. (Right) By the time it has reached its activity temperature, a lizard has turned light. This color change reduces heat gained from the sun by 23 percent.

Color often identifies the sex and social status of an individual, as in these bluehead wrasse (*Thalassoma bifasciatum*). (Left) Adult male (top) and female. (Right) A female bluehead wrasse in the process of transforming to a male.

Three species of salamanders form a mimicry complex in eastern North America. The red eft (*Notophthalmus viridescens*, top left) and red salamander (*Psuedotriton ruber*, top right) have skin toxins that deter predators. The red-backed salamander (*Plethodon cinereus*, middle left) is not protected by toxins, but predators confuse the erythristic form of that species (middle right) with the toxic species. The experiment described in the text used the mountain dusky salamander (*Desmognathus ochropheus*, bottom left) as a palatable control.

The gular fans of lizards are used in social displays. Color, size, and shape identify the species and sex of an individual. (All the lizards in these photographs are males.) (Top left) Carolina anole, *Anolis carolinensis*, from Florida. (Top right) knight anole, *Anolis equestris*, from Cuba. (Bottom left) *Anolis grahami* from Jamaica. (Bottom right) *Anolis chrysolepis* from Brazil.

The colors of birds result from a combination of structural characteristics and pigments. The blue of a blue jay (left) is produced by small particles in the feathers that scatter blue light back toward the viewer. The red of a cardinal and many other birds is produced by a caretenoid pigment, zoonerythrin.

 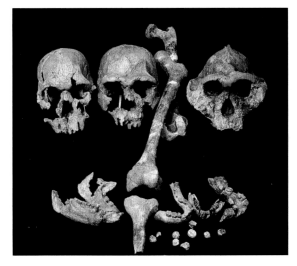

Skulls of hominids. (Left) Skulls of a gracile australopithicine (top) and a robust form. (Right) Skulls of *Homo habilis* (left), *Homo erectus* (center), and a robust australopithecine (right). Limb bones and mandibles are also shown.

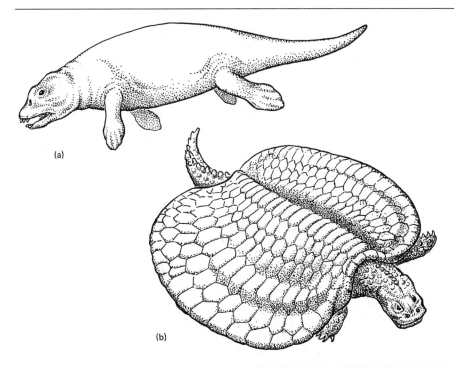

Figure 13–29 Placodonts were slow-swimming animals that probably fed on mollusks. *Placodus* (a) was relatively unspecialized, but *Henodus* (b) had dermal armor that was almost as extensive as a turtle's.

(a)

(b)

Cretaceous. Two lineages of plesiosaurs can be distinguished, and these appear to have evolved side by side, indicating a considerable ecological separation between them (Figure 13–30). One lineage is composed of long-necked animals with small heads, whereas the other contained short-necked animals with long skulls. Both had heavy, rigid trunks and (unlike nothosaurs) appear to have rowed through the water with limbs that functioned like oars and may also have acted as hydrofoils, increasing the efficiency of swimming (reviewed by Storrs 1993). **Hyperphalangy,** the addition of joints to the toes, increased the size of the paddles, and some plesiosaurs had as many as 17 phalanges per digit. In both types of plesiosaurs the nostrils were located high on the head just in front of the eyes.

The long-necked plesiosaurs reached their zenith in *Elasmosaurus,* which lived in the late Cretaceous. The lineage can be traced back to *Plesiosaurus* in the early Jurassic. That form had 35 cervical vertebrae. The *Elasmosaurus* line is characterized by a progressive increase in the number of cervical vertebrae and a reduction in the size of the head. Not only did the number of cervical vertebrae increase, but individual vertebrae became longer. *Microcleidus* of the middle Jurassic had 39 or 40 cervical vertebrae, and *Elasmosaurus* had 76. Even in the early forms the

body was not well streamlined, and as the neck became longer, the streamlining became even poorer. The size of the paddles relative to the size of the body decreased from the Jurassic to the Cretaceous. Clearly, the *Elasmosaurus* line of plesiosaurs were not rapid swimmers. They may have used an ambush strategy to capture fish (Massare 1987, 1988).

The short-necked plesiosaurs followed a completely different evolutionary pathway, leading to an increasingly streamlined body form. The neck became shorter and the paddles larger. These were probably speedier swimmers than the long-necked plesaiosaurs, and might have captured their prey by pursuit in the manner of seals and sea lions (Massare 1988). In a sense, the short-necked plesiosaurs were converging on the morphological adaptations of ichthyosaurs, although they never attained the perfection of streamlining seen in derived members of that group.

Ichthyosaurs The ichthyosaurs ("fish reptiles") were the most specialized of the aquatic tetrapods of the Mesozoic (Figure 13–31). In many aspects of their body form they resemble porpoises (small toothed whales). Ichthyosaurs had a dorsal fin that was supported only by stiff tissue, not by bone, and the upper lobe of the caudal fin similarly lacked skeletal

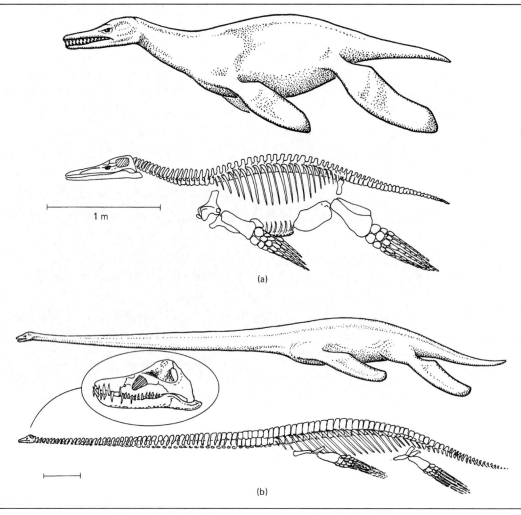

Figure 13–30 Two radically different sorts of plesiosaurs had evolved by the Jurassic: (a) *Polycotylus*, a short-necked form; (b) *Elasmosaurus*, a long-necked form. (Body outlines modified from D. M. S. Watson, 1951, *Paleontology and Modern Biology*, Yale University Press, New Haven, CT; skeletons modified from J. Piveteau, 1955, *Traité de Paléontologie*, volume 5, Mason et Cie, Paris, France.)

support. (The vertebral column of derived ichthyosaurs bent sharply downward into the ventral lobe of the caudal fin.) We know of the presence of these soft tissues because many ichthyosaur fossils in fine-grained sediments near Holzmoden in southern Germany contain an outline of the entire body preserved as a carbonaceous film.

Ichthyosaurs had both forelimbs, and hind limbs (unlike cetaceans, which retain only the forelimbs). The limbs of ichthyosaurs were modified into paddles by hyperphalangy and **hyperdactyly** (the addition of extra digits). Fossil ichthyosaurs with embryos in the body cavity indicate that these animals gave birth to living young. One fossil appears to be an individual that died in the process of giving birth, with a young ichthyosaur emerging tailfirst as do baby porpoises.

The stomach contents of ichthyosaurs, preserved in some specimens, include cephalopods, fish, and an occasional pterosaur. Ichthyosaurs had large heads with long, pointed jaws that were armed with sharp teeth in most forms, although a few ichthyosaurs were toothless. The large eyes were supported by a ring of sclerotic bones. Triassic ichthyosaurs were poorly streamlined, and may have

Figure 13–31 Evolutionary change in ichthyosaurs led to increasingly streamlined animals: (a) *Cymbospondylus* from the Triassic; (b) *Opthalmosaurus* from the late Jurassic. (c) The dorsal fin and the upper lobe of the caudal fin of ichthyosaurs were stiff tissue, not supported by bone.

used anguilliform locomotion. The greater streamlining of later forms may have been associated with the development of carangiform locomotion and rapid pursuit of prey like that of extant tunas. The Jurassic was the high point of ichthyosaur diversity; they were less abundant in the early Cretaceous and only a single genus remained in the late Cretaceous.

Ichthyosaurs became extinct before the end of the period. Even ichthyosaurs, which were the fastest swimmers among the Mesozoic marine reptiles, were probably not as swift as extant toothed whales, and the diversification of fast, agile fish in the Cretaceous may have contributed to the decline of marine reptiles (Massare 1988).

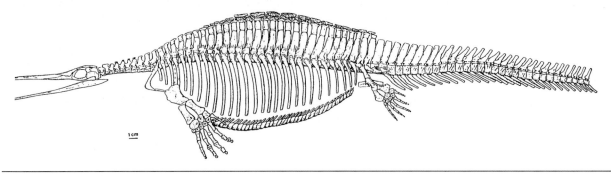

Figure 13–32 *Hupehsuchus*, skeletal reconstruction based primarily on the holo-type. (Figure and caption from R. L. Carroll and Z. Dong, 1991, *Philosophical Transactions of the Royal Society, London* 331:131–153.) Note the dermal armor on the back and the well-developed gastralia (ventral ribs).

Hupehsuchia Fossils of two aquatic reptiles from the Triassic of China, *Hupehsuchus* and *Nanchangosaurus*, have been described as a previously unknown lineage of Mesozoic diapsids (Carroll and Dong 1991). The phylogenetic affinities of these animals are unclear. They share derived characters with ichthyosaurs, but these may represent convergence. *Hupehsuchus* was about 2 meters long. Its distinctive features include a deep, laterally compressed body with stout limbs, each bearing five toes, an elongate snout that lacks teeth, and a long tail (Figure 13–32). The ribs are heavy, and ventral ribs (gastralia) are present. The dorsal midline of the body, from the neck to the base of the tail, is covered by bony plates.

The biology of an animal like *Hupehsuchus* is hard to imagine. It probably swam with lateral undulations, and the articulating surfaces of the neural arches indicate that the greatest amount of movement would have occurred in the posterior trunk and the tail. The heavy ribs and gastralia would have counteracted the natural buoyancy of an animal with air-filled lungs, but the dermal plates on the back were well above the center of gravity and would have threatened to turn *Hupehsuchus* upside-down. The absence of teeth indicates that *Hupehsuchus* had a specialized mode of feeding. One possibility is that it was a suspension feeder and had plates of baleen that strained invertebrates from the water as it swam, although not all features of its morphology are consistent with this interpretation.

The Lepidosauria

The lepidosaurs are distinguished by derived characters of the skeleton and the soft anatomy (Benton 1985a). Among these, the long bones are capped by articulating surfaces (epiphyses) of bone or calcified cartilage, the postparietal and tabular bones have been lost from the skull, the kidney has a sexual segment, and the structures of the pituitary and adrenal glands are distinctive. Perhaps the most interesting derived character of the lepidosaurs is determinate growth. That is, growth stops when the cartilaginous plate that separates the ends of the long bones and the epiphyses becomes fully ossified. This modification of the pattern of continuous growth that is seen, for example, in crocodilians and turtles may be associated with specialization on insects as food, and the importance of not growing too large to be efficient predators of insects (Carroll 1977, 1987). Two major groups of lepidosaurs can be distinguished, the Sphenodontidae (represented by the living tuatara) and the Squamata (lizards, snakes, and amphisbaenians).

Sphenodontidae The correct phylogenetic allocation of the living tuatara (*Sphenodon punctatus* and *S. guntheri*) and their fossil relatives has been a source of confusion for more than a century. In many books this animal is still listed as a rhynchocephalian. The order Rhynchocephalia was named by Albert Günther in the 1860s when he restudied the tuatara, which had previously been considered an agamid lizard. The striking anatomical differences between tuatara and lizards seemed to Günther to warrant separating the groups at the ordinal level.

Subsequent study has changed this view: Derived characters of tuatara and their fossil relatives link those animals to lizards. Also, differences between the sphenodontids and the other diapsids grouped with them as rhynchosaurs have become apparent and have weakened the basis for regarding rhynchocephalians as a natural group. The sphenodontids appear to be the sister group of the lizards + snakes + amphisbaenians.

Some of the distinctive characteristics of sphenodontids are found in their teeth, which are fused to the summit of the jaws and are not replaced during the lifetime of an animal. In the upper jaw a row of teeth on the palate parallels the teeth in the maxilla. When the lower jaw closes, its teeth fit between the two rows of teeth in the upper jaw. Tuatara chew their food very thoroughly, repeatedly crushing the food item and using the tongue to move it from one side of the jaws to the other. After 20 or 30 minutes of this processing, the body contents of an insect such as a grasshopper have been squeezed out, and the exoskeleton looks like a used tube of toothpaste.

Sphenodontids were diverse through the early Mesozoic, and some sphenodontids are known from the Cretaceous, but there is no fossil record from the Cenozoic. Fossils of sphenodontids from Triassic deposits in Great Britain include both carnivorous and herbivorous forms. The only living sphenodontids are the two species of tuatara, which occur on about 20 islands off the coast of New Zealand.

Squamata The squamates show numerous derived characters of the cranial and postcranial skeleton and soft tissues (Benton 1985a). The most conspicuous of these is the loss of the lower temporal bar and the quadratojugal bone that formed part of that bar (Figure 13–33). This modification is part of a suite of structural changes in the skull that contribute to the development of a considerable degree of kinesis. The living tuatara shows the ancestral condition for squamates, with the quadratojugal linking the jugal and the quadrate bones to form a complete lower temporal arch. (This fully diapsid condition is not characteristic of all sphenodontids, however; some of the Mesozoic forms did not have a complete lower temporal arch.)

Early lizards are not well known. The fossil genera *Paliguana* and *Palaeagama* from the late Permian and early Triassic of South Africa are probably not lizards, but they do show changes in the structure of the skull that probably parallel the changes that occurred in early squamates. The gap between the quadrate and jugal widened, and the complexly interdigitating suture between the frontal and parietal bones on the roof of the skull became straighter and more like a hinge. Additional areas of flexion evolved at the front and rear of the skull and in the lower jaw of some lizards. These changes were accompanied by the development of a flexible connection at the articulation of the quadrate bone with the squamosal that provided some mobility to the quadrate. This condition, known as **streptostyly,** increases the force the pterygoideus muscle can exert when the jaws are nearly closed (Smith 1980).

In snakes the flexibility of the skull was increased still further by loss of the second temporal bar, which was formed by a connection between the postorbital and squamosal bones (Figure 13–33). A further increase in the flexibility of the joints between other bones in the palate and the roof of the skull produced the extreme flexibility of snake skulls. The role of skull kinesis in the ecology of snakes is discussed further in Chapter 15.

The third group of squamates has a completely different sort of skull specialization. The amphisbaenians are small, legless, burrowing animals. They use their heads as rams to construct tunnels in the soil, and the skull is heavy with rigid joints between the bones. Their specialized dentition allows them to bite small pieces out of large prey.

Marine Lizards: The Mosasaurs

The mosasaurs were varanoid lizards, closely related to the living monitor lizards of the genus *Varanus*. They differ from living varanoids in their enormous size (more than 10 meters for the largest species) and their specializations for marine life. Mosasaurs are known only from the late Cretaceous, but they were a large group with nearly 20 genera. The body and tail of mosasaurs were long, the tail was laterally flattened, and the limbs were modified as flippers (Figure 13–34). The skulls of mosasaurs were very like those of varanoid lizards—elongate with sharp teeth and a flexible joint in the lower jaw between the dentary and splenial. One genus, *Globidens,* had teeth with hemispherical crowns, as its name suggests. It probably fed on mollusks. Several fossils of ammonites (free-swimming mollusks related to the squid

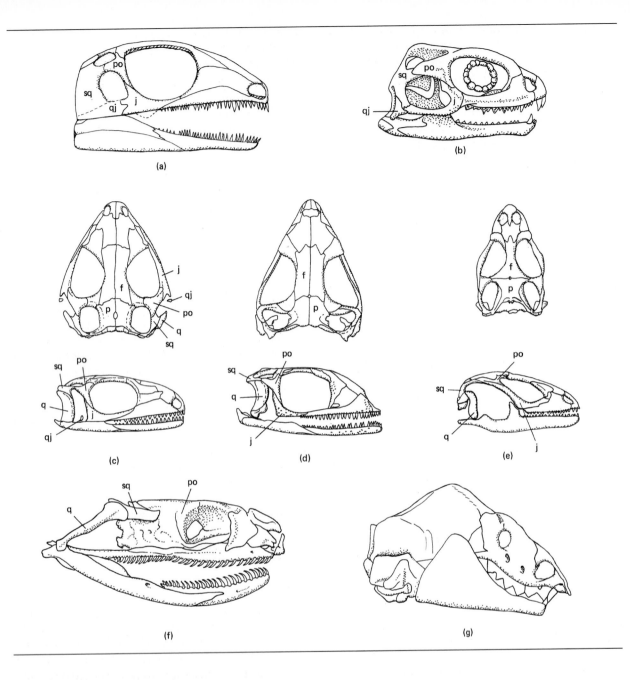

(a)

(b)

(c)

(d)

(e)

(f)

(g)

and nautilus) have been found bearing a pattern of tooth marks that suggests they had been bitten by another species of mosasaur, *Prognathodon overtoni*.

Mosasaurs probably lived in shallow seas. Hundreds of well-preserved specimens have been found, but none shows a trace of embryos within the body cavity. That situation contrasts with the discovery of several examples of ichthyosaurs with embryos, and suggests that despite their other specializations for marine life, the mosasaurs did not develop viviparity, but instead they retained the ancestral squamate characteristic of laying eggs.

Other Terrestrial Vertebrates of the Late Mesozoic

There is a tendency to look on the Jurassic and Cretaceous as the Age of Dinosaurs and to forget that

Figure 13–33 Modifications of the diapsid skull in lepidosauromorphs. Fully diapsid forms like the Permian *Petrolacosaurus* (a) retain two complete arches of bone that define the temporal fenestrae. This condition is seen in living tuatara, *Sphenodon* (b). Lizards have achieved a kinetic skull by developing a gap between the quadrate and quadratojugal and by simplifying the suture between the frontal and parietal bones as shown by the modern collared lizard *Crotaphytus* (e). Probable transitional stages allowing increasing skull kinesis that occurred in a nonsquamate lineage are illustrated by *Paliguana* (c) and *Palaeagama* (d). In snakes (f) skull kinesis is further increased by loss of the upper temporal arch. Amphisbaenians (g), which use their heads for burrowing through soil, have specialized akinetic skulls. f, frontal; j, jugal; p, parietal; po, postorbital; q, quadrate; qj, quadratojugal; sq, squamosal. ([a] From R. R. Reisz, 1977, *Science* 196:1091–1093; [b,c] from A. S. Romer, 1966, *Vertebrate Paleontology*, 3d edition, University of Chicago Press, Chicago, IL; [f,g] from C. Gans, 1974, *Biomechanics: An Approach to Vertebrate Biology*, Lippincott, Philadelphia, PA.)

there was a very considerable nondinosaur terrestrial fauna as well as variations in the kinds of nonavian dinosaurs in different habitats (Lucas 1981). To a certain extent the nonavian dinosaurs do form a separate unit. The large carnosaurs were probably the only animals capable of preying on adults of the large herbivores. Nonetheless, there must have been interactions between nonavian dinosaurs and nondinosaurs. As far as we know, all nonavian dinosaurs reproduced by laying eggs, and their eggs were a food source that could be exploited by small predators. The Nile monitor lizard today is a predator on eggs of crocodiles, and it is likely that Cretaceous monitor lizards had a similar taste for eggs. Terrestrial crocodiles may have joined monitor lizards in raiding nests.

One of the distinctive features of nonavian dinosaurs is their large size. The smallest nonavian dinosaur skeletons known indicate a total adult length of about half a meter. Even that is large in comparison to living squamates; most lizards are smaller than 20 centimeters, and only crocodilians approach the bulk of even moderate-size nonavian dinosaurs. There were, of course, adaptive zones available for smaller vertebrates in the Jurassic and Cretaceous, and these were filled as they are now by squamates, turtles, amphibians, and mammals. Among these were the animals that might have stolen eggs from nests and served as food for juvenile nonavian dinosaurs.

Although nonavian dinosaurs are so impressive and distinctive that there is a tendency to think of them as inhabiting a world of their own, they were in reality a part of an ecosystem that included many vertebrates that would not appear strange to us today. As Benton (1985b) has pointed out, "The assemblage of tetrapod families that includes all modern tetrapod groups—frogs, salamanders, lizards, snakes, turtles, crocodilians, birds, and mammals—arose in the late Triassic, and increased in diversity through the Jurassic and early Cretaceous. During the Cretaceous the diversity of these modern groups was approximately equal to that of . . . dinosaurs and pterosaurs. In the latest Cretaceous the diversity of the modern groups rose dramatically and became twice that of the dinosaurs and pterosaurs, long before the terminal Cretaceous extinction event."

A striking feature of the Mesozoic world was the absence of large mammals, and the occupation of that adaptive zone by large archosaurs. Figure 13–35 shows the relative abundance of different groups of vertebrates at two Cretaceous fossil localities. The Lance locality appears to have been a wooded swampy habitat with large streams and some ponds. The Bug Creek locality was probably downstream from such a swamp, in the delta of a major waterway. In both localities nonavian dinosaurs are a minor component of the community in terms of species diversity, although one nonavian dinosaur is the equivalent of a great many smaller animals in terms of biomass. (Some paleontologists believe that the Bug Creek locality is of Paleocene rather than Cretaceous age [see Archibald (1989) for a review].) If this is true, the presence of dinosaurs and Cretaceous mammals at this site indicates either that these groups survived the Cretaceous–Tertiary extinction or that during the Paleocene fossils were eroded from older sediments and redeposited in the creek beds.)

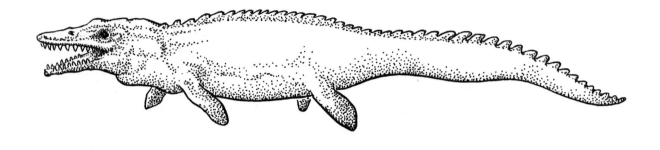

Figure 13–34 Mosasaurs were marine forms closely related to modern monitor lizards. Reconstruction of *Tylosaurus*, about 10 meters long.

Late Cretaceous Extinctions

After thriving and dominating the terrestrial habitat for 150 million years, nonavian dinosaurs disappeared near the end of the Cretaceous. The extinctions that occurred at the Cretaceous–Tertiary transition are an example of a recurrent biological phenomenon—mass extinction. In fact, the Cretaceous–Tertiary extinction was relatively small as mass extinctions go, but it is the most widely known because the disappearance of nonavian dinosaurs has captured the attention of both scientists and the public.

Biologists have struggled for years to explain the rapid change in faunal composition that occurred at the end of the Mesozoic, and there are nearly as many different hypotheses as there are authors. The catastrophism school of thought has enjoyed some vogue with the view that the giant archosaurs were wiped out by some geological or cosmic disaster. The currently popular hypothesis in the catastrophism sweepstakes is the suggestion that the impact of a comet or an asteroid with Earth ignited worldwide firestorms, or injected enough dust into the atmosphere to blot out the sun, thereby suppressing photosynthesis and leading to the extinctions of animals.

A high concentration of iridium and the presence of quartz crystals that show the effect of enormous force (known as shocked quartz) in some deposits at the end of the Cretaceous provide the basis for the inference of an asteroid impact. The Chicxulub crater on the Yucatan coast of Mexico appears to be of the correct age (64.4 ± 0.5 million years) for this event. The Manson crater in Iowa was originally thought to be the same age as Chicxulub, leading to the suggestion that two or more objects might have hit Earth in rapid succession, much like the multiple fragments of comet Shoemaker–Levy that struck Jupiter in 1994. However, it now appears

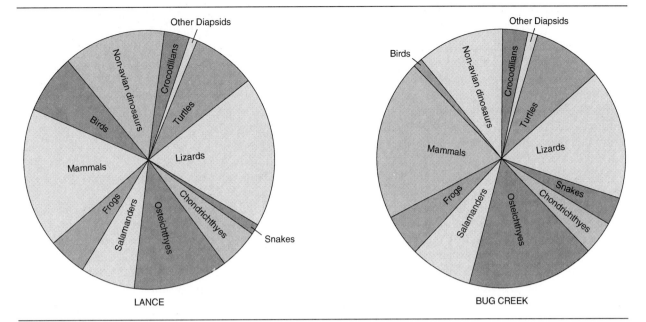

Figure 13–35 Relative abundance of genera at late Cretaceous fossil sites. Although dinosaurs were the most spectacular terrestrial vertebrates during the Mesozoic, they were surrounded by a large assortment of animals that were not very different from living species. Mammals were well represented at both sites. (Calculated from data in R. Estes and P. Berberian, 1970, *Breviora*, No. 343.)

that the Manson crater is nearly 10 million years older than Chicxulub (Izett et al. 1993).

In general, most geologists favor the hypothesis that a catastrophe, probably the impact that produced the Chicxulub crater, was responsible for the extinctions that occurred at the end of the Cretaceous. In contrast, although most paleontologists agree that impacts have occurred, they believe that normal biological, climatic, and geological processes provide the most plausible explanation for those faunal changes (Hallam 1987, Kerr 1988).

Evaluation of the two hypotheses requires consideration of what we can reasonably expect from the fossil record, and of events in the late Cretaceous preceding the Chicxulub impact. The catastrophism scenario predicts that all major extinctions would have occurred nearly instantaneously—within a week, or a year or two at the most—whereas the gradualist interpretation considers that the extinctions stretched over thousands or even tens of thousands of years. That's a substantial difference, but it cannot be distinguished in the fossil record. Dating events at the end of the Cretaceous involves large uncertainties: 1 to 2 million years for correlations of

marine and terrestrial deposits and 0.5 to 4 million years for dates based on rates of decay of radioactive elements (Benton 1990). Thus, dates can help to determine the sequence of events, but not their duration.

Our inability to measure the duration of events applies to the fossil record of all kinds of life— marine invertebrates, terrestrial plants, and terrestrial vertebrates. A second difficulty is particularly acute for studies of nonavian dinosaurs. Only in western North America is there a fossil record of dinosaurs that extends from the late Cretaceous across the Cretaceous–Tertiary boundary into the Paleocene (Archibald 1989, Dodson and Tatarinov 1990). Thus, we are basing inferences about worldwide changes on just one area.

Fossil evidence indicates that dinosaurs were probably declining before the end of the Cretaceous. The Judith River formation is about 75 million years old, and the Hell Creek formation is at the very end of the Cretaceous, 65 million years ago. The Judith River formation has yielded about 32 genera of nonavian dinosaurs, whereas the Hell Creek formation contains only about 18 genera. Ornithischian dinosaurs were most affected—ceratopsids fell from six genera to three, and hadrosaurs from seven genera to one. (No dinosaur fossils are found in the upper 2 or 3 meters of the Hell Creek formation, immediately before the iridium layer that is presumed to mark the Chicxulub impact, but fossils of other animals are also rare in this part of the deposit and paleontologists suspect that the absence of dinosaur fossils reflects a change in the conditions of deposition or preservation rather than the extinction of dinosaurs.)

The end of the Cretaceous (the Maastrichtian epoch, which extended from about 72 million years ago to 65 million years ago) shows a pattern of declining diversity of nonavian dinosaurs. Of the 13 families present at the start of the Maastrichtian, three disappeared in the early Maastrichtian and two more in the middle Maastrichtian. Thus, all the controversy about the extinction of dinosaurs revolves around eight families containing 10 genera and 12 species.

The average duration of species of nonavian dinosaurs in the fossil record is 2 to 3 million years, and genera last 5 to 6 million years. The limited diversity of Maastrichtian dinosaur faunas contrasts with totals of 40 to 80 species for equivalent time periods earlier in the late Cretaceous. What seems to have happened to nonavian dinosaurs near the end of the Cretaceous was not so much a massive increase in extinctions as a reduction in the appearance of new dinosaurs to replace those that became extinct (Padian and Clemens 1985, Jablonski 1986).

Most paleontologists believe that this pattern of declining diversity beginning several million years before the Chicxulub impact is more consistent with gradual extinction produced by geological and biological process than with an abrupt catastrophe. Hypotheses based on broad-scale changes in climate have been proposed repeatedly. The Mesozoic was characterized by stable temperatures with little variation from day to night, from summer to winter, and from north to south. Geological evidence indicates that climates became cooler and more variable at the end of the Cretaceous. Casper, Wyoming, is an important dinosaur fossil locality, and a reconstruction of its Mesozoic temperature regime suggests that the long-term temperature extremes probably lay between 11 and 34°C, with a mean of 22°C. Clearly, the climate has changed since then: The present temperature extremes at Casper are –27 to +40°C, with a mean of 8°C.

In addition to these changes in climate, geological events at the end of the Cretaceous were changing the land surface and altering terrestrial habitats. A reduction in sea level of 150 to 200 meters drained the shallow inland seas that had filled the centers of the continents during the late Cretaceous. Rivers cut their way down toward the new sea level, forming valleys instead of meandering across broad floodplains. Late Cretaceous nonavian dinosaurs appear to have been concentrated in river and floodplain habitats, and these changes would have reduced the habitat available to them (Schopf 1983). The floodplains were probably routes for north–south migration, and their interruption by valleys running east–west might have disrupted migratory patterns. Furthermore, the population sizes of large herbivorous dinosaurs and their theropod predators may have represented a delicate balance of predator–prey ratios and population densities (Farlow 1993). Perhaps changes in topography that restricted the movements of predators and their prey tilted the scales toward excessive predation or produced population densities that were too low to sustain normal social behavior and reproduction. In that scenario, an extraterrestrial impact, if it had any effect at all, would be just the last straw.

Summary

The major groups of tetrapods in the Mesozoic were members of the diapsid ("two arches") lineage. This group is distinguished particularly by the presence of two fenestrae in the temporal region of the skull that are defined by arches of bone. The archosauromorph lineage of diapsids contains the most familiar of the Mesozoic tetrapods, the nonavian dinosaurs.

Two major groups of nonavian dinosaurs are distinguished, the Saurischia and Ornithischia.

The saurischians included the sauropod dinosaurs—enormous herbivorous quadrupedal forms like *Apatosaurus* (formerly *Brontosaurus*), *Diplodocus,* and *Brachiosaurus*—and the theropods, which were bipedal carnivores. The carnosaurs (of which *Tyrannosaurus rex* is the most familiar example) were large theropods that probably preyed on the large sauropods. Other theropods were smaller: The ornithomimids were probably very like ostriches, and some had horny beaks and lacked teeth. The deinonychosaurs were fast-running predators. Ornithomimids probably seized small prey with hands that had three fingers armed with claws, whereas the deinonychosaurs probably were able to prey on dinosaurs larger than themselves. They may have hunted in packs and used the enormous claw on the second toe to slash their prey. Birds had evolved by the Jurassic: *Archaeopteryx*, the earliest bird known, is very like small coelurosaurs.

The ornithischian dinosaurs were herbivorous, and many had horny beaks on the snout and batteries of specialized teeth in the rear of the jaw. The ornithopods (duck-billed dinosaurs) and pachycephalosaurs (thick-headed dinosaurs) were bipedal, and the stegosaurs (plated dinosaurs), ceratopsians (horned dinosaurs), and ankylosaurians (armored dinosaurs) were quadrupedal. Although the saurischians and ornithischians represent independent radiations and had different ecological specializations, they share many morphological features that appear to reflect the mechanical constraints of being very large terrestrial animals.

Nonavian dinosaurs and birds are part of the archosaur lineage of the archosauromorph diapsids. Crocodilians and the extinct phytosaurs are also archosaurs, as are the pterosaurs. The phytosaurs were very like crocodilians in general body form, but the external nares of phytosaurs were located immediately anterior to the eyes instead of at the tip of the snout as they are in crocodilians. Phytosaurs were abundant in the Triassic, but appear to have been replaced by crocodilians in the Jurassic. Crocodilians were most diverse in the Cretaceous, but large species survived well into the Cenozoic, and at least one species may have been 15 meters long (as large as *Tyrannosaurus rex*). Only 21 species of crocodilians survive today. None is larger than 6 or 7 meters, and most are smaller than 4 meters. Pterosaurs were remarkably diverse in the Mesozoic; some species were the size of a sparrow and others were larger than the largest extant birds. Specializations of the skulls of pterosaurs suggest that they had a wide variety of diets, including preying on insects and small terrestrial vertebrates, eating fish, crushing snails, and sieving small crustaceans from the water with a comb-like array of fine teeth. Less familiar

archosauromorphs include the rhynchosaurs, specialized herbivorous tetrapods of the Triassic, and the lizard-like Prolacertiformes.

The lepidosauromorph lineage was less diverse in the Mesozoic than the archosauromorphs. The placodonts, plesiosaurs, and ichthyosaurs were marine mollusk eaters or fish eaters. The Younginiformes of the late Permian were mostly lightly built terrestrial animals about 50 centimeters long. The champsosaurs are a problematic group of crocodile-like animals about 2 meters in length. The surviving lepidosauromorphs are all members of the lepidosaur lineage. The sphenodontids are stocky terrestrial forms, about 50 centimeters long. The tuatara (*Sphenodon punctatus* and *S. guntheri*) are the only extant sphenodontids, but the group was more diverse in the Mesozoic. Lizards, snakes, and amphisbaenians compose the squamate lineage. The basic diapsid skull form has been extensively modified among squamates, particularly in relation to feeding habits (lizards and snakes) and the use of the head for burrowing (amphisbaenians).

The phylogenetic relationship of crocodilians and birds allows us to draw inferences about some aspects of the biology of nonavian dinosaurs. Characters that are shared by crocodilians and birds are probably ancestral for nonavian dinosaurs. Social behavior and parental care are the norm for crocodilians and birds, and increasing evidence suggests that nonavian dinosaurs, too, had elaborate social behavior and that at least some species cared for the young.

The mode of temperature regulation of extinct diapsids cannot be deduced from a phylogenetic approach because crocodilians are ectotherms, whereas birds are endotherms. The biophysical relationships between organisms and their environments are very sensitive to body size, and generalizations about nonavian dinosaurs as a group are not valid. The largest nonavian dinosaurs would have had stable body temperatures as a result of their great thermal inertia (gigantothermy), and the high metabolic rates that some paleontologists have postulated for them would have led to lethal overheating.

The Mesozoic was marked by a series of faunal changes. The extinctions at the end of the Cretaceous were the most dramatic of these, partly because the nonavian dinosaur fauna of the time was so spectacular, but they are not unique. The replacement of rhynchosaurs by nonavian dinosaurs in the late Triassic was another change of great magnitude, and it may be associated with the decline of the *Dicroidium* flora and its replacement by taller plants, including bennettitaleans. In the middle Cretaceous the diversity of sauropod dinosaurs declined while the ornithischians burgeoned. This faunal change was contemporaneous with the decline of bennettitaleans and the appearance of flowering plants (angiosperms). In this context of shifting faunas and floras throughout the Mesozoic, explanations of the late Cretaceous extinctions that depend on worldwide catastrophes are less persuasive than hypotheses based on the gradual changes in sea level, topography, and climate that are shown in the fossil record of the late Mesozoic.

References

Abler, W. L. 1992. The serrated teeth of tyrannosaurid dinosaurs and biting structures in other animals. *Paleobiology* 18:161–183.

Alexander, R. M. 1989. *Dynamics of Dinosaurs and Other Extinct Giants*. Columbia University Press, New York, NY.

Archibald, J. D. 1989. The demise of the dinosaurs and the rise of the mammals. Pages 48–57 in *The Age of Dinosaurs,* edited by K. Padian and D. J. Chure. The Paleontological Society, University of Tennessee Press, Knoxville, TN.

Benton, M. J. 1983a. The age of the rhynchosaur. *New Scientist* 98(1352):9–13.

Benton, M. J. 1983b. Large-scale replacements in the history of life. *Nature* 302:16–17.

Benton, M. J. 1983c. Dinosaur success in the Triassic: a noncompetitive ecological model. *The Quarterly Review of Biology* 58:29–55.

Benton, M. J. 1985a. Classification and phylogeny of diapsid reptiles. *Zoological Journal of the Linnaean Society (London)* 84:97–164.

Benton, M. J. 1985b. Mass extinctions among non-marine tetrapods. *Nature* 316:811–814.

Benton, M. J. 1990. *Vertebrate Paleontology.* HarperCollins Academic, London, UK.

Bonaparte, J. F., and M. Vince. 1979. El hallazgo del primer nido de dinosaurios triasicos (Saurischia, Prosauropoda). Triásico superior de Patagonia, Argentina. *Ameghiana* 16:173–182.

Buffrénil, V. de, J. O. Farlow, and A. de Ricqles. 1986. Growth and function of *Stegosaurus* plates: evidence from bone histology. *Paleobiology* 12:450–458.

Caple, G., R. P. Balda, and W. R. Willis. 1983. The physics of leaping animals and the evolution of preflight. *American Naturalist* 121:455–476.

Carroll, R. L. 1977. The origin of lizards. Pages 359–396 in *Problems in Vertebrate Evolution,* edited by S. M. Andrews, R. S. Miles, and A. D. Wells. Academic, London, UK.

Carroll, R. L. 1987. *Vertebrate Paleontology and Evolution.* Freeman, New York, NY.

Carroll, R. L., and Z. Dong. 1991. *Hupehsuchus,* an enigmatic aquatic reptiles from the Triassic of China, and the problem of establishing relationships. *Philosophical Transactions of the Royal Society, London* 331:131–153.

Charig, A. 1972. The evolution of the archosaur pelvis and hindlimb: an explanation in functional terms. In *Studies in Vertebrate Evolution,* edited by K. A. Joysey and T. S. Kemp. Winchester, Piscataway, NJ.

Charig, A. 1979. *A New Look at the Dinosaurs.* Mayflower, New York, NY.

Colbert, E. H. 1993. Feeding strategies and metabolism in elephants and sauropod dinosaurs. *American Journal of Science* 293-A:1–19.

Coombs, W. P., Jr. 1990. Behavior patterns of dinosaurs. Pages 32–42 in *The Dinosauria,* edited by D. B. Weishampel, P. Dodson, and H. Osmólska. University of California Press, Berkeley, CA.

Dodson, P. 1993. Comparative craniology of the Ceratopsia. *American Journal of Science* 293-A:200–234.

Dodson, P., and L. P. Tatarinov. 1990. Dinosaur extinction. Pages 55–62 in *The Dinosauria,* edited by D. B. Weishampel, P. Dodson, and H. Osmólska, University of California Press, Berkeley, CA.

Farlow, J. O. 1987. Speculations about the diet and digestive physiology of herbivorous dinosaurs. *Paleobiology* 13:60–72.

Farlow, J. O. 1990. Dinosaur energetics and thermal biology. Pages 43–55 in *The Dinosauria,* edited by D. B. Weishampel, P. Dodson, and H. Osmólska. University of California Press, Berkeley, CA.

Farlow, J. O. 1993. On the rareness of big, fierce animals: speculations about the body sizes, population densities, and geographic ranges of predatory mammals and large carnivorous dinosaurs. *American Journal of Science* 293-A:167–199.

Farlow, J. O., C. V. Thompson, and D. E. Rosner. 1976. Plates of the dinosaur *Stegosaurus:* forced convection heat loss fins? *Science* 192:1123–1125.

Feduccia, A. 1993. Evidence from claw geometry indicating arboreal habits of *Archaeopteryx*. *Science* 259:790–793.

Feduccia, A., and H. B. Tordoff. 1979. Feathers of *Archaeopteryx:* asymmetric vanes indicate aerodynamic function. *Science* 203:1021–1022.

Feduccia, A., and R. Wild. 1993. Birdlike characters in the Triassic archosaur *Megalancosaurus*. *Naturwissenschaften* 80:564–566.

Ferguson, M. W. J. 1985. Reproductive biology and embryology of the crocodilians. Pages 329–491 in *Biology of the Reptilia,* volume 14, edited by C. Gans, F. Billett, and P. F. A. Maderson, Wiley, New York, NY.

Galton, P. M. 1990. Basal Sauropodamoprha—prosauropods. Pages 320–344 in *The Dinosauria,* edited by D. B. Weishampel, P. Dodson, and H. Osmólska, University of California Press, Berkeley, CA.

Gauthier, J. 1986. Saurischian monophyly and the origin of birds. Pages 1–55 in *The Origin of Birds and the Evolution of Flight,* edited by K. Padian. *Memoirs of the California Academy of Sciences,* No. 8.

Gauthier, J., and K. Padian. 1985. Phylogenetic, functional, and aerodynamic analyses of the origin of birds and their flight. Pages 185–197 in *The Beginning of Birds, Proceedings of the International* Archaeopteryx *Conference, Eichstätt 1984,* edited by M. K. Hecht, J. H. Ostrom, G. Viohl, and P. Wellnhofer. Jura Museum, Eichstätt, West Germany.

Hallam, A. 1987. End-Cretaceous mass extinction event: argument for terrestrial causation. *Science* 238:1237–1242.

Hazlehurst, G. A., and J. M. V. Rayner. 1992. Flight characteristics of Triassic and Jurassic Pterosauria: an appraisal based in wing shape. *Paleobiology* 18:447–463.

Hopson, J. A. 1975. The evolution of cranial display structures in hadrosaurian dinosaurs. *Paleobiology* 1:21–43.

Hopson, J. A., and L. B. Radinsky. 1980. Vertebrate paleontology: new approaches and new insights. *Paleobiology* 6:250–270.

Horner, J. R. 1982. Evidence of colonial nesting and site fidelity among ornithischian dinosaurs. *Nature* 297:675–676.

Horner, J. R. 1984. The nesting behavior of dinosaurs. *Scientific American* 241(4):130–137.

Horner, J. R. 1994. Comparative taphonomy of some dinosaur and extant bird colonial nesting grounds. Pages 116–123 in *Dinosaur Eggs and Babies,* edited by K. Carpenter, K. F. Hirsch, and J. R. Horner. Cambridge University Press, Cambridge, UK.

Horner, J. R., and P. Makela. 1979. Nest of juveniles provides evidence of family structure among dinosaurs. *Nature* 282:296–298.

Izett, G. A., W. A. Cobban, J. D. Obradovich, and M. J. Kunk. 1993. The Manson impact structure: $^{40}Ar/^{39}Ar$ age and its distal impact ejecta in the Pieere Shale in southeastern South Dakota. *Science* 262:729–732.

Jablonski, D. 1986. Mass extinctions: new answers, new questions. Pages 43–61 in *The Last Extinction,* edited by L. Kaufman and K. Mallory. M.I.T. Press, Cambridge, MA.

Jenkins, F. A., Jr. 1993. The evolution of the avian shoulder joint. *American Journal of Science* 293-A:253–267.

Jensen, J. A. 1985. Three new sauropod dinosaurs from the Upper Jurassic of Colorado. *Great Basin Naturalist* 45:697–709.

Jones, D. R., and G. Shelton. 1993. The physiology of the alligator heart: Left aortic flow patterns and right-to-left shunts. *Journal of Experimental Biology* 176:247–269.

Kerourio, P. 1981. Nouvelles observations sur le mode de nidification et de ponte chez les dinosauriens du Cretace terminal du Midi de la France. *Comptes Rendu Sommaire des Séances de la Societe geologique de France* No. 1, pp. 25–28.

Kerr, R. A. 1988. Was there a prelude to the dinosaurs' demise? *Science* 239:729–730.

Kirkland, J. I. 1994. Predation of dinosaur nests by terrestrial crocodilians. Pages 124–133 in *Dinosaur Eggs and Babies,* edited by K. Carpenter, K. F. Hirsch, and J. R. Horner. Cambridge University Press, Cambridge, UK.

Lang, J. W. 1986. Male parental care in mugger crocodiles. *National Geographic Research* 2:519–525.

Lang, J. W. 1989. Social behavior. Pages 102–117 in *Crocodiles and Alligators,* edited by C. A. Ross. Facts on File, New York, NY.

Lillywhite, H. B. 1991. Sauropods and gravity. *Natural History* December 1991, p. 33.

Lucas, S. G. 1981. Dinosaur communities of the San Juan Basin: a case for lateral variations in the composition of Late Creta-

ceous dinosaur communities. Pages 337–393 in *Advances in San Juan Basin Paleontology,* edited by S. G. Lucas, J. K. Rigby, Jr., and B. S. Kues. University of New Mexico Press, Albuquerque, NM.

Massare, J. A. 1987. Tooth morphology and prey preference of Mesozoic marine reptiles. *Journal of Vertebrate Paleontology* 7:121–137.

Massare, J. A. 1988. Swimming capabilities of Mesozoic marine reptiles: implications for methods of predation. *Paleobiology* 14:187–205.

McIntosh, J. S. 1989. The sauropod dinosaurs: a brief survey. Pages 85–99 in *The Age of Dinosaurs,* Short Courses in Paleontology, No. 2, edited by K. Padian and D. J. Chure. The Paleontological Society, University of Tennessee Press, Knoxville, TN.

Moratalla, J. J., and J. E. Powell. 1994. Dinosaur nesting patterns. Pages 37–46 in *Dinosaur Eggs and Babies,* edited by K. Carpenter, K. F. Hirsch, and J. R. Horner. Cambridge University Press, Cambridge, UK.

Norell, M. A., J. M. Clark, D. Demberelyin, B. Rhinchen, L. M. Chiappe, A. R. Davidson, M. C. McKenna, P. Altangerel, and M. J. Novacek. 1994. A theropod dinosaur embryo and the affinities of the Flaming Cliffs dinosaur eggs. *Science* 266:779–782.

Ostrom, J. H. 1974. Archaeopteryx and the evolution of flight. *Quarterly Review of Biology* 49:27–47.

Ostrom, J. H. 1986. The cursorial origin of avian flight. Pages 73–81 in *The Origin of Birds and the Evolution of Flight,* edited by K. Padian. *Memoirs of the California Academy of Sciences,* No. 8.

Padian, K., and W. A. Clemens. 1985. Terrestrial vertebrate diversity: episodes and insights. Pages 41–96 in *Phanerozoic Diversity Patterns,* edited by J. W. Valentine. Princeton University Press, Princeton, NJ.

Padian, K., and J. M. V. Rayner. 1993. The wings of pterosaurs. *American Journal of Science* 293-A:91–166.

Paladino, F. V., M. P. O'Connor, and J. R. Spotila. 1990. Metabolism of leatherback turtles, gigantothermy, and thermoregulation of dinosaurs. *Nature* 344:858–860.

Peterson, A. 1985. The locomotor adaptations of *Archaeopteryx:* glider or cursor? Pages 99–103 in *The Beginning of Birds, Proceedings of the International* Archaeopteryx *Conference, Eichstätt 1984,* edited by M. K. Hecht, J. H. Ostrom, G. Viohl, and P. Wellnhofer. Jura Museum, Eichstätt, West Germany.

Pooley, A. C., and C. Gans. 1976. The Nile crocodile. *Scientific American* 234:114–124.

Rayner, J. M. V. 1988. The evolution of vertebrate flight. *Biological Journal of the Linnean Society* 34:269–287.

Romer, A. S. 1966. *Vertebrate Paleontology,* 3d edition. University of Chicago Press, Chicago, IL.

Ruben, J. 1991. Reptilian physiology and the flight capacity of *Archaeopteryx. Evolution* 45:1–17.

Ruben, J. 1993. Powered flight in *Archaeopteryx:* response to Speakman. *Evolution* 47:935–938.

Schopf, T. J. M. 1983. Extinction of the dinosaurs: a 1982 understanding. Pages 415–422 in *Large Body Impacts and Terrestrial Evolution,* edited by L. T. Silver and P. Schultz. Geological Society of America Special Paper 190.

Seymour, R. S. 1979. Dinosaur eggs: gas conductance through the shell, water loss during incubation and clutch size. *Paleobiology* 5:1–11.

Smith, K. K. 1980. Mechanical significance of streptostyly in lizards. *Nature* 283:778–779.

Spotila, J. R., M. P. O'Connor, P. Dodson, and F. V. Paladino. 1991. Hot and cold running dinosaurs: body size, metabolism and migration. *Modern Geology* 16:203–227.

Storrs, G. W. 1993. Function and phylogeny in sauropterygian (Diapsida) evolution. *American Journal of Science* 293-A:63–90.

Thulborn, T. 1990. *Dinosaur Tracks*. Chapman and Hall, London, UK.

Troyer, K. 1984. Microbes, herbivory and the evolution of social behavior. *Journal of Theoretical Biology* 106:157–169.

Unwin, D. M. 1987. Pterosaur locomotion: joggers or waddlers? *Nature* 327:13–14.

Unwin, D. M., and N. Bakhurina. 1994. *Sordes pilosus* and the nature of the pterosaur flight apparatus. *Nature* 371:62–64.

Weishampel, D. B. 1981. Acoustic analyses of potential vocalization in lambeosaurine dinosaurs (Reptilia: Ornithischia). *Paleobiology* 7:252–261.

Yalden, D. W. 1985. Forelimb function in *Archaeopteryx*. Pages 91–97 in *The Beginning of Birds, Proceedings of the International Archaeopteryx Conference, Eichstätt 1984*, edited by M. K. Hecht, J. H. Ostrom, G. Viohl, and P. Wellnhofer. Jura Museum, Eichstätt, West Germany.

Geography and Ecology of the Mesozoic 14

D uring the Mesozoic, Pangaea reached its greatest development—all of the Earth's land surface had coalesced into a single continent that stretched from the South Pole to the North Pole. As the continental blocks merged, plants and animals that had evolved in isolation met and interacted. Some forms became extinct, whereas others spread across the globe.

Terrestrial environments changed substantially during the Mesozoic, and some of the changes in plant growth form may have been associated with the appearance of long-necked sauropod dinosaurs that could browse on all but the very tallest kinds of trees. Hadrosaurs and ceratopsians, too, were specialized herbivores, and the trail of destruction left by a herd of these animals might have created a suitable habitat for the growth of angiosperms (flowering plants) which appear in the fossil record in the Mesozoic.

Pangaea, the World Continent

At the beginning of the Triassic, the entire land area of Earth was concentrated in the supercontinent Pangaea, which straddled the Equator. The maximum development of Pangaea occurred at approximately the middle/late Triassic boundary. The southern part of Pangaea, areas that now form Antarctica and Australia were close to the South Pole, and the northern part of modern Eurasia was within the Arctic Circle (Figure 14–1), but even these high latitudes were relatively warm. Large amphibians are found in Triassic deposits from Australia, Antarctica, Greenland, and Spitzbergen, and coal deposits in both the northern and southern hemispheres point to moist climates. In contrast, low and middle latitudes were probably dry, either seasonally or year-round until the late Cretaceous and early Tertiary, when coal deposits in middle latitudes suggest that the climate had become wetter (Wilson et al. 1994, Manspeizer 1994).

Despite the continuity of land from the extreme north of Pangaea to its southern tip, Triassic floras and faunas show regional characters that probably reflect differences in patterns of rainfall and seasonal temperature extremes in areas far from the sea. The fragmentation of Pangaea began in the Jurassic with the separation of Laurasia and Gondwana by a westward extension of the Tethys Sea (Figure 14–2) and continued through the remainder of the Mesozoic. By the late Cretaceous, the continents had nearly reached their current positions (Figure 14–3).

Terrestrial Ecosystems

The Mesozoic is marked by a series of large-scale changes in flora and fauna beginning with the Permian/Triassic event, which extended over a period of about 25 million years, and culminating in the Cretaceous/Tertiary transition. Terrestrial ecosystems had achieved an essentially modern food web by the end of the Permian. Plants grew in communities that contained mixtures of species, they were

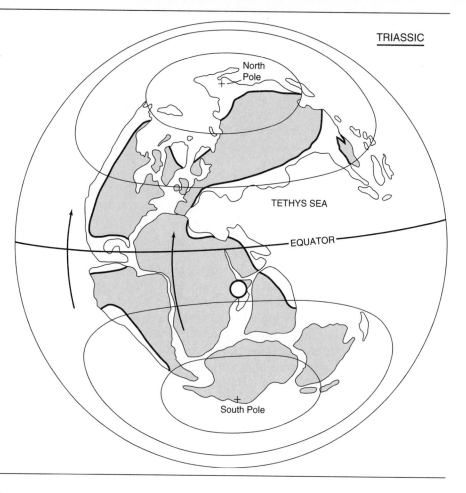

Figure 14–1 Location of continental blocks in the Triassic. The arrows indicate the general northward drift of Pangaea during the early Mesozoic. Epicontinental seas (shown by shading) were confined to the margins of continents.

consumed by herbivorous insects and vertebrates, and carnivorous invertebrates and vertebrates preyed on the herbivores (DiMichele and Hook 1992). The kinds of plants and animals in Permian ecosystems were quite different from those in modern ecosystems, however. During the Mesozoic angiosperms (flowering plants) and mammals appeared and diversified, and ecosystems of the early Tertiary have close modern analogues (Wing and Dieter–Sues 1992).

Mesozoic vegetation included familiar modern groups of gymnosperms such as conifers (pines and other cone-bearing trees), ginkgophytes (relatives of the living ginkgo), as well as cycads. Conifers ranged from small bushes to tall forest trees, and many have structural characteristics that suggest they lived in relatively dry habitats. Cycads, too, appear to have lived in dry regions. Bennettitalean cycads had thick stems, from half a meter to three meters tall. Many species had recessed stomata that

would have limited transpiration of water, and hairs on the leaf surfaces that would have reflected sunlight. They were probably characteristic of open habitats, and the thick external coverings of their trunks might have protected them from fires.

Angiosperms appear in the fossil record in the Cretaceous, although extrapolations of the time of origin of angiosperms based on analysis of DNA suggest that they might have originated as early as the Paleozoic (Martin et al. 1989). By the middle of the Cretaceous angiosperms were well established, especially in middle latitudes. Northern Laurasia and southern Gondwana continued to be dominated by conifers and ferns. A late Cretaceous deposit in Wyoming that had been buried by volcanic ash allowed Wing and his colleagues to reconstruct the spatial relations of the plant community as well as to determine the kinds of plants that had been present (Wing et al. 1993). Angiosperms accounted for 61 percent of the species of plants, but only 12 percent

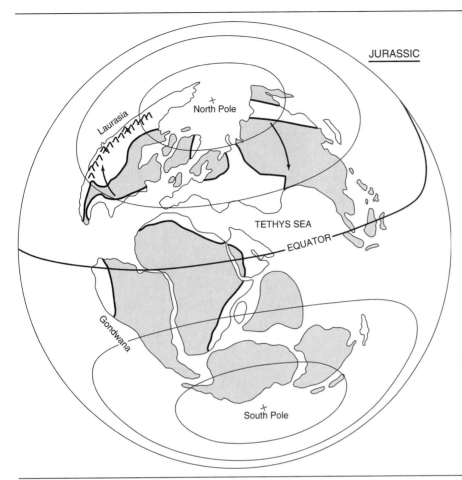

Figure 14–2 Location of continental blocks in the Jurassic. The northward drift of the continents continued, and Laurasia began to rotate as shown by the arrows, ripping North America from its contact with South America and diminishing the size of the Tethys Sea. Epicontinental seas (shown by shading) were more extensive than in the Triassic. Toward the end of the Jurassic the Nevadan orogeny produced the ancestral Sierra Nevada Mountains (△△△).

of the plant cover. In contrast, stream-side sites of the same age were dominated by angiosperms. Thus, generalizations about the changing composition of vegetation during the Mesozoic may reflect a large influence of local variation in the species present at the particular sites that have been studied. Clearly, ferns and gymnosperms remained significant elements of some floras through the entire Mesozoic. Angiosperms appear to have enjoyed their greatest success in disturbed habitats.

At the start of the Triassic, most herbivorous vertebrates were quadrupedal and their feeding was probably restricted to plants within a meter of the ground. A plant could escape these herbivores by growing out of their reach. Seed ferns, cycads, and seedlings of all kinds of plants were available to these browsers, but adult trees were safe.

The appearance of prosauropod dinosaurs in the late Triassic changed that situation. The long necks of these animals allowed them to reach much

higher than other vertebrate herbivores, especially when they stood erect. In Chapter 13 we considered this and other specializations of herbivorous nonavian dinosaurs from the dinosaurs' point of view: The ability to reach upward broadened the range of food available to them. But what did this change mean to plants? From a plant's perspective, the advent of long-necked and bipedal herbivores meant that the browse height increased from about 1 meter above the ground to about 4 meters. No longer could plants grow out of the reach of browsers in just a few years; now they were exposed to browsing for many years, perhaps even for their entire lives. This change in herbivores would have had profound effects on the life history of plants. It might have shifted the balance of costs and benefits away from investing energy in rapid growth (to outgrow browsers) and toward putting that energy into production of structures such as spines (to deter vertebrate herbivores). It could have changed the opti-

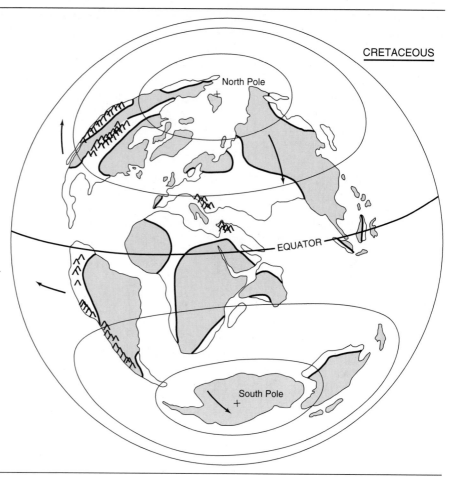

Figure 14–3 Location of continental blocks in the Cretaceous. The final breakup of Laurasia and Gondwana occurred in the Cretaceous. The northern continental blocks continued the rotation they began in the Jurassic, and were approaching their current positions by the end of the Cretaceous. The southern blocks moved apart (arrows). Note that India was still in the Southern Hemisphere and Australia, Papua New Guinea, and New Zealand were well south of their current positions. Extensive epicontinental seas (indicated by shading) covered large parts of North and South America, Eurasia, and Africa.

CRETACEOUS

North Pole

EQUATOR

South Pole

mal age for reproduction, and the optimal size of seeds.

Herbivores are not entirely a negative factor in the lives of plants, because animals that eat seeds carry them away from the parent plant and deposit them in new locations. An edible covering is the bribe a plant offers to animals to induce them to disperse its seeds, and both gymnosperms and angiosperms include species that rely on animals for seed dispersal.

By the middle Jurassic, sauropod dinosaurs could probably browse at heights up to 10 meters if they stood on four legs, and could have reached still higher if they were able to stand erect by using the tail as a counterbalance. Weighing up to 100,000 kilograms, sauropods could also topple even large trees by walking over them, just as elephants do now. The peg-like or spatulate teeth of sauropods show few signs of wear, suggesting that they swallowed twigs and leaves with relatively little oral processing and

relied on a gastric mill to crush their food. Stegosaurs, the second most abundant kind of vertebrate herbivores, were probably limited to feeding close to ground level. They had horny beaks, and their teeth do show signs of wear, suggesting that they fed on plants that may have been fibrous enough to require chewing.

By the Cretaceous, Laurasia and Gondwana were separated by the Tethys Sea, and differences appear in the ecosystems of the northern and southern continents. Tetanurae (see Figure 13–13) became the dominant carnivores in Laurasia, whereas in South America ceratosaurs were dominant and tetanurans were rare. African dinosaur faunas are poorly known, and initial discoveries of carnivores have revealed unexpected similarities to Laurasia (Sereno et al. 1994). Sauropods appear to have continued as the dominant vertebrate herbivores in South America and probably in Africa, hadrosaurs were rare, and ceratopsians were absent. In contrast, hadrosaurs and ceratopsians dominate the faunas of late Cretaceous herbivores in the Northern Hemisphere. Hadrosaurs occurred in both North America and Eurasia, whereas ceratopsians were apparently restricted to western North America. Both groups have horny bills and extensive dentitions with closely packed, interlocking teeth and would have been able to chew tough plants such as ferns and cycads. These herbivores, which probably lived in migratory or nomadic herds that may have numbered in the thousands of individuals, could have had profound effects on the vegetation of North America in the late Cretaceous. The increase in the relative abundance of angiosperms in disturbed habitats is approximately coincident with the increase in abundance and diversity of hadrosaurs and ceratopsians in the late Cretaceous (Wing and Dieter-Sues 1992). The spread of angiosperms may have been facilitated by herds of dinosaurs if fast-growing angiosperms were able to colonize habitats after the passage of a group of hadrosaurs or ceratopsians left a community of ferns and cycads crushed and torn.

Are Mass Extinctions Periodic?

Large-scale extinctions, such as the one that occurred at the end of the Cretaceous, are inherently dramatic, and in the mid 1980s a suggestion that extinctions occur in a regular cycle appeared to support the hypothesis that impacts by extraterrestrial objects cause mass extinctions. An examination of the marine fossil record led Raup and Sepkoski (1984, 1986) to suggest that major extinctions occur at intervals of approximately 26 million years. Astrophysicists seized on this hypothesis, and sought a source of extraterrestrial objects that could bombard Earth at 26 million year intervals.

Popular attention has moved on to other sensations, but the possible periodicity of extinctions remains a source of heated controversy among scientists. The suggestion that extinctions have occurred at regular intervals sounds like a hypothesis that can be tested, but statistical analysis is anything but simple. In the first place, statistical analyses of cycles are inherently complex. Added to that problem is the uncertainty of dating geological deposits, which grows larger as one moves back in time. Furthermore, the span of time encompassed by the hypothesis is relatively brief, and the analysis is based on only nine extinctions. That is a small sample, especially considering the difficulties of dating the actual events and in carrying out statistical tests.

One approach to difficult statistical analyses is to build your own computer-generated data set on the basis of an assumption about the pattern of events (for example, that they were randomly distributed in time), and then to compare those computer-generated data with the actual observations. This process, known to statisticians as "boot strapping," allows you to see how often the results you observe in nature would occur by chance. Several computer simulations of this sort have shown that mathematical models based on random events can produce a spacing of extinctions that appears cyclic. From 5 to 35 percent of the extinction sequences generated by these models were at least as close to a 26 million-year cycle as the actual data (Hoffman 1989).

Scientists usually accept a probability of .05 (i.e., a 5 percent chance of error) as the break-point for accepting or rejecting a hypothesis. The hypothesis is provisionally accepted when the chance of error is less than 5 percent, and it is rejected when the chance of error exceeds 5 percent. Thus, the 5 to 35 percent chance of error predicted by the computer simulations suggests that the hypothesis of a 26 million-year cycle of extinctions should be rejected.

References

DiMichele, W. A., and R. W. Hook. 1992 (rapporteurs). Paleozoic terrestrial ecosystems. Pages 205–325 in *Terrestrial Ecosystems Through Time*, edited by A. K. Behrensmeyer, J. D. Damuth, W. A. DiMichele, R. Potts, H. Dieter-Sues, and S. L. Wing, University of Chicago Press, Chicago, IL.

Hoffman, A. 1989. Changing paleontological views on mass extinction phenomena. Pages 1–18 in *Mass Extinctions:*

Processes and Evidence, edited by S. K. Donovan, Belhaven Press, London, UK.

Manspeizer, W. 1994. The breakup of Pangea and its impact on climate: Consequences of Variscan–Alleghanide orogenic collapse. Pages 169–185 in *Pangea: Paleoclimate, Tectonics, and Sedimentation During Accretion, Zenith, and Breakup of a Supercontinent,* edited by G. D. Klein, Geological Society of America, Special Paper 288.

Martin, W., A. Gierl, and H. Saedler. 1989. Molecular evidence for pre-Cretaceous angiosperm origins. *Nature* 339:46–48.

Raup, D. M., and J. J. Sepkoski. 1984. Periodicities of extinctions in the geologic past. *Proceedings National Academy Sciences USA* 81:801–805.

Raup, D. M., and J. J. Sepkoski. 1986. Periodic extinction of families and genera. *Science* 231:833–836.

Sereno, P. C., J. A. Wilson, H. C. E. Larsson, D. B. Dutheil, and H.-D. Sues. 1994. Early Cretaceous dinosaurs from the Sahara. *Science* 266:267–271.

Wilson, K. M, D. Pollard, W. W. Hay, S. L. Thompson, and C. N. Wold. 1994. General circulation model simulations of Triassic climates: preliminary results. Pages 91–116 in *Pangea: Paleoclimate, Tectonics, and Sedimentation During Accretion, Zenith, and Breakup of a Supercontinent,* edited by G. D. Klein, Geological Society of America, Special Paper 288.

Wing, S. L., and H. Dieter–Sues 1992 (rapporteurs). Mesozoic and early Cenozoic terrestrial ecosystems. Pages 327–416 in *Terrestrial Ecosystems Through Time,* edited by A. K. Behrensmeyer, J. D. Damuth, W. A. DiMichele, R. Potts, H. Dieter–Sues, and S. L. Wing, University of Chicago Press, Chicago, IL.

Wing, S. L., L. J. Hickey, and C. C. Swisher, 1993. Implications of an exceptional fossil flora for late Cretaceous vegetation. *Nature* 363:342–344.

THE LEPIDOSAURS: TUATARA, LIZARDS, AMPHISBAENIANS, AND SNAKES

15

We noted in our discussion of the fauna of the Mesozoic in Chapter 13 that although terrestrial environments were dominated by nonavian dinosaurs, many of the animals present would not look strange in the modern world. Mammals, birds, and squamates all originated and diversified in the Mesozoic, and their overshadowing by dinosaurs is as much a matter of our perception as of biological reality. Lizards and especially snakes are elements of the diapsid lineage that have had their greatest flowering in the Cenozoic, and the extant species of squamate diapsids outnumber the species of mammals.

Some aspects of the biology of squamates probably give us an impression of the ancestral way of life of amniotic tetrapods, although many of the structural characteristics of lizards and snakes are derived. One of the derived characteristics of lizards is determinate growth; that is, increase in body size stops when the growth centers of the long bones ossify. This mechanism sets an upper limit to the size of individuals of a species, and may be related to the specialization of most lizards as predators of insects. The predatory behavior of lizards ranges from sitting in one place and ambushing passing insects to seeking food by traversing the home range in an active, purposeful way. Broad aspects of the biology of lizards are correlated with these foraging modes, including morphology, exercise physiology, and social behavior. The anatomical specializations of snakes are associated with their elongate body form and include modifications of the jaws and skull that allow them to subdue and swallow large prey.

The Lepidosaurs

The lepidosaur lineage of diapsids includes the Squamata, which is the second largest group of extant tetrapods (Tables 15–1 and 15–2 and Figures 15–1 and 15–2). Squamates are defined by a variety of derived characters (see Chapter 13), of which determinate growth may have the most general significance. Crocodilians and turtles continue to grow all through their lives, although the growth rates of adults are much slower than those of juveniles. This pattern of growth can be viewed as *functionally* determinate in that the largest individuals are only half again the size of the smallest sexually mature adults, but squamates (and birds and mammals) have a *structural* type of determinate growth. The cartilaginous epiphyseal plates of the long bones are the sites at which growth occurs as cells proliferate. Growth continues while the epiphyseal plates are composed of cartilage, and stops completely when the epiphyses fuse to the shafts of the bones, obliterating the cartilaginous plates. The development of determinate growth in lepidosaurs may initially have been associated with the insectivorous diet that is presumed to have been characteristic of early lepidosaurs. Lizard-size animals can prey on insects without requiring the sorts of morphological or ecological specializations that are necessary for large insect-eating vertebrates.

The Sphenodontidae, represented now only by the two species of tuatara of New Zealand, is the sister group of the Squamata. Squamates have diversified greatly since their origin in the Triassic. Within the squamates, lizards can be distinguished in colloquial terms but not phylogenetically, because snakes and amphisbaenians are derived from lizards. Thus "lizards" is a paraphyletic group (not including all descendants). Nonetheless, lizards, snakes, and amphisbaenians are largely distinct in ecology and behavior, and a colloquial separation is useful in considering them.

The Radiation of Sphenodontids and the Biology of Tuatara

The sphenodontids were a diverse group in the Mesozoic. Triassic forms were small, with body lengths of only 15 to 35 centimeters. Most of the Triassic sphenodontids had teeth that were fused to the jaws (**acrodont**) like the teeth of the extant tuatara (*Sphenodon*), but others had teeth attached to the jaw largely by their lateral surfaces (**pleurodont**), like those of some lizards. The teeth of the Triassic sphenodontids suggest that both insectivorous and herbivorous diets were represented in the group. Sphenodontids from the Jurassic and Cretaceous included small terrestrial forms and also marine animals as much as 1.5 meters long. Some had transversely broadened teeth and were probably herbivorous.

The two species of *Sphenodon,* known as tuatara, are the only extant sphenodontids. ("Tuatara" is a Maori word, and does not add an "s" to form the plural.) Tuatara have derived characters that distinguish them from the Mesozoic forms (Whiteside 1986). Tuatara formerly inhabited the North and South islands of New Zealand, but the advent of humans and their associates (cats, dogs, rats, sheep, and goats) exterminated them on the mainland. Now populations are found on only about 30 small islands off the coast. Tuatara have been fully protected in New Zealand since 1895, but only one species, *Sphenodon punctatus*, was recognized. In fact, a second species of tuatara, *S. guntheri,* had been described in 1877. *Sphenodon guntheri* is much less common than *S. punctatus*. The only surviving population of *S. guntheri* is a group of fewer than 300 adults living on 1.7 hectares of scrub on the top of North Brother Island. The lack of fit among the phylogenetic status of tuatara and the legislation designed to protect them left a crack that *S. guntheri* nearly fell through (Daugherty et al. 1990). Because this species was not recognized as being phylogenetically distinct from *S. punctatus* it received no special protection. The 300 animals on North Brother Island were regarded as not very important from a conservation perspective compared to the large populations of tuatara (*S. punctatus*) on some of the other islands. Probably it was only the presence of a lighthouse that was staffed until 1990 by resident keepers who deterred illegal landings and poaching that saved the tuatara. The East Island population of *S. guntheri* (the only other population of the species) became extinct during this century. This example illustrates the crucial role that taxonomy plays in conservation—a species must be recognized before it can be protected. In other words, bad taxonomy can kill (May 1990).

Adult tuatara are about 60 centimeters long. They are nocturnal and in the cool, foggy nights that characterize their island habitats they cannot raise their body temperatures during activity by basking in the sun as lizards do. Body temperatures from 6 to 16°C have been reported for active tuatara, and these are low compared to most lizards. During the day, tuatara do bask in the sun and raise their body temperatures to 28°C or higher. Tuatara feed largely on invertebrates, with an occasional frog, lizard, or seabird for variety. The jaws and teeth of tuatara produce a shearing effect during chewing: The upper jaw contains two rows of teeth, one on the maxilla and the other on the palatine bones. The teeth of the lower jaw fit between the two rows of upper teeth, and the lower jaw closes with an initial vertical movement, followed by an anterior sliding movement. As the lower jaw slides, the food item is bent or sheared between the triangular cusps on the teeth of the upper and lower jaws.

Tuatara live in burrows that they may share with nesting seabirds. The burrows are spaced at intervals of 2 or 3 meters in dense colonies, and tuatara may use vocalizations in their social interactions. Male tuatara bite each other when establishing dominance and bite females during mating.

The ecology of tuatara rests to a large extent on exploitation of the resources provided by colonies of seabirds. Tuatara feed on the birds, which are most vulnerable to predation at night. In addition, the quantities of guano produced by the birds, scraps of

Table 15–1 Characteristics of lineages of extant lepidosaurs. The table follows the sequence of lineages shown in Figure 15–1.

Sphenodontidae: 2 species of tuatara, now restricted to islands off the coast of New Zealand.

Iguania

    Corytophanidae: 9 species of small to medium-size (10–20 cm) arboreal Neotropical lizards. They have long tails, laterally flattened bodies, and some (*Corytophanes, Laemanctus*) have crests on their heads that may make it hard for predators to recognize them as lizards when they are seen as silhouettes against a patch of light.

    Crotaphytidae: The collared and leopard lizards. These are 6 species of medium-size North American desert lizards that prey on other lizards.

    Hoplocercidae: 10 species of terrestrial and arboreal lizards with a geographic range extending from Panama to Brazil.

    Iguanidae: Only 31 species remain in the current classification of this family. They are large (to >1.5 m) terrestrial and arboreal herbivorous lizards. The Neotropical green iguana (*Iguana iguana*) is the most familiar member of the family, which also includes the black and ground iguanas of Central America and the West Indies, the Galapagos marine and land iguanas, and the Fijian iguanas. The North American chuckwalla (*Sauromalus*) is an iguanid. This rock-dwelling lizard is dorsoventrally flattened, and seeks shelter from predators in rock crevices, where it inflates its lungs to wedge itself in place.

    Opluridae: 7 species of small (10 cm), mostly terrestrial lizards from Madagascar and the Comoro Archipelago.

    Phrynosomatidae: About 117 species of North American lizards. The group takes its name from the horned lizards (*Phrynosoma*), but more than half the species are in the genus *Sceloporus* (spiny swifts), which range from southern Canada to Panama. Most phrynosomatids are terrestrial, and some are specialized to enter rock crevices (dorsoventrally flattened) or to live on loose sand (fringes of scales on the toes, valvular nostrils that exclude sand, earless or with skin folds and elongated scales that prevent sand grains from entering the ears). Some species of *Sceloporus* and *Urosaurus* are arboreal.

    Polychrotidae: About 250 species of primarily small (10 cm) South American lizards, most of them in the genus *Anolis*. *Anolis carolinensis*, the green anole, has a geographic range that extends northward to North Carolina. Most species of *Anolis* are arboreal.

    Tropiduridae: More than 200 species of small to medium-size (10–20 cm) South American lizards that occupy regions from sea level to the high Andes, including deserts, rainforests, and grasslands.

Acrodonta

    Chamaeleonidae: 90 species of primarily arboreal lizards, but including a few grassland and terrestrial species, found in Africa and Madagascar and extending into southern Spain and along the west coast of the Mediterranean. Chameleons have the laterally flattened bodies characteristic of many arboreal lizards, and additional specializations, including zygodactyl feet, prehensile tails, eyes that swivel to provide a 360-degree field of view, and a projectile tongue. The leaf chameleons (*Brookesia, Rhampholeon*) are as small as 25 mm, whereas some species of *Chamaeleo* grow to more than 60 cm. (Note that the family name and the genus are spelled with an *ae*, but the common name "chameleon" is spelled with an *e*.)

    Leiolepididae: 14 species of mainly herbivorous terrestrial lizards found in Southeast Asia, India, the Middle East, and Saharan Africa. *Leiolepis* and some species of *Uromastyx* are medium-size (20 cm), and other *Uromastyx* approach a meter in length.

    Agamidae: This family, centered in Asia, contains some 300 species of small to large (10 cm to 1 m) lizards that extend through the Middle East into Africa and along the Indoaustralian Archipelago into Australia. Small agamids such as *Calotes* are arboreal and have the same body form as arboreal corytophanids. The Australian lizard *Moloch* is a spiny, dorsoventrally flattened ant-eating species reminiscent of the North American horned lizards, and *Hydrosaurus* is a large herbivore like the Neotropical green iguana that lives in trees overhanging rivers.

Gekkota

    Gekkonidae: Geckos, with some 700 species and a geographic distribution that includes every continent except Antarctica, are one of the largest families of lepidosaurs. Extant species range in size from very small (30 mm) to medium-size (30 cm).

    Pygopodidae: The 34 species of pygopods are found only in Australia. They are elongate, nearly limbless terrestrial lizards.

Amphisbaenia

    Bipedidae: 3 species of round-headed amphisbaenians (*Bipes*) found in Mexico. This lineage is unique among amphisbaenians in retaining well-developed forelimbs that are used to help penetrate the soil surface. Once underground, *Bipes* use the head for burrowing.

    Trogonophiidae: 6 species of round-headed amphisbaenians that use an oscillating movement of the head in digging. Trogonophiids are found in north Africa and western Asia.

    Amphisbaenidae: 140 species of round-headed, spade-snouted, and keel-headed amphisbaenians found in southern Europe, western Asia, Africa, North and South America, and the West Indies. *(Continued)*

Table 15–1 *(Continued)*

Scincomorpha

    Dibamidae: 2 genera of small to medium (5–20 cm) limbless, burrowing lizards. *Dibamus* is found in Indomalaysia, whereas *Anelytropsis* is from Mexico.

    Gymnophthalmidae: About 150 species of small (less than 6 cm) lizards that live in the leaf litter of Neotropical forests. Limb reduction is widespread in this lineage.

    Teiidae: About 80 species of active terrestrial lizards ranging from nearly the Canadian border in North America to central Argentina, and including the West Indies. Teids range from small insectivorous species of *Ameiva*, *Cnemidophorus*, and *Kentropyx* to species such as the tegus (*Tupinambis*) and the caiman lizard (*Dracena*) that can grow to a meter or more.

    Lacertidae: About 150 species of small to medium-size terrestrial lizards from the Old World with a range of body forms very like those seen among teiids. Lacertids are found over all of Europe, Africa, and Asia. Like the teiids, lacertids are mostly terrestrial, and include some large predators as well as a great number of smaller species.

    Xantusiidae: 19 species of tiny to small (3–10 cm), secretive lizards from southwestern North America, Central America, and Cuba. All have the eyelids fused into a transparent scale that covers the eye and gives the group its common name, spectacled lizards.

    Scincidae: Some 700 species make this one of the most species-rich lineages of lepidosaurs. Skinks occur on all of the continents except Antarctica. Nearly all are very small to medium-size (5–20 cm) terrestrial lizards, and many show limb reduction. Most are insectivorous, but some of the large (40 cm) species of the Australian blue-tongued skinks (*Tiliqua*) consume plant material, and the Solomon Islands arboreal skink (*Corucia zebrata*) has specializations of the gut that promote fermentative digestion of plant material by symbiotic microorganisms.

    Cordylidae: 42 species of small to medium-size terrestrial or rock-dwelling lizards from sub-Saharan Africa. *Cordylus* is a rock dweller and has the dorsoventrally flattened body typical of lizards that seek shelter in crevices. It is heavily armored, and many species have exceedingly sharp spines along the body margins and the tail. Skull kinesis is discussed in the text in terms of the feeding mechanisms employed by lizards, but *Cordylus* uses skull kinesis in a different way: By contracting its jaw muscles, *Cordylus* raises the braincase relative to the lower jaw, and it uses this mechanism to wedge itself into crevices. *Platysaurus* is a small rock dweller without the heavy body armor of *Cordylus*, and *Chamaesaura* is an elongate, nearly limbless lizard. Most cordylids are viviparous, except for *Platysaurus*, which lays eggs.

    Gerrhosauridae: The 27 species of gerrhosaurids are found in sub-Saharan Africa and on Madagascar. This lineage shows parallels to the cordylids: *Gerrhosaurus* has heavy body armor, and *Tetradactylus* is elongate and has reduced limbs. *Angolosaurus skoogi* lives in shifting sand dunes in the Namib Desert, where it eats seeds that blow in from outside the dunes and buries itself in sand to avoid extreme temperatures. Gerrhosaurids are egg layers.

Anguimorpha

    Xenosauridae: This small family includes 7 species of medium-size terrestrial lizards. *Shinisaurus* is found in China and *Xenosaurus* in Mexico. *Shinisaurus* is a semiaquatic lizard that lives along stream banks and forages in the water, catching fish and aquatic invertebrates. *Xenosaurus* lives in the moist leaf litter of high-altitude cloud forests. Both genera are viviparous.

    Anguidae: The 90 species of anguids occur in North and South America and in Europe, the Middle East, and southern China. All anguids have body armor, and a fold along the side of the body allows the trunk to expand and contract as they breathe. Most are terrestrial, foraging in leaf litter, and four genera are legless: *Anniella* (California and Baja California) is a small species (10–20 cm) that spends the day underground and emerges at night to hunt on the surface. *Anguis* (Europe), *Ophisaurus* (North America, Europe, and Asia), and *Ophioides* (South America) are limbless surface dwellers that forage in leaf litter and dense vegetation. *Anguis* is small, but *Ophisaurus* grows to a trunk length of 0.5 m, with a tail at least as long as its trunk.

    Varanidae: The 31 species in this group include the single species of *Lanthanotus*, the Bornean earless monitor, and 30 species of monitor lizards (*Varanus*). *Lanthanotus* is an elongate lizard with a trunk 15–20 cm long. It is secretive and semiaquatic, spending the day in a burrow and foraging at night on both land and water. Monitor lizards range in size from *Varanus brevicauda*, which has a total length of only 10 cm, to the Komodo dragon, *V. komodoensis*, which can exceed 3 m in length and may weigh 75 kg. All monitors are active predators, and the larger species patrol a large home range searching for food in holes and beneath rocks and logs. Some species are moderately arboreal. Varanids occur in Africa, Asia, and the East Indies, but about half the species are limited to Australia.

    Helodermatidae: The 2 species of helodermatids are the only poisonous lizards. These are large (25- to 40-cm trunk length), heavy-bodied lizards found in the southwestern United States and Mexico.

    Serpentes: See Figure 15–2 and Table 15–2.

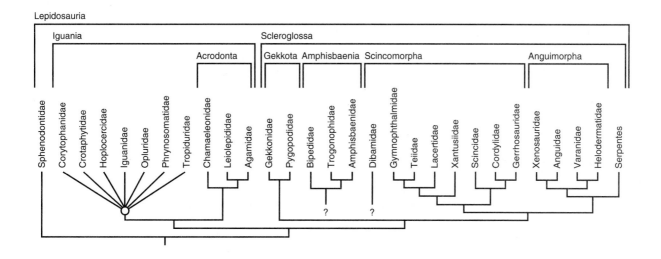

Lepidosauria

Iguania · Scleroglossa

Acrodonta · Gekkota · Amphisbaenia · Scincomorpha · Anguimorpha

Sphenodontidae · Corytophanidae · Crotaphytidae · Hoplocercidae · Iguanidae · Opluridae · Phrynosomatidae · Tropiduridae · Chamaeleonidae · Leiolepididae · Agamidae · Gekkonidae · Pygopodidae · Bipedidae · Trogonophidae · Amphisbaenidae · Dibamidae · Gymnophthalmidae · Teiidae · Lacertidae · Xantusiidae · Scincidae · Cordylidae · Gerrhosauridae · Xenosauridae · Anguidae · Varanidae · Helodermatidae · Serpentes

Figure 15–1 Phylogenetic relationships of extant lepidosaurs. The circle indicates that branching sequences are unresolved within the Iguania, and the correct placement of the Amphisbaenia and Dibamidae is unknown. (Modified from various sources, including R. Estes and G. Pregill [editors], 1988, *Phylogenetic Relationships of the Lizard Families*, Stanford University Press, Stanford, CA; D. Cundall, V. Wallach, and D. A. Rossman, 1993, *Zoological Journal of the Linnean Society* 109:235–273; and David Hillis and Alan Savitzky, personal communication.)

the food they bring to their nestlings, and the bodies of dead nestlings attract huge numbers of arthropods that are eaten by tuatara. These arthropods are largely nocturnal and must be hunted when they are active. Thus the crepuscular (= occurring at dusk and dawn) activity of tuatara and the low body temperatures that result from being active at those times of day are probably specializations that stem from the association of tuatara with colonies of nesting seabirds. This pattern of behavior and thermoregulation probably does not represent even the ancestral condition for sphenodontids, and there is no reason to interpret it as being ancestral for lepidosaurs or diapsids.

The Radiation of Squamates

The fossil record of lizards is largely incomplete through the middle of the Mesozoic, but late Jurassic deposits in China and Europe include members of most of the groups of extant lizards. The major

groups of lizards had probably diverged by the end of the Jurassic. The oldest snake that is adequately known is *Dinilysia* from the late Cretaceous of South America. Amphisbaenians are known from the Paleocene, but had probably originated earlier.

Lizards

The approximately 3000 species of lizards range in size from diminutive geckos only 3 centimeters long to the Komodo monitor lizard, which is 3 meters long at maturity and weighs some 75 kilograms. A reconstruction of the skeleton of a fossil monitor lizard, *Megalania prisca*, from the Pleistocene of Australia is 5.5 meters long, and in life the lizard might have weighed more than 1000 kilograms. About 80 percent of extant lizards weigh less than 20 grams as adults and are insectivorous. Spiny swifts and japalures (Figure 15–3a and b) are examples of these small, generalized insectivores. Other small lizards have specialized diets: The North American horned lizards and the Australian moloch (Figure 15–3e and f) feed on ants. Most geckos (Figure 15–3g) are noc-

Table 15–2 Characteristics of lineages of extant snakes. The table follows the sequence of lineages shown in Figure 15–2.

Scolecophidia
  Leptotyphlopidae: About 100 species of tiny (10 cm) to small fossorial snakes found in Africa, southwestern Asia, South and Central America, and southwestern North America.
  Typhlopidae: About 150 species of small to medium-size fossorial snakes with reduced eyes. Typhlopids are found on all continents except North America and Antarctica.
  Anomalepididae: 18 species of small (20 cm) to medium-size (75 cm) fossorial snakes from Central and South America.

Alethinophidia
  Anomochilidae: 2 species of small (less than 40 cm) burrowing snakes from the Malayan Peninsula and East Indies.
  Uropeltidae: About 40 species of small to medium-size fossorial snakes (20–70 cm) from India and Sri Lanka.
  Cylindrophiidae: 8 species of burrowing snakes (25–85 cm) from Sri Lanka, Burma, Indochina, and the East indies.
  Aniliidae: A single species of fossorial snake from northern South America.
  Xenopeltidae: One species of medium-size fossorial snake found in the East Indies.
  Loxocemidae: A single medium-size species of semifossorial snake from southern Mexico and Central America.
  Pythonidae: 21 species, many of of which are large (2 m) to enormous (10 m or more). Found in Asia, Africa, and Australia.
  Boinae: 22 species of mainly large to enormous terrestrial (*Boa*), semiaquatic (*Eunectes*), and arboreal (*Corallus*, *Epicrates*) snakes found from Mexico through subtropical South America and the West Indies.
  Erycinae: 15 species of medium-size semifossorial snakes. *Eryx* and *Gongylophis* are found from Africa and India to central Asia. *Calabaria* occurs only in West Africa, and *Charina* and *Lichanura* occur in western North America.
  Boyleriidae: 2 species of medium-size boa-like snakes known only from Round Island, near Mauritius in the Indian Ocean.
  Tropidophiidae: 20 species of small, terrestrial boa-like snakes from Central America, northern South America, and the West Indies.
  Acrochordidae: Two species of medium-size to large (about 2 m) aquatic snakes from southern Asia, the East Indies, and northern Australia.

Colubroidea
  Atractaspidae: 60 species of small to medium-size African and Asian snakes. The atractaspids are secretive, living in leaf litter and spending time underground. They have elongated fangs on the maxillae, sometimes preceded by several smaller teeth.
  Elapidae: About 250 species of venomous snakes with hollow fangs near the front of relatively immobile maxillae. Elapids occur on all continents except Antarctica, and are most diverse in Australia. Sea snakes are elapids.
  Viperidae: About 200 species of medium size to large (2 m) venomous snakes in which the maxilla is rotated about its attachment to the prefrontal, allowing the fang to rest horizontally when the mouth is closed. True vipers (about 60 species) are found in Eurasia and Africa; pit vipers are found in the New World and in Asia. Viperids are absent from Australia and Antarctica.
  Colubridae: About 1600 species of tiny to very large snakes, found on all continents except Antarctica. Many colubrids have glands that secrete venom that kills prey, but they lack hollow teeth specialized for injecting venom.

turnal, and many species are closely associated with human habitations.

Lizards are adaptable animals that have occupied habitats ranging from swamp to desert and even above the timberline in some mountains. Many species are arboreal, and the most specialized of these are frequently laterally flattened and often have peculiar projections from the skull and back that help to obscure their outline. The Old World chameleons (Chamaeleonidae) are the most specialized arboreal lizards (Figure 15–3h). Their **zygo-**dactylous (*zygo* = joined, *dactyl* = digit) feet grasp branches firmly, and additional security is provided by the prehensile tail. The tongue and hyoid apparatus are specialized, and the tongue can be projected forward more than a body's length to capture insects that adhere to its sticky tip. This feeding mechanism requires good eyesight, especially the ability to gauge distances accurately so that the correct trajectory can be employed. The chameleon's eyes are elevated in small cones and are independently movable. When the lizard is at rest, the eyes

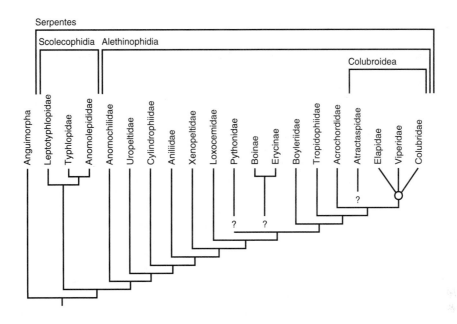

Figure 15–2 Phylogenetic relationships of extant snakes. Note the lack of resolution of branching sequences involving the Pythonidae and Boinae + Erycinae. Branching sequences within the Colubroidea are also unresolved. (Modified from D. Cundall, V. Wallach, and D. A. Rossman, 1993, *Zoological Journal of the Linnean Society* 109:235–273; A. G. Kluge, 1993, *Zoological Journal of the Linnean Society* 107:293–351; and David Hillis and Alan Savitzky personal communication.)

swivel back and forth giving the lizard a view of its surroundings. When an insect is spotted, both eyes fix on it, and a cautious stalk brings the lizard within shooting range.

Most large lizards are herbivores. Many iguanas (family Iguanidae) are arboreal inhabitants of the tropics of Central and South America. Large terrestrial iguanas occur on islands in the West Indies and the Galapagos Islands, probably because the absence of predators has allowed them to spend a large part of their time on the ground. Smaller terrestrial herbivores like the black iguanas (Figure 15–3c) live on the mainland of Mexico and Central America, and still smaller relatives such as the chuckwallas and desert iguanas range as far north as the western United States. Many species of lizards live on beaches, but few extant species actually enter the water. The marine iguana of the Galapagos Islands is an exception. The feeding habits of the marine iguana are unique. It feeds on seaweed, diving 10 meters or

more to browse on algae growing below the tide mark.

An exception to the rule of herbivorous diets for large lizards is found in the monitor lizards (family Varanidae). Varanids are active predators that feed on a variety of vertebrate and invertebrate animals, including birds and mammals (Figure 15–3j). Few lizards are capable of capturing and subduing such prey, but varanids have morphological and physiological characteristics that make them effective predators of vertebrates. The Komodo monitor lizard is capable of killing adult water buffalo, but its normal prey is deer and feral goats (Auffenberg 1981). (Large monitor lizards were widely distributed on the islands between Australia and Indonesia during the Pleistocene and may have preyed on pygmy elephants that also lived on the islands [Diamond 1987].) The hunting methods of the Komodo monitor are very similar to those employed by mammalian carnivores, showing that

Figure 15–3 Parallel and convergent evolution of body forms among lizards. Small, generalized insectivores: (a) spiny swift, *Sceloporus*, Phrynosomatidae. (b) japalure, *Calotes*, Agamidae. Herbivores: (c) black iguana, *Ctenosaura*, Iguanidae; (d) mastigure, *Uromastyx*, Leiolepididae. Ant specialists: (e) horned lizard, *Phrynosoma*, Phrynosomatidae; (f) spiny devil, *Moloch*, Agamidae. Nocturnal lizards: (g) Tokay gecko, *Gekko*, Gekkonidae. Arboreal lizards: (h) African chameleon, *Chamaeleo*, Chamaeleonidae. Legless lizards: (i) North American glass lizard, *Ophisaurus*, Anguidae. Large predators: (j) monitor lizard, *Varanus*, Varanidae.

a simple brain is capable of complex behavior and learning. In the late morning a Komodo monitor waits in ambush beside the trails deer use to move from the hilltops, where they rest during the morning, to the valleys, where they sleep during the afternoon. The lizards are familiar with the trails the deer use and often wait where several deer trails converge. If no deer pass the lizard's ambush, it moves into the valleys, systematically stalking the thickets where deer are likely to be found. This purposeful hunting behavior, which demonstrates familiarity with the behavior of the prey and with local geography, is in strong contrast to the opportunistic seizure of prey that characterizes the behavior of many lizards.

Limb reduction has evolved repeatedly among lizards (Greer 1991), and every continent has one or more families with legless, or nearly legless, species (Figure 15–3i). Leglessness in lizards is usually associated with life in dense grass or shrubbery in which a slim, elongate body can maneuver more easily than a short one with functional legs. Some legless lizards crawl into small openings among rocks and under logs, and a few are subterranean.

Amphisbaenians

The amphisbaenians include about 150 species of extremely fossorial squamates (*fossor* = a digger). Most amphisbaenians are legless; however, the three species in the Mexican genus *Bipes* have well-developed forelegs that they use to assist entry into the soil but not for burrowing underground. The skulls of amphisbaenians are used for tunneling, and they are rigidly constructed. The dental structure is also distinctive: Amphisbaenians possess a single median tooth in the upper jaw—a feature unique to this group of vertebrates. The median tooth is part of a specialized dental battery that makes amphisbaenians formidable predators, capable of subduing a wide variety of invertebrates and small vertebrates. The upper tooth fits into the space between two teeth in the lower jaw and forms a set of nippers that can of bite out a piece of tissue from a prey item too large

(e)

(f)

(g)

(h)

(i)

(j)

Figure 15–3 *(Continued)*

for the mouth to engulf as a whole. The earliest amphisbaenian known is a fossil from the late Cretaceous (Wu et al. 1993).

The skin of amphisbaenians also is distinctive (Figure 15–4). The **annuli** (rings) that pass around the circumference of the body are readily apparent from external examination, and dissection shows that the integument is nearly free of connections to the trunk. Thus, it forms a tube within which the body of the amphisbaenian can slide forward or backward. The separation of the trunk and skin is employed during the concertina locomotion that all amphisbaenians use underground. Integumentary muscles run longitudinally from annulus to annulus. Their contraction causes the skin over the area of muscular contraction to telescope and to buckle outward, anchoring that part of the amphisbaenian against the walls of its tunnel. Next, contraction of muscles that pass anteriorly from the vertebrae and ribs to the skin slide the trunk forward within the tube of integument. Amphisbaenians can move backward along their tunnels with the same mechanism by contracting muscles that pass posteriorly from the ribs to the skin. (The name "Amphisbaenia" is derived from the Greek roots *amphi* [double] and *baen* [walk] in reference to the ability of amphisbaenians to move forward and backward with equal facility.) A similar type of rectilinear (straight-line) locomotion is used by some heavy-bodied snakes, but the telescoping ability of the skin of snakes is generally restricted to the lateroventral portions of the body, whereas the skin is loose around the entire circumference of the body of amphisbaenians.

The burrowing habits of amphisbaenians make them difficult to study. Three major functional categories can be recognized: Some species have blunt heads; the rest have either vertically keeled or horizontally spade-shaped snouts. Blunt-snouted forms burrow by ramming their heads into the soil to compact it (Figure 15–4b). Sometimes an oscillatory rotation of the head with its heavily keratinized scales is used to shave material from the face of the tunnel. Shovel-snouted amphisbaenians ram the end of the tunnel, then lift the head to compact soil into the roof (Figure 15–4c). Wedge-snouted forms ram the snout into the end of the tunnel and then use the snout or the side of the neck to compress the material into the walls of the tunnel (Figure 15–4d–f). In parts of Africa representatives of the three types occur together and share the subsoil habitat. The unspecialized blunt-headed forms live near the surface where the soil is relatively easy to tunnel through, and the specialized forms live in deeper, more compact soil. The geographic range of the unspecialized forms is greater than that of the specialized ones, and in areas in which only a single species of amphisbaenian occurs it is a blunt-headed species.

The relationship between unspecialized and specialized burrowers is puzzling. One would expect that the specialized forms with their more elaborate methods of burrowing would replace the unspecialized ones, but this has not happened. The explanation may lie in the conflicting selective forces on the snout. On one hand, it is important to have a snout that will burrow through soil, but on the other hand, it is also important to have a mouth capable of tackling a wide variety of prey. The specializations of the snout that make it an effective structure for burrowing appear to reduce its effectiveness for feeding. The blunt-headed amphisbae-

nians may be able to eat a wider variety of prey than the specialized forms can. Thus, in loose soil where it is easy to burrow, the blunt-headed forms may have an advantage. Only in soil too compact for a blunt-headed form to penetrate might the specialized forms find the balance of selective forces shifted in their favor.

Snakes

The 2500 species of snakes range in size from diminutive burrowing species that feed on termites and grow to only 10 centimeters to the large constrictors, which approach 10 meters in length (Siegel et al. 1987). The Scolecophidia includes three families of small burrowing snakes with shiny scales and reduced eyes. Traces of the pelvic girdle remain in most species, but the braincase is snake-like. Burrowing snakes in the families Aniliidae and Uropeltidae use their heads to dig through soil, and the bones of their skulls are solidly united. The sole xenopeltid, the sunbeam snake of southeastern Asia, is a ground-dwelling species that takes its common name from its highly iridescent scales. The Boinae are mostly New World snakes, whereas the pythons are found in the Old World. The anaconda, a semiaquatic species of boa from South America, is considered the largest extant species of snake. It probably approaches a length of 10 meters, and the reticulated python of southeast Asia is nearly as large. Not all boas and pythons are large, however; some secretive and fossorial species are considerably less than 1 meter long as adults. The wart snakes in the family Acrochordidae are entirely aquatic; they lack the enlarged ventral scales that characterize most terrestrial snakes, and they have difficulty moving on land.

The Colubroidea includes most of the extant species of snakes, and the family Colubridae alone contains two-thirds of the extant species. The diversity of the group makes characterization difficult. Colubroids have lost all traces of the pelvic girdle, they have only a single carotid artery, and the skull is very kinetic. Many colubroid snakes are venomous, and snakes in the families Elapidae and Viperidae have hollow fangs at the front of the mouth that inject extremely toxic venom into their prey.

The body form of even a generalized snake such as the kingsnake (Figure 15–5a) is so specialized that little further morphological specialization is associ-

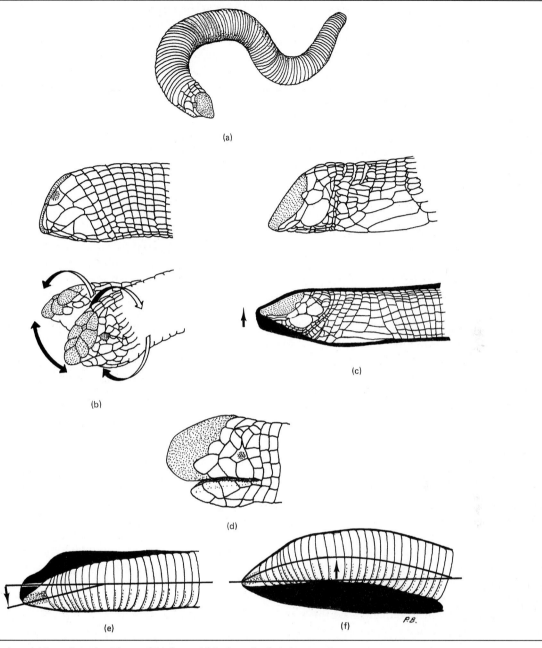

Figure 15–4 Amphisbaenians (a, *Monopeltis*, from Africa) are legless burrowing squamates; (b) blunt-snouted (*Agamodon*, from Africa); (c) shovel-snouted (*Rhineura* from Florida); (d) wedge-snouted (*Anops*, Brazil). ([b], [c], [e], and [f] modified from C. Gans, 1974, *Biomechanics*, Lippincott, Philadelphia, PA.)

ated with different habits or habitats. Kingsnakes are constrictors, and they crawl slowly, poking their heads under leaf litter and into holes that might shelter prey. Chemosensation is an important means of detecting prey for these snakes. Snakes have forked tongues, with widely separated tips that can move independently. When the tongue is projected, the tips are waved in the air or touched to the ground. Then the tongue is retracted, and chemical stimuli are transferred to the paired vomeronasal organs.

(See Halpern [1992] for more details.). The forked shape of the tongue of snakes (which is seen also among the Amphisbaenia, Lacertiformes, and Varanoidea) may allow them to sample two points in space simultaneously. In that manner, these squamates may be able to detect gradients of chemical stimuli and localize objects (Schwenk 1994).

Nonconstrictors such as the whipsnake (Figure 15–5b) move quickly and are visually oriented. They forage by crawling rapidly, frequently raising the head to look around. Many arboreal snakes are extremely elongated and frequently have large eyes (Figure 15–5c). Their length distributes their weight and allows them to crawl over even small twigs without breaking them. Burrowing snakes, at the opposite extreme of snake body form, are short and have blunt heads and very small eyes (Figure 15–5d). The head shape assists in penetrating soil, and a short body and tail create less friction in a burrow than would the same mass in an elongate body. Vipers, especially forms like the African puff adder (Figure 15–5e), are heavy-bodied with broad heads.

The sea snakes (Figure 15–5f) are derived from terrestrial elapids. Sea snakes are characterized by extreme morphological specialization for aquatic life: The tail is laterally flattened into an oar, the large ventral scales are reduced or absent in most species, and the nostrils are located dorsally on the snout and have valves that exclude water. The lung extends back to the cloaca and apparently has a hydrostatic role in adjusting buoyancy during diving as well as a respiratory function. Oxygen uptake through the skin during diving has been demonstrated in sea snakes. *Laticauda* are less specialized than other sea snakes, and may represent a separate radiation into the marine habitat. They retain enlarged ventral scales and emerge onto land to bask and to lay eggs. The other sea snakes are so specialized for marine life that they are helpless on land, and these species are viviparous. The locomotor specializations of snakes reflect differences in their morphology associated with different predatory modes (discussed in the following section) and the properties of the substrates on which they move (Box 15–1).

Snake skeletons are delicate structures that do not fossilize readily. In most cases we have only vertebrae, and little information has been gained from the fossil record about the origin of snakes. The earliest fossils known are from Cretaceous deposits and seem to be related to boas. Colubrid snakes are first known from the Oligocene, and elapids and viperids appeared during the Miocene.

Functionally snakes are extremely specialized legless lizards. They may have reached this specialization from a fossorial stage. Differences in the eyes of lizards and snakes have been interpreted as evidence of that transition. In lizards the eye is focused by distorting the lens, thus changing its radius of curvature, whereas in snakes the lens is moved in relation to the retina to bring objects into focus. The morphology of the retina of snakes also differs from that of lizards. Snakes have no fovea centralis, their retinal cells lack colored oil droplets, and there is a unique ophidian double cone. These differences in the eyes may indicate that snake ancestors passed through a stage in which they were so specialized for burrowing that the eyes had nearly been lost. Among the most specialized burrowing snakes and lizards, the eyes are very reduced and probably capable of little more than distinguishing between light and darkness. According to this hypothesis, the eye reevolved when snakes reentered an aboveground niche, but the structural details of the original lizard eye were not exactly duplicated.

A fossorial stage is not the only plausible explanation of the origin of snakes, however. Both epigean (ground surface) and aquatic origins have been suggested. Legs are not particularly useful to a small predatory squamate in a number of habitats. Dense vegetation entangles the legs as an animal tries to draw them forward, and there is no space to use legs in small openings such as cracks in rocks. Possibly the initial radiation of snakes took advantage of these sorts of microhabitats. Extant lizards show an array of elongate forms with reduced legs, and Mesozoic lizards were probably at least equally diverse. Alternatively, fossils of some elongate varanid-like lizards with reduced limbs from late Cretaceous marine deposits in Israel suggest that an aquatic stage is an alternative to the hypothesis of the evolution of snakes from fossorial lizards. *Ophiomorus* and *Pachyrachis* were both more than a meter long and had more than 100 presacral vertebrae. Both lack forelimbs and the pectoral girdle. *Ophiomorus* had well-developed remnants of the rear limb and pelvic girdle, but in *Pachyrachis* they were much reduced (Carroll 1987).

The specializations of snakes compared to legless lizards appear to reflect two selective pressures—locomotion and predation. Elongation of the body is characteristic of snakes. The reduction in body diameter associated with elongation has been accompanied by some rearrangement of the internal anatomy of snakes. The left lung is reduced or entirely absent, the gallbladder is posterior to the liver, the right kidney is anterior to the left, and the gonads may show similar displacement.

Legless lizards face problems in swallowing prey. The primary difficulty is not the loss of limbs, because few lizards use the legs to seize or manipulate food. The difficulty stems from the elongation that is such a widespread characteristic of legless forms. As the body lengthens, the mass is redistributed into a tube with a smaller diameter. As the mouth gets smaller, the maximum size of the prey that can be eaten also decreases, and an elongate animal is faced with the difficulty of feeding a large body through a small mouth. Most legless lizards are limited to eating relatively small prey, whereas snakes have morphological specializations that permit them to engulf prey considerably larger than the body diameter (see the following section). This difference may be one element in the great evolutionary success of snakes in contrast to the limited success of legless lizards and amphisbaenians.

Ecology and Behavior of Squamates

The past quarter century has seen an enormous increase in the number and quality of field studies of the ecology and behavior of snakes and lizards. As a result, our understanding of how these ectothermal amniotes function has been greatly broadened. Studies of lizards have been particularly fruitful, in large measure because many species of lizards are conspicuous and active during the day. Field observations of animals are difficult at best, and species that have characteristics that make them relatively easy to study are overrepresented in the literature. Much less is known about species with cryptic habits. The discussions that follow rely on studies of particular species, and it is important to remember that no one species or family is representative of lizards or snakes as a group.

Foraging and Feeding

The methods that snakes and lizards use to find, capture, subdue, and swallow prey are diverse and are important in determining the interactions among species in a community. Astonishing specializations have evolved: Blunt-headed snakes with long lower jaws that can reach into a shell to winkle out a snail, nearly toothless snakes that swallow bird eggs intact and then slice them open with sharp ventral processes (hypapophyses) on the neck vertebrae, and chameleons that project their tongues to capture insects or small vertebrates on the sticky tips are only a sample of the diversity of feeding specializations of squamates.

Studies of the ecology of snakes have emphasized the morphological specializations associated with different dietary habits (for examples, see Pough 1983), whereas similar studies of lizards have focused on correlations among physiological and behavioral characteristics (summarized by Huey and Bennett 1986).

Feeding Specializations of Snakes The entire skull of a snake is much more flexible than the skull of a lizard. The snake skull contains eight links, with joints between them that permit rotation (Figure 15–7). This number of links gives a staggering degree of complexity to the movements of the ophidian skull, and to make things more complicated, the linkage is paired—each side of the head acts independently. Furthermore, the pterygoquadrate ligament and quadratosupratemporal ties are flexible. When they are under tension they are rigid, but when they are relaxed they permit sideward movement as well as rotation. All of this results in a considerable degree of three-dimensional movement in a snake's skull.

The mandibles of lizards are joined in a symphysis, but in snakes they are attached only by muscles and skin and can spread sideward and move forward or back independently. Loosely connected mandibles and flexible skin in the chin and throat allow the jaw tips to spread so that the widest part of the prey passes ventral to the articulation of the jaw with the skull.

Swallowing movements take place slowly enough to be observed easily (Figure 15–8). A snake usually swallows prey head first, perhaps because that approach presses the limbs against the body, out of the snake's way. Small prey may be swallowed tail first or even sideward. The mandibular and ptery-

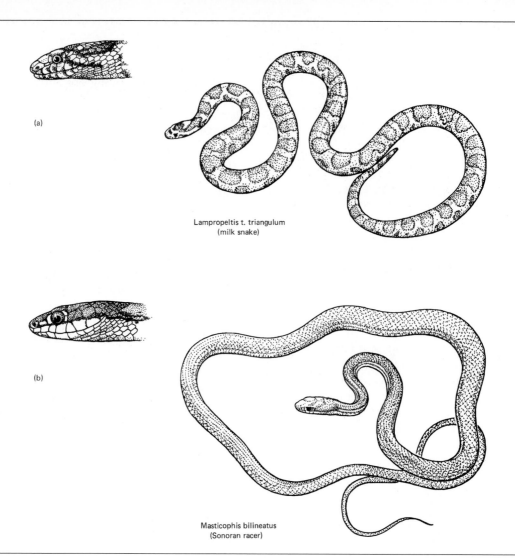

(a)

Lampropeltis t. triangulum
(milk snake)

(b)

Masticophis bilineatus
(Sonoran racer)

Figure 15–5 Body forms of snakes: (a) Constrictors such as the kingsnake (*Lampropeltis*) crawl slowly, poking their heads into holes and under logs. They are often nocturnal and probably rely on chemoreception to detect prey that is hidden from sight. (b) Active, visually oriented snakes like the whipsnakes (*Masticophis*) forage by moving rapidly across the ground with their heads raised. They are diurnal and have large eyes. Vision is probably an important sensory mode for these snakes. (c) Arboreal snakes such as *Leptophis* are often elongate; they probably rely mainly on vision to detect prey in their three-dimensional habitat. (d) Burrowing snakes like *Typhlops* have small, rounded or pointed heads with little distinction between head and neck, short tails, and smooth, often shiny scales. Their eyes are greatly reduced in size. (e) Vipers, especially the African vipers like the puff adder (*Bitis arietans*), have large heads and stout bodies that accommodate large prey. (f) Sea snakes such as *Laticauda* have a tail that is flattened from side to side and valves that close the nostrils when they dive.

goid teeth of one side of the head are anchored in the prey and the opposite jaw is advanced. The mandibles also are protracted independently or together and grip the prey ventrally. Once it has reached the esophagus, the prey is forced toward the stomach by contraction of the neck muscles.

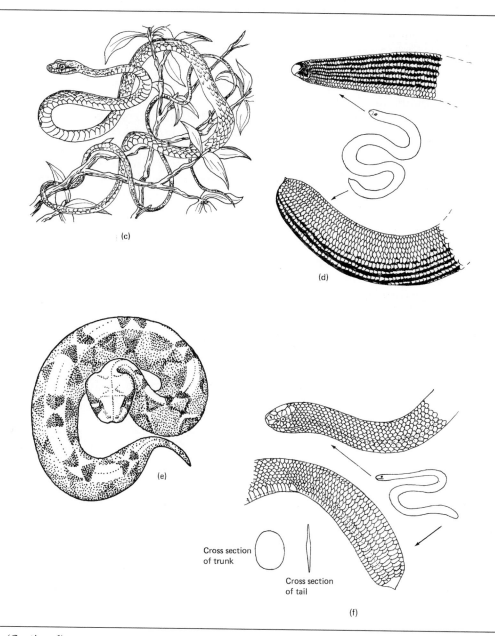

(c)

(d)

(e)

Cross section
of trunk

Cross section
of tail

(f)

Figure 15–5 *(Continued)*

Most species of snakes seize prey and swallow it as it struggles. The risk of damage to the snake during this process is a real one, and various features of snake anatomy seem to give some protection from struggling prey. The frontal and parietal bones of a snake's skull extend downward, entirely enclosing the brain and shielding it from the protesting kicks of prey being swallowed. Possibly the kinds of prey that can be attacked by snakes without a specialized feeding mechanism are limited by the snake's ability to swallow the prey without being injured in the process.

Constriction and venom are predatory specializations that permit a snake to tackle large prey with little risk of injury to itself. Constriction is characteristic of the boas and pythons as well as a number of colubrid snakes. Despite travelers' tales of animals crushed to jelly by a python's coils, the process of

Box 15–1 The Way of a Snake

Snakes commonly employ four types of locomotion. In **lateral undulation** (Figure 15–6a) the body is thrown into a series of curves. The curves may be irregular, as shown in the illustration of a snake crawling across a board dotted with fixed pegs. Each curve presses backward; the pegs against which the snake is exerting force are shown in solid color. The lines numbered 1 to 7 are at 3-inch intervals, and the position of the snake at intervals of 1 second is shown.

Rectilinear locomotion (Figure 15–6b) is used primarily by heavy-bodied snakes. Alternate sections of the ventral integument are lifted clear of the ground and pulled forward by muscles that originate on the ribs and insert on the ventral scales. The intervening sections of the body rest on the ground and support the snake's body. Waves of contraction pass from anterior to posterior, and the snake moves in a straight line. Rectilinear locomotion is slow, but it is effective even when there are no surface irregularities strong enough to resist the sideward force exerted by serpentine locomotion. Because the snake moves slowly and in a straight line it is inconspicuous, and rectilinear locomotion is used by some snakes when stalking prey.

Concertina locomotion (Figure 15–6c) is used in narrow passages such as rodent burrows that do not provide space for the broad curves of serpentine locomotion. A snake anchors the posterior part of its body by pressing several loops against the walls of the burrow and extends the

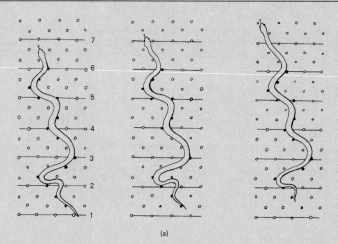

(a)

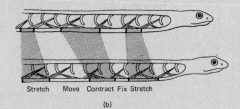

Stretch   Move   Contract   Fix   Stretch

(b)

Figure 15–6 Locomotion of snakes: (a) lateral undulation; (b) rectilinear; (c) concertina; (d) sidewinding.

front part of its body. When the snake is fully extended it forms new loops anteriorly and anchors itself with these while it draws the rear end of its body forward.

Sidewinding locomotion (Figure 15–6d) is used primarily by snakes that live in deserts where windblown sand provides a substrate that slips away during serpentine locomotion. A sidewinding snake raises its body in loops, resting its weight on two or three points that are the only parts of the body in contact with the ground. The loops are swung forward through

Continued

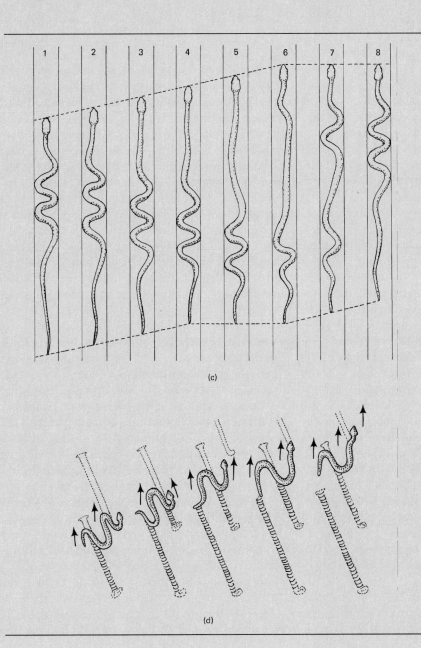

(c)

(d)

Continued

Box 15–1 *(Continued)*

the air and placed on the ground, the points of contact moving smoothly along the body. Force is exerted downward; the lateral component of the force is so small that the snake does not slip sideward. This downward force is shown by imprint of the ventral scales in the tracks.

Because the snake's body is extended nearly perpendicular to its line of travel, sidewinding is an effective means of locomotion only for small snakes that live in habitats with few plants or other obstacles.

constriction involves very little pressure. A constrictor seizes prey with its jaws and throws one or more coils of its body about the prey. The loops of the snake's body press against adjacent loops, and friction prevents the prey from forcing the loops open. Each time the prey exhales, the snake takes up the slack by tightening the loops slightly. Two hypotheses have been proposed to explain the cause of death from constriction. The traditional view holds that prey suffocates because it cannot expand its thorax to inhale. Another possibility is that the increased internal pressure interferes with, and eventually stops, the heart (Hardy 1994).

Snakes that constrict their prey must be able to throw the body into several loops of small diameter to wrap around the prey. Constrictors achieve these small loops by having short vertebrae and short trunk muscles that span only a few vertebrae from the point of origin to the point of insertion. Contraction of these muscles produces sharp bends in the trunk that allow constrictors to press tightly against their prey. However, the trunk muscles of snakes are also used for locomotion, and the short muscles of constrictors produce several small-radius curves along the length of the snake's body. That morphology limits the speed with which constrictors can move, because rapid locomotion by snakes is accomplished by throwing the body into two or three broad loops. This is the pattern seen in fast-moving species such as whipsnakes, racers, and mambas. The muscles that produce these broad loops are long, spanning many vertebrae, and the vertebrae also are longer than those of constrictors.

In North America fast-moving snakes (colubroids) first appear in the fossil record during the Miocene, a time when grasslands were expanding. Constrictors, largely erycines, predominated in the snake fauna of the early Miocene, but by the end of that epoch the snake fauna was composed primarily of colubroids. Alan Savitzky (1980) has suggested that fast-moving colubroid snakes had an advantage over slow-moving boids in the more open habitats that developed during the Miocene, and that the radiation of colubroids involved a complex interaction between locomotion and feeding. Rodents were probably the most abundant prey available to the snakes, and rodents are dangerous animals for a snake to swallow while they are able to bite and scratch. Constriction provided a relatively safe way for boids to kill rodents, but the long vertebrae and long trunk muscles that allowed colubroids to move rapidly through the open habitats of the Miocene would have prevented them from using constriction to kill their prey.

Savitzky suggested that colubroids initially used venom to immobilize prey. Duvernoy's gland, found in the upper jaw of extant colubrid snakes, is homologous to the venom glands of viperids and elapids. In many of these snakes Duvernoy's gland produces a toxic secretion that immobilizes prey. (Some extant colubrids have venom that is dangerously toxic to animals as large as humans.) Thus, the evolution of venom that could kill prey may have been a key feature that allowed Miocene colubroid snakes to dispense with constriction and become morphologically specialized for rapid locomotion in open habitats. The presence of Duvernoy's gland appears to be an ancestral character for colubroid snakes, as this hypothesis predicts. Some colubrids, including the ratsnakes (*Elaphe*), gophersnakes (*Pituophis*), and kingsnakes (*Lampropeltis*), have lost the venom-producing capacity of the Duvernoy's gland, and these are the groups in which constriction has been secondarily developed as a method of killing prey.

In this context, the front-fanged venomous snakes (Elapidae and Viperidae) are not a new devel-

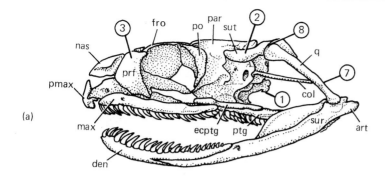

(a)

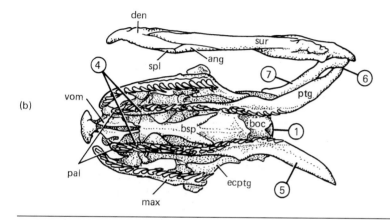

(b)

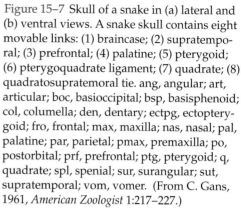

Figure 15–7 Skull of a snake in (a) lateral and (b) ventral views. A snake skull contains eight movable links: (1) braincase; (2) supratemporal; (3) prefrontal; (4) palatine; (5) pterygoid; (6) pterygoquadrate ligament; (7) quadrate; (8) quadratosupratemoral tie. ang, angular; art, articular; boc, basioccipital; bsp, basisphenoid; col, columella; den, dentary; ectpg, ectopterygoid; fro, frontal; max, maxilla; nas, nasal; pal, palatine; par, parietal; pmax, premaxilla; po, postorbital; prf, prefrontal; ptg, pterygoid; q, quadrate; spl, spenial; sur, surangular; sut, supratemporal; vom, vomer. (From C. Gans, 1961, *American Zoologist* 1:217–227.)

opment, but instead represent alternative specializations of an ancestral venom delivery system. Given the ancestral nature of venom for colubroid snakes, one would expect that different specializations for venom delivery would be represented in the extant snake fauna, as indeed they are. A variety of snakes have enlarged teeth (fangs) on the maxillae. Three categories of venomous snakes are recognized (Figure 15–9): Opisthoglyphous, proteroglyphous, and solenoglyphous. This classification is descriptive, and represents convergent evolution by different phylogenetic lineages.

Opisthoglyphous (*ophistho* = behind, *glyph* = hollowed) snakes have one or more enlarged teeth near the rear of the maxilla with smaller teeth in front. In some forms the fangs are solid; in others there is a groove on the surface of the fang that may help to conduct saliva into the wound. Several African and Asian opisthoglyphs can deliver a dangerous or even lethal bite to large animals, including humans, but their primary prey is lizards or birds, which are often held in the mouth until they stop struggling and are then swallowed.

Proteroglyphous snakes (*proto* = first) include the cobras, mambas, coral snakes, and sea snakes in the Elapidae. The hollow fangs of the proteroglyphous snakes are located at the front of the maxilla, and there are often several small, solid teeth behind the fangs. The fangs are permanently erect and relatively short.

Solenoglyphous (*solen* = pipe) snakes include the pitvipers of the New World and the true vipers of the Old World. In these snakes the hollow fangs are the only teeth on the maxillae, which rotate so that the fangs are folded against the roof of the mouth when the jaws are closed. This folding mechanism permits solenoglyphous snakes to have long fangs that inject venom deep into the tissues of the prey. The venom, a complex mixture of enzymes and other substances (Table 15–3), first kills the prey and then speeds its digestion after it has been swallowed.

Snakes that can inject a disabling dose of venom into their prey have evolved a very safe prey-catch

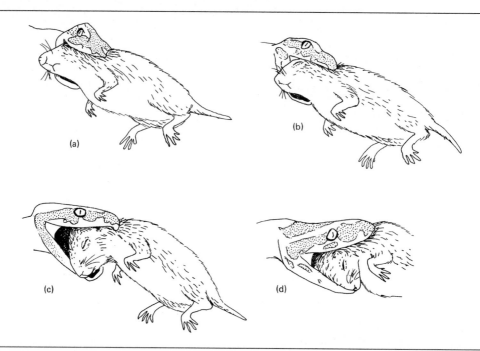

Figure 15–8 Snakes use a combination of head movements and protraction and retraction of the jaws to swallow prey. (a) Prey grasped by left and right jaws at the beginning of the swallowing process. (b) The upper and lower jaws on the right side have been protracted, disengaging the teeth from the prey. (c) The head is rotated counterclockwise, moving the right upper and lower jaws over the prey. The recurved teeth slide over the prey like the runners of a sled. (d) The upper and lower jaw on the right side are retracted, embedding the teeth in the prey and drawing it into the mouth. Notice that the entire head of the prey has been engulfed by this movement. Next the left upper and lower jaws will be advanced over the prey by clockwise rotation of the head. The swallowing process continues with alternating left and right movements until the entire body of the prey has passed through the snake's jaws. (Modified from T. H. Frazetta, 1966, *Journal of Morphology* 118:217–296.)

ing method. A constrictor is in contact with its prey while it is dying and runs some risk of injury from the prey's struggles. A solenoglyphous snake needs only to inject venom and allow the prey to run off to die. Later the snake can follow the scent trail of the prey to find its corpse. This is the prey-capture pattern of most vipers, and experiments have shown that a viper can distinguish the scent trail of a mouse it has bitten from trails left by uninjured mice.

Several features of the body form of vipers allow them to eat larger prey in relation to their own body size than can most nonvenomous snakes. Many vipers, including rattlesnakes, the jumping viper, and the African puff adder and Gaboon viper, are very stout snakes. The triangular head shape that is usually associated with vipers is a result of the out-

ward extension of the rear of the skull, especially the quadrate bones. The wide-spreading quadrates allow passage through the mouth, and even a large meal makes little bulge in the stout body and thus does not interfere with locomotion. Vipers have specialized as relatively sedentary predators that can prey even on quite large animals. The other family of terrestrial venomous snakes, the cobras, mambas, and their relatives, are primarily slim-bodied snakes.

Foraging Behavior and Energetics of Lizards The activity patterns of lizards span a range from extremely sedentary species that spend hours in one place to species that are in nearly constant motion. Field observations of the tropidurid lizard *Leiocephalus schreibersi* and the teiid *Ameiva chrysolaema* in the

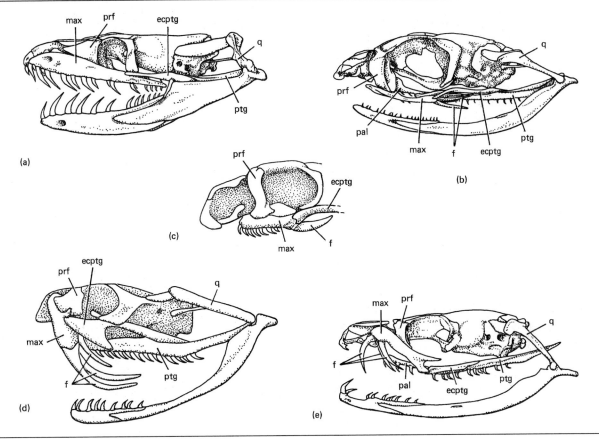

Figure 15–9 Dentition of snakes: (a) aglyphous (without fangs), African python, *Python sebae*; (b, c) opisthoglyphous (fangs in the rear of the maxilla), African boomslang, *Dispholidus typus*, and Central American false viper, *Xenodon rhabdo-cephalus*; (d) solenoglyphous (fangs on a rotating maxilla), African puff adder, *Bitis arietans*; (e) proteroglyphous (permanently erect fangs at the front of the maxilla), African green mamba, *Dendroaspis jamesoni*. The fangs of solenoglyphs (d) are erected by an anterior movement of the pterygoid that is transmitted through the ectopterygoid and palatine to the maxilla, causing it to rotate about its articulation with the prefrontal, thereby erecting the fang. Some opisthoglyphs, especially *Xenodon* (c), have the same mechanism of fang erection. ecptg, ectopterygoid; fro, frontal; max, maxilla; pal, palatine; par, parietal; pmax, premaxilla; prf, prefrontal; ptg, pterygoid; q, quadrate; sut, supratemporal.

Dominican Republic revealed two extremes of behavior. *Leiocephalus* rested on an elevated perch from sunrise to sunset and was motionless for more than 99 percent of the day. Its only movements consisted of short, rapid dashes to capture insects or to chase away other lizards. These periods of activity never lasted longer than 2 seconds, and the frequency of movements averaged 9.6 per hour. In contrast, *Ameiva* were active for only 4 or 5 hours in the middle of the day, but they were moving more than 70 percent of that time, and their velocity averaged one body length every 2 to 5 seconds.

The same difference between the two species was seen in a laboratory test of spontaneous activity: *Ameiva* was more than 20 times as active as *Leiocephalus*. In fact, the teiids were as active in exploring their surroundings as small mammals tested in the same apparatus. A xantusiid lizard tested in the laboratory apparatus had a pattern of spontaneous activity that fell approximately midway between that of the teiid and the tropidurid. Thus, a spectrum of spontaneous locomotor activity is apparent in lizards that extends from species that are nearly motionless through species that move at intermediate rates to species that are as active as mammals.

For convenience the extremes of the spectrum are frequently called sit-and-wait predators and widely foraging predators, respectively, and the intermediate condition has been called a cruising forager. Other field studies have shown that this spectrum of locomotor behaviors is widespread in lizard faunas. In North America, for example, spiny swifts (*Sceloporus*) are sit-and-wait predators, many skinks (*Eumeces*) appear to be cruising foragers, and whiptail lizards (*Cnemidophorus*) are widely foraging predators. The ancestral locomotor pattern for lizards may have been that of a cruising forager, and both sit-and-wait predation and active foraging may represent derived conditions. (A spectrum of foraging modes is not unique to lizards; it probably applies to nearly all kinds of mobile animals, including fishes, mammals, birds, frogs, insects, and zooplankton.)

The ecological, morphological, and behavioral characteristics that are correlated with the foraging modes of different species of lizards appear to define many aspects of the biology of these animals (Huey and Pianka 1981). For example, sit-and-wait predators and widely foraging predators consume different kinds of prey and fall victim to different kinds of predators. They have different social systems, probably emphasize different sensory modes, and differ in some aspects of their reproduction and life history.

These generalizations are summarized in Table 15–4 and are discussed in the following sections. However, a weakness of this analysis must be emphasized: Sit-and-wait species of lizards (at least, the ones that have been studied most) are primarily iguanians, whereas widely foraging species are mostly scincomorphs. That phylogenetic split raises the question of whether the differences we see between sit-and-wait and widely foraging lizards are really the consequences of the differences in foraging behavior, or if they are ancestral characteristics of iguanian versus scincomorph lizards. If that is the case, their association with different foraging modes may be misleading. In either case, however, the model presented in Table 15–4 provides a useful integration of a large quantity of information about the biology of lizards; it represents a hypothesis that will be modified as more information becomes available.

Lizards with different foraging modes use different methods to detect prey: Sit-and-wait lizards normally remain in one spot from which they can survey a broad area. These motionless lizards detect the movement of an insect visually and capture it with a quick dash from their observation site. Sit-and-wait lizards may be most successful in detecting and capturing relatively large insects like beetles and grasshoppers. Active foragers spend most of their time on the ground surface, moving steadily and

Table 15–3 **Components of the venoms of squamates.**

Compound	Occurrence	Effect
Proteinases	All venomous squamates, especially viperids	Tissue destruction
Hyaluronidase	All venomous squamates	Increases tissue permeability, hastens the spread of other constituents of venom through the tissues
L-Amino acid oxidase	Many snakes, but absent from vipers, elapids, and sea snakes	Attacks a wide variety of substrates, causes great tissue destruction
Cholinesterase	High in elapids, may be present in sea snakes, low in viperids	Unknown, it is not responsible for the neurotoxic effects of elapid venom
Phospholipases	All venomous squamates	Attacks cell membranes
Phosphatases	All venomous squamates	Attacks high-energy phosphate compounds such as ATP
Basic polypeptides	Elapids and sea snakes	Blocks neuromuscular transmission

poking their snouts under fallen leaves and into crevices in the ground. They apparently rely largely on chemical cues to detect insects, and they probably seek out local concentrations of patchily distributed prey such as termites. Widely foraging species of lizards appear to eat more small insects than do lizards that are sit-and-wait predators. Thus, the different foraging behaviors of lizards lead to differences in their diets, even when the two kinds of lizards occur in the same habitat.

The different foraging modes also have different consequences for the exposure of lizards to their own predators. A lizard that spends 99 percent of its time resting motionless is relatively inconspicuous, whereas a lizard that spends most of its time moving is easily seen. Sit-and-wait lizards are probably most likely to be discovered and captured by predators that are active searchers, whereas widely foraging lizards are likely to be caught by sit-and-wait predators. As a result of this difference, foraging modes may alternate at successive levels in the food chain: Insects that move about may be captured by lizards that are sit-and-wait predators, and those lizards may be eaten by widely foraging predators. Insects that are sedentary are more likely to be discovered by a widely foraging lizard, and that lizard may be picked off by a sit-and-wait predator.

The body forms of sit-and-wait lizard predators may reflect selective pressures different from those that act on widely foraging species. Sit-and-wait lizards are often stout-bodied, short-tailed, and cryptically colored. Many of these species have dorsal patterns formed by blotches of different colors that probably obscure the outlines of the body as the lizard rests motionless on a rock or tree trunk. Widely foraging species of lizards are usually slim and elongate with long tails, and they often have patterns of stripes that may produce optical illusions as they move. However, one predator-avoidance mechanism, the ability to break off the tail when it is seized by a predator (**autotomy**), does not differ among lizards with different foraging modes (Box 15–2).

What physiological characteristics are necessary to support different foraging modes? The energy requirements of a dash that lasts for only a second or two are quite different from those of locomotion that is sustained nearly continuously for several hours. Sit-and-wait and widely foraging species of lizards differ in their relative emphasis on the two metabolic pathways that provide the ATP that is used for activity and in how long that activity can be sustained. Sit-and-

wait lizards move in brief spurts, and they rely largely on anaerobic metabolism to sustain their movements. (See Chapter 4 for a discussion of aerobic and anaerobic metabolism.) These lizards quickly become exhausted when they are forced to run continuously on a treadmill, whereas widely foraging species can sustain activity for long periods without exhaustion.

Those differences in locomotor behavior are associated with differences in the oxygen transport systems of the lizards: Widely foraging species of lizards have larger hearts and more red blood cells in their blood than do sit-and-wait species. As a result, each beat of the heart pumps more blood and that blood carries more oxygen to the tissues of a widely foraging species of lizard than a sit-and-wait species.

Sustained locomotion is probably not important to a lizard that makes short dashes to capture prey or to escape from predators, but sprint speed might be vitally important in both those activities. As one would predict, the sprint speed of the sit-and-wait lizards is generally greater than that of the widely foraging species.

The continuous locomotion of widely foraging species of lizards is energetically expensive. Measurements of energy expenditure of lizards in the Kalahari showed that the daily energy expenditure of a widely foraging species averaged 150 percent of that of a sit-and-wait species. However, the energy that the widely foraging species invested in foraging was more than repaid by its greater success in finding prey. The daily food intake of the widely foraging species was 200 percent that of the sit-and-wait predator. As a result, the widely foraging species had more energy available to use for growth and reproduction than did the sit-and-wait species, despite the additional cost of its mode of foraging.

Diet, Nutrition, and Captive Husbandry

Although we know a great deal about the foraging behavior and food habits of lizards, we know relatively little about their nutritional requirements. Free-ranging lizards probably consume a variety of different kinds of invertebrates. The prey itself, the food in the prey's gut, and even soil or plant material that is consumed accidentally with the prey are potential sources of vitamins and minerals. Animals in captivity usually have less diverse diets than those of wild animals, and a diet that lacks essential nutri-

Table 15–4 Ecological and behavioral characteristics associated with the foraging modes of lizards. Foraging modes are presented as a continuum from sit-and-wait predators to widely foraging predators. In most cases data are available only for species at the extremes of the continuum. See the text for details.

Character	Sit-and-Wait	Cruising Forager	Widely Foraging
Foraging Behavior			
Movements/hour	Few	Intermediate	Many
Speed of movement	Low	Intermediate	Fast
Sensory modes	Vision	Vision and olfaction	Vision and olfaction
Exploratory behavior	Low	Intermediate	High
Types of prey	Mobile, large	Intermediate	Sedentary, often small
Predators			
Risk of predation	Low	?	Higher
Types of predators	Widely foraging	?	Sit-and-wait and widely foraging
Body form			
Trunk	Stocky	Intermediate?	Elongate
Tail	Often short	?	Often long
Physiological characteristics			
Endurance	Limited	?	High
Sprint speed	High	?	Intermediate to low
Aerobic metabolic capacity	Low	?	High
Anaerobic metabolic capacity	High	?	Low
Heart mass	Small	?	Large
Hematocrit	Low	?	High
Energetics			
Daily energy expenditure	Low	?	Higher
Daily energy intake	Low	?	Higher
Social behavior			
Size of home range	Small	Intermediate	Large
Social system	Territorial	?	Not territorial
Reproduction			
Mass of clutch (eggs or embryos) relative to mass of adult	High	?	Low

Source: Based on data from L. J. Vitt and J. D. Congdon, 1978, *American Naturalist* 112:595–608; R. B. Huey and E. R. Pianka, 1981, *Ecology* 62:991–999; W. E. Magnusson et al., 1985 *Herpetologica* 41:324–332; R. B. Huey and A. F. Bennett, 1986, pages 82–98 in *Predator–Prey Relationships*, edited by M. E. Feder and G. V. Lauder, University of Chicago Press, Chicago, IL.

ents can depress reproduction even if it does not produce visible abnormalities.

Proper nutrition is a central part of captive breeding programs for endangered species, and ingenious ways are being developed to ensure that the diets of animals in these programs provide essential elements. Metabolic bone disease can result from a diet with a ratio of calcium to phosphorus less than 1.5:1, but many insects have much lower calcium: phosphorus ratios—around 0.5:1. Mary Allen, at the National Zoo in Washington, developed a method of ensuring that insects provide these minerals in the proper ratio (Allen and Oftedal 1989, Allen et al. 1993). Crickets were fed diets that contained up to 12 percent calcium. After 48 hours dense material was visible in the intestinal tracts of crickets that had received diets containing at least 8 percent calcium (Figure 15–11). These crickets had whole-body calcium:phosphorus ratios of 1.7:1.

Geckos were fed crickets that had received high- and low-calcium food. X-rays show that animals fed high-calcium crickets had well mineralized bone, whereas lizards fed low-calcium crickets had poorly mineralized bone and some of their bones had fractures (Figure 15–12). High-calcium crickets are now fed to amphibians, reptiles, birds, and mammals at the National Zoo.

Social Behavior and Reproduction

Squamates employ a variety of visual, auditory, chemical, and tactile signals in the behaviors they use to maintain territories and to choose mates. Iguanians use mainly visual signals, some gekkotans use vocalization, and scincomorphs, anguimorphs, and snakes use pheromones extensively. The various sensory modalities employed by animals have biased the amount of information we have about the behaviors of different species. Because humans are primarily visually oriented, we perceive the visual displays of other animals quite readily. The auditory sensitivity of humans is also acute, and we can detect and recognize vocal signals that are used by other species. However, the olfactory sensitivity of humans is low and we lack a vomeronasal system, so we are unable to perceive most chemical signals used by squamates. One result of our sensory biases has been a concentration of behavioral studies on organisms that use visual signals. As a result of this focus, the extensive repertoires of visual displays of iguanian lizards figure largely in the literature of behavioral ecology, but much less is known about the chemical and tactile signals that are probably important for other lizards and for snakes.

The social behaviors of squamates appear to be limited in comparison to those of crocodilians, but many species show dominance hierarchies or territoriality. The signals that are used in agonistic encounters between individuals are often similar to those used for species and sex recognition during courtship. Parental attendance at a nest during the incubation period of eggs occurs among squamates, but extended parental care of the young is unknown.

Territorial Behavior and Courtship Iguanian lizards employ primarily visual displays during social interactions. The polychrotid genus *Anolis* includes some 200 species of small to medium-size lizards that occur primarily in tropical America. Male *Anolis* have gular fans, areas of skin beneath the chin that can be distended by the hyoid apparatus during visual displays. (See color insert.) The brightly colored scales and skin of the gular fans of many species of *Anolis* are conspicuous signaling devices, and they are used in conjunction with movements of head and body.

Figure 15–13 shows the colors of the gular fans of eight species of *Anolis* that occur in Costa Rica. No two species have the same combination of colors on their gular fans, and thus it is possible to identify a species solely by seeing the colors it displays. In addition, each species has a behavioral display that consists of raising the body by straightening the forelegs (called a pushup), bobbing the head, and extending and contracting the gular fan. The combination of these three sorts of movements allows a complex display. The three movements can be represented graphically by an upper line that shows the movements of the body and head and a lower line that shows the expansion and contraction of the gular fan. This representation is called a **display action pattern.** No two display action patterns are the same, so it would be possible to identify any of the eight species of *Anolis* by seeing its display action pattern.

The behaviors that territorial lizards use for species and sex recognition during courtship are very much like those employed in territorial defense— pushups, head bobs, and displays of the gular fan. A territorial male lizard is likely to challenge any lizard in its territory, and the progress of the interaction depends on the response it receives. An aggressive response indicates that the intruder is a male and stimulates the territorial male to defend its territory, whereas a passive response from the intruder identifies a female and stimulates the territorial male to initiate courtship. These behaviors are illustrated by the displays of a male *Anolis carolinensis* shown in Figures 15–14 and 15–15. The first response of a territorial lizard to an intruder is the assertion–challenge display shown at the top of Figure 15–14. The dewlap is extended, and the lizard bobs at the intruder. In addition the nuchal (neck) and dorsal crests are slightly raised, and a black spot appears behind the eye. The next stage depends on the sex of the intruder and its response to the challenge from the territorial male (Figure 15–15). If the intruder is a male and does not retreat from the initial challenge, both males become more aggressive. During aggressive posturing (middle panel in Figure 15–14) the males orient laterally to each other, the nuchal and dorsal crests are fully erected, the body is compressed laterally, the black spot behind the eye darkens, and the throat is swelled. All of these postural changes make the lizards appear larger and presumably more formidable to the opponent. If the

Box 15–2 Caudal Autotomy: Your Tail or Your Life

Autotomy (self-amputation) of appendages is a common predator-escape mechanism among invertebrates and vertebrates. The tail is the only appendage that vertebrates are known to autotomize, and the capacity for caudal autotomy is developed to some degree among salamanders, tuatara, lizards, and a few amphisbaenians, snakes, and rodents (Arnold 1988). In most cases autotomy is followed by regeneration of a new tail.

The caudal autotomy of squamates occurs either at distinctive fracture planes that are found in all but the four to nine anteriormost caudal vertebrae or between vertebrae. The caudal muscles are segmental, and pointed processes from adjacent segments interdigitate. The caudal arteries have sphincter muscles just anterior to each fracture site and the veins have valves. Autotomy appears to be an active process that requires contraction of the caudal muscles, bending the tail sharply to one side and initiating separation. The vertebral centrum ruptures, and the processes of the caudal muscles separate. The arterial sphincter muscles contract and the venous valves close, preventing loss of blood. An autotomized tail twitches rapidly for several minutes, and its violent writhing can distract the attention of a predator while the lizard itself scurries to safety (Figure 15–10). The tails of some juvenile skinks are bright blue, and in experiments lizards with these colorful tails were more effective at using autotomy to escape from predatory snakes than were lizards with tails that had been painted black. Of course, an autotomized tail receives no blood flow and its muscular activity is sustained by anaerobic metabolism. The anaerobic metabolic capacity of lizard tail muscles appears to be substantially greater than that of limb muscles.

The point of autotomy is normally as far posterior on the tail as possible. When the tail of a lizard is seized with forceps, autotomy usually occurs through the plane of the vertebra immediately anterior to the point at which the tail was being held, thereby minimizing the amount of tail lost. When the tail is regenerated the vertebrae are replaced by a rod of cartilage that does not contain fracture planes. Consequently, future autotomy must occur anterior to the regenerated portion of the tail. Some geckos adjust the point of autotomy according to their body temperature: Autotomy occurs closer to the body when the lizard is cold than when it is warm. When a lizard is cold it cannot run as fast as it can when it's warm, and perhaps the longer segment of tail left behind by a cold lizard occupies the predator's attention long enough to allow the lizard to reach safety.

Leaving your tail in the grasp of a predator is certainly better than being eaten, but it is not free of costs. The tail acts as a signal of status among some species of lizards, and a lizard that autotomizes a large part of its tail may fall to a lower rank in the dominance hierarchy as a result. Losing the tail also affects the energy balance of a lizard: For example, the rate of growth of juvenile lizards that have autotomized their tails is reduced while their tails are being regenerated. Many lizards store fat in the tail, and the females mobilize this energy while they are depositing yolk in eggs. Sixty percent of the total fat storage was located in the tails of female geckos (*Coleonyx brevis*). Autotomy of the tail by gravid female geckos resulted in their producing smaller clutches of eggs than lizards with tails. Some lizards, especially the small North American skink *Scincella lateralis*, eat their autotomized tails if they can, thereby recovering the lost energy.

intruder is a receptive female, the territorial male initiates courtship (bottom panel in Figure 15–14).

The differences in color and movement that characterize the dewlaps and display action patterns of *Anolis* are conspicuous to human observers, but do the lizards also rely on them for species identification? Indirect evidence suggests that the lizards probably do use gular fan color and display action patterns for species identification. For example, examination of communities of *Anolis* that contain

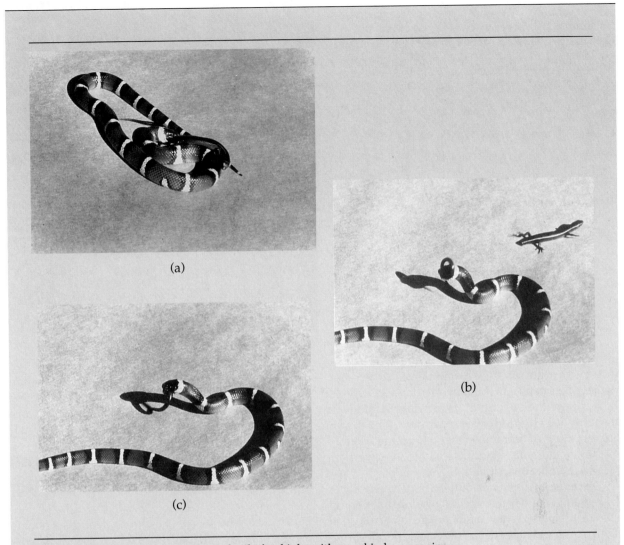

Figure 15–10 The freshly autotomized tail of a skink writhes and jerks, engaging the attention of a predatory kingsnake while the lizard escapes. In this sequence of photographs a kingsnake seizes a skink by the tail (a). The skink autotomizes its tail and runs off (b), leaving the snake struggling to swallow the wriggling tail (c). (Photographs by Benjamin E. Dial.)

many species show that the differences in the colors of the gular fans and in the display action patterns is greatest for those species that encounter each other most frequently.

Pheromonal communication, mediated by the vomeronasal organ, probably occurs in several lineages of lizards, primarily scincomorphs and anguimorphs, although chemical cues may be more important for some iguanians than has been realized. A review of the subject can be found in Simon (1983). Territorial male *Sceloporus* and other phrynosomatid lizards rub secretions from their femoral glands on objects in their territories (see Mason 1992). These secretions contain protein and sometimes lipids. Exploratory behavior by lizards, including iguanians, involves touching the tongue

Figure 15–11 X-ray of crickets fed a diet containing 12 percent calcium for 72 hours. The calcium in the gastrointestinal tract appears as a radiopaque structure along the midline of the cricket. (From M. E. Allen and O. T. Oftadal, 1989, *Journal of Zoo and Wildlife Medicine* 20:26–33. Photograph courtesy of Mary E. Allen.)

to the substrate, and the vomeronasal organ may detect pheromones in the femoral gland secretions. In addition, the secretions absorb light strongly in the ultraviolet portion of the spectrum. Some lizards are sensitive to ultraviolet light (Fleishman et al. 1993), and femoral gland secretions may be both visual and olfactory signals.

Experimental studies of the use of pheromones have employed several species of North American skinks (*Eumeces*). Male and female broad-headed skinks (*E. laticeps*) can detect the cloacal odors of conspecifics, and males detect skin odors. Male skinks (but not females) distinguish between the cloacal odors of male and female conspecifics, and males will follow scent trails of females but will not follow the trails of other males. Female skinks do not follow the scent trails of either sex. The discriminatory ability of male skinks is quite precise—they can distinguish the cloacal odors of conspecific females from the scents of females of other closely related species (Cooper and Vitt 1986).

Territoriality, the relative importance of vision compared to olfaction, and foraging behavior appear to be broadly correlated among lizards. The elevated perches from which sit-and-wait predators survey their home ranges allow them to see both intruders

and prey, and they dash from the perch to repel an intruder or to catch an insect. In contrast, widely foraging lizards are almost entirely nonterritorial, and olfaction is as important as vision in their foraging behavior. These lizards spend most of their time on the ground, where their field of vision is limited and they probably have little opportunity to detect intruders.

Reproduction and Parental Care A receptive female snake or lizard often raises its tail and bends it to the side, signaling its willingness to mate. The male responds by pressing his cloacal region against the cloacal region of the female. Male lizards often secure a grip with their jaws on the neck or shoulders of the female, but snakes rarely use their jaws during mating. Male squamates have paired copulatory organs called **hemipenes.** The name stems from the mistaken belief that the two organs were used together. In fact, only the hemipenis on the side nearest the female is used. Each hemipenis is a hollow, closed-end structure like the finger of a glove, and some squamates have hemipenes that are bifurcated at the tips. In their retracted state, the hemipenes lie in the base of the tail just posterior to the cloaca with their closed ends pointing caudally.

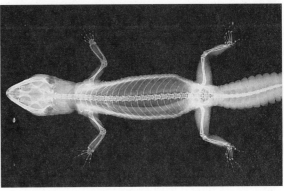

Figure 15–12 X-rays of leopard geckos (*Eublepharis macularis*). The lizard in (a) was fed low-calcium crickets, whereas the lizard in (b) received high-calcium crickets. The lizard on the low-calcium cricket diet has a less mineralized skeleton (compare the ribs in the two images), and fractures can be seen in left ulna. (Photographs courtesy of Mary E. Allen.)

The walls of the hemipenes contain large sinuses, and the hemipenes are erected by filling the sinuses with blood. As the organ becomes enlarged, the propulsive muscle squeezes the hemipenis inside out. It emerges from the posterior edge of the cloaca of the male and is inserted into the cloaca of the female. The hemipenes of many squamates have spines on their bases that probably help to secure them within the cloaca of the female during the transfer of sperm, a process that may require several minutes or even longer. When copulation is finished, blood drains from the sinus and a retractor muscle pulls it back into its resting position.

Squamates show a range of reproductive modes from oviparity (development occurs outside the body of the female and is supported entirely by the yolk) to viviparity (the eggs are retained in the oviducts and development is supported by transfer of nutrients from the mother to the fetuses). Intermediate conditions include retention of the eggs for a time after they have been fertilized, and the production of precocial young that were nourished primarily by material in the yolk. Oviparity is assumed to be the ancestral condition, and viviparity has evolved at least 45 times among lizards and 35 times among snakes (Blackburn 1982, 1985; Shine 1985).

Viviparous squamates have specialized chorioallantoic placentae, and in the Brazilian skink *Mabuya heathi* more than 99 percent of the mass of the fetus results from transport of nutrients across the placenta (Blackburn et al. 1984).

Viviparity is not evenly distributed among lineages of squamates. Nearly half of the origins of viviparity in the group have occurred in the family Scincidae, whereas viviparity is unknown in teiid lizards and occurs in only two genera of lacertids. Viviparity has advantages and disadvantages as a mode of reproduction. The most commonly cited benefit of viviparity is the opportunity it provides for a female snake or lizard to use her own thermoregulatory behavior to control the temperature of the embryos during development. This hypothesis is appealing in an ecological context, because a relatively short period of retention of the eggs by the female might substantially reduce the total amount of time required for development, especially in a cold climate. However, a test of the egg-retention hypothesis by Robin Andrews and Barbara Rose produced unexpected results (Andrews and Rose 1994). The striped plateau lizard, *Sceloporus virgatus*, is found in the Sierra Madre of Mexico and the mountain ranges of southeastern Arizona and south-

Figure 15–13 Species-typical displays of *Anolis* lizards. Eight species of *Anolis* from Costa Rica can be separated into three groups on the basis of the size and color pattern of their gular fans. *Simple* fans are unicolored, *compound* fans are bicolored, and *complex* fans are bicolored and very large. The display action pattern of each species is shown graphically beneath the drawing of the lizard. The horizontal axis is time (the duration of these displays is about 10 seconds) and the vertical axis is vertical height. Solid line shows movements of the head; dashed line indicates extension of the gular fan. (Modified from A. A. Echelle et al., 1971, *Herpetologica* 27:221–288.)

western New Mexico. Female plateau lizards retain their eggs longer than do females of many other species of *Sceloporus*, and retain their eggs still longer if the summer rainy season is late and the environment is too dry for successful egg development. Andrews and Rose found that the thermoregulatory behavior of gravid female lizards did not keep their eggs warmer than they would have been in a nest—the average body temperature of gravid female lizards (25.4°C) was very similar to the average temperature of nest sites (25.2°C). Furthermore, eggs that were retained by females developed more slowly than eggs that were deposited in nests. Eggs that were laid at the normal time hatched on September 3rd, whereas eggs that were retained for 30 days hatched on September 16th. The difference in hatching date means that hatchlings from eggs that had been retained for 30 days would have had about two weeks less time to feed and grow before entering hibernation. That might be a substantial handicap, because hatchling plateau lizards can increase their body mass by 40% in two weeks.

Although egg retention slows embryonic development of plateau lizards, it might accelerate development for species that live in climates where nest sites would be cold. Furthermore, the disadvantage of viviparity, in terms of its effect on reproductive output, may be lower in cold regions than in warm ones. Lizards in warm habitats may produce more than one clutch of eggs in a season, but that is not possible for a viviparous species because development takes too long. However, in a cold climate lizards are not able to produce more than one clutch of eggs in a breeding season anyway, and viviparity would not reduce the annual reproductive output of a female lizard. Phylogenetic analyses of the origins of viviparity suggest that it has evolved most often in cold climates, as this hypothesis predicts, but other origins appear to have taken place in warm climates, and more than one situation favoring viviparity among squamates appears likely.

Viviparity has costs as well as benefits. The agility of a female lizard is substantially reduced when her embryos are large. Experiments have shown that pregnant female lizards cannot run as fast as nonpregnant females and that snakes find it easier to capture pregnant lizards than nonpregnant ones. Females of some species of lizards become secretive when they are pregnant, perhaps in response to their vulnerability to predation. They reduce their activity and spend more time in hiding places. This behavioral adjustment may contribute to the reduction in body temperature seen in pregnant females of some species of lizards, and it probably reduces their rate of prey capture as well.

In general, large species of squamates produce more eggs or fetuses than do small species, and within one species large individuals often have more offspring in a clutch than do small individuals. Both phylogenetic and ecological constraints play a role in determining the number of young produced, however. All geckos have a clutch size of either one or two eggs, and all *Anolis* produce only one egg at a time. Lizards with stout bodies usually have clutches that are a greater percentage of the body mass of the mother than do lizards with slim bodies. The division between stout and slim bodies approximately parallels the division between sit-and-wait predators and widely foraging predators. It is tempting to infer that a lizard that moves about in search of prey finds a bulky clutch of eggs more hindrance than a lizard that spends 99 percent of its time resting motionless. However, some of the divisions among modes of predatory behavior, body form, and relative clutch mass also correspond to the phylogenetic division between iguanian and scincomorph lizards, and as a result it is not possible to decide which characteristics are ancestral and which may be derived.

All-female (**parthenogenetic**) species of squamates have been identified in six families of lizards and one snake. The phenomenon is particularly widespread in the teiids (especially *Cnemidophorus*)

Figure 15–14 Displays by a male *Anolis carolinensis*. Top: Assertion–challenge display. Middle: Aggressive posturing between males. Bottom: Courtship. Note the extension of the dewlap, the species-typical head bob, and the absence of the dorsal and nuchal crests and the eyespot. (From D. Crews, 1978, in *Behavior and Neurology of Lizards*, edited by N. Greenberg and P. D. MacLean, U. S. Department of Health, Education, and Welfare, National Institutes of Mental Health, Rockville, MD.)

and lacertids (*Lacerta*) and occurs in several species of geckos. Parthenogenetic species are known or suspected to occur among chameleons, agamids, xantusiids, and typhlopids. However, parthenogenesis is probably more widespread among squamates than this list indicates because parthenogenetic species are not conspicuously different from bisexual species. Parthenogenetic species are usually detected when a study undertaken for an entirely different purpose reveals that a species contains no males.

Confirmation of parthenogenesis can be obtained by obtaining fertile eggs from females raised in isolation, or by making reciprocal skin grafts between individuals. Individuals of bisexual species usually reject tissues transplanted from another individual because genetic differences between them lead to immune reactions. Parthenogenetic species, however, produce progeny that are genetically identical to the mother, so no immune reaction occurs and grafted tissue is retained.

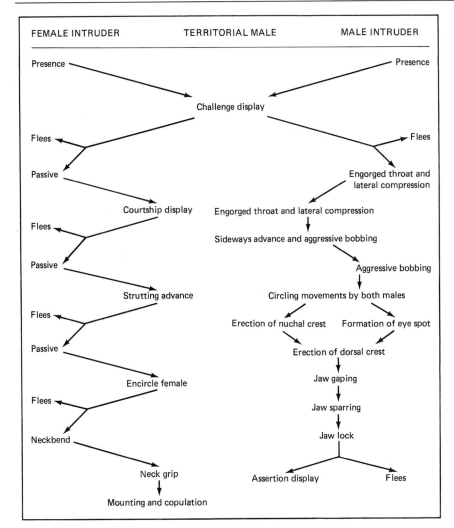

Figure 15–15 Normal sequence of behaviors for a territorial male *Anolis carolinensis* confronting an intruding male or female anole. A territorial male challenges any intruder, and the response of the intruder determines the subsequent behavior of the territorial male. (From D. Crews, 1975, *Herpetologica* 31:37–47.)

The chromosomes of lizards have allowed the events that produced some parthenogenetic species to be deciphered. Many parthenoforms appear to have had their origin as interspecific hybrids. These hybrids are diploid (2*n*) with one set of chromosomes from each parental species. For example, the diploid parthenogenetic whiptail lizard *Cnemidophorus tesselatus* is the product of hybridization between the bisexual diploid species *C. tigris* and *C. septemvittatus* (Figure 15–17). Some parthenogenetic species are triploids (3*n*). These forms are usually the result of a backcross of a diploid parthenogenetic individual to a male of one of its bisexual parental species or, less commonly, the result of hybridization of a diploid parthenogenetic species with a male of a bisexual species different from its parental species. A parthenogenetic triploid form of *C. tesselatus* is apparently the result of a cross between the parthenogenetic diploid *C. tesselatus* and the bisexual diploid species *C. sexlineatus*.

It is common to find the two bisexual parental species and a parthenogenetic species living in overlapping habitats. Parthenogenetic species of *Cnemidophorus* often occur in habitats like the floodplains of rivers that are subject to frequent disruption. Disturbance of the habitat may bring together closely related bisexual species, fostering the hybridization that is the first step in establishing a parthenogenetic species. Once a parthenogenetic species has become established, its reproductive potential is twice that of a bisexual species because every individual of a parthenogenetic species is capable of pro-

Box 15–3 Designer Lizards

A female lizard has only a certain amount of energy to devote to reproduction, and that energy can be used to produce a few large eggs (i.e., eggs with a large quantity of yolk) or several smaller eggs. All else being equal, large eggs produce large hatchlings and small eggs produce small hatchlings, so the question is Are large hatchlings more likely to survive than small hatchlings? And if large hatchlings do survive better, is the difference in survival great enough to make up for producing fewer individual hatchlings?

This question has been addressed in studies of the side-blotched lizard, *Uta stansburiana* (Sinervo et al. 1992). Female side-blotched lizards produce several clutches annually and mature in one year. The lizards normally lay small eggs in the first clutch and larger eggs in subsequent clutches. Does that shift in egg size reflect a survival advantage for large hatchlings later in the season?

To answer that question, Sinervo and his colleagues had to compare the survival of large and small hatchlings at different times of year, but the natural variation in the size of eggs is small and it would be hard to detect differences in survival, even though a very small difference could have a significant effect in evolutionary time. To increase his chances of seeing an effect of hatchling size on survival, Sinervo created artificially small hatchlings by using a hypodermic needle and syringe to remove yolk from some eggs. In

a different experiment (Sinervo and Licht 1991) he devised a method to produce giant hatchlings by surgically removing some ova from the ovary before yolk had been deposited. In that situation, the total amount of yolk available was divided among fewer eggs, and each egg was larger than normal. These methods allowed Sinervo to produce hatchlings that ranged from 50 percent to 150 percent of the normal size (Figure 15–16).

Studies at two field sites showed that female *Uta stansburiana* could maximize the number of surviving offspring from their first clutches by producing small eggs, whereas larger eggs would maximize survival of offspring in later clutches. That change in optimum egg size as the summer progresses probably reflects a changing ecological setting for the hatchlings. Hatchlings from the first clutch of eggs emerge into a world without hatchling *Uta*—only their siblings and hatchlings from other first clutches compete with them for living space. In that situation, large body size may not confer an advantage, so a female can achieve the maximum number of surviving young by producing a large clutch of small eggs. Subsequent clutches don't have it so easy—they must establish home ranges in places that already have resident juvenile *Uta* from earlier clutches. In that situation, large hatchlings may be more likely to insert themselves successfully into the existing social sys-

ducing young. Thus, when a flood or other disaster wipes out most of the lizards, a parthenogenetic species can repopulate a habitat faster than a bisexual species.

Parental care has been recorded for more than 100 species of squamates (Shine 1988). A few species of snakes and a larger number of lizards remain with the eggs or nest site. Some female skinks remove dead eggs from the clutch. Some species of pythons brood their eggs: The female coils tightly around the eggs, and in some species muscular contractions of

the female's body produce sufficient heat to raise the temperature of the eggs to about 30°C, which is substantially above air temperature. One unconfirmed report exists of baby pythons returning at night to their empty eggshells, where their mother coiled around them and kept them warm. Little interaction between adult and juvenile squamates has been documented. In captivity female prehensile-tailed skinks (*Corucia zebrata*) have been reported to nudge their young toward the food dish, as if teaching them to eat. Prehensile-tailed skinks, which occur only on

Figure 15–16 Hatchling side-blotched lizards, *Uta stansburiana*. The normal size for hatchlings (center) compared to gigantized (left) and miniaturized (right) individuals. (Photograph courtesy of Barry Sinervo.)

tem, and females achieve the maximum number of surviving young by producing a smaller clutch that contains large eggs.

In some situations the naturally occurring egg size was not the best option for a female. For example, at the Los Banos field site in 1989 the average size of eggs in natural clutches was about 0.4 grams, but the maximum number of surviving offspring would be produced by eggs that weighed about 0.55 grams. In other words, naturally occurring variation was insufficient to produce eggs of the optimum size for survival.

Reintroduction of captive-bred juveniles to their natural habitats is a frequent goal of management programs for endangered species. Each captive-bred juvenile can represent thousands of hours of painstaking care, and tens or hundreds of thousands of dollars of investment. Clearly, it is desirable to maximize the chances that the juveniles will survive after they are released, and studies like Sinervo's are producing information that can be applied to management programs (Sinervo, 1994). Perhaps it will be possible to use these techniques to produce hatchlings of the sex and size that will maximize the success of reintroductions (Pough 1994).

the Solomon Islands, are herbivorous and viviparous.

Free-ranging baby green iguanas have a tenuous social cohesion that persists for several months after they hatch. The small iguanas move away from the nesting area in groups that may include individuals from several different nests. One lizard may lead the way, looking back as if to see that others are following. The same individual may return later and recruit another group of juveniles. During the first 3 weeks after they hatch, juvenile iguanas move up into the forest canopy and are seen in close association with adults. During this time the hatchlings probably ingest feces from the adults, thereby inoculating their guts with the symbiotic microbes that facilitate digestion of plant material (Troyer 1982). After their fourth week of life, the hatchlings move down from the forest canopy into low vegetation, where they continue to be found in loosely knit groups of two to six or more individuals that move, feed, and sleep together. This association might provide some protection from predators, and if the hatchlings continue to eat fecal material, it is another opportunity to ensure that each lizard has received its full complement of gut microorganisms.

Thermoregulation and the Ecology and Behavior of Squamates

The extensive repertoire of thermoregulatory mechanisms employed by ectotherms allows many species of lizards and snakes to keep body temperature within a range of a few degrees during the part of a day when they are active. Many species of lizards have body temperatures between 33 and 38°C while they are active (the **activity temperature range**), and snakes often have body temperatures between 28 and 34°C. Avery (1982) has summarized the information about body temperatures of squamates in the field.

These activity temperature ranges have been the focus of much research: Field observations show that thermoregulatory activities may occupy a considerable portion of an animal's time. Less obvious, but just as important, are the constraints that the need for thermoregulation sets on other aspects of the behavior and ecology of squamates. For example, some species of lizards and snakes are excluded from certain habitats because it is impossible to thermoregulate. In temperate regions the activity season lasts only during the months when it is warm and sunny enough to permit thermoregulation; at other times of the year snakes and lizards hibernate. Even during the activity season, time spent on thermoregulation may not be available for other activities. Avery (1976) proposed that lizards in temperate regions show less extensive social behavior than do tropical lizards because thermoregulatory behavior in cool climates requires so much time.

The physiological advantages associated with a stable body temperature were discussed in Chapter 4. The body tissues of an organism are the site of a tremendous variety of biochemical reactions proceeding simultaneously and depending on one another to provide the proper quantity of the proper substrates at the proper time for reaction sequences. Each reaction has a different sensitivity to temperature, and regulation is greatly facilitated when temperature variation is limited. Thus, coordination of internal processes may be a major benefit of thermoregulation for squamates. If the temperature stability that a snake or lizard achieves by thermoregulation is important to its physiology and biochemistry, one would expect to find that the internal economy of an animal functions best within its activity temperature range, and that is often the case. Examples of physiological processes that work best at temperatures within the activity range can be found at the molecular, tissue, system, and whole-animal levels of organization.

Organismal Performance and Temperature

How effectively does an organism carry out its normal activities at different temperatures? Does a difference in body temperature have a measurable effect on how fast a lizard can run, for example, or is running speed independent of temperature? These questions can be addressed with ecological analyses of organismal performance; that is, with measurements of the effectiveness of a process at different temperatures. Much information is available for squamates about the effects of temperature on processes as diverse as intracellular chemical reactions, recovery from disease, growth rates, and predatory success. Temperature profoundly affects the ability of squamates to carry out different activities, but not all activities are affected in parallel ways, and behavioral shifts (that is, qualitative rather than quantitative changes) are seen in some cases. Furthermore, the relationship between body temperature and physiological processes is a two-way interaction: Physiological capacities change in response to changes in body temperature, but under some conditions the sequence of cause and effect is reversed and squamates manipulate their body temperatures in response to internal conditions such as feeding status or pregnancy.

The wandering garter snake (*Thamnophis elegans vagrans*) provides examples of the effects of body temperature on a variety of physiological and behavioral functions (Stevenson et al. 1985). Wandering garter snakes are diurnal, semiaquatic inhabitants of lakeshores and stream banks in western North America. They hunt for prey on land and in water, and feed primarily on fishes and amphibians. Chemosensation is an important mode of prey detection for snakes and is accomplished by flicking the tongue. Scent molecules are transferred from the

(a)

(b)

(c)

(d)

(e)

Figure 15–17 The apparent sequence of crosses leading to the formation of diploid and triploid unisexual species of *Cnemidophorus*. Hybridization of the bisexual diploid species (a) *C. tigris* and (b) *C. septemvittatus* produced a unisexual diploid form with half of its genetic complement derived from each of the parental species (an allodiploid). This parthenogenetic form is called (c) *C. tesselatus*. Hybridization between a diploid *C. tesselatus* and a male of the bisexual species (d) *C. sexlineatus* produced a unisexual triploid form with its genetic complement derived from three different parental species (an allotriploid). This parthenogenetic triploid form is also called (e) *C. tesselatus*. Thus, *C. tesselatus* consists of clones of both diploid and triploid lineages, although taxonomists will probably treat these two forms as separate species in the near future. (Photographs by C. J. Cole and C. M. Bogert, American Museum of Natural History, courtesy of C. J. Cole.)

tips of the forked tongue to the epithelium of the vomeronasal organ in the roof of the mouth. The garter snakes are diurnal; they spend the night in shelters, where their body temperatures fall to ambient levels (4 to 18°C), and emerge in the morning to bask. During activity on sunny days the snakes maintain body temperatures between 28 and 32°C.

Stevenson and his associates measured the effect of temperature on the speed of crawling and swimming, the frequency of tongue flicks, the rate of digestion, and the rate of oxygen consumption of the snakes (Figure 15–18). Crawling, swimming, and tongue flicking are elements of the foraging behavior of garter snakes, and the rates of digestion and oxygen consumption are involved in energy utilization. The ability of garter snakes to crawl and swim was severely limited at the low temperatures they experience during the night when they are inactive. At 5°C snakes often refused to crawl and at 10°C they were able to crawl only 0.1 meter per second and could swim only 0.25 meter per second. The speed of both types of locomotion increased at higher temperatures. Swimming speed peaked near 0.6 meter per second at 25 and 30°C, and crawling speed increased to an average of 0.8 meter per second at 35°C. The rate of tongue flicking increased from less than 0.5 flick per second at 10°C to about 1.5 flicks per second at 30°C. The rate of digestion increased slowly from 10 to 20°C and more than doubled between 20 and 25°C. It did not increase further at 30°C, and dropped slightly at higher temperatures. The rate of oxygen consumption increased steadily as temperature rose to 35°C, which was the highest temperature tested because higher body temperatures would have been injurious.

All five measures of performance by garter snakes increased with increasing temperature, but the responses to temperature were not identical. For example, swimming speed did not increase substantially above 20°C, whereas crawling speed continued to increase up to 35°C. The rate of digestion peaked at 25 to 30°C and then declined, but the rate of oxygen consumption increased steadily to 35°C. More striking than the differences among the functions, however, is the apparent convergence of maximum performance for all of the functions on temperatures between 28.5 and 35°C. This range of temperatures is close to the body temperatures of active snakes in the field on sunny days (28 to 32°C). Anywhere within that range of body temperatures, snakes would be able to crawl, swim, and tongue flick at rates that are at least 95 percent of their maximum rates.

The relationship between the body temperatures of active garter snakes and the temperature sensitivity of various behavioral and physiological functions reported by Stevenson and his colleagues is probably common for squamates. That is, in most cases the body temperatures they maintain during activity are the temperatures that maximize organismal performance. However, at least two types of variation complicate the picture of squamate thermoregulation: changes in behavior that accompany changes in body temperature and changes in thermoregulation in response to the physiological status of an animal.

Behavioral Changes A change in the body temperature of a squamate may be accompanied by a qualitative change in behavior instead of by the graded levels of performance shown by garter snakes. For example, *Agama savignyi* is an agamid lizard that lives in desert areas of the Middle East. It shows a pronounced temperature sensitivity of sprint speed: At a body temperature of 18°C it can run only 1 meter per second, but at 34°C it runs about 3 meters per second. *Agama savignyi* lives in open habitats where it may be some distance from shelters that could provide protection from predators. Clearly, the lizards are better able to run to a shelter when they are warm than when they are cool, and they displayed two types of defensive behavior, depending on their body temperature. At body temperatures between 18 and 26°C most *A. savignyi* did not try to run from a predator; instead, they leaped from the substrate and attempted to bite. However, at body temperatures of 30°C or above, most lizards ran away. This sort of qualitative shift in behavior at different body temperatures may be a widespread response among squamates to the effects of body temperature on their ability to carry out certain activities.

Effects of Nutritional Status and Bacterial Infections Several internal states of squamates and other ectotherms influence body temperature. A thermophilic (*thermo* = heat, *philo* = loving) response after feeding is widespread: Individuals with food in the gut maintain higher body temperatures than do individuals without food. A higher body temperature accelerates digestion and increases digestive efficiency and water uptake, so a warm animal digests

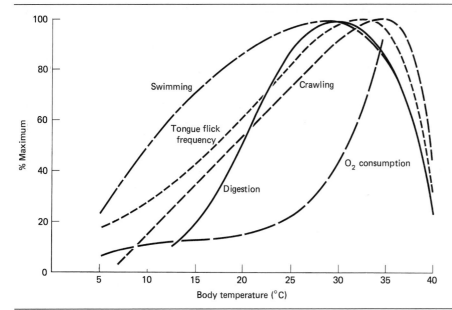

Figure 15–18 Effect of temperature on several activities by the wandering garter snake, *Thamnophis elegans vagrans.* The horizontal axis shows the body temperature of the snakes and the vertical axis shows the percentage of maximum performance achieved at each temperature. (From R. D. Stevenson et al., 1985, *Physiological Zoology* 58:46–57.)

its food more rapidly and assimilates a higher proportion of the energy and water present in the food. Conversely, fasting animals regulate their body temperatures at low levels that reduce their metabolic rates and conserve their stored energy.

Behavioral fever is another common response of ectotherms: Individuals infected by bacteria change their thermoregulatory behavior and maintain body temperatures several degrees higher than those of uninfected controls. These behavioral fevers have been demonstrated in arthropods, fishes, frogs, salamanders, turtles, and lizards. The release of prostaglandin E_1, which acts on thermoregulatory centers of the anterior hypothalamus, appears to be the immediate cause of both the behavioral fevers of ectotherms and the physiological fevers of endotherms. Survival is enhanced by fever, apparently because bacterial growth is limited by a reduction in the availability of iron at higher temperatures.

Reproductive Status Pregnancy affects thermoregulation by squamates. The rate of embryonic development of squamates is strongly affected by temperature, and one of the major advantages of viviparity is thought to be the opportunity it provides for the mother to control the temperature of embryos during development. The body temperatures of female squamates during pregnancy may be different from the temperatures they would normally maintain. For example, pregnant female spiny swifts (*Sceloporus*

jarrovi) had an average body temperature of 32.0°C, whereas male lizards in the same habitat had an average body temperature of 34.5°C (Beuchat 1986). The female lizards changed their thermoregulatory behavior after they had given birth, and the average body temperature of postparturient female lizards was 34.5°C, like that of the males. The low body temperatures of pregnant lizards were unexpected because one can easily think of reasons why giving birth as early in the year as possible would be advantageous for the lizards.

That line of reasoning suggests that female lizards should maintain higher-than-normal body temperatures during pregnancy, or at least they should not reduce their body temperatures. Contrary to that prediction, the body temperatures maintained by pregnant squamates appear to converge toward 32°C whether the normal body temperature for the species is higher or lower. If the body temperature of pregnant female lizards is a compromise between the thermal requirements of the mother and the best temperature for embryonic development, this convergence might indicate that temperatures near 32°C are particularly favorable for embryonic development (Beuchat and Ellner 1987). Experiments have shown that surprisingly small differences in the incubation temperature of lizard and snakes eggs can produce profound morphological, physiological, and behavioral differences in the hatchlings.

Temperature and the Ecology of Squamates

Squamates, especially lizards, are capable of very precise thermoregulation, and microhabitats at which particular body temperatures can be maintained may be one of the dimensions that define the ecological niches of lizards. The five most common species of *Anolis* on Cuba partition the habitat in several ways (Figure 15–19). First, they divide the habitat along the continuum, from sunny to shady: Two species (*A. lucius* and *A. allogus*) occur in deep shade in forests, one (*A. homolechis*) in partial shade in clearings and at the forest edge, and two (*A. allisoni* and *A. sagrei*) in full sun. Within habitats in the sun-shade continuum, the lizards are separated by the substrates they use as perch sites. In the forest *A. lucius* perches on large trees up to 4 meters above the ground, whereas *A. allogus* rests on small trees within 2 meters of the ground. *A. homolechis,* which does not share its habitat with another common species of *Anolis,* perches on both large and small trees. In open habitats *A. allisoni* perches more than 2 meters above the ground on tree trunks and houses, and *A. sagrei* perches below 2 meters on bushes and fenceposts.

Some species of lizards do not thermoregulate. Lizards that live beneath the tree canopy in tropical forests often have body temperatures very close to air temperature (that is, they are thermally passive), whereas species that live in open habitats thermoregulate more precisely. The relative ease of thermoregulation in different habitats may be an important factor in determining whether a species of lizard thermoregulates or allows its temperature to vary with ambient temperature.

The distribution of sunny areas is one factor that determines the ease of thermoregulation. Sunlight penetrates the canopy of a forest in small patches that move across the forest floor as the sun moves across the sky. These patches of sun are the only sources of solar radiation for lizards that live at or near the forest floor, and the patches may be too sparsely distributed or too transient to be used for thermoregulation. In open habitats sunlight is readily available, and thermoregulation is easier. The difference in thermoregulatory behavior of lizards in open and shaded habitats can be seen even in comparisons of different populations within a species. For example, *Anolis sagrei* occurs in both open and forest habitats on Abaco Island in the Caribbean. Lizards in open habitats bask and maintain body temperatures between 32 and 35°C from about 8:30 in the morning through about 5:00 in the afternoon. Lizards in the forest do not bask, and their body temperatures vary from a low of 24°C to a high of 28°C over the same period.

The task of integrating thermoregulatory behavior with foraging is relatively simple for sit-and-wait foragers such as *Anolis.* These lizards can readily change their balance of heat gain and loss by making small movements in and out of shade or between calm and breezy perch sites while they continue to scan their surroundings for prey. Widely foraging lizards may have more difficulty integrating thermoregulation and predation. They are continuously moving between sun and shade and in and out of the wind, and their body temperatures are affected by their foraging activity. These lizards sometimes have to stop foraging to thermoregulate, resuming foraging only when they have warmed or cooled enough to return to their activity temperature range.

Body size is yet another variable that can affect thermoregulation. An example of the interaction of body size, thermoregulation, and foraging behavior is provided by three species of teiid lizards (*Ameiva*) in Costa Rica. *Ameiva* are widely foraging predators that move through the habitat, pushing their snouts beneath fallen leaves and into holes. The largest of the three species, *A. leptophrys,* had an average body mass of 83 grams, the middle species, *A. festiva,* weighed 32 grams, and the smallest, *A. quadrilineata,* weighed 10 grams (Figure 15–20). The study site was at the side of a road and the adjacent forest. The three species foraged in different parts of this habitat: *A. quadrilineata* spent most of its time in the short vegetation at the edge of the road, *A. festiva* was found on the bank beside the road, and *A. leptophrys* foraged primarily beneath the canopy of the forest (Figure 15–21). The different foraging sites of the three species may reflect differences in thermoregulation that result from the variation in body size.

The thermoregulatory behavior of the three species was the same: A lizard basked in the sun until its body temperature rose to 39 to 40°C, then moved through the mosaic of sun and shade as it searched for food. The body temperatures of the lizards dropped as they foraged, and lizards ceased foraging and resumed basking when their body temperatures had fallen to 35°C. Thus, the time that a lizard could forage depended on how long it took for its body temperature to cool from 39 or 40°C to 35°C. The rate of cooling for *Ameiva* in the shade was

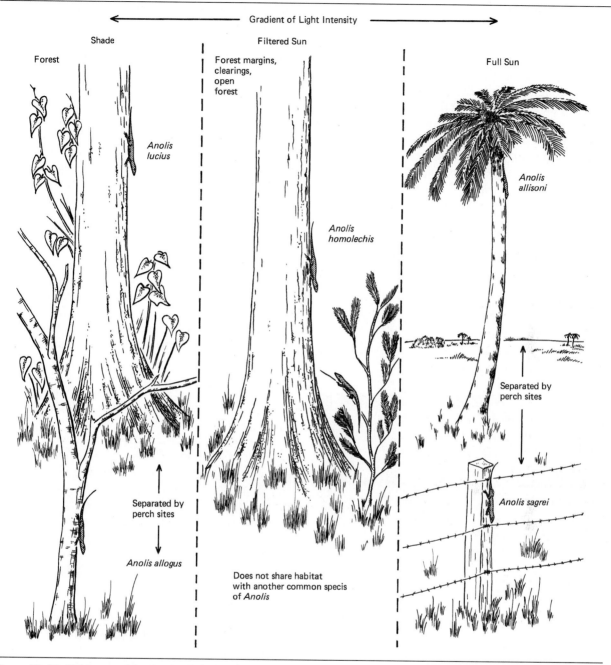

Shade

Forest

Filtered Sun

Forest margins,
clearings,
open
forest

Full Sun

*Anolis
lucius*

*Anolis
homolechis*

*Anolis
allisoni*

Separated by
perch sites

Separated by
perch sites

Anolis allogus

Does not share habitat
with another common specis
of *Anolis*

Anolis sagrei

Figure 15–19 Habitat partitioning by Cuban species of *Anolis*. See text for explanation. (Based on R. Ruibal, 1961, *Evolution* 15:98–111.)

inversely proportional to the body size of the three species: *A. quadrilineata* cooled in 4 minutes, *A. festiva* in 6 minutes, and *A. leptophrys* in 11 minutes (Figure 15–22a). That relationship appears to explain part of the microhabitat separation of the three species: The smallest species, *A. quadrilineata*, cools so rapidly that it may not be able to forage effectively in shady microhabitats, whereas *A. leptophrys* cools slowly and can forage in the shade beneath the forest canopy. The species of intermediate body size, *A. festiva*, uses the habitat with an intermediate amount of shade.

The slow rate of cooling of *A. leptophrys* may explain why it is able to forage in the shade, but field observations indicate that its foraging is actually restricted to shade; it emerges from the forest only to bask. Does some environmental factor prevent *A. leptophrys* from foraging in the sun? The answer to that question may lie in the way the body temperatures of the three species increase when they are in open microhabitats. Body size profoundly affects the equilibrium temperature of an organism in the sun. A lizard warms by absorbing solar radiation, and as it gets warmer its heat loss by convection, evaporation, and reradiation also increases. When the rate of heat loss equals the rate of heat gain the body temperature does not increase further. Large lizards reach that equilibrium at higher body temperatures than do small ones. Computer simulations of the heating rates of the three *Ameiva* in sun showed that *A. quadrilineata* and *A. festiva* would reach equilibrium at body temperatures of 37 to 40°C, but *A. leptophrys* would continue to heat until its body temperature reached a lethal 45°C (Figure 15–22b). This analysis suggests that *Ameiva leptophrys* would die of heat stress if it spent more than a few minutes in a sunny microhabitat, but the two smaller species of *Ameiva* would not have that problem.

Thus, as a result of the biophysics of heat exchange, the large body size of *A. leptophrys* apparently allows it to forage in shaded habitats (because it cools slowly), but prevents it from foraging in sunny habitats (because it would overheat). Field observations of the foraging behavior of hatchling *A. leptophrys* emphasize the importance of the heat exchange in the foraging behavior of these lizards. Hatchling *A. leptophrys* forage in open habitats like *A. festiva* and *A. quadrilineata* rather than under the forest canopy like adult *A. leptophrys*. That is, the juveniles of the large species of *Ameiva* behave like adults of the smaller species, probably because of the importance of body size and heat exchange in determining the mirohabitats in which lizards can thermoregulate.

The difference in the use of various microhabitats by these three species of lizards looks at first

(a) (b)

(c)

Figure 15–20 Three species of *Ameiva* that forage in different habitats: (a) *Ameiva leptophrys*, which has an average adult mass of 83 grams; (b) *Ameiva festiva*, average adult mass 32 grams; (c) *Ameiva quadrilineata*, average adult mass 10 grams. ([a] Courtesy of Daniel H. Janzen; [b] and [c] photographed by Michael Hopiak, courtesy of the Cornell University Herpetology Collection.)

glance like an example of habitat partitioning in response to interspecific competition for food. That is, because all three species eat the same sort of prey, they could be expected to concentrate their foraging efforts in different microhabitats to reduce competition. However, this analysis of the thermal requirements of the lizards suggests that interspecific competition for food is, at most, a secondary factor.

If competition for food were important, one would not expect to find hatchling *A. leptophrys* foraging in the same microhabitat as adult *A. quadrilineata*, because the similarity in size of the two lizards would intensify competition for food. The hypothesis that competition for food determines the microhabi-

tat distribution of the animals predicts that the forms most similar in body size should be mostly widely separated in the habitat. In contrast, the hypothesis that energy exchange with the environment is critical in determining where a lizard can forage predicts that species of similar size will live in the same habitat, and that is approximately the pattern seen. Apparently, the physical environment (radiant energy) is more important than the biological environment (interspecific competition for food) in determining the microhabitat distributions of these lizards. That conclusion reflects the broad-scale ecological significance of the morphological and physiological differences between ectotherms and endotherms, a theme that is developed in Chapter 16.

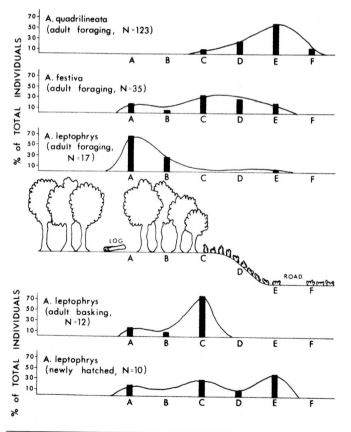

Figure 15–21 Foraging sites of three species of *Ameiva* in Costa Rica. The histograms show the number of individuals of the three species seen in each of six locations: A, a small clearing in the forest; B, immediately inside the forest edge; C, outside the edge of the forest; D, midway between the edge of the forest and open area; E, low vegetation beside a road; F, low vegetation in a large open area without trees. (From P. E. Hillman, 1969, *Ecology* 50:476–481.)

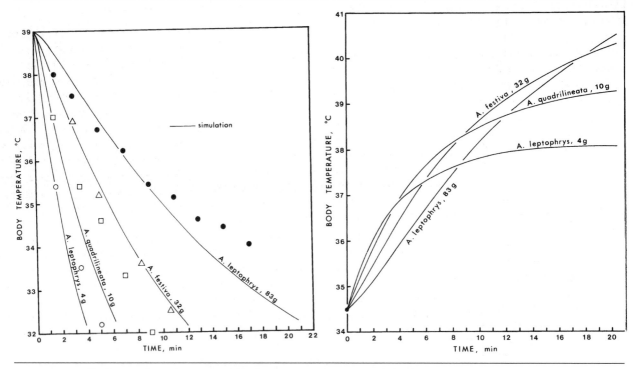

Figure 15–22 Cooling and heating rates of the three sympatric species of *Ameiva*. The solid lines are computer simulations of heating and cooling rates, and the symbols in (a) show the close correspondence of measured body temperatures to those predicted by the computer simulation. The largest species, *Ameiva leptophrys*, heats and cools more slowly than the smaller species. If it remained in the sun its body temperature would rise above 40°C. The smaller species heat and cool more rapidly than the large species and reach temperature equilibrium at lower body temperatures. [(a) From P. E. Hillman, 1969, *Ecology* 50:476–481; (b) courtesy of Peter E. Hillman.]

Summary

The extant lepidosaurs include the squamates (lizards, amphisbaenians, and snakes) and their sister group, the Sphenodontidae. The lepidosaurs, with nearly 6000 species, form the second largest group of extant tetrapods. The two species of tuatara of the New Zealand region are the sole extant sphenodontids. They are lizard-like animals, about 60 centimeters long, with a dentition and jaw mechanism that give a shearing bite. Sphenodontids were diverse in the Mesozoic and included terrestrial insectivorous and herbivorous species as well as a marine form.

Lizards range in size from tiny geckos only 3 centimeters long to the Komodo monitor lizard, which reaches a length of 3 meters. The Iguania and Gekkota are composed largely of stout-bodied lizards and show considerable diversity of body form. The Scincomorpha and anguimorpha are elongate and show less morphological diversity than do iguanians.

Differences in ecology and behavior parallel the phylogenetic divisions: Many iguanians are sit-and-wait predators that maintain territories and detect prey and intruders by vision. Iguanians often employ colors and patterns in visual displays during courtship and territorial defense. Many scincomorph and anguimorph lizards are often widely foraging predators that detect prey by olfaction and do not maintain territories. Pheromones are important in the social behaviors of many of these lizards.

Amphisbaenians are specialized burrowing squamates. Their skulls are solid structures that they use for tunneling through soil. Many amphisbaenians have blunt heads, and others have vertically keeled or horizontally spade-shaped snouts. The dentition of amphisbaenians appears to be specialized for nipping small pieces from prey too large to be swallowed whole. The skin of amphisbaenians is loosely attached to the trunk, and amphisbaenians slide backward or forward inside the tube of their skin as they move through tunnels with concertina locomotion.

Snakes are derived from lizards. Repackaging the body mass of a vertebrate into a serpentine form has been accompanied by specializations of the mechanisms of locomotion (serpentine, rectilinear, concertina, and sidewinding), prey capture (constriction and the use of venom), and swallowing (a highly kinetic skull).

Many squamates have complex social behaviors associated with territoriality and courtship, but parental care is only slightly developed. Fertilization is internal, and viviparity has evolved 80 or more times among squamates. Thermoregulation is another important behavior of squamates, and various activities are influenced by body temperature. The ecological niches of some lizards may be defined in part by the microhabitats needed to maintain particular body temperatures. Feeding status, pregnancy, and bacterial infections can change the thermoregulatory behavior of squamates, causing an affected individual to maintain a higher or lower body temperature than it otherwise would.

References

Allen, M. E., and O. T. Oftedal. 1989. Dietary manipulation of the calcium content of feed crickets. *Journal of Zoo and Wildlife Medicine* 20:26–33.

Allen, M. E., O. T. Oftedal, and D. E. Ullrey. 1993. Effect of fdietary calcium concentration on mineral composition of fox geckos (*Hemidactylus garnoti*) and Cuban tree frogs (*Osteopilus spetentrionalis*). *Journal of Zoo and Wildlife Medicine* 24:118–128.

Andrews, R. M., and B. R. Rose. 1994. Evolution of viviparity: constraints on egg retention. *Physiological Zoology* 67:1006–1024.

Arnold, E. N. 1988. Caudal autotomy as a defense. Pages 235–273 in *Biology of the Reptilia*, volume 16, edited by C. Gans and R. B. Huey, Liss, New York, NY.

Auffenberg, W. 1981. *The Behavioral Ecology of the Komodo Monitor*. University Presses of Florida, Gainesville, FL.

Avery, R. A. 1976. Thermoregulation, metabolism, and social behaviour in Lacertidae. Pages 245–259 in *Morphology and Biology of Reptiles*, edited by A. d'A. Bellairs and C. B. Cox. Academic, London, UK.

Avery, R. A. 1982. Field studies of reptilian thermoregulation. Pages 93–166 in *Biology of the Reptilia*, volume 12, edited by C. Gans and F. H. Pough. Academic, London, UK.

Beuchat, C. A. 1986. Reproductive influence on the thermoregulatory behavior of a live-bearing lizard. *Copeia* 1986:971–979.

Beuchat, C. A., and S. Ellner. 1987. A quantitative test of life history theory: thermoregulation by a viviparous lizard. *Ecological Monographs* 57:45–60.

Blackburn, D. G. 1982. Evolutionary origins of viviparity in the Reptilia, I: Sauria. *Amphibia–Reptilia* 3:185–205.

Blackburn, D. G. 1985. Evolutionary origins of viviparity in the Reptilia, II: Serpentes, Amphisbaenia, and Ichthyosauria. *Amphibia–Reptilia* 6:259–291.

Blackburn, D. G., L. J. Vitt, and C. A. Beuchat. 1984. Eutherian-like reproductive specializations in a viviparous reptile. *Proceedings of the National Academy of Science USA* 81:4860–4863.

Carroll, R. L. 1987. *Vertebrate Paleontology and Evolution*. Freeman, New York, NY.

Cooper, W. E., Jr., and L. J. Vitt. 1986. Interspecific odour discriminations among syntopic congeners in scincid lizards (Genus *Eumeces*). *Behaviour* 97:1–9.

Daugherty. C. H., A. Cree, J. M. Hay, and M. B. Thompson. 1990. Neglected taxonomy and continuing extinctions of tuatara (Spendonod). *Nature* 347:177–179.

Diamond, J. M. 1987. Did Komodo dragons evolve to eat pygmy elephants? *Nature* 326:832.

Fleishman, L. J., E. R. Loew, and M. Leal. 1993. Ultraviolet vision in lizards. *Nature* 365:397.

Greer, A. E. 1991. Limb reduction in squamates: identification of the lineages and discussion of the trends. *Journal of Herpetology* 25:116–173.

Halpern, M. 1992. Nasal chemical senses in reptile: structure and function. Pages 423–523 in *Biology of the Reptilia*, volume 18, edited by C. Gans and D. Crews. University of Chicago Press, Chicago, IL.

Hardy, D. L., Sr. 1994. A re-evaluation of suffocation as the cause of death during constriction. *Herpetological Review* 25:45–47.

Huey, R. B., and A. F. Bennett. 1986. A comparative approach to field and laboratory studies in evolutionary biology. Pages 82–98 in *Predator–Prey Relationships*, edited by M. E. Feder and G. V. Lauder. University of Chicago Press, Chicago, IL.

Huey, R. B., and E. R. Pianka. 1981. Ecological consequences of foraging mode. *Ecology* 62:991–999.

Mason, R. T. 1992. Reptilian pheromones. Pages 114–228 in *Biology of the Reptilia*, volume 18, edited by C. Gans and D. Crews. University of Chicago Press, Chicago, IL.

May, R. M. 1990. Taxonomy as destiny. *Nature* 347:129–130.

Pough, F. H. (editor). 1983. Adaptive radiation within a highly specialized system: the diversity of feeding mechanisms of snakes. *American Zoologist* 23:339–460.

Pough, F. H. 1994. Zoo–academic research collaboration: How close are we? *Herpetologica* 49:500–508.

Savitzky, A. H. 1980. The role of venom delivery strategies in snake evolution. *Evolution* 34:1194–1204.

Schwenk, K. 1994. Why snakes have forked tongues. *Science* 263:1573–1577.

Shine, R. 1985. The evolution of viviparity in reptiles: an ecological analysis. Pages 606–694 in *Biology of the Reptilia*, volume 15, edited by C. Gans and F. Billett. Wiley, New York, NY.

Shine, R. 1988. Parental care in reptiles. Pages 275–329 in *Biology of the Reptilia*, volume 16, edited by C. Cans and B. R. Huey. Liss, New York, NY.

Siegel, R., J. T. Collins, and S. S. Novak. 1987. *Snakes: Ecology and Evolutionary Biology*. Macmillan, New York, NY.

Simon, C. A. 1983. A review of lizard chemoreception. Pages 119–133 in *Lizard Ecology: Studies of a Model Organism,* edited by R. B. Huey, E. R. Pianka, and T. W. Schoener. Harvard University Press, Cambridge, MA.

Sinervo, B. 1994. Experimental manipulations of clutch size and offspring size in lizards: mechanistic, evolutionary, and conservation considerations. In *Captive Management and Conservation of Amphibians and Reptiles,* edited by J. B. Murphy, J. T. Collins, and K. Adler, Contributions to Herpetology, Society for the Study of Amphibians and Reptiles, Oxford, UK.

Sinervo, B., and P. Licht. 1991. Hormonal and physiological control of clutch size, egg size, and egg shape in side-blotched lizards (*Uta stansburiana*): constraints on the evolution of life histories. *Journal of Experimental Zoology* 257:252–264.

Sinervo, B., P. Doughty, R. B. Huey, and K. Zamudio. 1992. Allometric engineering: a causal analysis of natural selection on offspring size. *Science* 258:1927–1930.

Stevenson, R. D., C. R. Peterson, and J. S. Tsuji. 1985. The thermal dependence of locomotion, tongue flicking, digestion, and oxygen consumption in the wandering garter snake. *Physiological Zoology* 58:46–57.

Troyer, K. 1982. Transfer of fermentative microbes between generations in a herbivorous lizard. *Science* 216:540–542.

Whiteside, D. I. 1986 The head skeleton of the Rhaetian sphenodontid *Diphydontosaurus avonis* gen. et sp. nov. and the modernizing of a living fossil. *Philosophical Transactions of the Royal Society (London) Series B* 312:379–430.

Wu, X.-C., D. R. Brinkman, A. P. Russell, Z.-M. Dong, P. J. Currie, L.-H. Hou, and G.-H. Cul. 1993. Oldest known amphisbaenian from the Upper Cretaceous of Chinese Inner Mongolia. *Nature* 366:57–59.

ECTOTHERMY: A LOW-COST APPROACH TO LIFE

16

Ectothermy is an ancestral character of vertebrates, but like many ancestral characters it is just as effective as its derived counterpart. Furthermore, the mechanisms of ectothermal thermoregulation are as complex and specialized as those of endothermy. In Chapter 4 we examined the behavioral and physiological mechanisms that ectotherms use to control their body temperatures, and in Chapter 15 we considered the effect of these thermoregulatory mechanisms on the ecology and behavior of lizards and snakes. Here we consider the consequences of ectothermy in shaping broader aspects of the life-style of fishes, amphibians, and reptiles. Ectothermal tetrapods have low energy requirements compared to endotherms, and this character allows them to be successful in environments in which food is in short supply, either seasonally or chronically. Within that general characterization, the success of various species is related to specializations that facilitate finding food or mates under difficult circumstances, or allow them to endure prolonged unfavorable periods. The conclusion from this examination is that success in difficult environments is as likely to reflect the ancestral features of a group as its specializations.

Vertebrates and Their Environments

Vertebrates manage to live in the most unlikely places. Amphibians live in deserts where rain falls only a few times a year and several years may pass with no rainfall at all; lizards live on mountains at altitudes above 4000 meters where the temperature falls below freezing nearly every night of the year and does not rise much above freezing during the day; birds and mammals live near the poles where the coldest temperatures on earth are recorded in winter and the sun barely appears above the horizon for weeks on end; and fishes live in the depths of the sea where sunlight never penetrates.

Of course, vertebrates do not seek out only inhospitable places to live—birds, lizards, mammals, and even amphibians can be found on the beaches at Malibu, sometimes running between the feet of surfers, and fishes cruise the shore. However, even this apparently benign environment is stressful to some animals, and examination of the ways that vertebrates have managed to live in stressful environments has provided much information about how they function as organisms, that is, how morphology, physiology, ecology, and behavior interact.

In some cases elegant adaptations allow specialized vertebrates to colonize stressful habitats. More common and more impressive than these specializations, however, is the realization of how minor are the modifications of the basic vertebrate body plan that allow animals to endure environmental temperatures from –70 to +70°C, or water conditions ranging from complete immersion in water to complete independence of liquid water. No obvious dif-

497

ferences distinguish animals from vastly different habitats—an arctic fox looks very much like a desert fox, and a lizard from the Andes Mountains looks like one from the Atacama Desert. The adaptability of vertebrates lies in the combination of minor modifications of their ecology, behavior, morphology, and physiology. A view that integrates these elements shows the startling beauty of organismal function of vertebrates.

Endotherms and ectotherms react somewhat differently to many environmental factors. The essence of the endothermal approach to life is the use of energy to maintain internal homeostasis. Endotherms can regulate their body temperatures and body fluid and salt concentrations with remarkable precision in the face of extreme fluctuations in their environment. Ectotherms are also capable of remarkable homeostasis, but the general characteristic of ectotherms is low rates of energy consumption (Chapter 4). In many cases ectotherms save energy by relaxing their limits of homeostasis, whereas endotherms expend energy to maintain homeostasis. This difference in the responses of endotherms and ectotherms to stressful environments has broad implications: By relaxing homeostasis ectotherms are able to do things that would not be feasible for an endotherm. On the other hand, the activity of ectotherms may be curtailed for long periods during unfavorable seasons because they are not able to maintain homeostasis. The energy requirements of endothermy are a substantial factor in many aspects of the ecology and behavior of birds and mammals, and inability to obtain enough energy may exclude endotherms from activity in certain habitats during some parts of the year: Some endotherms migrate or hibernate to avoid conditions in which they cannot get enough energy to maintain homeostasis.

Characteristics of Environments

The life of a vertebrate is energetically expensive; homeostasis and mobility require ATP (adenosine-5'-triphosphate), which is derived from metabolism of food. It follows that extreme environments are those in which energy is limited or the cost of homeostasis and mobility is high, and the most difficult situations are those in which both conditions apply. That qualitative generalization applies to ectotherms

and endotherms, although energy is likely to become limiting for endotherms before ectotherms because endothermal life is more energetically expensive than ectothermal life.

Habitats in which food (energy) is scarce abound. In ecological terms all animals are consumers. That is, animals depend on the primary production of plants: Herbivorous animals eat plants and carnivorous animals eat herbivores. Thus, the path of energy flow through an ecological community begins with the fixation of carbon by plants, and one measure of the availability of energy in a habitat is the primary production of the plants in the habitat (Table 16–1). Tropical forests and estuarine marine environments are the most productive habitats on Earth, producing as much as 4000 or 5000 grams of carbon per square meter annually. At the opposite extreme, deserts and the open sea produce less than one-tenth as much carbon as do tropical forests and estuaries, and the annual production of arctic habitats is still lower.

The primary production of a habitat tells how much energy is potentially available to animals but does not necessarily reflect the accessibility of the energy. A tropical forest has an enormous production of new plant tissue every year, but much of that material is leaves that are in the canopy of the forest, which is 20 meters or more above the ground. Those leaves are not accessible to herbivores that cannot climb trees. Furthermore, animals have many habitat requirements besides a source of energy: They need water, shelter from the elements, places to escape predators, microclimates that are suitable for reproduction and for rearing young, and so on. Some vertebrates use their mobility to occupy habitats that do not have all the resources they need. For example, small birds in deserts can fly to water sources that may be many kilometers away, whereas mammals of the same body size are less mobile and must subsist on the water resources they can find in a restricted area.

The availability of resources that will sustain an individual is not enough to allow a species to occupy a habitat: Reproduction is necessary for a population to be self-sustaining. This requirement means that the population density must be sufficient for males and females to locate each other during the breeding season, and that the habitat must provide conditions suitable for reproduction and the growth of young. Elaborate behavioral modifications that permit the sexes to find each other are seen in some

Table 16–1 Net primary production of carbon and average plant biomass for major ecosystems

	Net Primary Production (dry g/m² year⁻¹)		Biomass per Unit Area (dry kg/m²)	
	Range	Mean	Range	Mean
Terrestrial habitats				
Tropical forests	1000–5000	2000	6–80	45
Temperate forest	600–2500	1300	6–200	30
Tropical savanna	200–2000	700	0.2–15	4
Temperate grassland	150–1500	500	0.2–5	1.5
Desert	10–250	70	0.1–4	0.7
Arctic	0–10	3	0–0.2	0.02
Aquatic habitats				
Estuaries	500–4000	2000	0.04–4	1
Lakes and streams	100–1500	500	0–0.1	0.02
Open ocean	2–400	125	0–0.005	0.003

Source: R. H. Whittaker, 1970, *Communities and Ecosystems*, Macmillan, New York, NY.

species of vertebrates, and juveniles sometimes do not live in the same habitat as adults.

Extreme conditions in many habitats are seasonal; periods of plenty may alternate with periods of great scarcity. These cycles are conspicuous in arctic regions, where the long days of summer allow plants to grow rapidly and winters bring all primary production to a halt. Other habitats have annual cycles, too; even some tropical forests have pronounced wet (monsoon) and dry seasons that greatly affect plant production. The responses of animals to seasonally extreme conditions are somewhat different from their responses to chronic shortages of energy. Hibernation and estivation are ways to remain in a habitat during difficult periods, and migration allows animals to move from habitat to habitat as the availability of energy waxes and wanes.

Ectotherms in Extreme Environments

The low energy requirements of ectotherms allow them to exploit many environments in which energy is scarce seasonally or at all times. Also, the ability of ectotherms to become torpid allows them to wait out periods when the environment may be too hot or too dry to permit activity. Temporary relaxation of homeostasis—allowing physiological variables to fluctuate more widely than usual—may permit organisms to occupy environments that would not otherwise be habitable.

The responses of ectotherms to stressful environments usually involve suites of characters that include behavior and physiology as well as body structure. Patterns of life history may be modified in ways that ensure that males and females can find each other to mate, or that facilitate the development of embryos and young. These phenomena are well illustrated by vertebrates in three environments: the deep sea, the desert, and freezing water.

The Deep Sea

Oceans cover 69 percent of the Earth's surface (Figure 16–1) and are classified into different depth zones. Two major life zones exist: the pelagic, where organisms live a free-floating or swimming existence, and the benthic, where organisms associate with the bottom. Solar light is totally extinguished at a depth of 1000 meters in the clearest oceans and at much shallower depths in the less clear coastal seas. As a result, large portions of the oceans are aphotic, or perpetually dark (about 75 percent by volume), interrupted only by the flashes and glow of bioluminescent organisms. A distinctive and bizarre array of deep-sea fishes has evolved in those zones.

The abundance, size, and species diversity of fish decreases as you descend into the ocean depths. These trends are not surprising, for all animals ulti-

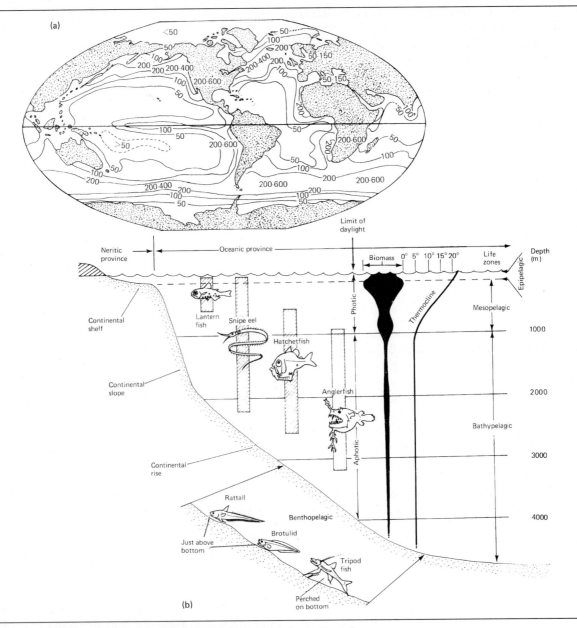

Figure 16–1 Life zones of the ocean depths. (a) Annual productivity at the ocean surface. Numbers are grams of carbon produced per square meter per year. Rich assemblages of deep-sea fishes occur where highly productive waters overlie deep waters. (b) Schematic cross section of the life zones within the deep sea. Various pertinent physical and biotic parameters are superimposed on the arbitrary life zones, as are the vertical ranges of several characteristic fish species, some of which migrate on a daily basis. ([a] Modified from H. Friedrich, 1973, *Marine Biology*, University of Washington Press, Seattle, WA; [b] modified from N. B. Marshall, 1971, *Explorations of the Life of Fishes*, Harvard University Press, Cambridge, MA.)

mately depend on plant photosynthesis, which is limited to the epipelagic regions. Below the epipelagic, animals must depend on descent of food from above—a rain of detritus from the surface into the deep sea. The decrease of fishes with depth is inevitable, for the amount of food available must diminish if it is consumed during descent. Sampling confirms this decrease in food: Surface plankton can reach biomass levels of 500 milligrams wet mass per cubic meter; at a depth of 1000 meters, where aphotic conditions commence, plankton decrease to 25 milligrams per cubic meter; at depths of 3000 to 4000 meters to 5 milligrams per cubic meter; and at 10,000 meters to 0.5 milligram per cubic meter.

Fish diversity parallels this decrease: About 750 species of deep-sea fishes are estimated to occupy the mesopelagic zones and only 150 species the bathypelagic regions. Deep seas that lie under areas of high surface productivity contain more and larger fish species than do regions that underlie less productive surface waters. High productivity occurs in areas of upwelling, where currents recycle nutrients previously removed by the sinking of detritus. In these places deep-sea fishes tend to be most diverse and abundant.

In tropical waters, photosynthesis continues throughout the year. Away from the tropics it is more cyclic, following seasonal changes in light, temperature, and sometimes currents. Diversity and abundance of deep-sea fishes decrease away from the tropics. Over 300 species of meso- and bathypelagic fishes occur in the vicinity of tropical Bermuda. In the entire Antarctic region, only 50 species have been described. The high productivity of Antarctic waters is restricted to a few months of each year. Apparently, the sinking of detritus through the rest of the year is insufficient to nourish a diverse assemblage of deep-sea fishes.

We emphasize that availability of food (energy) is the most formidable environmental problem that deep-sea fishes encounter and, indeed, many of their specific characteristics may have been selected by food scarcity. Another variable factor is hydrostatic pressure, whereas low temperature and the absence of light are constants.

With increased depth each species is further removed from a primary source of food. In general, mesopelagic fishes and invertebrates, which inhabit depths from 100 to 1000 meters, migrate vertically. At dusk they ascend toward the surface, only to descend again near dawn apparently following

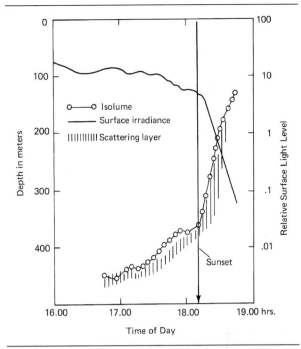

Figure 16–2 Upward migration of mesopelagic fishes as indicated by changes in depth of deep scattering layer. Sonar signals readily reflect off the swim bladders of mesopelagic fishes, and aggregations of them produce an acoustic scattering layer. At sunset the intensity of light at the surface decreases (right axis) as does the light penetrating the sea. Plotting the depth of a single light intensity (isolume, read depth on left axis) against time illustrates the progressively shallower depth at which a given light intensity occurs during dusk. The light intensity chosen for this example falls between that of starlight and full moonlight, and it corresponds closely with the upper boundary of the deep scattering layer, which rises rapidly with the upward migration of mesopelagic fishes. (Modified from E. M. Kampa and B. P. Boden, 1957, *Deep-Sea Research* 4:73–92.)

light-intensity levels (Figure 16–2). Several benefits and costs probably result from this behavior. By rising at dusk, mesopelagic fishes enter a region of higher productivity where food is more concentrated. However, they also increase their exposure to predators. Furthermore, ascending mesopelagic fishes are exposed to temperature increases that can exceed 10°C. The energy costs of maintenance at these higher temperatures can double, or even triple. In contrast, daytime descent into cooler waters lowers metabolism, which conserves energy and reduces the chance of predation because fewer predators exist at increased depth.

Bathypelagic Fishes It is less certain that bathypelagic fishes undertake daily vertical migrations. There is little metabolic economy from vertical migration within this aphotic zone because temperatures are uniform (about 5°C). Furthermore, the cost and time of migration over the several thousand meters from the bathypelagic region to the surface would probably outweigh the energy gained from invading the rich surface waters. Instead of migrating, bathypelagic fishes are specialized to live a less active life than those of their meso- and epipelagic counterparts.

Pelagic deep-sea fish display a series of structures related to their particular life-styles. For example, eye size and function correlate with depth. Mesopelagic fishes have large eyes, pelagic fishes have smaller eyes, and benthic fishes vary in eye size. The retina contains a high concentration of visual pigment, the photosensitive chemical that absorbs light in the process of vision. The visual pigments of deep-sea fishes are most efficient in absorbing blue light, which is the wavelength of light that is most readily transmitted through clear water. Many deep-sea fishes and invertebrates are emblazoned with startling bioluminescent designs formed by photophore organs that emit blue light.

Deep-sea fishes have less dense bone and less skeletal muscle than do fishes from shallower depths. Surface fishes have strong ossified skeletons and large red muscles especially adapted for continuous cruising. Mesopelagic fishes, which swim mostly during vertical migration, have a more delicate skeleton and less axial red muscle. In bathypelagic fishes, the axial skeleton and the mass of muscles are greatly reduced and locomotion is limited (Figure 16–3).

The jaws and teeth of deep-sea fishes are often enormous in proportion to the rest of the body. If a fish rarely encounters potential prey, it is probably important to have a mouth large enough to engulf nearly anything it does meet and a gut that can extend to accommodate a meal (Figure 16–4). Increasing the chances of encountering prey is also important to survival in the deep sea. Rather than searching for prey through the blackness of the depths, the ceratioid anglerfishes dangle a bioluminescent bait in front of them. The bioluminescent lure is believed to mimic the movements of zooplankton and to lure fishes and larger crustaceans to the mouth. Prey is sucked in with a sudden opening of the mouth, snared in the teeth, and then swallowed. The jaws of many anglerfishes expand and

Figure 16–3 Reduction in the amount of bone in oceanic fishes that live at increasing depths as seen in x-rays. Upper: surface-dwelling jackfish (Carangidae); middle: mesopelagic, vertically migrating lanternfish (Myctophidae); bottom: mesopelagic, deep-living bristlemouth (Gonostomatidae). Note also the change in the relative size of the eye.

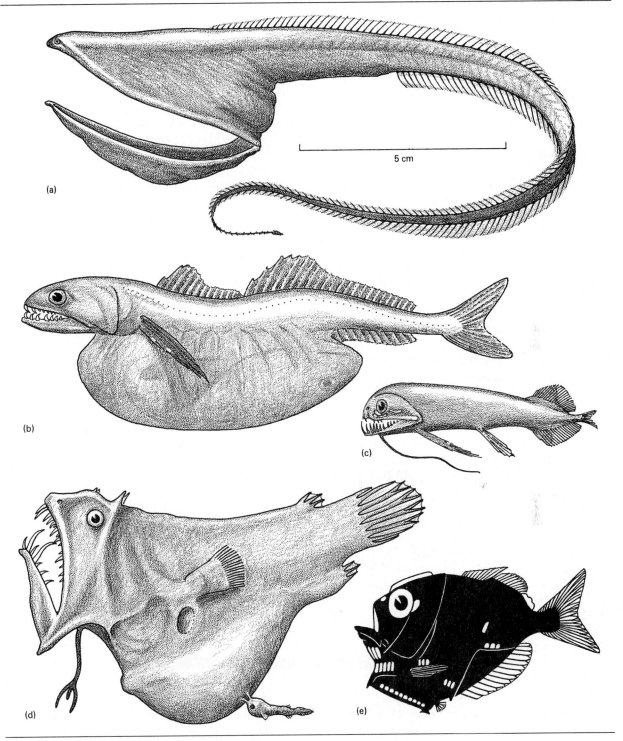

Figure 16–4 Deep-sea fishes showing their large mouths and distendable guts: (a) pelican eel, *Eurypharynx pelecanoides;* (b) deep-sea perch, *Chiasmodus niger,* its belly distended by a fish bigger than itself; (c) stomiatoid, *Aristostomias grimaldii;* (d) female anglerfish, *Liophryne argyresca,* with a parasitic male attached to her belly; (e) hatchet fish, *Sternoptyx diaphana.* The light areas are luminous photophores.

the stomach stretches to accommodate prey larger than the predator. Thus deep-sea fishes, like most teleosts, show major specializations in locomotor and feeding structures. Unlike surface teleosts, however, the scarcity of food has selected for structures that minimize the costs of foraging and maximize the capture of prey.

Most features of the biology of deep-sea fishes are directly related to the scarcity of food, but in some cases the relationship is indirect. For example, most species are small (the average length is less than 5 centimeters) and individuals of a species are not abundant. In so vast a habitat the number of anglerfishes (ceratioids) when most concentrated does not exceed 150,000 individuals per cubic mile. This estimate represents all ceratioid species. The density of females in the most common species is typically less than one female per cubic mile. Imagine trying to find another human under similar circumstances! Yet to reproduce, each fish must locate a mate and recognize it as its own species.

Distinctive bioluminescent patterns characterize the males and females of many bathypelagic fishes, and these patterns may be used in sex and species recognition. Female anglerfishes possess a bioluminescent lure that is species specific in appearance. Although it is used to lure prey, the bait is probably also used to attract males. Visual detection of other individuals much beyond 40 or 50 meters is not possible, however, because light does not travel far in water and other senses must be used. Scent trails are used by many deep-sea fishes. The females secrete a pheromone, and males usually have enlarged olfactory organs. Sensing the pheromone during searching movements, males swim upstream to an intimate encounter.

The life history of ceratioid anglerfishes dramatizes how selection adapts a vertebrate to its habitat. The adults typically spend their lives in aphotic regions below 1000 meters. Fertilized eggs, however, rise to the surface, where they hatch into larvae. The larvae remain mostly in the upper 30 meters where they grow, and later descend to the aphotic region. Descent is accompanied by metamorphic changes that differentiate females and males. During metamorphosis, young females descend to great depths, where they feed and grow slowly, reaching maturity after several years. Many species resemble a large mouth accompanied by a stomach, a description that certainly fits their appearance.

Female anglerfishes feed throughout their lives, whereas males feed only during the larval stage. Metamorphic changes in males prepare them for a different future, for their function is reproduction, literally by lifelong matrimony. The body may elongate and axial red muscles develop. The males cease eating, and the enlarged liver apparently provides energy for an extended period of swimming. The olfactory organs of males enlarge at metamorphosis and the eyes continue to grow. These changes suggest that adolescent males spend a brief, but active, free-swimming period concentrated on finding a female. The search must be precarious, for males must search vast, dark regions for a single female while running a gauntlet of other deep-sea predators. In the young adults there is an unbalanced sex ratio; often more than 30 males exist for every female. Apparently, at least 29 of those males cannot expect to locate a virgin female.

Having found a female, a male does not want to lose her, and he ensures a permanent union by attaching himself as a parasite to the female. When a male finds a female, he bites into her flesh and attaches himself firmly. Preparation for this encounter begins during metamorphosis when the male's teeth degenerate and strong tooth-like bones develop at the tips of the jaws. A male remains attached to the female for life, and in this parasitic state he grows and his testes mature. Monogamy prevails in this pairing, for females invariably have but one attached male. As humans we consider this life-style bizarre, and, indeed, it is unknown among other vertebrates. But it has been successful—some 200 ceratioid species exist. The lesson these unique fishes provide is that vertebrate life is a plastic venture capable of adapting to extreme conditions. In the ceratioids, two features stand out: efficient energy utilization and reproduction. Actually, adaptations in all vertebrates relate to these two goals, but they are not often painted in such bold relief.

Ectotherms in the Heat: Deserts

Deserts can be produced by various combinations of topography, air movements, and ocean currents, but whatever their cause, deserts have in common a scarcity of liquid water. That characteristic is at the root of many of the features of deserts that make them difficult places for vertebrates to live. The dry air characteristic of most deserts seldom contains

Figure 16–5 The desert tortoise, *Gopherus agassizi*. (a) An adult tortoise. (b) A tortoise entering its burrow. (Photographs by R. Bruce Bury, U.S. Fish & Wildlife Service.)

enough moisture to form clouds that would block solar radiation during the day or radiative cooling at night. As a result, the daily temperature excursion in deserts is large in comparison to that of more humid areas. Scarcity of water is reflected by sparse plant life and a correspondingly low primary productivity in desert communities. Food shortages may be chronic and exacerbated by seasonal shortages and unpredictable years of low production when the usual pattern of rainfall does not develop.

Not all deserts are hot; indeed, some are distinctly cold—most of Antarctica and the region of Canada around Hudson Bay and the Arctic Ocean are deserts. The low-latitude deserts north and south of the equator are hot deserts, however, and it is in these low-latitude deserts that vertebrates encounter the most difficult problems of desert life.

The scarcity of rain contributes to the low primary production of deserts and also means that sources of liquid water for drinking are usually unavailable to small animals that cannot travel long distances. These animals obtain water from the plants or animals they eat, but plants and insects have ion balances that differ from those of vertebrates. In particular, potassium is found in higher concentrations in plants and insects than it is in vertebrate tissues, and excreting the excess potassium can be difficult if water is too scarce to waste in the production of large quantities of urine.

The low metabolic rates of ectotherms alleviate some of the stress caused by scarcity of food and water, but many desert ectotherms must temporarily extend the limits within which they regulate body temperatures or body fluid concentrations, become inactive for large portions of the year, or adopt a combination of these responses. Tortoises, lizards, and anurans from deserts illustrate these phenomena.

Balancing Water and Energy Budgets The largest ectothermal vertebrates in the deserts of North America are tortoises. The Bolson tortoise (*Gopherus flavomarginatus*) of northern Mexico probably once reached a shell length of a meter, although predation by humans has apparently prevented any tortoise in recent times from living long enough to grow that large. The desert tortoise (*G. agassizi*) of the southwestern United States is smaller than the Bolson tortoise, but it is still an impressively large turtle (Figure 16–5). Adults can reach shell lengths approaching 50 centimeters and may weigh 5 kilograms or more. A study of the annual water, salt, and energy budgets of desert tortoises in Nevada shows the difficulties they face in that desert habitat (Nagy and Medica 1986).

Desert tortoises construct shallow burrows that they use as daily retreat sites during the summer and deeper burrows for hibernation in winter. The tortoises in the study area emerged from hibernation in spring, and aboveground activity extended through the summer until they began hibernation again in November. Doubly labeled water was used to measure the energy expenditure of free-ranging tortoises (Box 16–1). Figure 16–7 shows the annual cycle of time spent above ground and in burrows by the tortoises, and the annual cycles of energy, water, and

Box 16–1 Doubly Labeled Water

This technique is widely employed in studies of the energy consumption of wild vertebrates because it is the only method of measuring metabolism without restraining the animal in some way. The method measures carbon dioxide production, which in turn can be used to estimate oxygen consumption. "Doubly labeled" refers to water that carries isotopic forms of oxygen and hydrogen: The oxygen atom has been replaced with its stable isotope oxygen-18 (^{18}O), and one hydrogen atom (H) has been replaced by its radioactive form, tritium (^{3}H). (Doubly labeled water is prepared by mixing appropriate quantities of water containing the hydrogen isotope [^{3}HHO] with water containing the oxygen isotope [$H_2 \, ^{18}O$]. When this mixture is diluted by the body water of an animal the concentrations of the isotopes are so low that few individual water molecules contain both isotopes.) A measured amount of doubly labeled water is injected into an animal and allowed to equilibrate with the body water for several hours (Figure 16–6a), and then a small blood sample is withdrawn. The concentrations of tritium and oxygen-18 in the blood are measured, and the total volume of body water can be calculated from the dilution of the doubly labeled water that was injected.

After that first blood sample has been withdrawn, the animal is released and recaptured at intervals of several days or weeks. Each time the animal is recaptured a blood sample is taken and the concentrations of tritium and oxygen-18 are measured. The calculation of the amount of carbon dioxide produced by the animal is based on the difference in the rates of loss of tritium and oxygen-18. Tritium, which behaves chemically like hydrogen, is lost as water—that is, as ^{3}HHO—whereas oxygen-18 is lost both as water ($H_2 \, ^{18}O$) and as carbon dioxide ($C^{18}OO$). Thus, the decline in the concentration of tritium in the blood of the animal is a measure of the rate of water loss, and the decline in the concentration of oxygen-18 is a measure of the rates of loss of carbon dioxide and water (Figure 16–6b). The difference between decrease in concentration of tritium and oxygen-18, therefore, is the rate of loss of carbon dioxide, and this is proportional to the rate of oxygen consumption. A full description of the use of doubly labeled water for metabolic studies can be found in Nagy (1983).

Figure 16–6 Doubly labeled water. (a) The reaction of carbon dioxide and water to produce carbonic acid, and the subsequent reconversion of carbonic acid to water and carbon dioxide produce an equilibrium of ^{18}O between H_2O and CO_2 when water labeled with the isotope is injected into a vertebrate. (b) Differential washout of hydrogen and oxygen isotopes in the body water of an animal that has been injected with doubly labeled water. (Modified from K. A. Nagy, 1988, in *Stable Isotopes in Ecological Research*, edited by J. Ehleringer, P. Rundall, and K. Nagy. Springer, New York, NY.)

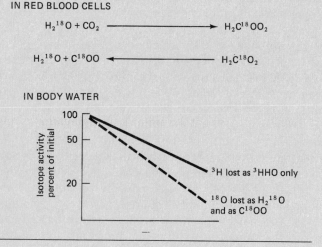

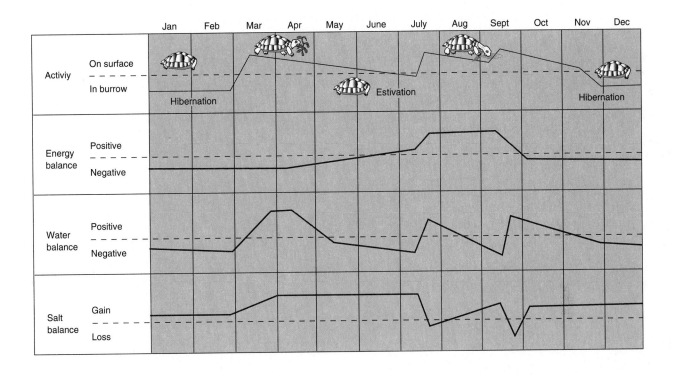

Figure 16–7 Annual cycle of desert tortoises. See text for details. (Based on K. A. Nagy and P. A. Medica, 1986, *Herpetologica* 42:73–92.)

salt balance. A positive balance means that the animal shows a net gain, whereas a negative balance represents a net loss. Positive energy and water balances indicate that conditions are good for the tortoises, but a positive salt balance means that ions are accumulating in the body fluids faster than they can be excreted. That situation indicates a relaxation of homeostasis and is probably stressful for the tortoises. The figure shows that the animals were often in negative balance for water or energy, and they accumulated salt during much of the year. Examination of the behavior and dietary habits of the tortoises through the year shows what was happening.

After they emerged from hibernation in the spring, the tortoises were active for about 3 hours every fourth day; the rest of the time they spent in their burrows. From March through May the tortoises were eating annual plants that had sprouted

after the winter rains. They obtained large amounts of water and potassium from this diet, and their water and salt balances were positive. The osmotic concentration of the tortoises' body fluids increased by 20 percent during the spring, indicating that they were osmotically stressed as a result of the high concentrations of potassium they were ingesting. Furthermore, the energy content of the plants was not great enough to balance the metabolic energy expenditure of the tortoises, and they were in negative energy balance. During this period the tortoises were using stored energy by metabolizing their body tissues.

As ambient temperatures increased from late May through early July, the tortoises shortened their daily activity periods to about 1 hour every sixth day. The rest of the time the tortoises spent estivating in shallow burrows. The annual plants died, and the tortoises shifted to eating grass and achieved positive energy balances. They stored this extra energy as new body tissue. The dry grass contained little water, however, and the tortoises were in negative water balance. The osmotic concentrations of their body fluids remained at the high levels they had reached earlier in the year.

In mid-July thunderstorms dropped rain on the study site, and most of the tortoises emerged from estivation. They drank water from natural basins, and some of the tortoises constructed basins by scratching shallow depressions in the ground that trapped rain water. The tortoises drank large quantities of water (nearly 20 percent of their body mass) and voided the contents of their urinary bladders. The osmotic concentrations of their body fluids and urine decreased as they moved into positive water balance and excreted the excess salts they had accumulated when water was scarce. The behavior of the tortoises changed after the rain: They fed every 2 or 3 days and often spent their periods of inactivity above ground instead of in their burrows.

August was dry, and the tortoises lost body water and accumulated salts as they fed on dry grass. They were in positive energy balance, however, and their body tissue mass increased. More thunderstorms in September allowed the tortoises to drink again and to excrete the excess salts they had been accumulating. Seedlings sprouted after the rain, and in late September the tortoises started to eat them.

In October and November the tortoises continued to feed on freshly sprouted green vegetation, but low temperatures reduced their activity and they were in slightly negative energy balance. Salts accumulated and the osmotic concentrations of the body fluids increased slightly. In November the tortoises entered hibernation. Hibernating tortoises had low metabolic rates and lost water and body tissue mass slowly. When they emerged from hibernation the following spring they weighed only a little less than they had in the fall. Over the entire year, the tortoises increased their body tissues by more than 25 percent, and balanced their water and salt budgets, but they did this by tolerating severe imbalances in their energy, water, and salt relations for periods that extended for several months at a time.

The ability to tolerate physiological imbalances is an important aspect of the ability of ectothermal vertebrates to occupy habitats where seasonal shortages of food or water occur. The chuckwalla (*Sauromalus obesus*) is an herbivorous iguanid lizard that lives in the rocky foothills of desert mountain ranges (Figure 16–8). The annual cycle of the chuckwallas, like that of the desert tortoises, is molded by the availability of water. The lizards face many of the same stresses that the tortoises encounter, but their responses are different: The lizards have nasal glands that allow them to excrete salt at high concentrations, and they do not drink rain water but instead depend on water they obtain from the plants they eat.

Figure 16–8 Chuckwalla, *Sauromalus obesus*. (Photograph by F. Harvey Pough.)

Two categories of water are available to an animal from the food it eats: free water and metabolic water. Free water corresponds to the water content of the food, that is, molecules of water (H_2O) that are absorbed across the wall of the intestine. Metabolic water is a by-product of the cellular reactions of metabolism. Protons are combined with oxygen during aerobic metabolism, yielding a molecule of water for every two protons. The amount of metabolic water produced can be substantial; more than a gram of water is released by metabolism of a gram of fat (Table 16–2). For animals like the chuckwalla that do not drink liquid water, free water and metabolic water are the only routes of water gain that can replace the water lost by evaporation and excretion.

Table 16–2 Quantity of water produced by metabolism of different substrates

Compound	Grams of Water/Gram of Compound
Carbohydrate	0.556
Fat	1.071
Protein	0.396 when urea is the end product
	0.499 when uric acid is the end product

Chuckwallas were studied at Black Mountain in the Mohave Desert of California (Nagy 1972, 1973). They spent the winter hibernating in rock crevices and emerged from hibernation in April. Individual lizards spent about 8 hours a day on the surface in April and early May (Figure 16–9). By the middle of May air temperatures were rising above 40°C and

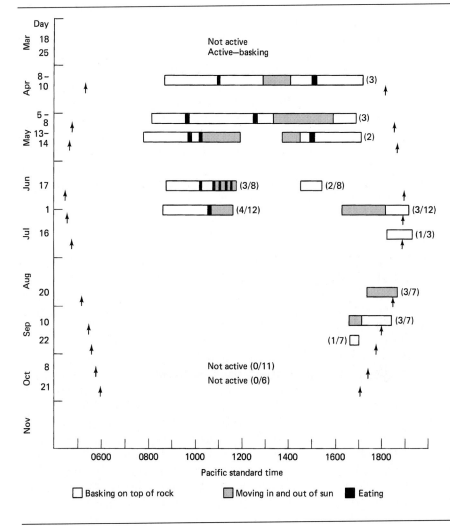

Figure 16–9 Daily behavior patterns in chuckwallas through their activity season. Arrows indicate sunrise and sunset. Numbers in parentheses for April and May indicate the number of animals whose behavior was recorded. Thereafter the fraction in parentheses indicates the number of lizards active and observed out of the number known to be present. (From K. A. Nagy, 1973, *Copeia* 1973:93–102.)

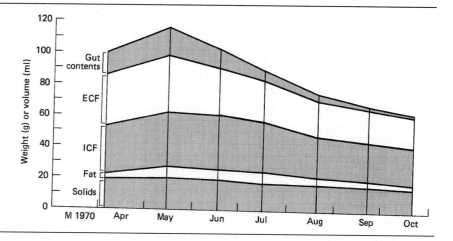

Figure 16–10 Seasonal changes in body composition of chuckwallas. The total body water is composed of the extracellular fluid (ECF; blood plasma, urine, and water in lymph sacs) and the intracellular fluid (ICF; the water inside cells). (From K. A. Nagy, 1972, *Journal of Comparative Physiology* 79:39–62.)

the chuckwallas retreated into rock crevices for about 2 hours during the hottest part of the day, emerging again in the afternoon. At this time of year annual plants that sprouted after the winter rains supplied both water and nourishment, and the chuckwallas gained weight rapidly. The average increase in body mass between April and mid-May was 18 percent (Figure 16–10). The water content of the chuckwallas increased faster than the total body mass, indicating that they were storing excess water.

By early June the annual plants had withered, and the chuckwallas were feeding on perennial plants that contained less water and more ions than the annual plants. Both the body masses and the water contents of the lizards declined. The activity of the lizards decreased in June and July: Individual lizards emerged in the morning or in the afternoon, but not at both times. In late June the chuckwallas reduced their feeding activity and in July they stopped eating altogether. They spent most of the day in the rock crevices, emerging only in the late afternoon to bask for an hour or so every second or third day. From late May through autumn the chuckwallas lost water and body mass steadily, and in October they weighed an average of 37 percent *less* than they had in April when they emerged from hibernation.

The water budget of a chuckwalla weighing 200 grams is shown in Table 16–3. In early May the annual plants it is eating contain more than 2.5 grams of free water per gram of dry plant material, and the lizard shows a positive water balance, gaining about 0.8 gram of water per day. By late May, when the plants have withered, their free water content has dropped to just under a gram of water per

gram of dry plant matter, and the chuckwalla is losing about 0.8 gram of water per day. The rate of water loss falls to 0.5 gram per day when the lizard stops eating.

Evaporation from the respiratory surfaces and from the skin accounts for about 61 percent of the total water loss of a chuckwalla. When the lizards stop eating, they also become inactive and spend most of the day in rock crevices. The body temperatures of inactive chuckwallas are lower than the temperatures of lizards on the surface. As a result of their low body temperatures, the inactive chuckwallas have lower rates of metabolism: They breathe more slowly and lose less water from their respiratory passages. Also, the humidity is higher in the rock crevices than it is on the surface of the desert, and this reduction in the humidity gradient between the animal and the air further reduces evaporation. Most of the remaining water loss by a chuckwalla occurs in the feces (31 percent) and urine (8 percent). When a lizard stops eating, it also stops producing feces and reduces the amount of urine it must excrete. The combination of these effects reduces the daily water loss of a chuckwalla by almost 90 percent.

The food plants were always hyperosmotic to the body fluids of the lizards and had high concentrations of potassium. Despite this dietary salt load, the osmotic concentrations of the body fluids of the chuckwallas did not show the variation seen in tortoises because the lizards' nasal salt glands were able to excrete ions at high concentrations. The concentration of potassium ions in the salt gland secretions was nearly ten times their concentration in urine. The formation of potassium salts of uric acid was the second major route of potassium excretion by the

Table 16-3 **Seasonal changes in the water balance of a 200-gram chuckwalla.**

	Early May	Late May	September
Food intake (g dry mass/day)	2.60	2.86	0.00
Water content of food (g/g dry mass)	2.53	0.96	—
Water gain (g/day)			
Free water	6.56	2.74	0.0
Metabolic water	0.68	0.68	0.20
Total water gain	7.24	3.41	0.20
Water loss (g/day)	6.41	4.26	0.52
Net water flux (g/day)	+0.81	−0.84	−0.32

Source: K. A. Nagy, 1972, *Journal of Comparative Physiology* 79:39–62.

lizards, and was nearly as important in the overall salt balance as nasal secretion. The chuckwallas would not have been able to balance their salt budgets without the two extrarenal routes of ion excretion, but with them they were able to maintain stable osmotic concentrations.

Both the chuckwallas and tortoises illustrate the interaction of behavior and physiology in responding to the stresses of their desert habitats. The tortoises lack salt-secreting glands and store the salt they ingest, tolerating increased body fluid concentrations until a rainstorm allows them to drink water and excrete the excess salt. Some of the tortoises constructed basins that collected rain water that they drank, whereas other tortoises took advantage of natural puddles. The chuckwallas were able to stabilize their body fluid concentrations by using their nasal glands to excrete excess salt, but they did not take advantage of rainfall to replenish their water stores. Instead, they became inactive, reducing their rates of water loss by almost 90 percent, and relying on energy stores and metabolic water production to see them through the period of drought.

Conditions for the chuckwallas were poor at the Black Mountain site during Nagy's study. Only 5 centimeters of rain had fallen during the preceding winter, and that low rainfall probably contributed to the early withering of the annual plants that forced the chuckwallas to cease activity early in the summer. Unpredictable rainfall is a characteristic of deserts, however, and the animals that live in them must be able to adjust to the consequences. Rainfall records from the weather station closest to Black Mountain showed that in 5 of the previous 10 years the annual total rainfall was about 5 centimeters.

Thus, the year of the study was not unusually harsh, and conditions are sometimes even worse—only 2 centimeters of rain fell during the winter after the study. However, conditions in the desert are sometimes good. Fifteen centimeters of rain fell in the winter of 1968, and vegetation remained green and lush all through the following summer and fall. Chuckwallas and tortoises live for decades, and their responses to the boom or bust conditions of their harsh environments must be viewed in the context of their long life spans. A temporary relaxation of the limits of homeostasis in bad years is an effective trade-off for survival that allows the animals to exploit the abundant resources of good years.

Desert Amphibians Permeable skins and high rates of water loss are characteristics that would seem to make amphibians unlikely inhabitants of deserts, but certain species are abundant in desert habitats. Most remarkably, these animals succeed in living in the desert *because* of their permeable skins, not despite them. Anurans are the most common desert amphibians, but tiger salamanders are found in the deserts of North America and several species of plethodontid salamanders occupy seasonally dry habitats in California.

The spadefoot toads are the most thoroughly studied desert anurans (Figure 16–11). They inhabit the desert regions of North America, including the edges of the Algodones Sand Dunes in southern California where the average annual precipitation is only 6 centimeters and in some years no rain falls at all. An analysis of the mechanisms that allow an amphibian to exist in a habitat like that must include consideration of both water loss and gain.

Figure 16–11 A desert spadefoot toad, *Scaphiopus multiplicatus.* (Photograph by David Dennis.)

The skin of desert amphibians is as permeable to water as that of species from moist regions. A desert anuran must control its water loss behaviorally by its choice of sheltered microhabitats free from solar radiation and wind movement. Different species of anurans utilize different microhabitats—a hollow in the bank of a desert wash, the burrow of a ground squirrel or kangaroo rat, or a burrow the anuran excavates for itself. All these places are cooler and wetter than exposed ground.

Desert anurans spend extended periods underground, emerging on the surface only when conditions are favorable. Spadefoot toads construct burrows about 60 centimeters deep, filling the shaft with dirt and leaving a small chamber at the bottom which they occupy. In southern Arizona the spadefoots construct these burrows in September, at the end of the summer rainy season, and remain in them until the rains resume the following July.

At the end of the rainy season when the spadefoots first bury themselves, the soil is relatively moist. The water tension created by the normal osmotic pressure of a spadefoot's body fluids establishes a gradient favoring movement of water from the soil into the toad. In this situation, a buried spadefoot can absorb water from the soil just as the roots of plants do. With a supply of water available, a spadefoot toad can afford to release urine to dispose of its nitrogenous wastes.

As time passes, the soil moisture content decreases and the soil moisture potential becomes more negative, until it equals the water potential of the spadefoot. At this point there is no longer a gradient allowing movement of water into the toad. When its source of new water is cut off, a spadefoot stops excreting urine and instead retains urea in its body, increasing the osmotic pressure of its body fluids. Osmotic concentrations as high as 600 milliOsmolal have been recorded in spadefoot toads emerging from burial at the end of the dry season. The low water potential produced by the high osmotic pressure of the spadefoot's body fluids may reduce the water gradient between the animal and the air in its underground chamber so that evaporative water loss is reduced. Sufficiently high internal osmotic pressures should create potentials that would allow spadefoot toads to absorb water from even very dry soil.

The ability to continue to draw water from soil enables a spadefoot toad to remain buried for 9 or 10 months without access to liquid water. In this situation its permeable skin is not a handicap to the spadefoot—it is an essential feature of the toad's biology. If the spadefoot had an impermeable skin, or if it formed an impermeable cocoon as some other amphibians do, water would not be able to move from the soil into the animal. Instead, the spadefoot would have to depend on the water contained in its body when it buried. Under those circumstances spadefoot toads would probably not be able to invade the desert because their initial water content would not see them through a 9-month dry season.

A different pattern of adaptation to arid conditions is seen in a few treefrogs. The African rhacophorid *Chiromantis xerampelina* and the South American hylid *Phyllomedusa sauvagei* lose water through the skin at a rate only one-tenth that of most frogs. *Phyllomedusa* has been shown to achieve this low rate of evaporative water loss by using its legs to spread the lipid-containing secretions of dermal glands over its body surface in a complex sequence of wiping movements. These two frogs are unusual also because they excrete nitrogenous wastes as salts of uric acid rather than as urea. This uricotelism provides still more water conservation.

Ectotherms in the Cold: Supercooling, Antifreeze, and Freeze Tolerance

Temperatures drop below freezing in the habitats of many ectothermal vertebrates on a seasonal basis, and some animals at high altitudes may experience freezing temperatures on a daily basis for a substantial part of the year. Ectotherms lack the metabolic capacities to maintain body temperatures above freezing under those conditions and instead they show one of two responses: freezing avoidance (by supercooling or the use of antifreeze compounds) or tolerance of freezing and thawing.

Cold Fish Body fluid concentrations of marine fishes are 300 to 400 milliOsmolal, whereas seawater has a concentration near 1000 milliOsmolal. The temperature at which water freezes is affected by its osmotic concentration: Pure water (0 milliOsmolal) freezes at 0°C, and increasing the osmotic concentration lowers the freezing point. The osmotic concentrations of the body fluids of marine fishes correspond to freezing points of –0.6 to 0.8°C, and the freezing point of seawater is –1.86°C. The temperature of Arctic and Antarctic seas falls to –1.8°C in winter, yet the fish swim in this water without freezing.

One of the early studies of freezing avoidance of fishes was conducted by P. F. Scholander and his colleagues in Hebron Fjord in Labrador (Scholander et al. 1957). In summer the temperature of the surface water at Hebron Fjord is above freezing, but the water at the bottom of the fjord is –1.73°C (Figure 16–12). In winter the temperature of the surface water falls to –1.73°C, like the bottom temperature. Several species of fishes live in the fjord, and some are bottom dwellers, whereas others live near the surface. These two zones present different problems to the

fishes: The temperature near the bottom of the fjord is always below freezing, but ice is not present because ice is lighter than water and remains at the surface. Surface-dwelling fishes live in water temperatures that rise well above freezing in the summer and drop below freezing in winter, and they are also in the presence of ice.

The body fluids of bottom-dwelling fish in Hebron Fjord have freezing points of –0.8°C year round. Because the body temperatures of these fishes are –1.73°C, the fish are supercooled. That is, the water in their bodies is in the liquid state despite the fact that it is below its freezing point. When water freezes, the water molecules become oriented in a crystal lattice. The process of crystallization is accelerated by nucleating agents that hold water molecules in the proper spatial orientation for freezing, and in the absence of nucleating agents pure water can remain liquid at –20°C. In the laboratory the fishes from the bottom of Hebron Fjord can be supercooled to –1.73°C without freezing, but if they are touched with a piece of ice, which serves as a nucleating agent, they freeze immediately. At the bottom of the fjord there is no ice, and the bottom-dwelling fishes exist year round in a supercooled state.

What about the fishes in the surface waters? They do encounter ice in winter when the water temperature is below the osmotically determined freezing point of their body fluids, and supercooling would not be effective in that situation. Instead, the surface-dwelling fishes synthesize antifreeze substances in winter that lower the freezing point of their body fluids to approximately the freezing point of the seawater in which they swim.

Antifreeze compounds are widely developed among vertebrates (and also among invertebrates and plants). Marine fishes have two categories of organic molecules that protect against freezing: glycoproteins with molecular weights of 2600 to 33,000, and polypeptides and small proteins, with molecular weights of 3300 to 13,000 (Hew et al. 1986, Davies et al. 1988). These compounds are extremely effective in preventing freezing. For example, the blood plasma of the Antarctic fish *Trematomus borchgrevinki* contains a glycoprotein that is several hundred times more effective than salt (sodium chloride) in lowering the freezing point. Apparently, the glycoprotein is absorbed on the surface of ice crystals and hinders their growth by preventing water molecules from assuming the proper orientation to join the ice crystal lattice.

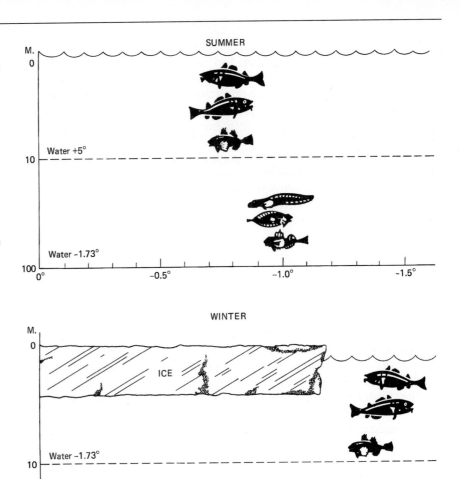

Figure 16–12 Water temperatures and distribution of fishes in Hebron Sound in summer and winter. The freezing point of the blood plasma of the fishes is indicated by the position of the symbol on the horizontal axis. Shallow-water fishes show a decrease of 0.7°C in the freezing point of their blood in winter, whereas deepwater fishes have the same freezing point all year. (From P. F. Scholander et al., 1957, *Journal of Cellular and Comparative Physiology* 79:39–62.)

Some terrestrial ectotherms rely on supercooling. Mountain-dwelling lizards such as Yarrow's spiny lizard (*Sceloporus jarrovi*), which lives at altitudes up to 3000 meters in western North America, are exposed to temperatures below freezing on cold nights, but sunny days permit thermoregulation and activity. These animals have osmotic concentrations of 300 milliOsmolal, which correspond to freezing points of –0.6°C, but they withstand substantially lower temperatures before they freeze. For example, spiny lizards supercooled to an average temperature of –5.5°C before they froze. At –3°C the

lizards had not frozen after 30 hours. The lizards spent the nights in rock crevices that were 5 to 6°C warmer than the air temperature, and the combination of this protection and their ability to supercool was usually sufficient to allow the lizards to survive. However, a few individuals at the highest altitudes were found frozen in their rock crevices during most winters.

Frozen Frogs Terrestrial amphibians that spend the winter in hibernation show at least two categories of responses to low temperatures. One group, which

Figure 16–13 Wood frog (*Rana sylvatica*): (a) at normal temperature; (b) frozen. (Photographs courtesy of J. and K. Storey.)

includes salamanders, toads, and aquatic frogs, buries deeply in the soil or hibernates in the mud at the bottom of ponds. These animals apparently are not exposed to temperatures below the freezing point of their body fluids, and as far as we know, they have no antifreeze substances and no capacity to tolerate freezing. However, other amphibians apparently hibernate close to the soil surface, and these animals are exposed to temperatures below their freezing points. Unlike fishes and lizards, these amphibians freeze at low temperatures, but they are not killed by freezing (Figure 16–13). These species can remain frozen at –3°C for several weeks, and they tolerate repeated bouts of freezing and thawing without damage. However, temperatures below –10°C are lethal.

Tolerance of freezing refers to the formation of ice crystals in the extracellular body fluids; freezing of the fluids inside the cells is apparently lethal. Thus, freeze tolerance involves mechanisms that control the distribution of ice, water, and solutes in the bodies of animals (Storey 1986). The ice content of frozen frogs is usually in the range 34 to 48 percent. Freezing of more than 65 percent of the body water appears to cause irreversible damage, probably because too much water has been removed from the cells.

Freeze-tolerant frogs accumulate low-molecular-weight substances in the cells that prevent intracellular ice formation. Wood frogs, spring peepers, and chorus frogs use glucose as a cryoprotectant, whereas gray treefrogs use glycerol. Glycogen in the liver appears to be the source of the glucose and glycerol. The accumulation of these substances is apparently stimulated by freezing, and is initiated within minutes of the formation of ice crystals. This mechanism of triggering the synthesis of cryoprotectant substances has not been observed for any other vertebrates or for insects.

Frozen frogs are, of course, motionless. Breathing stops, the heartbeat is exceedingly slow and irregular or may cease entirely, and blood does not circulate through frozen tissues. Nonetheless, the cells are not frozen and they have a low level of metabolic activity that is maintained by anaerobic metabolism. The glycogen content of frozen muscle and kidney cells decreases, and concentrations of lactic acid and alanine (two end products of anaerobic metabolism) increase.

The ecological significance of freeze tolerance in some species of amphibians is unclear. The four species of frogs so far identified as being freeze tolerant all breed relatively early in the spring, and shallow hibernation may be associated with early emergence in the spring. Being among the first individuals to arrive at the breeding pond may increase the chances for a male of obtaining a mate, and it gives larvae the longest possible time for development and metamorphosis, but it also entails risks. Frequently, frogs and salamanders move across snowbanks to reach the breeding ponds, and they enter ponds that are still partly covered with ice. A cold snap can lead to the entire surface of the pond freezing again, trapping some animals under the ice and others in shallow retreats under logs and rocks around the pond. Freeze tolerance may be important to these animals even after their winter hibernation is over.

The Role of Ectothermal Tetrapods in Terrestrial Ecosystems

In the previous sections we have illustrated some of the ways that an ectothermal approach to life allows animals to exploit habitats or adaptive zones that would be difficult or impossible for an endotherm to occupy. The key to this ability of ectotherms is the low energy requirement of ectothermy; an ectotherm simply requires less energy on an hourly, daily, or annual basis than does an endotherm of the same body size.

However, the amount of energy required by an ectotherm and an endotherm is not the only difference between them; equally significant is what they do with that energy once they have it. Endotherms expend more than 90 percent of the energy they take in to produce heat to maintain their high body temperatures. Less than 10 percent, often as little as 1 percent, of the energy a bird or mammal assimilates is available for net conversion (that is, increasing the species' biomass by growth of an individual or production of young). Ectotherms do not rely on meta-

bolic heat. The solar energy they use to warm their bodies is free in the sense that it is not drawn from their food. Thus, most of the energy they ingest is converted into the biomass of their species. Values of net conversion for amphibians and reptiles are between 30 and 90 percent (Table 16–4). As a result of that difference in how energy is used, a given amount of chemical energy invested in an ectotherm produces a much larger biomass return than it would have from an endotherm. A study of salamanders in the Hubbard Brook Experimental Forest in New Hampshire showed that, although their energy consumption was only 20 percent that of the birds or small mammals in the watershed, their conversion efficiency was so great that the annual increment of salamander biomass was equal to that of birds or small mammals. Similar comparisons can be made among lizards and rodents in deserts. (For details, see Pough [1980, 1983].)

Body size is another major difference between ectotherms and endotherms that directly affects their roles in terrestrial ecosystems. Ectotherms are smaller than endotherms, partly because the energetic cost of endothermy is very high at small body sizes. As body mass decreases the mass specific cost of living

Table 16–4 **Efficiency of biomass conversion by ectotherms and endotherms. These are net conversion efficiencies calculated as (energy converted/energy assimulated) × 100.**

Ectotherms		Endotherms	
Species	Efficiency	Species	Efficiency
Red-backed salamander *Plethodon cinereus*	48	Kangaroo rat *Dipodomys merriami*	0.8
Mountain salamander *Desmognathus ochrophaeus*	76–98	Field mouse *Peromyscus polionotus*	1.8
Panamanian anole *Anolis limifrons*	23–28	Meadow vole *Microtus pennsylvanicus*	3.0
Side-blotched lizard *Uta stansburiana*	18–25	Red squirrel *Tamiasciurius hudsonicus*	1.3
Hognose snake *Heterodon contortrix*	81	Least weasel *Mustela rixosa*	2.3
Python *Python curtus*	6–33	Savanna sparrow *Passericulus sandwichensis*	1.1
Adder *Vipera berus*	49	Marsh wren *Telmatodytes palustris*	0.5
Average of 12 species	50	Average of 19 species	1.4

Source: F. H. Pough, 1980, *The American Naturalist* 115:92–112.

(energy per gram) for an endotherm increases rapidly, becoming nearly infinite at very small body sizes (Figure 16–14). This is a finite world, and infinite energy requirements are just not feasible. Thus, energy requirements, among other factors, apparently set a lower limit to the body size that is possible for an endotherm. The mass specific energy requirements of ectotherms also increase at small body sizes, but because the energy requirements of ectotherms are about one-tenth those of endotherms of the same body size, an ectotherm can be about an order of magnitude smaller than an endotherm. A mouse-size mammal weighs about 20 grams, and few adult birds and mammals have body masses less than 10 grams. The very smallest species of birds and mammals weigh about 3 grams, but many ectotherms are only one-tenth that size (0.3 gram). Amphibians are especially small—20 percent of the species of salamanders and 17 percent of the species of anurans have adult body masses less than 1 gram, and 65 per-

cent of salamanders and 50 percent of anurans are smaller than 5 grams. Squamates are generally larger than amphibians, but 8 percent of the species of lizards and 2 percent of snakes weigh less than 1 gram. The largest amphibians and squamates weigh substantially less than 100 kilograms, whereas more than 10 percent of the species of extant mammals weigh 100 kilograms or more (Figure 16–15).

Small amphibians and squamates occupy key positions in terms of energy flow through an ecosystem: Because they are so small, they can capture tiny insects and arachnids that are too small to be eaten by birds and mammals. Because they are ectotherms, they are efficient at converting the energy in the food they eat into their own tissues. As a result, the small ectothermal vertebrates in terrestrial ecosystems can be viewed as repackaging energy into a form that avian and mammalian predators can exploit. In other words, if you put yourself in the position of a shrew or a bird searching for a meal in the Hubbard

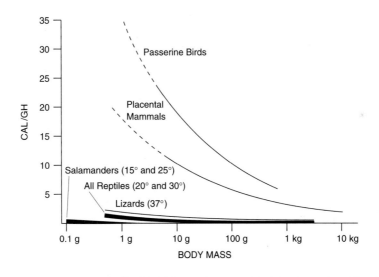

Figure 16–14 Resting metabolic rate as a function of body size for terrestrial vertebrates. (This semilogarithmic presentation emphasizes the dramatic increase in mass-specific metabolic rate at small body sizes—compare it to the log–log graph in Figure 4–13.) Metabolic rates for salamanders are shown at 15 and 25°C as the lower and upper limits of the darkened area, and data for all reptiles combined are shown at 20 and 39°C. The curve for anurans falls within the all-reptiles area, and the relationship for nonpasserine birds is similar to that for placental mammals. Dotted portions of the lines for birds and mammals show hypothetical extensions into body sizes below the minimum for adults of most species of birds and mammals. (From F. H. Pough 1980 *American Naturalist* 115:92–112. Reprinted by permission of the University of Chicago Press. © 1980 by The University of Chicago.)

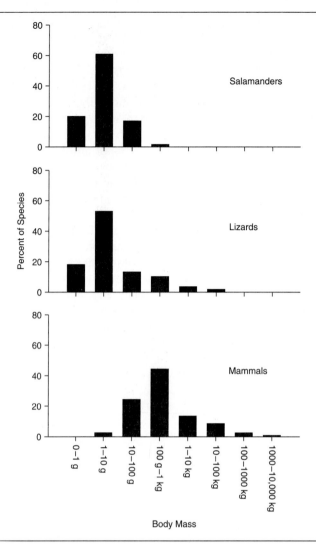

Figure 16–15 Adult body masses of amphibians and reptiles. (Based on data in F. H. Pough, 1983, pages 141–188 in *Behavioral Energetics,* edited by W. P. Aspey and S. I. Lustick, Ohio State University Press, Columbus, OH; and R. M. May, 1988, *Science* 241:1441–1449.)

Brook Forest, you would be well advised to eat salamanders. In this context, frogs, salamanders, lizards, and snakes occupy a position in terrestrial ecosystems that is important both quantitatively (in the sense that they constitute a substantial energy resource) and qualitatively (in that ectotherms exploit food resources that are not available to endotherms).

In a very real sense small ectotherms can be thought of as living in a different world from that of endotherms. As we saw in the case of the three species of *Ameiva* lizards in Costa Rica (Chapter 15), interactions with the physical world may be more important in shaping the ecology and behavior of small ectotherms than are biological interactions such as competition. In some cases these small vertebrates may have their primary predatory and competitive interactions with insects and arachnids rather than with other vertebrates. For example, orb-web spiders and *Anolis* lizards on some Caribbean islands are linked by both predation (adult lizards eat spiders and spiders may eat hatchling lizards) and competition (lizards and spiders eat many of the same kinds of insects). When lizards were removed from experimental plots the abundance of insect prey increased and the spiders consumed more prey and survived longer (Schoener and Spiller 1987).

Thus, ectothermy and endothermy represent fundamentally different approaches to the life of a terrestrial vertebrate, each with its own advantages and disadvantages. An appreciation of ectotherms and endotherms requires understanding the functional consequences of the differences between them. Ectothermy is an ancestral character of vertebrates, but it is a very effective way of life in modern ecosystems.

Summary

Ectotherms do not use chemical energy from the food they eat to maintain high body temperatures. The results of that ancestral vertebrate characteristic are far-reaching for extant ectothermal vertebrates, and ectotherms and endotherms represent quite different approaches to vertebrate life.

Because of their low energy requirements, ectotherms can colonize habitats in which energy is in short supply, either seasonally (deserts) or chronically (the deep sea). Ectotherms are able to extend some of their limits of homeostasis to tolerate high or low body temperatures, and high or low body water contents when doing so allows them to survive in difficult conditions.

When food is available, ectotherms are efficient at converting the energy it contains into their own body tissues for growth or reproduction. Net conversion efficiencies of ectotherms average 50 percent of the energy assimilated compared to an average of 1.4 percent for endotherms.

Ectotherms can be smaller than endotherms because their mass-specific energy requirements are low, and many ectotherms weigh less than a gram, whereas most endotherms weigh more than 10 grams. As a result of this difference in body size many small ectotherms, such as salamanders, frogs, and lizards, eat prey that is too small to be consumed by endotherms. The efficiency of energy conversion by ectotherms and their small body sizes lead to a distinctive role in modern ecosystems, one that is in many respects quite different from that of terrestrial ectotherms. Understanding these differences is an important part of understanding the organismal biology of terrestrial ectothermal vertebrates.

References

Davies, P. L., C. L. Hew, and G. L. Fletcher. 1988. Fish antifreeze proteins: physiology and evolutionary biology. *Canadian Journal of Zoology* 66:2611–2617.

Hew, C. L., G. K. Scott, and P. L. Davies. 1986. Molecular biology of antifreeze. Pages 117–123 in *Living in the Cold: Physiological and Biochemical Adaptation,* edited by H. C. Heller, X. J. Musacchia, and L. C. H. Wang. Elsevier, New York, NY.

Nagy, K. A. 1972. Water and electrolyte budgets of a free-living desert lizard, *Sauromalus obesus. Journal of Comparative Physiology* 79:39–62.

Nagy, K. A. 1973. Behavior, diet and reproduction in a desert lizard, *Sauromalus obesus. Copeia* 1973:93–102.

Nagy, K. A. 1983. The doubly labeled water (^{3}HH^{18}O) method: a guide to its use. *UCLA Publication 12-1417.* University of California, Los Angeles, CA.

Nagy, K. A., and P. A. Medica. 1986. Physiological ecology of desert tortoises in southern Nevada. *Herpetologica* 42:73–92.

Pough, F. H. 1980. The advantages of ectothermy for tetrapods. *The American Naturalist* 115:92–112.

Pough, F. H. 1983. Amphibians and reptiles as low-energy systems. Pages 141–188 in *Behavioral Energetics: Vertebrate Costs of Survival,* edited by W. P. Aspey and S. I. Lustick. Ohio State University Press, Columbus, OH.

Schoener, T. W., and D. A. Spiller. 1987. Effect of lizards on spider populations: manipulative reconstruction of a natural experiment. *Science* 236:949–952.

Scholander, P. F., L. Van Dam, J. W. Kanwisher, H. T. Hammel, and M. S. Gordon. 1957. Supercooling and osmoregulation in Arctic fish. *Journal of Cellular and Comparative Physiology* 49:5–24.

Storey, K. B. 1986. Freeze tolerance in vertebrates: biochemical adaptation of terrestrially hibernating frogs. Pages 131–138 in *Living in the Cold: Physiological and Biochemical Adaptations,* edited by H. C. Heller, X. J. Musacchia, and L. C. H. Wang. Elsevier, New York, NY.

PART IV

TERRESTRIAL ENDOTHERMS: BIRDS AND MAMMALS

Birds and mammals are the vertebrates with which people are most familiar, partly because many species are large and diurnal, and partly because birds and mammals have colonized nearly every terrestrial habitat on Earth. The success of birds and mammals in many habitats is related to their endothermy, which allows them to be active at night and in cold weather. Those are conditions in which terrestrial ectotherms find it difficult or impossible to thermoregulate and, consequently, are inactive.

Flight dominates the biology of birds: Most of the morphological features of birds are directly or indirectly related to the requirements of flight, and many of the distinctive aspects of their behavior and ecology stem from the mobility that flight provides. Migration, for example, is a particularly avian characteristic because, of all the terrestrial vertebrates, birds are best able to move long distances.

No one feature of mammals characterizes the group and dominates its biology as flight does for birds, but sociality comes close. Many features of the biology of mammals are related to their interactions with other individuals of their species, ranging from the period of dependence of young on their mother to life-long alliances between individuals that affect their social status and reproductive success in a group.

Humans differ from other vertebrates in the extent to which they have come to dominate all of the habitats of Earth, and in their effects, direct and indirect, on other vertebrates. In this portion of the book we explore the evolution of birds and mammals, the adaptive zones opened to them by their distinctive characteristics, and the origin and radiation of humans as an example of vertebrate evolution.

CHARACTERISTICS OF BIRDS: SPECIALIZATIONS FOR FLIGHT

17

The linear progression of a book does not lend itself to following the simultaneous evolution of several different phylogenetic lineages. At this point we must turn our attention back to the conditions in the middle of the Mesozoic that were described in Chapter 14 and consider the diversification of another group of diapsids, the birds, from their origin in or before the Jurassic. *Archaeopteryx* is the oldest fossil of a bird that is known, but by the time it lived in the Jurassic it was probably an archaic relic that existed contemporaneously with more derived birds. The debate about the role of flight in the origin of birds was described in Chapter 13. Despite our uncertainty about the route by which birds took to the air, the demands of flight have clearly shaped many aspects of the morphology of birds. In this chapter we consider the relationships between the body forms of birds and the physical and biological requirements of flight.

Birds as Flying Machines

In some respects birds are variable: Beaks and feet are specialized for different modes of feeding and locomotion, the morphology of the intestinal tract is related to dietary habits, and wing shape reflects flight characteristics. Despite that variation, however, the morphology of birds is more uniform than that of mammals, and much of this uniformity is a result of the specialization of birds for flight.

Consider body size as an example: Flight imposes a maximum body size on birds. The muscle power required to take off increases by a factor of 2.25 for each doubling of body mass. That is, if species B weighs twice as much as species A, it will require 2.25 times as much power to fly at its minimum speed. If the proportion of the total body mass allocated to flight muscles is constant, the muscles of a large bird must work harder than the muscles of a small bird. In fact, the situation is still more complicated because the power output is a function of both muscular force and wing beat frequency. Large birds have lower

wing beat frequencies than small birds for mechanical and aerodynamic reasons. As a result, if species B weighs twice as much as species A, it will develop only 1.59 times as much power from its flight muscles, although it needs 2.25 times as much power to fly. Therefore, large birds require longer takeoff runs than small birds, and ultimately a body mass is reached at which any further increase in size would move a bird into a realm in which its flight muscles were not able to provide enough power to take off.

Calculations of this maximum size from aerodynamic principles suggest that it lies near 12 kilograms and that estimate corresponds reasonably well with the observed body masses of birds. The mute swan weighs about 12 kilograms, and the trumpeter swan is about 17 kilograms. The largest flying bird known was a giant condor that had a wingspan estimated to be 7 meters and a possible body mass of 20 kilograms. Large pterosaurs also had body masses estimated to be in the region of 20 kilograms.

Flightless birds are spared the mechanical constraints associated with producing power for flight,

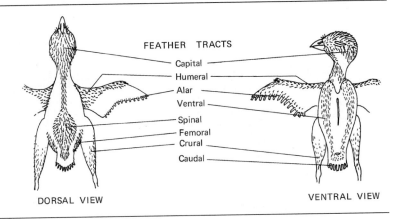

Figure 17–1 Feather tracts of a typical songbird.

FEATHER TRACTS

Capital
Humeral
Alar
Ventral
Spinal
Femoral
Crural
Caudal

DORSAL VIEW

VENTRAL VIEW

but they still do not approach the body sizes of mammals. The largest extant bird is the flightless ostrich, which weighs about 150 kilograms, and the largest bird known, one of the extinct elephantbirds, weighed an estimated 450 kilograms. In contrast, the largest mammal, the blue whale, weighs more than 135,000 kilograms. If one restricts the comparison to quadrupedal, terrestrial mammals, the elephant weighs some 5000 kilograms.

The structural uniformity of birds is seen even more clearly if their body shapes are compared to those of other diapsids such as the dinosaurs. There are no quadrupedal birds, for example, nor any with horns or bony armor. Even those species of birds that have become secondarily flightless retain many ancestral characters and the constraints that are associated with them. In this chapter we consider the body form and function of birds, especially in relation to the requirements of flight.

Feathers and Flight

Feathers develop from pits or follicles in the skin, generally arranged in tracts or **pterylae,** which are separated by patches of unfeathered skin, the **apteria** (Figure 17–1). Some species, such as ratites, pen-

guins, and mousebirds, lack pterylae, and the feathers are uniformly distributed over the skin.

For all their structural complexity, feathers are remarkably simple and uniform in chemical composition. More than 90 percent of a feather consists of beta-keratin, a protein related to the keratin of scales and to the hair and horn of mammals. About 1 percent of a feather consists of lipids, about 8 percent is water, and the remaining fraction consists of small amounts of other proteins and pigments, such as melanin. The colors of feathers are produced by structural characters and pigments. (See the color insert.)

Basic Types of Feathers

Ornithologists usually distinguish five types of feathers: (1) contour feathers, including typical body feathers and the flight feathers (remiges and rectrices); (2) semiplumes; (3) down feathers of several sorts; (4) bristles; and (5) filoplumes. **Contour feathers** (Figure 17–2) include a short, tubular base, the **calamus,** which remains firmly implanted within the follicle until molt occurs. Distal to the calamus is a long, tapered **rachis,** which bears closely spaced side branches called **barbs,** the lowermost of which externally mark the division between calamus and rachis. The barbs on either side of the rachis constitute a sur-

Figure 17–2 Typical contour feather (a wing quill) showing its main structural features. The inset and electron micrograph show details of the interlocking mechanism of the proximal and distal barbules. (From A. M. Lucas and P. R. Stettenheim, 1972, *Avian Anatomy and Integument,* Agriculture Handbook 362, United States Department of Agriculture, Washington, DC; and H. E. Evans, 1982, in *Diseases of Cage and Aviary Birds,* 2d edition, edited by M. L. Petrak, Lea & Febiger, Philadelphia, PA. Photograph by Alan Pooley.)

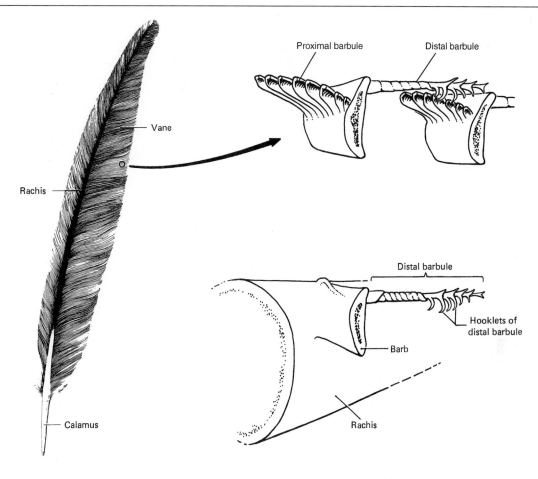

Vane

Rachis

Calamus

Proximal barbule   Distal barbule

Distal barbule

Hooklets of
distal barbule

Barb

Rachis

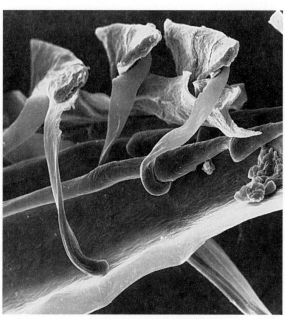

Figure 17–3 A red-tail hawk, showing the extreme development of slotting in the outer primaries. (Photograph © Fred Tilly, The National Audubon Society Collection.)

face called a **vane.** Vanes may be symmetrical or asymmetrical. The proximal portions of the vanes of a feather have a downy or plumulaceous texture, being soft, loose, and fluffy. This gives the plumage of a bird its excellent properties of thermal insulation. The more distal portions of the vanes have a pennaceous or sheet-like texture, firm, compact, and closely knit. This exposed part provides an airfoil, protects the downy undercoat, sheds water, reflects or absorbs solar radiation, and may have a role in visual or auditory communication. The barbules are structures that maintain the pennaceous character of the feather vanes. They are arranged in such a way that any physical disruption to the vane is easily cor-

rected by the bird's preening behavior, in which the bird realigns the barbules by drawing its slightly separated bill over them.

The **remiges** (wing feathers, singular **remex**) and **rectrices** (tail feathers, singular **rectrix**) are large, stiff, mostly pennaceous contour feathers that are modified for flight. For example, the distal portions of the outer primaries of many species of birds are abruptly tapered or notched, so that when the wings are spread the tips of these primaries are separated by conspicuous gaps or slots (Figure 17–3). This condition reduces the drag on the wing and, in association with the marked asymmetry of the outer and inner vanes, allows the feather tips to twist as the

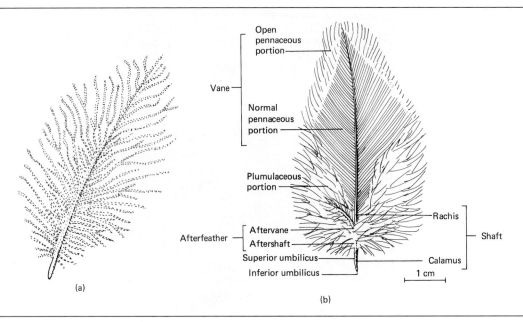

Figure 17–4 Comparison of a semiplume (a) and body contour feather (b).

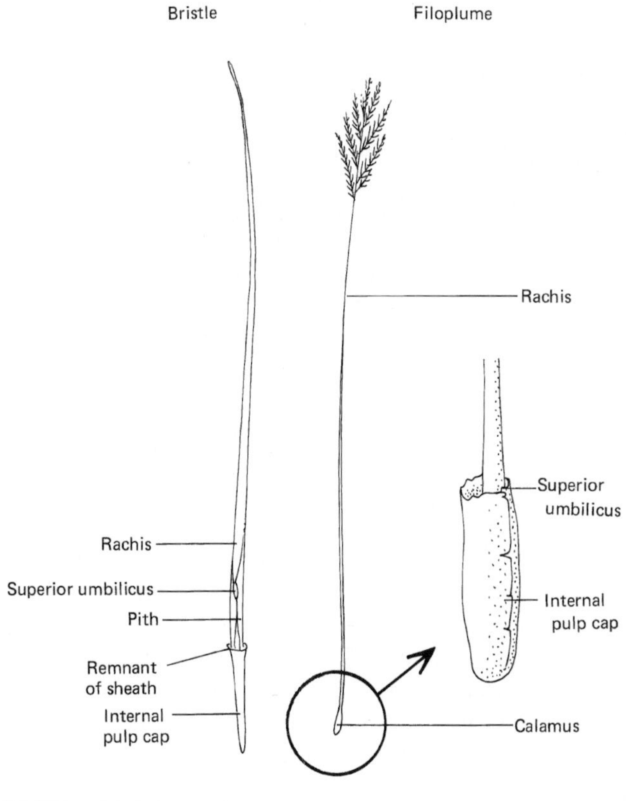

Figure 17–5 Comparison of a bristle (left) and a filoplume (right).

Bristle

Filoplume

Rachis

Superior umbilicus

Internal pulp cap

Calamus

Rachis

Superior umbilicus

Pith

Remnant of sheath

Internal pulp cap

wings are flapped and to act somewhat as individual propeller blades (see Figure 17–6).

Semiplumes are feathers intermediate in structure between contour feathers and down feathers. They combine a large rachis with entirely plumulaceous vanes and can be distinguished from down feathers by the fact that the rachis is longer than the longest barb (Figure 17–4). Semiplumes are mostly hidden beneath the contour feathers. They provide thermal insulation and help to fill out the contour of a bird's body.

Down feathers of various types are entirely plumulaceous feathers in which the rachis is shorter than the longest barb or entirely absent. Down feathers provide insulation for adult birds of all species. In addition, natal down, which is structurally simpler than adult down, provides an insulative covering on many birds at hatching or shortly thereafter. Natal downs usually precede the development of the first contour feathers, and down feathers are associated with apteria (the spaces between the contour feather tracts). Definitive downs are those that develop with the full body plumage. Uropygial gland downs are associated with the large sebaceous gland found at the base of the tail in most birds. The papilla of the gland usually bears a tuft of modified, brush-like down feathers that aid in transferring the oily secretion from the gland to the bill to provide waterproof dressing to the plumage.

Powder down feathers, which are difficult to classify by structural type, produce an extremely fine, white powder composed of granules of keratin. The powder, which is shed into the general plumage, is nonwettable and is therefore assumed to provide another kind of waterproof dressing for the contour feathers. All birds have powder down, but it is best developed in herons.

Bristles are specialized feathers with a stiff rachis and barbs only on the proximal portion or none at all (Figure 17–5). Bristles occur most commonly around the base of the bill, around the eyes, as eyelashes, and on the head or even on the toes of some birds. The distal rachis of most bristles is colored dark brown or black by melanin granules. The melanin not only colors the bristles but also adds to their strength, resistance to wear, and resistance to photochemical damage. Bristles and structurally intermediate feathers called semibristles screen out foreign particles from the nostrils and eyes of many birds and act as tactile sense organs and possibly as aids in the aerial capture of flying insects, as for example, the long bristles at the edges of the jaws in nightjars and flycatchers.

Filoplumes are fine, hair-like feathers with a few short barbs or barbules at the tip (Figure 17–5). In some birds, such as cormorants and bulbuls, the filoplumes grow out over the contour feathers and contribute to the external appearance of the plumage, but usually they are not exposed. Filoplumes are sensory structures that aid in the operation of other feathers. Filoplumes have numerous free nerve endings in their follicle walls, and these nerves connect to pressure and vibration receptors around the follicles. Apparently, the filoplumes transmit information about the position and movement of the contour feathers via these receptors. This sensory system probably plays a role in keeping the contour feathers in place and adjusting them properly for flight, insulation, bathing, or display.

Aerodynamics of the Avian Wing Compared to Fixed Airfoils

Unlike the fixed wings of an airplane, the wings of a bird function both as an airfoil (lifting surface) and as a propeller for forward motion. The avian wing is admirably suited for these functions, consisting of a light, flexible airfoil (Figure 17–6). The primaries, inserted on the hand bones, do most of the propelling when a bird flaps its wings, and the secondaries along the arm provide lift. Removal of flight feathers from the wings of doves and pigeons shows that when only a few of the primaries are pulled out the bird's ability to fly is greatly altered, but a bird can still fly when as much as 55 percent of the total area of the secondaries has been removed.

A bird also has the ability to alter the area and shape of its wings and their positions with respect to the body. These changes in area and shape cause corresponding changes in velocity and lift that allow a bird to maneuver, change direction, land, and take off. Moreover, a bird's wing is not a solid structure like a conventional airfoil such as an airplane wing,

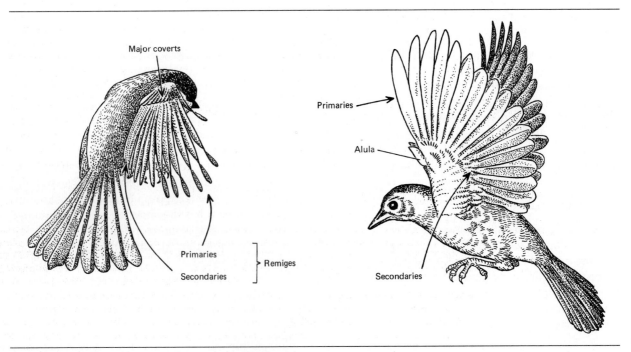

Figure 17–6 Drawings from high-speed photographs show the twisting and opening of the primaries during flapping flight. (From A. C. Thompson, 1964, *A New Dictionary of Birds*, McGraw-Hill, New York, NY.)

but allows some air to flow through and between the feathers.

Obviously, the aerodynamic properties of a bird's wing in flight—even in nonflapping flight—are vastly more complex than those of a fixed wing on an airplane or glider. Nevertheless, it is instructive to consider a bird's wing in terms of the basic performance of a fixed airfoil. (See Norberg [1985] and Rayner [1988] for reviews.) Although a bird's wing actually moves forward through the air, it is easier to think of the wing as stationary with the air flowing past. The flow of air produces a force, which is usually called the **reaction.** It can be resolved into two components: the **lift,** which is a vertical force equal to or greater than the weight of the bird, and the **drag,** which is a backward force opposed to the bird's forward motion and to the movement of its wings through the air.

When the leading end of a symmetrically streamlined body cleaves the air, it thrusts the air equally upward and downward, reducing the air pressure equally on the dorsal and ventral surfaces. No lift results from such a condition. There are two ways to modify this system to generate lift. One is to increase the **angle of attack** of the airfoil, and the other is to modify its surface configuration.

When the contour of the dorsal surface of the wing is convex and the ventral surface is concave (a **cambered airfoil**) the air pressure against the two surfaces is unequal because the air has to move farther and faster over the dorsal convex surface relative to the ventral concave surface (Figure 17–7). The result is reduced pressure over the wing, or lift. When the lift equals or exceeds the bird's body weight, the bird becomes airborne. The camber of the wing varies in birds with different flight characteristics and also changes along the length of the wing. Camber is greatest close to the body and decreases toward the wingtip. This change in camber is one of the reasons why the proximal part of the wing generates greater lift than the distal part.

If the leading edge of the wing is tilted up so that the angle of attack is increased, the result is increased lift up to an angle of about 15 degrees, the **stalling angle.** This lift results more from a decrease in pressure over the dorsal surface than from an increase in pressure below the airfoil. If the smooth flow of air over the wing becomes disrupted, the airflow begins to separate from the wing because of the increased air turbulence over the wing. The wing is then stalled. Stalling can be prevented or delayed by the use of slots or auxiliary airfoils on the leading edge of the main wing. The slots help to restore a smooth flow of air over the wing at high angles of attack and at slow speeds. The bird's **alula** has this effect, particularly during landing or takeoff (Figure 17–7). Also, the primaries act as a series of independent, overlapping airfoils, each tending to smooth out the flow of air over the one behind.

Another characteristic of an airfoil has to do with wingtip vortexes—eddies of air resulting from outward flow of air from under the wing and inward flow from over it. This is **induced drag.** One way to reduce the effect of these wingtip eddies and their drag is to lengthen the wing so that the tip vortex disturbances are widely separated and there is proportionately more wing area where the air can flow smoothly. Another solution is to taper the wing, reducing its area at the wingtip where induced drag is greatest. The ratio of length to width is called the **aspect ratio.** Long, narrow wings have high aspect ratios and high lift-to-drag (L/D) ratios. High-

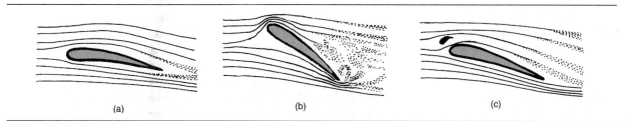

(a)　　　　　　　　　(b)　　　　　　　　　(c)

Figure 17–7 Diagram of airflow around a cambered airfoil. At a low angle of attack (a) the air streams smoothly over the upper surface of the wing and creates lift. When the angle of attack becomes steep (b), air passing over the wing becomes turbulent, decreasing lift enough to produce a stall. A wing slot (c) helps to prevent turbulence by directing a flow of rapidly moving air close to the upper surface of the wing.

performance sailplanes and albatrosses, for example, have aspect ratios of 18:1 and L/D ratios in the range of about 40:1.

Wing loading is another important considera-tion. This is the mass of the bird divided by the wing area. The lighter the loading, the less is the power needed to sustain flight. Small birds usually have lighter wing loading than do large birds, but wing loading is also related to specializations for powered versus soaring flight. The comparisons in Table 17–1 illustrate both of these trends. Small species such as hummingbirds, barn swallows, and mourning doves have lighter wing loading than large species such as the peregrine, golden eagle, and mute swan; yet the 3-gram hummingbird, a powerful flier, has a heavier wing loading than the more buoyant, sometimes soaring barn swallow, which is more than five times heavier. Similarly, the rapid-stroking peregrine has a heavier wing loading than the larger, often soaring golden eagle.

Flapping Flight

Flapping flight is remarkable for its automatic, unlearned performance. A young bird on its maiden flight uses a form of locomotion so complex that it defies precise analysis in physical and aerodynamic terms. The nestlings of some species of birds devel-op in confined spaces such as burrows in the ground or cavities in tree trunks in which it is impos-sible for them to spread their wings and practice flapping before they leave the nest. Despite this seeming handicap, many of them are capable of fly-ing considerable distances on their first flights. Div-ing petrels may fly as far as 10 kilometers the first

time out of their burrows. On the other hand, young birds reared in open nests frequently flap their wings vigorously in the wind for several days before fly-ing—especially large birds such as albatrosses, storks, vultures, and eagles. Such flapping may help to develop muscles, but it is unlikely that these birds are learning to fly; however, a bird's flying abilities do improve with practice for a period after it leaves the nest.

There are so many variables involved in flap-ping flight that it becomes difficult to understand exactly how it works. A beating wing is flexible and porous and yields to air pressure, unlike the fixed wing of an airplane. Its shape, wing loading, camber, sweepback, and the position of the individual feath-ers all change remarkably as a wing moves through its cycle of locomotion. This is a formidable list of variables, and it is no wonder that flapping flight has not yet fully yielded to explanation in aerodynamic terms; however, the general properties of a flapping wing can be described (see Figure 17–8).

We can begin by considering the flapping cycle of a small bird in flight. A bird cannot continue to fly straight and level unless it can develop a force or thrust to balance the drag operating against forward momentum. The flapping of the wings, especially the wingtips (primaries), produces this thrust, whereas the inner wings (secondaries) are held more nearly stationary with respect to the body and gen-erate lift. It is easiest to consider the forces operating on the inner and outer wing separately. Most of the lift, and also most of the drag, comes from the forces acting on the inner wing and body of a flying bird. The forces on the wingtips derive from two motions that have to be added together. The tips are moving forward with the bird, but at the same time they are also moving downward relative to the bird. The wingtip would have a very large angle of attack and would stall if it were not flexible. As it is, the forces on the tip cause the individual primaries to twist as the wing is flapped downward (see Figure 17–6) and to produce the forces diagrammed in Figure 17–8c.

The forces acting on the two parts of the bird combine to produce the conditions for equilibrium flight shown. The positions of the inner and outer wings during the upstroke (dotted lines) and down-stroke reveal that vertical motion is applied mostly at the wingtips (Figure 17–8a). Thus, the inner wing (the secondaries) acts as if the bird were gliding to gener-ate the forces shown in Figure 17–8b, and the outer wing generates the force shown in part c. Most of the

Table 17–1 **Wing loading of some North Amer-ican birds**

Species	Body Mass (g)	Wing Area (cm^2)	Wing Loading (g/cm^2)
Ruby-throated hummingbird	3.0	12.4	0.24
Barn swallow	17.0	118.5	0.14
Mourning dove	130.0	357.0	0.36
Peregrine falcon	1,222.5	1,342.0	0.91
Golden eagle	4,664.0	6,520.0	0.71
Mute swan	11,602.0	6,808.0	1.70

Source: E. L. Poole, 1938, *Auk* 55:511–517.

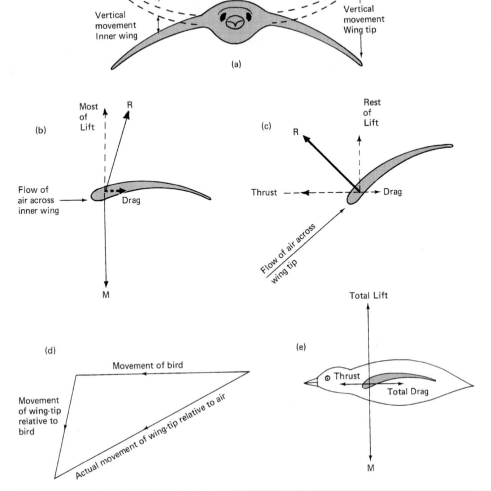

lift to counter the pull of gravity (M) is generated by the inner wing and body. Canting of the wing tip (the primaries) during the downstroke (Figure 17–8c) produces a resultant force (R) that is directed forward. The movement of the wingtip relative to the air is affected by the forward motion of the bird through the air (Figure 17–8d). As a result of this motion and the canting of the wing tips during the downstroke, the flow of air across the primaries is different from the flow across the secondaries and the body (Figure 17–8e). When flight speed through the air is constant, the forces acting on the inner wing and the body and on the outer wing combine to produce a set of summed vectors in which thrust exceeds total drag and lift at least equals the body mass.

As the wings move downward and forward on the downstroke, which is the power stroke, the trailing edges of the primaries bend upward under air pressure, and each feather acts as an individual propeller biting into the air and generating thrust. Contraction of the **pectoralis major,** the large breast muscle, produces the downstroke during level flapping flight. During this downbeat, the thrust is greater than the total drag, and the bird accelerates. In small birds, the return stroke, which is upward and backward, provides little or no thrust and is mainly a passive recovery stroke. The bird decelerates during the recovery stroke.

For larger birds with slower wing actions, the time of the upstroke is too long to spend in a state of deceleration. A similar situation exists when any bird takes off: It needs thrust on both the downstroke and the upstroke. Thrust on the upstroke is produced by bending the wings slightly at the wrists and elbow and by rotating the humerus upward and backward. This movement causes the upper surfaces of the twisted primaries to push against the air and to produce thrust as their lower surfaces did in the downstroke. In this type of flight the wingtip describes a rough figure eight through the air. As speed increases the figure-eight pattern is restricted to the wingtips.

A powered upstroke results mainly from the contraction of the **supracoracoideus,** a deep muscle underlying the pectoralis major and attached directly to the keel of the sternum. It inserts on the dorsal head of the humerus by passing through the foramen triosseum, formed where the coracoid, furcula, and scapula join (Figure 17–9). In most species of birds the supracoracoideus is a relatively small, pale muscle with low myoglobin content, easily fatigued. In species that rely on a powered upstroke for fast, steep takeoffs, for hovering, or for fast aerial pursuit, the supracoracoideus is relatively larger. The ratio of weights of the pectoralis major and the supracoracoideus is a good indication of a bird's reliance on a powered upstroke; such ratios vary from 3:1 to 20:1. The total weight of the flight muscles is also indicative of the extent to which a bird depends on powered flight. Strong fliers such as pigeons and falcons have breast muscles comprising more than 20 percent of body weight, whereas in some owls, which have very light wing loading, the flight muscles make up only 10 percent of total weight.

A flying bird increases its speed by increasing the amplitude of its wing beats, but the frequency of wing beats remains nearly constant at all speeds during level flight. Large birds have slower wing beat frequencies than those of small birds, and strong fliers usually have slower beat frequencies than

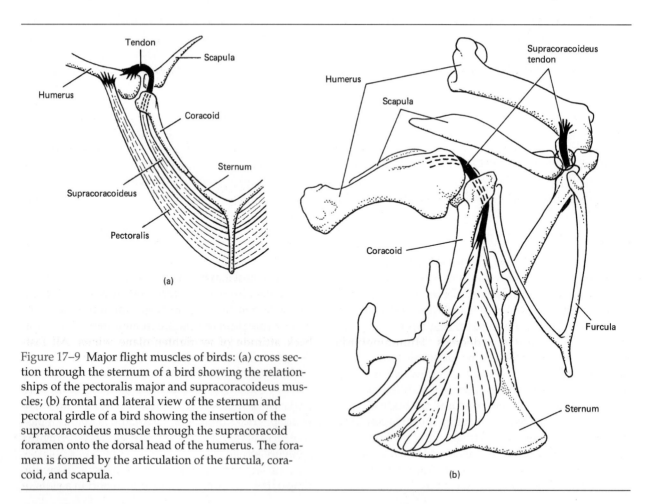

Figure 17–9 Major flight muscles of birds: (a) cross section through the sternum of a bird showing the relationships of the pectoralis major and supracoracoideus muscles; (b) frontal and lateral view of the sternum and pectoral girdle of a bird showing the insertion of the supracoracoideus muscle through the supracoracoid foramen onto the dorsal head of the humerus. The foramen is formed by the articulation of the furcula, coracoid, and scapula.

those of weak fliers. The frequency of respiration of many birds during flight appears to have a constant relationship to the wing beat frequency. Some birds, especially those that have low wing beat frequencies, breathe once per wing beat cycle, whereas birds with high wing beat frequencies have breathing cycles that span several wing beats. In general, inspiration appears to occur during or at the end of a wing upstroke, with expiration at the end of a downstroke.

A brief description of the flight of hummingbirds serves to indicate an extreme of specialization of flapping flight. Hummingbirds can hover and also fly backward. The hand bones form most of the wing skeleton of hummingbirds; the forearm and upper arm are very short. In addition, there is little articulation in the elbow and wrist joints, so that the entire wing is essentially an inflexible framework that can be moved very freely and in almost any direction at the shoulder joint. Thus, the entire wing functions essentially as a variable-pitch propeller. The wings beat in a horizontal plane as the bird hovers, and both the upstroke and the downstroke are powered (Figure 17–10). The supracoracoideus muscle is large in relation to the pectoralis, and the entire flight muscle mass is large in relation to body size, comprising 30 percent of total body weight. Hovering requires very high energy utilization because no lift is generated by forward movement through the air.

Wing Structure and Flight Characteristics

Wings may be large or small in relation to body size, resulting in light wing loading or heavy wing loading. They may be long and pointed, short and rounded, highly cambered or relatively flat, and the width and degree of slotting are additional important characteristics.

Depending on whether a bird is primarily a powered flier or a soaring form, the various segments of the wing (hand, forearm, upper arm) are lengthened to different degrees. Hummingbirds have very fast, powerful wing beats, requiring maximum propulsive force from the primaries. The hand bones of hummingbirds are longer than the forearm and upper arm combined. Most of the flight surface is formed by the primaries, and hummingbirds have only six or seven secondaries. Frigate birds are marine species with long, narrow wings specialized for powered flight as well as for gliding and soaring. All three segments of the forelimb are about equal in

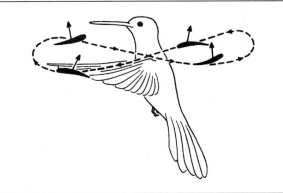

Figure 17–10 Diagrammatic representation of the wing movements of a hummingbird in hovering flight.

length. The soaring albatrosses have carried lengthening of the wing to the extreme found in birds: The humerus or upper arm is the longest segment, and there may be as many as 32 secondaries in the inner wing (Figure 17–11).

Ornithologists recognize four structural and functional types of wings (Figure 17–12). Birds that live in forests and woodlands where they must maneuver around obstructions have **elliptical wings.** These wings have a low aspect ratio, tend to be highly cambered, and usually have a high degree of slotting in the outer primaries. These features are generally associated with slow flight and a high degree of maneuverability. Although some species with elliptical wings, notably upland gamebirds such as pheasants and grouse, have fast takeoff speeds, they maintain rapid flight only for short distances.

Many birds that are aerial foragers, make long migrations, or have a heavy wing loading that is related to some other aspect of their lives, such as diving, have **high-speed wings.** These wings have a moderately high aspect ratio, taper to a pointed tip, have a flat profile (little camber), and often lack slots in the outer primaries. In flight they show the swept-back attitude of jet fighter plane wings. All fast-flying birds have converged on this form.

Seabirds, particularly those such as albatrosses and shearwaters that rely on **dynamic soaring,** have long, narrow, flat wings lacking slots in the outer primaries. Some albatrosses have aspect ratios of 18:1 and lift-to-drag ratios similar to those of high-performance sailplanes. Dynamic soaring is possible only where there is a pronounced vertical wind gradient, with the lower 15 or so meters of air being

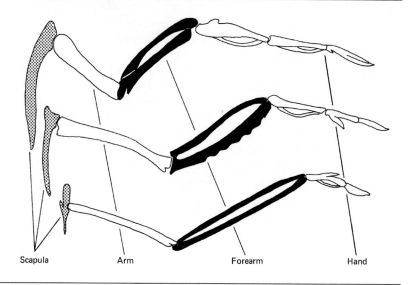

Figure 17–11 Comparison of the relative lengths of the proximal, middle, and distal elements of the wing bones of a hummingbird (top), frigate bird (middle), and albatross (bottom) drawn to the same size.

Scapula   Arm   Forearm   Hand

slowed by friction against the ocean surface. Furthermore, dynamic soaring is feasible only in regions where winds are strong and persistent, such as in the latitudes of the Roaring Forties. This is where most albatrosses and shearwaters are found. Starting from the top of the wind gradient, an albatross glides downwind with great increase in ground speed (kinetic energy). Then, as it nears the surface, it turns and gains altitude while gliding into the wind. Because the bird flies into wind of increasing speed as it rises, its loss of airspeed is not as great as its loss of ground speed, and consequently it does not stall until it has mounted back to the top of the wind gradient, where the air velocity becomes stable. At that point, the bird has converted much of its kinetic energy to potential energy and it turns downwind to repeat the cycle.

Large species of albatrosses weigh 10 to 12 kilograms and have wing spans greater than 3 meters, making them among the largest flying birds. Because their pectoral muscles are relatively small and weak, they can take off only by running or paddling and flapping into a strong wind, or by launching forth from the brink of a precipice.

The slotted **high-lift wing** is a fourth type. It is associated with static soaring typified by vultures, eagles, storks, and some other large birds. This wing has an intermediate aspect ratio between the elliptical wing and the high aspect ratio wing, a deep camber, and marked slotting in the primaries. When the bird is in flight the tips of the primaries turn markedly upward under the influence of air pressure and body weight. Static soarers remain airborne mainly by seeking out and gliding in air masses that are rising at a rate faster than the bird's sinking speed. Hence, a light wing loading and maneuverability (slow forward speed and small turning radius) are advantageous. Broad wings provide the light wing loading, and the highly developed slotting enhances maneuverability by responding to changes in wind currents with changes in the positions of individual feathers instead of movements of the entire wing. A bird cannot soar in tight spirals at high speed, and flying slowly with enough lift to prevent stalling requires a high angle of attack. The deeply slotted primaries apparently make the combination of low speed and high lift possible. The distal, emarginated portion of each primary produces lift by acting as a separate high aspect ratio airfoil set at a high angle of attack. This design reduces induced drag.

In regions where topographic features and meteorological factors provide currents of rising air, static soaring is an energetically cheap mode of flight. By soaring rather than flapping, a large bird the size of a stork can decrease by a factor of 20 or more the energy required for flight per unit of time, whereas the saving is only one-tenth as much for a small bird such as a warbler. It is little wonder, then, that most large land birds perform their annual migrations by soaring and gliding as much of the time as possible, and some condors and vultures cover hundreds of kilometers each day soaring in search of food.

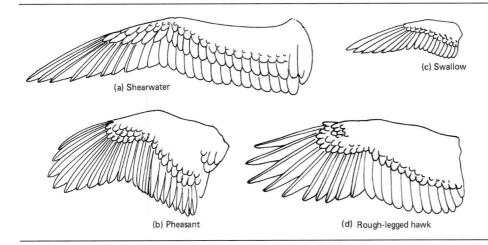

Figure 17–12 Comparison of four basic types of bird wings: (a) high aspect ratio; (b) elliptical; (c) high speed; (d) slotted.

(a) Shearwater

(c) Swallow

(b) Pheasant

(d) Rough-legged hawk

Body Form and Flight

Many aspects of the morphology of birds appear to have been molded by aerodynamic forces. (See Raikow [1985] and Rayner [1988] for reviews.) Feathers, for example, provide lift and streamlining during flight. Feathers are light, yet they are strong and resilient for their weight. Of course, flying is not the only function of feathers—they also provide the insulation necessary for endothermy and their colors and shapes function in crypsis and display.

Structural modifications can be seen in several aspects of the structure of birds. The avian skeleton is not lighter in relation to the total body mass of a bird than is the skeleton of a mammal of similar size, but the distribution of mass is different. Many bones are pnuematized (air-filled), and the skull is especially light, but the leg bones of birds are heavier than those of mammals. Thus, the total mass of the skeleton of a bird is similar to that of a mammal, but more of a bird's mass is concentrated in its hind limbs.

Characteristics of some of the organs of birds reduce body mass. For example, birds lack urinary bladders, and most species have only one ovary (the left). The gonads of both male and female birds are usually small; they hypertrophy during the breeding season and regress when breeding has finished.

Power-producing features are equally important components of the ability of birds to fly. The pectoral muscles of a strong flier may account for 20 percent of the total body mass. The power output per unit mass of the pectoralis major of a turtledove during level flight has been estimated to be 10 to 20 times that of most mammalian muscles. Birds have large hearts and high rates of blood flow and complex lungs that use cross-current flows of air and blood to maximize gas exchange and to dissipate the heat produced by high levels of muscular activity during flight. The brains of birds are similar in size to the brains of rodents, and the forebrain and cerebellum are well developed. Birds rely heavily on visual information and the optic lobes are especially large. The sense of smell is not well developed in most birds and the olfactory lobes are correspondingly small.

Streamlining

Birds are the only vertebrates that move fast enough in air for wind resistance and streamlining to be important factors in their lives. Many passerine birds are probably able to fly 50 kilometers per hour or even faster when they must, although their normal cruising speeds are lower. Ducks and geese can fly at 80 or 90 kilometers per hour, and peregrine falcons reach speeds as high as 200 kilometers per hour when they dive on prey. Fast-flying birds have many of the same structural characters as those seen in fast-flying aircraft. Contour feathers make smooth junctions between the wings and the body and often between the head and body as well, eliminating sources of turbulence that would increase wind resistance. The feet are tucked close to the body during flight, further improving streamlining.

At the opposite extreme, some birds are slow fliers. Many of the long-legged, long-necked wading birds, such as spoonbills and flamingos, fall in this category. Their long legs trail behind them as they fly and their necks are extended. They are far from streamlined, although they may be strong fliers.

Skeleton and Muscles

The hollow, air-filled (pneumatic) bones of birds (Figure 17–13) are probably an ancestral character of the archosaur lineage, not a derived character of birds. Not all birds have pneumatic bones. In general, pneumatization of bones is better developed in large birds than in small ones. Diving birds (penguins, grebes, loons) have little pneumaticity, and the bones of diving ducks are less pneumatic than those of nondivers.

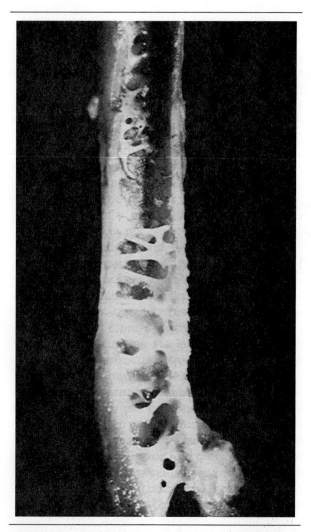

Figure 17–13 The hollow bones of birds are reinforced by struts.

The distribution of pneumaticity among the bones of the skeleton also varies. The skull is pneumatic in nearly all birds, although the kiwi, a flightless bird, lacks air spaces in the skull. The sternum, pectoral girdle, and humerus are pneumatic and are part of a system of interconnected air sacs that allow a one-way flow of air through the lungs (discussed in the following section). Pneumaticity extends through the rest of the appendicular skeleton of some birds, even into the phalanges.

Except for the specializations associated with flight, the skeleton of a bird is very much like that of a small, bipedal archosaur (Figure 17–14). The pelvic girdle of birds is elongated, and the ischium and ilium have broadened into thin sheets that are firmly united with a **synsacrum** that is formed by the fusion of 10 to 23 vertebrae. The long tail of ancestral diapsids has been shortened in birds to about five free caudal vertebrae and a **pygostyle** formed by the fusion of the remaining vertebrae. The pygostyle supports the tail feathers (rectrices). The thoracic vertebrae are joined by strong ligaments that are often ossified. The relatively immobile thoracic vertebrae, the synsacrum, and the pygostyle in combination with the elongated, roof-like pelvis produce a nearly rigid vertebral column. Flexion is possible only in the neck, at the joint between the thoracic vertebrae and the synsacrum, and at the base of the tail. The rigid trunk is balanced on the legs. The femur projects anteriorly and its articulation with the tibiotarsus and fibula is close to the center of gravity of the bird.

The wings are positioned above the center of gravity. The sternum is greatly enlarged compared to other vertebrates, and (except in flightless birds) it bears a keel from which the pectoralis and supracoracoideus muscles originate (Figure 17–9). Strong-flying birds have well-developed keels and their flight muscles are large. The scapula extends posteriorly above the ribs and is supported by the coracoid, which is fused ventrally to the sternum. Additional bracing is provided by the clavicles, which, in most birds, are fused at their distal ends to form the **furcula** (wishbone).

The relative size of the leg and flight muscles of birds is related to their primary mode of locomotion. Flight muscles comprise 25 to 35 percent of the total body mass of strong fliers such as hummingbirds and swallows. These species have small legs and the leg muscles account for as little as 2 percent of the body mass. Predatory birds such as hawks and owls use their legs to capture prey. In these species the

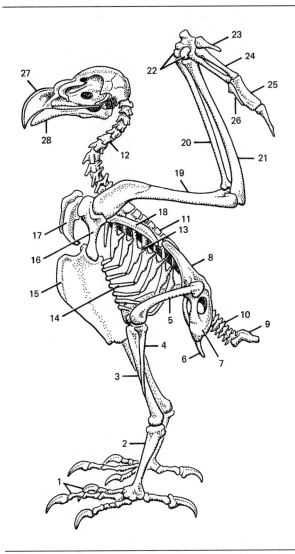

Figure 17–14 Skeleton of an eagle. 1, Toes; 2, tarsus; 3 and 4, tibiotarsus formed by the fused tibia (3) and fibula (4); 5, femur; 6, pubis; 7, ischia; 8, ilium; 9, pygostyle; 10, caudal vertebrae; 11, thoracic vertebrae; 12, cervical vertebrae; 13, uncinate process of rib; 14, sternal rib; 15, sternum; 16, coracoid; 17, furcula; 18, scapula; 19, humerus; 20, radius; 21, ulna; 22, carpal bones; 23, first digit; 24, metacarpal; 25, second digit; 26, third digit; 27, upper mandible; 28, lower mandible. (From J. Dorst, 1974, *The Life of Birds*, Columbia University Press, New York, NY.)

flight muscles make up about 20 percent of the body mass and the limb muscles are 10 percent. Swimming birds—ducks and grebes, for example—have an even division between limb and flight muscles, and the combined mass of these muscles may be 30

to 60 percent of the total body mass. Birds such as rails, which are primarily terrestrial and run to escape from predators, have limb muscles that are larger than their flight muscles.

Muscle fiber types and metabolic pathways also distinguish running birds from fliers. The familiar distinction between the light meat and dark meat of a chicken reflects those differences. Fowl, especially domestic breeds, rarely fly, but they are capable of walking and running for long periods. The dark color of the leg muscles reveals the presence of myoglobin in the tissues and indicates a high capacity for aerobic metabolism in the limb muscles of these birds. The white muscles of the breast lack myoglobin and have little capacity for aerobic metabolism. The flights of fowl (including wild species such as pheasants, grouse, and quail) are of brief duration and are used primarily to evade predators. The bird uses an explosive takeoff, fueled by anaerobic metabolic pathways, followed by a long glide back to the ground. Birds that are capable of strong sustained flight have dark breast muscles with high aerobic metabolic capacities.

The skulls of most birds consist of four bony units that can move in relation to each other. This skull kinesis is important in some aspects of feeding. The upper jaw flexes upward as the mouth is opened, and the lower jaw expands laterally at its articulation with the skull (Figure 17–15). The flexion of the upper and lower jaws increases the bird's gape in both the vertical and horizontal planes and probably assists in swallowing large items.

Heart, Lungs, and Gas Exchange

The heart of birds is composed of morphologically distinct left and right atria and ventricles. The separation of the ventricle into left and right halves is an archosaurian character that is seen in crocodilians as well as in birds. However, birds lack the capacity possessed by crocodilians to shunt blood between the pulmonary and systemic circulation. In birds, as in mammals, the blood must flow in series through the two circuits.

The respiratory system of birds is unique among extant vertebrates (Figure 17–16). The air sacs that occupy much of the dorsal part of the body and extend into the pneumatic spaces in many of the bones provide a system in which airflow through the lungs is unidirectional instead of tidal (in and out) as it is in mammalian lungs. The air sacs are poorly vas-

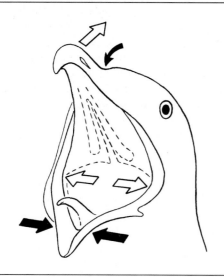

Figure 17–15 Skull and jaw kinesis of birds. A yawning herring gull (*Larus argentatus*) shows the kinetic movements of the skull and jaws that occur during swallowing. White arrows show the positions of outward flexion and black arrows show inward flexion. (From P. Bühler, 1981, in *Form and Function in Birds,* volume 2, edited by A. S. King and J. McLelland, Academic, New York, NY.)

cularized, and gas exchange occurs only in the parabronchial lung. The combined volume of the air sacs is about nine times the volume of the parabronchial lung itself. The flow of air through this extensive respiratory system may help to dissipate the heat generated by high levels of muscular activity during flight.

Air flows through the parabronchial lung in the same direction during both inspiration and exhalation (Figure 17–17). During inspiration the volume of the thorax increases, drawing air through the bronchus and into the posterior thoracic and abdominal air sacs and the parabronchial lung. Simultaneously, air from the parabronchial lung is drawn into the clavicular and anterior thoracic sacs. On expiration the volume of the thorax decreases, air from the posterior thoracic and abdominal sacs is forced into the parabronchial lung, and air from the clavicular and thoracic sacs is forced out through the bronchus.

Gas exchange takes place in a network of tiny, blind-end air capillaries that intertwine with equally fine blood capillaries. Air flow and blood flow pass in opposite directions, and the capillary pairs compose a cross-current exchange system. This arrangement is more effective at exchanging lung gases than is the mammalian lung, and the efficiency of gas exchange in the parabronchial lung may allow birds to breathe at higher altitudes than mammals (Box 17–1). Birds do sometimes penetrate to very high altitudes, either as residents or in flight. For example, radar tracking of migrating birds shows that they sometimes fly as high as 6500 meters, the alpine chough lives at altitudes around 8200 meters on Mount Everest, and bar-headed geese pass directly over the summit of the Himalayas at altitudes of 9200 meters during their migrations.

Vocalization is a second function of the respiratory systems of many vertebrates, including birds. However, the vocal organ of birds, the **syrinx** (plural syringes), is unique. The syrinx lies at the base of the trachea where the two bronchi diverge. Several of the cartilaginous rings that support the trachea and bronchi are modified and surrounded by muscles. The anatomy of the syrinx varies among birds—some are purely tracheal, some involve both the trachea and the bronchi, and others are purely bronchial. Two membranes are associated with the syrinx, one in the base of each bronchus. Songbirds have five to nine (usually seven) pairs of syringeal muscles, parrots and lyrebirds have three pairs, falcons have two, and most other birds have only a single pair of syringeal muscles. Storks, vultures, ratites, and most pelicans lack syringeal muscles.

The mechanism of avian song production is an area of controversy. One view holds that sound is produced by vibration of the syringeal membranes and that modulations of that sound are the result of changes in the vibration of the membranes (Greenewalt 1968, 1969). An alternative hypothesis maintains that pulsatile activity of the abdominal muscles and resonance in the trachea are important in modulating the amplitude and frequencies produced by the syringeal membranes (Gaunt et al. 1982, Nowicki 1987).

Feeding, Digestion, and Excretion

The digestive systems of birds show some differences from those of other vertebrates, although the distinctions are less pronounced than the differences in the respiratory systems. With the specialization of the forelimbs as wings, which largely precludes their having any role in prey capture, birds have concentrated their predatory mechanisms in their beaks and

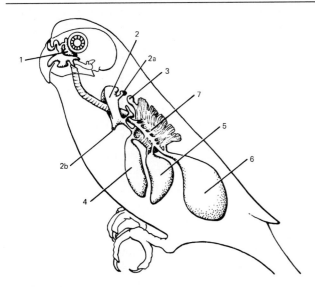

Figure 17–16 The lung and air sac system of the budgerigar; only the left side is shown. 1, Infraorbital sinus; 2, clavicular air sac; 2a, axillary diverticulum to the humerus; 2b, sternal diverticulum; 3, cervical air sac; 4, cranial thoracic air sac; 5, caudal thoracic air sac; 6, abdominal air sacs; 7, parabronchial lung. (From H. E. Evans, 1982, in *Diseases of Cage and Aviary Birds*, 2d edition, edited by M. L. Petrak, Lee & Febiger, Philadelphia, PA.)

feet. The absence of teeth prevents birds from doing much processing of food in the mouth, and the gastric apparatus takes over some of that role.

The Esophagus and Crop   Birds often gather more food than they can process in a short period, and the excess is held in the esophagus. Many birds have a **crop,** an enlarged portion of the esophagus that is specialized for temporary storage of food. The crop of some birds is a simple expansion of the esophagus, whereas in others it is a unilobed or bilobed structure (Figure 17–19). An additional function of

the crop is transportation of food for nestlings. When the adult returns to the nest it regurgitates the material from the crop and feeds it to the young. In doves and pigeons the crop of both sexes produces a nutritive fluid (crop milk) that is fed to the young. The milk is produced by fat-laden cells that detach from the squamous epithelium of the crop and are suspended in an aqueous fluid. Crop milk is rich in lipids and proteins, but contains no sugar. Its chemical composition is similar to that of mammalian milk, although it differs in containing intact cells. The proliferation of the crop epithelium and the for-

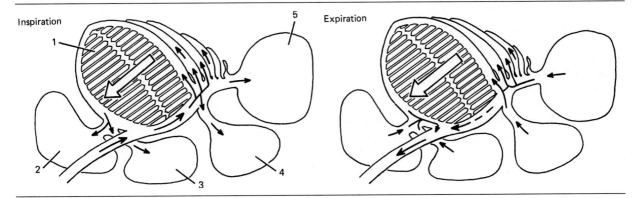

Figure 17–17 Pattern of airflow during inspiration and expiration. Note that air flows through the parabronchial lung during both phases of the respiratory cycle. 1, Parabronchial lung; 2, clavicular air sac; 3, cranial thoracic air sac; 4, caudal thoracic air sac; 5, abdominal air sacs. (From P. Scheid, 1982, in *Avian Biology,* volume 6, edited by J. R. King and K. C. Parkes, Academic, New York, NY.)

Box 17–1 High-Flying Birds

Birds regularly fly at altitudes higher than human mountain climbers can ascend without using auxiliary breathing apparatus, and the ability of birds to sustain activity at high altitudes is a result of the morphological characteristics of their pulmonary systems. Mammals have difficulty breathing at high altitudes because of the low oxygen pressure in the air. The atmosphere is most dense at the surface of the earth (where the entire weight of the atmosphere is pressing down on it), and it becomes increasingly less dense at higher altitudes. At sea level atmospheric pressure is 760 millimeters of mercury (760 torr in the units of the International System). The composition of dry air by volume is 79.02 percent nitrogen and other inert gases, 20.94 percent oxygen, and 0.04 percent carbon dioxide. These gases contribute to the total atmospheric pressure in proportion to their abundance, so the contribution of oxygen is 20.94 percent of 760 torr, or 159.16 torr. The pressure exerted by an individual gas is called the partial pressure of that gas. The rate and direction of diffusion of gas between the air in the lungs and the blood in the pulmonary capillaries is determined by the difference in the partial pressures of the gas in the blood and in the lungs. Oxygen diffuses from air in the lungs into blood in the pulmonary capillaries because oxygen has a higher partial pressure in the air than in the blood, whereas carbon dioxide diffuses in the opposite direction because its partial pressure is higher in blood than in air.

At higher altitudes the atmospheric pressure is lower. At 7700 meters, the atmospheric pressure is only 282 torr, and the partial pressure of oxygen in dry air is about 59 torr. As a result of the low atmospheric pressure at this altitude the driving force for diffusion of oxygen into the

blood is small. (The actual pressure differential is reduced even below this figure because air in the lungs is saturated with water, and water vapor contributes to the total pressure in the lungs. The vapor pressure of water at 37°C is 47 torr. Thus, the partial pressure of oxygen in the lungs is 20.94 percent × (282 − 47 torr) = 49 torr.)

The tidal ventilation pattern of the lungs of mammals means that the partial pressure of oxygen in the pulmonary capillaries can never be higher than the partial pressure of oxygen in the expired air. The best that a tidal ventilation system can accomplish is to equilibrate the partial pressures of oxygen in the pulmonary air and in the pulmonary circulation. In fact, failure to achieve complete mixing of the gas within the pulmonary system means that oxygen exchange falls short even of this equilibration, and blood leaves the lungs with a partial pressure of oxygen slightly lower than the partial pressure of oxygen in the exhaled air. The cross-current blood flow system in the parabronchial lung of birds ensures that the gases in the air capillaries repeatedly encounter a new supply of deoxygenated blood (Figure 17–18, left side of diagram).

When blood enters the system (on the left side of the diagram) it has the low oxygen pressure of mixed venous blood (P_v). The blood entering the leftmost capillary is exposed to air that has already had much of its oxygen removed farther upstream. Nonetheless, the low oxygen pressure of the mixed venous blood ensures that even in this part of the parabronchus, oxygen uptake can occur. Blood flowing through capillaries farther to the right in the diagram is exposed to higher partial pressures of oxygen in the parabronchial gas and takes up correspondingly more oxygen. The oxygen pressure of the blood that flows out of the lungs

mation of crop milk is stimulated by prolactin, as is the lactation of mammals. A nutritive fluid produced in the esophagus is fed to hatchlings by greater flamingos and by the male emperor penguin.

The hoatzin, a South American bird, is the only avian species known to employ foregut fermentation (Figure 17–20). Hoatzins are herbivorous—green leaves make up more than 80 percent of the diet.

(P_a) is the result of mixing of blood from all the capillaries. The oxygen pressure of the mixed arterial blood is higher than the partial pressure of oxygen in the exhaled air.

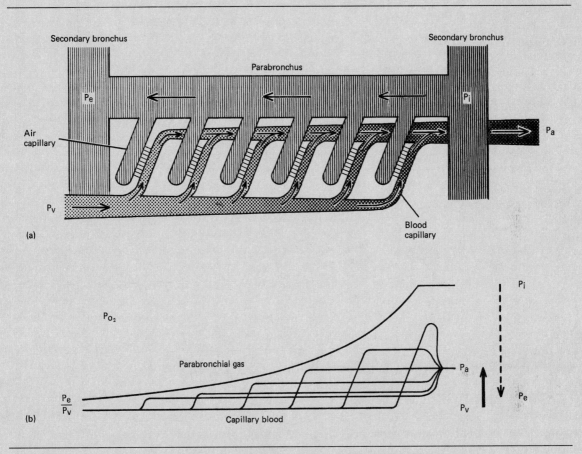

Figure 17–18 Diagram of gas exchange in a cross-current lung. Air flows from right to left in this diagram and blood flows from left to right. P_e = oxygen pressure in the air exiting the parabronchus, P_v = oxygen pressure in the mixed venous blood entering the blood capillaries, P_a = oxygen pressure in the blood leaving the blood capillaries P_i = oxygen pressure in the air entering the parabronchus. Top: General pattern of air and blood flow through the parabronchial lung. Bottom: Diagrammatic representation of cross-current gas exchange. (From P. Scheid, 1982, in *Avian Biology*, volume 6, edited by J. R. King and K. C. Parkes, Academic, New York, NY.)

More than a century ago, naturalists observed that hoatzins smell like fresh cow manure, and a study of the crop and lower esophagus revealed that volatile fatty acids are the source of the odor (Grajal et al. 1989). Bacteria and protozoa like those found in the rumen of cows (Box 21–1) break down the plant cell walls, and bacterial extracts from the hoatzin's crop were as effective as those from a cow's rumen in

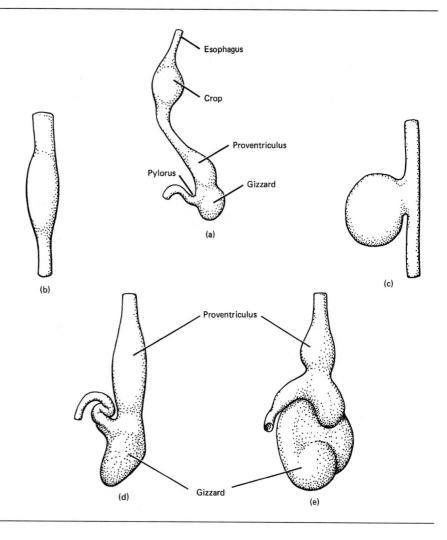

Figure 17–19 Anterior digestive tract of birds. (a) The relationship among the parts. The relative sizes of the proventriculus and gizzard vary in relation to diet. Carnivorous and fish-eating birds like the great cormorant have a relatively small crop (b) and gizzard (c), whereas seed eaters and omnivores like the peafowl have a large crop (d) and muscular gizzard (e). (From J. McLelland, 1979, *Form and Function in Birds*, edited by A. S. King and J. McLelland, Academic, London, UK.)

digesting plant material. Volatile fatty acids produced by the process of fermentation are absorbed by the gut.

The Gastric Apparatus The form of the stomachs of birds is related to their dietary habits. Carnivorous and piscivorous (fish-eating) birds need expansible storage areas to accommodate large volumes of soft food, whereas birds that eat insects or seeds require a muscular organ that can contribute to the mechanical breakdown of food. Usually, the gastric apparatus of birds consists of two relatively distinct chambers, an anterior glandular stomach (**proventriculus**) and a posterior muscular stomach (**gizzard**). The proventriculus contains glands that secrete acid and digestive enzymes. The proventriculus is especially large in species that swallow large items such as fruit or fish intact (Figure 17–19).

The gizzard has several functions, including storage of food while the chemical digestion that was begun in the proventriculus continues, but its most important function is the mechanical processing of food. The thick, muscular walls of the gizzard squeeze the contents, and small stones that are held in the gizzards of many birds help to grind the food. In this sense the gizzard is performing the same function that is performed by the teeth of mammals. The pressure that can be exerted on food in the gizzard is intense. A turkey's gizzard can grind up two dozen walnuts in as little as 4 hours, and can crack hickory nuts that require 50 to 150 kilograms of pressure to break under experimental conditions.

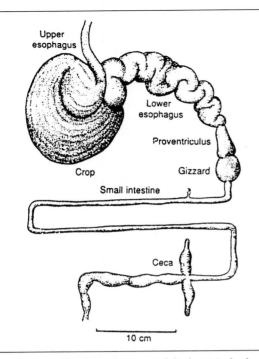

Figure 17–20 The digestive tract of the hoatzin looks very much like that of a ruminant mammal (see Figure 21–14). The muscular crop and the anterior esophagus are greatly enlarged. Cornified ridges on the inner surface of the crop probably grind the contents of the crop, reducing the particle size. This process is analagous to a cow chewing its cud, but has the advantage that mechanical procesing and fermentation occur in the same structure. (From A. Grajal, S. D. Strahl, R. Parra, M. G. Dominguez, and A. Neher, 1989, *Science* 245:1236–1238.)

The Intestine, Ceca, and Cloaca The small intestine is the principal site of chemical digestion as enzymes from the pancreas and intestine break down the food into small molecules that can be absorbed across the intestinal wall. The mucosa of the small intestine is modified into a series of folds, lamellae, and villi that increase its surface area. The large intestine is relatively short, usually less than 10 percent of the length of the small intestine. Passage of food through the intestines of birds is quite rapid: Transit times for carnivorous and fruit-eating species are in the range of a few minutes to a few hours. Passage of food is slower in herbivores and may require a full day. Birds generally have a pair of ceca at the junction of the small and large intestines. The ceca are small in carnivorous, insectivorous, and seed-eating species, but they are large in herbivorous and omnivorous species such as cranes, fowl, ducks, geese, and the ostrich. Symbiotic microorganisms in the ceca apparently ferment plant material.

The cloaca temporarily stores waste products while water is being reabsorbed. The precipitation of uric acid in the form of urate salts frees water from the urine, and this water is returned to the bloodstream. Species of birds that have salt-secreting glands can accomplish further conservation of water by reabsorbing some of the ions that are in solution in the cloaca and excreting them in more concentrated solutions through the salt glands (Chapter 4). The mixture of white urate salts and dark fecal material that is voided by birds is familiar to anyone who has washed an automobile.

The Hind Limbs and Locomotion

Unlike most tetrapods, birds usually are specialized for two or more different modes of locomotion: bipedal walking or swimming with the hind limbs and flying with the forelimbs.

Walking, Hopping, and Perching

Terrestrial locomotion may involve walking or running, supporting heavy bodies, hopping, perching, climbing, wading in shallow water, or supporting the body on insubstantial surfaces such as snow or floating vegetation. We will consider cursorial adaptations first, because birds evolved from bipedal dinosaurs and because the principles of cursorial adaptation have been well worked out in the quadrupedal mammals. Modifications usually associated with running in quadrupeds are (1) a progressive increase in the lengths of the distal limb elements relative to the proximal ones, (2) a decrease in the area of the foot surface that makes contact with the ground, and (3) a reduction in the number of toes. All three of these cursorial trends are expressed to varying degrees in running birds; however, problems of balance are more critical for bipeds than for quadrupeds, and these problems may have restricted the evolution among bipeds of some of the cursorial adaptations found in quadrupeds.

Because the center of gravity must lie over the feet of a biped to maintain balance, a reduction in the

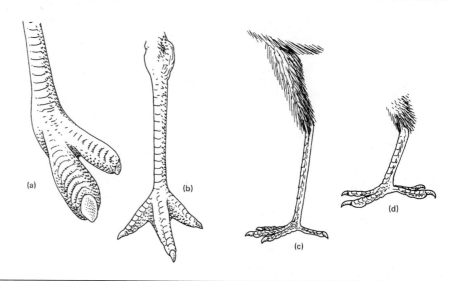

Figure 17–21 Avian feet with various specializations for terrestrial locomotion: (a) ostrich, with only two toes; (b) rhea, with three toes; (c) secretary bird, with a typical avian foot; (d) roadrunner, with zygodactyl foot. (Not drawn to scale.)

surface area in contact with the ground can be achieved only by some sacrifice in stability. No bird has reduced the length and number of toes in contact with the ground to the extent that the hooved mammals have, but the large, fast-running ostrich has only two toes on each foot and many other cursorial species have only the three forward-directed toes in contact with the ground (Figure 17–21).

Weight-bearing characteristics such as large, heavy leg bones arranged in vertical columns as supports of great body mass are well known in large mammals such as elephants. No surviving birds show specializations of this kind, but these features are seen in some of the large, flightless terrestrial birds of earlier times. The extinct elephantbirds (Aepyornithiformes) of Madagascar and moas (Dinornithiformes) of New Zealand were herbivores that evolved on oceanic islands in the absence of large carnivorous mammals and survived there until contact with humans in the post-Pleistocene period (Figure 17–22). The giant, carnivorous *Diatryma* (Gruiformes) and related species were successful continental forms in the Americas until the appearance of large mammalian carnivores. Among the large, flightless terrestrial birds, only some of the fleet-footed cursors such as the rheas and ostrich have managed to survive in the presence of large placental carnivores.

Hopping, a succession of jumps with the feet moving together, is a special form of pedal locomotion found mostly in perching, arboreal birds. It is most highly developed in the passerines, and only a few nonpasserine birds regularly hop. Many passerines cannot walk, and hopping is their only mode of terrestrial locomotion. Some groups of passerines have a more terrestrial mode of existence and these birds possess a walking gait as well as the ability to hop. Larks, pipits, starlings, and grackles are examples. The separation between walking and hopping passerines cuts across families. For example, in the family Corvidae, the ravens, crows, and rooks are walkers, whereas the jays and magpies are hoppers.

The most specialized avian foot for perching on branches is one in which all four toes are free and mobile, of moderate length, and with the hind toe well developed, lying in the same plane as the forward three, and opposable to them (**ansiodactyl**). Such a foot produces a firm grip and is highly developed in the passerine birds. The **zygodactylous** condition, with two toes forward and opposable to two extending backward is characteristic of birds such as parrots and woodpeckers, which climb or perch on vertical surfaces.

The tendons that flex the toes of a perching bird can lock the foot in a tight grip so that the bird does not fall off its perch when it relaxes or goes to sleep. The plantar tendons, which insert on the individual phalanges of the toes, slip in grooves and sheaths that are positioned anterior to the knee joint and posterior to the ankle joint in such a way that the weight of the resting bird tightens the tendons and

Figure 17–22 Large, flightless, terrestrial birds: (a) elephantbird of Madagascar and (b) *Diatryma*, compared with (c) the cursorial ostrich.

curls the toes around the perch. Thus, muscular contraction is not required to hold the toes closed. Furthermore, the tendons lying underneath the toe bones have hundreds of minute, rigid, hobnail-like projections that mesh with ridges on the inside surface of the surrounding tendon sheath. The projections and ridges lock the tendons in place in the sheaths and help to hold the toes in their grip around the branch.

Birds that wade are specialized for walking at various depths and over various types of bottoms. Every shoreline community has its complement of short-legged, intermediate, and long-legged waders, each exploiting a somewhat different wading and feeding niche. A similar stratification of species can often be seen in mixed species aggregations of herons in a southern swamp such as the Florida Everglades (Figure 17–23). If the bottom substrate is soft mud, wading birds require feet with longer toes, as seen in herons, or with partial webbing between short toes (flamingos, avocets, and others).

Birds that walk on insubstantial surfaces usually have long toes (Figure 17–24). The jacanas or lily-trotters, which walk on floating vegetation, have the longest toes and claws relative to their body size of any birds. Certain sandgrouse have especially short and broad claws and densely feathered toes, the whole foot forming a broad base for the bird when walking in loose sand. The winter plumage of north-

ern grouse has similar modifications of the feet for walking on snow.

Some birds have poorly developed pedal locomotion. Highly aerial species such as swifts, swallows, and some goatsuckers use their feet for little more than clinging to a perch, whereas very specialized aquatic forms such as loons and grebes have their legs positioned so far back to maximize propulsion through the water that they cannot walk upright on land. Penguins stand and walk upright, but their legs are so short that they cannot move very fast. They have developed a mode of tobogganing over snow and sometimes use their modified wings as well as their legs for propulsion, one of the few cases in which the forelimbs are used by birds for nonaerial locomotion on land.

Climbing

Birds climb on tree trunks or other vertical surfaces by using their feet, tails, beaks, and, rarely, their forelimbs. Several distantly related groups of birds have independently acquired specializations for climbing and foraging on vertical tree trunks. Species such as woodpeckers and woodcreepers, which use their toes as supports, begin foraging near the base of a tree trunk and work their way vertically upward, head first, clinging to the bark with strong feet on short legs. The tail is used as a prop to brace the body

Figure 17–23 The water depths at which wading birds forage are determined in part by the length of their legs. From left to right this figure shows species that feed in water of progressively greater depth: green heron, little blue heron, reddish egret, common egret, and great blue heron.

against the powerful pecking exertions of head and neck, and the pygostyle and free caudal vertebrae in these species are much enlarged and support strong, stiff tail feathers. A similar modification of the tail is found in certain swifts that perch on cave walls and inside chimneys.

Nuthatches and similarly modified birds climb on trunks and rock walls in both head-upward and head-downward directions while foraging, and in these species, which do not use their tails for sup-

port, the claw on the hallux is larger than those on the forward-directed toes and is strongly curved.

Although a few nestling birds clamber about with their wings, only the young hoatzin of South America has evolved a special modification of the forelimbs for climbing. Hoatzins take a long time to begin to fly—60 to 70 days. Until they can fly, they clamber through vegetation using large claws on the first and second digits of the wing that are moved by special muscles. Later in life the claws fall off, and

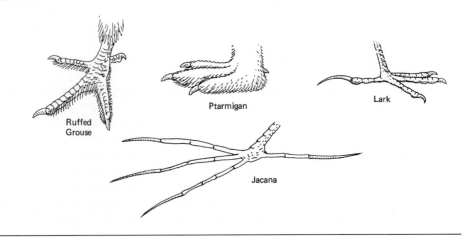

Figure 17–24 Specializations of the feet of birds that walk on insubstantial surfaces like snow (ruffed grouse, ptarmigan), sand (lark), or lily pads (jacaña).

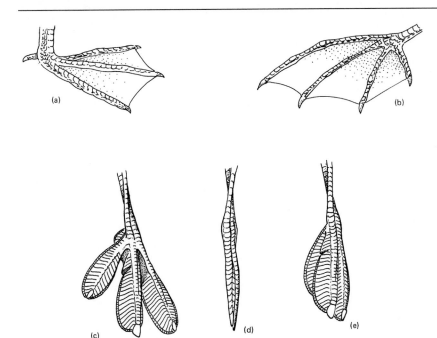

Figure 17–25 Webbed and lobed feet of some aquatic birds: (a) duck, showing partial webbing; (b) cormorant, showing totipalmate webbing. The lobed foot of a grebe showing how it is rotated during a stroke: (c) position of toes during backward power stroke in side view; (d) front view; (e) side view of the rotated foot during the forward recovery stroke. (From R. T. Peterson, 1978, *The Birds*, 2d edition, Time-Life Books, New York, NY.)

the wing of the adult assumes a typical avian condition, but even the adults are weak fliers and they continue to use their wings to help in climbing among the dense branches of their tropical, swampy habitat. These characteristics of hoatzins may be related to their herbivorous diet. The large crop associated with foregut fermentation of plant material takes up a great deal of space in the trunk, and the sternum is reduced. As a result, there is less area for attachment of flight muscles than in other birds of the same size.

Swimming on the Surface

Although no birds have become fully aquatic like the ichthyosaurs and cetaceans, nearly 400 species are specialized for swimming. Nearly half of these aquatic species also dive and swim under water.

Modifications of the hind limbs are the most obvious avian specializations for swimming. Other changes include a wide body that increases stability in water, dense plumage that provides buoyancy and insulation, a large preen gland, producing oil that waterproofs the plumage, and structural modifications of the body feathers that retard penetration of water to the skin. The legs are near the rear of a bird's body where the mass of leg muscles interferes least with streamlining and where the best control of steering can be achieved.

The feet of aquatic birds are either webbed or lobed (Figure 17–25). Webbing between the three forward toes (palmate webbing) has been independently acquired at least four times in the course of avian evolution. Totipalmate webbing of all four toes is found in pelicans and their relatives.

Lobes on the toes have evolved convergently in several phylogenetic lines of aquatic birds. There are two different types of lobed feet. Grebes are unique in that the lobes on the outer sides of the toes are rigid and do not fold back as the foot moves forward. A grebe rotates its foot 90 degrees so that the inner side points forward, the toes with their lobes slicing through the water like knife blades for an efficient recovery stroke with minimum drag. A simpler mechanism for the recovery stroke occurs in all the other lobe-footed swimmers where the lobes are flaps that fold back against the toes during forward movement through water and flare open to present a maximum surface on the backward stroke.

Diving and Swimming Under Water

The transition from a surface swimming bird to a subsurface swimmer has occurred in two fundamentally different ways: either by further specialization of a hind limb already adapted for swimming or by modification of the wing for use as a flipper under water. Highly specialized foot-propelled divers have evolved independently in grebes, cormorants, loons, and the extinct Hesperornithidae. All these families except the loons include some flightless forms (Figure 17–26). Wing-propelled divers have evolved in the Procellariiformes (the diving petrels), the Sphenisciformes (penguins), and the Charadriiformes (auks and related forms). Only among the waterfowl are there both foot-propelled and wing-propelled diving ducks, but none of these species is as highly modified for diving as specialists such as the loons or auks. The water ouzels or dippers are passerine birds that dive and swim under water with great facility using their small, round wings, but they lack any other morphological specializations.

Other morphological, behavioral, and physiological modifications are important in the evolution of diving birds, such as buoyancy and oxidative metabolism under water. Diving birds overcome buoyancy by reducing the volume of their air sacs, having bones with reduced pneumaticity, expelling air from their plumage before submerging, and in the case of some penguins, by swallowing small stones that act as ballast. Like other diving air-breathers, birds constrict their peripheral blood flow, reduce heart rate, and otherwise lower their metabolic rate while under water. They also tend to have a large blood volume with a high oxygen-carrying capacity and muscles that are especially rich in myoglobin, they are able to tolerate high carbon dioxide levels in blood, and they can obtain considerable energy from anaerobic metabolism. These fea-

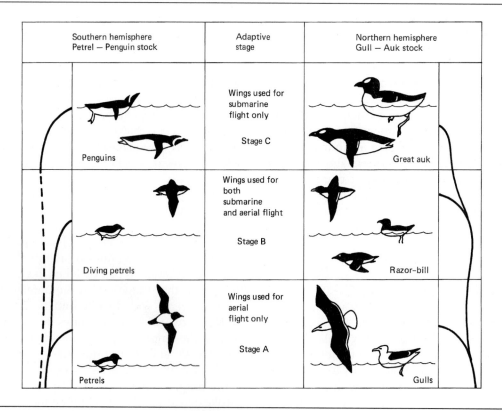

Figure 17–26 Parallel evolution of swimming and diving birds in the Southern Hemisphere (Procellariiformes and Sphenisciformes) and the Northern Hemisphere (Charadriiformes). (From R. W. Storer, 1971, in *Avian Biology*, volume 1, edited by D. S. Farner and J. R. King, Academic, New York, NY.)

tures allow diving times from 1 to 3 minutes, with a maximum recorded survival time of 15 minutes.

The Sensory Systems

A bird moves rapidly through three-dimensional space and requires a continuous flow of sensory information about its position and the presence of obstacles in its path. Vision and hearing are the senses best suited to provide this sort of information on a rapidly renewed basis, and birds have well-developed visual and auditory systems. Their predominance is reflected in the brain: The optic lobes are large and the tectum is an important area for processing visual and auditory information. Olfaction is relatively unimportant for most birds and the olfactory lobes are small. The cerebellum, which coordinates body movements, is large. The cerebrum is less developed in birds than it is in mammals and is dominated by the corpus striatum. Many aspects of the behavior of birds are relatively stereotyped in comparison to the more plastic behavioral responses of mammals, and that difference may reflect the greater development of the neopallium of mammals.

Vision

The eyes of birds are large—so large that the brain is displaced dorsally and caudally and in many species the eyeballs meet in the midline of the skull. The eyes of some hawks, eagles, and owls are as large as the eyes of humans. In its basic structure the eye of a bird is like that of any other vertebrate, but the shape varies from a flattened sphere to something approaching a tube (Figure 17–27). An analysis of the optical characteristics of birds eyes suggests that these differences are primarily the result of fitting large eyes into small skulls. The eyes of a starling are small enough to be contained within the skull, whereas the eyes of an owl bulge out of the skull. An owl would require an enormous, unwieldy head to accommodate a flat eye like that of a starling. The tubular shape of the owl's eye allows it to fit into a reasonable size skull.

Although the basic structures of vertebrate eyes are similar (Chapter 3), the methods of focusing vary (Figure 17–28). Mammalian eyes have spherical lenses that account for most of the bending of light rays that focuses them on the retina. The focus is adjusted to accommodate nearby or distant objects by contracting or relaxing muscles within the ciliary body that change the shape of the lens. In birds, both the cornea and the lens contribute to focus, and accommodation is produced by changing the curvature of both structures. Muscles within the ciliary body change the shape of the lens, and a second set of muscles that are associated with the cornea change its curvature.

The pecten is a conspicuous structure in the eye of birds (Figure 17–27). It is formed exclusively of blood capillaries surrounded by pigmented tissue and covered by a membrane; it lacks muscles and

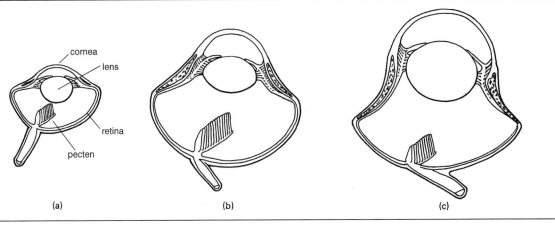

Figure 17–27 Variation in the shape of the eye of birds: (a) flat, typical of most birds; (b) globular, found in most falcons; (c) tubular, characteristic of owls and some eagles.

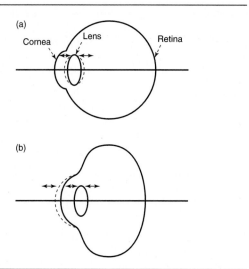

Figure 17–28 Accommodation mechanisms of vertebrate eyes. (a) Mammalian eye: contraction and relaxation of muscles associated with the ciliary bodies change the shape of the lens—more spherical for close vision, flatter for distant objects. (b) Bird eye: in addition to the change in shape of the lens, a second set of muscles anchored around the corneal margin change the curvature of the cornea. (Modified from G. R. Martin, 1987, *Nature* 328:383.)

nerves. The pecten arises from the retina at the point where the nerve fibers from the ganglion cells of the retina join to form the optic nerve. In some species of birds the pecten is small, but in other species the pecten extends so far into the vitreous humor of the eye that it almost touches the lens. The function of the pecten remains uncertain after 200 years of debate. Possible functions for the pecten in vision have been proposed (reduction of glare, a mirror to reflect objects above the bird, production of a stroboscopic effect, a visual reference point), but none seems very likely. The pecten is formed largely by blood capillaries, and its highly vascularized structure suggests that it may provide nutrition for the retinal cells and help to remove metabolic waste products that accumulate in the vitreous humor.

Oil droplets are found primarily in the cone cells of the avian retina, as they are in other vertebrates. The droplets range in color from red through orange and yellow to green; red droplets occur only in birds and turtles. Different color oil droplets are associated with various types of cone cells, and these cells are localized in different areas of the retina. For example, the central dorsal part of the retina of the

pigeon has a predominance of photoreceptors with red droplets and the ventral and lateral parts of the retina have cells with yellow droplets. The oil droplets act as filters, absorbing some wavelengths of light and transmitting others. The function of the oil droplets is unclear and is certainly complex because the various colors of droplets are combined with different kinds of photoreceptor cells and different visual pigments. Birds like gulls, terns, gannets, and kingfishers that must see through the surface of water have a preponderance of red droplets. Aerial hawkers of insects (swifts, swallows) have predominantly yellow droplets.

Hearing

In birds, as in other diapsids, the columella and its cartilaginous extension, the extracolumella, transmit vibrations of the tympanum to the oval window of the inner ear. The cochlea of birds has the same basic structure as that of other vertebrates (Chapter 3), but it appears to be specialized for fine distinctions of the frequency and temporal pattern of sound. The cochlea of a bird is about one-tenth the length of the cochlea of a mammal, but it has about ten times as many hair cells per unit of length. The space above the basilar membrane (the scala vestibuli) is nearly filled in birds by a folded, glandular tegmentum. This structure may damp sound waves, allowing the ear of a bird to respond very rapidly to changes in sounds.

The openings of the external auditory meatus are small, only a few millimeters in diameter in most birds, and are covered with feathers that may ensure laminar airflow across the opening during flight. The columellar muscle inserts on the columella. On contraction this muscle draws the columella away from the oval window, decreasing the sound energy transmitted to the inner ear. This mechanism may protect the ear against the noise of wind during flight and it may also help to tune the auditory system. In starlings contraction of the columellar muscle increases the effect of the middle ear as a filter for sounds of different frequencies. The sounds that most readily pass the filter are those in the range of greatest auditory sensitivity, which, in most birds, corresponds to the frequency range of their own vocalizations.

Localization of sounds in space can be difficult for small animals such as birds. Large animals localize the source of sounds by comparing the time of arrival, intensity, or phase of a sound in their left and right ears, but none of these methods is very effective when the distance between the ears is small. The pneumatic construction of the skulls of birds may allow them to use sound that is transmitted through the air-filled passages between the middle ears on the two sides of the head to increase their directional sensitivity. If this is true, internally transmitted sound would pass from the middle ear on one side to the middle ear on the other side and reach the *inner* surface of the contralateral tympanum. Here it would interact with the sound arriving on the external surface of the tympanum via the external auditory meatus. The vibration of each tympanum would be the product of the combination of pressure and phase of the internal and external sources of sound energy, and the magnitude of the cochlear response would be proportional to the difference in pressure across the tympanic membrane. Localization of sound by owls is apparently further aided by asymmetry of the external auditory openings (Box 17–2).

The sensitivity of the auditory system of birds is approximately the same as that of humans, despite the small size of birds' ears. Most birds have tympanic membranes that are large in relation to the size of the head. A large tympanic membrane enhances auditory sensitivity, and owls (which have especially sensitive hearing) have the largest tympani relative to their head size among birds. Sound pressures are amplified during transmission from the tympanum to the oval window of the cochlea because the area of the oval window is smaller than the area of the tympanum. The reduction ratio for birds ranges from 11 to 40. High ratios suggest sensitive hearing, and the highest values are found in owls; songbirds have intermediate ratios (20 to 30). (The ratio is 36 for cats and 21 for humans.) The inward movement of the tympanum as sound waves strike it is opposed by air pressure within the middle ear, and birds show a variety of features that reduce the resistance of the middle ear. The middle ear is continuous with the dorsal, rostral, and caudal air cavities in the pneumatic skulls of birds. In addition to potentially allowing sound waves to be transmitted to the contralateral ear, these interconnections increase the volume of the middle ear and reduce its stiffness, thereby allowing the tympanum to respond to faint sounds. Owls have more spacious middle ear cavities and more extensive communication of the middle ear with the air spaces of the skull than do other birds.

Olfaction

The sense of smell is well developed in some birds and poorly developed in others; a review of the olfactory systems of birds can be found in Bang and Wenzel (1985). The size of the olfactory bulbs is a rough indication of the sensitivity of the olfactory system. Relatively large bulbs are found in ground nesting and colonial nesting species, species that are associated with water, and carnivorous and piscivorous species of birds. Some birds use scent to locate prey. The kiwi, for example, has nostrils at the tip of its long bill and finds earthworms underground by smelling them. Turkey vultures follow airborne odors of carrion to the vicinity of a carcass, which they then locate by sight. Sponges soaked in fish oil and placed on floating buoys attracted shearwaters, fulmars, albatrosses, and petrels even in the dark.

Olfaction probably plays a role in the orientation and navigation abilities of some birds. The tube-nosed seabirds (albatrosses, shearwaters, fulmars, and petrels) nest on islands, and when they return from foraging at sea they approach the islands from downwind. Homing pigeons use olfaction (as well as other mechanisms) to navigate (Ioalè et al. 1990). The well-developed nasal bulbs of colonial nesting species of birds suggest the possibility that olfaction is used for social functions such as recognition of individuals, but this has never been demonstrated.

Other Senses

Birds use a variety of cues for navigation, and many of these are extensions of the senses we have discussed into regions beyond the sensitivity of mammals. For example, birds can detect polarized light when it falls on the area of the retina that normally receives skylight, but other parts of the retina cannot detect polarization. Ultraviolet light (light with a wavelength of less than 400 nanometers) is not detected by most mammals, but pigeons are more sensitive to ultraviolet light in the region of 350 nanometers than they are to light at any wavelength in the visible part of the spectrum.

Some birds are sensitive to very small differences in air pressure. Experiments showed that pigeons were able to detect the difference in air pressure between the ceiling and floor of a room, and a pigeon flying upward at a rate of 4 centimeters per second would be able to detect a change in its alti-

Box 17–2 Not Hearing Straight: Ear Asymmetry of Owls

Owls are acoustically the most sensitive of birds. At frequencies up to 10 kilohertz (10,000 cycles per second), the auditory sensitivity of an owl is as great as that of a cat. Owls have large tympanic membranes, large cochleae, and well-developed auditory centers in the brain. Some owls are diurnal, others crepuscular (active at dawn and dusk), and some are entirely nocturnal. In an experimental test of their capacities for acoustic orientation, barn owls were able to seize mice in total darkness (Konishi 1973). If the mice towed a piece of paper across the floor behind them, the owls struck the rustling paper instead of the mouse, showing that sound was the cue they were using.

A distinctive feature of many owls is the facial ruff that is formed by stiff feathers. The ruff acts as a parabolic sound reflector, focusing sounds with frequencies above 5 kilohertz on the external auditory meatus and amplifying them by 10 decibels. The ruffs of some owls are asymmetric, and that asymmetry appears to enhance the ability of owls to locate prey. When the ruffs were removed from the barn owls in Konishi's experiments, the owls made large errors in finding targets.

Asymmetry of the aural system of owls goes beyond the feathered ruff. The skull itself is markedly asymmetric in many owls (Figure 17–29), and these are the species with the greatest auditory sensitivity (Norberg 1977). The asymmetry ends at the external auditory meatus; the middle and inner ears of owls are bilaterally symmetric. The asymmetry of the external ear openings of owls assists with localization of prey in the horizontal and vertical axes (Norberg 1978). The time at which sounds are received by the two ears can tell the horizontal direction of the source, and owls are capable of detecting differences of a few hundredths of a millisecond in the arrival times of a sound at the left and right ears. The vertical direction of a sound source can be determined by the differential sensitivity of the two ears to sounds coming from above and below the level of the owl. The soft tissue and feathers surrounding the face of an owl produce

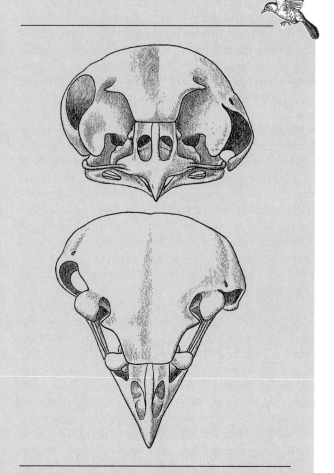

Figure 17–29 The skulls of many owls show pronounced asymmetry in the position of the external auditory meatus that assists in localization of sound. (From R. Å. Norberg, 1978, *Philosophical Transactions of the Royal Society of London B* 282:325–410.)

a situation of reversed asymmetry in which the actual directional sensitivity of the ears is the reverse of what would be expected from examination of the skull. The left ear, which opens low on the side of the head, is most sensitive to sounds that originate *above* the level of the owl's head, and the right ear is sensitive to sounds coming from *below* the level of the head. This auditory asymmetry applies only to sounds with frequencies above 6000 hertz; sensitivity to lower frequencies is bilaterally symmetrical.

tude of 4 millimeters. This sensitivity may be useful during flight, and may also play a role in allowing birds to anticipate changes in weather patterns that are important in the timing of migration.

Another kind of sensory information that may tell birds about the weather is infrasound—very low frequency sounds (less than 10 cycles per second) that are produced by large-scale movements of air. Thunderstorms, winds blowing across valleys, and other geophysical events produce infrasound that is propagated over thousands of miles. The lower-frequency limit of human hearing is around 20 hertz, but birds can hear well into the region of infrasound.

Pigeons, for example, can detect frequencies as low as 0.05 hertz (3 cycles per *minute*).

The magnetic field of the earth provides a cue that could be used for orientation if an animal were able to perceive it, and considerable evidence suggests that birds are able to detect magnetic fields. The orientation of several species of birds during their migratory periods can be adjusted in predictable ways by placing them in an artificial magnetic field. The mechanism by which the magnetic field is detected is not known, but deposits of magnetite recently described in the heads of pigeons may be involved (Wiltschko and Wiltschko 1988).

Summary

Feathers are the distinguishing character of birds, and flight is the distinctive mode of avian locomotion. In many respects the morphology of birds is shaped by the demands of flight. Flapping flight is a more complicated process than flight with fixed wings like those of aircraft, but it can be understood in aerodynamic terms. Feathers compose the aerodynamic surfaces responsible for lift and propulsion during flight and they also provide streamlining. There are many variations and specializations within the four basic types of wings, but in general high speed and elliptical wings are used in flapping flight, whereas high aspect ratio and high-lift wings are used for soaring and gliding. The hollow birds of bones, probably an ancestral character of the archosaur lineage, combine lightness and strength. The air sacs, some of which fill several of the pneumatic bones, create a one-way flow of air through the lung that enhances oxygen uptake, and the air sacs probably also help to dissipate the heat produced by muscles during flight.

Some parts of the skeletons of birds are light in relation to the sizes of their bodies, and some of this lightness has been achieved by modification of the skull. Several skull bones that are separate in other diapsids are fused in birds, air sacs are extensively developed, and teeth are absent. The function of teeth in processing food has been taken over by the muscular gizzard, which is part of the stomach of birds. The gizzard is well developed in birds that eat hard items such as seeds and it may contain stones that probably assist in grinding food. Vision is the primary sense for most birds, and visual capacities extend to detection of ultraviolet and polarized light. Hearing is also important and the auditory system of birds is capable of very precise discriminations of the frequencies and temporal patterns of sound. Sensitivity extends downward to include sounds with frequencies lower than 10 cycles per second. Sensitivity to magnetism has been demonstrated for several species of birds, but the mechanisms involved are unknown. Olfactory sensitivity is variably developed among birds: Some species use scent to locate food and for navigation, but other species appear not to use olfaction at all.

References

Bang, B. G., and B. M. Wenzel. 1985. Nasal cavity and olfactory system. Pages 195–225 in *Form and Function in Birds*, volume 3, edited by A. S. King and J. McLelland. Academic, London, UK.

Gaunt, A. S., S. L. L. Gaunt, and R. M. Casey. 1982. Syringeal mechanics reassessed: evidence from *Streptopelia. Auk* 99:474–494.

Grajal, A. S., D. Strahl, R. Parra, M. G. Dominguez, and A. Neher. 1989. Foregut fermentation in the hoatzin, a Neotropical leaf-eating bird. *Science* 245:1236–1238.

Greenewalt, C. H. 1968. *Bird Song: Acoustics and Physiology.* Smithsonian Institution Press, Washington, DC.

Greenewalt, C. H. 1969. How birds sing. *Scientific American* 221:126–139.

Ioalè, P., M. Nozzolini, and F. Papi. 1990. Homing pigeons do extract directional information from olfactory stimuli. *Behavioral Ecology and Sociobiology* 26:301–306.

Konishi, M. 1973. How the owl tracks its prey. *American Scientist* 61:414–424.

Norberg, R. Å. 1977. Occurrence and independent evolution of bilateral ear asymmetry in owls and implications for owl taxonomy. *Philosophical Transactions of the Royal Society of London* B280:375–408.

Norberg, R. Å. 1978. Skull asymmetry, ear structure and function, and auditory localization in Tengmalm's owl, *Aegolius funereus* (Linné). *Philosophical Transactions of the Royal Society of London* B282:325–410.

Norberg, U. M. 1985. Flying, gliding, and soaring. Pages 129–158 in *Functional Vertebrate Morphology*, edited by M. Hildebrand, D. M. Bramble, K. F. Liem, and D. B. Wake. Harvard University Press, Cambridge, MA.

Nowicki, S. 1987. Vocal tract resonances in oscine bird sound production: evidence from birdsongs in a helium atmosphere. *Nature* 325:53–55.

Raikow, R. J. 1985. Locomotor system. Pages 57–147 in *Form and Function in Birds*, volume 3, edited by A. S. King and J. McLelland. Academic, London, UK.

Rayner, J. M. V. 1988. Form and function in avian flight. Pages 1–66 in *Current Ornithology*, volume 5, edited by R. J. Johnston. Plenum, New York, NY.

Wiltschko, W., and R. Wiltschko. 1988. Magnetic orientation in birds. Pages 67–121 in *Current Ornithology*, volume 5, edited by R. F. Johnston. Plenum, New York, NY.

THE ECOLOGY AND BEHAVIOR OF BIRDS

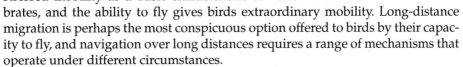

18

I n the preceding chapter we examined the functional relationships among structural characteristics of birds and the physical and biological requirements of flight. In this chapter we consider some of the consequences of flight for the biology of birds. We have stressed mobility as a basic characteristic of vertebrates, and the ability to fly gives birds extraordinary mobility. Long-distance migration is perhaps the most conspicuous option offered to birds by their capacity to fly, and navigation over long distances requires a range of mechanisms that operate under different circumstances.

A second characteristic of birds is diurnality; most species are active only by day. We know that diurnality is not necessarily related to flight, because bats are nocturnal, but the diurnal activity of birds has made them popular and accessible subjects for biological field studies. Two important areas of modern biology, optimal foraging theory and the behavioral ecology of mating systems, have drawn heavily on studies of birds for data that can be generalized to other vertebrates. Both types of studies emphasize the role of behavior in enhancing the fitness of individuals, and the study of reproductive biology, in particular, includes examples in which individual fitness may be maximized by cooperating with other individuals. These altruistic behaviors are normally directed toward relatives of the individual engaging in the behavior, as predicted by the theory of inclusive fitness explained in Chapter 1.

The Evolution of Birds

The fossil record of birds, which begins with *Archaeopteryx* in Jurassic sediments from Germany, is extensive and has been the subject of several reviews (Martin 1983a,b; Olson 1985; Cracraft 1986; Lockley et al. 1992). Fossils of true flying birds are known from Cretaceous deposits in Europe, Asia, and North and South America, and two extant orders of birds, the Charadriiformes (shorebirds) and Procellariiformes (tube-nosed seabirds), can be traced back to the Cretaceous. By the early Tertiary birds

were diverse, and several extant orders and families were represented.

Early Birds

Archaeopteryx was probably a late-surviving relict that was contemporaneous with more typically avian birds. It had teeth, claws on the fingers, and a long tail. If it were not for the presence of feathers, *Archaeopteryx* could readily be classified as a small dinosaur. However, several lines of evidence suggest that *Archaeopteryx* was capable of flight (Chapter 13):

Its skeletal proportions were similar to those of some extant flying birds, the number of primaries and secondaries were identical to those of extant birds, and the asymmetry of its flight feathers is like that seen in extant flying birds, and the furcula was large. On balance, these characteristics are more consistent with the view that *Archaeopteryx* was a flying bird than with Ostrom's earlier proposal that *Archaeopteryx* represented a transitional, preflight stage of avian evolution (Ostrom 1974, 1979; Martin 1983b; Rayner 1988).

The next birds known after *Archaeopteryx* are from early Cretaceous deposits in China (*Sinornis*, Figure 18–1) and Spain (*Iberomesornis* and *Concornis*). *Sinornis* and *Concornis* were the size of sparrows, and *Iberomesornis* was about twice as large. All had pectoral girdles that were more derived than that of *Archaeopteryx*, with strut-like coracoids, a furcular process, and an ulna that was longer than the humerus. In addition, all had a pygostyle instead of a long tail. *Sinornis* and *Concornis* had a keeled sternum. The vertebral column contained only 11 dorsal vertebrae, compared to 14 for *Archaeopteryx*. The short trunk and tail of *Sinornis* shifted the center of mass toward the forelimbs, as in extant birds, rather than toward the hind limbs as in cursorial terrestrial archosaurs, and many of its derived characters are associated with flight (Sereno and Chenggang 1992). For example, its wrist could bend backward sharply, as in extant birds, so the wing could be tucked against the body. These early Cretaceous birds retained ancestral characters as well, including teeth in the jaws and separate metatarsal bones in the hand.

Slightly younger is *Ambiortus*, which was found in Cretaceous deposits in Mongolia. *Ambiortus* had a well-developed keel on the sternum, and its coracoid, scapula, and furcula were typically avian. A third early Cretaceous bird, *Enaliornis*, was found in England. *Enaliornis*, which is placed in the order Hesperornithiformes, was a specialized diving bird that was probably flightless.

Two flightless birds are known from the late Cretaceous, *Patagopteryx* and the remarkable *Mononykus olecranus*. *Mononykus* means "one claw," and refers to the peculiar structure of the forelimb, which is short and ends in a stout claw (Altangerel et al. 1993a,b). In other respects, *Mononykus* shows derived features of birds, including a keeled sternum, a fibula that was reduced to a small spike of bone, and fused metacarpals.

Figure 18–1 *Sinornis*, a flying bird from early Cretaceous lakebed deposits in China. (From P. C. Sereno and R. Chenggang, 1992, *Science* 255:845–848. © AAAS 1992.)

In addition to these fossils, an abundance of fossilized feather impressions and tracks indicates that by the early Cretaceous birds inhabited both the Northern and Southern hemispheres. The differences among the fossil species known—flying and flightless birds and a foot-propelled diver—combined with the wide geographic distribution of birds in the Cretaceous suggest that more of the evolution of birds took place in the Jurassic than has previously been appreciated. Avian evolution is an area of lively controversy, and views of the phylogenetic relationships of birds are likely to change substantially as additional material is studied.

Martin (1983a) has suggested that a split occurred early in the evolution of birds, producing two lineages—the Sauriurae and Ornithurae (lizard-tails and bird-tails, respectively). The sauriurine lineage included *Archaeopteryx* and a group of plesiomorphic birds known as Enantiornithes (opposite birds). The anterior thoracic vertebrae were fused in all sauriurines, and an anteromedial process of the scapula abutted these vertebrae. The enantiornithians were flying forms with well-developed wings, but the keel on the sternum was not large and the coracoids may have been important sites of muscle origins. The enantiornithians had a bird-like wing skeleton that included a fused carpometacarpus, but the metatarsal morphology was so dinosaur-like that their identification as birds was initially received with skepticism. Enantiornithians are known from fossil deposits in both the Northern and Southern hemispheres, and the material from Argentina is particularly diverse. Enantiornithians were apparently predators, and one form had raptorial claws like those of extant hawks.

Most enantiornithians were crow-size to heron-size, but *Alexornis* was smaller, only the size of a sparrow. *Gobipteryx*, an enantiornithian known from skulls found in late Cretaceous deposits in Mongolia, is unique among Cretaceous birds in lacking teeth. In the same deposits that yielded fossils of adult *Gobipteryx*, the Polish Mongolian Expedition found fossilized eggs, some of which contained well-preserved skeletons with skulls very like those of adult *Gobipteryx*. The skeletons of these embryos were well developed, and it is likely that the chicks were precocial at hatching.

The other half of the Mesozoic avian dichotomy proposed by Martin is the Ornithurae, which includes all the extant birds. The earliest ornithurines known are the Hesperornithiformes (western birds). *Enaliornis* establishes the presence of hesperornithiforms in the early Cretaceous, and they were well established by the late Cretaceous. The hesperornithiforms were medium-size to large flightless birds that were specialized for foot-propelled diving (Figure 18–2). The body and neck were elongate, the

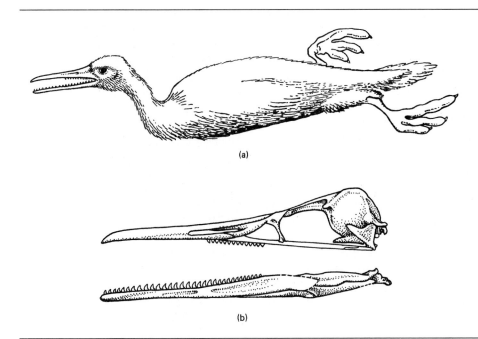

(a)

(b)

Figure 18–2 The hesperornithiforms were flightless, toothed birds. (a) Restoration of *Hesperornis*; (b) skull of *Parahesperornis*. Note the teeth in the maxilla and dentary bones. [(a) From A. Feduccia, 1980, *The Age of Birds*, Harvard University Press, Cambridge, MA; (b) from L. D. Martin, 1983, in *Perspectives in Ornithology*, edited by A. H. Brush and G. A. Clark, Jr., Cambridge University Press, Cambridge, UK.]

sternum lacked a keel, and the bones were not pneumatic. Teeth remained in the maxilla and dentary. Feathers preserved with two specimens of *Parahesperornis* were plumulaceous and the birds may have had a furry appearance somewhat like that of the extant kiwis. The feet were placed far posteriorly on the body (a position that is characteristic of many foot-propelled diving birds) and the toes had lobes like those seen in extant grebes. The femur and tibiotarsus were locked in place and could not be rotated under the body. As a result, hesperornithiforms would not have been able to walk on land and probably pushed themselves along by sliding on their stomachs. The lateral placement of the feet would have made it possible for hesperornithiforms to exert force directly backward during swimming and diving without an upward component that would have tended to drive them toward the surface. Pachyostosis (an increased density of bone) gave hesperornithiforms a high specific gravity that would have facilitated diving. Coprolites (fossilized feces) found in association with *Hesperornis* and *Baptornis* contain the remains of small fishes.

Another toothed bird, *Ichthyornis*, is known from the same late Cretaceous deposits in Kansas that contain *Hesperornis*, but the ichthyornithiforms were flying birds with a well-developed keel on the sternum. Several species of *Ichthyornis* have been named, mostly on the basis of differences in size. In general, ichthyornithiforms were the size of gulls and terns, and they may have had similar habits. By the late Cretaceous these toothed birds were probably oceanic relicts like extant frigate birds.

The Evolution of Derived Orders and Families of Birds

Two extant orders of birds have been identified in the late Cretaceous, the Charadriiformes (wading birds) and perhaps the Procellariiformes (tube-nosed seabirds). The charadriiforms are abundant and diverse in deposits from New Jersey and Wyoming, whereas the presence of Procellariiformes is deduced from the presence of only two bones in the New Jersey deposits.

Most orders of birds had evolved by the end of the Eocene some 40 to 50 million years ago. Groups associated with deposits of Eocene age include ostriches (Struthioniformes), rheas (Rheiformes), the recently extinct elephantbirds of Madagascar (Aepyornithiformes), penguins (Sphenisciformes), the

tube-nosed marine birds (Procellariiformes), the kingfisherlike birds (Coraciiformes), woodpeckers (Piciformes), and the perching birds (Passeriformes). The first diurnal birds of prey (Falconiformes) had already appeared in the Paleocene. Five other orders first appear in the fossil record around the end of the Eocene or the beginning of the Oligocene: doves and pigeons (Columbiformes), cuckoos (Cuculiformes), goatsuckers (Caprimulgiformes), swifts (Apodiformes), and trogons (Trogoniformes), so that by the beginning of the Oligocene at least 26 orders of birds had become differentiated.

A major radiation of avian families occurred during the Tertiary (Feduccia 1995). The Eocene was the epoch of greatest diversification of birds. More surviving families arose in that period than at any other time. Most of these families consist of additional water birds and nonpasserine forest dwellers. A second radiation occurred in the Miocene and included a few additional families of water birds, but mostly land-dwelling passerines that were adapted to drier, less forested environments. Most families of birds had evolved by the end of the Miocene, and many still-existing genera and some species were present by the Pliocene. Birds formed complex ecological communities by the middle of the Tertiary (Warheit 1992).

Phylogeny of Extant Birds

Phylogenetic relationships among extant birds are poorly known and are the subject of continuing controversy. Several views of these relationships and the difficulties inherent in their study can be found in reviews by Cracraft (1986), Olson (1985), Sibley et al. (1988), Raikow (1985), Houde (1986, 1987), and Sibley and Ahlquist (1990).

The lack of consensus about the phylogeny of extant birds makes it impossible to provide a cladogram that represents a widely accepted hypothesis of evolutionary relationships. Charles Sibley and his colleagues (Sibley et al. 1988, Sibley and Ahlquist 1990) have presented a cladistic analysis based on comparisons of DNA. Their classification of passerine birds has been well received, but their analysis of the relationships of nonpasserine birds is more controversial.

One of the major controversies in bird phylogeny centers on the relationships of a group of flightless birds known as ratites. Extant ratites include ostriches (Africa), rheas (South America), emus and cassowaries (Australia), and kiwis (New Zealand). All of these land masses were part of the southern supercontinent Gondwana, and it has been suggested that the ratites arose from a single flightless ancestor that was widely distributed on Gondwana. By this hypothesis the ratites form a monophyletic group and their current geographic distribution reflects the breakup of Gondwana in the late Mesozoic and early Cenozoic.

An alternative view proposes that the similarities of extant ratites are ancestral characters that were found in many lineages of birds. The ratites and another group of birds, the tinamous of Central and South America, share a paleognathous palatal structure. This palate is characterized by long prevomers that extend posteriorly to articulate with the palatines and pterygoids and by large basipterygoid processes that articulate with the pterygoids. The paleognathus palate is distinguished from the neognathus palate that is characteristic of other birds. Paleognathous palatal characters are seen in toothed birds such as *Hesperornis* and in the Cretaceous toothless bird *Gobipteryx,* and they are present in early stages of the embryonic development of neognathus birds. Thus, the paleognathus condition of ratites and tinamous could have evolved by neoteny from a neognathus form. This possibility means that the paleognathus birds are not necessarily a monophyletic lineage. Furthermore, the discovery of birds related to ostriches in Paleocene and Eocene deposits in North America and Europe casts doubt on the assumption that the current distribution of ratites is the result of events associated with the breakup of Gondwana (Houde 1986).

The phylogeny and zoogeography of ratites are of considerable significance in current biochemical studies of the phylogeny of birds, because Sibley and Ahlquist based the calibration of the DNA molecular clock on the assumption that the separation of the lineages of ratites was caused by the breakup of Gondwana. If the ratites do represent two or more independent origins, or if the origins of some ratites were in the Northern Hemisphere rather than in Gondwana, the calibration of the DNA clock is erroneous and conclusions about relationships among other groups of birds that were based on that calibration are weakened.

The most commonly used classification of birds derives from Storer (1971). Because it is the system used by most references, you will need to be familiar with it to read the ornithological literature, and we

present it in Table 18–1. The Sibley and Ahlquist (1990) hypothesis of phylogenetic relationships is presented in Figure 18–3 and Table 18–2.

Birds as Model Organisms

Birds might almost have been designed as the ideal vertebrate animals for biologists to study. They are diverse (nearly 9700 species), widespread, conspicuous, and largely diurnal. Most birds are visually oriented, and respond to stimuli such as colors, patterns, and movements that humans also are able to perceive. A worldwide corps of amateur ornithologists has helped to assemble the huge amount of information we have about the life history and population ecology of birds.

Studies of birds have contributed to our understanding of vertebrates, especially in areas such as ecology, morphology, and behavior (Konishi et al. 1989). These topics reveal fascinating examples of the interdependence of structure and function that emphasize the importance of broadly integrative studies in organismal biology. In this chapter we focus on aspects of feeding, reproduction, and behavior.

Divergence and Convergence in Feeding

Birds show morphological, physiological, and behavioral specializations associated with feeding on diverse sources of food. Modifications of the beak are often associated with dietary specializations.

Beaks and Tongues The presence of a horny beak in place of teeth is not unique to birds—we have noted the same phenomenon in turtles, rhynchosaurs, dinosaurs, pterosaurs, and the dicynodonts—but the diversity of beaks among birds is remarkable. The range of morphological specializations of beaks defies complete description, but some categories can be recognized (Figure 18–4).

Insectivorous birds such as warblers, which find their food on leaf surfaces, usually have short, thin, pointed bills that are adept at seizing insects, whereas aerial sweepers such as swifts, swallows, and nighthawks, which catch their prey on the wing, have short, weak beaks and a wide gape. Kinesis of the lower jaw substantially increases the

Table 18-1 Classification of extant birds to the level of orders. The geographic regions represent the distribution of the entire order; families within an order often have smaller distributions. Compare the groupings suggested by this traditional classification with those shown in Figure 18–3.

Classification	Approximate Number of		Geographic Region
	Families	Species	
Palaeognathae			
Tinamiformes (tinamous)	1	47	Neotropics
Rheiformes (rheas)	2	2	Neotropics
Struthioniformes (ostriches)	1	1	Africa
Casuariiformes (emus and cassowaries)	2	4	Australia, New Guinea
Dinornithiformes (kiwis)	1	3	New Zealand
Neognathae			
Podicipediformes (grebes)	1	21	Worldwide
Sphenisciformes (penguins)	1	17	Southern Hemisphere
Procellariiformes (albatrosses, shearwaters, petrels)	1	115	Worldwide
Pelecaniformes (tropicbirds, boobies, gannets, cormorants, pelican, frigatebirds)	5	63	Worldwide
Anseriformes (screamers and waterfowl)	4	161	Worldwide
Phoenicopteriformes (flamingos)	1	6	Worldwide, except Australia
Ciconiiformes (herons, bitterns, whale-head stork, storks, ibises, spoonbills)	5	110	Worldwide
Falconiformes (condors, hawks, eagles, kites, falcons, caracaras)	3	304	Worldwide
Galliformes (curassows, guans, chachalacas, megapodes, guineafowl, pheasants, quail, grouse, turkeys)	5	183	Worldwide
Gruiformes (rails, coots, sungrebes, kagu, sunbittern, roatelos, buttonquail, cranes, limpkin, trumpeters, seriemas, bustards)	9	196	Worldwide
Charadriiformes (shorebirds, plovers, sandpipers, gulls, jaegers, skuas, skimmers, terns, auks, murres, puffins, sandgrouse)	11	366	Worldwide

gape of the nightjar; the distance between the two rami of the lower jaw increases from 12.5 millimeters when the mouth is closed to 40 millimeters when it is opened. Stiff feathers at the corners of the mouth further increase the insect-trapping area. It is always the case that one can find exceptions to generalizations about the correspondence between morphology and behavior, and the bills of insectivorous birds reveal many of these. For example, some hornbills use their heavy, simitar-shaped beaks to pick up termites and small insects from the ground, and flycatchers, which snap up insects in flight, have hooked beaks.

Many carnivorous birds, such as gulls, ravens, crows, and roadrunners, use their heavy pointed beaks to kill their prey, but most hawks, owls, and eagles kill prey with their talons and employ their beaks to tear off pieces small enough to swallow. Falcons stun prey with the impact of their dive, and then bite the neck of the prey to disarticulate the cervical vertebrae. The true falcons (the genus *Falco*) have a structure called a tomial tooth that aids in the process. This tooth actually is a sharp projection from the upper mandible that matches a corresponding notch on the bottom mandible. Shrikes, a group of predatory passerine birds that employ neck biting to kill prey, also have a tomial tooth. Fish-eating birds such as cormorants and pelicans have beaks with a sharply hooked tip that is used to seize fish, and mergansers have long, narrow bills with a series of serrations along the sides of the beak in addition to a hook at the tip. Darters and anhingas have harpoon-like bills that they use to impale fish. The massive beak and jaw

Table 18-1 *(Continued)*

| Classification | Approximate Number of | | Geographic Region |
	Families	Species	
Gaviiformes (loons)	1	5	New World, Eurasia
Columbiformes (doves, pigeons)	1	310	Worldwide
Psittaciformes (parrots)	1	358	Pantropical and Australia
Coliiformes (mousebirds)	1	6	Africa
Musophagiformes (turacos)	1	23	Africa
Cuculiformes (cuckoos, hoatzin)	6	143	Worldwide
Strigiformes (owls)	2	186	Worldwide
Caprimulgiformes (nightjars, poorwills, frogmouths, oilbird)	6	105	Worldwide
Apodiformes (swifts, hummingbirds)	3	422	Worldwide
Trogoniformes (trogons, quetzals)	1	39	Pantropical except Australasia
Coraciiformes (kingfishers, todies, motmots, bee-eaters, rollers, hoopoes, hornbills)	14	218	Worldwide
Piciformes (jacamars, barbets, honeyguides, toucans, woodpeckers)	5	355	Worldwide
Passeriformes (perching birds, including the songbirds)			
Tyranni (broadbills, pittas, asities, New Zealand wrens, tyrant flycatchers, cotingas, manakins, woodcreepers, ovenbirds, antbirds, tapaculos)	13	1151	Pantropical
Passeres			
Crows and related forms	14	1101	Worldwide
Thrushes and related forms	4	610	Worldwide
Nuthatches, wrens, and related forms	11	1195	Worldwide
Larks, sparrows, finches, and related forms	6	1651	Worldwide
Wood warblers, tanagers, blackbirds, and related forms	3	781	New World, Eurasia

Based on F. B. Gill, 1989, *Ornithology*, Freeman, New York, NY, and personal communication. The numbers of families and species are based on C. G. Sibley and B. L. Monroe, Jr., 1990, *Distribution and Taxonomy of Birds of the World*, Yale University Press, New Haven, CT.

apparatus of the gigantic predatory bird *Diatryma* was probably used to kill and dismember prey (Box 18–1).

Spoonbills have flattened bills with broad tips that they use to create currents that lift prey into the water column where it can be seized. Many aquatic birds strain small crustaceans or plankton from water or mud with bills that incorporate some sort of filtering apparatus. Dabbling ducks have bills with horny lamellae that form crosswise ridges and their tongues also have horny projections. The tongue and bill are densely invested with sensory corpuscles and form a filter system that allows ducks to scoop up a billfull of water and mud, filter out the prey, and allow the debris to escape. Flamingos have a similar system. The bill of a flamingo is sharply bent and the anterior part is held in a horizontal position when the flamingo lowers its head to feed. The lower jaw of a flamingo is smaller than the upper jaw, and it is the upper jaw that moves during feeding while the lower jaw remains motionless. (This reversal of the usual vertebrate pattern is possible because of the kinetic skull of birds.)

Seeds contain the energy and nutrients that plants have invested in reproduction, and seeds are usually protected by hard coverings (husks) that must be removed before the nutritious contents can be eaten. Specialized seed-eating birds use one of two methods to husk seeds before swallowing them. One group holds the seed in its beak and slices it by making fore-and-aft movements of the lower mandible, whereas birds in the second group hold the seed against ridges on the palate and crack the husk by exerting an upward pressure with their

Table 18-2 Sexual dimorphism in the song control regions of the brains of birds. The average ratio of the volumes of five SCRs in males compared to females (Male:Female) parallels the difference in the sizes of the song repertories of males and females.

	Zebra Finch	Canary	Chat	Bay Wren	Buff-Breasted Wren
SCR volume ratio	4.0:1.0	3.1:1.0	2.3:1.0	1.3:1.0	1.3:1.0
Song repertoire	Males only	Males>>>females	Males>>females	Males = females	Males = females

Source: Modified from E. A. Brenowitz, A. P. Arnold, and R. N. Levin, 1985, *Brain Research* 343:104–112.

robust lower mandible. After the husk has been opened, both kinds of birds use their tongues to remove the contents. Other birds have different specializations for eating seeds: Crossbills extract the seeds of conifers from between the scales of the cones, using the diverging tips of their bills to pry the scales apart. Woodpeckers, nuthatches, and chickadees may wedge a nut or acorn into a hole in the bark of a tree and then hammer at it with their sharp bill until it cracks.

The production of fruits is a strategy of seed dispersal for plants, and birds are important dispersal agents. The bright colors of many fruits advertise their presence or ripeness and attract birds that eat them. The pulp surrounding the seed is digested during its passage through the bird's intestine, but the seeds are not damaged. The birds may discard the seed before eating the pulp, or the seeds may pass through the gut to be voided with the feces. In some cases the chemical or mechanical stress of its passage through the digestive system of a bird facilitates subsequent sprouting by the seed. Fruit-eating birds do not have to penetrate a hard covering, but the very size of the fruit may make it hard to swallow. Skull kinesis can be important for fruit eaters; the widening of the gape as the mouth opens may allow them to swallow large items.

Many birds use their beaks to probe for hidden food. Long-billed shorebirds probe in mud and sand to locate worms and crustaceans. These birds display a form of skull kinesis in which the flexible zone in the upper jaw has moved toward the tip of the beak, allowing the tip of the upper jaw to be lifted without opening the mouth (Figure 18–5). This mechanism enables long-billed waders to grasp prey under the mud.

Tongues are an important part of the food-gathering apparatus of many birds. Woodpeckers drill holes into dead trees and then use their long tongues to investigate passageways made by wood-boring insects. The tongue of the green woodpecker, which extracts ants from their tunnels in the ground, extends four times the length of its beak. The hyoid bones that support the tongue are elongated and housed in a sheath of muscles that passes around the outside of the skull and rests in the nasal cavity (Figure 18–6). When the muscles of the sheath contract, the hyoid bones are squeezed around behind the skull, and the tongue is projected from the bird's mouth. The tip of a woodpecker tongue has barbs that impale insects and allow them to be pulled from their tunnels. Nectar-eating birds such as hummingbirds and sunbirds also have long tongues and a hyoid apparatus that wraps around the back of the skull. The tip of the tongue of nectar-eating birds is divided into a spray of hair-thin projections, and capillary force causes nectar to adhere to the tongue.

Dietary Specializations The morphology of birds' bills is sometimes closely correlated with methods of prey capture or dietary specialization. For example, the spoonbills, *Platalea*, are related to storks and herons. Like their relatives, spoonbills feed on aquatic animals including fishes, amphibians, and crayfish, but they differ from storks and herons in bill structure and in feeding methods.

Storks and herons have long, pointed bills. They locate prey visually while wading in shallow water, and they seize prey items in their open bills. This method of hunting depends on being able to see individual prey items under water, and a heron may extend its wings, shading a patch of water to reduce the reflection from the water's surface. Spoonbills also hunt in this way, but they have a second method of capturing prey that is effective even in murky water (Weihs and Katzir 1994).

Spoonbills take their name from the shape of their beak, which is broadened from side to side,

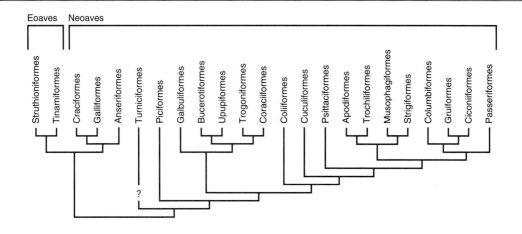

Figure 18–3   A hypothesis of the phylogenetic relationships of extant birds, from C. G. Sibley, and J. E. Ahlquist, 1990, *Phylogeny and Classification of Birds*, Yale University Press, New Haven, CT. This analysis, which is based on comparisons of DNA, suggests groupings different from those shown in Table 18–1. EOAVES: Struthioniformes (ostrich, rheas, cassowaries, emus, kiwis); Tinamiformes (tinamous). NEOAVES: Craciformes (guans, chachalacas, curassows, megapodes); Galliformes (pheasants, quail, grouse, guineafowl); Anseriformes (screamers, ducks, swans, geese); Turniciformes (buttonquail); Piciformes (honeyguides, woodpeckers, barbets, toucans, toucanets); Galbuliformes (jacamars, puffbirds); Bucerotiformes (hornbills); Upupiformes (hoopoes, wood hoopoes, scimitar-bills); Trogoniformes (trogons); Coraciiformes (rollers, motmots, todies, kingfishers, bee-eaters); Coliiformes (colies); Cuculiformes (cuckoos, coucals, hoatzin, anis, roadrunners); Psittaciformes (parrots, macaws); Apodiformes (swifts); Trochiliformes (hermits, hummingbirds); Musophagiformes (turacos, plantain-eaters, go-away-birds); Strigiformes (owls, owlet-nightjars, frogmouths, oilbird, potoos, nightjars, nighthawks, whip-poor-wills); Columbiformes (pigeons, doves); Gruiformes (sunbittern, bustards, cranes, limpkin, trumpeters, seriemas, kagu, rails, gallinules, coots, mesites, monias, roatelos); Ciconiiformes (sandgrouse, seedsnipe, woodcock, snipe, sandpipers, curlews, jacanas, thick-knees, avocets, stilts, plovers, lapwings, jaegers, skuas, skimmers, gulls, terns, auks, murres, puffins, guillemots, osprey, eagles, Old World vultures, hawks, kites, harriers, secretary-bird, falcons, caracaras, grebes, tropicbirds, boobies, gannets, anhingas, darters, cormorants, shags, herons, egrets, bitterns, hamerkop, flamingos, ibises, spoonbills, shoebill, pelicans, New World vultures, condors, storks, openbills, adjutant, jabiru, frigatebirds, penguins, loons, storm-petrels, shearwaters, petrels, albatrosses); Passeriformes (the arrangement of passerine birds is close to that shown in Table 18–1).

especially near the tip (Figure 18–8). The dorsal surface of the beak is curved, and the ventral surface is flat. In clear water a spoonbill wades forward slowly, seizing prey in its beak just as a heron does. In murky water, however, a spoonbill uses a different method of foraging, sweeping its beak through the water just a few centimeters above the bottom. The curved profile of the bill creates a vortex in the water that pulls small objects, including prey animals, off the bottom and up into the water column where they are seized on the next sweep.

Bill morphology can affect the choice of food items on a very fine scale. The African black-bellied finch, *Pyrenestes ostrinus*, includes individuals with two different types of bills—large and small (Smith 1987). There is no sexual dimorphism in bill size, and

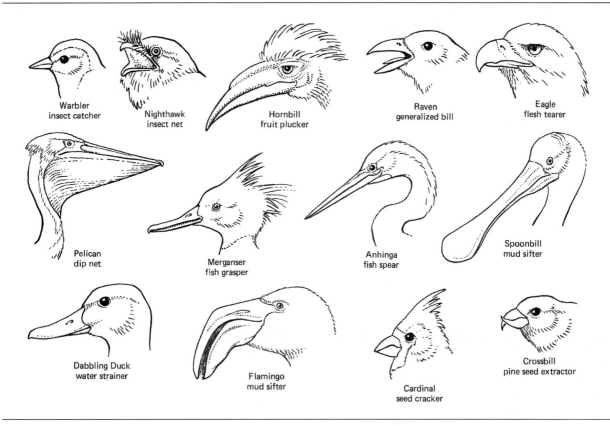

Figure 18–4 Examples of specializations of the beaks of birds.

no evidence of assortative mating. That is, a large-billed individual is equally likely to be a male or female, and is equally likely to mate with an individual of the opposite bill morph. A mixed pair produces young with large and small bills, but no intermediates.

Finches are seed eaters, and bill size affects the ease with which they can eat different kinds of seeds. The African black-bellied finch eats the seeds of different species of sedges, and a range of seed sizes is available. Individuals of the large-billed morph are able to crack hard seeds more rapidly than small-billed individuals can, but small-billed birds consume soft seeds more rapidly than large-billed individuals can. These differences in seed handling time are reflected in the diets of the two morphs. Large-billed individuals are nearly three times as likely as small-billed birds to have hard seeds in their crop, and small-billed birds are three times as likely to have soft seeds (Figure 18–9). Thus, both bill morphs of *Pyrenestes ostrinus* specialize on the sizes of seeds they can consume most effectively.

Foraging Behavior of Birds: The Concept of Optimization

Refined morphological specializations, such as the correspondence of beak shapes and diet, demonstrate such appropriate design that we are intuitively led to hypothesize the existence of similar perfection in physiological and behavioral adaptations to the environment. A general hypothesis is implicit in much of modern biology—that over evolutionary time natural selection has favored organisms with genotypes providing them with characteristics that solve environmental problems in an optimal way for survival and reproduction. In the past quarter century this general hypothesis has seen vigorous testing and modification, in large part by ecologists working with **optimal foraging theory** (OFT) (Krebs et al. 1983, Stephens and Krebs 1987). Numerous other possible optimalities have been investigated, primarily those involving behavioral characteristics of animals, all borrowing from economics the procedures of cost/benefit analysis. However, the concept

Figure 18–5 Long-billed wading birds that probe for worms and crustaceans in soft substrates can raise the tips of the upper bill without opening their mouth. (From P. Béhler, 1981, in *Form and Function in Birds*, volume 2, edited by A. S. King and J. McLelland, Academic, New York, NY.)

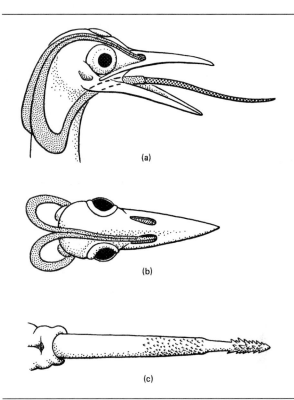

Figure 18–6 Hyoid apparatus of a woodpecker, showing the mechanism for protrusion of the tongue beyond the tip of the mandibles. The tongue itself is about the length of the bill, and it can be extended well beyond the tip of the beak by muscles that move the elongated hyoid apparatus. The detail of the tongue shows the barbs on the tip that impale prey.

of optimization in biology remains controversial, and OFT is criticized by some ecologists as strongly as it is defended by others (Pierce and Ollason 1987, Stearns and Schmid–Hempel 1987).

The technique of OFT allows us to compare the predictions of a theoretical model to actual observations in the natural world. If the model and nature resemble one another, we have probably correctly considered the essential elements of the organism's adaptations in constructing the model; if they do not match well, the nature of the mismatch is a guide to other factors that must be considered. OFT has been used to investigate what animals feed on, where they go to feed, and how they search for food. It is possible to describe the rules animals use to make decisions to eat one food item and ignore another, or to hunt here and not there.

Most food resources used by birds occur in patches, and the food supply within a patch generally diminishes with time (seeds become depleted or insects take evasive action after becoming aware of a foraging predator). A foraging bird must decide how long to feed in one patch before moving to the next, taking into account that no food intake at all occurs when it is travelling between patches.

Theoretically a forager should behave as though it estimates the average rate of its food intake in the environment in which it is feeding and compares its current rate of food intake with the average, including the cost of moving from one patch to another. When the rate of food capture in a patch falls to the environmental average, it is time to switch. This

hypothesis is known as the **marginal value theorem** (Figure 18–10). The graph shows the rate of prey capture for a predator foraging in patches where the prey density is high or low. As the bird continues to forage in a patch, the rate of prey capture decreases. After a period of feeding (t_1 or t_2) the rate of prey capture in the patch will have decreased until it equals the average capture rate for the entire habitat, shown by the dotted horizontal line. Optimal foraging theory predicts that a predator will move to a new patch when its capture rate has dropped to the average for the habitat. The graph shows that a predator behaving in that manner will forage longer (t_2) in a patch where prey is abundant than in a patch where prey is scarce (t_1), thereby maximizing its energy intake during the time it spends foraging.

Most field studies of this hypothesis have been unsuccessful because the researcher has little way of

Box 18–1 Giant Predatory Birds

One does not usually think of birds as being frightening, although some eagles have wingspans of 2 meters or more. However, the extinction of dinosaurs at the end of the Mesozoic left empty the adaptive zone that had been filled by bipedal carnivores, and giant flightless birds appear to have filled this role from the Paleocene until the Pleistocene (Marshall 1994).

The earliest of these dinosaur analogs were the diatrymas (terror cranes), which are known from Paleocene and Eocene deposits in North America and Europe. The diatrymas were 2 meters tall and had massive legs and toes with enormous claws. The head was huge—nearly as large as that of a horse—and had an enormous, hooked beak (Figure 18–7). A mechanical analysis of the feeding apparatus of *Diatryma* suggests that the birds were equipped to be ferocious predators: "Whatever Diatryma ate, it could bite it hard" (Witmer and Rose 1991). Indeed, the skull and jaws of *Diatryma* show the thick mandibular rami and evidence of massive adductor muscles.

These features are unlike those of any extant bird (perhaps fortunately for us), but they are seen in hyaenas where they are associated with the ability to crush bones. Perhaps *Diatryma*, like hyaenas, was a scavenger as well as a predator and was able to crush bones to eat the marrow inside.

South America was isolated from the northern continents in the early Cenozoic, and another lineage of giant predatory birds, the phorusrachids, evolved on that continent. The phorusrachids, which were 1.5 to 2.5 meters tall, were more lightly built than the diatrymas and were probably faster runners. Like the diatrymas, the phorusrachids had huge beaks and powerful claws. Phorusrachids are known from South America from the Oligocene to the end of the Pliocene; they disappeared about the time of the great faunal interchange between North and South America. However, a phorusrachid is known from Pleistocene deposits in Florida, apparently representing a genus that moved north across the Central American land bridge.

Figure 18–7 Terror crane, *Diatryma*, attacking a condylarth.

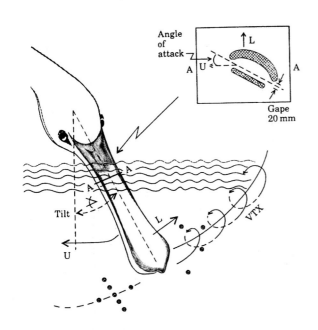

Figure 18–8 When a spoonbill sweeps its beak through the water, the curved upper surface and flat lower surface create vortex currents (shown by spiraling arrows marked VTX) that lift prey from the bottom. In this drawing the bill is being swept from right to left (indicated by the arrow marked U). The sweeping motion generates a lift (L). The line A—A indicates the position of the cross section of the beak shown in the inset. (From D. Weihs and G. Katzir, 1994, *Animal Behaviour* 47:649–654; courtesy of D. Weihs and G. Katzir.)

knowing the food abundance in a patch. On the other hand, laboratory tests of foraging have been rewarding. By creating patches with known prey numbers and varying not only those numbers but the time required to change patches, researchers have shown that there is a remarkable correspondence between the predictions of the model and the behavior of foraging birds.

What food to eat depends on the net energy content of the food after the energy costs of search and handling (the energy required to husk or kill) have been subtracted. OFT assumes that it is advantageous for a predator to maximize its energy intake per unit of time (Figure 18–11). The rate of energy intake for prey of different sizes depends on both the energy content of the prey and the time it takes to handle (subdue and swallow) the prey item. Large prey individuals contain more energy than small individuals, but they also require longer handling times. The graph shows this relationship for a wagtail eating dungfly larvae. Larvae that are 7 millimeters long provide the greatest energy return per second of handling time. Because food patches in nature are composed of several types of edible material, a foraging organism must decide whether to eat an item or ignore it. Optimal foraging theory predicts

that a low net energy item should be rejected if the predator could expect to find an energetically superior tidbit soon after rejecting the lower-quality item. Even if a feeding patch is rich in inferior food items, theoretically they should be completely ignored if there is a greater net energy gain from rejecting them in favor of some higher-quality item also in the patch. However, when high-quality items are scarce, the wise forager eats what is at its bill.

These predictions of what prey should be eaten and what should be ignored are excellent beginnings for laboratory experiments, and they have also been tested successfully in the field. J. D. Goss–Custard (1977) studied the European redshank (*Tringa totanus*). These sandpipers feed on polychaete worms by probing in estuarine intertidal mud with their long bills. Such habitats often contain large numbers of several species of these worms but little else that appeals to a redshank. Goss–Custard was able to study redshanks where only two species of worm occurred in significant numbers, a large species with a patchy distribution and an ubiquitous small species. As food items for the redshank these two species seem to differ only in the greater quantity of nutritious flesh in the larger species. The rate at which large worms were captured increased as the

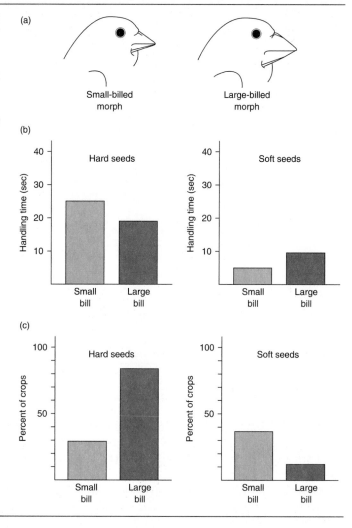

Figure 18–9 The African black-bellied seed eater. (a) The two bill morphs, large and small. (b) The small-billed morph takes longer than the large-billed morph to eat hard seeds, but eats soft seeds faster than the large-billed morph can. (c) The small-billed morph is most likely to have soft seeds in its crop, whereas the large-billed morph is most likely to have hard seeds. (Based on T. B. Smith, 1987, *Nature* 329:717–719.)

density of large worms in a patch increased (Figure 18–12a). When the rate of capture of large worms was low the birds ate many small worms, but they ignored small worms when they were able to capture large worms (Figure 18–12b). That is, they ignored less profitable prey items (small worms) when more profitable prey items (large worms) were readily available. As the rate of capture of large worms increased the redshanks became more and more selective no matter what the density of small worms was, as the theoretical predictions indicate they should.

Not all studies of the prey optimization of foragers have matched theoretical predictions. Several of these inconsistencies can be resolved by recognizing that energy content is not the only value of food. Nutritional characteristics also affect the value of a

given food. Trace nutrients or the relative proportions of carbohydrate, fat, and protein may add to the value of a food item, or toxins may detract from its value.

Perhaps the greatest value of OFT has been in sharpening our perspective of the trade-offs necessary in adaptation to the complex environments in which animals live. However, the skepticism about OFT expressed by Pierce and Ollason (1987) is shared by many ecologists. Few behaviors match our simplistic models; animals must make compromises among simultaneous and conflicting demands. Adaptations also may be greatly affected by phylogenetic constraints. These complications must be remembered when the results of OFT studies appear to demonstrate that animals are behaving optimally.

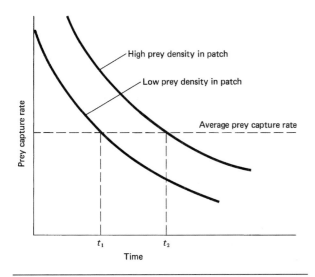

Figure 18–10 Marginal value theorem of optimal foraging.

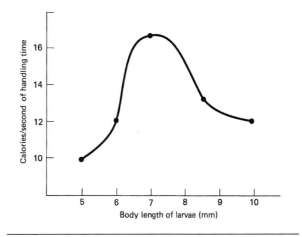

Figure 18–11 Rate of energy intake for prey of different sizes. (Data from N. B. Davies, 1977, *Journal of Animal Ecology* 46:37–57.)

Social Behavior and Reproduction

Vision and hearing are the major sensory modes of birds as they are of humans, and one result of this correspondence has been the important role played by birds in behavioral studies. Most birds are active during the day, and they are relatively easy to observe. A tremendous amount of information has been accumulated about the behavior of birds under natural conditions, and this background has contributed to the design of experimental studies in the field and in the laboratory.

The activities associated with reproduction are among the most complex and conspicuous behaviors of birds, and much of our understanding of the evolution and function of the mating systems of vertebrates is derived from studies of birds. Classic work in avian ethology, such as Konrad Lorenz's studies of imprinting and Niko Tinbergen's demonstration of innate responses of birds to specific visual stimuli, has formed a basis for current studies of behavioral ecology (Box 18–2).

Vocalization and Visual Displays

Birds use colors, postures, and vocalizations for species, sex, and individual identification. Studies of

birdsong have contributed greatly to our understanding of communication by vertebrates, and important general concepts such as species specificity in signals and innate predisposition to learning were first developed in studies of birdsong. Studies of the neural basis of song are leading to a close integration of behavior and neurobiology. (See Konishi [1985] for a review.)

Birdsong has a specific meaning that is distinct from a birdcall. The song is usually the longest and most complex vocalization produced by a bird. In many species songs are produced only by mature males, and only during the breeding season. Song is a learned behavior that is controlled by a series of song control regions (SCRs) in the brain. During the period of song learning, which occurs early in life (Box 18–2), new neurons are produced. These neurons connect a part of the SCR that is associated with song learning to a region that sends impulses to nerve cells that control the vocal muscles (Nordeen and Nordeen 1988). Thus, song learning and song production are closely linked in male birds.

The SCRs are under hormonal control, and in many species of birds the SCRs of males are larger than those of females and have more and larger neurons and longer dendritic processes (Nottebohm and Arnold 1976). The vocal behavior of female birds varies greatly across taxonomic groups: In some species females produce only simple calls, whereas in other species the females engage with males in complex song duets. The SCRs of females of the lat-

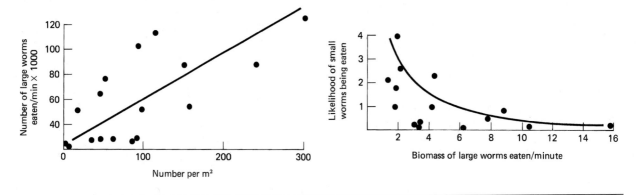

Figure 18–12 Redshanks feeding on polychaete worms illustrate the behavior expected from an optimally foraging predator. (Data from J. D. Gross–Custard, 1977, *Animal Behaviour* 25:10–29.)

ter species are very similar in size to those of males (Table 18–2). The function of the SCR in female birds of species in which females do not vocalize has been unclear, but recent experiments suggest that it plays a role in species recognition. When the SCR of female canaries was inactivated, the birds no longer distinguished the vocalizations of male canaries from those of sparrows.

A birdsong consists of a series of notes with intervals of silence between them (Figure 18–14). Changes in frequency (frequency modulation) are conspicuous components of the songs of many birds, and we have noted that the avian ear may be very good at detecting rapid changes in frequency (Chapter 17). Birds often have more than one song type, and some species may have repertoires of several hundred songs.

Birdsongs identify the particular species of bird that is singing, and they often show regional dialects. These dialects are transmitted from generation to generation as young birds learn the songs of their parents and neighbors. In the indigo bunting, one of the best studied species, song dialects that were characteristic of small areas persisted up to 15 years, which is substantially longer than the life of an individual bird (Payne et al. 1981). Birdsongs also show individual variation that allows birds to recognize the songs of residents of adjacent territories and to distinguish the songs of these neighbors from those of intruders. Male hooded warblers remember the songs of neighboring males and recognize them as individuals when they return to their breeding sites

in North America after spending the winter in Central America (Godard 1991).

The songs of male birds identify their species, sex, and occupancy of a territory. Territorial males respond to playbacks of the songs of other males with vocalizations, aggressive displays, and even attacks on the speaker. These behaviors repel intruders, and broadcasting recorded songs in a territory from which a territorial male has been removed delays the occupation of the vacant territory by a new male.

Visual displays are frequently associated with songs; for example, a particular body posture that displays colored feathers may accompany singing. Male birds are often more brightly colored than females and have feathers that have become modified as the result of sexual selection. In this process, females mate preferentially with males that have certain physical characteristics. As a result of that response by females, those physical characteristics contribute to the reproductive fitness of males, even though they may have no useful function in any other aspect of the ecology or behavior of the animal. The colorful speculum on the secondaries of male ducks, the red epaulets on the wings of male red-winged blackbirds, the red crowns on kinglets, and the elaborate tails of male peacocks are familiar examples of specialized areas of plumage that are involved in sexual behavior and display. It has been suggested that the bright colors of male birds may be an indication of good nutritional status or resistance to parasites, and thus could provide a basis for females to evaluate the merits of several potential mates, although this hypothesis is still controversial (Hamilton and Zuk 1982, Slagsvold and Lifjeld 1992, Petrie 1994).

Some species of birds use brightly colored objects to attract females; male bowerbirds, for example, decorate their bowers with feathers from other birds, shells, or shiny bits of glass and metal. A particularly dramatic example of this behavior is provided by Archbold's bowerbird (*Archboldia papuensis*) of New Guinea (Frith and Frith 1990). Blue is an especially popular color for bowerbird ornaments, and male *Archboldia* collect the display plumes from the male King of Saxony bird of paradise (*Pteridophora alberti*). During the mating season, the male bird of paradise grows a single long feather from above each eye. These plumes look like thin wires with squares of blue plastic fastened to them at intervals. The Friths found that the bowers of several *Archboldia* were decorated with three to six *Pteridophora* plumes, which occupied a central position in the bower mat. When the Friths moved the plumes to the edges of the bower, the bower owner promptly returned them to their conspicuous location.

Conspicuous or aerodynamically cumbersome feathers can make a male bird vulnerable to capture by visually guided predators, and the bright colors and special adornments of the breeding season are often discarded for a more sober, even cryptic, appearance during the rest of the year. Thus, the male African standard-wing nightjar has specially elongated and flagged second primaries that are used in flight displays during courtship, but these feathers probably slow the male's flight and make it easier for an aerial predator to capture him (Figure 18–15). As soon as courtship is over, the male bites off the projecting parts of the feathers, leaving the stubs in the wings. The pattern of molting is so arranged that the primaries are not replaced until just before the next breeding season. This pattern of molting differs from that of all other caprimulgids and from that of female standard-wing nightjars. The usual pattern for caprimulgids is to begin molt in the spring with the outermost primary, and to move sequentially through the primaries to the tenth, and this is the pattern followed by the female standard-wing nightjar. In the case of the males, however, molt begins in the center of the wing with the fifth and sixth primaries and ends with the tenth and the stump of the second. That sequence leaves just enough time for the second

primary to grow to its full length at the beginning of the next courtship season.

Vocalizations are not the only sounds that birds use in courtship; nonvocal sounds are produced by the feathers of some species. The drumming of male grouse in the spring is a familiar example of a nonvocal sound that plays a role in courtship. Sounds are often produced as a by-product of the beating of a bird's wings in flight, and only slight modification in the shapes of primaries or tail feathers is needed to produce the characteristic whistling and buzzing sounds made by certain kinds of ducks, bustards, and hummingbirds when they fly. Such sounds may be used in territorial advertisement or as individual location signals among birds flying at night or in heavy fog. Other species of birds have undergone more specific modification of their flight feathers to produce sounds used in displays. Among the tropical American manakins one finds not only narrowed and stiffened primaries involved in the production of sounds during displays, but also secondaries with thickened, club-like shafts that apparently act like castanets to produce clicks when the wings are moving (Figure 18–16). Other species, including goatsuckers, owls, doves, and larks, clap their wings together in flight, producing characteristic sounds associated with courtship or territorial defense.

Mating Systems and Parental Investment

The mating systems of vertebrates are believed to reflect the distribution of food, breeding sites, and potential mates. These resources affect individuals of the two sexes differently, and some types of resource distributions give one sex the opportunity to achieve multiple matings by controlling access to the resources. Studies of birds have contributed very largely to the development of theories of sexual strategies (Emlen and Oring 1977, Oring 1982).

The energy cost of reproduction for males of most species of vertebrates is probably lower than the cost for females. Sperm are small and cheap to produce compared to eggs that must supply the nutrients required for embryonic development. Furthermore, a male does not necessarily have any commitment beyond insemination, whereas a female must at least carry the eggs until they are deposited, and often is involved in brooding eggs and caring for the young as well. Courtship and territorial displays may be energetically expensive (see Chapter 11), but

Box 18–2 Training Bird Brains

The process known as imprinting has played a prominent role in studies of bird behavior. Imprinting is a special kind of learning that occurs only during a restricted period in ontogeny called the critical period. Once imprinting is established, it is permanent and cannot be reversed. A flock of geese were imprinted on the famous ethologist Konrad Lorenz. The geese followed Lorenz around as if he were a mother goose.

The young of precocial bird species learn the characteristics that identify their parents in the hours immediately after hatching. Young ducks, for example, will imprint on an object that moves and makes a noise. Normally this object would be a parent, but in experimental situations young ducks will imprint on other animals (including humans) or on inanimate objects such as a ticking clock that is trundled along on a cart.

Most birds learn their own species' song early in life by hearing a parent bird sing. Studies of zebra finches show that the song-learning interval corresponds to period during which new neurons are rapidly incorporated into the song control region of the brain. The images and vocalizations that birds learn early in life form the basis for their social and reproductive behavior as adults. Birds that are cross-fostered by adults of a different species, particularly males, subsequently attempt to mate with the foster parent's species, and birds that are hand-reared by humans identify their keepers as sexual partners when they are adults.

The confusion of species-identification by birds that have imprinted on a foster parent or a keeper can be disastrous for programs in which endangered species are reared in captivity and then released. Young birds must recognize appropriate mates if they are to establish a breeding population, and captive-rearing programs go to great lengths to ensure that the young birds are properly imprinted. Hatchling California condors, for example, are reared in enclosed incubators and fed by a technician who inserts her hand into a rubber glove modeled to look like the head of an adult condor (Figure 18–13a).

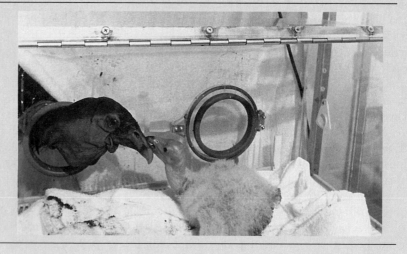

Figure 18–13 Captive husbandry of endangered species of birds. (a) A hatchling California condor in its incubator with the model condor head used to feed it. (b) A bald ibis. (Photographs: [a] San Diego Zoo; [b] David Hosking.)

Still more training may be necessary to produce captive-reared young that can survive after they have been released. The bald ibis (Figure 18–13b) is an example of how complicated a process this can be. The geographic range of the bald ibis once extended from the Middle East and North Africa north to Switzerland and Germany but the wild population has dwindled to fewer than 300 birds in a reserve in Morocco. Bald ibises flourish in captivity, and there are more than 800 captive individuals. This species seems ideal for reintroduction—it is prolific, and its disappearance from the wild seems to have been caused by human predation rather than by pollution or loss of habitat. Yet two attempts to establish populations by releasing captive-reared birds failed.

The reason for the failures seems to have been the absence of normal social behaviors in the captive-reared birds. Bald ibises are social birds with extended parental care, and it seems that juveniles learn appropriate behaviors from adults. For some reason, this did not occur in captivity. An attempt is now underway to instruct young bald ibises in these social skills—human foster parents are hand-rearing the birds, teaching them to find their way to fields where they can forage, to recognize predators and other dangers such as automobiles, and to engage in mutual preening, which is an important social behavior.

This example illustrates the importance of basic biological information to applied programs in organismal biology, such as captive husbandry and reintroduction. Organisms are enormously complicated, and successful management of endangered species requires integrating information about their ecology, physiology, and behavior.

Figure 18–13 *(Continued)*

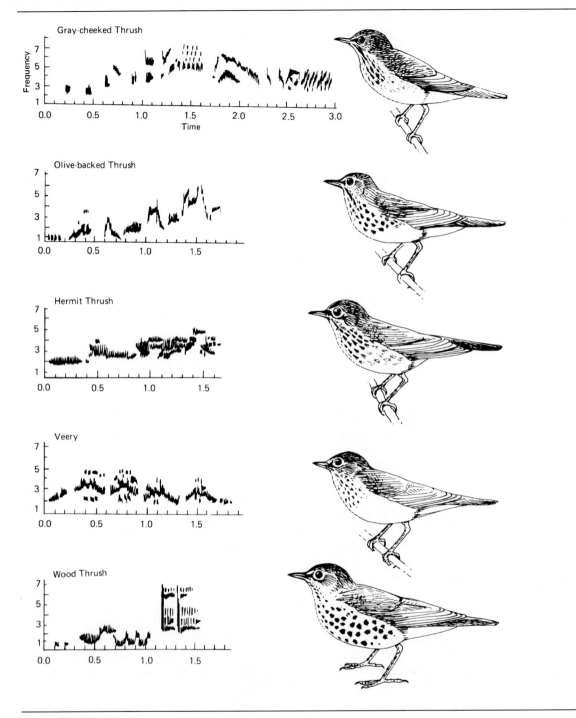

Figure 18–14 Difference in the songs of some species of thrushes from eastern North America. The sonographs are plots of sound frequency in kilohertz versus time in seconds. They show differences in loudness (amplitude modulation) by varying degrees of darkness and change in frequency (frequency modulation) by upward or downward slopes within a syllable. (From W. C. Dilger, 1956, *Auk,* 73:313–353; and R. C. Stein, 1956, *Auk* 73:503–512.)

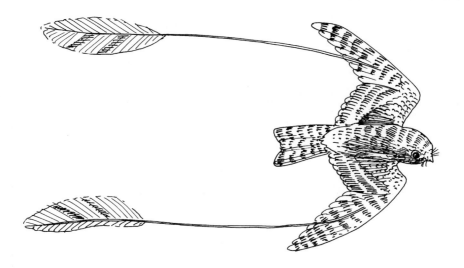

Figure 18–15 Male standard-wing nightjar, showing the elongated second primaries that are used in an aerial courtship display.

even if that is generally true, most male vertebrates can potentially mate more often than females. A male is ready to mate again very quickly after inseminating a female, but a female must ovulate and yolk a new clutch of eggs before she can attempt a second mating. Because of this disparity in the costs of reproduction for males and females, the routes to maximizing reproductive success may be different for the two sexes. For males the most productive strategy may be to mate with as many females as possible, whereas a female may maximize her success by devoting time to careful choice of the best male and to care of her young.

The extent to which males can achieve multiple matings depends on ecological conditions and especially on the availability of food and nest sites, which are the resources most needed by females. Theoretically, a male could increase his opportunities to mate by defending these resources—excluding other males and mating with all the females in the area. His ability to do that will depend on the spatial distribution of resources. If food and nest sites are more or less evenly distributed through the habitat, it is unlikely that a male could control a large enough area to monopolize many females. Under those conditions, all males will have access to the resources and to the females. On the other hand, if resources are clumped in space with barren areas between the patches, the females will be forced to aggregate in

the resource patches and it will be possible for a male to monopolize several females by defending a patch. Males that are able to defend good patches should attract more females than males defending patches of lower quality.

The temporal distribution of breeding females is also important in determining the potential for a male to achieve multiple matings. A male can mate with only one female at a time, so an important consideration is the number of receptive females relative to the number of breeding males at any moment. This ratio is called the **operational sex ratio** (OSR). If the actual ratio of males to females in a population is 1:1 and if all the females become receptive at the same time, there would be one receptive female for each breeding male and the OSR would be 1:1. In that situation, males have little opportunity to monopolize a large number of females and the variance in mating success among males will be low. That is, most males will mate with one female and relatively few males will have no matings or more than one mating. On the other hand, if females become receptive over a period of weeks, the number of breeding males at any given time will be larger than the number of females and the OSR will be greater than 1:1. In that situation it is possible for one male to mate with many females over the course of the breeding season. As the OSR rises above 1:1, competition between males for mates increases, the

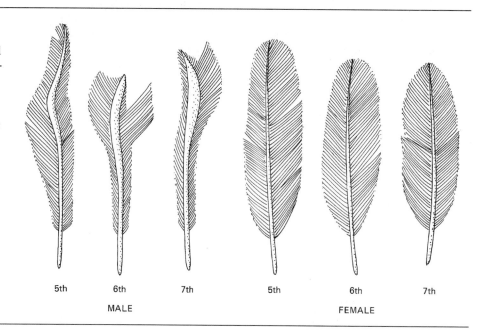

Figure 18–16 Sound-producing feathers: Secondaries of the male (left) and female (right) South American manakin. The shafts of the male's feathers are thickened and produce sounds when the wings are moved in display.

5th 6th 7th   5th 6th 7th
MALE   FEMALE

variation in reproductive success of individual males increases, and sexual selection is likely to become more intense. These are the elements that contribute to determining the mating strategies of males and females.

Social vertebrates exhibit one of two broad categories of mating systems—monogamy or polygamy. Monogamy (*mono* = one, *gamy* = marriage) refers to a pair bond between a single male and a single female. The pairing may last for part of a breeding season, an entire season, or for a lifetime. Polygamy (*poly* = many) refers to a situation in which an individual has more than one mate in a breeding season.

Polygamy can be exhibited by males, females, or both sexes. In polygyny (*gyn* = female) a male mates with more than one female, whereas in polyandry (*andr* = male) a female mates with two or more males. Promiscuity is a mixture of polygamy and polyandry in which both males and females mate with several different individuals.

Monogamy is the dominant social system of birds. Both parents in monogamous mating systems usually participate in caring for the young, and 93 percent of the species of birds that produce altricial young (which require extensive parental care) are monogamous compared to 83 percent monogamy among species with precocial young.

Promiscuity is the second most common mating system for birds, accounting for 6 percent of the extant species of birds. Two percent of the species of

birds are polygynous and only 0.4 percent are polyandrous. Despite their relative rarity, promiscuous, polygynous, and polyandrous species of birds have been extensively studied because these unusual mating systems can reveal much about the mechanisms of sexual selection and evolution.

Monogamy Monogamy occurs in so many species of birds in so many different ecological conditions that no one mechanism is likely to explain its prevalence. Resource distribution and the degree of parental care appear to be two factors that are frequently important in monogamy. When nest sites and food are evenly distributed through a habitat, a male or female cannot control access to these resources. If neither sex has the opportunity to monopolize additional members of the opposite sex by controlling resources, monogamy is the reproductive strategy that maximizes the fitness of individuals. When the territory quality of one male is much like that of all other males, a female can probably maximize her reproductive success by pairing with an unmated male. Perhaps a more important incentive for monogamy for many species of birds is the need for attendance by both parents to raise a brood to fledging. Dramatic examples include situations in which continuous nest attendance by one parent is necessary to protect the eggs or chicks from predators while the other parent forages for food. This situation is commonly observed in seabirds that nest in dense colonies that sometimes include mixtures of two or more species. In the absence of an attending parent, neighbors raid the nest and kill the eggs or chicks. The male and female alternate periods of nest attendance and foraging, and some species engage in elaborate displays when the parents switch duties (Figure 18–17). A third situation that could make monogamy advantageous would be a sex ratio that deviates widely from 1:1. When one sex is in short supply, individuals of that sex may be the resource that individuals of the other sex defend. For example, female ducks suffer higher mortality than males and the sex ratios of ducks are biased toward males. As a result of the shortage of female ducks, competition between males for mates is intense. A male duck pairs with a female several months before the breeding season begins and defends her against other males.

Polygyny When an individual male can control or gain access to several females, the male can increase his reproductive success by mating with more than one female. In **resource defense polygyny** males control access to females by monopolizing critical resources such as nest sites or food that have patchy distributions. A male that stakes out its territory in a high-quality patch can attract many females. For this system to work, a female must benefit from mating with a male that already has one mate. That is, the reproductive fitness of a female must be greater as a secondary mate on a high-quality territory than it would be as a primary mate on a territory of lower quality. Red-winged blackbirds are a familiar example of resource defense polygyny (Orians 1980). Male blackbirds arrive at their marshy breeding areas before females and compete for territories. When the females arrive, they have a choice among a variety of territories of different quality, each defended by a male blackbird. A female should choose to mate polygynously if the difference in quality of the territories is large enough so that she will raise more young than she would by mating monogamously with a male on a poorer territory. This hypothesis is known as the **polygyny threshold model** (Orians 1969).

In **male dominance polygyny** males are not defending females, nor are they defending a resource that females require. Instead, males compete for females by establishing patterns of dominance or by demonstrating their quality through displays. This type of reproductive system is typical of situations in which the male is not involved in parental care and no potential exists for controlling resources or mates. Birds with precocial young in rich habitats often show male dominance polygyny, although this mating system is not limited to such species. The sizes of male territories in male dominance polygyny and the degree of aggregation of males are not set directly by the resources in the habitat, as is the case with resource defense polygyny. Instead, the distribution of males is determined by the sizes of the home ranges of females. Aggregations of many males in a small area are called **leks**. The prairie chicken of western North America is a well-studied lekking species (Wiley 1973). During the breeding season male prairie chickens congregate in traditional lek sites. Each male occupies a small territory (from 13 to 100 square meters in area) in the lek, and within this territory performs a courtship display that includes elaborate postures (Figure 18–18). Two colorful sacks that are outgrowths from the esophagus are filled with air and project through the breast feathers. Air is expelled from these sacks with a popping sound. Females visit the leks and copulate with a single male. The central sites appear to be the most favored and the 10 percent of the males in the most central sites obtain 75 percent of the matings.

In **harem defense polygyny** males control access to females that are gregarious and form groups. This mating system is rare among birds, but well known among mammals—harems of females guarded by a male are typical of some bats, many species of pinnipeds (seals, sea lions, and the walrus), and some ungulates (deer and elk). The first well-documented example of harem defense polygyny among birds was described for the Montezuma oropendola (*Psarocolius montezuma*). These oropendolas are large Neotropical passerine birds; females weigh about 250 grams and males weigh about 520 grams. The females breed colonially and one tree may contain 30 or more nests. The aggressive activities of the males are not directed to defending any resource in the environment that the females require; instead, the males defend groups of females. High-ranking males exclude low-ranking males from the area of the colony, and mating success of males is correlated with the time they spend in the colony.

Male Incubation and Polyandry Incubation and care of the young by both parents is considered to be the ancestral condition for birds, and it occurs in the vast majority of the extant species of birds. Howev-

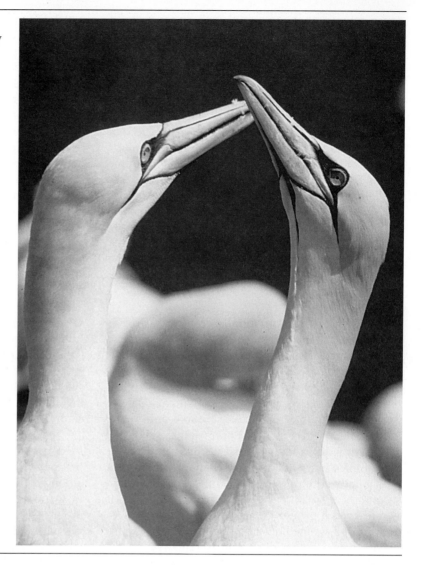

Figure 18–17 Nest exchange display of northern gannets. (Photograph by Mary Tremaine, courtesy of the Cornell Laboratory of Ornithology.)

er, biparental care is not always required, and situations in which it is possible for one parent to care for the eggs and young allow the development of the polygynous mating strategies we have discussed. Males of polygynous species of birds do not participate in parental care, and all the tasks of incubation and rearing the young are performed by the female. Less commonly it is the female that is emancipated and the male that assumes parental responsibilities. The rarity of this situation probably reflects the relative parental investments of the two sexes. A male bird can leave a newly laid clutch of eggs and breed again as soon as he locates a receptive female, whereas a female who leaves her nest must ovulate and yolk a new clutch of eggs before she can obtain a second breeding. Thus, a female has more to lose by abandoning her eggs that does a male.

Despite the imbalance of energy investment in reproduction by males and females, there are a few situations in which the ability of a female bird to increase her fitness by multiple breedings equals or exceeds that of males, and these are the cases in which polygamy is balanced between males and females (promiscuity) or favors females (polyandry). Several species of charadriiforms show this pattern of breeding, including jacanas, stints, sandpipers, and phalaropes.

In **rapid multiple clutch polygamy** both sexes have an opportunity to increase their fitness by multiple breedings in rapid succession, and both males and females incubate eggs. Temminck's stint (*Calidris temminckii*), a charadriiform that occurs in northern Europe, provides one of the best-studied examples (Hildén 1975). Male stints establish territories and display to attract females. Every female breeds in rapid succession with two males on different territories, and every male breeds with two females. The first clutch of eggs is incubated by the male, and the second clutch is incubated by the female. The evolution of this reproductive system appears to depend on the ready availability of resources that allow single-parent care to lead to successful fledging.

In polyandrous mating systems females control or gain access to multiple males. **Resource defense polyandry,** like its counterpart resource defense polygyny, is based on the ability of one sex to control access to a resource that is critical for the other sex. This pattern of breeding among birds seems to be typical of situations in which the cost of each reproductive effort for the female is low (because food is abundant and a clutch contains only a few small eggs) and the probability of successful fledging is small. Spotted sandpipers (*Actitis macularia*) provide an example of this mating strategy (Oring 1982). Predation on sandpiper nests is high, and the resource that female sandpipers control is replacement clutch-es for males that have lost their clutches to predators. Male spotted sandpipers form territories and incubate the eggs. A female spotted sandpiper mates with a male and remains in his territory at least until she has laid three eggs. After that, she may move away to breed with other males, leaving parental care to the male, or she may remain in the territory. If she remains she may or may not participate in parental care. A female that has moved away and bred with other males may return and breed again with the original male if their first clutch is destroyed.

Female access polyandry has been described for the gray phalarope (*Phalaropus fulicarius*). This shorebird occurs in flocks with large home ranges that include feeding areas separated by hundreds of meters. Defense of a resource is not possible, but females initiate courtship and may limit access to males by interactions among themselves (Kistchinski 1975). The male phalarope incubates the eggs, driving the female away after she has finished laying.

In **cooperative polyandry** a single female and a group of males form a communal breeding unit in which all males have an opportunity to mate. Females are able to maintain a male group because of the advantages to males in cooperating as a breeding unit. In the Galapagos hawk (*Buteo galapagoensis*), for example, female mortality is higher than male mortality and the OSR on Santiago Island is 2.3

males per female. The habitat is saturated with territories, and territories remain stable for periods exceeding the life spans of individual birds. Some females do not breed, apparently because they cannot obtain territories. Breeding groups persist from year to year, changing only with the death of a member. Monogamous males have higher reproductive success than males that are part of a breeding group (an average of 1.0 versus 0.67 fledged young per year on Santiago Island), but joining a group may help males to establish territories initially, and some polyandrous males appear to become monogamous after several years. Consequently, the best chance for a male Galapagos hawk to increase its lifetime fitness may be initially to become a member of a breeding group (Faaborg et al. 1980). The Tasmanian native hen (*Tribonyx mortierii*) shows a variation on this pattern: The sex ratio is about 1.5 males per female, and the breeding population consists of pairs and trios. A trio is usually composed of two brothers and an unrelated female. The trios form when the birds are about a year old, and they remain together for life. Trios have larger territories than pairs, and trios have larger clutches and fledge more young per year than do pairs.

The conspicuousness of birds and the relative ease with which they can be studied has made them a mainstay of sociobiological research. The diversity of avian mating systems and the correlations between ecological conditions and certain types of mating systems have contributed largely to our current understanding of vertebrate behavior. Recent work has begun to emphasize the roles of individual experience and of lability of breeding systems. If environmental conditions determine the relative advantages of different mating systems, how should organisms respond to variation in these conditions? One possible response is a flexible mating system that responds in ecological time to changes in ecological conditions. Investigation of the short-term causes and consequences of variation in avian mating systems is emerging as an area of increasing importance for both ornithologists and behaviorists (Oring 1982).

Oviparity, Nesting, and Brooding Eggs

Elaborate and diverse behaviors are associated with egg laying and parental care. Nest preparation by birds runs the gamut from nothing more than the fairy tern's selection of a branch on which it balances its egg, to the multiroom communal nests of weaver birds, which are used by generation after generation. Incubation provides heat for the development of eggs and the presence of a parent is a deterrent to many predators. However, some birds leave their eggs for periods of days while they forage, and brood parasites deposit their eggs in the nests of other species of birds and play no role in brooding or rearing their young.

Oviparity In contrast to the diversity of mating strategies of birds, their mode of reproduction is limited to laying eggs. No other group of vertebrates that contains such a large number of species is exclusively oviparous. Why is this true of birds?

Constraints imposed on birds by their specializations for flight are often invoked to explain the failure of birds to evolve viviparity, but those arguments are not particularly convincing when one remembers that bats have successfully combined flight and viviparity. Furthermore, flightlessness has evolved in at least 15 families of birds, but none of these flightless species has evolved viviparity.

Oviparity is presumed to be the ancestral reproductive mode for diapsids, and it is retained by both extant groups of archosaurs, the crocodilians and the birds. However, viviparity has evolved nearly 100 times in the other major lineage of extant diapsids, the lepidosaurs (Chapter 15), so the capacity for viviparity is clearly present in diapsids. A key element in the evolution of viviparity among lizards and snakes appears to be the retention of eggs in the oviducts of the female for some period before they are deposited. This situation occurs when the benefits of egg retention outweigh its costs. For example, the high incidence of viviparity among snakes and lizards in cold climates may be related to the ability of a female ectotherm to speed embryonic development by thermoregulation. A lizard that basks in the sun can raise the temperature of eggs retained in her body, but after the eggs are deposited in a nest the mother no longer has any control over their temperature and rate of development. Birds are endotherms and brood their eggs, thereby controlling their temperature after the eggs are laid. Thus egg retention provides no thermoregulatory advantage for a bird.

Broad aspects of the biology of birds may create an unfavorable balance of costs and benefits of egg retention, thereby making it unlikely that any lineage of birds would take the first step in an evolutionary process that has repeatedly led to viviparity among snakes and lizards. If birds are viewed as being specialized for the production of one relatively large egg at a time and for complex egg incubation and parental care, the potential advantages of egg retention are greatly diminished and the costs of decreased fecundity and increased risk of maternal mortality are increased (Blackburn and Evans 1986). Perhaps it is this balance of costs and benefits rather than any single factor that is responsible for the retention of the ancestral reproductive mode by all extant birds. The same line of reasoning probably can be applied to crocodilians, which construct nests and care for their young, and it can be extended with caution to speculations about the reproductive mode of dinosaurs.

Nesting Construction of nests is an important aspect of avian reproduction because nests provide protection for the eggs from such physical stresses as heat, cold, or rain and from predators. Bird nests range from shallow holes in the ground to enormous structures that represent the combined efforts of hundreds of individuals over many generations (Figure 18–19). The nests of passerines are usually cup-shaped structures composed of plant materials that are woven together. Swifts use sticky secretions from buccal glands to cement material together to form nests, and grebes, which are marsh-dwelling birds, build floating nests from the buoyant stems of aquatic plants. A review of bird nests can be found in Collias and Collias (1984).

Most birds nest individually, but some lineages are exceptions: Only 16 percent of passerines nest in colonies, but 98 percent of seabirds are colonial nesters (Wittenberger and Hunt 1985, Kharitonov and Siegel–Causey 1988). Nesting colonies of some species of penguins, petrels, gannets, gulls, terns, and auks contain hundreds of thousands of individuals. Colonies are smaller in most other groups of birds; colonies of herons, storks, doves, swifts, and passerines contain a few tens of nests. Colonial nesting offers advantages and disadvantages. A colony is a concentration of potential prey that may attract predators, but the density of nesting birds may provide a degree of protection. In many colonies the nests are located two neck lengths apart, and an intruder is menaced from all sides by snapping beaks. Centrally placed nests may be better protected against predators than nests on the periphery of the colony.

Mixed colonies of two or more species of seabirds occur, but at least some of these may be transitional situations in which one species is in the process of displacing the other. For example, on the eastern coast of North America the greater black-backed gull (*Larus marinus*) is extending its range southward, apparently in response to the abundance of food available in garbage dumps. As it moves south, it is invading the breeding colonies of herring gulls (*Larus argentatus*). Great black-backed gulls, which are larger than herring gulls and breed earlier in the year, appear to be displacing herring gulls from some of their traditional breeding sites.

Incubation The megapodes, known as mound birds, bury their eggs in sand or soil and rely on heat from the sun or rotting vegetation for incubation, and the Egyptian plover buries its eggs in sand, but all other birds are believed to brood their eggs using metabolic heat. Some species of birds begin incubation as soon as the first egg is laid and others wait until the clutch is complete. Starting incubation immediately may protect the eggs, but it means that the first eggs in the clutch hatch while the eggs that were deposited later are still developing, forcing the parents to divide their time between incubation and gathering food for the hatchlings. Furthermore, the eggs that hatch last produce young that are smaller than their older nestmates and these young probably have less chance of surviving to fledge. Most passerines, as well as ducks, geese, and fowl, do not begin incubation until the next-to-last or last egg has been laid.

Prolactin, secreted by the pituitary gland, suppresses ovulation and induces brooding behavior, at least in those species of birds that wait until a clutch is complete to begin incubation. The insulating properties of feathers that are so important a feature of the thermoregulation of birds become a handicap during brooding when it is necessary for the parent to transfer metabolic heat from its own body to the eggs. Prolactin plus estrogen or androgen stimulates the formation of brood patches in female and male birds, respectively. These brood patches are areas of bare skin on the ventral surface of a bird. The feathers are lost from the brood patch and blood vessels proliferate in the dermis, which may double in thickness and give the skin a spongy texture. Not all birds develop brood patches, and in some species only the female has a brood patch, although the male may share in incubating the eggs. Ducks and geese create brood patches by plucking the down

Figure 18–19 The nests of birds show great diversity. Some nests are no more than shallow depressions, whereas other birds build elaborate structures. The piping plover (a), like many shorebirds, lays its eggs in a depression scraped in the soil, whereas the bald eagle (b) constructs an elaborate nest that is used year after year. Coots (c) build floating nests, and the Australian mallee fowl (d) scrapes together a pile of sand in which it buries its eggs. Heat from the sun warms the eggs, and the male mallee fowl adds and removes sand to keep the temperature stable.
([a] © Allan D. Cruickshank from National Audubon Society; [b] © Joan Baron; [c] © Bruce W Heinemann; [d] © Jen and DES Bartlett.)

feathers from their breasts; they use the feathers to line their nests. Some penguins lay a single egg that they hold on top of their feet and cover with a fold of skin from the belly, which envelopes the egg as it rests on the bird's feet.

The temperature of eggs during brooding is usually maintained within the range 33 to 37°C, and some eggs can withstand periods of cooling when the parent is off the nest (see Drent [1975] for details).

Tube-nosed seabirds (Procellariiformes) are known for the long periods that adults spend away from the nest during foraging. Fork-tailed storm petrels (*Oceanodroma furcata*) lay a single egg in a burrow or rock crevice. Both parents participate in incubation, but the adults forage over vast distances and both parents may be absent from the nest for periods of hours or even for several days at a time. The mean period of parental absence was 11 days (during an incubation period that averaged 50 days) for storm petrels studied in Alaska, and eggs were exposed to ambient temperatures of 10°C while the parents were away. Experimental studies showed that storm petrel eggs were not damaged by being cooled to 10°C every 4 days (Vleck and Kenagy 1980). The pattern of development of chilled eggs was like that of eggs incubated continuously at 34°C, except that each day of chilling added about one day to the total time required for the eggs to hatch.

Parent birds turn the eggs as often as several times an hour during incubation, and individual eggs are moved back and forth between the center and the edge of the clutch. Temperature variation exists within a nest, and shifting the eggs about may ensure that they all experience approximately the same average temperature. In addition, turning the eggs may help to prevent premature fusion of the chorioallantoic membrane with the inner shell membrane. Embryos attain a stable orientation during incubation—that is, the same side is usually uppermost. This position is dictated by the asymmetric distribution of embryonic mass, and the process of turning the egg allows the embryo to assume its equilibrium position. Apparently this mechanism assures that when the chorioallantoic and inner shell membranes fuse (approximately midway through incubation), the embryo is in a position that will facilitate hatching.

Incubation periods are as short as 10 to 12 days for some species and as long as 60 to 80 days for others. In general, large species of birds have longer incubation periods than small species, but ecological factors also contribute to determining the length of the incubation period. The effect of parental absence in slowing development has already been mentioned, and the amount of time a parent is absent may depend on its foraging success. A high risk of predation may favor rapid development of the eggs. Among tropical tanagers, species that build open-topped nests near the ground are probably more vulnerable to predators than are species that build similar nests farther off the ground. The incubation periods of species that nest near the ground are short (11 to 13 days) compared to those of species that build nests at greater heights (14 to 20 days). Species of tropical tanagers with roofed-over nests have still longer incubation periods—17 to 24 days.

The inorganic part of eggshells contains about 98 percent crystalline calcite, $CaCO_3$, and the embryo obtains about 80 percent of its calcium from the egg-

shell. An organic matrix of protein and mucopolysaccharides is distributed through the shell and may serve as a support structure for the growth of calcite crystals. Eggshell formation begins in the isthmus of the oviduct. Two shell membranes are secreted to enclose the yolk and albumen, and carbohydrate and water are added to the albumen by a process that involves active transport of sodium across the wall of the oviduct followed by osmotic flow of water. The increased volume of the egg contents at this stage appears to stretch the egg membranes taut. Organic granules are attached to the egg membrane, and these **mammillary bodies** appear to be the sites of the first formation of calcite crystals (Figure 18–20). Some crystals grow downward from the mammillary bodies and fuse to the egg membranes, and other crystals grow away from the membrane to form cones. The cones grow vertically and expand horizontally, fusing with crystals from adjacent cones to form the palisade layer. Changes in the ionic composition of the fluid surrounding the egg during shell formation lead to an increase in the concentrations of magnesium and phosphorus and a change in the pattern of crystallization in the surface layers of the shell.

The eggshell is penetrated by an array of pores that allow oxygen to diffuse into the egg and carbon dioxide and water to diffuse out (Figure 18–21). Pores occur at the junction of three calcite cones, but only 1 percent or less of those junctions form pores; the rest are fused shut. Pores occupy about 0.02 percent of the surface of an eggshell. The morphology of the pores varies in different species of birds: Some pores are straight tubes, whereas others are branched. The openings of the pores on the surface of the eggshell may be occluded to varying degrees with organic or crystalline material. Additional information about the structure and function of avian eggs can be found in Carey (1983).

Water evaporates from an egg during development and the loss of water creates an air cell at the blunt end of the egg. The embryo penetrates the membranes of this air cell with its beak 1 or 2 days before hatching begins, and ventilation of the lungs begins to replace the chorioallantoic membrane in gas exchange. Pipping, the formation of the first cracks on the surface of the eggshell, follows about half a day after penetration of the air cell, and actual emergence begins half a day later. Shortly before hatching the chick develops a horny projection on its upper mandible. This structure is called the egg tooth, and it is used in conjunction with a hypertrophied muscle on the back of the neck (the hatching muscle) to thrust vigorously against the shell. The egg tooth and hatching muscle disappear soon after the chick has hatched. In those species of birds that delay the start of incubation until all the eggs have been laid, an entire clutch nears hatching simultaneously. Hatching may be synchronized by clicking sounds that accompany breathing within the egg, and both acceleration and retardation of individual eggs may be involved. A low-frequency sound produced early in respiration, before the clicking phase is reached, appears to retard the start of clicking by advanced embryos. That is, the advanced embryos do not begin clicking while other embryos are still producing low-frequency sounds. Subsequently, clicking sounds or vocalizations from advanced embryos appear to accelerate late embryos. Both effects were demonstrated by Vince (1969) in experiments with bobwhite quail eggs. She found that she could accelerate the hatching of a late egg by 14 hours when she paired it with an early egg that had started incubation 24 hours sooner, and the presence of the late egg delayed hatching of the early egg by 7 hours.

Parental Care

The ancestral form of reproduction in the archosaur lineage appears to consist of the deposition of eggs in a well-defined nest site, attendance at the nest by one or both parents, hatching of precocial young, and a period of association between the young and one or both parents. All of the crocodilians that have been studied conform to this pattern, and evidence is increasing that at least some dinosaurs remained with their nests and young.

Extant birds follow these ancestral patterns, but not all species produce precocial young. Instead, hatchling birds show a spectrum of maturity that extends from precocial young that are feathered and self-sufficient from the moment of hatching to altricial forms that are naked and entirely dependent on their parents for food and thermoregulation (Figure 18–22 and Table 18–3). The most precocial birds at hatching are the megapodes, and these show the most ancestral form of nesting, burying their eggs in nests made from mounds of soil and vegetation very much like those of crocodilians. Newly hatched megapodes scramble to the surface already feathered and capable of flight. Most precocial birds are covered with down at hatching and can walk, but are not able to fly.

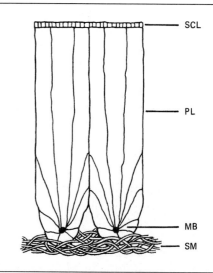

Figure 18–20 Diagram of the crystal structure of an avian eggshell. Crystallization begins at the mammillary bodies (MB) and crystals grow into the outer shell membrane (SM) and upward to form the palisade layer (PL). Changes in the chemical composition of the fluid surrounding the growing eggshell are probably responsible for the change in crystal form in the surface crystalline layer (SCL). (From C. Carey, 1983, in *Current Ornithology*, volume 1, edited by R. F. Johnston, Plenum, New York, NY.)

The distinction between precocial and altricial birds extends back to differences in the amount of yolk originally in the eggs and includes differences in the relative development of organs and muscles at hatching, and the rates of growth after hatching (Table 18–4). Robert Ricklefs (1979) has proposed that the physical maturity of tissues at hatching, especially skeletal muscles, can be used to subdivide the growth patterns of birds. Mature tissues may not be capable of rapid growth, and these mature tissues may set the limits to the growth of other tissues. Ricklefs emphasized the relationship between mode of development and food supply, and David Winkler and Jeffrey Walters (1983) have pointed to a strong phylogenetic influence on development modes. Most differences in developmental mode occur between orders, not within them, even when species within an order differ substantially in their ecology.

After altricial young have hatched they are guarded and fed by one or both parents. In some species the parents are assisted by nest helpers (Box 18–3). Adults of some species of birds carry food to nestlings in their beaks, but many species swallow

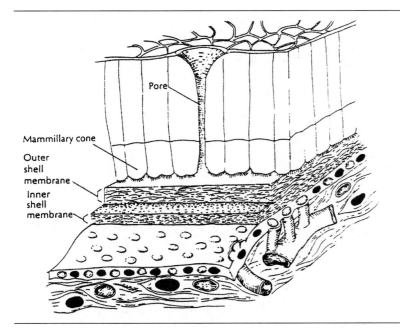

Figure 18–21 A diagram of the structure of an eggshell. Pore canals penetrate the calcified region, allowing oxygen to enter and carbon dioxide and water to leave the egg. The gases are transported to and from the embryo via blood vessels in the chorioallantoic membrane. (Source: Frank B. Gill, *Ornithology*, Freeman, New York, NY.)

food and later regurgitate it to feed the young. Hatchling altricial birds respond to any disturbance that might signal the arrival of a parent at the nest by gaping their mouths widely. The sight of an open mouth appears to stimulate a parent bird to feed it, and the young of many altricial birds have brightly colored mouth linings. Ploceid finches have covered nests, and the mouths of the nestlings of some species are said to have luminous spots that have been likened to beacons showing the parents where to deposit food in the gloom of the nest.

The duration of parental care is variable: The young of small passerines fledge (leave the nest) about 2 weeks after hatching and are cared for by their parents for an additional one to three weeks. Larger species of birds such as the tawny owl spend a month in the nest and receive parental care for an additional 3 months after they have fledged, and the young of the wandering albatross requires a year to become independent of its parents.

Migration and Navigation

The mobility that characterizes vertebrates is perhaps most clearly demonstrated in their movements over enormous distances. These displacements, which may cover half the globe, require both endurance and the ability to navigate. Other verte-

brates migrate, even over enormous distances, but migration is best known among birds.

Long-Distance Movements of Vertebrates

Four major kinds of long-distance movements by vertebrates can be distinguished, although the categories blend into each other. **Dispersal movements** are universal among animals, even among species in which individuals typically occupy limited home ranges or territories. These dispersal movements of individuals, often by juveniles or nonreproductive adults, may be related to intraspecific competition for resources and can result in expansion of the species' geographic range. Thus, overcrowding in the center of the range may lead some individuals to move to new areas where they may establish new populations.

Nomadism is a more or less irregular or random movement by a population of individuals (usually organized in herds, flocks, or schools) to favorable feeding or breeding areas that are of unpredictable occurrence in time or space. Nomadism allows animals to exploit local and temporary sources of food. Such movements are especially characteristic of ungulate mammals living in environments subject to harsh and variable conditions such as Arctic tundras, arid grasslands, and deserts. Most nomadic mam-

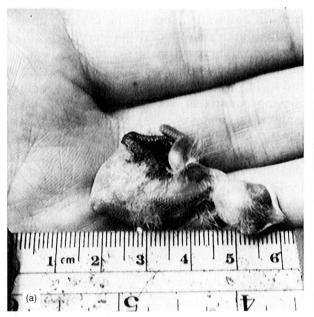

Figure 18–22 Altricial chicks (a) such as that of the tree swallow *(Tachycinta bicolor)* are entirely naked when they hatch and unable even to stand up. Precocial species (b) such as the snowy plover *(Charadius alexandrinus)* are covered with down when they hatch and can stand erect and even walk. The plover in this photograph has just hatched; it retains the egg tooth on the tip of its bill, whereas the tree swallow chick is 5 days old. The dark color on the leg of the swallow is ink, used to identify individual hatchlings for a study of parental care. (Photographs by David Winkler.)

Table 18–3   **Maturity of birds at hatching.**

Precocial: eyes open, covered with feathers or down, leave nest after 1 or 2 days
1. Independent of parents: megapodes
2. Follow parents, but find their own food: ducks, shorebirds
3. Follow parents and are shown food: quail, chickens
4. Follow parents and are fed by them: grebes, rails

Semiprecocial: eyes open, covered with down, able to walk but remain at nest and are fed by parents: gulls, terns

Semialtricial: covered with down, unable to leave nest, fed by parents
1. Eyes open: herons, hawks
2. Eyes closed: owls

Altricial: eyes closed, little or no down, unable to leave nest, fed by parents: passerines

Source: Modified from M. M. Nice, 1962. *Transactions of the Linnaean Society of New York* 8:1–211.

mals have favored calving grounds to which females return on a regular basis. Some pelagic fishes and a few species of birds in boreal forests and in austral deserts are also nomadic.

Emigration, also referred to as **invasion** or **irruption,** is another kind of movement characteristic of species in unpredictable environments. These are irregular movements of large numbers of individuals into areas where the species is not usually found. Irruptions result from periods of favorable feeding and breeding conditions that allow popula-

Table 18–4   **Comparison of altricial and precocial birds.**

Amount of yolk in eggs	precocial > altricial
Amount of yolk remaining at hatching	precocial > altricial
Size of eyes and brain	precocial > altricial
Development of muscles	precocial > altricial
Size of gut	altricial > precocial
Rate of growth after hatching	altricial > precocial

Box 18–3 Built-in Babysitters: Nest Helpers

A peculiar feature of the reproductive biology of more than 200 species of birds is the existence of nest helpers that provide care to offspring that are not their own. Most examples of nest helpers occur in Australia or the tropics, and few are known from Europe or North America. The mating systems of species with helpers vary from monogamy (the most common situation) through species with multiple breeders of either sex. Most species that have nest helpers are territorial, but some are colonial. Helper systems are characterized by regular involvement of the helpers in feeding and care of the young. Helpers defer their own breeding for one or more years while they assist in raising the offspring of other birds.

The peculiarity of helper systems lies in the expenditure of time and energy by helpers in caring for young that are not genetically their own. This altruistic behavior would appear to reduce the fitness of helpers, and that paradox has stimulated many studies. The concept of **kin selection** has contributed substantially to understanding helper systems. Stated in simplified form, this hypothesis proposes that an individual can increase its fitness by providing assistance to a related individual because relatives share alleles of common descent, and these alleles (not individuals) are the units of inheritance. Thus, the **inclusive fitness** of an individual consists of (1) its own reproductive success, plus (2) the additional reproductive success of relatives that results from the altruistic behavior of the individual multiplied by the fraction of alleles shared with each relative, minus (3) any decrease in the reproductive success of the individual that results from its altruistic behavior.

The hypothesis of kin selection predicts that nest helpers will be related to the individuals they help, and this is often the case. For example, almost 50 percent of the cases of nest helpers among Florida scrub jays involved birds helping their own parents, and 25 percent involved birds

helping a parent and a stepparent. Birds helping their own siblings accounted for 20 percent of the nest helpers, and less than 5 percent of the examples involved helping entirely unrelated individuals (Woolfenden 1975, 1981). However, this pattern is not universal, and other examples of nest helpers involve more complicated genetic relationships among the participants, including cases where the helpers are unrelated to the individuals they help.

Nest helpers do help; nests with helpers almost always fledge more young than nests without helpers. Thus, kin selection could produce some benefit to the helpers, but why would the helpers not increase their fitness still more by breeding themselves? In other words, why *do* helpers help? Also, why do helpers *ever* help nonrelatives?

Studies of these questions have stimulated much discussion and various points of view; a summary can be found in Oring (1982). Several general hypotheses have been proposed:

1. A shortage of breeding territories, nest sites, or potential mates may make it difficult for young birds to breed. Helping to raise younger siblings may be the best way to mark time until an opportunity to breed presents itself.
2. Becoming a nest helper may be a way to gain access to a territory and, eventually, to a mate.
3. Some components of parental care are learned by experience, and birds that act as helpers for one or more breeding seasons may fledge more young when they do reproduce as a result of the experience they have gained.

These hypotheses are not mutually exclusive—they all may apply to some species—and hypotheses 2 and 3 suggest that some advantage could be gained from helping even unrelated individuals.

tions to increase in numbers until food gives out. The mass exodus that follows is usually accompanied by high mortality. Norwegian lemmings normally dwell in alpine tundra, but occasionally make irruptive movements down fjords to the sea. In the boreal forest, crossbills and nutcrackers, which respond to superabundant but irregular production of spruce nuts, and also goshawks and snowy owls, which depend on snowshoe hares and lemmings for food, are additional examples of species that show emigration.

Migration, on the other hand, refers to a regular, predictable movement between two home ranges or territories, typically between a summer breeding area and a winter, nonbreeding area. It is important to emphasize, however, that true migrations are separated only by definition from emigration and nomadism. All sorts of gradations and combinations occur, even within a single species. Arctic caribou are at times nomadic, but they also make regular migratory movements to and from winter and summer feeding grounds. Some segments of the snowy owl and goshawk populations are regularly migratory, but occasional large-scale invasions bring them far south of their usual wintering ranges.

Migrations often involve movements over thousands of kilometers, especially in the case of birds nesting in northern latitudes, some marine mammals, sea turtles, and fishes. Short-tailed shearwaters, for example, make an annual migration between their breeding range in southern Australia and the North Pacific that requires a round trip of more than 30,000 kilometers (Figure 18–23). Gray whales make round trips of 9000 kilometers or more between their feeding grounds in the Bering Sea and calving areas in bays on the coast of Baja California. The 600-kilometer migration by North American caribou probably is the longest distance regularly traveled by a land mammal. Less mobile vertebrates such as amphibians may migrate a kilometer or more from nonbreeding habitats to breeding areas.

Migratory Movements of Birds

Few migratory movements are as dramatic as those of birds, and in no other group has migration been so well studied. Migration is a widespread phenomenon among birds—about 40 percent of the bird species in the Palearctic are migratory, and an estimated total of some 5 billion birds migrate from the Palearctic every year. Data accumulated from recaptures of banded birds have established the origins, destinations, and migratory pathways for many species. A remarkable feature of most of these migrations is their relatively recent origin: Migrations are responses to seasonal changes in the availability of resources. These variations in the resource base are, in turn, the result of seasonal cycles in the climate, and worldwide patterns of climate have changed frequently during the past 2 million years. Current avian migratory patterns are probably no more than 15,000 years old (Moreau 1972).

Many birds return each year to the same migratory stopover sites, just as they may return to the same breeding and wintering sites year after year. Migrating birds may be concentrated at high densities at certain points along their traditional migratory routes. For example, species that follow a coastal route may be funnelled to small points of land, such as Cape May, New Jersey, from which they must initiate long overwater flights. At these stopovers, migrating birds must find food and water to replenish their stores before they venture over the sea, and they must also avoid the predators that congregate at these sites. Development of coastal areas for human use has destroyed many important resting and refueling stations for migratory birds. The destruction of coastal wetlands has caused serious problems for migratory birds on a worldwide basis. Loss of migratory stopover sites may remove a critical resource from a population at a particularly stressful stage in its life cycle.

The Advantages of Migration The high energy costs of migration must be offset by energy gained as a result of moving to a different habitat. The normal food sources for some species of birds are unavailable in the winter, and the benefits of migration for those species are starkly clear. Other species may save energy mainly by avoiding the temperature stress of northern winters. In other cases the main advantage of migration may come from breeding in high latitudes in the summer when the long days provide more time to forage than the birds would have if they remained closer to the equator.

Competition may also play a role in migration. Direct evidence of competition for food resources is lacking, but indirect evidence was provided by a study of birds that migrate between the Neotropics and North America (Cox 1968). The extent of differentiation of the beak is considered to be a measure of the difference in feeding niches of birds. Birds that

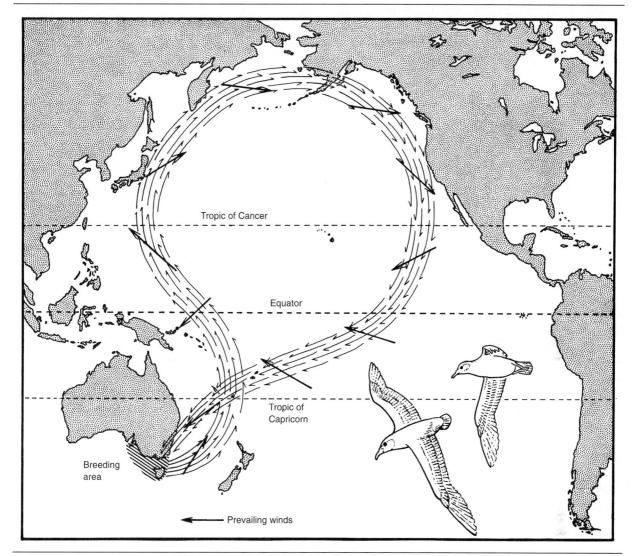

Figure 18–23 The migratory path of the short-tailed shearwater from its Australian breeding area to its northern range takes advantage of the prevailing winds in the Pacific region to reduce the energy cost of migration. (From A. J. Marshall and D. L. Serventy, 1956, *Proceedings of the Zoological Society of London* 127:489–510.)

have bills of different sizes and shapes are usually assumed to be exploiting different sources of food even when they forage in the same place, whereas birds with bills of the same size and shape are assumed to eat the same kinds of food and must forage in different habitats to avoid competition. During the breeding season, when food requirements are high, competition should be more intense between species with bills of similar shape than between species with bills of different shapes. Cox compared the bill morphology of resident and migrant tropical birds: If competition is irrelevant to migration, the similarity of bill shape should be the same for resident and migrant species. If competition is important in determining migration, one would predict that species with similar bill shapes would migrate, because they are less able to separate their feeding niches ecologically.

Cox found that variation in bill form was greatest in resident tropical birds, and decreased as the proportion of migratory species in a group increased (Figure 18–24). This result is consistent with the predictions of the hypothesis that competition is an important force in migration. Cox proposed that species of birds that are not ecologically separated by bill morphology avoid competition for food during the breeding season by migrating northward.

Physiological Preparation for Migration Migration is the result of a complex sequence of events that integrate the physiology and behavior of birds. Fat is the principal energy store for migratory birds, and birds undergo a period of heavy feeding and premigratory fattening (**Zugdisposition,** migratory preparation) in which fat deposits in the body cavity and subcutaneous tissue increase tenfold, ultimately reaching 20 to 50 percent of the nonfat body mass. Fat is metabolized rapidly when migration begins, and many birds migrate at night and eat during the day. Even diurnal migrants divide the day into periods of migratory flight (usually early in the day) and periods of feeding. In addition, pauses of several days to replenish fat stores are a normal part of migration. *Zugdisposition* is followed by **Zugstimmung** (migratory mood), in which the bird undertakes and maintains migratory flight. In caged birds, which are prevented from migrating, this condition results in the well-known phenomenon of **Zugunruhe** (migratory restlessness).

Preparation for migration must be integrated with environmental conditions, and this coordination appears to be accomplished by the interaction of internal rhythms with an external stimulus. Daylength is the most important cue for *Zugdisposition* and *Zugstimmung* for birds in north temperate regions. Northward migration in spring is induced by increasing daylength (Figure 18–25). The direction in which migratory birds orient during *Zugunruhe* depends on their physiological condition. In this experiment, photoperiod was manipulated to bring one group of indigo buntings into their autumn migratory condition at the same time that a second group of birds was in its spring migratory condition. When the birds were tested under an artificial planetarium sky, the birds in the spring migratory condition oriented primarily in a northeasterly direction (Figure 18–25b), whereas birds in the fall migratory condition oriented in a southerly direction (Figure 18–25c). Dark circles show the mean nightly headings pooled for several observations for each of six birds in the spring migratory condition and five birds in the fall condition.

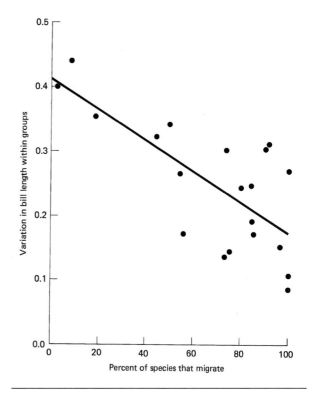

Figure 18–24 Birds that winter in Central America show an inverse relationship between the amount of variation in beak length within a group and the percentage of species in that group that migrate to North America during the breeding season. Variation in beak length is expressed as coefficient of variation for each group. (Data from G. W. Cox, 1968, *Evolution* 22:180–192.)

After breeding, many species of birds enter a refractory period in which they are unresponsive to long daylengths, their gonads regress to the nonbreeding condition, and they molt. Decreasing photoperiods in autumn may accelerate the southward migration. Many birds are again refractory to photoperiod stimulation for several weeks after the autumnal migration and require an interval of several weeks of short photoperiods before they can again be stimulated by long photoperiods.

Underlying the responses of birds to changes in daylength is an endogenous (internal) rhythm. This circannual (about a year) cycle can be demonstrated by keeping birds under constant conditions. Fat deposition and migratory restlessness coincide in most species and alternate with gonadal develop-

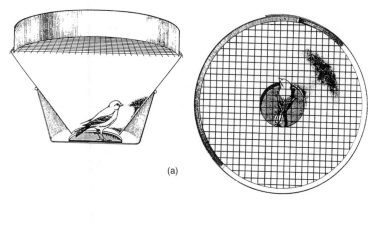

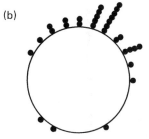
(a)

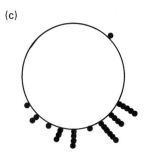

Figure 18–25 Direction in which migratory birds orient during *Zugunruhe*. (a) The birds were tested in circular cages that allow a view of the sky. The bird stands on an ink pad, and each time it hops onto the sloping wall it leaves a mark on the blotting paper that lines the cage. (b) In spring, the birds oriented toward the north, and (c) in autumn toward the south. (Source: [a] Illustration by Adolph E. Brotman from "The Stellar-Orientation System of a Migratory Bird" by Stephen To Emlen, in *Scientific American*, August 1975 pp 102–111; reprinted with permission; [b,c] data from S. T. Emlen, 1969, *Science* 165:716–718.)

(b)

(c)

SPRING CONDITION

AUTUMN CONDITION

ment and molt as they do in wild birds. When the rhythms are free-running (that is, when they are not cued by external stimuli), they vary between 7 and 15 months. In other words, the birds' internal clocks continue to run, but in the absence of the cue normally provided by changing daylength the internal rhythms drift away from precise correspondence with the seasons.

Orientation and Navigation

The seasonal migrations of vertebrates that cover thousands of kilometers and, especially, their ability to return regularly to the same locations, year after year, pose an additional question—How do they find their way? Various hypotheses propose mechanisms to explain how animals navigate on these journeys and the explanations fall into two general categories: (1) Long-distance migration is an extension of the tendency to explore territory beyond the local home range, learning to recognize landmarks as one goes along or, (2) the ability to home through unfamiliar territory results from an internal navigation system. There are no clear answers, but experiments reveal that many vertebrates can find their way home when they are displaced and that a number of different mechanisms and sensory modalities are involved. A general review of bird migration and navigation can be found in Alerstam (1990).

The homing pigeon has become a favorite experimental animal for studies of navigation. As long as people have raised and raced pigeons, it has been known that birds released in unfamiliar territory vanish from sight flying in a straight line, usually in the direction of home. How do pigeons accomplish this feat? There is no complete answer yet, but experiments have shown that navigation by homing pigeons (and presumably by other vertebrates as well) is complex and is based on a variety of sensory cues. On sunny days pigeons vanish toward home and return rapidly to their lofts. On overcast days vanishing bearings are less precise and birds more often get lost. These observations led to the idea that pigeons use the sun as a compass.

Of course, the position of the sun in the sky changes from dawn to dusk. That means that a bird must know what time of day it is to use the sun to tell direction, and this time-keeping ability requires some sort of internal clock. If that hypothesis is correct, it should be possible to fool a bird by shifting its clock forward or backward. For example, if one turns on the lights in the pigeon loft 6 hours before sunrise every morning for about 5 days, the birds will become accustomed to that artificial sunrise, and at any time during the day they will assume that the time is 6 hours later than it actually is. When those birds are released they will judge direction by the sun, but their internal clocks will be wrong by 6 hours. As a consequence of that error, the birds should fly off on courses that are displaced by 90° from the correct course for home. Clock-shifted pigeons react differently under sunny and cloudy skies, indicating that pigeons have at least two mechanisms for navigation (Figure 18–26). Each dot in these plots shows the direction in which a pigeon vanished from sight when it was released in the center of the large circle. The home loft is straight up in each diagram. The solid bar extending outward from the center of each circle is the average vector sum of all the individual vanishing points. When they are able to see the sun, control birds that have been kept on the normal photoperiod orient predominantly in the direction of home (Figure 18–26a). Birds that have had their photoperiods shifted 6 hours fast disappear on bearings westward of the true direction of home (Figure 18–26b). However, when the sun is obscured by clouds, the birds cannot use it for navigation and must rely on other mechanisms. Under those conditions both control and clock-shifted birds orient correctly toward home (Figure 18–26c, d).

Polarized light is another cue vertebrates use to determine directions. In addition, some vertebrates have been shown to detect ultraviolet light and to sense extremely low-frequency sounds, well below the frequencies humans can hear. Those sounds are generated by ocean waves and air masses moving over mountains and can signal a general direction over thousands of kilometers, but their use as cues for navigation remains obscure. The senses that have been shown to be involved in navigation by pigeons do not end here. Pigeons can also navigate by recognizing airborne odors as they pass over the terrain. Even magnetism is implicated: On cloudy days pigeons wearing small magnets on their heads have their ability to navigate disrupted, but on sunny days magnets have no effect. When it is clear, pigeons apparently rely on their sun compass and magnetic cues are ignored.

Results of this sort are being obtained with other vertebrates as well, and lead to the general conclusion that a great deal of redundancy is built into navigation systems. Apparently, there is a hierarchy of useful cues. For example, a bird that relies on the sun and polarized light to navigate on clear days could switch to magnetic direction sensing on heavily overcast days. For both conditions, it might use local odors and recognition of landmarks as it approaches home.

Many birds migrate only at night. Under these conditions a magnetic sense of direction might be important (Able and Able 1990, 1993; Wiltschko et al. 1993). Several species of nocturnally migrating birds use star patterns for navigation. Apparently, each bird fixes on the pattern of particular stars and uses their motion in the night sky to determine a compass direction. As is the case for sun compass navigation, an internal clock is required for this sort of celestial navigation, and artificially changing the time setting of the internal clock produces predictable changes in the direction in which a bird orients.

Despite numerous studies the complexities of navigation mechanisms of vertebrates are far from being fully understood and much controversy surrounds some hypotheses. The built-in redundancy of the systems makes it difficult to devise experiments that isolate one mechanism. When it is deprived of the use of one sensory mode, an animal is likely to have one or several others it can use instead. This redundancy itself, and the remarkable sophistication with which many vertebrates navigate, show the importance of migration in their lives.

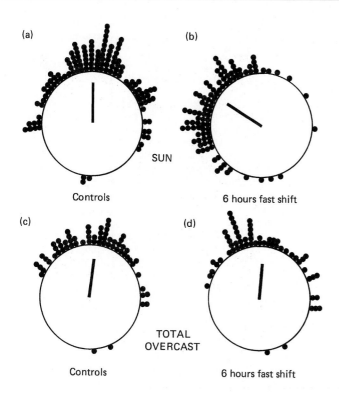

(a)

(b)

SUN

Controls

6 hours fast shift

(c)

(d)

TOTAL
OVERCAST

Controls

6 hours fast shift

Figure 18–26 Orientation of clock-shifted pigeons under sunny and cloudy skies. The line shows the average direction chosen by the birds. (From W. T. Keeton, 1969, *Science* 165:922–928.)

Summary

The fossil record reveals that birds radiated in the Jurassic. *Archaeopteryx,* the earliest fossil bird known, was probably a late-surviving relict that was contemporaneous with more advanced birds. It retained a large number of ancestral characters and, except for the presence of feathers, was very like a small dinosaur. True flying birds were widespread and diverse by the Cretaceous. Extant orders of birds had evolved by the end of the Eocene, but the phylogenetic relationships among the groups of extant birds are poorly understood.

The ecology and behavior of birds are directly influenced by their ability to fly. The mobility of birds allows them to exploit food supplies that have patchy distributions in time and space and to feed and reproduce in different areas. Migration is the most dramatic manifestation of the mobility of birds, and some species travel tens of thousands of kilo-meters in a year. Migrating birds use a variety of cues for navigation, including the position of the sun, polarized light, the Earth's magnetic field, and infrasound.

Many of the complex social behaviors of birds are associated with reproduction, and birds have contributed greatly to our understanding of the relationship between ecological factors and the mating systems of vertebrates. All birds are oviparous, perhaps because egg retention is the first step in the evolution of viviparity and the specializations of the avian way of life do not make egg retention advantageous for birds. Monogamy, with both parents caring for the young, is the most common mating system for birds, but polygyny (a male mating with several females) and polyandry (a female mating with several males) also occur.

References

Able, K. P., and M. A. Able. 1990. Calibration of the magnetic compass of a migratory bird by celestial rotation. *Nature* 347:378–380.

Able, K. P., and M. A. Able. 1993. Daytime calibration of magnetic orientation in a migratory bird requires a view of skylight polarization. *Nature* 364:523–525.

Alerstam, T. 1990. *Bird Migration.* Cambridge University Press, Cambridge, UK.

Altangerel, P., M. A. Norell, L. M. Chiappe, and J. M. Clark. 1993a. Flightless bird from the Cretaceous of Mongolia. *Nature* 362:623–626.

Altangerel, P., M. A. Norell, L. M. Chiappe, and J. M. Clark. 1993b. Correction: flightless bird from the Cretaceous of Mongolia. *Nature* 363:188.

Blackburn, D. G., and H. E. Evans. 1986. Why are there no viviparous birds? *American Naturalist* 128:165–190.

Carey, C. 1983. Structure and function of avian eggs. Pages 69–103 in *Current Ornithology,* volume 1, edited by Richard F. Johnston. Plenum, New York, NY.

Collias, N. E., and E. C. Collias. 1984. *Nest Building and Bird Behavior.* Princeton University Press, Princeton, NJ.

Cox, G. W. 1968. The role of competition in the evolution of migration. *Evolution* 22:180–192.

Cracraft, J. A. 1986. The origin and early diversification of birds. *Paleobiology* 12:383–399.

Drent, R. 1975. Incubation. Pages 333–420 in *Avian Biology,* volume 5, edited by D. S. Farner, J. R. King, and K. C. Parkes. Academic, Orlando, FL.

Emlen, S. T., and L. W. Oring. 1977. Ecology, sexual selection, and the evolution of mating systems. *Science* 197:215–223.

Faaborg, J., Tj. deVries, C. B. Patterson, and C. R. Griffin. 1980. Preliminary observations on the occurrence and evolution of polyandry in the Galapagos hawk *(Buteo galapagoensis). Auk* 97:581–590.

Feduccia, A. 1995. Explosive evolution in Tertiary birds and mammals. *Science* 267:637–638.

Frith, C. B., and D. W. Frith. 1990. Archbold's bowerbird *Archboldia papuensis* (Ptilonorhynchidae) uses plumes from King of Saxony bird of paradise *Pteridophora alberti* (Paradisaeidae) as bower decoration. *Emu* 90:136–137.

Godard, R. 1991. Long-term memory of individual neighbors in a migratory songbird. *Nature* 350:228–229.

Goss–Custard, J. D. 1977. Optimal foraging and size selection of worms by redshank, *Tringa totanus. Animal Behaviour* 25:10–29.

Hamilton, W. D., and M. Zuk. 1982. Heritable true fitness and bright birds: a role for parasites? *Science* 218:384–387.

Hildén, O. 1975. Breeding system of Temminck's stint, *Calidris temminckii. Ornis Fennica* 52:117–146.

Houde, P. 1986. Ostrich ancestors found in the Northern Hemisphere suggest new hypothesis of ratite origins. *Nature* 324:563–565.

Houde, P. 1987. Critical evaluation of DNA hybridization studies in avian systematics. *Auk* 104:17–32.

Kharitonov, S. P., and D. Siegel-Causey. 1988. Colony formation in seabirds. Pages 223–272 in *Current Ornithology,* volume 5, edited by R. F. Johnston. Plenum, New York, NY.

Kistchinski, A. A. 1975. Breeding biology and behaviour of the grey phalarope, *Phalaropus fulicarius,* in east Siberia. *Ibis* 117:285–301.

Konishi, M. 1985. Birdsong: from behavior to neuron. *Annual Review of Neurosciences* 8:125–170.

Konishi, M. S. T. Emlen, R. E. Ricklefs, and J. C. Wlngfield. 1989. Contributions of bird studies to biology. *Science* 246:465–472.

Krebs, J. R., D. W. Stephens, and W. J. Southerland. 1983. Perspectives in optimal foraging. Pages 165–216 in *Perspectives in Ornithology,* edited by A. H. Brush and G. A. Clark, Jr. Cambridge University Press, Cambridge, UK.

Lockley, M. G., S. Y. Yang, M. Matsuka, F. Fleming, and S. K. Lim. 1992. The track record of Mesozoic birds: evidence and implications. *Philosophical Transactions of the Royal Society (London) Series B* 336:113–134.

Marshall, L. G. 1994. Terror birds of South America. *Scientific American* 270(2):90–95.

Martin, L. D. 1983a. The origin and early radiation of birds. Pages 291–338 in *Perspectives in Ornithology,* edited by A. H. Brush and G. A. Clark, Jr. Cambridge University Press, Cambridge, UK.

Martin, L. D. 1983b. The origin of birds and of avian flight. Pages 105–129 in *Current Ornithology,* volume 1, edited by Richard F. Johnston. Plenum, New York, NY.

Moreau, R. E. 1972. *The Palearctic–African Bird Migration Systems.* Academic, New York, NY.

Nordeen, K. W., and E. J. Nordeen. 1988. Projection neurons within a vocal motor pathway are born during song learning in zebra finches. *Nature* 334:149–151.

Nottebohm, F., and A. P. Arnold. 1976. Sexual dimorphism in vocal control areas of the songbird brain. *Science* 194:211–213.

Olson, S. L. 1985. The fossil record of birds. Pages 79–238 in *Avian Biology,* volume 8, edited by D. S. Farner, J. R. King, and K. C. Parkes. Academic, Orlando, FL.

Orians, G. H. 1969. On the evolution of mating systems in birds and mammals. *American Naturalist* 103:589–603.

Orians, G. H. 1980. *Some Adaptations of Marsh-Nesting Blackbirds.* Princeton University Press, Princeton, NJ.

Oring, L. W. 1982. Avian mating systems. Pages 1–92 in *Avian Biology,* volume 6, edited by D. S. Farner, J. R. King, and K. C. Parkes. Academic, New York, NY.

Ostrom, J. H. 1974. *Archaeopteryx* and the origin of flight. *Quarterly Review of Biology* 49:27–47.

Ostrom, J. H. 1979. Bird flight: how did it begin? *American Scientist* 67:46–56.

Payne, R. B., W. L. Thompson, K. L. Fiala, and L. L. Sweany. 1981. Local song traditions in indigo buntings: cultural transmission of behavior patterns across generations. *Behaviour* 77:199–221.

Petrie, M. 1994. Improved growth and survival of offspring of peacocks with more elaborate trains. *Nature* 371:598–599.

Pierce, G. J., and J. G. Ollason. 1987. Eight reasons why optimal foraging theory is a complete waste of time. *Oikos* 49:111–118.

Raikow, R. J. 1985. Problems in avian classification. Pages 187–212 in *Current Ornithology,* volume 2, edited by Richard F. Johnston. Plenum, New York, NY.

Rayner, J. M. V. 1988. The evolution of vertebrate flight. *Biological Journal of the Linnean Society* 34:269–287.

Ricklefs, R. E. 1979. Adaptation, constraint, and compromise in avian postnatal development. *Biological Review* 54:269–290.

Sereno, P. C., and R. Chenggang. 1992. Early evidence of avian flight and perching: new evidence from the Lower Cretaceous of China. *Science* 255:845–848.

Sibley, C. G., and J. E. Ahlquist. 1990. *Phylogeny and Classification of Birds*. Yale University Press, New Haven, CT.

Sibley, C. G., J. E. Ahlquist, and B. L. Monroe. 1988. A classification of the living birds of the world, based on DNA–DNA hybridization studies. *Auk* 105:409–423.

Slagsvold, T., and J. T. Lifjeld. 1992. Plumage color is a condition-dependent sexual trait in male pied flycatchers. *Evolution* 46:825–828.

Smith, T. B. 1987. Bill size polymorphism and intraspecific niche utilization in an African finch. *Nature* 329:717–719.

Stearns, S. C., and P. Schmid–Hempel. 1987. Evolutionary insights should not be wasted. *Oikos* 49:118–125.

Stephens, D. W., and J. R. Krebs. 1987. *Foraging Theory*. Princeton University Press, Princeton, NJ.

Storer, R. W. 1971. Classification of birds. Pages 1–18 in *Avian Biology*, volume 1, edited by D. S. Farner, J. R. King, and K. C. Parkes. Academic, New York, NY.

Vince, M. A. 1969. Embryonic communication, respiration, and the synchronization of hatching. Pages 233–260 in *Bird Vocalizations*, edited by R. A. Hinde. Cambridge University Press, Cambridge, UK.

Vleck, C. M., and G. J. Kenagy. 1980. Embryonic metabolism of the fork-tailed storm petrel: physiological patterns during prolonged and interrupted incubation. *Physiological Zoology* 53:32–42.

Warheit, K. L. 1992. A review of the fossil seabirds from the Tertiary of the North Pacific: plate tectonics, paleoceanography, and faunal change. *Paleobiology* 18:401–424.

Weihs, D., and G. Katzir. 1994. Bill sweeping in the spoonbill, *Platalea leucordia*: evidence for a hydrodynamic function. *Animal Behaviour* 47:649–654.

Wiley, R. H. 1973. Territoriality and non-random mating in the sage grouse, *Centrocercus urophasianus*. *Animal Behaviour Monographs* 6:87–169.

Wiltschko, W., U. Munro, H. Ford, and R. Wiltschko. 1993. Red light disrupts magnetic orientation of migratory birds. *Nature* 364:525–527.

Winkler, D. W., and J. R. Walters. 1983. The determination of clutch size in precocial birds. Pages 33–68 in *Current Ornithology*, volume 1, edited by Richard F. Johnston. Plenum, New York, NY.

Witmer, L. M., and K. D. Rose. 1991. Biomechanics of the jaw apparatus of the gigantic Eocene bird *Diatryma*: implications for diet and mode of life. *Paleobiology* 17:95–120.

Wittenberger, J. F., and G. L. Hunt, Jr. 1985. The adaptive significance of coloniality in birds. Pages 1–78 in *Avian Biology*, volume 8, edited by D. S. Farner, J. R. King, and K. C. Parkes. Academic, Orlando, FL.

Woolfenden, G. E. 1975. Florida scrub jay helpers at the nest. *Auk* 92:1–15.

Woolfenden, G. E. 1981. Selfish behavior by Florida scrub jay helpers. Pages 257–260 in *Natural Selection and Social Behavior*, edited by R. D. Alexander and D. W. Tinkle. Chiron, New York, NY.

19
THE SYNAPSIDA AND THE EVOLUTION OF MAMMALS

A gain we must backtrack, this time to the end of the Paleozoic (Chapters 9 and 10) to find the origins of the final lineage of vertebrates, the synapsids. The synapsids actually had their first radiation (eupelycosaurs and caseasaurs) and second radiation (nonmammalian therapsids) in the Paleozoic, before the radiations of the diapsids we have already discussed. However, the third radiation of the synapsid lineage (mammals) reached its peak in the Cenozoic. Nonetheless, through the late Paleozoic and early Mesozoic the synapsid lineage was becoming increasingly mammal-like and mammals and nonavian dinosaurs both appeared on the scene in the Triassic.

In Chapter 4 we discussed the paradox of the origin of endothermy: The two components of endothermal thermoregulation are a high metabolic rate that produces heat and insulation that retains that heat in the body, and neither characteristics is advantageous without the previous existence of the other. We suggested that the solution of the paradox might lie in the evolution of high rates of metabolism for some purpose other than thermoregulation. One possible benefit of a high rate of metabolism would be to enhance the locomotor endurance of predatory synapsids, and it is gratifying to see in the fossil record just the sorts of structural changes in the limbs and rib cage of synapsids that might be expected to accompany an increasingly active form of locomotion and foraging. The three groups of extant mammals—monotremes, marsupials, and placentals—had evolved by the late Mesozoic, and they were accompanied by several groups of mammals that are now extinct.

Terrestrial Vertebrates of the Late Paleozoic

The synapsid lineage was the first group of amniotes to radiate widely in terrestrial habitats. During the late Carboniferous and the first half of the Permian the basal synapsids were the most abundant terrestrial vertebrates, and from the mid-Permian into the Triassic nonmammalian therapsids were the top carnivores in the food web. Many eupelycosaurs and early therapsids weighed more than a kilogram, but most of the synapsid lineages had disappeared by the late Triassic, and the surviving forms were smaller and perhaps nocturnal.

The appearance of mammalian skeletal characters among synapsids can be linked to increasing capacity for locomotor activity. The legs became longer and were progressively held more nearly under the body with accompanying changes in the pelvic girdle. The teeth and the jaw muscles became increasingly differentiated and probably permitted more effective oral breakdown. These structural characters suggest that derived carnivorous synapsids were active predators, and their foraging behavior may have required a high aerobic metabolic capacity (see Chapter 4). Changes in the structure of the ribs suggest a shift from a reptilian to a mammalian method of lung ventilation. The metabolic rates of derived carnivorous synapsids were probably sufficient for some degree of endothermal thermoregulation, and the early mammals appear to have been nocturnal. The sensory requirements of nocturnal activity may have led to an increase in the sizes of the brains of early mammals compared to their diurnal contemporaries, the lizards and sphenodontids.

The Synapsid Skull

In ancestral amniotes the skull roof and cheek were solid, and the jaw muscles lay within this cover of dermal bones. This is the **anapsid** (without arches) skull condition. Among the earliest amniotes was a late Carboniferous group known as limnoscelids. These anapsids had an open suture between the skull roof and the cheek region. It seems likely that the edges of the bones bordering this slit—the supratemporal and postorbital above and the squamosal below—were mechanically advantageous sites for attachment of the jaw muscles. The edges of a temporal fenestra may provide a more secure origin for jaw muscles than does a flat surface. If some of the adductor muscles of the lower jaw found a particularly favorable origin on this region, selection could have favored the enlargement of the gap, giving a progressively larger area for muscle attachment and ultimately forming the temporal fenestra characteristic of the synapsids (Figure 19–1).

Eupelycosaurs and Caseasaurs

The Synapsida includes two major lineages, the Caseasauria and the Eupelycosauria (Figure 19–2). The eothyridid caseasaurs were small, carnivorous animals from the early Permian that represent the most plesiomorphic synapsid skull patterns known. The caseids were larger and were herbivorous. The premaxilla of caseasaurs is tilted forward to form a pointed rostrum that is particularly prominent in caseids (Figure 19–3). The Caseasauria is the sister taxon of the Eupelycosauria, which includes all remaining synapsids.

The most generalized eupelycosaurs were the varanopsids and ophiacodontids. Some of these were large animals; *Ophiacodon major* was 3 meters long and may have weighed 200 kilograms. The ophiacodontids appear to have been semiaquatic fish eaters. Their heads were long and slender and each jaw contained 40 or more small, sharp teeth. The morphological specializations of eupelycosaurs are closely associated with differences in diet. The ophiacodontids, and to a still greater extent the haptodonts and sphenacodontids, became increasingly specialized for predation, whereas the edaphosaurids were herbivores. Carnivorous eupelycosaurs had long, narrow skulls (Figure 19–3). Their dentition was **heterodont,** with large canine-like teeth followed by smaller teeth. Herbivorous synapsids had short, broad skulls. *Edaphosaurus* (Edaphosauridae) had **homodont** dentition (teeth of uniform size and shape) with peg-like teeth.

Carnivorous Eupelycosaurs

Sphenacodontids probably preyed on a variety of small and large animals, including other sphenacodontids as well as the herbivorous caseids and edaphosaurids. The progressive changes that can be seen in the structure of the jaws and teeth apparently reflect selection for more effective predation (Figure 19–3). The anterior teeth became larger and better suited for grasping and killing large prey.

Other changes in the form of the sphenacodontid skull reflect the mechanical requirements of the specialized dentition. The enlarged maxillary teeth seen in *Dimetrodon* have deep roots in the maxilla, and in the evolution of this lineage the maxilla gradually increased in depth. As the maxilla grew downward, the palate, which was flat ancestrally, became arched. The significance of this arch is twofold: First, an arched palate is mechanically stronger than a flat one. Second, the arch of the palate provides space for air to pass over prey held in the mouth. The arched palate of sphenacodontids was the first step toward the development of a separation of the mouth and nasal passages seen in some therapsids and in mammals. The reflected lamina of the angular bone that first appears among sphenacodontids becomes more prominent in therapsids and assumes an important role in the evolution of the mammalian middle ear (Box 19–1).

The postcranial skeleton of sphenacodontids also changed in ways that appear to reflect selection

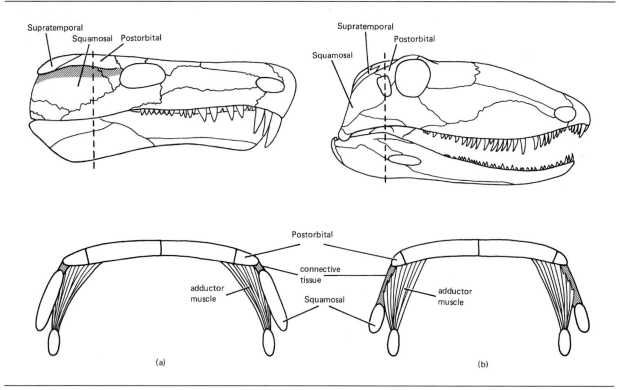

Figure 19–1 Hypothetical origin of the synapsid temporal fenestra from a suture between the cheek and roof of an anapsid skull: (a) *Limnoscelis,* anapsid; (b) *Ophiacodon,* synapsid. (Modified from T. S. Kemp, 1982, *Mammal-Like Reptiles and the Origin of Mammals,* Academic, London, UK; and A. S. Romer, 1966, *Vertebrate Paleontology,* 3d edition, University of Chicago Press, Chicago, IL.)

Figure 19–2 Phylogenetic relationships of the Synapsida. This diagram shows the probable relationships among the major groups of synapsids. Extinct lineages are marked by a dagger (†). The numbers indicate derived characters. **1.** Synapsida: Lower temporal fenestra present, plus additional features of the skull. **2.** Eupelycosauria: Snout deeper than it is wide, frontal bone forming a large portion of the margin of the orbit, and other skull characters. **3.** Sphenacodontia: A reflected lamina on the angular bone, retroarticular process of the articular bone turned downward, and additional features of the skull. **4.** Therapsida: Reflected lamina of angular deeply notched, upper canine enlarged, maxilla extended dorsally, plus other skull characters. **5.** Theriodontia: Coronoid process on dentary, quadrate and quadratojugal reduced and lacking a sutural connection to the squamosal, and other skull features. **6.** Eutheriodontia: Temporal fossa completely open dorsally, plus other features of the skull. **7.** Cynodontia: Postcanine teeth with anterior and posterior accessory cusps and small cusps on the inner side, partial bony secondary palate, plus other characters of the skull and post-cranial skeleton. (Based on T. S. Kemp, 1983, *Zoological Journal of the Linnean Society* 77:353–384; M. J. Benton (editor), 1988, *The Phylogeny and Classification of the Tetrapods,* Special Volume No. 35B, The Systematics Association, Oxford University Press, Oxford, UK; R. L. Carroll, 1988, *Vertebrate Paleontology and Evolution,* Freeman, New York, NY; M. J. Benton, 1990, *Vertebrate Paleontology,* HarperCollins Academic, London, UK; M. J. Benton (editor), 1993, *The Fossil Record 2,* Chapman & Hall, London, UK; J. A. Hopson, 1994, pages 190–219 in *Major Features of Vertebrate Evolution,* edited by D. R. Prothero and R. M. Schoch, Short Courses in Paleontology No. 7, Paleontological Society and University of Tennessee Press, Knoxville, TN; and J. A. Hopson, personal communication.)

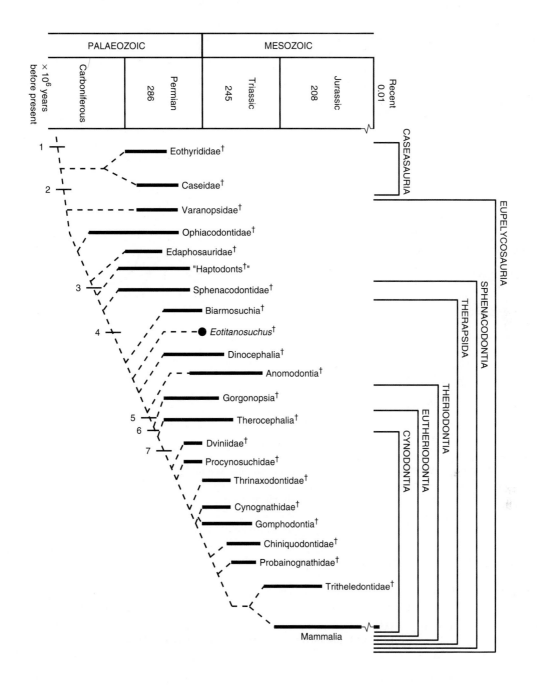

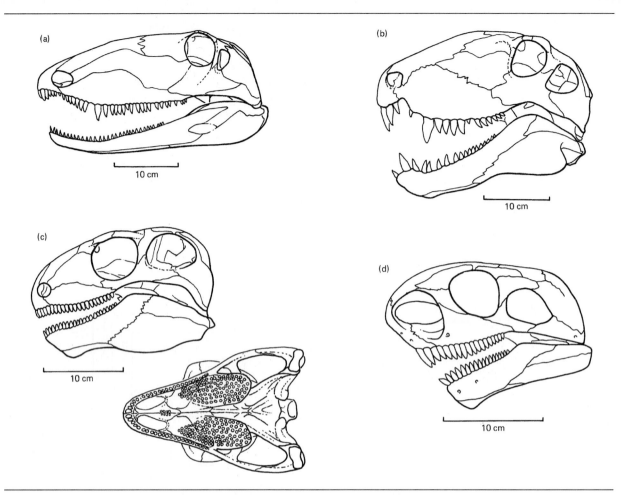

Figure 19–3 Among early synapsids, carnivorous forms had long, narrow skulls (a, *Ophiacodon*, an ophiacodontid; b, *Dimetrodon*, a sphenacodontid). Their dentition was heterodont, with well-developed canine-like teeth. Herbivores *Edaphosaurus* (c) and *Cotylorhynchus* (d) had short, broad skulls. *Edaphosaurus* had homodont dentition with peg-like teeth; *Cotylorhynchus* was more heterodont. Palatal teeth of *Edaphosaurus* bit against similar teeth in the lower jaw. (Modified from A. S. Romer, 1966, *Vertebrate Paleontology*, 3d edition, University of Chicago Press, Chicago, IL.)

for increased effectiveness of predation (Figure 19–4). The legs, although they were still held out horizontally from the body in the ancestral pattern, were longer and slimmer in the derived sphenacodontids than in earlier forms, suggesting increasingly active search and pursuit of prey. Reduction of the intercentra and development of slanting (instead of horizontal) articulations between adjacent vertebrae may suggest that lateral undulation of the vertebral column during locomotion was decreased in these predators.

A remarkable feature of some sphenacodontids and edaphosaurids was the elongation of the neural spines of the trunk region (Figure 19–5). A *Dimetrodon* 3 meters long had spines that rose as much as 1.5 meters above the vertebrae and supported a crest of skin that may have been a temperature-regulating device. These crested synapsids probably relied on absorbing solar energy to raise their body temperatures to optimal activity levels, just as living reptiles do. Marks of blood vessels on the spines indicate that the tissue they supported

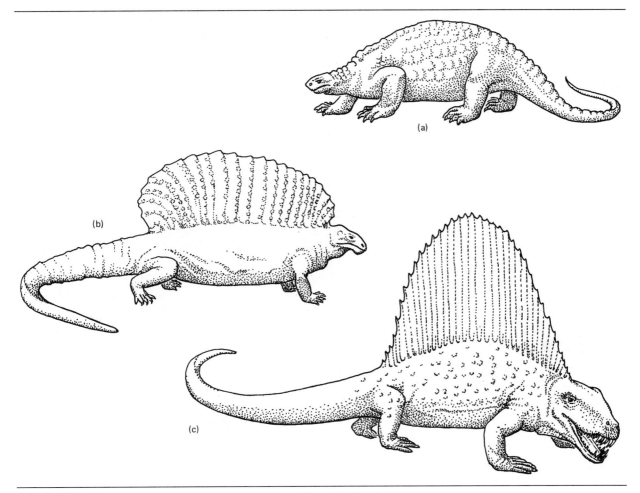

Figure 19–4 Body forms of herbivorous and carnivorous synapsids: (a) *Coty-lorhynchus*, a caseid; (b) *Edaphosaurus*, an edaphosaurid; (c) *Dimetrodon*, a sphenacodontid.

was heavily vascularized, and fossils of *Dimetrodon* usually have the spines aligned like pickets in a fence, suggesting that in life the spines were covered by connective tissue and skin that held the spines in place after death.

The potential blood flow through the crests of these sailback synapsids, indicated by the extent of vascularization, so greatly exceeds any reasonable metabolic requirement for such a tissue that it seems likely that the animals shunted blood through the crest in response to their thermoregulatory requirements. In the morning a *Dimetrodon* could orient its body perpendicular to the sun's rays and allow a large volume of blood to flow through the crest where the blood would be warmed by the sun and the heat carried back into the animal's body. When a

Dimetrodon was warm enough, blood flow through the crest could be restricted, and the heat would be retained within the body.

One mathematical model of thermoregulation by *Dimetrodon* suggests that the crest could have increased both the rate of heating in the morning and the maximum body temperature that could be achieved (Table 19–1). The effect of the crest might have been particularly significant for large species such as *Dimetrodon limbatus* and *D. grandis*. The rates of heating calculated for these animals are two or three times faster with the crest than without it, and the maximum body temperatures they could achieve with a crest (36 to 37°C) are 5°C higher than those they would have reached without a crest (Haack 1986). However, mathematical models of energy

Figure 19–5 *Dimetrodon grandis* (American Museum of Natural History, #315862, photograph by Julius Kirschner, Courtesy of the Department of Library Services.)

exchange are sensitive to the assumptions on which they are based, and a different analysis concluded that the presence or absence of a crest would have had little effect on the rate of heating (Tracy et al. 1986).

Herbivorous Synapsids: The Edaphosaurids and Caseids

While the carnivores evolved a body form that made them more effective predators, two lineages of synapsids were becoming specialized herbivores. Although they are only distantly related, the body forms of caseids and edaphosaurids were similar to each other (Figure 19–4). They were heavy-bodied animals with short, sturdy legs sprawled out from the trunk. The skull was small in proportion to the body, and the dentition was specialized for crushing and shredding vegetation rather than for killing and tearing apart prey.

The facial region of caseids and edaphosaurids was short (Figure 19–3). This morphology increases the biting force at the front of the jaws, although it

does so at the cost of a reduction in the speed of jaw closure. For an herbivore, however, force is more important than speed. Plant material is difficult to chew, and the jaw articulation of herbivorous vertebrates is frequently displaced from the horizontal plane of the tooth row. This displacement may increase the leverage of the jaw muscles, and it also produces a horizontal displacement between the upper and lower rows of teeth when the jaws close. The grinding action that occurs as the teeth in the lower jaw move past those in the upper jaw may be helpful in chewing plants.

The dentition of caseids was heterodont with the largest teeth in the front of the jaws. The posterior teeth had cutting edges that were formed by many small cusps like the teeth of some extant lizards. The earliest edaphosaurid, *Ianthasaurus* of the late Carboniferous, was small with a skull only 7 centimeters long. Its teeth were sharply pointed, and it may have been an insectivore. In contrast, *Edaphosaurus* (a Permian form) was about 2 meters long. It had large bulbous teeth of uniform size along the margins of the jaws, and teeth on the palate that occluded with teeth

Table 19–1 **Effect of the crest on the heating rates of _Dimetrodon_.**

Species	Estimated Body Mass (kg)	Estimated Time Required to Raise Body Temperature (hr)			
		5°C		10°C	
		With Crest	Without Crest	With Crest	Without Crest
D. milleri	50	2	4	4	6.5
D. limbatus	140	4	8	8	—
D. grandis	250	3	10.5	11	—

Source: S. R. Haack, 1986, _Paleobiology_ 12:450–458.

on the inner surface of the lower jaws. The neural spines of _Ianthasaurus_ and _Edaphosaurus_ are elongate, giving them a dorsal crest like that of _Dimetrodon_, except that edaphosaurid spines had horizontal projections, rather like stubby yards on the masts of a square-rigged sailing ship.

Therapsida and Theriodontia

From the mid-Permian to the end of the Triassic there was a flourishing fauna of mammal-like animals that are grouped under the general name of nonmammalian therapsids (Kemp 1982, 1988a; Hotton 1991; Hopson and Barghusen 1986; Hopson 1991, 1994). Like the earlier synapsids, therapsids radiated into herbivorous and carnivorous forms. Some of the herbivorous nonmammalian therapsids were large, heavy-bodied, slow-moving animals that probably congregated in herds as antelopes do. Other herbivorous forms were small and superficially very like rodents. The carnivorous nonmammalian therapsids were more agile animals than the large herbivores. The pattern of increasing effectiveness as terrestrial predators, which we have traced from the earliest anthracosaurs through eupelycosaurs, continued in the carnivorous nonmammalian therapsids.

The success of nonmammalian therapsids was noteworthy. Late Permian fossil deposits all over the world indicate that the ecosystems of the time were dominated by these animals. The remarkable similarity of faunas in different parts of the world at the end of the Paleozoic testifies not only to the continuity of landmasses and climates, but also to the very high degree to which nonmammalian therapsids were successful in exploiting the resources of their environments.

The transition from the Permian to the Triassic was a period of great extinction, actually greater in magnitude than the better-known extinction of many diapsids at the end of the Mesozoic. The known genera of tetrapods represented by fossils dropped from 200 in the late Permian to 50 in the early Triassic. The pattern of extinction was not regular; some old groups persisted while seemingly more derived forms disappeared. Among the herbivores, only the tusked dicynodonts survived into the Triassic in abundance. The most derived groups of carnivores, the therocephalians and nonmammalian cynodonts, survived the Permo–Triassic transition and diversified in the early Triassic.

Early Nonmammalian Therapsids: Biarmosuchians, _Eotitanosuchus_, Dinocephalians, and Anomodonts

Therapsids such as _Biarmosuchus_ (late Permian) differed from sphenacodontid eupelycosaurs in several ways that appear to be related to increasing specialization as predators (Hopson 1994). The temporal fenestra of these nonmammalian therapsids is larger than it was in sphenacodontids, and a portion of the the jaw-closing muscles appears to have passed out through the fenestra to attach to the lateral surface of the skull roof. The upper canines are longer than those of sphenacodontids, and the entire skull is more rigid than it was in earlier synapsids. The postcranial skeleton of nonmammalian therapsids also shows changes from the sphenacodontid condition—the pectoral and pelvic girdles are less massive, and the limbs are more slender. The shoulder joint appears to have allowed more freedom of movement of the forelimb.

Biarmosuchians are the most plesiomorphic nonmammalian therapsids known, and are relatively small (skull lengths less than 10 centimeters), but *Eotitanosuchus*, which is known only from a skull (even the lower jaw is missing), had a head the size of a very large dog. In the late Permian a group of large nonmammalian therapsids, the dinocephalians, dominated land faunas worldwide. The earliest dinocephalians were up to 3 meters long, and showed resemblances to carnivorous eupelycosaurs (Figure 19–6). Later dinocephalians had heavy bodies, short limbs, and long tails. Dentition was concentrated at the front of the mouth: The canines and incisors were enlarged and the postcanine teeth reduced in size and number. The long jaws would have produced a large kinetic force as they snapped closed, embedding the anterior teeth deep in the flesh of their prey. Then the interdigitating upper and lower incisors would have permitted the animal to tear off bite-size pieces.

The herbivorous dinocephalians were huge, clumsy-looking animals (Figure 19–7). Some, like *Moschops*, were nearly 3 meters long and more than 1.5 meters at the shoulder. These herbivores had teeth that decreased gradually in size posteriorly. Teeth in the upper and lower jaws interdigitated and vegetation was torn into pieces between adjacent teeth. The dinocephalians show a thickening of the roof of the skull that may indicate head-butting behavior associated with intraspecific combat. *Moschops* had a skull roof 10 centimeters thick. This bony shield would have transmitted the force of impact around the sides of the skull to the occipital condyle, which was directly in line with the shield (Barghusen 1975). Thus the impact would be transmitted along the neck and dissipated by the massive shoulders and trunk.

Dinocephalians flourished in the early part of the late Permian, but had disappeared before the end of that period. Other herbivorous nonmammalian

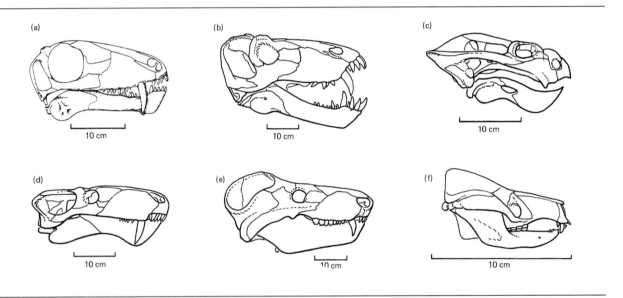

Figure 19–6 Skulls of nonmammalian therapsids. Early carnivores like (a) the biarmosuchian *Biarmosuchus* and (b) the dinocephalian *Titanophoneus* had skulls that closely resemble those of sphenacodontid eupelycosaurs. The herbivorous dicynodonts (c, *Dicynodon*) had lost most or all incisors and cheek teeth, retaining only the tusk-like canines. Gorgonopsids (d, *Scymnognathus*) and cynognathid cynodonts (e, *Cynognathus*) were carnivores specialized for attacking large prey. The tritylodontid cynodonts (f, *Oligokyphus*) were herbivores with a very rodent-like dentition, including a diastema between the incisors and cheek teeth. (Modified from A. S. Romer, 1966, *Vertebrate Paleontology*, 3d edition, University of Chicago Press, Chicago, IL; and R. L. Carroll, 1988, *Vertebrate Paleontology and Evolution*, Freeman, New York, NY.)

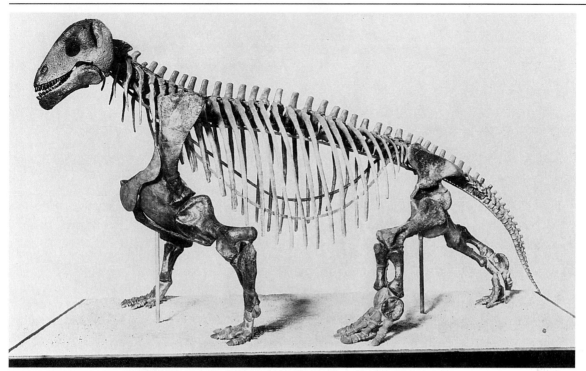

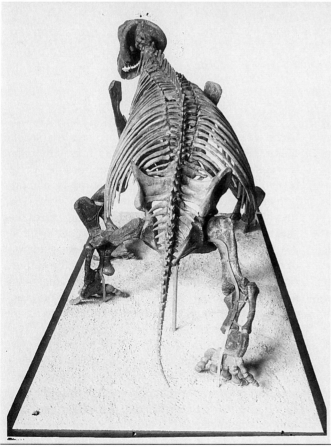

Figure 19–7 The dinocephalian *Moschops capensis* (AMNH 5552). Note the thick skull roof, and massive neck, shoulder girdle, and forelimbs. (Courtesy of the American Museum of Natural History.)

Figure 19–8 A reconstruction of the early Triassic dicynodont *Lystrosaurus*. (By Gregory Paul, from G. King, 1990, *The Dicynodonts, A Study in Paleobiology,* Chapman & Hall, London, UK.)

therapsids that appeared at the same time persisted until the late Triassic. These were the anomodonts, a collection of herbivores that were in many respects the most successful of all therapsids. In the late Permian they dominated terrestrial faunas in diversity and in the number of specimens found as fossils. The most diverse and lasting of the anomodonts were the animals known as dicynodonts (King 1990).

A distinctive feature of dicynodonts was the extreme specialization of the skull for an herbivorous diet. In most forms all of the teeth were lost, except for the upper canines, which were retained as a pair of tusks (Figure 19–8). The jaws are presumed to have been covered with a horny beak like that of turtles. A beak has some advantages over teeth as a structure for grinding plant material. Unlike a row of teeth, a horny beak can provide a continuous cutting surface, and it can be replaced continuously as it is worn away. The structure of the jaw articulation of dicynodonts permitted extensive fore-and-aft movement of the lower jaw, shredding the food between the two cutting plates.

Theriodontia and Eutheriodontia

In the world of the late Permian and early Triassic, the major predators were carnivorous nonmammalian therapsids or theriodonts—gorgonopsians, therocephalians, and nonmammalian cynodonts. Elaboration of the coronoid process of the dentary bone at the rear of the lower jaw, a feature that characterizes theriodonts, provided additional space for insertion of jaw muscles. Accompanying changes in the skull roof opened the temporal fossa completely in the eutheriodonts (therocephalians and nonmammalian cynodonts), providing still more space for jaw muscles.

Gorgonopsians were lightly built animals the size of coyotes or wolves (*Lycaenops*, Figure 19–9). Their canine teeth were long and blade-like, with serrated margins, and the incisors were well developed (Figure 19–6). The postcanine teeth were small and gorgonopsians probably used the anterior teeth to kill and tear chunks of flesh from their prey. The long tail characteristic of eupelycosaurs was greatly shortened in the gorgonopsians, and the lateral

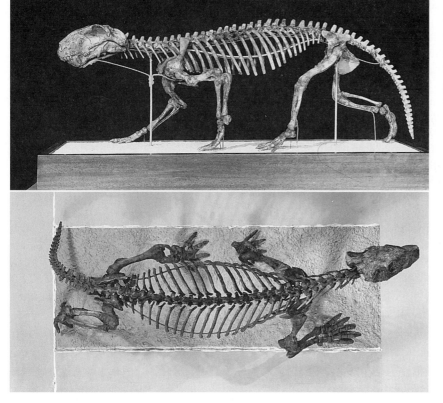

Figure 19–9 The gorgonopsian *Lycaenops ornatus* (AMNH 2240). The hind limbs appear to have been carried nearly vertically, but the forelimbs extended outward from the body. (Courtesy of the American Museum of Natural History.)

undulation of the trunk that was characteristic of the locomotion of early synapsids may have been abandoned by gorgonopsians. The forelimbs were probably still held in a sprawled posture, but the hind limbs could apparently be brought beneath the body. This arrangement would allow the hind limbs to provide thrust for rapid locomotion.

The Cynodonts

Cynodonts, the last of the major groups of therapsids, appeared in the late Permian and have persisted as mammals to the present. By the end of the Triassic nonmammalian cynodonts included some very mammal-like groups and had given rise to the mammals themselves. A marked reduction in body size characterized the cynodont lineages. Some early cynodonts were as big as large dogs, but by the middle Triassic, other nonmammalian cynodonts had skulls only 70 millimeters long, and in the latest Triassic the total body length of the earliest mammals was less than 100 millimeters.

Both carnivorous and herbivorous forms are represented among the nonmammalian cynodonts. The carnivores include the early Triassic cynognathids, the middle Triassic chiniquodontids, and the late Triassic tritheledontids. These animals had well-developed canine teeth and incisors, like the earlier groups of carnivorous therapsids, but the postcanine teeth of derived cynodonts also were well developed and had small accessory cusps anterior and posterior to the main cusp. Traces of wear on the surface of the postcanine teeth of *Cynognathus* indicate that serrations on the cusps of the teeth produced a tearing and cutting action. This arrangement would have allowed cynognathids to use the cheek teeth to cut off bite-size pieces of meat, and perhaps to slice the chunks into smaller pieces before they were swallowed.

Some features of the soft anatomy of nonmammalian therapsids can be inferred from traces left on the bones. The smooth surfaces of the premaxilla, maxilla, and dentary adjacent to the teeth suggest that nonmammalian cynodonts had muscular lips like those of early mammals. The lips of nonmammalian therapsids could have functioned in conjunction with the cusped cheek teeth and, probably, a muscular tongue to manipulate food as it was chewed.

In contrast to nonmammalian cynodonts, the cheek teeth of carnivorous dinocephalians and gorgonopsids are poorly developed, and these animals apparently used their incisors to nip off bits of flesh. A pair of bony plates in the roof of the mouth formed a partial secondary palate that was probably completed by soft tissue. This structure separated the passageway for air from the mouth cavity.

Surprisingly, some of these changes in the jaws may have reflected selective forces that acted initially on the ability of nonmammalian cynodonts to hear. The jaw muscles of nonmammalian cynodonts showed progressive differentiation of the superficial masseter muscle in association with the development of a pronounced coronoid process of the dentary bone (Figure 19–10). These changes probably improved the biting force of the jaws, and also allowed a reduction in the size of the articular and quadrate bones that lay posterior to the dentary bone in the lower jaw. The articular and quadrate of nonmammalian cynodonts are homologous with the malleus and incus of the middle ear of extant mammals (Box 19–1), and it is likely that they already played a role in sound transmission (Crompton 1985). If this was the case, lighter bones would be more effective than heavy bones at transmitting vibrations picked up by the tympanic membrane, but a reduction in the size of the bones would have reduced the strength and load-bearing capacity of the jaw joint. Thus, the reorganization of the jaw muscles and the development of the coronoid process of the dentary bone seen in nonmammalian cynodonts could have been advantageous because they reduced the forces applied to the jaw joint during biting, allowing the articular and quadrate bones to be small. In this analysis, selective pressures to improve hearing are seen as being the important factors in the reorganization of the mammalian jaw.

However, the improvement in hearing would have some cost associated with it. A weak jaw would have limited the ability of nonmammalian cynodonts to chew their food: The canines and incisors could puncture and tear prey, but the cheek teeth could not process food effectively, because those chewing motions (especially unilateral chewing) would exert stress on the jaw joint. Ultimately the cynodont lineage evolved a new load-bearing jaw

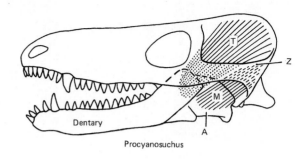

Procyanosuchus

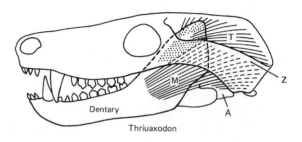

Thriuaxodon

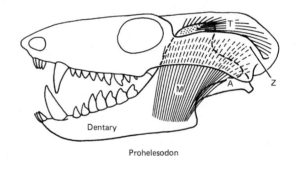

Prohelesodon

Figure 19–10 In nonmammalian cynodonts the dentary bone increased in size relative to the other bones of the lower jaw. As this happened, the masseter muscle, one of the major jaw-closing muscles, progressively changed its insertion from the angular bone to the dentary. A, angular bone; Z, zygomatic arch; M, masseter muscle; T, temporalis muscle. (From T. S. Kemp, 1982, *Mammal-like Reptiles and the Origin of Mammals*. Academic, London, UK.)

joint between the dentary and squamosal bones that was strong enough to resist the forces exerted during strong biting. This is the mammalian type of jaw suspension. A classic example of an evolutionary intermediate is provided by the trithelodontid *Diarthrog-*

Box 19–1 The Evolution of the Mammalian Middle Ear

More than a century ago embryological studies demonstrated that the malleus and incus of the middle ear of mammals were homologous with the articular and quadrate bones that formed the ancestral jaw joint of amniotes. More recently the transition has been traced in fossils from its beginning in basal synapsids through nonmammalian therapsids to early mammals (Allin 1975).

The jaw joint of basal synapsids was formed by the articular and quadrate bones, which were sturdy structures that resisted the compressive forces applied to them during biting. A key to the change in the jaw joint, and the reduction in the size of the articular and quadrate, was a change in the mechanics of the jaw in the synapsid lineage that reduced the stress on the articulation (Bramble 1978).

The transition from the nonmammalian to the mammalian condition can be visualized by comparing the posterior half of the skull of *Thrinaxodon*, a nonmammalian cynodont (Figure 19–11a), with that of *Didelphis*, the Virginia opossum (Figure 19–11b). The jaw joint between the articular and quadrate of *Thrinaxodon* was also part of the auditory apparatus. The reflected lamina of the angular bone is a distinctive feature of the jaw of synapsids. In nonmammalian therapsids the reflected lamina was probably the principal support of the tympanic membrane, and vibrations were transmitted to the stapes via the articular and quadrate. The mammalian jaw joint is formed by the dentary and squamosal bones. The tympanum and middle ear ossicles lie behind the jaw joint and are much reduced in size, but they retain the same relation to each other as they had in nonmammalian cynodonts. The lower jaw of a fetal mammal viewed from the medial side shows that the angular (tympanic), articular (malleus), and quadrate (incus) develop in the same positions they have in the nonmammalian cynodont skull (Figure 19–11c). The homolog of the angular, the tympanic bone, supports the tympanum of mammals. The retroarticular process of the articular bone is the manubrium of the malleus, and the ancestral jaw joint persists as the articulation between the malleus and incus.

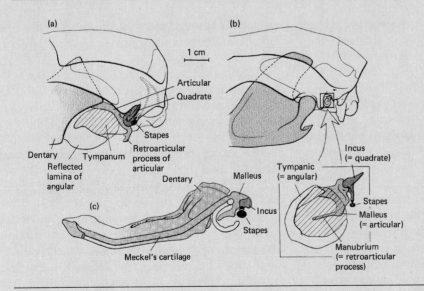

Figure 19–11 Origin of the middle ear bones of mammals: (a) *Thrinaxodon*, a nonmammalian cynodont; (b) *Didelphis*, the Virginia opossum; (c) embryonic mammal. (From A. W. Crompton and F. A. Jenkins, Jr., 1979, pages 59–73 in *Mesozoic Mammals: The First Two-Thirds of Mammalian History*, edited by J. A. Lillegraven, Z. Kielan-Jaworowska, and W. A. Clemens, University of California Press, Berkeley, CA.)

nathus. This animal gets its name from its double jaw joint (*di* = two, *arthro* = joint, *gnath* = jaw). It has both the ancestral articular-quadrate joint and, next to it, the mammalian dentary-squamosal articulation.

The herbivorous nonmammalian cynodonts grouped in Figure 19–2 as Gomphodontia include the early Triassic diademodontids, the middle Triassic traversodontids, and the late Triassic tritylodontids. As the names suggest, features of their teeth are particularly significant in these groups. The upper and lower cheek teeth developed matching surfaces associated with crushing, cutting, and grinding plant material.

The arrangement of teeth in the jaws of gomphodonts was also quite mammalian. For example, the tritylodontid *Oligokyphus* had a large diastema separating the anterior teeth from the cheek teeth (Figure 19–6). The postcanine teeth were multirooted like the cheek teeth of mammals, and had crescentic cusps on their surfaces forming opposing pairs of cutting edges suitable for shredding plant material. *Oligokyphus* was about 50 centimeters in total length, and nearly half of that was tail.

The postcranial skeleton of the nonmammalian cynodonts was also verging on the mammalian condition. The structure of the pelvis and femur indicates that the hind limb was held almost directly beneath the body and moved in a parasagittal plane (parallel to the long axis of the body). The ability to shift between a sprawling and an upright posture that was present in gorgonopsians had been lost in cynodonts. The forelimbs of cynognathids and diademodontids appear to have retained a widespread position, but changes in the shape of the scapula suggest that the speed of the stride was increased. In the tritylodontid *Oligokyphus* the shoulder girdle was very like that of mammals, and the forelimb probably moved in a parasagittal plane. Taken together, these changes indicate increased power and agility in the nonmammalian cynodonts.

Running and Breathing

Skeletal changes of this sort are consistent with the hypothesis that synapsids at the nonmammalian cynodont level were developing a respiratory system more like that of mammals than reptiles. That would be a significant development because locomotion, respiration, and metabolism are closely linked. Early tetrapods walked with lateral undulations of the trunk that were produced by alternating unilateral contractions of the thoracic muscles. That is, contraction of the muscles on the left side of the trunk bent the trunk to the left, and was followed by contraction of muscles on the right side of the trunk. This ancestral method of locomotion is retained in salamanders and lizards.

Lizards inhale and exhale by contracting the trunk muscles to change the volume of the thoracic cavity. As the rib cage expands, pressure in the lungs is reduced below atmospheric pressure, and air flows in. When the rib cage contracts, pressure in the lungs is increased above atmospheric pressure and air is forced out. The muscles that change the volume of a lizard's thoracic cavity during breathing are the same muscles that produce lateral bending during locomotion, and the two activities are incompatible—a lizard cannot breathe and walk simultaneously. When a green iguana is walking, the lateral bending of the trunk increases pressure in the lobe of the lung on the concave side, and reduces pressure on the convex side (Carrier 1987). As a result, air is forced back and forth between the lobes of the lung, but the total lung volume does not change substantially and little air is inhaled or exhaled (Figure 19–12). Respiratory gas exchange nearly ceases when a lizard walks or runs, and resumes when it stops. Behavioral and physiological characteristics of lizards appear to be related to this conflict between locomotion and respiration: Lizards typically move in a stop-and-go fashion. That is, they move a short distance, then stop and breathe, and move again. When lizards are forced to sustain locomotion they rely on anaerobic metabolism, which does not depend on delivery of oxygen to the tissues (Chapter 4).

Mammals, in contrast to lizards, flex the vertebral column vertically during locomotion. The expansion and contraction of the thoracic cavity produced by this movement actually enhance filling and emptying the lungs, as shown in outline by Figure 19–13. Galloping dogs and horses take one breath per stride, and they run at the same stride frequency at all speeds. (They increase the length of each stride to go faster.) As the vertebral column bends, the lungs are compressed, pressure in the lungs rises, and air is exhaled. As the vertebral column straightens, the thoracic cavity expands, pressure in the lungs drops below atmospheric pressure and air is drawn into the lungs. (Studies of breathing by trotting dogs have revealed a triphasic process that involves an interaction between the viscera, espe-

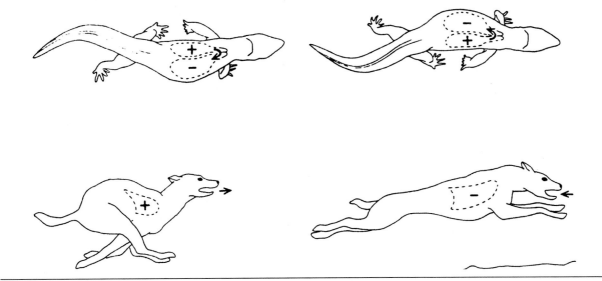

Figure 19–12 The effect of axial bending on lung volume of a running lizard and a galloping dog. The bending axis of the thorax of the lizard is between the right and left lungs. As the lizard bends laterally, the lung on the concave side is compressed and air pressure in that lung increases (shown by +) while air pressure in the lung on the convex side is reduced (shown by –). Air may be pumped between the lungs (arrow), but little or no air will move in or out of the animal. In contrast, the bending axis of the thorax of a galloping mammal is dorsal to the lungs. As the vertebral column bends, the volume of the thoracic cavity decreases, and pressure in the lungs rises (shown by +), pushing air out of the lungs (arrow). When the vertebral column straightens, the volume of the thoracic cavity increases, pressure in the lungs falls (shown by –), and air is pulled into the lungs (arrow). (From D. R. Carrier, 1987, *Paleobiology* 13:326–341.)

cially the liver, and the diaphragm [Bramble and Jenkins 1993]. Inertial movements of the viscera are modulated by contractions of the diaphragm and force air in and out of the lower [diaphragmatic] lobes of the lungs, whereas pressure exerted by the scapula on the chest wall as the foot hits the ground ventilates the upper [apical] lobes.)

The diaphragm, a sheet of muscle at the posterior end of the thoracic cavity of mammals, is essential to the high rates of metabolism that mammals achieve during activity. In its resting state the diaphragm is convex anteriorly (i.e., toward the thoracic cavity), and it flattens when it contracts. Thus, contraction of the diaphragm increases the volume of the thoracic cavity and lowers pressure in the lungs so that air flows inward (inspiration). Relaxation of the diaphragm allows it to resume its convex shape, decreasing the volume of the thoracic cavity, raising pressure in the lungs and forcing air out (expiration). Cutting the nerves that stimulate con-

traction of the diaphragm of rats had no effect on their resting rates of oxygen consumption, but almost entirely abolished the increase in oxygen consumption that normally accompanies activity (Ruben et al. 1987). Aerobic scope (the difference between the rate of oxygen consumption at rest and during activity) dropped by 80 percent—from 1.66 milliliters of oxygen per gram per hour to less than 0.5 milliliters of oxygen per gram per hour (Figure 19–13). This information helps us to interpret the significance of changes in the skeletons of nonmammalian cynodonts in the Triassic.

We've mentioned progressive movement of the limbs to a more vertical orientation during the evolution of nonmammalian therapsids. Furthermore, nonmammalian cynodonts such as *Thrinaxodon* had an abrupt change in the size and shape of ribs in the lumbar region of the trunk (Figure 19–14), suggesting that these animals might have had a muscular diaphragm like that of mammals.

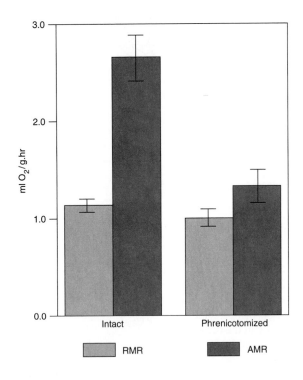

Figure 19–13 Mean resting metabolic rate (RMR) and active metabolic rate (AMR) of intact rats (left) and phrenectomized rats in which the nerves innervating the diaphragm have been cut (right). The bars show standard errors of the means. The intact rats have far greater capacity to increase the rate of oxygen consumption during activity than do the phrenectomized rats. (From J. A. Ruben et al., 1988, *Paleobiology* 13:54–59.)

With elimination of lateral bending of the vertebral column during locomotion and a diaphragm that could modulate inertial movements of the viscera, nonmammalian cynodonts may have been capable of relatively high rates of ventilation during activity. Details of the structure of their nasal passages appear to support that speculation. High rates of ventilation present a problem for animals that breathe air: The surfaces of the lungs where gas exchange takes place are thin because gas must be exchanged very rapidly—a red cell is in the gas-exchanging portion of the lung circulation for less than a second. These thin tissues are susceptible to drying, and to minimize water loss the inhaled air must be warmed to lung temperature and saturated with water vapor before it reaches the gas-exchange surfaces. This process occurs as the air passes over warm, moist surfaces in the nasal passages.

The higher the rate of ventilation, the more elaborate are the structures a vertebrate needs to warm and humidify the air it inhales. Lizards have low rates of ventilation, and the simple walls of their nasal passages are adequate to treat the air before it reaches their lungs. Birds and mammals have higher rates of ventilation than lizards, and require structures that increase the surface area of the nasal passages.

The turbinals of mammals are thin sheets of bone located in the nasal passages. Three groups of turbinals can be distinguished on the basis of the bones to which they attach (Figure 19–16). The nasoturbinals and ethmoturbinals are in the dorsal and posterior part of the nasal cavity, whereas the maxilloturbinals are anterior and ventral. The naso- and ethmoturbinals are covered with sensory (olfactory) epithelia, and are protected from respiratory currents by a transverse lamina (TL in Figure 19–15). These structures are associated with olfaction—when you sniff hard, air is drawn over them and the sensory cells respond to odoriferous molecules. The maxilloturbinals are covered with nonsensory respiratory mucosa, and are responsible for warming and humidifying inhaled air. They lie in the nasal passages where air passes over them with each breath. The elaborate convolutions of the maxilloturbinals produce the large surface area that is needed by an animal with a high rate of ventilation.

Ridges in the nasal passages of cynodonts suggest that turbinals may have been present in these animals (Hillenius 1992). So, we have evidence suggesting that nonmammalian cynodonts had mammal-like limb movements and a mammal-like diaphragm. These features are compatible with breathing and running at the same time. If nonmammalian cynodonts also had maxilloturbinals, it is plausible that they had high rates of ventilation and high rates of metabolism. The presence of so many mammal-like morphological characters in nonmammalian cynodont therapsids may indicate that their exercise physiology and thermoregulation were also basically mammalian.

The First Mammals

The origin of mammals is often treated as if it were a discrete event that happened suddenly sometime in the late Triassic or early Jurassic, but this is an

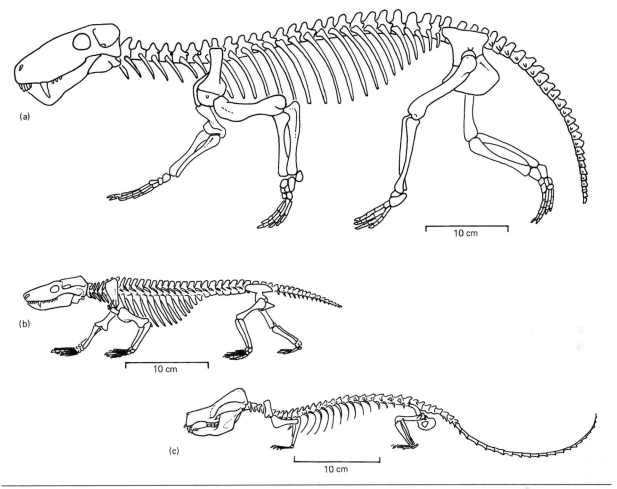

Figure 19–14 Postcranial skeletons of nonmammalian therapsids. *Lycaenops* (a), a gorgonopsid, had ribs that extended the full length of the trunk like those of extant lizards. The nonmammalian cynodonts *Thrinaxodon* (b) and *Oligokyphus* (c) showed a reduction of the size in the ribs in the lumbar region like that seen in mammals, suggesting the presence of a muscular diaphragm. (Modified from A. S. Romer, 1966, *Vertebrate Paleontology*, 3d edition, University of Chicago Press, Chicago, IL.)

oversimplified view. The animals we know as mammals are the products of an evolutionary lineage that stretches back to the division between diapsids and synapsids in the Carboniferous. The anatomical structures that characterize mammals evolved in mosaic fashion through the Permian and Triassic, and their original functions were not necessarily the same as those they have today. The biology of animals in the synapsid lineage was also changing during the late Paleozoic and early Mesozoic. We have traced the evidence for increasing locomotor agility and increased specialization of the jaws and teeth in nonmammalian cynodonts, and the most derived nonmammalian cynodonts are little different from the earliest mammals.

Tritheledontids: The Sister Group of Mammals?

The Tritheledontidae (also known as the Diarthrognathidae) is a poorly known group of small nonmammalian cynodonts from the late Triassic and early Jurassic of South Africa. Tritheledontids may be the lineage most closely related to mammals (Figure

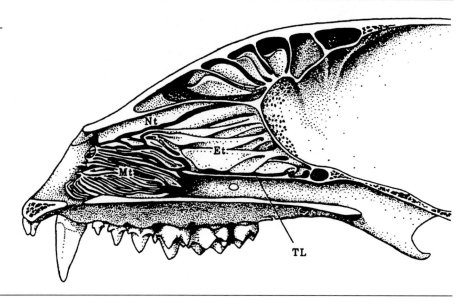

Figure 19–15 The snout of a raccoon showing the nasal passages and turbinals. Eth, ethmoturbinal; Mt, maxilloturbinal; Nt, nasoturbinal; TL, transverse lamina. (From W. J. Hillenius, 1992, *Paleobiology* 18:17–29.)

19–16). *Diarthrognathus*, the best known of the tritheledontids, possessed the mammalian dentary-squamosal joint and also the nonmammalian articular-quadrate joint, and its skull and teeth showed derived mammalian characters: The postorbital and prefrontal bones were absent and the zygomatic arch was slender. The postcranial skeleton of *Diarthrognathus* has not been described, but it is essentially identical to that of early mammals (J. A.

Hopson, personal communication). Furthermore, tritheledontids are the only nonmammalian cynodonts known in which the teeth were covered with prismatic enamel like that of mammals.

Thus, the tritheledontids may be the sister group of mammals, but that relationship is by no means certain and another group might be more closely related to mammals than are the tritheledontids. The tritylodontids have derived mammalian

Figure 19–16 Phylogenetic relationships of the Mammalia. This diagram shows the probable relationships among the major groups of mammals. Extinct lineages are marked by a dagger (†). The numbers indicate derived characters. **1.** Mammalia: A well-developed jaw joint formed by the squamosal and dentary bones, four lower incisors, plus specializations of the braincase and ear. **2.** "Holotheres": Molars with a triangular pattern of cusps. **3.** Tribosphenia: New internal cusp (protocone) on upper molars that occludes into a basin on lower molars. **4.** Theria: Characters of the braincase, a vertical tympanic membrane. Note that the placement of the Monotremata and Multituberculata in this cladogram differs slightly from that of Figure 21–1, reflecting differences in the sources on which the two figures are based. These differences emphasize that the relationships shown in cladograms are hypotheses, and are subject to test and change. (Based on T. S. Kemp, 1983, *Zoological Journal of the Linnean Society* 77:353–384; M. J. Benton (editor), 1988, *The Phylogeny and Classification of the Tetrapods,* Special Volume No. 35B, The Systematics Association, Oxford University Press, Oxford, UK; R. L. Carroll, 1988, *Vertebrate Paleontology and Evolution,* Freeman, New York, NY; M. J. Benton, 1990, *Vertebrate Paleontology,* HarperCollins Academic, London, UK; M. J. Benton (editor), 1993, *The Fossil Record 2,* Chapman & Hall, London, UK; J. A. Hopson, 1994, pages 190–219 in *Major Features of Vertebrate Evolution,* edited by D. R. Prothero and R. M. Schoch, Short Courses in Paleontology No. 7, Paleontological Society and University of Tennessee Press, Knoxville, TN; and J. A. Hopson, personal communication.)

characters in the skull and postcranial skeleton, but their dentition is that of highly specialized herbivores, whereas early mammals are carnivores. We have placed the tritheledontids as the sister group of mammals, largely on the basis of derived mammalian features of the skull and postcranial skeleton. Nevertheless, a case can be made for giving that position to the tritylodontids (Kemp 1983, 1988b; Rowe 1988). The main problem with identifying the sister group of mammals is that many aspects of morphology show extreme convergence. Specialists differ in their opinions about which characters are homologous and which are convergent, and these differences of interpretation lead to different phylogenies.

Early Mammals

Part of the difficulty in identifying the sister group of mammals is a reflection of the fragility and small size of the skeletons of these small animals. Most fossil material consists only of scraps, and the postcranial skeletons of several groups are unknown. The fossils are often so small that they must be examined under a microscope, and they are found by sifting and washing tons of dirt and rock. In fact, searching for fossils of early mammals is very much like searching for gold.

The oldest mammal, *Adelobasileus cromptoni* (about 225 million years old), is known only from an isolated braincase from the late Triassic of Texas. *Sinoconodon*, from the early Jurassic of China, is the

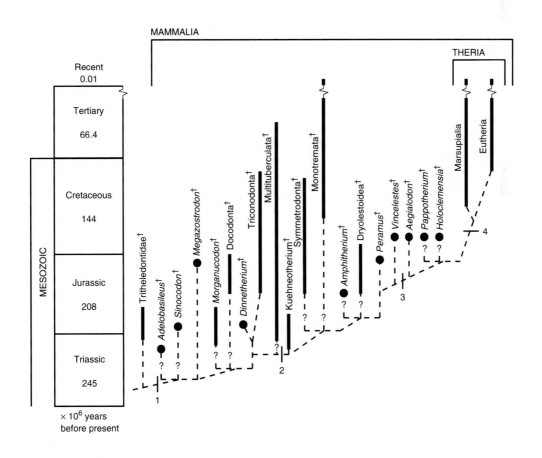

most plesiomorphic mammal about which we have much information. It presents a mosaic of derived and ancestral characters—a robust dentary-squamosal jaw joint (derived) plus a reduced quadrate-articular joint (ancestral), double-rooted postcanine teeth (derived) plus multiple replacements of the incisors and canines (ancestral), and apparent replacement of molars and imprecise occlusion between upper and lower molars (ancestral).

Mammals beyond *Sinoconodon* have precise molar occlusion that is produced by an interlocking arrangement of the upper and lower molars that brings them together in a consistent manner. Precise occlusion makes it possible for the cusps on the teeth to cut up food very thoroughly, creating a large surface area for digestive enzymes to act on, and thereby promoting rapid digestion. The increasing intricacy of tooth occlusion through the nonmammalian cynodont lineage suggests that the efficiency of chewing was increasing in the synapsid lineage, possibly in parallel with the increasing energy demands of higher metabolic rates. The loss of multiple tooth replacement in the mammalian lineage may be related to the demands of food processing. Reptiles, which do not have precise tooth occlusion, replace their teeth many times during their lives, but this system would not work for a mammal because only a fully erupted tooth occludes properly with its counterpart in the upper jaw.

Caroline Pond (1977) pointed out that precise tooth occlusion would be difficult to achieve in the growing jaw of a juvenile animal, and suggested that lactation is the mammalian solution to this problem. Milk does not require teeth to process, and the molar teeth of mammals do not appear until their jaws have nearly completed growth. Furthermore, mammalian molars are not replaced during an individual's life. Of course, if one set of molars must endure for a lifetime, they'd better be tough enough to withstand a lifetime's wear. Mammalian teeth display a variety of derived features, singly and in combination, that increase their durability (Janis and Fortelius 1988).

Megazostrodon had molar teeth with three cusps arranged in a line from anterior to posterior; the middle cusp was the largest. When the jaws closed, the external faces of the cusps of the lower molars slipped up past the inner faces of the cusps of the upper molars and produced a shearing action like the blades of a pair of scissors. The postcranial skeleton of *Megazostrodon* had many derived features

(Figure 19–17). The joint between the atlas and axis was apparently capable of as much rotation as that of extant mammals, and the joint between the atlas and the double occipital condyles allowed flexion and extension. The cervical, thoracic, and lumbar vertebrae were well differentiated, and the cervicothoracic flexure indicates that the head and neck were held in an upright posture. The lumbar vertebrae lacked ribs, as they do in extant mammals. The anterior neural spines were angled posteriorly, and the posterior spines were angled anteriorly. Three transitional vertebrae are evident in the posterior thoracic region. This pattern of vertebrae in mammals is associated with flexion and extension of the trunk in a vertical plane in contrast to horizontal flexion. The pelvic girdle of *Megazostrodon* was fully mammalian: It had a narrow, rod-like ilium that was directed anterodorsally, a large obturator foramen, and a reduced pubis. The form of the limbs suggests that *Megazostrodon* was an active forager, with climbing abilities similar to those of many small mammals today (Jenkins and Parrington 1976).

The lineage that included *Morganucodon*, the docodonts, *Dinnetherium*, and the triconodonts also was characterized by molar teeth with three cusps arranged in a line. The main cusp of the lower molars (a) occluded between the main cusp (A) and the anterior accessory cusp (B) of the corresponding upper molar (Figure 19–18).

A different pattern of occlusion characterizes the "holotheres." The cusps are arranged in a trian-

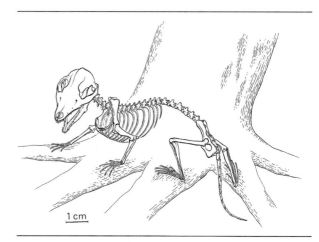

Figure 19–17 Reconstruction of *Megazostrodon,* an early Jurassic mammal. (From F. A. Jenkins, Jr., and F. R. Parrington, 1976, *Philosophical Transactions of the Royal Society (London) Series B* 273:387–431.)

gle instead of in a line. The apex of the triangle on the lower molars points out (toward the lips), whereas the apex on the upper molars points in. The main cusps of the lower (a) and upper (A) molars occlude between adjacent teeth on the opposite jaw. The Mesozoic fossil record of mammals consists mostly of teeth, and when we speak of the evolution of Mesozoic mammals we're mostly talking about the evolution of the crown structure of their teeth. The basic structure of reversed triangles, first seen in *Kuehneotherium* in the early Jurassic, is present with little change in the symmetrodonts (middle Jurassic to late Cretaceous). *Amphitherium*, from the middle Jurassic, had developed an enlarged heel on the lower molar bearing a fourth cusp (d) which overlapped the adjoining tooth and formed a basin that received the main cusp (A) of the upper molar. This structure was elaborated in *Peramus*, which also had an inward extension of the upper molar that was enlarged in early Cretaceous "holotheres" to form a new internal cusp. More derived Mesozoic mammals such as *Aegialodon* and *Pappotherium* had **tribosphenic** molars that combined a shearing action between cusps with a mortar-and-pestle crushing between cusps and basins (Figure 19–19). Mammals with this tribosphenic type of molar are known as the Tribosphenia (Figure 19–16). The enormous diversity of molars seen among extant mammals (marsupials and eutherians) is derived from this condition.

Different interpretations of some details of mammalian evolution are plausible (e.g., Kemp 1988b, Rowe 1988), and two groups of mammals do not fit easily into a phylogeny based on molar structure—the multituberculates because their teeth are so specialized, and the monotremes because their teeth are so reduced. The multituberculates were a diverse and successful lineage of mammals that included nearly 50 genera and survived for 100 million years, from the late Jurassic into the Oligocene. Multituberculates are known mainly from Laurasia, but teeth that have been identified as those of multituberculates have been found in early Cretaceous deposits in Morocco and late Cretaceous sites in Argentina. Some multituberculates were the size of small mice, and others grew as large as a woodchuck. The skulls of multituberculates have a rodent-like appearance that is produced by a pair of long incisors separated by a diastema from the cheek teeth (Figure 19–20). The lower premolars of most multituberculates were specialized as large shearing blades (**plagiaulacoid** premolars); similar premolars are found among some living marsupials. Multituberculates were probably omnivorous, much like extant rodents that include both plant and animal material in their diet.

The phylogeny of the Monotremata (duck-billed platypus and the echidnas) has been difficult to assess because the duck-billed platypus has extremely reduced teeth and echidnas lack teeth altogether. Fossil monotremes with teeth have been found in Cretaceous and Miocene deposits in Australia. The teeth are not true tribosphenic molars, and opinions about their homologies vary (Kielan–Jaworowska et al. 1987, Archer et al. 1992, 1993). We have followed Hopson's (1994) conclusion that the crown patterns of the teeth of these early monotremes are derived from the pattern of tooth cusps seen in very early "holotheres."

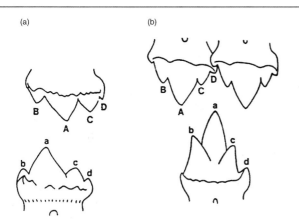

(a)   (b)

Figure 19–18 Two types of dentition seen in early mammals. (a) The upper and lower molars of *Morganucodon*, seen from the lingual side, show the triconodont pattern of tooth occlusion. The main cusp of the lower molar (a) occludes between the main cusp (A) and the anterior accessory cusp (B) of the corresponding upper molar. (b) The upper and lower molars of *Kuehneotherium*, seen from the lingual side, illustrate the "holothere" pattern. The main cusp of the lower molar (a) occludes between two adjacent upper molars. (From J. A. Hopson, 1994, pages 190–219 in *Major Features of Vertebrate Evolution*, edited by D. R. Prothero and R. M. Schoch, Short Courses in Paleontology No. 7, Paleontological Society and University of Tennessee Press, Knoxville, TN.)

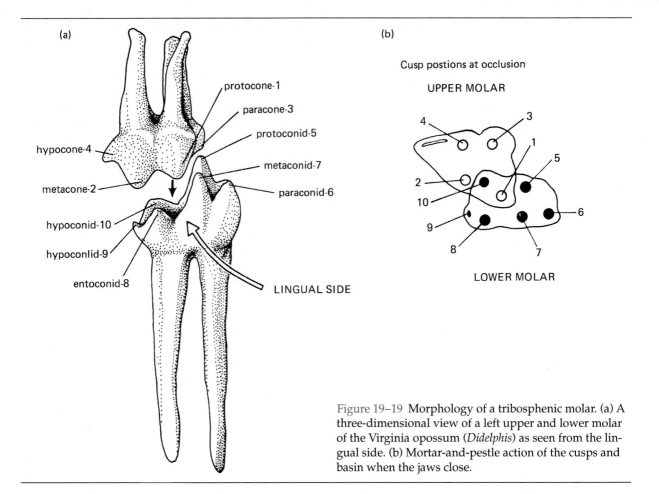

(a)

protocone-1
paracone-3
protoconid-5
hypocone-4
metaconid-7
metacone-2
paraconid-6
hypoconid-10
hypoconlid-9
entoconid-8

LINGUAL SIDE

(b)

Cusp postions at occlusion

UPPER MOLAR

4   3
1
5
2
10
9
6
8   7

LOWER MOLAR

Figure 19–19 Morphology of a tribosphenic molar. (a) A three-dimensional view of a left upper and lower molar of the Virginia opossum (*Didelphis*) as seen from the lingual side. (b) Mortar-and-pestle action of the cusps and basin when the jaws close.

The Adaptive Zone of Early Mammals

The structure of the teeth of Mesozoic mammals suggests that they occupied at least three general feeding niches: The shearing dentition of triconodonts probably indicates that they preyed on other vertebrates, whereas the interdigitating cusps of the teeth of symmetrodonts may have been suited to puncturing and crushing the exoskeletons of insects. Multituberculates were the most diverse of the Mesozoic mammals, and appear to have included forms that may have been primarily insectivorous as well as species that were probably omnivorous or herbivorous.

Mammals were far from being the only small tetrapods with these dietary habits in the Triassic and Jurassic. Nonavian dinosaurs and rhynchosaurs were substantially larger than any of the Mesozoic mammals, but the early mammals shared their habitats with other synapsids—nonmammalian cynodonts such as the chiniquodontids (probably carnivorous) and traversodontids (probably herbivorous)—and with the diapsids that were continuing their radiation—lizards (mostly insectivorous) and sphenodontids (insectivorous and herbivorous). Some ecological separation among these groups may have been related to body size: The chiniquodontids and traversodontids were larger than most of the other forms, but mammals, lizards, and sphenodontids were all about the same size. The ecological distinctions among these groups may have rested in the times of day during which they were active—the diapsids during the day and the mammals at night.

Changes in the auditory apparatus and in the size of the brain of mammals compared to nonmammalian cynodonts and diapsids suggest that the mammals were active at dusk or at night, because the time of day an animal is active deter-

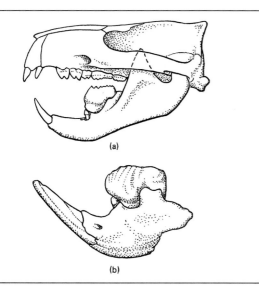

Figure 19–20 Multituberculates: (a) skull of ptilodontid multituberculate; (b) fragment of the lower jaw of the multituberculate *Mesodma formosa* showing the plagaulacoid premolar. (From W. A. Clemens, 1979, in *Mesozoic Mammals: The First Two-Thirds of Mammalian History*, edited by J. A. Lillegraven, Z. Kielan-Jarowoska, and W. A. Clemens, University of California Press, Berkeley, CA.)

mines which sensory systems are important. Vision is a very effective sense during the day, and has the advantage of revealing a large amount of information about an animal's surroundings. For example, it is readily apparent visually that some objects are closer to the observer than others, or that some are approaching and others are moving away. These spatial relationships are difficult to perceive in the dark, when an animal must depend primarily on olfaction or hearing. Those senses require that spatial relationships be inferred by comparing the intensity of stimuli over intervals of time: A sound that becomes louder may mean that the source of the noise is approaching, whereas a sound that grows fainter means that the source is moving away. Similarly, an odor that changes in intensity may mean that its source is approaching or receding.

Making sequential comparisons of the intensity of olfactory and auditory stimuli over time may require more neural integration in the brain than do visual comparisons, and the size of the brain in early mammals has been interpreted as evidence of nocturnal activity. The volume of the braincase of the earliest mammals appears to have been four or five times the volume of the braincases of nonmammalian cynodonts such as *Probainognathus*. The olfactory bulbs and cerebral hemispheres were large (Kielan-Jaworowska 1986). Jerison (1973) suggested that this increase in the size of the brain may be correlated with the ability of early mammals to process more sensory information than could nonmammalian cynodonts, especially information from the nose and the inner ear. That interpretation is speculative, but it is consistent with what is known of the structure of the earliest mammals. If the large size of the brains of early mammals was a consequence of the need to integrate olfactory and auditory information about their environment, as Jerison believes, the nocturnal habits of early mammals may have been a key factor in the subsequent radiation of mammals as large-brained, social vertebrates (see Chapter 23).

Summary

The synapsid lineage is characterized by a fenestra on the side of the skull bordered dorsally by the squamosal and postorbital bones and ventrally by the jugal bone. The first synapsids were the eupelycosaurs and caseasaurs of the late Carboniferous and early Permian. Many eupelycosaurs were large animals, and some may have weighed as much as 250 kilograms. Progressive changes in the jaws and teeth of sphenacodontid eupelycosaurs indicate increasingly effective predatory capacities. The girdles became less massive and the limbs slimmer in later forms, suggesting the development of a more active mode of foraging. From the mid-Permian through the Triassic these trends continued in the nonmammalian therapsids, particularly the nonmammalian cynodonts. By the end of the Triassic, derived nonmammalian cynodonts had a mosaic of mammalian and nonmammalian characters. A substantial reduction in body size accompanied the development of mammalian characters, and early mammals were only 100 millimeters long and probably weighed less than 50 grams.

Most information about these mammals comes from fossilized teeth, which are the most durable parts of the skeleton. Early mammals have traditionally been divided into two major lineages on the basis of tooth structure: Triconodont molars had three cusps arranged in a straight anteroposterior line, whereas holothere molars had three cusps forming a triangle. Currently, it seems most likely that the three groups of living mammals—the monotremes, marsupials, and eutherians—are all derived from the "holothere" lineage.

The tooth structures of Mesozoic mammals suggest that they included insectivorous, carnivorous, omnivorous, and herbivorous species. Contemporaries of the mammals included diapsids (lizards and sphenodontids) that had similar body sizes and dietary habits. Mesozoic mammals were probably nocturnal, using heat produced by metabolism and insulation provided by hair to sustain body temperatures several degrees above ambient temperature. The large brains of early mammals (compared to those of their diapsid contemporaries) may have been associated with the ability to process olfactory and auditory information.

References

Allin, E. F. 1975. Evolution of the mammalian middle ear. *Journal of Morphology* 147:403–437.

Archer, M., F. A. Jenkins, Jr., S. J. Hand, P. Murray, and H. Godthelp. 1992. Description of the skull and non-vestigial dentition of a Miocene platypus (*Obdurodon dicksoni* n. sp.) from Riversleigh, Australia and the problem of monotreme origins. Pages 15–27 in *Platypus and Echidnas,* edited by M. Augee. Royal Zoological Society of New South Wales, Sydney, Australia.

Archer, M., P. Murray, S. Hand, and H. Godthelp. 1993. Reconsideration of monotreme relationships based on the skull and dentition of the Miocene *Obdurodon dicksoni.* Pages 75–94 in *Mammal Phylogeny: Mesozoic Differentiation, Multituberculates, Monotremes, Early Therians, and Marsupials,* edited by F. S. Szalay, M. J. Novacek, and M. C. McKenna. Springer, New York, NY.

Barghusen, H. R. 1975. A review of fighting adaptations in dinocephalians (Reptilia, Therapsida). *Paleobiology* 1:295–311.

Bramble, D. M. 1978. Origin of the mammalian feeding complex: models and mechanisms. *Paleobiology* 4:271–301.

Bramble, D. M., and F. A. Jenkins, Jr. 1993. Mammalian locomotor-respiratory integration: implications for diaphragmatic and pulmonary design. *Science* 262:235–240.

Carrier, D. R. 1987. The evolution of locomotor stamina in tetrapods: circumventing a mechanical constraint. *Paleobiology* 13:326–341.

Crompton, A. W. 1985. Origin of the mammalian temporomandibular joint. Pages 1–18 in *Development of Temporomandibular Joint Disorders,* edited by D. S. Carlson, J. McNamara, and K. A. Ribbens. Monograph 16, Craniofacial Growth Series, University of Michigan Press, Ann Arbor, MI.

Haack, S. C. 1986. A thermal model of the sailback pelycosaur. *Paleobiology* 12:450–458.

Hillenius, W. J. 1992. The evolution of nasal turbinates and mammalian endothermy. *Paleobiology* 18:17–29.

Hopson, J. A. 1991. Systematics of the nonmammalian Synapsida and implications for patterns of evolution in synapsids. Pages 635–693 in *Origins of the Higher Groups of Tetrapods,* edited by H.-P. Schultze and L. Trueb. Cornell University Press, Ithaca, NY.

Hopson, J. A. 1994. Synapsid evolution and the radiation of non-eutherian mammals. Pages 190–219 in *Major Features of Vertebrate Evolution,* edited by D. R. Prothero and R. M. Schoch. Short Courses in Paleontology No. 7. Paleontological Society and University of Tennessee Press, Knoxville, TN.

Hopson, J. A., and H. R. Barghusen. 1986. An analysis of therapsid relationships. Pages 83–106 in *The Ecology and Biology of Mammal-Like Reptiles,* edited by N. Hotton III, P. D. MacLean, J. J. Roth, and E. C. Roth. Smithsonian Institution Press, Washington, DC.

Hotton, N., III. 1991. The nature and diversity of synapsids: prologue to the origin of mammals. Pages 598–634 in *Origins of the Higher Groups of Tetrapods,* edited by H.-P. Schultze and L. Trueb. Cornell University Press, Ithaca, NY.

Janis, C. M., and M. Fortelius. 1988. On the means whereby mammals achieve increased functional durability of their dentitions, with special reference to limiting factors. *Biological Review* 63:197–230.

Jenkins, F. A., Jr., and F. R. Parrington. 1976. The postcranial skeletons of the Triassic mammals *Eozostrodon, Megazostrodon,* and *Erythrotherium. Philosophical Transactions of the Royal Society, Series B* 173:387–431.

Jerison, H. J. 1973. *Evolution of Brain and Intelligence.* Academic, New York, NY.

Kemp, T. S. 1982. *Mammal-like Reptiles and the Origin of Mammals.* Academic, London, UK.

Kemp, T. S. 1983. The relationships of mammals. *Zoological Journal of the Linnaean Society* 77:353–384.

Kemp, T. S. 1988a. Interrelationships of the Synapsida. Pages 1–22 in *The Phylogeny and Classification of the Tetrapods,* volume 2, edited by M. J. Benton. Clarendon, Oxford, UK.

Kemp, T. S. 1988b. A note on the Mesozoic mammals, and the origin of therians. Pages 23–29 in *The Phylogeny and Classification of the Tetrapods*, volume 2, edited by M. J. Benton. Clarendon, Oxford, UK.

Kielan–Jaworowska, Z. 1986. Brain evolution in Mesozoic mammals. Pages 21–34 in *Vertebrates, Phylogeny, and Philosophy*, edited by K. M. Flanagan and J. A. Lillegraven. Contributions to Geology, Special Paper 3. University of Wyoming, Laramie, WY.

Kielan–Jaworowska, Z., A. W. Crompton, and F. A. Jenkins, Jr. 1987. The origin of egg-laying mammals. *Nature* 326:871–873.

King, G. 1990. *The Dicynodonts: A Study in Paleobiology*. Chapman & Hall, London, UK.

Pond, C. M. 1977. The significance of lactation in the evolution of mammals. *Evolution* 31:177–199.

Rowe, T. 1988. Definition, diagnosis, and origin of Mammalia. *Journal of Vertebrate Paleontology* 8:241–264.

Ruben, J. A., A. F. Bennett, and F. L. Hisaw. 1987. Selective factors in the origin of the mammalian diaphragm. *Paleobiology* 13:54–59.

Tracy, C. R., J. S. Turner, and R. B. Huey. 1986. A biophysical analysis of possible thermoregulatory adaptations in sailed pelycosaurs. Pages 195–206 in *The Ecology and Biology of Mammal-like Reptiles*, edited by N. Hotton III, P. D. MacLean, J. J. Roth, and E. C. Roth. Smithsonian Institution Press, Washington, DC.

20 GEOGRAPHY AND ECOLOGY OF THE CENOZOIC

The role of Earth history in shaping the evolution of vertebrates is difficult to overestimate. The positions of continents have affected climates and the ability of vertebrates to migrate from region to region. The geographical continuity of Pangaea allowed synapsids to spread to nearly all parts of the world in the late Paleozoic. However, by the late Mesozoic Pangaea no longer existed as a single entity. Epicontinental seas extended across the centers of North America and Eurasia, and the southern continents were separating from the northern continents and from each other. Continental drift in the late Mesozoic and early Cenozoic moved most of the landmasses of the Earth from the tropics into higher latitudes. This movement and changes in patterns of ocean currents resulted in a cooling trend leading to a series of ice ages that began in the Pleistocene and continue to the present. We are currently living in a relatively ice-free interglacial period. Both the fragmentation of the land masses and the changes in climate during the Cenozoic have been important factors in the evolution of mammals.

Continental Geography During the Cenozoic

The breakup of Pangaea in the Jurassic was initiated by the movement of North America to open the ancestral Atlantic Ocean. Rifts began to open in Gondwana, and India moved northward on its separate oceanic plate (Figure 20–1), eventually to coalesce with Eurasia. The collision of the Indian and Eurasian plates produced the Himalayas, the highest mountain range in today's world.

South America, Antarctica, and Australia separated from Africa as a single continental unit during the middle and late Cretaceous. Land connections persisted across this platform during the Paleocene and perhaps the early Eocene. In the middle to late Eocene, Australia separated from Antarctica and, like India, drifted northward. Intermittent land connections between South America and Antarctica were retained until quite recently via the Scotia Island arc.

In the Northern Hemisphere a land bridge between Alaska and Siberia—the trans-Bering Bridge—broadly connected North America to Asia at high but relatively ice-free latitudes (Figures 20–1). Intermittent connections also persisted between eastern North America and Europe via Greenland and Scandinavia from the Cretaceous to the early Eocene. However, migrations of mammals between North America and Europe via this route appear to have been prominent only during the early Eocene. The connection apparently broke during the Eocene, and continued migrations between North America and the Old World must have occurred via the Bering land bridge linking Siberia and Alaska. Other tectonic movements also influenced Cenozoic climates. For example, the uplift of mountain ranges such as the Rockies, the Andes, and the Himalayas in the middle Tertiary cast rain shadows along their western flanks, resulting in the replacement of woodlands by grasslands.

Continental Movements and Vegetational Changes

An appreciation of the fact that the major land masses were moving away from the equatorial region and toward the poles is the key to understanding the changes in climatic conditions that accompanied the radiation of Cenozoic mammals. The passage of landmasses over the polar region, Antarctica in the early Tertiary and Greenland in the later Tertiary, allowed the formation of polar ice caps and cooling of the temperate regions. These events culminated in the periods of glaciation ("ice ages") of the Pleistocene. The story of Cenozoic mammal evolution is also the story of the temperate regions of the Northern Hemisphere becoming cooler and drier with accompanying changes in vegetation (for example, the replacement of lush tropical-like forests with woodland and grassland).

The warm and humid conditions typical of the Jurassic and early to middle Cretaceous changed toward the end of the Cretaceous when widespread moderate cooling took place. Yet the world of the early Tertiary still reflected the "hothouse" world of the Mesozoic: tropical-like forests were found in high latitudes, and even within the Arctic circle. A mammalian fossil assemblage from the early Eocene of Ellesmere Island (Canada) shows the presence of mammals resembling (although not closely related to) the tree-dwelling primates and flying lemurs of present-day Southeast Asia. The peak of climatic warming in the higher latitudes occurred around the start of the middle Eocene (Figure 20–2). From this high point, the higher-latitude regions started to cool, with a rather precipitous drop in mean annual temperature in the late Eocene, plunging the Earth into the start of the icy world of the later Tertiary.

What caused this dramatic change in earth's temperatures? Present studies of continental movements and the reconstruction of ancient ocean currents indicate that the middle Eocene was when Australia broke away from Antarctica and Greenland broke away from Norway. Cold polar bottom water was formed with the isolation of land masses over the poles, and ocean circulation carried the cold

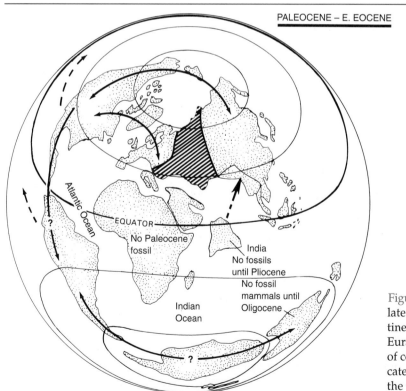

PALEOCENE – E. EOCENE

Figure 20–1 Continental positions in the late Paleocene and early Eocene. An epicontinental sea (crosshatching) extended across Eurasia. Dashed arrows show the direction of continental drift, and solid arrows indicate the major land bridges mentioned in the text.

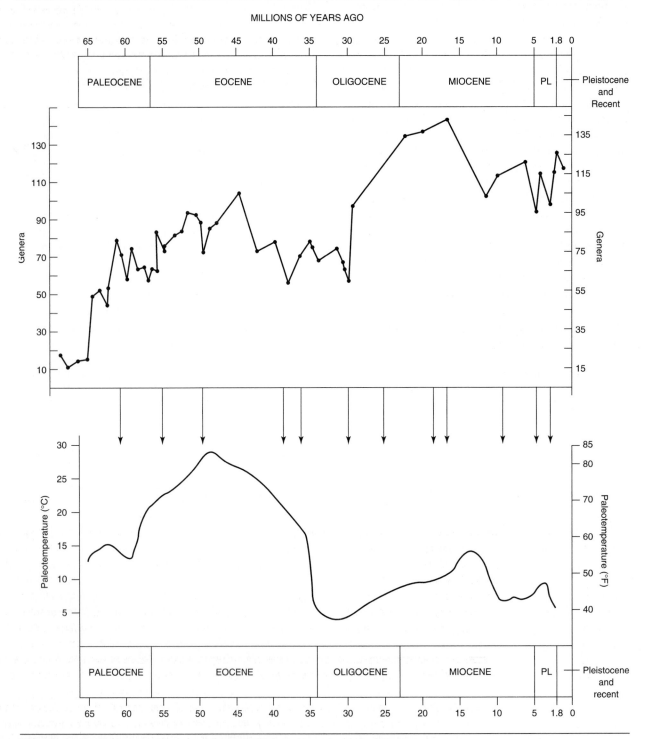

Figure 20–2 Numbers of mammalian genera in North America and paleotempera-
tures in the North Sea during the Cenozoic. Note that the peaks in total genera and
in temperature substantially coincide. The arrows indicate times of major falls in
sea levels. (Modified from R. K. Stucky, 1990, *Current Mammalogy* 2:375–432; and
C. M. Janis, 1993, *Annual Review of Ecology and Systematics* 24:467–500.)

water toward the equator, cooling the temperate latitudes. The Antarctic ice cap probably formed around this time, although the Arctic ice cap formed somewhat later, as we shall see.

Following a rather cool Oligocene, temperatures started to rise in higher latitudes in the early Miocene, reaching a second peak in the middle Miocene (but nowhere near as high as in the middle Eocene). This Miocene was now drier than the Eocene, and the combination of warmth and dryness promoted the spread of grasslands and the evolution of mammals adapted to a savanna-type of habitat. The mammals inhabiting North America at this time showed great similarity to those of present-day East Africa, although they were not necessarily closely related to them.

This Miocene warming may be due to the opening of Drake's Passage between Antarctica and South America, which isolated the cold polar water around Antarctica. However, expansion of the Antarctic ice cap in the late Miocene once again brought cooling to the higher latitudes, a trend that has persisted to the present day with occasional remissions.

Note that changes in the diversity of mammals paralleled changing temperatures at higher latitudes throughout the Cenozoic, with peaks seen in the mid-Eocene and mid-Miocene. This pattern makes sense when we think of today's world, in which a much greater diversity of mammals exists in the tropical forests than in the temperate woodlands. It may seem paradoxical that the diversity appears to be greater at the mid-Miocene peak than at the warmer mid-Eocene peak. Several phenomena probably contribute to this paradox. Some of these factors are artifacts of the fossil record, whereas others probably represent real ecological differences. First, there is the "Pull of the Recent." That is, the nearer we get to the present day, the less likely it is that fossils have been eroded and lost forever, so the *apparent* diversity of fossils in young deposits is greater than in old ones. Second, there were more large mammals in the Miocene than there had been in the Eocene, and large teeth and bones have a greater chance of preservation than small ones, again biasing the fossil record. Finally, the changing Cenozoic climates had produced a fractionation of habitat types by the Miocene. Tropical forests were still found at the equator, but other kinds of habitats (such as woodland and grassland) were found at higher latitudes, and these diverse habitats probably supported a great diversity of mammals.

The Pleistocene Ice Ages

The extensive episodic continental glaciers that characterize the Pleistocene were events that had not been a feature of the world since the Paleozoic. These Ice Ages had an important influence not only on Cenozoic mammal evolution in general, but also on our own evolution and even on our present civilizations. The world can still be considered to be in the grip of an Ice Age, but at the present time we inhabit a warmer, interglacial, period.

For example, at present the volume of ice on earth constitutes about 26 million cubic kilometers. During glacial episodes in the Pleistocene there could be as much as 77 million cubic kilometers of ice, perhaps even more. An enormous volume of water was, and still is, locked up in glaciers and polar ice caps. The melting of the glaciers of the last glacial episode at the end of the Pleistocene, around 12 thousand years ago, caused the sea level to rise by about 140 meters (almost half the height of the Empire State Building) to its present relatively stable condition. If the present-day glaciers were to melt, sea level would rise at least another 50 meters, submerging most of the world's coastal cities. No wonder present-day environmentalists are concerned about the possibilities of global warming caused by human activities!

Today glaciers cover 10 percent of the earth's land surface, mostly in polar regions but also on high mountains. At times in the Pleistocene an ice mass that was probably between 3 and 4 kilometers thick covered as much as 30 percent of the land, and extended southward in North America to 38°N latitude (southern Illinois, Figure 20–3). A similar ice sheet covered northern Europe. However, the area covering much of Alaska, Siberia, and Beringia (Pleistocene land which is now underwater as the Bering Straits) was free of ice cover, and had a unique high-latitude, savanna-like fauna, mixing southern elements, such as lions and antelope, with northern elements, such as reindeer, that persist in Arctic latitudes today.

These continental glaciers advanced and retreated several times during the Pleistocene. (The Southern Hemisphere was less affected because the southern continental landmasses were further from the poles than the northern ones at this time, as they are today.) Table 20–1 shows the major episodes of glaciation, but we now know that many (perhaps a dozen or more) minor ones were interspersed between these major ones.

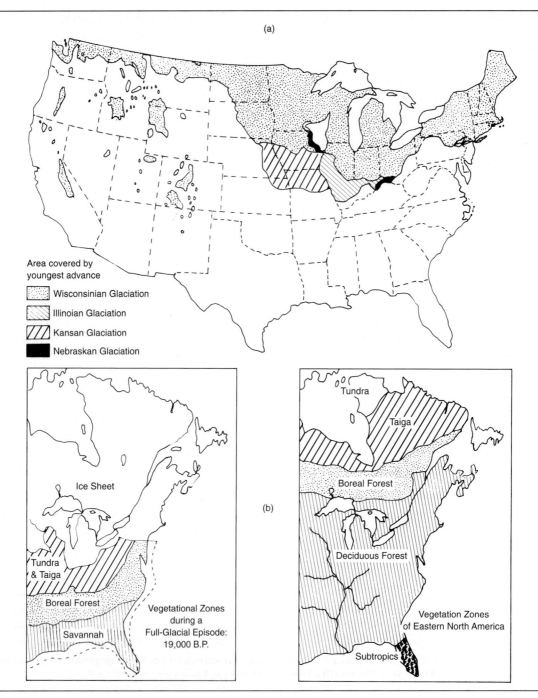

Figure 20–3 Pleistocene glaciation. (a) Extent of glaciation in North America. Several advances of the continental glacier are shown. In the west montane glaciers advanced and retreated. (b) Effect of glacial periods on the position of biomes in North America. Biomes are shifted southward during periods of glacial advance (left), and extend northward duing interglacial periods like the present (right).

Continental glaciation had a greater effect on world climates than just ice covering the high latitudes. Although popular books depict mammoths struggling to free themselves from ice, glaciers advance slowly enough for animals to migrate toward the equator. Drying of the ice-free portions of the Earth due to the volume of water tied up in glaciers was at least as important for mammalian evolution as the glaciers themselves. Many of the equatorial areas that today are covered by lowland rain forests were then much drier, even arid. Even today's relatively mild interglacial period is apparently drier than other interglacial periods in the Pleistocene. For example, remains of hippos are found in what is now the Sahara Desert.

With each glacial episode the forests in the Amazonian and Congo basins contracted into several isolated refugia separated by savannas. During interglacial periods, the forests again spread over the basins, and the savannas contracted and were fractionated. As a result of these alternating expansions and contractions, plant and animal populations were repeatedly isolated and remixed. During isolations one species might be separated into several populations. The effect of glaciation on equatorial aridity, and the resultant contraction of moist habitats, may explain the high species diversity that now exists in these areas (Terborgh 1992).

What caused these episodes of glaciation? A long-standing theory suggests that the amount of solar radiation impinging on the Earth (solar insolation) varies enough to affect the Earth's climate. In the 1930s the Yugoslavian astronomer Milutin Milankovitch proposed that episodes of glaciation are initiated by the fortuitous (or perhaps unfortunate!) combination of several small variations of the passage of the Earth's orbit around the sun, and the relative position of the Earth to the sun. Three cycles interact here, each with its own characteristic periodicity (time elapsed between the extremes of the cycle): (1) the Earth's elliptical orbit around the sun (with a periodicity of 100,000 years), (2) the tilt of the Earth's rotational axis (with a periodicity of 40,000 years), and (3) the precession (wobble) of Earth's rotational axis (with a periodicity of 26,000 years (Imbrie and Berger 1984).

Each of these orbital properties produces different effects. Change in tilt and precession modify the distribution of sunlight with respect to season and latitude, but not total global insolation, whereas changes in the Earth's orbit result in minute changes

Table 20–1 **Glacial and interglacial stages from two northern hemisphere regions.***

Age[†] (millions of years ago)	Midcontinent of North America	European Alps
0.03	Wisconsin—late *Interstadial II*	Würm—late
0.06	Wisconsin—middle *Interstadial I*	Würm—early *Riss-Würm*
0.12	Wisconsin—early *Sagamonian*	Riss *Mindel-Riss*
0.50	Illinoian *Yarmouthian*	Mindel *Günz-Mindel*
1.30	Kansan *Aftonian*	Günz *Donau-Günz*
1.30	Nebraskan	Donau

*Interglacials in italics
[†]Age refers to beginning of each glacial episode in millions of years before the present.

in global insolation. Normally these properties are cycling out of phase, like discordant keys played on a piano, but every so often they line up together like keys making a chord. Milankovitch suggested that the critical factor that leads to a glacial episode is a change in the amount of summer insolation at high latitudes. It appears that glacial episodes get their start not from the world as a whole getting colder year round, but from cool summers that prevent the melting of winter ice. In contrast, the winters during glacial periods may have been warmer than those of the present day.

It is important to realize that these Milankovitch cycles must have been in existence throughout Earth's history. However, in the Cenozoic it was only after the formation of the Arctic ice cap in the Pleistocene (possibly as early as the Pliocene) that there existed sufficient build-up of polar ice to plunge the Northern Hemisphere into an Ice Age.

Cenozoic Mammals and Vicariance Biogeography

As the geography of the world changed during the Cenozoic to assume its present character, so the types of mammals changed to fit the changing patterns of climate and vegetation. Following their origin in the latest Triassic and their persistence through the Mesozoic at small body size and low levels of morphological diversity, mammals diversified in the Cenozoic. Thus the Cenozoic is commonly known as the Age of Mammals, even though the time elapsed since the start of the Cenozoic represents only about a third of the time spanned by the total history of mammals. The radiation of mammals is almost certainly related to the extinction of the dinosaurs, which left the world free of large-sized tetrapods, providing a window of opportunity for other groups. Large terrestrial birds, including herbivores like the ostrich and carnivores (now all extinct), also diversified, and in the early Tertiary there was a modest radiation of terrestrial crocodiles.

The radiations of mammals took place in two major waves, one in the continuing hothouse world of the early Tertiary and one more adapted to the later Tertiary ice house. We call the first wave of mammals "archaic" types, a pejorative term that really refers to our own perspective from the comfort of the Recent. Archaic mammals are those that did not leave members of their lineages surviving until the present day. (Members of most present-day orders did not make their first appearance until the Eocene.) The early archaic mammalian faunas have a very different aspect from the ones that followed. They appear to have been composed mainly of small to medium-sized generalized mammals with a predominance of arboreal over terrestrial types. Larger, specialized predators did not appear until the late Paleocene, and herbivores with teeth suggestive of a completely herbivorous diet did not appear until the late Eocene. It may be the case that forests bounced back so successfully after the extinction of herbivorous dinosaurs that they dominated terrestrial habitats, so that it was not until the climatic changes of the Eocene produced more diverse habitats that mammals began to radiate into the terrestrial niches apparent today (reviewed in Janis 1993).

During the early Cenozoic the continents had not moved as far from their equatorial positions as they are today, but the major landmasses were perhaps more isolated than at present. Australia had broken free from the other southern continents by the mid-Eocene, but North and South America did not come into contact until the Pliocene. India, having broken away from Africa in the Mesozoic, did not make contact with Asia until the Miocene, when its impact resulted in the uplift of the Himalayan

mountain ranges. Africa, floating northward, did not make contact with Eurasia until the late Oligocene or early Miocene, closing off the original east–west expanse of the Tethys Sea to form the now-enclosed Mediterranean basin. Europe was divided from Asia by a north–south epicontinental sea, the Turgai Straits (the remnants of which can be seen in existing bodies of water such as the Black Sea) until the Eocene/Oligocene boundary.

The radiation of mammals was concurrent with this fractionation of the continental masses. Various mixtures of the basal mammalian stocks were isolated on different continents, and hence from each other. This apparent movement and separation of ancestral stocks from each other, by the Earth's physical processes rather than by their own movements is termed **vicariance.** Much of the difference in the distribution of present-day mammals (for example, the isolation of the majority of marsupials in Australia) can be ascribed to what we would term **vicariance biogeography.**

Other patterns of mammal distribution can be explained by **dispersal,** which reflects movements of the animals themselves, usually by the relatively slow (in biological time) spread of populations rather than the long-distance movements of individual animals. Dispersal can result in extinctions as well as radiations. For example, when the Turgai Straits dried up, mammals that had previously inhabited Asia flooded into Europe and some uniquely European mammals (mainly early types) became extinct. This episode of Eocene/Oligocene extinctions was so dramatic that it is known as The Great Dying.

Today the mammals of the Northern Hemisphere (Holarctica) and of Africa, Madagascar, South America, and Australia are strikingly different from each other. These four geographic groupings fall into three major faunal provinces: (1) a Laurasian fauna, including Africa (often considered as a separate Ethiopian fauna south of the Sahara), consisting almost exclusively of eutherians; (2) a South American, or New World tropical fauna, containing a mixture of eutherians and metatherians; and (3) and Australian fauna, containing monotremes, metatherians, and a sprinkling of eutherians.

However, the fauna of Africa used to be even more distinct from the fauna of the rest of Holarctica before Africa collided with Eurasia in the Oligocene/Miocene. Similarly, the South American fauna was more distinct from that of Holarctica before South America was connected with North America in the Pliocene (as we shall discuss later). Even the North American fauna was much more distinct from that of the rest of Holarctica for most of the Tertiary than it is today. Much of the present North American fauna (e.g., deer, bison, and rodents such as voles) crossed over from Eurasia in the Pliocene and Pleistocene via the Bering land bridge. Today only Madagascar and Australia retain distinctly different faunas. It is unfortunate that we have so little fossil evidence from the early Cenozoic of India, as it, too, must have harbored a distinctive early mammalian fauna.

Convergent Evolution of Mammalian Feeding Types

The interaction of mammalian feeding specializations with other aspects of morphology and with behavior and ecology is strikingly illustrated by the convergent and parallel evolution of Cenozoic mammals. Figure 20–4 shows examples of convergence. Specialized herbivores include the extinct litopterns (Figure 20–4a) of South America, the Holarctic artiodactyls (Figure 20–4b) of Eurasian origin, and the horse-like perissodactyls (most of the evolution of which took place in North America) are strikingly similar in many aspects (Figure 20–5). Even the Australian diprotodonts such as the kangaroo (Figure 20–4c) demonstrate striking convergences in jaws, teeth, and feeding mechanisms.

The carnivorous mammals show similar convergences. For example, the true wolf (Figure 20–4d) in the Holarctic, the marsupial Tasmanian wolf (Figure 20–4e) in Australia and Tasmania, and the extinct marsupial borhyaenids (Figure 20–4f) of South America have similar body morphology and tooth form, although only the two wolves have the long legs that distinguish fast-running predators.

Mammals specialized for feeding on ants and termites (myrmecophagy) include the giant anteater in South America (Figure 20–4g), the aardvark in Africa (Figure 20–4h), the pangolin in Asia, Africa, and Europe (Figure 20–4i), and the spiny anteater in Australia (Figure 20–4j). Ants and termites are social insects and employ group defense. They often build impressive earthen nests containing thousands of individuals, some of which are strong-jawed soldiers. The convergent myrmecophagous specializations of these unrelated or distantly related mam-

mals include a reduction in the number and size of teeth, changes in jaw and skull shape and strength, and forelimbs modified for digging (see Figure 21–14). Some mammals with a carnivore ancestry have adopted a myrmecophagous lifestyle and show similar, though less extensive, specializations. These species include a marsupial anteater, the Australian numbat, and an African hyaena, the aardwolf.

Sometimes similar forms evolved under what appear to be much less isolated conditions. Thus hares and rabbits (Figure 20–4k) evolved in the Holarctic where other small herbivores, the rodents, also occurred. Other forms evolved in one area but migrated and survived only in another. Examples are horses and rhinoceroses (Figure 20–4l,m), which originated in North America but now are found as native animals only in Asia and Africa. Isolated landmasses also saw the evolution of unique forms that seem to have had few close ecological counterparts. These include the sirenians (Figure 20–4n), which originated in Africa and spread to North and South America and throughout the Indian and Western Pacific Oceans, and the extinct desmostylians (Figure 20–4o), which were large freshwater or shallow-water marine forms that are thought to have evolved on the coast of Africa and dispersed to the Pacific coasts of North America and Asia.

Early Tertiary Mammals in Laurasia

The first Tertiary mammals, from the early Paleocene in North America, were of three major types. Ancestral marsupials appear to have been omnivorous and arboreal, like modern opossums. A second group were small placentals, mostly shrew-like terrestrial "protoeutherians" or basal ungulates ("condylarths"), from which most extant mammals have evolved. A third group, the Multituberculata, were mostly small, arboreal rodent-like animals.

The diversity of marsupials declined as that of eutherians rose in North America in the late Cretaceous and the Paleocene. Why were marsupials so successful in North America in the middle Cretaceous and then much less successful in the Paleocene? Their decline in North America is especially puzzling in the light of their spread across the entire Northern Hemisphere in the Eocene. The earliest fossils of marsupials currently known from Europe are from the late Paleocene. African marsupial fossils are known from the Paleocene–Eocene boundary, and the first indisputably marsupial Asian fossil to be discovered comes from the Oligocene (summarized by Benton 1985).

One explanation of the decline of marsupials in North America suggests that the earliest eutherians originated and first radiated in Eurasia, whereas the marsupials first radiated in North America, in isolation from eutherians. Perhaps the arrival of placental mammals in North America via the Bering land bridge from Asia upset the community balance between marsupials and the other Cretaceous mammals. This argument implies, however, that marsupials cannot compete with placentals for the same or closely related adaptive zones. As we shall see, this alleged inferiority of marsupials probably does not exist.

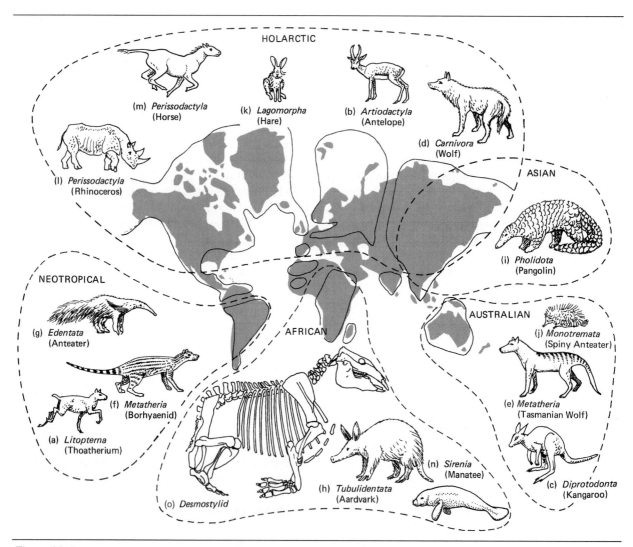

Figure 20–4 Radiation and convergence of mammals evolving in isolation during the Cenozoic. Superimposed on a map of the recent relative positions of the continents (shaded) are the land areas (solid lines) thought to have existed during the early Eocene. Note that these areas do not occupy the relative positions they may have had during the Eocene. Isolating distances may be underestimated by this projection. Mammals that evolved during various periods of the Cenozoic (not just the Eocene) are illustrated with the landmasses on which they probably originated. Mammals of Northern Hemisphere origin are grouped as Holarctic.

Tertiary Mammals of Australia and South America

Most of us are aware that the mammals of Australia differ from those of Asia, Europe, Africa, and North America. The mammals of South America also differ from those of the Northern Hemisphere, but less so than do the Australian mammals (Keast 1972). These two continents provide particularly clear examples of the effects of vicariance biogeography. Although both continents appear to have been invaded by animals that moved southward, the timing of those movements in relation to the changing connections between continents illustrates the role that chance events play in biogeography and evolution.

Tertiary Mammals of South America The paleontological record of South America is extensive and shows a history of connection, separation, and reconnection to North America. The effects of these geological events can be traced in the changing fauna of South American mammals. South America was isolated from North America by a seaway between Panama and the northwestern corner of South America from the late Jurassic until the end of the Tertiary. In the Pliocene, about 2.5 million years ago, the Central American land bridge was established between North and South America and animals from the two continents were free to mix for the first time in more than 100 million years (Marshall et al. 1982). Faunal interchange by island hopping or rafting commenced in the late Miocene as the two American landmasses drew ever nearer to each other (summarized by Janis 1993). The result was a massive exchange of elements between two radically different groups of animals that enriched the faunas of both continents (Mares and Genoways 1982).

Three major groups of Tertiary mammals can be distinguished in South America (Patterson and Pascual 1972). The earliest inhabitants—marsupials, edentates, and "condylarths"—appear to have been derived from North American forms that entered South America in the late Cretaceous before the connection between North and South America was broken, or moved from island to island while the seaway was still narrow. Equally important were the groups of North American mammals that failed to colonize South America at this time, notably insectivores, carnivores, and rodents. We have no explanation of why these groups did not succeed in populating South America, but their absence is important because it left adaptive zones open and marsupials radiated into them during the isolation of South America in the Tertiary.

The result of these radiations was the evolution in South America of a varied and complex fauna of marsupials, edentates, and ungulates. These animals showed extensive convergence with mammals in North America and the Old World in their general body form and ways of life (see Savage 1986 for restorations of the South American mammals). The resemblances between the mammals evolving in parallel on North and South America in the Tertiary were only general, however, not exact. Enough differences existed so that most North and South American forms that came into contact in the Pliocene (after the Central American land bridge arose) did so with little evidence of direct competition or displacement.

The South American marsupials present in the Cretaceous radiated into adaptive zones that were occupied in North America by insectivores, carnivores, multituberculates, and rodents. The caenolestid lineage included forms with teeth like multituberculates and others that were even more rodent-like. Argyrolagids had long hind legs and tails reminiscent of such hopping placental mammals as kangaroo rats and jerboas. Most spectacular of the marsupials of South America were the borhyaenids. Some borhyaenids were small generalized predators with head-plus-body lengths up to 75 centimeters, whereas *Thylacosmilus* was the size of a mountain lion (1.5 meters long) and converged on the skull features of the saber-tooth cats of the Northern Hemisphere. All the borhyaenids were relatively short-legged animals that probably stalked or ambushed their prey rather than pursuing it. The adaptive zone for large cursorial predators in Tertiary South America was apparently occupied by the psilopterids and phorusrachids, flightless birds standing 1.5 meters or taller (Marshall 1994).

The edentates are one of the success stories of South American mammals. Armadillos were the first group of South American edentates to appear in the fossil record, and they became very diverse in the Miocene. In the late Pliocene and Pleistocene the armadillo *Pampatherium* was a large as a rhinoceros. A second group of armored edentates related to armadillos, the glyptodonts (Figure 20–6b), appeared in the Eocene and radiated in the Miocene and Pliocene. Glyptodonts were encased in bone, and resembled a Volkswagen beetle in size and

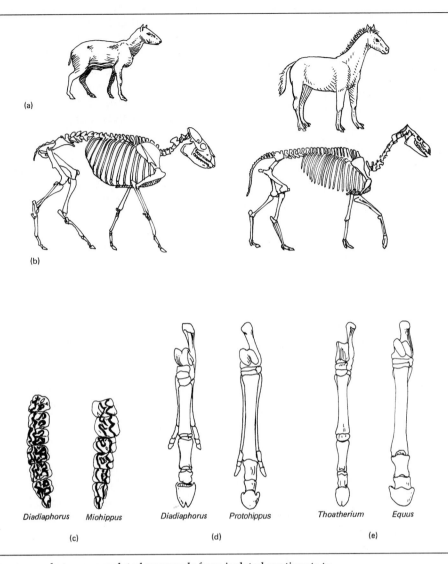

Figure 20–5 Convergence between unrelated mammals from isolated continents to grazing and sustained running over hard ground. South American litopterns (left) and North American (Holarctic) perissodactyls (right) (not to scale). (a) Reconstructions of the Miocene South American litoptern *Diadiaphorus* and a late Miocene and Pliocene North American and Asian three-toed horse, *Hypohippus*. (b) Skeletons of *Diadiaphorus* and *Hypohippus*. (c) Occlusal surface of the cheek teeth of *Diadiaphorus* and comparable developments in perissodactyls from the mid-Oligocene to Miocene *Miohippus*. (d) Frontal view of the left hind feet of the three-toed forms *Diadiaphorus* and *Protohippus* from the Miocene. (e) Frontal views of the hind foot of the Miocene one-toed litoptern *Thoatherium* and the Pleistocene one-toed horse *Equus*. The litopterns reached this degree of specialization much earlier than the horses, which did not show similar reduction of the toes until millions of years later.

shape; a carapace covered the body, a separate bony casque protected the head, and bony rings encircled the tail. In one species the tail terminated in a spiked knob like a medieval mace. In the Pliocene some glyptodonts achieved lengths in excess of 4 meters. Large flightless birds, the brontornithids, with massive beaks and heavy legs might have been specialized predators of glyptodonts.

Figure 20–6 Some extinct mammals of the Cenozoic. (a) *Phenacodus*, a Paleocene and Eocene herbivore of Europe and North America; (b–e) South American mammals: (b) one of the armored glyptodonts, mid-Eocene to Pleistocene edentates; (c) tapir-like *Pyrotherium*, a late Eocene amblypod; (d) rhinoceroslike *Toxodon*, a Pliocene and Pleistocene notoungulate; (e) *Andrewsarchus*, a late Eocene giant from Mongolia with a skull over 1 meter long, is the largest carnivorous land mammal known. Not to the same scale.

A third group of successful edentates, the ground sloths, is well represented by fossils. Early members appeared in the Oligocene, and by the Pleistocene some forms were the size of elephants. *Nothrotherium*, one of the megatheriid sloths that spread into North America, was more than 3 meters tall, and a South American ground sloth, *Megatherium*, reached 6 meters. These animals were tall enough to rest on their hind legs and browse on tree branches. Dung and even mummified body tissues of ground sloths have been found in caves in arid regions of North and South America, sometimes in association with human artifacts. Two other groups of edentates, the tree sloths and the anteaters, spread from South America to North America, but have only sparse fossil records on either continent.

The third group of early colonizers of South America was the "condylarths" and their descendents, the unique endemic South American ungulates. These animals radiated into several orders, many of them containing extremely large animals. The litopterns were cursorial herbivores ranging in size from small, delicately built adianthids through the middle-size proterotheriids to the large macrauchenids. The proterotheriids were horse-like in their general body form and showed the same pattern of reduction of the digits. They lacked the high-crowned (hypsodont) teeth of horses, however, suggesting that they were browsers, not grazers. The macraucheniids, some of which stood 1.5 meters at the shoulder, were camel-like but their nostrils were located high on the top of the head. Some paleontologists interpret this character as indicating that macraucheniids had an elongated proboscis.

The astrapotheres and pyrotheres were two orders of large, heavy-bodied herbivores with small proboscises and tusks (Figure 20–6c). Pyrotheres appeared in the Eocene and increased progressively in size until the middle Oligocene when they became extinct. Astrapotheres, also of Eocene origin, persisted through the Miocene.

The notoungulates showed an early dichotomy; two groups (Typoptheria and Hegetotheria) were rodent-like and rabbit-like herbivores. Some of these were quite large: *Hegetotherium* of the Oligocene and Miocene was the size of a small dog, and *Typotherium* was the size of a bear. The order Toxodontia contained large rhinoceros- or hippopotamus-like animals. *Toxodon*, an abundant animal in the Pleistocene, was nearly 3 meters in length (Figure 20–6d).

A second category of mammals made their way to South America in the late Eocene and early Oligocene, perhaps by rafting from Africa. These were placental mammals and included the caviomorph rodents (guinea pigs and Patagonian hares) and platyrrhine primates (New World monkeys). The caviomorph rodents diversified in the Miocene and Pliocene to include the largest rodents that have ever lived. *Telicomys*, a dinomyid ("terrible mouse"), was the size of a small rhinoceros.

A third group, arriving in the latest Miocene, were the procyonid carnivores (raccoons, coatis, and their relatives) originating from North America. These animals made their way south, probably by island-hopping.

A striking feature of the South America fauna in the middle Tertiary was the variety of large herbivorous mammals (Mares and Genoways 1982). In South America seven groups of ungulates fell into this category, plus five families of edentates and two families of rodents. In North America, in contrast, just three groups of ungulates (artiodactyls, perissodactyls, and proboscideans) plus the ground sloths (which had invaded from South America by island hopping during the Miocene) filled the large herbivore adaptive zone.

What happened when the establishment of a land bridge in the Pliocene allowed these two very different faunas to mix? The results were less dramatic than one might have expected. Animals from North America moved southward, and South American forms moved northward. On each continent the newcomers and the native fauna appeared to coexist; for the most part the immigrants enriched the existing fauna rather than displacing it. However, a later disparity exists. Although Pleistocene extinction affected the largest mammals on both continents, the southern immigrants to North America were more profoundly affected than North American forms. The northern immigrants in South America were more successful, and today about half the generic diversity of South American mammals consists of forms with a North American origin. The southern species that persist in North America are mostly confined to Central America and the southern United States.

Several groups of large South American mammals, including the astrapotheres and some notoungulates, disappeared in the Miocene and early Pliocene before the land bridge formed. These extinctions were probably associated with the Miocene orogeny that lifted the Andean Cordillera to its present height. The rain shadow that formed on the eastern side of the mountains created a drier and more open habitat. The loss of some native South American forms at this time was balanced by the evolution of new ones, particularly ground sloths and glyptodonts. The native South American mammals that were present at the end of the Pliocene largely persisted through the Pleistocene. The dramatic extinctions of South American mammals occurred at the end of the Pleistocene. Large species, both natives and immigrants, were most affected. These faunal changes in South America seem to be part of a worldwide pattern, because similar extinctions occurred in North America and Africa at the same time and may be related to human activities.

Other marsupials have fared well in South America. In the late Pliocene, when the land connection to North America was established, marsupials comprised 19 percent of the genera of mammals in South America. Despite the loss of the borhyaenids, marsupials now include 17 percent of the genera of Recent mammals in South America, an insignificant decline. One familiar South American marsupial invader, the Virginia opossum, has extended its range northward into southern Canada

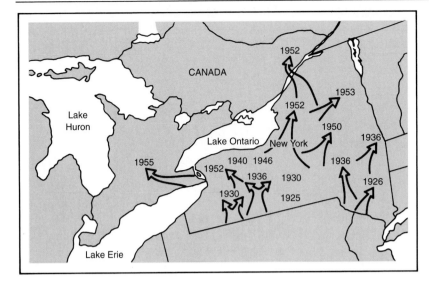

Figure 20–7 Northward invasion of the opossum (*Didelphis virginiana*) in eastern North America. Capture dates are placed over localities from which specimens were first obtained. Direct introductions into northern areas made during the early 1900s were always unsuccessful. Subsequently, naturally occurring invasions into these same northern areas by expanding, locally adapted populations have been very successful. Little or no northward expansion of the opossum's range has occurred since 1955. (Data from W. J. Hamilton, 1958, *Cornell University Experiment Station Memoirs* 354:1–48; and R. L. Peterson, 1966, *The Mammals of Eastern Canada*, Oxford University Press, Toronto, Canada.)

(Figure 20–7). Thus, the history of South America provides no evidence to support the simplistic notion of a general inferiority of marsupials when they are in competition with placental mammals (Kirsch 1977).

Cenozoic Isolation of Australia   The mammalian fauna of Australia has always been composed almost entirely of marsupials rather than placentals (Rich and van Tets 1985). The earliest marsupial fossils currently known from Australia are of early Eocene age, and there are abundant and diverse remains from the Miocene onward. A single early Eocene molar is the only record of a terrestrial Australian placental until late in the Cenozoic (Godthelp et al. 1992).

South America, Antarctica, and Australia were still close together in the late Mesozoic and early Cenozoic, and Australia was probably populated by marsupials that moved from South America across Antarctica, which was warm and moist until about 35 million years ago (Kerr 1987, Eaton 1993, Janis 1993). Marsupial fossils from the Cretaceous are known from Peru and Bolivia in South America, and a fossil South American marsupial was found in Eocene deposits in Antarctica in 1982.

Once marsupials reached Australia, they enjoyed the advantages of long-term isolation, and a radiation of diverse types followed. Kangaroos and wallabies, the wombats and koala, marsupial mice and marsupial moles, bandicoots, and the probably extinct Tasmanian wolf evolved to fill a variety of niches with food habits ranging from complete herbivory to carnivory (Archer and Clayton 1984).

Bats are the only major group of placentals that diversified in Australia before humans arrived. Rodents founded colonies in Australia by island hopping from Southeast Asia in the late Miocene and on several occasions since, but these incursions probably had little overall effect on the Australian marsupials. A far greater threat has been the prehistoric invasion by humans and dogs, which already has led to the near or complete extinction of the Tasmanian wolf and endangers numerous other forms.

Chance in Evolution

The remarkable convergences of the mammalian faunas of North America, South America, and Australia clearly show the role that chance events play in evolution. For example, an adaptive zone exists for predators, and the group that radiates to fill that zone depends in large measure on what phylogenetic lineage is in the right place at the right time. Thus, in North America placental canids gave rise to a variety of dog-like carnivores ranging in size from foxes to wolves, whereas in South America and Australia it was borhyaenid and dasyurid marsupials,

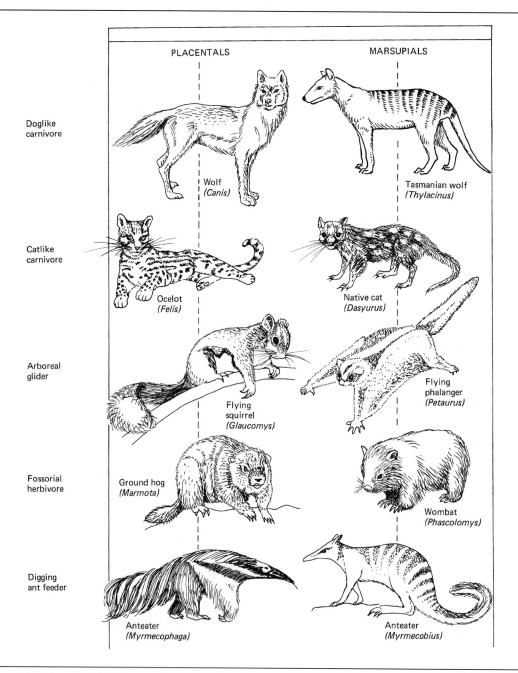

Figure 20–8 Convergences in body form and habits between placental and marsupial mammals. (From G. G. Simpson, G. S. Pittendrigh, and L. H. Tiffany, 1957, *Life*, Harcourt, Brace, Jovanovich, New York, NY.)

respectively, that produced dog-like predators (Figure 20–8). The same comparisons can be made between other mammalian adaptive zones: North American artiodactyls, South American litopterns, edentates, and ground sloths, and Australian kangaroos and giant wombats were all medium-size or large herbivores. The North American lagomorphs, the South American typotheres, and the Australian quokkas were small, rabbit-like herbivores. Not all the parallels are complete: The North American saber-tooth cats and the South American saber-tooth borhyaenids were convergent, but the Australian marsupial lion (*Thylacoleo*), although it may have been a cat-like carnivore, lacked the enormous canines of the saber-tooth forms.

The picture that emerges is of a mammalian body plan that can evolve into a wide variety of specialized forms when given the opportunity. The events that create an opportunity are largely the results of chance and ecology. The coincidence in time between the formation or breaking of connections between landmasses and the presence or absence of particular phylogenetic lineages at those times determines which lineages will dominate a particular adaptive zone on a particular landmass. The importance of chance and vicariance is not limited to mammals, but it is clearly demonstrated by a comparison of the mammalian faunas of the Northern and Southern hemispheres.

References

Archer, M., and G. Clayton (editors). 1984. *Vertebrate Zoogeography and Evolution in Australasia*. Hesperian, Carlisle, Australia.

Benton, M. J. 1985. First marsupial fossil from Asia. *Nature* 318:313.

Eaton, J. G. 1993. Marsupial dispersal. *National Geographic Research and Exploration* 9:436–443.

Godthelp, H., M. Archer, and R. Citelli. 1992. Earliest known Australian Tertiary mammal fauna. *Nature* 356:514–516.

Imbrie, J., and A. Berger (editors). 1984. *Milankovitch and Climate Change*. Elsevier, Amsterdam, Netherlands.

Janis, C. M. 1993. Tertiary mammal evolution in the context of changing climates, vegetation, and tectonic events. *Annual Review of Ecology and Systematics* 24:467–500.

Keast, A. L. 1972. Comparisons of contemporary mammal faunas of southern continents. Pages 433–501, in *Evolution, Mammals, and Southern Continents*, edited by A. Keast, F. C. Erk, and B. Glass. State University of New York Press, Albany, NY.

Kerr, R. A. 1987. Ocean drilling details steps to an icy world. *Science* 236:912–913.

Kirsch, J. A. W. 1977. The six-percent solution: second thoughts on the adaptedness of the Marsupialia. *American Scientist* 65:276–288.

Mares, M. A., and H. H. Genoways (editors). 1982. *Mammalian Biology in South America*, volume 6, Special Publication Series. Pymatuning Laboratory of Ecology, University of Pittsburgh, Pittsburgh, PA.

Marshall, L. G. 1994. The terror birds of South America. *Scientific American* 270(2):90–95.

Marshall, L. G., S. D. Webb, J. J. Sepkoski, and D. M. Raup. 1982. Mammalian evolution and the great American interchange. *Science* 215:1351–1357.

Patterson, B., and R. Pascual. 1972. The fossil mammal fauna of South America. Pages 274–309, in *Evolution, Mammals and Southern Continents*, edited by A. Keast, F. C. Erk, and B. Glass. State University of New York Press, Albany, NY.

Rich, P. V., and G. F. van Tets. 1985. *Kadimakara, Extinct Vertebrates of Australia*. Pioneer Design Studio, Lilydale, Australia.

Savage, R. J. G. 1986. *Mammalian Evolution: An Illustrated Guide*. Facts on File Publications, New York, NY.

Terborgh, J. 1992. *Diversity and the Tropical Rainforest*. Scientific American Library, New York, NY.

21 CHARACTERISTICS OF MAMMALS

The diversification of therian mammals in the Cenozoic has continued some trends seen among Paleozoic and Mesozoic synapsids (Chapter 19). In particular, the specialization of the jaws and teeth that allowed more effective processing of food has continued in mammals. A useful perspective on mammalian diversity can be gained by examining their feeding specializations. The morphology of the skulls, jaws, and teeth of mammals is clearly related to dietary habits, and very similar specializations have evolved convergently or in parallel in distinct evolutionary lineages in different parts of the world. This situation reflects the fragmentation of landmasses during the Cenozoic (Chapter 20) which left different combinations of mammalian stocks on different continents. In this chapter we provide an overview of the diversity of extant mammals.

The Major Lineages of Mammals

Mammals alone widely exploit the resources of earth, from pole to pole, mountaintop to deep sea, and even the night sky. Only birds and arthropods approach such variety. In this chapter we survey the major groups of mammals and emphasize some characteristics of extant mammals, especially the structure and function of the integument, the feeding system, the nervous system, and the reproductive system. Two other salient characteristics of mammals, their endothermy and sociality, are treated in detail in Chapters 22 and 23.

Current opinion favors a monophyletic origin of mammals as depicted by Figure 21–1 and Table 21–1. The most likely sister group of mammals is one of the tritheledontids, a derived cynodont clade (Chapter 19). Mammalian higher systematics is in a state of flux (Benton 1988, Novacek 1993). A widely repeated phylogeny of mammals (Novacek 1992) is presented here with modification and some cladogram

node justifications. But it is important to remember that cladograms are hypotheses of relationships; they are always subject to revision in the light of new information or new interpretations of existing material. The systematics of the mammalian clade is perhaps the prime example. There are 20 orders of extant mammals with unresolved relationships between many of the orders.

Derived mammals are divisible into three major groups, separated on the basis of reproductive biology. The Prototheria, or monotremes, survive as about six species isolated in Australia and New Guinea (Griffiths 1979). They are grouped in two lineages: the echidnas and the duck-billed platypuses. Prototherians lay eggs that are incubated and hatched outside the reproductive tract of the females. Nevertheless, echidnas and platypuses are hairy, endothermic, milk-producing vertebrates with only the dentary in the lower jaw, and thus qualify as mammals.

The remaining two groups of extant mammals are closely related but have had separate evolutionary histories since the earliest Cretaceous. The 250 or so species of Metatheria (marsupials) are distinguished by their short gestation periods, tiny, feebly developed offspring, and (in many but not all species) a protective pouch, the **marsupium.** This pouch lies over the mammary glands of the female and the young crawl into it immediately after birth to feed and complete development. The extant marsupials are restricted to the Australian region and the New World.

The eutherians, or placental mammals, which include about 3800 species, are born at a more advanced state of development than are marsupials. Some are able to run or swim by their mother's side within minutes of birth, and most of the species with long periods of postnatal development are weaned sooner after birth than are marsupial young.

These differences between the reproductive patterns of Metatheria and Eutheria reflect different evolutionary responses to the stresses of the environment (Lillegraven 1974). Eutherians produce advanced young with superior survival potential, but at high and prolonged cost to the mother. The metatherian female invests much less before the birth of her young and can quickly recycle her reproductive apparatus to produce subsequent broods if appropriately stimulated by environmental conditions (Parker 1977).

Eutherians are the dominant mammals of recent times. Over one-third of the extant species of eutherian mammals are insectivores and bats. Because of their small size and nocturnal habitats, these mammals are not well known to the general public. The Insectivora, which feed on the flesh of soft-bodied invertebrates or pierce the exoskeletons of insects to eat their soft parts, may provide us with models of the life of the earliest mammals. The winged Chiroptera (bats) generally feed on flying insects or on fruit and nectar. The major groups of bats use echlocation for navigation and prey capture. By far the most familiar small mammals are species of Rodentia (mice, voles, squirrels, and their many relatives). These small, gnawing, plant-eating eutherians are among the most successful of mammals. They are also humanity's greatest mammalian competitors and constant, if uninvited, companions.

Domesticated mammals are found among the species of plant-eating Artiodactyla and Perissodactyla, upon whom we depend for food, fiber, and labor, and the flesh-eating Carnivora, from which we gain cooperative assistance, protection, and endless hours of enjoyment.

The fish-, squid-, and plankton-eating Cetacea include the largest vertebrates ever known to have lived on earth. The primates, with their generally catholic tastes in food, are human's closest relatives. As such they are a perennial attraction and an important guide to understanding our own aptitudes and deficiencies.

Although we will draw examples from other lineages in our discussion of mammals, these six groups—Insectivora, Chiroptera, Primates, Rodentia, Carnivora, and Artiodactyla—receive most of our attention.

The Mammalian Integument

In many ways the outside covering of mammals is the key to their unique way of life and contains many of the unique, derived characters of the group. We have emphasized that endothermy is an energetically expensive process; a bird or mammal may use nearly all the energy in the food it eats merely to keep itself warm. Clearly, reducing the heat lost to the environment can be an important factor in the energy balance of an endothermal homeotherm. Part of this reduction is accomplished behaviorally (for example, by seeking a sheltered microclimate) and part by physiological mechanisms such as directing blood flow away from the body surface. In either case, the characteristics of an animal's integument are major factors that affect its heat exchange with the environment. Much of the ability of mammals to live in harsh climates is attributable to properties of their integument. The skin, with its hair, glands, and sense cells influences every aspect of mammalian life.

The variety of mammalian integuments is enormous. Some small rodents have exceedingly delicate epidermis only a few cells thick. Human epidermis varies from a few dozen cells thick over much of the body to over a hundred cells thick on the palms and soles. Elephants, rhinoceroses, hippopotamuses, tapirs, and pigs were once classified as pachyderms because their epidermis is several hundreds of cells thick. The texture of the external surface of the epidermis varies from smooth (in fur-covered skins and the hairless skin of cetaceans) to rough, dry, and crinkled (many hairless terrestrial mammals). The tail of many rodents is covered by epidermal scales very like those of lizards.

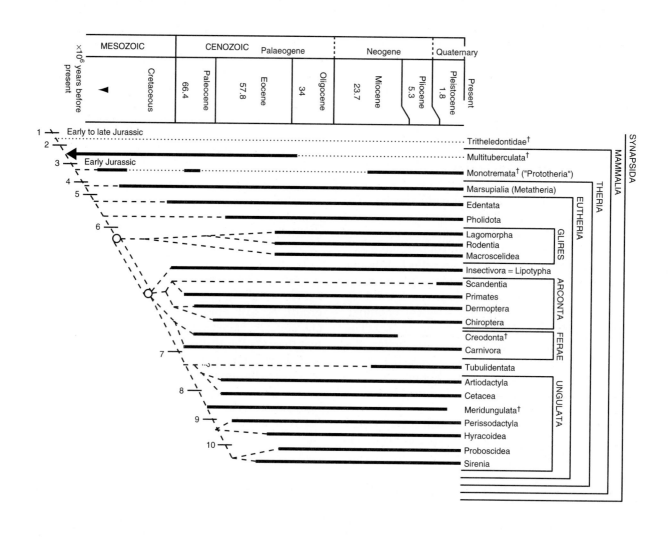

Integumentary Derivatives

The versatility of the mammalian integument lies in its derivatives: growing, replaceable hair; lubricant- and oil-producing holocrine (sebaceous) glands; apocrine and exocrine glands that secrete volatile substances, water, and ions; the unique mammary glands; scales, nails, claws, hoofs, and horns. The integument is also a contributing factor in the growth of antlers. Not every species has each class of integumentary derivative, but the various types often occur in unrelated mammals, attesting to their functional significance. The derivatives of the mam-

malian skin are derived from epidermis, but are supported, nourished, and innervated through the dermis. Figure 21–2 shows the superficial layer of the skin, the **epidermis,** which is formed by dead cells that are derived from an underlying germinal layer resting on the deeper still, relatively cell-poor **dermis.** Beneath that, the **hypodermis** is a final layer rich in subcutaneous fat cells. Specialized structures of the skin include guard hairs, undercoat or wool hairs, sebaceous glands, apocrine glands, and sweat glands. Sensory nerve endings include free nerve endings (probably pain receptors), beaded nerve nets around blood vessels, Meissner's corpuscles

1. Prismatic enamel on teeth. **2.** Dentary replaces surangular in jaw articulation, action of molars involves lateral movement of the lower jaw. **3.** Triangular (tribosphenic) pattern of main molar cusps. **4.** Theria: Loss of eggshell, multiple ear ossicles, vertical tympanic membrane, characters of braincase including double walled construction, expanded neopallium in brain, characters of vertebrae and long bones. **5.** Chorioallantoic placenta, long gestation. **6.** Stapes stirrup-shaped (i.e., penetrated by a fenestra). **7.** Clavicle reduced or absent, bunodont crowns on molars, hoofs often substituted for claws or nails. **8.** Expanded auditory capsule roof. **9.** Mastoid bone not visible on external surface of skull. **10.** Orbits anteriorly placed, bilophodont teeth. (Based on M. J. Novacek,

1993, *Journal of Mammalian Evolution* 1:3–30; and M. J. Novacek, 1992, *Nature* 356:121–125; with additional material from M. C. McKenna, 1975, pages 21–46 in *Phylogeny of the Primates*, edited by W. P. Luckett and F. S. Szalay, Plenum, New York, NY; J. F. Eisenberg, 1981, *The Mammalian Radiations*, The University of Chicago Press, Chicago, IL; T. S. Kemp, 1982, *Mammal-like Reptiles and the Origin of Mammals*, Academic, London, UK; T. S. Kemp, 1983, *Zoological Journal of the Linnaean Society* 77:353–384; M. S. Novacek and A. R. Wyss, 1987, *Cladistics* 2:257–287; Z. Kielan–Jaworowska, A. W. Crompton, F. A. Jenkins, Jr., 1987, *Nature* 326:871–873; and F. S. Szalay, M. J. Novacek and M. C. Mckenna, editors, 1993, *Mammal Phylogeny*, Vols. 1 and 2, Springer, New York, NY.)

Figure 21–1 Phylogenetic relationships of Mammals. The diagram shows the probable relationships among the major groups of mammals. Extinct lineages are marked by a dagger (†). The numbers indicate derived characters that distinguish the lineages. The circles indicate that the relationships of the lineages at that node cannot yet be defined, and the lineages that lie between characters 6 and 7 are all included in that uncertainty. Note that the placement of the Multituberculata in this figure differs slightly from that of Figure 19–16, reflecting differences in the sources on which the figures are based. We have retained the difference to emphasize that the relationships shown in cladograms are hypotheses, not facts.

(touch receptors), Pacinian corpuscles (pressure receptors), nerve terminals around hair follicles, and warmth and cold receptors. Vascular plexuses (intertwined blood vessels) of the skin are involved in thermoregulation.

Hair Hair has a variety of functions in a mammal, including concealing an animal from predators and signaling to other members of its species, but the basic function of hair in nearly all mammals is insulation. Prominent features of the pelage of extant mammals are its growth, replacement, color, and mobility. A hair grows from a deep invagination of the germinal layer of the epidermis called the hair follicle (Figure 21–3). A hair is composed of keratin, pigments, and, in some cases, tiny encapsulated air bubbles. These chemical substances are common to all tetrapods, but their organization and chemistry make hair uniquely mammalian. The keratin of hair is of two types: soft and hard. **Soft keratin,** found in the core or medulla of a hair shaft, is similar to that of the epidermis but is often interrupted by masses of trapped air bubbles. **Hard keratin** in the cortex covering the medulla and the sheath that encapsulates the hair surface results from a gradual fusion of living epidermal cells into a hard, chemically inert

homogeneous mass that forms stiff hair, bristles, and quills. (Bird feathers and leg scales, as well as the scales of snakes and lizards, are hard keratin.) When the soft keratin portion is dominant, a supple pelage such as wool is produced.

The color of hair depends on the quality and quantity of melanin injected into the forming hair by melanocytes at the base of the hair follicle. Black hair has dense melanin deposits in both cortex and medulla; the white hair of humans has an unpigmented cortex but pigmented medulla. When the medullary portion of the hair shaft is absent, a fine, often blond or reddish pelage (depending on the melanin) results. The color patterns of mammals are built up by the colors of individual hairs. Most mammals are countershaded (dark above and lighter beneath), a combination that is inconspicuous under most conditions, whereas other species (skunks, for example) that rely on advertising their noxious properties to deter predators have strongly contrasting patterns that are visible even at low levels of illumination.

Because exposed hair is nonliving, it wears and bleaches. Replacement occurs by growth of an individual hair or by **molting,** in which old hairs are replaced. Human hair lengthens about one-third of a

Table 21–1 **Classification of extant mammals and approximate numbers of species. The geographic regions represent the distribution of the entire group; families within an order often have smaller distributions. Extinct forms are indicated by a dagger (†).**

Classification	Species	Major Examples
Prototheria		
Allotheria†		
Multituberculata†		
Ornithodelphia		
Monotremata	3	Spiny anteaters, duck-billed platypus, 2 to 10 kg.
Theria		
Metatheria		
Marsupialia	280	Opossums, bandicoots, koala, wombats, kangaroos, wallabies; primarily 5 g to 5 kg, some Australian forms are large (90 kg); Nearctic, Neotropical, and Australian regions.
Eutheria		
Edentata	30	Anteaters, sloths, armadillos; 20 g to 33 kg; Nearctic and Neotropical regions.
Pholidota	7	Pangolins; 2 to 33 kg; Ethiopian and Oriental regions.
Lagomorpha	69	Pikas, rabbits, hares; 180 g to 7 kg; worldwide except Antarctica, introduced in Australia by humans.
Rodentia	1814	Rats, mice, squirrels, agouti, capybara; 7 g to over 50 kg; worldwide except Antarctica.
Macroscelidea	15	Elephant shrews; 25 to 500 g; Ethiopian region with one species in Morocco and Algeria.
Insectivora (=Lipotypha)	390	Hedgehogs, moles, shrews; 2 g to 1 kg; worldwide except Antarctica.
Scandentia	16	Tree shrews; 400 g; Oriental region.
Primates	235	Lemurs, monkeys, great apes, humans; 85 g to over 275 kg; primarily Oriental, Ethiopian, and Neotropical regions, humans are now worldwide.
Dermoptera	2	Flying lemurs; 1 to 2 kg; Oriental region.
Chiroptera	986	Bats; 4 g to 1.4 kg; worldwide except Antarctica.
Carnivora	274	Dogs, bears, raccoons, weasels, hyaenas, cats, sea lions, walrus, seals (these last three are often assigned to the Pinnipedia); 70 g to 760 kg, some marine forms over 100 kg; worldwide.
Tubulidentata	1	Aardvark; 64 kg; Ethiopian region.
Artiodactyla	213	Swine, hippopotamuses, deer, giraffe, antelope, sheep, goats, cattle; 2.5 to 3600 kg; worldwide except Antarctica (introduced into Australia and New Zealand by humans).
Cetacea	80	Porpoises, toothed whales, baleen whales; 20 to 200,000 kg; worldwide in oceans and in some rivers and lakes in Asia, South America, northern America and Eurasia.
Perissodactyla	17	Horses, tapirs, rhinoceroses; 200 to 3600 kg; Africa, southern and central Asia, central and northern South America.
Hyracoidea	7	Hyraxes; 4 kg; Ethiopian and southern Saharan regions.
Proboscidea	2	Elephants; 4500 to 7000 kg; Ethiopian and Oriental regions.
Sirenia	4	Dugongs, manatees; 140 to over 1000 kg; coastal waters and estuaries of all tropical and subtropical oceans except the eastern Pacific (in the Atlantic drainage they enter rivers).

Sources: J. A. Hopson, 1970, *Journal of Mammalogy* 51:1–9; M. C. McKenna, 1975, in *Phylogeny of the Primates,* edited by W. P. Luckett and F. S. Szalay, Plenum, New York, NY; T. A. Vaughn, 1986, *Mammalogy,* 3d edition, Saunders College, Philadelphia; J. F. Eisenberg, 1981, *The Mammalian Radiations,* University of Chicago Press, Chicago, IL; R. M. Nowak and J. L. Paradiso, 1991, *Walker's Mammals of the World,* 5th edition, Johns Hopkins University Press, Baltimore, MD; S. Anderson and J. K. Jones, Jr., 1984, *Orders and Families of Recent Mammals of the World,* Wiley–Interscience, New York, NY; M. J. Novacek and A. R. Wyss, 1987, *Cladistics* 2:257–287.

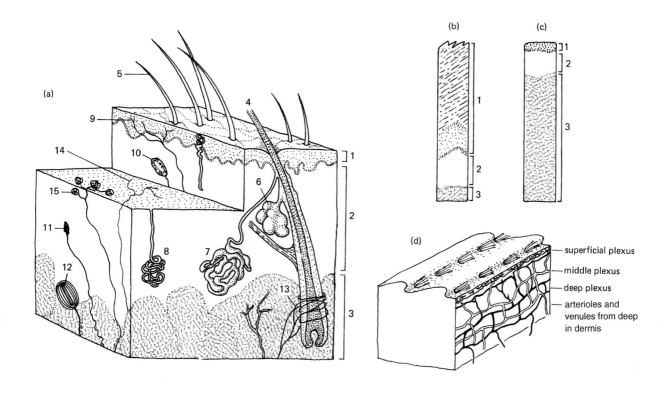

Figure 21–2 Structure of mammalian skin. (a) Composite of skin and appendages showing: (1) layers and surface configuration of the epidermis, (2) the relatively cell-poor dermis, (3) the hypodermis rich in subcutaneous fat cells, (4) guard hairs, (5) undercoat or wool hairs, (6) sebaceous gland, (7) apocrine gland, (8) sweat gland, (9) free nerve endings, (10) beaded nerve nets around blood vessels, (11) Meissner's corpuscles, (12) Pacinian corpuscles, (13) nerve terminals around a hair follicle, (14) heat, and (15) cold receptors. (b) Thick epidermis of the skin on the sole of the human foot. (c) Thin, hairy skin and subcutaneous tissue from the human thigh (numbering as in part a). (d) Canine skin showing three of the vascular plexuses involved in thermoregulation. Note also the numerous hair shafts emerging from a single hair follicle complex, a characteristic of furred mammals. (Modified from various sources, especially A. W. Ham and D. W. Cormack, 1973, *Histology*, 8th edition, Lippincott, Philadelphia, PA; and R. J. Harrison and E. W. Montagna, 1973, *Man*, 2d edition, Appleton-Century-Crofts, New York, NY.)

millimeter per day during active growth. When it becomes quiescent, the follicle produces a bulbous anchor that firmly locks the hair in place. Most mammals have pelage that grows and rests in seasonal phases. Molting usually occurs once or twice a year. Only by this mechanism can mammals like the snowshoe hare and ermine change from a summer camouflage of brown to a winter coat of white. Skins with quiescent pelage are the ones of value in the fur trade, because only then are the hairs tightly bound to the dermis.

Fur consists of closely placed hairs, often produced by multiple hair shafts arising from a single complex root (Figure 21–2d). Its insulating effect depends on its ability to trap air within the fur coat, and this is proportional to the length of the hairs.

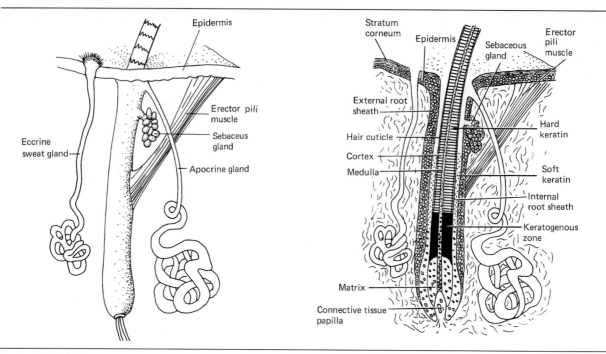

Figure 21–3 The structure of mammalian hair and associated glands. Composite view of hair follicle (left) and surrounding structures with dermis dissected away and (right) a longitudinal section to show the internal structure of a typical hair. (Modified from various sources, especially A. W. Ham and D. W. Cormack, 1973, *Histology*, 8th edition, Lippincott, Philadelphia, PA; and R. J. Harrison and E. W. Montagna, 1973, *Man*, 2d edition, Appleton-Century-Crofts, New York, NY.)

Mammals from polar and temperate regions have longer coats in winter than in summer. The spacing of hairs is also important because it determines the size of entrapped air spaces; winter coats are denser than summer pelage.

Most hairs lie at an angle to the skin surface. Attached about midway along the length of each hair follicle is a bundle of smooth muscle, the **erector pili** (Figure 21–3). This muscle pulls the hair into a nearly perpendicular orientation to the skin. The result is to increase the depth of the pelage, and thus trap a larger volume of air. A curious side effect noticeable in near-naked mammals such as humans are the dimples (goose pimples) on the skin's surface over the insertion of contracted erector pili muscles. Cold stimulates a general contraction of the erector pili via the sympathetic nerves, as do other stressful conditions such as fear and anger. The significance of these behaviorally induced erections of the pelage appears to lie in increasing the apparent size and, therefore the apparent strength of the excited mammal. Often such displays are limited to specific

regions, such as the hackle and tail hair of dogs. Displays such as those of a porcupine are impressive and are clearly understood by a wide range of other animals.

Glandular Structures Secretory structures of the skin develop from the epidermis. **Sebaceous glands** open into the neck of a hair follicle and often lie in the space bounded by the hair shaft, the superficial epidermis, and the erector pili muscle. When the muscle contracts, the sebaceous gland is squeezed and its oily contents flow into the follicle and thence onto the skin surface (Figure 21–3). Sebaceous glands may also be found in hairless areas of the mouth and lips, the end of the penis, and around the vulva, the nipples of the mammary glands, and the edges of the eyelids. The oils lubricate and waterproof the hair and skin. Specialized sebaceous glands of sheep produce lanolin.

A second type of gland common to many mammals is the **apocrine gland.** These glands open near but not into the hair follicles, and also occur in skin that lacks hair. Apocrine glands are tubular epidermal invaginations that extend deep into the dermis. The exact functions of apocrine secretions are not well understood. Although the secretion from human apocrine glands is odorless, bacteria on the body surface convert it to an odorous product. The various musk and scent glands that are so characteristic of mammals are thought to be enlarged, modified apocrine glands. **Mammary glands** closely resemble apocrine glands in development, structure, and mode of secretion.

Mammary glands (Figure 21–4) develop as paired ridges of hypertrophied ventral epithelium of the fetus that run from the axillary to the inguinal regions. This mammary ridge differentiates into discrete localized thickenings in different positions. Manatees (Sirenia) have one pair of axillary mammae, primates have one pair of thoracic mammae, some artiodactyls have abdominal mammae, and perissodactyls have inguinal mammae. The number of mammae that ultimately develop roughly corresponds to the number of young born in a litter. Certain opossum-like marsupials are reported to have nearly 20 mammae. In male mammals further development of the mammae does not occur after birth. In females the thickened epithelium produces numerous elongate cords that branch and proliferate into the dermis. Hollow sacs develop at the terminal ends of these cords and become continuous with channels that develop through the cords of epithelium to produce a highly branched duct system. Under hormonal stimulation the terminal sacs become more numerous, enlarge greatly, and the cells of the sac walls begin to secrete milk. Milk is a water-based solution of proteins (primarily the phosphoprotein calcium caseinate), the sugar lactose, and lipids in the form of several different fatty acids combined in suspended droplets. The fat droplets often are capped by or contain cell fragments, indicating a mode of secretion similar to that of apocrine glands, which also lose portions of cells during secretion. The proportions of these primary constituents vary from species to species (Table 21–2). The concentrations of the major components are far above those of the maternal blood and, in general, growth of a newborn mammal is positively correlated with the energy content of the milk.

Milk of the hooded seal contains more than 60 percent fat, and is the richest milk described for any mammal (Oftedal et al. 1988). Hooded seal pups are weaned four days after birth, at which point they have doubled their initial body mass from an average of 22.0 kilograms at birth to 42.6 kilograms at weaning (Bowen et al. 1985). The increased body mass is mostly blubber. This is the shortest period of lactation known for any mammal, and may reflect the ecological circumstances in which hooded seals give birth. In March and April when young hooded seals are born, floating pack ice is rapidly being reduced to small, drifting pieces by spring storms and rising temperatures. Under these conditions, a female seal cannot be certain that she will find her pup again if she leaves it to seek food. Thus, the rapid fattening of the pups may enable them to survive on stored fat until they have grown large enough to take to the sea themselves.

Pregnancy hormones, important coordinators of behavior, physiology, and environment, stimulate the mammary glands into a physiological state capable of lactation, but milk does not flow until some time after suckling has begun. Monotremes lack highly specialized openings for the mammae, but other mammals have nipples that are complexly innervated epidermal organs. Certain artiodactyls have analogous but much larger structures called teats.

Similar in structure to the apocrine and mammary glands and often grouped with them are the **sweat** or **eccrine glands.** The coiled tubules of sweat glands are never associated with hair follicles but occur on the hairless surfaces that come in contact with the substrate: soles of feet and prehensile tails. Secretions maintain skin pliability, increase traction, and enhance sensitivity of skin pressure receptors. Most mammals, particularly those without underfur, have sweat glands widely distributed over the body surface.

Sweat glands are also used in thermoregulation. Water secreted by the sweat gland is forced onto the skin surface by special contractile elements of the gland, evaporating and cooling the skin, which, in turn, cools the blood flowing in the subcutaneous vascular beds. When perspiration is copious, considerable amounts of sodium and chloride are lost and must be replaced in the diet. Some sweat glands respond to stimuli other than heat. In conjunction with nearby apocrine glands, these sweat glands contribute to odor production under conditions of stress and excitement.

Claws, Nails, Hoofs, Horns, and Antlers Some integumentary appendages are involved in locomotion,

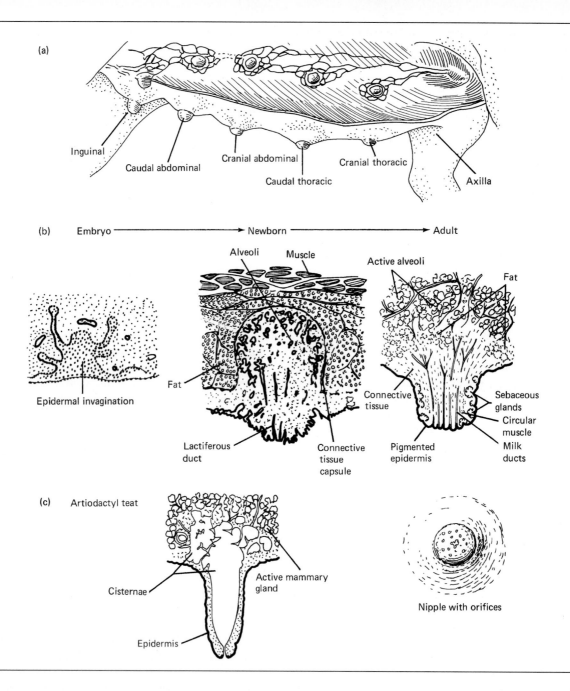

(a)

Inguinal   Caudal abdominal   Cranial abdominal   Caudal thoracic   Cranial thoracic   Axilla

(b)   Embryo ⟶ Newborn ⟶ Adult

Alveoli   Muscle   Active alveoli   Fat

Fat   Connective tissue   Sebaceous glands   Circular muscle   Milk ducts

Epidermal invagination

Lactiferous duct   Connective tissue capsule   Pigmented epidermis

(c)   Artiodactyl teat

Cisternae   Active mammary gland

Epidermis

Nipple with orifices

offense, defense, or display. Claws, nails, and hoofs are accumulations of keratin that protect the terminal phalanx of the digits. Claws are digital appendages of the skin (Figure 21–5). The claws of most animals are exposed and wear against the ground, but cat-like carnivores have retractable claws. Muscular tension on the flexor digitorum longus tendon causes the third phalanx to rotate on the end of the second phalanx. This rotation exposes the claw and stretch-

es the dorsal elastic ligament. Relaxation of the flexor digitorum muscle allows the elastic ligament to retract the claw. The hoof, characteristic of ungulates, is illustrated by that of the horse. Horseshoes are used to retard wear of the hoof on the hard substrates that humans impose on domestic horses. They can be nailed directly to the hoof because the hoof is dead keratin like your fingernails. In horses the flexor digitorum longus tendon returns the hoof

Figure 21–4 Mammary glands. (a) The positions of nipples in eutherians as illustrated by a dissection of the dog. Only the axillary mammae fail to develop in these large-litter-bearing canids. (b) Stages in the development of the human mammary gland resemble those of apocrine and eccrine glands. An embryo 15 centimeters in length has a simple proliferation and invagination of epidermal cells (left). A newborn infant (middle) has the basic elements of a nipple, multiple lactiferous ducts, and nonsecretory alveoli. The tip of the mature, lactating primate breast shows the addition of circular constrictor muscles, sebaceous glands, and swollen active alveoli which vent their product via multiple openings on the nipple. (c) The artiodactyl teat delivers a much larger volume of milk in a shorter time via the development of large milk-holding cisternae adjacent to the elongate teat, which has a single wide terminal opening. This system probably represents an anti-predator adaptation to a nomadic mammalian life on open grasslands.

to a proper angle after it has been bent by the weight of the body. The human fingernail is a simpler structure than either the retractable claw or the hoof, but was derived from ancestral claws.

Horns and antlers are characteristic of many large ungulates (Bubenik and Bubenick 1990). Their primary roles appear to be social recognition, sexual display, and jousting between males, although they may also be used for defense.

Horns may be formed entirely by epidermal tissue, as in the nasal horns of a rhinoceros (Figure 21–6a), which have no contribution from the bony elements of the skull. Thus rhinoceros horns can be cut off without major direct effect on the animal. The horns slowly grow back, but the process takes many years during which the calves of dehorned mothers are susceptible to predation by hyenas (Berger and Cunningham 1994).

In contrast to rhinoceros horns, the short knobby horns of a giraffe (Figure 21–6b) are composed of both skin and bone. The skin covering the horns of a giraffe is scarcely differentiated from that elsewhere on the body. The North American pronghorn antelope is the only extant mammal with bifurcated horns (Figure 21–6c). The fork is composed entirely of epidermal tissue, which covers a simple projection of the frontal bone known as the **os cornu** (horn bone). The pronghorn is unique also in being the only mammal that sheds the epidermal horny sheaths. The members of the artiodactyl family Bovidae have evolved a great variety of horns based on a common developmental plan (Figure 21–6d; Janis 1982). True horns are supported by a well-developed *os cornu* that is sheathed by a nonshedding, nonrenewable horn of hard keratin (Figure 21–6e). Complex shapes are characteristic of the horns of many species of antelopes (Figure 21–6f).

Antlers are a derived character of deer (Cervidae). They are used for intraspecific aggression, especially in ritualized head wrestling matches during the breeding season. The ancestral character state for deer was to have tusks formed by the canine teeth (Figure 21–7a). Derived cervids developed solid, bony antlers in conjunction with tusks, as seen in the muntjac (Figure 21–7b). The most derived cervids lack canines altogether, and the males of many species have enormous, complex antlers, such as those of the caribou (Figure 21–7c). Female caribou, unlike the females of other species of deer, have antlers, although they are smaller than the antlers of males. Antlers are replaced annually. The integument forms the velvet that nourishes and protects the growing antler (Figure 21–7d). When the antler matures, the integument dies and is shed, exposing the bone. The complex junction between the antler and the pedicel of the skull eventually weakens and the antler is shed.

Table 21–2 **Composition of milk of various species (grams per liter).**

Species	Proteins	Sugars	Fats	Inorganic Salts
Porpoise	110	13	460	6
Whale	95	4	200	10
Cat	92	50	35	11
Pig	74	32	45	10
Cow	35	45	40	9
Elephant	32	72	190	6
Human	11	75	35	3
Human blood plasma	ca. 8	0.1	0.7	ca. 1

Source: *The New Larousse Encyclopedia of Animal Life*, revised edition, 1980, Larousse, New York, NY.

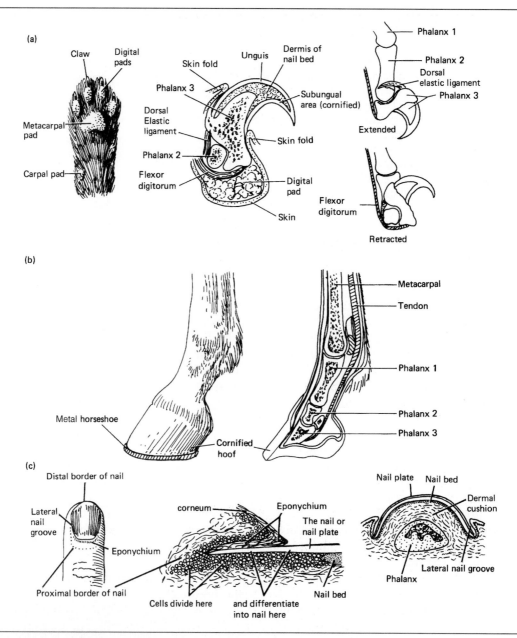

Figure 21–5 Skin appendages associated with terminal phalanges. (a) Retractable claws. Left: Hair and thick epidermal pads associated with the base of the claws. Center: Cross section of a claw showing its close relationships with the blood vessels, dermis, and bone of the third (terminal) phalanx. Right: Claw retraction mechanism characteristic of cats. (b) The hoof of a horse. Left: Normal appearance of the hoof of a shod horse. (Horse shoes are devices used to minimize wear of the hoof on unnaturally (human produced) hard and abrasive surfaces.) Right: Longitudinal section of lower foot showing relationship of phalanges to hoof. (c) The human nail. Left: Distinct regions on a nail correspond to the regional specializations of the epidermis associated with the nail (Center). Right: A cross section of the end of a finger shows the close association of the nail with the dermis and terminal phalanx of the digit.

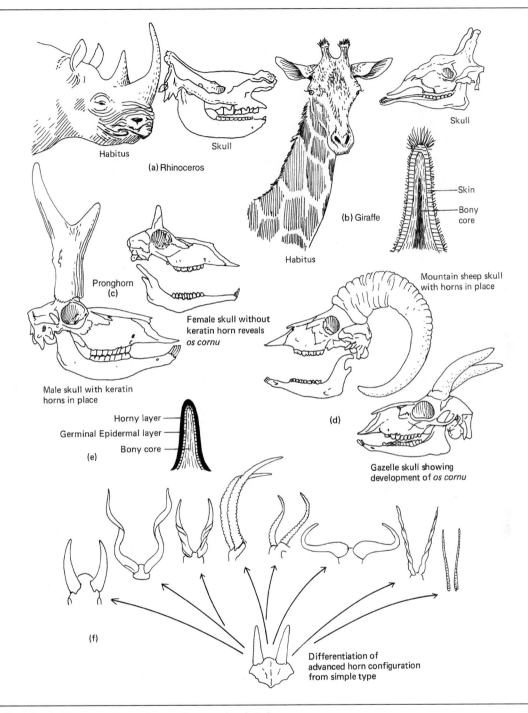

Figure 21–6 Cephalic appendages: horns. See text for details.

The Oligocene ancestors of the deer-like ungulates had long, stabbing upper canines but no antlers. During the Miocene, forms evolved that had both stabbing canines and modest antlers, as do Asian muntjacs. Not until the Pliocene did tuskless antlered deer become widespread. In studies of the fighting behavior of muntjacs, Cyrille Barrette (1977) has found that tusks are used offensively, whereas

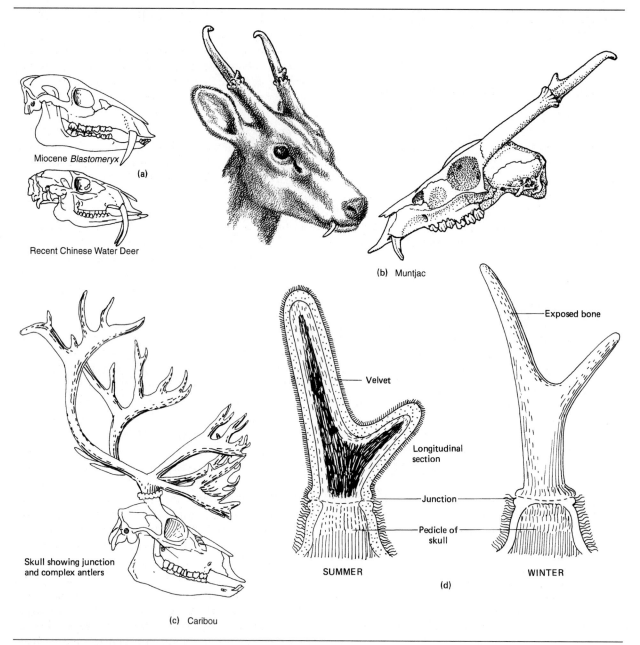

(a) Miocene *Blastomeryx*

Recent Chinese Water Deer

(b) Muntjac

Skull showing junction and complex antlers

(c) Caribou

Velvet

Longitudinal section

Junction

Pedicle of skull

SUMMER

Exposed bone

WINTER

(d)

Figure 21–7 Cephalic appendages: antlers. See text for details.

antlers may be offensive when used to stab and defensive when used as shields against blows from an opponent. Males engage in head wrestling and attempt to twist and shove their opponent off balance so that the tusks can be used. Winning a tusk-and-antler fight rewards the successful male muntjac with access to receptive females and successful reproduction. A possible first stage in the evolution of antlers would be defensive shielding against tusk blows. The larger the protoantler, the more effective the shield, and at some point in their evolution the effectiveness of antlers could have made tusks obsolete. Subsequently, additional functions have probably been added to the roles played by antlers, and sexual selection seems likely to be among them. This is another example of the maxim that the evolutionary origin of a character may be quite different from its current use.

Mammalian Food and Feeding Specializations

An efficient means of obtaining and processing food to sustain high energy needs is absolutely essential to an endotherm. Mammals meet these needs in various ways, as seen in their trophic (nutritional) adaptations. The trophic apparatus includes the teeth and jaws, the muscles involved in chewing, the alimentary canal, and even locomotion and social behavior.

Mammalian Teeth

Each tooth is attached in a jaw socket by a type of bone known as **cement,** hard and rigid like other bone, but wearing rapidly when exposed. The core of a tooth is hollow and filled with nerves, blood vessels, and cells that produce **dentine,** which forms the inner layer of the tooth. Dentine contains more mineralized material and less organic matter than bone and is harder, heavier, and more resistant to wear than cement. The exposed part of the tooth is encased in **enamel,** a unique substance that is the hardest, heaviest, and most friction-resistant tissue evolved by vertebrates. Unlike both cement and dentine, enamel is ectodermal, not mesodermal, in origin. Enamel is totally acellular and is nearly devoid of organic matter, being composed of large, uniformly oriented hydroxyapatite crystals.

The teeth of a generalized mammal are **heterodont** and are designated from anterior to posterior on each side of each jaw as **incisors** (usually two or three, but up to five for marsupials), **canines** (never more than one), **premolars** (generally two to four), and **molars** (variable, but often three). Most mammals have two sets of teeth during an animal's lifetime, a condition known as **diphyodonty** (*di* = two, *phyo* = grow, produce). The lacteal (milk or deciduous) dentition appears first in a generally regular order from anterior to posterior. The molars are adult teeth that function throughout life, but ontogenetically they are posterior, unreplaced members of the lacteal dentition. In many species the premolars have taken a form very similar to that of the molars, an evolutionary phenomenon known as molarization of the premolars. It is often difficult to distinguish premolars from molars in an intact skull, especially in herbivores.

Because digestion is basically a chemical process, increasing the area for contact and penetration of digestive enzymes allows rapid chemical breakdown of food. Most vertebrates use their teeth to catch, grasp, or crop food but not to process it. Mammals are different: The anterior teeth (incisors and canines) of mammals retain this basic function, but the posterior teeth (premolars and molars) are modified to masticate foodstuffs. During mastication, the food is repeatedly worked between interlocking occlusion surfaces of the upper and lower dentition by the action of the mobile, sensitive tongue and a characteristic mammalian facial structure, the cheeks, which assist in retaining food in the mouth when chewing. The food thus becomes a loose pulp that readily absorbs gastric juice in the stomach.

Dentition

The skulls of extant mammals can be arranged to illustrate a functional classification of recent eutherians based on their trophic specializations, especially their teeth (Figure 21–8). Such an arrangement is not phylogenetic; instead, it provides an ecological perspective of mammalian trophic diversity. The earliest therian mammals had a dentition similar to that found in relatively unspecialized insectivores extant today. The hedgehog, *Erinaceus,* has a dentition with piercing cusps on most of the teeth (Figure 21–8a). More specialized is the battery of sharp teeth of the mole, *Scalopus* (Figure 21–8b), which are used for piercing and holding worms and insects.

Myrmecophagous forms (*myrme* = ant, *phago* = to eat) like the armadillo (*Dasypus*), the anteater (*Cyclopes*), and the giant anteater (*Myrmecophaga*) feed on ants and termites in their nests (Figure 21–8c to e). Hence, in numerous unrelated mammalian groups, flat crushing teeth or no teeth at all are to be found, a peculiarity coupled with long, mobile, and worm-like tongues with exceptional protrusibility. Enlarged salivary glands produce a viscous, sticky secretion that coats the tongue. Myrmecophagous mammals also have elongate snouts and digging specializations.

Omnivorous mammals from many lineages retain some piercing and ripping cusps in the anterior teeth but have flat, broad crushing cusps posteriorly (Figure 21–8f to h). Because of their tough cel-

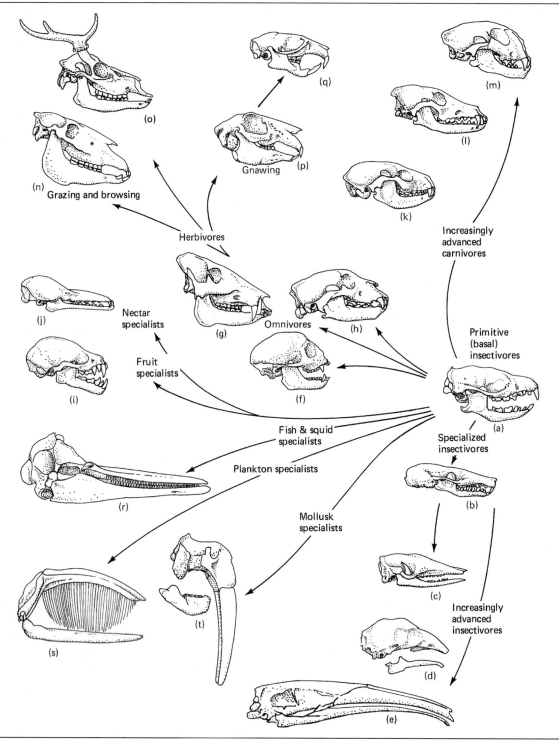

Figure 21–8 Feeding specializations of teeth and skulls of mammals: (a) hedgehog, (b) mole, (c) armadillo, (d) anteater, (e) giant anteater, (f) marmoset, (g) peccary, (h) bear, (i) fruit-eating bat, (j) nectar-eating bat, (k) raccoon, (l) coyote, (m) mountain lion, (n) the grazing horse, (o) deer, (p) jackrabbit, (q) woodrat, (r) porpoise, (s) right whale, (t) walrus. See text for details.

lulose cell walls, many plant tissues are resistant to efficient digestion. Grinding plant matter between rough surfaces is the usual method of disrupting cell walls. The cheek teeth of these omnivores are **brachyodont** (*brachy* = short) and **bunodont** (*buno* = a hill or mound), meaning that the cusps are rounded. These teeth process soft animal material that can be ground (not sliced) into digestible pieces and plant material like soft roots, tubers, and berries.

A similar dentition with much-enlarged anterior biting teeth occurs in the fruit-eating bat, *Artebius* (Figure 21–8i), which bites chunks from fruit and crushes the pulp for its juices with the broad flat posterior teeth. Another bat, *Choeronycterus* (Figure 21–8j), uses a long tongue to extract nectar from flower and has greatly reduced dentition.

Herbivores of two very different types show similarities in the great relative size of the flat grinding cheek teeth and the development of a gap (a **diastema**) between these cheek teeth and the anterior food-procuring teeth. A diastema positions the cutting apparatus away from the face and allows the snout to penetrate narrow openings or reach close to the ground during feeding. Both rabbits (Lagomorpha; Figure 21–8p) and rodents (Rodentia; Figure 21–8q) have a greatly enlarged pair of incisors in both the upper and the lower jaw. These incisors are used to gnaw through hard plant coverings to reach tender material inside, or for nibbling grasses and shrubs, and they continue to grow through life. Behind the incisors, a long diastema separates the gnawing apparatus from the plant-crushing cheek teeth. A soft fold of cheek skin is puckered inward across the diastema while gnawing and closes off the mouth from flying particles of bark and wood. Lagomorph incisors are completely encased in enamel (except where it wears off at the tips) and there is a second set of small incisors immediately behind the first. The incisors of rodents, however, have enamel only on their anterior surfaces. The enamel is the hardest part of the tooth and wears more slowly than the dentine behind it, producing a self-sharpening chisel edge.

Some ungulates, such as horses and zebras (Figure 21–8n), retain a full complement of upper and lower incisors. Artiodactyls such as cows and deer (Figure 21–8o) usually lack upper incisors and upper canines. They have canines that are similar in shape and function to incisors and have migrated forward to join the incisors in a single functional cutting edge. The lower incisors and incisiform canines

bite against a strongly cornified (calloused) palatal plate.

Carnivores that feed on flesh cut or sheared from the carcasses of prey have developed the fourth upper premolar and first lower molar into a scissors-like pair of shearing blades known as the **carnassial apparatus** (Figure 21–8k to m).

Marine habits have produced some highly specialized forms, such as the fish-trap dentition of the porpoise, *Delphinus*, the toothless plankton-straining right whale, *Eubalaena*, and the mollusk-crushing dentition of the walrus, *Odobenus* (Figure 21–8r to t). Porpoises are toothed whales (Cetacea, Odontoceti). Most toothed whales feed on fishes or squid, and their teeth consist of a long series of nearly identical sharp cones (Figure 21–8r). This dental pattern is convergent with the fish-trap type of dentition already described for gars, ichthyosaurs, and crocodilians. A second type of specialized dentition of cetaceans is represented by the baleen or whalebone whales (Cetacea, Mysticeti) of which the right whale is an example (Figure 21–8s). The teeth of mysticetes have been replaced by sheets of a fibrous, stiff, horn-like epidermal derivative known as baleen which extend downward from the upper jaw. Baleen whales are filter feeders, straining small organisms from the water with their baleen sieves. The ten species of baleen whales include the largest mammal (and the largest vertebrate), the blue whale, which may reach a mass of 160,000 kilograms.

Types of Mammalian Teeth

Mammalian teeth are structures with well-defined growth characteristics. Because the texture and quality of mammalian foods differ enormously, cheek teeth are variously specialized. Generalized mammalian dentitions have brachyodont cheek teeth with rectangular crowns that do not protrude much above the gums. The occlusal (grinding) surface is not smooth, but is marked by one to several sharp cusps that interlock (occlude) when the jaws close. This cusp pattern is referred to as **tubercular** (*tuber* = a knob or hump) and permits both piercing and crushing of the food.

Most teeth have a discrete period of growth during ontogeny, and growth is not resumed if the tooth is subsequently worn or broken. An immature incisor tooth has an open pulp cavity, indicating that it is growing (Figure 21–9a), whereas a mature incisor has a narrow, canal-like pulp cavity that restricts nutrient supply and growth (Figure 21–9b). The incisor teeth of rodents and rabbits have a persistently open pulp cavity, grow throughout life, and have enamel only on the outer face (Figure 21–9c). A typical molar tooth has multiple roots (Figure 21–9d). The low, rounded cusps identify the tooth as bunodont. If the cusps of the cheek teeth fuse into ridges, useful in grinding plant material, a **lophodont** tooth is formed (Figure 21–9c). Two further specializations for herbivory are evident in the dentition of the horse: The crowns stand high above the gums (the **hypsodont** condition, *hyps* = high), and the occlusal surfaces of the cheek teeth show complex ridges composed of dentine adjacent to ridges of enamel and areas of exposed cement. As a hypsodont tooth wears, the various hardnesses of tooth material wear differentially to maintain an uneven grinding surface (Figure 21–9f and g). Similar specializations can be seen in the dentition of cattle (Artiodactyla). In artiodactyls, however, the ridges are all derived from elongate longitudinal cusps and are termed **selenodont** (*selen* = a crescent moon; Figure 21–9h). As these complex teeth wear down, the difference in hardness of the cement, enamel, and dentine produces a rough, self-sharpening surface. Almost all imaginable combinations of cusp pattern and tooth height have evolved as specializations associated with diet (Janis and Fortelius 1988).

Carnivores must be able to shear or cut away chunks of flesh from a large carcass. Once obtained, such pieces are readily digested without much mastication. Hence, the occlusal surfaces of shearing teeth (carnassials) in carnivore dentition are sharp and knife-like and present little grinding surface. These teeth of derived carnivores are called **sectorial** (*sect* = a knife), and they slide past one another like scissor blades to cut flesh.

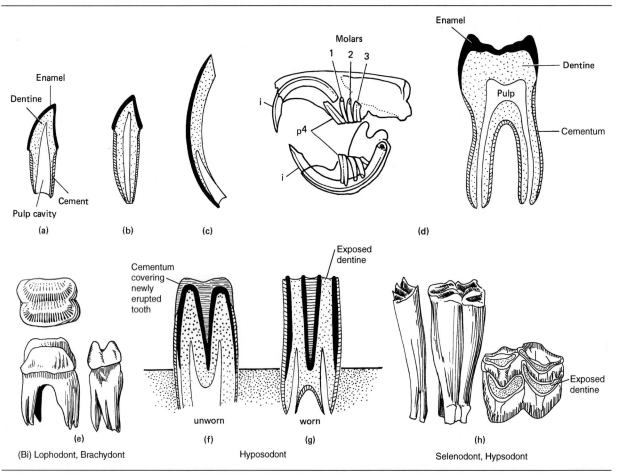

Figure 21–9 The structure of mammalian teeth. See text for details.

Figure 21–10 shows the canine, premolar, and molar teeth of several species of carnivores that have different diets. Canids have diverse diets and only moderately specialized dentition. The anterior cheek teeth of dogs are sectorial, and a carnassial apparatus is present, but well-developed molars with tubercular cusps follow these slicing teeth and allow dogs to crush bones for their marrow. This explains why a dog often holds a bone between its paws and gnaws with the bone far back in the corner of its mouth. When meat is on the bone, the dog does not place the bone as far back in its mouth and uses the carnassials to slice off the meat. In contrast, crushing mastication is of so little importance to cats, which are exclusively flesh-eaters, that only a vestigial first molar occurs in the upper jaw. The carnassial apparatus is the major part of the cheek dentition of cats.

Similar dentitions may be developed independently using different elements. The posterior crushing elements of the badger are the result of enlargement of the first molar, whereas in the mongoose both molars 1 and 2 are retained.

Not all carnivores eat meat, and some are specialized herbivores. Cheek teeth seen in occlusal view show that the premolars are molarized and the molars are large in the omnivorous raccoon, the herbivorous lesser panda, and the bamboo-eating giant panda.

Nontrophic Functions of Teeth

Many wild swine and some domestic breeds have enlarged upper and lower canines. They are used for rooting and digging in the soil for the storage roots

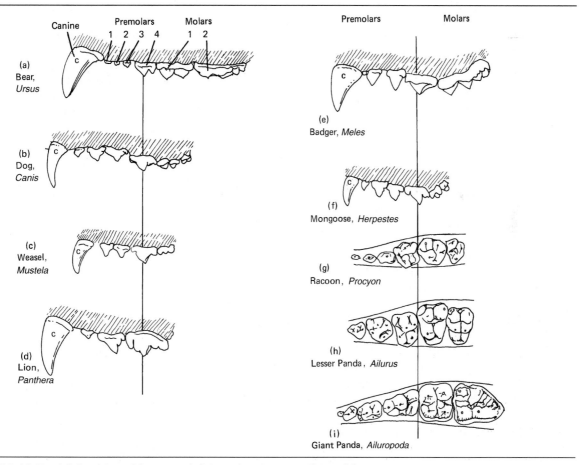

Figure 21–10 Partial dentition of the upper left jaw of various members of the order Carnivora. The vertical lines locate the main cusp of premolar 4, which forms the carnassial apparatus when it is present. (After B. Peyer, 1968, *Comparative Odontology.* University of Chicago Press, Chicago, IL.)

and tubers of plants. In addition, the tusks are larger in the males of many species than in the females. In some species these tusks have lost their digging function and must be considered sexually dimorphic characters, probably connected with aggressive and/or sexual display.

The canines of the walrus (Pinnipedia) are massive (Figure 21–8t). Walruses are marine and crawl out of the sea onto drifting ice or onto the shore using their tusks as levers to lift their massive bodies from the water. Major functions of the tusks may be in social communication as well as ice picks for climbing. Other well-known tusks are those of elephants. Here the incisors of the upper jaw form the tusk, which is pure dentine (ivory). Both sexes of the African elephant have tusks, but female Indian elephants are usually tuskless. Tusks are used in offense and defense, and to hold down branches pulled into reach by the trunk for browsing.

In many primates, the canines are enlarged and are larger in males than in females. By rolling back the very flexible lips, these primates present a fierce aggressive display. It has been suggested that the disproportionately large roots of human canines are a remnant of teeth that once functioned in these displays.

Mammalian Mastication

Teeth, of course, are only part of a mammal's adaptation to a specialized diet. Other skeletomuscular specializations allow them to use teeth effectively in mastication. Many variations in mammalian mastication can be resolved by the study of five pairs of muscles, their sites of origin and insertion, and the articulation between the mandible and the skull (Figure 21–11a). Depression of the lower jaw requires little muscular force because gravity does most of the work. The **digastric** muscles (having two fleshy parts; *di* = two, *gaster* = belly) originate near the back of the skull and insert on the inner ventral borders of each side of the mandible. Each digastric passes posterior to the angle of the jaw, where the muscle constricts to a ligamentous neck, dividing it into the two bellies that give it its name. Contraction of the digastric appears to aid forceful opening of the mouth.

Mastication is a complex activity that involves several pairs of muscles (Figure 21–11a). The **temporalis** originates on the skull roof and inserts on the coronoid process of the mandible. In addition,

another muscle, the **masseter,** originates on the zygomatic arch and inserts on the posterior half of the lateral surface of the mandible. The fibers run obliquely from the arch posteriorly and down to the mandible. The masseter allows the lower jaw to be moved laterally in a grinding rotary path relative to the upper jaw (Figure 21–11c). A final set of muscles of mastication is hidden from superficial view. The **pterygoideus** muscles originate on the base of the skull posterior to the palate; the fibers run laterally and obliquely to insert on the medial surface of the angle of the mandible (Figure 21–11b). Contraction of the muscles of one side pulls the mandible toward the midline. Acting together, they protrude the lower jaw.

The shape of the lower jaw (mandible) and of the mandibular condyle and its fossa, as well as the orientation of the fibers in these masticatory muscles, clearly reflects specialization for different functions. Some mammalian jaws have the condylar processes in or very near the occlusal plane of the teeth (Figure 21–12a). These jaws close like scissors when the strongly developed temporalis and masseter muscles contract. The posterior teeth come into contact before the anterior ones. Other mammals have condylar processes high above the occlusal plane (Figure 21–12b) and the jaws close in a nearly parallel fashion, all the occlusal surfaces approaching each other at about the same time. The former orientation is ideal for slicing and cutting, the latter for crushing, and with the addition of lateral motion, permits grinding along the whole the tooth row.

Because carnivores generally attack, hold, and kill with well-developed canine teeth, it is not surprising that they usually have large temporalis muscles (Table 21–3) attached to a prominent coronoid process. The mandibular condyles of carnivores are cylindrical bars set deeply in fossae that allow no anterior–posterior movement and little lateral shifting of the lower jaws. These features help prevent dislocation of the jaw by struggling prey. In contrast, both the temporalis muscles and the coronoid process are reduced in highly specialized herbivores, but the ramus of the jaw is very deep, to accommodate insertion of the large masseter muscles needed to apply force to the cheek teeth during mastication. The mandibular condyles of herbivores are of various shapes and orientations, permitting anterior–posterior and lateral movement. The human jaw, condyles, and musculature are intermediate between these extremes.

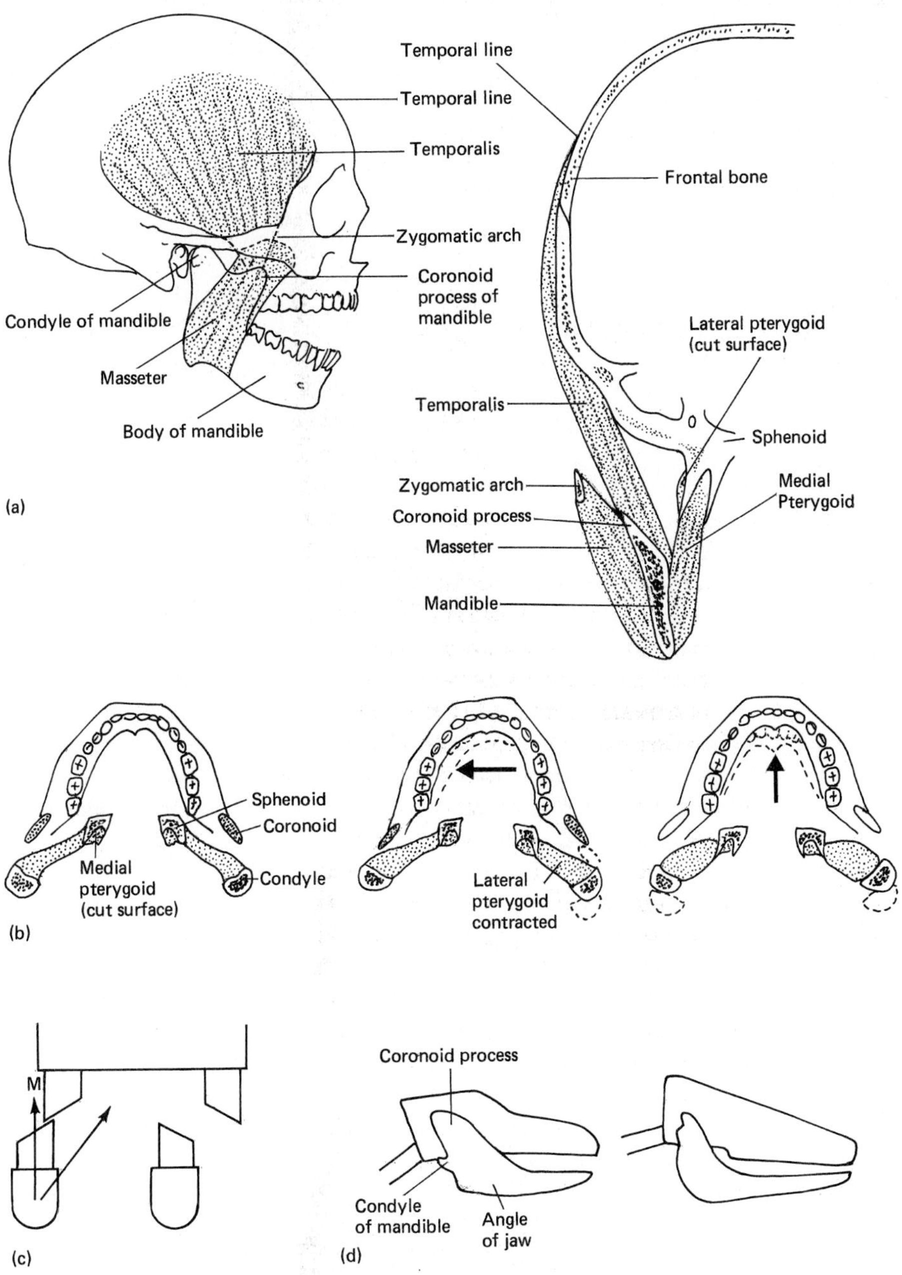

Figure 21–11 Masticatory muscles and jaw movements. See text for details.

Coevolution of Ungulates and Grasses

Grazing herbivores, both invertebrates (primarily insects) and vertebrates, can have a major impact on vegetation. Plants have reacted to these attacks in a multitude of ways, including the evolution of chemical and mechanical defenses. **Allelochemicals** are compounds that are distasteful or toxic to herbivores and are deposited by plants in otherwise edible tissues. Mechanical defenses include thorns, spines, and structures in the cell walls. Fossil evidence demonstrates the antiquity of the interactions of climate, the resulting vegetation, and its defenses, and the responses of plant eating mammals to these changing situations (Chapter 20; also see Collinson and Hooker 1991).

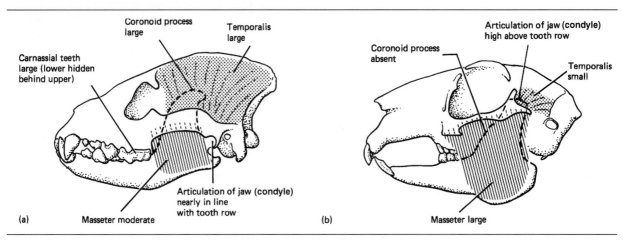

Figure 21–12 The shape of the jaws is influential in producing the differing jaw actions of carnivore (a) and herbivore (b).

The mechanical defenses of grasses provide an excellent illustration of plant–animal coevolution. The 700 genera and perhaps 9000 species of grasses are worldwide in occurrence and are considered the height of flowering plant evolution. The first undoubted fossil grasses are from the early Tertiary. Several features of grasses evolved simultaneously with the evolution and radiation of herbivorous mammals, and many descendants of this herbivore radiation today depend on grasses.

Reconstruction of ancestral characters suggests that the first grasses were low-growing tufts. Instead of growing from the tip of the stem as other plants do, grasses grow from the base of the leaf blade. A grass, therefore, elevates the oldest, nongrowing part of its foliage, and this can be removed without stop-ping growth. Most grazing mammals cannot graze close enough to the ground to remove all the leaf tissue, yet from the earliest radiations of herbivores the muzzle has shown elongation permitting deeper and deeper cropping of plant tufts (Figure 21–13). Grasses have repeatedly evolved nearly stemless growth forms with growing portions very close to the ground, out of reach of even the most specialized grazers.

Plant cells are encased in a rigid cell wall. To obtain the cell protoplasm, an herbivore must rupture this cell wall and expose the cell contents. Grasses have made this as unrewarding as possible by incorporating crystalline silica into the fibrous cell wall. Silica is a hard mineral, and chewing enough of it will grind teeth to useless, flat nubbins. It is probably a defense against both vertebrate and invertebrate herbivores (Vicari and Bazely 1993, Hochuli 1993).

Cellulose and lignin are common constituents of all plant cell walls. Cellulose is a complex carbohydrate closely related to starch, yet it cannot be digested by any multicellular animal. Grasses often have greatly elongated cells and thickened cell walls, making the amount of directly digestible foodstuffs a small proportion of the total leaf. It seems improbable that vertebrates should evolve complex specializations to enable them to feed on plants with such tooth-destroying texture and low nutritional value, yet the Artiodactyla, Perissodactyla, Sirenia, Hyracoidea, Proboscidea, Rodentia, Lagomorpha, Primates, and the Marsupialia contain many species that eat grasses. Why should this be so?

Table 21–3 **Weights of the jaw-closing muscles of some mammals (weight of each muscle as a percentage of the total).**

	Temporalis	Masseter	Pterygoideus
Carnivores			
Tiger	48	45	7
Bear	64	30	7
Dog	67	23	10
Herbivores			
Zebra	10	50	40
European bison	10	60	30
Horse	11	57.5	31.5

Source: R. McN. Alexander, 1968, *Animal Mechanics,* University of Washington Press, Seattle, WA.

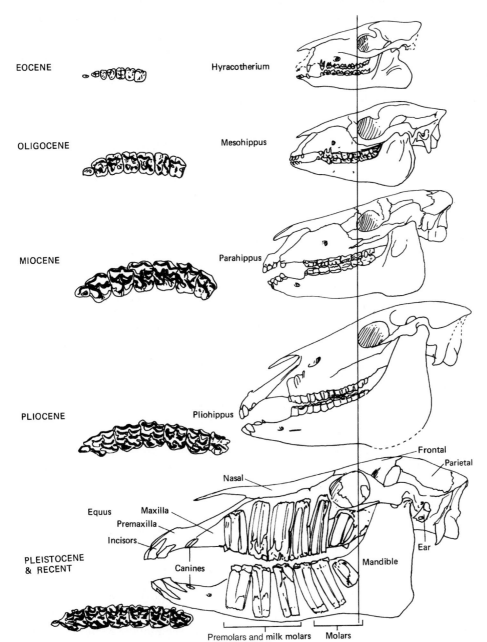

EOCENE   Hyracotherium

OLIGOCENE   Mesohippus

MIOCENE   Parahippus

PLIOCENE   Pliohippus

Frontal

Parietal

Nasal

Equus   Maxilla

Premaxilla

Incisors

Ear

PLEISTOCENE
& RECENT

Mandible

Canines

Premolars and milk molars   Molars

Figure 21–13 Development of hypsodont dentition in horse-like perissodactyls. Note that this is not a phylogenetic lineage, but it illustrates changes that took place in a number of lineages that became specialized to feed on abrasive vegetation during the Miocene and Pliocene. The molars became broader and flatter, increasing their total surface area, and the premolars became identical in form to the molars. The pattern of enamel, dentine, and cement on the occlusal surfaces of the teeth became increasingly complex, producing a self-sharpening grinding surface. The teeth became increasingly hyposodont (long in the root to crown dimension) with long-persisting open roots that allowed continued growth to replace wear. The skull of the modern horse *Equus* is dissected around the base of the teeth to show the extreme hyposodonty. (Modified from B. Peyer, 1968, *Comparative Odontology*, University of Chicago Press, Chicago, IL.)

The whole of the Mesozoic was warm by present climatic standards, and the climate became increasingly moist as the period progressed. During the Cretaceous, ferns and cycads similar to those now living in subtropical South America were distributed from the paleo-equator to nearly 90°N and S latitude. The late Cretaceous was a period of gradual withdrawal of the shallow seas from much of the land and a decided cooling and drying of the Earth's climate. Except for a brief period of warming between the Paleocene and Eocene, this cooling and increased seasonality has continued for 70 million years. The angiosperms, present since the Jurassic, radiated and were followed in short order by the radiation of mammals. By the end of the Miocene, even angiosperm forests had been somewhat diminished, and savannas followed by treeless grasslands began their global spread. Today, as they have since the Pliocene, grasses dominate the worldwide terrestrial flora (Janis 1993). Survival of some phylogenetic lineages may have depended on grasses as the predominant food source despite the relative difficulty in eating and absorbing them.

Specializations of the Digestive System of Herbivorous Mammals

Grazing mammals have teeth with large, rough occlusal surfaces that can rupture the thick cell walls of leaves and grasses. Additional specializations include an overall increase in the size of the teeth, increased molarization of the premolars, and development of complex folds of enamel, dentine, and cement that result in all three substances being exposed on the occlusal surface simultaneously. To deal with the exceedingly abrasive silica, the cheek teeth became elongate, permitting them much longer service before they were worn away. Ultimately, the cheek teeth of various lineages of herbivorous mammals evolved either of two specializations: In some herbivores, tooth eruption patterns allow a basically diphyodont mammal to be functionally polyphyodont (having several successive replacement teeth). Examples are the serially replacing teeth of manatees and one rock wallaby, the narbalek. As the teeth of these mammals wear down they migrate forward along the jaws and finally are shed. New teeth erupt posteriorly to replace them. Elephants have a different trick up their sleeve. While they have a total complement of only six teeth in each jaw (three deciduous premolars and three replacement molars), each tooth is large enough to fill the entire jaw space. As a tooth wears down, another one moves in from the back to replace it. When the last (sixth) tooth has been worn down, no extra ones can arise. Some small herbivores (lagomorphs and rodents) have continuously growing cheek teeth that renew themselves from below as they are worn away from above.

Several clades of mammals have independently evolved a complex forestomach that allows symbiotic microorganisms to convert the cellulose and lignin of plant cell walls into digestible nutrients. The most derived stomach anatomy of mammals is

the **ruminant** digestive system, and the animals that possess it (camels, giraffes, antelope, cattle, sheep, goats, and deer) are called ruminants or cud-chewing mammals. Similar but less complex specializations of the stomach are found in a variety of other animals, including kangaroos, colobus, langur, and proboscis monkeys, some rodents, and hippos, but these forms do not chew the cud. In contrast, other mammals (perissodactyls, howler monkeys, hoalas, wombats, elephnats, some rodents, and lagomorphs) have specialized hindguts (caecum and/or colon) that provide space where gut microbial symbionts can ferment plant compounds to digestible end products (Box 21–1).

Cursorial Specializations of Ungulates and Their Predators

Grasses flourish in open habitats that place a premium on efficient, rapid locomotion over generally flat terrain. With nowhere to hide, most large grazing animals run to escape predators, and many species migrate long distances to find grass at the proper stage of growth. Such locomotion is termed **cursorial** and is characteristic of most large herbivores, represented today primarily by the ungulates (Perissodactyla and Artiodactyla).

The ancestral foot posture of mammals is the **plantigrade** type (Figure 21–15), in which the sole of the foot rests flat on the ground and the entire foot skeleton supports the weight of the body. In the case of the hind foot illustrated, support extends from the calcaneus to the terminal phalanges. Several mammals, especially stealthy or moderately cursorial (running) carnivores, support their weight only on the phalanges. This **digitigrade** posture reduces friction and increases the length or reach of the stride by an amount equal to the length of the vertical metapodial and the mesopodial elements.

Efficiency in cursorial locomotion is promoted by a long stride length, and speed is further enhanced by increasing the number of strides per unit of time. The ultimate in cursorial adaptations is the **unguligrade** posture achieved independently by numerous different ungulate clades. Here the weight of the body is supported entirely by the hoof-clad terminal phalanx or phalanges, and the effective stride length includes the contribution of the second and third phalanges. Increased efficiency is fostered by reducing the distal mass, and therefore the inertia

Box 21–1 Herbivores, Microbes, and the Ecology of Digestion

Plant food is much more abundant than animal food, and it does not run away, but the energy content of plant material is generally lower than that of animal tissues. The protein content of leaves and stems is usually low, and the protein is enclosed by a tough carbohydrate cell wall. Although specialized teeth can rupture the leaves and expose the cells, only enzymes (cellulases) can break through the cellulose protecting the cytoplasm. However, no multicellular animal has the ability to synthesize cellulases. Thus efficient use of plants as food requires indirect attack on the cell walls by enzymes produced by microorganisms that live as symbionts in the guts of herbivorous animals. The complex gut morphologies characteristic of herbivorous vertebrates provide fermentation chambers that promote the action of these microorganisms.

The many separate evolutions of fermentative digestion among vertebrates have resulted in distinctly different solutions to the problems posed by plants as food (Janis 1976). Two different fermentative systems have evolved among ungulates. Horses and other perissodactyls are examples of **monogastric, (= caecalid = hindgut)** fermenters. These have a simple stomach and an enormous caecum—a closed-end sac at the junction of the small and large intestines (Figure 21–14, left). Other ungulate hindgut fermenters include rhinoceroses, tapirs, and elephants. Hindgut fermentation is also known among birds, lizards, turtles, and fishes. Even omnivorous and carnivorous mammals such as humans and dogs obtain measurable amounts of nutrients from hindgut fermentation.

Cows are examples of **ruminant (= foregut)** fermenters, in which the nonabsorptive forestomach is divided into three chambers that store and process food, followed by a fourth chamber in which digestion occurs (Figure 21–14, right). Some other ruminants have only three-chambered stomachs, the forestomach having but two divisions.

In typical mammalian fashion, hindgut fermenters chew their food as they eat and digestion is initiated by enzymes in the saliva and continued when the food reaches the stomach, which is acid. The stomach is not disproportionately large, and the partly digested mass of food moves into the small intestine rather rapidly as new food is eaten. Absorption of nutrients occurs in the small intestine. At the end of the small intestine, finely ground particles of food (the small solids portion of the chyme—Chapter 3) pass into the caecum. Larger food particles move through the large intestine and are passed as feces. In the caecum microorganisms attack the intact plant cell walls, releasing nutrients and forming volatile fatty acids. The products of this fermentation and the symbiotic microorganisms growing on them pass into the large intestine and some of the material is absorbed across the intestinal wall.

Foregut fermenters (ruminants) swallow partly chewed food into the first chamber of the forestomach, the **rumen,** which is enormously enlarged. Here the food is moistened and churned, mixing thoroughly with the symbiotic microorganisms that live in the rumen. Large particles of food float on top of the rumen fluid and are passed to the **reticulum,** a blind-end sac with honeycomb partitions in its walls. Here small masses (**cuds**) of moist plant material are formed. These cuds are regurgitated when the animal is at rest, chewed again, and swallowed into the rumen. Only after material is finely ground does it pass from the rumen through the **omasum** and into the acid stomach (**abomasum**), where it is processed by the usual digestive enzymes of vertebrates. These enzymes act on the plant material, on the products of fermentation from the rumen, and on the accumulation of microorganisms themselves, rupturing their cell walls and releasing the protein and carbohydrate they contain. (This makes the use of the term symbiosis to describe the ruminant/rumen protist relationship questionable.) From

Continued

Box 21–1 *(Continued)*

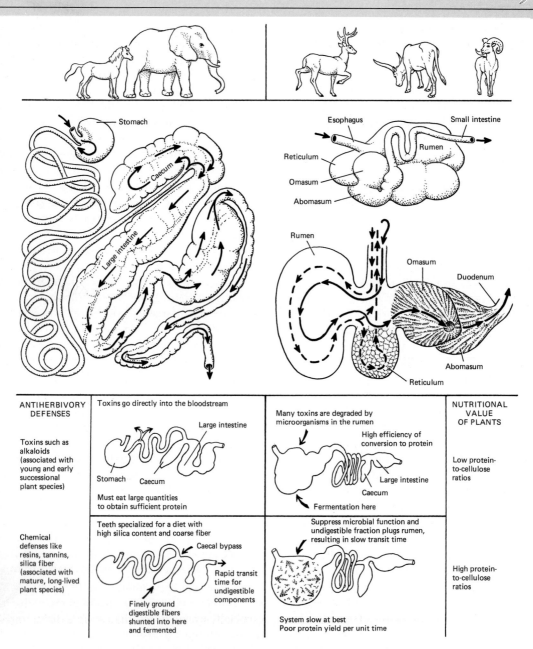

Figure 21–14 Monogastric and digastric digestive systems. (Left) The monogastric system: Fermentation occurs in the enlarged caecum and large intestine. (Right) The digastric (ruminant) system: Food passes slowly through the four chambers of the stomach of a ruminant.

the acid stomach the chyme passes into the small intestine, where the products of both microbial digestion and acid digestion are absorbed.

Distinct advantages and disadvantages are associated with each of the two kinds of fermentative digestion. Foregut fermentation can be extremely efficient because the microorganisms attack the plant material *before* it reaches the small intestine, where most absorption takes place. In contrast, the food has already passed the small intestine of a hindgut fermenter before it is mixed with microorganisms in the caecum. Furthermore, the microorganisms from the rumen are digested in the acid last chamber of the stomach, and the material that moves into the intestine contains all the substances the microorganisms have synthesized as well as the material released from the plants by fermentation. A hindgut fermenter does not digest the microorganisms expelled from the caecum, so this potential source of energy and nutrients is not exploited. An additional advantage of foregut fermentation is the role microorganisms in the rumen play in detoxifying chemical compounds that would be harmful to a vertebrate. A hindgut fermenter receives no such benefit and must absorb allelochemicals from plants into its bloodstream and transport them to its liver for detoxification.

On the other hand, a hindgut fermentation system processes material rapidly. Food moves through the gut of a horse in 30 to 45 hours, compared to 70 to 100 hours for a cow. Hindgut fermentation works well with food that has relatively high concentrations of protein, because a large volume of food can be processed rapidly. The system is not efficient at extracting energy from the food, but by processing a large volume of food rapidly a horse can obtain a large quantity of energy in a short time. In addition, a hindgut system is effective when the food contains a large proportion of indigestible material, such as silica, resins, or tannins, because these compounds pass quickly through the gut without entering the caecum. In contrast, a foregut system is slow because food cannot pass out of the rumen until it has been ground into very fine particles. Ruminants do not do well on diets that contain high levels of tannins or resins because these compounds suppress microbial function in the rumen, and plants with high silica contents break down so slowly that they impede the movement of food out of the rumen.

These differences in digestive physiology are reflected in the ecology of foregut and hindgut fermenters. Hindgut fermenters can survive on very low-quality food such as straw as long as it is available in large quantities. Consequently, perissodactyls can survive on dry pastures in regions of seasonal drought, whereas ruminants cannot process low-quality food fast enough to subsist.

Some rodents and lagomorphs with well-developed caecal fermentation produce feces from which relatively little of the products of fermentation has been absorbed. These animals regularly eat some of the feces they deposit in their burrows (**coprophagy**), thereby retrieving the products of microbial fermentation as well as the protein and carbohydrates synthesized by the microbes. Lagomorphs produce special moist fecal pellets at night which they subsequently eat. Young koalas feed for an extended period on the feces of their mother. The end result of all gut fermentation is the gleaning of more protein and energy from plant material, especially from grasses, than simple direct digestion would produce.

of the limbs, by concentrating muscles in the upper limbs and using tendons to transmit muscular forces to the slim lower limbs. The muscles of the upper limb insert very close to the joint over which they act, increasing their effective speed at the cost of reduced force. Elastic ligaments running across the joints of the limbs are stretched when those joints are bent by the weight of the body and they snap back when the weight is released, straightening the joint and returning much of the energy stored in them. Behavioral characteristics also aid long-distance travel. The stride length of the gait used in steady travel coincides with the natural oscillation period of a pendulum with dimensions similar to those of the limb. Thus, little additional energy is spent in reversing the motion of the limb at the extremes of its excursion—gravity and elastic recoil do much of the work.

Such extensive locomotor specializations have obviously had an effect on the carnivores that prey on grassland herbivores. Three families of the Carnivora have members that are specialized to prey on large grazing animals: Hyaenidae (hyaenas), Canidae (dogs and wolves), and Felidae (cats). Both hyenas and dogs hunt in packs and use stamina and

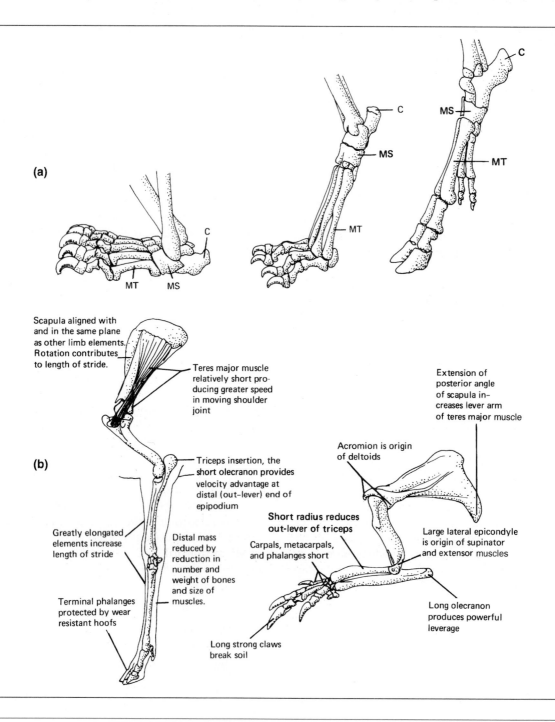

(a)

(b)

Scapula aligned with and in the same plane as other limb elements. Rotation contributes to length of stride.

Teres major muscle relatively short producing greater speed in moving shoulder joint

Extension of posterior angle of scapula increases lever arm of teres major muscle

Acromion is origin of deltoids

Triceps insertion, the short olecranon provides velocity advantage at distal (out-lever) end of epipodium

Short radius reduces out-lever of triceps

Large lateral epicondyle is origin of supinator and extensor muscles

Greatly elongated elements increase length of stride

Distal mass reduced by reduction in number and weight of bones and size of muscles.

Carpals, metacarpals, and phalanges short

Terminal phalanges protected by wear resistant hoofs

Long strong claws break soil

Long olecranon produces powerful leverage

cooperation to exhaust and kill large prey. Many of the specializations for speed and efficiency seen in ungulates have analogs in cursorial carnivores, especially the canids. Few felids run down their prey. Cats stalk by vision and when close to their prey (usually less than 100 meters) they attack with a brief burst of speed followed by a spring. Only the lion hunts in groups, and its social system is accordingly complex. The cheetah is very different from other cats and is distinct from all other mammals in being a solitary, cursorial predator. Although it rarely runs for distances greater than 600 meters, it is the fastest terrestrial animal, having been clocked at 112 kilometers per hour.

Evolution of the Mammalian Nervous and Sensory Systems

The basic structure of vertebrate nervous systems and the regions of the generalized vertebrate brain, all of which are retained in mammals, have been reviewed (Chapter 3). Two characteristics of the mammalian brain that set it apart from those of other vertebrates are (1) the evolution of a superficial mat of gray (nuclear) matter from the ancestral condition in which the nuclei were buried within the white matter, and (2) the specialization of that coat of gray matter in the telencephalon of mammals to form the **neopallium.**

Thus the mammalian brain is enlarged primarily by the evolution of extensive layers of neuron cell bodies on the surface of the ancestral brain stem. These new layers are especially thick over the telencephalon of the anteriormost part of the brain. This trend is evident from endocasts of Mesozoic mammals. These small nocturnal creatures evolved from

lineages whose nervous systems have left convincing evidence of diurnal activity with a dependence on bright light vision (Jerison 1973). Which of the five classic senses the early mammals inherited— sight, smell, touch, taste, and hearing—might best have replaced the bright light vision so important as a distance sense to the synapsid and therapsid ancestors of mammals?

The Distance Senses of Early Mammals

One might first suppose that evolution would have reworked the visual system given the extensive correlates of vision, brain, and behavior already inherent as a distance sense in the mammalian genotype. However, there is one major limitation to the information-rich visual system that mammals inherited. Cones have relatively low photosensitivity; they are at least two orders of magnitude less sensitive than rods. An eye adapted for acute vision required the brightness of daylight to function effectively, but early mammals are believed to have been nocturnal. Thus nocturnal mammal-like tetrapods would have required different visual systems, or even a different sensory modality than their synapsid ancestors to obtain and process information from a distance.

Vision The most obvious adaptations of an early mammal to the new sensory demands of a life at night would be those of a nocturnal eye, that is, one with a pure rod retina. Extant members of the Insectivora seem to support this hypothesis, and many otherwise specialized mammals also have a predominantly nocturnal eye. However, the quality of image formed by the eyes of nocturnal vertebrates is poor. If these eyes are representative of those of early

Figure 21–15 Contrasting specializations of the limbs of mammals. (a, left to right) Plantigrade, digitigrade, and unguligrade postures. C, calcaneus; MS, mesopodial elements; MT, metapodial elements. (b) The skeletal elements of the left forelimb of a deer (cursorial) and the armadillo (fossorial) compared. In many ways, specializations of fossorial (digging) limbs are the opposite of those for efficient cursorial limbs. The length of reach when digging (equivalent to length of stride) is usually unimportant so that the limbs are generally short. Power, however, is very important. Thus the muscles and their bony attachments are enlarged and generally insert far from the joint over which they act, enhancing their leverage. The distal mass of the limb is often increased by broadening and flattening the bony elements. (From various sources, including M. Hildebrand, 1988, *Analysis of Vertebrate Structure*, 3d edition, Wiley, New York, NY.)

mammals, the usefulness of vision as a distance sense would not have been great.

Acute vision does occur in several mammalian lineages, and it is clear that many independent alterations of the ancestral mammalian eye have occurred. The overwhelming majority of mammals appears to have retinas that are dominated by rods, although rudimentary color discrimination is widespread (Jacobs 1993). Primates are the only mammals with well-developed trichromatic color discrimination comparable to that found in other vertebrates. This kind of color vision is produced by three visual pigments that have different absorption spectra and, consequently, are sensitive to different wavelengths of light. A trichromatic retina is not the only way to see color, however. Certain diurnal mammals like squirrels appear to have good dichromatic color vision, which is based on cones with two visual pigments. The cones of mammals show such distinct differences from the cones of other vertebrates that they probably are not homologous structures. Mammalian cones probably evolved from pure rod retinas during the first major postarchosaurian radiation of mammals. Only recently, during the Cenozoic radiations of diurnal mammals, have retinal systems similar to those of diapsids evolved in some mammals. The vast majority of mammals have rather poor visual acuity. Elephants cannot distinguish two objects separated by less than 10 minutes and 15 seconds of arc on the retina. Chimpanzees, however, can distinguish points separated by only 26 seconds of arc. The evidence indicates that early mammals probably had a visual system with low acuity that was not appropriate for a primary distance sense.

Touch Mechanoreception of their immediate surrounding environment may well have been important to early mammals. Pits in the rostral region of fossil skulls suggest that some synapsids might have had vibrissae that detected obstacles near the face. However, the distance sensitivity of vibrissae is limited.

Taste and Smell Taste can be discounted as contributing significantly to the expansion of the mammalian brain. It probably was the major sensory modality involved in the coevolution of chemical defenses of plants and vertebrate herbivores. Nevertheless, it requires direct application of the taste buds to the environment and this greatly limits its value as a distance sense.

Olfaction, on the other hand, detects substances borne by air currents. Thus, unlike touch and taste stimuli, olfactory stimuli can originate at a distance. Olfactory stimuli are not blocked by most natural obstacles, and olfaction functions in daylight and at night. A very small amount of stimulant is often detectable. Endocasts of early mammal brains indicate the presence of enlarged olfactory regions. However, olfactory information is not transmitted rapidly, and the direction that scent particles travel depends on the vagaries of air currents. An additional problem with olfaction as a primary distance sense is its poor resolution of spatial information about the source. Although olfaction has great significance to vertebrates, especially to mammals, it is a rather poor candidate for a rapid nocturnal distance sense to replace diurnal vision.

Hearing Audition has several advantages as a distance sense compared to vision and olfaction. Sound is not readily blocked by obstacles in the environment as is light and it is transmitted more directionally and much faster than odors. The mammalian ear shows the importance of audition. The phylogenetically oldest portion of the auditory system (the inner ear) is derived and elaborate in even the earliest mammals as are the newer structures (the tympanic cavity and middle ear). The outer portions (the pinna and auditory tube) are unique to mammals. The combined effect of these specialized structures is exceptional frequency discrimination, broad sensitivity to various intensities of sound, and precision of directionality. Cranial casts of early mammals show that auditory areas of the brain were large.

The Evolution of Mammalian Brain Size

Why did the earliest mammals have brain volumes four or five times those of other tetrapods and why do extant mammals have brain volumes four to five times those of the earliest mammals? Active organisms require a three-dimensional method of perception of the world around them. In diapsids, which are predominantly diurnal, vision provides this information. We have pointed out, however, that vision would not have been an adequate sense for early mammals, which were nocturnal. Smell and hearing appear to have been more likely senses for obtaining information about the world, and casts made of the inside of brain cases of early mammals indicate that the regions associated with those sensory modalities were indeed enlarged. Three-dimensional information about the world must be extracted from scent or sound largely by comparing the strength of a stimulus from one moment to the next, or as the head is turned from one side to the other. This means having the ability to monitor incoming information, store it, and then to compare it with subsequent information. Evaluations of positions in three-dimensional space are based on the differences in signals received through time. Thus, processing information about the world would have required integrative ability in the central nervous system (CNS) of an early mammal.

This integrative ability might have had selective value in its own right. It should permit more complex and particularly more flexible behavior patterns. The ability to make associations between past events and a present situation is the basis of learning. Selection for this sort of integrative ability might be the force that produced a second four- to fivefold increase in mammalian average relative brain size that occurred during the Cenozoic. Most archaic mammals had brains smaller than those of Recent mammals. As mammals evolved, the average brain size increased as did the range of relative brain sizes (Figure 21–16).

Integrative and associative capacity appears to be best developed among humans, although it would be rash to underestimate the complexity of the neural capacities of other mammals. Successes in communicating with chimpanzees and gorillas and studies of toothed cetaceans suggest that an integrative and associative capacity is not uniquely human.

The features that we proudly consider attributes of human intelligence may be approached more closely by other vertebrates than we realize because their features, like ours, evolved through processes that affected all mammals. Nobel laureate Konrad Lorenz (1977) said: "I . . . consider human understanding in the same way as any other phylogenetically evolved function which serves the purposes of survival, that is, as a function of a natural physical system interacting with a physical external world." In Chapter 24 we examine more closely the precise physical external world of the ancestors of humans to gain insight into the evolution of our species.

Specialization of the Auditory System: Echolocation

Many mammals are more sensitive than humans to one or another sensory modality: The olfactory sensitivity of dogs is probably the most familiar example of the sensory capacity of a mammal that exceeds our own. Equally impressive is the use that bats and cetaceans have made of hearing as a distance sense for navigation. An examination of echolocation illustrates the way in which a sensory capacity that is ancestral for mammals can be elaborated, and how environment and phylogeny interact.

Several derived mammals emit sounds above 20 kilohertz (20,000 cycles per second), called **ultrasound** because it is above the range of normal human sensitivity (the 10 octaves between 20 hertz and 20 kilohertz). Elephants emit sounds below this range (**infrasound**) in connection with group movements and reproductive receptivity. When infant

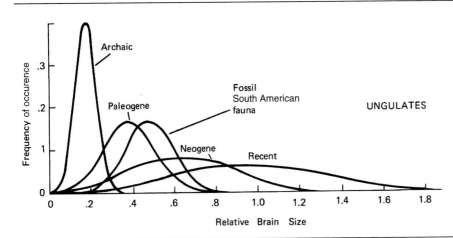

Figure 21–16 Changes in the relative brain size of ungulates of the Northern Hemisphere and South America during the Cenozoic. The Paleogene includes species from the Paleocene through the Oligocene epochs, the Neogene species from the Miocene through the Pleistocene epochs. (After H. J. Jerison, 1973, *Evolution of the Brain and Intelligence*, Academic, New York, NY. Also see L. B. Radinsky, 1978, *American Naturalist*, 112:815–831.)

rodents stray outside their nest, they emit ultra-sounds that stimulate adults to retrieve them. Some adult rodents, a few marsupials, dermopterans, pinnipeds, many insectivores, microchiropteran bats, and odontocete cetaceans emit such sounds throughout their life as part of a sound-based distance sensing system of **echolocation.** The most thoroughly studied echolocating mammals are the microchiropteran bats and the toothed cetaceans.

Bat Echolocation Because of their nocturnal and secretive habits, bats have until recently been little understood. How do bats avoid obstacles under conditions where vision can be of little or no use? Between 1793 and 1799, Lazzaro Spallanzani, an Italian naturalist and physiologist, elegantly tested the auditory contribution to obstacle avoidance of bats:

> I had two slender conical tubes soldered from very thin brass plate and introduced them with their thinner end in front into the ears and auditory openings of a bat. The tubes were externally covered with pitch, which served to fill up the space between tube and the deep concavity of the external ear and to attach the tube to the ear. In this way the air had no passage to the internal ear other than through the tubes. The animal (which could see) showed no influence of this impediment during its flight. But when I closed the tubes with pitch so that the air could no longer enter the auditory duct, the animal did not fly at all, or its flights were short and uncertain, and it frequently fell. This experiment which is so decisively in favor of hearing has been repeated with equal results both in blinded bats and in seeing ones.

Spallanzani also captured bats from the belfry of the cathedral at Pavia, removed their eyes, released them, and 4 days later captured some of these bats at the same roost. These blinded bats were as full of insects as any sighted bat, indicating that blinded bats could use some sense other than vision to capture insects on the wing and return to a home roost. Spallanzani thought that they heard the buzz of the insects' wings and that obstacle avoidance was based on a bat's ability to hear the sounds of its wings and body as they were reflected from objects. At the time of Spallanzani's studies there was a general ignorance of the properties of sound; that sound existed below and above the range of human audition was alien to scientific thought. Unfortunately this surprisingly modern, experimentally tested hypothesis was eclipsed by the scientific politics of the day. The eminent and influential French anatomist Baron

Georges Cuvier derided both the experiments and conclusions of the Italian without providing any new experimental evidence.

At the beginning of the twentieth century, several researchers, often unaware of Spallanzani's work, independently concluded from experiments that ears were important for bat navigation. In 1938, G. W. Pierce and D. R. Griffin, with newly developed high-frequency acoustic detectors, reported that bats emit intense ultrasonic sound with a spitting motion. In the following years Griffin and R. Galambos at Harvard and S. Dijkgraaf at Utrecht independently discovered that flying bats produce ultrasonic cries that echo back from obstacles and that bats have ears that are sensitive to these sound frequencies. They also rediscovered Spallanzani and, much to their credit, brought his 150-year-old work to the attention of the twentieth century (Dijkgraaf 1960). Subsequently, Griffin and Galambos showed that basically the same echolocation behavior is used by bats for navigation and for capturing food (Griffin et al. 1960).

The weight gained by bats that had been allowed to feed for a known period of time in a swarm of tiny insects showed that they must average as many as 10 mosquito or 14 fruit fly captures per minute. High-speed photography has shown two separate catches in 0.5 second. By use of electronic transducers capable of detection of high-frequency sound, a three-phase hunting pattern has been defined (Figure 21–17).

The initial search phase of the little brown bat (*Myotis lucifugus*) is characterized by fairly straight flight and the emission of ten or so pulsed sounds separated by silent periods of more than 50 milliseconds. Each of the 10 pulses in a call is about 2 milliseconds in duration, and each pulse constitutes a downward sweep of frequencies starting at about 85 kilohertz and ending near 35 kilohertz. These bat calls are therefore frequency modulated (FM). Other bats use different pulse lengths and frequencies that vary from family to family of bats; some produce constant frequency calls that either terminate in a short downward FM sweep or, like the calls of some FM species, include several simultaneous harmonics. Despite our inability to hear their high-frequency cries, many bats produce sounds of extraordinary loudness. Sound intensities higher than 200 dynes per square centimeter (as loud as a jet airplane) have been measured 5 centimeters from the mouths of some bats.

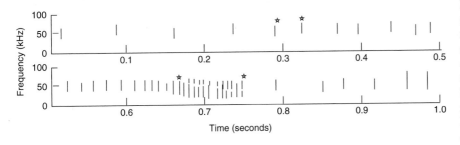

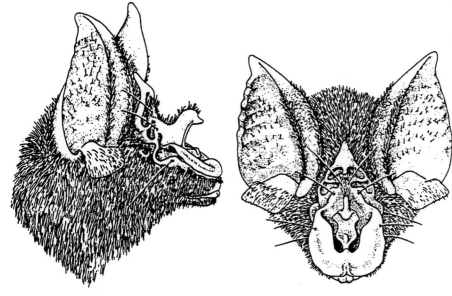

Figure 21–17 Bat echolocation sounds and modifications of the nose to broadcast them. The sound spectrograph analysis is of the frequency-modulated (FM) pulses emitted by the little brown bat, *Myotis lucifugus*, during an interception maneuver. Frequency in kilohertz is plotted against time during the continuous 1-second record. Filled stars indicate typical loud pulses near the time of detection of the target; open stars indicate the onset and completion of the terminal buzz just before capture of prey. (After M. S. Gordon et al., 1982, *Animal Physiology, Principles and Adaptations*, 4th edition, Macmillan, New York, NY.)

The second phase of the hunting pattern begins as a bat detects an insect. Fruit flies and mosquitoes can be detected from a distance of about a meter. The interval between pulses shortens, the silent intervals falling to less than 10 milliseconds for *M. lucifugus*. One hundred cries per second, each lasting only 0.5 to 1 millisecond, are typical as the bat alters its flight path to intercept its prey.

Finally, the terminal phase of the hunt is characterized by a buzz-like emission of ultrasound. The intervals between pulses are less than 10 milliseconds, the pulse duration is about 0.5 millisecond, and the frequencies drop to 25 to 30 kilohertz. The exact details vary from species to species, but the general behaviors are consistent. When the bat is within a few millimeters of the prey, it often scoops with a wing or with the membrane between its legs and pulls the insect toward the mouth.

To accomplish these amazing feats, bats must hear and recognize high-frequency echoes bounced off the bodies of insects, determine target direction and distance with great accuracy, and be able to orient, approach, and capture the insect even though it may be moving. How do they do it?

The bat larynx is typically mammalian. Only in mammals is the larynx an organ capable of producing sounds of complex frequency and temporal modulation and variation. (Birds use a different anatomical mechanism.) Situated just posterior to the hyoid bone, the larynx is a box of cartilages and muscles encircling the esophageal end of the trachea. All inspired and expired air passes through its confines. The central opening of the larynx can be closed by one or two paired folds of tissue that oppose each other across the tracheal opening. The deepest of these folds are the vocal cords; they have thickened edges and considerable associated musculature. Sounds are produced by forcing air through the slit between these tensed sheets of tissue. Modulation of frequency and loudness is accomplished by alteration of the tension on the vocal cords and entire larynx, the amount of air expelled per unit time through the structure, and sometimes by extralaryngeal resonating chambers. Bats produce echolocation sounds without major modification of this basic mammalian larynx, although the whole structure is enlarged and the fleshy folds of the vocal cords are exceptionally thin. The production of such brief, rapidly repeated and precisely patterned sounds seems a Gargantuan task for the tiny muscles controlling the vocal cords, but large muscles would contract and relax too slowly to produce rapid pulses.

The sounds produced by the larynx are emitted through the open mouth or the nose, depending on the family of bat. Mouth calls have a wide angle of dispersion (180 degrees or more). The noses of those bats that use them to broadcast are complex structures (Figure 21–17). They have epidermal flaps and a nostril spacing that concentrates and focuses the sound in a narrow cone (less than 90 degrees) in front of the bat, much like a megaphone. The calls travel through the air in radially expanding waves at about 34 centimeters per millisecond. Because of the dispersion pattern, the amount of sound energy striking a target decreases as the square of the distance traveled. A small object intercepts very little sound energy and thus can reflect very little. As the echo is reflected back toward the bat, its energy continues to diminish as the square of the distance.

Thus, the returned sound—despite its initial loudness—is exceedingly faint. In addition, only those wavelengths in the emitted call that are approximately equal to or shorter than the diameter of the reflecting object will be returned. Despite these problems, bats can detect and locate remarkably small objects. Little brown bats can detect wires 1 millimeter in diameter from a distance of 2 meters and wires only 0.08 millimeter in diameter from shorter distances. Even irregularities on surfaces can be located. Fish-eating bats apparently locate fish by detecting ripples on the water surface, and bats use echolocation to find the cracks in rocks where they cling while they sleep.

Many features of an echo convey information. An object's size is indicated by the frequencies in the echo; large objects reflect longer wavelengths (lower frequencies) than small ones. The extremely high fre-

quencies of the emitted calls are necessary to detect very small objects. The character of the reflecting surface is indicated by the character of the echo. A smooth, hard surface such as the exoskeleton of a beetle returns a sharp echo, whereas a blurred echo indicates a rough surface like the body of a moth. The time required for an echo to return is directly proportional to the distance from the bat to the target, and the change in return time between successive calls can indicate the relative movement of the bat and its target. As a bat approaches a target, the call repetition rate increases, giving the bat more and more precise information about the target's location.

Several features of the morphology and neurology of the auditory system of echolocating bats contribute to the sensitivity of their hearing and their ability to process the information contained in echoes. The tympanic membranes and ear ossicles are small and light, and are easily set into motion. Contraction of the middle ear muscles briefly damps the sensitivity of the ear as each cry is emitted; thus, the bat does not deafen itself. A padding of blood sinuses, fat, and connective tissue isolates the bony labyrinth of the inner ear from the rest of the skull and reduces direct bone conduction of sound into the inner ear.

Perception of the direction of a returning echo is aided by the large, complex pinnae and by a neural mechanism known as contralateral inhibition. Stimulation of cells sensitive to a particular frequency in the inner ear on one side of the head produces a transient desensitization of the cells that respond to the same frequency in the ear on the other side of the head. The effect of that desensitization is to increase the contrast between the intensity of sound perceived by the two ears and thus to permit more precise determination of the direction of an echo.

One question that has long puzzled scientists is how an individual bat can discriminate between the echoes of its own call and those of other bats. When thousands of bats fly out of a roost in the evening the din must be enormous. One mechanism that probably contributes to a bat's ability to recognize the echoes of its own calls is a brief neural sensitization following emission of a call to sounds of the same wavelengths that were emitted. The sensitive period begins about 2 milliseconds after the end of the call and lasts about 20 milliseconds. Because sound travels 34 centimeters per millisecond in air, this timing means that a bat is especially sensitive to echoes of its own call from objects at distances between 30 centimeters and 4 meters. This same mechanism probably helps a bat on its final approach to prey.

The behavior of some insects that are potential prey for bats indicates that a degree of coevolution has occurred. Some noctuid moths are sensitive to the ultrasonic sounds produced by hunting bats. Low-intensity ultrasonic sounds cause the moths to fly away from the source, whereas loud sounds cause them to stop flying and drop to the ground. Some arctiid moths produce ultrasonic sounds themselves, and these sounds should be audible to bats. This moth family includes species that contain chemical compounds that make them unpalatable, and it has been suggested that their ultrasonic emissions advertise their distastefulness to bats, just as the bright colors and conspicuous displays of other insects advertise their unpalatability to predators that hunt by vision.

Cetacean Echolocation Many mammals live in close association with water. Some are aquatic, and a few are truly pelagic. The most specialized aquatic mammals are the members of the Cetacea—the whales and porpoises. Anatomical, biochemical, and embryological evidence suggests they evolved from the same group of Paleocene ancestors as did the artiodactyls. The demands of aquatic life have so entirely reshaped cetaceans that there is little external indication of this relationship.

The two lineages of extant cetaceans are the Mysticeti (the baleen whales) and the Odontoceti (the toothed whales). They appear to have radiated independently in the Eocene from a third lineage of ancestral toothed whales, all of which are now extinct. Baleen whales, which filter small organisms from the water, are not known for their use of echolocation in finding their prey, although they produce intricate vocalizations in the range of human hearing that probably are related to their complex social behavior. These are the whales whose songs are available on recordings. Calculations of the sound energy emitted suggests that some of their songs travel at least hundreds of miles through the sea. Certainly human instrumentation can track individual animals from such distances (Amato 1993). It is not inconceivable that all the mysticete whales of a particular species in the same ocean could be in indirect vocal communication with each other.

The toothed whales share the same general modifications of body form for aquatic life that are seen in baleen whales, but are otherwise different animals. The toothed whales are much smaller than baleen whales. Although the largest species, the sperm whale, reaches a length of 18 meters and a mass of 53,000 kilograms, most odontocetes are considerably smaller. Many toothed whales live close to shore, and there are even some purely freshwater species in the Amazon and Orinoco rivers of South America, the Ganges River in India, and Lake Tung'ing in China. Most toothed whales eat fishes or squid and their distributions are limited to areas of high productivity where food is abundant. Frequently, productive waters are also murky; thus many odontocetes must hunt in water in which vision is limited. Other species, such as the sperm whale which feeds on squid, hunt at extreme depths where sunlight does not penetrate. Dives of 1000 meters are probably routine for sperm whales. Odontocete evolution has solved the sensory problems in these habitats by an echolocation system as sophisticated as that of bats (Au 1993).

The difficulties inherent in any echolocation system are complicated for cetaceans by the acoustical properties of water. Sound travels about five times as fast in water as in air, and therefore the wavelength of any sound frequency is five times longer in water than in air. Because objects reflect only those wavelengths equal to or shorter than their diameters, cetaceans must use exceedingly high-frequency sounds to produce wavelengths short enough for the detection of small objects. Not surprisingly, the echolocation clicks of the bottlenose porpoise include frequencies from 20 to 220 kilohertz and much of the sound energy is in frequencies higher than the echolocation sounds of bats.

A second problem faced by mammals in water is the difficulty of matching the acoustic impedance of the ear to that of water. Sound energy is not transmitted well across an air–water interface. About 99.9 percent of the energy that reaches such an interface is reflected and only 0.1 percent crosses the boundary. Terrestrial vertebrates face the same problem because the inner ear is water filled. The middle ear converts sound energy transmitted through air to mechanical movement of bones, and thence to displacement of the fluid in the inner ear. The system does not work underwater, however, because a new water–air interface is created at the tympanic membrane and most of the sound energy is reflected from the tympanum back into the water.

A related problem is the efficiency of transmission of sound energy from water into body tissues, especially fat and muscle. A very high proportion of the acoustic energy that impinges on a body surface submersed in water is absorbed and propagates through the tissues, echoing back and forth to produce a diffuse buzz in the inner ear that conveys no directional information. Cetaceans have apparently solved these twin problems by abandoning the middle ear as a sound receiving and transmitting organ. The inner ear is isolated from sound propagated through the body tissues, and sound is conducted to the inner ear via a special fat body that extends from the lower jaw to the auditory bulla. The bulla is composed of extremely dense bone, which does not transmit sound readily and is separated from the rest of the skull by soft, sound-absorbing tissues. Thus the inner ear is isolated from sound approaching from other parts of the body. Kenneth Norris first proposed that the intramandibular fat body receives sound energy through an acoustic window formed by an area of very thin bone on the side of the mandible (Figure 21–18; see Norris 1974). Sound penetrates this thin bone readily and enters the fat body, which serves as a waveguide conducting the sound energy back to the bulla. The fat body termi-nates on the bulla itself where the bone of the bulla is thin. Norris claimed that the sound a cetacean perceives is not received via the middle ear but instead comes from the lower jaw. A test supported this hypothesis: Sound from a speaker beamed at the acoustic window of the lower jaw caused a large increase in nerve cell activity in the auditory region of the brain, but sound impinging on the side of the head did not.

Sound production underwater also poses problems not experienced by animals in air. Diving animals submerge with a limited supply of air, and it is probably undesirable to waste any of the oxygen supply by bubbling air out of the mouth in the process of generating sounds. Norris has suggested that the site of sound production in cetaceans has shifted from the larynx to the complex system of nasal passages and diverticulae that extend upward from the pharynx to the blowhole, the single external naris of odontocetes, which is located on the top of the head. Air is forced back and forth in these passages to produce sound energy which is reflected forward from the broad anterior surface of the skull and focused by an acoustic lens formed by the oil-

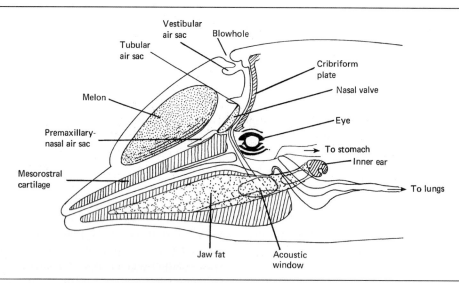

Figure 21–18 Proposed odontocete echolocation sound production and reception apparatus shown for the bottle-nosed dolphin (Tursiops). Sound produced by air shuttled between air sacs through the nasal valve is focused by the oil of the melon into a forwardly directed beam. Some sound may also be guided by the mesorostral cartilage. Returned sound is channeled through the mandibular (jaw) fat bodies and especially the acoustic window on the lower jaw to the otherwise isolated and fused middle and inner ears. (Modified after K. S. Norris, 1966, in *Evolution and Environment*, edited by E. T. Drake, Peabody Museum Centenary Volume, Yale University, New Haven, CT.)

filled **melon.** This fat body is a characteristic feature of odontocete cetaceans, and it is responsible for the external shape of the head. The skull itself occupies a relatively small portion of the head. The melon of the sperm whale is the source of sperm oil that was especially valued for the clear smokeless flame it produced when burned in oil lamps. As much as 30 barrels of oil can be obtained from the melon of a large sperm whale.

Sound passes from one medium to another as it leaves a cetacean's head. Because the acoustic coupling between oil and water is good, little of the sound energy is lost by reflection as it would be in an air-to-water transmission, but the sound waves may be bent. It seems likely that the melon serves as a flexible lens, changing its shape from moment to moment to beam the sound energy in different directions. The melon of some cetaceans changes shape very conspicuously during echolocation. The echolocation sounds of the bottlenose porpoise are broadband clicks (20 to 220 kilohertz). The highest frequencies in the click are emitted in a narrow beam directly forward and on a level with the animal's head. The lower-frequency sounds are dispersed in wider vertical and horizontal arcs. In addition, an echolocating porpoise moves its head around as if scanning with its sound beam.

The ability of porpoises to detect objects and to distinguish between similar objects is remarkable. Many of Norris's early studies were conducted with a bottlenose dolphin named Alice. Although she was frequently described affectionately by Norris and his associates as the world's dumbest dolphin, when she was blindfolded, Alice was able to distinguish between two steel balls, one 57 millimeters in diameter and the other 64 millimeters, from a distance of 2 meters in 1.5 seconds (Norris 1974). Blindfolded she could tell the difference between different species of fish or between fresh fish and day-old fish, and she was able to pick up her vitamin pill (5 millimeters in diameter) from the concrete bottom of her tank.

In their natural surroundings porpoises use their echolocation abilities to navigate and to find food. The lower-frequency sounds that are beamed in a broad arc would reflect from large objects and are probably the important components of navigation, whereas the high-frequency sounds that are concentrated in a narrow beam are probably used to locate prey. The best sound-reflecting surface in fishes is the gas-filled swim bladder (Figure 8–5), and it is prob-

ably this structure that odontocetes most easily detect. It has been suggested that the reduced air bladders of some fast-swimming open-water fishes such as tunas make them harder for cetaceans to detect by echolocation.

Mammalian Reproduction

Juvenile mammals depend on one and sometimes both parents for periods of weeks, months, or years after birth. This unavoidable association with parents and siblings is an important part of the sociality that is such a distinctive feature of mammals (Chapter 22). The dependence of young mammals on their parents is a result of the reproductive specializations of mammals. Understanding mammalian reproductive anatomy, physiology, and ecology is pivotal to understanding mammals.

Modes of Reproduction

Unique to mammals is the evolution of lactation, whereby the newborn mammal is supplied with food (as milk) by the mother (Pond 1975). A major evolutionary phenomenon within mammals has been the refinement of viviparity into two distinct processes in which food for a developing embryo is supplied continuously by the mother before birth and (in the most derived clade) a chorioallantoic placenta. As might be expected from knowledge of their ancestry, mammals show a gradation from oviparity to viviparity.

Monotremes The platypus and echidna retain their heritage as egg layers (Figure 21–19a and b). The ovaries are bigger than in other mammals for they produce eggs with large amounts of yolk. As in lizards there are two oviducts, each opening into the urogenital sinus. The ovulated eggs are fertilized prior to entry into the uterus. There they begin to differentiate and are coated with a leathery, mineralized shell. Although both oviducts are present in monotremes, only the left duct functions; as in birds, the right oviduct is reduced in size. Unlike birds, however, a recognizable shell gland secretion section is absent. There is no distinct separation between tube, isthmus, and uterus (Figure 21–19b).

The platypus generally lays two eggs, which are incubated in a nest. Echidnas lay a single egg and

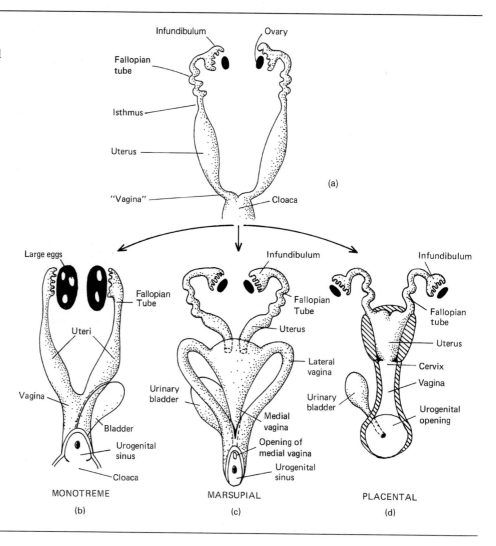

Figure 21–19 Representative female reproductive tracts of hypothetical ancestral and major clades of living mammals: (a) hypothetical ancestral mammal with tract similar to that of a lizard; (b) egg-laying monotreme; (c) marsupial showing complicated vaginal structures; (d) placental mammal as exemplified by an advanced primate and shown in partial longitudinal section.

carry it about in a metatherian-like pouch. A shell tooth, like that of birds, allows the monotreme young to break out of its shell.

Marsupials The reproductive tract of female marsupials combines structural features found in both monotremes and eutherians (Figure 21–19c and d). Following copulation, sperm pass from the vagina through the uterus and into the fallopian tubes, where fertilization takes place. Secretions of the oviduct assist sperm in reaching the egg. Although the penis of mammals is a single organ, in many marsupials its distal end is forked to fit into each lateral vagina.

The amount of yolk present in a marsupial egg, although greater than the yolk in a placental mam-

mal's egg, nevertheless is limited. A yolk sac placenta is often formed, but actual implantation, if it occurs at all, is brief. Gestation lasts for 13 days in the opossum and only for several weeks even in the large kangaroos. The degree of maturity at birth is probably best visualized by considering that the newborn of a large kangaroo, with adult size equaling that of a human, is only 2 to 3 centimeters long. Usually, the medial vagina serves as the birth canal. Following birth the embryos crawl from the vaginal region into the marsupial pouch (marsupium), where they grasp a mammary gland teat with their mouth. The teat swells so that the baby cannot drop off. Usually, more zygotes develop into embryos than there are teats. In the opossum, for example, there are about 13 nipples, but between 20 and 40

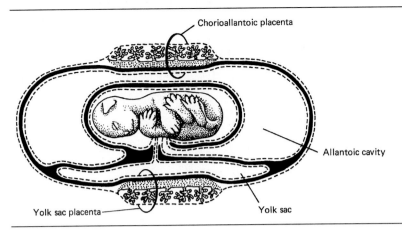

Chorioallantoic placenta

Allantoic cavity

Yolk sac placenta

Yolk sac

Figure 21–20 Two basic types of placental structures as seen in a transitional stage of an implanted embryo of the cat. Both a yolk sac and a chorioallantoic placenta are present in this stage. The chorioallantoic placenta grows outward and takes over the function of the earlier, primitive yolk sac placenta. (Modified after W. W. Ballard, 1964, *Comparative Anatomy and Embryology*, Ronald Press, New York, NY.)

eggs are ovulated, fertilized, and gestated. This apparent waste may reflect the difficulty that each newborn faces in finding and entering the marsupium and securing a teat.

Placentals In placentals the paired oviducts are fused to form a common vagina and sometimes a common large uterus, as in the human female (Figure 21–19d). In many placentals (some bats, rodents, and carnivores) the uteri remain distinct as right and left halves, or are partly fused to form a T-shaped uterus. In all placentals the **fallopian tubes** are small, short, and separate. As a rule, fertilization of the egg takes place in the fallopian tube after the sperm travels through the cervix, uterus, and up the tubes. The fertilized egg is propelled into the uterus by contractions of the fallopian tubes and implants on the uterine wall, where it develops the embryonic–maternal connections called the **placenta.** As embryonic–maternal dependency evolved in mammals the importance of yolk diminished and eutherian eggs are small.

The placenta is a special vascular extraembryonic structure derived from the extraembryonic membranes of the amniotic egg (Figure 21–20). Its development and maintenance represent major energy investments by the mother, especially early in the pregnancy. Placentae are either nondeciduous or deciduous. In the nondeciduous placenta all the fetal and uterine membrane linings remain intact throughout gestation. As a result, six distinct layers separate the fetal blood from the maternal blood. In a deciduous placenta varied degrees of erosion of the membranes occur and the number of layers decreases. In humans, only two layers remain; and in rabbits

and some rodents, such as the guinea pig, the fetal and maternal bloodstreams are separated only by the fetal endothelial lining of the capillaries. The hypothesis that reducing the number of membranes across the fetal–maternal barrier improves the efficiency of exchange sounds plausible, but has little evidence to support it (Mossman 1987). Perhaps of greater importance is the total area and the amount of vascular penetration associated with the fetal–maternal or placental junctions. Variation of placental morphology among eutherians defies simple description (see Mossman 1987, for details), but in most species the placental exchange structures between fetus and mother appear to act as a countercurrent exchange mechanism (Figure 21–21). This anatomical arrangement is functionally similar to the countercurrent exchange arrangement in the gills of teleost fishes (Chapter 8). Thus, the fetal capillary blood flow is opposite to the maternal capillary blood flow, and in this manner oxygen and carbon dioxide, nutrients, and wastes are supplied to and removed from the fetus in an efficient manner.

In contrast to birds and monotremes, the placentals do not invest energy in laying down chemically specialized yolk for embryonic development because they supply energy via the placenta. However, the placenta itself represents a large amount of energy invested by the mother in the growth of a specialized tissue. Eutherian embryos are afforded the protection of the womb and develop at the high body temperatures typical of a mammal. Nevertheless, adulthood comes slowly and parental postnatal nurture and protection are crucial to survival. The true ecological value of mammalian viviparity is therefore hard to ascribe to any single feature (Lille-

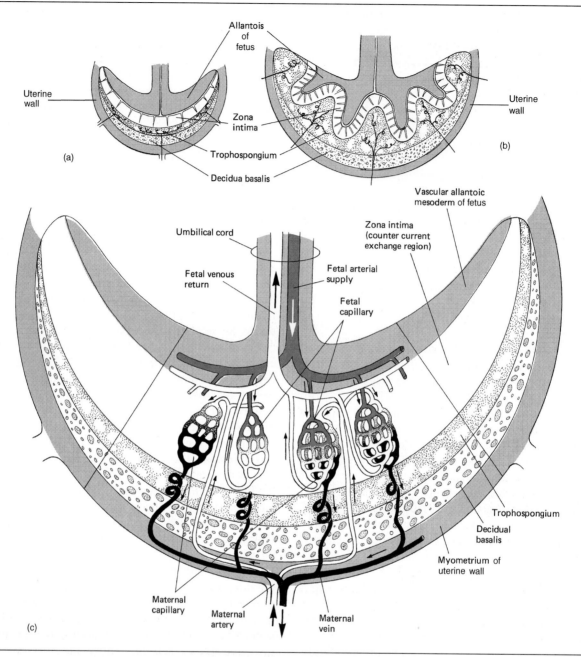

Figure 21–21 Generalized placental blood flow showing the counter current structure. Exchange occurs across the capillaries of fetus and of the mother in the zona intima. Attachment of the placenta to the uterus occurs at the decidua basalis. The simpler anatomy of (a) is typical of a labyrinthine-type placental arrangement as found in carnivores, rodents, chiropterans, and lagomorphs. The highly lobulated placenta of (b) is typical of most ungulates, cetaceans, and primates and increases the exchange area by folding of the zona intima. The cross section in (c) reveals the architecture of the maternal circulation and the fetal circulation; the stippled area (the trophospongium) contains the coiled maternal venous return vessels. (Modified after H. W. Mossman, 1987, *Vertebrate Fetal Membranes,* Rutgers University Press, New Brunswick, NJ.)

graven et al. 1987). It correlates, however, with the facts that parents can direct more attention to fewer but larger offspring, and that the offspring may learn complex behaviors during the period of parental care. The diverse social behaviors of mammals, many of which involve interactions between parents and their offspring, are discussed in Chapter 23.

Timing and the Hormonal Control of Reproduction

The sex of a vertebrate is usually a matter of inheritance (Chapter 3). Whether marsupial or placental, the long-term behavioral and physiological roles of mammalian parents with respect to their offspring are precisely defined and essential for reproductive success. A female mammal must be finely tuned physiologically and behaviorally if her offspring are to survive. In species where the males participate in reproduction after insemination, they too must coordinate sex-specific behaviors with their mates and offspring. Reproductive cycles and their synchronism with resource-determining environmental conditions are controlled by the release of sex hormones (Blum 1986).

Reproductive Cycles The ability to adjust reproductive resource expenditure to appropriate environmental conditions is made possible by reproductive cycling. A single reproductive cycle of the female mammal is called an **estrous cycle.** Specific reproductive events are associated with each of four periods of the cycle, the durations of which vary greatly in different mammals (Figure 21–22). **Proestrus** starts each reproductive cycle and is typified by initial growth of an ovarian follicle and its secretion of estrogen. **Estrus** begins with **ovulation** and modification of the follicle into the corpus luteum, which secretes progesterone. In many mammals estrus is the period of maximum female receptivity to the sexual overtures of males. **Metestrus** is a period of regression of the corpus luteum, with a declining secretion of progesterone. **Diestrus** is a variable period of quiescence before the next proestrus. In many primates metestrus is terminated by a menses, a discharge of blood and cellular debris from the endometrial lining of the uterus. Menstruation in the human coincides with the onset of diestrus and is associated with low levels of all sex hormones in the bloodstream (Figure 21–22). The reproductive cycle in those species with a menses is known as the **menstrual cycle.**

In many mammals diestrus is prolonged throughout the year and females are receptive to males only once a year. Such species are **monestrous. Polyestrous** species commence proestrus following a short diestrus and are sexually receptive several times during a breeding season or cycle continuously throughout the year. Except for some of the most derived primates, most female mammals respond to copulation attempts by males only during estrus, when ovulation takes place. The chance for fertilization of a ripe egg is thus maximized. The hypothalamus and the pituitary gland, collectively termed the **pituitary axis** (Chapter 3), are involved in the initiation of estrus (Figures 21–22 and 21–23).

Proestrus commences as follicular development and secretion of estrogen begin. Estrogen has multiple effects in the estrus cycle. First, it causes the uterine lining to grow and to enrich its vascularization. This process prepares for the possible implantation of a fertilized egg. At approximately this stage the reproductive cycle can be considered to pass into estrus. Follicular growth reaches the stage where ovulation occurs and the ripe ovum is shed into the coelom. The remaining follicle cells enlarge and form the **corpus luteum.** This ovarian structure manufactures not only estrogen but also progesterone, which prepares the uterus for implantation by inducing uterine secretions and inhibiting contractions of its smooth muscle. The high levels of estrogen and progesterone feed back on the pituitary axis, inhibiting its secretion of estrus initiating hormones. If pregnancy does not occur, the corpus luteum begins to degenerate and reduced secretion of both estrogen and progesterone results in collapse of the thick uterine linings. In some derived primates the uterine lining sloughs off, forming the menstrual flow. This process is assisted by the release of vasodilators such as histamine and prostaglandins, which promote endometrial hemorrhage (Heap and Flint 1984). Metestrus can therefore be considered to pass into diestrus as the uterine lining begins to collapse.

In the male, as in the female, gonadotropins secreted by the pituitary axis have as their target organs the gonads. These pituitary hormones stimulate the production of sperm in the seminiferous tubules of the testes and cause the interstitial cells of the testes to secrete the steroid hormone testosterone, which is essential to the maturation of sperm. Testosterone, like estrogen, has a feedback effect on the pituitary axis that can reduce gonadotropic output. Despite this negative feedback, testosterone is

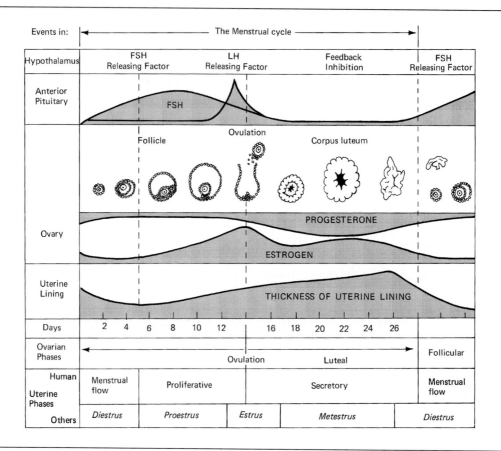

The Menstrual cycle

Events in:					
Hypothalamus	FSH Releasing Factor	LH Releasing Factor	Feedback Inhibition	FSH Releasing Factor	

Days	2 4 6 8 10 12	16 18 20 22 24 26	
Ovarian Phases	Ovulation	Luteal	Follicular
Human Uterine Phases	Menstrual flow	Proliferative	Secretory	Menstrual flow
Others	*Diestrus*	*Proestrus*	*Estrus*	*Metestrus*	*Diestrus*

Figure 21–22 Relationships among reproductive structures and hormones during the menstrual cycle in the human female. Correspondence between the human menstrual cycle and the estrus cycle typical of most mammals is indicated at the bottom (uterine phases). The height of the shaded regions is proportional to the circulating blood levels of the indicated hormones. FSH, follicle stimulating hormone; LH, luteinizing hormone. (Modified after R. P. Shearman, 1972, *Human Reproductive Physiology,* Blackwell Scientific, Oxford, UK.)

relatively inactive in depressing the function of the pituitary, and a delicate balance is achieved: Sperm and testosterone production proceed continuously. In several mammals, such as humans, a male reproductive cycle is therefore usually not described. In many mammal species, however, breeding activity and spermatogenesis are seasonal and synchronous with estrus (Short 1984a).

If implantation occurs following copulation, the female estrous cycle is interrupted and metestrus is prolonged. The ovarian corpus luteum does not degenerate, but continues to produce high levels of estrogen and progesterone. Maintenance of the corpus luteum past its usual nonpregnant functional period is achieved by the secretion of hormones from the placenta. Placental hormone secretion is not inhibited by high estrogenic levels in the blood, and a continued increase in blood levels of estrogen and progesterone follows. As the placenta grows, it becomes a new source of estrogen and progesterone (Figure 21–24). These placental secretions of estrogenic substances replace that of the corpus luteum and assure that the uterus retains the vascularization and that supplies nutrients to the developing embryo.

Just prior to birth of the embryo (parturition) the hypothalamic peptide hormone oxytocin is secreted from the posterior pituitary gland of the mother (Figure 21–23). This hormone stimulates contractions of the uterus (onset of labor), which lead to expulsion of

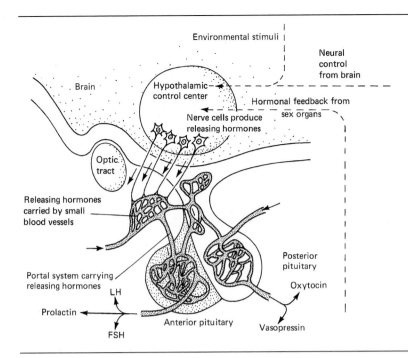

Environmental stimuli

Neural control from brain

Brain

Hypothalamic control center

Hormonal feedback from sex organs

Nerve cells produce releasing hormones

Optic tract

Releasing hormones carried by small blood vessels

Portal system carrying releasing hormones

LH

Prolactin

FSH

Anterior pituitary

Posterior pituitary

Oxytocin

Vasopressin

Figure 21–23 Relationships of the pituitary axis, which is composed of the hypothalamus and pituitary gland, in the neurohumoral control of reproductive cycles. Note that blood flow to the posterior pituitary lobe may or may not pass through the portal system (Modified after R. P. Shearman, 1972, *Human Reproductive Physiology*, Blackwell Scientific, Oxford, UK.)

the fetus and its placenta. In human females naturally occurring **endorphins,** nervous system secretions unique to vertebrates with actions similar to morphine, are released at triple the normal level during labor.

Studies of sheep and cattle have demonstrated that the fetus may play a role in the initiation of birth. Normal function of the fetal pituitary axis and adrenal glands is required for normal birth to occur. Cortisol secreted by the fetal adrenal glands causes increased activity of enzymes that convert progesterone to estrogen in the placenta. The result is a decrease in progesterone production and an increase in estrogen. The higher levels of estrogen probably stimulate production of prostaglandin in the fetal membranes and in the lining of the uterus. In turn, prostaglandin increases the excitability of the uterus and promotes the uterine contractions that expel the fetus. This mechanism by which the fetus gives the signal for delivery has so far been demonstrated only in ruminants, but it may be more widespread among mammals. There is evidence for increased fetal adrenal growth just prior to delivery in primates.

Lactation Postnatal growth of the young mammal is supported by the unique, derived character of the Mammalia: production of maternal milk in the mammary glands. Parental care is variously elaborated into many other areas of the life of young mammals. Lactation, like the estrous cycle and pregnancy, is largely regulated by hormones and usually requires the presence of suckling young.

The high levels of circulating estrogen and progesterone during pregnancy induce enlargement of the mammary tissues and the accumulation of fats and connective tissue. Milk production does not take place until parturition has occurred. An abrupt decline in estrogen and progesterone levels occurs when their source during pregnancy, the placenta, is expelled in the afterbirth. The hypothalamus, thus released from feedback inhibition, secretes yet another unique pituitary releasing hormone into the pituitary portal system (Figures 21–23 and 21–25). This factor instigates secretion of prolactin by the anterior pituitary gland. Prolactin stimulates milk production, but does not itself cause the delivery of milk. Suckling by the newborn infant stimulates nerve receptors in the nipple, and this information is transmitted to the hypothalamus and on to the posterior pituitary. Oxytocin, the hormone involved in the induction of labor, is released and causes milk ejection by inducing contractions of the smooth muscles of the mammary gland, and milk flow starts a short time after the beginning of suckling. In cursorial ungulates such as artiodactyls, in which nursing

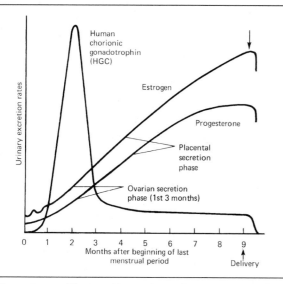

Figure 21–24 Changes in circulating hormone levels during a human pregnancy as indicated by their presence in urine. (Redrawn from R. P. Shearman, 1972, *Human Reproductive Physiology*, Blackwell Scientific, Oxford, UK.)

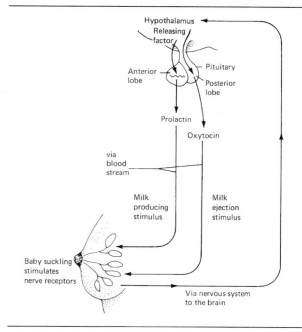

Figure 21–25 Role of the pituitary axis in lactation. Note that milk production and secretion are promoted by the anterior pituitary lobe hormone oxytocin. Both of these processes are initiated via neural connection from the nipple to the hypothalamus, a complex reflex which is stimulated by suckling. (Modified after R. P. Shearman, 1972, *Human Reproductive Physiology*, Blackwell Scientific, Oxford, UK.)

could expose mothers and young to predators, milk is immediately available from a cistern-like udder. Suckling by the young maintains lactation. In many mammals, estrus is not initiated during the period of lactation. This delay seems to occur in species that are energy limited. In other species, however, estrus quickly begins again. In humans there is variation among individuals, and among populations. In some mothers, menstruation is delayed for three to four months during lactation; in others, the menstrual cycle commences soon after birth. In food-limited hunter–gatherer human populations, females may not resume menstrual cycling until the cessation of nursing two to four years after birth.

Ecological Aspects of Reproduction by Metatherians and Eutherians

Marsupial gestation seldom lasts more than 30 days, and the fetus does not usually implant in the uterus until the last few days of gestation. Instead, a shell-like membrane surrounds the embryo and most of prenatal development depends on consumption of yolk (**lecithotrophy**). The embryo is born at an early stage of development, and unless it finds its way into the marsupium and a mammary teat, its survival is impossible. In most placentals the embryo is born at a comparatively advanced stage of development, and although dependence on parental care is often prolonged, the young may be quickly weaned and begin to cope directly with the world. Humans are one of the notable exceptions to the general mammalian pattern of early weaning and independence. The studies of Geoffrey Sharman (1976) and his colleagues in Australia and subsequent work by other researchers have shown that although marsupials and placentals have invested approximately the same amount of energy in an offspring by the time it is weaned, the patterns of energy investment are different for the two groups. Marsupial development is largely a postnatal cost (lactation), whereas placental development is largely a prenatal cost (placentation). This difference may have important ecological and evolutionary implications in times of energy stress.

The placental reproductive mode involves a large maternal investment in intrauterine structures to assure development. When food is scarce, it is usually the mother who suffers because the embryo's demands continue in spite of external environmental conditions. In placental mammals natural abortion during food shortage is rare. In many rodent populations, for example, limited food resources may restrict the number of zygotes implanted but does not affect the number of fetuses aborted. However, in marsupials unfavorable food resources result in what might be termed marsupium abortion, for lactation declines. The young, therefore, suffer directly and can be entirely abandoned. After parturition, placentals, of course, can also abandon young and do, but the energy already invested in reproduction is high. The marsupial reproductive plan may be more successful than that of placentals when environmental conditions at the time of birth are poor, for less energy has been invested in prenatal development (Parker 1977).

Timing of Reproduction Although the investment in a fetus at birth is greater for a placental than for a marsupial, both placentals and marsupials have ways of coping with varying, harsh, and unpredictable conditions. One such response takes the form of timing copulations so they result in births during the season of the year when food is most plentiful. To accomplish this, copulation should occur during an estrous cycle when the gestation period typical of the species will produce newborn in spring or summer in temperate regions or during the rainy season in the tropics. The length of gestation in placental mammals varies greatly, and generally correlates with size. Most large placentals, for example, carry embryos from six months to a year or even longer. As a result, copulations that take place in late summer or autumn result in births the following spring or summer when food is adequate to sustain lactation and the weaning of young.

Embryonic Diapause Several species of marsupials and placentals use **delayed implantation** or **embryonic diapause** to assure that young are born at favorable periods of the year or to avoid the birth of a second litter before the first litter has been weaned. Although details of the mechanisms for inducing embryonic diapause are not fully understood, two modes of control appear to be involved, one biological and the other environmental.

Several species of carnivores halt development of the new, unattached blastocyst immediately following fertilization. Growth depends on uterine

secretions, and at this stage of pregnancy uterine secretions are controlled by hypothalamic and pituitary outputs. Thus, the growth of the blastocyst can be arrested via central nervous system control via the hypothalamus. This mechanism does not work after implantation of the blastocyst on the uterine wall when fetal growth becomes solely dependent on the passage of nutrients across the placental barrier. Martens, skunks, and badgers breed in the summer and employ embryonic diapause to prolong gestation. The diapause may last 10 months, assuring that birth occurs during the spring of the following year. Diapause is apparently triggered by changes in day length, and results in a cessation of uterine thickening and secretions associated with estrus. The diapause may result from melatonin release and ultimate suppression of the secretion of progesterone by the corpus luteum following ovulation.

Many species of bats in temperate zones employ embryonic diapause, but as a consequence of their winter hibernation. Lowering the body temperature slows all physiological processes, including the rate of fetal growth. However, the California leaf nose bat slows the rate of embryonic development and spreads it over a 9-month gestation period without diapause. The mechanism in this case is unknown (Short 1984b). Bats feed on nocturnal flying insects, which are unavailable in cold weather. Thus, the delayed implantation shown by bats clearly fits the pattern of adjusting the time of birth to correspond with a season of abundant food.

Among rodents and insectivores diapause serves a different purpose—the temporal separation of litters to avoid competition for resources. Small polyestrous forms can enter estrus every four to five days (Table 21–4), have gestation lengths of around 20 days, and become pregnant even when feeding the previous litter. Weaning a litter usually requires a month. Arrival of a second litter before the first litter has been weaned would strain the mother's ability to provide enough milk. To avoid this energetic dilemma, diapause maintained by lactation delays the development of the second litter until the first litter has finished suckling. The hormonal mechanisms are not totally clear, but suckling apparently inhibits the secretion of estrogen from the ovary via stimulation of the pituitary axis.

Embryonic diapause reaches its zenith in the large macropods of Australia. The red kangaroo, which inhabits the climatically unpredictable arid desert regions of Australia, can have an arrested

Table 21–4 **Length of the gestation period and estrous cycle in some mammals (days).**

Nonprimate Species	Gestation	Estrous Cycle	Primates	Estrous Gestation	Cycle
Mouse	20	4–5	Lemur	120–140	
Rat	22	4–5	Tarsier	180	24
Rabbit	28	*	Wooly monkey	139	
Cat	63	18	Rhesus monkey[†]	150–180	27
Dog	63	235	Chacma baboon[†]	180–190	20–36
Lion	110	21	Langur[†]	170–190	
Goat	150	20	Gibbon[†]	210	
Domestic cattle	278	21	Orangutan[†]	220–270	
Horse	340	20	Chimpanzee[†]	216–260	36
Dolphin	360		Gorilla[†]	250–290	
African elephant	660	42	Human[†]	267 avg.	28

*Continuous; induced ovulation

[†]Old World monkeys and the great apes menstruate externally. Rhesus monkeys and humans show maximal behavioral changes during menstruation; other species have maximal change during estrus.

Source: Modified from R. J. Harrison and W. Montagna, 1973, *Man*, 2d edition, Appleton Century Crofts, New York; and V. Hayssen, A. van Tienhoven, and A. van Tienhoven, 1993, *Asdell's Patterns of Mammalian Reproduction: A Compendium of Species Specific Data*, Cornell University Press, Ithaca, NY.

blastocyst in the uterus, an immature suckling joey tucked away in its marsupium, and a nearly weaned juvenile that still needs occasional sips of milk to supplement its newly acquired diet of vegetation. This suite of reproductive interactions is induced by the joey, whose suckling produces nervous stimulation of the hypothalamus, causing the pituitary hormone release that diminishes and reduces follicular development and subsequent ovulation. The output of estrogen and progesterone by the corpus luteum is suppressed also by the increased prolactin release brought about by suckling. The red kangaroo represents the utmost in opportunistic reproduction: As long as environmental conditions are suitable, the young can develop for long periods in the pouch before they are weaned. If conditions deteriorate, which can happen quickly and unpredictably in the Australian desert, a female can abandon her joeys with no delay in producing the next litter. The diapausing blastocysts immediately begin to grow and are born in about 30 days (Table 21–4). The red kangaroo is a prime example of biological control of diapause, where the proximal cause is the presence of a suckling joey. But even in this example, the ultimate control remains the environment and its vagaries.

Summary

The first mammals were small and probably insectivorous. Because of their ability to maintain a relatively constant internal body temperature, they were probably active at night and in cool seasons—times when ectotherms are inactive.

A cladistic classification of mammals retains the three widely recognized mammal groupings of the Prototheria (the egg-laying monotremes), Metatheria (the marsupials), and the Eutheria (the placental mammals). Mammalian diversity is astonishing, and mammals have radiated into almost every habitat on earth.

A few selected aspects of the adaptive radiation of derived mammals give an appreciation for the versatility of mammalian structure and function. The ways mammals interact with their environments, satisfy their nutritional demands, sense the world around them, and reproduce are areas of particular importance.

The mammalian integument is a unique structure that plays a major role in mammalian homeostasis, especially thermoregulation, and in mammalian social behavior. Its derivatives include hair, scales, claws, hoofs, horns, and glands that secrete substances ranging from milk to pheromones. Its sensory transducers provide vital information about the environment.

The specializations of mammalian teeth are basic to mammalian evolution. Jaw, muscle, and locomotor anatomy also reflect feeding adaptations. A case can be made for plant evolution having stimulated changes in herbivore evolution that, in turn, affected carnivore evolution.

Evolutionary changes in mammals also occurred in the evolution of the brain, especially the increased size of the cerebral hemispheres in many, but not all, lineages. Mammals evolved from earlier synapsids that could process three-dimensional visual reconstructions of the world. The original sensory demands made upon a nocturnal mammal-like therapsid may have acted as an evolutionary stimulus to brain enlargement. Auditory specialists that echolocate (bats and cetaceans) demonstrate the analytical power of the mammalian nervous system.

Reproduction stands at the base of the complex adaptations of mammals. The placenta and lactation provide energy under neurohormonal control, but the reproductive strategies of the Metatheria (marsupials) and Eutheria (placentals) are quite different. Marsupial reproduction appears to reduce the energy investment in the embryo and concentrate the maternal energy contribution in lactation, whereas placental reproduction emphasizes energy investment in utero.

References

Amato, I. 1993. A sub surveillance network becomes a window on whales. *Science* 261:549–550.

Au, W. W. L. 1993. *The Sonar of Dolphins.* Springer, New York, NY.

Barrette, C. 1977. Fighting behavior of muntjac and the evolution of antlers. *Evolution* 31:169–176.

Benton, M. J. 1988. The relationships of the major group of mammals: new approaches. *Trends in Ecology and Evolution* 3:40–45.

Berger, J., and C. Cunningham. 1994. Horns, hyenas, and black rhinos. *National Geographic Research and Exploration* 10:241–244.

Blum, V. 1986. *Vertebrate Reproduction.* Springer, New York, NY.

Bowen, W. D., O. T. Oftedal, and D. J. Boness. 1985. Birth to weaning in 4 days: remarkable growth in the hooded seal, *Cystophora cristata. Canadian Journal of Zoology* 63:2841–2846.

Bubenik, G. A., and A. B. Bubenik. 1990. *Horns, Pronghorns, and Antlers.* Springer, New York, NY.

Collinson, M. E., and J. J. Hooker. 1991. Fossil evidence of interactions between plants and plant-eating mammals. *Philosophical Transactions of the Royal Society, London Series B* 333:197–208.

Dijkgraff, S. 1960. Spallanzani's unpublished experiments on the sensory basis of object perception in bats. *Isis* 51:9–20.

Griffin, D. R., F. A. Webster, and C. R. Michael. 1960. The echolocation of flying insects by bats. *Evolution* 8:141–154.

Griffiths, M. 1979. *Biology of the Monotremes.* Academic, New York, NY.

Heap, R. B., and A. P. F. Flint. 1984. Pregnancy. Pages 153–194 in *Reproduction in Mammals.* volume 3, *Hormonal Control of Reproduction,* edited by C. R. Austin and R. V. Short. Cambridge University Press, Cambridge, UK.

Hochuli, D. F. 1993. Does silica defend grasses against invertebrate herbivores? *Trends in Ecology and Evolution* 8:418–419.

Jacobs, G. H. 1993. The distribution and nature of colour vision among the mammals. *Biological Review* 68:413–471.

Janis, C. M. 1976. The evolutionary strategy of the Equidae and the origins of rumen and cecal digestion. *Evolution* 30:757–774.

Janis, C. M. 1982. Evolution of horns in ungulates: ecology and paleoecology. *Biological Review* 57:261–318.

Janis, C. M. 1993. Tertiary mammal evolution in the context of changing climates, vegetation, and tectonic events. *Annual Review of Ecology and Systematics* 24:467–500.

Janis, C. M., and M. Fortelius. 1988. On the means whereby mammals achieve increased functional durability of their dentitions, with special reference to limiting factors. *Biological Review* 63:197–230.

Jerison, H. J. 1973. *Evolution of the Brain and Intelligence.* Academic Press, New York, NY.

Lillegraven, J. A. 1974. Biological considerations of the marsupial–placental dichotomy. *Evolution* 29:707–722.

Lillegraven, J. A., S. D. Thompson, B. K. McNab, and J. L. Patton. 1987. The origin of eutherian mammals. *Biological Journal of the Linnaean Society* 32:281–336.

Lorenz, K. 1977. *Behind the Mirror: A Search for a Natural History of Human Knowledge,* 1st American edition, translated by R. Taylor. Harcourt Brace Jovanovich, New York, NY.

Mossman, H. W. 1987. *Vertebrate Fetal Membranes.* Rutgers University Press, New Brunswick, NJ.

Norris, K. S. 1974. *The Porpoise Watcher.* George J. McLeod, Toronto, Canada.

Novacek, M. J. 1992. Mammalian phylogeny: shaking the tree. *Nature* 356:121–125.

Novacek, M. J. 1993. Reflections on higher mammalian phylogenetics. *Journal of Mammalian Evolution* 1:3–30.

Oftedal, O. T., D. J. Boness, and W. D. Bowen. 1988. The composition of hood seal *(Cystophora cristata)* milk: an adaptation for postnatal fattening. *Canadian Journal of Zoology* 66:318–322.

Parker, P. 1977. An ecological comparison of marsupial and placental patterns of reproduction. Pages 273–286, in *Biology of the Marsupials,* edited by B. Stonehouse and D. Gillmore. University Park Press, Baltimore, MD

Pond, C. M. 1975. The significance of lactation in the evolution of mammals. *Evolution* 31:177–199.

Sharman, G. B. 1976. Evolution of viviparity in mammals. Pages 32–70 in *Reproduction in Mammals,* volume 6, *The Evolution of Reproduction,* edited by C. R. Austin and R. V. Short. Cambridge University Press, Cambridge, UK.

Short, R. V. 1984a. Oestrous and menstrual cycles. Pages 115–152 in *Reproduction in Mammals,* volume 6, *The Evolution of Repro-duction,* edited by C. R. Austin and R. V. Short. Cambridge University Press, Cambridge, UK.

Short, R. V. 1984b. Species differences in reproductive mechanisms. Pages 24–61 in *Reproduction in Mammals,* volume 4, *Reproductive Fitness,* edited by C. R. Austin and R. V. Short. Cambridge University Press, Cambridge, UK.

Vicari, M., and D. R. Bazely. 1993. Do grasses fight back? The case for antiherbivore defenses. *Trends in Ecology and Evolution* 8:137–141.

ENDOTHERMY: A HIGH-ENERGY APPROACH TO LIFE

22

Endothermy is a derived character of mammals and birds. The two lineages evolved endothermy independently, but the costs and benefits are the same for both. Endothermy is a superb way to become relatively independent of many of the stresses of the physical environment, especially cold. Birds and mammals can live in the coldest habitats on Earth, provided that they can find enough food. That qualification expresses the major problem of endothermy: It is energetically expensive. Endotherms need a reliable supply of food. As a result, the conspicuous interactions of endotherms are often with their biological environment—predators, competitors, and prey—rather than with the physical environment as was often the case for ectotherms. Because energy intake and expenditure are important factors in the daily lives of endotherms, calculations of energy budgets can help us to understand the consequences of some kinds of behavior.

When all efforts at homeostasis are inadequate, endotherms have two more methods of dealing with harsh conditions: (1) Birds and large species of mammals can migrate to areas where conditions are more favorable. (2) Many species of small mammals and some birds can become torpid. This response, a temporary drop in body temperature, conserves energy and prolongs survival at the cost of abandoning the benefits of homeothermy.

Costs and Benefits

Endothermy has both benefits and costs compared to ectothermy. On the positive side, endothermy allows birds and mammals to maintain high body temperatures when solar radiation is not available or is insufficient to warm them—at night, for example, or in the winter. The thermoregulatory capacities of birds and mammals are astonishing; they can live in the coldest places on Earth. On the negative side, endothermy is energetically expensive. We have pointed out that the metabolic rates of birds and mammals are nearly an order of magnitude greater than those of amphibians and reptiles. The energy to sustain those high metabolic rates comes from food, and endotherms need more food than do ectotherms.

Of course, a host of other differences distinguish the ecology and behavior of endotherms and ectotherms, and these also can be considered costs or benefits of the different methods of thermoregulation. In this chapter we concentrate on how endotherms use energy, the ways in which endotherms thermoregulate in cold and in hot environments, and how endotherms control their water gains and losses. These topics are intimately related, because the high metabolic rates of endotherms are associated with more precise homeostatic control of their internal environment than is necessary for many ectotherms. For example, rising body temperature can be countered by evaporative cooling (sweating or panting), but too much evaporative cooling depletes water stores and leads to other problems. Body size plays a large role in determining the stresses to which endotherms are subjected and the responses that are possible for them.

The same sorts of habitats that are stressful for ectotherms (Chapter 16) are also difficult for endotherms, although not always for the same reasons. Cold temperatures, for example, make ectothermal thermoregulation difficult or impossible and may present a risk of freezing. Endotherms have sufficient insulation and thermogenic (heat-producing) capacity to survive low temperatures, but they need a plentiful and regular supply of food to do that. Indeed, their high energy requirements appear to shape several aspects of the biology of endotherms, such as the relationship among body size, diet, and home range discussed in the next chapter. The role that energy gain and use play in the day-to-day lives of endotherms can be illustrated by an energy budget.

Energy Budgets of Vertebrates

An understanding of the costs of living can be obtained by constructing an energy budget for an animal. An energy budget, like a financial budget, shows income and expenditure but uses units of energy as currency. The energy costs of different activities can be evaluated. This quantitative approach to the study of ecology and behavior offers the promise of understanding some of the reasons why animals behave as they do.

A Daily Energy Budget: The Vampire Bat

Studies of vampire bats (*Desmodus rotundus*) by Brian McNab (1973) have revealed a clear-cut relationship between energy intake, energy expenditure, and the species' geographic range. These bats of the suborder Microchiroptera inhabit the Neotropics and are specialized to feed exclusively on blood (Figure 22–1). Their daily pattern of activity is simple: They spend about 22 hours in their caves, fly out at night to a feeding site, and return after they have fed. Typically, a vampire flies about 10 kilometers round-trip at 20 kilometers per hour to find a meal. Thus, a bat spends half an hour per day in flight. The remaining time outside the cave may be spent in feeding.

The vampire's food is convenient for energetic calculations because of the relatively constant caloric content of blood. The information needed to construct an energy budget is the following:

I = ingested energy (blood). The blood a bat drinks must provide the energy needed for all of its life processes: maintenance, activity, growth, and reproduction.

E = excreted energy. As in any animal, not all of the food ingested is digested and taken up by the bat. The energy contained in the feces and urine is lost.

$I - E$ = assimilated energy. This is the energy actually taken into the bat's body.

M = metabolism. This can be subdivided into M_i (metabolism while the bat is inside the cave) and M_o (nonflight metabolism while the bat is outside the cave).

A = cost of activity, a half hour of flight per day.

B = biomass increase. This term is the profit a bat shows in its energy budget. It may be stored as fat or used for growth or for reproduction (production of gametes, growth of a fetus, or nursing a baby).

In its simplest form, the energy budget is

$$\text{energy in} = \text{energy out} \pm \text{biomass change}$$

The biomass term appears as $\pm$ because an animal metabolizes some body tissues when its energy expenditure exceeds its energy intake. (This is what every dieter hopes to do to lose weight.) Translating this general equation into the terms defined gives

$$I - E = M_i + M_o + A + B$$

All these terms can be measured and expressed as kilojoules per bat per day (kJ/bat · day). These calculations are based on McNab's studies and apply to a Brazilian vampire bat weighing 42 grams.

Ingested energy: In a single feeding a vampire can consume 57 percent of its body mass in blood, which contains 4.6 kJ/g. Thus the ingested energy is

$$42 \text{ g} \times 57\% \times 4.6 \text{ kJ/g blood} = 110.2 \text{ kJ}$$

Excreted energy: A vampire excretes 0.24 g urea in the urine plus 0.95 g of feces daily. Urea contains

Figure 22–1 The geographic range of the vampire bat, *Desmodus rotundus* (shaded area), closely approximates the 10°C isotherm for the minimum average temperature during the coldest month of the year (dashed line) at the northern limit of its range (in Mexico) and the southern limit (in Uruguay, Argentina, and Chile). The positions of the 10°C isotherm and the altitudinal range of the bats in the Andes Mountains are not known and are indicated by question marks. (Based on B. K. McNab, 1973, *Journal of Mammalogy* 54:131–144.)

10.5 kJ/g and the feces contain 23.8 kJ/g. Thus the excreted energy is

$$0.24 \text{ g urea} \times 10.5 \text{ kJ/g} + 0.95 \text{ g feces} \times 23.8 \text{ kJ/g} = 2.5 \text{ kJ} + 22.6 \text{ kJ} = 25.1 \text{ kJ}$$

Assimilated energy: The energy the bat actually assimilates from blood equals the energy ingested (110.2 kJ) minus that excreted (25.1 kJ). Thus a vampire's energy intake is 85.1 kJ/day:

$$110.2 \text{ kJ} - 25.1 \text{ kJ} = 85.1 \text{ kJ}$$

Metabolism: In a tropical habitat, 20°C is a reasonable approximation of the temperature a bat experiences both inside and outside the cave. While at rest in the laboratory at 20°C a vampire's meta-

bolic rate is 3.8 mL O_2/g · hr. The terms for metabolism can be calculated and converted to joules using the energy equivalent of oxygen (20.1 J/mL O_2):

$$M_i = 42 \text{ g} \times 3.8 \text{ mL } O_2/\text{g} \cdot \text{hr} \times 20.1 \text{ J/mL } O_2 \times 22 \text{ hr/day} = 70.6 \text{ kJ}$$
$$M_o = 42 \text{ g} \times 3.8 \text{ mL } O_2/\text{g} \cdot \text{hr} \times 20.1 \text{ J/mL } O_2 \times 1.5 \text{ hr/day} = 4.8 \text{ kJ}$$

Activity: The metabolism of a bat flying at 20 km/hr is three times its resting metabolic rate (3×3.8 mL O_2/g · hr = 11.4 mL O_2/g · hr). The cost of the round trip from the cave to the feeding site is

$$A = 42 \text{ g} \times 11.4 \text{ mL } O_2/\text{g} \cdot \text{hr} \times 20.1 \text{ J/mL } O_2 \times 0.5 \text{ hr} = 4.8 \text{ kJ}$$

Biomass change: The quantities calculated so far are fixed values that the bat cannot avoid. The biomass change is a variable value. If the assimilated energy is greater than the fixed costs, this energy profit can go to biomass increase. Fixed costs that exceed the assimilated energy are reflected as a loss of biomass. For the situation described there is an energy profit:

$$I - E = M_i + M_o + A \pm B$$
$$110.2 \text{ kJ} - 25.1 \text{ kJ} = 70.6 \text{ kJ} + 4.8 \text{ kJ} + 4.8 \text{ kJ} \pm B$$
$$B = 4.9 \text{ kJ/bat} \cdot \text{day}$$

These calculations show that vampires can live and grow under the conditions assumed. What happens if we change some of the assumptions? McNab points out that the northern and southern limits of the geographic range of vampires conform closely to the winter isotherms of 10°C. That is, the minimum temperature outside the cave during the coldest month of the year is 10°C; the bats do not occur in regions where the minimum temperature is lower. Is this coincidence, or is 10°C the lowest temperature the bats can withstand? Calculating an energy budget for a vampire under these colder conditions provides an answer.

Caves have very stable temperatures that usually do not vary from summer to winter. We will assume that temperature remains constant at 20°C in the cave our imaginary vampires inhabit. In that case only the conditions a bat encounters outside the cave are altered. Because of limitations of stomach capacity ingestion cannot increase beyond 57 percent of body mass, the value assumed in the previous calculation. Therefore, we need recalculate only M_o, A, and B.

Metabolism outside: At 10°C a bat must increase its metabolic rate to maintain its body temperature, and laboratory measurements indicate the resting metabolic rate increases to 6.3 mL O_2/g · hr:

$$M_o = 42 \text{ g} \times 6.3 \text{ mL } O_2/\text{g} \cdot \text{hr} \times 20.1 \text{ J/mL } O_2 \times 1.5 \text{ hr} = 7.9 \text{ kJ}$$

Activity: The cost of activity will not change, because the metabolic rate of the bat during flight (11.4 mL O_2/g · hr) is higher than the resting metabolic rate needed to keep it warm (6.3 mL O_2/g · hr). Only the term M_o changes, increasing from 4.8 to 7.9 kJ, and the sum of the energy costs becomes 83.3 kJ/bat · day.

Because the assimilated energy remains at 85.1 kJ/bat day, only 1.8 kJ is available for biomass increase. The assumptions in these calculations introduce a degree of uncertainty, and probably 1.8 kJ is not different from 0 kJ. At 10°C a bat uses all its energy staying alive. Thus a vampire bat could live under those conditions, but it would have no energy surplus for growth or reproduction. If the temperature outside the cave were lower than 10°C, the bat would have a negative energy balance and would lose body mass with each meal. The agreement between our calculations and the actual geographic distribution of the bats suggests that energy may be a biologically significant factor in limiting their northward and southward spread.

Additional calculations reveal more about the selective forces that shape the life of vampire bats. A bat's stomach can hold a volume of blood equal to 57 percent of its body mass, but a bat cannot fly with that load. The maximum flight load is 43 percent of the body mass. Before it can take off to start the flight back to its cave, therefore, a bat must reduce the weight gained from its meal. Vampires do this by rapidly excreting water. Within 2 minutes after it begins to eat, a vampire starts to emit a stream of dilute urine. Experiments reveal that a vampire produces urine at a maximum rate of 0.24 milliliter per gram body mass · hour. Thus in the hour and a half the bat may spend in feeding, it could excrete as much as 15 grams of water—more than enough to allow it to fly (Busch 1988).

Although rapid excretion of water solves the bat's immediate problem, it introduces another. The bat is left with a stomach full of protein-rich food that will yield a large amount of urea. To excrete this urea, the bat needs water to form urine. By the time a vampire gets back to its cave it is facing a water shortage instead of a water excess. Unlike many mammals, vampires seldom drink water but depend instead on blood for their water requirements. Like other mammals adapted to conditions of water scarcity, vampire bats have kidneys capable of producing very concentrated urine to conserve water (see Table 4–4). As a result of its unusual ecology and behavior, a vampire bat can be considered to live in a desert of its own making in the midst of a tropical forest.

Endotherms in the Cold: The Arctic

As the energy budget for the vampire bat revealed, endotherms expend most of the energy they consume just keeping themselves warm even in the moderate conditions of tropical and subtropical climates. Nonetheless, endotherms have proven themselves very adaptable in extending their thermoregulatory responses to allow them to inhabit even arctic and antarctic regions (Bech and Reinertsen 1989, Davenport 1992). Not even small body size is an insuperable handicap to life in these areas: Redpolls and chickadees that weigh only 10 grams overwinter in central Alaska.

Aquatic life in cold regions places still more stress on an endotherm. Because of the high heat capacity and conductivity of water, an aquatic animal may lose heat at 50 to 100 times the rate it would if it were moving at the same speed through air. Even a small body of water is an infinite heat sink for an endotherm; all of the matter in its body could be converted to heat without appreciably raising the temperature of the water. How, then, do endotherms manage to exist in such stressful environments?

Increased Heat Production Versus Decreased Heat Loss

There are potentially two solutions to the problems of endothermal life in cold environments and the special problems of aquatic endotherms in particular. A stable body temperature could be achieved by increasing heat production or by decreasing heat loss. On closer examination the option of increasing heat production does not seem particularly attractive. Any significant increase in heat production would require an increase in food intake. This scheme poses obvious ecological difficulties in terrestrial arctic and antarctic habitats where primary production is extremely low, especially during the coldest parts of the year. For most polar animals the quantities of food necessary would probably not be available, and a number of studies have shown that metabolic rates of most polar endotherms are similar to those of related species from temperate regions.

Because they lack the option of increasing heat production significantly, conservation of heat within the body is the primary thermoregulatory mechanism of polar endotherms. Insulative values of pelts from arctic mammals are two to four times as great as those from tropical mammals. In arctic species insulative value is closely related to the length of the fur (Figure 22–2). Small species such as the least weasel and the lemming have fur only 1 or 1.5 centimeters long. Presumably, the thickness of their fur

is limited because longer hair would interfere with the movement of their legs. Large mammals (caribou, polar and grizzly bears, dall sheep, and arctic fox) have hair 3 to 7 centimeters long. There is no obvious reason why their hair could not be longer; apparently, further insulation is not needed. The insulative values of pelts of short-haired tropical mammals are similar to those measured for the same hair lengths in arctic species. Long-haired tropical mammals, like the sloths, have less insulation than arctic mammals with hair of similar length.

A comparison of the lower critical temperatures of tropical and arctic mammals illustrates the effectiveness of the insulation provided (Figure 22–3). **Lower critical temperature** is the environmental temperature at which energy utilization must rise above its basal level to maintain a stable body temperature. A number of tropical mammals have lower critical temperatures between 20 and 30°C. As air temperatures fall below their lower critical temperatures, the animals are no longer in their thermoneutral zones and must increase their metabolic rates to maintain normal body temperatures. For example, a tropical raccoon has increased its metabolic rate approximately 50 percent above its standard level at an environmental temperature of 25°C.

Arctic mammals are much better insulated; even small species like the least weasel and the lemming have lower critical temperatures in still air that are between 10 and 20°C, and larger mammals have thermoneutral zones that extend well below freezing (Chappell 1980). The arctic fox, for example, has a lower critical temperature of –40°C, and at –70°C (approximately the lowest air temperature it ever encounters) has elevated its metabolic rate only 50 percent above its standard level. Under those conditions the fox is maintaining a body temperature approximately 110°C above air temperature. Arctic birds are equally impressive. An arctic jay has a lower critical temperature below 0°C in still air and can withstand –70°C with a 150 percent increase in its metabolic rate, whereas an arctic gull, like the arctic fox, has a lower critical temperature near –40°C and can withstand –70°C with a modest increase in metabolism.

Clearly, hair or feathers can provide superb insulation for a terrestrial endotherm. These external insulative coverings are of limited value to aquatic animals, however, because when air trapped between hairs is displaced by water the coverings lose most of their insulative value. The insulation of

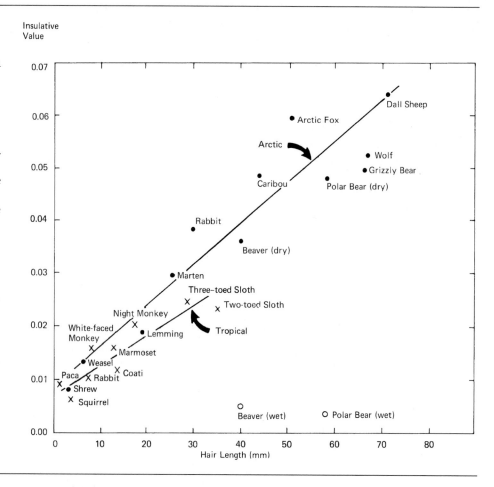

Figure 22–2 The insulative values of the pelts of arctic mammals (•) are proportional to the length of the hair. Pelts from tropical mammals (×) have approximately the same insulative value as those of tropical mammals at short hair lengths, but long-haired tropical mammals like sloths have less insulation than arctic mammals with hair of the same length. Immersion in water greatly reduces the insulative value of hair, even for such semi-aquatic mammals as the beaver and polar bear (•). (Modified from P. F. Scholander et al., 1950, *Biological Bulletin* 99:237–258.)

polar bear hair falls almost to zero when it is wet, and even seal hair loses much of its insulative value (Figure 22–2). In water, fat is a far more effective insulator than hair, and aquatic mammals have thick layers of blubber. This blubber forms the primary layer of insulation; skin temperature is nearly identical to water temperature and there is an abrupt temperature change through the blubber. Its inner surface is at the animal's core body temperature.

The insulation provided by blubber is so effective that pinnipeds and cetaceans require special heat-dissipating mechanisms to avoid overheating when they engage in strenuous activity, or venture into warm water or onto land. This heat dissipation is achieved by shunting blood into capillary beds in the skin outside the blubber layer and into the flippers, which are not covered by blubber. Selective perfusion of these capillary beds enables a seal or porpoise to adjust its heat loss to balance its heat production. When it is necessary to conserve energy, a countercurrent heat exchange system in the blood vessels supplying the flippers is brought into operation; when excess heat is to be dumped, blood is shunted away from the countercurrent system into superficial veins.

The effectiveness of the insulation of marine mammals is graphically illustrated by the problems experienced by the northern fur seal (*Callorhinus ursinus*) during its breeding season. Northern fur seals are large animals; males attain body masses in excess of 250 kilograms. Unlike most pinnipeds, fur seals have a dense covering of fur, which is probably never wet through. They are inhabitants of the North Pacific. For most of the year they are pelagic, but during summer they breed on the Pribilof Islands in the Bering Sea north of the Aleutian Peninsula. Male fur seals gather harems of females on the shore. Here they must try to prevent the females from straying, chase away other males, and copulate with willing females.

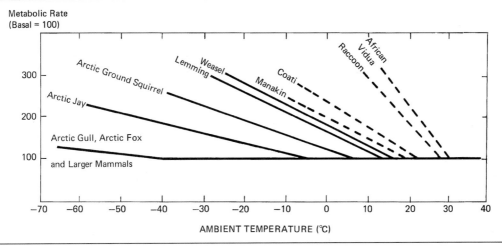

Figure 22–3 Lower critical temperatures for birds and mammals. Solid lines, arctic birds and mammals; dashed lines, tropical birds and mammals. The basal metabolic rate for each species is considered to be 100 units to facilitate comparisons among species. The metabolic rate of a species rises above its basal rate at the lower critical temperature for the species. Because of their effective insulation, arctic birds and mammals can maintain resting metabolic rates at lower environmental temperatures than tropical species, and show smaller increases in metabolism (i.e., flatter slopes) below the lower critical temperature. (Modified from P. F. Scholander et al., 1950, *Biological Bulletin* 99:237–258.)

George Bartholomew and his colleagues have studied both the behavior of the fur seals and their thermoregulation (Bartholomew and Wilke 1956). Summers in the Pribilof Islands (which are near 57°N latitude) are characterized by nearly constant overcast and air temperatures that rise only to 10°C during the day. These conditions are apparently close to the upper limits the seals can tolerate. Almost any activity on land causes the seals to pant and to raise their hind flippers (which are abundantly supplied with sweat glands) and wave them about. If the sun breaks through the clouds, activity suddenly diminishes—females stop moving about, males reduce harem guarding activities and copulation, and the adult seals pant and wave their flippers. If the air temperature rises as high as 12°C, females, which never defend territories, begin to move into the water. Forced activity on land can produce lethal overheating.

At that time, seal hunters herded the bachelor males from the area behind the harems inland preparatory to killing and skinning them. Bartholomew recorded one drive that took place in the early morning of a sunny day while the air temperature rose from 8.6°C at the start of the drive to 10.4°C by the end. In 90 minutes the seals were driven about 1 kilometer with frequent pauses for rest. "The seals were panting heavily and frequently paused to wave their hind flippers in the air before they had been driven 150 yards from the rookery. By the time the drive was half finished most of the seals appeared badly tired and occasional animals were dropping out of the pods [groups of seals]. In the last 200 yards of the drive and on the killing grounds there were found 16 'roadskins' (animals that had died of heat prostration) and in addition a number of others prostrated by overheating." The average body temperature of the roadskins of this drive was 42.2°C, which is 4.5°C above the 37.7°C mean body temperature of adults not under thermal stress.

Fur seals can withstand somewhat higher temperatures in water than they can in air because of the greater heat conduction of water, but they are not able to penetrate warm seas. Adult male fur seals apparently remain in the Bering Sea during their pelagic season. Young males and females migrate into the North Pacific, but they are not found in waters warmer than 14 to 15°C, and they are most abundant in water of 11°C. Their inability to regulate body temperature during sustained activity and their sensitivity to even low levels of solar radiation and moderate air temperatures probably restrict the location of potential breeding sites and their movements during their pelagic periods. Summers in the Pribilofs are barely cool enough to allow the seals to breed there. An increase in summer temperature associated with global warming might drive the seals from their traditional breeding grounds.

Migration to Avoid Stressful Conditions

Every environment has unfavorable aspects for some species, and these unfavorable conditions are often seasonal, especially in latitudes far from the Equator. The primary cause of migrations is usually related to seasonal changes in climatic factors such as temperature or rainfall. In turn, these conditions influence food supply and the occurrence of suitable breeding conditions.

We can consider the costs and benefits of migrating by considering two kinds of animals that represent extremes of body size. The baleen whales are the largest animals that have ever lived, and hummingbirds are among the smallest endotherms; yet both whales and hummingbirds migrate.

Whales

The annual cycle of events in the lives of the great baleen whales is particularly instructive in showing how migration relates to the use of energy and how it correlates with reproduction in the largest of all animals. Most baleen whales summer in polar or subpolar waters of either the Northern or the Southern Hemisphere, where they feed on krill or other crustaceans that are abundant in those cold, productive waters. For three or four months each year a whale consumes a vast quantity of food that is converted into stored energy in the form of blubber and other kinds of fat. During this same time pregnant female whales nurture their unborn young, which may grow to one-third the length of their mothers before birth.

Near the end of summer the whales begin migration toward tropical or subtropical waters where the females bear their young and nurse them for a period of time before making the reverse migration. During this winter sojourn in warm waters,

some of the whales also mate. The young grow rapidly on the rich milk provided by their mothers, and by spring the calves are mature enough to travel with their mothers back to arctic or antarctic waters. The calves are weaned about the time they arrive in their summer quarters. From a bioenergetic and trophic point of view the remarkable feature of this annual cycle is that virtually all of the energy required to fuel it comes from ravenous feeding and fattening during the three or four months spent in polar seas. Little or no feeding occurs during migration or during the winter period of calving and nursing. Energy for all these activities comes from the abundant stores of blubber and fat.

The gray whale (*Eschrichtius robustus*) of the Pacific Ocean has one of the longest and best known migrations (Figure 22–4). The summer feeding waters are in the Bering Sea and the Chukchi Sea north of the Bering Strait in the Arctic Ocean. A small segment of the population moves down the coast of Asia to Korean waters at the end of the Arctic summer, but most gray whales follow the Pacific Coast of North America, moving south to Baja California and adjacent parts of western Mexico. They arrive in December or January, bear their young in shallow, warm waters, and then depart northward again in March. Some gray whales make an annual round-trip of at least 9000 kilometers and the adults eat little or nothing for the eight months they are away from their northern feeding grounds.

The amount of energy expended by a whale in this annual cycle is phenomenal. The basal metabolic rate of a gray whale with a fat-free body mass of 50,000 kilograms is approximately 979,000 kilojoules per day. If the metabolic rate of a free-ranging whale, including the locomotion involved in feeding and migrating, is about three times the basal rate (a typical level of energy use for mammals), the whale's average daily energy expenditure is around 2,937,000 kilojoules. Body fat contains 38,500 kilojoules per kilogram, so the whale's daily energy expenditure is equivalent to metabolizing over 76 kilograms of blubber or fat per day. Assuming an energy content of 20,000 kilojoules per kilogram for krill and a 50 percent efficiency in converting the gross energy intake of food into biologically usable energy, the energy requirement for existence is equivalent to a daily intake of 294 kilograms of food.

To exist for 245 days during the migration and calving in midlatitudinal waters without eating, the whale must metabolize a minimum of 18,375 kilo-

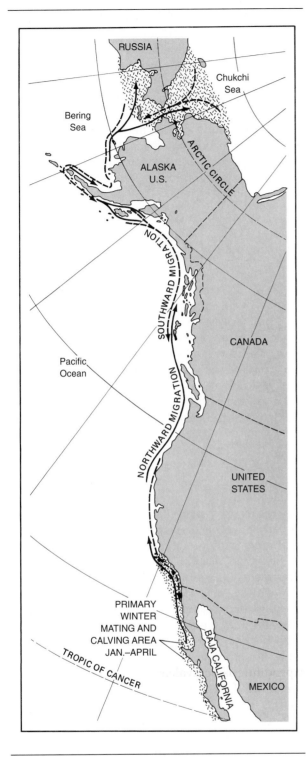

Figure 22–4 Migratory route of the gray whale between the Arctic Circle and Baja California.

grams of fat. Accumulating that amount of fat in 120 days of active feeding in arctic waters at a conversion efficiency of 50 percent requires the consumption of 70,438 kilograms of krill, or 586 kilograms per day. The total food intake per day on the feeding grounds to accommodate the whale's daily metabolic needs plus energy storage for the migratory period would be not less than 294 + 586 = 880 kilograms of krill per day.

This is a minimum estimate for females because the calculations do not include the energetic costs of the developing fetus or the cost of milk production. Nor does it include the cost of transporting 20,000 kilograms of fat through the water. But a large whale can do all this work and more without exhausting its insulative blanket of blubber because nearly half the total body mass of a large whale consists of blubber and other fats.

Why does a gray whale expend all this energy to migrate? The adult is too large and too well insulated ever to be stressed by the cold arctic and subarctic waters that do not vary much from 0°C from summer to winter anyway. It seems strange for an adult whale to abandon an abundant source of food and go off on a forced starvation trek into warm waters that may cause stressful overheating. The advantage probably accrues to the newborn young, which, though relatively large, lacks an insulative layer of blubber. If the young whale were born in cold northern waters it would probably have to utilize a large fraction of its energy intake (milk produced from its mother's stored fat) to generate metabolic heat to regulate its body temperature. That energy could otherwise be used for rapid growth. Apparently, it is more effective, and perhaps energetically more efficient, for the mother whale to migrate thousands of kilometers into warm waters to give birth and nurse in an environment where the young whale can invest most of its energy intake in rapid growth.

Hummingbirds

At the opposite end of the size range, hummingbirds are the smallest endotherms that migrate. Ornithologists have long been intrigued by the ability of the ruby-throated hummingbird (*Archilochus colubris*), which weighs only 3.5 to 4.5 grams, to make a nonstop flight of 800 kilometers during migration across the Gulf of Mexico from Florida to the Yucatan Peninsula. For many years it seemed impossible that such a small bird with such a high resting metabolic rate could store enough energy for such a long flight.

Measurements of body fat, oxygen consumption during flight, and speed of flight have finally solved the riddle. Like most migratory birds, ruby-throated hummingbirds store subcutaneous and body fat by feeding heavily prior to migration. A hummingbird with a lean mass of 2.5 grams can accumulate 2 grams of fat. Measurements of the energy consumption of a hummingbird hovering in the air in a respirometer chamber in the lab indicate an energy consumption of 2.89 to 3.10 kilojoules per hour. Hovering is energetically more expensive than forward flight, so these values represent the maximum energy used in migratory flight. Even so, two grams of fat produces enough energy to last for 24 to 26 hours of sustained flight. Hummingbirds fly about 40 kilometers per hour, so crossing the Gulf of Mexico requires about 20 hours. Thus, by starting with a full store of fat, they have enough energy for the crossing with a reserve for unexpected contingencies such as a headwind that slows their progress. In fact, most migratory birds wait for weather conditions that will generate tailwinds before they begin their migratory flights, thereby further reducing the energy cost of migration.

Torpor as a Response to Low Temperatures and Limited Food

We have stressed the high energy cost of endothermy because the need to collect and process enough food to supply that energy is a central factor in the lives of many endotherms. In extreme situations environmental conditions may combine to overpower a small endotherm's ability to process and transform enough chemical energy to sustain a high body temperature through certain critical phases of its life. For diurnally active birds, long cold nights during which there is no access to food can be lethal, especially if the bird has not been able to feed fully during the daytime. Cold winter seasons usually present a dual problem for resident endotherms: (1) the necessity to maintain high body temperature against a greatly increased temperature difference between its internal temperature and the ambient temperature, and (2) a relative scarcity of food. In

response to such problems, some birds and mammals have mechanisms that permit them to avoid the energetic costs of maintaining a high body temperature under unfavorable circumstances by entering a state of torpor (adaptive hypothermia). By entering torpor an endotherm is giving up many of the advantages of endothermy, but in exchange it realizes an enormous saving of energy. Thus, endotherms enter torpor only when they would face critical shortages of energy or water if they remained at normal body temperature.

Physiological Adjustments During Torpor

When an endotherm becomes torpid profound changes occur in a variety of physiological functions (Heller 1987). Although body temperatures may fall very low during torpor, temperature regulation does not entirely cease. In **deep torpor** an animal's body temperature drops to within 1°C or less of the ambient temperature, and in some cases (bats, for example) extended survival is possible at body temperatures just above the freezing point of the tissues. Arctic ground squirrels actually allow the temperature of parts of their bodies to supercool as low as –2.9°C (Barnes 1989). Oxidative metabolism and energy use are reduced to as little as one-twentieth of the rate at normal body temperatures. Respiration is slow, and the overall breathing rate can be less than one inspiration per minute. Heart rates are drastically reduced and blood flow to peripheral tissues is virtually shut down, as is blood flow posterior to the diaphragm. Most of the blood is retained in the core of the body. In this respect, the cardiovascular adjustments that occur during torpor are like those seen in diving animals (Jones et al. 1988).

Deep torpor is a comatose condition much more profound than the deepest sleep. Voluntary motor responses are reduced to sluggish postural changes, but some sensory perception of powerful auditory and tactile stimuli and ambient temperature changes is retained. Perhaps most dramatically, a torpid animal can arouse spontaneously from this state by endogenous heat production that restores the high body temperature characteristic of a normally active endotherm. Some endotherms can rewarm under their own power from the lowest levels of torpor, whereas others must warm passively with an increase in ambient temperature until some threshold is reached at which arousal starts.

There are varying degrees of torpor, from the deepest states of hypothermia to the lower range of body temperatures reached by normally active endotherms during their daily cycles of activity and sleep. Nearly all birds and mammals, especially those with body masses under 1 kilogram, undergo **circadian temperature cycles.** These cycles vary from 1 to 5°C or more between the average high-temperature characteristic of the active phase of the daily cycle and the average low-temperature characteristic of rest or sleep (Aschoff 1982). Small birds (sunbirds, hummingbirds, chickadees) and small mammals (especially bats and rodents) may drop their body temperatures during quiescent periods from 8 to 15°C below their regulated temperature during activity. Even a bird as large as the turkey vulture (about 2.2 kilograms) regularly drops its body temperature at night. When all these different endothermic patterns are considered together, no really sharp distinction can be drawn between torpor and the basic daily cycle in body temperature that characterizes most small to medium-size endotherms.

Body Size and the Occurrence of Torpor

Species of endotherms capable of deep torpor are found in a number of groups of mammals and birds. All three subclasses of mammals include species capable of torpor. The echidna, the platypus, and several species of small marsupials possess various patterns of hypothermia, but the phenomenon is most diverse among placentals, particularly among bats and rodents. Certain kinds of hypothermia have also been described for some insectivores, particularly the hedgehog, some primates, and some edentates. Deep torpor, contrary to popular notion, is not known for any of the carnivores, despite the fact that some of them den in the winter and remain inactive for long periods. Among birds, torpor occurs in some of the goatsuckers or nightjars and in hummingbirds, swifts, mousebirds, and some passerines (sunbirds, swallows, chickadees, and others). Other species, including larger ones like turkey vultures, show varying depths of hypothermia at rest or in sleep but are not in a semicomatose state of deep torpor.

Torpor and body size are closely related, especially among mammals (French 1986). The largest mammals that undergo deep torpor are marmots, which weigh about five kilograms. The limitation on body size reflects a balance between the energy expenditure at normal body temperatures and during torpor and the time and energy spent in entry into torpor and in arousal. Torpor is not as energetically advantageous for a large animal as for a small one. In the first place, the energetic cost of maintaining high body temperature is relatively greater for a small animal than for a large one and, as a consequence, a small animal has more to gain from becoming torpid. Second, a large animal cools off more slowly than a small animal and does not lower its metabolic rate as rapidly.

Furthermore, large animals have more body tissue to rewarm on arousal, and their costs of arousal are correspondingly larger than those of small animals (Figure 22–5). An endotherm weigh-

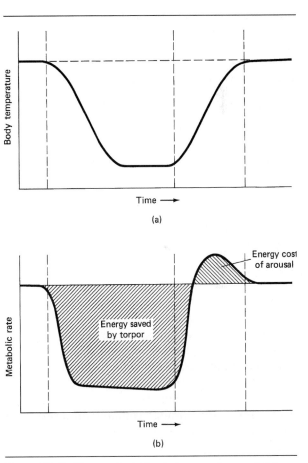

Figure 22–5 Changes in (a) body temperature and (b) metabolic rate (energy consumption) during torpor. A decrease in metabolic rate precedes a fall in body temperature to a new set point. An increase in metabolism produces the heat needed to return to normal body temperatures; the metabolic rate during arousal briefly overshoots the resting rate.

ing a few grams, such as a little brown bat or a hummingbird, can warm up from torpor at the rate of about 1°C per minute, and be fully active within 30 minutes or less depending on the depth of hypothermia. A 100-gram hamster requires more than 2 hours to arouse, and a marmot takes many hours. Entrance into torpor is slower than arousal. Consequently, daily torpor is feasible only for very small endotherms; there would not be enough time for a large animal to enter and arouse from torpor during a 24-hour period. Moreover, the energy required to warm up a large mass is very great. Oliver Pearson calculated that it costs a 4-gram hummingbird only 0.48 kilojoule to raise its body temperature from 10 to 40°C. That is 1/85 of the total daily energy expenditure of an active hummingbird in the wild. By contrast, a 200-kilogram bear would require 18,000 kilojoules to warm from 10 to 37°C, the equivalent of a full day's energy expenditure. The smaller potential savings and the greater costs of arousal make daily torpor impractical for any but small endotherms.

Medium-size endotherms are not entirely excluded from the energetic savings of torpor, but the torpor must persist for a longer period to realize a saving. For example, ground squirrels and marmots enter prolonged torpor during the winter **hibernation** when food is scarce. They spend several days at very low body temperatures (in the region of 5°C), then arouse for a period before becoming torpid again (Box 22–1). Still larger endotherms would have such large total costs of arousal (and would take so long to warm up) that torpor is not feasible for them even on a seasonal basis. Bears in winter dormancy, for example, lower their body temperatures only about 5°C from normal levels, and metabolic rate decreases about 50 percent. Even this small drop, however, amounts to a large energy saving through the course of a winter for an animal as large as a bear. That small reduction in body temperature, plus the large fat stores bears accumulate before retreating to their winter dens, is sufficient to carry a bear through the winter.

Energy Relations of Daily Torpor

Studies of daily torpor in birds have emphasized the flexibility of the response in relation to the energetic stress faced by individual birds. Susan Chaplin's work with chickadees provides an example (Chaplin 1974). These small (10- to 12-gram) passerine birds are winter residents in northern latitudes, where they regularly experience ambient temperatures that do not rise above freezing for days or weeks (Figure 22–9).

Chaplin found that in winter chickadees around Ithaca, New York, allow their body temperatures to drop from the normal level of 40 to 42°C that is maintained during the day to 29 to 30°C at night. This reduction in body temperature permits a 30 percent reduction in energy consumption. The chickadees rely primarily on fat stores they accumulate as they feed during the day to supply the energy needed to carry them through the following night. Thus the energy available to them and the energy they utilize at night can be estimated by measuring the fat content of birds as they go to roost in the evening and as they begin activity in the morning. Chaplin found that in the evening chickadees had an average of 0.80 gram of fat per bird. By morning the fat store had decreased to 0.24 gram. The fat metabolized during the night (0.56 gram per bird) corresponds to the metabolic rate expected for a bird that had allowed its body temperature to fall to 30°C.

Chaplin's calculations show that this torpor is necessary if the birds are to survive the night. It would require 0.92 gram of fat per bird to maintain a body temperature of 40°C through the night. That is more fat than the birds have when they go to roost in the evening. If they did not become torpid, they would starve before morning. Even with torpor, they use 70 percent of their fat reserve in one night. They do not have an energy supply to carry them far past sunrise and chickadees are among the first birds to start to forage in the morning. They also forage in weather so foul that other birds, which are not in such precarious energy balance, remain on their roosts. The chickadees must reestablish their fat stores each day if they are to survive the next night.

Hummingbirds, too, may depend on the energy they gather from nectar during the day to carry them through the following night. These very small birds (4 to 10 grams) have extremely high energy expenditures and yet are found during the summer in northern latitudes and at high altitudes. An example of the lability of torpor in hummingbirds was provided by studies of nesting broad-tailed hummingbirds at an altitude of 2900 meters near Gothic, Colorado (Calder and Booser 1973). Ambient tem-

Box 22–1 Waking Up Is Hard Work: The Cost of Arousal

Hibernation is an effective method of conserving energy during long winters, but hibernating animals do not remain at low body temperatures for the whole winter. Periodic arousals are normal, and these arousals consume a large portion of the total amount of energy used by hibernating mammals (French 1986). An example of the magnitude of the energy cost of arousal is provided by Lawrence Wang's study of the Richardson's ground squirrel (*Spermophilus richardsonii*, Figure 22–6) in Alberta, Canada (Wang 1978).

The activity season for ground squirrels in Alberta is short: They emerge from hibernation in mid-March and adult squirrels reenter hibernation 4 months later, in mid-July. Juvenile squirrels begin hibernation in September. When the squirrels are active they have body temperatures of 37 to 38°C, and their temperatures fall as low as 3 to 4°C when they are torpid. Figure 22–7 shows the body temperature of a juvenile male ground squirrel from September through March; periods of torpor alternate with arousals all through the winter. Hibernation began in mid-September with short bouts of torpor followed by rewarming. At this time the temperature in the burrow was about 13°C. As the winter progressed and the temperature in the burrow fell, the intervals between arousals

Figure 22–6 Richardson's ground squirrel, *Spermophilus richardsonii*. (Photograph by Gail R. Michener.)

lengthened and the body temperature of the torpid animal declined. By December, January, and February the burrow temperature had dropped to 0°C and the periods between arousals averaged 14 to 19 days. In late February the periods of torpor became shorter, and in early March the squirrel emerged from hibernation.

Continued

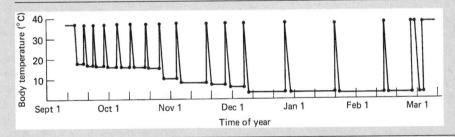

Figure 22–7 Record of body temperature during a complete torpor season for a Richardson's ground squirrel. Torpor cycles are initially short and become longer as winter progresses, then shorten again as spring approaches. (From L. C. H. Wang, 1978, *Strategies in Cold: Natural Torpidity and Thermogenesis*, edited by L. C. H. Wang and J. W. Hudson, Academic, New York, NY.)

Box 22–1 *(Continued)*

Figure 22–8 Record of body temperature during a single torpor cycle for a Richardson's ground squirrel. (From L. C. H. Wang, 1978, *Strategies in Cold: Natural Torpidity and Thermogenesis*, edited by L. C. H. Wang and J. W. Hudson, Academic, New York, NY.)

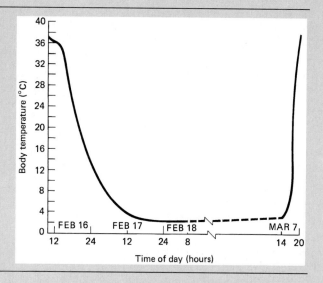

A torpor cycle consists of entry into torpor, a period of torpor, and an arousal (Figure 22–8). Entry into torpor began shortly after noon on February 16, and 24 hours later the body temperature had stabilized at 3°C. This period of torpor lasted until late afternoon on March 7, when the squirrel started to arouse. In 3 hours the squirrel warmed from 3 to 37°C. It maintained that body temperature for 14 hours, and then began entry into torpor again.

These periods of arousal account for most of the energy used during hibernation (Table 22–1). The energy costs associated with arousal include the cost of warming from the hibernation temperature to 37°C, the cost of sustaining a body temperature of 37°C for several hours, and the metabolism above torpid levels as the body temperature slowly declines during reentry into torpor. For the entire hibernation season the combined metabolic expenditures for those

peratures drop nearly to freezing at night, and hummingbirds become torpid when energy is limiting. Calder and Booser were able to monitor the body temperatures of nesting birds by placing an imitation egg containing a temperature-measuring device in the nest. These temperature records showed that hummingbirds incubating eggs normally did not become torpid at night. The reduction of egg temperature that results from the parent bird's becoming torpid does not damage the eggs, but it slows development and delays hatching. Presumably, there are advantages to hatching the eggs as quickly as possible, and as a result brooding hummingbirds expend energy to keep themselves and their eggs warm through the night, provided that they have the energy stores necessary to maintain the high metabolic rates needed.

On some days bad weather interfered with foraging by the parent birds, and as a result they apparently went into the night with insufficient energy supplies to maintain normal body temperatures. In this situation the brooding hummingbirds did become torpid for part of the night. One bird that had experienced a 12 percent reduction in foraging time during the day became torpid for 2 hours, and a second that had lost 21 percent of its foraging was torpid for 3.5 hours. Thus, torpor can be a flexible response that integrates the energy stores of a bird with environmental conditions and biological requirements such as brooding eggs (Calder 1994).

Table 22–1 Use of energy during differen phases of the torpor cycle by Richardson's ground squirrel.

Month	Percentage of Total Energy per Month			
	Torpor	Warming	Intertorpor Homeothermy	Reentry
July	8.5	17.2	56.5	17.8
September	19.2	15.2	49.9	15.7
November	20.8	23.1	43.1	13.0
January	24.8	24.1	40.0	11.1
March	3.3	14.0	76.4	6.3
Average for season	16.6	19.0	51.6	12.8

Source: L. C. H. Wang, 1978, pages 109–145 in *Strategies in Cold: Natural Torpidity and Thermogenesis*, edited by L. C. H. Wang and J. W. Hudson, Academic, New York, NY.

three phases of the torpor cycle account for an average of 83 percent of the total energy the squirrel uses.

Surprisingly, we have no clear idea of why a hibernating ground squirrel undergoes these arousals that increase its total winter energy expenditure nearly fivefold. Ground squirrels do not store food in their burrows, so they are not using the periods of arousal to eat. They do urinate during arousal, and it is possible that some time at a high body temperature is necessary to carry out physiological or biochemical activities such as resynthesizing glycogen, redistributing ions, or synthesizing serotonin. Arousal may also allow a hibernating animal to determine when environmental conditions are suitable for emergence. Whatever their function, the arousals must be important because the squirrel pays a high energy price for them during a period of extreme energy conservation.

Endotherms in the Heat: Deserts

Hot, dry areas place a more severe physiological stress on endotherms than do the polar conditions we have already discussed. The difficulties endotherms encounter in deserts result from a reversal of their normal relationship to the environment. Endothermal thermoregulation is most effective when an animal's body temperature is higher than the temperature of its environment. In this situation heat flow is from the animal to its environment, and thermoregulatory mechanisms achieve a stable body temperature by balancing heat production and heat loss. Very cold environments merely increase the gradient between an animal's body temperature and the environment. The example of arctic foxes with lower critical temperatures of –40°C illustrates the success that endotherms have had in providing sufficient insulation to cope with enormous gradients between high core body temperatures and low environmental temperatures.

In a desert the gradient is not increased; it is reversed. Desert air temperatures can climb to 40 or 50°C during summer, and the ground temperature may exceed 60 or 70°C. Instead of losing heat to the environment, an animal is continually absorbing heat, and that heat plus metabolic heat must somehow be dissipated to maintain the animal's body

Figure 22–9 The black-capped chickadee, *Parus atricapillus*, is a small bird that winters in cold climates. (Photograph © Gregory K. Scott)

temperature in the normal range. It can be a greater challenge for an endotherm to maintain its body temperature 10°C below the ambient temperature than to maintain it 100°C above ambient.

Evaporative cooling is the major mechanism an endotherm uses to reduce its body temperature. The evaporation of water requires approximately 2427 kilojoules/kilogram. (The exact value varies with temperature.) Thus, evaporation of a liter of water dissipates 2427 kilojoules, and evaporative cooling is a very effective mechanism as long as an animal has an unlimited supply of water. In a hot desert, however, where the thermal stress is greatest, water is a scarce commodity and its use must be carefully rationed. Calculations show, for example, that if a kangaroo rat were to venture out into the desert sun, it would have to evaporate 13 percent of its body water per hour to maintain a normal body temperature. Most mammals die when they have lost 10 to 20 percent of their body water, and it is obvious that, under desert conditions, evaporative cooling is of limited utility except as a short-term response to a critical situation.

Unable to rely on evaporative cooling, endotherms have evolved a number of other responses that have allowed a diverse assemblage of birds and mammals to inhabit deserts. The mechanisms they use are complex and involve combinations of ecological, behavioral, morphological, and physiological mechanisms that act together to enhance the effectiveness of the entire system. As a start toward unraveling some of these complexities, we can categorize three major classes of responses of endotherms to desert conditions as follows:

1. Some endotherms manage to avoid desert conditions by behavioral means. They live in deserts but are rarely exposed to the full stress of desert life.
2. Other endotherms have relaxed the limits of homeostasis. They manage to survive in deserts by tolerating greater ranges of variation in characters such as body temperature or body water content than normal.
3. Specializations such as torpor in response to shortages of food or water and a reduced standard metabolic rate (and, consequently, a reduction in the amount of metabolic heat an animal must dissipate) are combined with some of the characters mentioned in 1 and 2 in some desert endotherms.

Large Mammals in Hot Deserts

Large animals, including humans, have specific advantages and disadvantages in desert life that are directly related to body size. A large animal has nowhere to hide from desert conditions. It is too big

Figure 22–10 Bactrian camels in the Gobi Desert. In the heat of the day, most of these camels have faced into the sun to reduce the amount of direct solar radiation they receive and have pressed against each other to reduce the heat they gain by convection and reradiation. (Photograph © George Holton)

to burrow underground, and few deserts have vegetation large enough to provide useful shade to an animal much larger than a jackrabbit. On the other hand, large body size offers some options not available to smaller animals. Large animals are mobile and can travel long distances to find food or water, whereas small animals may be limited to home ranges only a few meters or tens of meters in diameter. Large animals have small surface/mass ratios and can be well insulated. Consequently, they absorb heat from the environment slowly. A large body mass gives an animal a large thermal inertia; that is, it can absorb a large amount of heat before its body temperature reaches dangerous levels.

The Camel The dromedary camel (*Camelus dromedarius*) is the classic large desert animal (Figure 22–10). There are authentic records of journeys in excess of 500 kilometers, lasting two or three weeks, during which the camels did not have an opportunity to drink. The longest trips are made in winter and spring when air temperatures are relatively low and scattered rainstorms may have produced some fresh vegetation that provides a little food and water for the camels.

Camels are large animals—adult body masses of dromedary camels are 400 to 450 kilograms for females and up to 500 kilograms for males. The camel's specializations for desert life are revealed by comparing the daily cycle of body temperature in a camel that receives water daily and one that has been deprived of water (Figure 22–11). The watered camel shows a small daily cycle of body temperature with a minimum of 36°C in the early morning and a maximum of 38°C in midafternoon. When a camel is deprived of water, the daily temperature variation triples. Body temperature is allowed to fall to 34.5°C at night and climbs to 40.5°C during the day.

The significance of this increased daily fluctuation in body temperature can be assessed in terms of the water that the camel would expend to prevent the 6°C rise by evaporative cooling. With a specific heat of 3.4 kilojoules/(kilogram · °C), a 6°C increase in body temperature for a 500-kilogram camel represents storage of 10,000 kilojoules of heat. Evaporation of a kilogram of water dissipates approximately 2427 kilojoules. Thus, a camel would have to evaporate slightly more than 4 liters of water to maintain a stable body temperature at the nighttime level; by tolerating hyperthermia during the day it can conserve that water.

In addition to the direct saving of water not used for evaporative cooling, the camel receives an indirect benefit from tolerating hyperthermia in a reduction of energy flow from the air to the camel's body. As long as the camel's body temperature is below air temperature a gradient exists that causes the camel to absorb heat from the air. At a body temperature of 40.5°C the camel's temperature is equal to that of the air for much of the day, and no net heat exchange takes place. Thus, the camel saves an additional quantity of water by eliminating the temperature gradient between its body and the air. The combined effect of these measures on water loss is illustrated

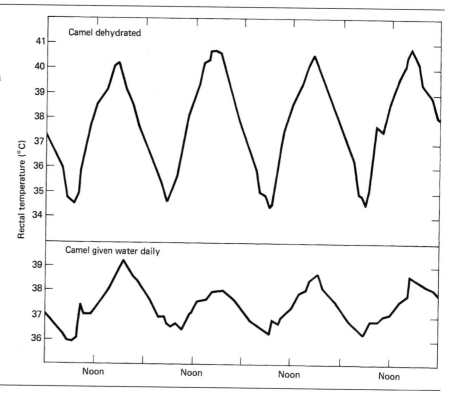

Figure 22–11 Daily cycles of body temperature for a dehydrated camel (top) and a camel with daily access to water (bottom). (From K. Schmidt–Nielsen et al., 1957, *American Journal of Physiology* 188:103–112.)

by data from a young camel (Table 22–2). When deprived of water the camel reduced its evaporative water loss by 64 percent and reduced its total daily water loss by half.

Behavioral mechanisms and the distribution of hair on the body aid dehydrated camels in reducing their heat load. In summer camels have hair 5 or 6 centimeters long on the back and up to 11 centimeters long over the hump. On the ventral surface and legs the hair is only 1.5 to 2 centimeters long. Early in the morning camels lie down on surfaces that have cooled overnight by radiation of heat to the night sky. The legs are tucked beneath the body and the ventral surface, with its short covering of hair, is placed in contact with the cool ground. In this position a camel exposes only its well-protected back and

sides to the sun and places its lightly furred legs and ventral surface in contact with cool sand, which may be able to conduct away some body heat. Camels may assemble in small groups and lie pressed closely together through the day. Spending a day in the desert sun squashed between two sweaty camels may not be your idea of fun, but in this posture a camel reduces its heat gain because it keeps its sides in contact with other camels (both at about 40°C) instead of allowing solar radiation to raise the fur surface temperature to 70°C or above.

Despite their ability to reduce water loss and to tolerate dehydration, the time eventually comes when even camels must drink. These large, mobile animals can roam across the desert seeking patches of vegetation produced by local showers and move

Table 22-2 **Daily water loss of a 250-kg camel.**

Condition	Feces	Urine	Evaporation	Total
	\multicolumn{4}{c}{Water Loss (L/day) by Different Routes}			
Drinking daily (8 days)	1.0	0.9	10.4	12.3
Not drinking (17 days)	0.8	1.4	3.7	5.9

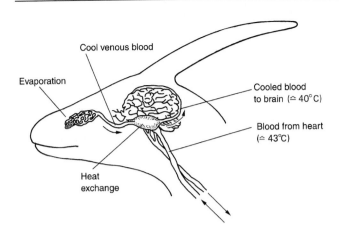

Figure 22–12 Countercurrent heat-exchange mechanism that may cool blood going to a gazelle's brain. (Modified from C. R. Taylor, 1972, in *Comparative Physiology of Desert Animals*, edited by G. M. O. Maloiy, Academic, London, UK.)

Cool venous blood

Evaporation

Cooled blood to brain ($\cong 40°C$)

Blood from heart ($\cong 43°C$)

Heat exchange

from one oasis to another, but when they drink they face a problem they share with other grazing animals: Water holes can be dangerous places. Predators frequently center their activities around water holes, where they are assured of water as well as a continuous supply of prey animals. Reducing the time spent drinking is one method of reducing the risk of predation, and camels can drink remarkable quantities of water in very short periods. A dehydrated camel can drink as much as 30 percent of its body mass in 10 minutes. (A very thirsty human can drink about 3 percent of body mass in the same time.)

The water a camel drinks is rapidly absorbed into its blood. The renal blood flow and glomerular filtration rate increase and urine flow returns to normal within a half hour of drinking. The urine changes from dark brown and syrupy to colorless and watery. Aldosterone stimulates sodium reabsorption, which helps to counteract the dilution of the blood by the water the camel has drunk. Nonetheless, dilution of the blood causes the red blood cells to swell as they absorb water by osmosis. Camel erythrocytes are resistant to this osmotic stress, but other desert ruminants have erythrocytes that can burst under hypotonic conditions. Bedouin goats, for example, have fragile erythrocytes, and the water a goat drinks is absorbed slowly from the rumen. Goats require two days to return to normal kidney function after dehydration.

Other Large Mammals Richard Taylor has investigated the temperature and water relations of several species of African antelope that live in arid grasslands or desert regions (Taylor 1972). These animals, which range in size from the 20-kilogram Thomson's gazelle and 50-kilogram Grant's gazelle (*Gazella thomsoni* and *G. granti*) to the 100-kilogram oryx (*Oryx beisa*) and 200-kilogram eland (*Taurotragus oryx*) utilize heat storage like the dromedary, but allow their body temperatures to rise considerably above the 40.5°C level recorded for the camel. Taylor recorded rectal temperatures of 45°C for the oryx and 46.5°C for the Grant's gazelle. (Thomson's gazelles, which do not penetrate into desert areas, began to pant at air temperatures above 42°C and used evaporative cooling to maintain body temperature below air temperature.)

Body temperatures above 43°C rapidly produce brain damage in most mammals, but Grant's gazelles maintained rectal temperatures of 46.5°C for as long as 6 hours with no apparent ill effects. These antelope keep brain temperature below body temperature by using a countercurrent heat exchange to cool blood before it reaches the brain. In ungulates the blood supply to the brain passes via the external carotid arteries (Figure 22–12). At the base of the brain these arteries break into a rete mirabile that lies in a venous sinus. The blood in the sinus is venous blood, returning from the walls of the nasal passages where it has been cooled by the evaporation of water. This cool venous blood thus cools the warmer arterial blood before it reaches the brain. A mechanism of this sort is widespread among mammals.

Unlike the dromedary camel, antelopes are apparently independent of drinking water even during summer. Taylor suggests that one of the mechanisms that permits this independence is behavioral. The leaves of a desert shrub, *Diasperma,* are an important part of the diet of the antelopes. During the day, when air temperature is high and humidity is low, these leaves contain about 1 percent water by weight. They are so dry that they disintegrate into powder when they are touched. At night, however, as air temperatures fall and relative humidity increases, the leaves take up water and after 8 hours have a water content of 40 percent. If antelopes ate the leaves at night rather than in the daytime, they might obtain enough water from their food to be independent of water holes.

Large animals such as those we have discussed illustrate one approach to desert life. Too large to escape the stresses of the environment, they survive by tolerating a temporary relaxation of homeostasis. Their success under the harsh conditions in which they live is the result of complex interactions between diverse aspects of their ecology, behavior, morphology, and physiology. Only when all of these features are viewed together does an accurate picture of an animal emerge.

Birds in Desert Regions

Although birds are relatively small vertebrates, the problems they face in deserts are more like those experienced by camels and antelope than like those of small mammals. Birds are predominantly diurnal and few seek shelter in burrows or crevices. Thus, like large mammals, they meet the stresses of deserts head on and face the antagonistic demands of thermoregulation in a hot environment and the need to conserve water.

Also like large mammals, birds are mobile. It is quite possible for a desert bird to fly to a mountain range on a daily basis to reach water. For example, mourning doves in the deserts of North America congregate at dawn at water holes, some individuals flying 60 kilometers or more to reach them. The normally high and labile body temperatures of birds give them an advantage in deserts that is not shared by mammals. With body temperatures normally around 40°C, birds face the problem of a reversed temperature gradient between their bodies and the environment for a shorter portion of each day than would a mammal. Furthermore, birds' body temperatures are normally variable, and birds tolerate moderate hyperthermia without apparent distress.

These are all preadaptations to desert life that are present in virtually all birds. Neither the body temperatures nor the lethal temperatures of desert birds are higher than those of related species from non-desert regions.

The mobility provided by flight does not extend to fledgling birds, and the most conspicuous adaptations of birds to desert conditions are those that ensure a supply of water for the young. Altricial fledglings, those that need to be fed by their parents after hatching, receive the water they need from their food. One pattern of adaptation in desert birds ensures that reproduction will occur at a time when succulent food is available for fledglings. In the arid central region of Australia, bird reproduction is precisely keyed to rainfall. The sight of rain is apparently sufficient to stimulate courtship, and mating and nest building commence within a few hours of the start of rain. This rapid response ensures that the baby birds will hatch in the flush of new vegetation and insect abundance stimulated by the rain.

A different approach, very like that of mammals, has been evolved in columbiform birds (pigeons and doves), which are widespread in arid regions. Fledglings are fed on pigeon's milk, a liquid substance produced by the crop under the stimulus of prolactin. The chemical composition of pigeon's milk is very similar to that of mammalian milk; it is primarily water plus protein and fat, and it simultaneously satisfies both the nutritional requirements and the water needs of the fledgling. This approach places the water stress on the adult, which must find enough water to produce milk as well as meeting its own water requirements.

Seed-eating desert birds with precocial young, like the sandgrouse found in the deserts of Africa and the Near East, face particular problems in providing water for their young. Baby sandgrouse begin to find seeds for themselves within hours of hatching, but they are unable to fly to water holes as their parents do and seeds do not provide the water they need. Instead, adult male sandgrouse transport water to their broods. The belly feathers, especially in males, have a unique structure in which the proximal portions of the barbules are coiled into helices. When the feather is wetted, the barbules uncoil and trap water. The feathers of male sandgrouse hold 15 to 20 times their weight of water, and the feathers of females hold 11 to 13 times their weight.

Male sandgrouse in the Kalahari Desert of southern Africa fly to water holes just after dawn and soak their belly feathers, absorbing 25 to 40 milliliters of water. Some of this water evaporates on the flight back to their nests, but calculations indicate that a male sandgrouse could fly 30 kilometers and arrive with 10 to 28 milliliters of water still adhering to its feathers. As the male sandgrouse lands, the juveniles rush to him, and seizing the wet belly feathers in their beaks, strip the water from them with downward jerks of their heads. In a few minutes, the young birds have satisfied their thirst and the male rubs himself dry on the sand.

Small Mammals in Hot Deserts

Rodents are the preeminent small mammals of arid regions. It is a commonplace observation that population densities of rodents may be higher in deserts than in moist situations. A number of features of rodent biology can be viewed as preadaptive for extending their geographic ranges into hot, arid regions. Among the most important of these preadaptations are the normally nocturnal habits of many rodents and their practice of living in burrows. A burrow provides ready escape from the heat of a desert, giving an animal access to a microenvironment within its thermoneutral zone while soil temperatures on the surface climb above 60°C. Rodents that live in burrows during the day and emerge to forage at night escape the desert heat so successfully that their greatest temperature stress may be cold. Because of the normal absence of clouds, deserts cool rapidly after sundown and many deserts are distinctly chilly at night during much of the year.

Although retreat to a burrow during the day provides direct escape from heat, it is not, by itself, a solution to the other major challenges of desert life—the chronic shortages of food and water. What the burrow does provide is the shelter and microclimate an animal needs in order to solve the other problems. We pointed out earlier that a kangaroo rat would reach its lethal limit of dehydration in less than 2 hours if it had to rely on evaporative cooling to maintain its body temperature at normal levels during the day. By retreating into its burrow, a kangaroo rat avoids that use of water. Indeed, the water savings of a burrow probably go beyond that. As a rodent in a burrow loses water by evaporation, the air in the burrow becomes humid and may approach saturation. At the same time of day, the relative humidity of air outside the burrow may be only 20 to 30 percent. The higher humidity of burrow air reduces an animal's evaporative water loss.

A further saving is achieved in some animals (including birds and lizards in addition to mammals) by a countercurrent water recycler in the nasal passages. The air an animal exhales leaves the nares at a temperature lower than that at which it left the lungs. The phenomenon has important implications in terms of energy and water balance.

A brief consideration of the respiratory cycle illustrates the mechanism involved. As air is inhaled, it passes over moist tissues in the nasal passages. The nasal passages themselves are narrow and the wall surface area is large. As the air passes over these moist surfaces, it is warmed and humidified so that when it enters the lungs it is saturated with water at the animal's core body temperature. Temperature equilibration and saturation with water vapor are essential to protect the lungs, which are delicate structures that would be damaged if they were exposed to dry air. As the relatively dry inhaled air passes over the moist tissues of the nasal passages, evaporation cools the walls of the nasal passages. When the warm, saturated air from the lungs is exhaled, water from the air condenses on the cool walls. This process of evaporation on inhalation and condensation on exhalation saves water. It must be emphasized that this countercurrent exchange of heat and energy is not an adaptation to desert life. It is an inevitable consequence of the anatomy and physiology of the nasal passages. However, a preliminary comparison of the nasal heat and water exchange of five species of rodents suggests that interspecific differences may be related to the aridity of the habitat (Welch 1984). Two desert rodents, the kangaroo rat (a heteromyid) and the Australian hopping mouse (a murid), were especially good at recovering water when they were breathing air of low humidity. In contrast, two murid rodents from moist habitats, the deer mouse and the house mouse, were good at recovering heat when they were breathing cold air.

The importance of the nasal countercurrent in water recovery can be illustrated by calculations (Table 22–3). During the day a kangaroo rat in its burrow inhales air that is at 30°C and 80 percent relative humidity. This air contains 24 milligrams of water per liter of air. (Saturated air, at 100 percent relative humidity, contains 30 milligrams of water per liter at 30°C.) The kangaroo rat must warm this air to core temperature (38°C) and add enough water to

Table 22–3   Water recycled by nasal countercurrent exchange in a kangaroo rat.

Condition	Inhaled Air				Exhaled Air			
	Temperature (°C)	RH (%)	Water Content (mg/L)	Water Added (mg/L)	Temperature (°C)	RH (%)	Water Content (mg/L)	Water Recovered (mg/L)
Daytime, in burrow	30	80	24	22	31	100	31	15
Night, on surface	15	20	2.5	43.5	14.5	100	12	34

raise the relative humidity at that temperature to 100 percent. At 38°C saturated air contains 46 milligrams water per liter, so the amount of water that must be evaporated in the nasal passages is 22 milligrams per liter of inhaled air.

Under the conditions we have assumed, a kangaroo rat exhales air at 31°C. That is, as the air travels out through the nasal passages it is cooled 7°C by contact with the walls. The exhaled air is still saturated with water, but at 31°C saturated air contains only 31 milligrams of water per liter instead of the 46 milligrams/liter it contained at 38°C. Thus, a kangaroo rat evaporates 22 milligrams of water per liter of air to saturate the air before it enters the lungs, and recovers 15 milligrams of water per liter of air on exhalation. The nasal countercurrent system reduces its respiratory water loss by nearly 70 percent from that which the kangaroo rat would experience if the air were exhaled at core body temperature. The savings amounts to 15 milligrams of water per liter of air under the burrow conditions we have assumed.

The water saving achieved at night when the rat is outside its burrow is still greater. The outside air is cool (15°C) and dry (20 percent relative humidity). Consequently, the kangaroo rat must evaporate 43.5 milligrams of water per liter of air to bring it to saturation at core temperature. The increased evaporation cools the kangaroo rat's nose to 14.5°C, and air exhaled at this temperature contains only 12 milligrams of water per liter. Thus, under nighttime conditions, a kangaroo rat recovers 74 percent of the water evaporated in the nasal passages, a saving of 34 milligrams of water per liter of inhaled air.

Evaporation of water from the respiratory passages is one major avenue of water loss, and water excreted with the urine and feces is another. Rodents in general have the ability to produce relatively dry feces and concentrated urine. The laboratory white rat, for example, can produce urine with twice the osmotic concentration humans can achieve. The dromedary camel has a urine-concentrating ability approximately equivalent to that of a rat, and so do dogs and cats. In desert rodents, such as kangaroo rats, sand rats, and jerboas, urine concentrations of 3000 to 6000 milliOsmoles/liter are commonly observed, and the current world champion urine concentrator appears to be the Australian hopping mouse, which can produce urine concentrations in excess of 9000 milliOsmoles/liter. As we pointed out in Chapter 4, high urine concentrations in mammals are associated with long loops of Henle in the kidney which enhance the countercurrent multiplier function.

As a result of their low evaporative water losses and ability to concentrate urine and produce relatively dry feces, many desert rodents are completely independent of liquid water. Their water loss has been reduced to the point at which they are able to obtain all the water they need from air-dried seeds. Part of this water comes from water actually contained in the food, and part from the water that is produced as the food is oxidized (metabolic water, Table 16–2).

The free water content of food depends in part on the relative humidity at which it is stored. We pointed out that *Diasperma* leaves have a water content of only 1 percent during the day but absorb water at night when the relative humidity rises, and it appears to be at night that the antelope eat them. Seeds show a similar variation in water content with humidity. Knut Schmidt–Nielsen found that the barley he fed his kangaroo rats contained less than 4 percent water when it was kept at 10 percent relative humidity, but when it was stored at 76 percent relative humidity, the water content rose to 18 percent. He reported that bannertail kangaroo rats (*Dipodomys spectabilis*) may store several kilograms of plant material in their burrows, where it is exposed

Figure 22–13 The antelope ground squirrel, *Ammospermophilus leucurus*, is a diurnal desert rodent. (Photograph by George A. Bartholomew and Mark A. Chappell.)

to the high relative humidity of the burrow atmosphere. The smaller Merriam's kangaroo rat (*D. merriami*) does not store much food in its own burrow, but sneaks into the burrows of bannertail kangaroo rats and may help itself to the food stored there. Gerbils in the Old World deserts also store seeds in their burrows. Air-dried seeds contained 4 to 7 percent water, whereas seeds taken from the gerbils' burrows contained 30 percent water. Thus, hoarding food in the burrow not only provides a hedge against food shortages but increases the amount of water available to an animal from the food.

Not all rodents that live in deserts are nocturnal. Ground squirrels are diurnal and are conspicuous inhabitants of deserts (Figure 22–13). They can be seen running frantically across the desert surface even in the middle of day. The almost frenetic activity of desert ground squirrels on intensely hot days is a result of the thermoregulatory problems that small animals experience under these conditions. Studies of the antelope ground squirrel (*Ammospermophilus leucurus*) at Deep Canyon, near Palm Springs, California, provide information about how the behavior of the squirrels is affected by the heat stress of the environment (Chappell and Bartholomew 1981a,b).

The heat on summer days at Deep Canyon is intense, and standard operative temperatures in the sun are as high as 70 to 75°C (Box 22–2). The squirrels are heat-stressed for most of the day. Standard operative temperature rises above the thermoneutral zone of ground squirrels within 2 hours after sunrise, and the squirrels follow a bimodal pattern of activity that peaks in midmorning and again in the late afternoon. Relatively few squirrels were active in the middle of the day. The body temperatures of antelope ground squirrels are labile, and body temperatures of individual squirrels varied as much as 7.5°C (from 36.1 to 43.6°C) during a day. The squirrels use his lability of body temperature to store heat during their periods of activity.

The high operative temperatures limit the time that squirrels can be active in the open to no more than nine to 13 minutes. The squirrels sprint furiously from one patch of shade to the next, pausing only to seize food or to look for predators. They minimize their exposure to the highest temperatures by running across open areas, and seek shade or their burrows to cool off. On a hot summer day a squirrel can maintain a body temperature below 43°C (the maximum temperature it can tolerate) only by retreating every few minutes to a burrow deeper than 60 centimeters where the soil temperature is 30 to 32°C. The body temperature of an antelope ground squirrel shows a pattern of rapid oscillations, rising while the squirrel is in the sun and falling when it retreats to its burrow (Figure 22–15). Ground squirrels do not sweat or pant; instead, they use this combination of transient heat storage and passive cooling in a burrow to permit diurnal activity. The strategy the antelope ground squirrel uses is basically the same as that employed by a camel—saving water by allowing the body temperature to rise until the heat can be dissipated passively. The difference between the two animals is a consequence of their difference in body size:

Box 22–2 How Hot Is It?

Measurements of environmental temperatures figure largely in studies of the energetics of animals, but the actual process of making the measurements is complicated, and no one measurement is necessarily appropriate for all purposes. You will recall from Chapter 4 that exchange of energy between animals and their environments involves radiation, convection, conduction, and evaporation. In addition, metabolic heat production contributes significantly to the body temperatures of endotherms. The thermal environment of an animal is determined by all of the routes of heat exchange operating simultaneously, and its body temperature includes the effect of metabolism as well. The question How hot is it? translates to What is the heat load for an animal in this environment? Answering that question requires integrating all the routes of energy exchange to give one number that represents the environmental heat stress. (It is easier to think of heat stress as coming from a hot environment and being represented by a risk of overheating [hyperthermia], but the same reasoning applies to a cold environment. In that situation the heat stress is loss of heat and the risk of hypothermia.)

Physiological ecologists have developed several measurements of the environmental heat stress on an organism, and Figure 22–14 illustrates four of these. The data come from a study of the thermoregulation of the antelope ground squirrel in a desert canyon in California (Chappell and Bartholomew 1981a).

The easiest measurement to make is the temperature of the air (T_a, frequently called **ambient temperature**). At Deep Canyon, California, in June air temperature rose from about 25°C at dawn to a peak above 50°C in late afternoon, and then declined. Air temperature is a factor in conductive and convective heat exchange. An

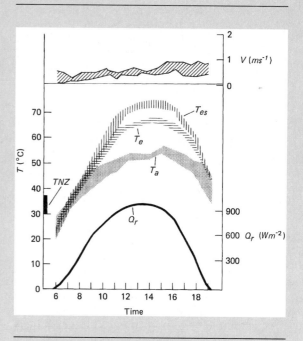

Figure 22–14 Ground-level meterological conditions in open sunlit areas at Deep Canyon, California, during June. Wind velocity (V) in meters per second, solar insolation (Q_r) in watts per square meter, effective environmental temperature (T_e) and standard operative temperature (T_{es}) in °C. The thermoneutral zone (TNZ) of ground squirrels is indicated. (From M. A. Chappell and G. A. Bartholomew, 1981, *Physiological Zoology* 54:81–93. Copyright 1981 by The University of Chicago. All rights reserved.)

animal gains heat by conduction and convection when the air temperature is warmer than the animal's surface temperature and loses heat when the air is cooler. Conductive heat exchange is usually small, but convection can be an important component of the overall energy budget of an organism. However, the magnitude

---

A camel weighs 500 kilograms and can store heat for an entire day and cool off at night, whereas an antelope ground squirrel weighs about 100 grams and heats and cools many times in the course of a day.

The tails of many desert ground squirrels are wide and flat, and the ventral surfaces of the tails are usually white. The tail is held over the squirrel's back with its white ventral surface facing upward. In this

of convective heat exchange depends on wind-speed as well as air temperature. Thus, measuring air temperature provides only part of the information needed to assess just one of the three important routes of heat exchange. Consequently, air temperature is not a very useful measure of heat stress.

If air temperature is unsatisfactory as a measure of environmental heat load because it makes only a small contribution to the overall energy exchange, perhaps a measurement of the major source of heat is what is needed. **Solar insolation** in this arid habitat is the major source of heat stress, and the magnitude of the insolation (Q_r) can be measured with a device called a pyranometer. Solar insolation rises from 0 at dawn to about 900 watts per square meter in midday, and falls to zero at sunset. This measurement provides information about how much solar energy is available to heat an animal, but that is still only one component of the energy exchange that determines the heat stress.

The **effective environmental temperature** (T_e) combines the effects of air temperature, ground temperature, solar insolation, and wind velocity. The effective environmental temperature is measured by making an exact copy of the animal (a mannikin), equipping it with a temperature sensor such as a thermocouple, and putting the mannikin in the same place in the habitat that the real animal occupies. Frequently, taxidermic mounts are used as mannikins: The pelt of an animal is stretched over a framework of wire or a hollow copper mold of the animal's body. Because the mannikin has the same size, shape, color, and surface texture as the animal, it responds the same way as the animal to solar insolation, infrared radiation, and convection. The equilibrium temperature of the mannikin is the temperature that a metabolically inert animal would have as a result of the combination of radiative and convective heat exchange. At Deep Canyon the temperatures of mannikins of antelope ground squirrels increased more rapidly than air temperatures and stabilized near 65°C from midmorning through late afternoon. The temperatures of the mannikins were 15°C higher than air temperature, showing that the heat load experienced by the ground squirrels was much greater than that estimated from air temperature alone.

Because the mannikin is a hollow shell—a pelt stretched over a supporting structure—it does not duplicate the thermoregulatory processes that have important influences on the thermoregulation of a real animal. Metabolic heat production increases the body temperature of a ground squirrel, evaporative water loss lowers body temperature, and changes in peripheral circulation and raising or lowering the hair change the insulation. The effects of these factors can be incorporated mathematically if the appropriate values for metabolism, insulation, and whole-body conductance are known. The result of this calculation is the **standard operative temperature** (T_{es}); an explanation of how to calculate T_{es} can be found in Bakken (1980). For the ground squirrels in our example the standard operative temperature was nearly 10°C higher than the effective temperature and about 25°C higher than the air temperature. For much of the day the T_{es} of a ground squirrel in the sun at Deep Canyon was 30°C or more above the squirrel's upper critical temperature of 43°C. Similar calculations can provide values for T_{es} in other microenvironments the squirrels might occupy—in the shade of a bush, for example, or in a burrow. This is the information needed to evaluate the behavior of the squirrels to determine if their activities are limited by the need to avoid overheating.

position it acts as a parasol, shading the squirrel's body and reducing the standard operative temperature. The tail of the antelope ground squirrel is relatively short, extending only halfway up the back, but the shade it gives can reduce the standard operative temperature by as much as 6 to 8°C. The Cape ground squirrel (*Xenurus inauris*) of the Kalahari Desert has an especially long tail that can be extend-

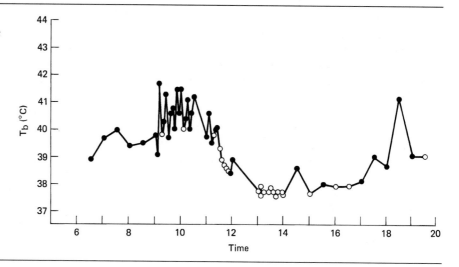

Figure 22–15 Short-term cycles of activity and body temperature of an antelope ground squirrel: •, active; ○, inactive. (From M. A. Chappell and G. A. Bartholomew, 1981, *Physiological Zoology*, 54:81–93. Copyright 1981 by The University of Chicago. All rights reserved.)

ed forward nearly to the squirrel's head. Cape ground squirrels use their tails as parasols (Figure 22–16), and observations of the squirrels indicated that the shade may significantly extend their activity on hot days.

The Use of Torpor by Desert Rodents

The significance of daily torpor as an energy conservation mechanism in small birds was illustrated ear-lier. Many desert rodents have the ability to become torpid. In most cases the torpor can be induced by limiting the food available to an animal. When the food ration of the California pocket mouse (*Perognathus californicus*) is reduced slightly below its daily requirements, it enters torpor for a part of the day. In this species even a minimum period of torpor results in an energy saving. If a pocket mouse were to enter torpor and then immediately arouse, the process would take 2.9 hours. Calculations indicate

(a)   (b)

Figure 22–16 Cape ground squirrel (*Xenurus inauris*) using its tail as a parasol: (a) The erected tail shades the dorsal surface of the animal; (b) the tail is held over the back of a horizontal squirrel, shading its head and body. (Photographs courtesy of Albert F. Bennett; from A. F. Bennett, R. B. Huey, H. John-Alder, and K. A. Nagy, 1984, *Physiological Zoology*, 57:57–62. Copyright 1984 by The University of Chicago. All rights reserved.)

that the overall energy expenditure during that period would be reduced 45 percent compared to the cost of maintaining a normal body temperature for the same period. In this animal, the briefest possible period of torpor gives the animal an energetic saving, and the saving increases as the time spent in torpor is lengthened.

The duration of torpor is proportional to the severity of food deprivation in the pocket mouse. As its food ration is reduced, it spends more time each day in torpor and conserves more energy. Adjustment of the time spent in torpor to match the availability of food may be a general phenomenon among seed-eating desert rodents. These animals appear to assess the rate at which they accumulate food supplies during foraging rather than their actual energy balance. Species that accumulate caches of food will enter torpor even with large quantities of stored food on hand if they are unable to add to their stores by continuing to forage. When seeds were deeply buried in the sand, and thus hard to find, pocket mice spent more time in torpor than they did when the same quantity of seed was close to the surface (Reichman and Brown 1979). This behavior is probably a response to the chronic food shortage that may face desert rodents because of the low primary productivity of desert communities and the effects of unpredictable variations from normal rainfall patterns, which may almost completely eliminate seed production by desert plants in dry years.

Summary

Endothermy is an energetically expensive way of life. It allows organisms considerable freedom from the physical environment, especially low temperatures, but it requires a large base of food resources to sustain high rates of metabolism. Energy budgets—calculations that quantify the energy expenditures and energy returns of specific activities—can provide insights about the energy implications of many behaviors of endotherms.

Endothermy is remarkably effective in cold environments; some species of birds and mammals can live in the coldest temperatures on Earth. The insulation provided by hair, feathers, or blubber is so good that little increase in metabolic heat production is needed to maintain body temperatures 100°C above ambient temperatures. In fact, some aquatic mammals, such as northern fur seals, are so well insulated that overheating is a problem when they are on land or in water warmer than 10 or 15°C.

Endothermy is most effective in situations in which an animal is warmer than its environment. In this situation metabolism and insulation are adjusted to balance heat production and heat loss. In hot environments the temperature gradient can be reversed—the environment is hotter than the organism—and this condition creates problems for endotherms. Evaporative cooling is effective as a short-term response to overheating, but it depletes the body's store of water and creates new problems. Small animals, rodents for example, can often avoid much of the stress of hot environments by spending the day underground in burrows and emerging only at night when it is cool. Larger animals have nowhere to hide and must meet the stress of hot environments head-on. Camels and other large mammals of desert regions relax their limits of homeostasis when they are confronted by the twin problems of high temperatures and water shortage: They allow their body temperatures to rise during the day and fall at night. This physiological tolerance is combined with behavioral and morphological characteristics that reduce the amount of heat that actually reaches their bodies from the environment.

Mobility is an important part of the response of large endotherms to both hot and cold environments: Seasonal movements away from unfavorable conditions (migration) or regular movements between scattered oases that provide water and shade are options available to medium-size or large mammals. The great mobility of birds makes these sorts of movements feasible even for relatively small species.

When the stresses of the environment overwhelm the regulatory capacities of an endotherm and resources to sustain high rates of metabolism are unavailable, many small mammals (especially rodents) and some birds enter torpor, a state of adaptive hypothermia. During torpor the body temperature is greatly reduced and the animal becomes inert. Periods of torpor can be as brief as a few hours (nocturnal hypothermia is widespread), or can last for many weeks. Mammals that hibernate (enter torpor during winter) arouse at intervals of days or weeks, warming to their normal temperature for a few hours and then returning to a torpid condition. Torpor conserves energy at the cost of forfeiting the benefits of endothermy.

The most remarkable feature of the ability of birds and mammals to live in diverse climates is not the specializations of arctic or desert animals, remarkable as they are, but the realization that only minor changes in the basic endothermal pattern are needed to permit existence over nearly the full range of environmental conditions on Earth.

References

Aschoff, J. 1982. The circadian rhythm of body temperature as a function of body size. Pages 173–188 in *A Companion to Animal Physiology*, edited by C. R. Taylor, K. Johansen, and L. Bolis. Cambridge University Press, Cambridge, UK.

Bakken, G. S. 1980. The use of standard operative temperature in the study of the thermal energetics of birds. *Physiological Zoology* 53:108–119.

Barnes, B. M. 1989. Freeze avoidance in a mammal: body temperatures below 0°C in an arctic hibernator. *Science* 244:1593–1595.

Bartholomew, G. A., and F. Wilke. 1956. Body temperature in the northern fur seal, *Callorhinus ursinus*. *Journal of Mammalogy* 37:327–337.

Bech, C., and R. E. Reinertsen (editors). 1989. *Physology of Cold Adaptation in Birds*. Plenum, New York, NY.

Busch, C. 1988. Consumption of blood, renal function, and utilization of free water by the vampire bat, *Desmodus rotundus*. *Comparative Biochemistry and Physiology* 90A:141–146.

Calder, W. A. 1994. When do hummingbirds use torpor in nature? *Physiological Zoology* 67:1051–1076.

Calder, W. A., and J. Booser. 1973. Hypothermia of broad-tailed hummingbirds during incubation in nature with ecological correlations. *Science* 180:751–753.

Chaplin, S. B. 1974. Daily energetics of the black-capped chickadee, *Parus atricapillus*, in winter. *Journal of Comparative Physiology* 89:321–330.

Chappell, M. A. 1980. Thermal energetics and thermoregulatory costs of small Arctic mammals. *Journal of Mammalogy* 1:278–291.

Chappell, M. A., and G. A. Bartholomew. 1981a. Standard operative temperatures and thermal energetics of the antelope ground squirrel *Ammospermophilus leucurus*. *Physiological Zoology* 54:81–93.

Chappell, M. A., and G. A. Bartholomew. 1981b. Activity and thermoregulation of the antelope ground squirrel *Ammospermophilus leucurus* in winter and summer. *Physiological Zoology* 54:215–223.

Davenport, J. 1992. *Animal Life at Low Temperature*. Chapman & Hall, New York, NY.

French, A. R. 1986. Patterns of thermoregulation during hibernation. Pages 393–402 in *Living in the Cold: Physiological and Biochemical Adaptations*, edited by H. C. Heller, X. J. Musacchia, and L. C. H. Wang. Elsevier, New York, NY.

Heller, H. C. (editor). 1987. Living in the cold. *Journal of Thermal Biology* 12, No. 2. (The entire issue is devoted to this topic.)

Jones, D. R., W. K. Milsom, and N. H. West (editors). 1988. The comparative physiology and biochemistry of cardiovascular, respiratory, and metabolic responses to hypoxia, diving, and hibernation. *Canadian Journal of Zoology* 66:3–200.

McNab, B. K. 1973. Energetics and distribution of vampires. *Journal of Mammalogy* 54:131–144.

Reichman, O. J., and J. H. Brown. 1979. The use of torpor by *Perognathus amplus* in relation to resource distribution. *Journal of Mammalogy* 60:550–555.

Taylor, C. R. 1972. The desert gazelle: a paradox resolved. Pages 215–217 in *Comparative Physiology of Desert Animals*, edited by G. M. O. Maloiy. *Symposia of the Zoological Society of London, No. 31*.

Wang, L. C. H. 1978. Energetic and field aspects of mammalian torpor: the Richardson's ground squirrel. Pages 109–145 in *Strategies in Cold: Natural Torpidity and Thermogenesis*, edited by L. C. H. Wang and J. W. Hudson. Academic, New York, NY.

Welch, W. R. 1984. Temperature and humidity of expired air: interspecific comparisons and significance for loss of respiratory heat and water from endotherms. *Physiological Zoology* 57:366–375.

Body Size, Ecology, and Sociality of Mammals

We have noted the increase in brain size that has occurred during the evolution of mammals, and suggested that part of the origin of the derived features of the mammalian brain might be sought in the nocturnal habits that are postulated for Mesozoic mammals. Relying on scent or hearing instead of vision to interpret their surroundings, ancestral mammals may have experienced selection for an increased ability to associate and compare stimuli received over intervals of time. The associative capacity resulting from changes in the brain during the evolution of mammals might also contribute to more complex behavior, and the association of mother and young during nursing could provide an opportunity to modify behavior by learning.

Indeed, social behaviors and interactions between individuals play a large role in the biology of mammals. These behaviors are modified by the environment, and clear-cut relationships between energy requirements, resource distribution, and social systems can often be demonstrated. In this chapter we consider some examples of those interactions that illustrate the complexity of the evolution of mammalian social behavior. In addition, we consider the social behavior of several species of primates. The social behavior of many primates is elaborate, but not necessarily more complex than that of some other kinds of mammals, including canids. However, primates have been the subjects of more field studies than have other mammals, with the result that we know a great deal about their social behavior, its consequences for the fitness of individuals, and even a little about the way some species of primates view their own social systems.

Social Behavior

Sociality means living in structured groups, and some form of group living is found among nearly all kinds of vertebrates. However, the greatest development of sociality is found among mammals, and much of the biology of mammals can best be understood in the context of what sorts of groups form, the advantages of group living for the individuals involved, and the behaviors that stabilize groups.

Mammals may be particularly social animals as a result of the interaction of several mammalian characteristics, no one of which is directly related to sociality, that in combination create conditions in which sociality is likely to evolve. Thus, the relatively large brains of mammals (which presumably facilitate complex behavior and learning), long gestation periods and (for some species) continued growth of the central nervous system after birth, prolonged association of parents and young, and high metabolic rates and endothermy (with the resulting high resource requirements) may be viewed as conditions that are conducive to the development of interdependent social units (Eisenberg 1981).

711

Of course, not all mammals are social. Mammalian characteristics do not necessarily lead to sociality: Solitary and social species are known among marsupials and placentals. Monotremes appear to be solitary, but the three living monotreme species provide too small a sample to form a basis for speculations about the phylogenetic origins of sociality among mammals. Of course, the social behavior of mammals does not operate in a vacuum; it is only one part of the biology of a species. Social behavior interacts with other kinds of behavior, such as food gathering, predator avoidance, and reproduction, with the morphological and physiological characteristics of a species, and with the distribution of the resources in the habitat. Our emphasis in this chapter is on that interaction, and we illustrate the interrelationships of behavior and ecology with examples drawn from both predators and their prey.

The social behavior of mammals is an area of active research, and our treatment is necessarily limited. Additional information can be found in reviews by Eisenberg (1981), Wittenberger (1981), Rubenstein and Wrangham (1986), Gittleman (1989), and Caro (1994). Some interesting examples of the costs and benefits of social behavior were briefly reviewed by Lewin (1987).

Population Structure and the Distribution of Resources

From an ecological perspective, the distribution of resources needed by a species is usually a major factor in determining its social structure. If resources are too limited to allow more than one individual of a species to inhabit an area, there is little chance of developing social groupings. Thus, the distribution of resources in the habitat and the amount of space needed by an individual to fill its resource requirements are important factors influencing the sociality of mammals.

Most animals have a **home range,** an area within which they spend most of their time and find the food and shelter they need. Home ranges are not defended against the incursions of other individuals (an area that is defended is called a **territory**), and the value of a home range probably lies in the familiarity of an individual animal with the locations of food and shelter. Many species of vertebrates employ a type of foraging known as trap lining, in which they move over a regular route and visit specific places where food may be available. For example, a mountain lion may carefully approach a burrow where a marmot lives, beginning its stalk long before it can actually see whether the marmot is outside its burrow, and a hummingbird may return to patches of flowers at intervals that match the rate at which nectar is renewed. This kind of behavior demonstrates a familiarity with the home range and with the resources that are likely to be available in particular places.

The **resource dispersion hypothesis** predicts that the size of the home range of an individual animal will depend primarily on two factors: (1) the resource needs of the individual, and (2) the distribution of resources in the environment. That is, individuals of species that require large quantities of a resource such as food should have larger home ranges than individuals of species that require less food. Similarly, the home ranges of individuals should be smaller in a rich environment than in one where resources are scarce. The resource dispersion hypothesis is a very general statement of an ecological principle. It applies equally well to any kind of animal and to any kind of resource. The resources usually considered are food, shelter, and access to mates. In Chapter 18 we considered the role of monopolization of resources by individuals in relation to the mating systems of birds, and here we discuss the role of resource dispersion in relation to home range size and sociality of mammals.

Body Size and Resource Needs

Studies of mammals have concentrated on food as the resource of paramount importance in determining the sizes of home ranges. Reviews by Gittleman and Harvey (1982), Clutton–Brock and Harvey (1983), Macdonald (1983), and McNab (1983) provide additional details. Food requirements are assumed to be equivalent to energy requirements, and we discussed the relationship between energy requirements and body size in Chapters 4 and 16.

You will recall that the energy consumption of vertebrates increases in proportion to body mass raised to a power that is usually between 0.75 and 1.0. If one assumes that energy requirements determine home range size, one can predict that home range size will also increase in proportion to the 0.75 or 1.0 power of body mass. That prediction appears to be correct in general, but perhaps wrong in detail (Figure 23–1). That is, the sizes of the home ranges of

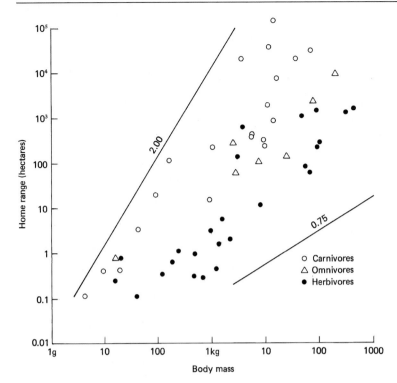

Figure 23–1 Home-range size of mammals as a function of body mass. The lines have slopes of 0.75 and 2.0, which illustrate allometric increases in home-range size on this double logarithmic scale. (From B. K. McNab, 1983, in *Advances in the Study of Mammalian Behavior*, edited by J. F. Eisenberg and D. G. Kleiman, Special Publication 7, The American Society of Mammalogists.)

mammals do increase with increasing body size, but the rates of increase (the slopes in Figure 23–1) are somewhat greater than expected. Home ranges appear to be proportional to body mass raised to powers between 0.75 and 2.0. This relationship between energy requirements and home range sizes suggests that energy needs are important in determining the size of the home range, but that additional factors are involved. One possibility is that the efficiency with which an animal can find and utilize resources decreases as the size of a home range increases. If that is true, the sizes of home ranges would be expected to increase with the body sizes of animals more rapidly than energy requirements increase with body size.

The failure of the resource dispersion hypothesis to predict the exact relationship between body size and the size of home ranges indicates that we have more to learn about how animals use the resources of their home ranges. So far we have been assuming that resources are distributed evenly throughout the home range, but that assumption overlooks the structural complexity of most habitats. What insights can be obtained from a more realistic consideration of how mammals gather food?

The Availability of Resources

Three factors seem likely to be important in determining the availability of food to mammals: what they eat, whether their food is evenly dispersed through the habitat or is found in patches, and how they gather their food. We will consider examples of each of these factors.

Dietary Habits Figure 23–1 shows that home range size increases with body size, and that dietary habits also affect the size of the home range. Herbivores have smaller home ranges than do omnivores of the same body size, and carnivores have larger home ranges than do herbivores or omnivores. For example, the home range of an elk (an herbivore) that weighs 100 kilograms is approximately 100 hectares. A bear (an omnivore) of the same body size has a home range larger than 1000 hectares, and a tiger (a carnivore) has a home range of more than 10,000 hectares.

The relationship between home range size and dietary habits of mammals probably reflects the abundance of different kinds of food. The grasses and leaves eaten by some herbivores are nearly ubiquitous, and a small home range provides all the food an individual requires. The plant materials (seeds and fruit) eaten by omnivores are less abundant than leaves and grasses, and different species of plants produce seeds and fruit at different seasons. Thus, a large home range is probably necessary to provide the food resources needed by an individual omnivore. The vertebrates that are eaten by carnivores are still less abundant, and a correspondingly larger home range is apparently needed to ensure an adequate food supply.

Distribution of Resources We have continued to assume that resources are evenly distributed through the habitat and that one part of a home range is equivalent to another part in terms of the availability of food. This assumption may be valid for some grazing and browsing herbivorous mammals, but it is clearly not true for mammals that seek out fruiting trees (which represent patches of food) or for any carnivorous mammal that preys on animals that occur in groups. The sizes of the home ranges of animals that utilize food that occurs in patches should reflect the quality of the habitat: Home ranges should be small if concentrations of food are abundant and large if concentrations of food are widely dispersed.

That relationship was well illustrated by a study of the home ranges of arctic foxes (*Alopex lagopus*) in Iceland (Hersteinsson and Macdonald 1982). The foxes live in social groups consisting of one male and two females plus the cubs from the current year. The home ranges of the individuals of a group overlap widely with each other, and there is very little overlap between the home ranges of members of different groups. The home ranges of the foxes are located along the coast and do not extend far into the uplands (Figure 23–2). Between 60 and 80 percent of the diet is composed of items the foxes find on the shore—the carcasses of seabirds, seals, and fishes, and invertebrates from clumps of seaweed washed up on the beach. Little food is available for foxes in the uplands. The foxes concentrate their foraging on the beach during the 3 hours before low tide, which is the best time for beachcombing. The foxes approach the shore carefully, stalking along gullies. They creep out on the beach carefully, apparently looking for birds that are resting or feeding. If birds are present, the foxes stalk them. If no birds are present, the foxes search the beach for carrion.

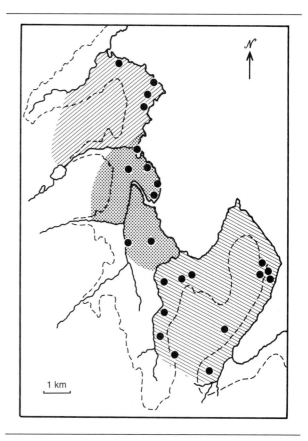

Figure 23–2 Map of the territorial boundaries of three groups of arctic foxes in Iceland. The 200-meter contour line is shown. Black dots mark the sites of dens used by the foxes. (From P. Hersteinsson and D. W. Macdonald, 1982, in *Telemetric Studies of Vertebrates*, edited by C. L. Cheeseman and R. B. Mitson, *Symposia of the Zoological Society of London, No. 49*, Academic, London, UK.)

Three groups of foxes were studied in detail, using radiotelemetry to follow the movements of individuals. The areas of the home ranges varied more than twofold, from 8.6 to 18.5 square kilometers (Table 23–1). The sizes of the home ranges were more similar when only the coastline was considered: Each territory included between 5.4 and 10.5 kilometers of coastline.

The coastline was, of course, the source of most of the food the foxes were eating, but not all areas of the coastline accumulated floating objects. The distribution of food on the beaches was patchy and depended on the directions of currents. As a result, some parts of the shore were more productive than others. The length of productive coastline occupied by each group of foxes was quite similar—from 5.4

Table 23–1 Home ranges of three groups of arctic foxes in Iceland.

	Group 1	Group 2	Group 3	Average
Total area (km^2)	10.3	8.6	18.5	12.5
Length of coastline (km)	5.6	5.4	10.5	7.2
Length of productive coastline (km)	5.6	5.4	6.0	5.7
Driftwood productivity (logs/year)	1800	1800	2100	1900

Source: P. Hersteinsson and D. W. Macdonald, 1982, pages 259–289 in *Telemetric Studies of Vertebrates,* edited by C. L. Cheeseman and R. B. Mitson, Symposia of the Zoological Society of London, No. 49, Academic, London, UK.

to 6.0 kilometers. Farmers in Iceland use driftwood to make fence posts, and the amount of driftwood that was harvested by the farmers from the coasts in the home ranges of the three groups of foxes varied only from 1800 to 2100 logs per year. Because both driftwood and carrion are moved by currents and deposited on the beaches, the harvest of driftwood by farmers probably reflects the harvest of carrion by the foxes. Thus, the sizes of the home ranges of the three groups appear to match the distribution of their most important food resource, and the productive areas of the home ranges of the three groups are very similar despite the twofold difference in total areas of their home ranges.

Group Size and Hunting Success It is readily apparent that the average size of the home range of a species of mammal can influence the social system of that species. Individuals of a species probably encounter each other frequently when home ranges are small, whereas individuals of species that roam over thousands of hectares may rarely meet. Thus, the distribution of resources in relation to the resource needs of a species is one of the factors that can set limits to the degree to which social groupings can occur. However, sociality may influence resource distribution if groups of animals are able to exploit resources that are not available to single individuals.

The influence of sociality on resource distribution may be seen among predatory animals that can hunt individually or in groups. Some species of prey are too large for an individual predator to attack, but are vulnerable to attack by a group of predators. For example, spotted hyenas (*Crocuta crocuta*) weigh about 50 kilograms. When hyenas hunt individually they feed on Thomson's gazelles (*Gazella thomsoni*, 15 kilograms) and juvenile wildebeest (*Connochaetes taurinus*, about 30 kilograms) (Figure 23–3). Howev-

er, when hyenas hunt in packs they feed on adult wildebeest (about 200 kilograms) and zebras (*Equus burchelli*, about 220 kilograms). Some species of prey have defenses that are effective against individual predators but less effective with groups of predators. For example, the success rate for solitary lions (*Panthera leo*) hunting zebras and wildebeest is only 15 percent, whereas lions hunting in groups of six to eight individuals are successful in up to 43 percent of their attacks. Groups of lions make multiple kills of wildebeest more than 30 percent of the time, but individual lions kill only a single wildebeest.

The relationship of sociality and body size of a predator to the size of its prey is shown in Figure 23–4: Social predators (defined in this study as those that hunt in groups of eight to ten individuals) attack larger prey than less social predators (average group sizes of 1.6 to 3.1 individuals), and these weakly social predators attack larger prey than do solitary predators (average group sizes of 1.0 to 1.3 individuals). Thus, one consequence of sociality for predatory mammals appears to be an increase in the potential food resources of an environment: Individual predators may be able to extend the range of prey species they can attack by hunting in groups.

Of course, the major disadvantage to hunting in a group is that there are more mouths to feed when you do get prey. The food requirement of a group of predators is the sum of the individual requirements of the members of the group, and the per capita amount of food obtained by hunting in a group would have to exceed that caught by a solitary hunter to make group hunting advantageous. Packer and Ruttan (1988) have reviewed factors that could contribute to the evolution of cooperative hunting.

The question is, do predators form groups *because* that allows them to hunt large prey, or *must*

Figure 23–3 Spotted hyenas (a) prey on small animals like the Thomson's gazelle (b) when they hunt individually. When they hunt in packs, they attack larger prey such as the wildebeest (c) and zebras (d). (Photographs by Sara Cairns.)

they hunt large prey because they live in groups for some other reason? A study of lions on the Serengeti Plains suggests that the second hypothesis is correct (Packer et al. 1990). Female lions are the only female felids to live in social groups. Female lions defend a group territory and protect their cubs from other groups of female lions. The high population densities that are characteristic of lions may have favored group defense of a territory. The presence of large prey makes it possible for lions to hunt in groups, but group hunting does not increase the amount of food available per lion—a lion hunting by herself can catch as much prey as her share of a group capture. Thus, groups form because of the advantages they provide in the social structure of the population of lions on the Serengeti, and group hunting is a by-product of that social structure.

A similar interpretation has been suggested for the formation of groups of male cheetahs (Caro 1994). Male cheetahs may live alone or form permanent coalitions of two or three individuals that live and hunt together. In Caro's study, these coalitions were often composed of littermates, and a coalition was more successful in occupying a territory than was a single male. Competition for territories was intense, and territorial disputes were an important source of mortality for male cheetahs. Cheetahs hunting singly concentrated on small prey such as Thomson's gazelles, whereas coalitions attacked larger prey such as wildebeest. Overall foraging success increased with group size for male cheetahs, but Caro concluded that the benefits of a coalition in holding a territory and controlling access to females was probably more important than its effect on food intake.

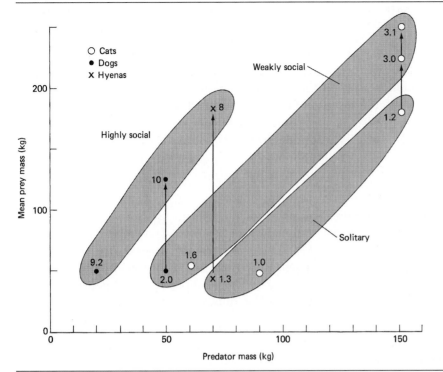

Figure 23–4 Size of prey in relation to predator mass for solitary predators and for predators that hunt in small or large groups. Vertical lines connect points for species that hunt in groups of variable size. (From B. K. McNab, 1983, in *Advances in the Study of Mammalian Behavior*, edited by J. F. Eisenberg and D. G. Kleiman, Special Publication 7, The American Society of Mammalogists.)

Advantages of Sociality

The advantages of cooperative hunting may provide a basis for sociality among some carnivorous mammals, but sociality is not limited to species of mammals that hunt in groups and the potential advantages of sociality are not confined to predatory behavior. Mammals may derive benefits from sociality in terms of avoiding predation and in reproduction and care of young.

Defenses Against Predators

One probable advantage of sociality is a reduction in the risk of predation for an individual that is part of a group compared to the risk for a solitary individual. The benefits of sociality in avoiding predation take many forms. A group of animals may be more likely to detect the approach of a predator than an individual would be, simply because a group has more eyes, ears, and noses to keep watch. Alternatively, an individual in a group may be able to devote a larger proportion of its time to feeding and less to watching for predators than a solitary individual can. Mammals that live in groups generally occur in

open habitats, whereas solitary species are usually found in forests, and that relationship may reflect in part the antipredator aspects of group living. It is important to note that the benefits of sociality in predator avoidance result from the reduction in the risk of predation for an *individual* that is part of a group, not for the species as a whole.

Sociality and Reproduction

Groupings of animals are important factors in mating systems and in care of young. In Chapter 18 we described mating systems in the context of avian biology, and most aspects of that discussion apply equally well to the mating systems of mammals. In the next section we extend the analysis by considering the specific relationships among body size, habitat, diet, antipredator behavior, and mating systems of several species of African ungulates.

The extensive period of dependence of many young mammals on their parents provides a setting in which many benefits of sociality can be manifested. Maternal care of the offspring is universal among mammals, and males of many species also play a role in parental care. Group living provides opportunities for complex interactions among adults and juveniles that involve various sorts of **alloparental behavior** (care provided by an individual that is not a parent of the young receiving the care). Collaborative rearing of young of several mothers is characteristic of lions and of many canids. Frequently, nonbreeding individuals join the mothers in protecting and caring for the young. Among dwarf mongooses (*Helogale undulata*) this kind of behavior extends to the care of sick adults, and reports of similar behaviors exist for mammals as diverse as elephants and cetaceans. Many social groups of mammals consist of related individuals, and these helpers may increase their inclusive fitness by assisting in rearing the offspring of their kin. (See Chapter 18 for a discussion of helpers.)

Body Size, Diet, and the Structure of Social Systems

The complex relationships among body size, sociality, and other aspects of the ecology and behavior of herbivorous mammals are illustrated by the variation in social systems of African antelopes (family Bovidae) (Jarman 1974, Leuthold 1977, Estes 1991). The smallest species of these bovids have adult weights of 3 to 4 kilograms (the dik-diks, *Madoqua*, and some duikers, *Cephalophus*), and the largest (the African buffalo, *Syncerus caffer*) weighs 900 kilograms (Figure 23–5). The smallest species are forest animals that browse on the most nutritious parts of shrubs, live individually or in pairs, defend a territory, and hide from predators (Table 23–2). The largest species (including the 300-kilogram eland, *Taurotragus oryx*, and the African buffalo) are grassland animals that graze unselectively, live in large herds, are migratory, and use group defense to deter predators. Species with intermediate body sizes are also intermediate in these ecological and behavioral characteristics. It seems likely that the correlated variation in body size, ecology, and behavior among these bovids reveals functional relationships among these aspects of their biology. How might such diverse features of mammalian biology interact?

The feeding habits of these antelope appear to provide a key that can be used to understand other aspects of their ecology and behavior. The diets of the different species are closely correlated with body size and the habitats in which the species live. In turn, those relationships are important in setting group size. The size of a group determines the distribution of females in time and space, and this is a major factor in establishing the mating system used by males of a species. Group size also plays an important role in determining the appropriate antipredator tactics for a species. Mating systems and antipredator mechanisms are central factors in the social organization of a species.

Body Size and Food Habits

Antelope are ruminants, relying on symbiotic microorganisms in the rumen to convert cellulose from plants into compounds that can be absorbed by the vertebrate digestive system (Box 21–1). The effectiveness of ruminant digestion is proportional to body size. This relationship exists because the volume of the rumen in species with different body sizes is proportional to body mass, whereas metabolic rates are proportional to the 0.75 power of body mass. The ecological consequence of this difference in allometric slopes is illustrated in Figure 23–6: A large ruminant has proportionately more capacity to process food than does a small ruminant. As one moves downward to animals of very small body size, the metabolic requirements become high in relation to the volume of rumen that is available to ferment plant material.

Because of this relationship, small ruminants must be more selective feeders than large ruminants. That is, a large ruminant has so much volume in its rumen that it can afford to eat large quantities of food of low nutritional value. It does not extract much energy from a unit volume of this food, but it is able to obtain its daily energy requirements by processing a large volume of food. Small ruminants, in contrast, must eat higher quality food and rely upon obtaining more energy per unit volume from the smaller volume of food that they can fit into their rumen in a day. In fact, 40 kilograms is the approximate lower limit of body size at which an unselective ruminant can balance its energy budget; species larger than 40 kilograms can be unselective grazers, whereas smaller species must eat only the most nutritious parts of plants (Van Soest 1982).

The species of antelopes in this example can be divided into four feeding categories:

Type I   species are very selective grazers and browsers. They feed preferentially on certain

(a)

(b)

(c)

Figure 23–5 The Bovidae includes species with a wide range of adult body sizes. The dik-dik (a) is among the smallest species, the impala (b) is medium-size, and the African buffalo (c) is one of the largest. ([a] Photograph by Jack Cranford; [b,c] photograph by Sara Cairns.)

species of plants, and they choose the parts of those plants that provide the highest-quality diet—new leaves (which have a higher nitrogen content than mature leaves) and fruit. Dik-diks and duikers fall in this category, and they have adult body masses between 3 and 20 kilograms.

Type II species are moderately selective grazers and browsers. They eat more parts of a plant than the type I species, and they may have seasonal changes in diet as they exploit the availability of fresh shoots or fruits on particular species of plants. Thomson's gazelle (*Gazella thomsoni*) and the impala (*Aepyceros melampus*) weigh 20 to 100 kilograms and have type II diets.

Type III species are primarily grazers that are unselective for species of grass, but selective for the parts of the plant. That is, they eat the leaves and avoid the stems. Hence, they are selecting for a growth stage: They avoid grass that is too short, because that limits food

Table 23–2 Correlations of the ecology and social systems of African ungulates.

Diet Type	Examples	Body Mass (kg)	Food Habits	Group Size	Mating System	Predator Avoidance
I	Dikdik, some duikers	3–20	Highly selective browser	1 or 2	Stable pair, territorial	Hide
II	Thompson's gazelle, impala	20–100	Moderately selective browser and grazer	2 to 100	Male territorial in breeding season, temporary harems	Flee
III	Wildebeest, hartebeest	100–200	Grazers, selective for growth stage	Large herds	Nomadic, temporary harems	Flee, hide in herd, threaten predator
IV	Eland, buffalo	300–900	Grazers, unselective	Large herds	Male hierarchy	Group defense

Source: Modified from P. J. Jarman, 1974, *Behaviour* 58:215–267.

intake, and also avoid grass that is too long (because it has too many stems that are low-quality food). Wildebeest (*Connochaetes taurinus*) and hartebeest (*Alcelaphus buselaphus*), which weigh about 200 kilograms, are type III feeders,

Type IV species are unselective grazers and browsers. They eat all species of plants and all parts of the plant. Eland (*Taurotragus oryx*, 300 kilograms) and buffalo (*Syncerus caffer*, 900 kilograms) are type IV species.

Food Habits and Habitat

The food habits of the different species of antelope are important in determining what sorts of habitats provided the resources they need. Selective feeding operates at three levels: the vegetation type, the species and individual groups of plants, and the parts of plants eaten. The type of vegetation present largely depends on the habitat—forests contain shrubs and bushes, whereas the plains are covered with grass. The resources needed by species with type I diets are found in forests where the presence of a diversity of species with different growth seasons ensures that fresh leaves and fruits will be available at all times of the year. Species with type II diets are found in habitats that are a mosaic of woodland and grassland, and type III species (which are primarily grazers) are found in savanna and grassland areas. Species with type II and type III diets may move from place to place in response to patterns of

rainfall. For example, wildebeest require grass that has put out fresh new growth, but that has not had time to mature. To find grass at this growth stage, wildebeest have extensive nomadic movements that follow the seasonal pattern of rain on the African plains.

Type IV feeders eat almost any kind of plant material, and they can find something edible in almost any habitat. They occupy a range of habitats, including grassland and brush, and do not have nomadic movements.

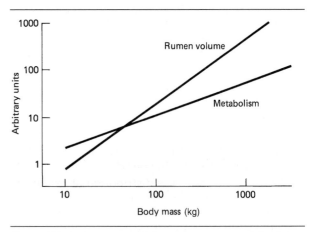

Figure 23–6 Illustration of the consequences of the different allometric relationships between rumen volume, which increases in proportion to body size (an allometric slope of 1), and energy requirement, which is proportional to metabolism (an allometric slope of 0.75). Both axes are drawn with logarithmic scales, and the scale of the vertical axis is in arbitrary units.

Habitat and Group Size

The habitats in which antelope feed and the types of food they eat set certain constraints on the sorts of social groupings that are possible. For example, species with type I diets live in forests and feed on scattered, distinct items. They eat an entire leaf or fruit at a bite and they must move between bites. A type I feeder completely removes the items it eats, so it changes the distribution of resources in its habitat (Figure 23–7). A second individual cannot feed close behind the first, because the food resources of an individual bush or shrub are entirely consumed by the first individual to feed there. As a result, the feeding behavior of species with type I diets makes it impossible for a group of animals to feed together. If one individual attempts to follow behind another to feed, the second animal must search to find food items overlooked by the individual ahead, and consequently it falls behind. Alternatively, it can move aside from the path of the first animal to find an area that has not already been searched. In either case, small animals in dense vegetation rapidly lose track of each other, and no cohesive group structure is maintained.

Instead, type I species are solitary or occur in pairs, and the individuals of a pair are only loosely associated as they feed. A type I diet places a premium on familiarity with a home range, because a tree or bush is a patch of food that must be visited repeatedly to harvest fruit or new leaves as they appear. Also, the resources represented by the patches of food can be defended against intruders, and species of antelope with type I diets are territorial. The resources of a territory do not change very much from one season to the next, and the only seasonal variation in the group size of species with type I diets results from the presence of a juvenile in its parents' territory for part of the year.

Species with type II and type III diets are less selective than species with type I diets, and their feeding has less impact on the distribution of resources. These species do not remove all of the food resource in an area, and other individuals can feed nearby. Type III feeders, in particular, graze as they walk—taking a bite of grass, moving on a few steps, and taking another bite. This mode of feeding changes the abundance of food, but not its distribution in space, and herds of wildebeest graze together, all moving in the same direction at the same speed and maintaining a cohesive group. Rainfall is a major determinant of

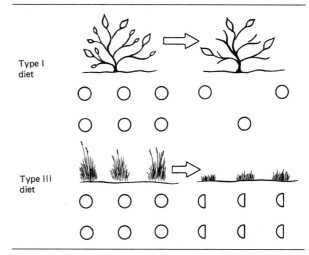

Figure 23–7 Diagrammatic illustration of the effect of feeding by a selective browser with a type I diet (a) and a grazer with a type III diet (b). The browser removes entire food items, thereby changing the distribution of food in the habitat, as well as the abundance of food. The grazer removes part of a grass clump, changing the abundance of food in the habitat but not its distribution. (From P. J. Jarman, 1974, *Behaviour* 58:215–267.)

the distribution of suitable food in the habitat of these species. The rainstorms that stimulate the growth of grass are erratic, and the patch sizes in which food resources occur are enormous—hundreds of square kilometers of new grass where rain fell are separated by hundreds of kilometers of old, dry grass that did not receive rain. Instead of having home ranges or territories, species with type II and type III diets move nomadically with the rains. Group sizes change as the distribution of resources changes, from half a dozen to 60 individuals for species with type II diets, and from herds of 300 or 400 to superherds of many thousands of individuals during the nomadic movements of wildebeests.

Species of bovids with type IV diets are so unselective in their choice of food that they can readily maintain large groups, and herds of buffalo number in the hundreds. Because these species can eat almost any kind of vegetation, the distribution of resources does not change seasonally and the size of the herds is stable.

Group Size and Mating Systems

The mating systems used by African antelope are closely related to the size of their social groups and the distribution of food because those are the major factors that determine the distribution of females and the potential for males to obtain opportunities to mate by controlling resources that females need. The females of species with type I diets are dispersed because the distribution of resources in the habitat does not permit groups of individuals to form. A male of a type I species can defend food resources, but individuals must disperse through the territory to feed and it is not feasible for a male to maintain a harem of females. Males of type I species pair with one female; the male defends its territory year round, the pair bond with an individual female appears to be stable, and offspring are driven out of the territory as they mature.

A group of individuals of type II species contains several males and females. The evenly distributed nature of food of these species makes it difficult for a male to monopolize resources. Only some of the males are territorial, and even this territoriality is manifested for only part of the year. A territorial male tries to exclude other males from its territory and to gather harems of females. Exclusive mating rights are achieved by holding a patch of ground and containing females within it, driving them back if they try to leave. These species have no long-term association between a male and a particular female.

Type III species are nomadic, and males establish territories only when the herd is stationary. During these periods male wildebeest hold harems of females within their territories, but the association between a male and females is broken when the herd moves on. However, mothers and their daughters maintain associations for 2 or 3 years. Unmated male wildebeest form bachelor herds with hierarchies, and individuals at the top of a bachelor hierarchy try to displace territorial males. If a territorial male is displaced, it joins a bachelor herd at the bottom of the hierarchy and must work its way up to the top before it can challenge another territorial male.

The social structure of buffalo herds differs in two respects from that of wildebeests: (1) Each herd includes many mature males that form a dominance hierarchy. The individuals near the top of the hierarchy court receptive females, but no territoriality or harem formation is seen. (2) The female membership of the herd is fixed, and this situation results in a degree of genetic relationship among all the members of a herd of buffalo. That genetic relationship among individuals creates situations in which kin

selection may be a factor in the social behavior of the African buffalo.

Mating Systems and Predator Avoidance

Prey species have a variety of ways to avoid predators, but only some will work in a given situation. In general, a prey species can (1) avoid detection by a predator, (2) flee after it has been detected but before the predator attacks, (3) flee after the predator attacks, or (4) threaten or attack the predator. Body size, habitat, group size, and the mating system all contribute to determining the risk of predation faced by a species and which predator avoidance methods are most effective.

Predators usually attack prey that are the same size as the predator or smaller. Small species of animals potentially have more species of predators than do large species. Species of antelopes with type I diets are small and, consequently, they are at risk from many species of predators. Furthermore, small antelopes may not be able to run fast enough to escape a predator after it has attacked. On the other hand, these small antelopes live in dense habitats, where they are hard to see. They are cryptically colored and secretive, and they rely on being inconspicuous to avoid detection by predators. If they are pursued, they may be able to use their familiarity with the geography of their home range to avoid capture.

Groups of animals are more conspicuous to predators than are individuals, but groups also have more eyes to watch for the approach of a predator. Species of antelopes with type II diets live in small groups in open habitats, where they can detect predators at a distance. These antelopes avoid predators by fleeing either before or after the predator attacks (Box 23–1). Small predators may be attacked by the antelope, but usually only when a member of the group has been captured, and this sort of defense is normally limited to a mother protecting her young; the rest of the group does not participate.

Species of antelope in the type III diet category are large enough to have relatively few predators, and in a group they may be formidable enough to scare off a predator. Wildebeest sometimes form a solid line and walk toward a predator: This behavior is effective in deterring even lions from attacking. Many predators of wildebeest focus their attacks on

Box 23–1 Unprofitable Prey?

A distinctive behavior—stotting, which consists of leaping vertically into the air—is used by some species of antelope when they are threatened by a predator. The function of stotting is unclear. It may be an alarm signal that alerts other individuals of the species to the presence of a predator, but the advantage to the individual that gives the warning is not clear. Altruistic behavior of this type is usually associated with kin selection, but the individuals in groups of antelope with type II diets are not closely related and do not show other types of altruistic behavior such as group defense of young. It has been suggested that some behaviors of prey species that had been considered to be altruistic alarm signals are really signals directed to the predator by fleet-footed prey.

Alarm signals are given by many other kinds of vertebrates. A familiar example is the white underside of the tail of deer (Figure 23–8). A deer that sees a predator at a distance does not immediately flee but stands watching the predator. It may flick its tail up and down, exposing the white ventral surface in a series of flashes. European hares stand erect on their hind legs when a fox in the open approaches within 30 meters of the hare. In this posture, the hare is readily visible to the fox.

This kind of behavior is not limited to mammals. For example, several related species of fleet-footed lizards that live in open desert habitats have dorsal colors that blend with the substrate on which they live, and a pattern of white with black bars on the underside of the tail. These lizards stand poised for flight as a predator approaches, looking back over their shoulder at the predator. The tail is curled upward, exposing the contrasting black and white pattern on its

Figure 23–8 Alarm signals to conspecifics or signals to a predator? The white-tail deer displays the white underside of its tail when it detects a predator. (Photograph © Stephen J. Krasemann.)

ventral surface, and waved from side to side. Is it possible that in behaviors of this sort the prey animal is signaling to the predator that it has been detected, and to attack it will be unprofitable because the prey is ready to flee? Does the prey's behavior deter pursuit by the predator?

This intriguing hypothesis requires experimental test of the prediction that predators are less likely to attack prey individuals that engage in these behaviors than individuals that do not signal their awareness of the predator's approach. However, if the hypothesis can be supported by experiments, some puzzling examples of apparently altruistic signals can be reinterpreted as behavior that benefits the individual giving the signal (Hasson 1991).

calves, and defense of a calf is usually undertaken only by the mother. Much of the antipredator behavior of wildebeest depends on the similarity of appearance of individuals in the herd to each other. Field observations have shown that the individuals

in a group of animals that are distinctive in their markings or behavior are most likely to be singled out and captured by predators.

One of the unavoidable events that makes a female wildebeest distinctive is giving birth to a calf, and the reproductive biology of the species has specialized features that appear to minimize the risks associated with giving birth. The breeding season and birth of wildebeests are highly synchronized. Mating occurs in a short interval, and as a consequence 80 percent of the births occur within a period of 2 or 3 weeks. Furthermore, nearly all of the births that will take place on a day occur in the morning in large aggregations of females, all giving birth at once. A female wildebeest who is slightly out of synchrony with other members of her group can interrupt delivery at any stage up to emergence of the calf's head so as to join the mass parturition. Presumably, this remarkable synchronization and control of parturition reflects the advantage of presenting predators with a homogeneous group of cows and calves rather than a group with only a few calves that could readily be singled out for attack.

Buffalo are formidable prey even for a pride of lions and they escape much potential predation simply as a result of their size. When buffalo are attacked, they engage in group defense, and if a calf is captured its distress cries bring many members of the group to its defense. This altruistic behavior probably represents kin selection, because the stability of the female membership of buffalo herds results in genetic relationships among the individuals.

Evolution of Social Behaviors of Antelope

It is tempting to view the continuum of increasing sociality from antelopes with type I diets through species with type IV diets as an evolutionary progression from simple social systems to complex ones. However, this view would almost certainly be wrong. The relationships among body size, diet, habitat selection, group size, mating system, and antipredator behavior form a web, not a ladder. If any one aspect of the biology of antelopes can be seen as limiting the range of possibilities for other features of their biology, body size is probably the key factor. The relationships among body size, rumen size, and metabolic requirements appear to define the four dietary types we have discussed and to determine what sorts of habitats supply the resources they require. In turn, the habitats in which different species of antelope live constrain the sorts of social systems they can exhibit.

If that is so, a progression from simple social systems to more complex ones is unlikely. Ruminant

digestion is most advantageous for medium-size animals, and the first ruminants probably weighed about 100 kilograms and had diets like those of impalas and gazelles (Van Soest 1982). If dietary habits and habitat selection were important influences on the ecology and behavior of the early ruminants, it is likely that the ancestral social system was some form of temporary territoriality and harem formation. The monogamous mating system seen in dik-diks and the huge herds of buffalo sharing some degree of kinship probably both represent derived behaviors.

Social Systems Among Primates

The phylogenetic relationship of humans to other primates has led some biologists to assume that these are the animals that should have the most elaborate social systems, and that the study of the social systems of primates will provide information about the evolution of human behavior. Both assumptions are controversial: Increasing information indicates that complex social systems exist among many kinds of vertebrates other than primates. Interpretation of primate behavior in the context of human evolution is fraught with difficulty and must be approached cautiously. Nonetheless, primates do have elaborate and complex social systems, and more long-term research has focused on the social systems of primates than on any other vertebrates.

The approximately 200 species of primates (Table 23–3) are ecologically diverse. They live in habitats ranging from lowland tropical rain forests, to semideserts, to northern areas that have cold, snowy winters. Some species are entirely arboreal, whereas others spend most of their time on the ground. Many are generalist omnivores that eat fruit, flowers, seeds, leaves, bulbs, insects, bird eggs, and small vertebrates, but many of the colobus (*Colobus*) and howler monkeys (*Alouatta*) are specialized folivores (leaf eaters) with sacculated stomachs in which bacteria and protozoans ferment cellulose, and some of the small prosimians and callithricid monkeys are insectivores.

R. W. Wrangham (1982) proposed that the social systems of primates can best be classified on the basis of the amount of movement of females that occurs between groups (Table 23–4). Four categories can be defined on this basis:

1. **Female transfer systems.** In species with this type of social organization most females move away from the group in which they were born to join another social group. Because of this migration of females among groups, the females in a group are not closely related to each other. In contrast, males often remain with their natal groups and associations of male kin may be important elements of the social behaviors of these species of primates. Male chimpanzees, for example, cooperate in defending their territories from invasion by neighboring males. The majority of species of primates with female transfer systems live in relatively small social groups.

2. **Nonfemale transfer systems.** Most females of these species spend their entire lives in the group in which they were born. Social relations among the females in a group are complex and are based on kinship. Males of these species emigrate from their natal group as adolescents and may continue to move among groups as adults. In some of these species a single male lives with a group of females until he is displaced by a new male, whereas in other species several males may be part of the group and maintain an unstable dominance hierarchy among themselves. Group size is usually larger for nonfemale transfer species than for species with female transfer.

3. **Monogamous species.** A single male and female form a pair, sometimes accompanied by juvenile offspring. These species of primates show little sexual dimorphism, the sexes share parental care and territorial defense, and the offspring are expelled from the parents' territory during adolescence.

4. **Solitary species.** These live singly or as females with their infants and juvenile offspring. Male prosimians maintain territories that include the home ranges of several females and exclude other males from their territories, whereas male orangutans do not defend territories, but instead repulse other males when a female within the male's home range comes into estrus.

Three ecological factors appear to be particularly important in shaping the social systems of primates, as they are for other vertebrates:

1. The defensibility of food resources appears to determine whether individuals will benefit from (a) not attempting to defend resources, (b) defending individual territories, or (c) forming long-term relationships with other individuals and jointly defending resources.

2. The distribution of food in time and in space may determine how large a group can be, and whether the group can remain stable or must break into smaller groups when food is scarce.

3. The risk of predation may determine whether individuals can travel alone or require the protection of a group, whether the benefit of the additional protection that is provided by a large versus a small group outweighs the added competition among individuals in a large group, and whether the presence of males is needed to protect young.

Life within a group of primates is a balance between competition and cooperation (Figure 23–9). Competition is manifested by aggression. Some aggression—for example, the defense of food, rest, sleeping sites, or mates—is closely linked to resources. Other types of aggression involve establishing and maintaining dominance hierarchies, which may be an indirect form of resource competition if high-ranking individuals have preferential access to resources.

Cooperation, too, is diverse. Grooming behavior in which one individual picks through the hair of another, removing ectoparasites and cleaning wounds, is the most common form of cooperation. Other types of cooperation include sharing food or feeding sites, collective defense against predators, collective defense of a territory or a resource within a home range, and the formation of alliances between individuals. Two-way, three-way, and even more complex alliances that function during competition within a group are common among primates.

Kinship and the concept of inclusive fitness play important roles in the interpretation of primate social behavior. A behavior must not decrease the fitness of the individual exhibiting the behavior if it is to persist in the repertoire of a species. Fitness is nearly impossible to demonstrate in wild populations, and behaviorists normally search for effects that are likely to be correlated with fitness, such as access to females (for males), interbirth interval (for females), or the probability that offspring will survive to reproductive age. Behaviors that increase these measures are assumed to increase fitness. The behaviors may directly benefit the individual displaying the behavior (personal fitness) or they may be costly to the personal fitness of the individual but sufficient-

Table 23-3 **Social organization of extant primates.**

Taxon	Social Organization
Lemuroidea Aye-aye (*Daubentonia*, 1 species) Lemurs (*Lemur* and 9 other genera, 18 species) Indri (*Indri*, 1 species) Sifaka (*Propithecus*, 2 species)	Largely solitary or monogamous
Lorisoidea Bushbabies (*Galago*, 8 species) Lorises (*Loris*, 1 species; *Nycticebus*, 2 species) Potto (*Perodicticus*, 1 species) Angwantibo (*Arctocebus*, 1 species)	Largely solitary
Tarsoidea Tarsiers (*Tarsius*, 3 species)	Solitary or monogamous

Ceboidea = Platyrrhines (New World monkeys)
Callimiconidae
    Goeldi's marmoset (*Callimico*, 1 species)
Callithrichidae
    Marmosets and tamarins (4 genera,
        15 species)
Cebidae
    Howler monkeys (*Alouatta*, 6 species)
    Spider monkeys (*Ateles, Brachyteles,
        Lagothrix*, 7 species)
    Capuchin monkeys (*Cebus*, 4 species)
    Squirrel monkeys (*Samiri*, 2 species)
    Owl monkeys (*Aotus*, 1 species)
    Uakaris (*Cacajao*, 2 species)
    Titis (*Callicebus*, 3 species)
    Sakis (*Chiropotes* and *Pithecia*, 6 species)
	Small groups with one resident male
	
	Largely monogamous pairs
	
	Monogamous pairs, or small to large groups

ly beneficial to its close relatives to offset the cost to the individual (inclusive fitness).

Social Relationships Among Primates

Four general types of relationships among individuals have been described in the social behavior of primates (For more details see Watts [1985], Richard [1985], Smuts et al. [1987], Dunbar [1988], Cheney and Seyfarth [1990], and Nishida [1990].)

Adult–Juvenile Associations Primates are born in a relatively helpless state compared to many mammals, and they are dependent on adults for unusually long periods. The relationship of a mother to her infant is variable within a species—some mothers are protective, whereas others are permissive. Permissive mothers often wean their offspring earlier than pro-

tective mothers and may have shorter intervals between the birth of successive offspring, although this relationship has not been observed in all species. The offspring of permissive mothers may suffer higher rates of mortality than the offspring of protective mothers, and the incompetence of some first-time mothers appears to lead to high mortality among firstborn offspring.

Older siblings often participate in grooming and carrying an infant, but they may also assault, pinch, and bite the infant while it is being fed or groomed by the mother. Allomaternal behavior provided by an adult female who is not the mother includes cuddling, grooming, carrying, and protecting an infant. Several factors seem to influence allomaternal behavior: Young infants are preferred to older ones, infants of high-ranking mothers receive more attention and less abuse than infants of low-ranking

Table 23–3 *(Continued)*

Taxon	Social Organization
Catarrhini (Old World monkeys and apes)	
Cercopithecoidea	
Cercopithecidae	Mostly small to large groups
Vervet monkey, guenons, and others	
(*Cercopithecus*, 17 species)	
Mangabeys (*Cercocebus*, 5 species)	
Macaques (*Malaca*, 12 species)	
Baboons (*Papio*, 4 species; *Theropithecus*, 1 species)	
Yellow or savanna baboon (*Papio cynocephalus*)	
Hamadryas baboon (*Papio hamadryas*)	
Drill (*Papio leucophaeus*)	
Mandrill (*Papio sphinx*)	
Gelada baboon (*Theropithecus gelada*)	
Colobus monkeys (*Colobus*, 7 species)	
Langurs (*Nasalis, Presbytis, Pygathrix,*	
Rhinopithecus, 20 species)	
Hominoidea (apes and humans)	
Hylobatidae	Monogamous
Gibbons (*Hylobates*, 9 species)	
Pongidae	
Orangutan (*Pongo*, 1 species)	Solitary
Panidae	
Gorilla (*Gorilla*, 1 species)	Small groups with a variable number of resident males
Chimpanzee (*Pan*, 2 species)	Closed social network containing several breeding males and females
Hominidae	Closed social network containing several breeding males and females
Human (*Homo*, 1 species)	

Source: Modified from B. B. Smuts et al., editors, 1987, *Primate Societies,* University of Chicago Press, Chicago, IL.

mothers, and siblings may participate more than unrelated females in allomaternal behavior (Nicolson 1987). Males of the monogamous New World primates participate extensively in caring for infants, carrying them for much of the day and sharing food with them, whereas the relationships of males of Old World primates with infants are more often characterized by proximity and friendly contact than by care.

Female Kinship Bonds The females of some species of semiterrestrial Old World primates live in groups that include several males and females. This social organization is typical of yellow baboons (*Papio cynocephalus*), several species of macaques (*Macaca*), and vervet monkeys (*Cercopithecus aethiops*). Females of these species remain for their entire lives in the troops in which they were born, whereas males migrate to other troops when they mature. The role of female kinship bonds is much smaller in female transfer systems because the females in a group are not closely related.

The females within a group form a dominance hierarchy and compete for positions in the hierarchy. Related females within a group are referred to as **matrilineages.** Females consistently support their female relatives during encounters with members of other matrilineages. The supportive relationship among females within a matrilineage is an important element of the social structure of a group. For example, when their female kin are nearby, young animals can dominate older and larger opponents from subordinate matrilineages. Furthermore, high-ranking females retain their position in the hierarchy even when age or injury reduces their fighting ability. An adolescent female yellow baboon normally attains a

Table 23–4 Characteristics of the social systems of primates.

System	Group Size	Number of Males in Group	Male Behavior	Examples
Female transfer	Small	One or many	Territorial, harems, sometimes male kinship groups	Chimpanzee, gorilla, howler monkeys, hamadryas baboons, colobus monkeys, some langurs
Nonfemale transfer	Large	One or several	Male hierarchy, whole group (males and females) may exclude conspecifics from food sources	Most cercopithecines: yellow baboons, mangabeys, macaques, guenon monkeys
Monogamous	Male and female, plus juvenile offspring	One	Both sexes participate in territorial defense and parental care	Gibbons, marmosets, tamarins, indri, titis
Solitary	Individual, or female plus juvenile male offspring	—	Range of male overlaps ranges of several females	Bushbabies, tarsiers, lorises, orangutans

Source: Based on R. W. Wrangham, 1982, pages 269–289 in *Current Problems in Sociobiology*, edited by King's College Sociobiology Group, Cambridge University Press, Cambridge, UK.

rank in the group just below that of her mother, and this inheritance of status provides stability in the dominance relationships among the females of a group. However, the social rank of the matrilineage is not fixed: Low-ranking female yellow baboons, with their female kin, may challenge higher-ranking individuals, and if they are successful their entire matrilineage may rise in rank within the group.

The female kinship bonds are clearly important elements of the social structure of nonfemale transfer systems, but the exact contribution of the long-term relationships among females to the fitness of individual females is not clear. In some troops high-ranking females are young when they first give birth and have short interbirth intervals and high infant survival, but those correlations are not present in all the troops that have been studied. Furthermore, female kinship bonds are manifested weakly if at all in female transfer systems, which include most species of apes and many species of monkeys.

Male–Male Alliances Male primates in nonfemale transfer species often form dominance hierarchies, but male rank depends mainly on individual attributes and is therefore less stable than female dominance systems based on matrilineage. Young adult males, which are usually recent immigrants from another group, have the greatest fighting ability and usually achieve the highest rank. Some older males achieve stable alliances with each other that enable them to overpower younger and stronger rivals in competition for opportunities to court receptive females. These males probably achieve greater mating success by engaging in these reciprocal alliances than they would achieve on the basis of their individual ranks in the hierarchy.

Cooperative relationships among males are most common in female transfer systems, because the males of these species remain in their natal group. As a result, kin relationships exist among the males in a group. In red colobus monkeys (*Colobus badius*), for

Figure 23–9 Social behaviors of yellow baboons: (a) male friend grooming a female baboon in estrus; (b) aggression among male baboons. (Photographs by Carol Saunders.)

example, only males born in the group appear to be accepted by the adult male subgroup, and the membership of this subgroup can remain stable for years. Adult males spend much of their time in close proximity to each other and cooperate in aggression against males of a neighboring group. Male chimpanzees (*Pan troglodytes*) spend more time together than do females and engage in a variety of cooperative behaviors, including greeting, grooming, and sharing meat. However, this apparent cooperation is simply a way of cementing relationships that are based on intense and sometimes violent competition over females. For example, Goodall (1986) reported the systematic killing of an entire group of males by the males of a neighboring group, which then took over the females in that community.

Male–Female Friendships Among Baboons Barbara Smuts's (1985) observations of a group of yellow baboons revealed that interactions between individual male and female baboons were not randomly distributed among members of the group. Instead, each female had one or two particular males called friends. Friends spent much time near each other and groomed each other often. These friendships lasted for months or years, including periods when the female was not sexually receptive because she was pregnant or nursing a baby. Male friends were solicitous of the welfare of their female friends and of their infants. Similar male–female friendships have been described in mountain gorillas (*Gorilla gorilla beringei*), gelada baboons (*Theropithecus gelada*),

hamadryas baboons (*Papio hamadryas*), rhesus macaques (*Macaca mulatta*), and Japanese macaques (*Macaca fuscata*).

The advantage of these friendships for a female appears to lie in the protection that males provide to the females and their offspring from predators and from other members of the group. The advantage for a male of friendship with a female is less apparent. If the female's offspring had been sired by the male, protecting it would contribute to the male's fitness. However, in Smuts's study of yellow baboons, only half the friendships between males and infants involved relationships in which the male was the likely sire of the infant. The other friendships involved males that had never been seen mating with the mother of the infant. The advantage of friendship for males may depend on long-term associations with females. Smuts noted that males who participated in a friendship with a female had a significantly increased chance of mating with that female many months later when she was again receptive.

How Do Primates Perceive Their Social Structure?

The summary of primate social structures presented above represents the results of tens of thousands of hours of observations of individual animals over periods of many years. Statistical analyses of interactions between individuals—grooming sessions, aggression, defense—reveal correlations associated with factors including age, personality, kinship, and rank. Do the animals themselves recognize those relationships?

That is a fascinating but difficult question, especially with studies of free-ranging animals, but observations are accumulating that suggest that primates probably do recognize different kinds of relationships among individuals. For example, when juvenile rhesus macaques are threatened by another monkey, they scream to solicit assistance from other individuals who are out of sight. The kind of scream they give varies depending on the intensity of the interaction (threat or actual attack) and the dominance rank and kinship of their opponent. Furthermore, a mother baboon appears to be able to interpret the screams of her juvenile and to respond more or less vigorously depending on the nature of the threat her offspring faces. When tape-recorded screams were played back to the mothers, the mothers responded most strongly to screams that were given during an attack by a higher-ranking opponent, less strongly to screams that were given in response to interactions with lower-ranking opponents, and least strongly to screams that were given in interactions with relatives.

In a similar experiment with free-ranging vervet monkeys, the screams of a juvenile were played back to three females, one of whom was the mother of the juvenile. The mother responded more strongly to the screams than did the other two monkeys, as one would expect if females can recognize the voices of their own offspring. However, the other two monkeys responded to the screams by looking toward the mother, suggesting that they were able not only to associate the screams with a particular juvenile, but also to associate that juvenile with its mother.

Observations of redirected aggression also suggest that some primates classify other members of a group in terms of matrilineage and friendships. When a baboon or macaque has been attacked and routed by a higher-ranking opponent, the victim frequently attacks a bystander who took no part in the original interaction. This behavior is known as redirected aggression, and the targets of redirected aggression are relatives or friends of the original opponent more frequently than would be expected by chance. Vervet monkeys show still more complex forms of redirected aggression: They are more likely to behave aggressively toward an individual when they have recently fought with one of that individual's close kin. Furthermore, an adult vervet is more likely to threaten a particular animal if that animal's kin and one of its own kin fought earlier that same day. This sort of feud is seen only among adult vervets, suggesting that it takes time for young animals to learn the complexities of the social relationships of a group.

These sorts of observations suggest that adult primates have a complex and detailed recognition of the genetic and social relationships of other individuals in their group. Furthermore, they may be able to recognize more abstract categories, such as *relative* versus *nonrelative*, *close relative* versus *distant relative*, or *strong friendship bond* versus *weak friendship bond*, that share similar characteristics independent of the particular individuals involved.

Summary

Sociality, the formation of structured groups, is a prominent characteristic of the behavior of many species of mammals. However, social behavior is only one aspect of the biology of a species, and social behaviors coexist with other aspects of behavior and ecology, including finding food and escaping from predators.

The size and geography of an animal's home range is related to the distribution and abundance of resources, the body size of the animal, and its feeding habits. Large species have larger home ranges than small species, and for any given body size the sizes of home ranges are in the order carnivores > omnivores > herbivores.

Social systems are related to the distribution of food resources and to the opportunities for an individual (usually, a male) to increase access to mates by controlling access to resources. Dietary habits, the structural habitat in which a species lives, and its means of avoiding predators are closely linked to body size and mating systems. These aspects of biology form a web of interactions, each influencing the others in complex ways.

The social systems of primates, especially cercopithecoid monkeys, have been the subjects of field studies and more information about social behavior under field conditions is available for primates than for other mammals. The social systems of primates are complex, but not unique among mammals. Some primates are solitary or monogamous, but others live in groups and display behaviors that suggest not just recognition of other individuals, but also recognition of the genetic and social relationships among other individuals. Studies of other kinds of mammals will probably reveal similar phenomena, and understanding the behavior of mammals requires a broad understanding of their ecology and evolutionary histories.

References

Caro, T. M. 1994. *Cheetahs of the Serengeti Plains*. University of Chicago Press, Chicago, IL.

Cheney, D., and R. Seyfarth. 1990. *How Monkeys See the World*. University of Chicago Press, Chicago, IL.

Clutton–Brock, T. H., and P. H. Harvey. 1983. The functional significance of variation in body size among mammals. Pages 632–663 in *Advances in the Study of Mammalian Behavior*, edited by J. F. Eisenberg and D. G. Kleiman. Special Publication 7, The American Society of Mammalogists, Lawrence, KS.

Dunbar, R. I. M. 1988. *Primate Social Systems*. Cornell University Press, Ithaca, NY.

Eisenberg, J. F. 1981. *The Mammalian Radiations*. University of Chicago Press, Chicago, IL.

Estes, R. D. 1991. *The Behavior Guide to African Mammals*. University of California Press, Berkeley, CA.

Gittleman, J. L. (ed). 1989. *Carnivore Behavior, Ecology, and Evolution*. Cornell University Press, Ithaca, NY.

Gittleman, J. L., and P. H. Harvey. 1982. Carnivore home-range size, metabolic needs, and ecology. *Behavioral Ecology and Sociobiology* 10:57–63.

Goodall, J. 1986. *The Chimpanzees of Gombe*. Harvard University Press, Cambridge, MA.

Hasson, O. 1991. Pursuit-deterrent signals: communication between prey and predator. *Trends in Ecology and Evolution* 6:325–329.

Hersteinsson, P., and D. W. Macdonald. 1982. Some comparisons between red and arctic foxes, *Vulpes vulpes* and *Alopex lagopus*, as revealed by radio tracking. Pages 259–289 in *Symposia of the Zoological Society of London*, No. 49, edited by C. L. Cheesman and R. B. Mitson. Academic, London, UK.

Jarman, P. J. 1974. The social organization of antelope in relation to their ecology. *Behaviour* 58:215–267.

Leuthold, W. 1977. *African Ungulates: A Comparative Review of their Ethology and Behavioral Ecology*. Springer, New York, NY.

Lewin, R. 1987. Social life: a question of costs and benefits. *Science* 236:775–777.

Macdonald, D. W. 1983. The ecology of carnivore social behavior. *Nature* 301:379–381.

McNab, B. K. 1983. Ecological and behavioral consequences of adaptation to various food resources. Pages 664–697 in *Advances in the Study of Mammalian Behavior*, edited by J. F. Eisenberg and D. G. Kleiman, Special Publication 7, The American Society of Mammalogists, Lawrence, KS.

Nicolson, N. A. 1987. Infants, mothers, and other females. Pages 330–342 in *Primate Societies*, edited by B. Smuts, D. Cheney, R. Seyfarth, R. Wrangham, and T. Struhsaker. University of Chicago Press, Chicago, IL.

Nishida, T. (ed) 1990. *The Chimpanzees of the Mahale Mountains. Sexual and Life History Strategies*. Tokyo University Press, Tokyo, Japan.

Packer, C., and L. Ruttan. 1988. The evolution of cooperative hunting. *The American Naturalist* 132:159–198.

Packer, C., D. Scheel, and A. E. Pusey. 1990. Why lions form groups: food is not enough. *The American Naturalist* 136:1–19.

Richard, A. F. 1985. *Primates in Nature*. Freeman, San Francisco, CA.

Rubenstein, D. I., and R. W. Wrangham (editors). 1986. *Ecological Aspects of Social Evolution: Birds and Mammals*. Princeton University Press, Princeton, NJ.

Smuts, B. 1985. *Sex and Friendship in Baboons*. Aldine, Hawthorne, NY.

Smuts, B., D. Cheney, R. Seyfarth, R. Wrangham, T. Struhsaker (eds). 1987. *Primate Societies*. University of Chicago Press, Chicago, IL.

Van Soest, P. J. 1982. *Nutritional Ecology of the Ruminant*. O & B Books, Corvallis, OR.

Watts, E. S. (ed). 1985. *Nonhuman Primate Models for Human Growth and Development*. Liss, New York, NY.

Wittenberger, J. F. 1981. *Animal Social Behavior*. Duxbury, Boston, MA.

Wrangham, R. W. 1982. Mutualism, kinship, and social evolution. Pages 269–289 in *Current Problems in Sociobiology*, edited by King's College Sociobiology Group. Cambridge University Press, Cambridge, UK.

24 HUMANS AS VERTEBRATES

Humans are primates, and complex social systems like those discussed in Chapter 23 are an ancestral character. The fossil record indicates that the first stone tools are at least 2.5 million years old. Tool cultures are clearly associated with the first appearance of the genus *Homo*, but may not be an exclusive characteristic of our genus. Molecular techniques of studying genetic relationships among organisms have recently been applied to human evolution and have challenged some prevailing views: The molecular studies suggest that the separation of humans from the African great apes was more recent than had been inferred from anatomical studies, and that the closest extant relatives of humans are chimpanzees. Furthermore, application of cladistic methods to human fossils reveals that humans retain more ancestral morphological character states than some of the extant great apes.

The growth of human populations, especially in the past 50,000 years, is a unique phenomenon in the history of vertebrates. Never before has a single species so dominated the resources of the Earth. The consequences of this domination, and the environmental changes that have resulted, have been catastrophic for many other species, especially of vertebrates. The first extinctions of other species of vertebrates that can be traced to humans may have occurred during the Pleistocene when humans first entered North America, although this hypothesis is controversial. Unquestionable and documented extinctions caused by humans can be traced back to the Age of Exploration, and the situation continues to worsen. In this chapter we trace the origin of humans and the spread of modern humans, and discuss the impact of humans on other vertebrates and the prospects and vital importance of reversing the increasing rate of extinction of species.

The Origin of Humans

Human beings are primates. We have descended from arboreal ancestors that lived in early Tertiary forests 65 million years ago. Our closest extant relatives are the chimpanzees and gorillas of Africa. What is the basis for these statements, which Darwin's Victorian contemporaries found so startling and upsetting? In the following sections we summarize the evidence and the inferences about human phylogeny and evolution.

Characteristics of Primates

Humans share many biological traits with animals variously called apes, monkeys, and prosimians. Using the principle of homology (Chapter 1), comparative anatomy demonstrates that all of these mammals can be grouped together as Primates. Obvious traits that characterize primates are (1) retention of the clavicle (which is reduced or lost in many mammalian lineages) as a prominent element of the pectoral girdle; (2) a shoulder joint allowing a high degree of limb movement in all directions and an elbow joint permitting rotation of the forearm; (3) general retention of five functional digits on the fore and hind limbs; (4) enhanced mobility of the digits, especially the thumb and big toes, which are usually opposable to the other digits; (5) claws modified into flattened nails; (6) sensitive tactile pads developed on the distal ends of the digits; (7) reduced snout and olfactory apparatus with most of the skull posterior to the orbits; (8) reduction in number of teeth compared to stem mammals but retention of simple molar cusp patterns; (9) complex visual apparatus with high acuity, color perception, and a trend toward development of forward-directed binocular eyes; (10) large brain relative to body size, in which the cerebral cortex is particularly enlarged; (11) trend toward derived fetal nourishment mechanisms; (12) only two mammary glands (some exceptions); (13) typically, only one young per pregnancy associated with prolonged infancy and preadulthood; and (14) trend toward holding the trunk of the body upright leading to facultative bipedalism (Szalay and Delson 1979). It is important to understand that several of the above characteristics can be identified as trends within the primates and thus may not be apparent in ancestral members of the lineage.

Most of these traits have long been attributed to an arboreal life. All of the basic modifications of the appendages can be seen as contributions to arboreal locomotion, as can the stereoscopic depth perception that results from binocular vision as well as the enlarged brain for neuromuscular coordination between visual perception and locomotory response. Most primates are arboreal, but some have become secondarily terrestrial and humans are the most terrestrial of all. Even so, many of the traits that are most distinctively human have been said to derive from earlier arboreal specializations. However, arboreality cannot be the entire basis for these primate characteristics. Squirrels provide a telling counter example; they are arboreal but show few of the specializations seen in primates (Cartmill 1974). Although squirrels do have a clavicle and good mobility of the shoulder and elbow joint, the range of movement is not as great as that of primates. The thumbs of squirrels are greatly reduced, there is no

opposability, and the remaining digits have long, sharp, recurved claws lacking tactile pads at their tips. Squirrels have large olfactory organs and snouts, laterally directed eyes, and no notable brain enlargement compared with fully terrestrial and diurnal rodents. Squirrel life histories are typical of small mammals: They produce large litters of fast-growing young. Nevertheless, squirrels are fully arboreal and can match or exceed the climbing skills of similar-size primates. If primate characters are not *required* by an arboreal mammal, what do they indicate about our lineage's early evolution?

Small marsupials and true (African) chameleons may give a clue, because they are also arboreal and share more convergences with ancestral primates than do squirrels. These three convergent taxa (stem primates, small marsupials, and chameleons) are all visually directed predators on arboreal insects. They all use grasping digits during cautious well-controlled pursuit of insects on slender branches. Apparently we may attribute many of the basic characters of primates not just to arboreal habits, but to very specialized ones—visual location and manual capture of arboreal insect prey.

Evolutionary Trends and Diversity in Primates

Table 23–4 presents a traditional classification of modern primates, and Figure 24–1 presents a simplified cladogram. It is generally agreed that the first primates evolved from a line of arboreal proto-insectivores not unlike the present-day tree shrews (Scandentia: Tupaiidae; Martin 1990). Most species of these small, agile southeast Asian mammals prefer tropical forest growth, running across the forest floor and climbing (especially on woody vines) high in the canopy. Tree shrews are slender, with long bushy tails and rather short limbs. They are thus more like squirrels than monkeys (Figure 24–2) and the name Tupaiidae comes from the Malay word for squirrel. The skull and teeth show most of the unspecialized characteristics of ancestral mammals, including nostrils at the end of a pronounced moist snout, a large gape, lateral eyes, and occipital condyles located at the extreme posterior of the skull with the head carried fully out in front of the vertebral column. Tree shrews have lost only one incisor and one premolar from the ancestral mammalian dental formula of three incisors, one canine, four premolars, and three molars. The cheek teeth also retain the ancestral tricuspid condi-

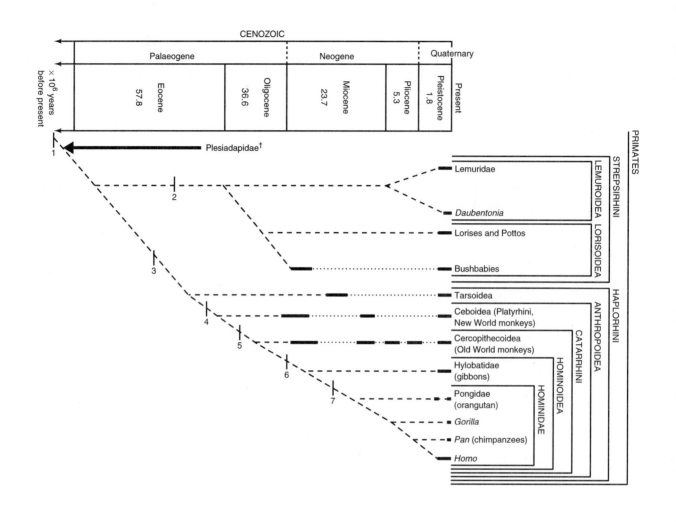

CENOZOIC

Palaeogene | Neogene | Quaternary

× 10⁶ years before present

| Eocene 57.8 | Oligocene 36.6 | Miocene 23.7 | Pliocene 5.3 | Pleistocene 1.8 | Present |

Plesiadapidae†

1

2

3

4

5

6

7

Lemuridae

Daubentonia

Lorises and Pottos

Bushbabies

Tarsoidea

Ceboidea (Platyrhini, New World monkeys)

Cercopithecoidea (Old World monkeys)

Hylobatidae (gibbons)

Pongidae (orangutan)

Gorilla

Pan (chimpanzees)

Homo

PRIMATES

STREPSIRHINI

LEMUROIDEA

LORISOIDEA

HAPLORHINI

ANTHROPOIDEA

CATARRHINI

HOMINOIDEA

HOMINIDAE

tion (Luckett 1980). Although the tree shrews, because of some of their specializations, are not thought to be in the direct lineage of primates, they offer the best extant model of the basal stock of our lineage.

The first primates appear at the Cretaceous–Tertiary boundary, and the lineage shows considerable diversity by the Eocene. The earliest primates were rather like tree shrews, perhaps with some development of prehensile fingers and toes, opposability of the first digits, and enlarged brains. In a short period of geological time, these **prosimians** spread over large portions of the Old and New Worlds and dif-

ferentiated into ecological forms ranging from insectivorous to vegetarian in their diets and from nocturnal to diurnal in their daily activity cycles (Figure 24–3). Most were primarily arboreal and retained ancestral features such as long snouts and long tails. They became more clearly identifiable as primates by changes leading to increased range of movement in their appendicular skeleton, the tendency for the eyes to move forward, and—most notably—larger brain capacity. The late Eocene prosimians possessed brains that were larger in relation to their body sizes than any other Eocene mammals. This allometric

1. Primates: Cheek teeth bunodont or brachydont; limbs plantigrade; a nail (instead of a claw) always present in extant forms, at least on the pollex (thumb). 2. Strepsirhini (lemurs and lorises): Cranium elongate; orbit and temporal fossa broadly confluent. 3. Haplorhini [tarsiers plus anthropoids]: Cranium short; orbit and temporal fossa separated ventrally by a postorbital plate. 4. Anthropoidea: Tubular external auditory meatus; broad interorbital bony pillar and prominent thickening [the glabella] medially above orbits distinct from brow ridges [the tori]; third premolar single cusped, bilaterally compressed; forth premolar bi-cuspid, slightly longer than broad; lower molars increase in size posteriorly, the third only slightly larger than the second, all with five cusps, the hypoconulid small; second upper molar larger than first or third; distinctive muscle, nerve and articular modifications of the humerus; characteristic ankle and thumb bones. 5. Catarrhini [Old World monkeys, apes, and humans]: Nasal aperture high and oval-shaped; premaxilla short and broad; upper incisors of similar size, the first spatulate and the second caniform; canine slender, high-crowned relative to length and sexually dimorphic; third and fourth upper premolars broad in buccolingual direction; upper molars similarly broad with four cusps, the hypocone small; molar enamel thin; tooth rows form a V-shape; palate shallow and longer than broad; distinctive characters of humerus, radius, and ulna. 6. Hominoidea [apes and humans]: Frontal bone wide posteriorly; cusps on upper premolars nearly uniform; third lower premo-lar low-crowned; modifications of the vertebral column foreshadowing those seen in Recent hominoids [see below]; scapula with elongated vertebral border and robust acromion; humeral head rounded, medially oriented, and longer than femoral head. Additional characters of hominoids absent in the problematic fossil *Proconsul* include enlarged sinuses; first upper incisor as broad as high; lower molars broad; palate deep; clavicle elongated; shaft of ulna bowed; radial head rounded; blade of ilium broadened; thoracic vertebrae protrude into thoracic cavity; five lumbar, four or five sacral, and six caudal vertebrae, no tail; various characteristics of muscle insertions; several derived features of bones of the appendages. 7. Hominidae [apes and humans]: Canines robust and long relative to their height; lower third molar robust, not laterally compressed; tooth rows form a U-shape. Additional characters of Recent hominids include mastoid process distinct; medial and lateral pterygoids of equal size; maxillary sinus enlarged; orbits higher than broad; lengthened premaxilla; second upper incisor spatulate; third and fourth lower premolars elongated; enamel on molars thick; ischial tuberosities absent; four or five lumbar vertebrae; several derived characters of trunk and arm muscles and arm bones. (Based on R. L. Cichon, 1983, pages 783–837 in R. L. Cichon and R. S. Corruccini, *New Interpretations of Ape and Human Ancestry*, Liss, New York, NY; and on data from various sources in E. Delson, editor, 1985, *Ancestors: The Hard Evidence*, Liss, New York, NY.)

Figure 24–1 Phylogenetic relationships of the primates. This diagram shows the probable relationships among the major living groups of primates. An extinct lineage possibly at the base of primate phylogeny is marked with a dagger (†).

enlargement was probably associated with the sophisticated neuromuscular coordination required for complicated arboreal insect-eating activities, complex visual functions, and control of a grasping, manipulative hand. These fossil primates are most closely related to extant lemurs, pottos, aye-ayes, lorises, and galagos. The first primates have long been grouped with these extant forms and the tarsiers as "prosimians." Today, like so many other groups composed of ancestral forms, the prosimians are considered paraphyletic. The species generally considered prosimian are not all closely related, but instead they retain ancestral primate features. They are restricted to parts of Africa and Southeast Asia and to the island of Madagascar, where the lemurs and their relatives continued to maintain a diversity of forms in the absence of predators and other competing primates. Today many of the lemurs are threatened with extinction, owing to destruction of their forest habitat (Jolly 1980). By the Oligocene the first simians had evolved in most parts of the world, and the prosimians began to disappear.

There is much disagreement about the evolutionary transition from the prosimians to the

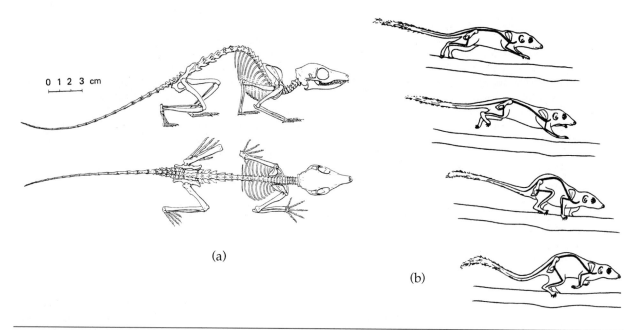

Figure 24–2 Skeleton of tree shrew (*Tupaia glis*) (a) in typical posture with flexed limbs and quadrupedal stance and (b) sequential phases of the bounding run used at top speed along a limb. Note the great flexibility of both axial and appendicular elements in locomotion, an ancestral characteristic tree shrews have in common with primates. This flexibility contrasts with that of many nonarboreal or cursorial mammals. [Modified from F. A. Jenkins, Jr. (editor), 1974, *Primate Locomotion*, Academic, New York, NY.]

Anthropoidea or higher primates (Cichon and Fleagle 1987). Fossil primates from the critical Oligocene and Miocene periods are rare. Most of the fossils from the Miocene are, in fact, already derived New World (platyrrhine) or Old World (catarrhine) monkeys or apes (hominoids). The morphological features of these groups and their fossil histories suggest that they evolved from a tarsier-like ancestor in Asia in the late Eocene. Today's tarsiers are small nocturnal arboreal insect eaters specialized for jumping between vertical stems and trunks in thickets and forests. One curious shared derived feature of these primates is the condition of the nose. The nostrils of tarsiers and all monkeys and apes are surrounded by smooth dry skin or hair rather than the moist, glandular skin of the nose seen in more plesiomorphic primates.

These derived primates invaded Africa where the characteristics of the Anthropoidea evolved. By the early Oligocene some of these primates had moved across the South Atlantic Ocean, which was much narrower than it is now. (Rodents ancestral to the South American guinea pigs and their relatives [caviomorphs] also arose in Africa and crossed the Atlantic at about the same time.)

All known platyrrhine anthropoids have been restricted to the American tropics since their first appearance in the fossil record in the late Oligocene. All catarrhines with the exception of *Homo sapiens* have been similarly restricted to the Old World. But Oligocene platyrrhines from Argentina resemble Egyptian catarrhines of the same period in a number of dental features, indicating close relationship of the two clades.

The New World monkeys do not share the derived features of Old World monkeys, apes, and humans, but have unique features of their own. As the group name indicates, platyrrhine monkeys have flat noses, with nostrils that are far apart and face outward. Most platyrrhines are smaller than catarrhines. The long tail of platyrrhines is an ancestral primate characters that was retained in the lin-

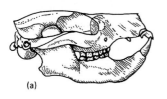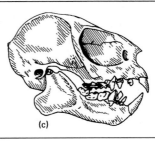

(a)

(b)

(c)

Figure 24–3 Reconstruction of fossil skulls of some early primates and primate-like mammals, showing variation in dentition with feeding habits: (a) *Plesiadapis*, a late Paleocene form, was rodent-like and questionably in the primate lineage; (b) *Notharctus*, an Eocene lemur, was more omnivorous and carnivorous, while (c) *Tetonius*, a derived Eocene tarsier-like form, had highly specialized teeth, suggesting a basically predatory mode of life. The large orbits of *Tetonius* clearly indicate nocturnal habits. Not to same scale. (Modified from A. S. Romer, 1966, *Vertebrate Paleontology*, 3d edition, University of Chicago Press, Chicago, IL.)

eage. The tail is specialized as a prehensile organ in a number of the medium and large platyrrhines. The pollex (thumb) is not fully opposable to the other fingers, and is reduced or absent in some species. However, the hallux (big toe) is fully opposable. Three premolars are present in each jaw. In addition to the typical South American monkeys, the platyrrhines also include the squirrel-like marmosets (callithrichids). Marmosets are specialized small primates, some with a highly developed taste for the edible gums exuded from the bark of certain trees. Marmoset toes, except the hallux, have secondarily claw-like nails that dig into bark and aid in climbing.

The catarrhines include the Old World monkeys, apes, and humans. We catarrhines have nostrils that are close together, open forward and downward, and have a smaller bony opening from the skull than is the case for platyrrhines. There is a trend toward large body size in our lineage, which culminates in the great apes and humans. Although catarrhines are mostly arboreal, there is a marked tendency toward upright carriage and secondary specializations for terrestrial locomotion. The tail is often short or absent, and tail prehensility has never evolved. The second premolar is always absent, giving a total of two incisors, one canine, two premolars, and three molars on each side of the upper and lower jaw. This condition is traceable to fossils in the Oligocene. The group consists of two clades: the Old World monkeys (Cercopithecoidea); and the apes and humans (Hominoidea), the latter including the gibbons

(Hylobatidae), the orangutan (Pongidae) and the African great apes, fossil forms such as the genus *Australopithecus*, and humans (Hominidae). The Cercopithecoidea constitute a spectrum of nearly 80 species of often highly colorful, sometimes impressively fierce and behaviorally complex vertebrates. They are almost universally social animals living in habitats that range from open grassy plains to the tops of mature rain forests and from the highlands of the Himalayas to intertidal mangrove swamps.

Evolution and Phylogeny of the Hominoidea

Apes and humans are placed in the Hominoidea. The fossil record for this clade is better than that of other lineages of primates, despite a gap between eight and four million years ago.

The hominoids are more diverse than are their nearest relatives, the cercopithecoid monkeys. There is a great range in body size (from Asian gibbons to the huge African gorillas) and in body covering (from the dense hair covering of gibbons to the sparse remnants of hair in humans). Gibbons move through the trees most frequently by brachiation (swinging from the underside of one branch to the underside of the next using the hands to grasp the branches), and are among the most versatile arboreal acrobats alive. Gibbons become bipedal when moving on the ground, holding their arms outstretched for balance as a tightrope walker uses a pole. The larger and more sluggish orangutans rarely swing by their arms, and prefer slow quadrupedal climbing among the branches of trees. The African gorillas and chimpanzees are more at home on the ground, where they most often progress quadrupedally, supported on their forelimbs by their knuckles. Although all modern hominoids can stand erect and walk to some degree on their hind legs, only humans among extant forms display an erect bipedal mode of striding locomotion involving a specialized structure of the pelvis and hind limbs, thereby freeing the forelimbs from obligatory functions of support, balance, or locomotion.

Hominoids are morphologically distinguished from other recent anthropoids, including their sister taxon the cercopithecoids, by a pronounced widening and dorsoventral flattening of the trunk relative to body length so that the shoulders, thorax, and hips have become proportionately broader than in monkeys. The clavicles are elongated, the iliac blades of the pelvis are broad, and the sternum is a broad structure, the bony elements of which fuse soon after birth to form a single flat bone. The shoulder blades of hominoids lie over a broad, flattened back, in contrast to their lateral position next to a narrow chest in monkeys (Figure 24–4) and most other quadrupeds. The pelvic and pectoral girdles of hominoids are relatively closer together because the lumbar region of the vertebral column is short. The caudal vertebrae have become reduced to vestiges in all recent hominoids, and normally no free tail appears postnatally. There is a tendency for the spinal column to arch and, especially in the lumbar region, to approach the center of gravity of the body when the trunk is held upright. Balance is assisted by the flat thorax, which places the center of gravity near the vertebral column. These and other anatomical specializations of the trunk are common to all hominoids and help to maintain the erect postures that these primates assume during sitting, vertical climbing, and walking bipedally (Campbell 1985). Secondary curves of the spine are the consequence of bipedal locomotion, and form in humans as soon as an infant learns to walk.

Especially in the last 2 million years, brain enlargement has been a dominant evolutionary force molding the shape of the hominoid skull, and has tended to increase the neurocranial vault more than its base. The skulls of hominoids also differ from those of other catarrhines by their extensive formation of sinuses, hollow air-filled spaces lined with mucous membranes that develop between the outer and inner surfaces of skull bones. Chimpanzees, gorillas, and humans share the derived character of large frontal sinuses.

In the identification of fossils, differences in the cheek teeth between hominoids and Old World monkeys are of paramount significance because many primate fossils are represented by individual teeth, or teeth in fragments of jawbone. The Old World monkeys have lower molars (M_1 and sometimes M_2) with four cusps, one at each corner of a rectangular crown; the anterior pair of cusps, like the posterior pair, are each connected by a ridge. Since the Miocene, hominoid lower molars have had crowns with five cusps, but in extant humans (although not most fossil relatives) this pattern is frequently obscured by crenulations or variations in the number

Figure 24–4 Cephalic view of chest and pectoral girdle (right half) of macaque (cercopithecoid monkey) and human (hominoid).

Macaque

Human

of cusps. When five cusps do appear the grooves between the cusps usually resemble the letter Y, with the open part of the Y embracing the hypoconid cusp (Figure 24–5). This molar pattern, called Y-5, has persisted among hominoids for more than 20 million years (Day 1986).

Figure 24–6 presents a phylogeny of the Hominoidea. The earliest indications of both the cercopithecoid and the hominoid lineages are fossil teeth and jaws found in the Egyptian Fayum deposits of the Oligocene, approximately 35 million years old. Fossils assigned to the genus *Propliopithecus* and others of similar structure, especially *Aegyptopithecus*, have ancestral characters of both clades of catarrhines. *Oligopithecus*, which occurs in the same deposit, has derived characters of the molars that suggest it is near the stock that gave rise to the cercopithecoid monkeys. If this hypothesis is correct, the division between the cercopithecoids and hominoids probably occurred near the Eocene–Oligocene boundary.

By the Miocene ancestral hominoids had diversified into a number of ecological types and had spread widely over the Old World, including Africa, Europe, and Asia. One genus, *Pliopithecus* (Figure 24–7), has a skull like that of gibbons but its appendicular and trunk skeletons have forelimbs and hind limbs about equal in length. It was not as specialized for brachiation and arboreal life as extant gibbons or the somewhat later *Oreopithecus* (Figure 24–8). *Oreopithecus*, an Italian fossil primate, resembles gibbons in its postcranial skeleton but it has a less specialized skull and its teeth lack the Y-5 configuration. Thus *Oreopithecus* may be either a cercopithecoid monkey or a pre-Y-5 hominoid, and may have been much like extant gibbons (Andrews 1992).

Of great interest are Miocene fossils variously referred to as dryopithecids and sivapithecids which demonstrate great ape/human characteristics. These primates occurred widely across Eurasia and Africa and have been assigned various generic names: *Dryopithecus*, *Proconsul*, *Kenyapithecus*, *Otavipithecus*, *Sivapithecus*, and *Ramapithecus* (*Proconsul* is usually synonymized with *Dryopithecus*, and *Ramapithecus* is often included in *Sivapithecus*, illustrating the lumping typical of some areas of primate paleontology). They are all closely related hominoids, but paleoclimatic and paleofaunistic evidence suggests they occupied a wide range of ecological niches (Andrews 1992).

Some early Miocene dryopithecids lived in heavily forested areas of Africa and later Eurasia. The later ramapithecines first appear in the fossil record 17 million years ago during the Miocene when the heavily forested areas of the Old World were giving way to mixed environments consisting of patches of forests, savanna woodlands, and open grasslands (Chapter 20). The kinds of mammalian fossils found with ramapithecines suggest that these primates lived on the edge of the forests and open areas and obtained food from both habitats. If this hypothesis is correct, some lineages of primates may have been evolving toward a terrestrial mode of locomotion, and it is among the Miocene forms of 13 to 14 million years ago (such as *Kenyapithecus* and *Otavipithecus*) that we are most likely to find the ancestors of the great apes and humans.

The oldest fossils that show great ape/human characteristics are some Miocene teeth and jaw fragments from the Siwalik Hills of Pakistan and India; Maboko Island and Fort Ternan, Kenya; Namibia; and localities in Europe and China. They span an enormous time period for primate remains, from about 17 to 9 million years ago. They are sometimes called the "god apes" because several of the generic

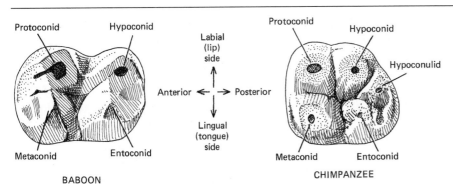

Figure 24–5 Right lower molar patterns of Old World monkey (baboon) and hominoid (chimpanzee) compared. Shading indicates low crown areas or grooves between major cusps and ridges illustrating the Y-5 configuration of hominoids.

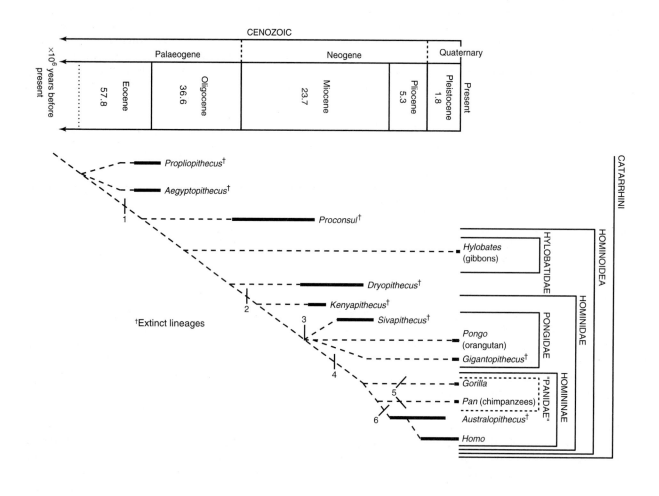

†Extinct lineages

names are derived from the names of Hindu gods. *Sivapithecus* and the possibly congeneric *Ramapithecus* are the best known, and were contemporary with other Miocene forms of sivapithecids. It is widely accepted that these are the probable ancestral stock from which the great apes and human primates evolved. The shape and relative size of the teeth are more like those of humans than like those of apes, but the lower tooth row is V-shaped unlike apes *or* humans.

The size of the teeth and jaws suggests animals about a meter or so in stature. The drastic change in dentition, with reduced canines and thick enamel molars, suggests to some workers that these primates may have fed on material that required crushing (thick enamel) and grinding (no interlocking canines to inhibit rotary chewing movements). Although many of their fossils are associated with fossils of browsing and grazing ungulates, indicating that these primates lived in savanna woodlands and at the edge of grasslands, the grinding teeth of these fossils are most like arboreal nut and hard fruit feeders among extant primates. The conclusion now seems to be that by Miocene time derived hominoid

1. Hominoidea: See node 6 in Figure 24–1. **2.** Hominidae: See node 7 in Figure 24–1. **3.** Ponginae (orangutan and fossil relatives): Zygomatic arch flattened, facing anteriorly or even slightly superiorly; loss of glabellar thickening; narrow interorbital distance; small incisive foramen, first upper incisor much larger than second; distinctive junction of premaxilla with maxilla and 18 additional derived characters. **4.** Homininae (African apes, australopithecines, and humans): Large sphenopalatine fossae; several characteristics of the nasopalatine region; presence of a fronto-ethmoid sinus; lower molar trigonids markedly reduced; several derived characters of limbs indicative of vertical climbing and incipient bipedal locomotion. **5.** African apes: Thin enamel caps on molars; extensive modification of the hands and less extensive modifications of the arms reflecting the occurrence of knuckle-walking. **6.** Australopithecines and humans: Progressive increase in cranial capacity; anterior shift in position of foramen magnum and occipital condyles; reduction in size and sexual dimorphism of canines; reduction in diastema posterior to canines; decreased molar length relative to width; increased molar crown height; distinctive enamel prism patterns; progressive reduction in length of forelimb relative to length of trunk; shortened ischium; broad ilium; pronounced lumbosacral curve in vertebral column; modifications of arm and leg muscles. (Based on R. L. Cichon, 1983, pages 783–837 in R. L. Cichon and R. S. Corruccini, *New Interpretations of Ape and Human Ancestry*, Plenum, New York, NY; and on data from various sources in E. Delson, editor), 1985, *Ancestors: The Hard Evidence*, Liss, New York, NY.)

Figure 24–6 Phylogenetic relationships of the Hominoidea. This diagram shows the probable relationships among the major groups of hominoids. Extinct lineages are marked by a dagger (†). The quotation marks indicate that the frequently used family name "Panidae" has a paraphyletic composition, since chimpanzees are thought to be the sister group of hominids.

populations in Asia, Europe, and Africa were approaching characteristics appropriate for ancestors of the great apes and humans. The closest extant relative to the sivapithecids, especially to *Ramapithecus*, appears to be the orangutan. It seems an inescapable conclusion that in numerous dental and associated cranial features humans retain ancestral characteristics of our clade while the lineages of great apes have each derived unique specializations. That is, the apes evolved from a more human-like lineage, not the other way around!

Origin and Evolution of Humans

Humans differ enough in their anatomy from the apes so that they have usually been placed in a separate family, called the Hominidae, although biochemical evidence suggests a much closer intrafamilial relationship between apes and humans. The jaw in the human clade is shorter in association with the shortening of the muzzle, certain teeth are smaller and the entire dentition is more uniform in size and shape. In particular, the canines lie on the same occlusal plane as the incisors and cheek teeth. Apes have a prominent diastema between the canines and incisors that accommodates occlusion of the jaws, whereas in humans all the teeth touch their adjacent members (Figure 24–9). The jaws of an ape are rectangular or U-shaped, with the four incisors lying at right angles to the canines and cheek teeth, which form nearly parallel lines along the jaw's two sides. The ancestors of the hominids, and the hominoids such as *Sivapithecus*, had an almost V-shaped jaw. Some of the earliest hominids had a U-shaped jaw. The jaw of members of the genus *Homo* is bow-shaped, with the teeth running in a curve that is widest at the back of the mouth. The human palate is prominently arched, whereas the ape palate is flatter between their parallel rows of cheek teeth.

Various evolutionary trends within the hominids can also be identified. The articulations of the skull with the vertebral column (**occipital condyles**) and the **foramen magnum** (for the passage of the spinal cord) shifted from the ancestral position at the rear of the braincase to a position under the braincase. This change balances the skull on top of the vertebral column, in association with an upright, vertical posture of the trunk. The braincase itself became greatly enlarged in association with an increase in forebrain size. By the end of the middle Pleistocene a prominent vertical forehead developed, in contrast to the sloping foreheads of the apes. The brow ridges and crests for muscular attachments on the skull became reduced in size in association with the reduction in size of the muscles that once attached to them. The human nose became a more

Figure 24–7 Reconstruction of *Pliopithecus*, a generalized Miocene ape.

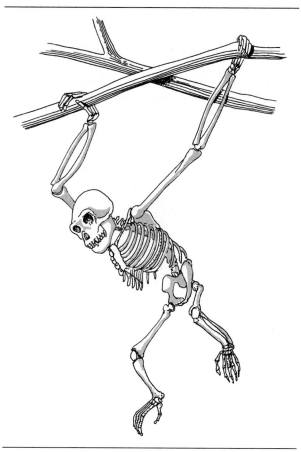

Figure 24–8 Reconstruction of *Oreopithecus*, a gibbon-like primate of the Miocene.

prominent feature of the face, with a distinct bridge and tip.

The most radical changes in the hominid post-cranial skeleton are associated with the assumption of a fully erect, bipedal stance: the S-shaped curvature of the vertebral column; the modification of the

pelvis and position of the **acetabulum** (socket in the hip for the ball joint with the femur) in connection with upright bipedal locomotion; and the lengthening of the leg bones and their positioning as vertical columns directly under the head and trunk. The feet of humans show drastic modification for

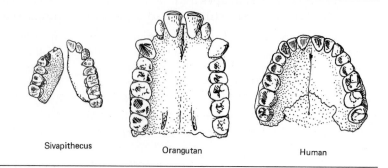

Figure 24–9 Upper jaw of *Sivapithecus*, an orangutan, and a human compared.

Sivapithecus   Orangutan   Human

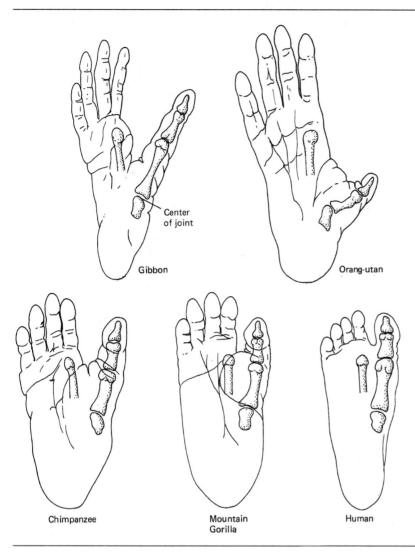

Figure 24–10 Feet of extant hominoids compared, showing some skeletal parts (metatarsals for digits in I and II and phalanges of digit I) in relation to foot form. Note difference in arboreal specializations of gibbon and orangutan, and the degrees of specialization for terrestrialism in chimpanzee, gorilla, and human.

Gibbon

Center of joint

Orang-utan

Chimpanzee

Mountain Gorilla

Human

bipedal striding locomotion, having become flattened except for a tarso-metatarsal arch with corresponding changes in the shapes and positions of the tarsals (ankle bones) and with close, parallel alignment of all five metatarsals and digits; the big toe is no longer opposable as in apes and monkeys (Figure 24–10), although it may still have had the capacity to diverge from the rest of the toes in early hominids.

It is important to emphasize that among mammals the bipedal, striding mode of terrestrial locomotion is unique to the human line of evolution and that it may have been a key change that made possible the evolution of other distinctively human traits, such as the perfection of tool-using hands, and thus indirectly probably stimulated the evolution of ever larger forebrains. In addition to changes in the foot, evolution of a bipedal, striding gait from the anthropoid quadrupedal mode required major reorganization in the structure and function of the pelvis and hind leg bones, and their associated muscles. Figure 24–11 contrasts the shape and orientation of the gorilla and human pelvis to show the postural differences between quadrupedal and bipedal locomotion. In the gorilla the ischium protrudes backward prominently and the long ilium extends forward and lateral to the vertebral column; during normal terrestrial locomotion the whole pelvis is tilted toward the horizontal. In humans the ischium is much shorter, the broadened ilium extends toward the midventral (abdominal) region, and the pelvis is held vertically during walking.

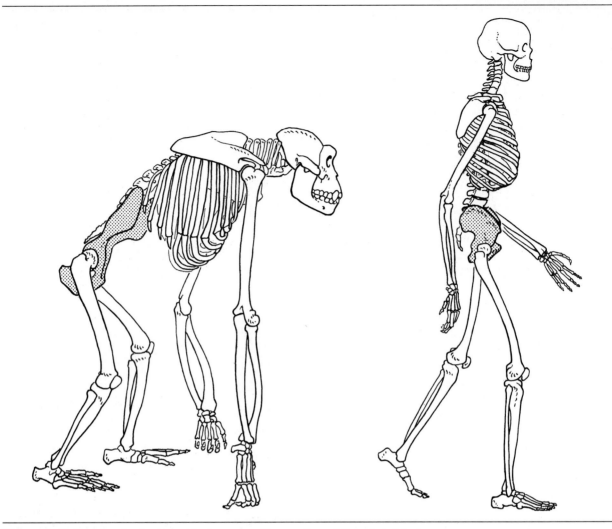

Figure 24–11 Shape and orientation during general locomotion of the pelvis (shaded) in gorilla and human.

What evolutionary events turned the human into an obligate terrestrial biped from ancestors that clearly were arboreal? How did our tool-using hands and our big brain with its associated implications for intelligence, derived social organization, and culture evolve?

The Miocene, like the Pleistocene (Chapter 20) after it, was a period of increasing cooling, drying, and seasonality of climate. Seasonality favors herbs and annual plants over woody perennials. As tropical forests yielded to more open woodland savannas and grasslands in the Miocene, arboreal vertebrates, including various clades of hominoids, came under selective forces favoring terrestrial life and upright

bipedalism. Some hominids of this time had dentition suited for grinding coarse substances and may have been omnivorous, foraging on seeds, nuts, and other small food items gleaned from the ground. They probably also scavenged small animal carcasses and preyed on vulnerable animals. This much of the story is inferred from the combination of ape-like and human-like traits found in various Miocene and early Pliocene hominoid fossils.

We can speculate further that once freed of locomotor requirements, the forelimbs and hands could respond to selective forces that would shape them into increasingly effective manipulative feeding organs. Anatomical changes could open new avenues for behavioral change. Increased survival rates of the young may have resulted because the foraging members of family groups could gather and carry quantities of food to less mobile nursing mothers and their dependent offspring. The prolongation of gestation, increase in intervals between births, decrease in number of young per pregnancy, lengthening of time to reach sexual maturity, and postreproductive survival of old individuals constitute unique changes in primate evolution culminating in the human line (Figure 24–12). All these behaviors are traits that favor the evolution of a well-developed social organization. Increase in brain capacity—permitting functions such as communication, eye–limb coordination, manual dexterity, memory, and other functions associated with the neocortex—would be favored by social cooperation in food gathering, hunting, or defense against powerful predators. The increases in body size, tool use, and brain size in the derived lines of hominids known from the late Pliocene and early Pleistocene coincide with evidence of the cooperative hunting of large mammals.

Before Darwin's time it was believed that the modern human species was only a few thousand years old, and by counting the genealogy of the Old Testament, it was thought possible to fix the date of Adam's creation in the Garden of Eden rather precisely. Even after the discovery of the first human fossil bones in Europe and later in Asia and Africa, it was long thought that true human beings (genus *Homo*) were no more than about 500,000 years old, roughly mid-Pleistocene in origin. Better methods of dating fossils have revealed that the original estimates of the age of some fossils were erroneously short. More important, older and older fossils with hominid characteristics have been uncovered, and this mosaic evolution makes it difficult to specify a particular fossil as the first true human.

A phylogenetic tree of the Hominoidea (Figure 24–6) indicates that the human, chimpanzee, and gorilla are more closely related to each other, in terms of recency of their common ancestor, than they are to orangutans or gibbons. These relationships among extant apes and humans were first determined on the basis of comparative anatomy by T.H. Huxley, who published his ideas in 1863. They have gained additional support and clarity from recent biochemical, serological, and cytological comparisons as well as from the fossil record of the earlier hominoids. The majority of studies, especially the molecular ones,

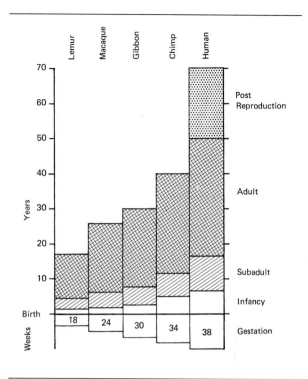

Figure 24–12 Progressive changes in the length of life phases and gestation in primates. Note the proportional increase in the length of the postjuvenile stages as gestation increases. Only humans have a significant postreproductive stage, although the stage has also occasionally been observed in chimpanzees.

support a close relationship between chimpanzees and humans, although some anatomical studies place humans and gorillas closer to each other.

The extant African apes have no known fossil record. At present we must depend on molecular studies of extant forms to gain insight into the history of our nearest surviving relatives. This evidence indicates that gorillas separated from the chimpanzee and human common ancestor between 8 and 10 million years ago. The gorillas were subsequently isolated by unknown phenomena into eastern and western populations about 3 million years ago. Sometime in the interim, perhaps as recently as 6 million years ago, chimpanzee and human lineages separated (Morell 1994). Two or possibly three species of chimpanzee exist today with geographic differences in genetic patterns similar to that found in gorillas. However, the differences in chimpanzee populations are thought to have occurred more recently, between about 2.5 and 1.6 million years ago.

A surprising conclusion from the emerging discipline of molecular evolution is that humans are more closely related to the extant great apes of Africa than morphologists or paleontologists had thought (Cichon and Corruccini 1983, Lewin 1984). For example, hybridization of nonrepeated DNA sequences of humans and chimpanzees indicate 98 percent identity. Electrophoretic measurement of genetically controlled polymorphic proteins (allozymes) in the six extant genera of hominoids show that the degree of genetic difference among human, chimpanzee, gorilla, and orangutan is no greater than that observed among species in the same genus in many other taxa of vertebrates, including most mammals. The difference between the gibbons (Hylobatidae) and other hominoids is only slightly greater than the average difference between species in the same genus among other vertebrates (Bruce and Ayala 1979). High-resolution analysis of chromosomal morphology also reveals close relatedness among the hominoids: Humans are most closely allied to chimpanzees, the gorillas are slightly more distantly related, and the orangutans still farther removed (Yunis and Prakash 1982). These sorts of data and calculations, based on the concept of the **molecular clock** (the supposedly unvarying rate of generic change within a clade over evolutionary time), suggest that *Homo* and the extinct but related genus *Australopithecus* last shared a common ancestor with the African apes no more than 5 to 10 million years ago, in the late Miocene to early Pliocene. This is a much more recent split than most biologists have previously supposed, and indicates a phase of rapid evolutionary change in the lineage leading to the specialized apes (Andrews 1992).

If this molecular time estimate is correct, a gap of only 1 to 5 million years occurs in the geological record before the first undoubted hominid fossils are found in mid-Pliocene deposits in Kenya and Ethiopia. About 8 million years ago, at the beginning of the essentially blank period for hominoid fossils in East Africa, tectonic forces split the region almost in two along a north to south feature still dramatically obvious today, the Rift Valley. New mountain boundaries and resulting changes in the ecological landscape appear to have divided an as yet unknown ancestral population of Homininae. The western population survived in moist forested lowlands and became the ancestors of the chimpanzees. The eastern population lived in the rain shadow of the new mountains where forests gave way to woodlands and woodlands to savannas. As conditions became dryer and more seasonal 2 to 3 million years ago, savannas became increasingly widespread in the rain shadow. The hominids isolated in the East became our direct ancestors (Coppens 1994).

Although isolated fragments as old as 5.6 million years are known from Africa, they are too small to yield much information. No interpretable hominid remains older than about 3.8 million years had been discovered until recently. A new species of fossil hominid based on 17 specimens, all but four of them teeth, represents the remains of a form that lies close to the divergence between our closest African ape relatives and the lineage to humans (White et al. 1994, Wood 1994). *Australopithecus ramidus*, as the fragments have been named, comes from Ethiopian sediments about 4.4 million years of age and indicative of a wooded habitat (WoldeGabriel et al. 1994). *A. ramidus* has many chimpanzee-like features, is the most ape-like hominid ancestor known, and lacks some of the derived traits shared by the next most recent fossil hominids and all later hominids. Tentative evaluation of the scant evidence, however, indicates that *A. ramidus* was bipedal (judging from the position of the foramen magnum), had a human rather than ape-like arm (especially the shoulder joint and elbow), and had incisiform canines with less sexual dimorphism than modern African apes, traits that are hallmarks of the human lineage.

A most remarkable specimen, a very substantial part of a single young adult female skeleton of the next most recent known hominid, *Australopithecus afarensis,* is known in popular literature by the nickname Lucy. Lucy was discovered by Donald Johanson in the desert of the Afar plain, also in Etheiopia, not far from the Red Sea, in a deposit dated at 3.2 million years (Johanson and Edey 1981). Lucy's species represents a line of near-*Homo* primates that stretches back into the Pliocene toward the separation of the pongid and hominid clades. She is an astonishing specimen in several respects. Lucy is the most complete pre-*Homo* hominid fossil ever found, consisting of more than 60 pieces of bone from the skull, lower jaw, arms, legs, pelvis, ribs, and vertebrae. Paleoanthropologists believe that she stood fully erect. Her overall body size was small. Young but fully grown when she died, Lucy was only about 1 meter tall and weighed perhaps 30 kilograms (Figure 24–13), but other finds indicate that males of her species were larger, averaging 1.5 meters tall and 45

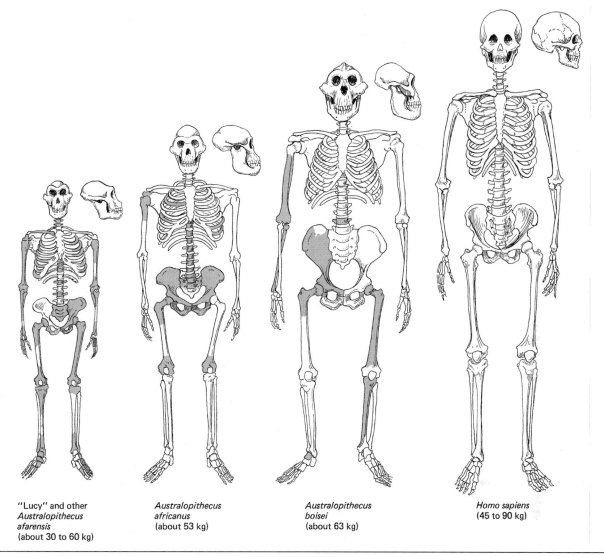

"Lucy" and other
*Australopithecus
afarensis*
(about 30 to 60 kg)

*Australopithecus
africanus*
(about 53 kg)

*Australopithecus
boisei*
(about 63 kg)

Homo sapiens
(45 to 90 kg)

Figure 24–13 Reconstructed skeletons of four hominid species, showing relative stature and stance. Shaded portions of the postcranial skeletons of the fossils show the parts actually known from specimens. Although almost all australopithecine skulls are fragmentary, sufficient material exists to reconstruct entire skulls with fair certainty. Too little of the most ancient australopithecine (*A. ramidus*) has been discovered to permit confident reconstruction.

kilograms (McHenry 1994). Although her hip and limb clearly indicate bipedalism, detailed analysis of the fossils and reconstruction of musculature demonstrate a bipedalism different from modern humans and retention of a partly arboreal anatomy (Berge 1994). Arboreality may also be reflected in her hand bones. Lucy and her kind may have slept and taken refuge in trees, bipedalism being important in the efficient harvesting of fruits in their open forest and woodland habitat (Hunt 1994). Her teeth and lower jaw are also clearly human-like. Her diet may have been rich in hard objects, probably fruits that would have been unevenly distributed in space and time (Andrews and Martin 1991). Unfortunately, we know little about the size and shape of Lucy's cranium. Other specimens of *A. afarensis* indicate a brain

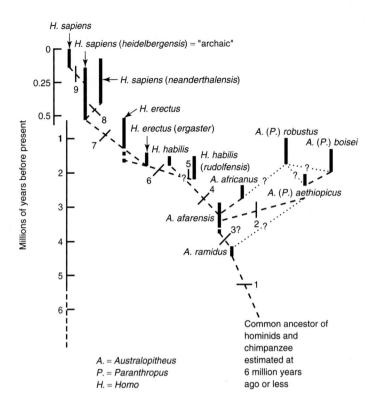

H. sapiens

H. sapiens (heidelbergensis) = "archaic"

H. sapiens (neanderthalensis)

H. erectus

H. erectus (ergaster)

A. (P.) robustus

A. (P.) boisei

H. habilis

H. habilis (rudolfensis)

A. africanus

A. (P.) aethiopicus

A. afarensis

A. ramidus

Millions of years before present

Common ancestor of hominids and chimpanzee estimated at 6 million years ago or less

A. = Australopitheus
P. = Paranthropus
H. = Homo

size of 380 to 450 cubic centimeters, quite close to that of modern chimpanzees and gorillas (Simons 1989, Lewin 1993).

Australopithecus afarensis has been found by Mary Leakey and her associates at Laetoli in Tanzania from an even earlier geological formation radiometrically dated between 3.6 and 3.8 million years in age. This is the only species of hominid known from that time. Thus the remarkable fossilized hominid footprints found at Laetoli of the same age were probably made by *A. afarensis*, proving the antiquity of bipedalism in hominid ancestry, long before the acquisition of a large brain. The analysis of these footprints in volcanic ash beds indicates that they do not differ substantially from modern human trails made on a similar substrate (White 1980). Indeed, the upright stance associated with bipedal striding and the consequent physical requirement to support the head squarely on top of the vertebral column may have been prerequisites that permitted the later evolution of a large forebrain.

Lucy and her kin were either contemporary with, or gave rise to, other hominids such as *Aus-*

tralopithecus africanus, another Pliocene hominid from East Africa, considered by some paleoanthropologists as ancestral to *Homo*. It now seems possible, however, that Lucy's species may have given rise to both the later *Australopithecus* and directly or indirectly to *Homo* (Figure 24–14).

The later australopithecines were distributed in East and South Africa from about 2.5 to 1 million years ago. The later australopithecines are represented by two different types: three large, *robust* forms with sagittal crests on the skulls and enormous grinding postcanine teeth (often placed in their own genus *Paranthropus*); and one smaller, more delicate, *gracile* form that was very *Homo*-like in build, had fully upright posture, and lacked sagittal crests and massive eye ridges (Johanson and Edey 1981, Delson 1987). (See the color insert.) *Australopithecus africanus* was the gracile form, and was perhaps close to the *Homo* lineage. *Australopithecus africanus* may have been a wide-ranging hunter and gatherer with an omnivorous diet. The three robust forms, *Australopithecus robustus* of South Africa and *A. aethiopicus* and *A. boisei* of East Africa, were terrestrial

1. Australopithecines and humans: See node 6 in Figure 24–6. 2. *"Paranthropus"*: An extensively pneumatized and projecting mastoid process incorporating on its expanded medial wall the occipitomastoid crest that provides a broad origin for the digastric muscle, an enlarged occipital–marginal venous sinus system, keystone-shaped nasal bones that extend beyond the fronto-maxillary suture to the glabella, more than 20 characters of the dentition. 3. *Australopithecus* and *Homo*: Tooth rows convex laterally, lower third premolar nearly bicuspid, hypoconulids of first and second lower molars small, mastoid not greatly enlarged, origin of digastric muscle restricted to a strong tendon with its base ossified as a juxtamastoid ("beside the mastoid") eminence, nasal bones triangular with the apex at the fronto-maxillary suture and distinctly beneath the glabella. 4. *Homo*: Increased mobility of the elbow joint and solidity of the knee joint (the reverse of the condition in early australopithecines, especially *A. afarensis*) indicating a shift from suspension or climbing to terrestrial bipedal locomotion, associated features of the femur and pelvis, reduced face size and projection associated with reduced robustness of the jaw and dentition, increase in brain size. 5. *Homo habilis*: Cheek teeth narrow in buccolingual direction, palate retracted under the nose, brain size increased by 45 percent over *Australopithecus africanus*. 6. *Homo erectus*: Distinguished on the basis of frontal proportions; parietal measurements; occipital curvature; the anatomy of the gleonoid cavity and the tympanic plate; straightness of the basicranium; the range of brain sizes; thickness of the cranial bones; a keeled, tent-shaped cranial vault; and a robust pelvis. 7. Archaic *Homo sapiens*: Basicranial flexion essentially modern with corresponding changes in facial prognathism and the angulation of the tympanic and petrous bones. 8. *Homo sapiens neanderthalensis*: Nose large relative to face; midface projection and large sinuses; long pubic bone; relatively short, stout limbs. 9. *Homo sapiens sapiens*: Long, curved parietal bones contribute to a spherical cranial vault and a short, high skull with its maximum dimension high on the head; a weak brow ridge but prominent, strongly developed chin; gracile (lightly built) long-boned postcranial skeleton. (Based on E. Delson, editor, 1985, *Ancestors: The Hard Evidence*, Liss, New York, NY.)

Figure 24–14 A hypothesis of the phylogenetic relationships within the Hominidae. Note change in time scale at 0.5 million years ago. Names in parentheses are sometimes considered full generic or subgeneric names (those that are capitalized) or species or subspecies names (those that are in lower case). Paleoanthropologists have claimed identifiable distinctiveness for all such indicated taxa. (Based primarily on B. Wood, 1994, *Nature* 371:280–281 and E. Delson, 1987, *Nature* 327:654–655.)

savanna-dwelling vegetarians, somewhat analogous to the forest-inhabiting gorillas. The gracile species either was replaced by *Homo* or, according to some theories, evolved into *Homo*.

A controversial form (or forms) described from fragmentary remains, existed in East Africa around 2 million years ago (Walker in Rasmussen 1993, Tattersall 1993). The actual taxonomic status of these highly variable remains, which have been called *H. habilis*, is much debated (Wood 1987). *Homo habilis* may have existed as early as 2 million years ago and have been present at the demise of some species of *Australopithecus*. The structure of the hand and wrist, and the long and powerful arms of both *Australopithecus* and *H. habilis* indicate arboreal capabilities (Susman and Stern 1982). The lower limbs of *H. habilis* are also much like Lucy's, suggesting the possibility that these small species of hominids (*Australopithecus afarensis* and *Homo habilis*) were differentiated by cranial capacity, not by body size or locomotor morphology. Cranial capacities of *Australopithecus afarensis* are the smallest so far known in the hominid fossil record and fall between 380 and 450 cubic centimeters in the three specimens measured; cranial capacities of *Homo habilis* are between 500 and 750 cubic centimeters (Campbell 1985, Falk 1985). Some paleoanthropologists believe that the brain of *A. afarensis* was still ape-like in appearance and had not undergone the cortex reorganization characteristic of *Homo*.

The earliest recognized simple stone tools are found throughout East Africa and date to 2.5 to 2.7 million years ago. They are called the Oldowan culture and they continue to appear in the geological record relatively unchanged for 1 million years. For some time most paleoanthropologists believed they were made by *Homo*, not *Australopithecus* (Lewin 1988, but see Susman 1988). Careful analysis of the structures of the modern human hand that allow the precise grasping necessary for stone tool manufacture and the evidence of these structures on the bones of the hand, especially the thumb (Susman 1994), may lead to a very different conclusion. On the basis of the only thumb morphology data available (Aiello 1994), the Oldowan tools could have been made by *Homo* or *Australopithecus robustus*, a small-brained species not generally considered in direct *Homo* ancestry. Perhaps all of the hominid species in existence 2 million years ago—both gracile and robust australopithecines and the newly emerged *Homo habilis*—might have been tool makers. Unfortunately, thumbs are rare among these fossils. It is interesting, however, that australopithecines thought to have become extinct 500,000 years before the earliest Oldowan tool finds (*A. afarensis*) do not have the hand bone characteristics of precision grip. Whatever occurred in the Oldowan tool-making industry, *A. africanus* disappeared from the fossil record around 2 million years ago. The two robust types persisted past the recorded existence of *H. habilis* to about 1 million years ago.

Perhaps as long as 2 million years ago, and definitely by 1.8 million years ago, a new hominid appears in the fossil record. *Homo erectus*, originally described as *Pithecanthropus erectus*, probably originated in East Africa, where it coexisted for at least several hundred thousand years with two of the robust australopithecines (Harris et al. 1988). *Homo erectus* was the first intercontinentally distributed human. It appears to have spread to Asia no later than 1.8 million years ago, and subsequently perhaps into Europe (Swisher et al. 1994). The species disappeared between 200,000 and 300,000 years ago. Three characteristics are especially important features of *Homo erectus*: substantially larger body size than earlier hominids (to 1.85 meters and at least 65 kilograms), a major increase in female size that reduced sexual dimorphism, and a brain volume overlapping that of extant humans. Chimpanzees have cranial capacities ranging from 280 to 500 cubic centimeters, gorillas from 350 to 750 cubic centimeters, australopithecines from 350 to 700 cubic centimeters, *Homo erectus* from 775 to 1100 cubic centimeters, and *Homo sapiens* from 950 to 2200 cubic centimeters.

The facial features of *H. erectus* were plesiomorphic—a massive prognathous (projecting) jaw, almost no chin, large teeth, sloping forehead, prominent bony eyebrow ridges, and broad, flat nose (Figure 24–15). *Homo erectus* made stone tools, and may have used fire. A cave near Zhoukoudian, China, about 50 kilometers southwest of Beijing was occu-

pied by *H. erectus* almost continuously for more than 200,000 years beginning at least 460,000 years ago (Wu and Lin 1983). From these and other sites it can be said that *H. erectus* appears to have changed gradually (unless fragments of unrecognized species have been lumped with *H. erectus*). Later individuals have even larger cranial capacities than earlier ones, but the change is not great (Rightmire 1990). Stone tools no more sophisticated than the Oldowan culture are the only tools earlier than about 1.4 million years ago. Novel tools, especially cleavers and so-called hand axes appear in East Africa about 1.4 million years ago. These later tools of African *H. erectus*, known as the Acheulean tool industry, changed very little in style for the next 1.2 million years, although the materials chosen changed and later samples have a much higher proportion of small tools than do early examples. The lack of dramatic advance in tool manufacture over this immense period is surprising, especially in light of the spread of these tools to Southwest Asia and Western Europe (Burenhult 1993).

Interesting ecological problems and possibilities attend the changes in the human lineage seen in *Homo erectus*. The notably larger body size of males in earlier species of our lineage implies a ranging pattern in which females foraged in smaller territories than males and the probability of a polygynous mating system (McHenry 1994). The substantial increase of body size and reduction in sexual dimorphism of *H. erectus* and subsequent forms of *Homo* imply behavioral changes. These new characteristics may be directly related to a significant increase in group mobility and ultimately to intercontinental distribution. Brain expansion relates to these phenomena in more than the obvious ways. Brain tissue is a metabolically expensive tissue both to grow and to maintain. Most of the growth of the brain occurs during embryonic development, and requires energy input from the mother. Thus, selective pressures for larger brains can be satisfied only in an environment that provides sufficient energy, especially to the pregnant or lactating female (Foley and Lee 1991). The evolution of larger brains may have required increased foraging efficiency (in part achieved through larger female size and mobility), high-quality foods in substantial quantities (in part achieved through the use of tools and fire), and a change in life history pattern that probably exaggerated the ancestral primate character of slow rates of pre- and postnatal development (thus lowering

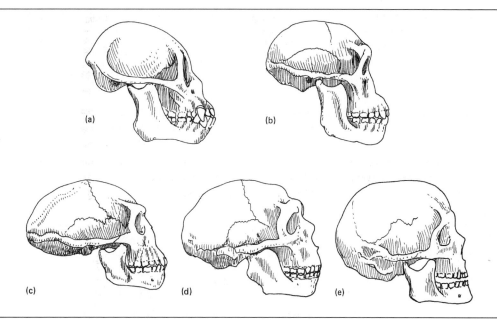

Figure 24–15 Reconstructed skulls of hominids and ancestral hominoid compared: (a) *Dryopithecus*; (b) *Australopithecus africanus*; (c) *Homo erectus*; (d) *Homo sapiens neanderthalensis*; (e) *Homo sapiens sapiens*.

daily energy demands and also a females' lifetime reproductive output).

Homo erectus is probably the sister group of our own species, *Homo sapiens*, which came into existence around 200,000 to 300,000 years ago. No derived characters have yet been defined that clearly distinguish the latest fossils assigned to *Homo erectus* from fossils believed to be early (archaic) *Homo sapiens*. The characters used to separate the two taxa are all measurements (continuous variables), not the presence or absence of a distinctive structure. On average, *Homo sapiens* have a slightly larger brain, a thicker and more robust skull, larger teeth, and less prognathism (projection of the jaw beyond the plane of the upper face) than *Homo erectus*. All of these characters show overlap between the two groups; the clearest distinction is the reduction in prognathism.

Did *Homo sapiens* evolve just once, or several times? That is, is it a monophyletic species? The "multiregional model" of modern human origins achieved prominence with the discovery that *Homo erectus* from Southeast Asia dated from the same time as the oldest *Homo erectus* from Africa (Swisher et al. 1994). This finding opened the question of where *H. erectus* first evolved—in Africa, or in Asia?

The lack of fossils of australopithecines or *H. habilis* anywhere but eastern and southern Africa suggests that an African origin of *H. erectus* is most likely. But what is the significance of regional variation in *H. erectus*? Some paleoanthropologists have proposed that *H. erectus* already showed the characteristics that would appear later in *H. sapiens* populations in the same regions. Thus, they suggested, *H. sapiens* may have evolved independently in different areas, each from an already distinctive local population of *H. erectus* (Walpoff in Mellars and Stringer 1989). If this is the case, *H. sapiens* would be a polyphyletic species. To make it a monophyletic lineage, it would have to be expanded to include the hominids now recognized as *H. erectus* or contracted to identify just one regional group. Perhaps it is fortunate that analysis of both fossil and modern material has not supported the multiregional hypothesis (Lahr 1994). The results of this analysis indicate that there was most probably a single African origin for all *Homo sapiens*.

Whatever is eventually understood about the transition to modern humans, it will undoubtedly be accompanied by shifts in the assignment of names to those fossils. This has already occurred with the most famous of fossil humans, the Neandertals. The first recognized discovery of fossils of *Homo sapiens* were from the Neander Valley in western Germany, and used to be called Neandertal Man (for a time considered a separate species, *Homo neanderthalensis*). The Neandertal features first appear in fossils about 100,000 years ago. Neandertals were 1.5 to 1.7 meters tall and had receding foreheads, prominent brow ridges, and weak chins, but their brains were as large or larger than those of modern-day *Homo sapiens*. They were stone tool makers, producing tools known as the Mousterian tool industry, with a well-organized society and advancing culture. The Neandertal people present a curious puzzle, because since their initial discovery in 1856 more modern-appearing forms of *Homo sapiens* have been unearthed dating to about 200,000 years before the appearance of the Neandertals.

It is now widely (but by no means universally) accepted that *H. sapiens* arose from *H. erectus* in Africa at least 300,000 years ago (Stringer and Andrews 1988, Lahr 1994). The early forms are known as **archaic** *H. sapiens* and share many features with *H. erectus*. About 200,000 years later a distinct population, probably originally in Europe, had evolved numerous derived characters that may be interpreted in part as progressively specializing them for life in a cold climate. (Similar traits are seen in some *H. sapiens sapiens* populations in cold regions.) Perhaps because of strong selection and restricted gene flow, the humans known as Neandertals became much more distinctive and derived in their peculiar characters than any extant population. They are thus called *H. sapiens neanderthalensis* by most anthropologists.

On the average, Neandertals appear to have been much stronger than extant humans. They had short limbs with attachments for massive muscles. Another striking difference between Neandertals and extant humans is the difference in the size of facial features: Neandertals had thick brow ridges, large noses, and broad midface regions. Neandertals had well-developed incisor and canine teeth, and the facial features of Neandertals may have been associated with structural modifications that supported these large teeth. The front teeth of Neandertals typically show very heavy wear, sometimes down to the roots. Were Neandertals processing tough food between their front teeth, or perhaps chewing hides to soften them as do some modern aboriginal peoples? Fossilized feces attributable to both archaic *H. sapiens* and *H. sapiens nean-*

derthalensis ranging from 300,000 to 50,000 years ago appear to indicate an exclusively meat diet (Kurtén 1986). Of course, the sample size is small and we have no way of knowing if such habits were merely seasonal.

Neandertals hunted wild horse, mammoth, bison, giant deer, and woolly rhinoceros from close range: Their Mousterian-style hunting tools were of the punching, stabbing, and hacking type. Although almost all skeletal remains of Neandertals show evidence of serious injury during life, one in five persons was over 50 years old at the time of death. It was not until after the Middle Ages that historic human populations again achieved this longevity. Whether or not the Neandertals had the capacity for complex speech remains controversial (Gibbons 1992), but they were the first humans known to bury their dead, often it appears after considerable ritual. Of special importance are burials at Shanidar cave in Iraq that include a variety of plants recognized in modern times for their medicinal properties.

The emerging picture of *H. sapiens neanderthalensis* is quite different from that proposed for much of the time since their discovery in 1856; they seem now to have been adept hunters who probably lived in a complex society. The Neandertals vanished between 40,000 and 34,000 years ago, long after modern *Homo sapiens sapiens* appeared. Whether the observed rapid shift from Neandertal to fully modern humans was a result of competitive exclusion, inbreeding of subspecies leading to the disappearance of one phenotype, or behavioral shifts leading to transformation of phenotypic characters to a modern morph is debated (Delson 1985, Gould 1988). Anatomically, modern humans, *Homo sapiens sapiens*, known in southern Europe as Cro-Magnon people, are distinguished by suites of subtle skeletal features equivalent to those distinguishing modern races. It is with the Cro-Magnons that an unbroken trail of evidence and history leads to the present.

Origins of Human Technology and Culture

Humans are tool users and tool makers *par excellence*. Some other animals also use tools to a limited extent, usually in rather stereotyped and instinctive ways. Egyptian vultures open ostrich eggs by picking up stones in their beaks and dropping them on the shells, a Galapagos finch holds twigs and cactus spines in its beak to probe for insects in holes or under bark, both baboons and chimpanzees use sticks and stones as weapons very much in the way that ancestral humans must have done, and chimpanzees use these same types of materials as tools in obtaining food. The use of roots, wooden clubs, and stones as hammers and anvils to crack five different species of nuts may be a culturally transmitted behavior in some West African chimpanzee populations (Boesch et al. 1994).

One of the remarkable conclusions from paleoanthropological findings is that the use of tools, chipped stones, and possibly modified bone and antler precedes the origin of the big-brained *Homo sapiens* by at least 1.5 million years. Modified stone tools are known from deposits that are 2.7 million years old. Later in the fossil record such tools are found in association with remains of *Homo habilis*, and the anatomy of some australopithecines is appropriate for a tool maker (Susman 1994). The use of tools by ancestral hominids may have been a major factor in the evolution of the modern *Homo* type of cerebral cortex, although tool use need not necessarily be associated with brain enlargement. Once the manufacture and use of tools began to increase fitness, selective pressures might have favored neural mechanisms promoting improved crafting and use of tools (Calvin 1994). In fact, the elaborate brain of *Homo sapiens* may be the consequence of culture as much as its cause (but see Susman 1988).

By at least 750,000 years ago *Homo erectus* had perfected stone tools and had also learned to control and use, but perhaps not to make, fire. Actual hearths in caves are not widely recognized before 500,000 years ago. With fire humans could cook their food, increasing its digestibility and nutritional value and preserving meat for longer periods than it would remain usable in a raw state; they could keep themselves warm in cold weather; they could ward off predators; and they could light up the dark to see, work, and socialize.

Neandertal people practiced ritual burial in Europe and the Near East at least 60,000 years ago, suggesting that religious beliefs had developed by that time. Not long afterward the cave bear became the focus of a cult in Europe. By 40,000 years ago, Cro-Magnon people began constructing their own dwellings and living in communities—things Neandertal peoples also probably did. The domestication of animals and plants, the development of agriculture, and the dawn of civilization were soon to follow.

The Human Race and the Future of Vertebrates

Some 45,000 species of extant vertebrates have been described by scientists, and new species are still being discovered, especially among oceanic and tropical fishes. Even in such well-studied groups as birds and mammals an occasional new species is still found. Extant species, of course, represent only a small fraction of the diversity and number of vertebrate species that have existed on Earth. Since the evolution of the first ostracoderms, hundreds of thousands of vertebrate species have evolved and then become extinct. Just as death is the natural end of the life of an individual, so extinction of a species is a natural result of the evolutionary process. What is new is the extent to which one species, *Homo sapiens*, has been responsible for recent extinctions and for endangering many forms of life. Past extinctions probably resulted mainly from climatic and geological changes that altered living conditions to the disadvantage of some species while favoring others. The emergence of human beings as a single dominant species on earth adds a new dimension to the problems of survival for other species (Kaufman and Mallory 1986).

The impact of humans on other vertebrates may have begun as early as the Pleistocene and has increased in historical and modern times as the worldwide population of humans has mushroomed (Figure 24–16). Several million years of hominoid evolution in Africa led to a population of *Homo* estimated at 125,000 during the middle of the Pleistocene. As the human population spread into Europe and Asia, it slowly increased to 1 million individuals during the next 700,000 years. Humans possibly underwent a drastic reduction in numbers during the last Pleistocene ice age 60,000 to 80,000 years ago, but then began to expand again (Gibbons 1993). By the end of the last glacial advance, about 10,000 years ago, hunting and food-gathering techniques had allowed humans to spread over most of the earth and to reach a level of more than 5 million persons. By 2000 years ago, developing civilizations in both the Old World and the Americas had further increased the world population to more than 100 million, although in the New World human populations were still counted in the few millions. Not until after 1800 did the human population reach 1 billion individuals; by 1950 it had more than doubled again to nearly 2.5

billion; by 1987 it was 5 billion; in 1994, 5.6 billion. By 2000 it will be more than 6 billion, unless catastrophe of unparalleled dimensions intervenes.

Humans and the Pleistocene Extinctions of Mammals

The number of genera of Cenozoic mammals reached peaks in the Miocene and early Pleistocene. Only 60 percent of the known fossilized Pleistocene genera are living now. The dramatic and recent decline in certain mammals is emphasized by calculating extinction rates of different mammalian orders (Figure 24–17). Extinction seems to have occurred mostly among terrestrial mammals. For example, the fissipeds, which are the closest terrestrial relatives of the marine pinnipeds, show a large increase in the rate of extinction, whereas the pinnipeds do not. For all mammals extinction increased from 25 genera per million years in the Miocene to 40 in the Pliocene and to over 200 in the Pleistocene.

What caused the extinctions? Some feel that delayed effects of the Miocene climatic change and the coming of the ice ages were the root causes; others think that the extinctions were largely caused by another phenomenon of the Pleistocene—the evolution of hunting humans with their newly manufactured tools and social skills. The latter view has been pursued intensively by P. S. Martin at the University of Arizona (Martin and Wright 1967, Martin 1973, Mosimann and Martin 1975, Diamond 1983, Martin and Klein 1984, Stuart 1991). What types of information lead to these conclusions? The most important points in favor of human causation are that the highest rates of extinction (1) were largely a postglacial event, and (2) occurred more frequently among large mammals than among small mammals. The climatic changes in postglacial time have left a more subtle record than the human spear points found in association with mammal remains and are consequently more difficult to evaluate. It is perhaps best to attribute the faunal changes to both human and climate factors.

The Human Factor Clearly, the high rates of mammalian extinction occurred primarily during the Wisconsin and, especially, in postglacial time (Figure 24–18). Examples of extinction are mastodons, mammoths, and ground sloths. Various camels and antelopes also became extinct, but their remains are rarely found in association with human artifacts. Martin contends that large mammals were preyed upon by invading nomadic humans, and archaeological data support this view. Presumably, large

YEARS AGO	CULTURAL STAGE	ASSUMED DENSITY PER SQUARE KILOMETER		TOTAL POPULATION (MILLIONS)
1,000,000	Lower Paleolithic		.00425	
35,000–40,000	Lower Mesolithic		.02	
10,000	Mesolithic		.04	
2,000–9,000	Village farming and Urban, Religious, and Government centers		1.0	133
160	Farming and Industrial		6.2	906
A.D. 2000	Technological		46.0	6,270

Figure 24–16 Human population growth and dispersion from lower Paleolithic time through different cultural stages projected to 2000. Each grid unit represents an assumed population density of one human per square kilometer. (Based on E. S. Deevey, 1960, *Scientific American* 203(3):194–204.)

mammals would be more susceptible to extinction by humans than would small mammals. The numbers of small- versus large-bodied genera do show differential survival (Figure 24–19) at the generic level.

D. H. Janzen (1983) proposed that humans had help in contributing to the megafauna demise from large contemporary carnivores. He agreed with Martin that humans migrated into an area that had abundant large herbivorous mammals and began extensive hunting. The presence of humans and their kills would have provided more carrion to the carnivores, whose population densities, Janzen pointed out, would have increased in response to the food supply. The combined predation pressure of increasing human and carnivore populations could have led to further decimation of the large herbivores. When the prey populations became critically low the nomadic hunters would have moved on and the populations of carnivores would subsequently decline because their resource base was gone.

Norman Owen–Smith (1987) suggested that the elimination of large herbivorous mammals (megaherbivores) by humans could have had a cascading effect on smaller species because grazing by the megaherbivores was a major factor in creating and maintaining spatial diversity of habitats. Once the megaherbivores were gone, habitats would have become more uniform and this change would have been detrimental for the small herbivores that were dependent on nutrient-rich and spatially diverse habitats.

A relation can be shown between the arrival of prehistoric humans in different regions and the extinction of large herbivorous mammals. It would appear that gradually a hunting technology developed in Africa during the middle Wisconsin glaciation and then spread northward and burst into the Western Hemisphere as the late Wisconsin glacier retreated. In the New World humans were present at least 25,000 years ago (Patrusky 1980, Adovasio and Carlisle 1986) and possibly more than 40,000 years ago. Evidence of extensive hunting of mammals began in the colder north, perhaps with a new wave of humans immigrating from Asia some 13,000 years ago (Mosimann and Martin 1975). No extinction of similar proportion has occurred in Africa, where human hunting apparently evolved side-by-side with the evolution of large mammals. Extinctions in northern Eurasia were moderate, staggered in time, and occurred more than 20,000 years after the appearance of humans with advanced stone tools

(Stuart 1991). Extinction of the marsupial megafauna of Australia also occurred (Gillespie et al. 1978, Martin and Klein 1984), but not until these large Australian mammals had coexisted with aboriginal humans for several thousand years. Thus, the model of overkill of large mammals by humans may not apply to megafaunal extinction in Australia.

The Climatic Factor Could human hunters have been the primary cause of extinction of large mammals in the Pleistocene? The evidence shows that prehistoric humans did hunt mammals, and probably often with devastating results. Fire was used to stampede large herds of ungulates over cliffs to their deaths— a potentially wasteful technique when it met with substantial success, for only a small portion of the carcasses could be utilized. Nevertheless, other scientists do not attribute the demise of the large mammals solely to human influence, but suggest that changing climates and their influence on habitat availability were as important as humans in produc-

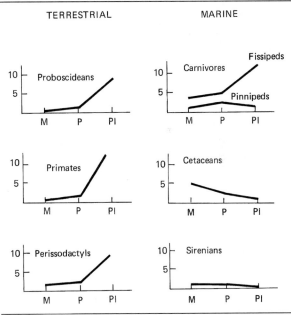

Figure 24–17 Generic extinction rates for various terrestrial and marine mammalian groups. Numbers are the estimated number of genera that became extinct per million years. How much of the change in mammalian fauna over this period is ascribable to the climatic changes rampant during these epochs (Chapter 20) and how much to the growth and dispersal of the predatory human population remains debated. M, Miocene; P, Pliocene; Pl, Pleistocene.

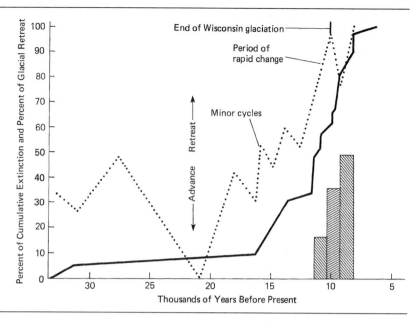

Figure 24–18 Coincidence between the extinction of North American mammals, glacial retreat, and the population of humans. Solid line represents the cumulative percentage extinction of 40 mammalian species; dashed line is the percentage of withdrawal of the late Wisconsin glaciation; histograms are arbitrary units of relative abundance of evidence for the occurrence of humans in North America. [From J. J. Hester, 1967, pages, 169–192 in P. S. Martin and H. E. Wright, Jr. (editors), *Pleistocene Extinctions: The Search for a Cause,* Yale University Press, New Haven, CT.]

ing extinctions. In their view, large mammals are more susceptible to environmental stress. Severe restrictions of forage and available water and elimination of migratory corridors to more suitable habitat would affect large mammals with their correspondingly large requirements more than smaller mammals. Thus, large North American mammals may well have been on the decline when intensive human hunting began.

R. D. Guthrie (in Martin and Klein 1984) has pointed out that most climatic theories of Pleistocene extinction are too ambiguous and diffuse to account for the final extinction event, however effective they are at depicting environmental stress. In the general trend of replacement of closed-canopy forests by more open, grass-dominated steppes and savannas that began in the early Tertiary, Guthrie sees increasing seasonality as being of primary significance. The climatic trend throughout the Cenozoic was toward a shorter growing season. At first this trend increased the mosaic patchwork of forest, scrub, and grassland over much of the world. This increasingly complex pattern of vegetation supported a large, diverse fauna.

Indeed, the spread of grasses promoted faunal diversification. Responding to the spread of grasslands, the generally large-bodied ungulates reached maximum diversity in the late Miocene and Pliocene, contributing significantly to the total number of mammalian genera known from those epochs.

But as the trend continued toward seasonal extremes of temperature and aridity the diversity of vegetation declined. The flora over broad expanses of continents changed from a mosaic of plant associations to a latitudinal and altitudinal zonation of broad bands of plant communities with low diversity. The lower overall diversity of floras and greater regional homogeneity reached its maximum after the Wisconsin glacial advance. The reduction of plant diversity may have eliminated many herbivores, especially larger forms, and restricted all but a few species to limited ranges.

The postglacial period beginning 15,000 to 10,000 years ago appears to have been more stressful to large mammal faunas than the previous 2.5 million years of the Pleistocene (Stuart 1991). Shorter growing seasons mean less plant productivity to support herbivores, even in the restricted ranges where they found suitable floras. Guthrie argues that the rapid advent of short growing seasons and heightened seasonality critically decreased the resources available to many species of large herbivorous mammals with long periods of growth to maturity and long gestation periods. These changes constituted environmental challenges to which they could not rapidly adapt physiologically or genetically. The fossil record attests to the resulting decrease in local faunal diversity, distributional ranges, and frequently to the resulting extinctions without the influence of human hunting.

Many large mammals persisted in some regions despite prehistoric human hunting pressure. In the New World camels and horses became extinct, but they survived in Europe. Others, like bison and elk, survived in the New World in spite of humans. The Egyptian civilization knew of mammoths, and other evidence indicates that the pharaohs and mammoths were contemporaries, although geographically widely separated (Rosen 1994). There is currently no final solution to the puzzle of Pleistocene extinctions; a combination of unique new climatic stress and human predation seems most plausible.

Human Settlement of Oceanic Islands

Recent analyses of fossil deposits from islands in the Pacific, Indian, Caribbean, and Mediterranean oceans indicate that the arrival of humans on these islands was accompanied by large-scale extinctions of native animals (James et al. 1987). Excavation of fossils preserved in a tube-like cave in a lava flow of Haleakala Volcano on the Hawaiian island of Maui has provided a record of faunal changes over the past 7750 years. This record shows that before humans reached the islands the bird fauna included three times as many native species as are present now. At least two-thirds of the native species of birds became extinct following settlement of the island by Polynesians between 1000 and 400 years ago. Similar results from other islands in the Pacific suggest that the total number of species of birds living in the recent past may have been much larger than the approximately 9000 species that now exist.

The Age of Exploration and Discovery

Since 1600, more than 200 extinctions have been documented among vertebrates, mostly birds and mammals. Many more other vertebrates have probably disappeared without a recorded history. All can be attributed directly or indirectly to the activities of humans. The history of extinctions among the endemic birds of the Indian Ocean Mascarene Islands off the African coast is a particularly dismal but revealing example. The island of Mauritius was apparently uninhabited by humans until discovered in 1505 by the Portuguese. The endemic dodo, a large, flightless bird related to pigeons, had become extinct by 1681, the victim of slaughter by sailors for

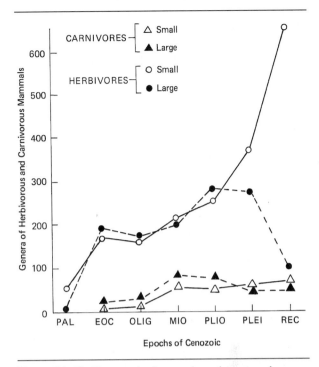

Figure 24–19 Changes in the number of genera of carnivorous and herbivorous mammals known from the epochs of the Cenozoic. Mammals of 20 kilograms adult body weight or less are considered small. The large herbivores that became extinct were primarily taxa that evolved after the end-Miocene extinction and radiated in the Plio-Pleistocene (Chapter 20; see C. Janis, 1993, *Annual Reviews of Ecology and Systematics* 24:467–500.

food and of predation on its eggs and young by introduced pigs and monkeys. A similar bird, the solitaire, had vanished on Reunion by 1750, and another, on Rodriguez, before 1800. In all, as many as 24 endemic birds have become extinct in the Mascarene Islands since 1600, more than 50 percent of the original avifauna. Today, many of the remaining endemic species of parrots, pigeons, and falcons number only a handful of individuals, so severe has been the human destruction of the Mascarene habitat.

The pattern of destruction of the vertebrate faunas of the Mascarene islands is characteristic of the effects of European exploration and discovery in the fifteenth through nineteenth centuries. Island endemics were brought to extinction either directly, through slaughter for food, or indirectly as the result of introductions of feral populations of domestic animals that altered island ecosystems. Subsequently, not only island but continental faunas and floras began to be affected as exploding human populations changed the face of the land.

Commercial Exploitation of Vertebrates

Explorers were soon followed by fur trappers, commercial hunters, fishermen, whalers, plume hunters, and a host of others who made their living by exploiting wild animal populations for food, leather, fur, oil, ivory, and other commercial products. Hunting was often done by companies with small armies of organized hunters. These enterprises have caused great destruction among vertebrate populations, but have actually brought few species to extinction because it usually becomes uneconomical to hunt out the last few hard-to-reach individuals of a species.

When Vitus Bering voyaged among the islands of the Bering Sea and North Pacific Ocean he found millions of fur seals breeding on the Pribilof Islands, and great numbers of sea otters spread over an even wider range (Figure 24–20). The furs of these marine mammals are among the most luxuriant known and were highly valued in China and Europe. Soon after the reports of the Bering expedition circulated the sealers and otter pelt purveyors began their work.

Between 1745 and 1867 Russian explorers and later the Russian–American Company shipped 260,790 sea otter pelts from Alaska. In 1757 alone the Company killed more than 7700 sea otters. From 1823 to 1830 fewer than 100 otters per year were taken, and the Russian–American Company shifted its activity to other resources.

For a time when so few pelts were received in trade people assumed the species had become extinct, but fortunately a few pairs managed to survive in isolated parts of the range. After complete protection in North America in 1911 and careful transplantation of individuals since 1965, the otter population has slowly increased. The wild population may now total 150,000. Sea otters are important predators in subtidal kelp beds and have an influence on the distribution and abundance of other species of animals and even plants in the community. Indeed, around Amchitka Island the sea otter shows signs of overpopulation and in southern parts of the range sea otters are in conflict with fishermen over a favorite food—abalone.

Between 1908 and 1910 Japanese seal hunters slaughtered four million fur seals in the Pribilofs. For-tunately, through diplomatic negotiations undertaken by the United States, the Fur Seal Treaty of 1911 was signed by Japan, Russia, Canada, and the United States. The first international agreement to protect a marine resource, it provides protection, management, and harvest by the United States with quotas of the harvested pelts going to the signatories. As a consequence, the northern fur seal population rebounded in numbers and has been harvested annually on a planned and scientifically managed basis for more than 50 years. Unfortunately, southern fur seal species were not so lucky, and they exist today only as remnant populations on a few islands scattered from southern California to Chile. The impact of fur seal populations on their ecosystem is also appreciable. Today's large populations are a serious economic and political problem for the Bering Sea fishermen because of their alleged efficiency in competing for the same seafood resources as humans.

A similar history can be related about the northern elephant seals, which come ashore to calve and breed on islands adjacent to southern California and western Mexico (Figure 24–21). They were slaughtered in the tens of thousands for oil and dog food, until there were only a few, perhaps as few as 20 individuals, left in the early 1930s. After protection by the governments of the United States and Mexico, the populations have increased, slowly at first and then more rapidly, until now there are more than 100,000 of these great, lumbering beasts. They are still increasing, having recently begun breeding on the California mainland south of San Francisco. However, biochemical studies reveal virtually no genetic variation in the population, apparently as a result of its having been reduced to so few individuals. The evolutionary consequences of this lack of genetic variation remain to be seen.

These three marine mammals are good examples of how well vertebrate populations can recover from greatly reduced numbers after heavy killing stops. A different story has to be told for the great whales. All of the large baleen whales, except the gray whale, which has responded to protection, are critically endangered as a consequence of the relentless use by the whaling nations of highly efficient modern techniques for capture and processing. The blue whale, perhaps the largest animal ever to have existed, escaped persecution during the early days of whaling. When whalers traveled in sailing ships and harpooned from open boats, blue whales simply were too fast to be caught and too heavy to be handled if

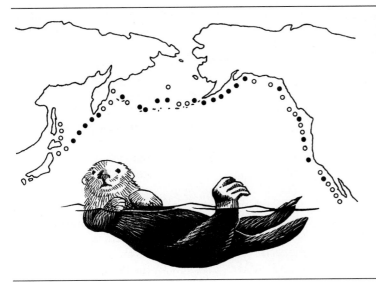

Figure 24–20 Original distribution of the sea otter (open circles) compared with its present relict occurrence (closed circles). (From V. Ziswiler, 1967, *Extinct and Vanishing Animals*, Springer, New York, NY.)

by luck one was harpooned. The modern catcher ship, armed with cannon-launched harpoons that explode inside the animal, proved to be the nemesis of the blue whale. International treaties have so far not been very effective in controlling whaling; some of the signatories of the treaties simply announce that despite an international ban on whaling to which they agreed, they will continue to hunt whales for what they call "research purposes." Molecular sleuthing has indicated that Japanese "research" whalers kill a variety of endangered whales that are then sold for food (Baker and Palumbi 1994). It is unclear whether the blue whale, fin whale, bowhead whale, and the right whales can ever recover from their pathetically reduced numbers.

Many avian species have been overharvested for commercial purposes. The passenger pigeon was probably brought to extinction as a result of commercial hunting, although the destruction of the nut-producing hardwood forests in the eastern United States contributed to its demise. Roosting and nesting in dense flocks of many millions of birds made the passenger pigeon especially vulnerable to shot and nets. In 1813, John James Audubon made a 55-mile trip along the Ohio from his home to Louisville, Kentucky. Wild passenger pigeons were migrating through the region in such numbers that "the light of noonday was obscured as by an eclipse." Audubon estimated the flock to be a mile wide with an average density of two birds per square yard. During a 3-hour period he calculated that more than 1,100,000,000 passenger pigeons must have flown by, and the flight lasted for three days.

Even if we reduce Audubon's estimate by half, the number of pigeons must have been truly astounding. Passenger pigeons were social nesters. The last great nesting occurred in 1878 near Petoskey, Michigan, and covered more than 40,000 hectares of forest; many millions of birds were present. Unfortunately, passenger pigeons were so tasty that the demand for them in urban markets seemed insatiable. The adult birds were shot, netted, and removed from nests by the millions. The New York City market alone would receive 100 barrels a day, week after week, without a drop in price. Chicago, St. Louis, Boston, and all the eastern cities, great and small, joined in the demand for pigeon flesh. The pigeons themselves fed on the seeds of oaks, hickory, and other nut-producing trees and also depended on the trees for nesting sites. As the eastern forests were demolished during the eighteenth and nineteenth centuries, the pigeons were forced westward and northward into marginal habitats, which probably were not favorable for the mass reproductive efforts that characterized the species. Today, much of the eastern United States is again covered by hardwood forest, but it is nearly all second-growth stands, different in species composition and habit from the late pre-Columbian forest. Some people still speculate about the disappearance of the passenger pigeon—that they all drowned in a great storm over Lake Michigan—but the true story is revealed by the

Figure 24–21 A breeding colony of elephant seals on Isla Ano Nuevo off the coast of California. A male (foreground) displays in front of a group of females and pups. (Photograph by F. Harvey Pough.)

statistics of the marketplace and human land use. The last bird died in the Cincinnati Zoo in 1914.

Extinction of a species or population results when the mortality of individuals is continually greater than the production of new individuals. Generally, when excessive mortality caused by humans is eliminated, even drastically reduced populations rebound to the limits imposed by factors other than human predation. However, this resilient capacity of vertebrate populations has limits that vary for different species and seem to be determined by a *critical minimum population size,* below which the species cannot recover because reproduction has been affected independently of mortality. In such cases, even if mortality can be greatly reduced, reproduction is still insufficient to result in a net increase in numbers and the species becomes extinct (Frankel and Soulé 1981).

Highly social and gregarious species seem to have large critical minimum population sizes, whereas widely dispersed, territorial species and island endemics may have very small critical minimum numbers. The passenger pigeon is an example of the first group. After the last great nesting in 1878 in Michigan, many millions of pigeons still existed despite the slaughter that occurred there, but even though significant human depredations on the passenger pigeons stopped after that time, the remaining flocks were never again able to reproduce in suf-

ficient numbers to replace the natural losses in the adult population, and the species had become extinct in the wild by the turn of the century. The small amount of recorded information on the bird's breeding habits indicates that mating and nesting may have been initiated by social stimulation involving communal displays of large numbers of birds closely packed together. Other species seem to have such a low critical population size that the concept has no application to them. Certainly, this would seem to be the case of the Laysan duck, which appears to have been reduced in 1930 to a single adult female and her clutch of fertile eggs. By the late 1950s there were again an estimated 580 to 740 ducks on Laysan Island, although the wild population has since dropped back to a smaller number of birds.

Several factors probably play a role in determining the critical minimum population size for a species. The population density may become too low for males to have much chance to find females, a possible problem now for some of the great whales, especially because their sound communication channels are becoming more and more jammed by shipping noise. Population density could become too low for adequate stimulation of courtship and mating, as indicated for the passenger pigeon. The population might become too small to counterbalance mortality from large numbers of surviving predators, parasites, or competitors. (For example, the rare Kirtland's warbler has had a problem with nest parasitism by the brown-headed cowbird in recent years.)

The total population size might become too small to survive a natural disaster such as a flood, fire, or cyclone, particularly if the population is localized. Remnant populations of island endemics are especially vulnerable to this factor. The last few surviving Laysan honeycreepers were blown off the island by a storm in 1923, and the same fate could be in store for the relict populations of the kestrel, pink pigeon, and parakeet on the island of Mauritius, which has a history of periodically being hit by cyclones.

Reduction of fecundity or increased susceptibility to diseases and parasites owing to genetic deterioration from inbreeding are other factors to consider when a species population becomes reduced to a very few individuals; the Laysan duck may be suffering from such problems at the present time. Reduced fecundity is sometimes characteristic of highly inbred captive populations of vertebrates (see Schonewald–Cox et al. 1983). Exactly which problems may be affecting an endangered species can be determined only by intensive study of the species and its environment, a costly and time-consuming process—and one in which we are already far behind.

The Impact of Agriculture and Land Development

Human use and abuse of land for agriculture, timbering, mining, and suburban living are the most detrimental of our activities to wild animal populations, primarily because of the inexorable destruction of natural habitat. Evolutionarily this would be comparable to the spread of grasslands in the Miocene or any of the great climatic changes that have precipitated biotic revolutions throughout the history of vertebrate life except for two critical factors: No natural habitat is unaffected or favored by twentieth-century practices and the rate at which habitats of every kind are being profoundly altered and destroyed is unprecedented. Although as yet only a few species of vertebrates have been directly destroyed and brought to total extinction as a result of human land uses, many have suffered major and perhaps irreversible reductions in distribution and density. No doubt many others will join the ranks of threatened or endangered wildlife in the next few decades as a result of the diminution and deterioration of natural habitats. The actual magnitude of species extinctions due to twentieth-century habitat alteration will not become clear until well into the twenty-first century.

Perhaps the earliest and most persistent conflict between humans and wild animals has been with other predators, particularly the large carnivores. Early humans were efficient and destructive predators in their own right, and with some justification they regarded the carnivores as competitors for the animals that humans hunted for food. Humans undoubtedly had to protect their kills from scavenging predators, and they surely robbed other predators of kills. Direct interactions between individual humans and carnivores must have been common in the Pleistocene. From these remote times human attitudes toward other large and powerful predators have been molded by fear and by the idea that other predators compete directly for available food. Except for isolated instances, neither belief is justified by objective analysis, but they still influence many contemporary minds.

With the domestication of livestock and pastoral nomadism as a way of life, humans began a systematic destruction of large carnivores that has continued and accelerated to the present. In North America, the timber wolf, mountain lion, black bear, and grizzly bear have been forced to retreat to more and more remote areas as ranchers and sheep herders arrived in the Midwest and West. The same can be said of the lion in Africa and Asia, of the tiger and wildcat in Asia, and of the jaguar in the Americas. Tillage and development of intensive farming beginning 10,000 years ago provided the final insults for large predators, which simply are not compatible with intensive, disruptive human uses of the land.

Unconscionably, the carnivores continue to be pursued by humans, and governments are still persuaded to spend millions of dollars a year on misguided and unjustified predator control programs. Predator control usually results in a need for rodent control, but rodents are much more difficult and costly to control than predators. It is ecologically and economically wise to accept the loss of a few pheasants and lambs killed by foxes and coyotes in exchange for the greater benefits of their impact on rodent and rabbit populations.

Other vertebrates have frequently been considered to be competitors of humans. The Boer farmers who settled in South Africa regarded the existing wild ungulate herds as competitors for their cattle, sheep, and goats, and systematically destroyed ungulates, as they did the Cape lion. The blue buck had been exterminated by 1800; the last of the quagga, a once-common zebra, disappeared in 1878; Burchell's zebra had been eliminated from all areas other than game parks by 1920; and three other ungulates, the bontebok, the white-tailed gnu, and the Cape mountain zebra, exist today only as remnant populations in wildlife reservations.

Human land use has transformed one-third to one-half of Earth's ocean and ice-free surface (Vitousek 1994). The loss of natural environments due to their conversion to intensive agriculture and other uses of land and water is by far the most important factor in the overall reduction of wild vertebrate populations. This will continue to be the overriding problem for the survival of many species in the closing years of the twentieth century and the beginning of the twenty-first. The cutting of the original eastern hardwood forests in North America for timber, firewood, and agricultural clearing had an effect on the passenger pigeon at least equal to the killing of these birds for food.

The prairie grasslands have been among the most disrupted natural communities in North America. This is especially true of the tall grass prairie, which was ploughed under and no longer exists. Even the short grass prairie is no longer truly natural. It has been invaded by many exotic species of grasses and weeds introduced through human agency and has been drastically altered by overgrazing and hoofed stock soil compaction of domestic livestock. Many of these western ranges are receiving additional insults from extensive mining processes, especially strip mining for coal.

Continued growth of the human population means that more and more land must be devoted to agriculture and other dramatically altering uses required to sustain unprecedented numbers of humans. Thus, more natural habitat is lost, and fewer places are left for wild animals, except for the very limited number of species that can adjust to the altered conditions created by humans. Most of the agriculturally productive areas of the world have been developed, but good agricultural lands are being reduced by suburban developments, businesses, roads, airports, and artificial reservoir and lakes.

Farmers worldwide are increasingly turning their attention to marginal lands. Deserts, for example, can be highly productive under irrigation.

Much of the water in such areas is fossil water trapped in isolated aquifers from prehistoric times when local climates were wetter. This water is no longer being replenished by regional rainfall. Thus where fossil water has been used it is becoming an increasingly scarce commodity. In some places irrigation followed by rapid evaporation into the dry atmosphere has increased the salinity of the soil, greatly reducing the harvest. One of the most critical areas of agricultural devastation in the late twentieth century is the tropical rain forests of South America, Africa, and Asia. When the forest is cut, the soils generally do not sustain intensive agriculture for more than a few years. Once abandoned these soils are frequently so changed in texture and composition that the original forest trees cannot come back; instead, there is a new plant community of heat- and drought-tolerant grasses and shrubs, which are poor pastures for livestock. Gone forever is the rich diversity of tree species and associated flora and the animals of these tropical forests (Figure 24–22). Genetic resources of potential value to agriculture and medicine are being lost along with complex ecosystems as tropical forest disappear (Myers 1984).

In addition to a loss of biodiversity, failure to keep large portions of some natural ecosystems intact may have serious consequences on climate, composition of the atmosphere, and on the distribution and quality of water, perhaps the most limiting resource for life. This has been dramatically demonstrated in recent years in the sub-Saharan regions of Africa (Soulé 1986). There is growing evidence, too, that cutting the Amazon tropical forest is having an impact on rainfall patterns in that region (Myers 1984). Deforestation can have consequences well beyond the actual area of forest that is cut (Figure 24–23). Clearcutting can create desert-like regions, and the runoff of heavy tropical rains from these barren areas can cause flash floods that destroy habitats many miles downstream. On the positive side, restoration projects can reestablish some tropical habitats when they are undertaken with the necessary biological and political skill (Janzen 1988).

Global consequences of human activity are also becoming apparent. The exchange of materials between terrestrial ecosystems and the atmosphere is increasingly dominated by human activities, and rising concentrations of many compounds have resulted from human biological or industrial processes (Mooney et al. 1987). Volumes have been written about the "greenhouse gases" (for example,

Woodward 1992), but many people remain unaware of our alterations of the global nitrogen cycle. Humanity fixes more nitrogen annually than is fixed by all natural pathways combined (Vitousek 1994), leading to eutrophication and other community altering phenomena. Acid precipitation, which results when oxides of sulfur and nitrogen produced by combustion of fossil fuels combine with water vapor in the atmosphere, has acidified soils and water systems in North America and Europe, especially the former Soviet bloc. Acid precipitation has also been discovered in China (Galloway et al. 1987, Schindler 1988).

Chlorofluorocarbons of the sort used to produce plastic foams, for dry cleaning, and in refrigeration systems drift into the stratosphere where ultraviolet radiation breaks down the chlorofluorocarbon molecules, releasing chlorine atoms as one of the products of the breakdown. In turn, the chlorine atoms react with ozone, which is the principal component of the atmosphere that blocks potentially damaging ultraviolet radiation from reaching Earth's surface. A single chlorine atom can catalyze the destruction of 100,000 molecules of ozone. Over the last decade the concentration of ozone in the stratosphere has dropped by 3 percent, and the decline is continuing. A major reduction in the concentration of ozone in the stratosphere could affect plant and animal life worldwide, and the growing hole in the ozone layer that appears annually over Antarctica (and over the Northern Hemisphere as well) is convincingly associated with the release of these fluorocarbons into the atmosphere (Kerr 1987, Bowman 1988).

Many other human practices threaten wildlife: the pet trade and the traffic of wild animals, the introduction of exotic plants and animals, the modification of waterways and water distribution by canals and dams, and chemical pollution (Kaufman and Mallory 1986). Humans have had, and continue to have, disastrous effects on other vertebrate species. The *Global 2000 Report* predicted that more species will become extinct between 1980 and the year 2000 than in previous recorded history. We humans are about to achieve this dubious milestone.

Human Concern for Vanishing Wildlife

Human impact on other forms of life has led to a new classification of species. The categories include "rare species," "unique species," "threatened species," and "endangered species." The International Union for the Conservation of Nature and Natural Resources, in cooperation with organizations such as the World Wildlife Fund and the Inter-

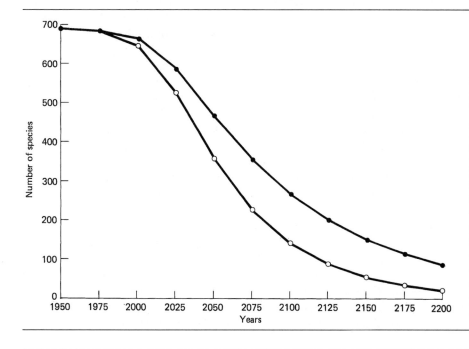

Figure 24–22 Predicted loss of mammalian species from tropical moist forests in the New World due to the reduction in extent of their habitat. The projected rate of deforestation is based on a continuation of the current rate. The upper curve represents the immediate loss of species expected. The lower curve is the final number of species expected at equilibrium if deforestation were to cease at that date. No date for reaching this equilibrium is indicated by these data. (Based on data from the Society for Conservation Biology.)

Figure 24–23 Neotropical forests: (a, b) undisturbed forest have a high diversity of plant and animal species; (c, d) after destruction the forests may be unable to regenerate their original diversity. (Photographs by John B. Heiser.)

national Council for Bird Preservation, attempts to keep current a catalog of specific details on all rare, severely threatened, or endangered species of plants and animals. The information is published in an ongoing series of *Red Data Books*, arranged taxonomically. Over 1000 species of vertebrates are now included. Increasingly in recent years governments have become concerned with such species too. Public Law 93-205, the United States Endangered Species Act of 1973, awaiting reauthorization at the time of this writing, gives authority to the Secretary of the Interior to designate endangered and threatened species (Table 24–1), to promote ways to preserve them from further diminution and, if possible, to

Table 24–1 Number of endangered species of vertebrates officially listed by the United States Department of the Interior in 1982/1987/1994.

Taxonomic Category	USA	Foreign	Total
Mammals	33/46/56	223/242/251	256/288/307
Birds	66/77/73	144/141/153	210/218/226
Reptiles	14/14/16	55/60/63	69/74/79
Amphibians	5/5/6	8/8/8	13/13/14
Fishes	32/43/63	11/11/11	43/54/74
Totals	150/185/214	441/462/486	591/647/700

Source: *Endangered Species Technical Bulletin,* VII:11, November 1982; XII:4, April 1987; and XIX:4, July/August 1994, U.S. Fish and Wildlife Service, Washington, DC.

increase their numbers. There are also important international agreements, such as the Convention on International Trade in Endangered Species of Wild Fauna and Flora. Many states and provinces also now have their own endangered species laws. Although the executive branch of the U.S. government has never been aggressive in putting the provisions of the Endangered Species Act into operation, public interest in the survival of endangered species has increased greatly.

Characteristics of Endangered Species

Are there predisposing characteristics of a species that are likely to lead to its becoming an endangered species? Birds and mammals that are restricted to islands are highly vulnerable, but there are many other variables that influence the survival potential of a species. By comparing the characteristics of threatened and endangered species with related and apparently thriving species, we can begin to identify the main factors involved (Table 24–2). Theoretically, one could check each species against this list of variables and come up with a statistical estimate of its chances for continued survival in the modern world. At present our knowledge of the vast majority of extant species is so minimal that the best we can do is to make a qualitative appraisal.

David Ehrenfeld (1970) constructed an interesting description of the hypothetical *most endangered animal:* "It turns out to be a large predator with a narrow habitat tolerance, long gestation period, and few young per litter. It is hunted for a natural product and/or for sport, but is not subject to efficient game management. It has a restricted distribution, but travels across international boundaries. It is intolerant of humans, reproduces in aggregates, and has nonadaptive behavioral idiosyncrasies." Although there really is no such animal, the polar bear comes close to approximating this model. Conversely, if one takes the opposite characteristics (those listed under "safe" in Table 24–2), a composite picture of a typical "wild" animal of the twenty-first century develops. Familiar existing approximations are the herring gull, house sparrow, gray squirrel, Virginia opossum, Norway rat, and carp.

In short, the result of human influence on environment and our wanton destruction of animal life is a greatly reduced diversity in faunas and floras. Increasingly, there is an emergence of a relatively few, superabundant, highly successful species that dominate ecosystems. Unless sufficiently large tracts of natural ecosystems are left intact to maintain some possibility of ecosystem balance, these organisms probably are going to interact with one another and with human crops in boom and bust oscillations of their population sizes.

Actions to Save Threatened Fauna and Flora

What can we do to prevent even more species from becoming rare, endangered, or extinct? Two cardinal requirements must be met if natural biotas and ecosystems are to remain intact for future genera-

Table 24–2   Characteristics of species that decrease or increase survival potential.

Endangered	Safe
Individuals of large size (cougar)	Individuals of small size (coyote)
Predator (hawk)	Grazer, insectivore, scavenger (turkey vulture)
Narrow habitat tolerance (orangutan)	Wide habitat tolerance (macaques)
Valuable fur, hide, oil (chinchilla)	Not a source of natural products and not exploited for research or pet purposes (gray squirrel)
Hunted for the market or hunted for sport where there is no effective game management (passenger pigeon)	Commonly hunted for sport in game managment areas (mourning dove)
Has a restricted distribution: island, desert watercourse, bog (Bahamas parrot)	Has broad distribution (herring gull)
Lives largely in international waters, or migrates across international boundaries (green sea turtle)	Has populations that remain largely within the territory(ies) of a specific country(ies) (snapping turtle)
Intolerant of the presence of humans (grizzly bear)	Tolerant of humans (black bear)
Species reproduction in one or two vast aggregates (West Indian flamingo)	Reproduction by solitary pairs or in many small- or medium-size aggregates (house sparrow)
Long gestation period; one or two young per litter, and/or extended maternal care (giant panda)	Short gestation period; more than two young per litter, and/or young become independent early and mature quickly (raccoon)
Has behavioral idiosyncrasies that are nonadaptive today (redheaded woodpecker—flies in front of cars)	Has behavior patterns that are particularly adaptive today (burrowing owl: highly tolerant of noise and low-flying aircraft; lives near the runways of airports)
Top-of-the-food-chain predator that is subjec to the effects of biological magnification of chemical pollutants (peregrine)	Lower trophic level predator that feeds on less contaminated prey (kestrel)

Source: D. W. Ehrenfeld, 1970, *Biological Conservation*, Holt, Rinehart, & Winston, New York, NY.

tions: (1) preservation of natural habitats and ecosystems on a large and representative scale, and, of paramount importance, (2) control of human population growth.

For the short term, the most immediately critical action is habitat preservation. Preservation must be of large blocks of natural ecosystems—10,000 to 20,000 square kilometers or more. These blocks must be large enough for the natural communities of plants and animals to remain self-perpetuating. Unfortunately, there is nothing approaching scientific understanding of how large is large enough, the immediate socioeconomic question that arises (Soulé 1986). The larger national parks and game preserves of the world show what can be done when natural areas are protected against major disruptive land use by humans.

The Kruger National Park was the first game park established in Africa; portions of it have been in preserve status since 1895; its present boundaries, encompassing more than 12,000 square kilometers of lowland bush-veld along the border with Mozambique, have been intact since 1926. Although situated squarely in one of the most critical sociopolitical trouble centers of the world and surrounded by lands that are heavily used by humans for agriculture and stock raising, the park remains relatively inviolate and is fenced around its entire perimeter. Kruger has lost only one species of plant or animal from its natural ecosystem, and it has been in existence long enough to give assurance that as long as it is properly cared for and preserved and if climate changes do not alter its habitats, it can continue to hold its living treasures indefinitely. One can see and study in Kruger more than 44 species of fish, 29 species of amphibians, nearly 100 species of squamates, more than 440 species of birds, and 114 species of mammals, including 16 species of large

mates, more than 440 species of birds, and 114 species of mammals, including 16 species of large ungulates, lions, leopards, cheetahs, hyaenas, and wild dogs.

Any conservation issue ultimately rests directly on a solution to the problems of the human population explosion. The human population has been increasing at an exponential rate for thousands of years, a pattern of growth that is unprecedented in other animal populations and that *cannot* continue indefinitely. An increase in the numbers of human beings leads to exponential rates of increase in the use of resources—food, fiber, fuel—and the production of waste products that pollute the global ecosystem: wastes from human food consumption, from use of fossil fuels, from fertilizing agricultural lands, and from synthetic chemical products and other sources. It is important to understand that these rates of use and waste frequently increase by exponents that are larger than the exponent of population increase. This is especially true as global communication (television most especially) exposes the overwhelming majority of humans to a Western life-style, raising worldwide expectations and aspirations. Worldwide per capita energy use has nearly tripled since 1850, and most of the increase has occurred in the last few decades (Vitousek 1994). Typically, the use of materials in a modern technological society increases four to five times faster than population growth, a rate far faster than in pretechnological societies.

The United Nations World Conference on Population in 1994 showed dramatically that many people throughout the world do not appreciate the disastrous consequences of human population growth and exponential rates of increase in the uses of land and resources. They assume that technology and production can keep pace with population growth. We are still wedded to the progressivist philosophy of the eighteenth and nineteenth centuries, a philosophy that followed in the wake of the scientific and industrial revolutions. It is the philosophy of the ever-expanding economy and TV commercials. "Progress is our most important product," where progress equates with material growth and expansion.

Technology, per se, need not be a negative force. Throughout this book, we have given examples of the direct application of technology to conservation and management of endangered species. The same technology that can be used to destroy the world can

very probably be used to help restore it, except in one instance: When a species is lost to extinction, it is lost forever. Many biologists and conservationists are not convinced that more technology is needed as much as a new frame of reference built on restraint—restraint on the use of resources and, above all, restraint on human procreation. A respect and reverence for all life, not just human life, must be realized before acceptance of such a world view can be expected to have an effect.

The world's human population is currently more than 5.6 billion persons, twice the number at the end of World War II. Presently, a quarter of a million humans are added each day, a population the size of New York City is added each month, and 94 million new people demand resources each year. Nine-tenths of today's humans live in what are known as the developing nations where the rates of population growth are highest. The situations in these nations is grave.

It has been estimated that no more than *two* billion humans can be indefinitely sustained on Earth. At that population, a reasonable quality of life (having about half the energy consumption per capita per year as United States citizens currently consume) could be sustainably provided for all (Pimentel et al. 1994). If these estimates are even vaguely correct then human population restraint and action toward a sustainable future must begin now. Allison Jolly (1985) has calculated that the total world population of all wild primates combined is less than that of any of the Earth's major cities. The entire extant populations of many species are no larger than that of a small town. Conditions are equally critical for many other vertebrates. Almost two decades ago Jolly (1980) expressed the problems we face even more painfully as biologists today:

> This realization has been painful. It began for me in Madagascar, where the tragedy of forest felling, erosion, and desertification is a tragedy without villains. Malagasy peasant farmers are only trying to change wild environments to feed their own families, as mankind has done everywhere since the Neolithic Revolution. The realization grew in Mauritius, where I watched the world's last five echo parakeets land on one tree and knew they will soon be no more. It has come through an equally painful intellectual change. I became a biologist through wonder at the diversity of nature. I became a field biologist because I preferred watching nature go its own way to messing it about with experiments. At last I understood that biol-

ogy, as the study of nature apart from man, is a historical exercise. From the Neolithic Revolution to its logical sequels of twentieth-century population growth, biochemical engineering of life forms, and nuclear mutual assured destruction, human mind has become the chief factor in biology. . . . the urgent need in [vertebrate] studies is conservation. It is sheer self-indulgence to write books to increase understanding if there will soon be nothing left to understand.

Summary

Evidence for the origin of *Homo sapiens* comes from the Cenozoic fossil record of primates and from comparative study of existing monkeys, apes, and human beings. The traits that characterize the Primates have evolved in relation to an arboreal, insect-grasping mode of existence; secondarily, some primates are terrestrial, humans most of all.

The first primates were probably much like the tree shrews, but the earliest primates for which there are substantial fossils (from the late Paleocene) were similar to the extant lemurs, pottos, and galagos. These prosimians quickly spread over the Old and New Worlds and differentiated into a variety of ecological forms. All were arboreal and some possessed somewhat larger brains in relation to body size than do most other mammals.

By the Oligocene two other distinct groups of higher primates had evolved: the platyrrhine monkeys of the New World tropics and the Old World catarrhine monkeys and apes. The anthropoid apes and human-like species, including *Homo sapiens*, are grouped in the Hominoidea. The fossil record for this clade is richer than for other lineages of primates. Many morphological features distinguish the hominoids from other catarrhines, and enlargement of the brain has been a major evolutionary force molding the shape of the hominoid skull, especially in the later part of human evolution.

The first known hominoids occur in the Oligocene, 35 million years ago. By the Miocene hominoids had diversified into a number of environments and had spread widely over Africa, Europe, and Asia. Fossils showing the derived characters of the great apes and humans have been found in China, Pakistan, India, Kenya, Turkey, and Hungary and date from about 17 to nine million years ago. Humans retain ancestral characters of these early forms that the apes have lost in the evolution of more derived characters.

A variety of hominid (human) fossils occur in late Pliocene and early Pleistocene deposits of Kenya and Ethiopia, some 4.4 million years old. The earliest, with small cranial volumes and often heavy features of skull, tooth, and jaw, are assigned to the genus *Australopithecus*. The first stone tools are found in East Africa dating from about 2.5 million years ago. As more early human-like fossils are unearthed, it becomes increasingly difficult to separate them from the genus *Homo*. Some fossils currently called *Homo habilis* are older than some forms of *Australopithecus*.

Homo erectus ranged across Africa and Eurasia from about 1.8 million years to 300,000 years ago. This hominid had a brain capacity approaching the lower range of *Homo sapiens*, made stone tools, and used fire. *Homo sapiens*, the only surviving species of the family Hominidae, came into existence around 200,000 to 300,000 years ago. By 40,000 to 60,000 years ago some populations of *Homo* were barely distinguishable in structure from extant humans. They had a well-organized society with a rapidly developing culture, especially obvious in their use of stone tools.

The biological success of human beings is measured by their universal distribution over the Earth and by the fact that humans derive some portion of their livelihood from every ecosystem and habitat in the biosphere. This all-pervasive status of *Homo sapiens* has had a dramatic impact on environments and on other animals, particularly on other vertebrates. Although extinction of species is a natural result of evolutionary processes through time, there is no precedent for the extent to which one species—*Homo sapiens*—has been responsible for extinctions of many other forms of life.

The first significant human impacts on other vertebrates may have occurred in prehistoric time. Extinction of many species of large vertebrates in the late Pleistocene may be the result of hunting by humans, although other predators and changes in climate may have been additional factors. Modern industrialized people have exerted drastic influences on populations of other animals. The influence of humans on vertebrate species within historical time has occurred in three overlapping phases: first, the period of exploration and discovery in the fifteenth through nineteenth centuries when many island endemics were brought to extinction; second, and following close on the heels of the explorers, were the fur trappers, commercial hunters, fishermen, and other harvesters who exploited wild vertebrate populations for commerce; third and most serious are the long-term effects of human agriculture and other land uses that destroy natural habitats. More species will have joined the ranks of endangered wildlife and will become extinct in the last two decades of the twentieth century than in all the centuries since 1600, owing to the diminution and deterioration of natural habitats. Such losses will be particularly severe in the biotically rich tropical forests of South America, Africa, and Asia.

Two cardinal requirements must be met if we are to save a significant vestige of natural biotas and ecosystems intact for future generations of humans to know: preservation of natural habitats and ecosystems on a large and representative scale, and control of human population growth. Ultimately, and soon, human beings must learn to accept a morality of restraints—restraint on the use of resources and, above all, restraint on human procreation.

References

Adovasio, J. M., and R. C. Carlisle. 1986. The first Americans, Pennsylvania pioneers. *Natural History* 95(12):20–27.

Aiello, L. C. 1994. Thumbs up for our early ancestors. *Science* 256:1540–1541.

Andrews, P. 1992. Evolution and environment in the Hominoidea. *Nature* 360:641–646.

Andrews, P., and L. Martin. 1991. Hominoid dietary evolution. *Philosophical Transactions of the Royal Society, London B* 334:199–209.

Baker, C. S., and S. R. Palumbi. 1994. Which whales are hunted? A molecular genetic approach to monitoring whaling. *Science* 265:1538–1539.

Berge, C. 1994. How did the australopithecines walk? A biomechanical study of the hip and thigh of Australopithecus afarensis. *Journal of Human Evolution* 26:259–273.

Boesch, C., P. Marchesi, and N. Marchesi. 1994. Is nut cracking in wild chimpanzees a cultural behavior? *Journal of Human Evolution* 26:325–338.

Bowman, K. P. 1988. Global trends in total ozone. *Science* 239:48–50.

Bruce, E. J., and F. J. Ayala. 1979. Phylogenetic relationships between man and the apes: electrophoretic evidence. *Evolution* 33:1040–1056.

Burenhult, G. (editor). 1993. *The First Humans (The Illustrated History of Humankind, Volume 1)*. Harper Collins, New York, NY.

Calvin, W. H. 1994. The emergence of intelligence. *Scientific American* 271(4):100–107.

Campbell, B. 1985. *Human Evolution: An Introduction to Man's Adaptations*, 3d edition. Aldine, Hawthorne, NY.

Cartmill, M. 1974. Rethinking primate origins. *Science* 1984:436–443.

Cichon, R. L., and R. S. Corruccini (editors). 1983. *New Interpretations of Ape and Human Ancestry*. Plenum, New York, NY.

Cichon, R. L., and J. G. Fleagle. 1987. *Primate Evolution and Human Origins*. Aldine, Hawthorne, NY.

Coppens, Y. 1994. East side story: the origin of humankind. *Scientific American* 270(5):88–95.

Day, M. H. 1986. *Guide to Fossil Man*, 4th edition. University of Chicago Press, Chicago, IL.

Delson, E. 1985. *Ancestors: The Hard Evidence*. Liss, New York, NY.

Delson, E. 1987. Evolution and palaeobiology of robust *Australopithecus*. *Nature* 327:654–655.

Diamond, J. M. 1983. Extinctions, catastrophic and gradual. *Nature* 304:396–397.

Ehrenfeld, D. W. 1970. *Biological Conservation*. Modern Biology Series. Holt, Rinehart & Winston, New York, NY.

Falk, D. 1985. Hadar AL 162-28 endocast as evidence that brain enlargement preceded cortical reorganization in hominid evolution. *Nature* 313:45–47. (See also *Nature* 321:536–537, 1986.)

Foley, R. A., and P. C. Lee. 1991. Ecology and energetics of encephalization in hominid evolution. *Philosophical Transactions of the Royal Society, London B* 334:223–232.

Frankel, O. H., and M. E. Soulé. 1981. *Conservation and Evolution*. Cambridge University Press, Cambridge, UK.

Galloway, J. N., Z. Dianwu, X. Jiling, and G. E. Likens. 1987. Acid rain: China, United States, and a remote area. *Science* 236:1559–1562.

Gibbons, A. 1992. Neandertal language debate: tongues wag anew. *Science* 256:33–34.

Gibbons, A. 1993. Pleistocene population explosions. *Science* 262:27–28.

Gillespie, R., D. R. Horton, P. Ladd, P. G. Macumber, T. H. Rich, R. Thorne, and R. V. S. Wright. 1978. Lancefield swamp and the extinction of the Australian megafauna. *Science* 200:1044–1048.

Gould, S. J. 1988. A novel notion of Neanderthal. *Natural History* 97(6):16–21.

Harris, J. M., F. H. Brown, M. G. Leakey, A. C. Walker, and R. E. Leakey. 1988. Pliocene and Pleistocene hominid-bearing sites from west of Lake Turkana, Kenya. *Science* 239:27–33.

Hunt, K. D. 1994. The evolution of human bipedality: ecological and functional morphology. *Journal of Human Evolution* 26:183–202.

James, H. F., T. W. Stafford, Jr., D. W. Steadman, S. L. Olsen, P. S. Martin, A. J. T. Jull, and P. C. McCoy. 1987. Radiocarbon dates on bones of extinct birds from Hawaii. *Proceedings of the National Academy of Sciences USA* 84:2350–2354.

Janzen, D. H. 1983. The Pleistocene hunters had help. *American Naturalist* 121:598–599.

Janzen, D. H. 1988. Tropical ecology and biocultural restoration. *Science* 239:243–244.

Johanson, D. C. and M. A. Edey. 1981. *Lucy, The Beginnings of Humankind*. Simon & Schuster, New York, NY.

Jolly, A. 1980. *A World Like Our Own: Man and Nature in Madagascar*. Yale University Press, New Haven, CT.

Jolly, A. 1985. *The Evolution of Primate Behavior*, 2d edition. Macmillan, New York, NY.

Kaufman, L., and K. Mallory. 1986. *The Last Extinction*. M. I. T. Press, Cambridge, MA.

Kerr, R. A. 1987. Has stratospheric ozone started to disappear? *Science* 237:131–132.

Kurten, B. 1986. *How to Deep-Freeze a Mammoth*. Columbia University Press, New York, NY.

Lahr, M. M. 1994. The multiregional model of modern human origins: a reassessment of its morphological basis. *Journal of Human Evolution* 26:23–56.

Lewin, R. 1984. DNA reveals surprises in human family tree. *Science* 226:1179–1182.

Lewin, R. 1988. A new tool-maker in the hominid record? *Science* 240:724–725.

Lewin, R. 1993. *Human Evolution*, 3d edition. Blackwell Scientific, Boston, MA.

Luckett, W. P. 1980. *Comparative Biology and Evolutionary Relationships of Tree Shrews*. Plenum, New York, NY.

Martin, P. S. 1973. The discovery of America. *Science* 179:969–974.

Martin, P. S., and R. G. Klein (editors). 1984. *Quaternary Extinctions, a Prehistoric Revolution*. University of Arizona Press, Tucson, AZ.

Martin, P. S., and H. E. Wright, Jr. (editors). 1967. *Pleistocene Extinctions: The Search for a Cause*. Yale University Press, New Haven, CT.

Martin, R. D. 1990. *Primate Origins and Evolution: A Phylogenetic Reconstruction*. Princeton University Press, Princeton, NJ.

McHenry, H. M. 1994. Behavioral ecology implications of early hominid body size. *Journal of Human Evolution* 27:77–87.

Mellars P., and C. B. Stringer, editors. 1989. *The Human Revolution*. Edinburgh University Press, Edinburgh, UK.

Mooney, H. A., P. M. Vitousek, and P. A. Matson. 1987. Exchange of materials between terrestrial ecosystems and the atmosphere. *Science* 238:926–932.

Morell, V. 1994. Will primate genetics split one gorilla into two? *Science* 265:1661.

Mosimann, J. E., and P. S. Martin. 1975. Simulating overkill by paleoindians. *American Scientist* 63:304–313.

Myers, N. 1984. *The Primary Source, Tropical Forests and Our Future*. Norton, New York, NY.

Owen–Smith, N. 1987. Pleistocene extinctions: the pivotal role of megaherbivores. *Paleobiology* 13:352–362.

Patrusky, B. 1980. Pre-Clovis man: sampling the evidence. *Mosaic* 11(5):2–10.

Pimentel, D. R. Harman, and M. Pacenza. 1994. Natural resources and an optimum human population. *Population and Environment* 15:347–369.

Rasmussen, D. T. (editor). 1993. *The Origin and Evolution of Humans and Humanness*. Jones & Bartlett, Boston, MA.

Rightmire, G. P. 1990. *The Evolution of* Homo erectus. Cambridge University Press, New York, NY.

Rosen, B. 1994. Mammoths in ancient Egypt? *Nature* 369:364.

Schindler, D. W. 1988. Effects of acid rain on freshwater ecosystems. *Science* 239:149–157.

Schonewald–Cox, C. M., S. M. Chambers, B. MacBride, and L. Thomas (editors). 1983. *Genetics and Conservation*. Benjamin–Cummings, Menlo Park, CA.

Simons, E. L. 1989. Human origins. *Science* 245:1343–1350.

Soulé, M. E. (editor). 1986. *Conservation Biology, the Science of Scarcity and Diversity*. Sinauer Associates, Sunderland, MA.

Stringer, C. B., and P. Andrews. 1988. Genetic and fossil evidence for the origin of modern humans. *Science* 239:1263–1268.

Stuart, A. J. 1991. Mammalian extinctions in the Late Pleistocene of northern Eurasia and North America. *Biological Reviews* 66:453–562.

Susman, R. L. 1988. Hand of Paranthropus robustus from member 1, Swartkrans: fossil evidence for tool behavior. *Science* 240:781–784.

Susman, R. L. 1994. Fossil evidence for early hominid tool use. *Science* 265:1570–1573.

Susman, R. L., and J. T. Stern. 1982. Functional morphology of Homo habilis. *Science* 217:931–934.

Swisher, C. C., G. H. Curtis, and T. Jacob. 1994. Age of the earliest known hominids in Java, Indonesia. *Science* 263:1118–1121.

Szalay, F. S., and E. Delson. 1979. *Evolutionary History of the Primates*. Academic, New York, NY.

Tattersall, I. 1993. *The Human Odyssey: Four Million Years of Human Evolution*. Prentice Hall, New York, NY.

Vitousek, P. M. 1994. Beyond global warming: ecology and global change. *Ecology* 75:1861–1876.

White, T. D. 1980. Evolutionary implications of Pliocene hominid footprints. *Science* 208:175–176.

White, T. D., G. Suwa, and B. Asfaw. 1994. *Australopithecus ramidus*, a new species of early hominid from Aramis, Ethiopia. *Nature* 371:306–312.

WoldeGabriel, G., T. D. White, and H. Buffetaut. 1994. Ecological and temporal placement of early Pliocene hominids at Aramis, Ethiopia. *Nature* 371:330–333.

Wood, B. 1987. Who is the "real" *Homo habilis? Nature* 327(6119): 187–188.

Wood, B. 1994. The oldest hominid yet. *Nature* 371:280–281.

Woodward, F. I., editor. 1992. *Global Climate Change: the Ecological Consequences*. Academic, New York, NY.

Wu, R. and S. Lin. 1983. Peking man. *Scientific American* 248:86–94.

Yunis, J. J., and O. Prakash. 1982. The origin of man: a chromosomal legacy. *Science* 215:1525–1530.

GLOSSARY

Technical terms, and some words with special biological meanings, are included in this glossary. In general, if a word is defined in the text and appears only on the page where it is defined and immediately adjacent pages, we have not attempted to define it again in the glossary. Similarly, if the definition in a standard dictionary is adequate, we have not included it here. However, certain words have slightly different meanings in different contexts. In those cases we have included the word in this glossary and limited the definition to the use of the word in this book.

abduction Movement away from the midventral axis of the body (*see also* adduction).

acetabulum Depression on the pelvic girdle that accommodates the head of the femur.

adduction Movement toward the midventral axis of the body (*see also* abduction).

allochthonous With an origin somewhere other than the region where found.

allopatry Situation in which two or more populations or species occupy mutually exclusive, but often adjacent, geographic ranges.

ammonotelic Excreting nitrogenous wastes primarily as ammonia.

amniotes Those vertebrates whose embryos possess an amnion, chorion, and allantois (i.e., turtles, lepidosaurs, crocodilians, birds, and mammals).

amphicoelous Vertebral centrum with both the anterior and posterior surfaces concave.

anadromous Migrating up a stream or river from a lake or ocean to spawn (of fishes).

angiosperm Most advanced and recently evolved of the vascular plants characterized by production of seeds enclosed in tissues derived from the ovary, the combination of ovary and/or seed being of major importance to many vertebrates as food. The success of the angiosperms has had important consequences for the evolution of terrestrial vertebrates.

aphotic "Without light"—for example, in deep-sea habitats or caves.

apocrine gland Type of gland in which the apical part of the cell from which the secretion is released breaks down in the process of secretion (*see also* holocrine gland).

apomorphic A character that is changed from its preexisting condition (*see also* autapomorphy).

aposematic Device (color, sound, behavior) used to advertise the noxious qualities of an animal.

arcade Curve or arch in a structure, such as the tooth row of humans.

archaic Form typical of an earlier evolutionary time. When capitalized, referring to the Archaeozoic period, previous to 2.5 billion years before the present.

archipterygium Fin skeleton, as in a lungfish, consisting of symmetrically arranged rays that extend from a central skeletal axis.

aural Of the external or internal ear or sense of hearing.

autapomorphy An attribute unique to one group of organisms.

autochthonous With an origin in the region where found.

benthic Living at the soil/water interface at the bottom of a body of water.

bilateral symmetry (bisymmetry) Characteristic of a body which can be divided into mirror-image halves by a single plane of space.

biomass Living organic material in a habitat (available as food for other species).

biome Biogeographic region defined by a series of spatially interrelated and characteristic life forms (e.g., tundra, mesopelagic zone, tropical rain forest, coral reef).

brachial Pertaining to the forelimb.

branchial Pertaining to the gills.

branchiomeric Segmentation of structures associated with, or derived from, the ancestral pharyngeal arches (*see also* metameric).

carapace Dorsal shell, as of a turtle.

catadromous Migrating down a river or stream to a lake or ocean to spawn (of fishes).

catastrophism Hypothesis of major evolutionary change as a result of unique catastrophic events of broad geographic and thus ecologic effect.

centrum Vertebral element formed in or around the notochord (plural, centra).

cephalic Pertaining to the head.

ceratotrichia Keratin fibers that support the web of the fins of chondrichthyes.

choana Internal nares (plural, choanae).

chondrification Formation of cartilage.

clade Phylogenetic lineage originating from a common ancestral taxon and including all descendants (*see also* grade).

cladistic Pertaining to the branching sequences of phylogenesis.

cladogram Branching diagram representing the hypothesized relationships of taxa.

cleidoic egg One independent of environment except for heat and gas (carbon dioxide, oxygen, water vapor) exchange. Characteristic of amniotes.

cline Change in a biological character along a geographic gradient.

coelom (celom) Body cavity, lined with tissue of mesodermal origin.

coevolution Complex biotic interaction through evolutionary time resulting in the adaptation of the interacting species to unique features of the life histories of the other species in the system.

cones Photoreceptor cells in the vertebrate retina that are differentially sensitive to light of different wavelengths.

conspecific Belonging to the same species as that under discussion (*see also* heterospecific).

convergent evolution Evolution of similar characters in phyletically distantly related or unrelated forms.

cosmine Form of dentine containing branching canals characteristic of the cosmoid scales of crossopterygian fishes and early dipnoans.

countershaded Color pattern in which the aspect of the body that is more brightly lighted (normally, the dor-

sal surface) is darker colored than the less brightly illuminated surface. The effect of countershading is to make an animal harder to distinguish from its background.

cranial Pertaining to the cranium or skull, a unique and unifying characteristic of all vertebrates.

demersal More dense than water and therefore sinking, as in the eggs of many fishes and amphibians.

detritus Particulate organic matter that sinks to the bottom of a body of water.

deuterostomy Condition in which the embryonic blastopore forms the anus of the adult animal; characteristic of chordates (*see also* protostomy).

double cone Type of retinal photoreceptor in which two cones share a single axon (*see also* cones).

durophagous Feeding upon hard material.

ecosystem Community of organisms and their entire physical environment.

ectoderm One of the embryonic germ layers, the outer layer of the embryo.

edentulous Lacking teeth.

endemism Property of being endemic (i.e., found only in a particular region).

endoderm Innermost of the germ cell layers of late embryos.

epicontinental sea (epeiric sea) Sea extending within the margin of a continent.

epigenetic Pertaining to an interaction of tissues during embryonic development that results in the formation of specific structures.

epiphysis (1) Pineal organ, an outgrowth of the roof of the diencephalon. (2) (plural, epiphyses) Accessory center of ossification at the ends of the long bones of mammals, birds, and some squamates. When the ossifications of the shaft (diaphysis) and epiphysis meet, further lengthwise growth of the shaft ceases. This process produces a determinate growth pattern.

epiphyte Plant that grows nonparasitically on another plant.

estivation (aestivation) Form of torpor, usually a response to high temperatures or scarcity of water.

estuarine Pertaining to, or formed in, a region where the freshwater of rivers mixes with the seawater of a coast.

euryhaline Capable of living in a wide range of salinities (*see also* stenohaline).

euryphagous Eating a wide range of food items; a food generalist (*see also* stenophagous).

eurythermal Capable of tolerating a wide range of temperatures (*see also* stenothermal).

eurytopy Capable of living in a broad range of habitats.

extraperitoneal Positioned in the body wall beneath the lining of the coelom (the peritoneum) in contrast to being suspended in the coelom by mesenteries.

fossorial Burrowing through the soil.

fovea centralis Area of the vertebrate retina containing only cone cells, where the most acute vision is achieved at high light intensities.

furcula Avian wishbone formed by the fusion of the two clavicles at their central ends.

geosyncline Portion of the Earth's crust that has been subjected to downward warping. Sediments frequently accumulate in geosynclines.

gestation Period during which an embryo is developing in the reproductive tract of the mother.

gill arch Assemblage of tissues associated with a gill; the term may refer to the skeletal structure only or to the entire epithelial muscular and connective tissue complex.

Gondwana Supercontinent that existed either independently or in close contact with all other major continental land masses throughout vertebrate evolution until the middle of the Mesozoic and was composed of all the modern Southern Hemisphere continents plus the subcontinent of India.

grade A level of morphological organization achieved independently by different evolutionary lineages (*see also* clade).

gymnosperms Group of plants in which the seed is not contained in an ovary—conifers, cycads, and ginkos.

hemal arch Structure formed by paired projections ventral to the vertebral centrum and enclosing caudal blood vessels.

hermaphroditic Having both male and female gonads.

heterocoelus Having the articular surfaces of the vertebral centra saddle-shaped, as in modern birds.

heterospecific Belonging to a different species from that under discussion (*see also* conspecific).

heterosporous plants Plants with large and small spores; the smaller give rise to male gametophytes and the larger to female gametophytes (equivalent to protogymnosperms).

heterotrophic Capable of using only organic materials as a source of energy.

holocrine gland Type of gland in which the entire cell is destroyed with the discharge of its contents (*see also* apocrine gland).

homologous Characters having a common origin.

homology Pertaining to the fundamental similarity of individual structures that belong to different species within a monophyletic group.

hydrosphere Free liquid water of the earth—oceans, lakes, rivers, etc.

hyperdactyly Increase in the number of digits.

hyperphalangy Increase in the number of bones in the digits.

hypertrophy Increase in the size of a structure.

hypotremate Having the main gill openings on the ventral surface and beneath the pectoral fins as in skates and rays (*see also* pleurotremate).

inguinal Pertaining to the groin.

interspecific Pertaining to phenomena occurring between members of different species.

intraspecific Pertaining to phenomena occurring between members of the same species.

isohaline Of the same salt concentration.

isostasy Condition of gravitational balance between segments of the Earth's crust or of return to balance after a disturbance.

isostatic movement Vertical displacement of the lithosphere due to changes in the mass over a point or region of the earth.

isotherm Line on a map that connects points of equal temperature.

leptocephalus larva Specialized, transparent, ribbon-shaped larva of tarpons, true eels, and their relatives.

lithosphere Crust of the Earth.

littoral Pertaining to the shallow portion of a lake, sea, or ocean where rooted plants are capable of growing.

lophophorate Pertaining to several kinds of marine animals that possess ciliated tentacles (lophophores) used to collect food (e.g., pterobranchs).

meninges Sheets of tissue enclosing the central nervous system. In mammals these are the dura mater, arachnoid, and pia mater.

mesenteries Membranous sheets derived from the mesoderm that envelop and suspend the viscera from the body wall within the coelom.

mesoblast Mesodermal cell.

mesoderm Central of three germ layers of late embryos.

metameric Pertaining to ancestral segmentation, used in reference to serially repeated units along the body axis.

monophyletic lineage A taxon composed of a common ancestor and all its descendants.

monophyly Relationship of two or more taxa having a common ancestor.

morph Genetically determined variant in a population.

morphotypic Type of classification based entirely on physical form.

neoteny Retention of larval or embryonic characteristics past the time of reproductive maturity (*see also* paedomorphosis and progenesis).

neural arch Dorsal projection from the vertebral centrum that, at its base, encloses the spinal cord.

neural crest Embryonic cells unique to vertebrate animals, associated with the neurectoderm but subsequently widely migrating to participate in the formation of many tissues and structures which are characteristic of the subphylum.

neurocranium Portion of the head skeleton encasing the brain.

niche Pertaining to the functional role of a species or other taxon in its environment—the ways in which it interacts with both the living and nonliving elements.

occipital Pertaining to the posterior part of the skull.

ontogenetic Pertaining to the development of an individual organism.

operculum Flap or plate of tissue covering the gills.

orogeny Process of crustal uplift or mountain building.

osseous Bony.

out-group Group of organisms that is related to but removed from the group under study. One or more outgroups are examined to determine which character states are evolutionary novelties (apomorphies).

paedomorphosis Condition in which a larva becomes sexually mature without attaining the adult body form. Paedomorphosis may be achieved by neoteny or by progenesis.

palatoquadrate Upper jaw element of primitive fishes and chondrichthyes, portions of which contribute to the palate, jaw articulation, and middle ear of other vertebrates.

Pangaea (Pangea) Single supercontinent that existed during the mid-Paleozoic and consisted of all modern continents apparently in direct physical contact with a minimum of isolating physical barriers.

paraphyletic Taxon that includes the common ancestor and some but not all of its descendants.

Phanerozoic Period since the Cambrian.

pharyngotremy Condition in which the pharyngeal walls are perforated by slit-like openings; found in chordates and hemichordates.

photophore Light-emitting organ.

phylogenetic Pertaining to the development of an evolutionary lineage (*see also* ontogenetic).

physoclistic Lacking a connection from the gut to the swim bladder as adults (of fishes).

physostomous Having a connection between the swim bladder and gut in adults (of fishes).

piloerection Contraction of muscles attached to hair follicles resulting in the erection of the hair shafts.

piscivorous Fish-eating.

placoid scales Primitive type of scale found in elasmobranchs and homologous with vertebrate teeth.

plastron Ventral shell, as of a turtle.

plate tectonics Theory of Earth history in which the lithosphere is continually being generated from the underlying core at specific areas and reabsorbed into the core at others resulting in a series of conveyor-like plates which carry the continents across the face of the Earth.

plesiomorphic The ancestral character from which an apomorphy is derived.

pleurotremate Having the main gill openings on sides of the body anterior to the pectoral fins as in sharks (*see also* hypotremate).

polymorphism Simultaneous occurrence of two or more distinct phenotypes in a population.

polyphyletic A taxon that does not contain the most recent common ancestor of all the subordinate taxa of the taxon.

portal system Portion of the venous system specialized for the transport of substances from the site of production to the site of action. A portal system begins and ends in capillary beds.

postzygapophysis Articulating surface on the posterior face of a neural arch.

prezygapophysis Articulating surface on the anterior face of a neural arch.

progenesis Accelerated development of reproductive organs relative to somatic tissue, leading to paedomorphosis.

Proterozoic Later part of the Precambrian, from about 1.5 billion years ago until the beginning of the Cambrian 500 million years ago (*see also* Phanerozoic).

protostomy Condition in which the embryonic blastopore forms the mouth of the adult animal (*see also* deuterostomy).

protraction Movement away from the center of the body (*see also* retraction).

protrusible Capable of being moved away (protruded) from the body.

refugium Isolated area of habitat fragmented from a formerly more extensive biome.

rete mirabile "Marvelous net," a complex mass of intertwined capillaries specialized for exchange of heat and/or dissolved substances between countercurrent flowing blood.

retraction Movement toward the center of the body (*see also* protraction).

rod Photoreceptor cell in the vertebrate retina specialized to function effectively under conditions of dim light.

rostrum Snout; especially an extension anterior to the mouth.

scapulocoracoid cartilage In elasmobranchs and certain primitive gnathostomes, the single solid element of the pectoral girdle.

scutes Scales, especially broad or inflexible ones.

serial Repeated, as are the body segments of vertebrates.

sinus Open space in a duct or tubular system.

sister group Group of organisms most closely related to the study taxa, excluding their direct descendants.

somite Member of a series of paired segments of the embryonic dorsal mesoderm of vertebrates.

speciose Refers to a taxon that contains a large number of species.

squamation Scaly covering of the body.

stenohaline Capable of living only within a narrow range of salinity of surrounding water; not capable of surviving a great change in salinity (*see also* euryhaline).

stenophagous Eating a narrow range of food items; a food specialist.

stenothermal Capable of living or of being active in only a narrow range of temperatures (*see also* eurythermal).

stratigraphy Classification, correlation, and interpretation of stratified rocks.

stratum Layer of material (plural, strata).

sympatry Occurrence of two or more species in the same area.

symphysis A joint between bones formed by a pad or disk of fibrocartilage that allows a small degree of movement.

symplesiomorphy Character shared by a group of organisms that is found in their common ancestor.

talonid Basin-like heel on a lower molar tooth, found in certain mammals.

tarsometatarsus Bone formed by fusion of the distal tarsal elements with the metatarsals in birds and some dinosaurs (*see also* tibiotarsus).

taxon Any scientifically recognized group of organisms.

thecodont teeth Teeth set in bony sockets in the jaw.

tibiotarsus Bone formed by fusion of the tibia and proximal tarsal elements in birds and some dinosaurs (*see also* tarsometatarsus).

troglodyte Organism that lives in caves.

trophic Pertaining to feeding and nutrition.

ureotelic Excreting nitrogenous wastes primarily as urea.

uricotelic Excreting nitrogenous wastes primarily as uric acid and its salts.

urogenital Pertaining to the organs, ducts, and structures of the excretory and reproductive systems.

vacuoles Membrane-bound spaces within cells containing secretions, storage products, etc.

viscera Internal organs of the coelom.

visceral skeleton Skeleton primitively associated with the pharyngeal arches, uniquely derived from the neural crest cells and forming in mesoderm immediately adjacent to the endoderm lining the gut.

zygapophysis Articular process of the neural arch of a vertebra (*see also* postzygapophysis and prezygapophysis).

zygodactylous Type of foot, in which the toes are arranged in two opposable groups.

SUBJECT INDEX

Numbers in boldface indicate illustrations; other illustrations may be on pages with text.

Crocodiles, 7, 394
  saltwater, 395
Crocodilians, 7, 392–396
  parental care, 420–**421**
  salt secretion by, 138
Crocodylotarsi, 391–396
Crocuta crocuta, 119, 715–**716**
Crop, 537–**540**
  milk, 537–538, 702
Crossopterygians, 241–242
Cryptobranchids, 311, 325
Cryptobranchus, 311, **313**
Cryptodires, 357, 361
Cuora, 360
Cupula, 233–234
Cursorial adaptations, 541–542,
  658–663
Cyanocitta cristata, 347
Cyclopes, 649, **650**
  didactylus, 289, **294**
Cyclostomata, 181, 183
Cyclotosaurus, 294, **296**
Cynodonts, 605–614
  see also Mammals
Cynognathus, **602**, 605
Cyprinodonts, 258
Cyprinus carpio, 262–263

Darwin's frog, 338
Dasyatidae, 215
Dasypus, 649, **650**
Dear enemy recognition, 315-318
Deep scattering layer, 208
Deer, 645, **648**, 723
Deinocheirus, 404
Deinonychosaurs, 404–405, **406, 407,
  408**, 423
  relationship to birds, 409–410
Deinonychus, 404–405, **406, 407, 408**,
  423, 427–429
Deinosuchus, 392
Delphinus, **650**, 651
Deme, 15
Dendrobates pumilio, 338, 341
Dendrobatids, 348, 349
Denticles
  of thelodonts, 175
Dentine, 76, 78, 649
Dermis, 75, 76–77, 638
Dermochelyids, 357, **358**
Dermochelys, 357, 361, 368
  as model dinosaur, 425–429
Desert
  ectotherms, 504–513
  endotherms, 697–702, 708–709
Desmodus rotundus, 684–686
Desmognathus, 326
  fuscus, 327
  ochrophaeus, 349
Determinate growth, 451
Deuterostomes, 46–47, 50, 53, 67

Development, 64–73
Devonian
  as Age of Fishes, 238–239
  climates, 279
  continental positions, 276, **277**
  origin of tetrapods and, 287
  terrestrial ecosystems, 164,
    276–279
  vertebrate origins and, 60
Diadectes, 299, 301, 304, **302**
Diadectomorphs, 298-299
Diademodontids, 608
Diapause, embryonic, 679–680
Diaphragm, 609, **610**
Diapsids, 387–440
  early, 304, **306, 389**
  nitrogen excretion by, 137–139
Diarthrognathus, 606–608, 612
Diasperma, 702, 704
Diastema, 651
Diatryma, 542, 564
Dicroidium, 390
Dicynodonts, **602**, 604
Diet
  habitat and, 720
  home range and, **713**–714
  of *Homo sapiens*, 752
  of human ancestors, 740
  see also Herbivory
Digestion, 86–88
  bird, 536–541
  fermentative, 659–661
Digitigrade, 658, **662**
Dik–diks, 718, **719**
Dimetrodon, 595, **598**–600, 601
Dinnetherium, 614
Dinocephalians, 602–**603**
Dinosaurs. *see* Nonavian dinosaurs
Diphyodonty, 649
Diploceraspis, 297, **300**
Diplodocoids, 401–**402**, 422
Diplodocus, 401, 422
Diploid, 14
Diplorhinans, 172–174
Dipnoans. *see* Lungfishes
Dipodomys, 704–705
Diseases of tortoises, 378
Display action pattern, 475, **480**
Distal convoluted tubules, 124,
  132–134
Diving by birds, 546–547
Docodonts, 614
Dodo, 757
Dominance hierarchy, 727, 728
  of tortoises, 368–369
Dorsal hollow nerve cord, 45,
  47
  development of, 68
Dorsal root nerves, 104
Doubly labeled water, 506
Draco, 407

Drag
  bird flight and, 527, 533–534
  fishes and, 227–232
Dromedary camel, 699–701
Dryopithecids, 739
Dsungaripterus, 396-398
Duck–billed platypus, 238, 671
Duikers, 718
Dunkleosteus, 197
Dwarf mongooses, 718

Ear
  inner, 112, 212
  middle, 110–112, **606**–608
  of bats, 668
  of birds, 548–549
  of cetaceans, 670
  of mammals, 664
  ossicles, 111–112, 607
  pinna, 109–110
Echidna, 627, **628**, 671–672
Echinotriton, 346
Echolocation, 665–671
Ectoderm, 66, 68
Ectothermy, 141–147, 497–519
Edaphosaurids, 600
Edaphosaurus, 595, **598**, 600–601
Edentates, 629–**631**
Edmontosaurus, 427–429
Eels
  American, 255
Egg size optimization, 484–485
Eggs
  bird, shells, 581–**582, 583**
  development and, 64–68
  isolecithal, 64
  macrolecithal, 64, 66
  mesolecithal, 64, 66
  nonavian dinosaur, 426–**427**
  oligolecithal, 64, 66
  salamander, 329–330
  turtle, 370
El Niño, 22
Eland, 701–702, 718, 720
Elaphe, 468
Elapids, 468-469
Elasmobranchs, 5, 198–**216**
  extant forms, 204–**216**
Elasmosaurus, 431, **432**
Elastin, 91
Electric discharge, 235– 238
  by skates, 215
Electric eel respiration, 221
Electrocytes, 235
Electroreception, 236–**238**
  by coelacanth, 272
  by sharks, 210–**211**
Elephant seal, 292, 759
Elephantbirds, 542
Elephants, 654, 658, 664, 665
  as dinosaur model, 425

Saurischians, 400–405
 posture of, 398–400
 see also Birds
Sauriurines, 555
Saurolophines, 416, **418**
Sauromalus obesus, **508**–511
Sauropods, 400–403
 ecology, 422–423
 nests of, 425, 426
Scales
 ganoid, 248
 of *Cladoselache*, 201
 placoid, 206
Scalopus, 649
Scaphiopus multiplicatus, 341
Sceloporus, 477
 jarrovi, 489, 514
 virgatus, 479–481
Scincella lateralis, 476
Scincomorphs, 472
Scolecophidians, 460
Scombroid fishes, 151
Sea otters, 758, **759**
Seals
 elephant, 292, 758, **760**
 fur, 688–690, 758
 hooded, 643
Sebaceous glands, 642
Secondary sex characters, 118–119
Sectorial teeth, 652
Segregation distortion, 14
Selection,
 directional, 14
 disruptive, 14
 genic, 14
 interdemic, 15
 kin, 585, 717–718, 724, 725–726
 modes of, 14
 species level, 15
 stabilizing, 14
 see also Natural selection
Selenodont, 652
Semicircular canals, 112
Senses, 104–112
 chemoreception, 209–212, 460, 479
 electroreception, 210–211, 236–**238**, 272
 lateral line, 72, 209, 212, **233**–**234**
 magnetic, 551
 of birds, 547–551
 of fishes, 232–238
 of mammals, 663–664
 of squamates, 460, 479
Sertoli cells, 120
Sex chromosomes, 118
Sex determination, 118–119
 by exogenous hormones, 372–373
 of mammals, 119
 of teleosts, 262–265
 of turtles, 370–371, 379

Sex ratio, 372, 574–575
Sex reversal, 262–265, **270**
Sexual dimorphism, 19, **20**
Seymouria, 298, **302**
Seymouriamorphs, 298, **302**
Shad, American, 255
Sharks, 5, 151
 basking, 208
 breeding by, 213
 bull, 129
 buoyancy of, 129
 edestoid, 202
 extant, 204–215
 feeding by, 208–209
 fossil, 198–204
 galeoid, 215
 great white, 212–213
 hammerhead, **207**, 212, 214
 hybodont, 203–204
 hyostylic jaw, **205**–206
 liver, 206–208
 megamouth, 209
 nutrition for development, 214
 Port Jackson, 129
 scales, 206
 schooling by, 214
 spiny dogfish, 214
 squaloid, 215
 whale, 208
 see also Chondrichthyes
Shearwaters, 531–532
Shivering, 147
Shunt
 heat, 145–146
 intracardiac, 366, **367**, 395
 systemic–pulmonary, 95
Siberia, 161
Sickle–cell anemia, 13
Sidewinding, 467
Silurian, 163, 164
 continents during, 161
 vertebrate origins and, 60
Simoedosaurus, 429
Sinoconodon, 613–614
Sinornis, 554
Sinuses, 738
Siphon sac, 213
Sirenids, 325
Sister group, 37
Sivapithecids (*Sivapithecus*), 739–741
Skates, 215, **216**
Skeleton, 80–84
 appendicular, 80, 83–84
 axial, 80–83
 of birds, 534–**535**
 of turtles, 359–361
 visceral arch, 80, 81
 see also skull
Skin. *see* Integument

Skinks, 476–**477**
 broad–headed, 478
 parental care by, 484
 prehensile–tailed, 484-485
 viviparity and, 479
Skull, 80–81
 anapsid, 359, 595
 bird, 535, **536**
 diapsid, 387, **388**
 snake, 463-465, **469**
 squamate, 435, **436**
 synapsid, 595, **596**
Snakes, 7, 460–463, **464–465**
 feeding by, 463–**470**
 origin of, 462–463
 sea, 138, 462
 venom, 468–470
 wandering garter, 486–488
 see also Squamates
Soaring
 dynamic, 531–532
 static, 532
Social behavior
 of plethodontids, 315–320
 of primates, 726–730
 of squamates, 475–478
 of turtles, 368–370
Social systems
 of primates, 724–726
Sociality
 and defense, 717
Solenoglyphous, 469, **471**
Solitaire, 757
Somatic nerves, 104
Somites, 68
Sordes pilosus, 398
Spadefoot toads, 319, **324**, 341, 511–512
Speciation, 15, 19–27
 allopatric, 21, 22
 glaciation and, 625
 sympatric, 21
Species concepts, 19–20
Specific dynamic action, 147
Sperm, 120, 675–676
Spermatophore, 325, **329**
Spermophilus richardsonii, **695**–697
Sphenacodontids, 595–600
Sphenodon, 452–455
 guntheri, 434
 punctatus, 434–435
Sphenodontids, 7, 452
 see also Tuatara
Sphyrna, **207**, 212
Spinal cord, 82, 100–101
Spinosaurids (*Spinosaurus*), 403
Spiral valve, 87
Spleen, 96
Spoonbills, 560–561, **565**
Spotted sandpipers, 577
Squalus, 214

Turtles, 7, 356–383
  big–headed, 366
  green, 367, 369, 374–**375**
  hatchlings, 371–374
  hawksbill, 358
  Kemp's ridley, 377–**380**
  leatherback, 357, **358**, 361, 367, 425–429
  loggerhead, 357, **358**, 375–376
  mud, 357
  musk, 366
  nitrogen excretion by, 137–139
  salt secretion by, 138
  sea, 374–376, 377–382
  snapping, 357
  soft–shelled, 357, 361, 365
  see also Tortoises
Tusks, 653–654
Tylotriton, 346
Tympanum, 73, 111, 112
Typhlomolge, 311, **313**
Typhlonectes, 322–325
Typotheres (*Typotherium*), 632
Tyrannosaurids (*Tyrannosaurus*), 403

Ultrafiltrate, 124, 132
Ultrasound, 665–656
Ultraviolet radiation, 352
Undina, 273
Ungulates, 655–658, 677–678
Unguligrade, 658, **662**
Uranoscopids, 236
Urea, 127–129, 130–131
  cycle, 130
Ureotelism, 130–131
Uric acid, 130, 137
Uricotelism, 130, 137, 139, 344, 513
Urine, 125–127, 130
  of desert rodents, 704
  of mammals, 131–137
  of vampire bats, 686
Urochordates, 47
Urocordylus, 297
Urostyle, 318, **322**
Uta stansburiana, 484–**485**
Uterus, 121, 671–673

Varanids, 413,457-458, **459**
Varanoids, 435
Variation
  genetic, 16
  individual, 17–18
  phenotypic, 16
Vascular networks, 77
Vascular plants. *see* Plants
Vasodilation
  and thermoregulation, 145, 146
Vasopressin. *see* Antidiuretic hormone
Vegetal pole, 64, 66
Veins, 89, 91
Velociraptor, 423
Venom, 87, 468–470, 472
Ventilation, 88–89
Ventral root nerves, 104
Vertebrae, 80, 82–83
  of early tetrapods, **294**
  of sarcopterygians, **294**
Vertebrates
  defined, 45
  environment of origin, 60
  theory of marine origin, 61
Vestibule of ear, 112
Vicariance biogeography, 626
Vieraella, 310
Villi, 87
Viperids, 468-470
Visceral arches, 73
Visceral nerves, 104
Vision, 105–108
  of birds, **547**–548
  of bolitoglossine salamanders, 312
  of fishes, 232
  of mammals, 663–664
  of sharks, 212
Viviparity, 121
  in caecilians, 322–325
  in mammals, 673–674
  in squamates, 479–481
  lecithotrophic, 214, 678
  placentotrophic, 214
Vocal cords, 667–668

Vocalizations
  by anurans, 331–336
  by birds, 536, 567–568, **572**
  by tortoises, 368
Vomeronasal organ, 109
Vultures
  Egyptian, 752
  turkey , 549

Walrus, **650**, 651, 654
Warblers, hooded, 568
Water
  life in, 220–221
Water relations
  of amphibians, 343–346
  of chuckwalla, 509–511
  of desert rodents, 703–705
  of desert tortoises, 505–508
Weberian apparatus, 255, **264**
Westlothiana, **304**
Whales, 669, 758-759
  gray, 690–**691**, 759
  migration, 690–691
  right, **650**, 651
  sperm, 669
Wildebeest, 715, 720, 722, 723–724
Wings
  of birds, 526–528, 531–532, **533**
Woodpecker tongue, 560, **563**

Xenacanthus, 198, 202
Xenopus laevis, 234, 340, 346

Yolk, 64, 120
  in mammalian eggs, 671–673
  sac, 66
Younginiforms (*Youngina*), 429
Yunnanosaurids, 400

Zebras, 715
Zugdisposition, 588
Zugstimmung, 588
Zugunruhe, 588, **589**
Zygodactyly, 456, 542

AUTHOR INDEX

Able, K. P., 590
Able, M. A., 590
Abler, W. L., 403
Adams, H. P., 378
Adovasio, J. M., 755
Agassiz, L., 365
Ahlberg, P. E., 283, 293
Ahlquist, J. E., 556, 557
Aiello, L. C., 750
Alberch, P., 290
Aldridge, R. J., 58
Alerstam, T., 589
Alexander, R. M., 422
Allan, J. D., 189, 261
Allen, M. E., 474
Allin, E. F., 607
Altangerel, P., 426, 554
Amaral, W. W., 292
Amato, I., 669
Andrews, P., 739, 746, 747, 752
Andrews, R. M., 479–481
Arano, B., 326
Archer, M., 615, 633
Archibald, J. D., 437, 439
Arcucci, A. B., 359
Armstrong, H. A., 58
Arnold, A. P., 567
Arnold, E. N., 476
Arntzen, J. W., 326
Aschoff, J., 693
Asfaw, B., 746
Au, W. W. L., 669
Auffenberg, W., 457
Avery, R. A., 144, 486
Avise, J. C., 374
Ayala, F. J., 746

Baker, C. S., 759
Bakhurina, N., 398
Bakken, G. S., 707
Balda, R. P., 412
Balding, D., 139
Balment, R. J., 100
Bambach, R. K., 278
Bandoni, R. J., 163
Bang, B. G., 549

Bardin, C. W., 119
Barghusen, H. R., 601, 602
Barnard, D. E., 315
Barnes, B. M., 692
Barrett, S. F., 278
Barrette, C., 647–648
Bartholomew, G. A., 145, 689, 705, 706–707
Bass, A. H., 236
Bastian, J., 236
Bazely, D. R., 656
Bech, C., 687
Behrensmeyer, A. D., 163
Beireter–Hahn, J., 75
Bemis, W. E., 241, 267, 272
Bennett, A. F., 150, 463, 609
Bennett, M. V. L., 236
Benton, M. J., 38, 58, 388, 389, 429, 434, 435, 437, 439, 627–628, 636
Berge, C., 746–747
Berger, A., 625
Berger, J., 645
Berril, N. J., 53
Berry, K., 378
Berven, K. A., 341
Beuchat, C. A., 345, 479, 489
Blackburn, D. G., 479, 579
Blake, R. W., 223
Blaustein, A. R., 39, 341, 352
Blieck, A., 57, 167
Block, B. A., 151
Blum, J. J., 155
Blum, V., 675
Bodznick, D., 238
Boesch, C., 753
Bogert, C. M., 142
Bonamo, P. M., 277
Bonaparte, J. F., 425
Bone, Q., 226
Boness, D. J., 643
Bongaarts, J., 41
Booser, J., 694–696
Booth, J., 369–370
Bowen, B. W., 374
Bowen, W. D., 643
Bowman, K. P., 763
Bramble, D. M., 607, 609

Bray, A. A., 165, 287
Breden, F., 13
Breder, C., 223
Briggs, D. E. G., 55, 58
Brinkman, D. R., 459
Britten, R. J., 25
Brodal, A., 183
Brodie, E. D. III, 349
Brodie, E. D. Jr., 349
Brothers, E. B., 268
Brower, L. P., 347
Brown, F. H., 750
Brown, J. A., 100
Brown, J. H., 709
Brown, M. B., 378
Brown, W. M., 24
Bruce, E. J., 746
Brussard, P. F., 344
Brust, D. G., 338
Bubenick, A. B., 645
Bubenick, G. A., 645
Buffetaut, H., 746
Buffrénil, V. de, 413
Bull, J. J., 118, 119
Bullock, T., 236
Burchfield, P., 380
Burenhult, G., 750
Burggren, W. W., 95, 241, 267, 366
Burian, 44
Burke, R. L., 381
Burke, V. J., 376
Burr, B. M., 261
Busch, C., 686

Caceres, J., 348
Caillouet, C. W. Jr., 380
Calder, W. A., 694–696
Caldwell, M. W., 177
Calvin, W. H., 753
Calzolai, R., 367
Campana, S. E., 268
Campbell, B., 738, 749
Caple, G., 412
Carey, C., 581
Carey, F. G., 151
Carlisle, R. C., 755
Carlsen, E., 41

Lahr, M. M., 751, 752
Lamarck, J. B. 11
Lamb, M. J., 118, 119
Lang, J. W., 420
Langner, G., 238
LaPointe, J. L., 378
Larsson, H. C. E., 448
Lauder, G. V., 239, 245, 246, 248, 255
Laufer, E., 64, 84
Laurent, P., 221
Laurin, M., 305, 359
Le Douarin, N., 70
Leakey, M. G., 748
Leakey, R. E., 750
Leal, M., 478
Lee, M. S. Y., 305, 359
Lee, P. C., 750
LeFeuvre, E. M., 213
Leong, A. S.–Y., 338
Lessels, C. M., 367
Leuthold, W., 718
Lewin, R., 25, 712, 746, 748, 749
Licht, P., 484
Liem, K. F., 239, 245, 246, 248
Lifjeld, J. T., 569
Likens, G. E., 763
Lillegraven, J. A., 637, 673–675
Lillywhite, H. B., 402
Lim, S. K., 553
Lin, S., 750
Lindsey, C. C., 223, 234
Linnaeus, C., 29
Linzell, J. L., 138
Litman, G. W., 98
Lloyd, E. A., 44, 45
Lockley, M. G., 553
Loew, E. R., 478
Lohmann, K. J., 375
Lorenz, K., 570, 665
Loveridge, J. P., 139
Lovtrup, S., 53
Lucas, S. G., 437
Luckett, W. P., 733–734
Lund, R., 202
Lundrigan, B. L., 347
Lutz, G. J., 318

MacBride, B., 761
Macdonald, D. W., 712, 714
Mace, G. M., 39
Macumber, P. G., 755
Magnuson, J. J., 226
Magnusson, W. E., 336, 341
Maisey, J. G., 179, 181
Makela, P., 423
Malkin, B., 348
Mallatt, J., 172, 183, 192
Mallory, K., 753, 763
Manspeizer, W., 445
Marchesi, N., 753
Marchesi, P., 753

Mares, M. A., 629, 632
Marks, S. B., 312
Marquez M., R., 380
Marsh, R. L., 334, 337
Marshall, L. G., 564, 629
Marshall, N. B., 226
Martin, A. P., 198
Martin, L. D., 553, 554, 555, 747
Martin, P. S., 754, 755, 756
Martin, R. D., 733
Martin, W., 446
Marx, J., 64, 67
Mason, R. T., 477
Massare, J. A., 431, 433
Mathewson, R. F., 209
Matoltsy, A. G., 75
Matson, P. A., 763
Matsuka, M., 553
May, R. M., 39, 452
Mayr, E., 19
Maze, J. R., 163
McCosker, J. E., 273
McCoy, P. C., 757
McDiarmid, R. W., 352
McElroy, C. T., 40
McHenry, H. M., 747, 750
McIntosh, J. S., 400
McKenna, M. C., 426
McKerrow, W. S., 276
McMahon, A. P., 64
McMahon, T. A., 155
McMenamin, D. L., 162
McMenamin, M. A. S., 162
McNab, B. K., 673–675, 684–686, 712
Medica, P. A., 505
Meeker, L. J., 357
Mellars, P., 751
Melton, D. A., 68
Mendel, G., 11
Meyer–Bahlburg, H. F. L., 119
Meylan, A. B., 358, 374
Michael, C. R., 666
Mikulic, D. G., 58
Milankovitch, M., 625
Miles, R. S., 172, 192, 198
Miller, K., 370-371
Milner, A. R., 283, 293
Milsom, W. K., 692
Milton, K., 348
Minnich, J. E., 138
Mittwoch, U., 119
Moni, R. W., 348
Monroe, B. L., 556
Mooney, H. A., 762
Moore, P. D., 164
Moos, M. Jr., 348
Moratalla, J. J., 426
Moreau, R. E., 586
Morell, V., 118, 745
Morrison, G., 129
Mosimann, J. E., 754, 755

Moss, S. A., 205, 208, 214
Mossman, H. W., 673
Moy–Thomas, J. A., 172, 192
Mrosovsky, N., 370, 379
Munro, U., 590
Murray, P., 615
Myers, C. W., 348, 349
Myers, N., 762

Nachtigall, W., 273
Naftolin, F., 118
Nagy, K. A., 505, 506, 509–511
Naylor, G. J. P., 198
Neher, A., 539
Neilson, J. D., 268
Nelson, C. E., 370
Nelson, J. S., 183, 247, 248, 258
Nicolls, G. D., 163
Nicolson, N. A., 727
Nielsen, C., 46
Nieuwenhuys, R., 209, 231
Niklas, K. J., 163
Nishida, T., 726
Nishikawa, K. C. B., 315
Nishimura, H., 125
Noden, D. M., 73
Norberg, U. M., 527, 550
Nordeen, E. J., 567
Nordeen, K. W., 567
Norell, M. A., 426, 554
Norman, J. R., 255
Norris, K. S., 670, 671
Northcutt, R. G., 53, 55, 58, 185, 208, 238
Nottebohm, F., 567
Novacek, M. J., 426, 636
Novak, S. S., 460
Nowicki, S., 536
Nozzolini, M., 549
Nusse, R., 64

O'Connor, M. P., 358, 426
O'Hara, R. K., 341
Obradovich, J. D., 439
Oftedal, O. T., 474, 643
Ohno, S., 118
Ollason, J. G., 563, 566
Olsen, S. L., 757
Olson, E. C., 181
Olson, S. L., 38, 553, 556
Orians, G. H., 575
Oring, L. W., 569, 577, 578, 585
Ostrom, J. H., 407, 409, 412, 554
Owen, R. E., 214
Owen–Smith, N., 755
Owens, D. W., 380

Pacenza, M., 767
Packard, G. C., 370-371
Packard, M. J., 370-371
Packer, C., 715, 716

Latin and Greek Lexicon

Many biological names and terms are derived from Latin (L) and Greek (G). Learning even a few dozen of these roots is a great aid to a biologist. The following terms are often encountered in vertebrate biology. The words are presented in the spelling and form in which they are most often encountered; this is not necessarily the original form of the word in its etymologically pure state.

An example of how a root is used in vertebrate biology can often be found by referring to the subject index. Remember, however, that some of these roots may be used as suffixes or otherwise embedded in technical words and will require further searching to discover an example of its use. Further information can be found in a reference such as E. C. Jaeger (1955), *A Source Book of Biological Names and Terms*, 3rd ed., Charles C Thomas, Springfield, Illinois.

a, ab (L) away from
a, an (G) not, without
acanth (G) thorn
actin (G) a ray
ad (L) toward, at, near
aeros (G) the air
aga (G) very much, too much
aistos (G) unseen
al, alula (L) a wing
allant (G) a sausage
alveol (L) a pit
amblyls (G) blunt, stupid
ammos (G) sand
amnion (G) a fetal membrane
amphi, ampho (G) both, double
amplexus (L) an embracing
ampulla (L) a jug or flask
ana (G) up, upon, through
anat (L) a duck
angio (G) a reservoir, vessel
ankylos (G) crooked, bent
anomos (G) lawless
ant, anti (G) against
ante (L) before
anthrac (G) coal
apat (G) illusion, error
aphanes (G) invisible, unknown
apo, ap (G) away from, separate
apsid (G) an arch, loop
aqu (L) water
arachne (G) a spider
arch (G) beginning, first in time
argenteus (L) silvery
arthr (G) a joint
ascidion (G) a little bag or bladder
aspid (G) a shield
asteros (G) a star
atri, atrium (L) an entrance-room
audi (L) to hear
austri, australis (L) southern
av (L) a bird
baen (G) to walk or step
bas (G) base, bottom

batrachos (G) a frog
benthos (G) the sea depths
bi, bio (G) life
bi, bis (L) two
blast (G) bud, sprout
brachi (G) arm
brachy (G) short
branchi (G) a fin
buce (L) the check
cal (G) beautiful
calie (L) a cup
capit (L) head
carn (L) flesh
caud (L) tail
cene, ceno (G) new, recent
cephal (G) head
cer, cerae (G) a horn
cerc (G) tail
chir, cheir (G) hand
choan (G) funnel, tube
chondr (G) grit, gristle
chord (G) guts, a string
chorio (G) skin, membrane
chrom (G) color
cloac (L) a sewer
coel (G) hollow
cornu (L) a horn
cortic, cortex (L) bark, rind
costa (L) a rib
cran (G) the skull
creta (L) chalk
cretio (L) separate, sift
crine (G) to separate
cten (G) a comb
cut, cutis (L) the skin
cyn (G) a dog
cytos (G) a cell
dactyl (G) a finger
de (L) down, away from
dectes (G) a biter
dendro (G) a tree
dent, dont (L) a tooth
derm (G) skin

desmos (G) a chain, tie, or band
deuteros (G) secondary
di, dia (G) through, across
di, diplo (G) two, double
din, dein (G) terrible, powerful
dir (G) the neck
disc (G) a disk
dory (G) a spear
draco (L) a dragon
drepan (G) a sickle
dromos (G) a running course
duct (L) a leading
dur (L) hard
e, ex (L) out of, from, without
echinos (G) a prickly being
eco, oikos (G) a house
ect (G) outside
edaphos (G) the soil or bottom
eid (G) form, appearance
elasma (G) a thin plate
eleutheros (G) free, not bound
elopos (G) a kind of sea fish
embolo (G) like a peg or stopper
embryon (G) a fetus
emys (G) a fresh-water turtle
end (G) within
enter (G) bowel, intestine
eos (G) the dawn or beginning
ep (G) on, upon
equi (L) a horse
ery (G) to drag or draw
erythr (G) red
eu, ev (G) good, true
eury (G) broad
extra (L) beyond, outside
falc (L) a sickle, scythe
fer (L) carrier of
fil (L) a thread
fundus (L) bottom, foundation
galeos (G) a shark
gallus (L) poultry
gaster (G) the belly
genos (G) birth